ECOLOGY AND BEHAVIOUR OF THE LADYBIRD BEETLES (COCCINELLIDAE)

ECOLOGY AND BEHAVIOUR OF THE LADYBIRD BEETLES (COCCINELLIDAE)

Edited by I. Hodek, H.F. van Emden and A. Honěk

A John Wiley & Sons, Ltd., Publication

This edition first published 2012 © 2012 by Blackwell Publishing Ltd

Blackwell Publishing was acquired by John Wiley & Sons in February 2007. Blackwell's publishing program has been merged with Wiley's global Scientific, Technical and Medical business to form Wiley-Blackwell.

Registered office: John Wiley & Sons, Ltd, The Atrium, Southern Gate, Chichester, West Sussex, PO19 8SQ, UK

Editorial offices: 9600 Garsington Road, Oxford, OX4 2DQ, UK
            The Atrium, Southern Gate, Chichester, West Sussex, PO19 8SQ, UK
            111 River Street, Hoboken, NJ 07030-5774, USA

For details of our global editorial offices, for customer services and for information about how to apply for permission to reuse the copyright material in this book please see our website at www.wiley.com/wiley-blackwell.

The right of the author to be identified as the author of this work has been asserted in accordance with the UK Copyright, Designs and Patents Act 1988.

Library of Congress Cataloging-in-Publication Data
Ecology and behaviour of the ladybird beetles (Coccinellidae) / edited by I. Hodek, H.F. van Emden, and A. Honěk.
      p. cm.
   Includes index.
   ISBN 978-1-4051-8422-9 (cloth)
   1. Ladybugs.   I. Hodek, Ivo.   II. Van Emden, Helmut Fritz.   III. Honěk, A. (Alois)
   QL596.C65E26 2012
   595.76'9–dc23
                        2011045545

A catalogue record for this book is available from the British Library.

Wiley also publishes its books in a variety of electronic formats. Some content that appears in print may not be available in electronic books.

Set in 9/11 pt PhotinaMT by Toppan Best-set Premedia Limited
Printed and bound in Malaysia by Vivar Printing Sdn Bhd

1   2012

To all those scientists
who have now passed on
and who laid the foundation
of our present knowledge of Coccinellidae,
particularly to Michael Majerus
who intended to be one of authors
of this volume.

CONTENTS

DETAILED CONTENTS

9. COCCINELLIDS AND SEMIOCHEMICALS, 444

Jan Pettersson

12. RECENT PROGRESS AND POSSIBLE FUTURE TRENDS IN THE STUDY OF COCCINELLIDAE, 520

Helmut F. van Emden and Ivo Hodek

APPENDIX: LIST OF GENERA IN TRIBES AND SUBFAMILIES, 526

Oldrich Nedvěd and Ivo Kovář

Colour plate pages fall between pp. 250 and pp. 251

CONTRIBUTORS

PIOTR CERYNGIER *Centre for Ecological Research, Polish Academy of Sciences, Dziekanow Lesny, 05-092 Lomianki, POLAND*

HELMUT F. VAN EMDEN *School of Biological Sciences, University of Reading, Whiteknights, Reading RG6 6AS UK*

EDWARD W. EVANS *Department of Biology, Utah State University, Logan, UT 84322 USA*

JAMES D. HARWOOD *Department of Entomology, University of Kentucky, Lexington, KY 40546-0091 USA*

IVO HODEK *Institute of Entomology, Academy of Sciences, CZ 37005 České Budějovice, CZECH REPUBLIC*

ALOIS HONĚK *Department of Entomology, Crop Research Institute, CZ 16106 Prague 6, CZECH REPUBLIC*

IVO KOVÁŘ *Emer. Scientist of the National Museum, Prague; Current address: Zichovec, CZECH REPUBLIC*

ERIC LUCAS *Département des Sciences Biologiques, Université du Québec à Montréal, C.P. 8888 Succ. Centre-ville, Montréal, Québec H3C 3P8 CANADA*

J.P. MICHAUD *Department of Entomology, Kansas State University, 1232 240th Ave., Hays, KS 67601 USA*

OLDRICH NEDVĚD *Faculty of Science, University of South Bohemia and Institute of Entomology, Academy of Sciences, CZ 37005 České Budějovice, CZECH REPUBLIC*

JAN PETTERSSON *Department of Ecology, Swedish University of Agricultural Sciences, Box 7044, SE-750 07 Uppsala, SWEDEN*

REMY L. POLAND *Department of Genetics, University of Cambridge, Cambridge CB2 3EH UK, recent address: Clifton College, 32 College Road, Clifton, Bristol, BS8 3JH, UK*

HELEN E. ROY *NERC Centre for Ecology and Hydrology, Crowmarsh Gifford, Oxfordshire OX10 8BB UK*

JOHN J. SLOGGETT *Maastricht Science Programme, Maastricht University, P.O. Box 616, 6200 MD Maastricht, THE NETHERLANDS*

PREFACE

For more than a decade, no volume on the general aspects of ladybirds has been published, although the predaceous major part of this coleopteran family represents an important component of the natural enemies of Sternorrhyncha (aphids, coccids, aleyrodids and psyllids). These sucking insects are among the most dangerous pests of crops, as under suitable conditions their populations increase exponentially, especially when parthenogenesis and viviparity occur.

Although classical biological control with coccinellids has recently been mostly abandoned, the increasing concern about chemical pest control has increased the need for modern types of biological control, mostly involving conservation and augmentation, within the framework of integrated pest management. For the success of these sophisticated methods of control, precise knowledge of behaviour and ecological relations is indispensable.

Such knowledge, particularly in the areas of ethology and molecular genetics, has accumulated in the last decade and reviewing these areas is an important novel contribution of this book. We hope that, at this stage, it will help to improve conservation and augmentation control, but also that it will stimulate further research.

The book was very unfortunately interrupted by the premature death of Professor Michael Majerus, Department of Genetics, Cambridge University, who was our proposed author for Chapters 2 and 8. We were extremely lucky to find very able successors, John Sloggett (Chapter 2), and Helen Roy and Remy Poland

(for parasites and pathogens in Chapter 8), who fulfilled their task with great success. Many thanks!

We should like to thank the other authors of chapters for contributing their expertise so readily and for reacting so positively and constructively to our suggestions for revision. We are also grateful for the expert help we have received with taxonomic nomenclature: Dr O. Nedvěd of the University of South Bohemia, Czech Republic, and from the UK Drs V.F. Eastop, C.H. Lyal and D.J. Williams (Natural History Museum), Drs R.T.V. Fox and S.L.Jury (University of Reading) and Professor D.L.J. Quicke (Imperial College). Dr Lyal deserves additional thanks for his advice in relation to Chapter 1.

We should also like to thank Dr Ward Cooper of Wiley-Blackwell for enthusiastically agreeing to publish our book and Mr Kelvin Matthews for his help with the publication process. We should also like to thank the other Wiley-Blackwell staff who have worked so helpfully and efficiently on the production of this book.

Ivo Hodek
České Budějovice, Czech Republic

Helmut F. van Emden
Reading, UK

Alois Honěk
Prague, Czech Republic

March 2012

INTRODUCTION

The reader may find the following information helpful in order to use the book more easily.

To keep the flow of the text unbroken and to save repetition and space, taxonomic affiliations of organisms and species authorities are given only in the Taxonomic Glossary (following section, while the Subject Index is at the end of book, as usual). Readers should also note that the Latin names as given in this glossary are used in the text and tables, and therefore may not be the same as the older Latin names given in the original papers cited. However, because the species names have usually not changed, and the older names are also listed in the glossary, any confusion will be avoided.

The very common generic names are always abbreviated as follows throughout the text of all chapters: *A.* for *Adalia*; *C.*, *Coccinella*; *Cer.*, *Ceratomegilla*; *Chil.*, *Chilocorus*; *Col.*, *Coleomegilla*; *Har.*, *Harmonia*; *Hip.*, *Hippodamia*; *P.*, *Propylea*.

Some phenomena are discussed in more than one chapter in the book, but from different angles; the reader's attention is directed to this by cross-references to the section number.

TAXONOMIC GLOSSARY

With common names where appropriate

Coleoptera: Family Coccinellidae

Synonyms [in square brackets] as well as currently valid names are listed in alphabetical order.

Adalia bipunctata (L.) – two spot ladybird [*Adalia fasciatopunctata* (Faldermann)]

Adalia conglomerata (L.)

Adalia decempunctata (L.) – ten spot ladybird

Adalia deficiens Mulsant

[*Adalia fasciatopunctata* (Faldermann)] = *Adalia bipunctata* (L.)

[*Adalia flavomaculata* (De Geer)] = *Lioadalia flavomaculata* (De Geer)

Adalia tetraspilota (Hope)

[*Adonia*] = *Hippodamia*

[*Adonia arctica* (Schneider)] = *Hippodamia arctica* (Schneider)

[*Adonia variegata* (Goeze)] = *Hippodamia variegata* (Goeze)

Afidenta misera (Weise) [*Afidenta mimetica* Dieke]

Aphidentula bisquadripunctata (Gyllenhal) [*Epilachna bisquadripunctata* (Gyllenhal)]

Afissula rana Kapur

Afissula sanscrita (Crotch)

Aiolocaria hexaspilota (Hope) [*Aiolocaria mirabilis* (Motschulsky)]

[*Aiolocaria mirabilis* (Motschulsky)] = *Aiolocaria hexaspilota* (Hope)

Alloneda dodecaspilota (Hope)

Anatis halonis Lewis

Anatis labiculata (Say) – fifteen spotted lady beetle

Anatis mali (Say) – eyespotted lady beetle

Anatis ocellata (L.) – eyed ladybird

Anatis quindecimpunctata (DeGeer)

Anegleis cardoni (Weise)

[*Anisolemnia dilatata* (F.)] = *Megalocaria dilatata* (F.)

Anisolemnia tetrasticta Fairmaire

Anisosticta bitriangularis (Say)

Anisosticta novemdecimpunctata (L.) – water ladybird

Anisosticta sibirica Bielawski

Aphidecta obliterata (L.) – larch ladybird

Apolinus lividigaster (Mulsant) [*Scymnodes lividigaster* (Mulsant)]

Axinoscymnus cardilobus Ren & Pang

[*Azya trinitatis* (Marshall)] = *Pseudoazya trinitatis* (Marshall)

Azya orbigera Mulsant

Brachiacantha quadripunctata Melsheimer

Brachiacantha ursina (F.)

Brumoides septentrionis (Weise)

Brumoides suturalis (F.) – three-striped lady beetle [*Brumus suturalis* (Mani)]

[*Brumus quadripustulatus* (L.)] = *Exochomus quadripustulatus* (L.)

[*Brumus suturalis* (Mani)] = *Brumoides suturalis* (F.)

Bulaea lichatschovi (Hummel)

Callicaria superba (Mulsant)

Calvia albida Bielawski

Calvia decemguttata (L.)

Calvia duodecimmaculata Gebler

Calvia muiri Timberlake [*Eocaria muiri* Timberlake]

Calvia quatuordecimguttata (L.) – cream-spot ladybird

Calvia quindecimguttata (F.)

Calvia shiva Kapur

Calvia shiva pasupati Kapur

Calvia shiva pinaki Kapur

Calvia shiva trilochana Kapur

Ceratomegilla barovskii kiritschenkoi (Semenov-Tian-Shanski) [*Spiladelpha barovskii kiritschenkoi* Semenov-Tian-Shanski]

Ceratomegilla notata (Laicharting)

Ceratomegilla undecimnotata (Schneider) [*Semiadalia undecimnotata* (Schneider)]

Cheilomenes lunata (F.)

Cheilomenes propinqua vicina (Mulsant) [*Cheilomenes vicina* (Mulsant), *Cydonia vicina nilotica* Mulsant]

[*Cheilomenes sexmaculata* (F.)] = *Menochilus sexmaculatus* (F.)

Cheilomenes sulphurea (Olivier)

[*Cheilomenes vicina*] = *Cheilomenes propinqua vicina* (Mulsant)

[*Chilocorus baylei* (Blackburn)] = *Chilocorus malasiae* Crotch

Chilocorus bijugus Mulsant

Chilocorus bipustulatus (L.) – heather ladybird

Chilocorus braeti Weise

Chilocorus cacti (L.)

Chilocorus circumdatus (Gyllenhal)

Chilocorus discoideus Crotch

Chilocorus distigma Klug

Chilocorus geminus Zaslavsky

Chilocorus hauseri Weise

Chilocorus hexacyclus Smith

Chilocorus infernalis Mulsant

Chilocorus inornatus Weise

Chilocorus kuwanae Silvestri

Chilocorus malasiae Crotch [*Chilocorus baylei* (Blackburn)]

Chilocorus nigripes Mader

Chilocorus nigritus (F.)

Chilocorus orbus Casey

Chilocorus quadrimaculatus (Weise)

Chilocorus renipustulatus (Scriba) – kidney-spot ladybird

Chilocorus rubidus Hope

Chilocorus similis (Rossi)

Chilocorus stigma (Say) – twice-stabbed lady beetle

Chilocorus tricyclus Smith

Chnootriba similis (Thunberg)

Cleobora mellyi (Mulsant) – Tasmanian ladybird

Clitostethus arcuatus (Rossi)

Clitostethus oculatus (Blatchley) [*Nephaspis oculatus* (Blatchley)]

Coccidophilus citricola Brèthes

Coccidula rufa (Herbst)

Coccidula scutellata (Herbst)

[*Coccinella algerica* Kovář] = *Coccinella septempunctata algerica* Kovář

Coccinella californica Mannerheim

Coccinella explanata Miyatake

Coccinella hieroglyphica L. – hieroglyphic ladybird

Coccinella leonina transversalis F. [*Coccinella repanda* Thunberg, *Coccinella transversalis* F.]

Coccinella luteopicta (Mulsant)

Coccinella magnifica Redtenbacher – scarce seven spot ladybird [*Coccinella divaricata* Olivier)]

Coccinella monticola Mulsant

Coccinella nigrovittata Kapur [*Tytthaspis trilineata* (Weise)]

Coccinella novemnotata Herbst – nine-spotted lady beetle

Coccinella quinquepunctata L. – five spot ladybird

Coccinella reitteri Weise

[*Coccinella repanda* Thunberg] = *Coccinella leonina* F.

Coccinella septempunctata L. – seven spot ladybird

Coccinella septempunctata algerica Kovář [*Coccinella algerica* Kovář]

Coccinella septempunctata brucki Mulsant

[*Coccinella sinuatomarginata* Faldermann] = *Coccinula sinuatomarginata* (Faldermann)

[*Coccinella transversalis* F.] = *Coccinella leonina transversalis* F.

Coccinella transversoguttata Faldermann – transverse lady beetle

Coccinella transversoguttata richardsoni Brown

Coccinella trifasciata L. – three-banded lady beetle

Coccinella undecimpunctata L. – eleven spot ladybird

[*Coccinella undecimpunctata aegyptiaca* Reiche] = *Coccinella undecimpunctata menetriesi* Mulsant

Coccinella undecimpunctata menetriesi Mulsant [*Coccinella undecimpunctata aegyptiaca* Reiche]

Coccinula crotchi (Lewis)

Coccinula quatuordecimpustulata (L.)

Coccinula redimita (Weise)

Coccinula sinensis (Weise)

Coccinula sinuatomarginata (Faldermann) [*Coccinella sinuatomarginata* Faldermann]

Coelophora biplagiata Swartz [*Lemnia biplagiata* (Swartz)]

Coelophora bissellata Mulsant

Coelophora duvaucelii (Mulsant)

Coelophora inaequalis (F.) – common Australian lady beetle

Coelophora mulsanti (Montrouzier)

Coelophora quadrivittata Fauvel

Coelophora saucia Mulsant

Coleomegilla maculata (DeGeer) – spotted lady beetle

Coleomegilla maculata fuscilabris (Mulsant)

Coleomegilla maculata lengi Timberlake – twelve
  spotted ladybeetle
Coleomegilla quadrifasciata (Schoenherr)
Cryptognatha simillima Sicard
Cryptognatha nodiceps Marshall
Cryptognatha signata Korschefsky
Cryptogonus ariasi (Mulsant)
Cryptogonus kapuri Ghorpade
Cryptogonus orbiculus (Gyllenhal)
Cryptogonus postmedialis Kapur
Cryptogonus quadriguttatus (Weise)
Cryptolaemus montrouzieri Mulsant – mealybug
  destroyer
Curinus coeruleus Mulsant – metallic blue lady beetle
Cycloneda ancoralis (Germar)
Cycloneda limbifer Casey
Cycloneda munda (Say) – polished lady beetle
Cycloneda polita Casey
Cycloneda sanguinea (L.) – blood-red lady beetle
[*Cydonia vicina nilotica* Mulsant] = *Cheilomenes
  propinqua vicina* (Mulsant)
Declivitata spp.
Delphastus catalinae (Horn)
Delphastus pusillus (LeConte)
Diomus austrinus Gordon
Diomus hennesseyi Fürsch
Diomus pumilio Weise
Diomus seminulus (Mulsant)
Diomus thoracicus F.
[*Eocaria muiri* Timberlake] = *Calvia muiri* (Timberlake)
Epilachna admirabilis Crotch
[*Epilachna bisquadripunctata* (Gyllenhal)] = *Aphidentula
  bisquadripunctata* (Gyllenhal)
[*Epilachna boisduvali* (Mulsant)] = *Henosepilachna
  boisduvali* Mulsant
Epilachna borealis (F.) – squash beetle
Epilachna canina (F.)
[*Epilachna 'chrysomelina'* (F.)]; used variably either for
  Henosepilachna argus (Geoffroy) or *Henosepilachna
  elaterii* (Rossi) or *Henosepilachna vigintioctopunctata*
  (F.)
[*Epilachna cucurbitae*] = *Henosepilachna sumbana*
  Bielawski
Epilachna defecta Mulsant
Epilachna dregei Mulsant – potato ladybird
Epilachna dumerili Mulsant
Epilachna eckloni Mulsant
[*Epilachna enneasticta* Mulsant] = *Henosepilachna
  enneasticta* (Mulsant)
Epilachna eusema (Weise)

Epilachna karisimbica Weise
Epilachna marginella (F.)
Epilachna marginicollis (Hope)
Epilachna mexicana (Guérin-Méneville)
Epilachna mystica Mulsant
Epilachna nigrolimbata Thomson
[*Epilachna niponica* Lewis] = *Henosepilachna niponica*
  (Lewis)
Epilachna paenulata (Germar)
[*Epilachna philippinensis* Dieke] = *Henosepilachna
  vigintisexpunctata* (Boisduval)
[*Epilachna pusillanima* Mulsant] = *Henosepilachna
  pusillanima* (Mulsant)
[*Epilachna pustulosa* Kôno] = *Henosepilachna pustulosa*
  (Kôno)
Epilachna quadricollis (Dieke)
[*Epilachna septima* Dieke] = *Henosepilachna septima*
  (Dieke)
[*Epilachna sparsa orientalis* Dieke] = *Henosepilachna
  vigintioctopunctata* (F.)
Epilachna undecimvariolata (Boisduval)
Epilachna varivestis Mulsant – mexican bean beetle
[*Epilachna vigintioctomaculata* Motschulsky] =
  Henosepilachna vigintioctomaculata (Motschulsky)
[*Epilachna vigintioctomaculata* Motschulsky] =
  Henosepilachna vigintioctomaculata (Motschulsky)
Epilachna vigintisexpunctata (Boisduval)
[*Epilachna yasutomii* (Katakura)] = *Henosepilachna
  yasutomii* Katakura
Epiverta chelonia (Mader)
Eriopis connexa (Germar)
Exochomus childreni Mulsant
[*Exochomus concavus* Fürsch] = *Parexochomus troberti
  concavus* (Fürsch)
Exochomus flavipes (Thunberg)
Exochomus flaviventris Mader
Exochomus fulvimanus Weise
[*Exochomus lituratus* (Gorham)] = *Priscibrumus
  lituratus* (Gorham)
[*Exochomus melanocephalus* (Zoubkoff)] = *Parexochomus
  melanocephalus* (Zoubkoff)
[*Exochomus nigromaculatus* (Goeze) = *Parexochomus
  nigromaculatus* (Goeze)
Exochomus quadripustulatus (L.) – pine ladybird
  [*Brumus quadripustulatus* (L.)]
[*Exochomus troberti* Mulsant] = *Parexochomus troberti*
  (Mulsant)
Halmus chalybeus (Boisduval) – steelblue lady beetle
  [*Orcus chalybeus* (Boisduval)]
[*Halyzia hauseri* (Mader)] = *Macroilleis hauseri* (Mader)

Halyzia sanscrita Mulsant
Halyzia sedecimguttata (L.) – orange ladybird
Halyzia straminea (Hope)
Halyzia tschitscherini Semenov
Harmonia antipoda (Mulsant in White) – antipodean ladybird
Harmonia axyridis (Pallas) – harlequin ladybird or Asian multi-colored ladybeetle
[*Harmonia breiti* Mader] = *Harmonia expalliata* Sicard
Harmonia conformis (Boisduval)
Harmonia dimidiata (F.) [*Leis dimidiata* (F.)]
Harmonia eucharis (Mulsant)
Harmonia expalliata Sicard [*Harmonia breiti* Mader]
Harmonia quadripunctata (Pontoppidan) – cream-streaked ladybird
Harmonia octomaculata (F.)
Harmonia sedecimnotata (F.)
Harmonia yedoensis (Takizawa)
Henosepilachna argus (Geoffroy) – bryony ladybird
Henosepilachna bifasciata (L.)
Henosepilachna boisduvali (Mulsant) [*Epilachna boisduvali* Mulsant]
Henosepilachna dodecastigma (Wiedemann)
Henosepilachna elaterii (Rossi) [*Epilachna chrysomelina* (F.)]
Henosepilachna enneasticta (Mulsant) [*Epilachna enneasticta* Mulsant]
Henosepilachna guttatopustulata (F.)
Henosepilachna indica (Mulsant)
Henosepilachna niponica (Lewis) [*Epilachna niponica* Lewis]
Henosepilachna ocellata (Redtenbacher)
Henosepilachna processa Li
Henosepilachna pusillanima (Mulsant) [*Epilachna pusillanima* Mulsant]
Henosepilachna pustulosa (Kôno) [*Epilachna pustulosa* Kôno]
Henosepilachna septima (Dieke) [*Epilachna septima* Dieke]
Henosepilachna sumbana Bielawski [*Epilachna cucurbitae* Richards, *Henosepilachna cucurbitae* Richards]
Henosepilachna vigintioctomaculata (Motschulsky) [*Epilachna vigintioctomaculata* Motschulsky]
Henosepilachna vigintioctopunctata (F.) [*Epilachna vigintioctopunctata* (F.), *Epilachna sparsa orientalis* Dieke]
Henosepilachna vigintisexpunctata (Boisduval) [*Epilachna philippinensis* Dieke]
Henosepilachna yasutomii Katakura [*Epilachna yasutomii* (Katakura)]

Hippodamia arctica (Schneider) [*Adonia arctica* (Schneider)]
Hippodamia caseyi Johnson
Hippodamia convergens Guerin – convergent ladybeetle
Hippodamia glacialis (F.) – glacial lady beetle
Hippodamia parenthesis (Say) – parenthesis lady beetle
Hippodamia quinquesignata (Kirby)
Hippodamia quinquesignata punctulata Le Conte [*Hippodamia quinquesignata ambigua* LeConte]
Hippodamia septemmaculata (DeGeer)
Hippodamia sinuata Mulsant
Hippodamia tredecimpunctata (L.) – thirteen spot ladybird
Hippodamia variegata (Goeze) – Adonis ladybird, variegated lady beetle [*Adonia variegata* (Goeze)]
Hyperaspis aestimabilis Mader
Hyperaspis bigeminata (Randall)
Hyperaspis binotata (Say)
Hyperaspis campestris (Herbst)
[*Hyperaspis congressis* Watson] = *Hyperaspis conviva* Casey
Hyperaspis conviva Casey [*Hyperaspis congressis* Watson]
Hyperaspis desertorum Weise
Hyperaspis lateralis Mulsant
Hyperaspis notata Mulsant
Hyperaspis pantherina Fürsch
Hyperaspis raynevali Mulsant
Hyperaspis reppensis (Herbst)
Hyperaspis senegalensis Mulsant
Hyperaspis senegalensis hottentotta Mulsant
Hyperaspis sphaeridioides Mulsant
Hyperaspis undulata (Say)
Illeis bielawskii Ghorpade
Illeis cincta (F.)
Illeis galbula (Mulsant) [*Leptothea galbula* (Mulsant)]
Illeis koebelei Timberlake
Jauravia quadrinotata Kapur
[*Leis dimidiata* (F.)] = *Harmonia dimidiata* (F.)
[*Leis*] = *Harmonia*
[*Lemnia biplagiata* (Swartz)] = *Coelophora biplagiata* Swartz
[*Leptothea galbula* (Mulsant)] = *Illeis galbula* (Mulsant)
Lindorus lophanthae (Blaisdell) [*Rhyzobius lophantae* (Blaisdell), *Rhyzobius lorophantae* (Blaisdell)]
Lioadalia flavomaculata (De Geer) [*Adalia flavomaculata* (De Geer)]
Macroilleis hauseri (Mader) [*Halyzia hauseri* (Mader)]
Macronaemia hauseri (Weise)

Megalocaria dilatata (F.) [*Anisolemnia dilatata* (F.)]

[*Menochilus quadriplagiatus* (Swartz)] = *Menochilus sexmaculatus* (F.)

Menochilus sexmaculatus (F.) [*Cheilomenes sexmaculata* (F.), *Menochilus quadriplagiatus* (Swartz)]

Micraspis allardi (Mulsant)

Micraspis discolor (F.)

Microweisea sp.

Mulsantina hudsonica (Casey)

Mulsantina picta (Randall)

Myrrha octodecimguttata (L.) – eighteen spot ladybird

Myzia oblongoguttata (L.) – striped ladybird

Myzia subvittata (Mulsant)

Neda marginalis Mulsant

Neocalvia anastomozans Crotch

[*Nephaspis oculatus* (Blatchley)] = *Clitostethus oculatus* (Blatchley)

Nephus bilucernarius (Mulsant)

Nephus bisignatus (Boheman)

Nephus flavifrons (Melsheimer) [*Scymnus flavifrons* Melsheimer, North America, not *Scymnus flavifrons* Blackburn, Australia]

Nephus guttulatus (LeConte) [*Scymnus guttulatus* LeConte]

Nephus includens (Kirsch)

Nephus kiesenwetteri (Mulsant) [*Scymnus kiesenwetteri* Mulsant]

Nephus ornatus (LeConte) [*Scymnus ornatus* LeConte]

Nephus quadrimaculatus (Herbst) [*Scymnus quadrimaculatus* (Herbst)]

Nephus redtenbacheri (Mulsant)

Nephus soudanensis (Sicard) [*Scymnus soudanensis* Sicard]

Oenopia billieti (Mulsant)

Oenopia conglobata (L.) [*Synharmonia conglobata* (L.)]

Oenopia kirbyi Mulsant

Oenopia lyncea (Olivier)

Oenopia sexareata (Mulsant)

[*Olla abdominalis* (Say)] = *Olla v-nigrum* (Mulsant)

Olla v-nigrum (Mulsant) – ash-gray lady beetle [*Olla abdominalis* (Say)]

Orcus australasiae (Boisduval)

[*Orcus chalybeus* (Boisduval)] = *Halmus chalybeus* (Boisduval)

Palaeoneda auriculata (Mulsant) [*Paleoneda miniata* (Hope)]

[*Paleoneda miniata* (Hope)] = *Palaeoneda auriculata* (Mulsant)

Pania luteopustulata Mulsant

Paranaemia vittigera (Mannerheim)

Parastethorus nigripes (Kapur) [*Stethorus loxtoni* Britton & Lee]

Parexochomus melanocephalus (Zoubkoff) [*Exochomus melanocephalus* (Zoubkoff))

Parexochomus nigromaculatus (Goeze) [*Exochomus nigromaculatus* (Goeze)]

Parexochomus troberti (Mulsant) [*Exochomus troberti* (Mulsant)]

Parexochomus troberti concavus (Fuersch) [*Exochomus concavus* Fuersch]

Pentilia insidiosa Mulsant

Pharoscymnus anchorago (Fairmaire)

Pharoscymnus numidicus (Pic)

Pharoscymnus ovoideus Sicard

Phymatosternus lewisii (Crotch)

Platynaspis luteorubra (Goeze)

Priscibrumus lituratus (Gorham) [*Exochomus lituratus* (Gorham)]

Priscibrumus uropygialis (Mulsant)

Propylea dissecta (Mulsant)

Propylea japonica (Thunberg)

Propylea quatuordecimpunctata (L.) – fourteen spot ladybird [*Propylaea quatuordecimpunctata* (L.)]

Pseudoazya trinitatis (Marshall) [*Azya trinitatis* (Marshall)]

[*Pseudoscymnus tsugae* Sasaji & McClure] = *Sasajiscymnus tsugae* (Sasaji & McClure)

[*Pseudoscymnus kurohime* (Mityake)] = *Sasajiscymnus kurohime* (Miyatake)

Psyllobora confluens (F.)

Psyllobora vigintiduopunctata (L.) – twenty two spot ladybird [*Thea vigintiduopunctata* (L.)]

Psyllobora vigintimaculata (Say) – twenty-spotted lady beetle

[*Pullus auritus* (Thunberg)] = *Scymnus auritus* Thunberg

[*Pullus mediterraneus* (F.)] = *Scymnus marinus* (Mulsant)

[*Pullus subvillosus* (Goeze)] = *Scymnus subvillosus* (Goeze)

Rhyzobius litura (F.)

[*Rhyzobius lophanthae* (Blaisdell)] = *Lindorus lophanthae* (Blaisdell)

[*Rhyzobius lorophanthae* (Blaisdell)] = *Lindorus lophanthae* (Blaisdell)

Rhyzobius ventralis (Erichson) – black lady beetle

Rodatus major (Blackburn)

Rodolia cardinalis (Mulsant) – vedalia beetle

Rodolia fumida Mulsant

Rodolia guerini (Crotch)

Rodolia iceryae Janson
Rodolia occidentalis Weise
Sasajiscymnus kurohime (Miyatake) [*Pseudoscymnus kurohime* (Miyatake)]
[*Sasajiscymnus ningshanensis* (Sasaji & McClure)] = *Scymnus ningshanensis* Yu & Yao
Sasajiscymnus tsugae (Sasaji & McClure) [*Pseudoscymnus tsugae* Sasaji & McClure]
[*Scymnodes lividigaster* (Mulsant)] = *Apolinus lividigaster* (Mulsant)
Scymnus abietis (Paykull)
[*Scymnus aeneipennis* Sicard] = *Zagloba aeneipennis* (Sicard)
Scymnus apetzi Mulsant
Scymnus ater Kugelann
Scymnus auritus Thunberg [*Pullus auritus* (Thunberg)]
Scymnus coccivora Ayyar
Scymnus creperus Mulsant
Scymnus dorcatomoides Weise
Scymnus flavifrons Blackburn (Australia)
[*Scymnus flavifrons* Melsheimer] (North America) = *Nephus flavifrons* (Melsheimer)
Scymnus frontalis (F.)
[*Scymnus guttulatus* LeConte] = *Nephus guttulatus* (LeConte)
Scymnus haemorrhoidalis Herbst
Scymnus hilaris Motschulsky
Scymnus hoffmanni Weise
Scymnus impexus Mulsant
Scymnus interruptus (Goeze)
[*Scymnus kiesenwetteri* Mulsant] = *Nephus kiesenwetteri* (Mulsant)
Scymnus lacustris LeConte
Scymnus levaillanti Mulsant
Scymnus loewii Mulsant – dusky lady beetle
Scymnus louisianae Chapin
Scymnus marginicollis Mannerheim
Scymnus marinus (Mulsant) [*Scymnus mediterraneus* Iablokoff-Khnzorian, *Pullus mediterraneus* (Iablokoff-Khnzorian)]
[*Scymnus mediterraneus*] = *Scymnus marinus* (Mulsant)
Scymnus morelleti Mulsant
Scymnus nigrinus Kugelann
Scymnus ningshanensis Yu & Yao [*Sasajiscymnus ningshanensis* (Sasaji & McClure)]
[*Scymnus ornatus* LeConte] = *Nephus ornatus* (LeConte)
Scymnus otohime Kamiya
Scymnus posticalis Sicard
Scymnus postpictus Casey

Scymnus pyrocheilus Mulsant
Scymnus quadrillum Motschulsky
[*Scymnus (Nephus) quadrimaculatus* (Herbst)] = *Nephus quadrimaculatus* (Herbst)
Scymnus rubromaculatus (Goeze)
Scymnus sinuanodulus Yu & Yao
Scymnus smithianus Silvestri
Scymnus soudanensis Sicard
Scymnus subvillosus (Goeze) [*Pullus subvillosus* (Goeze)]
Scymnus suturalis Thunberg
Scymnus syriacus (Marsuel)
Scymnus tardus Mulsant
[*Semiadalia undecimnotata* (Schneider)] = *Ceratomegilla undecimnotata* (Schneider)
Serangium parcesetosum Sicard
[*Sidis*] = *Nephus*
[*Spiladelpha barovskii kiritschenkoi* Semenov-Tian-Shanski] = *Ceratomegilla barovskii kiritschenkoi* (Semenov-Tian-Shanski)
Sospita vigintiguttata (L.)
Stethorus bifidus Kapur
Stethorus gilvifrons (Mulsant)
Stethorus japonicus Kamiya
[*Stethorus loxtoni* Britton & Lee] = *Parastethorus nigripes* (Kapur)
Stethorus madecassus Chazeau
[*Stethorus picipes* Casey] = *Stethorus punctum picipes* Casey
[*Stethorus punctillum* Weise] = *Stethorus pusillus* (Herbst)
Stethorus punctum (LeConte) – spider-mite destroyer
Stethorus punctum picipes Casey [*Stethorus picipes* Casey]
Stethorus pusillus (Herbst) [*Stethorus punctillum* (Weise)]
Stethorus tridens Gordon
Stethorus vegans (Blackburn)
Subcoccinella vigintiquatuorpunctata (L.) – twenty four spot ladybird
[*Synharmonia conglobata* (L.)] = *Oenopia conglobata* (L.)
Synona obscura Poorani, Ślipiński & Booth
Synonycha grandis (Thunberg)
Thalassa saginata Mulsant
[*Thea vigintiduopunctata* (L.)] = *Psyllobora vigintiduopunctata* (L.)
Tytthaspis sedecimpunctata (L.) – sixteen spot ladybird
[*Tytthaspis trilineata* (Weise)] = *Coccinella nigrovittata* Kapur
[*Verania*] = *Micraspis*

Vibidia duodecimguttata (Poda)
Zagloba aeneipennis (Sicard) [*Scymnus aeneipennis* Sicard]

OTHER ORGANISMS

In alphabetical order within each taxon, and with Family given in italics within brackets after the authority

OTHER INSECTS

Dermaptera

Anechura harmandi (Burr) *(Forficulidae)* – hump earwig
Forficula auricularia L. *(Forficulidae)* – European earwig

Hemiptera: Heteroptera

Calocoris norvegicus (Gmelin) *(Miridae)* – strawberry bug
Campylomma verbasci (Meyer) *(Miridae)* – mullein bug
Caternaultiella rugosa (Schoutenden) *(Plataspidae)*
Eurygaster integriceps Puton *(Pentatomidae)* – sunn pest or corn bug
Geocoris punctipes (Say) *(Lygaeidae)* – big-eyed bug
Hyaliodes vitripennis (Say) *(Miridae)*
Lygus hesperus (Knight) *(Miridae)* – western plant bug
Lygus lineolaris (Palidot de Beauvois) *(Miridae)* – tarnished plant bug
Lygus *(Miridae)*
Nabis (*Reduviolus*) *americoferus* Carayon *(Nabidae)* – common damsel bug
Nysius huttoni White *(Lygaeidae)* – wheat bug
Orius insidiosus (Say) *(Anthocoridae)* – insidious flower bug
Podisus maculiventris (Say) *(Pentatomidae)* – spined soldier bug
Pyrrhocoris apterus (L.) *(Pyrrhocoridae)* – firebug
Sidnia kinbergi (Stål) *(Miridae)*

Hemiptera: Auchenorrhyncha

Homalodisca vitripennis (Germar) *(Cicadellidae)* – glassy-winged sharpshooter

Nilaparvata lugens (Stål) *(Delphacidae)* – brown planthopper
Philaenus spumarius (L.) *(Cercopidae)* – common froghopper or meadow spittlebug

Hemiptera: Sternorrhyncha: Aphidoidea

(Synonyms are in square brackets)

Acyrthosiphon caraganae (Cholodkovsky) *(Aphididae)*
Acyrthosiphon ignotum Mordvilko *(Aphididae)*
Acyrthosiphon kondoi Shinji *(Aphididae)* – blue alfalfa aphid
[*Acyrthosiphon nipponicum* (Essig & Kuwana)] = *Neoaulacorthum nipponicum* (Essig & Kuwana) *(Aphididae)*
Acyrthosiphon pisum (Harris) *(Aphididae)* – pea aphid
Adelges cooleyi (Gillette) *(Adelgidae)* – Cooley spruce gall adelgid
Adelges laricis Vallot *(Adelgidae)* – larch adelgid
Adelges nordmannianae (Eckstein) *(Adelgidae)*
[*Adelges nusslini* (Boerner)] = *Adelges nordmannianae* (Eckstein) *(Adelgidae)*
Adelges piceae (Ratzeburg) *(Adelgidae)* – balsam woolly adelgid
Adelges tsugae Annand *(Adelgidae)*– hemlock woolly adelgid
Aphis carduella Walsh *(Aphididae)*
[*Aphis cirsiiacanthoidis* Boerner] = *Aphis fabae cirsiiacanthoidis* Scopoli *(Aphididae)*
Aphis craccivora Koch *(Aphididae)* – cowpea aphid or groundnut aphid
Aphis cytisorum cytisorum Hartig *(Aphididae)*
Aphis fabae Scopoli *(Aphididae)* – black bean aphid
Aphis fabae cirsiiacanthoidis Scopoli *(Aphididae)*
Aphis farinosa J.F.Gmelin *(Aphididae)*
Aphis glycines Matsumura *(Aphididae)* – soybean aphid
Aphis gossypii Glover *(Aphididae)* – cotton aphid or melon aphid
Aphis hederae Kaltenbach *(Aphididae)*
[*Aphis helianthi* Monell] = *Aphis carduella* Walsh *(Aphididae)*
Aphis jacobaeae Schrank *(Aphididae)*
[*Aphis laburni* Kaltenbach] = *Aphis cytisorum cytisorum* Hartig *(Aphididae)*
Aphis nerii Boyer de Fonscolombe *(Aphididae)* – oleander aphid
Aphis pomi De Geer *(Aphididae)* – green apple aphid
Aphis punicae Passerini *(Aphididae)*

Aphis sambuci L. *(Aphididae)* – elder aphid

Aphis spiraecola Patch *(Aphididae)* – spiraea aphid or green citrus aphid

Aphis spiraephaga F.P. Müller *(Aphididae)*

Aphis spiraephila Patch *(Aphididae)*

Aphis urticata J.F. Gmelin *(Aphididae)*

[*Aulacorthum magnoliae* (Essig & Kuwana)] = *Neoaulacorthum magnoliae* (Essig & Kuwana) *(Aphididae)*

Aulacorthum solani (Kaltenbach) *(Aphididae)* – glasshouse potato aphid or foxglove aphid

Betulaphis brevipilosa Boerner *(Aphididae)*

Betulaphis quadrituberculata (Kaltenbach) *(Aphididae)*

Brachycaudus helichrysi (Kaltenbach) *(Aphididae)* – leaf-curling plum aphid

Brachycaudus persicae (Passerini) *(Aphididae)* – black peach aphid

Brachycaudus prunicola (Kaltenbach) *(Aphididae)*

Brachycaudus tragopogonis (Kaltenbach) *(Aphididae)*

Brevicoryne brassicae (L.) *(Aphididae)* – cabbage aphid

Callipterinella calliptera (Hartig) *(Aphididae)*

Capitophorus elaeagni (Del Guercio) *(Aphididae)*

Cavariella konoi Takahashi *(Aphididae)*

Ceratovacuna lanigera Zehntner *(Aphididae)* – sugar cane woolly aphid

Cervaphis quercus Takahashi *(Aphididae)*

Chaetosiphon fragaefolii (Cockerell) *(Aphididae)* – strawberry aphid

Chaitophorus capreae (Mosley) *(Aphididae)*

Chaitophorus leucomelas Koch *(Aphididae)*

[*Chaitophorus versicolor* Koch] = *Chaitophorus leucomelas* Koch *(Aphididae)*

Chromaphis juglandicola (Kaltenbach) *(Aphididae)* – walnut aphid

Cinara palaestinensis Hille Ris Lambers *(Aphididae)*

Delphiniobium junackianum (Karsch) *(Aphididae)*

Diuraphis noxia (Kurdjumov) *(Aphididae)* – Russian wheat aphid

Drepanosiphum platanoidis (Schrank) *(Aphididae)* – sycamore aphid

Dysaphis crataegi (Kaltenbach) *(Aphididae)* – hawthorn–parsnip aphid

Dysaphis devecta (Walker) *(Aphididae)* – rosy leaf-curling aphid

Dysaphis plantaginea (Passerini) *(Aphididae)* – rosy apple aphid

Elatobium abietinum (Walker) *(Aphididae)* – spruce aphid

Eriosoma lanigerum (Hausmann) *(Aphididae)* – woolly apple aphid

Eucallipterus tiliae (L.) *(Aphididae)* – lime aphid

Euceraphis betulae (Koch) *(Aphididae)*

Euceraphis punctipennis (Zetterstedt) *(Aphididae)*

Hyalopterus pruni (Geoffroy) *(Aphididae)* – mealy plum aphid

Hyperomyzus carduellinus (Theobald) *(Aphididae)*

Hyperomyzus lactucae (L.) *(Aphididae)* – blackcurrant–sowthistle aphid

Laingia psammae Theobald *(Aphididae)*

Liosomaphis berberidis (Kaltenbach) *(Aphididae)*

Lipaphis pseudobrassicae (Davis) *(Aphididae)* – turnip aphid or mustard aphid

[*Longiunguis donacis* (Passerini)] = *Melanaphis donacis* (Passerini) *(Aphididae)*

Macrosiphoniella artemisiae (Boyer de Fonscolombe) *(Aphididae)*

Macrosiphoniella sanborni (Gillette) *(Aphididae)* – chrysanthemum aphid

Macrosiphum albifrons Essig *(Aphididae)* – lupin aphid

Macrosiphum euphorbiae (Thomas) *(Aphididae)* – potato aphid

[*Macrosiphum ibarae* (Matsumura)] = *Sitobion ibarae* (Matsumura) *(Aphididae)*

Macrosiphum rosae (L.) *(Aphididae)* – rose aphid

Megoura viciae Buckton *(Aphididae)* – vetch aphid

Melanaphis donacis (Passerini) *(Aphididae)*

Metopolophium dirhodum (Walker) *(Aphididae)* – rose–grain aphid.

Metopolophium festucae (Theobald) *(Aphididae)* – fescue aphid

Microlophium carnosum (Buckton) *(Aphididae)*

Microsiphoniella artemisiae (Gillette) *(Aphididae)*

Mindarus abietinus Koch *(Aphididae)*

Myzocallis boerneri Stroyan *(Aphididae)*

Myzocallis carpini (Koch) *(Aphididae)*

Myzocallis castanicola Baker *(Aphididae)*

Myzocallis coryli (Goetze) *(Aphididae)* – hazel aphid or filbert aphid

Myzus cerasi (F.) *(Aphididae)* – cherry blackfly

Myzus persicae (Sulzer) *(Aphididae)* – peach–potato aphid

Myzus persicae nicotianae Blackman *(Aphididae)*

Neoaulacorthum magnoliae (Essig & Kuwana) *(Aphididae)*

Neoaulacorthum nipponicum (Essig & Kuwana) *(Aphididae)*

Neomyzus circumflexus (Buckton) *(Aphididae)*

Neophyllaphis podocarpi Takahashi *(Aphididae)*

Periphyllus californiensis (Shinji) *(Aphididae)*

Periphyllus lyropictus (Kessler) *(Aphididae)*

Periphyllus testudinaceus (Fernie) *(Aphididae)*

Phorodon humuli (Schrank) *(Aphididae)* – damson–hop aphid

Phyllaphis fagi (L.) *(Aphididae)*

Phylloxera glabra (von Heyden) *(Phylloxeridae)*

Pineus pini (Macquart) *(Adelgidae)*

Pseudoregma alexanderi (Takahashi) *(Aphididae)*

Pseudoregma bambucicola (Takahashi) *(Aphididae)*

Pterocallis alni (deGeer) *(Aphididae)*

Rhopalosiphum maidis (Fitch) *(Aphididae)* – corn leaf aphid

Rhopalosiphum padi (L.) *(Aphididae)* – bird cherry–oat aphid

Schizaphis graminum (Rondani) *(Aphididae)* – greenbug

Schizolachnus pineti (F.) *(Aphididae)*

Schizolachnus piniradiatae (Davidson) *(Aphididae)*

Sitobion akebiae (Shinji) *(Aphididae)*

Sitobion avenae (F.) *(Aphididae)* – grain aphid

Sitobion ibarae (Matsumura) *(Aphididae)*

Symydobius oblongus (von Heyden) *(Aphididae)*

Thelaxes dryophila (Schrank) *(Aphididae)*

[*Therioaphis maculata* (Buckton)] = *Therioaphis trifolii* (Monell) *(Aphididae)* – spotted alfalfa aphid

Therioaphis trifolii (Monell) *(Aphididae)* – spotted alfalfa aphid

Toxoptera aurantii (Boyer de Fonscolombe) *(Aphididae)* – black citrus aphid or tea aphid

Toxoptera citricidus (Kirkaldy) *(Aphididae)* – brown citrus aphid

[*Toxoptera graminum* (Rondani)] = *Schizaphis graminum* (Rondani) *(Aphididae)* – greenbug

Tuberculatus annulatus (Hartig) *(Aphididae)*

Tuberolachnus salignus (J.F. Gmelin) *(Aphididae)* – willow aphid

Uroleucon aeneum (Hille Ris Lambers) *(Aphididae)*

Uroleucon ambrosiae (Thomas) *(Aphididae)*

Uroleucon cichorii (Koch) *(Aphididae)*

Uroleucon cirsii (L.) *(Aphididae)*

Uroleucon compositae (Theobald) *(Aphididae)*

Uroleucon formosanum (Takahashi) *(Aphididae)*

Uroleucon jaceae (L.) *(Aphididae)*

Uroleucon nigrotuberculatum (Olive) *(Aphididae)*

Vesiculaphis caricis (Fullaway) *(Aphididae)*

Hemiptera: other Sternorrhyncha

Abgrallaspis cyanophylli (Signoret) *(Diaspididae)*

Acutaspis umbonifera (Newstead) *(Diaspididae)*

Agonoscena pistaciae Burckhardt et Lauterer *(Psyllidae)*

Aleurodicus cocois (Curtis) *(Aleyrodidae)*

Aleurodicus dispersus Russell *(Aleyrodidae)*– spiralling whitefly

Aleurotuba jelinekii (Frauenfeld) *(Aleyrodidae)*

Aleyrodes proletella (L.) *(Aleyrodidae)*

Aonidiella aurantii (Maskell) *(Diaspididae)* – California red scale

Aonidiella orientalis (Newstead) *(Diaspididae)*

Aonidimytilus albus (Cockerell) *(Diaspididae)* – cassava scale

Aspidiotus destructor Signoret *(Diaspididae)*– coconut scale

Aspidiotus nerii Bouché *(Diaspididae)*

Asterolecanium sp. *(Asterolecaniidae)*

Aulacaspis tegalensis (Zehntner) *(Diaspididae)* – sugar cane scale

Aulacaspis tubercularis Newstead *(Diaspididae)*

Bemisia tabaci (Gennadius) *(Aleyrodidae)*

[*Bemisia argentifolii* Bellows] = *Bemisia tabaci* Gennadius *(Aleyrodidae)*

Chionaspis alnus Kuwana *(Diaspididae)*

Chionaspis salicis (L.) *(Diaspididae)*

Chrysomphalus aonidum (L.) *(Diaspididae)* – Florida red scale

Chrysomphalus bifasciculatus Ferris *(Diaspididae)*

Coccus hesperidum L.*(Coccidae)* – soft brown scale

Coccus viridis (Green) *(Coccidae)* – green coffee scale or soft green scale

Dactylopius opuntiae (Cockerell) *(Dactylopiidae)*

Diaphorina citri Kuwayama *(Psyllidae)* – Asian citrus psyllid

Diaspidiotus perniciosus (Comstock) *(Diaspididae)*

Dysmicoccus *(Pseudococcidae)*

Eriococcus coriaceus Maskell *(Eriococcidae)*

Eulecanium caraganae Borchsenius *(Coccidae)*

Ferrisia virgata (Cockerell) *(Pseudococcidae)* – striped mealybug

Hemberlesia lataniae (Signoret) *(Diaspididae)*

Heteropsylla cubana Crawford *(Psyllidae)*

Icerya purchasi Maskell *(Monophlebidae)*– cottony cushion scale

Lepidosaphes beckii (Newman) *(Diaspididae)* – citrus mussel scale

Lepidosaphes cornutus Ramakrishna Ayyar *(Diaspididae)*

Lepidosaphes ulmi (L.) *(Diaspididae)* – mussel scale or oystershell scale

Maconellicoccus hirsutus (Green) *(Pseudococcidae)* – pink hibiscus mealybug

Matsucoccus feytaudi Ducasse *(Matsucoccidae)*
Matsucoccus josephi Bodenheimer & Harpaz
 (Matsucoccidae)
Matsucoccus matsumurae (Kuwana) *(Matsucoccidae)*
Melanaspis glomerata (Green) *(Diaspididae)*
Monophlebulus pilosior (Maskell) *(Monophlebidae)*
Orthezia urticae (L.) *(Ortheziidae)*
Parlatoria blanchardi (Tergioni Tozzetti) *(Diaspididae)*
Phenacoccus herreni Cox & Williams *(Pseudococcidae)*
Phenacoccus madeirensis Green *(Pseudococcidae)*
Phenacoccus manihoti Matile-Ferrero *(Pseudococcidae)*
 – cassava mealybug
Phoenicococcus marlatti (Cockerell) *(Phoenicococcidae)*
Pinnaspis buxi (Bouché) *(Diaspididae)*
Planococcus citri (Risso) *(Pseudococcidae)* – citrus
 mealybug
Planococcus minor (Maskell) *(Pseudococcidae)*
Pseudochermes fraxini (Kaltenbach) *(Eriococcidae)*
Pseudococcus cryptus Hempel *(Pseudococcidae)*
Pseudococcus maritimus (Ehrhorn) *(Pseudococcidae)*
Pseudococcus viburni (Signoret) *(Pseudococcidae)*
Psylla alni (L.) *(Psyllidae)*
Psylla jucunda Tuthill *(Psyllidae)*
Psylla mali (Schmidberger) *(Psyllidae)* – apple sucker
Psylla ulmi Forster *(Psyllidae)*
Psylla uncatoides (Ferris et Clyver) *(Psyllidae)*
Pulvinaria psidii Maskell *(Coccidae)* – green shield scale
Pulvinaria urbicola (Cockerell) *(Coccidae)*
Pulvinaria vitis (L.) *(Coccidae)* – woolly vine scale
Rastrococcus invadens Williams *(Pseudococcidae)*
Saissetia coffeae (Walker) *(Coccidae)* – helmet scale
Saissetia oleae (Olivier) *(Coccidae)* – olive scale
Siphoninus phillyreae (Haliday) *(Aleyrodidae)*
Trialeurodes vaporariorum (Westwood) *(Aleyrodidae)*
 – glasshouse whitefly
Unaspis citri (Comstock) *(Diaspididae)* – citrus snow scale
Unaspis euonymi (Comstock) *(Diaspididae)*
Unaspis yanonensis Kuwana *(Diaspididae)*

Thysanoptera

Thrips tabaci Lindeman *(Thripidae)* – onion thrips or
 tobacco thrips

Neuroptera

Chrysopa Leach *(Chrysopidae)*
Chrysopa oculata Say *(Chrysopidae)* – golden-eyed
 lacewing

Chrysopa perla (L.) *(Chrysopidae)*
Chrysopa sinica (Tjeder) *(Chrysopidae)*
Chrysoperla carnea (Stephens) *(Chrysopidae)* – green
 lacewing
Chrysoperla plorabunda Fitch *(Chrysopidae)*
Chrysoperla rufilabris (Burmeister) *(Chrysopidae)*

Lepidoptera

Acraea encedon (L.) *(Nymphalidae)*
[*Anagasta kuehniella* (Zeller)] = *Ephestia kuehniella*
 Zeller *(Pyralidae)*
Arctia *(Arctiidae)* – tiger moths
Choristoneura pinus Freeman *(Tortricidae)* – jack pine
 budworm
Danaus plexippus (L.) *(Nymphalidae)* – monarch
 butterfly
Ephestia kuehniella Zeller *(Pyralidae)* – Mediterranean
 flour moth
Galleria mellonella (L.) *(Pyralidae)* – wax moth
Helicoverpa armigera (Hübner) *(Noctuidae)* – cotton
 bollworm
Helicoverpa zea (Boddie) *(Noctuidae)* – corn earworm
Heliothis virescens (F.) *(Noctuidae)* – tobacco budworm
Hyphantria cunea (Drury) *(Arctiidae)* – fall webworm
Ostrinia nubilalis (Hübner) *(Crambidae)* – European
 corn borer
Pectinophora gossypiella (Saunders) *(Gelechiidae)* – pink
 bollworm
Phthorimaea operculella (Zeller) *(Gelechiidae)* – potato
 tuber moth
Pieris rapae (L.) *(Pieridae)* – small cabbage white
 butterfly
Plutella xylostella (L.) *(Plutellidae)* – diamondback
 moth
Sitotroga cerealella (Olivier) *(Gelechiidae)* – Angoumois
 grain moth
Spodoptera litura (F.) *(Noctuidae)* – common
 cutworm
Trichoplusia ni (Hübner) *(Noctuidae)* – cabbage looper
 or ni moth
Tyria *(Arctiidae)* – cinnabar moths

Diptera

Anopheles quadrimaculatus Say *(Culicidae)*
Aphidoletes aphidimyza (Rondani) *(Cecidomyiidae)*
Boettcheria latisterna Parker *(Sarcophagidae)*

Chetogena claripennis (Macquart) *(Tachinidae)*

Chrysotachina slossonae (Coquillett) *(Tachinidae)*

Contarinia nasturtii (Kieffer) *(Cecidomyiidae)* – swede midge

Cryptochaetum iceryae (Williston) *(Cryptochaetidae)*

Culex quinquefasciatus Say *(Culicidae)* – southern house mosquito

[*Degeeria luctuosa* (Meigen)] = *Medina luctuosa* (Meigen) *(Tachinidae)*

[*Doryphorophaga doryphorae*] = *Myiopharus doryphorae* (Riley) *(Tachinidae)*

Drosophila melanogaster Meigen *(Drosophilidae)*

Episyrphus balteatus (DeGeer) *(Syrphidae)* – marmalade hover fly

Euthelyconychia epilachnae (Aldrich) *(Tachinidae)*

[*Exoristoides slossonae* Coquillett] = *Chrysotachina slossonae* (Coquillett) *(Tachinidae)*

Helicobia rapax (Walker) *(Sarcophagidae)*

Lydinolydella metallica Townsend *(Tachinidae)*

[*Lypha slossonae* (Coquillett)] = *Chrysotachina slossonae* (Coquillett) *(Tachinidae)*

Medina collaris (Fallén) *(Tachinidae)*

Medina funebris (Meigen) *(Tachinidae)*

Medina luctuosa (Meigen) *(Tachinidae)*

Medina melania (Meigen) *(Tachinidae)*

Medina separata (Meigen) *(Tachinidae)*

Megaselia (Phoridae)

Myiopharus doryphorae (Riley) *(Tachinidae)*

[*Paradexodes epilachnae* Aldrich] = *Euthelyconychia epilachnae* (Aldrich) *(Tachinidae)*

Phalacrotophora Enderlein *(Phoridae)*

Phalacrotophora berolinensis Schmitz *(Phoridae)*

Phalacrotophora beuki Disney *(Phoridae)*

Phalacrotophora decimaculata Liu *(Phoridae)*

Phalacrotophora delageae Disney *(Phoridae)*

Phalacrotophora fasciata (Fallén) *(Phoridae)*

Phalacrotophora indiana Colyer *(Phoridae)*

Phalacrotophora nedae (Malloch) *(Phoridae)*

Phalacrotophora philaxyridis Disney *(Phoridae)*

Phalacrotophora quadrimaculata Schmitz *(Phoridae)*

Polich␣ta unicolor (Fallén) *(Tachinidae)*

Pseudebenia epilachnae Shima & Han *(Tachinidae)*

Ravinia errabunda (Wulp) *(Sarcophagidae)*

[*Sarcophaga helicis* Townsend] = *Helicobia rapax* (Walker) *(Sarcophagidae)*

[*Sarcophaga latisterna* (Parker)] = *Boettcheria latisterna* Parker *(Sarcophagidae)*

[*Sarcophaga reinhardii* Hall] = *Ravinia errabunda* (Wulp) *(Sarcophagidae)*

Strongygaster triangulifera (Loew) *(Tachinidae)*

Hymenoptera

[*Aminellus* Masi] = *Cowperia* Girault *(Encyrtidae)*

[*Aminellus sumatraensis* Kerrich] = *Cowperia sumatraensis* (Kerrich) *(Encyrtidae)*

Anagyrus Howard *(Encyrtidae)*

Anagyrus australiensis (Howard) *(Encyrtidae)*

Anagyrus kamali Moursi *(Encyrtidae)*

Anagyrus lopezi (De Santis) *(Encyrtidae)*

Anagyrus pseudococci (Girault) *(Encyrtidae)*

Anastatus Motschulsky *(Eupelmidae)*

[*Anisotylus* Timberlake] = *Homalotylus* Mayr *(Encyrtidae)*

Aphanogmus Thomson *(Ceraphronidae)*

Aphidius colemani Viereck *(Braconidae)*

Aphidius eadyi Starý *(Braconidae)*

Aphidius ervi Haliday *(Braconidae)*

Apis mellifera L. *(Apidae)* – European honey bee

Aprostocetus Westwood *(Eulophidae)*

Aprostocetus esurus (Riley) *(Eulophidae)*

Aprostocetus neglectus (Domenichini) *(Eulophidae))*

[*Atritomellus* Kieffer] = *Dendrocerus* (Megaspilidae)

Austroterobia Girault *(Pteromalidae)*

Aximopsis Ashmead *(Eurytomidae)*

Azteca instabilis (Smith) *(Formicidae)*

Baryscapus Foerster *(Eulophidae)*

Baryscapus thanasimi (Ashmead) *(Eulophidae)*

Brachymeria carinatifrons Gahan *(Chalcididae)*

Catolaccus Thomson *(Pteromalidae)*

Centistes scymni Ferrière *(Braconidae)*

Centistes subsulcatus (Thomson) *(Braconidae)*

Centistina nipponicus (Belokobylskij) *(Braconidae)*

Cerchysiella Girault [*Zeteticontus* Silvestri] *(Encyrtidae)*

Chartocerus subaeneus (Foerster) *(Signiphoridae)*

Cheiloneurus carinatus Compere *(Encyrtidae)*

Cheiloneurus cyanonotus Waterston *(Encyrtidae)*

Cheiloneurus liorhipnusi (Risbec) *(Encyrtidae)*

Cheiloneurus orbitalis Compere *(Encyrtidae)*

Chrysocharis johnsoni Subba Rao *(Eulophidae)*

Chrysonotomyia appannai (Chandy Kurian) *(Eulophidae)*

Coccidoctonus trinidadensis Crawford *(Encyrtidae)*

Conura Spinola *(Chalcididae)*

Conura paranensis (Schrottky) *(Chalcididae)*

Conura petioliventris (Cameron) *(Chalcididae)*

Conura porteri (Brèthes) *(Chalcididae)*

Cowperia Girault *(Encyrtidae)*

Cowperia areolata (Walker) *(Encyrtidae)*

Cowperia indica (Kerrich) *(Encyrtidae)*

Cowperia punctata Girault *(Encyrtidae)*

Cowperia subnigra Li *(Encyrtidae)*

Cowperia sumatraensis (Kerrich) *(Encyrtidae)*

Crematogaster Lund *(Formicidae)*

Crematogaster lineolata (Say) *(Formicidae)*

Dendrocerus (Megaspilidae)

Dendrocerus ergensis (Guesquiere) *(Megaspilidae)*

Dibrachys cavus (Walker) *(Pteromalidae)*

Dinocampus Foerster *(Braconidae)*

Dinocampus coccinellae (Schrank) *(Braconidae)*

[*Dinocampus nipponicus* Belokobylskij] = *Centistina nipponicus* (Belokobylskij) *(Braconidae)*

[*Dinocampus terminatus* (Nees)] = *Dinocampus coccinellae* (Schrank) *(Braconidae)*

Dolichoderus bidens L. *(Formicidae)*

[*Echthroplectis* Foerster] = *Homalotylus* Mayr *(Encyrtidae)*

Elasmus Westwood *(Eulophidae)*

Encarsia sophia (Girault & Dodd) *(Aphelinidae)*

[*Epidinocarsus lopezi* (De Santis)] = *Anagyrus lopezi* (De Santis) *(Encyrtidae)*

Eretmocerus mundus Mercet *(Aphelinidae)*

Eupelmus Dalman *(Eupelmidae)*

Eupelmus urozonus Dalman *(Eupelmidae)*

Eupelmus vermai (Bhatnagar) *(Eupelmidae)*

[*Eupteromalus* Kurdjumov] = *Trichomalopsis* Crawford *(Pteromalidae)*

Formica L. *(Formicidae)*

Formica obscuripes Forel *(Formicidae)*

Formica polyctena Foerster *(Formicidae)* – European red wood ant

Formica rufa L. *(Formicidae)* – wood ant

Formica rufibarbis F. *(Formicidae)* – red-barbed ant

Gelis Thunberg *(Ichneumonidae)*

Gelis agilis (F.) *(Ichneumonidae)*

[*Gelis instabilis* (Förster)] = *Gelis agilis* (F.) *(Ichneumonidae)*

Gelis melanocephalus (Schrank) *(Ichneumonidae)*

[*Hemaenasioidea* Girault] = *Homalotylus* Mayr *(Encyrtidae)*

Homalotyloidea dahlbomii (Westwood) *(Encyrtidae)*

Homalotylus Mayr *(Encyrtidae)*

Homalotylus affinis Timberlake *(Encyrtidae)*

Homalotylus africanus Timberlake *(Encyrtidae)*

Homalotylus agarwali Anis & Hayat *(Encyrtidae)*

Homalotylus albiclavatus (Agarwal) *(Encyrtidae)*

Homalotylus albifrons (Ishii) *(Encyrtidae)*

Homalotylus albitarsus Gahan *(Encyrtidae)*

Homalotylus aligarhensis (Shafee & Rizvi) *(Encyrtidae)*

Homalotylus balchanensis Myartseva *(Encyrtidae)*

Homalotylus brevicauda Timberlake *(Encyrtidae)*

[*Homalotylus californicus* (Say)] = *Homalotylus terminalis* (Say) *(Encyrtidae)*

Homalotylus cockerelli Timberlake *(Encyrtidae)*

Homalotylus ephippium (Ruschka) *(Encyrtidae)*

Homalotylus eytelweinii (Ratzeburg) *(Encyrtidae)*

Homalotylus ferrierei Hayat, Alam & Agarwal *(Encyrtidae)*

Homalotylus flaminius Dalman *(Encyrtidae)*

Homalotylus hemipterinus (De Stefani) *(Encyrtidae)*

Homalotylus formosus Anis & Hayat *(Encyrtidae)*

Homalotylus himalayensis Liao *(Encyrtidae)*

Homalotylus hybridus Hoffer *(Encyrtidae)*

Homalotylus hyperaspicola Tachikawa *(Encyrtidae)*

Homalotylus hyperaspidis Timberlake *(Encyrtidae)*

Homalotylus hypnos Noyes

Homalotylus indicus (Agarwal) *(Encyrtidae)*

Homalotylus latipes Girault *(Encyrtidae)*

Homalotylus longicaudus Xu & He *(Encyrtidae)*

Homalotylus longipedicellus (Shafee & Fatma) *(Encyrtidae)*

Homalotylus mexicanus Timberlake *(Encyrtidae)*

Homalotylus mirabilis (Brèthes) *(Encyrtidae)*

Homalotylus mundus Gahan *(Encyrtidae)*

Homalotylus nigricornis Mercet *(Encyrtidae)*

Homalotylus oculatus (Girault) *(Encyrtidae)*

Homalotylus pallentipes (Timberlake) *(Encyrtidae)*

Homalotylus platynaspidis Hoffer *(Encyrtidae)*

Homalotylus punctifrons Timberlake *(Encyrtidae)*

Homalotylus quaylei Timberlake *(Encyrtidae)*

Homalotylus rubricatus Sharkov *(Encyrtidae)*

Homalotylus scutellaris Tan & Zhao *(Encyrtidae)*

Homalotylus scymnivorus Tachikawa *(Encyrtidae)*

Homalotylus shuvakhinae Trjapitzin & Triapitsyn *(Encyrtidae)*

Homalotylus similis Ashmead *(Encyrtidae)*

Homalotylus sinensis Xu & He *(Encyrtidae)*

Homalotylus singularis Hoffer *(Encyrtidae)*

Homalotylus terminalis (Say) *(Encyrtidae)*

Homalotylus trisubalbus Xu & He *(Encyrtidae)*

Homalotylus turkmenicus Myartseva *(Encyrtidae)*

Homalotylus vicinus Silvestri *(Encyrtidae)*

Homalotylus yunnanensis Tan & Zhao *(Encyrtidae)*

Homalotylus zhaoi Xu & He *(Encyrtidae)*

Inkaka quadridentata Girault *(Pteromalidae)*

Iridomyrmex Mayr *(Formicidae)*

Isodromus niger Ashmead *(Encyrtidae)*

Lasius claviger (Roger) *(Formicidae)*

Lasius japonicus Santschi *(Formicidae)* – Japanese ant

Lasius niger (L.) *(Formicidae)* – black ant or garden ant

Lasius umbratus (Nylander) *(Formicidae)*

[*Lepidaphycus* Blanchard] = *Homalotylus* Mayr *(Encyrtidae)*

Linepithema humile (Mayr) *(Formicidae)* – Argentine ant

[*Lygocerus* Foerster] = *Dendrocerus (Megaspilidae)*

Lysiphlebus fabarum (Marshall) *(Braconidae)*

Lysiphlebus testaceipes (Cresson) *(Braconidae)*

[*Mendozaniella* Brethes] = *Homalotylus* Mayr *(Encyrtidae)*

Merismoclea rojasi De Santis *(Pteromalidae)*

Mesopolobus Westwood *(Pteromalidae)*

Mesopolobus secundus (Crawford) *(Pteromalidae)*

[*Mestocharis lividus* Girault)] = *Pediobius foveolatus* (Crawford) *(Eulophidae)*

Metastenus Walker *(Pteromalidae)*

Metastenus caliginosus Szelényi *(Pteromalidae)*

Metastenus concinnus Walker *(Pteromalidae)*

Metastenus indicus Sureshan & Narendran *(Pteromalidae)*

Metastenus sulcatus (Dodd) *(Pteromalidae)*

Metastenus townsendi (Ashmead) *(Pteromalidae)*

Microctonus Wesmael *(Braconidae)*

Monomorium minimum (Buckley) *(Formicidae)* – little black ant

Myrmica ruginodis Nylander *(Formicidae)*

Myrmica rugulosa Nylander *(Formicidae)*

[*Neoaenasioidea* Agarwal] = *Homalotylus* Mayr *(Encyrtidae)*

[*Neotainania* Husain & Agarwal] = *Uga* Girault *(Chalcididae)*

[*Nobrimus* Thomson] = *Homalotylus* Mayr *(Encyrtidae)*

Nothoserphus Brues *(Proctotrupidae)*

Nothoserphus admirabilis Lin *(Proctotrupidae)*

Nothoserphus aequalis Townes *(Proctotrupidae)*

Nothoserphus afissae (Watanabe) *(Proctotrupidae)*

Nothoserphus boops (Thomson) *(Proctotrupidae)*

Nothoserphus debilis Townes *(Proctotrupidae)*

Nothoserphus epilachnae (Pschorn-Walcher) *(Proctotrupidae)*

Nothoserphus fuscipes Lin *(Proctotrupidae)*

Nothoserphus mirabilis Brues *(Proctotrupidae)*

Nothoserphus partitus Lin *(Proctotrupidae)*

Nothoserphus scymni (Ashmead) *(Proctotrupidae)*

Nothoserphus townesi Lin *(Proctotrupidae)*

Omphale Haliday *(Eulophidae)*

[*Omphale epilachni* Singh & Khan] = *Chrysonotomyia appannai* (Chandy Kurian) *(Eulophidae)*

Ooencyrtus azul Prinsloo *(Encyrtidae)*

Ooencyrtus bedfordi Prinsloo *(Encyrtidae)*

Ooencyrtus camerounensis (Risbec) *(Encyrtidae)*

Ooencyrtus distatus Prinsloo *(Encyrtidae)*

Ooencyrtus epilachnae Annecke *(Encyrtidae)*

[*Ooencyrtus epulus* Annecke] = *Ooencyrtus camerounensis* (Risbec) *(Encyrtidae)*

Ooencyrtus polyphagus (Risbec) *(Encyrtidae)*

Ooencyrtus puparum Prinsloo *(Encyrtidae)*

Ooencyrtus sinis Prinsloo *(Encyrtidae)*

Oomyzus Rondani *(Eulophidae)*

Oomyzus mashhoodi (Khan & Shafee) *(Eulophidae)*

Oomyzus scaposus (Thomson) *(Eulophidae)*

Oomyzus sempronius (Erdoes) *(Eulophidae)*

Ophelosia bifasciata Girault *(Pteromalidae)*

Oricoruna Boucek *(Pteromalidae)*

Ophelosia crawfordi Riley *(Pteromalidae)*

Oricoruna orientalis (Crawford) *(Pteromalidae)*

Pachyneuron Walker *(Pteromalidae)*

Pachyneuron albutius Walker *(Pteromalidae)*

Pachyneuron altiscuta Howard *(Pteromalidae)*

Pachyneuron chilocori Domenichini *(Pteromalidae)*

[*Pachyneuron concolor* Förster] = *Pachyneuron muscarum* (L.) *(Pteromalidae)*

Pachyneuron muscarum (L.) *(Pteromalidae)*

[*Pachyneuron siculum* Delucchi] = *Pachyneuron muscarum* (L.) *(Pteromalidae)*

Pachyneuron solitarium (Hartig) *(Pteromalidae)*

[*Pachyneuron syrphi* (Ashmead)] = *Pachyneuron albutius* Walker *(Pteromalidae)*

Parachrysocharis Girault *(Eulophidae)*

Paratrechina Motschulsky *(Formicidae)*

Pediobius Walker *(Eulophidae)*

Pediobius amaurocoelus (Waterston) *(Eulophidae)*

[*Pediobius epilachnae* (Rohwer)] = *Pediobius foveolatus* (Crawford) *(Eulophidae)*

Pediobius foveolatus (Crawford) *(Eulophidae)*

[*Pediobius mediopunctata* (Waterston)] = *Pediobius foveolatus* (Crawford) *(Eulophidae)*

Pediobius nishidai Hansson *(Eulophidae)*

[*Pediobius simiolus* (Takahashi)] = *Pediobius foveolatus* (Crawford) *(Eulophidae)*

[*Perilitus americanus* Riley] = *Dinocampus coccinellae* (Schrank) *(Braconidae)*

[*Perilitus coccinellae* (Schrank)] = *Dinocampus coccinellae* (Schrank) *(Braconidae)*

Perilitus rutilus (Nees) *(Braconidae)*

Perilitus stuardoi Porter *(Braconidae)*

[*Perilitus terminatus* (Nees)] = *Dinocampus coccinellae* (Schrank) *(Braconidae)*

Pheidole megacephala (F.) *(Formicidae)* – big-headed ant

Phygadeuon subfuscus Cresson *(Ichneumonidae)*

[*Pleurotropis* Foerster] = *Pediobius* Walker *(Eulophidae)*

Pnigalio agraules (Walker) *(Eulophidae)*

Praon volucre (Haliday *(Braconidae)*
Pristomyrmex pungens Mayr *(Formicidae)*
Prochiloneurus aegyptiacus (Mercet) *(Encyrtidae)*
Prochiloneurus nigriflagellum (Girault) *(Encyrtidae)*
Pseudocatolaccus Masi *(Pteromalidae)*
Quadrastichus ovulorum (Ferrière) *(Eulophidae)*
[*Scymnophagus* Ashmead] = *Metastenus* Walker
    (Pteromalidae)
[*Scymnophagus mesnili* Ferrière] = *Metastenus
    concinnus* Walker *(Pteromalidae)*
Sigmoepilachna indica Khan, Agnihotri & Sushil
    (Eulophidae)
Solenopsis invicta Buren *(Formicidae)* – red imported
    fire ant
[*Syntomosphyrus taprobanes* Waterston] = *Oomyzus
    scaposus* (Thomson) *(Eulophidae)*
Syrphoctonus tarsatorius (Panzer) *(Ichneumonidae)*
Tamarixia radiata (Waterston) *(Eulophidae)*
Tapinoma nigerrimum (Nylander) *(Formicidae)*
Tetramorium caespitum (L.) *(Formicidae)*
Tetrastichus Haliday *(Eulophidae)*
[*Tetrastichus coccinellae* Kurdjumov] = *Oomyzus
    scaposus* (Thomson) *(Eulophidae)*
Tetrastichus cydoniae Risbec *(Eulophidae)*
Tetrastichus decrescens Graham *(Eulophidae)*
Tetrastichus epilachnae (Giard) *(Eulophidae)*
[*Tetrastichus melanis* Burks] = *Oomyzus scaposus*
    (Thomson) *(Eulophidae)*
Tetrastichus orissaensis Husain & Khan *(Eulophidae)*
[*Tetrastichus sexmaculatus* Chandy Kurian] = *Oomyzus
    scaposus* (Thomson) *(Eulophidae)*
[*Thomsonina* Hellen] = *Nothoserphus* Brues
    (Proctotrupidae)
Trichogramma Westwood *(Trichogrammatidae)*
Trichogramma evanescens Westwood
    (Trichogrammatidae)
Trichomalopsis Crawford *(Pteromalidae)*
Trichomalopsis acuminata (Graham) *(Pteromalidae)*
Trichomalopsis dubia (Ashmead) *(Pteromalidae)*
[*Tripolycystus* Dodd] = *Metastenus* Walker
    (Pteromalidae)
Uga Girault [*Neotainania* Husain & Agarwal]
    (Chalcididae)
Uga colliscutellum (Girault) *(Chalcididae)*
Uga coriacea Kerrich *(Chalcididae)*
Uga digitata Qian & He *(Chalcididae)*
Uga hemicarinata Qian & Li *(Chalcididae)*
Uga javanica Kerrich *(Chalcididae)*
Uga menoni Kerrich *(Chalcididae)*
Uga sinensis Kerrich *(Chalcididae)*
Wasmannia auropunctata (Roger) *(Formicidae)*

[*Watanabeia* Masner] = *Nothoserphus* Brues
    (Proctotrupidae)

Coleoptera (other than Coccinellidae)

Agelastica coerulea (Baly) *(Chrysomelidae)* – alder leaf
    beetle
Agonum dorsale (Pontoppidan) *(Carabidae)*
Chrysomela populi L. *(Chrysomelidae)* – poplar leaf
    beetle
Chrysophtharta bimaculata (Olivier) *(Chrysomelidae)*
    – eucalyptus leaf beetle
Diabrotica virgifera LeConte *(Chrysomelidae)* – western
    corn rootworm
Dicranolaius bellulus (Guérin-Méneville) *(Melyridae)*
Dytiscus sp. *(Dytiscidae)*
Galeruca interrupta arminiaca Weise *(Chrysomelidae)*
Galerucella lineola (F.) *(Chrysomelidae)*
Galerucella pusilla Duftschmid *(Chrysomelidae)*
    – golden loosestrife beetle
Galerucella sagittariae Gyllenhal *(Chrysomelidae)*
    – cloudberry beetle
Harpalus pennsylvanicus (De Geer) *(Carabidae)*
Hypera postica (Gyllenhal) *(Curculionidae)* – alfalfa
    weevil
Laricobius nigrinus Fender *(Derodontidae)*
Leptinotarsa decemlineata (Say) *(Chrysomelidae)* –
    Colorado potato beetle
[*Melasoma populi* (L.)] = *Chrysomela populi* L.
    (Chrysomelidae)
Melolontha sp. *(Scarabaeidae)*
Plagiodera versicolora (Laicharting) *(Chrysomelidae)*
Platynus dorsalis (Pontopiddan) *(Carabidae)*
Pterostichus melanarius (Illiger) *(Carabidae)*
[*Pyrrhalta luteola* (Muller)] = *Xanthogaleruca luteola*
    (Muller) *(Chrysomelidae)*
Sitona discoideus Gyllenhal *(Curculionidae)*
Stenotarsus rotundus Arrow *(Endomychidae)*
Tachyporus sp. *(Staphylinidae)*
Xanthogaleruca luteola (Muller) *(Chrysomelidae)* – elm
    leaf beetle

PARASITIC MITES

Coccipolipus Husband *(Podapolipidae)*
Coccipolipus africanae Husband *(Podapolipidae)*
Coccipolipus arturi Haitlinger *(Podapolipidae)*
Coccipolipus benoiti Husband *(Podapolipidae)*
Coccipolipus bifasciatae Husband *(Podapolipidae)*

Coccipolipus cacti Husband *(Podapolipidae)*
Coccipolipus camerouni Husband *(Podapolipidae)*
Coccipolipus chilocori Husband *(Podapolipidae)*
Coccipolipus cooremani Husband *(Podapolipidae)*
Coccipolipus epilachnae Smiley *(Podapolipidae)*
Coccipolipus hippodamiae (McDaniel & Morrill) *(Podapolipidae)*
Coccipolipus macfarlanei Husband *(Podapolipidae)*
Coccipolipus micraspisi Husband *(Podapolipidae)*
Coccipolipus oconnori Husband *(Podapolipidae)*
Coccipolipus solanophilae (Cooreman) *(Podapolipidae)*
Hemisarcoptes Lignieres *(Hemisarcoptidae)*
Hemisarcoptes cooremani Thomas *(Hemisarcoptidae)*
Leptus ignotus (Oudemans) *(Erythraeidae)*

ARACHNIDA (other than parasitic mites)

Agistemus longisetus Gonzalez *(Stigmaeidae)*
Amblyseius andersoni (Chant) *(Phytoseidae)*
Amblyseius fallacis (Garman) *(Phytoseidae)*
Amphitetranychus viennensis (Zacher) *(Tetranychidae)* – hawthorn spider mite or fruit tree spider mite
Araneus diadematus (Clerck) *(Araneidae)*
Bryobia praetiosa Koch *(Tetranychidae)* – clover mite
Bryobia rubrioculus (Scheuten) *(Tetranychidae)* – brown mite or brown apple mite
Cheiracanthium *(Miturgidae)*
Clubiona reclusa O.P.- Cambridge *(Clubionidae)*
[*Metatetranychus ulmi* (Koch)] = *Panonychus ulmi* (Koch) *(Tetranychidae)*
Misumenops tricuspidatus (F.) *(Thomisidae)*
Panonychus mori Yokovama *(Tetranychidae)*
Panonychus ulmi (Koch) *(Tetranychidae)* – red spider mite or European red mite
Tetranychus evansi Baker & Pritchard *(Tetranychidae)*
Tetranychus lintearius Dufour *(Tetranychidae)* – gorse spider mite
Tetranychus mcdanieli McGregor *(Tetranychidae)*
[*Tetranychus telarius* (L.)] = *Tetranychus urticae* Koch *(Tetranychidae)*
Tetranychus urticae Koch *(Tetranychidae)* – two-spotted spider mite
Thanatus Koch *(Philodromidae)*

NEMATODES

[*Coccinellimermis* Rubtzov] = *Hexamermis* Steiner *(Mermithidae)*

Hexamermis Steiner *(Mermithidae)*
Howardula Cobb *(Allantonematidae)*
Mermis Dujardin *(Mermithidae)*
Parasitilenchus coccinellinae Iperti & van Waerebeke *(Allantonematidae)*

MOLLUSCA

Arion ater (L.) *(Arionidae)* – European black slug
Arion hortensis (Férussac) *(Arionidae)* – garden slug
Deroceras reticulatum (Müller) *(Limacidae)* – grey field slug or grey garden slug
Tandonia budapestensis (Hazay) *(Milacidae)* – Budapest slug

AVES

Acrocephalus schoenobaenus (L.) *(Sylviidae)* – sedge warbler
Alauda arvensis L. *(Alaudidae)* – sky lark
Anthus campestris (L.) *(Motacillidae)* – tawny pipit
Anthus pratensis (L.) *(Motacillidae)* – meadow pipit
Anthus trivialis (L.) *(Motacillidae)* – tree pipit
Apus apus (L.) *(Apodidaae)* – common swift
Calandrella cinerea (Gmelin) *(Alaudidae)* – red-capped lark
Coturnix coturnix L. *(Phasianidae)* – quail
Coturnix japonicus Temminck & Schlegel *(Phasianidae)* – Japanese quail
Cuculus canorus L. *(Cuculidae)* – common cuckoo
Cyanistes caeruleus (L.) *(Paridae)* – blue tit
Delichon urbica (L.) *(Hirundinidae)* – house martin
Dendrocopos medius (L.) *(Picidae)* – middle spotted woodpecker
Ficedula parva (Bechstein) *(Muscicapidae)* – red-headed flycatcher
Hippolais icterina (Vieillot) *(Sylviidae)* – icterine warbler
Hirundo rustica L. *(Hirundinidae)* – barn swallow
Luscinia luscinia (L.) *(Muscicapidae)* – thrush nightingale
Luscinia svecica (L.) *(Muscicapidae)* – bluethroat
Melanocorypha calandra (L.) *(Alaudidae)* – Calandra lark
Motacilla flava L. *(Motacillidae)* – blue-headed wagtail
Muscicapa striata (Pallas) *(Muscicapidae)* – spotted flycatcher
Oenanthe oenanthe (L.) *(Muscicapidae)* – northern wheatear

Parus major L. *(Paridae)*
Passer domesticus (L.) *(Passeridae)* – house sparrow
Passer montanus (L.) *(Passeridae)* – tree sparrow
Phoenicurus ochruros (Gmelin) *(Muscicapidae)* – black redstart
Phoenicurus phoenicurus (L.) *(Muscicapidae)* – common redstart
Phylloscopus collybita (Vieillot) *(Phylloscopidae)* – common chiffchaff
Phylloscopus trochilus (L.) *(Phylloscopidae)* – willow warbler
Saxicola rubicola (L.) *(Muscicapidae)* – common stonechat
Sitta europaea L. *(Sittidae)* – wood nuthatch
Sturnus vulgaris L. *(Sturnidae)* – common starling
Sylvia atricapilla (L.) *(Sylviidae)* – blackcap
Sylvia borin (Boddaert) *(Sylviidae)* – garden warbler
Sylvia communis Latham *(Sylviidae)* – common whitethroat
Sylvia curruca (L.) *(Sylviidae)* – lesser whitethroat
Sylvia nisoria (Bechstein) *(Sylviidae)* – barred warbler
Turdus philomelos (Turton) *(Turdidae)* – song thrush
Turdus merula L. *(Turdidae)* – Eurasian blackbird

FUNGI

Alternaria Nees *(Pleosporaceae)*
Beauveria bassiana (Balsamo) Vuillemin *(Cordycipitaceae)*
Cladosporium Link *(Davidiellaceae)*
Erisyphe polygoni DC. *(Erisyphaceae)* – powdery mildew
Erysiphe cichoracearum DC *(Erisyphaceae)*
Hesperomyces Thaxter *(Laboulbeniaceae)*
Hesperomyces chilomenis (Thaxter) Thaxter *(Laboulbeniaceae)*
Hesperomyces coccinelloides (Thaxter) Thaxter *(Laboulbeniaceae)*
Hesperomyces hyperaspidis Thaxter *(Laboulbeniaceae)*
Hesperomyces virescens Thaxter *(Laboulbeniaceae)*
Isaria farinosa (Holmskiold ex S. F. Gray) Fries *(Cordycipitaceae)*
Isaria fumosorosea Wize *(Cordycipitaceae)*
Lecanicillium lecanii (Zimmerman) Zare & Gams *(Cordycipitaceae)*
[*Lecanicillium longisporum* (Zimmerman) Zare & Gams – as 'Vertalec'] = *Lecanicillium lecanii* (Zimmerman) Zare & Gams *(Cordycipitaceae)*
Metarhizium anisopliae (Metschnikoff) Sorokin *(Clavicipitaceae)*

Microsphaera alphitoides Griffon & Maublanc *(Erisyphaceae)*
Microsphaera pulchra Cooke & Peck *(Erisyphaceae)*
Neotyphodium lolii Latch *et al.* *(Clavicipitaceae)*
Nosema coccinellae Lipa *(Nosematidae)*
Nosema epilachnae Brooks, Hazard & Becnel *(Nosematidae)*
Nosema henosepilachnae Toguebaye & Marchand *(Nosematidae)*
Nosema hippodamiae Lipa & Steinhaus *(Nosematidae)*
Nosema tracheophila Cali & Briggs *(Nosematidae)*
Nosema varivestis Brooks, Hazard & Becnel *(Nosematidae)*
Tubulinosema hippodamiae Bjornson, Le, Saito & Wang *(Nosematidae)*
Oidium monilioides (Nees) Link *(Erisyphaceae)*
[*Paecilomyces farinosus* (Holmskiold ex S. F. Gray) Brown & Smith] = *Isaria farinosa* (Holmskiold ex S. F. Gray) Fries *(Cordycipitaceae)*
[*Paecilomyces fumosoroseus* (Wize) Brown & Smith] = *Isaria fumosorosea* Wize *(Cordycipitaceae)*
Pandora neoaphidis (Remaudière & Hennebert) Humber *(Entomophthoraceae)*
Phyllactinia moricola (Hennebert) Homma *(Erisyphaceae)*
Podosphaera leucotricha (Ellis & Everhart) E.S. Salmon *(Erisyphaceae)* – apple powdery mildew
Podosphaera tridactyla (Wallroth) de Bary *(Erisyphaceae)* – plum powdery mildew
Puccinia Pers. *(Pucciniaceae)*
Saccharomyces fragilis Kudrjawzew *(Saccharomycetaceae)*
Sphaerotheca castagnei Léverrier *(Erisyphaceae)* – hop powdery mildew
Sphaerotheca cucurbitae (Jaczewski) Z.Y. Zhao *(Erisyphaceae)*
Sphaerotheca pannosa (Wallroth) de Bary *(Erisyphaceae)* – rose powdery mildew
[*Verticillium lecanii* (Zimmerman) Viegas] = *Lecanicillium lecanii* (Zimmerman) Zare & Gams *(Cordycipitaceae)*

PROTISTS

Anisolobus indicus Haldar, Ray & Bose *(Gregarinidae)*
Brustiospora indicola Kundu & Haldar *(Brustiophoridae)*
Gregarina Dufour *(Gregarinidae)*
Gregarina barbarara Watson *(Gregarinidae)*

Gregarina californica Lipa *(Gregarinidae)*
Gregarina chilocori Obata *(Gregarinidae)*
Gregarina coccinellae Lipa *(Gregarinidae)*
Gregarina dasguptai Mandal, Rai, Pranhan, Gurung, Sharma, Rai & Mandal *(Gregarinidae)*
Gregarina fragilis Watson *(Gregarinidae)*
Gregarina hyashii Sengupta & Haldar *(Gregarinidae)*
Gregarina katherina Watson *(Gregarinidae)*
Gregarina ruszkowskii Lipa *(Gregarinidae)*
Gregarina straeleni Theodorides & Jolivet *(Gregarinidae)*

BACTERIA

Alcaligenes paradoxus Davis *(Alcaligenaceae)*
Bacillus thuringiensis Berliner *(Bacillaceae)*
Flavobacterium Bergey *(Flavobacteriaceae)*
Rickettsia da Rocha-Lima *(Rickettsiaceae)*
Spiroplasma Saglio, Lhospital, Lafleche, Dupont, Bove, Tully & Freundt *(Spiroplasmataceae)*
Wolbachia Hertig *(Rickettsiaceae)*

PLANTS

Abies balsamea (L.) Miller *(Pinaceae)* – balsam fir
Abutilon theophrasti Medikus *(Malvaceae)* – Chinese jute or Indian Mallow
Acacia Miller *(Fabaceae)*
Acalypha ostryaefolia Riddell *(Euphorbiaceae)*
Aconitum L. *(Ranunculaceae)*
Aframomum melegueta (Roscoe) K. Schum *(Zingiberaceae)* – grains of paradise or melegueta pepper
Agropyron desertorum (Link) Schultes *(Poaceae)* – crested wheatgrass
Alnus japonica (Thunberg) Steudel *(Betulaceae)*
Amaranthus hybridus L. *(Amaranthaceae)*
Ambrosia artemisiifolia L. *(Asteraceae)* – hogbrake
Anethum graveolens L. *(Apiaceae)* – dill
Artemisia tridentata Nuttall *(Asteraceae)* – sagebrush
Artemisia vulgaris L. *(Asteraceae)* – wormwood
Atriplex sagittata Borkhausen *(Chenopodiaceae)*
Benthamidia florida (L.) Spach *(Cornaceae)*
Berberis vulgaris L. *(Berberidaceae)* – common barberry
Beta vulgaris L. *(Chenopodiaceae)* – sugar beet and garden beet

Betula populifolia Marshall *(Betulaceae)*
Brassica campestris L. *(Brassicaceae)*
Brassica napus L. *(Brassicaceae)* – rape
Brassica napus L. subsp. *oleifera* DC Metzger *(Brassicaceae)* – oil seed rape
Brassica nigra (L.) Koch *(Brassicaceae)* – black (or brown) mustard
Brassica oleracea L. Italica group *(Brassicaceae)* – broccoli
Brassica oleracea L. *(Brassicaceae)*
Calotropis procera (Aiton) Aiton *(Apocynaceae)*
Caltha palustris L. *(Ranunculaceae)*
Cannabis sativa L. *(Cannabaceae)* – Indian hemp
Carduus crispus L. *(Asteraceae)*
Carpinus caroliniana Walter *(Betulaceae)* – American hickory
Caulophyllum robustum Maximowicz *(Berberidaceae)*
Centaurea jacea L. *(Asteraceae)*
Centrosema pubescens Bentham *(Fabaceae)*
Chamerion angustifolium (L.) Holub *(Onagraceae)* – fireweed
Chenopodium L. *(Chenopodiaceae)*
Cirsium arvense (L.) Scopoli *(Asteraceae)* – creeping thistle
Cirsium kagamontanum Nakai *(Asteraceae)*
Cirsium kamtschaticum De Candolle *(Asteraceae)*
Citrus sinensis (L.) Osbeck *(Rutaceae)* – sweet orange
Clematis L. *(Ranunculaceae)*
Coriandrum sativum L. *(Apiaceae)* – coriander
Cosmos *(Asteraceae)*
Cotoneaster integerrima Medikus *(Rosaceae)*
Cotoneaster tomentosus Lindley *(Rosaceae)*
Crotalaria striata De Candolle *(Fabaceae)*
Cucurbita maxima Duchesne *(Cucurbitaceae)* – squash or pumpkin
Cymbopogon citratus (De Candolle) Stapf *(Poaceae)* – lemon grass
Dendrocalamus giganteus Munro *(Poaceae)* – giant bamboo
Elytrigia repens (L.) Nevski *(Poaceae)* – couch grass
Erythrina corallodendron L. *(Fabaceae)*
Euonymus japonicus Thunberg *(Celastraceae)*
Euphorbia L. *(Euphorbiaceae)*
Eurotia ceratoides Meyer *(Chenopodiaceae)*
Genista L. *(Fabaceae)*
Glochidion ferdinandi (Müller Argiovensis) Bailey *(Euphorbiaceae)*
Helianthus annuus L. *(Asteraceae)* – sunflower
Hibiscus L. *(Malvaceae)*
Juglans cinerea L. *(Juglandaceae)* – butternut

Juniperus virginiana L. *(Cupressaceae)* – red cedar or Virginian pencil cedar

Lapsana communis L. *(Asteraceae)* – nipplewort

Leontopodium alpinum Cassini *(Asteraceae)* – edelweiss

Leucaena Bentham *(Fabaceae)*

Ligustrum L. *(Oleaceae)*

Lolium multiflorum Lamarck *(Poaceae)* – Italian rye grass

Lolium perenne L. *(Poaceae)* – rye grass

Lonicera periclymenum L. *(Caprifoliaceae)* – honeysuckle

Lupinus luteus L. *(Fabaceae)* – yellow lupin

Lupinus mutabilis Sweet *(Fabaceae)*

Malus pumila Miller *(Rosaceae)* – apple

Manihot esculenta Crantz *(Euphorbiaceae)*

Manikara zapota (L.) P. Royen *(Sapotaceae)* – sapodilla

Medicago sativa L. *(Fabaceae)* – alfalfa (or often lucerne in UK)

Mercurialis annua L. *(Euphorbiaceae)*– annual mercury

Momordica charantia L. *(Cucurbitaceae)* – bitter melon

Morus australis Poir. *(Moraceae)*

Nerium oleander L. *(Apocynaceae)* – oleander

Nitraria L. *(Nitrariaceae)*

Onoclea sensibilis L. *(Onocleaceae)*

Oryzopsis hymenoides (Roemer & Schultes) Piper *(Poaceae)* – Indian ricegrass

Paederia foetida L. *(Rutaceae)*

Pastinaca sativa L. *(Apiaceae)* – parsnip

Phaseolus vulgaris L. *(Fabaceae)* – French bean or haricot bean

Physalis alkekengi L. *(Solanaceae)* – Chinese lantern

Picea schrenkiana Fischer & Meyer *(Pinaceae)*

Pinus armandii Franchet *(Pinaceae)* – Chinese white pine

Pinus sylvestris L. *(Pinaceae)* –Scots pine

Pisonia L. *(Nyctaginaceae)*

Pisum sativum L. *(Fabaceae)* – garden pea

Pittosporum tobira Aiton *(Pittosporaceae)*

Populus L. *(Salicaceae)*

Prunus avium (L.) L. *(Rosaceae)* – gean or wild cherry

Prunus cerasus L. *(Rosaceae)* – morello cherry or amarelle cherry

Prunus persica (L.) *(Rosaceae)* – peach or nectarine Batsch

Pterostyrax hispidus Siebold & Zuccarini *(Styracaceae)*

Pyracantha coccinea M.H. Roemer *(Rosaceae)* – firethorn

Quercus rubra L. *(Fagaceae)* – red oak

Raphanus sativus L. *(Brassicaceae)* – radish

Robinia pseudoacacia L. *(Fabaceae)*

Rosa multiflora Thunberg *(Rosaceae)*

Rubus occidentalis L. (Rosaceae)

Salix L. *(Salicaceae)*

Sambucus nigra L. *(Caprifoliaceae)* – elder

Sambucus racemosa L. subsp. *sieboldiana* (Miquel) H. Hara *(Caprifoliaceae)*

Schizopepon bryoniaefolius Maximowicz *(Cucurbitaceae)*

Senecio jacobaea L. *(Asteraceae)* – ragwort

Senecio vulgaris L. *(Asteraceae)* – groundsel

Sinapis alba L. *(Brassicaceae)* – white (or yellow) mustard

Solanum japonense Nakai *(Solanaceae)*

Solanum nigrum L. *(Solanaceae)* – black nightshade

Solanum tuberosum L.*(Solanaceae)* – potato

Solidago canadensis L. *(Asteraceae)*

Spartium junceum L. *(Fabaceae)* – Spanish broom

Spiraea (Rosaceae)

Styrax L. *(Styracaceae)*

Symphoricarpos rivularis Sucksdorf *(Caprifoliaceae)*

Talinum triangulare (Jacquin) Willdenow *(Portulacaceae)*

Tamarix L. *(Tamaricaceae)*

Tanacetum vulgare L. *(Asteraceae)* – tansy

Taraxacum officinale Weber *(Asteraceae)* – dandelion

Trichosanthes kirilowii Maximowicz *(Cucurbitaceae)*

Tripleurospermum maritimum (L.) Koch *(Asteraceae)*

Triticum L. *(Poaceae)* – wheat

Tsuga Carrière *(Pinaceae)*

Typha latifolia L. *(Typhaceae)* – bullrush

Ulmus (Ulmaceae)

Urtica dioica L. *(Urticaceae)* – stinging nettle

Verbascum thapsus L. *(Scrophulariaceae)* – Aaron's rod

Vicia faba L. *(Fabaceae)* – broad bean or faba bean

Vicia sativa L. *(Fabaceae)* – vetch or tare

Vigna unguiculata (L.) Walpers subsp. *cylindrica* (L.) Verdcourt *(Fabaceae)*

Withania somnifera (L.) Dunal *(Solanaceae)*

Zea mays L. *(Poaceae)* – maize (corn in USA)

Zingiber officinale Roscoe *(Zingiberaceae)* – ginger

Chapter 1

PHYLOGENY AND CLASSIFICATION

Oldrich Nedvěd[1] and Ivo Kovář[2]

[1] Faculty of Science, University of South Bohemia & Institute of Entomology, Academy of Sciences, CZ 37005 České Budějovice, Czech Republic

[2] Emer. Scientist of the National Museum, Prague; Current address: Zichovec, Czech Republic

1.1 POSITION OF THE FAMILY

1.1.1 The Cerylonid complex

The family Coccinellidae includes approximately 6000 described species in some 360 genera and 42 tribes. Coccinellids belong to the superfamily **Cucujoidea** of the Coleoptera suborder Polyphaga, and the family is a member of the phylogenetic branch frequently referred to as the **Cerylonid complex** or series of families, which is composed of Alexiidae, Cerylonidae, Coccinellidae, Corylophidae, Discolomatidae, Endomychidae (*s. lat.* including Mychotheninae, Eidoreinae and Merophysiinae) and Latridiidae (Crowson 1955, Lawrence & Newton 1995). Bothrideridae were added later (Pal & Lawrence 1986). Monophyly of the Cerylonid series was based on morphological characters (Slipinski & Pakaluk 1992) and confirmed by parsimony analysis of molecular data (Hunt et al. 2007, Robertson et al. 2008).

Phylogenetic relations between the families and subfamilies included in the Cerylonid complex are rather complicated and not fully resolved (Slipinski & Pakaluk 1992). Early morphological studies supported a clade Endomychidae plus Corylophidae as the sister group of Coccinellidae (Crowson 1955, Sasaji 1971a). Affinities have been proposed between Endomychidae and Coccinellidae (Pakaluk & Slipinski 1990; Burakowski & Slipinski 2000) due to the presence of a characteristic basal (= median) lobe of male genitalia, pseudotrimerous tarsi and absence of coronal suture on the head of larvae (except in Epilachninae); Eupsylobiinae (Endomychidae) and Coccinellidae (Pakaluk & Slipinski 1990) due to the long penis and coccinellid-like tegmen; Mycetaeinae (Endomychidae) and Coccinellidae (number of abdominal spiracles, open middle coxal cavity, Kovář 1996; distinct pronotum with sublateral carina, hidden mesotrochantin, Tomaszewska 2000); Alexiidae (= Sphaerosomatidae) and Coccinellidae (Slipinski & Pakaluk 1992); Endomychidae plus Alexiidae and Coccinellidae (Pakaluk and Slipinski 1990); or between Corylophidae and Coccinellidae (Sasaji 1971a; anterior tentoria separated, frontoclypeal suture absent, common type of antenna and tarsus, Tomaszewska 2005).

1.1.2 Sister families

Tomaszewska (2000) argued monophyly of the Endomychidae plus Coccinellidae from the common characters of procoxal cavity externally open, tarsal formula 4-4-4, abdomen with five pairs of functional spiracles and median lobe without additional struts. However, the polyphyletic nature of the family Endomychidae has been known for a long time (Slipinski & Pakaluk 1992) and was confirmed with molecular analyses by Robertson et al. (2008) that grouped the subfamily Anamorphinae with the Corylophidae, and the rest with the Coccinellidae. Later analysis by Tomaszewska (2005), based on adult and larval characters, failed to confirm a sister relationship of the Endomychidae and Coccinellidae.

The comprehensive **molecular study** of beetles by Hunt et al. (2007) placed the family Alexiidae or the pair of Alexiidae plus Anamorphinae (Endomychidae) as the sister group of the Coccinellidae. Molecular analysis of the Cerylonid complex by Robertson et al. (2008) proposed the subfamily Leiestinae (Endomychidae) or the complex of Endomychidae (part) plus (Corylophidae plus Anamorphinae (Endomychidae)) as sister groups of the Coccinellidae.

1.1.3 Feeding habits

In either case, the sister group of Coccinellidae, and their common ancestor, were probably **fungivorous** (Giorgi et al. 2009). Feeding on hemipterans (mainly Sternorrhyncha) has evolved predominantly in coleopteran lineages that contain fungus feeders (Derodontidae, Silvanidae, Laemophloeidae, Nitidulidae, Endomychidae, Anthribidae) and whose ancestors were fungus feeders (Coccinellidae) or sap feeders (Scarabaeidae: Cetoniinae) (Leschen 2000).

1.1.4 Monophyly of Coccinellidae

The **monophyly** of the family Coccinellidae was based on morphological synapomorphies, and has repeatedly been supported by molecular phylogenetic analyses (Hunt et al. 2007, Robertson et al. 2008, Giorgi et al. 2009), even early ones (Howland & Hewitt 1995) that did not place the family in the Cucujiformia. The first to recognize the morphological synapomorphy (although the term did not exist at that time) of the family, the **siphonal structure of penis**, was Verhoeff (1895).

1.2 CHARACTERISTICS OF THE FAMILY

The family Coccinellidae may be distinguished from the rest of the cerylonid complex by the combination of the following **adult characters**: (i) five pairs of abdominal spiracles, (ii) tentorial bridge absent, (iii) anterior tentorial branches separated, (iv) frontoclypeal suture absent, (v) apical segment of maxillary palps never needle-like, (vi) galea and lacinia separated, (vii) mandible with reduced mola, (viii) procoxal cavities open posteriorly, (ix) middle coxal cavities open outwardly, (x) meta-epimeron parallel sided, (xi) femoral lines present on abdominal sternite 2, (xii) tarsal formula 4-4-4 or 3-3-3, tarsal segment 2 usually strongly dilated below, (xiii) male genitalia with tubular curved sipho, distally embraced by the tegmen. **Larvae** are armed with setae and setose processes (three pairs on each abdominal segment), antennae of one to three segments, not over three times as long as wide, frontoclypeal suture absent (except some Epilachnini). Pupae of coccinellids are of the type **pupa adectica obtecta** – all appendages are glued to the body by exuvial fluid. The pupa is attached to the substrate by the tip of the abdomen. Coccidophagy is a synapomorphy for the family, other feeding habits, including mycophagy, are secondary (Giorgi et al. 2009).

Liere and Perfecto (2008) suggested that the evolution of coccinellid morphology was influenced by relationships with ants. The result of a parallel evolution in several subfamilies is a mosaic of various grades of characters and/or apparent similarities, which mosaic may considerably obscure the true phylogenetic relationships within the family. One of these characters may be the enlarged eye canthus, formerly considered an apomorphy of the Chilocorinae, but which probably evolved independently in the Platynaspidini (Slipinski 2007). Mouthpart morphology was used in determining diet and host specificity in Coccinellidae (Klausnitzer 1993, Klausnitzer & Klausnitzer 1997), but it is also constrained phylogenetically (Samways et al. 1997). Major larval types were defined by LeSage (1991). Pupal morphology was used for phylogenetic purposes separately from other characters (Phuoc & Stehr 1974).

1.3 CHANGES IN THE CLASSIFICATION OF SUBFAMILIES

1.3.1 Morphologically based classifications

Six subfamilies of Coccinellidae were recognized by Sasaji (1968, 1971b): Sticholotidinae, Coccidulinae, Scymninae + Chilocorinae, Coccinellinae + Epilachninae. However, his knowledge of non-Japanese taxa was limited. Korscheffsky (1931) classified the genus *Lithophilus* in a separate subfamily on account of the tetramerous structure of tarsus. Klausnitzer (1969, 1970, 1971) distinguished this subfamily (as Tetrabrachinae) from its sister group Coccidulinae based mainly on six sclerites on the pronotum of larvae as apomorphy, while he considered adult tetrameric tarsi as a plesiomorphy. Kovář (1973) added this seventh subfamily (as Lithophilinae) to Sasaji's phylogenetic tree, but later reassigned it to tribal level within the Coccidulinae and proposed another subfamily, the Ortaliinae (Kovář 1996). Chazeau et al. (1990) retained Sasaji's six subfamilies with slightly rearranged tribes.

The **monophyly** of several of the six subfamilies proposed by Sasaji has been disputed. The higher classification of Coccinellidae appears to suffer from the presence of para- and polyphyletic taxa (Vandenberg 2002). This problem is most conspicuous in the formerly basal subfamilies Sticholotidinae and Coccidulinae, and in the fauna of poorly studied regions.

1.3.2 Split of Sticholotidinae

In morphology based studies, Duverger (2003), Slipinski and Tomaszewska (2005) and Vandenberg and Perez-Gelabert (2007) proposed dividing up the highly diverse subfamily **Sticholotidinae**. In *Sticholotis* and allies, the terminal segment of the maxillary palp has a long obliquely oriented distal sensory surface on one side of the tapered apex, similar to the large sensory surface of the securiform (axe-shaped) palps of derived 'true' ladybirds; while in *Sukunahikona* and allies the sensory surface is small, oval and positioned distally, suggesting a basal position for this group. The subfamily Sticholotidinae also formed an unresolved polytomy in cladistic analysis by Yu (1994).

The subfamily Sticholotidinae was established by Sasaji (1968) with four tribes, whilst Sticholotidinae *sensu* Kovář (1996) contained 10 tribes. However, the phylogenetic relations among the constituent tribes remained obscure for a long time, in spite of an attempt at clarification by Gordon (1977). These Sticholotidinae *s. lat.* were characterized (Kovář 1996), among other traits, by the shape of the apical maxillary palp segment which differs from the rest of Coccinellidae (narrowed apically, conical or barrel shaped). However, this applies to the recently re-established **Microweiseinae** (*sensu* Slipinski 2007), while in the Sticholotidinae *s. str.* the last segment is more or less enlarged, approaching the typical securiform shape found in the other Coccinellidae.

These two lineages differ substantially, such as in the shape of the metendosternite, presence or absence of anterolateral carinae of the pronotum, and also in characters of both male and female genitalia. Microweiseinae retained their basal position, while Sticholotidinae *s. str.* appeared among more derived subfamilies. Slipinski (2007) was even stricter, dividing the entire family into only two subfamilies, Microweiseinae and Coccinellinae that contained, in his approach, all other usually separate subfamilies as tribes.

1.3.3   Monophyly of other subfamilies

The traditional **Coccidulinae** constitutes another polyphyletic group which suffers from a paucity of serious global study (Pope 1988, Vandenberg 2002). Coccidulinae was a paraphyletic group in respect to the Scymninae + Chilocorinae in cladistic analysis by Yu (1994) based on adult characters. The genera *Bucolus* and *Cryptolaemus* were variously classified in the past as either Coccidulinae or Scymninae. Pope (1988) proposed combining the tribes of Coccidulinae and Scymninae into a single subfamily. On the other side, Chinese specialists (Pang et al. 2004, Ren et al. 2009) distinguished a higher number of subfamilies: Sticholotinae (*s. lat.*), Scymninae, Ortaliinae, Hyperaspinae, Aspidimerinae, Chilocorinae, Coccidulinae, Lithophilinae, Coccinellinae and Epilachninae.

1.3.3.1   Contribution of immature stages

Savoiskaya and Klausnitzer (1973) regarded the **larval** armature of the **thorax and abdomen** as of considerable taxonomical significance. Savoiskaya (1969) proposed the two tribes Tytthaspidini and Bulaeini as

independent, based mainly on morphology of larvae, while the tribe Coccinellini was considered polyphyletic (Savoiskaya & Klausnitzer 1973) only because different genera of this tribe have variable abdominal structures. Cladistic analysis based solely on larval characters would indicate not only paraphyletic and polyphyletic Chilocorinae, Coccidulinae and Scymninae, but also a generally unacceptable topology of the tree (Yu 1994).

1.3.4   Molecular analyses

In their molecular cladistic study, based on cytochrome oxidase I (COI) and internal transcribed spacers (ITS1), Cihakova and Nedvěd (unpublished) found that the subfamily Coccinellinae is monophyletic with the Halyziini at the base. The three subfamilies Chilocorinae, Scymninae and Coccidulinae formed a second closely related group, with the Chilocorini and Platynaspidini not grouped together. COI discriminated well at the intergeneric, interspecific and intraspecific levels (Palenko et al. 2004). On the other hand, Schulenburg et al. (2001) found the extremely long and variable ITS1 sequence to be unsuitable for phylogenetic reconstruction at the subfamily level. A small total **genome size** (0.19–0.99 pg DNA) is typical for many coccinellids, excluding *Exochomus* (Chilocorinae; 1.71 pg) (Gregory et al. 2003).

Although molecular studies usually prove stable for studying higher taxa (e.g. families of Coleoptera, Hunt et al. 2007), the relationships they suggest for lower levels (subfamilies and tribes) are sometimes dubious because of low sample sizes and the gene used. When only four coccinellid genera (*Adalia*, *Coccinella*, *Calvia* and *Exochomus*) were analysed, different outgroups and different tree-searching algorithms yielded strikingly different topologies (Mawdsley 2001). A cladogram based on combining 16S and 12S mitochondrial rDNA, used for determining genetic distances of 8 coccinellid species (Tinsley & Majerus 2007), gave ambiguous results. Even in larger studies, unexpected clades are reported, with only Coccinellinae remaining monophyletic (Hunt et al. 2007). The subfamily Chilocorinae is polyphyletic in the study by Robertson et al. (2008), because the Platynaspidini were grouped with the Hyperaspidini (Scymninae). In the same study, the Scymninae is polyphyletic, the tribe Ortaliini were grouped with the Epilachninae and the Stethorini with the Coccinellinae.

1.3.4.1 Alternative molecular methods

Zhang et al. (1999) used isoenzyme analysis to reveal the relationship between five species of Coccinellidae and Zhang and Zheng (2002) used RAPD analysis to reveal the relationship between six species of Coccinellinae, both with reliable results. Phylogenetic distance was also measured indirectly through the transmission efficiency of male-killing bacteria, which has been found to be perfect within the genus from which the bacterium was isolated, but was lower in less related species of coccinellids (Tinsley & Majerus 2007).

1.3.5 Rejection of the monophyly of subfamilies

A comprehensive molecular phylogenetic analysis of the family is still in progress, but the preliminary results (Giorgi et al. 2009) rejected the monophyly of four of the six subfamilies proposed by Sasaji. The **Sticholotidinae** were split into two unrelated groups, where the Sukunahikonini, Microweiseini and Serangiini formed a basal clade, while the Sticholotidini grouped together with the Exoplectrini and lay higher up the tree. The subfamily **Chilocorinae** might be limited to the Chilocorini + Telsimiini, while the Platynaspidini tend to group with the Scymninae. The **Scymninae** may also include the **Ortaliinae** and parts of the Coccidulinae; however, the genera *Bucolus* and *Cryptolaemus* should fall into the Coccidulinae *s. str.* The **Coccidulinae**, besides being added near or to the Scymninae, lose the Exoplectrini, which group either with the Sticholotidinae *s. str.* or the Chilocorinae *s. str.* The monophyly of **Epilachninae** was not rejected, but the sample was too small to support it. The **Coccinellinae** appeared as the only stable monophyletic subfamily. Both Coccinellinae and Epilachninae were monophyletic in a smaller study with 16 species and one gene (COI) (Fu & Zhang 2006). Genera *Epilachna* and *Henosepilachna* were mixed together; tribe Halyziini was embedded deeply in Coccinellini.

Similarly, in another recent molecular phylogenetic reconstruction (Magro et al. 2010) of the family, Coccinellinae remained the single subfamily supported as monophyletic. Although there was a relatively good sample of genes (six, both mitochondrial and nuclear, both ribosomal and protein), this study suffers from low taxon sampling, except for the subfamily Coccinel-linae. Monophyly of other subfamilies cannot be either rejected or supported. However, a noteworthy finding was the placement of the tribes Tytthaspidini and Halyziini well inside the tribe Coccinellini. The concept of these three tribes should be re-examined.

1.4 CHARACTERISTICS OF THE SUBFAMILIES AND TRIBES

1.4.1 Proposed classification

Based on the above suggestions for reclassifying the higher taxonomy of the family and on molecular evidence, we propose a preliminary classification (see Fig. 1.1) of Coccinellidae with nine subfamilies: (Microweiseinae, (Coccinellinae, (Epilachninae, (Sticholotidinae, Exoplectrinae, Chilocorinae, (Scymninae, Coccidulinae), Ortaliinae)))).

1.4.2 Microweiseinae

The basal subfamily – **Microweiseinae** – is composed of the tribes **Carinodulini, Sukunahikonini, Microweiseini** and **Serangiini** which, except for an apomorphic much simplified mandible (with single apical tooth and no mola), share many ancestral characters as are seen on the metendosternite with a broad and very short furcate stalk bearing slender anterolateral arms. Male tegmen asymmetrical with no or only slightly differentiated basal lobe. Maxillary palps with long slender apical segment with small sensoric area. Larvae densely granulate bearing single minute seta on each granule, without defence gland openings.

1.4.3 Sticholotidinae

Narrowly defined **Sticholotidinae** *s. str.* is composed of the tribes Shirozuellini (syn. Ghanini), Cephalo-scymnini, **Plotinini, Limnichopharini, Sticholotidini** and **Argentipilosini**. It is rather heterogeneous and displays both primitive and derived characters. Compactly articulated antenna with well-developed spindle–shaped 1–4 segmented club bearing concentration of short sensory setae on mesal surface of last antennomere. Maxillary palps geniculate, terminally pointed. Larvae broadly fusiform with lateral setose processes, finely granulate and densely pubescent,

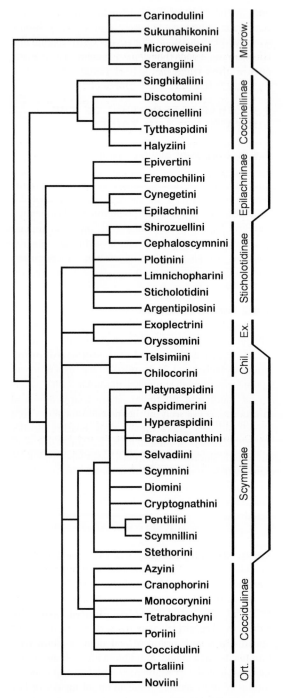

Figure 1.1 Proposed phylogenetic relationships between subfamilies and tribes compared with the classification by Kovář (1996) in right bars. Ort., Ortaliinae; Chil., Chilocorinae; Ex., Exoplectrinae; Microw., Microweiseinae. Tree drawn by Phy.fi online utility (Fredslund 2006).

abdominal segments 1–8 with defence gland openings. Pupa with urogomphi.

Among the tribes included in the Sticholotidinae *s. str.* an oriental tribe, the **Shirozuellini,** and the **Cephaloscymnini** from the New World are close to each other and more primitive than the rest in having an only slightly derived prosternum and a six-segmented abdomen (although in the Cephaloscymnini the last segment is reduced). However, the head capsule is largely derived in both, somewhat resembling that of the Chilocorinae. The remaining four tribes share a derived five-segmented abdomen.

1.4.4   Coccinellinae

A great number of characters are more or less universal and shared by particular tribes of the subfamily **Coccinellinae**. However, it seems that the **female genital plate** of the 'handle and blade' type is the only true **synapomorphy** of the Coccinellinae. Other characters may be shared with diverse tribes of other subfamilies. Mandible with double apical tooth and molar tooth. Maxillary palp is securiform. Abdomen with six ventrites. Colour pattern may be very striking and aposematic. Eggs are spindle shaped, laid in clusters in upright position, similarly to Epilachninae, while they are laid singly on their side in other subfamilies. Larvae are elongate fusiform with variable setose processes, often with conspicuous colour pattern.

The Neotropical tribe **Discotomini** combines the normal carnivorous type of mouthparts, dorsally shifted antennal insertions as usual in the Epilachninae, and the serrate type of antennae. The oriental **Singhikaliini** is a single tribe within the Coccinellinae represented by pubescent species. The peculiar colour pattern of *Singhikalia* resembles that of the tribe Noviini (Ortalinae). Fuersch (1990) moved the Singhikaliini into the Coccidulinae. Jadwiszczak (1990) supported their original position in the Coccinellinae as proposed by Miyatake (1972).

The tribes **Coccinellini, Tytthaspidini** (syn. Bulaeini) and **Halyziini** (syn. Psylloborini) form a major, species-rich part of the subfamily Coccinellinae, containing the species referred to as 'ladybirds' or 'ladybugs' in the narrow sense.

Although only the tribes Coccinellini and Halyziini are usually considered distinct, the third (Tytthaspidini) should also be treated as a distinct tribe because of the parallel development of certain important characters, especially the **frons** and **pronotum**. Both plesiomorphic and apomorphic types of frons occur here in parallel. Thus, the apomorphic frons type is homoplastic between the Tytthaspidini and the other tribes.

The plesiomorphic type of pronotum is quadrate with anterior corners which are never pointed, and anterior margin slightly emarginate, not or partly concealing the head; this character state is shared by certain, usually elongate members of all three tribes. However, the apomorphic state of a trapezoidal pronotum which is strongly emarginate anteriorly is predominant in the Halyziini and less frequently observed in the Coccinellini and Tytthaspidini.

In tribe **Halyziini** the eye canthus does not divide the eye (plesiomorphy), with the eye facets sometimes coarse, and a simple flagellum on the median lobe of the male genitalia. The mandible, with several small teeth arranged in a row, is an adaptation to mycetophagy. A similar mandible and a non-typical feeding habit occur also in some genera of the **Tytthaspidini** (*Tytthaspis*, *Bulaea*, *Isora*). The peculiar **simple type of male genitalia** in the above genera and the common type of **colour pattern** are striking. Due to the position of Tytthaspidini and Halyziini well inside the tribe Coccinellini in some molecular analyses (Magro et al. 2010) and similarity of some genera of tribe Coccinellini in their morphology and life history, the concept of these tribes must be widened (see Appendix: List of Genera). Differentiation of several parallel lineages within the numerous genera of the tribe **Coccinellini** may be expected to be identified in the future.

1.4.5 Epilachninae

The subfamily **Epilachninae** is characterized by synapomorphic features in the organization of the **mouthparts** in both adults and larvae, adapted to their phytophagous habit. The mandible possesses multidentate terebra but lacks a basal tooth (mola). The galea of the maxilla is large, round to transversely oval, and the mentum converges anteriorly. Maxillary palp of adults strongly securiform. Antennae are inserted in a depression on inner side of eye. Eggs are laid in clusters in upright position, similarly to Coccinellinae, while they are laid singly on their side in other subfamilies. **Larvae** are oval, armed with dense, **spinose senti** but without gland openings. Head with epicranial stem and frontal arms V-shaped.

The Epilachninae include four tribes, i.e. Epivertini, Epilachnini, Cynegetini and Eremochilini. The main part of the subfamily is formed by two tribes (Epilachnini and Cynegetini), both rich in species although the latter tribe is limited mostly to the Neotropical and Afrotropical regions. The two tribes differ mainly in the presence (in **Cynegetini**, syn. Madaini) or absence (in **Epilachnini**) of the anterior fovea of the elytral epipleura. They both have apomorphic positioning of the antennal insertions, which are dorsal and placed at or behind the level of the anterior margins of the eyes.

The tribe **Epivertini**, represented by the single Chinese species *Epiverta chelonia*, displays the plesiomorphic position of the antennal insertions, although the Cassidinae-like shape of the adult body is highly derived. The Neotropical tribe **Eremochilini** is characterized by cylindrical body form and quite hypognathous head with opistognathous mandibles and without labrum.

1.4.6 Exoplectrinae

For a long time the tribe Exoplectrini was considered a true member of the subfamily Coccidulinae. However, Gordon (1994) grouped it with the tribe Oryssomini and established the subfamily **Exoplectrinae**. Molecular analyses place the tribe Exoplectrini either near the Sticholotidinae *s. str.* or near the Chilocorinae *s. str.*, but never near any part of Coccidulinae. This leads us to treat this taxon on the subfamily level. However, morphological characterization of the subfamily is difficult. The adults share a finely expanded, anteriorly emarginate clypeus with the Azyini (Coccidulinae). Although the structures associated with the insertions of various movable parts of the body in the Exoplectrinae are poorly developed, they are very striking in the Azyini. Apomorphic states in the **Exoplectrini** opposite to Azyini are also evident in the antenna (with large lobate scape) and wide apical segment of the maxillary palp. All steps in the development of coccinellid eyes may be seen within genera of the Exoplectrinae. Some genera possess a compact and asymmetrical antennal club. Abdomen has five ventrites. The Exoplectrinae tribe has a Gondwanan origin. They are suspected to be not only predators but also plant feeders (Gordon 1985).

1.4.7 Chilocorinae

In the phylogeny of the three tribes until now placed in the subfamily **Chilocorinae**, Sasaji (1968) proposed the sequence from primitive to advanced forms from **Telsimiini** through **Platynaspidini** to **Chilocorini**. They all show a complete fusion of the clypeus with the eye canthus, together with distinctly ventral insertions of the antennae. The two former tribes do not occur in the New World. Chilocorini occur in the Old World, and one homogeneous line of Chilocorini genera has developed independently in the New World, the monophyly of which is suggested by the synapomorphic absence of a basal marginal line on the pronotum, a character shared by all the American species. Larvae of the **Telsimiini** are covered with waxy exudations like those of the Scymnini. Adults are pubescent. Adult **Chilocorini** are often smooth, and the **larvae** possess **very long setose processes** for mechanical defence. They are mostly coccidophagous, and the mandibles have a single apex.

The **Platynaspidini** share with the Chilocorinae *s. str.* (i.e. Telsimiini and Chilocorini) the **enlarged eye canthus**, dividing the eye and covering the antennal insertion. Molecular analysis, however, groups them with the Scymninae. Adult Platynaspidini are covered by long hairs; the **larvae** are very wide, have **short pubescence** and lack long setose projections. They are aphidophagous and myrmecophilous.

1.4.8   Ortaliinae

Two tribes (Noviini and Ortaliini), that combine some characters of the Coccidulinae and Scymninae and some Coccinellinae and even Epilachninae, have been placed in the separate subfamily **Ortaliinae** (Kovář 1996). Their body is robust, oval to rounded, discontinuous, of medium to large size. Pubescence is simple and short. The eyes are large to strikingly enlarged, prominent anteroventrally at the sides, with the eye facets small to minute. The antennal insertions are placed between the eyes. The apex of the mandible is bifid; a basal tooth is present. The pronotum is trapezoidal, emarginate anteriorly with the anterior corners widely rounded and the posterior corners not pointed. The elytral epipleura are not foveolate for reception of legs, and rather broad. The colour pattern, if present, is simple, not strongly aposematic, resembling that of certain Epilachninae.

In the tribes **Noviini** and **Ortaliini** there are some noteworthy common trends: enlarged eyes, shortened antenna and shortened elytral epipleuron, broadened

genital plate and spermatheca in the female, along with a certain peculiarity in larval organization: the lateral setose projections are nipple-shaped, and there are abundant clavate setae at the apex of the tibiotarsus. **Larvae** are covered by **poor waxy secretions**. The **Noviini** differ from the **Ortaliini** in the derived state of the antenna (7–8 segments, enlarged scape), the vertical basisternal lobes of the prosternum, and in the externally angulate legs and trimerous tarsi. *Ortalia* feeds on Psylloidea, larvae of some species are myrmecophagous.

1.4.9   Coccidulinae

The subfamily **Coccidulinae** has been considered to have the most primitive body organization of the Coccinellidae without the basal Microweiseinae. Six tribes were placed in the Coccidulinae by Chazeau et al. (1989) in their systematic survey: Tetrabrachyni, Coccidulini, Sumniini, Exoplectrini, Noviini, and Azyini. Fuersch (1990) added the Monocorynini and Singhikaliini (now Coccinellinae) to the Coccidulinae. Gordon (1974, 1994) excluded the Oryssomini from the Cranophorini and placed near the Exoplectrini, while Fuersch (1990) transferred the Oryssomini directly from the Sticholotidinae to the Coccinellinae. The tribe **Noviini** has been a **permanent member** of the **Coccidulinae** in many authoritative works since Sasaji (1968). To make Coccidulinae a natural group, some taxonomic changes in the position of several tribes are necessary. A combination of morphological and molecular studies suggests the following tribes as probably true members of the subfamily Coccidulinae: Tetrabrachini, Monocorynini, Coccidulini, Cranophorini, Poriini and Azyini. All tribes are of Gondwanan origin.

The body organization of Coccidulinae is rather simple and contains a set of plesiomorphic character states: the **antennae** are **long**, 10–11 segmented, with the basal flagellomeres slender and the club more or less striking, always with a well-developed apical segment (distinction from Scymninae); the pronotum is commonly quadrate with the anterior corners broadly rounded and the anterior margin slightly emarginate, so that in several tribes the head is partly covered by it; the posterior corners are more or less pointed. The subfamily may be characterized by a slightly convex, more or less elongate, moderately discontinuous body shape, well developed and sometimes double pubescence and elytral punctation. Eye facets are coarse. **Larvae** are fusiform, dorsal and lateral surfaces with diverse **setose processes**, many covered by at least sparse **waxy secretions**.

In the **Cranophorini** the pronotum is corylophid-like, where the head is completely hidden under the anterior lobe of the pronotum. Basal depressions on the pronotum are present only in the **Monocorynini** and they have also a large, quite compact antennal club with some resemblance to that of ants. The epipleuron is usually broad and entire without distinct foveae, but in the Azyini the epipleuron is modified, with the anterior fovea strikingly margined on the outer side even if the fovea is shallow.

The hypothesis that the **Tetrabrachini** is a distinct tribe rests solely on their unusual adaptations, such as tetramerous tarsi facilitating soil dwelling and six sclerites on pronotum of larvae. The characters of the **Azyini** are derived; the structures associated with the insertion of various movable parts of the body are very striking. The clypeus is derived, and the antenna (especially the small scape) and the apical segment of the maxillary palp (also small) remain plesiomorphic in the Azyini.

Some genera, endemic in the Australian region, i.e. *Cryptolaemus*, *Bucolus*, *Bucolinus* (but not *Scymnodes*) were considered to fall in the Scymnini, perhaps due their scymnoid or platynaspoid appearance. However, they are not true Scymnini having a plesiomorphic long antenna with club of the Coccidulinae type (well-developed terminal segment). The elytral pubescence of *Cryptolaemus* does not have vortices and the size is larger than in most Scymnini. However, **larvae** of *Cryptolaemus* and *Bucolus* are covered by an **extensive waxy secretion** like Scymnini. *Bucolus* (and included *Bucolinus*) has bilobate tibiae like *Azya*. These genera may be included in the Azyini. The tribe **Poriini** possesses fine faceted eyes, reduced prosternum, and metallic colouration. The nominal tribe **Coccidulini** seems to be a likely candidate for further splitting into several lineages.

1.4.10 Scymninae

Sasaji (1968) defined the subfamily **Scymninae** as composed of the tribes Scymnini, Stethorini, Hyperaspini (correctly Hyperaspidini), Aspidimerini and Ortaliini. Later, Sasaji (1971a, b) further added the

Scymnillini and Cranophorini (now considered Coccidulinae). The tribes **Cryptognathini** and **Selvadiini** were established by Gordon (1971, 1985) as members of the Scymninae, the separation of Brachiacanthini (syn. Brachiacanthadini) from Hyperaspidini has been proposed by Duverger (1989), and finally Gordon (1999) separated **Diomini**.

Scymninae have a rather compact, minute to small-sized body. The subfamily is distinguished from the **Coccidulinae** by **short antennae** and a weakly securiform to parallel-sided apical segment of the **maxillary palps**. The shape of the antenna is very important in the study of the phylogeny of Coccinellidae, but not always strictly applied though it is decisive. The reduced apical segment of the antennal club, often combined with distal flagellomeres being gradually broadened, provides a clear synapomorphy for the Scymninae. The **head** has an arched frons, rather large eyes with inner orbits usually parallel and facets that are usually fine. The **pronotum** is apomorphic, of the trapezoidal type, emarginate anteriorly, and never conceals the head, with anterior angles that are narrowly rounded to pointed and sides that frequently descend ventrally. The **larvae** have simple setose processes and are typically covered in **waxy exudations**.

The above listed characters made it possible to exclude the tribe **Cranophorini** and genera *Cryptolaemus* and *Bucolus* (now considered Coccidulinae) and tribe **Ortaliini** (Ortaliinae) from the Scymninae.

The narrowing and shortening of the **elytral epipleuron** and the broadening of the **mesocoxal distance** distinguish the subfamily Scymninae from the **Chilocorinae**. Complete fusion of the **clypeus with the eye canthus** together with distinctly ventral insertions of the antennae clearly distinguished Chilocorinae from typical Scymninae, but is no longer applicable after including **Platynaspidini** into Scymninae. The long and narrow eye canthus in front of the eyes is present in all groups of Scymninae, while the derived shortened type occurs in some genera of Scymnini and Hyperaspidini.

Probably the most primitive tribe of Scymninae is the **Stethorini** in having somewhat coarsely faceted eyes, a primitive abdomen (six ventrites), plesiomorphic male and female genitalia and (apart from the anteriorly lobate prosternum, partly concealing the mouthparts) no special structure for reception of moveable parts of the body. Stethorini is separated from other tribes of Scymninae because the **clypeus** is **not emarginate** around the antennal bases. The Stethorini have a peculiar feeding specialization (acarophagy), and worldwide distribution.

Plesiomorphic states of the prosternal carina and spermatheca are present in two Neotropical tribes, the Scymnillini and Pentiliini. These tribes have an apparently five-segmented abdomen and an autapomorphic development of the prosternum, which in its most apomorphic state forms an entire anterior lobe partially concealing the mouthparts. The elytral epipleuron is strongly foveolate in the **Pentiliini** and there is an unusual reduction of the apophysis of the ninth sternite in the **Scymnillini**.

A group of tribes Aspidimerini, Brachiacanthini, Selvadiini and Hyperaspidini possesses many derived characters including the particular states of the insertion of movable body parts, and the characters of genitalia of both sexes. The penultimate antennal segment is strikingly elongate. The exclusively Oriental tribe **Aspidimerini** remains relatively primitive in the degree of development of the eyes, male genitalia and also the spermatheca, as well as being completely pubescent. They have extremely short antenna and maxillary palpus. The **Hyperaspidini** have derived states in the above-mentioned characters, like loss of pubescence and usually bare genitalia. The body is not perfectly limuloid. The larvae have a wide body. The **Brachiacanthini** and **Selvadiini** have a combination of plesiomorphic and apomorphic states; both are restricted to the New World. The **Platynaspidini** (formerly Chilocorinae) may be related to these tribes.

The **Scymnini** have rather large, finely faceted, laterally not prominent pubescent eyes with parallel inner orbits, the pronotum is truly trapezoidal with very narrow anterior angles and a deeply emarginate anterior margin. The epipleuron is very narrow with its inner edge not reaching the epipleural apex. The body is **pubescent**. **Larvae** are covered by extensive **waxy secretions**.

1.5   FUTURE PERSPECTIVES

While coccinellid subfamilies are more or less worldwide in distribution, many tribes are restricted to particular biogeographical regions, and this has resulted in alternative classifications (Vandenberg 2002). Some derived groups of species are often classified under a separate name, leaving the rest of related

species as a paraphyletic assemblage. Vandenberg (2002) proposes these sets of genera to be reunited or paraphyletic genera to be split to achieve a balanced classification.

Knowledge of the relationships of subfamilies and tribes within the family will allow us better to discriminate general patterns and specific cases, as recommended by Sloggett (2005). It will enable us to name precisely some compared groups – e.g. contrast is often incorrectly emphasized between **aphidophagous** (in fact almost exclusively Coccinellini) and **coccidophagous** coccinellids (in fact a heterogeneous assemblage from the Chilocorinae, Scymninae plus Coccidulinae clade).

REFERENCES

Burakowski, B. and S. A. Slipinski. 2000. The larvae of Leiestinae with notes on the phylogeny of Endomychidae (Coleoptera, Cucujoidea). *Ann. Zool.* 50: 559–573.

Chazeau, J., H. Fuersch and H. Sasaji. 1989. Taxonomy of Coccinellidae. *Coccinella* 1: 6–8.

Chazeau, J., H. Fuersch and H. Sasaji. 1990. Taxonomy of Coccinellidae (corrected version). *Coccinella* 2: 4–6.

Crowson, R. A. 1955. *The Natural Classification of the Families of Coleoptera.* Nathaniel Lloyd, London. 187 pp.

Duverger, C. 1989. Contribution a l'étude des Hyperaspinae. Première note (Coleoptera, Coccinellidae). *Bull. Soc. Linn. Bordeaux* 17: 143–157.

Duverger, C. 2001. Contribution à la connaissance des Hyperaspinae. 2ème note. *Bull. Soc. Linn. Bordeaux* 29: 221–228.

Duverger, C. 2003. Phylogénie des Coccinellidae. *Bull. Soc. Linn. Bordeaux* 31: 57–76.

Fredslund, J. 2006. PHY·FI: fast and easy online creation and manipulation of phylogeny color figures. *BMC Bioinformatics* 7: 315.

Fu, J. and Y.C. Zhang. 2006. Sequence analysis of mtDNA-COI gene and molecular phylogeny on twenty seven species of coccinellids. *Entomotaxonomia* 28: 179–186. (In Chinese with English abstract.)

Fuersch, H. 1990. Valid genera and subgenera of Coccinellidae. *Coccinella* 2: 7–18.

Giorgi, J. A., N. J. Vandenberg, J. V. McHugh et al. 2009. The evolution of food preferences in Coccinellidae. *Biol. Control* 51: 215–231.

Gordon, R. D. 1971. A generic review of the Cryptognathini, new tribe, with a description of a new genus (Coleoptera: Coccinellidae). *Acta Zool. Lilloana* 26: 181–196.

Gordon, R. D. 1974. A review of the Oryssomini, a new tribe of the Neotropical Coccinellidae (Coleoptera). *Coleopt. Bull.* 28: 145–154.

Gordon, R. D. 1977. Classification and phylogeny of the New World Sticholotidinae (Coccinellidae). *Coleopt. Bull.* 31: 185–228.

Gordon, R. D. 1985. The Coccinellidae (Coleoptera) of America north of Mexico. *J. New York Entomol. Soc.* 93: 1–912.

Gordon, R. D. 1994. South American Coccinellidae (Coleoptera). Part III: definition of Exoplectrinae Crotch, Azyinae Mulsant, and Coccidulinae Crotch; a taxonomic revision of Coccidulini. *Rev. Bras. Ent.* 38: 681–775.

Gordon R. D. 1999. South American Coccinellidae (Coleoptera). Part VI: a systematic revision of the South American Diomini, new tribe (Scymninae). *Annales Zoologici (Warszawa)* 47 (Suppl.1): 1–219.

Gregory, T. R., O. Nedvěd and S. J. Adamowicz. 2003. C-value estimates for 31 species of ladybird beetles (Coleoptera: Coccinellidae). *Hereditas* 139: 121–127.

Howland, D. E. and G. M. Hewitt. 1995. Phylogeny of the Coleoptera based on mitochondrial cytochrome oxidase I sequence data. *Insect Mol. Biol.* 4: 203–215.

Hunt, T., J. Bergsten, Z. Levkanicova et al. 2007. A comprehensive phylogeny of beetles reveals the evolutionary origins of a superradiation. *Science* 318: 1913–1916.

Iablokoff-Khnzorian, S. M. 1982. *Les Coccinelles Coléoptères-Coccinellidae Tribu Coccinellini des regions Palearctique et Orientale*. Boubée, Paris. 568 pp.

Jadwiszczak, A. 1990. On the systematic position of the genus *Subepilachna* Bilawski (Coleoptera, Coccinellidae). *Coccinella* 2: 56.

Klausnitzer, B. 1969. Zur Kenntnis der Larve von *Lithophilus connatus* (Panzer) (Col., Coccinellidae). *Entomologische Nachrichten* 13: 33–36.

Klausnitzer, B. 1970. Zur Stellung der Lithophilinae unter besonderer Berücksichtigung larvaler Merkmale (Col., Coccinellidae). *Proc. XIII. Int. Congr. Ent. Moskow* 1: 155.

Klausnitzer, B. 1971. Über die verwandtschaftlichen Beziehungen der Lithophilinae und Coccidulini (Col., Coccinellidae). *Dtsch. Ent. Z. N. F.* 18: 145–148.

Klausnitzer, B. 1993. Zur Nahrungsökologie der mitteleuropäischen Coccinellidae (Col.). *Jber. naturwiss. Ver. Wuppertal* 46, 15–22.

Klausnitzer, B. and H. Klausnitzer. 1997. *Marienkäfer (Coccinellidae). 4. überarbeitete Auflage*. Die Neue Brehm-Bücherei Bd. 451, Westarp Wissenschaften Magdeburg. 104 pp.

Korscheffsky, R. 1931. *Coleopterorum catalogus, pars 118, 120. Coccinellidae I, II*. 224 and 435 pp. Berlin.

Kovář, I. 1973. Taxonomy and morphology of adults. *In* I. Hodek (ed). *Biology of Coccinellidae*. Academia, Prague, pp.15–35.

Kovář, I. 1996. Phylogeny. *In* I. Hodek and A. Honěk (eds). *Ecology of Coccinellidae*. Kluwer Academic Publishers, Netherlands. pp. 19–31.

Kovář, I. 2007. Coccinellidae. *In* I. Löbl and A. Smetana (eds). *Catalogue of Palearctic Coleoptera. Vol. 4*. Apollo Books, Stenstrup. pp. 568–631.

Lawrence, J. F. and A. F. Newton, Jr. 1995. Families and subfamilies of Coleoptera (with selected genera, notes, references and data on family-group names). *In* J. Pakaluk and S. A. Slipinski (eds). *Biology, Phylogeny, and Classification of Coleoptera. Papers Celebrating the 80th Birthday of Roy A. Crowson*. Muzeum i Instytut Zoologii PAN, Warszawa. pp. 775–1006.

LeSage, L. 1991. Coccinellidae (Cucujoidea). *In* F. W. Stehr (ed). *Immature Insects*. Volume 2. Kendall/Hunt Publishing Company, Dubuque, Iowa. pp. 485–494.

Leschen, R. A. B. 2000. Beetles feeding on bugs (Coleoptera, Hemiptera): repeated shifts from mycophagous ancestors. *Invert. Taxon.* 14: 917–929.

Liere, H. and I. Perfecto. 2008. Cheating on a mutualism: indirect benefits of ant attendance to a coccidophagous coccinellid. *Environ. Entomol.* 37: 143–149.

Magro, A., E. Lecompte, F. Magne, J. L. Hemptinne and B. Crouau-Roy. 2010. Phylogeny of ladybirds (Coleoptera: Coccinellidae): are the subfamilies monophyletic? *Mol. Phyl. Evol.* 54: 833–848.

Mawdsley, J. R. 2001. Mitochondrial cytochrome oxidase I DNA sequences and the phylogeny of Coccinellidae (Insecta : Coleoptera : Cucujoidea). *J. New York Entomol. Soc.* 109: 304–308.

Miyatake, M. 1972. A new Formosan species belonging to the genus Singhicalia Kapur, with proposal of a new tribe (Coleoptera: Coccinellidae). *Trans. Shikoku Entomol. Soc. Matsuyama* 11: 92–98.

Özdikmen, H. 2007. New replacement names for three preoccupied ladybird genera (Coleoptera: Coccinellidae). *Mun. Ent. Zool.* 2: 25–28.

Pakaluk, J. and S. A. Slipinski. 1990. Review of Eupsilobiinae (Coleoptera: Endomychidae) with descriptions of new genera and species from South America. *Revue Suisse de Zoologie* 97: 705–728.

Pal, T. K. and J. F. Lawrence. 1986. A new genus and subfamily of mycophagous Bothrideridae (Coleoptera: Cucujoidea) from the Indo-Australian Region, with notes on related families. *J. Austr. Entomol. Soc.* 25: 185–210.

Palenko, M. V., D. V. Mukha and I. A. Zakharov. 2004. Intraspecific and interspecific variation of the mitochondrial gene of cytochrome oxidase I in ladybirds (Coleoptera: Coccinellidae). *Rus. J. Gen.* 40: 148–151.

Pang, H., S. X. Ren, T. Zeng and X. F. Pang. 2004. Biodiversity and their utilization of Coccinellidae in China. Guangzhou, Science and Technology press of Guangdong. 168 pp. (In Chinese.)

Phuoc, D. T. and F. W. Stehr. 1974. Morphology and taxonomy of the known pupae of Coccinellidae (Coleoptera) of North America, with a discussion of phylogenetic relationships. *Contributions of the AEI* 10 (6): 125.

Poorani, J. 2003. A new species of the genus *Synonychimorpha* Miyatake (Coleoptera: Coccinellidae) from South India. *Zootaxa* 212: 1–6.

Pope, R. D. 1988. A revision of the Australian Coccinellidae (Coleoptera). Part I. Subfamily Coccinellinae. *Invert. Taxon.* 2: 633–735.

Ren, S., X. Wang, H. Pang, Z. Peng and T. Zeng. 2009. *Colored Pictorial Handbook of Ladybird Beetles in China.* Science Press, Beijing. 336 pp. (In Chinese.)

Robertson, J. A., M. F. Whiting and J. V. McHugh. 2008. Searching for natural lineages within the Cerylonid Series (Coleoptera: Cucujoidea). *Mol. Phyl. Evol.* 46: 193–205.

Samways, M. J., R. Osborn and T. L. Saunders. 1997. Mandible form relative to the main food type in ladybirds (Coleoptera: Coccinellidae). *Biocontrol Sci. Tech.* 7: 275–286.

Sasaji, H. 1968. Phylogeny of the family Coccinellidae (Coleoptera). *Etizenia*, Ocassional Publications of the Biological Laboratory, Fukui University 35: 1–37.

Sasaji, H. 1971a. Phylogenetic positions of some remarkable genera of the Coccinellidae (Coleoptera), with an attempt of the numerical method. *Memoirs of the Faculty of Education, Fukui University, Series II (Natural Science)* 21: 55–73.

Sasaji, H. 1971b. *Coccinellidae. Fauna Japonica.* Acad. Press of Japan, Japan. 340 pp.

Savoiskaya, G. I. 1969. On revealing of new taxonomical categories of the coccinellids. *Vest. sel'-khoz. Nauki, Alma-Ata,* 9: 101–106. (In Russian.)

Savoiskaya, G. I. and B. Klausnitzer. 1973. Morphology and taxonomy of the larvae with keys for their identification. *In* I. Hodek (ed). *Biology of Coccinellidae.* Academia, Prague. pp. 36–55.

von der Schulenburg, J. H. G., J. M. Hancock, A. Pagnamenta et al. 2001. Extreme length and length variation in the first ribosomal internal transcribed spacer of ladybird beetles (Coleoptera : Coccinellidae). *Mol. Biol. Evol.* 18: 648–660.

Slipinski, A. 2007. *Australian Ladybird Beetles (Coleoptera: Coccinellidae): Their Biology and Classification.* Australian Biological Resources Study, Canberra. 286 pp.

Slipinski, A. and W. Tomaszewska. 2002. *Carinodulinka baja,* new genus and a new species of Carinodulini from Baja California (Coleoptera, Coccinellidae). *Annales Zoologici* 52: 489–492.

Slipinski, A. and W. Tomaszewska. 2005. Revision of the Australian Coccinellidae (Coleoptera). Part 3. Tribe Sukunahikonini. *Austr. J. Entomol.* 44: 369–384.

Slipinski, S. A. and Pakaluk, J. 1992. Problems in the classification of the Cerylonid series of Cucujoidea (Coleoptera). *In* M. Zunino, X. Belles and M. Blas (eds). *Advances in Coleopterology.* European Association of Coleopterology, Barcelona. pp. 79–88.

Sloggett, J. J. 2005. Are we studying too few taxa? Insights from aphidophagous ladybird beetles (Coleoptera: Coccinellidae). *Eur. J. Entomol.* 102: 391–398.

Tinsley, M. C. and M. E. N. Majerus. 2007. Small steps or giant leaps for male-killers? Phylogenetic constraints to male-killer host shifts. *BMC Evol. Biol.* 7: 238.

Tomaszewska, K. W. 2000. Morphology, phylogeny and classification of adult Endomychidae (Coleoptera: Cucujoidea). *Annales Zoologici* 50: 449–558.

Tomaszewska, K. W. 2005. Phylogeny and generic classification of the subfamily Lycoperdininae with a re-analysis of the family Endomychidae (Coleoptera, Cucujoidea). *Annal. Zool.* 55: 1–172.

Ukrainsky, A. S. 2006. Five new replacement ladybird (Coleoptera: Coccinellidae) generic names. *Russian Entomological Journal* 15: 399–400.

Vandenberg, N. J. 2002. Coccinellidae Latreille 1807. *In* R. H. Arnett, Jr, M. C. Thomas, P. E. Skelley and J. H. Frank (eds). *American Beetles,* Volume 2. CRC Press, Boca Raton. pp. 371–389.

Vandenberg, N. J. and D. E. Perez-Gelabert. 2007. Redescription of the Hispaniolan ladybird genus *Bura* Mulsant (Coleoptera: Coccinellidae) and justification for its transfer from Coccidulinae to Sticholotidinae. *Zootaxa* 1586: 39–46.

Verhoeff, C. 1895. Beiträge zur vergleichenden Morphologie des Abdomens der Coccinelliden und über die Hinterleibsmuskulatur von Coccinella, zugleich ein Versuch, die Coccinelliden anatomisch zu begründen und natürlich zu gruppieren. *Arch. Naturg.* 61: 1–80.

Yu, G. 1994. Cladistic analyses of the Coccinellidae (Coleoptera). *Entomologica Sinica* 1: 17–30.

Zhang, Y. and Z. Zheng. 2002. RAPD analysis and its application to taxonomy of six species of lady beetles. *J. Northwest University (Natural Science Edition)* 32: 409–412.

Zhang, Y., Z. Zheng, J. Yang and S. An. 1999. A comparative study of esterase isozymes in five species of coccinellids (Coleoptera: Coccinellidae) and its application in taxonomy. *Entomotaxonomia* 21: 123–127.

Chapter 2

GENETIC STUDIES

John J. Sloggett[1] and Alois Honěk[2]

[1] Maastricht Science Programme, Maastricht University, P.O. Box 616, 6200 MD Maastricht, The Netherlands
[2] Department of Entomology, Crop Research Institute, CZ 16106 Prague 6, Czech Republic

2.1 INTRODUCTION

Coccinellids have been used as genetic models since the beginning of the 20th century. Interest arose initially because of the genetically polymorphic colour patterns possessed by a number of species, and historically the majority of genetic work has been on coccinellid colour pattern polymorphism. However, studies of coccinellid genetics have by no means been limited to colour pattern. A considerable number of studies exist on coccinellid cytology and on the genetics of other traits. More recent molecular genetic work has examined not only the evolution of specific gene regions, but has been used to make evolutionary and ecological inferences about diverse aspects of coccinellid biology including sexual and population biology and phylogenetic relationships. In this chapter we begin by discussing whole genome and cytogenetic aspects of coccinellid biology (Sections 2.2 and 2.3), before moving on to discuss the vast body of work on coccinellid colour pattern (2.4), and genetics studies of non-colour pattern traits (2.5). Finally, we consider the rapidly burgeoning number of molecular studies of coccinellids and what they tell us, not only about the coccinellid genome, but also more generally about coccinellid biology (2.6).

2.2 GENOME SIZE

Genome size, measured as haploid nuclear DNA content or **C-value**, has been studied in 31 species of coccinellid, primarily within the Coccinellinae, but overall representing six subfamilies and eight tribes (Gregory et al. 2003): this makes the Coccinellidae the third best studied beetle family after the Tenebrionidae and Chrysomelidae. The C-values of coccinellids vary between 0.19 pg and 1.71 pg, with a mean of 0.53 pg. By comparison with the two other beetle families, the Coccinellidae appear to be **more variable in genome size**, although their mean C-value is between that of tenebrionids and chrysomelids. There appears to be **very little intraspecific variation** in genome size. Most C-value variation occurs between coccinellid subfamilies, although even congeneric species can exhibit up to two-fold variation. Genome size is not correlated with body size, nor does it appear to be related to chromosome number, but larger genomes do appear to be associated with longer development. In comparison with fast-developing aphidophagous species, two slower developing coccidophagous species have some

of the largest coccinellid genomes. Due to the small number of species from many coccinellid subfamilies and ecotypes included, most of the conclusions of the C-value study remain tentative and require further verification (Gregory et al. 2003).

2.3 CHROMOSOMES AND CYTOLOGY

2.3.1 Chromosome numbers and banding

The **chromosomal complements** of something under 200 coccinellid species are now known (Smith & Virkki 1978, Lyapunova et al. 1984, Rozek & Holecova 2002). The reported **diploid number of chromosomes varies between 12 and 28** (Smith 1960, 1962a, Yadav & Gahlawat 1994). The distribution of species' chromosome numbers varies across subfamilies and tribes (Fig. 2.1); the most common number, comprising a diploid number of 18 autosomes and a Xy_p sex chromosome pair, is that which is often considered ancestral for the Coleoptera (Smith & Virkki 1978, Lyapunova et al. 1984).

Studies on chromosome numbers have been supplemented by an increasing amount of work on **chromosome structure and banding**. Differential staining of chromosomal regions gives banding patterns that allow different chromosomes of equal size to be uniquely identified, as well as potentially identifying changes in chromosome structure such as inversions. Most work has focused on chromosome **C-banding** patterns, which arise from intense staining of **heterochromatic** (genetically inactive) chromosome regions. In ladybirds, as in other beetles, these are mainly associated with regions near the centromere of a chromosome and in the short arms of chromosomes (Ennis 1974, Drets et al. 1983, Maffei et al. 2000, 2004, Rozek & Holecova 2002, Beauchamp & Angus 2006). A number of other banding techniques have also been investigated (Ennis 1974, 1975, Maffei & Pompolo 2007). Homology of chromosome banding patterns has been used to infer a monophyletic origin for the coccinellids *Chilocorus orbus*, *Chil. tricyclus* and *Chil. hexacyclus* (Ennis 1974, 1975, 1976).

2.3.2 Sex determination

The commonest form of sex determination system in coccinellids is the so-called **Xy parachute (Xy$_p$)**

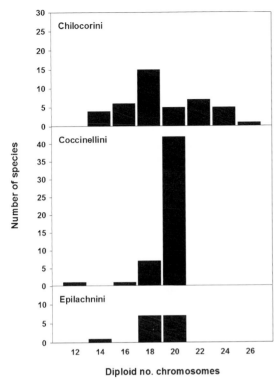

Figure 2.1 Diploid chromosome numbers of coccinellid species in the tribes Chilocorini, Coccinellini and Epilachnini. Female diploid chromosome numbers are used. From data in Smith & Virkki (1978), amended to take into account current views on the taxa included in these tribes.

system: females are XX, while males are Xy. The y chromosome is very small (which is why it is typically written as a lower case rather than an upper case Y) and it is paired with the X chromosome at metaphase I in a 'parachute' configuration, hence the name (Stevens 1906, Smith 1950). In some coccinellid species the y chromosome has been completely lost, giving an **XO system**: females possess two X chromosomes and males one. The absence of a y chromosome in such species suggests that, more generally in the Coccinellidae, it is the ratio of X chromosomes to haploid sets of autosomes (A) [2X:2A = female; 1X:2A = male] that determines sex rather than the occurrence of specific genes on the y chromosome which code for 'maleness' (Majerus 1994): both systems are known from other insects (Sanchez 2008).

In other species the X chromosome has fused with an autosome to form a **neo-XY** chromosome system: this is typical for the Chilocorini, although it also occurs in other groups (Smith & Virkki 1978).

Maffei and colleagues have investigated the special mechanism by which Xy_p sex chromosomes are associated during mitosis and meiosis. Their studies indicate that, depending on species, nucleolar proteins are either synthesized within or imported into the sex bivalent for this purpose. In the former case **fluorescent *in situ* hybridization (FISH)** techniques showed that **ribosomal DNA (rDNA)** genes mapped to the sex vesicle in *Olla v-nigrum* (Maffei et al. 2001a). In the latter case, in *Cycloneda sanguinea*, rDNA genes mapped to autosomes outside the sex vesicle; however, silver staining revealed that nucleolar material, which is typically associated with rDNA, was also associated with the sex chromosomes (Maffei et al. 2001b, 2004).

2.3.3 Supernumerary (B) chromosomes

A number of coccinellid groups contain species in which some individuals possess one or more **supernumerary chromosomes**, often called **B-chromosomes**, in addition to their normal chromosome complement. Populations containing such individuals are generally polymorphic, containing both individuals with and without B-chromosomes (e.g. Smith & Virkki 1978, Maffei et al. 2000, Tsurusaki et al. 2001). In a few known cases, such as in *Chilocorus stigma* and *Chil. rubidus*, all individuals bear at least one B-chromosome (Smith & Virkki 1978). The number of B-chromosomes carried by individuals can also vary. Tsurusaki et al. (2001) recorded between zero and four in one member of the *Henosepilachna* (= *Epilachna*) *vigintioctopunctata* species complex. The highest number of B-chromosomes recorded from a single individual is 13 from a male *Chil. rubidus* (Smith & Virkki 1978).

B-chromosomes are derived from the other chromosomes of the species that possess them or, more rarely, from the chromosomes of a related species through hybridization events (Jones & Rees 1982, Camacho et al. 2000). In *Henosepilachna* (= *Epilachna*) *pustulosa* some males possess a **supernumerary y chromosome**, probably derived from y chromosome duplication. During meiosis, both y chromosomes join with the X, producing a Xyy parachute, and ultimately gametes with either an X chromosome or two y chromosomes (Tsurusaki et al. 1993). Similar super-numerary y chromosomes are also known from *Coccinella quinquepunctata* (Lyapunova et al. 1984). Some of the B-chromosomes of coccinellids appear to be restricted to or to predominate in only one sex and might also have **originated from sex chromosomes** (Maffei et al. 2000, Tsurusaki et al. 2001). The formation of a neo-XY sex determination system from the fusion of an X chromosome with an autosome is also likely to give rise to B-chromosomes, arising both from the former y chromosome and from a centric fragment lost during the fusion (Smith & Virkki 1978). More generally B-chromosomes may arise as a result of **chromosomal fusion** or from the **loss of genetically inert, heterochromatic chromosomal arms** (Smith & Virkki 1978). In *Chil. stigma*, which is polymorphic for a number of chromosomal fusions (2.3.4), the mean number and range of B-chromosomes increases with an increase in the frequency of fusions across populations (Smith & Virkki 1978). In this species, as throughout the Chilocorini, unfused chromosomes have one fully heterochromatic arm. Centric fusion occurs between the other **euchromatic** parts of chromosomes, leaving the lost heterochromatic arms as supernumerary chromosomes (Smith 1959). Heterochromatic arms can also sometimes be lost and become B-chromosomes without chromosomal fusion (Smith & Virkki 1978).

In many cases after their genesis, supernumerary chromosomes are lost in subsequent generations. Based on the number of chromosomal fusions observed in *Chil. stigma*, and thus the number of related B-chromosomes expected to be present, only 22% of potential B-chromosomes survive as such in this species (Smith & Virkki 1978). In other organisms, both positive and negative direct effects on fitness have been recorded for B-chromosomes, as well as biased transmission rates to offspring that are in excess of the rates expected under simple Mendelian inheritance (Jones & Rees 1982, Camacho et al. 2000); however, no such studies yet exist for coccinellids. Henderson (1988) found a **negative correlation between male B-chromosome frequency and the proportion of males** in populations of British *Exochomus quadripustulatus*. He concluded that this correlation was a consequence of a third factor that affected both B-chromosome frequency and sex ratio. This conclusion was based on the observation that other coccinellids with a neo-XY sex determining system like that of

E. quadripustulatus showed similar sex ratio biases across populations without possessing B-chromosomes (Henderson & Albrecht 1988). It is perhaps worth noting that the studies pre-date detailed investigations on bacterial male-killing in coccinellids (Chapter 8), although none of the species studied by Henderson and Albrecht have been shown to possess male-killers (G.D.D. Hurst, personal communication). It is also now clear that B-chromosomes sometimes **directly affect the sex-ratio of the offspring** of individuals carrying them, although unlike in *E. quadripustulatus*, generally increasing the heterogametic sex (Beladjal et al. 2002, Werren & Stouthamer 2003, Underwood et al. 2005). Nonetheless the possibility that the correlation between *E. quadripustulatus* B-chromosome frequency and population sex ratio results from a direct relationship cannot be ruled out.

2.3.4 Cytogenetic changes, intraspecific cytogenetic variation and speciation

Relatively little **speciation** in the Coccinellidae appears to be directly related to cytogenetic changes. Most sub-families and tribes exhibit a relatively limited variation in chromosome number and structure; consequently the majority of speciation in these groups is unlikely to be linked to cytological differences. The most variable group is the Chilocorini where diploid numbers of chromosomes can vary between 14 and 26 (Smith 1959, Lyapunova et al. 1984; Fig. 2.1). Studies of this group in North America demonstrate that the relationship between cytology and reproductive isolation is complex, and that changes in cytology do not necessarily lead to reproductive isolation (Smith 1959, Smith & Virkki 1978).

Chilocorus stigma exhibits up to six intraspecific **chromosomal fusion polymorphisms** in a genome fundamentally of diploid number 26 (Smith 1959, 1962b, Smith & Virkki 1978). The frequency of fusions increases westwards and northwards, through the range of *Chil. stigma*, which covers most of North America except the extreme west (Fig. 2.2). In the east, from Florida to New York and Connecticut, no individuals exhibit chromosomal fusions. However, moving northwestwards, three fusions appear sequentially in Maine, eastern Ontario and central Ontario (Smith 1959, 1962b). Three further fusions have been found in more northerly parts of Ontario (Smith & Virkki

Figure 2.2 Geographic distribution of chromosome fusions in *Chilocorus stigma*. Circles show the average percentage of the maximum number of six fusions per individual (black). Thus the number of fusions increases to the north and west. From data in Smith & Virkki (1978).

1978). The clear distributional pattern of fusions suggests they are adaptive (Smith 1957, 1959, 1962b), although the nature of this adaptation is unknown. During chromosome pairing, individuals that are heterozygotic for fusions form trivalents, comprising one fusion chromosome paired with two homologous unfused ones. Fusion homozygotes form ring bivalents. In one case two fusions are semi-homologous (i.e. both contain one shared predecessor chromosome, in each cased fused to a different second chromosome): in this case heterozygotes for these fusions form a chain-of-four configuration. Meiosis involving all these configurations produces chromosomally balanced gametes (Smith 1966, Smith & Virkki 1978), making **fusion heterozygotes fully fertile**; thus *Chil. stigma* forms a **single species**.

In western North America, a parallel situation exists with respect to chromosomal fusions. Three monophyletic *Chilocorus* species, known as the *cyclus* complex, exhibit an increasing number of chromosomal fusions moving northwards. The most southerly species, *Chil. orbus*, has a diploid number of 22 chromosomes (two fusions). The more northerly *Chil. tricyclus* has 20 chromosomes (three fusions) and *Chil. hexacyclus* has 14 chromosomes (six fusions) (Smith 1959). However, unlike the chromosomal forms of *Chil. stigma*, the three species exhibit high levels of **hybrid sterility** as in heterozygote trivalents the chromosomes are often orientated incorrectly (Smith 1959, 1966). Thus while some gene flow occurs between species in the restricted areas where they overlap, it is limited (Smith 1966). It is unclear why parallel processes of chromosomal evolution have given rise to a single polymorphic species in the east but a complex of three separate species in the west, although a number of possibilities have been suggested. Smith (1959) indicated that *Chil. stigma* might represent an earlier stage than the *cyclus*-complex in the evolution of new species through cytogenetic change. This seems unlikely, and two further, non-exclusive possibilities were proposed by Smith (1966). The first is that *Chil. stigma* fusions are advantageous in the heterozygous form, maintaining the polymorphism in this species in a balanced state. The second is that fusions in the western *cyclus* complex evolved in isolated marginal populations in which they rapidly became homozygous; *cyclus* complex hybrids therefore lost the ability to form correctly orientated trivalents.

From these two examples, particularly that of the chromosomally polymorphic *Chil. stigma*, we can conclude that complete **reproductive isolation rarely arises as a consequence of cytogenetic changes alone**. Nonetheless, such changes can certainly play a role. In a cytogenetic study of the *Henosepilachna* (= *Epilachna*) *vigintioctomaculata* species complex, Tsurusaki et al. (1993) found that the species complex fell into two groups: group A, comprising *Henosepilachna* (= *Epilachna*) *vigintioctomaculata*, possessed relatively small heterochromatic segments in their chromosomes, whereas group B, comprising *H. pustulosa*, *Henosepilachna* (= *Epilachna*) *niponica* and *Henosepilachna* (= *Epilachna*) *yasutomii* possessed long heterochromatic segments in seven autosomes. Hybrids between group A and B members exhibited high mortality during embryogenesis. Tsurusaki et al. hypothesize that in hybrids replication of the different chromosomes will occur at a highly heterogeneous rate, because of the widely differing amounts of heterochromatin in the chromosomes from the two parental species. Consequently in the early stages of embryogenesis, the rate of cell division, which is probably maternally determined at this stage, is poorly synchronized with chromosome replication, leading to low embryonic survival (Fig. 2.3). Tsurusaki et al. thus provide a **cytogenetic basis for reproductive isolation** between group A and B species. It is worth noting, however, that even in this example additional factors appear to play a role, most notably incompatibility

between the female genital tract and allospecific sperm (Katakura 1986, Katakura & Sobu 1986). More generally, much work is still needed on the non-cytogenetic causes of reproductive isolation between closely related species, which overall remain poorly understood.

2.4 COLOUR PATTERN VARIATION

Coccinellid colouration has long attracted the attention of professional and amateur entomologists, probably because its high variation provided the opportunity of 'taxonomic description' of morphs to authors that did not come in contact with new species. These efforts peaked in the 1930s with Mader's (1926–37) catalogue which named both existing and theoretically possible morphs of Central European species.

A second and ultimately more important reason to study colour pattern variability was the insight it gave into evolutionary processes. Colour pattern studies first flourished at the time of the development of the synthetic theory of evolution. The authors hoped to observe the speciation process in action. 'If the genes that are unlike in two of these low ranking groups (subspecies and species) can be ascertained, there is some chance of obtaining a picture of the evolution process which produced the dissimilarity. This measure is of particular interest when it relates to those characters which competent taxonomists regard as the specific distinctions' (Shull 1949).

Conspicuous colour pattern polymorphisms played an important part in the debate surrounding natural selection and evolution. Early researchers believed that co-existing colour patterns were selectively neutral, and that their frequencies were consequently subject to genetic drift. Later studies showed that, in fact, they were under strong selection, which determined relative morph frequencies. However, the relative importance of selection and drift in the maintenance of conspicuous polymorphisms remained hotly debated for a long time (Ford 1964). In many polymorphic species, including ladybirds, there was consequently great interest not only in the genetic determination of alternative forms, but (i) their temporal and spatial variability, (ii) their ecological significance and (iii) the ultimate selective factors underlying the polymorphism. An enormous body of work on ladybirds was accomplished in the 20th century: its results are still significant for current biology and remain far from complete, even in well-investigated species (Majerus 1994).

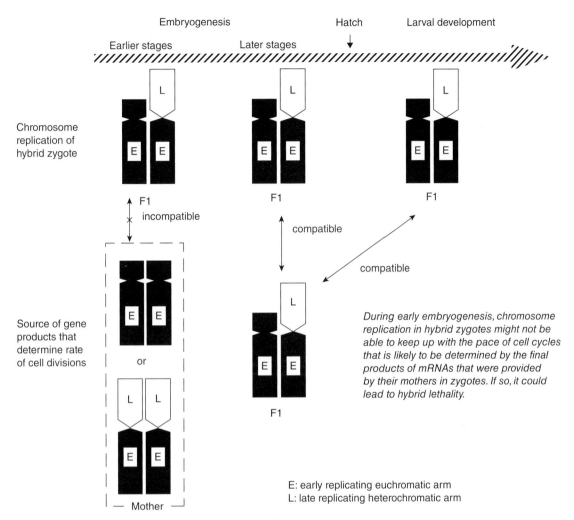

Figure 2.3 A hypothetical model of cytogenetic incompatibility between *Henosepilachna vigintioctomaculata*, with chromosomes with a small amount of heterochromatin, and species B members of the *Henosepilachna vigintioctomaculata* species complex, with chromosomes with a large amount of heterochromatin causing late replication (L). Redrawn from Tsurusaki et al. 1993, with permission.

2.4.1   The nature of colour patterns

The **elytral pattern** usually consists of dark spots on a light background (light or **non-melanic** morphs) or light spots on a dark background (dark or **melanic** morphs). The light areas (usually yellowish, reddish to brownish) are coloured by **carotenes**, the dark areas by **melanins** (Cromartie 1959). The spots always appear in predetermined positions and differ in presence, size and shape (Zimmermann 1931, Timofeeff-Ressovsky et al. 1965). The best method of describing variation of spot presence in 'light' morphs is probably that of Schilder (Schilder & Schilder 1951/2, Schilder 1952/3), illustrated in Fig. 2.4.

Several studies have concerned **morphological 'laws' of spot organization** and the ontogeny of the pattern. In particular species, not all possible combinations of spots are realized (Zarapkin 1938b, Filippov

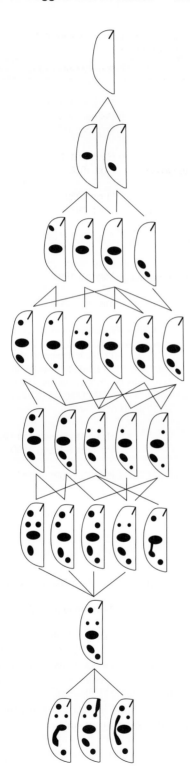

Figure 2.4 A classification of variation in elytral spot pattern using the example of *Hippodamia variegata*. The beetles are classified according to the cumulative number of spots and fusions. Each spot or fusion is a 'unit of melanic surface' and their sum roughly characterizes the degree of melanization. The horizontal series are defined by the particular number of spots and fusions, for example, the elytra shown in the lower series (bottom row) have 6 spots + 1 fusion (= 7 units of melanic surface), the elytron above has only 6 spots (= 6 units of melanic surface). Modified from Schilder & Schilder (1951/2).

1961). Principles of spot pattern variation were investigated by Smirnov (1927, 1932). In some species there exist several typical sequences in which particular spots appear after adult eclosion. This sequence of melanin deposition in particular spots was studied in *Adalia bipunctata* (Zarapkin 1930, 1938a, Majerus 1994). After spots are pigmented, the process of pigment deposition may continue so that the individuals become completely melanic. This process can occur within a few days, as in *Olla v-nigrum* (Vandenberg 1992), or over several months, as in *Anatis labiculata* (Majerus 1994).

Individuals can also differ in **scutum colouration**, which is also variable in pattern and degree of melanization. Particular patterns are associated with differing elytral patterns, so that light-spotted morphs have a light pronotum with dark marks of different shape while melanized individuals have dark elytra and pronota (Lusis 1932, Majerus 1994, Blehman 2007, 2009). The colour pattern of the head (frons) may also vary and the colouration is frequently typical of the sex (Table 2.1). Moreover the ventral side, thorax and abdomen also vary in the degree of melanization.

2.4.2 Genetic determination of colour patterns

Among many coccinellid species polymorphic for colour patterns, few have been the subject of detailed examination. Experimental studies are available for 15 species which have pronounced but not very complicated colour patterns and that are abundant and easy to collect and rear (Table 2.1). The studies were mostly accomplished in the first half of the 20th

Table 2.1 Experimental evidence of inheritance of and environmental effects of colour patterns: reports of crossing experiments and genotype x environment interactions of particular species, and their references.

Subject of the paper	Reference
Adalia bipunctata	
Melanic morphs dominant over light morph	Schroeder 1909
Multi-spot light morphs dominant over a melanic morph	Palmer 1911
Inheritance of multi-spot and melanic morphs in relation to typical (single spotted) morph	Palmer 1917
Inheritance of typical and dominant melanic morphs	Hawkes 1920
Allelomorphic series of three morphs, two melanic morphs dominant to typical (single spotted) morph	Fiori 1928
Dominance in allelomorphic series of morphs	Lusis 1928
Inheritance and dominance of 12 elytral colour morphs and pronotum pattern	Lusis 1932
Experiments on local differences in inheritance of colour morphs	Majerus 1994
Morph *tigrina* dominant over light *typica* and recessive to melanic morphs	Zakharov 1996
Adalia decempunctata	
Dominance of morphs in allelomorphic series	Lusis 1928
Inheritance of an allelic series, multi-spotted morph dominant to chequered and melanic morphs	Majerus 1994
Experiments on local variation in inheritance of colour morphs	Majerus 1994
Inheritance of slow and fast (dominant) deposition of carotenoids and melanin on elytra	Majerus 1994
Aphidecta obliterata	
Sex differences in head colouration on frons	Witter & Amman 1969
Sex differences in elytral colour pattern	Eichhorn & Graf 1971
***Calvia* sp.**	
Allelomorphic series and order of dominance of three morphs (originally considered separate species)	Lusis 1971
Coelophora inaequalis	
Inheritance of colour morphs	Hales 1976; Houston 1979
Eight allelic morphs of which non-melanic is dominant	Houston & Hales 1980
Coelophora quadrivittata	
Inheritance of three allelomorphs with pale morph dominant over spotted and striped morphs	Chazeau 1980
Coccinella septempunctata	
Low temperature during pre-imaginal development increases spot size, in three populations of Japan	Okuda et al. 1997
***Henosepilachna elaterii* (=*Epilachna chrysomelina*)**	
Selection of strains with different size of spots	Tennenbaum 1931
Spot size is influenced by selection as well as temperature	Zimmermann 1931
Increasing development temperature decreases spot size, different reaction in three geographically distant populations	Timofeeff-Ressovsky 1932, 1941
Inheritance of the size and form of an elytral spot	Tennenbaum 1933
Harmonia axyridis	
Inheritance of four alleles of elytral colour determination	Tan & Li 1934
Inheritance of six elytral colour and pattern morphs and elytral ridge	Hosino 1936
Inheritance of *forficula* and *transversifascia* morphs	Hosino 1939
Inheritance of *aulica* and *gutta* morphs	Hosino 1940a
Establishing allelomorphic series and order of dominance of seven morphs	Hosino 1940b
Rare melanic morph *distincta* described and order of its dominance determined	Hosino 1941
Inheritance of alleles of *succinea* morph with different numbers of spots	Hosino 1942
Inheritance of three alleles of *axyridis* and two alleles of *aulica* morphs	Hosino 1943a
Inheritance of *succinea* and *axyridis* morphs	Hosino 1943b

(Continued)

Table 2.1 (*Continued*)

Subject of the paper	Reference
Sixteen alleles determining colour polymorphism and order of their dominance	Tan 1946
Genetic variation of spot size in *succinea* morph	Hosino 1948
Distinguishing alleles of elytral colour pattern morphs	Komai 1956
Decreasing developmental temperature increases spot size in *succinea* morph	Sakai et al. 1974
Inheritance of rare *supercilia* and *interduo* morphs	Tan & Hu 1980
Sexual differences in labrum and prosternum colour	McCornack et al. 2007
Crosses between phenotypes illustrated	Seo et al. 2007
Temperature influences melanization in *succinea* morph	Michie et al. 2010
Hippodamia sinuata	
Inheritance of spotted and spotless (dominant) morph	Shull 1943
Hippodamia convergens	
Inheritance of spotted and spotless (dominant) morph	Shull 1944
Hippodamia quinquesignata	
Inheritance of spotted and spotless (dominant) morph	Shull 1945
Olla v-nigrum	
Inheritance of spotted and melanic (dominant) morphs	Vandenberg 1992
Propylea japonica	
Inheritance of three elytral colour and pattern morphs	Miyazawa & Ito 1921
Propylea quatuordecimpunctata	
Sexual differences in head (frons) and pronotum colour	Rogers et al. 1971
Polygenic inheritance of spot size	Majerus 1994

century and have been reviewed by Komai (1956). Table 2.1 shows that fewer crossing experiments have been made recently, except for the enormous but unfortunately largely unpublished work of the late Mike Majerus (Majerus 1994). His results indicate that the study of colour patterns requires further refinement, even in the most intensively studied species like *A. bipunctata*.

The published data on crossing experiments led to several generalizations. In most cases it was demonstrated that the morphs are determined by a series of **multiple alleles localized at the same locus**. However, species in which colour morphs are determined by two or more non-allelic genes might also exist (Komai 1956). Further ideas were proposed by Ford (1964) and advocated by Majerus (1994), who consider it more likely that such a multiple allele effect, as has been demonstrated in coccinellids, is caused by a sufficiently close juxtaposition of the loci of the genes so that crossing-over is most unlikely to separate them. Ford designated such a complex of loci as a '**supergene**'. Minor variation of **spot size** and shape is

controlled **polygenically**, for example in *Henosepilachna elaterii* (= *Epilachna chrysomelina*) (Timofeeff-Ressovsky 1932, Timofeeff-Ressovsky et al. 1965) or *Propylea quatuordecimpunctata* (Majerus 1994). For individual species, the **number of allelomorphs per locus** roughly correlates with the amount of experimental work invested in its analysis. Thus the number of established alleles for elytral colour pattern exceeded 22 in *Harmonia axyridis* (Tan 1946, Komai 1956) while for *A. bipunctata*, Majerus (1994) greatly increased this number above the 12 determined by Lusis (1928, 1932, Fig. 2.5). Patterns determined by particular alleles are frequently similar and difficult to distinguish. An example is the *succinea* group of morphs in *Har. axyridis* (light elytra with up to nine black spots), where morphs with particular numbers of dark spots are determined by particular alleles. Consequently some 15 alleles, each with a particular expression of the *succinea* morph in a homozygous state, may exist (Tan 1946). On the other hand, **modifiers** may vary the patterns determined by particular alleles. This became evident in crossing *A. bipunctata* males and females originating from distant geographic populations. Unexpected colour morphs appeared in the progeny, probably because the mixing of different sets of modifiers resulted in the normal dominance relationship breaking down (Majerus 1994). Heterozygous forms further increase intraspecific variation in colour morphs. In many species and populations, alleles determining melanic morphs are dominant to those determining light morphs. This led Tan (Tan 1946, Tan & Hu 1980) to formulate the '**mosaic dominance**' hypothesis which supposes that the heterozygote develops black pigmentation on any part of the elytron which is black in either homozygote (Tan 1946). Dominance of the melanic form is not typical for all species of coccinellids, however. This is because the order of dominance is determined by genetic environment of the allele for colour morph. Consequently species and populations exist where light morphs are dominant. Thus the allele determining the light morph is dominant over that for the melanic morph in *Hippodamia* (Shull 1949) and *Coelophora* (Houston 1979, Chazeau 1980).

2.4.3 Geographic variation

Geographic variation in morph frequencies was first systematically studied in *A. bipunctata* and *A. decempunctata* (Dobzhansky 1924b), and then in other species including *Anatis ocellata*, *Anisosticta novemdecimpunctata*, *Coccinella magnifica* (= *C. divaricata*), *C. septempunctata*, *C. transversoguttata*, *C. quinquepunctata*, *Coccinula quatuordecimpustulata*, *Hippodamia* (= *Adonia*) *variegata* and *Oenopia* (= *Synharmonia*) *conglobata* (Dobzhansky 1925, 1933, Dobzhansky & Sivertzew-Dobzhansky 1927). For all species, **geographic centres exist for less heavily pigmented populations**, with more distant areas inhabited by more heavily pigmented ones. Moreover, such centres for the different species roughly coincide in geographical location. The centre for light morphs lies in Central Asia for the eastern hemisphere and in California for the western hemisphere. The proportion of pigmented morphs increases radially in all directions from each 'light' centre.

Fine-scale patterns of geographic differentiation of proportions of colour morphs in local populations have been particularly well studied in two species,

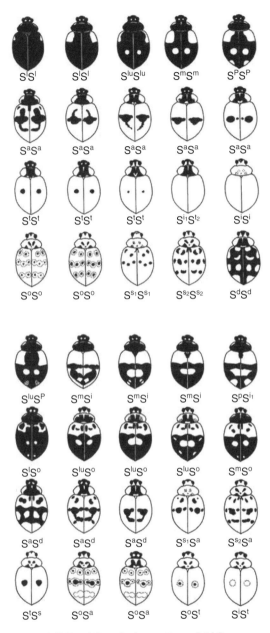

Figure 2.5 Variability of colour pattern of *Adalia bipunctata*. Homozygous morphs (top four rows) are determined by 12 alleles arranged in the order of dominance starting from top left. Some heterozygotes are shown in the bottom four rows. After Lusis (1932).

A. bipunctata and *Har. axyridis*. Proportions of light morphs (mostly *typica* with one black spot on each of the reddish elytra) and melanic morphs (with one, two or three reddish spots on each black elytron) were studied in **local populations of *A. bipunctata***. In areas with an oceanic climate, the proportion of melanic morphs may vary from 0–97% in the British Isles (Creed 1966, 1971a, b), 2–86% in the Netherlands (Brakefield 1984a, b, c), 0–83% in Norway (Bengtson & Hagen 1975, 1977), 6–88% in central Italy (Scali & Creed 1975) and 9–88% in some areas of the Baltic sea coast near St. Petersburg (Zakharov & Sergievsky 1978, 1980). Other populations generally have low proportions of melanics, for example 8–21% in Germany (Klausnitzer & Schummer 1983), 0–20% in Bohemia (Honek 1975), 0–15% on the Baltic coast of southern Finland (Mikkola & Albrecht 1988) and 3% in Stockholm (Zakharov & Shaikevich 2001). As one moves to Central Asia, the populations become more polymorphic than in Europe and the proportions of melanic morphs in local populations of Central Asia vary from 1–96% (Lusis 1973).

Several studies have attempted to find a **correlation between local environmental conditions and the proportion of melanics**. In Great Britain, melanic populations are concentrated in industrial areas (Creed 1966, 1971b) and around sources of air pollution in rural areas (Creed 1974). The same applies in the Baltic region (Lusis 1961, Zakharov & Sergievsky 1978, Mikkola & Albrecht 1988). The proportion of melanics also varies with distance from the sea, in central Italy (Scali & Creed 1975) and Norway (Bengtson & Hagen 1977): it is greatest near the sea, and decreases moving inland. However, contrasting trends were observed in the Netherlands (Brakefield 1984a, b, c). Local differences were observed in Ukraine, where Emetz (1984) in the course of eight years observed an increasing proportion of melanics (from 0 to 68%) on a burned site, while in the surrounding oak forest the proportion of melanics remained zero.

Variation in the proportions of particular **colour morphs in local *Har. axyridis* populations** have been studied by a number of authors (Dobzhansky 1924a, 1933, Komai et al. 1950, Komai & Hosino 1951, Komai 1956, Komai & Chino 1969, Vorontsov & Blehman 2001, Korsun 2004, Blehman 2007). The melanic morph *axyridis* is limited naturally to western Siberia where its distribution approximately coincides with the Jenisei river basin. Dobzhansky (1924a) and

Vorontsov and Blehman (2001) supposed that this morph occurs as a subspecies, probably differentiated from other populations by genetic changes at several loci. This hypothesis was supported by Blekhman (2008), who demonstrated the near-absence of a dominant elytral ridge allele (see also 2.5.1) in populations of *axyridis* morphs in western Siberia, as well as in more recent molecular genetic analyses (Blekhman et al. 2010). In the rest of the natural range of *Har. axyridis* (Pacific coast of Russia, China, Korea and Japan) all morphs occur together, but their proportions vary geographically. These populations consist of a mixture of light *succinea* morphs, melanic *axyridis*, *spectabilis* and *conspicua* morphs and a number of rare morphs. In Japan there is a marked cline, with decreasing proportions of *succinea* and increasing proportions of melanics, as one moves from the north to the south. The proportion of morphs is different in the continental populations of eastern Siberia, Korea and northern China where the proportions of melanic *spectabilis* and *conspicua* are higher than in Japan.

Other species have been less intensively studied. Studies of the geographic variation of *C. septempunctata* have concerned the size of spots. The maximum spot size occurs in ssp. *brucki* living on the Pacific coasts of Russia, Korea and Japan; while the centre of occurrence of small-spotted populations is in Central Asia. (Dobzhansky & Sivertzew-Dobzhansky 1927). Local trends for decreasing spot size in populations of warm areas compared to those of cooler areas were established by Tolunay (1939) in Turkey. The *Coccinella* species of North America are more variable. *Coccinella transversoguttata richardsoni* has several morphs that occur together, but whose frequency varies among local populations. By contrast, in *C. monticola*, *C. novemnotata* and *C. hieroglyphica*, particular morphs dominate in a particular area (Brown 1962). *Hip. variegata* can have 0–6 spots on each elytron. Six forms are found more frequently than the other 21 minority forms. The dominant forms are the same in different geographic populations of Central Europe (Schilder 1928, Strouhal 1939, Balthasarova-Hrubantova 1950, Schilder & Schilder 1951/2), but other spot forms occur in different proportions.

Variation in the extent of melanization between local populations of several species living in two relatively close localities in Central Asia, the warm Chuiskaya region and the cooler and overcast areas around Lake Issyk Kul, was described by Kryltzov (1956).

Melanization was on average greater in the Issyk Kul than the Chuiskaya region. The species exhibit either increases in spot size (*Psyllobora* (= *Thea*) *vigintiduopunctata, Bulaea lichatschovi, Coccinula sinuatomarginata, C. septempunctata, Coccinula quatuordecimpustulata*) or in the number of spots and confluences between them (*P. quatuordecimpunctata, Hippodamia tredecimpunctata, Hip. variegata*). The changing degree of melanization with climatic conditions is less pronounced in species with a dark ground colour than in species with a light ground colour (Kryltzov 1953).

Variation in the proportion of morphs on particular host plant/aphid systems represents a kind of micro-geographic variation. While no differences were established for populations of *A. bipunctata* and *A. decempunctata* on more than 20 host plants (Honek et al. 2005), significant variation was established between populations of *Har. axyridis* on coniferous and broad leaved plants (Komai & Hosino 1951). However, this variation could well be confounded by the presence of a sibling species, *Har. yedoensis*, which colonizes only conifers (Sasaji 1980).

2.4.4 Temporal variation

Temporal variation in the proportion of morphs at the same locality includes seasonal variation and long term changes over a span of years. **Seasonal variation** in morph frequency was first recorded in *A. bipunctata* by Meissner (1907a, b) and studied in detail by Timofeeff-Ressovsky (1940). During the 1930s, in Potsdam near Berlin he observed a regular increase in the proportion of melanic morphs over the summer breeding period and a decrease in the proportion of melanics during overwintering. The contrasting processes maintained an average proportion of melanics in the population. However, several studies from other geographic areas revealed no similar significant differences in the proportion of melanics early and late in the season (Lusis 1961, Bengtson & Hagen 1975, Honek 1975, Zakharov & Sergievsky 1980, Klausnitzer & Schummer 1983, Majerus and Zakharov 2000, Honek et al. 2005). No seasonal change was also found in *A. decempunctata* (Honek et al. 2005). Significant seasonal trends in the proportions of *Har. axyridis* colour morphs were the opposite of *A. bipunctata*: an increase in the light *succinea* morph through the vegetative season and lower melanic mortality in the winter was observed in populations of northern

China and Japan (Osawa & Nishida 1992, Wang et al. 2009). In central China, the frequency of the *succinea* morph was 48% in the spring, decreased to 22% in the summer and increased to 59% again in the autumn, while dark morphs exhibited a reciprocal change (Tan 1949). However, even in *Har. axyridis*, the temporal variation is area-specific and Kholin (1990) found no seasonal variation in populations of the Maritime Province of Russia.

In addition to seasonal fluctuations, **long-term changes** also have been observed in coccinellids. A decrease in the proportion of melanic *A. bipunctata* was observed in industrial areas of Birmingham in Great Britain from the 1960s onwards. From an initial 40–50%, the proportion of melanics decreased, between 1960 and the late 1970s, to about 10% and remained at this low level (Creed 1971a, Brakefield & Lees 1987). A similar decrease may have occurred during this period at Potsdam, where there were about 37–59% of melanics in the 1930s (Timofeeff-Ressovsky 1940), 15% in 1973 (Creed 1975) and only 5% in 1981 (Honek et al. 2005). Elsewhere the proportion of melanics has risen with increased industrial pollution. A large increase was observed in Gatchina, near St. Petersburg, where the proportion of melanics increased from 9% in the 1930s to 43% in 1975 (Zakharov & Sergievsky 1978). The proportion probably also increased in central Italy (Bologna) from 41% in 1926 (Fiori 1928) to 56% in 1974 (Scali & Creed 1975). The proportion of the melanic morph *sublunata* substantially increased at several localities in Central Asia, for example in Tashkent, from 47% in 1908 to 96% in 1972 (Lusis 1973). Examples of no change were observed in populations with low melanic frequencies (Klausnitzer & Schummer 1983). In *Har. axyridis* only slight long term changes were observed in a population in central Japan where the frequency of *succinea* was stable (43%) between 1912 and 1920 but decreased by 10% by the mid-1940s (Komai et al. 1950). In southern Slovakia, a population of *Hip. variegata* was sampled from 1937 to 2009 (Strouhal 1939, Balthasarova-Hrubantova 1950, Honek unpubl.): the dominant morphs remained the same and their frequencies varied little over the entire 70-year period.

2.4.5 Significance and evolution

The main interest of authors studying intraspecific colour polymorphism has been to explain its significance and origin. Because the patterns, body size and repellency/toxicity of coccinellid species differ substantially, it is likely that **the explanation will not be the same for all**. Majerus (1994) suggested that the significance of colour patterns would differ for (i) 'smaller species', (ii) the Chilocorini, (iii) host plant/habitat generalists and (iv) host plant/habitat specialists. For each group of species, light and melanic morphs have different significances and particular patterns may serve as a 'warning' in poisonous species and their mimics; 'cryptic colouration' could be adapted to particular environments, to signal for mates or as a means of thermoregulation.

In *A. bipunctata* four explanations of differences in local morph proportions were proposed: industrial melanism, (micro)climatic selection, assortative mating and Müllerian mimicry. The first two explanations are in fact complementary and well documented in some areas at least (Sergievsky & Zakharov 1981, 1983, 1989, Majerus & Zakharov 2000); the latter two represent distinct factors in the evolution of *A. bipunctata* polymorphism. **Industrial melanism** and its evolution were best documented in Britain (Creed 1966, 1971a) and northern Russia (Majerus & Zakharov 2000, Zakharov 2003), where a positive relationship between **smoke (small particle) pollution** of the air and the proportion of melanics observed exists (Creed 1971b, 1975). There was close correspondence between the proportion of melanics and the level of smoke pollution, but a less significant one or none with **sulphur dioxide pollution** (Lees et al. 1973, Brakefield & Lees 1987). However, industrial melanism does not exist in some areas, such as Bohemia and eastern Germany, where pollution was very high in the past (Honek 1975, Klausnitzer & Schummer 1983, Honek et al. 2005). This apparent discrepancy may be explained by another factor in melanism, **climate**. There is only a weak negative relationship between the proportion of melanics and **temperature**, but more important is a negative relationship between melanic proportion and **duration of sunshine** (Nefedov 1959, Lusis 1961). A statistically significant relationship was demonstrated for British populations where the correlation was $r = -0.59$ (Benham et al. 1974); after omitting sites for which sunshine hours were only estimated, it rose to $r = -0.75$ (Muggleton et al. 1975). Brakefield (1984a, b, c) also found a significant, negative correlation between sunshine hours and melanic frequency. However, the correlation between sunshine

hours and melanism may not apply on a very local scale in Britain (Creed 1975), to populations of warm Mediterranean areas (Scali & Creed 1975) and to populations of Central Asia with up to 96% of melanics (Lusis 1973) where sunshine is perhaps in excess. Climatic factors may explain the variation in frequency of melanic morphs in areas where industrial pollution fails to explain the situation. This applies to Norway where a coastal climate, i.e. the joint effects of temperature, humidity and sunshine, favours the occurrence of melanics (Bengtson & Hagen 1975, 1977). The authors found the best correlation ($r = 0.93$) was between percent melanics and an '**index of oceanity**', which combines the effects of temperature and rainfall. In the Netherlands, besides the expected negative correlation between melanic proportion and sunshine hours ($r = -0.45$, $p < 0.05$), there was also a correlation with humidity and the index of oceanity. The latter correlation was very high but, in contrast to Norway, a negative one ($r = -0.90$) (Brakefield 1984a, b, c). This suggests caution in interpreting correlations between climatic data and melanic proportions.

For the localities where seasonal differences in the proportion of melanic morphs of *A. bipunctata* exist, i.e. Germany (Timofeeff-Ressovsky 1940, Timofeeff-Ressovsky & Svirezhev 1966, 1967), Great Britain (Creed 1966, 1975) and the Netherlands (Brakefield 1985a), the mechanism of change probably comprises **balanced selection**, favouring melanics during the breeding period and non-melanics during the winter. One may calculate the selection coefficients s against black morphs in winter and t against red morphs in the summer (Creed 1975). This value is a quantitative measure of the intensity of selection which indicates the proportional reduction of the gametic contribution of a genotype compared to the favoured genotype. The average coefficients were $s = 0.52$ and $t = 0.33$ for the Potsdam population in the 1930s, and lower $s = 0.24$ and $t = 0.09$ for Birmingham in the 1960s. The magnitude of s increases with the lowest mean monthly temperature, and the values of t increased with the highest monthly maximum temperature (Creed 1975).

A **mating** advantage for melanic morphs in the breeding period of spring and summer may result from their **thermal properties**. The elytra of red morphs have a greater reflectance than those of the melanic morphs, and correspondingly absorb less radiant energy. As a consequence, dark animals exposed to sunlight have a greater temperature excess over ambient (by 2.1°C) and their initial rates of heating are 50% greater than non-melanics (Brakefield & Willmer 1985). The magnitude of temperature excess also increases with body size. The higher body temperatures of melanic individuals may increase activity during the breeding period. Melanic individuals then copulate more readily than the red ones. Lusis (1961) counted the proportion of red and dark copulating individuals on several days when the weather was convenient for breeding. The proportion of reds in copula at the time of the census ($23.0 \pm 6.5\%$ of the all red individuals recorded during the count) was significantly lower than the proportion of dark individuals ($31.8 \pm 7.1\%$). The same was true for Dutch populations, where melanics also gained a copulatory advantage (Brakefield 1984c). Melanic morphs also dispersed earlier to breeding sites in the spring, mated earlier and emerged from pupae earlier (Brakefield 1984b) which may all be a consequence of their thermal properties. Colour morphs of *Har. axyridis* also differ in several characters determining fitness including developmental rate at some stages, predation activity, longevity and fecundity (Soares et al. 2001, 2003).

In *A. bipunctata* **sexual selection and assortative mating** (preference for the mate of a particular morph) have also been supposed to maintain the variation in morph proportions observed in natural populations (O'Donald & Muggleton 1979). Some experiments indicated a **female preference** for males of the melanic morph *quadrimaculata*. This preference was observed in females regardless of their colour. It was frequency dependent and its intensity increased as the proportion of *quadrimaculata* males decreased (Majerus et al. 1982). After 14 generations of selection, O'Donald et al. (1984) and Majerus et al. (1986) were able to increase the degree of this preference in several isofemale lines. However, later studies revealed that even the combined effect of sexual and climatic selection cannot provide a general explanation for observed variation in morph frequency. In several British populations females had no mating preferences, or the melanic advantage was not frequency dependent (Kearns et al. 1990). Moreover, female preferences for melanic males in artificially selected lines disappeared after prolonged maintenance of these strains in the laboratory, and an attempt at selecting new isofemale

lines with similar preferences failed (Kearns et al. 1992). Similar preferences in both sexes for melanic mates were found in *C. septempunctata* (Srivastava & Omkar 2005) while the reverse was observed in Japanese *Har. axyridis* where females preferentially choose non-melanic males in spring, but non-melanics are less successful at mating than melanics in summer (Osawa & Nishida 1992). In this species, colour morph preference may be confounded by body size since in non-melanic males the mating individuals were larger than unmated individuals, while in melanic males there was no difference (Ueno et al. 1998).

Müllerian mimicry has also been proposed for maintaining colour polymorphism. Species whose adults have bright red or yellow colouration on the upper surface are thought to be aposematically coloured to prevent attacks by visual predators, mostly birds. The repellency is a consequence of previous negative experience of the predator with a distasteful prey, which has been demonstrated a number of times for coccinellids (e.g. Whitmore & Pruess 1982) (also Chapter 9). Several species of similar colouration may represent a Müllerian complex, with predators not distinguishing between the individual component species. The negative experience of the first attack on a member of this complex becomes generalized and all its members are protected in the future. Brakefield (1985b) hypothesized that polymorphic and relatively non-toxic *A. bipunctata* and *A. decempunctata* were mimicking two Müllerian models, the red *C. septempunctata* and black *Exochomus quadripustulatus*, both of which are highly distasteful and toxic to predators (Marples et al. 1989, Marples 1993). More recently, it has also been suggested that intraspecific variation in colour pattern might be an 'honest indicator' of chemical defensive strength (Bezzerides et al. 2007).

2.5   THE INHERITANCE OF OTHER TRAITS

2.5.1   Morphological characters: wing polymorphism

In addition to studies of colour pattern, there has been a diversity of other genetic studies on other biological characteristics of coccinellids. Studies of **morphological characters** include those of a transverse ridge occurring on the hind part of the elytra of *Har. axyridis*, which is polymorphic for this character. The ridge is thought to be encoded by a single dominant gene (Hosino 1936) and, although the function of the ridge is unknown, geographic variation in its occurrence appears to be clinal, suggesting it is under selection (Komai 1956, Sasaji 1980, Blekhman 2008, 2.4.3). The majority of morphological studies have been on **wing polymorphism** in coccinellids, which has been intensively studied in other insect groups (Zera & Denno 1997). In some species, such as *Subcoccinella vigintiquatuorpunctata* and *Rhizobius* spp., genetically controlled wing reduction occurs commonly in natural populations (Pope 1977, Hammond 1985). A more recent focus of such studies has been *A. bipunctata*, in which naturally occurring wingless individuals are very rare (Majerus & Kearns 1989, Marples et al. 1993). In this species, winglessness is controlled by a single recessive allele, although the extent of its development is determined by genetic background, modifier genes and environmental influences, notably temperature (Marples et al. 1993, Ueno et al. 2004, Lommen et al. 2005, 2009). Genetically controlled flightlessness has also been found in *Har. axyridis*: in this case the ladybirds appear morphologically normal, but structural modifications are present in the wing muscles (Tourniaire et al. 2000). Flightless strains of *Har. axyridis* have been used as a means of biological control, being largely unable to disperse away from the target crop (Gil et al. 2004, Seko et al. 2008, but see Seko & Miura 2009); a similar biological control function has been proposed for flightless *A. bipunctata* (Lommen et al. 2008; see also Chapter 11).

For wingless morphs of *A. bipunctata*, the **pattern of wing development** is similar to that of fully winged individuals, but is slower, making the wing discs smaller at pupation. Additionally the wings are truncated at adulthood, although this does not appear to arise from apoptosis (cell death). Expression of the gene *Distal-less* is limited to the proximal anterior margin of the wing discs in the larvae of wingless *A. bipunctata* morphs, while in winged morphs it is expressed all round the wing disc margin. The expression of this gene may determine the extent of winglessness (Lommen et al. 2009). Two other genes, *vestigial* and *scalloped*, have also been shown to play a role in ladybird wing development similar to the roles they play in *Drosophila melanogaster*. Interestingly, *scalloped* also plays a role in pupal ecdysis. It has been suggested

that disruption of *vestigial* and *scalloped* expression could be used as a means to produce wingless *Har. axyridis* for biological control purposes (Ohde et al. 2009).

2.5.2 Life history characters: heritability, selection experiments and genetic trade-offs

The majority of **life history characters**, such as body size, developmental and reproductive characteristics, and the requirement for diapause before reproduction, exhibit complex polygenic inheritance. There are two main approaches used to study genetic variation in such characters. **Heritability studies** give a measure of the contribution of genetic variation to total observed phenotypic variation in a species. For example, Ueno (1994a) reared *Har. axyridis* from different male–female parental combinations to estimate the heritabilities of body size and developmental characters; because males were mated to more than one female (although females were singly mated), he was able to make these estimates separately for the sexes. The estimates were moderate, varying between 0.24 and 0.56 of total phenotypic variance resulting from genetic variance. An alternative approach is to apply **directional selection** to a particular character, in order to demonstrate underlying genetic variation in the character. Thus, for example, it proved possible to select against the tendency for an obligate diapause before oviposition in *C. septempunctata* with eggs collected after short pre-ovipositional period, resulting in a general decline in the proportion of females requiring a diapause before reproduction (Hodek & Cerkasov 1961). This suggested that variation in the requirement for diapause in this species had an underlying genetic component (Chapter 6).

Linked to these sorts of studies are estimates of **genetic correlations** and **trade-offs** between traits. Selection on one particular trait may have correlated effects on other traits (**pleiotropy**) if the genes involved affect both traits. The traits may be obviously related: for example, in his study of *Har. axyridis*, Ueno (1994a) found high positive genetic correlations between body weight and body length (i.e. ladybirds that are longer are also heavier). Of much greater interest are the relationships between less closely related traits, since these give an idea of how genetic constraints can mould the overall phenotype of an organism. In his study, Ueno (1994a) found a possible negative correlation between body size and developmental time and a positive one between body size and developmental rate: thus, a longer developmental time and a slower developmental rate were apparently linked to smaller body size. This finding is in part supported by a study in which fast development was selected for in *Hippodamia convergens*: the fast developing larvae consumed more prey per unit time and exhibited a higher feeding efficiency. In selected lines there was a negative correlation between developmental time and body weight, even though body weight was apparently not affected by intense selection (Rodriguez-Saona & Miller 1995).

Of particular interest has been the relationship between traits related to the ability to exploit different types of food. Here, genetic trade-offs are seen as providing a basis for **dietary breadth and specialization**, with specialists being seen as having evolved very high efficiency on one type of food at the expense of an ability to exploit other types. The general focus has been on the suitability of food mediated by its allelochemical and nutrient content. Interestingly the majority of evidence from both phytophagous and aphidophagous coccinellids supports a view that, although genetic variation exists in the ability to process particular food species, genetic trade-offs in their nutritional suitability are either very weak or non-existent in ladybird consumers (Ueno et al. 1999, 2003, Ueno 2003, Fukunaga & Akimoto 2007; see also Rana et al. 2002, Sloggett 2008).

2.6   MOLECULAR GENETIC STUDIES

The first molecular genetic studies of the Coccinellidae were published in Japan in the 1970s (Sasaji & Ohnishi 1973a, b, Sasaji & Hisano 1977, Kuboki 1978); however, it is only in the last 20 years that molecular techniques for studying the Coccinellidae have come into regular use, mirroring the wider availability and lower cost of a diversity of methodological approaches. With the exception of the work discussed earlier on chromosomes (2.3.2) and wing development (2.5.1), there is little work providing insight into the biochemical, regulatory and developmental pathways connecting specific genes to particular phenotypic characteristics of coccinellids. However, coccinellids have been used as **models for the evolution of specific genomic regions** and **molecular markers**

have played a significant role in ecological and evolutionary studies of coccinellid biology.

A useful feature of coccinellids is that they can be **non-destructively sampled** for molecular genetic studies. Both the adults and larvae of coccinellids exude alkaloid-bearing haemolymph as a chemical defence, a process known as reflex bleeding (Happ & Eisner 1961, Kendall 1971; Chapters 8 and 9). The reflex blood contains enough protein for isozyme or allozyme analysis (Kuboki 1978, Ransford 1997) and it is also possible to amplify DNA from reflex blood using the polymerase chain reaction (PCR) (Karystinou et al. 2004). It is therefore not necessary to kill or damage live coccinellids in order to obtain samples for genetic analysis. This is a particularly valuable attribute for behavioural studies. For example, in a study of mating behaviour, sperm competition and paternity, Ransford (1997) was able to genotype adult *A. bipunctata* using allozymes from reflex blood extracts. This left the beetles alive for subsequent breeding and use in mating experiments.

2.6.1   Sequence evolution

2.6.1.1   Mitochondrial DNA and the inference of evolutionary history

Many studies of evolutionary history utilize **mitochondrial DNA (mtDNA)** as a molecular marker, because of its ease of PCR amplification, rarity of recombination and uniparental cytoplasmic inheritance (i.e. only through the female line). These factors have made mtDNA attractive when compared to nuclear molecular markers for phylogenetics and studies of genetic structure and phylogeography (Harrison 1989, Simon et al. 2006). Additionally more recently mtDNA has been proposed as a universal means for identifying species using a specific characteristic 'barcode' DNA sequence (Hebert et al. 2003, Hajibabaei et al. 2007). The universal utility of mtDNA has been questioned on a number of grounds, however (Ballard & Whitlock 2004, Hurst & Jiggins 2005). One objection has been that **mtDNA is often not selectively neutral**, as should be the case for molecular markers, because it is effectively **linked to endosymbiotic bacteria** such as male-killers, which are also inherited cytoplasmically and are likely to be under strong selection. In consequence interpretations of population structure or evolutionary history based on mtDNA are confounded by selection (Turelli et al.

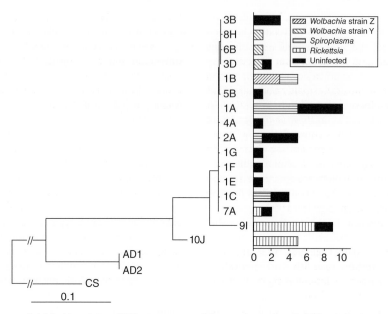

Figure 2.6 Phylogeny of *Adalia bipunctata* mtDNA sequences and the number and male-killing infection status of individuals bearing that mitotype. In the phylogeny CS, AD1 and AD2 are the mitotypes from one specimen of *Coccinella septempunctata* and two of *Adalia decempunctata*; all other terminal branches are *A. bipunctata* mitotypes. Scale bar = 0.1 substitutions per nucleotide for tree branch lengths. Redrawn from Schulenburg et al. (2002), with permission.

1992, Johnstone & Hurst 1996, Hurst & Jiggins 2005). The wide prevalence of male-killing bacteria in aphidophagous coccinellids (Chapter 8) has made them particularly suitable for studying this issue. Male-killing symbionts clearly affect host fitness, increasing female fitness at the cost of the dead males (Chapter 8). The consequent increase in infected daughters in a population will lead to a related increase in the mtDNA variant associated with the male-killer; furthermore because male-killers are rarely transmitted perfectly between generations, so that a proportion of uninfected individuals is produced, the mtDNA variant will also reach significant levels in uninfected individuals (Johnstone & Hurst 1996).

Schulenburg et al. (2002) investigated **mtDNA diversity and sequence divergence in relation to male-killing** in *A. bipunctata*. This coccinellid was known to harbour four different male-killers throughout its range: a *Rickettsia*, a *Spiroplasma* and two strains of *Wolbachia*. Schulenburg et al. sequenced two variable mtDNA regions (parts of the cytochrome oxidase subunit I (COI) and NADH dehydrogenase 5 (ND5)

genes) from 52 infected and uninfected ladybirds from nine populations extending from Britain to Russia. They then compared the distributions of different mitochondrial variants to the distributions of the male-killers across individuals and populations. Of a total of 10 mitotypes found associated with male-killing bacteria, nine were found to occur with only one bacterial type (Fig. 2.6). There was no significant differentiation of mitotypes from different geographic regions. Instead, **mitotypes were differentiated according to bacterial infection**, particularly between ladybirds with different such infections; differentiation between uninfected ladybirds and those with *Spiroplasma* and *Wolbachia* infections was much lower. A higher degree of mitotype differentiation among *Rickettsia*-infected individuals supported the idea that this bacterium had been present in *A. bipunctata* for longer than the others, allowing for greater subsequent mitochondrial sequence divergence.

Schulenburg et al.'s findings very clearly support the view that the distribution of DNA mitotypes is strongly correlated with the occurrence of the different male-killing symbionts: mtDNA variants and cytoplasmic male-killers are inherited together as a single unit and selection on one will result in a 'hitch-hiking' effect on the other. Building on this earlier study, Jiggins & Tinsley (2005) addressed the question of **selection on mitochondrial and nuclear DNA**. Concentrating exclusively on *Rickettsia*-bearing *A. bipunctata*, they showed that three different strains of the bacterium were associated with three different mtDNA variants. The occurrence of the three haplotypes varied geographically and, more equivocally, temporally. Tests on the haplotype frequency and genealogy of the mtDNA variants indicated that **mtDNA was under balancing selection** (i.e. selection maintaining stable frequencies of the mitotypes). Jiggins & Tinsley compared nuclear and mitochondrial genes by sequencing a nuclear gene (the glucose-6-phosphate dehydrogenase gene (*g6pd*)) in individuals from many of the same populations as those studied by Schulenburg et al. (2002) and comparing their data to Schulenburg et al.'s mtDNA data. In contrast to the mtDNA data, the **nuclear *g6pd* data was consistent with neutral evolution**. Furthermore, the gene diversity of mtDNA was much higher than that expected under neutrality when compared to *g6pd*. Jiggins & Tinsley also found strong differences in haplotype frequencies in infected and uninfected beetles: they were therefore able to conclude that selection was acting on the mtDNA through the male-killer rather than directly on the mtDNA; the latter would be more likely to result in similar haplotype frequencies in infected and uninfected beetles.

It is clear that in aphidophagous coccinellids the presence of cytoplasmically inherited male-killing symbionts and their linkage to mtDNA means that the assumptions of mtDNA selective neutrality required are violated. Hurst & Jiggins (2005) discuss the implications of this for studies using mtDNA. First, the **mtDNA diversity will be affected through selection on the symbiont**, confounding the use of mtDNA diversity in phylogeographic studies of population history and demography. Second, selection maintaining different symbionts in different populations, and the mitotypes associated with them, may lead to **increased estimates of population differentiation**, even if migration occurs between populations. Third, in the case of very closely related species, hybridization can lead to transfer of endosymbionts and associated mitotypes from one species to another;

if positive selection occurs for the symbiont in the new species, this can lead to **homogenization of the mtDNA of the two species.** This will make them indistinguishable in phylogenies, attempts to delimit species boundaries or when using mtDNA-based barcoding techniques. It is worth noting that none of these problems are limited to coccinellids: Hurst & Jiggins point out that symbiotic or parasitic inherited endosymbionts are very common in arthropods, meaning problems in the use of mtDNA markers may be extremely widespread.

Until now, much of the work using mtDNA in coccinellids has been phylogenetic (2.6.2.2). Most of the species in these phylogenies are too distantly related for there to be serious problems in the construction of a phylogeny. In most known cases, even sibling coccinellid species do not produce viable, fertile offspring (e.g. Zaslavskii 1963, Sasaji 1980, Ireland et al. 1986, Kobayashi et al. 2000, but see Sasaji et al. 1975); consequently cases of mtDNA introgression are likely to be very rare. Nonetheless, a lack of mtDNA neutrality could compromise calculations of times since species diverged based on the assumption of a neutral molecular clock (e.g. Palenko et al. 2004). The implications for population genetic and phylogeographic research of using mtDNA as a marker are more serious. However, allozymes or, more recently, microsatellites have generally been preferred to mtDNA for such purposes in studies of the Coccinellidae (see 2.6.2.3), and although mtDNA has been used in some studies (e.g. Blekhman et al. 2010), this is often in concert with additional nuclear molecular genetic data (e.g. Marin et al. 2010, C.E. Thomas et al. 2010). As our knowledge of the coccinellid genome increases, the potential of nuclear markers for population and evolutionary studies is becoming better characterized. For example, Jiggins & Tinsley's work suggests that the *g6pd* sequence is a suitable marker for population or phylogeographic studies; an increasing number of nuclear genomic regions have now also been characterized in molecular phylogenetic studies (2.6.2.2).

2.6.1.2   The ITS1 region

The **first ribosomal internal transcribed spacer (ITS1)** region is a non-coding region between the 18S and 5.8S ribosomal RNA genes. It is relatively fast-evolving due to its non-coding nature, and is easily

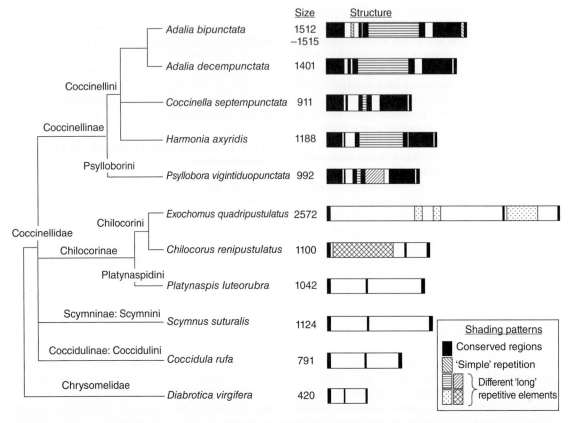

Figure 2.7 Summary of ITS1 sequence data from the Coccinellidae. From left to right, phylogenetic tree of ITS1 sequences, ITS1 sequence size in bp, and the structure of the ITS1 region. Redrawn from Schulenburg et al. (2001), with permission.

amplified from diverse taxa by PCR, due to the conserved gene regions lying on either of it, and the high copy number of rDNA arrays within the genome (Hillis & Dixon 1991). These characteristics have made it of value in phylogenetic and population studies of a diversity of taxa including insects (e.g. Schloetterer et al. 1994, Roehrdanz et al. 2003, Kawamura et al. 2007). The high number of copies of rDNA scattered across the nuclear genome of organisms means that heteroplasmy can occur (i.e. different rDNA copies have different sequences, e.g. Vogler & DeSalle 1994, Harris & Crandall 2000). Nonetheless, many taxa are believed to exhibit a high degree of rDNA copy homogeneity, as a result of concerted evolution across rDNA repeats (Hillis & Dixon 1991).

The ITS1 region has been particularly well studied in the Coccinellidae, by Schulenburg et al. (2001): they

sequenced the ITS1 regions of 10 coccinellid species in the subfamilies Coccinellinae, Chilocorinae, Scymninae and Coccidulinae. Both the **ITS1 size and interspecific size variation were extremely large**, with sizes ranging between 791 and 2572 base pairs (bp) (Fig. 2.7), compared to values of less than 550 bp for other polyphagous beetles. Even within tribes size variation remained high, being over 1.6-fold in the Coccinellini and 2.3-fold in the Chilocorini. The **ITS1 sequences of different species were strongly divergent**, and clearly homologous sequences could only be identified across all species at three very small regions of about 20 bp or less. Homology was higher within the subfamily Coccinellinae, comprising six regions of in total about 550 bp: it was possible to use these regions for phylogenetic reconstruction within the Coccinellinae. *Adalia bipunctata* specimens from

Britain, Germany and Russia exhibited minimal inter-individual ITS1 variation, this being limited to three nucleotide changes and two copy number differences in simple repetitive elements of two or three nucleotides. There was no indication of ITS1 heteroplasmy (within-individual variation between different ITS1 copies) in any of the coccinellid species studied.

Six coccinellid species exhibited **large units, of 26–212 bp, that were repeated more than once** within their ITS1 sequences (Fig. 2.7). Within the Coccinellinae, there was a high degree of sequence similarity in a given type of repeated element both intra- and interspecifically. This was not invariably the case for all repeated elements, however: sequence divergence varied both intraspecifically and, when elements were homologous, interspecifically. Schulenburg et al. (2001) argue that these long repetitive elements have played a significant role in the evolution of the large size of the coccinellid ITS1 region and the interspecific size variation exhibited by coccinellids. The repetition observed indicates that the evolution of new repeats is a common occurrence in the coccinellid ITS1 region, while a high level of sequence divergence in homologous repetitive elements means that many older repetitions may not be detected due to the effects of high nucleotide substitution rates. Schulenburg et al. suggest that the observed interspecific size variability in the ITS1 region potentially makes it an ideal molecular marker for coccinellid species in ecological studies (see also 2.6.2.1).

2.6.2 Molecular studies of coccinellid biology

Molecular studies now play a considerable role in studies of coccinellid ecology and evolutionary biology, and have already provided significant insights into the evolutionary history of members of the Coccinellidae, their population structure and behaviour. The areas where molecular studies have been or are likely to be of greatest value can be broadly divided into four areas: (i) molecular identification of species, populations or strains; (ii) phylogenetic studies; (iii) studies of population genetics and phylogeography, and (iv) studies of paternity and sperm competition. Clearly there is some overlap between these areas: for example, sequence data from studies aimed at providing molecular identification of coccinellid species may often be suitable for use in phylogenetic work (cf. Hajibabaei et al. 2007). A

number of different methodological approaches have been used, which are summarized in Table 2.2, although this is not an exhaustive list of the potential techniques available. The reader is referred to textbooks by Avise (2004), Beebee and Rowe (2008) and Freeland et al. (2011) for further background on the use of molecular genetic techniques in ecological and evolutionary studies.

2.6.2.1 Species, population and strain identification

Markers specific to species, populations or strains possess great potential in ecological work. Species-specific markers could be used to identify problematic life-history stages such as eggs or first instar larvae, which may otherwise require further rearing for unambiguous identification (Schulenburg et al. 2001). They are also likely to be of particular value in gut content analysis of potential predators of ladybirds, to identify ladybird prey consumed in the field; in particular field studies of intraguild predation may benefit from this approach (Gagnon et al. 2005, Weber & Lundgren 2009; Chapter 10). Population- or strain-specific markers could potentially have utility in the monitoring of the fate of released strains in an area where a natural population of the same species is already present, in competition experiments between strains or in the monitoring of multiple laboratory colonies with different origins for cross-contamination (Roehrdanz 1992, Roehrdanz & Flanders 1993). Although as yet there has been relatively little work published using genetic markers any of these ways, a number of studies have focused on the methodological approaches that would be most appropriate for such work.

Early work showed that both the **Restriction Fragment Length Polymorphism (RFLP)** and **Random Amplification of Polymorphic DNA (RAPD)** methods could be used to generate species-specific electrophoretic banding or fragmentation patterns (Roehrdanz 1992, Roehrdanz & Flanders 1993). RFLP analysis was carried out on a PCR-amplified 1200–1300 base pair mitochondrial region spanning parts of the 12S and 16S rDNA genes and the region between them. Using this method it was possible to distinguish *C. septempunctata*, *C. transversoguttata*, *Hip. variegata* and *P. quatuordecimpunctata* (Roehrdanz 1992). The proportion of shared restriction fragments varied from 97% between the congeneric *C. septempunctata* and

Table 2.2 Coccinellid studies using molecular genetic markers. Studies marked with an asterisk were preliminary studies that demonstrated that a particular methodological approach was suitable for a given use in coccinellid studies; however the method was not further utilized in such work.

Methodological approach	Description	Uses in coccinellid studies	References
Isozyme/ allozyme analysis	Variants of the same enzyme are separated by gel electrophoresis; they are visualized by adding the specific substrate with which the enzyme reacts and a stain specific to the product produced when the enzyme and substrate are present together. Allozymes are enzyme variants that are the product of a single genetic locus; isozymes are enzymes that that perform the same reaction, but occur at multiple loci.	Phylogenetics and phylogeography (species delimitation)	Sasaji & Ohnishi 1973b; Kuboki 1978; Sasaji & Nishide 1994
		Analysis of population genetic structure	Eggington 1986; Krafsur et al. 1992, 1995, 1996a, 1996b, 1997, 2005; Steiner & Grasela 1993; Coll et al. 1994; Krafsur & Obrycki 1996; Obrycki et al. 2001
		Paternity analysis	Ransford 1997
Restriction Fragment Length Polymorphism (RFLP) analysis	Detects genomic variation by cutting up the genome using restriction enzymes and using electrophoresis to examine the size of the resulting fragments. Polymorphisms at restriction enzyme cutting sites, as well as DNA sequence length polymorphisms, result in differing fragment lengths which can be separated using electrophoresis.	Species identification	Roehrdanz 1992*
		Analysis of population genetic structure	Haddrill 2001*
		Paternity analysis	Haddrill 2001*
Amplified Fragment Length Polymorphism (AFLP) analysis	Total genomic DNA is cut up using restriction enzymes and adaptor sequences joined to the ends of the resulting DNA fragments. A subset of the fragments is amplified using PCR with primers complementary to the adaptor and part of the fragment. The amplified fragments are visualized by electrophoresis.	Analysis of population genetic structure	Haddrill 2001*
		Paternity analysis	Haddrill 2001*
Random Amplification of Polymorphic DNA (RAPD) analysis	Uses PCR with short, relatively unspecific primers (typically 8–12 nucleotides) is to amplify small (300–2000 base pair) fragments of genomic DNA which are then resolved by electrophoresis. The resultant patterns of amplified DNA fragments, from a number of such primers, can be used to generate a characteristic DNA profile for a species, population or strain of an organism.	Species, population and strain identification	Roehrdanz 1992*; Roehrdanz & Flanders 1993*

Table 2.2 (*Continued*)

Methodological approach	Description	Uses in coccinellid studies	References
Microsatellite analysis	Uses tandemly repeated sequences of typically one to six nucleotides (e.g. CACACACA . . .), which are amplified using PCR primers designed to bind either side of the repeated sequence. The number of times the sequence is repeated, and thus the length of the microsatellite region, is frequently highly variable intraspecifically, making microsatellites suitable markers for population genetic analysis; in combination they can give a near-unique individual specific DNA fingerprint, making them of value in studies of individual reproductive success and paternity.	Species identification (single locus)	A. Thomas et al. 2010
		Analysis of population genetic structure or phylogeography	Haddrill 2001; Haddrill et al. 2008; Lombaert et al. 2010
		Paternity analysis	Haddrill et al. 2008
Inter-Simple Sequence Repeat (ISSR) analysis	Uses length polymorphism in genomic regions between microsatellite loci, which are amplified using primers complementary to two neighbouring microsatellites. As ISSRs are often more conserved than microsatellites they are used in phylogeographic studies, rather than studies of individuals.	Phylogeography (species delimitation)	Marin et al. 2010
Use of DNA sequences	PCR and sequencing of specific gene regions; the sequence is used directly in analyses	Phylogenetics	Howland & Hewitt 1995; Kobayashi et al. 1998; Palenko et al. 2004; Hunt et al. 2007; Robertson et al. 2008; Weinert 2008; Giorgi et al. 2009; Magro et al. 2010; Sloggett et al. in press
		Phylogeography (including species delimitation)	Kobayashi et al. 2000, 2011; Blekhman et al. 2010; Marin et al. 2010; C.E. Thomas et al. 2010
Use of specific PCR primers	Target-specific primers designed on the basis of sequence data. PCR of samples, followed by electrophoresis. Presence of a PCR amplification band indicates target DNA present.	Species identification	Gagnon et al. 2005*; Harwood et al. 2007
DNA microarray analysis	Labelled amplified DNA is simultaneously probed with many distinct gene- or organism-specific DNA oligonucleotides arranged together as microscopic spots on a solid matrix. Hybridization to the probes is detected through labelling, and conclusions drawn on the basis of which probes have hybridized to the DNA.	Species identification	Pasquer et al. 2009*

C. transversoguttata to 25% between the *Coccinella* species and *Hip. variegata*, suggesting a phylogenetic signal. The RAPD method was able to distinguish not only the four species, but intraspecifically between laboratory colonies with widely different geographic origins. Combinations of three or four specific primers were sufficient to completely differentiate the colonies, although many more primers were tested to find the most suitable (Roehrdanz 1992, Roehrdanz & Flanders 1993). With RAPD markers, as with RFLPs, although *C. septempunctata* and *C. transversoguttata* were distinct, they were more similar than *C. septempunctata*, *Hip. variegata* and *P. quatuordecimpunctata* were to each other. Although the authors suggested a number of uses for such markers, neither the RAPD nor RFLP methods appear to have been used in any practical application. The use of RAPD markers has in general declined, in large part due to the relatively low reproducibility of results with small variations in initial PCR conditions; both methods are now considered somewhat outmoded by comparison with the use of DNA sequences, either directly or, more often, to design PCR primers or probes specific to the entity being studied.

The direct use of DNA sequences for species identification has found its fullest expression in the idea of **DNA barcoding**, whereby short standardized sequences are used to distinguish species (Hebert et al. 2003, Hajibabaei et al. 2007). The barcode sequence of an unknown specimen, amplified using conserved primers, is compared to a library of such sequences obtained from specimens of known species identity; the specimen can then be identified if its sequence closely matches one in the library. A 650 bp fragment of the 5′ end of the mitochondrial COI gene has been adopted as a universal standard barcode region for animals. A number of such barcodes already exist for coccinellids in the GenBank database (http://www.ncbi.nlm.nih.gov/Genbank/); however, coccinellid barcodes have not yet been used in ecological studies, although it seems likely they will be used in the future. The reservations of Hurst & Jiggins (2005) about the potential for mtDNA to cross species boundaries must be borne in mind if very closely related species are studied using COI barcodes, although the risk is quite small (2.6.1.1). A potential alternative to using mtDNA barcodes is the use of the nuclear ITS1 sequences (Schulenburg et al. 2001; 2.6.1.2). These genomic regions exhibit high evolutionary rates in

coccinellids, making them species specific; furthermore they appear to be relatively homogeneous, both within and between individuals of a species.

Sequence data from identified target individuals can also be used to design **specific PCR primers or probes**. Gagnon et al. (2005) designed four sets of **species-specific primers** to amplify parts of the ITS1 regions of *C. septempunctata*, *P. quatuordecimpunctata* and *Har. axyridis* and the COI region of *Coleomegilla maculata*. Overall, the species-specificity of the primers was high, although there were two cases of cross-priming to other coccinellid species. After electrophoresis of the PCR product, it was possible to detect a PCR band from a target species' DNA when in a 10% mixture with that of another species, mimicking the proportion of prey DNA expected to occur in an intraguild predator. PCR of a **single microsatellite marker** specific to *A. bipunctata* has recently been used to identify *Har. axyridis* intraguild predation of this species (A. Thomas et al. 2010).

Microarray chips with COI DNA probes were used by Pasquer et al. (2009) to differentiate a number of different beneficial insects, including the coccinellids *A. bipunctata*, *C. septempunctata*, *Chilocorus nigritus*, *Cryptolaemus montrouzieri* and *Rhizobius lophanthae*. The chip, carrying a number of distinct **species-specific probes**, was exposed to fluorescence-labelled, PCR-amplified DNA from a single insect. The insect species was identified by which probes **hybridized** to the DNA. Pasquer et al. used at least four different probes for each beneficial insect species they wished to identify, to avoid a higher error rate consequent on the use of single probes (through false positives or negatives). The authors point out that, particularly for a species like *A. bipunctata* with a very high level of polymorphism in the COI gene (see 2.6.1.1), reliable identification can only be accomplished through the use of multiple probes. It seems likely that, as our knowledge of coccinellid genomes and markers increases, microarray chips will be increasingly used to identify not only species, but specific populations and strains as well.

It is worth noting that the **methods described in this section for coccinellids can be equally applied to the organisms with which they interact**. Laboratory- and field-based work already exists in which molecular techniques have been used to identify or quantify **prey in coccinellid diets**, (Weber & Lundgren 2009; Chapter 10). A less obvious application of molecular genetic methods is as a means to identify and study **coccinellid symbionts, pathogens and parasites**. Molecular methods are routinely used in the identification of male-killing endosymbionts (Chapter 8), but hold great and as yet untapped potential as an investigative tool with which to study other parasitic and pathogenic organisms of coccinellids, which in many cases have been poorly studied.

2.6.2.2   **Phylogenetics** (see also Chapter 1)

Phylogenetic studies of the Coccinellidae, like all such studies, originally utilized phenotypic characteristics, particularly **morphology**, and until recently our view of coccinellid phylogeny and consequent taxonomy has been almost exclusively morphologically based (e.g. Sasaji 1968, Fuersch 1996, Kovar 1996). The advent of molecular methods has not only made it possible to re-address questions about overall broad scale relationships between subfamilies and tribes, but also to address in detail smaller scale questions about the exact relationships between closely related species within the larger groupings. Although a number of methods have been shown to provide phylogenetic information, including studies of isozyme variation in closely related species (Sasaji & Ohnishi 1973b, Kuboki 1978, Sasaji & Nishide 1994), RFLP and RAPD studies (Roehrdanz 1992, Roehrdanz & Flanders 1993), the vast majority of molecular phylogenetic studies of coccinellids have been based on **DNA sequence data**. They include studies addressing the monophyly of the Coccinellidae and its position within the Coleoptera (Howland & Hewitt 1995, Hunt et al. 2007, Robertson et al. 2008), relationships between coccinellid subfamilies and tribes (Giorgi et al. 2009, Magro et al. 2010) and the phylogeny of closely related genera or species (Kobayashi et al. 1998, Palenko et al. 2004, Sloggett et al. in press).

Studies focused on **closely related genera or species** have been exclusively limited to mitochondrial COI sequence data. This sequence appears to exhibit the optimum level of divergence for phylogenetic analysis at the level of distinct species within genera or very closely related genera (Kobayashi et al. 1998, Palenko et al. 2004), although the issue of mtDNA linkage to endosymbionts must be considered for sibling species (Hurst & Jiggins 2005; 2.6.1.1). Above this taxonomic level, the phylogenetic utility of the sequence declines.

Mawdsley (2001) found that a COI sequence phylogeny of *C. septempunctata*, *A. bipunctata* and *Calvia quatuordecimguttata* from the Coccinellinae and *E. quadripustulatus* from the Chilocorinae varied markedly in its topology depending on the outgroup and tree-searching algorithm used. Sloggett et al. (in press) in comparing a number of genera of Coccinellini, found that the inferred amino acid sequences were more informative than DNA sequences for reconstruction of the relationship among genera; DNA sequence information was likely confounded by a high degree of homoplasy (i.e. identical bases in different species not occurring by common descent). In attempting a COI-based phylogeny of the Coleoptera, Howland and Hewitt (1995) found sequence variation within beetle families in some cases to be a great as that between beetle families, and concluded that a more conserved sequence was necessary. Studies of the phylogenetic relationships of the whole family Coccinellidae or of subfamilies or tribes within the Coccinellidae have used more slowly evolving sequences, particularly 18S ribosomal DNA (18S rDNA) and other rDNA regions (Hunt et al. 2007, Robertson et al. 2008, Giorgi et al. 2009, Magro et al. 2010); unlike studies of close relatives they have also all used more than one gene region.

DNA sequence-based studies of the **relationships of the Coccinellidae to other beetle groups** all agree that the Coccinellidae is a monophyletic group. Studies at **the subfamilial and tribal level** within the Coccinellidae have indicated that some traditionally recognized groups are paraphyletic, that is they do not include all the descendants of the most recent common ancestor and thus do not form natural groups. Giorgi et al. (2009) suggest that this is the case for the Sticholotidinae, Chilocorinae, Scymninae and Coccidulinae, although the Coccinellinae was supported as monophyletic. Their conclusions are broadly mirrored by Magro et al. (2010), who also considered the Epilachninae as a paraphyletic group. In the near future we are likely to have a much improved view of the relationships at all taxonomic levels within the group, which is currently being intensively investigated as part of the US National Science Foundation Partnerships for Enhancing Expertise in Taxonomy (PEET) and Assembling the Tree of Life (AToL) initiatives. For this work a broad range of gene regions are being used (mitochondrial 12S rDNA, 16S rDNA, COI and COII; nuclear 18S rDNA, 28S rDNA and histone

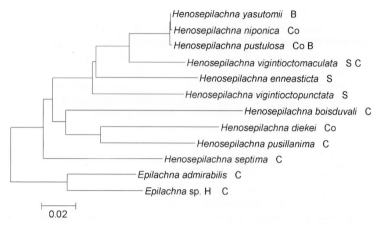

Figure 2.8 Phylogenetic tree of twelve Epilachninae, showing plant host shifts in this phytophagous group. The tree was constructed using 1000 bp cytochrome oxidase I (COI) mitochondrial sequences. Scale bar = 0.02 substitutions per nucleotide for tree branch lengths. Host plants: B, Berberidaceae; Co, Compositae; S, Solanaceae; C, Cucurbitaceae. Redrawn from Kobayashi et al. (1998) taking into account revised taxonomy, with permission.

H3 genes). Some of these regions are slowly evolving (18S and 28S), whereas others are fast evolving regions (12S, 16S, COI, COII): in this way it is hoped to provide complementary data on both older and more recent evolutionary events (J.A. Giorgi & J.V. McHugh, pers. comm.).

There are also a number of phylogenetic studies that have not aimed purely at providing phylogenetic clarity as an end in itself, but rather have aimed to address a variety of questions about the **evolution of other characteristics in a phylogenetic context** (Kobayashi et al. 1998, Weinert 2008, Giorgi et al. 2009, Sloggett et al. in press). For example, Giorgi et al. (2009) have addressed how **food preferences** have evolved within the Coccinellidae by mapping feeding preferences onto a DNA sequence phylogeny of subfamilies and tribes. From this they were able to conclude that coccidophagy is ancestral, with most other feeding preferences having evolved directly from coccidophagy. At a lower phylogenetic level, Kobayashi et al. (1998) used COI sequences to examine host plant shifts in the Epilachninae (Fig. 2.8). Sloggett et al. (in press) used COI sequences to construct a phylogeny of Coccinellini bearing different **chemical defences**. Using the phylogeny, they were able to show that the myrmecophilous *C. magnifica*, which is unusual amongst *Coccinella* species in possessing convergine–hippodamine chemical defences, is derived within the

Coccinella genus and is not in fact more closely related to other species with the same alkaloids, as had previously been suggested. They concluded that this change of chemical defence type was linked to the preference of *C. magnifica* for living with ants, which is also unique within the genus.

An interesting further development of the phylogenetic technique is provided by Weinert (2008), who studied **evolution of the endosymbiotic bacteria *Wolbachia* and *Rickettsia***, which can both cause male-killing in coccinellids. In her work, Weinert utilized DNA sequences and consequent phylogenies of both ladybird hosts and the bacteria that they harbour. Thus, she was able to show that both types of bacteria have invaded the Coccinellidae multiple times and that bacterial infection of new coccinellid species is most common when they are closely related to the ancestral host (i.e. decreases with increasing genetic distance). Nonetheless, within individual *Rickettsia* and *Wolbachia* clades that were exclusively coccinellid symbionts, only the *Rickettsia* clade exhibited close congruence with host phylogeny (Fig. 2.9).

2.6.2.3   Population genetic and phylogeographic studies

Population genetic studies are concerned with allele distributions and change, and **phylogeographic**

(a)

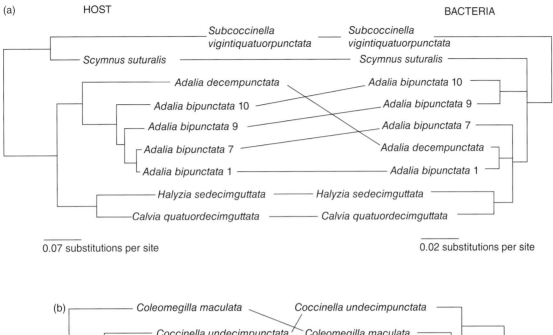

(b)

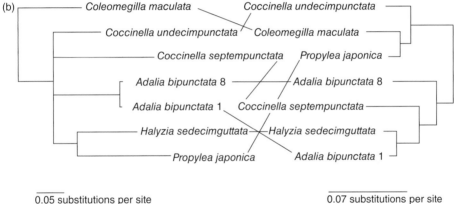

Figure 2.9 Tanglegrams of coccinellid host phylogeny with (a) an individual clade of *Rickettsia* coccinellid endosymbionts and (b) an individual clade of *Wolbachia* coccinellid endosymbionts. Scale bars represent substitutions per nucleotide for tree branch lengths. Connecting lines are drawn to connect bacteria strains to the host species they infect (or mitochondrial haplotypes within the host species *Adalia bipunctata*). While members of the *Rickettsia* clade exhibit close congruence with host phylogeny, members of the *Wolbachia* clade do not; consequently, a higher amount of horizontal transmission between more distantly related hosts can be inferred for members of the *Wolbachia* clade. Redrawn from Weinert (2008), with permission.

studies are concerned with the role that historical and biogeographical factors play in the current genetic structure of populations and species. Molecular studies across and between populations generally contain overlapping population genetic and phylogeographic components and therefore they are considered together here.

Although the majority of pre-molecular population genetic and evolutionary studies were concerned with colour pattern polymorphism (2.4), work away from this subject included chromosomal studies (2.3.4), and investigations of inbreeding (Lusis 1947) and hybridization (reviewed in Komai 1956). Arguably the most significant studies in the context of population genetic architecture were those of Lusis (1947), who investigated **inbreeding** in *A. bipunctata*. This species was found to exhibit very high levels of **inbreeding depression** consistent with a high number of deleterious recessive alleles occurring naturally in the heterozygous form. Similar results were obtained more recently by Morjan et al. (1999) for *P. quatuordecimpunctata* introduced to North America. The results suggest that mating between close relatives in the field is rare. Hurst et al. (1996), addressing this issue, showed that the low proportion of developed but dead eggs in field-collected *A. bipunctata* egg clutches was more similar to that of outbred than inbred clutches from the laboratory, the latter exhibiting very high egg mortality. Unfortunately, Hurst et al. did not consider the effect of multiple mating and sperm mixis (2.6.2.4). With multiple paternity of field-collected egg clutches, clutches would not be expected to exhibit equivalent high mortality to those from single inbred laboratory matings of virgin beetles, even if inbred matings do sometimes occur naturally.

More recently, the population genetics of *A. bipunctata* populations in Cambridge in the UK and Paris in France have been investigated using 10 polymorphic **microsatellite** loci (Haddrill 2001). This study found a **deficit of heterozygotes** at a number of loci examined in both populations, suggesting population substructuring or inbreeding. Estimates of inbreeding were particularly high, with mean relatedness in the two populations being 0.31 and 0.37, i.e. greater than that between half siblings. The results are particularly curious, given the earlier evidence of high inbreeding depression in *A. bipunctata*. A possible explanation is that the ladybirds sampled were part of the second generation that year: their parents may not have had to move too far to find food and mates after emergence, leading to matings between related individuals. It is worth noting that Hurst et al.'s (1996) inbreeding study, described above, used first generation eggs to estimate inbreeding in the field. Estimates of *A. bipunctata* **genetic differentiation** between the populations were low but significant, comparable to those obtained for *A. bipunctata* in earlier allozyme studies (Eggington 1986, Krafsur et al. 1996a). This is consistent with a relatively high level of gene flow between populations of this species, as well as with a large effective population size.

Prior to Haddrill's (2001) study, the majority of work on coccinellid genetic architecture had used **allozymes**. Most notably, in a long series of papers, Krafsur, Obrycki and co-workers described extensive allozyme-based population genetic analyses of native and introduced aphidophagous ladybirds of North America: these include *Hip. convergens*, *Col. maculata*, *A. bipunctata*, *C. septempunctata*, *P. quatuordecimpunctata*, *Hip. variegata* and *Har. axyridis* (Krafsur et al. 1992, 1995, 1996a, b, 1997, 2005, Krafsur & Obrycki 1996, Obrycki et al. 2001). These studies, which used a large number of putative allozyme loci (27–52, depending on species), showed that measures of genetic variation were similar to those for other beetles and that measures of population differentiation were generally low. Considered together with ecological studies, work relating to population genetic structure is consistent with populations of aphidophagous coccinellids being **near-panmictic**. This is unsurprising for a group of beetles that frequently move between habitats seeking aphids, as well as dispersing to and from overwintering sites on a yearly basis. While aphidophagous coccinellids might remain in the same place for more then one generation, as suggested by Haddrill for *A. bipunctata*, yearly movements between habitats will ultimately lead to genetic mixing between subpopulations. It is worth noting, however, that the aforementioned studies were all on aphidophagous habitat and dietary generalist species; more sedentary specialists might exhibit a lower degree of genetic mixis (Sloggett 2005). Similarly it is unclear to what extent these results could be extended to ladybirds with other diets, such as coccidophagous or phytophagous species.

A central aim of the allozyme studies was to investigate **genetic diversity in relation to the successful establishment of exotic biocontrol agents**. Measures of genetic diversity were similar for both native and exotic species, leading the authors to conclude that there is no obvious relationship between genetic diversity and successful colonization of new areas. More recent work by Lombaert et al. (2010) has thrown more light on the establishment by exotic ladybirds outside their native range, in a phylogeographic study examining the **origins of invasive *Har. axyridis* populations**. The authors used microsatellites and modern Bayesian computational techniques to compare different colonization scenarios. For each step of the worldwide spread of *Har. axyridis* (spread into eastern North America, western North America, Europe, South America and Africa), the Bayesian analysis gave a highest probability scenario for the

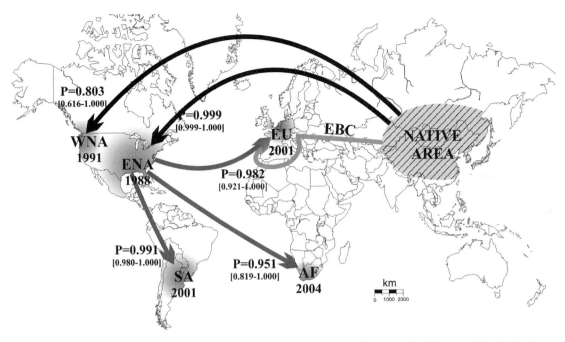

Figure 2.10 The most likely *Harmonia axyridis* invasion routes into eastern North America (ENA), western North America (WNA), South America (SA), Europe (EU) and Africa (AF), based on a study using microsatellites and Bayesian statistics. For each outbreak, the arrow indicates the most likely invasion pathway and the associated posterior probability value (P), with 95% confidence intervals in brackets. The years that invasive populations were first observed are also indicated. EBC is the European biocontrol strain, originally collected in 1982, which probably forms an additional source for European populations along with with ladybirds from eastern North America. From Lombaert et al. (2010), courtesy of the authors.

population origin (Fig. 2.10). The results showed that the first established non-native population, in eastern North America, acted as a 'bridgehead' source for the majority of the other non-native populations that subsequently established (Fig. 2.10). It may be inferred that some evolutionary change occurred in the eastern North American population that made *Har. axyridis* a much more successful colonizer of new regions than before.

Another phylogeographic use of molecular markers is to **delimit cryptic species** within complexes of close relatives that, on morphological grounds, can appear to be a single species. Work in this area has been carried out on phytophagous Epilachninae, where cryptic species can be specialized on different host plants. Very early on in the history of coccinellid molecular studies Kuboki (1978) used isozymes as a way of delimiting species within the *Henosepilachna*

vigintioctomaculata complex and examining the phylogenetic relationships between them. More recently, using a combination of breeding work, karyological studies and mitochondrial COI sequence analysis, Kobayashi et al. (2000) were able to show that the species *Henosepilachna* (= *Epilachna*) *vigintioctopunctata* was divided into two cryptic species. Conversely molecular studies can also show that species separated on morphological grounds, are in fact, a single entity. Marin et al. (2010), using crossing experiments, morphological analysis, analysis of COI sequences and **inter-simple sequence repeat (ISSR)** markers, showed that *C. septempunctata*, including the putative *C. septempunctata brucki* and *C. algerica*, were a single species and not a complex. The results from ISSR markers in this work are of particular interest. Due to linkage to endosymbiotic bacteria such as male killers, which are known in *C. septempunctata* (Majerus &

Hurst 1997), mtDNA may be a much less reliable genetic marker for such studies (2.6.1.1).

2.6.2.4   Reproductive success, paternity and sperm competition

Distinctive genetic markers are especially valuable in studies which aim to measure the success of particular strains or individuals in subsequent generations. The most important use of coccinellid genetic markers in this context has been in studying sperm competition; that is, to **deduce patterns of paternity when females mate multiply** and sperm from different males compete to fertilize a female's eggs, as is typically the case for coccinellids (Hodek & Ceryngier 2000; Chapter 3). If males carrying different genetic markers are used in experiments, the offspring can then be genotyped to deduce which male is the father.

Studies of sperm competition and paternity in coccinellids have included the Coccinellini *A. bipunctata*, *A. decempunctata* and *Har. axyridis* and the Epilachnini *Epilachna varivestris* and *Henosepilachna pustulosa* (Webb & Smith 1968, Nakano 1985, de Jong et al. 1993, 1998, Ueno 1994b, Ransford 1997, Haddrill et al. 2008); molecular studies have, however, been limited to *A. bipunctata*. The first studies of *A. bipunctata* were non-molecular, using its **polymorphic colour pattern** as a genetic marker. Virgin non-melanic females (i.e. homozygous recessive for colour pattern) were mated to non-melanic and melanic males (the latter being homozygote or heterozygote for the dominant melanic gene). By rearing the larvae to adulthood and scoring their colour pattern, it was possible to work out which male had fathered the offspring. Initial experiments, in which virgin females were mated to one male of each phenotype, concluded that *A. bipunctata* exhibited almost complete last male sperm precedence (de Jong et al. 1993). In a later series of experiments, however, de Jong et al. (1998) found much greater **variation in the proportion of eggs fertilized** by a melanic male, both when the female had previously been singly or multiply mated to non-melanic males. This was supported by evidence from the field, which showed that matings between heterozygote melanic and non-melanic individuals rarely resulted in a 1:1 proportion of the colour forms in the offspring, as would be expected with last male sperm precedence. In mating treatments consisting of two single matings, de Jong et al. (1998) observed a positive

correlation between the proportion of eggs fertilized by the second male and the **ratio of the duration of the second mating relative to the first**. They considered that manipulation of the first male's sperm by the second male was more successful the longer a male mated for; however, they did not discuss in detail by what mechanism sperm displacement occurred.

Colour pattern markers have been also used also to study sperm competition in *Har. axyridis* (Ueno 1994b) and, in the light of that study, to estimate how many males a female mates with in the field (Ueno 1996). However, there are some problems associated with the use of colour pattern markers. It is necessary to carry out time- and resource-intensive rearing of large numbers of offspring so that the colour patterns can be scored: de Jong et al. (1998) reared a total of 3108 *A. bipunctata* eggs to adults in their laboratory experiments and a further 5077 individuals from eggs obtained from field-collected adults. During the rearing process, it is important to ensure that there is no potential for selection to occur between larvae of different genotypes, which could potentially lead to biased estimates of morph frequencies: this in itself makes the rearing process more arduous, requiring the experimenter to avoid problems commonly encountered in the mass rearing of ladybirds such as crowding, food shortage or cannibalism. Furthermore, additional rearing may be required to verify parental genotypes: de Jong et al. (1993, 1998) had to mate all experimental melanic males to additional non-melanic virgin females, and rear resulting offspring to adult to verify whether the males were homozygote or heterozygote melanics.

Noting these types of problem, Ransford (1997) used an **allozyme**-based approach to study sperm competition in *A. bipunctata*. He used four allozyme variants at the isocitrate dehydrogenase 1 (*Idh1*) locus, of which one was common (92%) and three were rare (≤5%). The genotypes of individual ladybirds were deduced using electrophoretic analysis of reflex blood extracts, leaving the ladybirds alive for breeding and mating experiments. Neonate offspring were used whole to score their genotype. Therefore it was not necessary to rear them through to adult, as with colour pattern markers. Matings of males heterozygous for the different allozyme variants to homozygous virgin females produced 1:1 distributions of the paternal allozyme variants in the offspring, indicating that the variants were equivalent in their fertilization success. Two experiments on sperm precedence were carried out. The first used homozygous laboratory-bred virgin females mated to two males homozygous for different alleles. The second used males with rare alleles mated to field collected, non-virgin females: in this case the rarity of the experimental male alleles meant that many field-collected females had not mated previously with such males. In both experiments very **high variation in the proportion of offspring sired by the last-mating male was found**; with the only correlate of this variation being **mating duration**. Interestingly, among offspring not sired by the last mating, the relative precedence of a male mating over earlier matings persisted even after a further mating had occurred. Ransford's work suggests that sperm mixing occurs, and that the relative abundance of the sperm of different males determines how likely they are to fertilize the eggs. Although Ransford's allozyme study of paternity represents a significant advance over colour pattern-based paternity studies, a substantial amount of laboratory breeding was still required to produce A. bipunctata males homozygous for low frequency *Idh1* alleles, which are exceedingly scarce naturally.

Also studying A. bipunctata, Haddrill et al. (2008) used three extremely variable **microsatellite** loci to study mating frequency and paternity. All larvae from two egg clutches laid by individual field-collected females were genotyped. The females were also genotyped for exclusion purposes and the number of paternal microsatellite alleles present in each clutch determined for the three loci. Using this method, Haddrill et al. were able to estimate that individual A. bipunctata egg clutches were fathered by 2.5–3.5 males. Additionally, laboratory paternity studies indicated that **all males mating with a female then father at least a few offspring**; this indicates that females possess a limited ability, if any, for post-copulatory mate discrimination, for example by rejecting the sperm of certain males. Like other researchers, Haddrill et al. also found that **longer matings resulted in higher paternity of subsequent egg clutches**.

The majority of work on A. bipunctata is consistent with the conclusions of Ransford (1997), that A. bipunctata exhibits a mechanism of **instantaneous mixing during sperm displacement**. As a male transfers sperm he will displace earlier sperm in direct proportion to their instantaneous occurrence in the spermatheca of the female, including any he has already transferred (Parker et al. 1990). Consequently the more sperm a male transfers, the higher the proportion of subsequent eggs he fertilizes. In A. bipunctata and A. decempunctata, males that mate for longer transfer more sperm due to a mating mechanism in which individual spermatophores are transferred cyclically, with males undergoing up to three cycles in a single mating. Males that mate for longer undergo more cycles and transfer more spermatophores (Ransford 1997). Although the cyclical *Adalia* mating mechanism does not appear to be shared by most coccinellids (e.g. see Obata & Johki 1991), at least in *Har. axyridis* the mechanism of sperm displacement appears to be similar (Ueno 1994b). The availability of molecular methods makes it much easier to extend studies to coccinellids that are not polymorphic for colour pattern; thus in the future the generality of instantaneous mixing during sperm displacement across the Coccinellidae can be established. Because microsatellites provide a unique genetic fingerprint for each individual studied, they possess great potential for investigating multiple mating in coccinellids and to compare levels of promiscuity across individuals, populations and species. They are particularly promising markers for studies of populations from the field (cf. Haddrill et al. 2008).

2.7 CONCLUSIONS

Coccinellid genetic studies possess as much potential for the future as they have manifested in the past. In particular, due to the wealth of ecological data already available on coccinellids, the high visibility of the beetles in the field and their relative ease of maintenance in the laboratory, coccinellids are well suited for studies uniting both ecological and molecular approaches. This has already been shown, for example, in work on male-killing and on sperm competition, but as more genomic information on coccinellids becomes available, the approach is likely to be extended to other areas (Chapter 12).

This potential is best illustrated by reference to that genetic phenomenon which remains the best known for the Coccinellidae, colour pattern polymorphism. In spite of the vast body of knowledge on the inheritance of coccinellid colour patterns and on the selective pressures acting on them, as yet **we know very little about the genetic and developmental pathways that underlie colour pattern production**. Such knowledge would undoubtedly enhance our understanding of colour pattern polymorphism and its evolution, especially when integrated with what is already known from work in the field and laboratory. Coccinellid colour patterns, through studies of their variability, inheritance and maintenance, were of great importance to our understanding of evolution in the twentieth century. Molecular studies of their genetics and development could ensure that they remain at the forefront of evolutionary research in the twenty-first.

ACKNOWLEDGEMENTS AND DEDICATION

We thank Andrew Davis, Arnaud Estoup, Adiano Giorgi, Penny Haddrill, Ivo Hodek, Greg Hurst, Frank Jiggins, Grit Kunert, Eric Lombaert, Joe McHugh, Hinrich Schulenburg, Remy Poland and Lucy Weinert for their assistance in writing this chapter and their comments on it.

This chapter is dedicated to the late Professor Mike Majerus (1954–2009). Through his own research, and through his influence on many of his students who have gone on to study various aspects of ladybird genetics themselves, Professor Majerus made an enormous contribution to our current understanding of ladybird genetics. This is reflected in the large number of papers cited here that were authored or co-authored by himself and/or his students.

AH's contribution to this chapter was supported by grant # 522/08/1300 of the Grant Agency of the Czech Republic.

REFERENCES

Avise, J. C. 2004. *Molecular Markers, Natural History and Evolution*. 2nd edition. Sinauer Associates, Sunderland, MA. 684 pp.

Ballard, J. W. O. and M. C. Whitlock. 2004. The incomplete natural history of mitochondria. *Mol. Ecol.* 13: 729–744.

Balthasarova-Hrubantova, E. 1950. Contribution a la connaissance de la variabilite de l'espece *Adonia variegata* Goeze. *Prirodovedny sbornik (Bratislava)* 5: 193–206.

Beauchamp, R. L. and R. B. Angus. 2006. A chromosomal analysis of the four British ladybirds of the subfamily Coccidulinae (Coleoptera: Coccinellidae). *Proc. Russian Entomol. Soc.* 77: 18–27.

Beebee, T. J. C. and G. Rowe. 2008. *An Introduction to Molecular Ecology*. 2nd edn. Oxford University Press, Oxford. 400 pp.

Beladjal, L., T. T. M. Vandekerckhove, B. Muyssen et al. 2002. B-chromosomes and male-biased sex ratio with paternal inheritance in the fairy shrimp *Branchipus schaefferi* (Crustacea, Anostraca). *Heredity* 88: 356–360.

Bengtson, S. A. and R. Hagen. 1975. Polymorphism in the two-spot ladybird *Adalia bipunctata* in western Norway. *Oikos* 26: 328–331.

Bengtson, S. A. and R. Hagen. 1977. Melanism in the two-spot ladybird *Adalia bipunctata* in relation to climate in western Norway. *Oikos* 28: 16–19.

Benham, B. R., D. Lonsdale and J. Muggleton. 1974. Is polymorphism in two-spot ladybird an example of non-industrial melanism? *Nature* 249: 179–180.

Bezzerides, A. L., K. J. McGraw, R. S. Parker and J. Husseini. 2007. Elytra color as a signal of chemical defense in the Asian ladybird beetle *Harmonia axyridis*. *Behav. Ecol. Sociobiol.* 61: 1401–1408.

Blehman, A. V. 2007. Variability of pronotum patterns in ladybird beetle *Harmonia axyridis* Pallas (Coleoptera, Coccinellidae). *Ekologicheskaya Genetika* 5: 25–36.

Blehman, A. V. 2009. *Intrapopulation and geographic variability of* Harmonia axyridis *Pall. in a complex of polymorphic characters*. PhD Thesis. Biological Faculty of the Moscow State University of M. V. Lomonosov. Moscow. 24 pp.

Blekhman, A. V. 2008. Population variation of elytral ridge occurrence in ladybirds *Harmonia axyridis* Pallas. *Russian J. Genet.* 44: 1351–1354.

Blekhman, A. V., I. I. Goryacheva and I. A. Zakharov. 2010. Differentiation of *Harmonia axyridis* Pall. according to polymorphic morphological traits and variability of the mitochondrial COI gene. *Moscow Univ. Biol. Sci. Bull.* 65: 174–176.

Brakefield, P. M. 1984a. Ecological studies on the polymorphic ladybird *Adalia bipunctata* in the Netherlands. I. Population biology and geographical variation of melanism. *J. Anim. Ecol.* 53: 761–774.

Brakefield, P. M. 1984b. Ecological studies on the polymorphic ladybird *Adalia bipunctata* in the Netherlands. II. Population dynamics, differential timing of reproduction and thermal melanism. *J. Anim. Ecol.* 53: 775–790.

Brakefield, P. M. 1984c. Selection along clines in the ladybird *Adalia bipunctata* in The Netherlands: a general mating advantage of melanics and its consequences. *Heredity* 53: 37–49.

Brakefield, P. M. 1985a. Differential winter mortality and seasonal selection in the polymorphic ladybird *Adalia bipunctata* (L.) in the Netherlands. *Biol. J. Linn. Soc.* 24: 189–206.

Brakefield, P M. 1985b. Polymorphic Müllerian mimicry and interactions with thermal melanism in ladybirds and a soldier beetle: a hypothesis. *Biol. J. Linn. Soc.* 26: 243–265.

Brakefield, P. M. and D. R. Lees. 1987. Melanism in *Adalia* ladybirds and declining air pollution in Birmingham. *Heredity* 59: 273–277.

Brakefield, P. M. and P. G. Willmer. 1985. The basis of thermal melanism in the ladybird *Adalia bipunctata*. Differences in reflectance and thermal properties between the morphs. *Heredity* 54: 9–14.

Brown, W. J. 1962. A revision of the forms of *Coccinella* L. occuring in America north of Mexico (Coleoptera, Coccinellidae). *Can. Entomol.* 94: 785–808.

Camacho, J. P. M., T. F. Sharbel and L. W. Beukeboom. 2000. B-chromosome evolution. *Phil. Trans. R. Soc. Lond.* B 355: 163–178.

Chazeau, J. 1980. On polymorphism in elytral colouration pattern in *Coelophora quadrivittata* (Coleoptera, Coccinellidae). *Entomol. Exp. Appl.* 27: 194–198.

Coll, M., L. G. de Mendoza and G. K. Roderick. 1994. Population structure of a predatory beetle: the importance of gene flow for intertrophic level interactions. *Heredity* 72: 228–236.

Creed, E. R. 1966. Geographic variation in the two-spot ladybird in England and Wales. *Heredity* 21: 57–72.

Creed, E. R. 1971a. Industrial melanism in the two-spot ladybird and smoke abatement. *Evolution* 25: 290–293.

Creed, E. R. 1971b. Melanism in the two-spot ladybird, *Adalia bipunctata*, in Great Britain. In: Creed, E. R. (ed) *Ecological Genetics and Evolution*. Blackwell, Oxford, pp. 134–151.

Creed, E. R. 1974. Two spot ladybirds as indicators of intense local pollution. *Nature* 249: 390–391.

Creed, E. R. 1975. Melanism in the two spot ladybird: the nature and intensity of selection. *Proc. R. Soc. Lond.* B 190: 135–148.

Cromartie, R. I. T. 1959. Insect pigments. *Annu. Rev. Entomol.* 4: 59–73.

Dobzhansky, T. 1924a. Die geographische und individuelle Variabilität von *Harmonia axyridis* Pall. in ihren Wechselbeziehungen. *Biol. Zbl.* 44: 401–421.

Dobzhansky, T. 1924b. Über geographische und individuelle Variabilität von *Adalia bipunctata* und *A. decempunctata*. *Russkoe Entomol. Obozr.* 18: 201–211.

Dobzhansky, T. 1925. Die paläarktischen Arten der Gattung *Coccinula* Dobzh. *Zool. Anz.* 64: 277–284.

Dobzhansky, T. 1933. Geographical variation in lady-beetles. *Am. Nat.* 67: 97–126.

Dobzhansky, T. and N. P. Sivertzew-Dobzhansky. 1927. Die geographische Variabilität von *Coccinella septempunctata*. *Biol. Zbl.* 47: 556–569.

Drets, M. E., E. Corbella, F. Panzera and G. A. Folle. 1983. C-banding and non-homologous associations. II. The 'parachute' Xy_p sex bivalent and the behavior of heterochromatic segments in *Epilachna paenulata*. *Chromosoma* 88: 249–255.

Eggington, E. 1986. *Electrophoretic Variation amongst British Coccinellidae*. Unpublished undergraduate project report, University of Cambridge, Cambridge, UK.

Eichhorn, O. and P. Graf. 1971. Sex-linked colour polymorphism in *Aphidecta obliterata* L. (Coleoptera, Coccinellidae). *Z. Angew. Entomol.* 67: 225–231.

Emetz, V. M. 1984. Peculiarities of population dynamics and structure of *Adalia bipunctata* L. (Coleoptera, Coccinellidae) on burnt places. *Zh. Obshch. Biol.* 65: 391–395.

Ennis, T. J. 1974. Chromosome structure in *Chilocorus* (Coleoptera: Coccinellidae). I. Fluorescent and giemsa banding patterns. *Can. J. Genet. Cytol.* 16: 651–661.

Ennis, T. J. 1975. Feulgen hydrolysis and chromosome banding in *Chilocorus* (Coleoptera: Coccinellidae). *Can. J. Genet. Cytol.* 17: 75–80.

Ennis, T. J. 1976. Chromosome structure in *Chilocorus* (Coleoptera: Coccinellidae). II. The asynchronous replication of constitutive heterochromatin. *Can. J. Genet. Cytol.* 18: 85–91.

Filippov, N. N. 1961. Regularities of abberative variability of elytra pattern in Coleoptera. *Zool. Zh.* 40: 372–385.

Fiori, A. 1928. Richerche sul comportamento ereditario di alcune varieta di *Adalia bipunctata*. *Boll. Lab. Zool. Portici* 22: 285–304.

Ford, E. B. 1964. *Ecological genetics*, 1st edition. Methuen, London. 335 pp.

Freeland, J. R., H. Kirk and S. D. Petersen. 2011. *Molecular Ecology*, 2nd edition. John Wiley and Sons, Chichester. 464 pp.

Fukunaga, Y. and S. Akimoto. 2007. Toxicity of the aphid *Aulacorthum magnoliae* to the predator *Harmonia axyridis* (Coleoptera: Coccinellidae) and genetic variance in the assimilation of the toxic aphids in *H. axyridis* larvae. *Entomol. Sci.* 10: 45–53.

Fuersch, H. 1996. Taxonomy of coccinellids. *Coccinella* 6: 28–30.

Gagnon, A.-E., G. E. Heimpel and J. Brodeur. 2005. Detection of intraguild predation between coccinellids using molecular analyses of gut-contents. *Proc. Int. Symp. Biol. Control Aphids and Coccids, Tsuruoka, Japan, September 2005*, 155–159.

Gil, L., A. Ferran, J. Gambier et al. 2004. Dispersion of flightless adults of the Asian lady beetle, *Harmonia axyridis*, in greenhouses containing cucumbers infested with the aphid *Aphis gossypii*: effect of the presence of conspecific larvae. *Entomol. Exp. Appl.* 112: 1–6.

Giorgi, J. A., J. A Forrester, S. A Slipinski et al. 2009. The evolution of food preferences in Coccinellidae. *Biol. Control*, 51: 215–231.

Gregory, T. R., O. Nedved and S. J. Adamovicz. 2003. C-value estimates for 31 species of ladybird beetles (Coleoptera: Coccinellidae). *Hereditas* 139: 121–127.

Haddrill, P R. 2001. *The Development and Use of Molecular Genetic Markers to Study Sexual Selection and Population Genetics in the Two-spot Ladybird*, Adalia bipunctata *(L.)*. Unpublished PhD thesis, University of Cambridge, Cambridge, UK.

Haddrill, P. R., D. M. Shuker, W. Amos, M. E. N. Majerus and S. Mayes. 2008. Female multiple mating in wild and laboratory populations of the two-spot ladybird, *Adalia bipunctata*. *Mol. Ecol.* 17: 3189–3197.

Hajibabaei, M., G. A. C. Singer, P. D. N. Hebert and D. A. Hickey. 2007. DNA barcoding: how it complements taxonomy, molecular phylogenetics and population genetics. *Trends Genet.* 23: 167–172.

Hales, D. F. 1976. Inheritance of striped elytral pattern in *Coelophora inaequalis* (F). *Aust. J. Zool.* 24: 273–276.

Hammond, P. M. 1985. Dimorphism of wings, wing-folding and wing-toiletry devices in the ladybird, *Rhyzobius litura* (F.) (Coleoptera: Coccinellidae), with a discussion of interpopulation variation in this and other wing-dimorphic beetle species. *Biol. J. Linn. Soc.* 24: 15–33.

Happ, G. M. and T. Eisner. 1961. Hemorrhage in a coccinellid beetle and its repellent effect on ants. *Science* 134: 329–331.

Harris, D. J. and K. A. Crandall. 2000. Intragenomic variation within ITS1 and ITS2 of freshwater crayfishes (Decapoda: Cambaridae): implications for phylogenetic and microsatellite studies. *Mol. Biol. Evol.* 17: 284–291.

Harrison, R. G. 1989. Animal mitochondrial DNA as a genetic marker in population and evolutionary biology. *Trends Ecol. Evol.* 4: 6–11.

Harwood, J. D., N. Desneux, H. J. S. Yoo et al. 2007. Tracking the role of alternative prey in soybean aphid predation by *Orius insidiosus*: a molecular approach. *Mol. Ecol.* 16: 4390–4400.

Hawkes, O. A. M. 1920. Observations on the life history, biology and genetics of the lady-bird beetle, *Adalia bipunctata* (Mulsant). *Proc. Zool. Soc. Lond.* 1920: 475–490.

Hebert, P. D. N., A. Cywinska, S. L. Ball and J. R. deWaard. 2003. Biological identifications through DNA barcodes. *Proc. R. Soc. Lond.* B 270: 313–321.

Henderson, S. A. 1988. A correlation between B chromosome frequency and sex ratio in *Exochomus quadripustulatus*. *Chromosoma* 96: 376–381.

Henderson, S. A. and J. S. M. Albrecht. 1988. Abnormal and variable sex ratios in population samples of ladybirds. *Biol. J. Linn. Soc.* 35: 275–296.

Hillis, D. M. and M. T. Dixon. 1991. Ribosomal DNA: molecular evolution and phylogenetic inference. *Quart. Rev. Biol.* 66: 411–453.

Hodek, I. and J. Cerkasov. 1961. Prevention and artificial induction of the imaginal diapause in *Coccinella septempunctata* L. (Col.: Coccinellidae). *Entomol. Exp. Appl.* 4: 179–190.

Hodek, I. and P. Ceryngier. 2000. Sexual activity in Coccinellidae (Coleoptera): a review. *Eur. J. Entomol.* 97: 449–456.

Honek, A. 1975. Colour polymorphism in *Adalia bipunctata* in Bohemia. *Entomol. Germ.* 1: 345–348.

Honek, A., Z. Martinkova and S. Pekar. 2005. Temporal stability of morph frequency in central European populations of *Adalia bipunctata* and *A. decempunctata* (Coleoptera, Coccinellidae). *Eur. J. Entomol.* 102: 437–442.

Hosino, Y. 1936. Genetical studies of the lady-bird beetle, *Harmonia axyridis* Pallas (Report II). *Japan. J. Genet.* 12: 307–320.

Hosino, Y. 1939. Genetical studies of the lady-bird beetle, *Harmonia axyridis* Pallas (Report III). *Japan. J. Genet.* 15: 128–138.

Hosino, Y. 1940a. Genetical studies of the lady-bird beetle, *Harmonia axyridis* Pallas (Report IV). *Japan. J. Genet.* 16: 155–163.

Hosino, Y. 1940b. Genetical studies of the pattern types of the lady-bird beetle, *Harmonia axyridis* Pallas. *J. Genet.* 40: 215–228.

Hosino, Y. 1941. Genetical studies of the lady-bird beetle, *Harmonia axyridis* Pallas (Report V). *Japan. J. Genet.* 17: 145–155.

Hosino Y. 1942. Genetical studies of the lady-bird beetle, *Harmonia axyridis* Pallas (Report VI). *Japan. J. Genet.* 18: 285–296.

Hosino, Y. 1943a. Genetical studies of the lady-bird beetle, *Harmonia axyridis* Pallas (Report VII). *Japan. J. Genet.* 19: 167–181.

Hosino, Y. 1943b. Genetical studies of the lady-bird beetle, *Harmonia axyridis* Pallas (Report VIII). *Japan. J. Genet.* 19: 258–265.

Hosino, Y. 1948. Genetical studies of the lady-bird beetle, *Harmonia axyridis* Pallas (Report IX). *Japan. J. Genet.* 23: 90–95.

Houston, K. J. 1979. Mosaic dominance in the inheritance of the color patterns of *Coelophora inaequalis* (F.) (Coleoptera, Coccinellidae). *J. Aust. Entomol. Soc.* 18: 45–51.

Houston, K. J. and D. F. Hales. 1980. Allelic frequencies and inheritance of colour pattern in *Coelophora inaequalis* (F.) (Coleoptera, Coccinellidae). *Aust. J. Zool.* 28: 669–677.

Howland, D. E. and G. M. Hewitt. 1995. Phylogeny of the Coleoptera based on mitochondrial cytochrome oxidase I sequence data. *Insect Mol. Biol.* 4: 203–215.

Hunt, T., J. Bergsten, Z. Levkanicova et al. 2007. A comprehensive phylogeny of beetles reveals the evolutionary origins of a superradiation. *Science* 318: 1913–1916.

Hurst, G. D. D. and F. M. Jiggins 2005. Problems with mitochondrial DNA as a marker in population, phylogeographic and phylogenetic studies: the effects of inherited symbionts. *Proc. R. Soc.* B 272: 1525–1534.

Hurst, G. D. D., J. J. Sloggett and M. E. N. Majerus. 1996. Estimation of the rate of inbreeding in a natural population of *Adalia bipunctata* (Coleoptera: Coccinellidae) using a phenotypic indicator. *Eur. J. Entomol.* 93: 145–150.

Ireland, H., P. W. E. Kearns and M. E. N. Majerus. 1986. Interspecific hybridisation in the Coccinellidae: some observations on an old controversy. *Entomol. Rec. J. Var.* 98: 181–185.

Jiggins, F. M. and M. C. Tinsley. 2005. An ancient mitochondrial polymorphism in *Adalia bipunctata* linked to a sex-ratio-distorting bacterium. *Genetics* 171: 1115–1124.

Johnstone, R. A. and G. D. D. Hurst. 1996. Maternally inherited male-killing microorganisms may confound interpretation of mitochondrial DNA variability. *Biol. J. Linn. Soc.* 58: 453–470.

Jones, R. N. and H. Rees. 1982. *B chromosomes*. Academic Press, New York. 266 pp.

de Jong, P. W., M. D. Verhoog and P. M. Brakefield. 1993. Sperm competition and melanic polymorphism in the 2-spot ladybird, *Adalia bipunctata* (Coleoptera, Coccinellidae). *Heredity* 70: 172–178.

de Jong, P. W., P. M. Brakefield and B. P. Geerinck. 1998. The effect of female mating history on sperm precedence in the two-spot ladybird, *Adalia bipunctata* (Coleoptera, Coccinellidae). *Behav. Ecol.* 9: 559–565.

Karystinou, A., A. P. M. Thomas and H. E. Roy. 2004. Presence of haemocyte-like cells in coccinellid reflex blood. *Physiol. Entomol.* 29: 94–96.

Katakura, H. 1986. Evidence for the incapacitation of heterospecific sperm in the female genital tract in a pair of closely related ladybirds. *Zool. Sci.* 3: 115–121.

Katakura, H. and Y. Sobu. 1986. Cause of low hatchability by the interspecific mating in a pair of sympatric ladybirds (Insecta, Coleoptera, Coccinellidae): incapacitation of alien sperm and death of hybrid embryos. *Zool. Sci.* 3: 315–322.

Kawamura, K., T. Sugimoto, K. Kakutani, Y. Matsuda and H. Toyoda. 2007. Genetic variation of sweet potato weevils, *Cylas formicarius* (Fabricius) (Coleoptera: Brentidae), in main infested areas in the world based upon the internal transcribed spacer-1 (ITS-1) region. *Appl. Entomol. Zool.* 42: 89–96.

Kearns, P. W. E., I. P. M. Tomlinson, P. O'Donald and J. C. Veltman. 1990. Non-random mating in the two-spot ladybird (*Adalia bipunctata*), I. A reassessment of the evidence. *Heredity* 65: 229–240.

Kearns, P. W. E., I. P. M. Tomlinson, J. C. Veltman and P. O'Donald. 1992. Non-random mating in the two-spot ladybird (*Adalia bipunctata*), II. Further tests for female mating preference. *Heredity* 68: 385–389.

Kendall, D. A. 1971. A note on reflex bleeding in the larvae of the beetle *Exochomus quadripustulatus* (L.) (Col.: Coccinellidae). *The Entomologist* 104: 233–235.

Kholin, S. K. 1990. Stability of the genetic polymorphism in colour of *Harmonia axyridis* Pall (Coccinellidae, Coleoptera) in Maritime Province, USSR. *Genetika-Moskva* 26: 2207–2214.

Klausnitzer, B. and R. Schummer. 1983. Zum Vorkommen der Formen von *Adalia bipunctata* L. in der DDR (Insecta, Coleoptera). *Entomol. Nachr. Berlin* 27: 159–162.

Kobayashi, N., K. Tamura, T. Aotsuka and H. Katakura. 1998. Molecular phylogeny of twelve Asian species of Epilachnine ladybird beetles (Coleoptera, Coccinellidae) with notes on the direction of host shifts. *Zool. Sci.* 15: 147–151.

Kobayashi, N., Y. Shirai, N. Tsurusaki et al. 2000. Two cryptic species of the phytophagous ladybird beetle *Epilachna vigintioctopunctata* (Coleoptera: Coccinellidae) detected by analyses of mitochondrial DNA and karyotypes, and crossing experiments. *Zool. Sci.* 17: 1159–1166.

Kobayashi, N., M. Kumagai, D. Minegishi et al. 2011. Molecular population genetics of a host-associated sibling species complex of phytophagous ladybird beetles (Coleoptera: Coccinellidae: Epilachninae). *J. Zool. Syst. Evol. Res.* 49: 16–24.

Komai, T. 1956. Genetics of ladybeetles. *Adv. Genet.* 8: 155–188.

Komai, T. and M. Chino. 1969. Observations on geographic and temporal variations in the ladybeetle *Harmonia*. I. *Proc. Japan Acad.* 45: 284–288.

Komai, T. and Y. Hosino. 1951. Contributions to the evolutionary genetics of the lady-beetle, *Harmonia*. II. Microgeographic variations. *Genetics* 36: 382–390.

Komai, T., M. Chino and Y. Hosino. 1950. Contributions to the evolutionary genetics of the lady-beetle, *Harmonia*. I. Geographic and temporal variations in the relative frequencies of the elytral pattern types and in the frequency of elytral ridge. *Genetics* 35: 589–601.

Korsun, O. V. 2004. Polymorphism in natural populations of lady-beetle *Harmonia axyridis* Pall. (Insecta, Coleoptera, Coccinellidae) in the Eastern Baikal District. In: *Problemy ekologii i racionalnogo ispolzovania nrirodnych resursov v Dalnevostchnom regione*. Materialy regionalnoj nauchno-prakticheskoj konferencii, 21–23 dekabrja 2004g Vol 1, Blagoveschensk, Izdatelstvo BGPU, 195–199.

Kovar, I. 1996. Phylogeny. *In* Hodek I and Honek A *Ecology of Coccinellidae*. Kluwer Academic Publishers, Dordrecht, pp. 19–31.

Krafsur, E. S. and J. J. Obrycki. 1996. Gene flow in the exotic 14-spotted ladybird beetle, *Propylea quatuordecimpunctata*. *Genome* 39: 131–139.

Krafsur, E. S., J. J. Obrycki and R. V. Flanders. 1992. Gene flow in populations of the seven-spotted lady beetle, *Coccinella septempunctata*. *J. Heredity* 83: 440–444.

Krafsur, E. S., J. J. Obrycki and P. W. Schaefer. 1995. Genetic heterozygosity and gene flow in *Coleomegilla maculata* De Geer (Coleoptera: Coccinellidae). *Biol. Control* 5: 104–111.

Krafsur, E. S., P. Nariboli and J. J. Obrycki. 1996a. Gene flow and diversity at allozyme loci in the twospotted lady beetle (Coleoptera: Coccinellidae). *Ann. Entomol. Soc. Am.* 89: 410–419.

Krafsur, E. S., J. J. Obrycki and P. Nariboli. 1996b. Gene flow in colonizing *Hippodamia variegata* ladybird beetle populations. *J. Heredity* 87: 41–47.

Krafsur, E. S., T. J. Kring, J. C. Miller et al. 1997. Gene flow in the exotic colonizing ladybeetle *Harmonia axyridis* in North America. *Biol. Control* 8: 207–214.

Krafsur, E. S., J. J. Obrycki and J. D. Harwood. 2005. Comparative genetic studies of native and introduced Coccinellidae in North America. *Eur. J. Entomol.* 102: 469–474.

Kryltzov, A. I. 1953. Zavisimost izmenchivosti melanizacii zhukov ot ich osnovnoi okraski. *Zool. Zh.* 32: 915–919.

Kryltzov, A. I. 1956. Geographical variability of lady-birds (Coleoptera, Coccinellidae) in North Kirgisia. *Entomol. Obozr.* 35: 771–781.

Kuboki, M. 1978. Studies on the phylogenetical relationship in the *Henosepilachna vigintioctomaculata* complex based on variation of isozymes (Coleoptera: Coccinellidae). *Appl. Entomol. Zool.* 13: 250–259.

Lees, D. R., E. R. Creed and J. G. Duckett. 1973. Atmospheric pollution and industrial melanism. *Heredity* 30: 227–232.

Lombaert, E., T. Guillemaud, J.-M. Cornuet et al. 2010. Bridgehead effect in the worldwide invasion of the biocontrol harlequin ladybird. *PLoS ONE* 5: e9743.

Lommen, S. T. E., P. W. de Jong and P. M. Brakefield. 2005. Phenotypic plasticity of elytron length in wingless two-spot ladybird beetles, *Adalia bipunctata* (Coleoptera: Coccinellidae). *Eur. J. Entomol.* 102: 553–556.

Lommen, S. T. E., C. W. Middendorp, C. A. Luijten et al. 2008. Natural flightless morphs of the ladybird beetle *Adalia bipunctata* improve biological control of aphids on single plants. *Biol. Control* 47: 340–346.

Lommen, S. T. E., S. V. Saenko, Y. Tomoyasu and P. M. Brakefield. 2009. Development of a wingless morph in the ladybird beetle, *Adalia bipunctata. Evol. Dev.* 11: 278–289.

Lusis, Ya. Ya. 1928. On the inheritance of colour and pattern in the lady-beetles *Adalia bipunctata* and *A. decempunctata. Izvestija Bjuro Genetiki Akademii Nauk SSSR* 6: 89–163.

Lusis, Ya. Ya. 1932. An analysis of the dominance phenomenon in the inheritance of the elytra and pronotum colour in *Adalia bipunctata*. 1. *Trudy Laboratorii Genetiki Akademii Nauk SSSR* 9: 135–162.

Lusis, Ya. Ya. 1947. Some rules of reproduction in *Adalia bipunctata* L. 1. Heterozygosity of lethal alleles in populations. *Doklady Akademii Nauk SSSR* 57: 825–828.

Lusis, Ya. Ya. 1961. On the biological meaning of colour polymorphism of lady-beetle *Adalia bipunctata* L. *Latvijas Entomologs* 4: 2–29.

Lusis, Ya. Ya. 1971. Experimental data on the three species of genus *Calvia* (Coleoptera, Coccinellidae) from Central Asia. *Latvijas Entomologs* 14: 3–29.

Lusis, Ya. Ya. 1973. Taxonomical relationships and geographical distribution of forms in the ladybird genus *Adalia* Mulsant. *Petera Stutski Valsts Universitates Zinatniskie raksti* 184: 1–123.

Lyapunova, E. A., N. N. Vorontsov, J. S. Yadav et al. 1984. Karyological investigations on seven species of coccinellid fauna of USSR (Polyphaga: Coleoptera). *Zool. Anz.* 212: 185–192.

Mader, L. 1926–1937. *Evidenz der paläarktischen Coccinelliden und ihrer Aberationen in Wort und Bild. I. Teil.* Wien, Troppau. 412 pp.

Maffei, E. M. D. and S. G. Pompolo. 2007. Evaluation of hot saline solution and restriction endonuclease techniques in cytogenetic studies of *Cycloneda sanguinea* L. (Coccinellidae). *Genet. Mol. Res.* 6: 122–126.

Maffei, E. M. D., E. Gasparino and S. G. Pompolo. 2000. Karyotypic characterization by mitosis, meiosis and C-banding of *Eriopis connexa* Mulsant (Coccinellidae: Coleoptera: Polyphaga), a predator of insect pests. *Hereditas* 132: 79–85.

Maffei, E. M. D., S. G. Pompolo, L. A. O. Campos and E. Petitpierre. 2001a. Sequential FISH analysis with rDNA genes and Ag-NOR banding in the lady beetle *Olla v-nigrum* (Coleoptera: Coccinellidae). *Hereditas* 135: 13–18.

Maffei, E. M. D., S. G. Pompolo, J. C. Silva-Junior et al. 2001b. Silver staining of nucleolar organizer regions (NORs) in some species of Hymenoptera (bees and parasitic wasps) and Coleoptera (lady-beetles). *Cytobios* 104: 119–125.

Maffei, E. M. D., S. G. Pompolo and E. Petitpierre. 2004. C-banding and fluorescent *in situ* hybridization with rDNA sequences in chromosomes of *Cycloneda sanguinea* Linnaeus (Coleoptera, Coccinellidae). *Genet. Mol. Biol.* 27: 191–195.

Magro, A., E. LeCompte, F. Magne, J.-L. Hemptinne and B. Crouau-Roy. 2010. Phylogeny of ladybirds (Coleoptera: Coccinellidae): are the subfamilies monophyletic? *Mol. Phyl. Evol.* 54: 833–848.

Majerus, M. E. N. 1994. *Ladybirds (New Naturalist Series).* HarperCollins, London. 367 pp.

Majerus, M. E. N. and G. D. D. Hurst. 1997. Ladybirds as a model system for the study of male-killing symbionts. *Entomophaga* 42: 13–20.

Majerus, M. E. N. and P. W. E. Kearns. 1989. *Ladybirds (Naturalists' Handbooks 10).* Richmond Publishing Co., Slough. 103 pp.

Majerus, M. E. N. and I. A. Zakharov. 2000. Does thermal melanism maintain melanic polymorphism in the two-spot ladybird, *Adalia bipunctata* (Coleoptera, Coccinellidae)? *Zh. Obshch. Biol.* 61: 381–392.

Majerus, M. E. N., P. O'Donald and J. Weir. 1982. Evidence for preferential mating in *Adalia bipunctata. Heredity* 49: 37–49.

Majerus, M. E. N., P. O'Donald, P. W. E. Kearns and H. Ireland. 1986. Genetics and evolution of female choice. *Nature* 321: 164–167.

Marin, J., B. Crouau-Roy, J.-L. Hemptinne, E. Lecompte and A. Magro. 2010. *Coccinella septempunctata* (Coleoptera, Coccinellidae): a species complex? *Zool. Scr.* 39: 591–602.

Marples, N. M. 1993. Toxicity assays of ladybirds using natural predators. *Chemoecology* 4: 33–38.

Marples, N. M, P. M. Brakefield and R. J. Cowie. 1989. Differences between the 7-spot and 2-spot ladybird beetles (Coccinellidae) in their toxic effects on a bird predator. *Ecol. Entomol.* 14: 79–84.

Marples, N. M., P. W. de Jong, M. M. Ottenheim, M. D. Verhoog and P. M. Brakefield. 1993. The inheritance of a wingless character in the 2spot ladybird (*Adalia bipunctata*). *Entomol. Exp. Appl.* 69: 69–73.

Mawdsley, J. R. 2001. Mitochondrial cytochrome oxidase I DNA sequences and the phylogeny of Coccinellidae (Insecta: Coleoptera: Cucujoidea). *J. New York Entomol. Soc.* 109: 304–308.

McCornack, B. P., R. L. Koch and D. W. Ragsdale. 2007. A simple method for in-field sex determination of the multicolored Asian lady beetle *Harmonia axyridis*. *J. Insect Sci.* 7(10): 1–12.

Meissner, O. 1907a. Die relative Häufigkeit der Varietäten von *Adalia bipunctata* L. in Potsdam, 1906, nebst biologischen Bemerkungen über diese und einige andere Coccinelliden. *Z. Wiss. Insektenbiol.* 3: 12–20; 39–45.

Meissner O. 1907b. Die relative Häufigkeit der Varietäten von *Adalia bipunctata* L. in Potsdam, 1907, nebst biologischen Bemerkungen über diese und einige andere Coccinelliden. *Z. Wiss. Insektenbiol.* 3: 309–313; 334–344; 369–374.

Michie, L. J., F. Mallard, M. E. N. Majerus and F. M. Jiggins. 2010. Melanic through nature or nurture: genetic polymorphism and phenotypic plasticity in *Harmonia axyridis*. *J. Evol. Biol.* 23: 1699–1707.

Mikkola, K. and A. Albrecht. 1988. The melanism of *Adalia bipunctata* around the Gulf of Finland as an industrial phenomenon (Coleoptera, Coccinellidae). *Ann. Zool. Fenn.* 25: 177–185.

Miyazawa, B. and K. Ito. 1921. On the hybrids between varieties of *Propylea japonica*. *Japan. J. Genet.* 1: 13–20.

Morjan, W. E., J. J. Obrycki and E. S. Krafsur. 1999. Inbreeding effects on *Propylea quatuordecimpunctata* (Coleoptera: Coccinellidae). *Ann. Entomol. Soc. Am.* 92: 260–268.

Muggleton, J., D. Lonsdale and B. R. Benham. 1975. Melanism in *Adalia bipunctata* L. (Col., Coccinellidae) and its relationship to atmospheric pollution. *J. Appl. Ecol.* 12: 465–471.

Nakano, S. 1985. Sperm displacement in *Henosepilachna pustulosa* (Coleoptera: Coccinellidae). *Kontyû* 53: 516–519.

Nefedov, N. I. 1959. K faune i izmenchivosti nekotorych vidov kokcinellid Kabardinsko-balkarskoi ASSR. *Uchonye Zapiski Kabardinsko-Balkarskogo Gosudarstvennogo Universiteta* 5: 131–145.

O'Donald, P. and J. Muggleton. 1979. Melanic polymorphism in ladybirds maintained by sexual selection. *Heredity* 43: 143–148.

O'Donald, P., M. Derrick, M. Majerus and J. Weir. 1984. Population genetic theory of the assortative mating, sexual selection and natural selection of the two-spot ladybird, *Adalia bipunctata*. *Heredity* 52: 43–61.

Obata, S. and Y. Johki. 1991. Comparative study on copulatory behaviour in four species of aphidophagous ladybirds. *In* L. Polgar, R. J. Chambers, A. F. G. Dixon and I. Hodek (eds). *Behaviour and Impact of Aphidophaga*. SPB Academic Publishing, The Hague, pp. 207–212.

Obrycki, J.J., E. S. Krafsur, C. E. Bogran, L. E. Gomez and R. E. Cave. 2001. Comparative studies of three populations of

the lady beetle predator *Hippodamia convergens* (Coleoptera: Coccinellidae). *Fla. Entomol.* 84: 55–62.

Ohde, T., M. Matsumoto, M. Morita-Miwa et al. 2009. *Vestigial* and *scalloped* in the ladybird beetle: a conserved function in wing development and a novel function in pupal ecdysis. *Insect Mol. Biol.* 19: 571–581.

Okuda, T., T. Gomi and I. Hodek. 1997. Effect of temperature on pupal pigmentation and size of the elytral spots in *Coccinella septempunctata* (Coleoptera, Coccinellidae) from four latitudes in Japan. *Appl. Entomol. Zool.* 32: 567–572.

Osawa, N. and T. Nishida. 1992. Seasonal variation in elytral colour polymorphism in *Harmonia axyridis* (the ladybird beetle), the role of non-random mating. *Heredity* 69: 297–307.

Palenko, M. V., D. V. Mukha and I. A. Zakharov. 2004. Intraspecific and interspecific variation of the mitochondrial gene of cytochrome oxidase I in ladybirds (Coleoptera: Coccinellidae). *Russian J. Genet.* 40: 148–151.

Palmer, M. A. 1911. Some notes on heredity in the coccinellid genus *Adalia* Mulsant. *Ann. Entomol. Soc. Am.* 4: 283–302.

Palmer, M. A. 1917. Additional notes on heredity and life history in the coccinellid genus *Adalia* Mulsant. *Ann. Entomol. Soc. Am.* 10: 289–302.

Parker, G. A., L. W. Simmons and H. Kirk. 1990. Analysing sperm competition data: simple models for predicting mechanisms. *Behav. Ecol. Sociobiol.* 27: 55–65.

Pasquer, F., M. Pfunder, B. Frey and J. E. Frey. 2009. Microarray-based genetic identification of beneficial organisms as a new tool for quality control of laboratory cultures. *Biocontrol Sci. Techn.* 19: 809–833.

Pope, R. D. 1977. Brachyptery and wing-polymorphism among the Coccinellidae (Coleoptera). *Syst. Entomol.* 2: 59–66.

Rana, J. S., A. F. G. Dixon and V. Jarosik. 2002. Costs and benefits of prey specialization in a generalist insect predator. *J. Anim. Ecol.* 71: 15–22.

Ransford, M. O. 1997. *Sperm competition in the 2-spot ladybird*, Adalia bipunctata. Unpublished PhD thesis, University of Cambridge, Cambridge, UK.

Robertson, J. A., M. F. Whiting and J. V. McHugh. 2008. Searching for natural lineages within the Cerylonid Series (Coleoptera: Cucujoidea). *Mol. Phyl. Evol.* 46: 193–205.

Rodriguez-Saona, C. and J. C. Miller. 1995. Life history traits in *Hippodamia convergens* (Coleoptera: Coccinellidae) after selection for fast development. *Biol. Control* 5: 389–396.

Roehrdanz, R. L. 1992. Application of PCR techniques for identification of parasites and predators. *Proc. Fifth Russian Wheat Aphid Conf., Great Plains Agric. Council Pub.* 142: 190–196.

Roehrdanz, R. L. and R.V. Flanders. 1993. Detection of DNA polymorphisms in predatory coccinellids using polymerase chain reaction and arbitrary primers (RAPD-PCR). *Entomophaga* 38: 479–491.

Roehrdanz, R. L., A. L. Szalanski and E. Levine. 2003. Mitochondrial DNA and ITS1 differentiation in geographical populations of northern corn rootworm, *Diabrotica barberi* (Coleoptera: Chrysomelidae): identification of distinct genetic populations. *Ann. Entomol. Soc. Am.* 96: 901–913.

Rogers, C. E., H. B. Jackson, R. D. Eikenbary and K. J. Starks. 1971. Sex determination in *Propylea 14-punctata* (Coleoptera, Coccinellidae), an important predator of aphids. *Ann. Entomol. Soc. Am.* 64: 957–959.

Rozek, M. and M. Holecova. 2002. Chromosome numbers, C-banding patterns and sperm of some ladybird species from Central Europe. *Folia Biol., Krakow* 50: 17–21.

Sakai, T., Y. Uehara and M. Matsuka. 1974. The effect of temperature and other factors on the expression of elytral pattern in lady beetle, *Harmonia axyridis* Pallas. *Bull. Fac. Agric. Tamagawa Univ.* 14: 33–39.

Sanchez, L. 2008. Sex-determining mechanisms in insects. *Int. J. Dev. Biol.* 52: 837–856.

Sasaji, H. 1968. Phylogeny of the family Coccinellidae. *Etizenia* 35: 1–37.

Sasaji, H. 1980. Biosystematics on *Harmonia axyridis*-complex (Coleoptera: Coccinellidae). *Mem. Fac. Educ. Fukui Univ. Ser. II (Nat. Sci.)* 30: 59–79.

Sasaji, H. and K. Hisano. 1977. Genetic study on esteraseallozymes at two loci in *Harmonia axyridis. Zool. Mag. (Tokyo)* 86: 540.

Sasaji, H. and K. Nishide. 1994. Genetics of esterase isozymes in *Harmonia yedoensis* (Takizawa) (Coleoptera: Coccinellidae). *Mem. Fac. Educ. Fukui Univ. Ser. II (Nat. Sci.)* 45: 1–13.

Sasaji, H. and E. Ohnishi. 1973a. Disc electrophoretic study of esterase in ladybirds (preliminary report). *Mem. Fac. Educ. Fukui Univ. Ser. II (Nat. Sci.)* 23: 23–32.

Sasaji, H. and E. Ohnishi. 1973b. Intraspecific enzyme polymorphisms in *Harmonia axyridis* and its sibling species, with their species specificity. *Zool. Mag. (Tokyo)* 82: 340.

Sasaji, H., R. Yahara and M. Saito. 1975. Reproductive isolation and species specificity in two ladybirds of the genus *Propylea. Mem. Fac. Educ. Fukui Univ. Ser. II (Nat. Sci.)* 25: 13–34.

Scali, V. and E. R. Creed. 1975. The influence of climate on melanism in the two-spot ladybird, *Adalia bipunctata*, in central Italy. *Trans. R. Entomol. Soc. Lond.* 127: 163–169.

Schilder, F. A. 1928. Zur Variabilität von *Adonia variegata* Goeze (Col. Coccinell.). *Entomol. Blätter* 24: 129–142.

Schilder, F. A. 1952/3. Neue Variationsstudien an Coccinelliden. *Wiss. Z. Martin-Luther-Univ. Halle-Wittenberg* 2: 143–163.

Schilder, F. A. and M. Schilder. 1951/2. Methoden der Phänoanalyse von Tieren. *Wiss. Z. Martin-Luther-Univ. Halle-Wittenberg* 1: 81–91.

Schloetterer, C., M.-T. Hauser, A. von Haeseler and D. Tautz. 1994. Comparative evolutionary analysis of rDNA ITS regions in *Drosophila. Mol. Biol. Evol.* 11: 513–522.

Schroeder, C. 1909. Die Erscheinungen der Zeichnungsvererbung bei *Adalia bipunctata* L. und ihren ab. *6-pustulata* L. und *4-maculata* Scop. *Z. Wiss. Insektenbiol.* 5: 132–134.

Schulenburg J. H. G. vd, J. M. Hancock, A. Pagnamenta et al. 2001. Extreme length and length variation in the first ribosomal internal transcribed spacer of ladybird beetles (Coleoptera: Coccinellidae). *Mol. Biol. Evol.* 18: 648–660.

Schulenburg J. H. G. vd, G. D. D. Hurst, D. Tetzlaff et al. 2002. History of infection with different male-killing bacteria in the two-spot ladybird beetle *Adalia bipunctata* revealed through mitochondrial DNA sequence analysis. *Genetics* 160: 1075–1086.

Seko, S. and K. Miura 2009. Effects of artificial selection for reduced flight ability on survival rate and fecundity of *Harmonia axyridis* (Pallas) (Coleoptera: Coccinellidae). *Appl. Entomol. Zool.* 44: 587–594.

Seko, S., K. Yamashita and K. Miura. 2008. Residence period of a flightless strain of the ladybird beetle *Harmonia axyridis* Pallas (Coleoptera: Coccinellidae) in open fields. *Biol. Control* 47: 194–198.

Seo, M. J., E. J. Kang, M. K. Kang et al. 2007. Phenotypic variation and genetic correlation of elytra colored patterns of multicolored Asian lady beetles, *Harmonia axyridis* (Coleoptera, Coccinellidae) in Korea. *Korean J. Appl. Entomol.* 46: 235–249.

Sergievsky, S. O. and I. A. Zakharov. 1981. Ekologicheskaya genetika populiacij *Adalia bipunctata*: koncepcija 'zhestkogo i gibkogo' polimorfizma. *Issledovania po Genetike* 9: 112–120.

Sergievsky, S. O. and I. A. Zakharov. 1983. Izuchenie geneticheskogo polimorfizma dvutochechnoi bozhei korovki *Adalia bipunctata* (L.) Leningradskoi oblasti. Soobschenie II. Sostav populiacij goroda Leningrada. *Genetika-Moskva* 19: 635–640.

Sergievsky, S. O. and I. A. Zakharov. 1989. Reakcija populjacij na stresovyje vozdejstvija: koncepcija dvustupentshatogo reagirovania. *In* A. V. Jablokov (ed.). *Ontogenez, Evolucia,Biosfera.* Nauka, Moskva. pp. 157–173.

Shull, A. F. 1943. Inheritance in lady beetles. I. The spotless and spotted elytra of *Hippodamia sinuata. J. Heredity* 34: 329–337.

Shull, A. F. 1944. Inheritance in lady beetles. II. The spotless pattern and its modifiers in *Hippodamia convergens* and their frequency in several populations. *J. Heredity* 35: 329–339.

Shull, A. F. 1945. Inheritance in lady beetles. III – Crosses between variants of *Hippodamia quinquesignata* and between this species and *H. convergens. J. Heredity* 36: 149–160.

Shull, A. F. 1949. Extent of genetic differences between species of *Hippodamia* (Coccinellidae). *Hereditas* S8: 417–428.

Simon, C., T. R. Buckley, F. Frati, J. B. Stewart and A. T. Beckenbach. 2006. Incorporating molecular evolution into phylogenetic analysis, and a new compilation of conserved polymerase chain reaction primers for animal mitochondrial DNA. *Annu. Rev. Ecol. Evol. Syst.* 37: 545–579.

Sloggett, J. J. 2005. Are we studying too few taxa? Insights from aphidophagous ladybird beetles (Coleoptera: Coccinellidae). *Eur. J. Entomol.* 102: 391–398.

Sloggett, J. J. 2008. Habitat and dietary specificity in aphidophagous ladybirds (Coleoptera: Coccinellidae): explaining specialization. *Proc. Neth. Entomol. Soc. Meet.* 19: 95–113.

Sloggett J. J., S. A. Eimer, G. D. D. Hurst et al. In press. Explaining the alkaloids of *Coccinella magnifica* (Coleoptera: Coccinellidae): phylogenetic and ecological insights. *Biol. J. Linn. Soc.*

Smirnov, E. 1927. Mathematische Studien über individuelle und Kongregationenvariabilität. Verhandlungen des 5. *Internationalen Kongress für Vererbungswissenschaft*, Vol. 2, Berlin, 1373–1392.

Smirnov, E. 1932. Zur vergleichenden Morphologie der Zeichnung bei den Coccinelliden. *Z. Wiss. Zool.* 143: 1–15.

Smith, S. G. 1950. The cytotaxonomy of Coleoptera. *Can. Entomol.* 82: 58–68.

Smith, S. G. 1957. Adaptive chromosomal polymorphism in *Chilocorus stigma*. *Proc. Genet. Soc. Canada* 2: 40–41.

Smith, S. G. 1959. The cytogenetic basis of speciation in Coleoptera. *Proc. X Int. Congr. Genet., Montreal 1958* 1: 444–450.

Smith, S. G. 1960. Chromosome numbers of Coleoptera. II. *Can. J. Genet. Cytol.* 2: 66–68.

Smith, S. G. 1962a. Cytogenetic pathways in beetle speciation. *Can. Entomol.* 94: 941–955.

Smith, S. G. 1962b. Tempero-spatial sequentiality of chromosomal polymorphism in *Chilocorus stigma* Say (Coleoptera: Coccinellidae). *Nature* 193: 1210–1211.

Smith, S. G. 1966. Natural hybridization in the coccinellid genus *Chilocorus*. *Chromosoma* 18: 380–406.

Smith, S. G. and N. Virkki. 1978. *Animal Cytogenetics. Vol. 3: Insecta 5. Coleoptera.* Gebrüder Borntraeger, Berlin. 366 pp.

Soares, A. O., D. Coderre and H. Schanderl. 2001. Fitness of two phenotypes of *Harmonia axyridis* (Coleoptera, Coccinellidae). *Eur. J. Entomol.* 98: 287–293.

Soares, A. O., D. Coderre and H. Schanderl. 2003. Effect of temperature and intraspecific allometry on predation by two phenotypes of *Harmonia axyridis* Pallas (Coleoptera, Coccinellidae). *Env. Entomol.* 32: 939–944.

Srivastava, S. and Omkar. 2005. Mate choice and reproductive success of two morphs of the seven spotted ladybird, *Coccinella septempunctata*. *Eur. J. Entomol.* 102: 189–194.

Steiner, W. W. M. and J. J. Grasela. 1993. Population-genetics and gene variation in the predator, *Coleomegilla maculata* (De Geer) (Coleoptera: Coccinellidae). *Ann. Entomol. Soc. Am.* 86: 309–321.

Stevens, N. M. 1906. Studies in spermatogenesis. II. *Carneg. Inst. Wash. Publ.* 36: 33–74.

Strouhal, H. 1939. Variationsstatistische Untersuchung an *Adonia variegata* Gze (Col. Coccinell.). *Z. Morphol. Ökol. Tiere* 35: 288–316.

Tan, C. C. 1946. Mosaic dominance in the inheritance of color patterns in the lady-bird beetle, *Harmonia axyridis*. *Genetics* 31: 195–210.

Tan, C. C. 1949. Seasonal variations of color patterns in *Harmonia axyridis*. *Hereditas* S8: 669–670.

Tan, C. C. and K. Hu. 1980. On two new alleles of the color pattern gene in the lady-beetle, *Harmonia axyridis* and further proof of the mosaic dominance theory. *Zool. Res.* 1: 277–285.

Tan, C. C. and J. C. Li. 1934. Inheritance of the elytral color patterns of the lady-bird beetle, *Harmonia axyridis* Pallas. *Am. Nat.* 68: 252–265.

Tennenbaum, E. 1931. Variabilität der Fleckengröse innerhalb der Palästinarasse von *Epilachna chrysomelina* F. *Naturwissenschaften* 14: 490–493.

Tennenbaum, E. 1933. Zur Vererbung des Zeichnungmusters von *Epilachna chrysomelina* F. *Biol. Zbl.* 53: 308–313.

Thomas, A., S. Philippou, R. Ware, H. Kitson and P. Brown. 2010. Is *Harmonia axyridis* really eating *Adalia bipunctata* in the wild? *In* D. Babendreier, A. Aebi, M. Kenis and H. Roy (eds) *Working Group 'Benefits and Risks of Exotic Biological Control Agents'. Proceedings of the first meeting at Engelberg (Switzerland), 6 – 10 September, 2009. IOBC/wprs Bull.* 58: 149–153.

Thomas, C. E., E. Lombaert, R. Ware, A. Estoup and L. Lawson Handley. 2010. Investigating global invasion routes of the harlequin ladybird (*Harmonia axyridis*) using mtDNA. *In* D. Babendreier, A. Aebi, M. Kenis and H. Roy *Working Group 'Benefits and Risks of Exotic Biological Control Agents'. Proceedings of the first meeting at Engelberg (Switzerland), 6 – 10 September, 2009. IOBC/wprs Bull.* 58: 155–157.

Timofeeff-Ressovsky, N. V. 1932. The genogeographical work with *Epilachna chrysomelina* F. *Proc. Sixth Int. Congr. Genet., Ithaca, NY, 1932* 2: 230–232.

Timofeeff-Ressovsky, N. V. 1940. Zur Analyse des Polymorphismus bei *Adalia bipunctata* L. *Biol. Zbl.* 60: 130–137.

Timofeeff-Ressovsky, N. V. 1941. Temperaturmodifikabilität der Zeichnungmusters bei verschiedenen Populationen von *Epilachna chrysomelina* F. *Biol. Zbl.* 61: 68–84.

Timofeeff-Ressovsky, N. V. and Yu. M. Svirezhev. 1966. Adaptation polymorphism in populations of *Adalia bipunctata* L. *Problemy Kibernetiki* 16: 137–146.

Timofeeff-Ressovsky, N. V. and Yu. M. Svirezhev. 1967. Genetic polymorphism in populations, an experimental and theoretical investigation. *Genetika-Moskva* 3: 152–166.

Timofeeff-Ressovsky, N. V., E. A. Timofeeva-Ressovskaya and I. K. Zimmermann. 1965. Experimental and systematic analysis of geographic variability and speciation in *Epilachna chrysomelina* F. (Coleoptera, Coccinellidae). *Trudy Inst. Biol. Akad. Nauk SSSR Uralskii Filial* 44: 27–63.

Tolunay, M. A. 1939. *Coccinella Septempunctata'nin* bazi Türkiye Populationlari üzerine Variation-Qualitik araştirmalar. *Yülesek Ziraat Enstitüsü Çalişmalarindan* 94: 1–116.

Tourniaire, R., A. Ferran, L. Giuge, C. Piotte and J. Gambier. 2000. A natural flightless mutation in the ladybird, *Harmonia axyridis. Entomol. Exp. Appl.* 96: 33–38.

Tsurusaki, N., S. Nakano and H. Katakura. 1993. Karyotypic differentiation in the phytophagous ladybird beetles *Epilachna vigintioctopunctata* complex and its possible relevance to reproductive isolation, with a note on supernumerary Y chromosomes found in *E. pustulosa. Zool. Sci.* 10: 997–1015.

Tsurusaki, N., Y. Shirai, S. Nakano et al. 2001. B-chromosomes in the Indonesian populations of a phytophagous ladybird beetle *Epilachna vigintioctopunctata. Spec. Publ. Japan Coleopt. Soc., Osaka* 1: 65–71.

Turelli, M., A. A. Hoffmann and S. W. McKechnie. 1992. Dynamics of cytoplasmic incompatibility and mtDNA variation in natural *Drosophila simulans* populations. *Genetics* 132: 713–723.

Ueno, H. 1994a. Genetic estimations for body size characters, developmental period and development rate in a coccinellid beetle, *Harmonia axyridis. Res. Popul. Ecol.* 36: 121–124.

Ueno, H. 1994b. Intraspecific variation of P2 value in a coccinellid beetle, *Harmonia axyridis. J. Ethol.* 12: 169–174.

Ueno, H. 1996. Estimate of multiple insemination in a natural population of *Harmonia axyridis*. Coleoptera: Coccinellidae). *Appl. Entomol. Zool.* 31: 621–623.

Ueno, H. 2003. Genetic variation in larval period and pupal mass in an aphidophagous ladybird beetle (*Harmonia axyridis*) reared in different environments. *Entomol. Exp. Appl.* 106: 211–218.

Ueno, H., Y. Sato and K. Tsuchida. 1998. Colour-associated mating success in a polymorphic ladybird beetle, *Harmonia axyridis. Funct. Ecol.* 12: 757–761.

Ueno, H., N. Fujiyama, K. Irie, Y. Sato and H. Katakura. 1999. Genetic basis for established and novel host plant use in a herbivorous ladybird beetle, *Epilachna vigintioctomaculata. Entomol. Exp. Appl.* 91: 245–250.

Ueno, H., N. Fujiyama, I. Yao, Y. Sato and H. Katakura. 2003. Genetic architecture for normal and novel host-plant use in two local populations of the herbivorous ladybird beetle, *Epilachna pustulosa. J. Evol. Biol.* 16: 883–895.

Ueno, H., P. W. de Jong and P. M. Brakefield. 2004. Genetic basis and fitness consequences of winglessness in the two-spot ladybird beetle, *Adalia bipunctata. Heredity* 93: 283–289.

Underwood, D. A., S. Hussein, C. Goodpasture et al. 2005. Geographic variation in meiotic instability in *Eucheira socialis* (Lepidoptera: Pieridae). *Ann. Entomol. Soc. Am.* 98: 227–235.

Vandenberg, N. J. 1992. Revision of the New World lady beetles of the genus *Olla* and description of a new allied genus (Coleoptera, Coccinellidae). *Ann. Entomol. Soc. Am.* 85: 370–392.

Vogler, A. P. and R. DeSalle. 1994. Evolution and phylogenetic information content of the ITS-l region in the tiger beetle *Cicindela dorsalis. Mol. Biol. Evol.* 11: 393–405.

Vorontsov, N. N. and A. V. Blehman 2001. Distribution and intraspecific structure of ladybeetle *Harmonia axyridis* Pall., 1773 (Coleoptera, Coccinellidae). *In* V. A. Krasilov (ed). *Evolucia, Ekologia, Vidoobrazovanie.* Materialy konferencii pamjati Nikolaja Nikolaevitsa Voroshilova (1934–2000). Izdatelskij otdel UNCDO, Moskva, pp. 150–156.

Wang, S., J. P. Michaud, R. Z. Zhang, F. Zhang and S. Liu. 2009. Seasonal cycles of assortative mating and reproductive behaviour in polymorphic populations of *Harmonia axyridis* in China. *Ecol. Entomol.* 34: 483–494.

Webb, R. E. and F. F. Smith. 1968. Fertility of eggs of Mexican bean beetles mated alternately with normal and apholate-treated males. *J. Econ. Entomol.* 61: 521–523.

Weber, D. C. and J. G. Lundgren. 2009. Assessing the trophic ecology of the Coccinellidae: their roles as predators and as prey. *Biol. Control* 51: 199–214.

Weinert, L. A. 2008. *Incidence, Diversity and Evolution of Rickettsia and Other Endosymbionts that Infect Arthropods.* Unpublished PhD thesis, University of Edinburgh, Edinburgh, UK.

Werren, J. H. and R. Stouthamer. 2003. PSR (paternal sex ratio) chromosomes: the ultimate selfish genetic elements. *Genetica* 117: 85–101.

Whitmore, R. W. and K. P. Pruess. 1982. Response of pheasant chicks to adult lady beetles (Coleoptera, Coccinellidae). *J. Kansas Entomol. Soc.* 55: 474–476.

Witter, J. A. and G. D. Amman. 1969. Field identification and sex determination of *Aphidecta obliterata*, an introduced predator of *Adelges piceae. Ann. Entomol. Soc. Am.* 62: 718–721.

Yadav, J. S. and S. Gahlawat. 1994. Chromosomal investigations on five species of ladybird beetles (Coccinellidae, Coleoptera). *Folia Biol., Krakow* 42: 139–143.

Zakharov, I. A. 1996. A study of elytra pattern inheritance in *Adalia bipunctata. Genetika-Moskva* 32: 579–583.

Zakharov, I. A. 2003. Industrial melanism and its dynamics in populations of the two-spot ladybird *Adalia bipunctata* L. *Uspekhi Sovremennoi Biologii* 123: 3–15.

Zakharov, I. A. and S. O. Sergievsky. 1978. Studies on the variation of *Adalia bipunctata* populations in Leningrad city and suburbs. *Genetika-Moskva* 14: 281–284.

Zakharov, I. A. and S. O. Sergievsky. 1980. Study of the genetic polymorphism of two-spot ladybird *Adalia bipunctata* (L.) populations in Leningrad district. I. Seasonal dynamics of polymorphism. *Genetika-Moskva* 16: 270–275.

Zakharov, I. A. and E. V. Shaikevich. 2001. The Stockholm populations of *Adalia bipunctata* (L) (Coleoptera, Coccinellidae): a case of extreme female-biased population sex ratio. *Hereditas* 134: 263–266.

Zarapkin, S. R. 1930. Über die gerichtete Variabilität der Coccinelliden I. Allgemeine Einleitung und Analyse der Ersten

Pigmentierungsetappe bei *Coccinella 10-punctata*. *Z. Morphol. Ökol. Tiere* 17: 719–736.

Zarapkin, S. R. 1938a. Über die gerichtete Variabilität der Coccinelliden. V. Die Reihenfolge der Fleckenentstehung auf den Elytren der *Coccinella 10-punctata* (*Adalia 10-punctata*) in der ontogenetischen Entwicklung. *Z. Morphol. Ökol. Tiere* 34: 562–572.

Zarapkin, S. R. 1938b. Über die gerichtete Variabilität der Coccinelliden VI. Biometrische Analysis der Gerichteten Variabilität. *Z. Morphol. Ökol. Tiere* 34: 573–583.

Zaslavskii, V. A. 1963. Hybrid sterility as a limiting factor in the distribution of allopatric species. *Doklady Akademii Nauk SSSR* 149: 470–471.

Zera, A. J. and R. F. Denno. 1997. Physiology and ecology of dispersal polymorphism in insects. *Annu. Rev. Entomol.* 42: 207–230.

Zimmermann, K. 1931. Wirkung von Selektion und Temperatur auf die Pigmentierung von *Epilachna chrysomelina* F. *Naturwissenschaften* 19: 768–781.

Chapter 3

LIFE HISTORY AND DEVELOPMENT

Oldrich Nedvěd[1] and Alois Honěk[2]

[1] Faculty of Science, University of South Bohemia and Institute of Entomology, Academy of Sciences, CZ 37005 České Budějovice, Czech Republic
[2] Department of Entomology, Crop Research Institute, CZ 16106 Prague 6, Czech Republic

Ecology and Behaviour of the Ladybird Beetles (Coccinellidae), First Edition. Edited by I. Hodek, H.F. van Emden, A. Honěk.
© 2012 Blackwell Publishing Ltd. Published 2012 by Blackwell Publishing Ltd.

3.1 INTRODUCTION

Coccinellids are **holometabolous**, i.e. they have a 'complete metamorphosis', and pass through the following stages: egg, larva, pupa and adult. Egg stage lasts 15–20% of the total preimaginal developmental time, larva 55–65% and pupa 20–25% (Honěk & Kocourek 1990, Dixon 2000).

3.2 EGG

3.2.1 Egg morphology

Coccinellid eggs are usually elongate, oval or elliptic. They vary in **colour** from almost transparent (*Scymnus louisianae*; Brown et al. 2003), light grey (*Stethorus*), yellowish (*Halyzia*) through bright yellow (most species) to dark orange (Chilocorini), sometimes greenish (Klausnitzer 1969b). They are laid either upright, attached (glued) to the substrate by the lower end (Coccinellinae) or lying on their side (Scymninae, Coccidulinae, Chilocorinae). The 'upper' or anterior pole bears a ring of **micropyles** (Fig. 3.1) – pores in the chorion (the egg shell) through which spermatozoa

from spermatheca (receptaculum seminis) can enter during oviposition, and for oxygen diffusion. *A. bipunctata* possesses 40–50 pores in two rings, *Platynaspis luteorubra* have clusters of pores at both ends of the egg. Chilocorini have trumpet shaped structures besides the tube like micropyles (Ricci & Stella 1988). The egg of *Scymnus sinuanodulus* has a rosette of 4–11 cup-shaped and 13–20 semicircular structures that are like micropyles (Lu et al. 2002).

The surface of the coccinellid egg (**chorion**, egg shell) is usually smooth, except for the eggs of Epilachninae which bear a polygonal sculpture (Klausnitzer 1969b; Fig. 3.1c), and *Rhyzobius* with a granular surface (Ricci & Stella 1988). The smooth surface may be soaked with an oily **excretion** provided by accessory glands of the mother. The excretion is colourless, or rarely red (as in *Calvia quatuordecimguttata*; Klausnitzer & Klausnitzer 1986) and contains defensive or signalling alkanes (Hemptinne et al. 2000; Chapter 9.).

The yolk of the egg contains many **nutrients**. During embryonic development of the *A. bipunctata* egg mass, lipid and glycogen contents decline strongly, while egg protein decline more slowly. Free carbohydrates decline early in egg development and increase

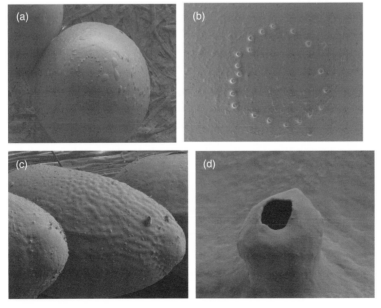

Figure 3.1 SEM photographs of ladybird eggs. (a, b) *Harmonia axyridis*, (c, d) *Cynegetis impunctata*. (a, c) upper parts of eggs with ring of micropyle openings; (b) ring of micropyles; (d) detail of a micropyle opening.

before hatching (Sloggett & Lorenz 2008). Energy per unit egg mass is lowest in the generalist *A. bipunctata*, which also has relatively large eggs, and is highest in the specialized *Anisosticta novemdecimpunctata*, which has small eggs.

A few days or hours before hatching the eggs become greyish, because the larvae are visible through the chorion. The first instar larvae of Coccinellinae and Epilachninae may possess special structures on head and prothorax called **egg teeth** which probably help the larva in hatching. Hatched larvae rest some time on the empty egg shells, and may eat them.

Ladybird eggs are defended chemically against predation, including that which is intra-guild, and they are suitable to varying degrees as food for other ladybirds (Rieder et al. 2008; Chapter 7).

3.2.2 Egg size

Insect species have a genetically fixed maximum number of ovarioles related to egg cluster size (see below) and/or a **fixed maximum size** of each single egg – traits expressed with ample food and other favourable environmental conditions. With a limited food supply, the size of eggs may be reduced and their number maintained, or decreased and their size maintained. Coccinellids adopt the second strategy (Stewart et a1. 1991a, b; Dixon & Guo 1993; see also 5.2.3). This is similar to birds, for example, where egg weight (E) across species increases in direct proportion (almost isometrically) to **body weight** (W) of the female and inversely to the **ovariole number** (O): $\log E = 0.83 \cdot \log(W/O) - 0.44$ (Stewart et al. 1991b).

The size of eggs is thus related to the average body size of the species; it is less than 0.4 mm in the smallest genera and over 2 mm in the largest genera. Eggs of *Scymnus louisianae* are 0.5 mm long and 0.2 mm wide (Brown et al. 2003). Adult *Clitostethus oculatus* are only 1.26 mm long and the eggs are 0.41 mm long and 0.18 mm wide, greenish to yellowish-white, with a reticulate chorion; whereas adult *Delphastus pusillus* are 1.39 mm long, the eggs are 0.43 mm long and 0.23 mm wide, white and smooth (Liu & Stansly 1996a). Eggs of *Stethorus pusillus* are 0.4 mm long, eggs of *A. bipunctata* 1.0 mm long, eggs of *C. septempunctata* 1.3 mm long, eggs of *Henosepilachna argus* 1.7 mm long, and eggs of *Anatis ocellata* 2.0 mm long (Klausnitzer & Klausnitzer 1986). Egg volume (V) in

microlitres may be calculated as: $V = LW^2\pi/6$, where L is length and W is the width of the egg (Takakura 2004).

The weight of the eggs of large species *C. septempunctata* (15.4 mg adult dry mass) and small species *P. quatuordecimpunctata* (3.7 mg adult dry mass) differed little (0.20 versus 0.18 mg) (Honěk et al. 2008). *C. septempunctata* produces a larger number of individually smaller eggs than *C. transversoguttata* (Kajita et al. 2009) which has the same body size. Similarly, *Har. axyridis* produces larger egg clusters with smaller eggs (0.28 mm³) than *Har. yedoensis* (0.36 mm³) (Osawa & Ohashi 2008). Species of similar adult size laying smaller eggs may incur higher costs per unit mass than species laying larger eggs. More specialized species reproducing at lower aphid densities may provide neonate larvae with more nutrients to facilitate the finding of an aphid colony (Sloggett & Lorenz 2008).

The **fitness** of young larvae is positively related to egg size. Starving first instar larvae survived substantially longer (using only each individual's **yolk reserves**) in large *Coelophora bissellata* (with egg volume 1.53 mm³) than did three other Malayan species with egg volumes of 0.81–1.20 mm³ (Ng 1988). However, small eggs have a higher **rate of embryonic development**, which decreases the risk of egg predation (Majerus 1994). Developmental time was 3.0 days in *Har. yedoensis* possessing larger eggs and 2.7 days in *Har. axyridis* with smaller eggs (Osawa & Ohashi 2008). There is a strong selection pressure on fast development and **hatching synchrony** in cannibalistic ladybirds. Minimum egg size is, however, constrained by the minimum size at which first instar larvae can find (by extended walking) and kill prey (Stewart et al. 1991b).

Small inter-individual variation in egg size was reported in *Har. axyridis*: mean egg mass was 0.246 mg, with the minimum average in offspring of individual females 0.196 and the maximum 0.278 (Prevolsek & Williams 2006). About 30 small eggs were laid compared with only about 20 large eggs. Intra-individual variation in egg size with varying food levels was minimal.

3.2.3 Cluster size

Eggs are either laid **singly** (Scymninae, Coccidulinae, Chilocorinae – mostly coccidophagous groups) or in clusters (**batch**, **clutch** – Coccinellinae – aphidopha-gous, Epilachninae –phytophagous) (Klausnitzer & Klausnitzer 1986). There are a few exceptions. In Chilocorinae, the coccidophagous *Orcus chalybeus* lay eggs in clusters (Thompson 1951); *Exochomus quadri-pustulatus* usually lay their eggs in small groups but sometimes also individually (Sengonca & Arnold 2003). Within Coccinellinae, *Synonycha grandis* lay **sparse clusters** with the eggs a few millimetres apart; *Megalocaria dilatata* lay eggs in two **rows** (Iablokoff-Khnzorian 1982). In the adelgid eating specialist *Aphidecta obliterata*, 20% of 'clusters' contain a single egg, and a maximum cluster size observed on spruce needles was five eggs (Timms & Leather 2007).

The eggs in clusters of Coccinellinae are typically tightly packed in an upright position. The cluster size is anatomically determined by the number of **ovari-oles** (3.5.5), which is similar in the left and right ovaries. The usual number of eggs per cluster in optimal conditions is equal to about half of the total ovariole number. This may be because the female lays eggs from one of the two ovaries at a time. Stewart et al. (1991a) proposed that about half the ovarioles are active in egg production at any time, while the rest are preparing for new oviposition. This inter-change of activity might enable more continuous egg production. The maximum number of eggs laid in one cluster in optimal conditions is close to the total number of ovarioles. However, dissections showed no significant correlation between ovariole number and egg cluster size in intraspecific data sets of *Har. axyridis* and *A. bipunctata* (Ware et al. 2008; 5.2.3).

Baungaard and Hämäläinen (1984) attest that the **maximum cluster size** is limited by the number of ovarioles in a single ovary, since only one fully developed egg within an ovariole can be laid, and all eggs in a cluster originate from the same ovary. They suggested that larger clusters are a result of laying a second cluster in very close proximity to the first one. However, Majerus (1994) reported egg clusters which contained numbers well in excess of the maximum numbers of ovarioles recorded for the respective species (*C. septempunctata* and *A. bipunctata*). He suggested that a cluster may include eggs from **both ovaries**, and that more than one egg may be deposited from one ovariole. Ware et al. (2008) reported egg clusters laid by an individual female that were bigger than the number of ovarioles in both ovaries together. We suggest that the count of ovarioles during dissection

may have been underestimated, because some of them had been emptied recently, therefore had not developed a second large maturing oocyte, and thus were overlooked.

In interspecific comparisons, there is a general linear correlation between the number of ovarioles and the cluster size (Dixon & Guo 1993). However, adult size is strongly determined by food availability during previ-ous larval development (Rhamhalinghan 1985) while cluster size is affected by food availability during the development of oocytes. **Oocyte development** and maturation takes a few days and there are several oocytes of different stages in each ovariole sequentially ordered in chambers or **follicles**, so that the female is able to lay a full clutch of eggs from both ovaries daily. The total **egg load** is usually split into two or more clusters. It is uncertain whether eggs of such small clusters come from one ovary only, or from a combina-tion of both.

Fois and colleagues (unpublished) found 5–72 eggs in clusters laid by a single female of *Har. axyridis*, the median and average being 28 eggs. This was a usual cluster size in this highly fertile species; well-fed young females laid two such clusters daily. However, the number of eggs laid per day was often lower than the total number of ovarioles of the female, which was 53–77 in this sample of *Har. axyridis*. The decrease in the number of eggs may have been due to **suboptimal feeding** with progressive **ageing**, and may vary with the **frequency of mating**. The average cluster weight of *Har. axyridis* is 5.6 mg (21 eggs) (Prevolsek & Wil-liams 2006).

The egg cluster size of *Har. axyridis* fed with eggs of *Ephestia kuehniella* is 29–32 and 28–39. Females fed with honey bee pollen oviposited only 8–14 eggs per cluster, while those fed with the pea aphid, *Acyrthosi-phon pisum*, laid 29–48 eggs per cluster (Berkvens et al. 2008a, b). A larval diet of unlimited aphids resulted in the largest clusters of eggs being laid by the resulting *Har. axyridis* and *A. bipunctata* females (Ware et al. 2008), irrespective of the constant number of ovari-oles. For more examples on the **effect of food** on cluster size see Table 3.1.

Egg cluster size is on average smaller (about 30 versus 40 eggs) for a short **photoperiod** (12 h light) than for a long one (16 h) in both the laboratory strain and wild population of the red *succinea* morph of *Har. axyridis* (Berkvens et al. 2008b), while it is larger (34 versus 28) in the melanic morph. The average cluster size in a melanic **colour morph** of *C. septempunctata*

Table 3.1 Egg cluster size (C), daily fecundity (D) and lifetime fecundity (F) of coccinellids. References for individual rows are numbered and listed below the table.

C	D	F	Prey, conditions	Ref.
Adalia bipunctata				
11–17	—	—	—	8
11–16 (1–47)	—	—	—	10
—	—	250	*Aphis fabae*	14
—	—	676	*Myzus persicae*	14
11–14	—	—	—	15
15 (2–43)	—	—	—	25
—	10	—	*Myzus persicae* on *Brassica napus*	28
—	4	—	*Myzus persicae* on *Sinapis alba*	28
—	7	—	*Myzus persicae* on *Vicia faba*	28
14	—	—		34
—	9.3	738	*Aphis fabae*	35
—	20.4	max. 1535	*Microlophium carnosum*	35
15 (3–25)	—	—	—	37
—	12–20	—	—	40
—	—	63	*Aphis fabae*	52
—	—	1011	*Phorodon humuli*	52
14–28	—	—	*Euceraphis betulae*, 25°C, 18L:6D	53
16–28	—	—	*Eucallipterus tiliae*, 25°C, 18L:6D	53
14–24	—	—	*Tuberculatus annulatus*, 25°C, 18L:6D	53
18–26	—	—	*Liosomaphis berberidis*, 25°C, 18L:6D	53
14–22	—	—	*Acyrthosiphon ignotum*, 25°C, 18L:6D	53
12–18	—	—	*Macrosiphoniella artemisiae*, 25°C, 18L:6D	53
12–21	—	—	*Cavariella konoi*, 25°C, 18L:6D	53
12–21	—	—	*Aphis fabae*, 25°C, 18L:6D	53
12–16	—	—	*Aphis farinosa*, 25°C, 18L:6D	53
0	—	—	*Aphis cirsiiacanthoidis*, 25°C, 18L:6D	53
0	—	—	*Aphis spiraephaga*, 25°C, 18L:6D	53
—	16	537	*Myzus persicae*, 25°C	63
—	—	600	—	95
—	—	39	*Acyrthosiphon caraganae*	97
—	—	264	*Aphis pomi*	97
—	—	94, 241	*Hyalopterus pruni*	97
—	—	161	*Rhopalosiphum padi*	97
30	—	—	—	108
31	20	—	*Acyrthosiphon pisum*, 22°C, 14L:10D, 30 days period; mean of individual maxima	115
—	16.7	978	19°C	51
—	22.5	835	23°C	51
—	25.9	759	27°C	51
—	18.3	501	pollen + *Ephestia kuehniella* eggs	51
—	23.0	992	*Acyrthosiphon pisum*	51
—	23.7	1079	*Myzus persicae*	51
—	28.1	796	*Acyrthosiphon pisum*	122
—	33.3	1864	pollen + *Ephestia kuehniella* eggs	122
—	11.3	890	pollen + *Artemia franciscana*	122
—	6.3	265	pollen + lyophilized artificial diet	122
—	10.8	468	pollen + lyoph. art. diet + *A. franciscana*	122

Table 3.1 (*Continued*)

C	D	F	Prey, conditions	Ref.
Adalia decempunctata				
11	—	—	—	21
Aiolocaria hexaspilota				
26 (3–74)	—	881 (784–1036)		47
Anatis ocellata				
—	—	300	—	58
Anegleis cardoni				
—	6.8	397	*Aphis gossypii*, 27°C, 14L:10D	84
—	6.3	328	*Aphis craccivora*, 27°C, 14L:10D	84
—	5.1	137	*Lipaphis pseudobrassicae*, 27°C, 14L:10D	84
Axinoscymnus cardilobus				
—	—	125	17°C	43
—	—	211	23°C	43
—	—	20	32°C	43
Brumoides suturalis				
—	—	142	*Ferrisia virgata* + *Planococcus minor*	29
—	—	182	*Ferrisia virgata*	29
—	—	18	*Phthorimaea operculella*	29
—	—	214	*Planococcus minor*	29
Callicaria superba				
—	—	237	—	49
Calvia decemguttata				
—	—	162–257	—	62
Calvia duodecimmaculata				
—	—	165	*Rhopalosiphum padi*	62
—	—	127	*Aphis pomi*	62
—	—	193	*Psylla mali*	62
Calvia quatuordecimguttata				
—	—	185	*Psylla alni*	62
—	—	247	*Psylla mali*	62
—	—	114–133	*Rhopalosiphum padi*	62
—	—	122 (106–142)	*Aphis pomi*	98
—	—	38 (28–46)	*Hyalopterus pruni*	98
—	—	158 (147–183)	*Psylla alni*	98
—	—	219 (198–243)	*Psylla mali*	98
18 (10–27)	—	142 (113–168)	*Psylla ulmi*	98
—	—	114 (98–128)	*Rhopalosiphum padi*	98
Ceratomegilla undecimnotata				
—	—	139	spring	47
—	—	73	summer	47
—	20	—	aphids	102
—	9–14	—	*Ephestia kuehniella* eggs	102
30	—	—	—	108

(*Continued*)

Table 3.1 (*Continued*)

C	D	F	Prey, conditions	Ref.
Chilocorus bipustulatus				
—	4.2	529	*Aspidiotus nerii*	114
Chilocorus nigritus				
—	3	—	*Aonidiella orientalis*	3
—	—	24	*Melanaspis glomerata*	22
—	—	370	*Aulacaspis tegalensis*	32
—	5	—	*Aspidiotus nerii*	36
—	—	57–93	—	50
—	—	151	*Parlatoria blanchardi*	72
—	3	564	*Abgrallaspis cyanophylli*, 20°C	88
—	6	1361	*Abgrallaspis cyanophylli*, 24°C	89
—	8	1008	*Abgrallaspis cyanophylli*, 26°C	88
—	6	872	*Abgrallaspis cyanophylli*, 30°C	88
—	3.5	432	*Aspidiotus nerii*, 26°C	88
—	—	81	*Aonidiella aurantii*, 27°C	50
—	—	86	*Diaspidiotus perniciosus*, 27°C	50
—	—	93	*Aspidiotus desctructor*, 27°C	50
—	—	87	*Aulacaspis tubercularis*, 27°C	50
—	—	72	*Chrysomphalus aonidum*, 27°C	50
—	—	79	*Hemberlesia lataniae*, 27°C	50
—	—	71	*Lepidosaphes cornutus*, 27°C	50
—	—	79	*Melanaspis glomerata*, 27°C	50
—	—	194	*Aonidiella aurantii*, 26°C	100
—	—	483	*Aspidiotus nerii*, 26°C	100
—	—	151	*Parlatoria blanchardii*, 27°C	72
—	—	292	*Aonidiella orientalis*, 24°C	3
—	—	57	*Aonidimytilus albus*, 27°C	3
—	—	121	*Aspidiotus desctructor*, 30°C	9
—	—	102	*Hemberlesia lataniae*, 30°C	9
—	—	136	*Melanaspis glomerata*, 30°C	9
—	—	370+	*Aulacaspis tegalensis*, 21°C	32
Clitostethus arcuatus				
—	—	181	*Siphoninus phillyreae*	125
Clitostethus oculatus				
—	3	—	*Bemisia tabaci*	64
—	0.6	52	20°C	92
—	1.0	81	26°C	92
—	0.7	33	31°C	92
Coccinella leonina transversalis				
16	—	—	—	74
—	—	915	*Aphis craccivora*	80
—	21	1400	lifetime mating	81
—	23	376	single mating	81
Coccinella novemnotata				
18	11.9	302	*Brevicoryne brassicae*, 25°C	24

Table 3.1 (*Continued*)

C	D	F	Prey, conditions	Ref.
Coccinella septempunctata				
25	18	—	15 mg *Acyrthosiphon pisum* daily	20
41	34	—	*Acyrthosiphon pisum* ad libitum	20
17	12.3	201	*Brevicoryne brassicae*, 25°C	24
34	—	—	in the laboratory, field collected	120
36	—	—	field observation	120
—	19	—	20 *Acyrthosiphon pisum* / day	26
—	32	—	30 *Acyrthosiphon pisum* / day	26
—	28 (23–37)	—	105 aphids / day	30
—	7 (3–10)	—	50 aphids / day	30
—	—	514–719	aphids	41
—	—	34–228	artificial diets	41
—	—	283–1182	aphids, 12–27 mg/day	46
—	—	0.3–50	artificial diet, 4.5–9.1 mg/day	46
—	—	1428 (268–2386)	*Acyrthosiphon pisum*, 25°C, 16L:8D	54
—	—	1286 (218–2291)	*Sitobion avenae*, 25°C, 16L:8D	54
—	—	626 (98–1528)	*Aphis fabae*, 25°C, 16L:8D	54
—	—	683 (134–1839)	*Aphis craccivora*, 25°C, 16L:8D	54
—	—	893	*Aphis spiraephila*	62
—	—	356	*Rhopalosiphum padi*	62
—	—	759	*Sitobion avenae*	62
—	—	513	*Schizaphis graminum*	62
—	—	476	*Aphis gossypii*	62
—	—	448	*Aphis pomi*	62
—	—	1061	*Aphis craccivora*	79
—	—	739	*Aphis gossypii*	79
—	—	203	*Aphis nerii*	79
—	—	1764	*Lipaphis pseudobrassicae*	79
—	—	1199	*Myzus persicae*	79
—	—	488	*Uroleucon compositae*	79
—	—	1061	*Aphis craccivora*	80
—	25	1764	lifetime mating	81
—	13	298	single mating	81
—	—	552	—	93
—	—	668	—	93
—	—	79–313	varied with parental age	105
—	—	705	melanic female + melanic male	106
—	—	396	typical female + typical male	106
50	—	—	—	108
	—	814	—	110
33–44	—	—	—	111
	22.4	287	*Aphis gossypii*, 25°C	118
15–84		718–1485 (max. 1953)	Tashkent	119
—	27	—	seasonal peak, alfalfa	127
Coccinella transversoguttata				
—	15	—	seasonal peak, alfalfa	127
Coccinella trifasciata				
10	—	—	—	108

(*Continued*)

Table 3.1 (*Continued*)

C	D	F	Prey, conditions	Ref.
Coccinella undecimpunctata				
21	12.4	371	*Brevicoryne brassicae*, 25°C	24
—	max. 40	85–135	Palestina	33
1–37	max. 89	451–746	Egypt	48
—	—	742–983	*Aphis craccivora*, single mated	132
Coelophora biplagiata				
20	—	—	—	99
Coelophora bissellata				
10	—	—	—	74
Coelophora inaequalis				
9	—	—	—	74
Coelophora saucia				
—	26 (max. 51)	785	16L:8D	76
—	16 (max. 40)	478	24L:0D	76
—	10 (max. 39)	311	8L:16D	76
—	2–3 (max. 50)	230–1044	varied with light colour	76
—	—	120	10 s mating	83
—	—	310	1 min mating	83
—	—	380	one hour mating	83
—	—	393	*Aphis craccivora*, newly emerged female	126
—	—	1980	*Aphis craccivora*, 20 days old female	126
—	—	240	*Aphis craccivora*, 60 days old female	126
—	—	1506	*Aphis craccivora*, 1 mating	129
—	—	2192	*Aphis craccivora*, 20 matings	129
Coleomegilla maculata				
—	2.6–3.0	—	pollen	23
—	1	—	*Leptinotarsa decemlineata* eggs	38
—	4	—	*Myzus persicae*	38
5–7	—	—	the first clutch	67
—	6.2	—	pollen + *Ephestia kuehniella* eggs	131
—	6.8	—	maize pollen	131
Delphastus catalinae				
—	34	—	—	103
Epilachna vigintisexpunctata				
24–28	—	—	—	94
Exochomus quadripustulatus				
—	2.6	91	*Acyrthosiphon pisum*	91
—	4.3	173	*Dysaphis plantaginea*	91
—	1–16	139	12/24°C	101
—	1–11	97	9/19°C	101

Table 3.1 (*Continued*)

C	D	F	Prey, conditions	Ref.
Harmonia axyridis				
—	5	257	*Pectinophora gossypiella* eggs	1
—	13	—	single pair in container	2
—	1.6–8.8	—	10–40 females per container	2
—	15	715	fresh eggs of *Sitotroga cerealella*	2
—	13	607	frozen eggs of *Sitotroga cerealella*	2
—	—	834	*Aphis fabae*	6
—	—	1536	*Diuraphis noxia*	6
29–32	—	—	*Ephestia* eggs	11
8–14	—	—	pollen	11
29–48	—	—	*Acyrthosiphon pisum*	12
28–39	—	—	*Ephestia* eggs	12
27	—	—	—	21
28 (5–72)	—	—	*Acyrthosiphon pisum*	27
—	—	751	*Aphis gossypii*, 14.5–18°C	39
—	—	164	artificial diet	42
—	25	3819	*Hyperomyzus carduellinus*, 25°C	44
—	14	945	*Hyperomyzus carduellinus*, 30°C	44
—	27	2310	*Myzus persicae*, 25°C	44
—	22	778	*Myzus persicae*, 30°C	44
—	16	—	*Aphis gossypii*	59
—	18	561	*Myzus persicae*, 25°C	63
—	—	400–800	6 weeks, 18°C	65
—	—	900–1300	6 weeks, 24°C	65
—	—	800–1100	6 weeks, 30°C	65
—	—	719	*Acyrthosiphon pisum*, 27°C	69
20	—	—	*Aphis spiraecola*	85
21	—	—	—	90
26	—	—	*Acyrthosiphon pisum*	86
12–52	20	100–200	*Hyalopterus pruni*	96
—	23	—	aphids	102
—	34–41	—	*Ephestia kuehniella* eggs	102
30	27 (max. 78)	1642 (703–2263)	*Aphis fabae*	107
23	—	—	alfalfa	111
31	—	—	weeds	111
47	19	—	*Acyrthosiphon pisum*, 22°C, 14L:10D, 30 days period; mean of individual maxima	115
39	—	455	frozen *Ephestia kuehniella* eggs	121
—	—	128	*Aphis glycines* susceptible soybean	68
—	—	22	*Aphis glycines* resistant soybean	68
—	10.3	—	flightless strain, 36. generation	130
—	19.3	—	control strain, 36. generation	130
—	—	888	*Ephestia kuehniella* eggs	136
—	—	823	*Schizaphis graminum*	136
Harmonia yedoensis				
24	—	401	frozen *Ephestia kuehniella* eggs	121

(*Continued*)

Table 3.1 (*Continued*)

C	D	F	Prey, conditions	Ref.
Henosepilachna sumbana				
21–32	—	—	—	94
Henosepilachna vigintioctopunctata				
26	—	—	—	94
Hippodamia convergens				
—	—	max. 1550	—	47
19	—	—	—	116
—	—	589–851	*Aphis craccivora*, single mated	132
Hippodamia tredecimpunctata				
10–40	—	107–407	USA	47
—	—	316	*Aphis gossypii*	62
—	—	345	*Aphis spiraephila*	62
—	—	158	*Brevicorine brasicae*	62
—	—	452	*Schizaphis graminum*	62
—	—	327–504	*Sitobion avenae*	62
Hippodamia variegata				
19	10.6	276	*Brevicoryne brassicae*, 25°C	24
—	21	842	*Myzus persicae*, 25°C	63
20	—	—	—	108
11–20	—	960 (789–1256)	*Dysaphis crataegi*, 25°C, 16L: 8D	61
Megalocaria dilatata				
2 × 14	—	—	—	47
Menochilus sexmaculatus				
—	—	861	multiple mating	13
—	—	70	single mating	13
—	44 (peak)	—	after 10 matings	75
—	55 (peak)	—	after 20 matings	75
—	55 (peak)	—	after 5 matings	75
—	—	1998	*Aphis craccivora*	80
—	32	1898	lifetime mating	81
—	24	1268	single mating	81
—	16–27	92–275	old females	82
—	16–33	214–555	old males	82
—	4–6	43–64	senescent males	82
—	22–30	767–1318	young females	82
—	23–30	1118–1464	young males	82
—	—	110	1 min mating	83
—	—	40	10 s mating	83
—	—	620	one hour mating	83
—	17.2	—	*Aphis craccivora*, 15L:9D, 18°C	109
—	9.9	—	*Aphis gossypii*, 15L:9D, 18°C	109
—	—	max. 2388	—	47
12	—	—	—	74
Micraspis discolor				
—	—	385	20°C	77
—	—	563	25°C	77
—	—	750	27°C	77
—	—	622	30°C	77

Table 3.1 (*Continued*)

C	D	F	Prey, conditions	Ref.
Nephus reunioni				
—	2.6 (max.)	177	*Planococcus citri*, 25°C	134
Nephus includens				
—	—	151	—	135
Oenopia conglobata				
5	—	—	—	21
8–15	—	—	—	47
—	—	439	*Aphis spiraephila*	62
—	—	643	*Rhopalosiphum padi*	62
—	—	327	*Aphis pomi*	62
—	—	349–453	Central Asia	119
Oenopia lyncea				
—	—	40	—	117
Olla v-nigrum				
27	—	—	—	108
17	5.7	72	*Acyrthosiphon pisum*, 15°C, 12L:12D	137
13	8.1	200	*Acyrthosiphon pisum*, 20°C, 12L:12D	137
13	12.4	269	*Acyrthosiphon pisum*, 25°C, 12L:12D	137
14	6.4	329	*Acyrthosiphon pisum*, 20°C, 18L:6D	137
Propylea dissecta				
—	—	150	interference with 4 females	71
—	—	336	single female	71
—	37 (peak)	—	after 10 matings	75
—	19 (peak)	—	after 20 matings	75
—	33 (peak)	—	after 5 matings	75
—	—	278–710	1 to multiple matings	78
—	—	942	*Aphis craccivora*	80
—	—	867	20 day old females after single mating	87
Propylea japonica				
—	—	644	—	47
7–9	—	—	—	56
Propylea quatuordecimpunctata				
1–24	20 (max. 65)	1308 (max. 1800)	England, laboratory	47
10	—		field collected in laboratory	120
6.4	—		field	120
—	—	386	*Myzus persicae* on Bt potatoes	55
—	—	345	*Myzus persicae* on non-Bt potatoes	55
—	—	278	*Aphis craccivora*	55
—	—	431	*Aphis craccivora* + *Myzus persicae* on Bt potatoes	55
—	—	119	*Aphis gossypii*	62
—	—	351	*Sitobion avenae*	62
—	—	211	*Aphis spiraephila*	62
12	—	—	—	108

(*Continued*)

Table 3.1 (*Continued*)

C	D	F	Prey, conditions	Ref.
Propylea quatuordecimpunctata × *Propylea japonica*				
—	—	149	—	47
Psyllobora confluens				
—	17	440	*Erysiphe cichoracearum*	18
Sasajiscymnus tsugae				
—	2.9	280 (max. 513)	*Adelges tsugae*, 25°C	45
Scymnus frontalis				
—	—	151	22°C	31
—	7.3	—	*Diuraphis noxia*, 26°C	73
Scymnus hoffmanni				
—	—	127	25°C	57
Scymnus interruptus				
—	—	88	20°C	133
—	—	402	24°C	133
Scymnus levaillanti				
—	4.9	428	20°C	113
—	5.9	456	25°C	113
—	8.3	393	30°C	113
—	2.6	83	35°C	113
Scymnus louisianae				
—	—	122	*Aphis glycines*, 23°C, 15:9 L:D	16
Scymnus marginicollis				
—	—	75	20–25°C	17
Scymnus marinus				
—	—	1.7	15°C	70
—	—	602	30°C	70
Scymnus sinuanodulus				
—	2.5	130 (max. 200)		66
Scymnus subvillosus				
—	—	99	*Hyalopterus pruni*, 5 aphids. per day	7
—	—	232	*Hyalopterus pruni*, 80 aphids per day	7
Scymnus syriacus				
—	—	588	*Aphis spiraecola* on *Citrus sinensis*	104
—	—	658	*Aphis spiraecola* on *Spirea* sp.	104
Serangium parcesetosum				
—	—	28	*Trialeurodes vaporariorum*	5
Serangium japonicum				
—	—	387	*Bemisia tabaci*, 20°C	124
Stethorus bifidus				
—	1	2–33	10 mites / day	19
—	2	24–67	20 mites / day	19
—	6	308–438	50 mites / day	19

Table 3.1 (*Continued*)

C	D	F	Prey, conditions	Ref.
Stethorus gilvifrons				
—	4.4	149	*Oligonychus coffeae*	123
Stethorus japonicus				
—	8.1	—	*Amphitetranychus viennensis*, 27°C: 16L8D	60
—	5.4	—	*Panonychus mori*, 27°C: 16L:8D	60
—	7.1	—	*Tetranychus urticae*, 27°C: 16L:8D	60
Stethorus pussilus				
—	—	54–60	—	4
—	—	122	Bt-maize	128
—	—	111	non-Bt-maize	128
Subcoccinella vigintiquatuorpunctata				
—	—	200–300	—	112
Synonycha grandis				
3–60	—	360	—	47

1, Abdel-Salam et al. 1997; 2, Abdel-Salam and Abdel-Baky 2001; 3, Ahmad 1970; 4, Alvarez-Alfageme et al. 2008; 5, Al-Zyoud et al. 2005; 6, Anonymous 1997; 7, Atlıhan and Güldal 2009; 8, Banks 1955, 1956; 9, Baskaran and Suresh 2006; 10, Baungaard and Hämäläinen 1981; 11, Berkvens et al., 2008a; 12, Berkvens et al., 2008b; 13, Bind 2007; 14, Blackman 1965; 15, Blackman 1967; 16, Brown et al. 2003; 17, Buntin and Tamaki 1980; 18, Cividanes et al. 2007; 19, Collyer 1964; 20, Dixon and Guo 1993; 21, Dobzhansky 1924; 22, Dorge et al. 1972; 23, Duan et al. 2002; 24, ElHag and Zaitoon 1996; 25, Ellingsen 1969; 26, Evans et al. 2004; 27, Fois and Nedvěd, unpublished; 28, Francis et al. 2001; 29, Gautam 1990; 30, Ghanim et al. 1984; 31, Gibson et al. 1992; 32, Greathead and Pope 1977; 33, Hafez in Iablokoff-Khnzorian 1982; 34, Hämäläinen and Markkula 1977; 35, Hariri 1966; 36, Hattingh and Samways 1994; 37, Hawkes 1920; 38, Hazzard and Ferro 1991; 39, He et al. 1994; 40, Hemptinne et al. 1992; 41, Hodek 1996a; 42, Hong and Park 1996; 43, Huang et al. 2008; 44, Hukushima and Kamei 1970; 45, Cheah and McClure 1998; 46, Chen et al. 1980; 47, Iablokoff-Khnzorian 1982; 48, Ibrahim 1955; 49, Iwata 1932; 50, Jalali and Singh 1989; 51, Jalali et al. 2009; 52, Kalushkov 1994; 53, Kalushkov 1998; 54, Kalushkov and Hodek 2004; 55, Kalushkov and Hodek 2005; 56, Kawauchi 1981; 57, Kawauchi 1985; 58, Kesten 1969; 59, Kim and Choi (1985); 60, Kishimoto 2003; 61, Kontodimas and Stathas 2005; 62, Kuznetsov 1975; 63, Lanzoni et al. 2004; 64, Liu et al. 1997; 65, Lombaert et al. 2008; 66, Lu and Montgomery 2001; 67, Lundgren and Wiedenmann 2002; 68, Lundgren et al. 2009; 69, McClure (1987); 70, M'Hamed and Chemseddine 2001; 71, Mishra and Omkar 2006; 72, Muralidharan 1994; 73, Naranjo et al. 1990; 74, Ng 1986; 75, Omkar and Mishra 2005b; 76, Omkar and Pathak 2006; 77, Omkar and Pervez 2002; 78, Omkar and Pervez 2005; 79, Omkar and Srivastava 2003; 80, Omkar et al. 2005a; 81, Omkar et al. 2005b; 82, Omkar et al. 2006b; 83, Omkar et al. 2006a; 84, Omkar et al. 2009; 85, Osawa 2005; 86, Perry and Roitberg 2005; 87, Pervez et al. 2004; 88, Ponsonby 2009; 89, Ponsonby and Copland 1998; 90, Prevolsek and Williams 2006; 91, Radwan and Lovei 1983; 92, Ren et al. 2002; 93, Rhamhalinghan 1986; 94, Richards and Filewood 1988; 95, Savoiskaya 1965; 96, Savoiskaya 1970; 97, Semyanov 1970; 98, Semyanov 1980; 99, Semyanov and Bereznaya 1988; 100, Senal 2006; 101, Sengonca and Arnold 2003; 102, Schanderl et al. 1988; 103, Simmons et al. 2008; 104, Soroushmehr et al. 2008; 105, Srivastava and Omkar 2004; 106, Srivastava and Omkar 2005; 107, Stathas et al. 2001; 108, Stewart et al. 1991b; 109, Sugiura and Takada 1998; 110, Sundby 1968; 111, Takahashi 1987; 112, Tanasijevic 1958; 113, Uygun and Atlıhan 2000; 114, Uygun and Elekcioglu 1998; 115, Ware et al. 2008; 116, Williams 1945; 117, Witsack 1971; 118, Xia et al. 1999; 119, Yakhontov 1958; 120 Honěk et al. 2008; 121, Osawa and Ohashi 2008, 122, Bonte et al. 2010, 123, Perumalsamy et al. 2010, 124, Yao et al. 2010, 125, Tavadjoh et al. 2010, 126, Omkar et al. 2010b, 127, Kajita and Evans 2010, 128, Li and Romeis 2010, 129, Omkar et al. 2010a, 130, Seko and Miura 2009, 131, Pilorget et al. 2010, 132, El-Heneidy et al. 2008, 133, Tawfik et al 1973, 134, Izhevsky and Orlinsky 1988, 135, Transfaglia and Viggiani 1972, 136, dos Santo et al. 2009, 137, Kreiter 1985.

in India was 8.9 eggs, while the same in typical morphs was 14.34 (Rhamhalinghan 1999).

3.2.4 Hatching rate

The percentage of eggs in a cluster that develop fully and from which larvae eventually hatch is called hatching rate, **egg viability**, or percentage fertility (100–progeny loss). It is often substantially less than 100%. The hatching rate differentiates fecundity (number of eggs per female) from **fertility** (number of viable progeny per female).

Some eggs are non-fertile due to low **sperm volume and quality**. Other eggs are fertilized but do not develop or fail to hatch due to **infection** by various bacteria. In most species of invertebrates, **male-killing** occurs during embryonic stages (early male-killing) and is associated with cytoplasmic bacteria, including *Wolbachia*, *Spiroplasma*, *Rickettsia*, *Flavobacterium* and gamma proteobacteria (Nakanishi et al. 2008; Chapter 8). These bacteria kill only, or mostly, male embryos, giving a hatching rate close to 50%. The female larvae of ladybirds, especially Coccinellinae, i.e. the sisters of the killed males, have a nutritive advantage over the females from uninfected clusters, because they **cannibalize** those undeveloped 'male' eggs (Majerus 1994).

Most authors do not discriminate between 'non-fertile' and 'non-developing' eggs and record the proportion of hatched eggs. Under optimal conditions this is high: 100% eggs of *C. septempunctata* and *C. leonina transversalis* fed with *Lipaphis pseudobrassicae*, *Myzus persicae* or *Aphis nerii* hatched (Gupta et al. 2006). There were 6% unhatched eggs in *P. quatuordecimpunctata* in England in the field (Banks 1956). They were consumed by newly born larvae. Laboratory progeny loss ranged from 5% in *C. leonina transversalis*, 6% in *P. dissecta*, through 10% in *Menochilus sexmaculatus* to 25% in *C. septempunctata* (Omkar et al. 2005) when the ladybirds were reared on a **suboptimal prey**, *Aphis craccivora*. In *A. bipunctata* fed with optimal prey, *Myzus persicae*, the hatching rate was 89%, associated with high **fecundity** (676 eggs per female), while both measures decreased on a suboptimal prey, *Aphis fabae* (56%, 250 eggs; Blackman 1965). The decrease of hatching rate of *C. septempunctata* among six prey species was clearly correlated with a decrease

of lifetime fecundity (Fig. 3.2; Omkar & Srivastava 2003).

Hatching rate varied with **temperature** in *Har. axyridis*. Between 18 and 24°C, the rate was 65–90%. At 30°C, it decreased to 30–65% (Lombaert et al. 2008). The highest temperature, 30°C, was lethal for most eggs of a population from the Czech Republic, regardless of humidity (O. Nedvěd, unpublished). In *Micraspis discolor* in India, the egg hatch rate was lowest (65%) at the lowest experimental temperature (20°C), highest (95%) at optimum temperature (27°C), and decreased (83%) at the highest temperature (30°C) (Omkar & Pervez 2002).

High **humidity** is favourable for embryonic development and larval hatching; 99% of the eggs of *Delphastus catalinae* hatched at 85% RH, while 85% hatched at 25% RH (Simmons et al. 2008).

There was no consistent difference in the hatching rate of *Har. axyridis* at long (16L:8D) and short (12L:12D) **photoperiod**s (Berkvens et al. 2008b). Wild populations hatched better than the laboratory strain. Nor in *Coelophora saucia* was the hatching rate very different (92–97%) under different photoperiods or under continuous light (Omkar & Pathak 2006).

Hatching rate was high (76–90%) when the eggs of *Menochilus sexmaculatus* were fertilized by young males (4–50 days), and low (12–25%) when fertilized by **old males** (60–110 days) (Omkar et al. 2006b). A similar pattern was found in *C. septempunctata* (Srivastava & Omkar 2004). Hatching rate of *A. bipunctata* remained high (80–90%) for 1 or 2 months old males and then decreased to 20–30% (Jalali et al. 2009a) and that of *Anegleis cardoni* increased from almost zero to more than 80% at a **female age** of 2 weeks (Omkar et al. 2009).

Multiple matings enhanced the total egg output and the percentage of hatching (Kesten 1969, Semyanov 1970, Omkar & Pervez 2005, Bind 2007). Hatching rate gradually decreased to 50% after the female of *P. dissecta* has been separated from the male (Omkar & Mishra 2005). However, the hatching rate was high in *Menochilus sexmaculatus*, *C. septempunctata* and *C. leonina transversalis* even when females mated only once (Omkar et al. 2005b). The level of **crowding** had a small detrimental effect on egg viability of *P. dissecta* (Mishra & Omkar 2006).

To describe **hatching synchrony** in individual species, three indices have been proposed (Perry &

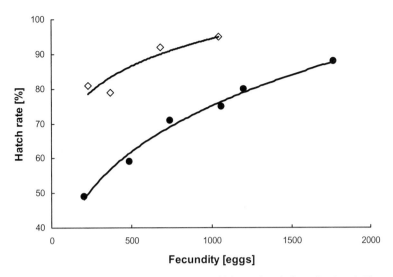

Figure 3.2 Relationship between average egg hatching rate (%) and lifetime female fecundity (eggs). The parallel increase in both parameters in *Coccinella septempunctata* (dots, lower line) was recorded simultaneously when six prey species were offered (after Omkar & Srivastava 2003). The increase in *Coelophora saucia* (diamonds, upper line) was recorded at (from the left) red, blue, yellow, and white light (after Omkar & Pathak 2006). Trend lines plotted as power functions.

Roitberg 2005): (i) total hatch time, related to batch size; (ii) the average interval between two sequential hatching larvae; (iii) the proportion of eggs per cluster, weighted by cluster size, that are vulnerable to sibling cannibalism by hatching later than the quiescent period of the first larva (141 min). The total duration of hatching of the first instar larva from an egg lasted 1 to 3.5 hours in *Har. axyridis*, and the total hatch time of the whole cluster (i) ranged from 61 to 203 minutes (Perry & Roitberg 2005). Hatching was synchronized within a cluster; the interval between hatching of successive larvae (ii) was 7–15 minutes. The proportion of eggs which showed a delayed hatch (iii) ranged from 0 to 7%.

The mean duration of hatching of larvae within the same batch (i) ranged between 1.01 hours in *Coelophora inaequalis* and 1.36 hours in *C. leonina transversalis* (Ng 1986). There were 45% (*Coelophora inaequalis*) to 83% (*C. leonina transversalis*) egg clusters which took longer to hatch than the duration between emergence of a larva and its cannibalization of eggs (a similar measure to (iii)).

Although hatching of ladybird larvae often occurs in daytime, most egg clusters (73%) of *P. dissecta*

hatched during the night (Mishra & Omkar 2004; Fig. 3.3).

3.2.5 Trophic eggs

Apart from viable hatching eggs, ladybirds also produce non-hatching eggs, which may be either **infertile** (not fertilized by sperm) or non-viable, where an embryo develops for some time but the larva does not emerge from the egg capsule. Non-hatching eggs are sometimes considered as trophic eggs that serve as the first food for the newly born sibling larvae. Egg **cannibalism** (5.2, Chapter 8) is undoubtedly advantageous to *A. bipunctata* larvae both in terms of faster development and increased survival (Roy et al. 2007). Perry and Roitberg (2005) showed that laying infertile eggs is also an active part of the maternal strategy in *Har. axyridis*. Females produced 56% more infertile eggs (23 versus 15% of the cluster) in low versus high food treatments; they manipulated the proportion of trophic eggs in favour of young first instar larvae that might have problems in finding essential prey (aphids). The oviposition sequence of infertile eggs within a cluster did not differ from random.

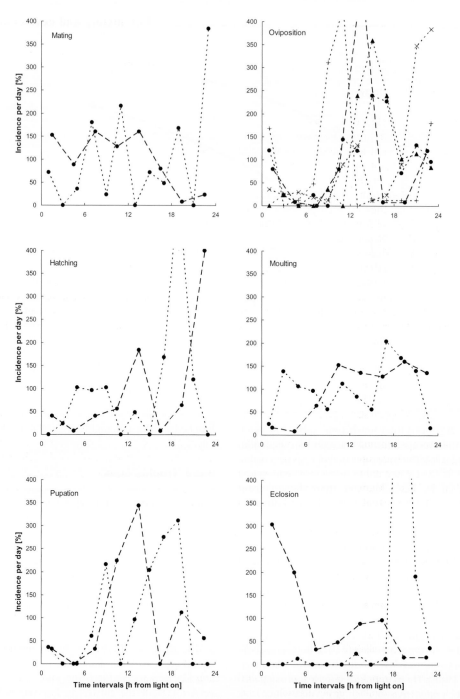

Figure 3.3 Circadian rhytmicity in life events of aphidophagous ladybird beetles (Coccinellini) in India. Dashed line and solid circles, *Menochilus sexmaculatus* (Omkar & Singh 2007); dotted line and open circles, *Propylea dissecta* (Mishra & Omkar 2004); dotted line and diagonal crosses, *Coccinella septempunctata*; dotted line and straight crosses: *Coccinella transversalis*; dotted line and triangles, *Propylea dissecta* (all three Omkar et al. 2004). Photophase: 0–12 h, scotophase: 12–24 h. All values are expressed in relative incidence, mean is 100%. Mating occurs more frequently around sunrise and around sunset; oviposition take place mainly during scotophase with two peaks (beginning and late scotophase); hatching from eggs occurs mainly during the second half of scotophase; larval moulting is more regularly distributed, with slight prevalence in scotophase; pupation occurs mainly during early or mid scotophase; adult emergence take place either during morning or during the second half of scotophase.

3.3 LARVA

3.3.1 Larval morphology

Larvae of most coccinellid tribes have an elongate body (Coccinellinae, Scymnini). It is ellipsoidal in the Hyperaspidini and Noviini and hemispherical in the Platynaspidini. The pronotum has two or four sclerotized plates (six in *Tetrabrachys*; Klausnitzer 1969a). The meso- and metanotum each have two plates, but in the Scymninae they are weakly developed or absent. The first eight abdominal segments bear six rows of characteristic structures as dorsal, dorsolateral and lateral pairs. These structures are well developed in the third and fourth instars and differ according to their shape, with seven types: seta, chalaza, tubercle, struma, parascolus, sentus and scolus (Gage 1920). These structures are important in taxonomy and identification. Larvae of Hyperaspidini and Platynaspidini are covered by chalazae and setae, the Scymnini and Noviini have tubercules, the Psylloborini and Tytthaspidini strumae, the Chilocorini senti and the Epilachninae scoli. Genera of Coccinellini have variable structures (Savoiskaya & Klausnitzer 1973). The more elaborate structures have a defensive role.

Larvae of most Scymninae, Telsimiini and Azyini (including *Cryptolaemus*) are covered with a white waxy exudation (filaments) as a defensive adaptation (Liere & Perfect 2008). Many larvae exhibit defensive reflex bleeding similar to that of adults, but the droplets exude from specific sutures or structures on the abdomen or thorax. Larvae of *Scymnus sinuanodulus* exhibit reflex bleeding of orange viscous droplets from thoracic tubercles (Lu et al. 2002). Larvae of third and fourth instars, namely of Coccinellinae, are brightly coloured.

The mandibles of larvae in the Hyperaspiidini, Platynaspidini, Stethorini, Scymnini and *Exochomus* have an apex ending with a single point; they have two apical teeth in the Coccinellini, Noviini and also in the larvae of *Chilocorus*, though in this genus the adults possess a mandible terminating in only a single point. In the phytophagous Epilachninae, mandibles are equipped with four or five large teeth. There is a row of smaller teeth or thick setae in the mycophagous Psylloborini and Tytthaspidini (Savoiskaya & Klausnitzer 1973). The larvae of *Scymnus levaillanti*, which employ pre-oral digestion, were more efficient in converting food to body mass than larvae of *Cycloneda sanguinea*, which use chewing and sucking (Isikber & Copland 2001).

The legs are elongate in the Epilachninae and Coccinellinae, but shorter in the Scymninae. The antennae are very short, 1–3 segmented; the first segment is short and wide (Savoiskaya & Klausnitzer 1973).

3.3.2 Instars

The individual substages during larval development separated by moulting (ecdysis) are called instars. The number of instars in the whole family is almost always four, independent of the species size, developmental conditions, etc. The constant and same number of instars in both aphidophagous and coccidophagous ladybirds, otherwise strongly different in many life history characteristics, was considered surprising by Dixon (2000). Fast development, typical and adaptive for aphidophagous Coccinellinae, would be better achieved by fewer instars, but their number seems to be phylogenetically constrained.

The last, fourth, instar eventually stops feeding and attaches to a substrate, forming the so-called prepupa. This pseudostage has sometimes been erroneously referred to as a fifth instar (Smith et al. 1999). Three instars were surprisingly reported in the coccidophagous *Hyperaspis campestris* (McKenzie 1932) and in an Egyptian population of *C. undecimpunctata* (while European populations of this species have four (Iablokoff-Khnzorian 1982)). Four larval instars were usually observed in *Clitostethus oculatus*, although a significant proportion of third instar larvae moulted directly to the pupal stage at 29°C or above (Ren et al. 2002). For these latter individuals, the third larval instar was prolonged to almost the length of the combined third and fourth instars of 'normal' beetles (6.1 days versus 6.3 days at 29°C, 3.4 days versus 4.9 days at 31°C and 4.7 days versus 5.0 days at 33°C).

However, a true higher number of instars, five, was reported for *Callicaria superba* (Iwata 1932), a large ladybird (12 mm) which feeds on Sternorrhyncha, Auchenorrhyncha and chrysomelid larvae. There is also a report of five larval instars in *Chil. nigritus* (Fitzgerald 1953, Chazeau 1981), a species feeding on coccids, aphids and whiteflies (Omkar & Pervez 2003). A small proportion of larvae of *Col. maculata* (Warren & Tadić 1967) and *Chil. bipustulatus* (Yinon 1969) went through five instars in the laboratory.

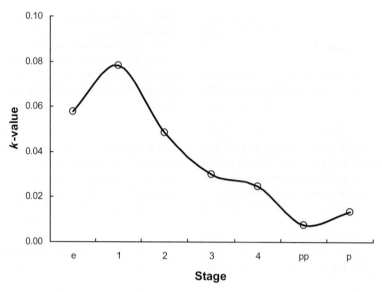

Figure 3.4. Stage specific mortality (k-values = $\log_{10}a_x - \log_{10}a_{x+1}$, where a_x is the initial number of individuals and a_{x+1} is the number of individuals surviving to the next stage) of *Propylea dissecta*. Average for five generations reared on five different aphid species. The first instar and eggs are most vulnerable to death, prepupa is the most resistant 'stage' (after Omkar & Pervez 2004).

A fifth larval instar occurred in 33% of the individuals of a Canadian population of the large invasive aphidophagous *Har. axyridis*. This extra instar had the same developmental time as ordinary fourth instars, but showed increased voracity and weight gain compared to the fourth larval instar, suggesting an increased fitness of those individuals and contributing to the invasiveness of the species (Labrie et al. 2006). A fifth instar was found to occur rarely also in European population of *Har. axyridis* at 20°C with an excess of food. Only females developed from these larvae (Ungerová, unpublished).

Moulting between instars occurs at any time of day, although in *P. dissecta* 61–77% of larvae moulted during the scotophase (Mishra & Omkar 2004).

During pre-adult development, mortality is highest in the first instar, possibly higher than in eggs, is least in prepupae and increases somewhat during the pupal stage (Fig. 3.4; Omkar & Pervez 2004b).

3.3.3 Development

The duration of larval development is species specific and strongly dependent on the ambient temperature

(3.6), and also on the quality and quantity of food (5.2).

Larval development of coccinellids is direct with no diapause. An exception is the fourth instar of the phytophagous *Epilachna admirabilis* which is sensitive to photoperiod (Takeuchi et al. 1999). The fourth instar larvae of *Scymnus abietis* have been observed very early in the spring, suggesting that they overwinter in this stage (Nedvěd & Kovář, unpublished).

In well-fed larvae, the first instar takes about 24% of the total developmental time, the second 17%, the third 19% and the fourth 40% (Honěk 1996). There was higher variation in these percentages in *A. bipunctata* (Obrycki & Tauber 1981) than in *Col. maculata* (Wright & Laing 1978). Even when the prepupa (8–11%) is not included, the last instar is always longer than any of the others.

Immune responses to microbial infection may cause developmental delay. The mean larval duration of *Serangium parcesetosum* was longest (22.5 days) when sprayed with medium and high dosages of *Beauveria bassiana*, intermediate (20 days) with a low dosage of *B. bassiana*, and lowest (18 days) for the blank control and the *Paecilomyces fumosoroseus* treatments (Poprawski et al. 1998). *B. bassiana* strain ATCC 44860 increased the development time of *Col. maculata* fed with Colorado potato beetle larvae, whereas the strain ARSEF 2991 reduced the development time (Todorova et al. 1996).

Development of phytophagous species may depend not only on the quality but also on the surface structure of the host plant. Larvae of the Mexican bean beetle, *Epilachna varivestis*, completed the first instar most quickly on pubescent soybean plants, whereas the duration of the third instar was shortest on glabrous plants. Larval mortality was 2.5–5 times greater on a densely pubescent isoline than on glabrous and normal soybean isolines (Gannon & Bach 1996).

The duration of pre-imaginal development differed among several isofemale lines (progeny of individual females, from several egg batches) of field-collected *Hip. convergens*, the values ranging from 223 to 273 degree days above the lower temperature threshold (Rodriguez-Saona & Miller 1999). Accordingly it is recommended that experiments on the effects of environmental factors on coccinellids should be (i) conducted with siblings (the progeny of one female) distributed into all treatments and (ii) repeated with the progeny of several unrelated females.

The duration of larval development further depends on **population density**. In *Chil. bipustulatus*, duration of development increased at higher density (Fomenko 1970). Rearing *P. dissecta* larvae at a moderate density of four larvae in a half litre beaker shortened developmental time and increased immature survival when compared with both single larvae and higher densities of larvae (Omkar & Pathak 2009). Mean developmental time of *Har. axyridis* in small (7 cm) Petri dishes was 13.7 days; in 15 cm Petri dishes it was 11.5 days; and 11.7 days in 0.5 l jars (Ungerová et al. 2010).

The effect of diverse environmental conditions on development may be mediated through the quality of prey. Duration of larval development of *P. japonica* was significantly longer if fed *Aphis gossypii* from cotton grown at elevated than at normal CO_2 concentrations (Gao et al. 2008).

3.3.4 Body size

When fed *ad libitum*, the body mass increases exponentially with the age of the larva, peaking in the late last instar larva. Due to the cessation of further weight gain at the end of the last larval instar (Ng 1991) the function of weight increase is sigmoid (Fig. 3.5).

Weight increases again in adults once they start to feed after emergence (Omkar et al. 2005a). The size of cuticular structures of successive instars increases in constant proportions. Weight increases exponentially, e.g. in *A. bipunctata* it is 0.04, 0.17, 0.52, 1.51 and 4.85 mg at the beginning of the first, second, third and fourth instar and prepupa, respectively (Jalali et al. 2009b).

The final size of the ladybird is largely determined during the fourth instar (Honěk 1996). For comparative purposes, the mean relative growth rate (RGR) may be calculated as RGR = dry weight gained during the feeding period / (length of feeding period in days · mean dry weight of predator during feeding period) (Waldbauer 1964). RGR values in *A. bipunctata* fed with *Myzus persicae* were 0.45, 0.70, 0.55 and 0.35 per day in the successive instars (Jalali et al. 2009b).

The relative food consumption of different larval instars is compared in 5.2. For example, Okrouhla et al. (1983) gave average values of 6, 11, 21 and 62% for the total food eaten by larvae of *Cheilomenes sulphurea*. In *A. bipunctata*, the percentages of food eaten by the four larval instars differed between food treatments. When fed with aphid *Myzus persicae*, they were 4, 9, 14 and 73% and with factitious food (a mixture of *Ephestia kuehniella* eggs and fresh bee pollen) they were 5, 8, 20 and 67% (Jalali et al. 2009b).

3.4 PUPA

3.4.1 Prepupal stage

The last (fourth) instar larva attaches itself by the tip of its abdomen ('anal organ', pygopod) to the substrate and prepares for pupation. The prepupae of particular species are as or more vulnerable to predation than their fourth instar larvae. Their physical defence structures are essentially the same, but they are practically **immobile** (Ware & Majerus 2008). Phorids attack host prepupae and parasitize them at the time of ecdysis to the pupal stage (Hurst et al. 1998). In life tables, the prepupa is often listed as one of the developmental stages, but strictly speaking it should be considered as a part of the **last instar larva**. For practical reasons, especially the differences between feeding larva and inactive prepupae, we recommend that these substages are separated in experiments.

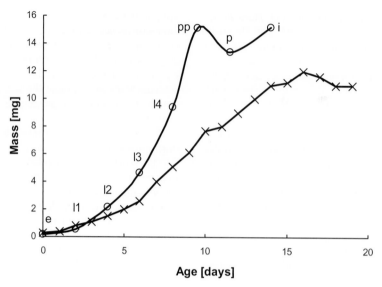

Figure 3.5 Growth curves of ladybirds. Crosses, growth of *Exochomus quadripustulatus* from egg to pupa at 25°C, fed with *Pulvinaria regalis* (after Sengonca & Arnold 2003). Circles, growth of *Coelophora biplagiata* from egg (e) through four larval instars (i), prepupa (pp), pupa (p) to adult (a); reared on *Aphis craccivora* at 27°C (after Omkar et al. 2005, mass was in fact double, but the values were reduced to a similar scale as the other species). Body mass increases exponentially from hatching to the fourth instar. This last instar stops to feed and grow and form a prepupa. The pupa is lighter than prepupa because the larval cuticle is shed, and some of the nutrients are spent through intensive metabolism. Part of the body water may be lost during the resting stage. Adults continue to grow, especially the females during egg production.

3.4.2 Pupal morphology

Pupae of coccinellids are of the type **pupa adectica obtecta**, which means that all appendages (antennae, limbs, wing pads) are glued to the body by exuvial fluid. The pupa is attached to the substrate (vegetation) by the tip of the abdomen. The **exuviae** (remains of the old cuticle) of the last larval instar are crumpled around the tip of abdomen in the Coccinellinae and Sticholotidinae. In the Chilocorinae, Scymninae and Ortaliinae the larval cuticle ruptures on the dorsal side, stretches and splits lengthwise, but is not shed and forms a **protective chamber** around the pupa. There are exceptions such as *Scymnus sinuanodulus* where the pupa is **naked** with the larval exuviae attached only to the last abdominal segment (Lu et al. 2002). In the Epilachninae, only the posterior third of the pupa is covered by the spiny larval exuviae. The pupal cuticle itself is either covered in long hairs (Epilachninae, Scymninae) or is apparently smooth, with only sparse very short glandular hairs (Coccinellinae).

The **size** of the pupa of particular species is highly dependent on prey quality and quantity during larval development, as well as on the temperature during development. The largest pupae develop at **medium temperatures**. The mean weight of pupae of *Exochomus quadripustulatus* was 11.9 mg at 9/19°C (thermoperiod), 12.5 mg at 12/24°C and 10.7 mg at a constant 25°C (Sengonca & Arnold 2003). The heaviest pupae of *Hip. convergens* were reared at 22°C as compared with 18, 26 and 30°C (Rodriguez-Saona & Miller 1999).

Har. axyridis larvae reared on *Acyrthosiphon pisum* produced heavier pupae (27.8 mg) than their siblings reared on the less **suitable prey** *Aphis craccivora* (20.8 mg) or on artificial diet (22.0 mg) (Ueno 2003). *Coleomegilla maculata* larvae fed with corn pollen gave much heavier pupae (13 mg) than individuals fed on the aphid *Rhopalosiphum padi* (7.3 mg) (Lundgren & Wiedenmann 2002). *Epilachna varivestis* reared on glabrous soybean plants had by 14–29% greater dry pupal weights than did individuals reared on the densely pubescent and normal isolines (Gannon & Bach 1996).

Male and female pupae cannot be distinguished by their external morphology. **Females** tend to be larger and heavier than **males** in the pupal stage: *Har. axyridis* females on average weighed 28.5 mg while males were 27.2 mg (Ueno 2003). Moreover, Ueno (2003) found slight intraspecific genetic variation (**family effect**) in pupal weights and an interaction of this genetic background with the type of prey. Based on these finding we suggest (i) strict use of siblings in all treatments of an experiment where effects of several environmental factors are to be studied, and (ii) the parallel use of several unrelated females/families for such experiments.

3.4.3 Timing of pupation

The development of pupae generally takes about 24% of the total pre-imaginal development time, i.e. less than in other Aphidophaga (Neuroptera: 33%, Syrphidae: 39%) (Honěk & Kocourek 1990). It may, however, exceed 40% in *Axinoscymnus cardilobus* (Huang et al. 2008) feeding on whiteflies. The **development rate** of pupae is linearly and positively related to **temperature** (3.6), and is usually **independent on the prey** quality and quantity in the preceding larval stage (Ahmad et al. 2006); by contrast, pupal and adult weight are dependent on previous feeding. Pupae of *Col. maculata* from larvae fed with pollen reached 14 mg and lasted 3.2 days while those fed on artificial diet weighed less (11 mg) and lasted for a comparable time (3.4 days) (Lundgren & Wiedenmann 2002). The complete pre-imaginal development of *Hip. convergens* on *Thrips tabaci* lasted 30 days, and 24 days on *Acyrthosiphon pisum*, while the duration of the pupal stage was not affected by the food (Schade & Sengonca 1998).

Under conditions of artificially elevated **CO_2 concentration**, the pupal development time was affected; both the larval and pupal durations of *Har. axyridis* fed *Aphis gossypii* were significantly shorter than at the normal concentration (Chen et al. 2005). A high level of **crowding** (35 larvae per beaker) shortened subsequent pupal development in *P. dissecta* (2.2 days in comparison to 3.4 days at low densities) (Omkar & Pathak 2009).

A relatively high incidence of pupation was observed during the late **photophase** (09:00–12:00 hours,

where 0 hours is the time the light was switched on) and early **scotophase** (12:00–15:00 hours) in *Menochilus sexmaculatus* (Omkar & Singh 2007; Fig. 3.3).

Mass-specific **respiration** rates in *Har. axyridis* increased with increases in pupal age (while it decreased in the other stages) (Acar et al. 2004). Below 10°C, CO_2 was produced primarily from non-oxygen-consuming reactions. CO_2 production gradually shifted more to oxygen-consuming reactions with increasing temperature.

The **abundance** of pupae is widely underestimated in the field because they are attached to the vegetation and thus unavailable for common **sampling methods** (sweeping, Kalushkov & Nedvěd 2005; beating, Kula & Nedvěd unpublished; sticky traps, Kaneko et al. 2006).

3.4.4 Places of pupation

The usual place for pupation is on the vegetation where the larva developed, so that the prepupae and pupae are exposed to **cannibalism** or **intraguild predation**. The prevalence of **parasitization** of ladybird pupae by *Phalacrotophora* flies increases strongly with the population density of the pupae, reaching up to 40% (Durska et al. 2003). Larvae of the genus *Chilocorus* aggregate before pupation (Fomenko 1970, Fomenko & Zaslavskii 1970).

In **aphidophagous** ladybirds, the place for pupation is mainly **foliage**, while some **coccidophagous** species pupate on the **bark** of branches and on tree trunks. **Cryptically** coloured pupae tend to be hidden on the underside surface of the leaves, while **aposematically** coloured ones may be exposed on the upper side and take advantage of basking in the sun (see thermal melanism, 3.4.6).

In central Honshu, hibernating and reproductive adults of *C. septempunctata* coexist in the same habitat during the mild winter. Although natural substrates were available, the beetles preferred to use **artificial substrates** such as metal cans (iron or aluminium), papers, and wooden materials discarded on the sunny slope as oviposition and pupation sites. They were warmed by solar radiation and served as **thermal microhabitats** (Ohashi et al. 2005). About 90% of the larvae of *Col. maculata* left the potato plant to pupate. They selected shelters that effectively reduced their intraguild predation by the lacewing *Chrysoperla rufilabris* (Lucas et al. 2000). The larvae and pupae of the ladybird *Thalassa saginata* develop inside **colonies of the ant** *Dolichoderus bidens* (Orivel et al. 2004).

3.4.5 Pupal defence

Pupae covered by the **larval skin** are protected **mechanically**, especially when the larva is spiny (Chilocorinae, Epilachninae) or covered by waxy exudations (Scymninae). The pupal cuticle also has its own **hairs**, which are especially long in the Epilachninae and Scymninae, and often have a **glandular** function, containing defensive compounds. Transparent viscous **droplets** are visible on the tips of the setae of the pupa of *Scymnus sinuanodulus* (Lu et al. 2002). The white pupa of *Clitostethus oculatus* has tiny clear droplets on the tips of the setae at high relative humidity. Larval exuviae together with whitefly debris cover the tip of the pupal abdomen. The yellowish pupae of *Delphastus pusillus* are covered by short hairs without any secretion, and several abdominal segments are enclosed within an unfragmented larval skin (Liu & Stansly 1996a).

The surface of the pupa of the phytophagous *Subcoccinella vigintiquatuorpunctata* bears glandular hairs producing three polyazamacrolide alkaloids. The secretion serves as a potent anti-predator defence against the predatory ant *Crematogaster lineolata* (Smedley et al. 2002). Comparative studies of the defensive chemistry of eggs, larvae, pupae and adults showed both qualitative and quantitative differences in **alkaloid** composition between the life stages of *Epilachna paenulata* (Camarano et al. 2006). The polyazamacrolide epilachnene, the principal component of the secretion of the pupal defensive hairs of *Epilachna varivestis*, proved to be a **deterrent** for *Har. axyridis* (Rossini et al. 2000). The oily droplets on the pupal integumental hairs of the squash beetle *Epilachna borealis* contain a mixture of tocopheryl acetates (Attygalle et al. 1996) and **polyazamacrolides** (Schroder et al. 1998). (more on semiochemicals in Chapter 9).

The pupa is not entirely immobile. If irritated by a predator or parasitoid, it quickly **raises** itself **upwards** several times. Pupae are better protected against predation than prepupae, but in most species they are still susceptible to attack. The **tough** pupal **integument** affords better protection than the soft larval skin (Ware & Majerus 2008).

3.4.6 Colouration and thermal melanism

Many coccinellid pupae are **cryptically coloured**, while in some large Coccinellinae the colours are bright and probably have a **warning** function. Species such as *C. septempunctata* and *Har. axyridis*, which mostly pupate exposed on the upper leaf surface, combine a bright orange background with black spots. The extent of **melanization** and the darkness of the orange-brown background increases in both species with decreasing **temperature** and increasing humidity. The pupa is light orange at 35°C and dark brown at 15°C in *C. septempunctata* (Hodek 1958, Okuda et al. 1997). In *C. septempunctata*, the percentage of the area from dorsal and lateral views which is black decreases linearly (from 45–55% to almost zero) with increasing temperature (from 20 to 36°C) (Rozsypalova 2007). Unlike the larvae and adults, which can control their body temperature behaviourally, the pupa relies on heating by the sun.

The **sensitive stage** for the determination of the extent of melanization is the late prepupa. Elements of the future **elytral pattern** may be visible on the pupal **wingpads**. In *Har. axyridis*, there is a similar trend as in *C. septempunctata*, but shifted to lower temperatures (17 to 33°C) (Ungerová & Nedvěd, unpublished). The extent of melanization of the adult is independent of the melanization of the pupa.

Blackening of the pupal case may also be caused by infection, **injury** or intoxication. This was observed in pupae of *C. septempunctata* whose larvae had been treated with azadirachtin (Banken & Stark 1997).

Although the colouration of ladybird pupae is less variable and less striking than that of adults, pupae can still be identified to species. For example, the colour and spotting of prepupae and pupae provide identification characters to distinguish two sympatric ladybirds, *Henosepilachna pusillanima* and *H. boisduvali* that have indistinguishable adult colouration (Nakano & Katakura 1999).

3.5 ADULT

3.5.1 Teneral development

Teneral development may include **sexual maturation** and the acquisition of **flight ability**. In many coccinellids, the activity of the follicular tissue in the **testes** starts in the pupa (Hodek & Landa 1971, Hodek 1996). After emerging from the pupa, females, and sometimes also males, show a **refractory period** of a few days in their mating behaviour. This was found to be equal in males and females of *A. bipunctata* (Hemptinne et al. 2001). Slight **protogyny** (the first mating of females taking place at earlier age) could theoretically occur when females mate before sexual maturity and store sperm, while males mate only after maturity. Mating of sexually non-mature females is common in many species with reproductive diapause (Chapter 6). Long refractory periods were observed in *Sasajiscymnus tsugae*: at 25°C when males matured at 19 days, whereas the female pre-oviposition period was 22.4 days (Cheah & McClure 1998). Omkar & Srivastava (2002) report **protandry** in Indian *C. septempunctata*. Males were ready to mate within 9 days at 27°C, while females took 11 days. Rana & Kakker (2000) reported an average pre-mating period in *C. septempunctata* of 6.4 days.

Temperature requirements for completing the teneral period may be calculated as for immature development (3.6) and may be prolonged by unsuitable prey (Hukushima & Kamei 1970) or prey scarcity (Kawauchi 1981).

The basic **sex ratio** in coccinellids is close to 50:50, except where there is an infection by a male-killing agent (Chapter 8). An increased proportion of males was found at high temperatures: in *Har. axyridis* at 30°C (62–82%) (Lombaert et al. 2008) and in *P. dissecta* at 35°C (62%) (Omkar & Pervez 2004b).

3.5.2 Wings and flight

The ability to **take-off** matures within a short period after the moult to adult. Adults of *C. septempunctata* began attempts to fly 40 hours after emergence at 26°C (Honěk 1990). Confinement of ladybirds in a limited space increased both the willingness to take-off and the **duration of flight** in *C. septempunctata* (Nedvěd & Hodek 1995).

There have been a few reports of the occasional occurrence of wingless or brachypterous individuals among normally winged ladybirds. Winglessness in *A. bipunctata* is determined by a single locus with the **wingless** allele recessive to the winged wildtype allele (Lommen et al. 2005) (Chapter 2). The occurrence of flightless ladybirds might increase biocontrol because flightless beetles have a longer **residence time** on the plants, even at lower prey density (Lommen et al. 2008). A homozygous **flightless** (but fully winged) strain of *Har. axyridis* was obtained by laboratory selection and used in biological control (Tourniaire et al. 2000). This strain had not only a longer residence time on aphid-infested plants but also a minimal potential to become invasive. Artificially, flightless ladybirds may be induced in a culture by limiting space during the emergence of adults from pupae individually placed into small chambers; this results in deformation of elytra and wings (N. Osawa, personal communication).

3.5.3 Pre-oviposition period

The number of days between female emergence and the laying of the first egg batch is called the pre-oviposition period. Its duration can be used to measure the intensity of diapause (Chapter 6). In dormant females, it can last weeks or even months.

Interspecific variation in the pre-oviposition period can be substantial. In **Coccinellinae**, it ranges between 0 and 10 days: in *Har. axyridis* it was 6–10 days, in *A. bipunctata* 3–8 days and in *Hip. variegata* it was surprisingly reported to be only 0–4 days (Lanzoni et al. 2004), which would indicate that some females of this species completed the development of their eggs while in the **pupal stage**. A short pre-oviposition period of 2–3 days was also found in *Anegleis cardoni* (Omkar et al. 2009). Adults of *C. septempunctata* fed on *Sitobion avenae* started mating 4–11 days after emergence, with an average **pre-mating period** of 6.4 days (Rana & Kakker 2000).

In **Scymninae**, the pre-oviposition period varies even among closely related species. In *Scymnus frontalis* it was 10–11 days (Gibson et al. 1992), while it was only two days in *S. louisianae* (Brown et al. 2003). In contrast, long pre-oviposition periods of 13 days at 31°C and 23 days at 20°C were observed in *Clitostethus oculatus* (Ren et al. 2002), and of 22 days at 25°C in *Sasajiscymnus tsugae* (Cheah & McClure 1998).

The pre-oviposition period generally increases at lower **temperatures**: it ranged from 20.5 days at a mean temperature of 15°C to 7.7 days at 30°C in *Scymnus frontalis* (Naranjo et al. 1990), from 9 days at 20° to 4 days at 35°C in *S. levaillanti* (Uygun & Atlıhan 2000) and from 24 days at 15°C and 3.5 days at 35°C in *S. marinus* (M'Hamed & Chemseddine 2001). **Food suitability** also modifies the pre-oviposition period: this was shorter in *Menochilus sexmaculatus* fed with *Aphis craccivora*, *Aulacorthum solani*, *Sitobion akebiae*, and *Myzus persicae* (7.3–8.0 days) than when fed with *A. gossypii* (11.6 days) (Sugiura & Takada 1998).

3.5.4 Size

Females are generally larger and heavier than males, although often not significantly so, due to high variability. The fresh weight of emerged **females** of *Anegleis cardoni* was constantly 1.16-times higher than that of **males** when reared on three different prey species, while the difference in weight within each of the two sexes between being fed the most and the least **suitable prey** species was 1.35 times (Omkar et al. 2009). Newly emerged females of *Har. axyridis* were 1.1 to 1.2-times heavier than males (Ungerová et al. 2010).

The positive relationship between male and female weight under the same conditions is linear after log–log transformation, giving the equation log W_M = –0.07 + 0.97 log W_F for *A. bipunctata* and W_M = –0.043 + 0.96 log W_F for *P. japonica* (Dixon 2000). **Crowding** of larvae in a small space reduced adult size of *P. dissecta*: 7 mg with single individuals, 11 mg at 4 individuals per beaker (0.7 l), and only 5 mg at 35 per beaker (Omkar & Pathak 2009). In small (7 cm) Petri dishes, the mean mass of *Har. axyridis* was 18.8 mg; in 15 cm Petri dishes 21.2 mg; in 0.5 l jars 25.1 mg (Ungerová et al. 2010).

Body length and width decreased as constant temperatures increased from 22 to 34°C in *Chil. nigritus*, but were highest at an alternating **temperatures** regime of 20/34°C (Ponsonby 2009). Females were only 1.03-times longer than males.

Apart from using body length and weight, the ventral body area (mm^2) may be calculated by measuring the body length and width and using the formula [½(body length)] × [½(body width)] (Obrycki et al. 1998, Giles et al. 2002, Phoofolo et al. 2007).

In interspecific comparisons among aphidophagous ladybirds, body size of a species is related to the body size and density of the prey. Small ladybirds can feed on small aphid species when these are in both high and low densities, but on large aphid species only at high densities, where young instars of the aphids are abundant. Large ladybirds cannot be sustained by low densities of small aphids due to the food limitation (Sloggett 2008).

3.5.5 Ovarioles

The ovarioles of Coccinellidae are of the **meroistic telotrophic** type. The number of ovarioles is species-specific, as well as dependent on the conditions during the larval development of the female (Table 3.2). There is a higher **number of ovarioles** in larger species; the number increases with body length by the power function with the exponent of 1.14 ± 0.02 (Fig. 3.6). Species with few ovarioles lay larger eggs than similar-sized species with many ovarioles (Stewart et al. 1991b; 5.2.3). The number of ovarioles has been reviewed from the viewpoint of taxonomy, evolution and fecundity by Rathour and Singh (1991). In species with a high number of ovarioles, there is a smaller number of chambers (follicles) with ripening oocytes (three in *C. septempunctata*) than in species with a small number of ovarioles (eight in *Stethorus* sp.) (Dobrzhansky 1926).

Intraspecific comparisons in *C. septempunctata* showed a positive linear relationship between the number of ovarioles and female **body weight** (Dixon & Guo 1993). The mean number of ovarioles in *C. septempunctata* females of 24 mg body weight was 80, while it was 139 in females weighing 31 mg (Rhamhalinghan 1985, 1986). In contrast, no correlation of the number of ovarioles with either **body length** or body weight of females was found in *Har. axyridis* (Nalepa et al. 1996, Osawa 2005).

Effects on the reproductive output of the resulting females of the **diet** provided to larvae have been reported. The number of ovarioles was 10% lower in *A. bipunctata* fed on the suboptimal prey *Aphis craccivora* than in females fed with the high quality prey *Acyrthosiphon pisum* (Ferrer et al. 2008). However, Ware et al. (2008b) reported no effect of larval diet on ovariole number in *Har. axyridis* and *A. bipunctata*, although maximum clutch size and oviposition rate were affected. Starvation does not alter the total number of ovarioles, it only changes the percentage of oosorptive and mature ovarioles (Osawa 2005). In those individuals of *C. septempunctata* with less than 100 ovarioles per female, almost all of the ovarioles (99.8%) were healthy and functional. By contrast, in individuals with a high ovariole number (102 to

Table 3.2 Number of ovarioles per ladybird female, female weight (mg) and size (mm). Indicated are ranges or mean numbers of ovarioles in both ovaries together, if directly referred to in the literature source in the right column, or double the number of ovarioles per one ovary if given so in the source. Size and fresh weight of each species are either original unpublished data measured by Nedvěd, or measured by Rhamhalinghan (four rows marked by asterisk *); averaged literature data on body length are given for other species. See also Fig. 3.6.

Species	Weight	Size	Ovarioles	Reference
Adalia bipunctata	—	—	45–51	Dobrzhansky (1926)
Adalia bipunctata	—	—	47	Ferrer et al. (2008)
Adalia bipunctata	13.5	4.9	43	Ferrer et al. (2008)
Adalia bipunctata	—	—	48	Ware et al. (2008)
Adalia decempunctata	10.7	4.6	48–52	Dobrzhansky (1926)
Adalia tetraspilota	—	4.5	34–48	Rathour & Singh (1991)
Afidenta misera	—	5.4	40–48	Rathour & Singh (1991)
Afissula rana	—	—	26	Rathour & Singh (1991)
Afissula sanscrita	—	—	18	Rathour & Singh (1991)
Aiolocaria hexaspilota	—	9.7	91	Dobrzhansky (1926)
Aiolocaria hexaspilota	—	9.7	88–96	Rathour & Singh (1991)
Alloneda dodecaspilota	—	6.8	32–44	Rathour & Singh (1991)
Anatis ocellata	60.7	8.5	56	Dobrzhansky (1926)
Anegleis cardoni	—	3.7	20	Rathour & Singh (1991)
Anisosticta novemdecimpunctata	—	3.5	24	Dobrzhansky (1926)
Apolinus lividigaster	—	4.0	14–26	Anderson (1981)
Callicaria superba	—	11.5	52–60	Rathour & Singh (1991)
Calvia albida	—	8.0	18–24	Rathour & Singh (1991)
Calvia decemguttata	27.3	6.6	29	Dobrzhansky (1926)
Calvia quatuordecimguttata	22.7	5.4	40	Dobrzhansky (1926)
Calvia quatuordecimguttata	22.7	5.4	40	Semyanov (1980)
Calvia shiva	—	—	28–36	Rathour & Singh (1991)
Calvia shiva pasupati	—	—	22	Rathour & Singh (1991)
Calvia shiva pinaki	—	—	18	Rathour & Singh (1991)
Calvia shiva trilochana	—	4.9	22	Rathour & Singh (1991)
Calvia sp.	—	—	24–28	Rathour & Singh (1991)
Ceratomegilla undecimnotata	44.0	6.6	42–62	Dobrzhansky (1926)
Chilocorus bijugus	—	—	40–48	Rathour & Singh (1991)
Chilocorus bipustulatus	—	3.5	25	Dobrzhansky (1926)
Chilocorus breiti	—	—	18–24	Rathour & Singh (1991)
Chilocorus hauseri	—	—	30–40	Rathour & Singh (1991)
Chilocorus nigritus	—	4.0	18	Rathour & Singh (1991)
Chilocorus rubidus	—	6.5	72–80	Rathour & Singh (1991)
Chilocorus sp.	—	—	32–40	Rathour & Singh (1991)
Chilocorus sp.	—	—	24–32	Rathour & Singh (1991)
Coccinella hieroglyphica	—	3.5	34	Dobrzhansky (1924)
Coccinella hieroglyphica	—	—	28	Dobrzhansky (1926)
Coccinella luteopicta	—	5.8	44–52	Rathour & Singh (1991)
Coccinella magnifica	—	7.4	69–74	Dobrzhansky (1926)
Coccinella septempunctata	—	—	96–119	Dobrzhansky (1926)
Coccinella septempunctata	44.4	7.2	102	Klausnitzer & Klausnitzer (1986)
Coccinella septempunctata	—	—	108–124	Rathour & Singh (1991)
Coccinella septempunctata	30.5*	—	129 (57–82 pairs)	Rhamhalinghan (1985)
Coccinella septempunctata	23.6*	—	86 (26–61 pairs)	Rhamhalinghan (1985)
Coccinella septempunctata	31.6*	—	149	Rhamhalinghan (1986)
Coccinella septempunctata	24.2*	—	74	Rhamhalinghan (1986)

(Continued)

Table 3.2 (*Continued*)

Species	Weight	Size	Ovarioles	Reference
Coccinella undecimpunctata	—	5.0	68	Dobrzhansky (1926)
Coccinula quatuordecimpustulata	—	3.5	20	Dobrzhansky (1926)
Coccinula redimita	—	4.0	20	Dobrzhansky (1926)
Coelophora bissellata	—	5.3	26–36	Rathour & Singh (1991)
Coelophora duvaucelii	—	—	36–44	Rathour & Singh (1991)
Cryptogonus ariasi	—	2.2	14	Rathour & Singh (1991)
Cryptogonus orbiculus	—	—	14	Rathour & Singh (1991)
Cryptogonus postmedialis	—	—	14	Rathour & Singh (1991)
Cryptogonus quadriguttatus	—	—	14	Rathour & Singh (1991)
Epilachna dumerili	—	—	24–32	Rathour & Singh (1991)
Epilachna marginicollis	—	—	32–36	Rathour & Singh (1991)
Epilachna mystica	—	—	44–52	Rathour & Singh (1991)
Exochomus quadripustulatus	9.7	4.3	26	Dobrzhansky (1926)
Halyzia sanscrita	—	—	38–44	Rathour & Singh (1991)
Halyzia straminea	—	7.3	56–64	Rathour & Singh (1991)
Harmonia axyridis	—	—	54–78	Dobrzhansky (1924)
Harmonia axyridis	34.7	6.5	64–76	Dobrzhansky (1926)
Harmonia axyridis	—	—	53–77	Fois & Nedvěd, unpubl.
Harmonia axyridis	—	—	62 (44–84)	Nalepa et al. (1996)
Harmonia axyridis	—	—	62	Ware et al. (2008)
Harmonia dimidiata	—	8.3	50–60	Rathour & Singh (1991)
Harmonia eucharis	—	8.0	48–164	Rathour & Singh (1991)
Harmonia quadripunctata	22.3	6.1	36–40	Dobrzhansky (1926)
Harmonia sedecimnotata	—	6.5	54–64	Rathour & Singh (1991)
Henosepilachna dodecastigma	—	—	56–60	Rathour & Singh (1991)
Henosepilachna indica	—	—	44–54	Rathour & Singh (1991)
Henosepilachna ocellata	—	—	58–64	Rathour & Singh (1991)
Henosepilachna processa	—	9.1	54–58	Rathour & Singh (1991)
Henosepilachna pusillanima	—	8.1	63	Dobrzhansky (1926)
Henosepilachna sp.	—	—	34	Rathour & Singh (1991)
Henosepilachna vigintioctomaculata	—	—	55	Dobrzhansky (1926)
Henosepilachna vigintioctomaculata	—	7.4	48–56	Rathour & Singh (1991)
Henosepilachna vigintioctopunctata	—	6.4	56–72	Rathour & Singh (1991)
Henosepilachna vigintioctopunctata	—	—	84	Dobrzhansky (1926)
Hippodamia septemmaculata	—	6.0	40–46	Dobrzhansky (1926)
Hippodamia tredecimpunctata	—	5.9	48–60	Dobrzhansky (1926)
Hippodamia variegata	10.3	4.8	45–49	Dobrzhansky (1926)
Hippodamia variegata	—	—	42–48	Rathour & Singh (1991)
Hyperaspis campestris	4.7	2.9	19–21	Dobrzhansky (1926)
Hyperaspis reppensis	—	2.8	17–20	Dobrzhansky (1926)
Illeis cincta	—	4.0	38–44	Rathour & Singh (1991)
Jauravia quadrinotata	—	2.2	18	Rathour & Singh (1991)
Megalocaria dilatata	—	13.0	40–52	Rathour & Singh (1991)
Menochilus sexmaculatus	—	5.0	40–52	Rathour & Singh (1991)
Micraspis allardi	—	4.5	28	Rathour & Singh (1991)
Myrrha octodecimguttata	11.3	4.3	16–17	Dobrzhansky (1926)
Myzia oblongoguttata	39.0	7.5	41–46	Dobrzhansky (1926)
Nephus redtenbacheri	—	1.6	12	Dobrzhansky (1926)
Oenopia billieti	—	3.9	24	Rathour & Singh (1991)
Oenopia conglobata	—	—	24	Dobrzhansky (1926)
Oenopia conglobata	11.0	4.6	24	Nedvěd, unpubl.
Oenopia kirbyi	—	3.3	24	Rathour & Singh (1991)

Table 3.2 (*Continued*)

Species	Weight	Size	Ovarioles	Reference
Oenopia sexareata	—	—	24	Rathour & Singh (1991)
Palaeoneda auriculata	—	10.5	84–88	Rathour & Singh (1991)
Pania luteopustulata	—	5.1	24–32	Rathour & Singh (1991)
Platynaspis luteorubra	—	3.0	19–22	Dobrzhansky (1926)
Priscibrumus uropygialis	—	—	24	Rathour & Singh (1991)
Propylea quatuordecimpunctata	13.0	4.1	23	Dobrzhansky (1926)
Psyllobora vigintiduopunctata	7.7	3.8	22–24	Dobrzhansky (1926)
Lindorus lophanthae	—	2.5	20	Stathas et al. 2002)
Rodolia guerini	—	—	40	Rathour & Singh (1991)
Rodolia sp.	—	—	24	Rathour & Singh (1991)
Scymnus ater	—	1.5	8	Dobrzhansky (1926)
Scymnus frontalis	3.1	2.5	12	Dobrzhansky (1926)
Scymnus haemorrhoidalis	1.3	2.1	10–12	Dobrzhansky (1926)
Scymnus interruptus	—	2.2	12	Dobrzhansky (1926)
Scymnus nigrinus	3.0	2.6	12	Dobrzhansky (1926)
Scymnus rubromaculatus	1.9	2.2	10–12	Dobrzhansky (1926)
Scymnus suturalis	0.9	1.8	14	Dobrzhansky (1926)
Stethorus pusillus	0.6	1.4	4	Dobrzhansky (1926)
Subcoccinella vigintiquatuorpunctata	13.3	3.8	33–35	Dobrzhansky (1926)
Tytthaspis sedecimpunctata	5.3	3.0	20–21	Dobrzhansky (1926)
Vibidia duodecimguttata	—	4.0	24	Dobrzhansky (1926)

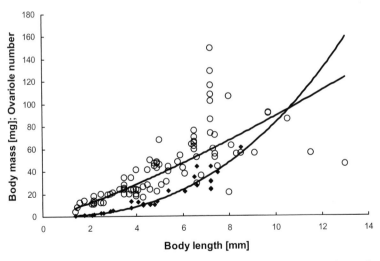

Figure 3.6 Relationships between body length (mm), body mass (mg, diamonds) and number of ovarioles (open circles) in Coccinellidae. Sources of the numbers of ovarioles are listed in Table 3.2. Fresh body masses are data measured by Nedvěd (unpublished), body length are either measured by Nedvěd or averaged literature data. Body mass increases with body length with the power function $M = 0.27 \cdot L^{2.46}$, number of ovarioles follow the power function (close to linear) $NO = 6 \cdot L^{1.14}$.

200 ovarioles per female), approximately 5.6% of the ovarioles were dysfunctional. The efficiency of the ovarioles (percentage of simultaneously active ones) was high in the summer generation and low in the winter generation. Crowding and competition among the ovarioles and inadequate nutrition were the factors affecting ovarian dysfunction (Rhamhalinghan 1986).

A five-stage rating system to describe ovarian development is based on the length of the terminal **follicle**, the number and shape of developing follicles in each ovariole, and the presence of yellow colour (**yolk**) in the terminal **oocyte** (Okuda et al. 1986, Phoofolo et al. 1995). Fully developed oocytes are **vitellinized** (yolk is deposited) and **chorionated** (covered by a chitinous envelope). A positive linear relationship has been observed between ovarian developmental rate and temperature. Oocytes gradually grow and ripen in individual ovarioles, and may then be laid synchronously from one or both ovaries. The number of eggs laid is positively correlated with the number of ovarioles, body length and body weight.

The rates of ovarian development and **oosorption** in predatory *Har. axyridis* are much higher than in herbivorous ladybirds. During **starvation**, the nutrients were found to be resorbed back from developing eggs. This oosorption mainly occurred during the intermediate developmental stage of the ovarioles, and the ovarian development and oosorption were asymmetric in the right and left ovaries in *Har. axyridis* (Osawa 2005).

3.5.6 Mating

3.5.6.1 Frequency and duration of mating

Ladybirds mate often and for a long time, and change partners (i.e. they are **promiscuous**). A **mating frequency** of about 20% at any given time was recorded for *A. bipunctata* in the field (Haddrill et al. 2007). The male uses his legs to hold himself onto the elytra of the female and is carried by her. Body shaking by males is essential for insemination. The spermatophore is formed in the bursa copulatrix (Obata 1988). Within 1 hour of mating the spermatophore is emptied, the envelope ejected from female's reproductive organs and is usually eaten by the female (Obata & Hidaka 1987).

The **mating duration** was recorded as 54 minutes with a range of 41–62 minutes in *C. septempunctata* (Rana & Kakker 2000). The duration was higher (51 minutes) when a melanic female mated with a melanic male of Indian *C. septempunctata* and lower (39 minutes) when a typical female mated with a typical male (Srivastava & Omkar 2005). Typical duration of mating of *Har. axyridis* in laboratory was between 2 and 3 hours (Awad & Nedvěd, unpublished). The mating posture in *Aiolocaria hexaspilota* may last several days (Iwata 1932). Manipulating the mating duration artificially had a dramatic influence on hatching rate in *Menochilus sexmaculatus* and *Coelophora saucia* (Omkar et al. 2006a). No eggs were fertilized after copulation lasting only 10 seconds, about 35% of eggs after a 1 minute copulation, and 80–90% after copulation lasting 1 hour or a natural long copulation.

Mating rhythmicity. Mating in *Menochilus sexmaculatus* (Omkar & Singh 2007) and *P. dissecta* (Mishra & Omkar 2004) in India occurred at all times of the day or night with a slight prevalence (66%) during the **photophase**.

The **willingness to mate** of both males and females of *P. dissecta* increased until their **age** had reached 10–30 days (Pervez et al. 2004). In *C. septempunctata* in India, females started to mate at 2 days and males at 4 days after emergence. Ten day old males and females were the most willing to mate (Srivastava & Omkar 2004). Mating of young (1–5 day old) adult *P. dissecta* lasted about 200 minutes; mating of those 20 days old lasted about 300 minutes (Pervez et al. 2004). The willingness to mate strongly increases when adults have no chance to mate. *Col. maculata* where the genders were isolated for only 1 day were 26 times more likely to mate than individuals kept in a mixed-sex group; isolating males had a stronger effect than isolating females (Harmon et al. 2008).

Multiple matings usually **enhance** the total egg output and percentage of eggs that hatch, even though one copulation is sufficient for permanent fertility of the female (except in some species such as *Stethorus pusillus* that lack a spermatheca; Putman 1955). Mating had a repeated stimulatory effect on the number of eggs laid in *A. bipunctata* (Semyanov 1970). The hatching rate in *Menochilus sexmaculatus* was 64% after multiple matings and 37% after only a single mating (Bind 2007). Egg hatching rate in this species decreased after separation of the female and male: hatching was reduced to 50% by 50, 40 and 30 days after the separation (following 5, 10 and 20 matings, respectively) (Omkar & Mishra 2005a). Results with *P. dissecta* (Omkar & Pervez 2005) were even more striking: 65% eggs hatched after a single mating, 72% after two, 78% after three and 93% after further multiple mating. However, the physical interference between individuals in a **limited space** had a small **detrimental effect** on egg viability of *P. dissecta*: 96% of the eggs laid by a single mated female hatched, while only 88% of the eggs hatched when laid by a female sharing a 9 cm Petri dish with four unmated males (Mishra & Omkar 2006).

Mating (or repeated mating) refusal by ladybird females has been observed frequently, but the causes and consequences of such behaviour have rarely been revealed. Two prominent hypotheses are that resistance is (i) a means of **avoiding costly** and **dangerous mating** and (ii) a means for the **selection of high-quality males**. Female refusal may be decreased by a nutritious **nuptial gift**, i.e. a spermatophore, provided by the male. The above hypotheses were investigated in *A. bipunctata*, in which females frequently display strong **copulation rejection** behaviour and ingest a **spermatophore** after copulation (Perry et al. 2009). Females deprived of food for 96 hours resisted mating more frequently and for longer periods than less starved females (starved for only 16 hours) and so they re-mated less frequently. The finding that starved females were more resistant suggests that mating is energetically costly, and that nuptial consumption of spermatophores does not offset these costs (Perry et al. 2009).

Female **longevity** decreased with increasing number of matings in both *Menochilus sexmaculatus* and *P. dissecta*, indicating there is a cost to mating. Thus very high mating activity, typical of many ladybirds, may have a deleterious effect on their fitness (Omkar & Mishra 2005a). Longevity decreased from 118 days for single, once-mated females to 76 days for repeatedly mated females of *Har. axyridis*. Due to this difference, once-mated females had higher lifetime fecundity (1326 eggs, hatchability 58%) than repeatedly mated ones (977 eggs, 51%; Fois et al. unpublished).

There was a transient benefit from **polyandry** (mating with several different males) in the reproductive success of *A. bipunctata* females. Fecundity and hatching rate were lower in females mated 10 times with one male than in females mated once with 10 different males (Haddrill et al. 2007). Promiscuous

females of *P. dissecta* were also more fecund and laid more viable eggs than monogamous ones, which had been confined in a cage with a single male. Amongst promiscuous females, those mated with several males (i.e. there was freedom of mate choice) had a significantly higher reproductive output than those mated daily with a new unmated male with no choice (Omkar & Mishra 2005a).

3.5.6.2 Sperm competition

Earlier studies suggested that most often it was the sperm of the last male (of many that copulated with the female) that fertilized the eggs (Ueno 1994, 1996). De Jong et al. (1993) provided evidence for almost complete second male sperm precedence in experiments with melanic and non-melanic *A. bipunctata* providing that the results were not obscured by female **rejection behaviour**. In a repeated experiment, however, there was a highly variable degree of paternity of the second male (de Jong et al. 1998).

Behavioural and molecular genetic data were used to examine how sperm from several males was used over time by females of *A. bipunctata*, and to link mating with **fertilization** (Haddrill et al. 2008). In the laboratory, the number of mates (males that copulated with the observed female) was usually similar to the number of fathers (males that passed their genes to the progeny), suggesting that females have little postcopulatory influence on the **paternity** of their eggs. Longer copulation resulted in a higher probability of paternity for any particular male, probably due to the transfer of larger numbers of sperm in multiple **spermatophores** (Haddrill et al. 2008).

3.5.6.3 Female choice and melanism

Hodek and Ceryngier (2000) regarded the finding that at least some coccinellid species do not mate at random as the most important among the sexual activities of coccinellids.

To find out whether ladybird females prefer a specific male phenotype, e.g. a colour morph in polymorphic species, **choice** and **no-choice mating tests** can be conducted. Also, observations can be made of **seasonal changes** in the frequency of colour morphs (individuals) and of pairs of different composition in the field.

Lusis (1961) observed that matings of *A. bipunctata* recorded near Riga and Moscow involved fewer matings between red females and red males and significantly more black males in pairs than would be expected with random mating. He hypothesized that the higher sexual activity of the melanic forms was due to their higher metabolic rate, as the result of higher absorbance of solar radiation. However, Creed (1975) did not observe this phenomenon in a population of *A. bipunctata* near Birmingham. Also in England, Muggleton (1979) found that mating preference was affected by the frequency of the different forms: regardless of colour, the rarer morph was preferred by females of both morphs. Majerus et al. (1982a) reported a preference for melanic males in another English population in 1981: while 34% of the non-mating males of *A. bipunctata* in North Staffordshire were melanics, the melanic proportion of mating males was 49%. For melanic females, however, the proportion was about 35% for both mating and non-mating ones. The authors (Majerus et al. 1982b) demonstrated a genetic basis for such preferential mating; differential competitiveness among males was not involved. However, the mechanism whereby females recognize melanic males remains unknown.

Female *Har. axyridis* expressed **visible mate preference**, by rejecting less-preferred phenotypes, and cryptically by **retaining eggs** for longer periods after mating with less-preferred males, apparently in order to replace their sperm later by that of a more-preferred male. Whereas **pair formation** was under female control, the **duration of copulation** was under male control. Males invested more time in mating with dissimilar females (Wang et al. 2009). Which males were preferred was not clear-cut, because there was a pleiotropic effect whereby female choice varied between the spring and autumn **generations**. The prevalence of melanics in spring decreased in summer because females preferred mating with the pale *succinea* morph (Osawa & Nishida 1992). In the summer generation, melanics were more successful.

Several studies based on mate-choice experiments have shown that most females seemed to prefer to mate with **melanic** morphs, especially the *conspicua* morph. Thus melanics dominate in laboratory cultures and **biocontrol stocks** although they remain relatively rare in the wild due to a set of selection pressures (Seo et al. 2008).

Two alternative hypotheses have been offered to explain why gender-specific reproductive behaviour may vary between generations: (i) **maternal factors** (epigenetic) that influence the expression of genes in progeny, and (ii) linkage disequilibria among **allele frequency** that cycle seasonally as a function of **assortative mating** (Wang et al. 2009).

In **field populations** of *Har. axyridis* in China, red phenotypes outnumbered melanics by 5:1 in the autumn, but melanics became equally abundant in the spring, suggesting that melanism is advantageous in winter, but costly in summer (Wang et al. 2009). In *A. bipunctata*, melanic forms were more abundant in the autumn than in the spring but not significantly so (Majerus & Zakcharov 2000, Honěk et al. 2005). The representation of melanic males in mating pairs observed in a **cage experiment** was higher than would be expected from random mating (Majerus et al. 1982a). Though it is still unknown how females recognise melanic males, as pointed out above, O'Donald and Majerus (1989) were able to show that **visual discrimination** by females was not involved. Specific cuticular alkanes (Hemptinne et al. 1998) may be responsible for the recognition, and higher body temperature and **activity** in melanic *A. bipunctata* under most conditions has been recorded (De Jong et al. 1996).

In the related *A. decempunctata*, the proportion of the three main colour morphs (*typica*, spotted; *decempustulata*, chequered; *bimaculata*, melanic) was found to be very stable both in space and time, with no environmental selection or female choice (Banbura & Majerus in Majerus 1998, Honěk et al. 2005).

Females of *Har. axyridis* uninfected with the ectoparasitic fungus *Hesperomyces virescens* were preferred by males as mating partners over **infected** ones (Nalepa & Weir 2007).

Melanic as well as typical individuals of *C. septempunctata* in India preferred to mate with melanic forms of the opposite sex. Mate choice was mainly determined by females and to a lesser degree by males (Srivastava & Omkar 2005). While Osawa (1994) suggested that the colour of the elytra of *Har. axyridis* is the most important factor in mate choice by females, Ueno (1994) stressed that size and activity were important.

Perry et al. (2009) investigated whether the extent of active choice/rejection of males by females depended on **male size** or whether unspecified rejection passively favoured large males that overwhelmed the females. They found that the extent of female resistance was independent of the size of the male, but that limited resistance resulted in a bias towards large males in copulations. A side effect of imperfect female resistance would be expected to result in **selection** for large male size. However, in all ladybirds, males are still slightly smaller than females, as also applies for the majority of animals, reflecting the higher energetic and nutritional cost of producing eggs in comparison with sperm.

3.5.6.4 Hybridization

Cases of spontaneous **interspecific mating** in the wild are rarely reported but may occur relatively often in a limited space in the laboratory. Cases of successful interspecific reproduction, i.e. **hybridization** are even rarer. Majerus (1997) listed the observations of interspecific matings in the field (five combinations, including taxonomically distant species) and laboratory (eight combinations, also with unrelated species). There were either no progeny after such mating or the progeny were apparently of the same species as the female, suggesting that the female was already mated with the same species male. The only **hybrid progeny** were reported for a couple consisting of an *A. bipunctata* female and an *A. decempunctata* male. In other hybridization combinations using previously isolated females, eggs showed signs of development but did not hatch (**embryonic mortality**), or larvae died in the first instar (*Anatis ocellata* × *An. labiculata*). A hybrid between *P. quatuordecimpunctata* and *P. japonica* had decreased fecundity (Iablokoff-Khnzorian 1982).

Reproductive isolation between two related *Henosepilachna* species is achieved by a combination of two mechanisms: (i) there is a choice for conspecific females by *Henosepilachna vigintioctomaculata* males; (ii) there is a difference in the intensity of male rejection between *H. vigintioctomaculata* females (strong) and *H. pustulosa* females (weak) (Matsubayashi & Katakura 2007). Host fidelity functioned as a strong **barrier** against gene flow between two other species *Henosepilachna niponica* and *Henosepilachna yasutomii*, but occasional interspecific mating has been achieved by addition of common host plant (Hirai et al. 2006).

3.5.7 Oviposition

3.5.7.1 Oviposition substrate

Ovipositing females have to select a suitable place for their eggs. This selection involves balancing several requirements which may sometimes even be conflicting: optimum **microclimate** for embryonic development, proximity of **food** for the hatched larvae, and **protection** from predators, parasites and competitors (5.4.1.3).

The Coccidulini lay eggs freely on the leaf surface (Klausnitzer & Klausnitzer 1986). *Megalocaria dilatata* (Coccinellini) lays eggs on the spines of plants and makes a barrier preventing an access to the egg cluster with a sticky secretion from the abdomen (Liu 1933). The aphidophagous generalist *A. bipunctata* laid more eggs (91%) on filter paper than on spruce needles (9%; Timms & Leather 2007), while the conifer adelgid-eating specialist *Aphidecta obliterata* laid more eggs on needles (77%). Many aphidophagous coccinellids lay eggs on the underside of **filter paper** when in laboratory containers, as in the field they lay eggs on the **underside** of the leaves on broad-leaved plants. In this way the coccinellids may reduce the risk of predation and rain washing off the eggs. Ladybirds specialised on conifer trees lay eggs on the **needles** or into bark crevices.

In Japan, *C. septempunctata* used for oviposition artificial insolated objects during the winter (Ohashi et al. 2005). *A. bipunctata* avoided laying eggs in patches with *Lasius niger* **ants** (Oliver et al. 2008).

Scymninae often hide their eggs in crevices of the substrate, or use an artificial **protection**. Females of *Sasajiscymnus kurohimae*, feeding on eusocial aphids, which have a soldier caste that defend their colonies, protect their eggs beneath undigested remnants of eaten aphids (Arakaki 1988). *Scymnus hoffmanni* similarly covers its eggs with cuticles of predated aphids (Kawauchi 1985). Eggs of *Scymnus louisianae* are laid predominately on the undersides of leaves, nestled among the leaf hairs (Brown et al. 2003).

Females of Chilocorinae (e.g. *Chil. rubidus*) lay one egg at a time under the scale of a larval prey coccid (Pantyukhov 1968). *Exochomus flavipes* have occasionally been observed laying eggs into a conspecific empty pupae from which a parasitoid had emerged (Geyer 1947).

Two whitefly predators, *Clitostethus oculatus* (Scymninae) and *Delphastus pusillus* (Microweiseinae) lay eggs singly or in groups of 2–4 on leaf surfaces, where whitefly eggs and nymphs are abundant (Liu & Stansly 1996a).

In **Epilachninae** that feed on plants as both adults and larvae, it is often difficult to distinguish oviposition preference from adult feeding preference, because oviposition is likely to occur at or near feeding sites. In laboratory assays, the distance between adult feeding scars and egg masses of *Henosepilachna niponica* was long (25 cm) and the females often placed eggs on artificial substrates rather than leaf discs (Fujiyama et al. 2008).

3.5.7.2 Oviposition rhythmicity

Many activities of ladybirds take place during the day, whereas they rest during the night (5.4.1.1). However, the oviposition activity of several species in India showed the opposite trend (Fig. 3.3). There, it was found that *C. septempunctata* females preferred to oviposit at the end of the scotophase and in the early photophase hours (21.00–1.00, where 0.00 means the beginning of the photophase), *P. dissecta* laid 86% eggs during the early half of scotophase (Mishra & Omkar 2004) or that they laid most eggs in the middle of the scotophase (15.00–17.00). *C. leonina transversalis* laid most by dusk, i.e. at the beginning of the scotophase (11.00–13.00) (Omkar et al. 2004). In *Menochilus sexmaculatus* (Omkar & Singh 2007), the peak of oviposition (62%) was attained during early scotophase (12:00–15:00).

3.5.7.3 Oviposition rate

The number of eggs laid per day, i.e. daily oviposition or reproductive rate, increases rapidly during several days after adult eclosion, to reach a maximum (which may equate to the number of ovarioles in both ovaries of the female) at about two or three weeks, and then slowly decreases during the remaining life span of the female, which might be several months.

These changes in oviposition rate (rapid increase followed by slow decrease) during adult life can be described by a **triangular fecundity function** (Dixon 2000). The pattern can be fitted by a third order power function (see fig. 2.11b in Dixon 2000) or by the Bieri model (Bieri et al. 1983, for its use see Lanzoni et al. 2004, and Fig. 3.7c). The Bieri model (Or = (a·(age − b))/ (exp(c·(age − b)))) provides a calculation for the time of **peak oviposition rate**. The age specific oviposition rate can also be fitted by a lognormal or other asymmetric distribution curve.

The increase and decrease of oviposition rate (**age specific fecundity**) can also be almost symmetric, as in *Anegleis cardoni*, where it was fitted simply by a second power curve (Omkar et al. 2009). The estimation of age at peak oviposition tends to be higher when fitted by this parabolic curve than when using an **asymmetric** distribution, and higher for treatments giving long (usually in better conditions) than short oviposition periods (less suitable conditions). Omkar & Mishra (2005) distinguished between short-lived females, that distributed their reproduction **uniformly** in their lifetime, and long-lived females which showed a high burst of reproductive activity followed by a gradual decline.

Egg production in *Scymnus louisianae* was found to decrease slowly, and it was rather equally distributed over a female lifespan of 80 days, though there was a rapid increase during 6 days after adult eclosion and a rapid drop during last 10 days of life (Brown et al. 2003, Fig. 3.7g). The females of *Exochomus quadripustulatus* laid eggs in an irregular pattern, with the number of eggs deposited within a single day ranging from 1 to 11 at a thermoperiod of 9/19°C and from 1 to 16 eggs at 12/24°C; the oviposition rate had a rather bimodal appearance (Sengonca & Arnold 2003; Fig. 3.7h). The rate of egg production of *Chil. nigritus* followed a cyclical pattern that lasted for approximately 22 days (Ponsonby & Copland 2007).

The daily oviposition rates of *C. septempunctata* and *P. quatuordecimpunctata* were not related to **female size** (Honěk et al. 2008).

Natality (m_x), which is the mean number of female offspring produced per surviving female during age interval x (Birch, 1948), is sometimes used instead of oviposition rate. The natality of *Clitostethus oculatus* fluctuated many times during the lifespan of females (Fig. 3.7b), and did not follow the typical triangular fecundity function referred to above (Ren et al. 2002).

3.5.7.4 Oviposition period

The duration of the period between the first and the last egg batch, the oviposition period, is a good measure of individual **fitness**. The number of days between the last oviposition and death of the female is called the **post-oviposition period**. It can be zero, when a female dies during its reproductive phase, or can last from a few days to weeks in **senescent** females. When a mean oviposition rate is calculated from the total lifetime fecundity of a female, only the reproductive (=oviposition) period should be used in the denominator.

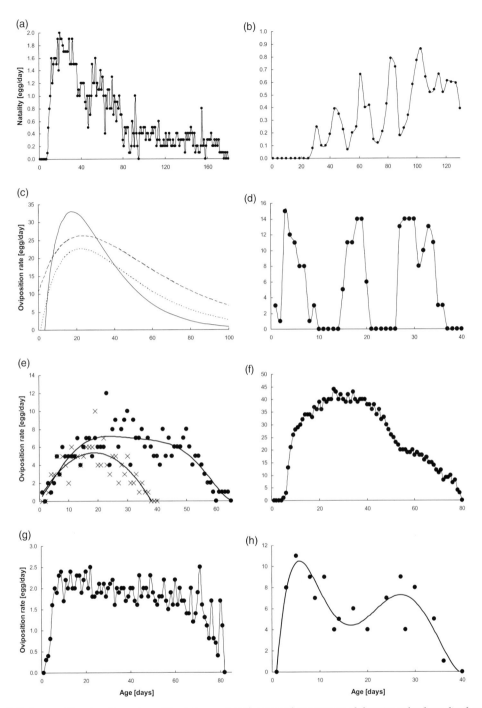

Figure 3.7 Age specific oviposition rates calling into question the general occurrence of the triangular fecundity function (Dixon 2000). (a) Triangular natality (m_x) with tail, *Axinoscymnus cardilobus* at 20°C fed with *Bemisia tabaci* (after Huang et al. 2008); (b) polymodal increasing natality of *Nephaspis oculatus* at 31°C fed with *Bemisia tabaci* and *B. argentifolii* (after Ren et al. 2002); (c) Bieri function calculated for *Harmonia axyridis* (solid line), *Hippodamia variegata* (dashed line) and *Adalia bipunctata* (dotted line) reared on *Myzus persicae* at 25°C (after Lanzoni et al. 2004); (d) polymodality in *Chilocorus nigritus* at 30°C, RH 62% and fed with *Abgrallaspis cyanophylli* (after Posonby & Copland 2007); (e) parabolic (second order power) function in *Anegleis cardoni* fed with *Lipaphis erysimi* (crosses) and long plateau of *Anegleis cardoni* fed with *Aphis craccivora* (dots; both after Omkar et al. 2009); (f) two levels in *Harmonia axyridis* at 25°C fed with *Aphis fabae* (after Stathas et al. 2001); (g) monotonous level (almost constant) in *Scymnus louisianae* fed with *Aphis glycines* (Brown et al. 2003); (h) bimodality in *Exochomus quadripustulatus* at thermoperiod 9/19°C fed with *Pulvinaria regalis* (Sengonca & Arnold 2003).

The length of the oviposition period is species-specific and decreases rapidly with increasing **temperature**, since **longevity** then decreases. The oviposition period of *Scymnus levaillanti* lasted 87, 76, 47 and 32 days at 20, 25, 30 and 35°C respectively (Uygun & Atlıhan 2000). No eggs were deposited at 15°C, even though the development thresholds for eggs and larvae were between 11–12°C. The oviposition periods of *Axinoscymnus cardilobus* lasted 109, 61, and 17 days at 17, 26 and 32°C, respectively (Huang et al. 2008).

The oviposition period may vary with **food** quality. The oviposition period was 22.4 days in *C. septempunctata* fed with the cereal aphid *Sitobion avenae* (Rana & Kakker 2000), but only 16 days when reared on *Brevicoryne brassicae* (ElHag & Zaitoon 1996). In *Anegleis cardoni*, it was twice as long (58 days) on *Aphis gossypii* than on *Lipaphis pseudobrassicae* (27 days) (Omkar et al. 2009).

While the average pre-oviposition and oviposition periods in melanic **colour morph** of *C. septempunctata* in India were 10.2 and 68.9 days, respectively, they were 14.6 and 76.9 days respectively in normal morphs (Rhamhalinghan 1986).

3.5.8 Fecundity

The total lifetime egg production (fecundity) is **species-specific**, increasing with the **size** of the ladybird, and dependent on **temperature** and on the quality and quantity of **food** both during the pre-imaginal development of the female and during reproduction. Fecundity is also influenced by **mating frequency**, **age** of the parents, **population density**, illumination, **photoperiod**, phenotype; it also changes with generation, etc. (Table 3.1). It is the product of daily fecundity (**oviposition rate**) and the duration of the **oviposition period**.

The relationship between food quantity and fecundity or fertility is called the **numerical response** (5.3). The shape of the relationship is usually a convex curve. In *Scymnus subvillosus* the numerical response was similar to its functional response that matched Holling's type II (Atlıhan & Güldal 2009). A twofold increase in **prey density** (*Aphis gossypii*) brought about a twofold increase in lifetime oviposition and mean oviposition rate in *C. septempunctata* (Xia et al. 1999).

The fecundity of *Anegleis cardoni* changed with **food suitability**: it was three times higher on *Aphis gossypii* than on *Lipaphis pseudobrassicae*, while the oviposition period was twice as long and the mean reproductive rate 1.34 times higher (Omkar et al. 2009). The fecundity of *C. septempunctata* was 35% higher when fed with *Aphis fabae* reared on a susceptible cultivar of *Vicia faba* than when the aphid had been fed on a resistant one (Shannag & Obeidat 2008).

A decrease in **prey density** (the scale *Abgrallaspis cyanophylli*) caused a significant but transient decline in egg production in *Chil. nigritus*. A heterogeneous prey population (mix of diverse developmental stages) elicited significantly higher levels of oviposition at similar host densities than homogeneous population (cohort of equally aged individuals), irrespective of the growth stage of the prey (Ponsonby & Copland 2007).

Of five generations of *Cer. undecimnotata* produced in experimental cages during a year, the greatest numbers of eggs were laid by females of the first and second **generations** (Katsoyannos et al. 1997).

Crowding has an adverse effect on fecundity. Individual females of *Har. axyridis* laid an average of 13 eggs per day, while females grouped as 10, 20, 30 and 40 per container laid 9, 6, 3 and 2 eggs per day, respectively (Abdel-Salam & Abdel-Baky 2001).

The parasitoid *Dinocampus coccinellae* has been reported as having a great effect on the fertility of mature female ladybirds. The **parasitoid** larva normally **castrates** the host ladybird and usually causes death. However, at relatively high temperatures (25°C), it was found that daily numbers of coccinellid eggs increased during two days after parasitization and then gradually decreased to stop 8–9 days after parasitization. Although mortality of the females was high (71%), half of those that survived began to oviposit again 12 days after the parasitoid larvae had emerged (Triltsch 1996).

Melanic females of *C. septempunctata* in India laid more eggs (302–412) than the females of the typical **colour morph** (278–379) (Rhamhalinghan 1986). In *Micraspis discolor* in India, fecundity was the lowest at the lowest experimental **temperature** of 20°C, highest at the optimum temperature, and decreased at the highest temperature of 30°C (Omkar & Pervez 2002) (3.6.4; Fig. 3.8b). In *Chil. nigritus*, no oviposition occurred at the low temperature of 18°C. The lifetime fecundity increased from 20 to 24°C and then

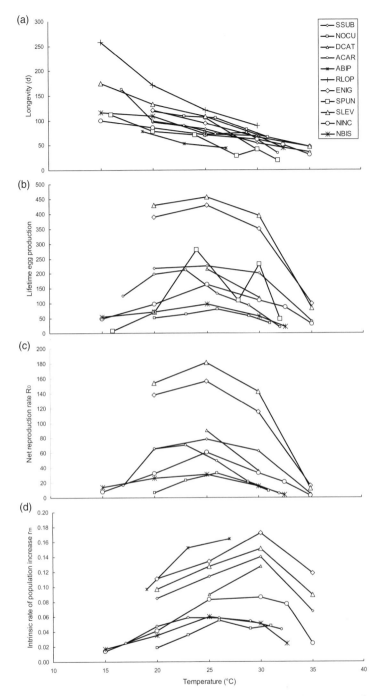

Figure 3.8 Temperature effects on selected parameters of adult life. (a) longevity; (b) lifetime egg production; (c) net reproduction rate R_0 (Andrewartha & Birch 1954); (d) intrinsic rate of population increase r_m (Andrewartha & Birch 1954). Data for 12 coccinellid species: *Adalia bipunctata*, ABIP (Jalali et al. 2009); *Axinoscymnus cardilobus*, ACAR (Huang et al. 2008); *Delphastus catalinae*, DCAT (Kutuk & Yigit 2007); *Diomus austrinus*, DAUS, (Chong et al. 2005); *Parexochomus nigromaculatus*, ENIG (Atlıhan & Özgökce 2002); *Clitostethus oculatus*, NOCU (Ren et al. 2002); *Nephus bisignatus*, NBIS (Kontodimas et al. 2007); *Nephus includens*, NINC (Kontodimas et al. 2007), *Lindorus lophantae*, RLOP (Stathas 2000); *Scymnus levaillanti*, SLEV (Uygun & Atlıhan 2000); *Scymnus subvillosus*, SSUB (Atlıhan & Chi 2008); *Stethorus pusillus*, SPUN (Roy et al. 2003).

decreased up to 30°C. No eggs were produced at 34°C. However, daily fecundity was highest at 30°C (Ponsonby 2009).

The **reproductive output** (percentage of body weight allocated to reproduction during 24 h) was similar (12.5%) in large *C. septempunctata* and small *P. quatuordecimpunctata* (Honěk et al. 2008).

3.5.9 Longevity

3.5.9.1 Voltinism

The number of generations per year and the lifespan of ladybirds depend on the **climatic conditions** in the region. However, many species, even in warmer climates, normally have a single generation (are **univoltine**), and live for one year including several months in **dormancy** (also Chapter 6).

In several species, a second overwintering has been reported in a minority of individuals. Thus there are records of a **second hibernation** in *Calvia quatuordecimgutta* (Kanervo 1946), in *P. quatuordecimpunctata* (Hariri 1966), in *Stethorus pusillus* (Putman 1955), in *Aiolocaria hexaspilota* (Iwata 1932, Savoiskaya 1970) and *C. septempunctata* (Sundby 1968). Savoiskaya (1970) reported that 15–20% of a *Har. axyridis* population in Kazakhstan may live and oviposit for 3 years.

Epilachna admirabilis is univoltine in northern Japan and hibernates as the final (fourth) instar larva. Some of the resulting adults then enter a second hibernation together with their own larvae (Katakura 1976).

Scymnus sinuanodulus regularly lives for **2 years** and the fecundity is the same in both years (Lu & Montgomery 2001). In *Pseudoscymnus tsugae* the mean longevity in the laboratory was 163 and 126 days for females and males, respectively, while some individuals of both sexes lived more than 300 days (Cheah & McClure 1998), suggesting the potential for a 2-year lifespan in the field.

While central European populations of *Cer. undecimnotata* are obligatorily univoltine (Ceryngier et al. 2004), at least a portion of the *Cer. undecimnotata* population in central Greece may complete two or more overlapping generations per year under artificial conditions in shaded outdoor cages (Katsoyannos et al. 1997; 6.2.8). *Hip. variegata* may complete seven generations between April and November under artificial conditions in outdoor cages in Greece. Adults of the

6th and 7th generations formed the hibernating population (Kontodimas & Stathas 2005).

Coccinella quinquepunctata, *C. septempunctata* and *C. magnifica* are potentially **polyvoltine** (Ceryngier et al. 2004). Heterogeneous voltinism (6.2.1.6) appears to be one of the factors responsible for the predominance of *C. septempunctata* in most habitats of the Palaearctic and for its successful invasion of the Nearctic Region (Hodek & Michaud 2008). The very successful invader *Har. axyridis* regularly has two generations per year, even in colder temperate countries (Brown et al. 2008), and unlike other ladybirds even reproduces late in the season.

The lifespan of the sequential generations of polyvoltine species usually decreases in the laboratory, probably due to inbreeding (6.2.1). Thus, in four successive generations of *Menochilus sexmaculatus* longevity was respectively 87, 89, 63 and 50 days in males, and 111, 101, 90 and 50 days in females (Hukushima & Kouyama 1974).

The overlap of generations in bivoltine species (such as *A. bipunctata* in Central Europe) enables the spread in the population of pathogens and parasites such as the mite *Coccipolipus hippodamiae* (Webberley et al. 2006; 8.3).

3.5.9.2 Effect of temperature, photoperiod and humidity

In coccinellids, as in other exotherms, **life span** shortens with increasing temperature. There are cases of a **linear decrease** of life span with temperature (3.6.4; Fig. 3.8a). For example, the linear decrease was clear in *Axinoscymnus cardilobus* reared at six temperatures: the longest longevity was 163 days at 17°C compared with only 34 days at 32°C (Huang et al. 2008). Adult *C. novemnotata* lived for 62, 48 and 21 days at 21, 27 and 32°C, respectively (McMullen 1967). *Olla v-nigrum* lived for 172 days at 15°C and only 51 days at 25°C (Kreiter & Iperti 1984). Males of Iranian population of *Stethorus gilvifrons* reared at various temperatures from 15 to 35°C had longest and shortest longevities of 18 and 13 days, and 17 and 9 days for females (Taghizadeh et al. 2008). The longevity of this coccinellid in Turkey also decreased with increasing temperature (20, 25, 30°C) and this was not affected by two different **photoperiod**s (16L:8D and 8L:16D) (Aksit et al. 2007). *Chil. nigritus* lived for 100–280 days at temperatures of 22–26°C, 65–115 days at 30°C and only 17–29 days at 34°C (Ponsonby 2009).

In contrast, in *Col. maculata* no such a clear tendency was recorded: longevity was 73, 74, 77, 45 and 80 days at 19, 21, 23, 25 and 27°C, respectively (Wright & Laing 1978). The mean adult longevities of *Delphastus catalinae* were 40 days at a constant 30°C and 42 days at **fluctuating temperatures** of 25/35°C (Kutuk & Yigit 2007).

The longevity of females of *Chil. nigritus* increased with increasing **relative humidity** (162 days at 40%, 220 at 80%), but net reproductive rate (R_0) and the maximum intrinsic rate of increase (r_{max}) were higher at the lower humidity (Senal 2006).

3.5.9.3 Effect of food

Adult longevity tends to be longer on a more suitable diet. Longevity of *Anegleis cardoni* was almost double when fed the more **suitable prey** *Aphis gossypii* than when fed the less suitable *Lipaphis pseudobrassicae* (Omkar et al. 2009). A diet of *Myzus persicae* increased the adult longevity and fecundity of *C. undecimpunctata* compared with one of *Aphis fabae* (Cabral et al. 2006). In contrast, however, the longevity of *Chil. nigritus* was almost twice when fed *Aspidiotus nerii* than *Abgrallaspis cyanophylli*, although the latter would appear the more suitable prey since it more than doubled fecundity (Ponsonby 2009). Moreover, the adult longevities of *Har. axyridis* and *A. bipunctata* females did not differ significantly when reared as larvae on different diets (either natural or artificial) and then on artificial diet as adults (120 to141 or 74 to 118 days, respectively; Ware et al. 2008). The longevity of females (59–76 days) and males (54–64 days) of *P. quatuordecimpunctata* was only little affected by seven aphid species provided as different food (Kalushkov & Hodek 2005).

The longevity of *Har. axyridis* fed with *Aphis glycines* reared on three susceptible varieties of soybean was greater (12.5 days) than on three **resistant varieties** (7.3 days) (Lundgren et al. 2009). In contrast, the adult longevity of *C. septempunctata* was not affected by whether *Aphis fabae* was reared on a susceptible (major) or on a partially resistant (79S4) cultivar of *Vicia faba* (Shannag & Obeidat 2008). No significant differences were reported in longevity or other characteristics of *Col. maculata* fed with **Bt-transgenic** corn pollen, non-Bt pollen, or *Schizaphis graminum* (Ahmad et al. 2006). Also *P. japonica* was not affected when fed with *Aphis gossypii* reared on either transgenic or non-transgenic cotton (Zhu et al. 2006).

The **quantity** of food has inconsistent effects on longevity. Increased prey (*Hyalopterus pruni*) consumption did not change the longevity of *Scymnus subvillosus*, but did result in a higher intrinsic rate of increase (Atlıhan & Güldal 2009). When the food provided to *Har. axyridis* was limited, mean longevity increased although mean reproductive life span and fecundity were reduced (Agarwala et al. 2008). However, there was a difference between groups of individuals, with one group showing a positive correlation and the other group a negative correlation between reproduction and longevity.

3.5.9.4 Effect of sexual activity

Males may have a longer or shorter longevity than females. Female *Psyllobora confluens* lived 46 days, but males 59 days, when fed with *Erysiphe cichoracearum* (Cividanes et al. 2007). In contrast, the mean longevity of female *Serangium parcesetosum* was longer (71 days) than that of males (60 days) (Al-Zyoud et al. 2005). There was no significant difference between the longevities of male and female *P. quatuordecimpunctata* (Kalushkov & Hodek 2005), but in *Chil. nigritus*, males are on average longer-lived than females (Ponsonby 2009).

Ageing trends were sex dependent in *P. dissecta*, with reproductive performance declining later in females than in males (Mishra & Omkar 2006).

There is a strong trade-off between the number of matings and longevity. Longevity decreased with an increasing number of matings in both *Menochilus sexmaculatus* and *P. dissecta* indicating a **cost of mating** (Omkar & Mishra 2005). Post-hibernation longevity was much shorter (76 days) in females of *Har. axyridis* that lived with male in limited space and mated regularly than in females mated before hibernation and then maintained without a male (107 days) and also much shorter than in virgin females (135 days) (Fois et al. unpublished).

3.6 TEMPERATURE AND DEVELOPMENT

Finally, we provide a general account of the more recent research on the main factor influencing coccinellid development, namely temperature. Earlier data have been reviewed by Honěk (1996). The life of exotherms, with their limited capacity for thermoregu-lation, is dependent on the external temperature. This restricts the range of conditions in which they can survive, determines the rate at which life processes proceed within this range and, together with body size (Gillooly et al. 2002), controls the fitness of the organism (Kingsolver & Huey 2008). Temperature determines the length of development of the immature stages, the length of the teneral period from adult ecdysis to first oviposition, the quantity and duration of oviposition and the length of life. The role of temperature in survival during hibernation and aestivation is described in 6.4.4.

3.6.1 Thermal constants

Growth and development of pre-adult stages occur only across a specific range of temperature. Within this range the **development time** and **rate of development** (a reciprocal of development time) vary with temperature. The **lower temperature threshold** is the temperature below which a particular stage of an animal cannot develop. After some time (3.6.4.1) it dies, probably because of the loss of correlation between physiological and behavioural functions. At the other extreme, the range of tolerated temperature is limited by the **upper temperature threshold** at which the animal dies from heat.

The relationship between the rate of development and temperature is linear only within the range of what are called the **ecologically relevant temperatures** at which most of the life activities of the animals take place. Close to the upper development limit, the relationship becomes non-linear, attains its highest point at the **temperature optimum** where development rate is highest (i.e. development time shortest), and it then decreases sharply as the upper lethal temperature limit is approached. The relationship between the rate of development and temperature thus typically has the shape of a right skewed peak, whose left slope also may divert from linearity near the lower development threshold (Gilbert & Raworth 1996). Several models which have been proposed to approximate this course of development rate with temperature, e.g. those of Lactin et al. (1995) and Briere et al. (1999) which enable the calculation of temperature optima and upper lethal temperatures, were reviewed by Kontodimas et al. (2004).

A **linear model** approximates the course of development rate in the ecologically relevant temperature range and enables the calculation of two thermal constants. The first of these is the above-mentioned **lower development threshold** (LDT; Honěk & Kocourek 1988), also known as the **basal temperature** Tb (Trudgill et al. 2005), and which is the temperature below which development ceases. The other constant is the **sum of effective temperatures** SET or, in other words, the thermal time T which is the number of day degrees [dd] above LDT for completion of a developmental stage). It has to be remembered that LDT is a virtual value, but it is convenient for predicting temperature effects under ecologically relevant temperatures. Under real conditions development is already seriously impaired below a temperature higher than the LDT by some 2–4°C.

3.6.2 Relationship between LDT and SET

Recent results on the effects of temperature on pre-imaginal development (Table 3.3) contain data for 44 populations of 25 coccinellid species. The results confirm the conclusion of the earlier review of Honěk (1996) that coccinellids are warm-adapted species with a relatively high LDT (mostly between 9–15°C; 17°C for tropical *Chil. nigritus*; Ponsonby 2009) and a low SET (200–320 dd for total development; but over 500 dd for small coccidophagous species; Table 3.3). For the whole family Coccinellidae, the average LDT calculated from Table 3.3 for eggs, larvae, pupae and total development, respectively, is 9.8, 9.3, 10.1 and 10.1°C and the average SET is 64, 167, 78 and 304 dd. Another plot of data for many species, both aphidophagous and coccidophagous, resulted in a LDT of about 10°C (Dixon et al. 1997).

The combination of a high LDT and a low SET guarantees a fast development at high temperatures, in contrast to cold adapted species whose LDT is low and SET high (Trudgill 1995). This adaptive covariance of thermal constants results in a **negative relationship** between LDT and SET (Honěk & Kocourek 1988), and this also holds for data presented in this chapter (Fig. 3.9). Interpretation of this correlation is difficult, because the negative slope of the regression of the LDT on the SET is both a consequence of biological variation and a statistical artefact. For a detailed discussion of this matter, see Honěk (1996).

Table 3.3 Thermal constants, lower development threshold LDT (°C) and sum of effective temperatures SET (dd) for the development of egg, larva, pupa and total pre-adult development. Calculated using a linear model of the development rate vs. temperature relationship and data of temperatures ≤30°C. Development time of the prepupae included in the larval stage.

	Egg		Larva		Pupa		Total		Reference
	LDT	SET	LDT	SET	LDT	SET	LDT	SET	
Adalia bipunctata	11.4	34.5	5.9	166.1	10.7	67.8	9.4	246	Olszak (1987)
*Adalia bipunctata**	7.8	52.0	7.7	180.2	9.4	86.1	8.4	314	Schuder et al. (2004)
*Adalia bipunctata**	6.2	56.5	9.7	119.4	9.5	81.2	9.2	252	Schuder et al. (2004)
*Adalia bipunctata**	10.4	44.0	11.4	114.0	11.7	71.1	11.2	232	Jalali et al. (2009a)
*Adalia bipunctata**	11.9	39.2	9.3	148.4	11.2	73.0	10.8	254	Jalali et al. (2009a)
*Adalia bipunctata**	11.5	39.8	13.3	108.8	12.1	67.6	12.6	218	Jalali et al. (2009)
*Adalia bipunctata**	—	—	—	—	—	—	10.3	268	Jalali et al. (2010)
*Adalia bipunctata**	—	—	—	—	—	—	10.4	266	Jalali et al. (2010)
Axinoscymnus cardilobus	8.8	63.7	8.9	128.5	9.5	114.7	9.3	302	Huang et al. (2008)
Calvia quatuordecimguttata†	6.5	52.6	7.5	171.6	8.5	64.3	7.7	288	LaMana & Miller (1995)
Calvia quatuordecimguttata†	5.8	55.9	7.1	174.1	8.2	68.3	7.3	294	LaMana & Miller (1995)
Calvia quatuordecimguttata†	8.3	45.3	8.2	166.0	9.4	60.8	8.5	271	LaMana & Miller (1995)
Calvia quatuordecimguttata†	7.2	52.4	7.7	173.9	8.3	66.5	7.5	292	LaMana & Miller (1995)
Chilocorus bipustulatus	—	—	—	—	—	—	11.1–13.0	475	Eliopoulos et al. (2010)
Chilocorus nigritus	13.9	88.3	16.2	241.8	13.4	79.0	15.6	363	Ponsonby & Copland (1996)
Clitostethus arcuatus	8.2	65.4	7.1	164.9	8.2	74.3	7.9	289	Mota et al. (2008)
Clitostethus oculatus	7.6	105.7	1.7	309.4	13.5	53.8	6.9	424	Ren et al. (2002)
Coccinella septempunctata	9.9	48.4	10.2	165.3	10.8	78.8	10.2	297	Katsarou et al. (2005)
Coccinella septempunctata	6.9	55.2	-2.1	270.6	0.9	90.8	0.7	405	Srivastava & Omkar (2003)
Delphastus catalinae	7.4	80.1	12.6	217.7	19.2	43.4	13.4	258	Kutuk & Yigit (2007)
*Diomus austrinus**	12.4	65.4	13.9	156.9	14.4	55.4	14.3	266	Chong et al. (2005)
*Diomus austrinus**	12.9	64.9	13.0	135.7	14.8	51.5	13.9	240	Chong et al. (2005)
Harmonia axyridis	10.2	47.0	10.9	162.9	10.9	69.2	10.8	279	LaMana & Miller (1998)
Harmonia axyridis	10.7	42.8	10.6	157.4	10.8	72.9	10.4	232	Schanderl et al. (1985)
Hippodamia convergens	8.7	50.4	10.8	121.9	11.3	50.0	10.6	224	Katsarou et al. (2005)
Hippodamia convergens	—	—	—	—	—	—	13.6	230	Rodriguez-Saona & Miller (1999)

(Continued)

Table 3.3 (Continued)

	Egg		Larva		Pupa		Total		Reference
	LDT	SET	LDT	SET	LDT	SET	LDT	SET	
Lindorus lophanthae	—	—	—	—	12.8	56.2	13.5	242	Cividanes & Gutierrez (1996)
Lindorus lophanthae*	8.7	114.0	8.0	240.4	8.3	90.0	8.2	444	Stathas (2000)
Lindorus lophanthae*	7.9	132.9	8.0	262.4	8.0	100.7	8.3	419	Stathas et al. (2002)
Nephus bisignatus	12.1	103.1	8.2	269.7	10.6	102.5	9.3	512	Kontodimas et al. (2004)
Nephus bisignatus	—	—	—	—	—	—	9.3	513	Kontodimas et al. (2007)
Nephus includens	11.8	101.9	10.4	196.8	11.1	103.8	11.0	403	Kontodimas et al. (2004)
Nephus includens	—	—	—	—	—	—	10.9	408	Kontodimas et al. (2007)
Nephus reunioni	16.4	100	8.4	274	9.2	116.0	11.9	469	Izhevsky & Orlinsky (1988)
Olla v-nigrum	9.0	46.2	11.9	133.1	12.5	45.8	11.4	220	Kreiter (1985)
Parexochomus nigromaculatus	11.5	57.7	7.9	140.4	6.6	89.7	8.8	284	Atlihan & Ozgokce (2002)
Propylea dissecta	1.2	58.7	4.2	176.2	3.2	65.4	3.6	299	Omkar & Pervez (2004)
Rodolia cardinalis	10.0	77.8	10.9	138.0	10.9	69.7	10.7	285	Grafton-Cardwell et al. (2005)
Sasajiscymnus tsugae	10.7	95.9	12.8	140.2	13.6	62.7	12.6	295	Cheah & McClure (1998)
Scymnus argentinicus	17.0	26.1	14.2	118.0	14.6	69.0	14.9	210	dos Santos (1992)
Scymnus levaillanti	11.8	64.6	8.2	142.5	10.7	88.8	10.0	296	Uygun & Atlihan (2000)
Scymnus levaillanti	6.3	77.4	15.1	65.9	6.3	72.4	11.3	201	Allawi (2006)
Scymnus marinus	13.7	28.1	—	—	—	—	10.6	657	M'Hamed & Chemseddine (2001)
Scymnus sinuanodulus	—	—	8.1	260.5	1.4	231.7	—	—	Lu & Montgomery (2001)
Scymnus subvillosus	9.8	66.1	5.5	136.4	11.9	138.0	5.5	329	Atlihan & Chi (2008)
Scymnus syriacus	16.0	41.1	7.9	119.1	7.9	102.4	11.5	236	Allawi (2006)
Stethorus japonicus‡	12.6	51.9	12.3	102.8	12.4	42.5	12.4	197	Mori et al. (2005)
Stethorus japonicus‡	12.8	50.4	12.4	99.9	12.4	42.3	12.5	193	Mori et al. (2005)
Stethorus pusillus	—	—	—	—	—	—	11.7	159	Raworth (2001)
Stethorus pusillus	11.5	61.5	11.4	102.5	11.0	52.1	11.4	215	Roy et al. (2002)

* different food type.
† different geographic populations.
‡ sex differences.

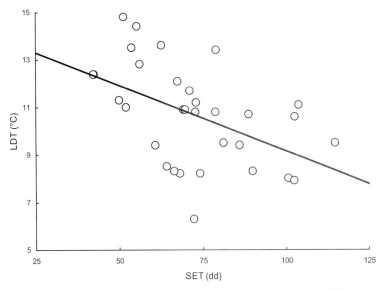

Figure 3.9 Lower development threshold LDT plotted against sum of effective temperatures SET in pupae of 32 species of coccinellid listed in Table 3.3 (data for three outlier species excluded). LDT = 14.7 − 0.0553 SET, R^2 = 0.239, P < 0.005.

3.6.3 Thermal window and development rate isomorphy

Two recent developments have provided rules that govern the temperature relationships of different species. A meta-analysis of data for many insect species has revealed that the span between the LDT and the optimum temperature for each individual species is about 20°C, regardless of whether the species are adapted to cold (i.e. have a low LDT) or warm conditions (Dixon et al. 2009). The existence of this **thermal window**, an intrinsic limit on thermal requirements, stresses the necessity of adaptation to the thermal conditions prevailing in the geographic area and in the ecological niche inhabited by the population. Another study (Jarosik et al. 2002) revealed that, in a population of any particular species, the LDT of the egg, larva and pupa is, in fact, identical and any experimentally established differences are probably caused by an observational bias. As a result, each developmental stage takes a constant proportion of the total development time and this proportion of development time does not change with temperature. This phenomenon was called **development rate isomorphy** and was tested with 426 populations belonging to 342 species of insects and mites. In a rigorous test, development

rate isomorphy was demonstrated for 243 (57%) of the populations and any violation of this principle in the rest of the species was very small. The existence of the 'thermal window' and 'development rate isomorphy' was tested and demonstrated to be true for a number of coccinellid species (Jarosik et al. 2002).

3.6.4 Other events affected by temperature

The prevailing temperature during pre-adult development influences adult **body size** but its effect is difficult to separate from the effect of the rate of food consumption. Recent results from *Diomus austrinus* (Chong et al. 2005) and *Aphidecta obliterata* (Timms & Leather 2008) were ambiguous, since temperature affected the body weight of males and females in a different way in the two species.

Temperature further determines the course of events of adult life. The thermal requirements for completing **teneral development**, i.e. the time from moult to adult to the first oviposition, can also be approximated by a linear relationship between development rate and temperature, but of course only if dormancy does not take place. The average threshold temperature LDT calculated from data in Table 3.4 (without the outlier at

Table 3.4 Thermal constants, lower development threshold LDT (°C) and sum of effective temperatures SET (dd) for teneral pre-oviposition period. Calculated using a linear model of development rate vs. temperature relationship, and data of temperatures ≤30°C.

Species	LDT	SET	Reference
*Adalia bipunctata**	10.9	69.1	Jalali et al. (2009)
*Adalia bipunctata**	1.3	142.4	Jalali et al. (2009)
Axinoscymnus cardilobus	7.8	127.4	Huang et al. (2008)
Clitostethus oculatus	5.1	317.3	Ren et al. (2002)
Delphastus catalinae	10.9	93.0	Kutuk & Yigit (2007)
Nephus bisignatus	10.0	102.1	Kontodimas et al. (2004)
Nephus includens	10.7	86.5	Kontodimas et al. (2004)
Olla v-nigrum	13.5	138.8	Kreiter (1985)
Scymnus subvillosus	10.9	93.0	Atlıhan & Chi (2008)

* different food type.

1.3°C LDT) is 9.5°C, which is similar to the LDT for immature stages. The average sum of effective temperature SET for teneral development is 127 dd.

Further important effects of temperature concern **reproduction**. Temperature influences **longevity**, the number of eggs laid and the distribution of oviposition through adult life. The data for nine species (Fig. 3.8) reveal a uniform trend. Mean longevity decreases monotonically with increasing temperature (Fig. 3.8a) but the production of eggs peaks at around 25°C (Fig. 3.8b). Using the distribution of mortality and oviposition in time, it is possible to calculate the **innate capacity for increase**, which also depends on temperature, of a coccinellid population (for details see e.g. Andrewartha & Birch 1954).

Parameters of offspring production, **net reproduction rate** R_0 (Fig. 3.8C) and **intrinsic rate of population increase** r_m (Fig. 3.8d) are also temperature sensitive and peak between 25–30°C as in *P. japonica*, where the highest r_m of 0.113 occurred at 25°C (Chi & Yang 2003). In *Stethorus pusillus*, the r_m followed a typical asymmetrical dome-shape pattern, as temperature increased, with maximum values of 0.196 per day at 30°C (Roy et al. 2003). Nevertheless, there exist warm-adapted species like *Stethorus gilvifrons* (Mulsant) whose fecundity, R_0 and r_m peak at 35°C (Taghizadeh

et al. 2008). While R_0 was highest (130) at 26°C and lower at both lower (22°C, 115) and higher temperatures (30°C, 82), r_m was greatest at the highest temperature in *Chil. nigritus* (Ponsonby 2009).

3.6.4.1 Tolerance to extreme temperatures

The **supercooling point** (SCP; Chapter 6) is naturally low for non-feeding stages, i.e. eggs and pupae, and high for larvae and adults. The mean SCP of *Har. axyridis* **eggs** was −27°C, and −21°C for **pupae**, while it was −14°C for larvae and −1.9°C for adults (Koch et al. 2004). The mean SCP of fresh eggs of *C. septempunctata* was −27°C and for older eggs it was −24.5°C, probably due to accumulation of metabolic water in the embryo (Nedvěd 1994). Wild-collected pupae of *C. septempunctata* in June and July had a SCP of −19°C. The mean SCP increased during larval development in *Cer. undecimnotata* (Nedvěd 1994): from −20°C in the second instar, through −16°C in the third , to −9°C in the fourth instar. In contrast, older and larger fourth instar larvae of *Exochomus quadripustulatus* had a lower SCP (−17°C) than smaller larvae of the same instar (−10°C) (Nedvěd 1994).

The **survival** of third and fourth instar larvae of *C. undecimpunctata* was higher than that of first and second instars when exposed to 6°C for a week. Larval survival declined sharply after 15 days. There was only 25% adult emergence from pupae chilled for 30 days. Adults survived the extended periods of **cold storage** better than the other developmental stages (Abdel-Salam & Abdel-Baky 2000).

Adults and second instar larvae of *Har. axyridis* were **heat stressed** at 35°C, and showed high mortality at 40°C, but these hot conditions were tolerated by all other stages (Acar et al. 2004). In our experiments with *Har. axyridis* (Fois & Nedvěd, unpublished), 100% **mortality** occurred in the egg and pupal stages at 35°C. The supercooling point (−17°C) and lower lethal temperature (−16.7°C) remained relatively constant for the overwintering populations in the outdoor hibernaculum. In contrast, the supercooling point and lower lethal temperature of the population overwintering indoors clearly increased as the winter progressed, from −18.5 to −13.2°C and −16.7 to −14.1°C, respectively (Berkvens et al. 2010).

The lower and upper thresholds for survival for 24 hours of *Delphastus catalinae* were around 0 and 40°C, respectively. Survival of their pupae was similar to that of adults (Simmons & Legaspi 2004). The first three instars of *Scymnus marinus* had an **optimum temperature** for development of 30°C; they also developed at 35°C but all died at 40°C. In contrast, the fourth instar larvae, prepupae and pupae of this species did develop at 40°C, although at a slower rate. The temperature optimum in these stages remained 30°C (M'Hamed & Chemseddine 2001).

In conclusion, coccinellids are average athletes in the wrestling match that Life has with Ambient Temperature. Within the range of performance of insects, coccinellids are no extremophiles; they do not live in particularly cold or warm conditions – or at least no such species have as yet been discovered.

ACKNOWLEDGEMENTS

AH's contribution to this chapter was supported by grant #522/08/1300 of the Grant Agency of the Czech Republic.

REFERENCES

Abdel-Salam, A. H. and N. F. Abdel-Baky. 2000. Possible storage of *Coccinella undecimpunctata* (Col., Coccinellidae) under low temperature and its effect on some biological characteristics. *J. Appl. Entomol.* 124: 169–176.

Abdel-Salam, A. H. and N. F. Abdel-Baky. 2001. Life table and biological studies of *Harmonia axyridis* Pallas (Col.: Coccinellidae) reared on the grain moth eggs of *Sitotroga cerealella* Olivier (Lep., Gelechiidae). *J. Appl. Entomol.* 125: 455–462.

Abdel-Salam, A. H., J. J. Ellington, M. A. El-Adl, A. M. Abou El-Naga and A. A. Ghanim. 1997. Longevity and fecundity of *Harmonia axyridis* adults reared on an artificial diet and pink bollworm eggs. *1st National Conference, Mansouri* 1: 50–57.

Acar, E. B., D. D. Mill, B. N. Smith, L. D. Hansen and G. M. Booth. 2004. Calorespirometric determination of the effects of temperature on metabolism of *Harmonia axyridis* (Col: Coccinellidae) from second instars to adults. *Environ. Entomol.* 33: 832–838.

Agarwala, B. K., H. Yasuda and S. Sato. 2008. Life history response of a predatory ladybird, *Harmonia axyridis* (Pallas) (Coleoptera: Coccinellidae), to food stress. *Appl. Entomol. Zool.* 43: 183–189.

Ahmad, R. 1970. Studies in West Pakistan on the biology of one nitidulid species and two coccinellid species (Coleoptera) that attack scale-insects (Hom., Coccoidea). *Bull. Entomol. Res.* 60: 5–16.

Ahmad, A., G. E. Wilde, R. J. Whitworth and G. Zolnerowich. 2006. Effect of corn hybrids expressing the coleopteran-specific Cry3Bb1 protein for corn rootworm control on aboveground insect predators. *J. Econ. Entomol.* 99: 1085–1095.

Aksit, T., I. Cakmak and G. Ozer. 2007. Effect of temperature and photoperiod on development and fecundity of an acarophagous ladybird beetle, *Stethorus gilvifrons*. *Phytoparasitica* 35: 357–366.

Allawi, T. F. 2006. Biological and ecological studies on *Scymnus syriacus* and *Scymnus levaillanti* (Coleoptera: Coccinellidae). *Eur. J. Entomol.* 103: 501–503.

Alvarez-Alfageme, F., N. Ferry, P. Castanera, F. Ortego and A. M. R. Gatehouse. 2008. Prey mediated effects of Bt maize on fitness and digestive physiology of the red spider mite predator *Stethorus punctillum* Weise (Coleoptera: Coccinellidae). *Transgenic Res.* 17: 943–954.

Al-Zyoud, F., P. Blaeser and C. Sengonca, 2005. Investigations on the biology and prey consumption of the predator *Serangium parcesetosum* Sicard (Coleoptera: Coccinellidae) by feeding on *Trialeurodes vaporariorum* Westwood (Homoptera: Aleyrodidae) as prey. *J. Plant Dis. Prot.* 112: 485–496.

Anderson, J. M. E. 1981. Biology and distribution of *Scymnodes lividigaster* (Mulsant) and *Leptothea galbula* (Mulsant) Australian Lady birds (Coleoptera: Coccinellidae). *Proc. Linn. Soc. N.S.W.* 105: 1–15.

Andrewartha, H. G. and L. C. Birch. 1954. *The distribution and abundance of animals*. The University of Chicago Press, Chicago.

Anonymous. 1997. Biological control of aphids in WANA using coccinellids. *Annual Report*, Icarda, Syria. p. 8.

Arakaki, N. 1988. Egg protection with faeces in the ladybeetle, *Pseudoscymnus kurohime* (Miyatake) (Coleoptera: Coccinellidae). *Appl. Entomol. Zool.* 23: 495–497.

Atlhan, R. and H. Chi. 2008. Temperature-dependent development and demography of *Scymnus subvillosus* (Coleoptera: Coccinellidae) reared on *Hyalopterus pruni* (Homoptera: Aphididae). *J. Econ. Entomol.* 101: 325–333.

Atlhan, R. and M. S. Özgökce. 2002. Development, fecundity and prey consumption of *Exochomus nigromaculatus* feeding on *Hyalopterus pruni*. *Phytoparasitica* 30: 443–450.

Atlhan, R. and H. Güldal. 2009. Prey density-dependent feeding activity and life history of *Scymnus subvillosus*. *Phytoparasitica* 37: 35–41.

Attygalle, A. B., S. R. Smedley, T. Eisner and J. Meinwald. 1996. Tocopheryl acetates from the pupal exocrine secretion of the squash beetle, *Epilachna borealis* (Coccinellidae). *Experientia* 52: 616–620.

Banken, J. A. O. and J. D. Stark. 1997. Stage and age influence on the susceptibility of *Coccinella septempunctata* (Coleoptera: Coccinellidae) after direct exposure to Neemix, a neem insecticide. *J. Econ. Entomol.* 90: 1102–1105.

Banks, C. J. 1955. An ecological study of Coccinellidae (Col.) associated with Aphis fabae Scop. on Vicia faba. *Bull. Entomol. Res.* 46: 561–587.

Banks, C. J. 1956. Observations on the behaviour and mortality in Coccinellidae before dispersal from egg-shells. *Proc. Roy. Entomol. Soc. London A* 31: 56–60.

Baskaran, R. K. and K. Suresh. 2006. Comparative biology and predatory potential of black beetle, *Chilocorus nigrita* (Fab.) on three scale insects. *J. Entomol. Res.* 30: 159–64.

Baungaard, J. and M. Hämäläinen. 1981. Notes on egg-batch size in *Adalia bipunctata* L. (Col.: Coccinellidae). *Ann. Entomol. Fenn.* 41: 25–27.

Berkvens, N., J. Bonte, D. Berkvens et al. 2008a. Pollen as an alternative food for *Harmonia axyridis*. *BioControl* 53: 201–210.

Berkvens, N., J. Bonte, D. Berkvens, L. Tirry and P. De Clercq. 2008b. Influence of diet and photoperiod on development and reproduction of European populations of *Harmonia axyridis* (Pallas) (Coleoptera: Coccinellidae). *BioControl* 53: 211–221.

Berkvens, N., J. S. Bale, D. Berkvens, L. Tirry and P. De Clercq. 2010. Cold tolerance of the harlequin ladybird *Harmonia axyridis* in Europe. *J. Insect Physiol.* 56: 438–444.

Bieri, M., J. Baumgartner, G. Bianchi, V. Delucchi and R. von Arx. 1983. Development and fecundity of pea aphid (*Acyrthosiphon pisum* Harris) as affected by constant temperatures and by pea varieties. *Mitt. Schweiz. Ent. Ges.* 56: 163–171.

Bind, R. B. 2007. Reproductive behaviour of a generalist aphidophagous ladybird beetle *Cheilomenes sexmaculata* (Coleoptera: Coccinellidae). *Internat. J. Trop. Insect Sci.* 27: 78–84.

Birch, L. C. 1948. The intrinsic rate of natural increase of an insect population. *J. Anim. Ecol.* 17: 15–26.

Blackman, R. L. 1965. Studies on specifity in Coccinellidae. *Annals Appl. Biol.* 56: 336–338.

Blackman, R. L. 1967. The effects of different aphid foods on Adalia bipunctata L. and Coccinella 7-punctata L. *Annals of Applied Biology* 59: 207–219.

Bonte, M., M. A. Samih and P. De Clercq. 2010. Development and reproduction of *Adalia bipunctata* on factitious and artificial foods. *BioControl* 55: 485–492.

Briere, J. F., P. Pracros, A. Y. LeRoux and J. S. Pierre. 1999. A novel model of temperature dependent development for arthropods. *Env. Entomol.* 27: 94–101.

Brown, G. C., M. J. Sharkey and D. W. Johnson. 2003. Bionomics of *Scymnus* (*Pullus*) *louisianae* J. Chapin (Coleoptera: Coccinellidae) as a predator of the soybean aphid, *Aphis glycines* Matsumura (Homoptera: Aphididae). *J. Econom. Entomol.* 96: 21–24.

Brown, P. M. J., H. E. Roy, P. Rothery et al. 2008. *Harmonia axyridis* in Great Britain: analysis of the spread and distribution of a non-native coccinellid. *BioControl* 53: 55–67.

Buntin, L. A. and G. Tamaki. 1980. Bionomics of *Scymnus marginicollis* (Coleoptera: Coccinellidae). *Can. Entomol.* 112: 675–680.

Cabral, S., A. O. Soares, R. Moura and P. Garcia. 2006. Suitability of *Aphis fabae*, *Myzus persicae* (Homoptera: Aphididae) and *Aleyrodes proletella* (Homoptera: Aleyrodidae) as prey for *Coccinella undecimpunctata* (Coleoptera: Coccinellidae). *Biol. Control* 39: 434–440.

Camarano, S., A. Gonzalez and C. Rossini. 2006. Chemical defense of the ladybird beetle *Epilachna paenulata*. *Chemoecology* 16: 179–184.

Ceryngier, P., J. Havelka and I. Hodek. 2004. Mating and activity of gonads in pre-dormant and dormant ladybirds (Coleoptera: Coccinellidae). *Invert. Reprod. Dev.* 45: 127–135.

Chazeau, J. 1981. La lutte biologique contre la conchinelle transparente du cocotier *Temnaspidotus destructor* (Signoret) aux Nouvelles Hebrides (Homoptera, Diaspididae). *Cahiers ORSTM, Serie Biologique* 44: 11–22.

Cheah, C. A. S. J. and M. S. McClure. 1998. Life history and development of *Pseudoscymnus tsugae* (Coleoptera: Coccinellidae), a new predator of the hemlock wooly adelgid (Homoptera: Adelgidae). *Env. Entomol.* 27: 1531–1536.

Chen, Z. H., E. Y. Chen and F. S. Yan. 1980. Effects of diets on the feeding and reproduction of *Coccinella septempunctata* L. *Acta Entomol. Sinica* 23: 141–148. (In Chinese with English summary.)

Chen, F. J., F. Ge and M. N. Parajulee. 2005. Impact of elevated CO_2 on tri-trophic interaction of *Gossypium hirsutum*, *Aphis gossypii*, and *Leis axyridis*. *Env. Entomol.* 34: 37–46.

Chi, H. and T. C. Yang. 2003. Two sex-life table and predation rate of *Propylea japonica* Thunberg (Coleoptera: Coccinellidae) fed on *Myzus persicae* Sulzer (Homoptera: Aphididae). *Env. Entomol.* 32: 327–333.

Chong, J. H., R. D. Oetting and L. S. Osborne. 2005. Development of *Diomus austrinus* Gordon (Coleoptera: Coccinellidae) on two mealybug prey species at five constant temperatures. *Biol. Control* 33: 39–48.

Cividanes, T. M. S., F. J. Cividanes and B. D. de Matos. 2007. Biology of *Psyllobora confluens* fed with *Erysiphe cichoracearum* fungus. *Pesquisa Agropecuaria Brasileira* 42: 1675–1679.

Cividanes, F. J. and A. P. Gutierrez. 1996. Modelling the age-specific per capita growth and reproduction of *Rhyzobius lophanthae* (Blaisd) (Coleoptera: Coccinellidae). *Entomophaga* 41: 257–266.

Collyer, E. 1964. Phytophagous mites and their predators in New Zealand orchards. *N. Z. J. of Agric. Res.* 7: 551–568.

Creed, E. R. 1975. Melanism in the two spot ladybird: the nature and intensity of selection. *Proc. R. Soc. London B.* 190: 135–148.

Dixon, A. F. G. 2000. *Insect Predator–Prey Dynamics: Ladybird Beetles and Biological Control*. Cambridge University Press, Cambridge, UK. 258 pp.

Dixon, A. F. G. and Y. Guo. 1993. Egg and cluster-size in ladybird beetles (Coleoptera, Coccinellidae) – the direct and indirect effects of aphid abundance. *Eur. J. Entomol.* 90: 457–463.

Dixon, A. F. G., J. L. Hemptinne and P. Kindlmann. 1997. Effectiveness of ladybirds as biological control agents: patterns and processes. *Entomophaga* 42: 71–83.

Dixon, A. F. G., A. Honěk, P. Keil et al. 2009. Relationship between the maximum and minimum thresholds for development in insects. *Funct. Ecol.* 23: 257–264.

Dobzhansky, F. G. 1924. Die weiblichen Generationsorgane der Coccinelliden als Artmerkmal betrachtet (Col.). *Entomol. Mitteilungsbl.* 13: 18–27.

Dobrzhansky, F. G. 1926. Genital apparatus of ladybirds as individual and group character. *Trans. Acad. Sci. USSR* 6, Series 20: 1385–1393; 1555–1586. (In Russian.)

Dorge, S. K., V. P. Dalaya and A. G. Pradhan. 1972. Studies on two predatory coccinellid beetles, *Pharoscymnus horni* Weise and *Chilocorus nigritus* Fab., feeding on sugarcane scales, *Aspidiotus glomerata* G. Labdev. *J. Sci. Technol. (B)* 10: 138–141.

Duan, J. J., G. Head, M. J. McKee et al. 2002. Evaluation of dietary effects of transgenic corn pollen expressing Cry3Bb1 protein on a non-target ladybird beetle, *Coleomegilla maculata. Entomol. Exp. Appl.* 104: 271–280.

Durska, E., P. Ceryngier and R. H. L. Disney. 2003. *Phalacrotophora beuki* (Diptera: Phoridae), a parasitoid of ladybird pupae (Coleoptera: Coccinellidae). *Eur. J. Entomol.* 100: 627–630.

ElHag, E. T. A. and A. A. Zaitoon. 1996. Biological parameters for four coccinellid species in Central Saudi Arabia. *Biol. Control* 7: 316–319.

El-Heneidy, A. H., A. A. Hafez, F. F. Shalaby and I. A. B. El-Din. 2008. Comparative biological aspects of the two coccinellid species; *Coccinella undecimpunctata* L. and *Hippodamia convergens* Guer. under laboratory conditions. *Egypt. J. Biol. Pest Control* 18: 51–59.

Ellingsen, I-J. 1969. Fecundity, aphid consumption and survival of the aphid predator Adalia bipunctata L. (Col., Coccinellidae). *Norsk. Entomol. Tidsskr.* 16: 91–95.

Eliopoulos, P. A., D. C. Kontodimas and G. J. Stathas. 2010. Temperature-dependent development of *Chilocorus bipustulatus* (Coleoptera: Coccinellidae). *Environ. Entomol.* 39: 1352–1358.

Evans, E. W., D. R. Richards and A. Kalaskar. 2004. Using food for different purposes: female responses to prey in the predator *Coccinella septempunctata* L. (Coleoptera: Coccinellidae). *Ecol. Entomol.* 29: 27–34.

Ferrer, A., A. F. G. Dixon and J. L. Hemptinne. 2008. Prey preference of ladybird larvae and its impact on larval mortality, some life-history traits of adults and female fitness. *Bull. Isectology* 61: 5–10.

Fitzgerald, D. V. 1953. Review of the biological control of coccids on coconut palms in the Seychelles. *Bull. Entomol. Res.* 44: 405–413.

Fomenko, R. B. 1970. The effect of aggregations of preimaginal stages of *Chilocorus bipustulatus* L. (Coleoptera, Coccinellidae) on the duration of their development. *Entomol. Obozr.* 49: 264–269. (In Russian.)

Fomenko, R. B. and V. A. Zaslavskii. 1970. Genetics of gregarious beaviour in *Chilocorus bipustulatus* L. (Coleoptera, Coccinellidae). *Dokl. Akad. Nauk SSSR* 192: 229–231. (In Russian.)

Francis, F., E. Haubruge, P. Hastir and C. Gaspar. 2001. Effect of aphid host plant on development and reproduction of the third trophic level, the predator *Adalia bipunctata* (Coleoptera: Coccinellidae). *Environ. Entomol.* 30: 947–952.

Fujiyama, N., C. Torii, M. Akabane and H. Katakura. 2008. Oviposition site selection by herbivorous beetles: a comparison of two thistle feeders, *Cassida rubiginosa* and *Henosepilachna niponica. Entomol. Exp. Appl.* 128: 41–48.

Gage, J. H. 1920. The larvae of the Coccinellidae. *Illinois Biol. Monogr.* 6: 232–294.

Gannon, A. J. and C. E. Bach. 1996. Effects of soybean trichome density on Mexican bean beetle (Coleoptera: Coccinellidae) development and feeding preference. *Env. Entomol.* 25: 1077–1082.

Gao, F., S. R. Zhu, Y. C. Sun et al. 2008. Interactive effects of elevated CO_2 and cotton cultivar on tri-trophic interaction of *Gossypium hirsutum, Aphis gossyppii*, and *Propylaea japonica. Env. Entomol.* 37: 29–37.

Gautam, R. D. 1990. Mass-multiplication technique of coccinellid predator, ladybird beetle (*Brumoides suturalis*). *Indian J. Agric. Sci.* 60: 747–750.

Geyer, J. W. C. 1947. A study of the biology and ecology of *Exochomus flavipes* Thunb. (Coccinellidae, Coleoptera). *J. Entomol. Soc. South Afr.* 9: 219–234; 10: 64–109.

Ghanim, A. E. B., B. Freier and T. Wetzel. 1984. Zur Nahrungsaufnahme und Eiablage von *Coccinella septempunctata* L. bei unterschiedlichem Angebot von Aphiden der Arten *Macrosiphum avenae* (Fabr.) und *Rhopalosiphum padi* (L.). *Arch. Phytopathol. Pflsch.* 20: 117–125.

Gibson, R. L., N. C. Elliott and P. Schaefer. 1992. Life history and development of *Scymnus frontalis* (Fabricius) (Coleoptera: Coccinellidae) on four species of aphid. *J. Kans. Entomol. Soc.* 65: 410–415.

Gilbert, N. and D. A. Raworth. 1996. Insects and temperature-a general theory. *Can. Entomol.* 128: 1–13.

Giles, K. L., R. D. Madden, R. E. Stockland, M. E. Payton and J. W. Dillwith. 2002. Host plants affect predator fitness via the nutritional value of herbivore prey: investigation of a plant–aphid–ladybeetle system. *BioControl* 47: 1–21.

Gillooly, J. F., E. L. Charnov, G. B. West, V. M. Savage and J. H. Brown. 2002. Effects of size and temperature on developmental time. *Nature* 417: 70–73.

Grafton-Cardwell, E. E., P. Gu and G. H. Montez. 2005. Effects of temperature on development of vedalia beetle, *Rodolia cardinalis* (Mulsant). *Biol. Contr.* 32: 473–478.

Greathead, D. J. and R. D. Pope. 1977. Studies on the biology and taxonomy of some *Chilocorus* spp. (Coleoptera: Coccinellidae) preying on *Aulacaspis* spp. (Hemiptera: Diaspididae) in East Africa, with the description of a new species. *Bull. Entomol. Res.* 67: 259–270.

Gupta, A. K., S. Srivastava, G. Mishra, K. Singh and Omkar. 2006. Survival, development and life tables of two congeneric ladybirds in aphidophagous guilds. *Insect Sci.* 13: 119–126.

Haddrill, P. R., D. M. Shuker, S. Mayes and M. E. N. Majerus. 2007. Temporal effects of multiple mating on components of fitness in the two-spot ladybird, *Adalia bipunctata* (Coleoptera: Coccinellidae). *Eur. J. Entomol.* 104: 393–398.

Haddrill, P. R., F. M. Waldron and B. Charles. 2008. Elevated levels of expression associated with regions of the Drosophila genome that lack crossing over. *Biol. Lett.* 4: 758–761.

Hämäläinen, M. and M. Markkula. 1977. Cool storage of *Coccinella septempunctata* and *Adalia bipunctata* (Col.: Coccinellidae) eggs for use in the biological control in greenhouses. *Ann. Agric. Fen.* 16: 132–136.

Hariri, G. E. 1966. Changes in metabolic reserves of three species of aphidophagous Coccinellidae (Coleoptera) during metamorphosis. *Entomol. Exp. Appl.* 9: 349–358.

Harmon, J. P., A. Hayden and D. A. Andow. 2008. Absence makes the heart grow fonder: Isolation enhances the frequency of mating in *Coleomegilla maculata* (Coleoptera: Coccinellidae). *J. Insect. Behav.* 21: 495–504.

Hattingh, V. and M. J. Samways. 1994. Physiological and behavioural characteristics of *Chilocorus* spp. (Coleoptera: Coccinellidae) in the laboratory relative to effectiveness in the field as biocontrol agents. *J. Econ. Entomol.* 87: 31–38.

Hawkes, O. A. M. 1920. Observations on the life history, biology and genetics of the ladybird beetle *Adalia bipunctata* (L.). *Proc. Zool. Soc. London* 78: 475–490.

Hazzard, R. V. and D. N. Ferro. 1991. Feeding responses of adult *Coleomegilla maculata* (Col.: Coccinellidae) to eggs of Colorado potato beetle (Col.: Chrysomelidae) and green peach aphids (Homoptera: Aphididae). *Environ. Entomol.* 20: 644–651.

He, J. L., E. P. Ma, Y. C. Shen, W. L. Chen and X. Q. Sun. 1994. Observations of the biological characteristics of *Harmonia axyridis* (Pallas) (Coleoptera: Coccinellidae). *J. Shanghai Agric. Coll.* 12: 119–124.

Hemptinne, J. L., A. F. G. Dixon and J. Coffin. 1992. Attack strategy of ladybird beetles (Coccinellidae): factors shaping their numerical response. *Oecologia* 90: 238–245.

Hemptinne, J. L., G. Lognay and A. F. G. Dixon. 1998. Mate recognition in the two spot ladybird beetle, *Adalia bipunctata*: role of chemical and behavioural cues. *J. Insect Physiol.* 44: 1163–1171.

Hemptinne, J. L., G. Lognay, C. Gauthier and A. F. G. Dixon. 2000. Role of surface chemical signals in egg cannibalism and intraguild predation in ladybirds (Coleoptera: Coccinellidae). *Chemoecology* 10: 123–128.

Hemptinne, J. L., A. F. G. Dixon and B. Adam. 2001. Do males and females of the two-spot ladybird, *Adalia bipunctata* (L.), differ in when they mature sexually? *J. Insect Behav.* 14: 411–419.

Hirai, Y., H. Kobayashi, T. Koizumi and H. Katakura. 2006. Field-cage experiments on host fidelity in a pair of sympatric phytophagous ladybird beetles. *Entomol. Exp. Appl.* 118: 129–135.

Hodek, I. 1958. Influence of temperature, relative humidity and photoperiodicity on the speed of development of *Coccinella septempunctata* L. *Acta. Soc. Entomol. Cechoslov. Praha* 55: 121–141.

Hodek, I. 1996a. Dormancy. *In* I. Hodek and A. Honěk (eds). *Ecology of Coccinellidae*. Kluwer Academic Publishers, Dordrecht. pp. 239–318.

Hodek I. 1996b. Food relationships. In: I. Hodek and A. Honěk (eds.): *Ecology of Coccinellidae*. Kluver Academic Publishers, Dordrecht. pp. 143–238.

Hodek, I. and P. Ceryngier. 2000. Sexual activity in Coccinellidae (Coleoptera): a review. *Eur. J. Entomol.* 97: 449–456.

Hodek, I. and V. Landa. 1971. Anatomical and histological changes during dormancy in two Coccinellidae. *Entomophaga* 16: 239–251.

Hodek, I. and J. P. Michaud. 2008. Why is *Coccinella septempunctata* so successful? (A point-of-view). *Eur. J. Entomol.* 105: 1–12.

Honěk, A. 1990. Seasonal changes in flight activity of *Coccinella septempunctata* L. (Coleoptera, Coccinellidae). *Acta Entomol. Bohemoslov.* 87: 336–341.

Honěk, A. 1996. Life history and development. *In* I. Hodek and A. Honěk (eds). *Ecology of Coccinellidae*. Kluwer Academic Publishers, Dordrecht. pp. 61–93.

Honěk, A. and F. Kocourek. 1988. Thermal requirements for development of aphidophagous Coccinellidae (Coleoptera), Chrysopidae, Hemerobiidae (Neuroptera), and Syrphidae (Diptera): some general trends. *Oecologia* 76: 455–460.

Honěk, A. and F. Kocourek. 1990. Temperature and development time in insects: a general relationship between thermal constants. *Zoologische Jahrbücher. Abteilung für Systematik und Ökologie der Tiere* 117: 401–439.

Honěk, A., A. F. G. Dixon and Z. Martinková. 2008. Body size, reproductive allocation, and maximum reproductive rate of two species of aphidophagous Coccinellidae exploiting the same resource. *Entomol. Exp. Appl.* 127: 1–9.

Honěk, A., Z. Martinková and S. Pekar. 2005. Temporal stability of morph frequency in central European populations of *Adalia bipunctata* and *A. decempunctata* (Coleoptera: Coccinellidae). *Eur. J. Entomol.* 102: 437–442.

Hong, O. K. and Y. C. Park. 1996. Laboratory rearing of the aphidophagous ladybeetle, *Harmonia axyridis*; yolk protein production and fecundity of the summer adult female. *Korean J. Appl. Entomol.* 35: 146–152.

Huang, Z., S. Ren and P. D. Musa. 2008. Effects of temperature on development, survival, longevity, and fecundity of the *Bemisia tabaci* Gennadius (Homoptera: Aleyrodidae) predator, *Axinoscymnus cardilobus* (Coleoptera: Coccinellidae). *Biol. Contr.* 46: 209–215.

Hukushima, S. and S. Kouyama. 1974. Life histories and food habits of *Menochilus sexmaculatus* Fabricius (Coleoptera: Coccinellidae). *Res. Bull. Fac. Agric. Gifu Univ.* 36: 19–29.

Hukushima, S. and M. Kamei. 1970. Effects of various species of aphids as food on development, fecundity and longevity of *Harmonia axyridis* Pallas (Coleoptera: Coccinellidae). *Res. Bull. Fac. Agric. Gifu Univ.* 29: 53–66.

Hurst, G. D. D., F. K. McMeechan and M. E. N. Majerus. 1998. Phoridae (Diptera) parasitizing *Coccinella septempunctata* (Coleoptera: Coccinellidae) select older prepupal hosts. *Eur. J. Entomol.* 95: 179–181.

Iablokoff-Khnzorian, S. M. 1982. *Les Coccinelles*. Société Nouvelle des Éditions Boubée, Paris. 568 pp.

Ibrahim, M. M. 1955. Studies on *Coccinella undecimpunctata aegyptiaca* Reiche. 2. Biology and life history. *Bull. Soc. Entomol. Egypte* 39: 395–423.

Isikber, A. A. and M. J. W. Copland. 2001. Food consumption and utilisation by larvae of two coccinellid predators, *Scymnus levaillanti* and *Cycloneda sanguinea*, on cotton aphid, *Aphis gossypii*. *BioControl* 46: 455–467.

Iwata, K. 1932. On the biology of two large lady-birds in Japan. *Trans. Kansai Ent. Soc.* 3: 13–26.

Izhevsky, S. S. and A. D. Orlinsky. 1988. Life history of the imported *Scymnus (Nephus) reunioni* (Col.: Coccinellidae) predator of mealybugs. *Entomophaga* 33: 101–114.

Jalali, M. A., L. Tirry, A. Arbab and P. De Clercq. 2010. Temperature-dependent development of the two-spotted ladybeetle, *Adalia bipunctata*, on the green peach aphid, *Myzus persicae*, and a factitious food under constant temperatures. *J. Insect Sci.* 10: article 124.

Jalali, M. A., L. Tirry and P. De Clercq. 2009a. Effects of food and temperature on development, fecundity and life-table parameters of *Adalia bipunctata* (Coleoptera: Coccinellidae). *J. Appl. Entomol.* 133: 615–625.

Jalali, M. A., L. Tirry and P. De Clercq. 2009b. Food consumption and immature growth of *Adalia bipunctata* (Coleoptera: Coccinellidae) on a natural prey and a factitious food. *Eur. J. Entomol.* 106: 193–198.

Jalali, S. K. and S. P. Singh. 1989. Biotic potential of three coccinellid predators on various diaspine hosts. *J. Biol. Control* 3: 20–23.

Jarosik, V., A. Honěk and A. F. G. Dixon. 2002. Developmental rate isomorphy in insects and mites. *Am. Nat.* 160: 497–510.

de Jong, P. W., S. W. S. Gessekloo and P. M. Brakefield. 1996. Differences in thermal balance, body temperature and activity between non-melanic and melanic two-spot ladybird beetles (*Adalia bipunctata*) under controlled conditions. *J. Exp. Biol.* 199: 2655–2666.

de Jong, P. W., P. M. Brakefield and B. P. Geerinck. 1998. The effect of female mating history on sperm precedence in the two-spot ladybird, Adalia bipunctata (Coleoptera, Coccinellidae). *Behavioral Ecology* 9: 559–565.

de Jong, P. W., M. D. Verhoog and P. M. Brakefield. 1993. Sperm competition and melanic polymorphism in the 2-spot ladybird, *Adalia bipunctata* (Coleoptera, Coccinellidae). *Heredity* 70, 172–178.

Kajita, Y., E. W. Evans and H. Yasuda. 2009. Reproductive responses of invasive and native predatory lady beetles (Coleoptera: Coccinellidae) to varying prey availability. *Environ. Entomol.* 38: 1283–1292.

Kajita, Y. and E. W. Evans. 2010. Alfalfa fields promote high reproductive rate of an invasive predatory lady beetle. *Biol. Invasions* 12: 2293–2302.

Kalushkov, P. 1998. Ten aphid species (Sternorrhyncha: Aphididae) as prey for *Adalia bipunctata* (Coleoptera: Coccinellidae). *Eur. J. Entomol.* 95: 343–349.

Kalushkov, P. and I. Hodek. 2004. The effects of thirteen species of aphids on some life history parameters of the ladybird *Coccinella septempunctata*. *BioControl* 49: 21–32.

Kalushkov, P. and I. Hodek. 2005. The effects of six species of aphids on some life history parameters of the ladybird *Propylea quatuordecimpunctata* (Coleoptera: Coccinellidae). *Eur. J. Entomol.* 102: 449–452.

Kalushkov, P. and O. Nedvěd. 2005. Genetically modified potatoes expressing Cry 3A protein do not affect aphidophagous coccinellids. *J. Appl. Entomol.* 129: 401–406.

Kalushkov, P. K. 1994. Longevity, fecundity and development of *Adalia bipunctata* (L.) (Col., Coccinellidae) when reared on three aphid diets. *Anz. Schad. Pflsch. Umweltschutz* 67: 6–7.

Kaneko, S., A. Ozawa, T. Saito, A. Tatara, H. Katayama and M. Doi. 2006. Relationship between the seasonal prevalence of the predacious coccinellid *Pseudoscymnus hareja* (Coleoptera: Coccinellidae) and the mulberry scale *Pseudaulacaspis pentagona* (Hemiptera: Diaspididae) in tea fields: Monitoring using sticky traps. *Appl. Entomol. Zool.* 41: 621–626.

Kanervo, V. 1946. Studien über die natürlichen Feinde des Erlenblattkäfers, *Melasoma aenea* L. (Col., Crysomelidae). *Annals Zool. Soc. Vanamo* 12: 206 pp. (In Finnish with German summary.)

Katakura, H. 1976. On the life cycle of *Epilachna admirabilis* (Coleoptera, Coccinellidae) in Sapporo, northern Japan, with special reference to its hibernation by adult stage. *Kontyu* 44: 334–336.

Katsarou, I., J. T. Margaritopoulos, J. A. Tsitsipis, D. C. Perdikis and K. D. Zarpas. 2005. Effect of temperature on development, growth and feeding of *Coccinella septempunctata* and *Hippodamia convergens* reared on the tobacco aphid, *Myzus persicae nicotianae*. *BioControl* 50: 565–580.

Katsoyannos, P., D. C. Kontodimas and G. J. Stathas. 1997. Phenology of *Hippodamia undecimnotata* (Col.: Coccinellidae) in Greece. *Entomophaga* 42: BP 283–293.

Kawauchi, S. 1981. The number of oviposition, hatchability and the term of oviposition of *Propylea japonica* Thunberg (Coleoptera, Coccinellidae) under different food condition. *Kontyu* 49: 183–191.

Kawauchi, S. 1985. Comparative studies on the fecundity of three aphidophagous coccinellids (Coleoptera: Coccinellidae). *Japan. J. Appl. Entomol. Zool.* 29: 203–209.

Kesten, U. 1969. Zur Morphologie und Biologie von *Anatis ocellata* (L) (Coleoptera: Coccinellidae). *Z. Angew. Entomol.* 63: 412–445.

Kim, G. H. and S. Y. Choi. 1985. Effects of the composition of artificial diets on the growth and ovarian development of an aphidivorous coccinellid beetle (*Harmonia axyridis* Pallas: Coccinellidae, Coleoptera). *Korean J. Entomol.* 15: 33–41.

Kingsolver, J. G. and R. B. Huey. 2008. Size, temperature, and fitness: three rules. *Evol. Ecol. Res.* 10: 251–268.

Kishimoto, H. 2003. Development and oviposition of predacious insects, *Stethorus japonicus* (Coleoptera: Coccinellidae), *Oligota kashmirica benefica* (Coleoptera: Staphylinidae), and *Scolothrips takahashii* (Thysanoptera: Thripidae) reared on different spider mite species (Acari: Tetranychidae). *Appl. Entomol. Zool.* 38: 15–21.

Klausnitzer, B. 1969a. Zur Kenntnis der Larve von *Lithophilus connatus* (PANZER) (Col., Coccinellidae). *Entomol. Nachrichten* 13: 33–36.

Klausnitzer, B. 1969b. Zur Unterscheidung der Eier mitteleuropäischer Coccinellidae. *Acta Entomol. Bohemoslov.* 66: 146–149.

Klausnitzer, B. and H. Klausnitzer. 1986. *Marienkäfer (Coccinellidae)*. A. Ziemsen Verlag, Wittenberg, Lutherstadt. 104 pp.

Koch, R. L., M. A. Carrillo, R. C. Venette, C. A. Cannon and W. D. Hutchison. 2004. Cold hardiness of the multicolored Asian lady beetle (Coleoptera: Coccinellidae). *Environ. Entomol.* 33: 815–822.

Kontodimas, D. C. and G. J. Stathas. 2005. Phenology, fecundity and life table parameters of the predator *Hippodamia variegata* reared on *Dysaphis crataegi*. *BioControl* 50: 223–233.

Kontodimas, D. C., P. A. Eliopoulos, G. J. Stathas and L. P. Economou. 2004. Comparative temperature-dependent development of *Nephus includens* (Kirsch) and *Nephus bisignatus* (Boehman) (Coleoptera: Coccinellidae) preying on *Planococcus citri* (Risso) (Homoptera: Psudococcidae): evaluation of a linear and various nonlinear models using specific criteria. *Env. Entomol.* 33: 1–11.

Kontodimas, D. C., P. G. Milonas, G. J. Stathas, L. P. Economou and N. G. Kavallieratos. 2007. Life table parameters of the pseudococcid predators *Nephus includens* and *Nephus bisignatus* (Coleoptera: Coccinellidae). *Eur. J. Entomol.* 104: 407–415.

Kreiter, S. 1985. *Étude bioécologique d'Olla v-nigrum (Mulsant) et essai de quantification de l'éficacité predatrice d'Adalia bipunctata (L.) contre les aphides en verger de pêchers (Coleoptera, Coccinellidae)*. Thése: Université de Droit, d'Économie et des Sciences d'Aix-Marseille. 326 pp.

Kreiter, S. and G. Iperti. 1984. Étude des potentialités biologiques et écologiques d'un prédateur aphidiphage *Olla*

V-nigrum Muls. (Coleoptera, Coccinellidae) en vue de son introduction en France. *109 Congr. Nat. Soc. Savantes Dijon* 2: 275–282.

Kutuk, H. and A. Yigit. 2007. Life table of *Delphastus catalinae* (Horn) (Coleoptera: Coccinellidae) on cotton whitefly, *Bemisia tabaci* (Genn.) (Homoptera: Aleyrodidae) as prey. *J. Plant Dis. Prot.* 114: 20–25.

Kuznetsov, V. N. 1975. Fauna and ecology of Coccinellidae of Ussuri. *Trudy Inst. Biopedol.* 28: 3–24. (In Russian.)

Labrie, G., E. Lucas and D. Coderre. 2006. Can developmental and behavioral characteristics of the multicolored Asian lady beetle *Harmonia axyridis* explain its invasive success? *Biol. Invasions* 8: 743–754.

Lactin, D. J., N. J. Holliday, D. L. Johnson and R. Craigen. 1995. Improved rate model of temperature-dependent development by arthropods. *Env. Entomol.* 24: 68–75.

LaMana, M. L. and J. C. Miller. 1995. Temperature-dependent development in a polymorphic lady beetle, *Calvia quatuordecimguttata* (Coleoptera: Coccinellidae). *Ann. Entomol. Soc. Am.* 88: 785–790.

LaMana, M. L. and J. C. Miller. 1998. Temperature-dependent development in an Oregon population of *Harmonia axyridis* (Coleoptera: Coccinellidae). *Env. Entomol.* 27: 1001–1005.

Lanzoni, A., G. Accinelli, G. G. Bazzocchi and G. Burgio. 2004. Biological traits and life table of the exotic *Harmonia axyridis* compared with *Hippodamia variegata* , and *Adalia bipunctata* (Col.: Coccinellidae). *J. Appl. Entomol.* 128: 298–306.

Li, Y. H. and J. Romeis. 2010. Bt maize expressing Cry3Bb1 does not harm the spider mite, *Tetranychus urticae*, or its ladybird beetle predator, *Stethorus punctillum*. *Biol. Control* 53: 337–344.

Liere, H. and I. Perfect. 2008. Cheating on a mutualism: Indirect benefits of ant attendance to a coccidophagous coccinellid. *Environ. Entomol.* 37: 143–149.

Liu, C. Y. 1933. Notes on the biology of two giant coccinellids in Kwangsi (*Caria dilatata* Fabr. and *Synonycha grandis* Thunb.) with special reference to the morphology of *Caria dilatata*. *Year Book Bur. Entomol., Hangchow* 1: 205–250.

Liu, T. X. and P. A. Stansly. 1996a. Morphology of *Nephaspis oculatus* and *Delphastus pusillus* (Coleoptera: Coccinellidae), predators of *Bemisia argentifolii* (Homoptera: Aleyrodidae). *Proc. Entomol. Soc. Wash.* 98: 292–300.

Liu, T. X., P. A. Stansly, K. A. Hoelmer and L. S. Osborne. 1997. Life history of *Nephaspis oculatus* (Coleoptera: Coccinellidae), a predator of *Bemisia argentifolii* (Homoptera: Aleyrodidae). *Ann. Entomo. Soc. Am.* 90: 776–782.

Lombaert, E., T. Malausa, R. Devred and A. Estoup. 2008. Phenotypic variation in invasive and biocontrol populations of the harlequin ladybird, *Harmonia axyridis*. *BioControl* 53: 89–102.

Lommen, S. T. E., P. W. De Jong and P. M. Brakefield. 2005. Phenotypic plasticity of elytron length in wingless two-spot ladybird beetles, *Adalia bipunctata* (Coleoptera: Coceinellidae). *Eur. J. Entomol.* 102: 553–556.

Lommen, S. T. E., C. W. Middendorp, C. A. Luijten et al. 2008. Natural flightless morphs of the ladybird beetle *Adalia bipunctata* improve biological control of aphids on single plants. *Biol. Control* 47: 340–346.

Lu, W. H. and M. E. Montgomery. 2001. Oviposition, development, and feeding of *Scymnus (Neopullus) sinuanodulus* (Coleoptera: Coccinellidae): a predator of *Adelges tsugae* (Homoptera: Adelgidae). *Ann. Entomol. Soc. Am.* 94: 64–70.

Lu, W. H., P. Souphanya and M. E. Montgomery. 2002. Descriptions of immature stages of *Scymnus (Neopullus) sinuanodulus* Yu and Yao (Coleoptera: Coccinellidae) with notes on life history. *Col. Bull.* 56: 127–141.

Lucas, E., D. Coderre and J. Brodeur. 2000. Selection of molting and pupation sites by *Coleomegilla maculata* (Coleoptera: Coccinellidae): avoidance of intraguild predation. *Environ. Entomol.* 29: 454–459.

Lusis, Ya. Ya. 1961. On the biological meaning of colour polymorphism of lady-beetle *Adalia bipunctata* L. *Latv. Entomol.* 4: 2–29. (In Russian with English summary.)

Lundgren, J. G. and R. N. Wiedenmann. 2002. Coleopteran-specific Cry 3Bb toxin from transgenic corn pollen does not affect the fitness of a nontarget species, *Coleomegilla maculata* DeGeer (Coleoptera: Coccinellidae). *Environ. Entomol.* 31: 1213–1218.

Lundgren, J. G., L. S. Hesler, K. Tilmon, K. Dashiell and R. Scott. 2009. Direct effects of soybean varietal selection and *Aphis glycines*-resistant soybeans on natural enemies. *Arthropod-Plant Interactions* 3: 9–16.

Majerus, M. E. N. 1994. *Ladybirds*. Harper Collins Publishers, London. 367 pp.

Majerus, M. E. N. 1997. Interspecific hybridisation in ladybirds (Col.: Coccinellidae). *Entomologist's Record* 109: 11–23.

Majerus, M. E. N. 1998. *Melanism. Evolution in action*. Oxford University Press, Oxford, 338 pp.

Majerus, M. E. N., P. O'Donald and J. Weir. 1982a. Evidence for preferential mating in *Adalia bipunctata*. *Heredity* 49: 37–49.

Majerus, M. E. N., P. O'Donald and J. Weir. 1982b. Female mating preference is genetic. *Nature (London)* 300: 521–523.

Majerus, M. E. N. and I. A. Zakharov. 2000. Does thermal melanism maintain melanic polymorphism in the two-spot ladybird *Adalia bipunctata* (Coleoptera: Coccinellidae)? *Zhurnal Obshchei Biologii* 61: 381–392.

Matsubayashi, K. W. and H. Katakura. 2007. Unilateral mate choice causes bilateral behavioral isolation between two closely related phytophagous ladybird beetles (Coleoptera: Coccinellidae: Epilachninae). *Ethol.* 113: 686–691.

McClure, M. S. 1987. Potential of the Asian predator, *Harmonia axyridis* Pallas (Coleoptera: Coccinellidae), to control *Matsucoccus resinosae* Bean and Godwin (Homoptera: Margarodidae) in the United States. *Environ. Entomol.* 16: 224–230.

McKenzie, H. L. 1932. The biology and feeding habits of *Hyperaspis lateralis* Mulsant. *Calif. Univ. Pubs. Ent.* 6: 9–20.

McMullen, R. D. 1967. The effects of photoperiod, temperature and food supply on rate of development and diapause in *Coccinella novemnotata*. *Can. Entomol.* 99: 578–586.

M'Hamed, T. B. and M. Chemseddine. 2001. Assessment of temperature effects on the development and fecundity of *Pullus mediterraneus* (Col. , Coccinellidae) and consumption of *Saissetia oleae* eggs (Hom. , Coccoida). *J. Appl. Entomol.* 125: 527–531.

Mishra, G. and Omkar. 2004. Diel rhythmicity of certain life events of a ladybird, *Propylea dissecta* (Mulsant). *Biol. Rhythm Res.* 35: 269–276.

Mishra, G. and Omkar. 2006. Conspecific interference by adults in an aphidophagous ladybird *Propylea dissecta* (Coleoptera: Coccinellidae): effect on reproduction. *Bull. Entomol. Res.* 96: 407–412.

Mori, K., M. Nozawa, K. Arai and T. Gotoh. 2005. Life-history traits of the acarophagous lady beetle, *Stethorus japonicus* at three constant temperatures. *BioControl* 50: 35–51.

Mota, J. A., A. O. Soares and P. V. Garcia. 2008. Temperature dependence for development of the whitefly predator *Clitostethus arcuatus* (Rossi). *BioControl* 53: 603–613.

Muggleton, J. 1979. Non-random mating in wild populations of polymorphic *Adalia bipunctata*. *Heredity* 42: 57–65.

Muralidharan, C. M. 1994. Biology and feeding potential of black beetle (*Chilocorus nigritus*), a predator on date palm scale (*Parlatoria blanchardii*). *Indian J. Agric. Sci.* 64: 270–271.

Nakanishi, K., M. Hoshino, M. Nakai and Y. Kunimi. 2008. Novel RNA sequences associated with late male killing in *Homona magnanima*. *Proc. Royal Soc. B-Biol. Sci.* 275: 1249–1254.

Nakano, S. and H. Katakura. 1999. Morphology and biology of a phytophagous ladybird beetle, *Epilachna pusillanima* (Coleoptera: Coccinellidae) newly recorded on Ishigaki Island, the Ryukyus. *Appl. Entomol. Zool.* 34: 189–194.

Nalepa, C. A., K. A. Kidd and K. R. Ahlstrom. 1996. Biology of *Harmonia axyridis* (Coleoptera: Coccinellidae) in winter aggregations. *Ann. Entomol. Soc. Am.* 89: 681–685.

Nalepa, C. A. and A. Weir. 2007. Infection of *Harmonia axyridis* (Coleoptera: Coccinellidae) by *Hesperomyces virescens* (Ascomycetes: Laboulbeniales): role of mating status and aggregation behavior. *J. Invertebrate Pathol.* 94: 196–203.

Naranjo, S. E., R. L. Gibson and D. D. Walgenbach. 1990. Development, survival, and reproduction of *Scymnus frontalis* (Coleoptera: Coccinellidae), an imported predator of Russian wheat aphid, at four fluctuating temperatures. *Ann. Entomol. Soc. Am.* 83: 527–531.

Nedvěd, O. 1994. *Cold Hardiness of Ladybirds* (Chladová odolnost slunéček.) PhD thesis, Institute of Entomology, České Budějovice. 116 pp. (In Czech.)

Nedvěd, O. and I. Hodek. 1995. Confinement stimulates trivial flight in *Coccinella septempunctata* (Coleoptera: Coccinellidae). *Eur. J. Entomol.* 92: 719–722.

Ng, S. M. 1986. Egg mortality of four species of aphidophagous Coccinellidae in Malaysia. *In* I. Hodek (ed). *Ecology of Aphidophaga*. Academia, Prague. pp 77–81.

Ng, S. M. 1988. Observations on the foraging behaviour of starved aphidophagous coccinellid larvae (Coleoptera: Coccinellidae). *In* E. Niemczyk and A. F. G. Dixon (eds). *Ecology and Effectiveness of Aphidophaga*. SPB Acad. Publ., The Hague. pp. 199–206.

Ng, S. M. 1991. Voracity, development and growth of larvae of *Menochilus sexmaculatus* (Coleoptera: Coccinellidae) fed on *Aphis spiraecola*. *In* L. Polgár, R. J. Chambers, A. F. G. Dixon and I. Hodek (eds). *Behaviour and Impact of Aphidophaga*. SPB Acad. Publ., The Hague. pp. 199–206.

Obata, S. 1988. Mating behaviour and sperm transfer in the ladybird beetle, *Harmonia axyridis* Pallas (Coleoptera: Coccinellidae). *In* E. Niemczyk and A. F. G. Dixon (eds). *Ecology and Effectiveness of Aphidophaga*. SPB Acad. Publ., The Hague. pp. 39–42.

Obata, S. and T. Hidaka. 1987. Ejection and ingestion of the spermatophore by the female ladybird beetle, *Harmonia axyridis* Pallas (Coleoptera: Coccinellidae). *Can. Entomol.* 119: 603–604.

Obrycki, J. J. and M. J. Tauber. 1981. Phenology of three coccinellid species: thermal requirements for development. *Ann. Entomol. Soc. Am.* 74: 31–36.

Obrycki, J. J., K. L. Giles and A. M. Ormord. 1998. Interactions between an introduced and indigenous coccinellid species at different prey densities. *Oecologia* 117: 279–285.

O'Donald, P. and M. E. N. Majerus. 1989. Sexual selection models and the evolution of melanism in ladybirds. *In* F. M. W. Feldman (ed). *Mathematical Evolutionary Theory*. Princeton University Press, Princeton, NJ. pp. 247–269.

Ohashi, K., Y. Sakuratani, N. Osawa, S. Yano and A. Takafuji. 2005. Thermal microhabitat use by the ladybird beetle, *Coccinella septempunctata* (Coleoptera: Coccinellidae), and its life cycle consequences. *Environ. Entomol.* 34: 432–439.

Okrouhla, M., S. Chakrabarti and I. Hodek. 1983. Developmental rate and feeding capacity in *Cheilomenes sulphurea* (Coleoptera: Coccinellidae). *Věst. Čs. Společ. Zool.* 47: 105–117.

Okuda, T., M. Hodková and I. Hodek. 1986. Flight tendency in *Coccinella septempunctata* in relation to changes in flight muscles, ovaries and corpus allatum. *In* I. Hodek (ed.). *Ecology of Aphidophaga*. Academia, Prague, W. Junk, Dordrecht. pp. 217–223.

Okuda, T., T. Gomi and I. Hodek. 1997. Effect of temperature on pupal pigmentation and size of the elytral spots in *Coc-*

cinella septempunctata (Coleoptera: Coccinellidae) from four latitudes in Japan. *Appl. Entomol. Zool.* 32: 567–572.

Oliver, T. H., I. Jones, J. M. Cook and S. R. Leather. 2008. Avoidance responses of an aphidophagous ladybird, *Adalia bipunctata*, to aphid-tending ants. *Ecol. Entomol.* 33: 523–528.

Olszak, R. W. 1987. The occurrence of *Adalia bipunctata* (L.) (Coleoptera, Coccinellidae) in apple orchards and the effect of different factors on its development. *Ekol. Polska* 35: 755–765.

Omkar and G. Mishra. 2005a. Evolutionary significance of promiscuity in an aphidophagous ladybird, *Propylea dissecta* (Coleoptera: Coccinellidae). *Bull. Entomol. Res.* 95: 527–533.

Omkar and G. Mishra. 2005b. Mating in aphidophagous ladybirds: costs and benefits. *J. Appl. Entomol.* 129: 432–436.

Omkar and S. Pathak. 2006. Effects of different photoperiods and wavelengths of light on the life-history traits of an aphidophagous ladybird, *Coelophora saucia* (Mulsant). *J. Appl. Entomol.* 130: 45–50.

Omkar and S. Pathak. 2009. Crowding affects the life attributes of an aphidophagous ladybird beetle, *Propylea dissecta*. *Bull. Insectology* 62: 35–40.

Omkar and A. Pervez. 2002. Influence of temperature on age-specific fecundity of the ladybeetle *Micraspis discolor* (Fabricius). *Insect Sci. Appl.* 22: 61–65.

Omkar and A. Pervez. 2003. Ecology and biocontrol potential of a scale-predator, *Chilocorus nigritus*. *Biocontrol Sci. Technol.* 13: 379–390.

Omkar and A. Pervez. 2004a. Comparative demographics of a generalist predatory ladybird on five aphid prey: a laboratory study. *Insect Sci.* 11: 211–218.

Omkar and A. Pervez. 2004b. Temperature-dependent development and immature survival of an aphidophagous ladybeetle, *Propylea dissecta* (Mulsant). *J. Appl. Entomol.* 128: 510–514.

Omkar and A. Pervez. 2005. Mating behavior of an aphidophagous ladybird beetle, *Propylea dissecta* (Mulsant). *Insect Sci.* 12: 37–44.

Omkar and S. K. Singh. 2007. Rhythmicity in life events of an aphidophagous ladybird beetle, *Cheilomenes sexmaculata*. *J. Appl. Entomol.* 131: 85–89.

Omkar and S. Srivastava. 2003. Influence of six aphid prey species on development and reproduction of a ladybird beetle, *Coccinella septempunctata*. *BioControl* 48: 379–393.

Omkar, G. Kumar and J. Sahu. 2009. Performance of a predatory ladybird beetle, *Anegleis cardoni* (Coleoptera: Coccinellidae) on three aphid species. *Eur. J. Entomol.* 106: 565–572.

Omkar, G. Mishra, S. Srivastava and A. K. Gupta. 2004. Ovipositional rhythmicity in ladybirds (Coleoptera: Coccinellidae): a laboratory study. *Biol. Rhythm Res.* 35: 277–287.

Omkar, G. Mishra, S. Srivastava, A. K. Gupta and S. K. Singh. 2005a. Reproductive performance of four aphidophagous ladybirds on cowpea aphid, *Aphis craccivora* Koch. *J. Appl. Entomol.* 129: 217–220.

Omkar, A. Pervez, G. Mishra, S. Srivastava, S. K. Singh and A. K. Gupta. 2005b. Intrinsic advantages of *Cheilomenes sexmaculata* over two coexisting *Coccinella* species (Coleoptera: Coccinellidae). *Insect Sci.* 12: 179–184.

Omkar, S. K. Singh and G. Mishra. 2010a. Multiple matings affect the reproductive performance of the aphidophagous ladybird beetle, *Coelophora saucia* (Coleoptera: Coccinellidae). *Eur. J. Entomol.* 107: 177–182.

Omkar, S. K. Singh and G. Mishra. 2010b. Parental age at mating affects reproductive attributes of the aphidophagous ladybird beetle, *Coelophora saucia* (Coleoptera: Coccinellidae). *Eur. J. Entomol.* 107: 341–347.

Omkar, K. Singh and A. Pervez. 2006a. Influence of mating duration on fecundity and fertility in two aphidophagous ladybirds. *J. Appl. Entomol.* 130: 103–107.

Omkar, S. K. Singh and K. Singh. 2006b. Effect of age on reproductive attributes of an aphidophagous ladybird, *Cheilomenes sexmaculata. Insect Sci.* 13: 301–308.

Orivel, J., P. Servigne, P. Cerdan, A. Dejean and B. Corbara. 2004. The ladybird *Thalassa saginata*, an obligatory myrmecophile of *Dolichoderus bidens* ant colonies. *Naturwissenschaften* 91: 97–100.

Omkar and S. Srivastava. 2002. The reproductive behaviour of an aphidophagous ladybeetle, *Coccinella septempunctata* Linnaeus. *Eur. J. Entomol.* 99: 465–470.

Osawa, N. 1994. The occurrence of multiple mating in a wild population of a ladybird beetle *Harmonia axyridis* Pallas (Coleoptera: Coccinellidae). *J. Ethol.* 12: 63–66.

Osawa, N. 2005. The effect of prey availability on ovarian development and oosorption in the ladybird beetle *Harmonia axyridis* (Coleoptera: Coccinellidae). *Eur. J. Entomol.* 102: 503–511.

Osawa, N. and T. Nishida. 1992. Seasonal variation in elytral colour polymorphism in *Harmonia axyridis* (the ladybird beetle): the role of non-random mating. *Heredity* 69: 297–307.

Osawa, N. and K. Ohashi. 2008. Sympatric coexistence of sibling species *Harmonia yedoensis* and *H. axyridis* (Coleoptera: Coccinellidae) and the roles of maternal investment through egg and sibling cannibalism. *Eur. J. Entomol.* 105: 445–454.

Pantyukhov, G. A. 1968. A study of ecology and physiology of the predatory beetle *Chilocorus rubidus* Hope (Coleoptera, Coccinellidae). *Zool. Zh.* 47: 376–386. (In Russian with English summary.)

Perry, J. C. and B. D. Roitberg. 2005. Ladybird mothers mitigate offspring starvation risk by laying trophic eggs. *Behav. Ecol. Sociobiol.* 58: 578–586.

Perry, J. C., D. M. T. Sharpe and L. Rowe. 2009. Condition-dependent female remating resistance generates sexual selection on male size in a ladybird beetle. *Anim. Behav.* 77: 743–748.

Perumalsamy, K., R. Selvasundaram, A. Roobakkumar, V. J. Rahman and N. Muraleedharan. 2010. Life table and predatory efficiency of *Stethorus gilvifrons* (Coleoptera: Coccinellidae), an important predator of the red spider mite, *Oligonychus coffeae* (Acari: Tetranychidae), infesting tea. *Exp. Appl. Acarol.* 50: 141–150.

Pervez, A., Omkar and A. S. Richmond. 2004. The influence of age on reproductive performance of the predatory ladybird beetle, *Propylea dissecta. J. Insect Sci.* 4: 22.

Phoofolo, M. W., J. J. Obrycki and E. S. Krafsur. 1995. Temperature-dependent ovarian development in *Coccinella septempunctata* (Coleoptera, Coccinellidae). *Ann. Entomol. Soc. Am.* 88: 72–79.

Phoofolo, M. W., K. L. Giles and N. C. Elliott. 2007. Quantitative evaluation of suitability of the greenbug, *Schizaphis graminum*, and the bird cherry-oat aphid, *Rhopalosiphum padi*, as prey for *Hippodamia convergens* (Coleoptera: Coccinellidae). *Biol. Control* 41: 25–32.

Pilorget, L., J. Buckner and J. G. Lundgren. 2010. Sterol limitation in a pollen-fed omnivorous lady beetle (Coleoptera: Coccinellidae). *J. Insect Physiol.* 56: 81–87.

Ponsonby, D. J. 2009. Factors affecting utility of *Chilocorus nigritus* (F.) (Coleoptera: Coccinellidae) as a biocontrol agent. *CAB Reviews: Perspectives in Agriculture, Veterinary Science, Nutrition and Natural Resources* 4: 1–20.

Ponsonby, D. J. and M. J. W. Copland. 1996. Effect of temperature on development and immature survival in the scale insect predator, *Chilocorus nigritus* (F.) (Coleoptera: Coccinellidae). *Biocontrol Sci. Technol.* 6: 101–109.

Ponsonby, D. J. and M. J. W. Copland. 1998. Environmental influences on fecundity, egg viability and egg cannibalism in the scale insect predator, *Chilocorus nigritus. BioControl* 43: 39–52.

Ponsonby, D. J. and M. J. W. Copland. 2007. Influence of host density and population structure on egg production in the coccidophagous ladybird, *Chilocorus nigritus* F. (Coleoptera: Coccinellidae). *Agric. Forest Entomol.* 9: 287–296.

Poprawski, T. J., J. C. Legaspi and P. E. Parker. 1998. Influence of entomopathogenic fungi on *Serangium parcesetosum* (Coleoptera: Coccinellidae), an important predator of whiteflies (Homoptera: Aleyrodidae). *Environ. Entomol.* 27: 785–795.

Prevolsek, J. and T. D. Williams. 2006. Individual variation and plasticity in egg size in multicolored Asian lady beetles (*Harmonia axyridis*). *Integr. Comp. Biol.* 46 Suppl. 1: E238–E238 (abstract).

Putman, W. L. 1955. Bionomics of *Stethorus punctillum* Weise in Ontario. *Can. Entomol.* 87: 9–33.

Radwan, Z. and G. L. Lovei. 1983. Aphids as prey for the coccinellid *Exochomus quadripustulatus. Entomol. Exp. Appl.* 34: 283–286.

Rana, J. S. and J. Kakker. 2000. Biological studies on 7-spot ladybird beetle, *Coccinella septempunctata* L. with cereal aphid, *Sitobion avenae* (F.) as prey. *Cereal Res. Comm.* 28: 449–454.

Rathour, Y. S. and T. Singh. 1991. Importance of ovariole number in Coccinellidae (Coleoptera). *Entomon* 16: 35–41.

Raworth, D. A. 2001. Development, larval voracity, and greenhouse releases of *Stethorus punctillum* (Coleoptera: Coccinellidae). *Can. Entomol.* 133: 721–724.

Ren, S. X., P. A. Stansly and T. X. Liu. 2002. Life history of the whitefly predator *Nephaspis oculatus* (Coleoptera: Coccinellidae) at six constant temperatures. *Biol. Contr.* 23: 262–268.

Rhamhalinghan, M. 1985. Intraspecific variations in ovariole number / ovary in *Coccinella septempunctata* L. (Coleoptera: Coccinellidae). *Ind. Zool.* 9: 91–97.

Rhamhalinghan, M. 1986. Seasonal variations in ovariole number/ovary in *Coccinella septempunctata* L. (Coleoptera: Coccinellidae). *Proc. Indian Natn. Sci. Acad.* 52: 619–623.

Rhamhalinghan, M. 1999. Variations in the realized fecundity of melanics of *Coccinella septempunctata* L. in relation to degree of elytral pigmentation and radiant heat level. *J.of Ins. Sci.* 12: 22–26.

Ricci, C. and I. Stella. 1988. Relationship between morphology and function in some Palearctic Coccinellidae. *In* E. Niemczyk and A. F. G. Dixon (eds). *Ecology and Effectiveness of Aphidophaga: Proceedings of an International Symposium, Teresin, Poland, 31 August–5 September, 1987.* SPB Academic Publishing, The Hague, The Netherlands. pp. 21–25.

Richards, A. M. and L. W. Filewood. 1988. The effect of agricultural crops and weeds on the bionomics of the pest species comprising the *Epilachna vigintioctopunctata* complex (Col., Coccinellidae). *J. Appl.Entomol.* 105: 88–103.

Rieder, J. P., T. A. S. Newbold, S. Sato, H. Yasuda and E. W. Evans. 2008. Intra-guild predation and variation in egg defence between sympatric and allopatric populations of two species of ladybird beetles. *Ecol. Entomol.* 33: 53–58.

Rodriguez-Saona, C. and J. C. Miller. 1999. Temperature dependent effects on development, mortality, and growth of *Hippodamia convergens* (Coleoptera: Coccinellidae). *Env. Entomol.* 28: 518–522.

Rossini, C., A. Gonzalez, J. Farmer, J. Meinwald and T. Eisner. 2000. Antiinsectan activity of epilachnene, a defensive alkaloid from pupae of Mexican bean beetles (*Epilachna varivestis*). *J. Chem. Ecol.* 26: 391–397.

Roy, M., J. Brodeur and C. Cloutier. 2002. Relationship between temperature and developmental rate of *Stethorus punctillum* (Coleoptera: Coccinellidae) and its prey *Tetranychus mcdanieli* (Acarina: Tetranychidae). *Env. Entomol.* 31: 177–187.

Roy, M., J. Brodeur and C. Cloutier. 2003. Effect of temperature on intrinsic rates of natural increase (rm) of a coccinellid and its spider mite prey. *BioControl* 48: 57–72.

Roy, H. E., H. Rudge, L. Goldrick and D. Hawkins. 2007. Eat or be eaten: prevalence and impact of egg cannibalism on two-spot ladybirds, *Adalia bipunctata. Entomol. Exp. Appl.* 125: 33–38.

Rozsypalova, A. 2007. *Polymorphism in Pupal Colouration in Ladybirds.* (Polymorfizmus zbarvení kukel slunéček.) Bachelor thesis, University of South Bohemia, České Budějovice. 23 pp. (In Czech.)

dos Santos, N. R. P., T. M. dos Santos-Cividanes, F. J. Cividanes, A. C. R. dos Anjos and L. V. R. de Oliveira. 2009. Biological aspects of *Harmonia axyridis* fed on two prey species and intraguild predation with *Eriopis connexa. Pesquisa Agropecuaria Bras.* 44: 554–560.

dos Santos, T. M. 1992. *Aspectos morphológicos e efeito da temperatura sobre a biologia de* Scymnus *(Pullus)* argentinicus *(Weise, 1906. (Coleoptera: Coccinellidae) alimentados com pulgão verde* Schizaphis graminum *(Rondani, 1852. (Homoptera: Aphididae).* Thesis, Escola superior de Agricultura de Lavras, Minas Gerais. 107 pp.

Savoiskaya, G. I. 1965. Biology and perspectives of utilisation of coccinellids in the control of aphids in south-eastern Kazakhstan orchards. *Trudy Inst. Zashch. Rast., Alma-Ata* 9: 128–156. (In Russian.)

Savoiskaya, G. I. 1970. Coccinellids of the Alma-Ata reserve. *Trudy Alma-Atin. Gos. Zapov.* 9: 163–187. (In Russian.)

Savoiskaya, G. I. and B. Klausnitzer, 1973. Morphology and taxonomy of the larvae with keys for their identification. In I. Hodek (ed.). *Biology of the Coccinellidae.* Prague & Hague, W. Junk, pp. 36–55.

Schade, M. and C. Sengonca. 1998. On the development, feeding activity and prey preference of *Hippodamia convergens* Guer.-Men. (Col., Coccinellidae) preying on *Thrips tabaci* Lind. (Thys., Thripidae) and two species of Aphidae. *Anz. Schadlingskunde Pflanzenschutz Umweltschutz* 71: 77–80.

Schanderl, H., A. Ferran and M. M. Larroque. 1985. Les besoins trophiques et thermiques des larves de la coccinelle *Harmonia axyridis* Pallas. *Agronomie* 5: 417–421.

Schanderl, H., A. Ferran and V. Garcia. 1988. L'élevage de deux coccinelles *Harmonia axyridis* et *Semiadalia undecimnotata* á l'aide d'oeufs d'*Anagasta kuehniella* tués aux rayon ultraviolets. *Entomol. Exp. Appl.* 49: 235–244.

Schroder, F. C., J. J. Farmer, S. R. Smedley, T. Eisner and J. Meinwald. 1998. Absolute configuration of the polyazamacrolides, macrocyclic polyamines produced by a ladybird beetle. *Tetrahedron Letters* 39: 6625–6628.

Schuder, I., M. Hommes and O. Larink. 2004. The influence of temperature on development of *Adalia bipunctata* (Coleoptera: Coccinellidae). *Eur. J. Entomol.* 101: 379–384.

Seko, T. and K. Miura. 2009. Effects of artificial selection for reduced flight ability on survival rate and fecundity of *Harmonia axyridis* (Pallas) (Coleoptera: Coccinellidae). *Appl. Entomol. Zool.* 44: 587–594.

Semyanov, V. P. 1970. Biological properties of *Adalia bipunctata* L. (Coleoptera: Coccinellidae) in conditions of

Leningrad region. *Zashch. Rast. Vredit. Bolez.* 127: 105–112. (In Russian.)

Semyanov, V. P. 1980. Biology of *Calvia quatuordecimguttata* L. (Coleoptera, Coccinellidae). *Entomol. Obozr.* 59: 757–763. (In Russian.)

Semyanov, V. P. and E. P. Bereznaya. 1988. Biology and prospects of using Vietnam's lady beetle *Lemnia biplagiata* (Swartz) for control of aphids in greenhouses. In E. Niemczyk and A. F. G. Dixon (eds), *Ecology and Effectiveness of Aphidophaga.* SPB Academic, The Hague. pp. 267–269.

Senal, D. 2006. *Avci böcek* Chilocorus nigritus *(Fabricius) (Coleoptera: Coccinellidae)'un bazi biyolojik ve ekolojik özellikleri ile doğaya adaptasyonu üzerinde aras, tirmalar.* PhD thesis, Department of Plant Protection, Institute of Natural and Applied Sciences, University of Çukurova, Turkey.

Sengonca, C. and C. Arnold. 2003. Development, predation and reproduction by *Exochomus quadripustulatus* L. (Coleoptera: Coccinellidae) as predator of *Pulvinaria regalis* Canard (Homoptera: Coccidae) and its coincidence with the prey in the field. *J. Plant Dis. Prot.* 110: 250–262.

Seo, M. J., G. H. Kim and Y. N. Youn. 2008. Differences in biological and behavioural characteristics of *Harmonia axyridis* (Coleoptera: Coccinellidae) according to color pattern of elytra. *J. Appl. Entomol.* 132: 239–247.

Shannag, H. K. and W. M. Obeidat. 2008. Interaction between plant resistance and predation of *Aphis fabae* (Homoptera: Aphididae) by *Coccinella septempunctata* (Coleoptera: Coccinellidae). *Ann. Appl. Biol.* 152: 331–337.

Simmons, A. M. and J. C. Legaspi. 2004. Survival and predation of *Delphastus catalinae* (Coleoptera: Coccinellidae), a predator of whiteflies (Homoptera: Aleyrodidae), after exposure to a range of constant temperatures. *Environ. Entomol.* 33: 839–843.

Simmons, A. M., J. C. Legaspi and B. C. Legaspi. 2008. Responses of *Delphastus catalinae* (Coleoptera: Coccinellidae), a predator of whiteflies (Hemiptera: Aleyrodidae), to relative humidity: oviposition, hatch, and immature survival. *Ann. Entomol. Soc. Am.* 101: 378–383.

Sloggett, J. J. 2008. Weighty matters: body size, diet and specialization in aphidophagous ladybird beetles (Coleoptera: Coccinellidae). *Eur. J. Entomool.* 105: 381–389.

Sloggett, J. J. and M. W. Lorenz. 2008. Egg composition and reproductive investment in aphidophagous ladybird beetles (Coccinellidae: Coccinellini): egg development and interspecific variation. *Physiol. Entomol.* 33: 200–208.

Smedley, S. R., K. A. Lafleur, L. K. Gibbons et al. 2002. Glandular hairs: pupal chemical defense in a non-native ladybird beetle (Coleoptera: Coccinellidae). *Northeastern Nat.* 9: 253–266.

Smith, K. M., D. Smith and A. T. Lisle. 1999. Effect of field-weathered residues of pyriproxyfen on the predatory coccinellids *Chilocorus circumdatus* Gyllenhal and *Cryptolaemus montrouzieri* Mulsant. *Austr. J. Exp. Agric.* 39: 995–1000.

Soroushmehr, Z., A. Sahragard and L. Salehi. 2008. Comparative life table statistics for the ladybeetle *Scymnus*

syriacus reared on the green citrus aphid, *Aphis spiraecola*, fed on two host plants. *Entomol. Sci.* 11: 281–288.

Srivastava, S. and Omkar. 2003. Influence of temperature on certain biological attributes of a ladybeetle *Coccinella septempunctata* Linnaeus. *Entomologica Sinica* 10: 185–193.

Srivastava, S. and Omkar. 2004. Age-specific mating and reproductive senescence in the seven-spotted ladybird, *Coccinella septempunctata*. *J. Appl. Entomol.* 128: 452–458.

Srivastava, S. and Omkar. 2005. Mate choice and reproductive success of two morphs of the seven spotted ladybird, *Coccinella septempunctata* (Coleoptera: Coccinellidae). *Eur. J. Entomol.* 102: 189–194.

Stathas, G. J. 2000. The effect of temperature on the development of the predator *Rhyzobius lophanthae* and its phenology in Greece. *BioControl* 45: 439–451.

Stathas, G. J., P. A. Eliopoulos, D. C. Kontodimas and J. Giannopapas. 2001. Parameters of reproductive activity in females of *Harmonia axyridis* (Coleoptera: Coccinellidae). *Eur. J. Entomol.* 98: 547–549.

Stathas, G. J., P. A. Eliopoulos, D. C. Kontodimas and D. T. Siamos. 2002. Adult morphology and life cycle under constant temperatures of the predator *Rhyzobius lophanthae* Blaisdell (Col., Coccinellidae). *Anz. Schädlingskunde* 75: 105–109.

Stewart, L. A., A. F. G. Dixon, Z. Ruzicka and G. Iperti. 1991a. Clutch and egg size in ladybird beetles. *Entomophaga* 36: 329–333.

Stewart, L. A., J. L. Hemptinne and A. F. G. Dixon. 1991b. Reproductive tactics of ladybird beetles – relationships between egg size, ovariole number and developmental time. *Func. Ecol.* 5: 380–385.

Sugiura, K. and H. Takada. 1998. Suitability of seven aphid species as prey of *Cheilomenes sexmaculata* (Fabricius) (Coleoptera: Coccinellidae). *Japanese J. Appl. Entomol. Zool.* 42: 7–14.

Sundby, R. A. 1968. Some factors influencing the reproduction and longevity of *Coccinella septempunctata* Linnaeus (Coleoptera: Coccinellidae). *Entomophaga* 13: 197–202.

Taghizadeh, R., Y. Fathipour and K. Kamali. 2008. Influence of temperature on life-table parameters of *Stethorus gilvifrons* (Mulsant) (Coleoptera: Coccinellidae) fed on *Tetranychus urticae* Koch. *J. Appl. Entomol.* 132: 638–645.

Takahashi, K. 1987. Differences in oviposition initiation and sites of lady beetles, *Coccinella septempunctata brucki* Mulsant and *Harmonia axyridis* (Pallas) (Coleoptera: Coccinellidae) in the field. *Japanese Journal of Applied Entomology and Zoology* 31, 253–254. (In Japanese.)

Takakura, K. 2004. Variation in egg size within and among generations of the bean weevil, *Bruchidius dorsalis* (Coleoptera, Bruchidae): effects of host plant quality and paternal nutritional investment. *Ann. Entomol. Soc. Am.* 97: 346–352.

Takeuchi, M., A. Shimizu, A. Ishihara and M. Tamura. 1999. Larval diapause induction and termination in a phytophagous lady beetle, *Epilachna admirabilis* Crotch (Coleoptera: Coccinellidae). *Appl. Entomol. Zool.* 34: 475–479.

Tanasijevic, N. 1958. Zur morphologie and biologie des luzerne marienkäfers *Subcoccinella vigintiquatuorpunctata* L. *Beiträge zur Entomologie* 8: 23–78.

Tavadjoh, Z., H. Hamzehzarghani, H. Alemansoor, J. Khalghani and A. Vikram. 2010. Biology and feeding behaviour of ladybird, *Clitostethus arcuatus*, the predator of the ash whitefly, *Siphoninus phillyreae*, in Fars province, Iran. *J. Insect Sci.* 10: 120.

Tawfik, M. F. S., S. Abul-Nasr and B. M. Saad. 1973. The biology of *Scymnus interruptus* Goeze (Coleoptera: Coccinellidae). *Bull. Soc. Entomol. Egypt* 57: 9–26.

Thompson, W. R. 1951. The specificity of host relations in predaceous insects. *Can. Entomol.* 1983: 262–269.

Timms, J. E. L. and S. R. Leather. 2007. Ladybird egg cluster size: relationships between species, oviposition substrate and cannibalism. *Bull. Entomol. Res.* 97: 613–618.

Timms, J. E. L. and S. R. Leather. 2008. How the consumption and development rates of the conifer specialist *Aphidecta obliterata* respond to temperature, and is better adapted to limited prey than a generalist? *Ann. Appl. Biol.* 153: 63–71.

Todorova, S. I., J. C. Cote and D. Coderre. 1996. Evaluation of the effects of two *Beauveria bassiana* (Balsamo) Vuillemin strains on the development of *Coleomegilla maculata lengi* Timberlake (Col, Coccinellidae). *J. Appl. Entomol.* 120: 159–163.

Tourniaire, R., A. Ferran, L. Giuge, C. Piotte and J. Gambier. 2000. A natural flightless mutation in the ladybird, *Harmonia axyridis*. *Entomol. Exp. Appl.* 96: 33–38.

Transfaglia, A. and G. Viggiani. 1972. Dati biologici sullo *Scymnus includens* Kirsch (Coleoptera: Coccinellidae). *Boll. Lab. Entomol. Agr. 'Philippo Silvestri'* 30: 9–18.

Triltsch, H. 1996. On the parasitization of the ladybird *Coccinella septempunctata* L (Col, Coccinellidae). *J. Appl. Entomol.* 120: 375–378.

Trudgill, D. L. 1995. Why do tropical poikilothermic organisms tend to have higher threshold temperature for development than temprate ones? *Funct. Ecol.* 9: 136–137.

Trudgill, D. L., A. Honĕk, D. Li and N. M. VanStraalen. 2005. Thermal time: concepts and utility. *Ann. Appl. Biol.* 146: 1–14.

Ueno, H. 1994. Itraspecific variation of P2 value in a coccinellid beetle, *Harmonia axyridis*. *J. Ethol.* 12: 169–174.

Ueno, H. 1996. Estimate of multiple insemination in a natural population of *Harmonia axyridis*. *Appl. Entomol. Zool.* 31: 621–623.

Ueno, H. 2003. Genetic variation in larval period and pupal mass in an aphidophagous ladybird beetle (*Harmonia axyridis*) reared in different environments. *Entomol. Exp. Appl.* 106: 211–218.

Ungerová, D., P. Kalushkov and O. Nedvĕd. 2010. Suitability of diverse prey species for development of *Harmonia axyridis* and the effect of container size. *IOBC Bulletin* 58: 165–174.

Uygun, N. and R. Atlıhan. 2000. The effect of temperature on development and fecundity of *Scymnus levaillanti*. *BioControl* 45: 453–462.

Uygun, N. and N. Z. Elekcioglu. 1998. Effect of three diaspididae prey species on development and fecundity of the ladybeetle *Chilocorus bipustulatus* in the laboratory. *BioControl* 43: 153–162.

Waldbauer, G. P. 1964. The consumption, digestion and utilization of solanaceous and non-solanaceous plants by larvae of the tobacco hornworm, *Protoparce sexta* (Johan) (Lepidoptera: Sphingidae). *Entomol. Exp. Appl.* 7: 253–269.

Wang, S., J. P. Michaud, R. Z. Zhang, F. Zhang and S. Liu. 2009. Seasonal cycles of assortative mating and reproductive behaviour in polymorphic populations of *Harmonia axyridis* in China. *Ecol. Entomol.* 34: 483–494.

Ware, R. L. and M. E. N. Majerus. 2008. Intraguild predation of immature stages of British and Japanese coccinellids by the invasive ladybird *Harmonia axyridis*. *BioControl* 53: 169–188.

Ware, R. L., B. Yguel and M. E. N. Majerus. 2008. Effects of larval diet on female reproductive output of the European coccinellid *Adalia bipunctata* and the invasive species *Harmonia axyridis* (Coleoptera: Coccinellidae). *Eur. J. Entomol.* 105, 437–443.

Warren, L. O. and M. Tadić. 1967. Biological observations in *Coleomegilla maculata* and its role as a predator of the fall webworm. *J. Econ. Entomol.* 60: 1492–1496.

Webberley, K. M., M. C. Tinsley, J. J. Sloggett, M. E. N. Majerus and G. D. D. Hurst. 2006. Spatial variation in the incidence of a sexually transmitted parasite of the ladybird beetle *Adalia bipunctata* (Coleoptera: Coccinellidae). *Eur. J. Entomol.* 103: 793–797.

Williams, J. L. 1945. The anatomy of the internal genitalia of some Coleoptera. *Proc. Entomol. Soc. Washington* 47: 73–91.

Witsack, W. 1971. Neufunde und zur Verbreitung von *Synharmonia lyncea* (Ol.), einem sehn seltenen Marienkäfer (Coccinellidae, Coleoptera). *Naturk. Jber. Mus. Heineanum (Halberstadt)* V/VI: 53–57.

Wright, E. J. and J. E. Laing. 1978. The effects on temperature on development, adult longevity and fecundity of *Coleomegilla maculata lengi* and its parasite, *Perilitus coccinellae*. *Proc. Entomol. Soc. Ontario* 109: 33–47.

Xia, J. Y., W. Van der Werf and R. Rabbinge. 1999. Temperature and prey density on bionomics of *Coccinella septempunctata* (Coleoptera: Coccinellidae) feeding on *Aphis gossypii* (Homoptera: Aphididae) on cotton. *Environ. Entomol.* 28: 307–314.

Yakhontov, V. V. 1958. Theoretical base for development of new method of biological control against pests. *Material of 1st conference on pathology of insects and biological control, Prague*, 455–479. (In Russian.)

Yao, S. L., Z. Huang, S. X. Ren, N. Mandour and S. Ali. 2010. Effects of temperature on development, survival, longevity, and fecundity of *Serangium japonicum* (Coleoptera: Coccinellidae), a predator of *Bemisia tabaci* Gennadius (Homoptera: Aleyrodidae). *Biocontrol Sci. Tech.* 21: 23–34.

Yinon, U. 1969. Food consumption of armored scale ladybeetle, *Chilocorus bipustulatus* (Coleoptera, Coccinellidae). *Entomol. Exp. Appl.* 12: 139–146.

Zhu, S. R., J. W. Su, X. H. Liu et al. 2006. Development and reproduction of *Propylaea japonica* (Coleoptera: Coccinellidae) raised on *Aphis gossypii* (Homoptera: Aphididae) fed transgenic cotton. *Zool. Stud.* 45: 98–103.

Chapter 4

DISTRIBUTION AND HABITATS

Alois Honěk

Department of Entomology, Crop Research Institute, CZ 16106, Prague 6, Czech Republic

Ecology and Behaviour of the Ladybird Beetles (Coccinellidae), First Edition. Edited by I. Hodek, H.F. van Emden, A. Honěk.
© 2012 Blackwell Publishing Ltd. Published 2012 by Blackwell Publishing Ltd.

4.1 INTRODUCTION

In the present epoch of molecular and mathematical approaches to biology we may ask, at the beginning of this chapter, why should one review and stimulate research of something as basic and simple as studies of coccinellid habitats? Our response is that habitat studies, i.e. the study of where coccinellids live, are indispensable. Knowledge of where the coccinellid lives provides basic information necessary for studies of ecology (determining species niches), ecophysiology (food and microclimatic preferences), biogeography (the factors limiting coccinellid distribution) and other disciplines. Habitat studies thus link natural reality with theory. Predictions of hypotheses derived from biological theories could of course be tested in the laboratory, if not solely on a computer. However, the final proof of the reality of such predictions can be done only after relating the ideas back to nature, and this means studying the presence and habits of coccinellids in natural habitats. Therefore the methods and results of habitat studies are of utmost importance. In this chapter we review the methods of collection and description of data concerning habitats, and interpret them in terms of the factors that cause variation in abundance while trying to keep a balance between earlier results and recent. Examples of coccinellid communities from economically important crops or intensively studied natural biotopes are provided and discussed at the end of this chapter.

The main difficulty of habitat studies is to define **what is the 'habitat'**. Man has a tendency to view his environment in terms of his own activities. Crops, gardens, forests or urban areas are automatically recognized as habitats. However, the vegetation, although uniform from our point of view, may be a mosaic of environments with often dramatically differing qualities. On the other hand, natural objects which appear different to our perception may be part of an integral habitat from the point of view of coccinellids. Two adjacent and evenly developed crops of different cereal species, two rows of garden trees each of a different species may represent a uniform habitat because of the absence of any difference that is important for the coccinellids. Since coccinellids can distinguish and select habitats with the qualities important to them, we too should aim to distinguish the boundaries of habitats and consider the environment from the 'point of view of a coccinellid'. Here we try to show some factors important for coccinellid decisions. The selection of these factors is based on our personal experience and is probably far from complete. However, we believe that considering at least these factors may contribute to understanding coccinellid distribution.

A habitat is populated by a specific **community of coccinellid species**. Here, the term 'community' is used in its broadest sense as a set of coccinellid individuals present in a given habitat at a particular time. We prefer this vague terminology over a detailed definition. Such a definition usually includes information concerning the previous history of the community (permanence in time, details of its origin) or the motivation of the animals that make up the community (hibernation, search for food). In the field, even the obvious causes are difficult to understand and including further criteria would only add a good deal of speculation. Perhaps this remark is superfluous since, in fact, use of the vast terminology attempting to classify communities is in decline, leaving room for new understanding of factors contributing to the structure of communities.

4.2 SAMPLING

Sampling is the initial phase in investigating the presence of coccinellids in a particular habitat. We may be interested in the proportion of different species in the entire family Coccinellidae (relative abundance), or in the absolute abundance (i.e. numbers per unit area). Maximizing the reliability of sampling is a prerequisite for making correct ecological conclusions and already at this stage of investigation there arise problems. Each sampling method only provides a differently biased estimate of relative and absolute abundance. The bias may arise from several sources: uneven distribution of species in the vegetation cover; different behaviour of coccinellids in relation to daily changes of light intensity, temperature and humidity; fluctuations of weather; different escape reactions of species and stages during the sampling procedure; different conspicuousness to the collector. Generally, it is easier to collect adults than larvae, and the first and second instars cannot be counted with any accuracy in the field. Here we review the different methods used for sampling coccinellids, how they are influenced by environmental factors and the efforts that have been made to standardize their results under different conditions, as well as attempts to recalculate 'absolute' from 'relative' abundance.

Several methods were tested for sampling from **herbaceous stands**. Collecting coccinellids by **hand-picking** from small plots ('quadrat sampling') was the sampling technique that gave the highest estimates of numbers per unit area, probably giving densities close to the true coccinellid abundance (Michels et al. 1997). True numbers of coccinellid adults and larvae were established by sampling out a $1\,m^2$ area fenced by a 40 cm plywood enclosure, until both plants and ground were clean of the insects. This method could be recommended for future studies as the yardstick of 'true abundance' of coccinellids for evaluating the efficiency of other methods. Lower abundances were obtained by 'removal sampling', which consisted of two 15-min periods of picking coccinellids from an unfenced area of $25\,m^2$. The differences between the results of both these methods were not significant for three (*Hip. convergens*, *Hip. sinuata*, *C. septempunctata*) species but not for *Col. maculata* (Michels et al.1997).

Sweeping with a sweep net is probably the most widely used method which measures an unknown function of coccinellid numbers and activity. It samples a narrow upper stratum of vegetation and the number of insects available for collecting is therefore dependent on plant stand height (Elliott et al. 1991). This relative method is also sensitive to the weather and to diurnal changes in coccinellid activity. Studies on the efficiency of this method have established a significant although sometimes low correlation between abundance established by sweeping and hand-picking. The magnitude of the error of the estimate of coccinellid abundance decreases and species presence increases with the number of sweeps in a non-linear convex manner (Elliott & Kieckhefer 1990).

Hand-shaking plants, an alternative to sweeping, is not used frequently, but is a convenient way of sampling wild herbaceous plants. Arefin and Ivliev (1988) compared its efficiency with sweeping in capturing different species (*C. septempunctata*, *Har. axyridis* and *P. quatuordecimpunctata*) and different developmental stages and found a high correlation between the results of both methods.

Visual counting on a transect (walking counts) is quick but weather sensitive; it is particularly convenient for recording highly visible brightly coloured adults, e.g. *C. septempunctata*. This method exploits the thermoregulatory behaviour of ladybirds – basking in the sunshine on cool days. The basking individuals are counted by a person who walks with the sun behind them. The method could be used for a quick comparison of coccinellid abundance in different crops and can supplement or replace sweeping (Honěk 1978a, 1982a, Lapchin et al. 1987, Iperti et al. 1988). Michels & Behle (1992) found that visual counting was the best method for estimating abundance of *Hippodamia* species in grain sorghum. Differences between observers could be removed, since there was a high correlation between the counts of those that worked in parallel (Frazer & Raworth 1985). By contrast, in strawberry crops where sweeping was not practicable, walking counts were influenced by temperature, solar radiation and time of day to the extent that they were of little practical use (Frazer & Raworth 1985).

Vacuum sampling by D-Vac has also been used for coccinellid studies. Like hand picking, it samples all development stages from the whole vegetation profile but its efficiency is limited by the influence of plant architecture. As with sweeping, efficiency is good if coccinellids sit at the top of the plants but low when they hide in the lower strata of the plant stand (Cosper et al. 1983). Ellington et al. (1984) found that the efficiency of D-Vac was superior to that of sweeping, as it collected more than twice the number of *Hippodamia* spp. adults than the latter method, but this was still only a fraction of the population counted by hand-picking. In stands of soybeans, D-Vac sampling was inferior to sweeping (Bechinski & Pedigo 1982).

Several studies have **compared the efficiency of sampling methods** in crop stands in relation to the crop and microclimate characteristics. For wheat, Elliott et al. (1991) proposed regression methods for converting the data from several sampling methods to absolute coccinellid numbers. Michels et al. (1997) compared hand-picking with timed sampling and sweeping. Hand-picking accurately estimated adult but underestimated larval density. Timed counts and sweeping results were correlated with the absolute density established by hand-picking. Regression models for converting the data of four coccinellid species to their absolute abundance included the number of tillers per 0.3 m of row, plant growth stage, height and number of aphids per tiller and these variables accounted for 89–93% of the variance. This study was further continued in alfalfa, where Elliott & Michels (1997) compared hand-picking, time-limited visual counting and sweeping; they proposed methods of converting the results of timed and sweep samples to absolute numbers of four coccinellid species. For sweeping and timed counting, plant height was included in the best regression models for adults. For larvae, the regression explained a low proportion of variance and included plant growth stage and aphid abundance. In sorghum crops, the efficiencies of hand-picking, timed sampling, sweeping and D-Vac sampling were compared and regression methods of calculating absolute abundance were established (Michels et al. 1996). In soybean crops, Arefin and Ivliev (1988) compared the efficiency of shaking the plant and sweeping, and found a high correlation between the results of both methods ($r^2 = 0.64$–0.98 for adults and larvae of *C. septempunctata, Har. axyridis* and *P. quatuor-decimpunctata*). However, the slope of the regression for abundance determined by sweeping differed from that found by shaking ($b = 0.16$ and 1.06, respectively). In cotton, Ellington et al. (1984) found that D-Vac sampling captured 34% and sweeping only 14% of the *Hippodamia* spp. adult population established by hand-picking.

Methods of sampling coccinellids from **trees and shrubs** have been less investigated than for herbs. Studies of tree faunas have been based on accessible parts of the crown, the lower stratum of large trees or whole small trees. Coccinellid faunas of upper parts of large trees remain uninvestigated. Three sampling methods have been used most frequently: (i) direct examination of a fixed number of twigs or leaves (Brown 2004, Michaud 2004, Oztemiz et al. 2008); (ii) collecting into a sheet by shaking (branch jarring) or beating (LaMana & Miller 1996, Brown & Schmitt 2001); and (iii) sweeping (Honěk & Rejmánek 1982, Honěk 1985b, Honěk et al. 2005). As yet no comparative studies of these methods are available. Sampling from trees and shrubs is made somewhat easier by the fact that their rigid architecture enables permanent localization of particular sampled twigs and leaves and counting prey. This enables precise counting coccinellid larvae whose distribution is largely determined by prey availability, but adults move in response to light and temperature conditions. Measuring leaf temperature is complex; it has been studied since the 1920s and is reviewed by Thofelt (1975). Microclimatic effects on coccinellid communities on trees have not yet been analyzed but are certainly important.

Besides sampling from plants, various **traps** can be used which collect coccinellid adults during flight. Such catches provide samples of adults of unknown origin, although the beetles probably mostly come from nearby stands, and are influenced by species-specific orientation (leading to different attractiveness of different kinds of trap), diurnal periodicity and flight ability.

Sticky cards are most frequently used because they catch large numbers of coccinellids. The numbers of adults stuck on the cards depends on the card colour. Yellow-coloured traps are better than red, green and white ones (Udayagiri et al. 1997) and are therefore generally used to study coccinellid movement within a field (Ives 1981a). Kokubu (1986) designed his experiments to estimate the numbers of immigrant and emigrant coccinellids in maize plots. He placed yellow sticky panels in parallel with the sides of the plot so that surfaces facing the plot sampled emigrant beetles while the opposite surface trapped immigrant ones. For *A. bipunctata, C. septempunctata* and *P. quatuordecimpunctata* there was, however, only a poor correlation between abundance on maize plants and trap catches. Similarly, Stephens and Losey (2004) found a poor correlation between sweeping, visual counts and sticky trap catches of *Har. axyridis, Col. maculata* and *C. septempunctata* in alfalfa. However, using sticky traps demonstrated different flight activity of *Col. maculata* and *C. septempunctata* in central and marginal parts of maize crops (Udayagiri et al. 1997, Bruck & Lewis 1998, Colunga-Garcia & Gage 1998) and were also useful in detecting long-term changes in abundance of coccinellid species that occurred before and after arrival of the invasive *Har. axyridis*. Yellow sticky traps can also be used for detecting the height which particular coccinellid species prefer when in flight (Parajulee & Slosser 2003).

In contrast to coloured cards, **window traps** (=impaction traps) collect a supposedly unbiased sample of flying coccinellids. The beetles bump into a transparent glass wall and fall into a collecting trough below. However, it remains unknown whether the glass wall is really 'invisible' to them. The species-selective capture by yellow sticky traps and window traps was demonstrated by Storck-Weyhermueller (1988); with both types of trap, the height and position above the ground influencing the catch of particular species. Boiteau et al. (1999) collected 21 species of coccinellids using window traps placed at heights of 0.8 to 14.3 m. *Hip. convergens, C. septempunctata* and *C. trifasciata* flew mostly near the ground while other species were less selective.

Malaise traps, tents catching flying insects that enter their interior through a gap and are led by phototaxis to collecting bottles, have been found to be convenient in detecting the peak of flying *C. septempunctata* during migration and non-migration flights. Sarospataki & Marko (1995) hung Malaise traps within a shrub stratum (0–2 m height), at the tree canopy stratum (12.5–14.5 m height) and above all the vegetation (25–27 m height). While the lower traps captured adults throughout the vegetative period, the traps placed above the vegetation captured adults only at the time of migration, in late July.

Although coccinellids are mostly day active, they are frequently caught in **light traps**. Trap efficiency is influenced by trap design and depends both on light source and the reflective surface behind the light. Light traps collect a mixture of species from coccinellid communities of different habitats near the trap and the samples have a great diversity (Honěk & Rejmánek 1982). Catches were used to compare the annual variation of species abundance (Honěk & Kocourek 1986) and long-term changes in the colour morph proportions of the different species (Honěk et al. 2005). Interpreting seasonal variation of catches in terms of number of generations per year (trivial flight) and flight to overwintering sites (migratory flight) needs background knowledge of species biology (Koch & Hutchinson 2003).

There certainly exist further methods used to capture other insect taxa but not applied to sampling coccinellids as yet. To use them the kind reader may stimulate her/his powers of imagination by reading standard reference books, e.g. Southwood and Henderson (2000).

Abundance of coccinellid individuals in a habitat is a dynamic state of balance between immigrating and emigrating individuals. The process of exchange of individuals at a place could be studied by **mark–recapture method**. A known number of individuals of a species are marked and released, then the individuals are sampled after some time and the proportion of released individuals in the total recaptured population is established. This method is convenient for study of abundant species and can determine total number of individuals at a place, their residence time, emigration rates, and distance of emigration from the place of their release. Calculation of these values was described for aphidophagous (van der Werf et al. 2000) and phytophagous coccinellids (Koji & Nakamura 2002). For marking individuals the authors use paint (van der Werf et al. 2000, Grez et al. 2005, Seko et al., 2008) or proteins, established after recapture by biochemical methods (Hagler & Naranjo 2004, Hagler & Jones 2010).

4.3   FACTORS DETERMINING THE COMPOSITION OF COMMUNITIES

The coccinellid community present at each site is a sample of the **local fauna** which is a total of all the species present in a geographic area. The **characteristics of a site** include the prey abundance and species, host plant, microclimate, surrounding landscape characteristics, as well as intraguild (Chapter 7) and other biotic factors. The **probability of being sampled** within a particular community is a function of the size of the coccinellid population. The probability of capturing rare species decreases with their rarity, and so their disappearance from samples for a time is no evidence of short-term extinction followed by recolonization. The estimated diversity of communities increases as sampling intensity increases to a maximum which is the full composition of the local fauna. This may be attained at any location if sampling effort is maximized and time is unlimited.

4.3.1   Local faunas

The relationship between the coccinellid faunas of particular geographic areas and the composition of their communities in particular habitats within these areas has still been little studied. There is probably no study that demonstrates geographic trends in abundance and diversity of species in particular habitats along a transect long enough to cross a range of geographic areas. Most studies have been performed in temperate regions with a similar diversity of coccinellid species, or limited to particular crops. It can be hoped that this section will stimulate further research.

4.3.1.1   Geographic differences

Unlike for other groups of organisms, for coccinellids there has been no study dividing the Earth's surface into a rectangular grid showing local variation in species diversity and abundance. Local lists of species for particular areas are not useful, because they are biased by different sizes of the areas and different sampling activity within them. Regardless of this limitation, it is clear that richness in genera and species is inversely related to the **geographical latitude** of an area. This may be illustrated by the example of faunas of areas of similar size and geography in Russia, sampled with comparable effort and separated by 15° of latitude. The fauna of the Magadan territory (60°N) contains 18 species from 9 genera (Ivliev et al. 1975) while that of the Primorskii Region (45°N) has 63 species from 29 genera (Kuznetsov 1975).

4.3.1.2 Invasion and extinction

The number of species in a geographical area may decrease through extinction, or increase through immigration by or artificial introduction of a new species. There is no record of any global **extinction** of a coccinellid species, but some species may become locally extinct from island faunas (Majerus 1994). However, the case of *C. undecimpunctata*, which invaded and spread in North America but now is nearly extinct after some 100 years of co-existence with the local fauna (Wheeler & Hoebke 2008), indicates that extinction could occur more frequently. More data are available for coccinellid introductions than for extinctions. Recent **introductions** of coccinellids into new areas have not been spontaneous but instead have followed human activity. Since the 1890s, 179 coccinellid species have been introduced deliberately or inadvertently to the USA and Canada, and 27 have become established. These species originally represented an insignificant fraction of the coccinellid species in the local communities, but this situation lasted only until the 1980s (Harmon et al. 2007). Since then immigrant species have come to replace indigenous species to an alarming extent, so that now introduced coccinellid species, mainly *C. septempunctata* and *Har. axyridis*, typically represent 60 to 80% of adult coccinellid communities (Gardiner et al. 2009). Mechanisms involved in the replacement by new species of the original fauna, and the fatal consequences for the latter thereof, include intraguild predation (Chapter 7) and 'marginalization' of native species – their displacement from preferred habitats to habitats that they had not occupied before. An example is provided by *C. septempunctata*, established in Utah (USA) between 1992–2001. Since then this species has increased in numbers while densities of native species, *Hip. convergens*, *Hip. quinquesignata*, *Hip. sinuata* and *C. transversoguttata*, have decreased. The decline of these native ladybirds mirrored a decline of pea aphid *Acyrthosiphon pisum*. As *C. septempunctata* depressed

prey availability for the adults of native species, these have shifted their foraging away from alfalfa to other crops and wild herbs. The reality of this scenario was confirmed by field experiments which revealed that native ladybirds were more sensitive to local aphid density than was *C. septempunctata* (Evans 2004). In eastern USA, Finlayson et al. (2008) compared composition of coccinellid communities on several habitats and in all places found prevalence of non-native over native species of coccinellids.

The limits that environmental conditions impose on the distribution of introduced species were demonstrated for *Har. axyridis*. At the northern boundary of its distribution in North America survival is limited by severe winter cold and this species only survives winter inside human houses. In northern marginal regions the distribution of human settlements coincides thus with distribution of this coccinellid species (Labrie et al. 2008).

4.3.1.3 Climatic changes

Recently a lot of attention has been paid to the consequences for coccinellids of the expected climatic change. Simulation, using data from the United Kingdom, revealed that warming of the climate could change the interaction between coccinellids and aphids in that the abundance of prey may decrease by 40–60% and the Julian date of their peak population could be advanced (Skirvin et al. 1997) (Fig. 4.1). As individual coccinellid species show a different response to prey abundance (Honěk et al. 2007), changes in prey number and their timing due to climate warming might change the composition of coccinellid communities. Other studies, however, throw a different light on the importance of local climate in determining the presence of coccinellid species. The accuracy of predictions of the geographic range of a species based on climatic characteristics was tested using data for 15 *Chilocorus* species introduced to various areas in tropical and subtropical zone, with the CLIMEX programme simulating the effect of climate on species distribution. This programme tests for a match between climate and biological characteristic of the species. Predictive models of distribution of the different *Chilocorus* species were based on the likelihood of their establishment with respect to their physiological characteristics and climatic tolerances. Model predictions were compared with data on the actual distribution of the species. The real distribution of four (27%) species was predicted

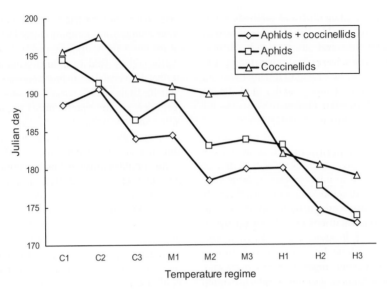

Figure 4.1 Simulation of timing (Julian day) of peak abundance of the grain aphid *Sitobion avenae* and its coccinellid predator *Coccinella septempunctata* in the UK under different scenarios of climatic change. Timing of maximum abundance was predicted using the model of Carter et al. (1982), three temperature regimes, cold (C), moderate (M) and hot (H) and three within-regime temperature modifications. The figure shows shortening of the vegetation period in terms of calendar time by c. 20 days for the *S. avenae* population alone, the *S. avenae* population subject to *C. septempunctata* predation, and the population of *C. septempunctata* (redrawn from Skirvin et al. 1997).

with 100% accuracy from climate data, but accuracy for the other species was limited by factors other than climate. Thus most predictions of change in the geographic distribution of species based on climatic data are likely to be wrong (Samways et al. 1999). Similarly a study of climatic limits for the geographic distribution of *Har. axyridis* showed that **climate** was **not** the **limiting factor**. Establishment and spread of this species was found to be likely in many regions, including ones which had not yet been invaded (Poutsma et al. 2008). In fact, a climate suitable for the establishment of *Har. axyridis* existed in many areas not occupied by this species for thousands of years before its recent nearly worldwide invasion.

4.3.2 Locality determinants

By analogy with a theatre performance, that of a coccinellid community at a particular site also involves 'scenery' which is the landscape, the host plant and the microclimate (the components dealt with in this chapter), and 'players' which comprise the prey,

conspecific and heterospecific members of the same trophic guild, as well as natural enemies. Except for prey which is dealt with here, the other 'players' are treated in Chapters 7 and 8. Here we present the factors that determine the presence or absence of a species in a particular habitat in the order of their importance which is prey, host plant, microclimate, and landscape.

4.3.2.1 Prey

Prey availability is the primary factor that brings together coccinellid assemblages; adults seek prey not only for their own welfare but as a resource for their progeny (Chapter 5). Assemblages are also formed because of satiation before hibernation (Chapter 6). It is still open for further research as to in which situations **abundance** of prey is **more important than** prey **species** in determining coccinellid presence in a habitat, and where the order of importance of these two determinants is reversed.

Coccinellid communities of particular habitats begin as assemblages of mobile adults which turn into assemblages of larvae after selective oviposition and selective larval survival. Larval communities thus usually are a copy of the composition of adult coccinellids. The **mechanisms whereby coccinellid assemble** and then oviposit have been reviewed by Evans (2003). Adult ladybirds are highly mobile in traversing the landscape (5.4.1.5), but become less active and produce more eggs as their rate of encounter with prey and aphid consumption increases, with the result that most eggs tend to be laid at sites of high aphid density. Females use their resources to produce eggs in modest numbers when prey consumption is limited. They may thus be prepared to lay some eggs quickly when they succeed in finding aphids in high numbers, but otherwise they may have little choice but to lay eggs at suboptimal sites. Upon locating patches of high prey density, females are faced with the decision of how long to remain. It is frequently thought that they become passively trapped at such patches until the aphid density there collapses. However, there is no such positive trapping. Oviposition has already terminated by the middle of the course of the prey population's development, and the cause is an **oviposition-deterring pheromone** (Ruzicka 1996, Klewer et al. 2007) secreted by the larvae from the anal disk that helps their locomotion (Laubertie et al. 2006). In this, the larvae are mainly responding to cues associated with presence of conspecific larvae (Yasuda et al. 2000, Hemptinne et al. 2000a). Oviposition-deterring pheromones may promote their departure from prey patches well before prey resources are exhausted. Females may also have an innate tendency to disperse regardless of local conditions.

Prey is an attractant and arrestant for immigrating adults. The minimum **prey density** capable of attracting adult *C. septempunctata* is below 10 aphids per square metre (Honěk 1980) and imigration may occur simultaneously with aphid immigration (Arefin & Ivliev 1988). At this low threshold density of prey, the coccinellids can just capture sufficient prey to maintain their body weight (Frazer & Gilbert 1976). A proportion of the adult population will leave the site but be replaced by new immigrants (Ives 1981a). Immigration and emigration rates depend on the **frequency of encounters** between coccinellids and their prey, which is how a positive correlation between the abundance of aphids and coccinellids arises. This was shown in alfalfa stands populated by different aphid and coccinellid species (Neuenschwander et al. 1975, Radcliffe et al. 1976, Honěk 1982a), in strawberry

plantations of British Columbia populated by the aphid *Chaetosiphon fragaefolii* and *C. californica* (Frazer & Raworth 1985), and bean stands infested with the aphid *Aphis craccivora* and populated with *Menochilus sexmaculatus* and *C. transversalis* (Agarwala & Bardhanroy 1999; Fig. 4.2). In tobacco stands, numbers of the aphid *Myzus persicae* and those of *Cer. undecimnotata*, *A. bipunctata*, *Hip. variegata* and *P. quatuordecimpunctata* were only loosely correlated because of a negative coccinellid response to the many aphids parasitized by hymenopteran parasitoids (Kavallieratos et al. 2004). However, evidence for a parallel increase in aphid and coccinellid abundance has been found to vary between coccinellid species. In cereal plots where aphid density was experimentally manipulated, numbers of *C. septempunctata* and *Hip. convergens* increased in parallel with aphid density while those of *Hip. tredecimpunctata*, *Col. maculata* and *Hip. parenthesis* were not significantly correlated (Elliott & Kieckhefer 2000). In soybean stands a positive relationship between the aphid *Aphis glycines* and coccinellid abundance was demonstrated for *Har. axyridis*, while with *C. septempunctata* increasing aphid abundance only prolonged residence time in experimental plots (Costamagna & Landis 2007). Only a few studies found that aphid and coccinellid abundance were not correlated (Sakuratani 1977). A positive correlation between the size of aphid colonies and the abundance of *A. bipunctata* was demonstrated in stands of three weeds, stinging nettle (*Urtica dioica*), scentless mayweed (*Tripleurospermum maritimum*) and wormwood (*Artemisia vulgaris*), where abundance of this ladybird and, in parallel, its dominance in coccinellid communities increased with aphid abundance and colony size (Honěk 1981).

Minor details of prey distribution may affect the decision by coccinellids to invade stands of a host plant. Coccinellids were found to aggregate more on grouped than on isolated aphid-infested plants of maize (Sakuratani et al. 1983) and alfalfa (Evans & Youssef 1992). The effect of aphid aggregation was also demonstrated in wheat stands where *A. bipunctata* preferred being on inflorescences (Hemptinne et al. 1988) where aphid populations are more densely aggregated than on leaves.

Recent studies have revealed mechanistic feedbacks on coccinellids of their effects on aphid behaviour. The **anti-predator defensive behaviour of prey** may reduce foraging by coccinellids and confound the density relationship between them. This behaviour

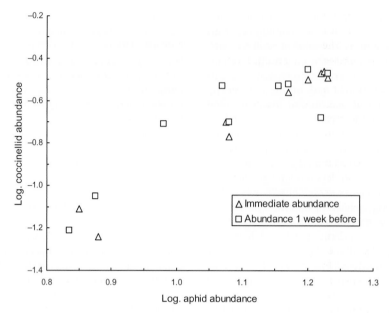

Figure 4.2 The cumulative abundance of adult *Menochilus sexmaculatus* and *Coccinella transversalis* in bean crops infested with *Aphis craccivora* in India. The figure shows the relationship between coccinellid and aphid abundance measured on a particular day (immediate abundance), and the abundance of coccinellids in relation to aphid abundance established 1 week earlier (redrawn from Agarwala & Bardhanroy 1999).

includes the **prey dropping** from the host plant, as happened in 60% of encounters between the aphid *Acyrthosiphon pisum* and *C. septempunctata* on alfalfa while only 7% of aphids drop spontaneously. The anti-predator dropping behaviour is specifically directed against coccinellid attack. Encountering heteropteran predators, *Nabis americoferus, Geocoris punctipes* and *Orius insidiosus*, resulted in only 14% drop of the aphids. Adult aphids had a significantly higher propensity to drop than nymphs but the density of aphids had no effect (Losey & Denno 1998). Also dropping of the aphid *Macrosiphum euphorbiae* from potato plants was increased by interaction with *Har. axyridis* (Narayandas & Alyokhin 2006). Dropping of aphids from host plant may lead to their predation on the ground, e.g. by carabids (Losey & Denno 1999). The efficiency of dropping as a protective strategy of the prey was demonstrated in *Har. axyridis*, whose predation of *A. pisum* doubled when dropping was artificially prevented (Francke et al. 2008). In practice, however, only a few encounters of coccinellids with aphid prey result in predation. Thus in *Hip. convergens* preying on *A. pisum* the ratio of encounters leading to

prey disturbance only to encounters resulting in prey consumption was 30:1 (Nelson & Rosenheim 2006). Whether similar mechanisms of predator–prey interaction are important in determining abundance of particular coccinellid species and affecting composition of coccinellid communities of particular habitats remains uncertain.

In addition to prey abundance, qualitative **taxonomic differences between prey species** are important for stenophagous coccinellids, and probably also for generalists. A review of prey specificity is given in 5.2. How might prey preferences influence selection of a particular habitat? In their review, Sloggett and Majerus (2000) address the concept of 'habitat preferences' from the point of view of diet specificity. They argue that limitations of consumption and competition for prey were probably important factor in the evolution of coccinellid preferences. Dietary specialization and associated preference for particular habitats has probably occurred in some lineages derived from generalist ancestors to avoid costs associated with migration between habitats and prey switching. Adaptation to particular food could be rather quick (Rana et al. 2002). In the generalist *A. bipunctata*, *A. pisum* is a preferred and *A. fabae* a non-preferred, but often consumed, aphid species. The initially poor performance of *A. bipunctata* improved during six generations of rearing on *A. fabae* prey, as development time decreased and adult weight and longevity increased. After six generations of selection, the strain by then adapted to *A. fabae* performed worse on *A. pisum* diet than the strain reared continually on that species. Thus the specialization on one kind of prey entails a trade-off in performance on another.

The 'innate' mechanisms leading to **preference for a particular species of prey** include perception of its chemical composition which may be **reflected** already **by its odour**. This was shown in females of *Cycloneda sanguinea* which were attracted to tomato plants infested by their preferred prey the aphid *M. euphorbiae* as opposed to plants infested by non-preferred prey, the mite *Tetranychus evansi* (Sarmento et al. 2007).

The **size** of the coccinellid relative to that of the prey is also a factor determining coccinellid preference. An analysis of the prey of ladybirds indicates that, the larger the ladybird species, the larger also the prey and/ or the greater the mobility (Dixon & Hemptinne 2001). The smallest ladybird species feed on mites, and the largest on caterpillars and beetle larvae. On a global scale, the size of coccinellid species inhabiting a geographic area is correlated with the size of the prey prevailing in this area. This means that the ratio of the numbers of species of aphidophagous and coccidophagous ladybirds in the Nearctic and Palearctic regions reflects the ratio of the number of species of aphids to the number of species of coccids in the two regions. Large coccinellid species, in particular the predators of large and/or active prey, also lay larger eggs than small species which are predators of small and/or slow moving prey (5.2.3; 5.2.4).

Variation of **quality acquired during prey development** may also influence coccinellid performance. The same prey fed different food may have different nutritive value for coccinellids. The aphid *A. pisum* reared on alfalfa had a 6.3-times greater content of myristic acid, a 2.7-times higher content of total fatty acids and a 1.7-times higher caloric content than *A. pisum* reared on beans. Individuals of *A. pisum* with as high a fat content fed to larvae of *Col. maculata* and *Hip. convergens* decreased their mortality and development time. This beneficial effect of high quality prey was significant when aphids were scarce, but not significant if food was in surplus (Giles et al. 2001, 2002).

Also the aphid *A. fabae* reared on hoary orache (*Atriplex sagittata*) was toxic to *A. bipunctata* while the same aphid species reared on other host plants was an acceptable (though not preferred) food for this coccinellid (Kalushkov 1998).

Oviposition follows successful predation, and so gives rise to the community of larvae. The number of eggs laid and the duration of oviposition are influenced by several factors. In the field, *C. septempunctata* start to oviposit after the aphid population has attained density of about one aphid per $300\,cm^2$ of leaf area, a threshold similar in different crops (Honěk 1980). In laboratory experiments with *A. bipunctata*, a species that prefers aggregated aphid populations, the minimum density for laying eggs was four-times higher at two aphids per $150\,cm^2$ leaf area, and females required at least 10 aphids per $150\,cm^2$ to achieve maximum oviposition (Hemptinne & Dixon 1991). In the field (an alfalfa crop), the threshold aphid density was not reached everywhere at the same time, and this caused a conspicuous asynchrony in ovariole maturation of *C. septempunctata* in different parts of the field (Honěk 1978b). As mentioned earlier in another context, limited prey consumption results in females producing only modest numbers of eggs given their restricted resources, and these eggs are frequently not laid and remain retained in the ovariole. As a consequence, the females are ready to lay eggs quickly when they succeed in finding aphids in high numbers (Evans 2003). In the field, the number of eggs deposited (Wratten 1973, Wright & Laing 1980, Ives 1981b, Kawauchi 1981, Ferran et al. 1984, Ghanim et al. 1984, Coderre et al. 1987) as well as egg cluster size (Agarwala & Bardhanroy 1999) is then frequently directly proportional (on a log–log plot) to the density of aphids. This is not only because of an increase in fertility but also because of an increasing number of coccinellid females (Neuenschwander et al. 1975, Turchin & Kareiva 1989, Ofuya 1991). Furthermore, females discriminate between preferred and nonpreferred prey and lay eggs preferentially in patches of suitable prey. Under laboratory conditions *A. bipunctata* laid 13 eggs /8 hours when provided with a suitable aphid, *A. pisum*, 11.4 eggs with the moderately suitable *A. fabae* and only 7.6 eggs when provided with the toxic aphid *Megoura viciae* (Frechette et al. 2006). The enhancing effect of prey abundance on oviposition is further magnified by suitable substrates available for oviposition. The availability of a suitable site was a significant factor in the selection of oviposition site and determination of cluster size in *Aphidecta obliterata* and *A. bipunctata* (Timms & Leather 2007).

4.3.2.2   Host plant

The prey of coccinellids is always encountered on its host plant, the effects of which on coccinellid attraction and oviposition are sometimes hard to separate (5.4). Plants, alone or in interaction with prey, are sources of **chemical cues** attracting coccinellid adults. Although the beetles' sensory capacities were thought to be poor, recent studies have revealed that their sense of smell is good (Pettersson et al. 2008, Hatano et al. 2009; and Chapter 9). Firstly, it may be the smell of intact plants that is detected. In the field, adult *C. septempunctata* were more abundant in barley plots containing the weeds creeping thistle (*Cirsium arvense*) and couch grass (*Elytrigia repens*) than in clean barley plots. In olfactometer experiments, adults then showed a significantly more positive response to odours of each of the two weeds than to that of barley alone (Ninkovic & Pettersson 2003). Similarly with *Col. maculata*, Zhu et al. (1999) observed significant electroantenogram responses to volatile compounds produced by intact maize plants at the three-leaf stage, i.e. the period when this crop is attractive to aphids. Odours emanating from host plants have also been found to stimulate coccinellid oviposition. Pieces of Eastern red cedar (*Juniperus virginiana*) wood split from the plant but not damaged by prey species attracted *Cycloneda munda*, *A. bipunctata*, *C. transversoguttata* and *Col. maculata* and initiated oviposition (Boldyrev et al. 1969). Shah (1983) demonstrated that *A. bipunctata* and *C. septempunctata* laid more eggs on the twigs of European barberry (*Berberis vulgaris*) than on the twigs of other woody plants, apple tree (*Malus pumila*) wild cherry (*Prunus avium*), sour cherry (*P. cerasus*), cotoneaster (*Cotoneaster tomentosus* and *C. integerrima*), common honeysuckle (*Lonicera periclymenum*) and common snowberry (*Symphoricarpos rivularis*). The aphids (*Acyrthosiphon pisum*) provided in this experiment to stimulate coccinellid oviposition were evenly dispersed over the experimental twigs but did not cause any damage on all species of experimental plants.

Other studies have demonstrated that **prey-induced plant chemicals** (5.4.1.2) can become arrestant or possibly attractant stimuli for adult coccinellids. *Hip. convergens* was attracted not only to radish leaves infested by *M. persicae* but also to radish leaves cleaned after previous colonization by this aphid (Hamilton et al. 1999). Also *C. septempunctata* females preferred not only odours of turnip, mustard and rape plants (*Brassica* spp.) actually populated with *M. persicae* but also odours of previously damaged and subsequently cleaned leaves (Girling & Hassall 2008). The adults of this species also responded positively to volatiles from barley plants infested or previously infested by the aphid *Rhopalosiphum padi*. while the volatiles emanating from uninfested plants or undisturbed aphids alone (placed on a filter paper and not releasing alarm pheromone) were not attractive (Ninkovic et al. 2001). Well known are the attractant and arrestant effects of **honeydew** (Carter & Dixon 1984, Pettersson et al. 2008). These roles of honeydew were shown in the open in experiments in which a synthetic mimic of honeydew was spread onto the crops to provide food for predators during the initial development of the aphid population. The honeydew mimic was either yeast autolysate or a suspension of the yeast *Saccharomyces fragilis* (as a source of amino acids) and sugar. Spraying the mimic on the fields increased the numbers of different coccinellid species in several crops (Hagen et al. 1971, Ben Saad & Bishop 1976, Nichols & Neel 1977, Evans & Richards 1997).

In contrast to adults, sensing by larvae is of aphid rather than host plant odours. This may help them to find prey on the host plant surface which is large relative to the size and movement capacities of larvae. Searching by larvae of *Hip. convergens* became more intensive after exposure to the odour of tobacco leaves infested with the aphid *Myzus persicae nicotianae* (Jamal & Brown 2001) while searching by *A. bipunctata* increased in response to the odour of crushed aphids, an odour made up mainly of aphid alarm pheromone (E)-β-farnesene (Hemptinne et al. 2000b; 9.3.2).

Of plant characteristics correlated with the presence of particular coccinellid species, the most obvious is **plant stature or architecture**. It is well established that some coccinellid species occur mainly in herbaceous stands, whilst others prefer shrubs or trees (4.4). The differences in coccinellid preference for plant type persist even when the host plants are populated by the same species of aphid (Iperti 1966; Fig. 4.3). The composition of coccinellid communities can even vary with plant species within stands consisting of plants of identical growth type, i.e. herbs or trees (Pruszynski & Lipa 1970). Although it is difficult to show conclusively that it is the particular species of plant that can be the real cause of coccinellid presence, it has been convincingly

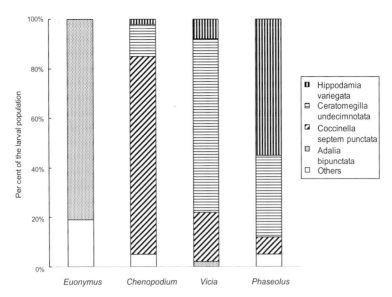

Figure 4.3 Proportion of larvae of different coccinellid species on four host plant species infested with the same aphid, *Aphis fabae*, in southern France (redrawn from Iperti 1966).

demonstrated in some cases. Among three crops, all infested by the aphid *M. persicae* at the same density, *C. transversoguttata* preferred sugar beet (*Beta vulgaris*) while its numbers on broccoli (*Brassica oleracea* Italica group) and radish (*Raphanus sativus*) were lower. *Scymnus marginicollis* preferred radish to broccoli and sugar beet (Tamaki et al. 1981). Sometimes species-specific host-plant effects become manifest only in combination with specific environmental conditions. In a study of herbaceous weeds, coccinellid communities were more and more dominated by *A. bipunctata* as the abundance of aphids increased. The only exception was the coccinellid community on the scentless mayweed (*Tripleurospermum maritimum*) growing on saline soils where *A. bipunctata* was replaced by *C. undecimpunctata*. The communities on stinging nettles (*Urtica dioica*) and wormwood (*Artemisia vulgaris*) were not affected by soil salinity (Honěk 1981). Preference for a particular host plant is sometimes affected by small details of its morphology. Floral architecture is critical in whether or not pollen and nectar of the host plant are accessible to predators. For *Col. maculata*, dill (*Anethum graveolens*) and coriander (*Coriandrum sativum*) both have a floral morphology that complements the head morphology of the adult ladybird (Patt et al. 1997).

That a host plant is suitable for coccinellids is indicated by the presence of larvae (5.2.2). Using this criterion LaMana and Miller (1996) found that, among common North American coccinellids, there were six species that preferred trees and five that preferred herbs. Although adults of all 11 species were present on both trees and herbs, larvae were found only on the preferred plant type. The difference between communities that occur on particular crops may be due to the timing of aphid presence. In Greece, *C. septempunctata* was the most abundant species on durum wheat, while *Hip. variegata* dominated cotton (Kavallieratos et al. 2002). This was probably because the monovoltine *C. septempunctata* dominated the early maturing durum wheat while populations of polyvoltine *Hip. variegata* increased in numbers with the course of the season to dominate in the late maturing cotton.

Co-existence of several coccinellid species on the same host plant is facilitated by them occupying different horizontal strata within the plant stand (4.3.2.3). Musser & Shelton (2003) found the adults of *Col. maculata* in the lower stratum of a maize crop while *Har. axyridis* was less selective (Fig. 4.4). The same vertical distribution of both species was found in field cage experiments by Hoogendoorn and Heimpel (2004). As vertical separation of different coccinellid species

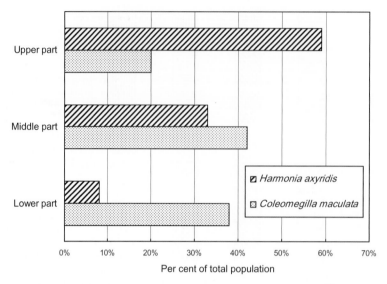

Figure 4.4 Differences in the vertical distribution (as proportion of the population of each species dwelling in the upper, middle and lower third of the plants) of adult *Harmonia axyridis* and *Coleomegila maculata* on a maize crop in the eastern USA (redrawn from Musser & Shelton 2003).

reduces competition between them, preferences for particular strata became exaggerated in cages that contained a mixture of both species; c. 70% of *Col. maculata* were in the bottom and lower sections of the maize plants and c. 80% of *Har. axyridis* were in the middle and top sections. In wheat, *A. bipunctata* preferred the inflorescences while *P. quatuordecimpunctata* preferred the lower parts of the plants (Hemptinne et al. 1988). In cotton, *Hip. convergens* and *Col. maculata* preferred leaves and terminals early in the season but later on the preferences differed and *Hip. convergens* was found on the fruits, whereas *Col. maculata* was found on both leaves and fruits (Cosper et al. 1983).

Coccinellid communities may also differ between **plants** of the same species but different **age** because ageing is accompanied by changes in size. Klausnitzer (1968) found a much higher incidence of *Myrrha octodecimguttata* adults (27%, *n* = 188) in crowns of old large trees than on young trees. By contrast, *Scymnus nigrinus*, the second species in order of abundance on the young trees was scarce in the crowns of old trees. Gumos and Wisniewski (1960) compared 10 and 40 year old pine stands and found a fourfold abundance of *Anatis ocellata* on the older trees. In general, coccinellid communities of young pines contained more

species with a broader ecological range than did those of the crowns of old pines which are inhabited by a lower number of stenotopic species. The succession of coccinellid species in the course of ageing of red pine stands infested with the aphid *Schizolachnus piniradiatae* in Canada was apparently brought about by a gradual decrease in prey density (Gagné & Martin 1968). *Coccinella transversoguttata* and *Scymnus lacustris* were dominant in younger stands, whereas *Mulsantina picta* and *Anatis mali* were most common in older plantations. In Israel, *Chil. bipustulatus* was more abundant in mature citrus groves (23–30 year old) than in young ones (7–9 year old) (Rosen & Gerson 1965). The causes of differences in attractivity for coccinellids other than tree size (e.g. microclimate) remain to be studied.

Effects of intraspecific **genetic differences** between host plants have been investigated mainly because of public concerns over food safety and the environmental hazards of transgenic and herbivore-resistant crops in general. No difference in the abundance of *C. septempunctata*, *Har. axyridis* and *Hip. convergens* were found between transgenic potatoes containing Cry3A gene for *Bacillus thuringiensis* (Bt) endotoxin and classical non-transgenic ones (Riddick et al. 2000; Fig. 4.5). Direct exposure to the Bt toxin expressed in transgenic plants had little effect on the activity and

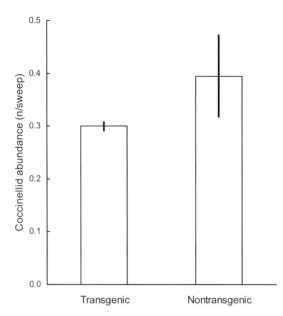

Figure 4.5 Cumulative abundance of adult *Coccinella septempunctata*, *Harmonia axyridis* and *Hippodamia convergens* on transgenic and nontransgenic potato. Average (±SE) results for 1994 and 1995. The differences between particular species and years were not significant (redrawn from Riddick et al. 2000).

abundance of *Menochilus sexmaculatus* (Dhillon & Sharma 2009). Differences in coccinellid abundance were only found (Torres & Ruberson 2005) when use of pesticides on transgenic crops ceased, and the increased numbers of predaceous Coccinellidae became part of an integrated pest management system (Obrycki et al. 2009). Selection of crops for resistance to phytophagous pests thus may result in decreasing abundance of pest populations with only minor effects on coccinellid assemblages so that the standard correlation between cocinellid and prey density (4.3.2.1) does not hold so strongly with host plant resistance. The synergism of lowering pest numbers as a consequence of breeding crops for resistance and simultaneous preservation of high coccinellid numbers in the crop stands may have positive consequences for using biological control (van Emden 2010). In stands of spring wheat cultivars resistant or susceptible to the aphid *Diuraphis noxia*, there were no differences in abundance of *C. septempunctata*, *C. transversoguttata*, *C. trifasciata* and *Hip. convergens* (Bosque-Perez et al. 2002). Francis et al. (2001) showed small differences

in oviposition and larval development of *A. bipunctata* fed the aphid *M. persicae* on bean (*Vicia faba*, glucosinolate free), rape (*Brassica napus*, low glucosinolate content) and white mustard (*Sinapis alba*, high glucosinolate content). However, the effect of host plant substances on coccinellid performance mediated via the prey is not always negligible. Thus endophytic fungi (endophytes) in the host plant may change plant and prey quality with effects on coccinellids. Mycotoxins from perennial ryegrass (*Lolium perenne*) infected with *Neotyphodium lolii* were transmitted by the aphid *R. padi* to *C. septempunctata*, whose development took longer and adult survival and fecundity were reduced (de Sassi et al. 2006) (5.2).

The host plant may further affect coccinellid communities by providing **supplementary food** (=non-prey food), pollen, nectaries or damaged fruits attractive to adults and/or enhancing larval performance (Lundgren 2009). Pollen is particularly important for larvae of some species. With *Col. maculata*, oviposition and larval density in maize stands increased after anthesis but the abundance of adults did not. Nevertheless, larvae fed only maize pollen showed extended development time and finally decreased adult weight and fecundity compared to beetles reared on aphids (Honěk 1978b, Lundgren & Wiedenmann 2004, Lundgren et al. 2004). Also Cottrell and Yeargan (1998a) reported an increase of larval populations of *Col. maculata* in plots with abundant pollen. Nectar, provided by flowers or extrafloral nectaries supplement coccinellid food. Host plants with more nectaries may be more attractive to adults but the presence of such supplementary food may decrease predation by them. Under experimental conditions there was a significant reduction in predation on the aphid *Aphis spiraecola* by *Har. axyridis* on apple shoots when a peach shoot with extrafloral nectaria was also provided (Spellman et al. 2006). Damaged fruits also provide additional sources of food for the beetles. *Harmonia axyridis* adults preferred damaged pumpkin, apple, grape and raspberry fruits over intact fruits, because of the higher sugar content, and not only as a water source (Koch et al. 2004).

Finally, **plant surfaces** are important in the searching and prey capture efficiency of both larvae and adults (5.4). Cottrell and Yeargan (1999) established that larvae of *Col. maculata* born on hophornbeam copperleaf *Acalypha ostryaefolia* were unable to leave because they could not pass over the glandular trichomes on the leaf petioles. The only way they could

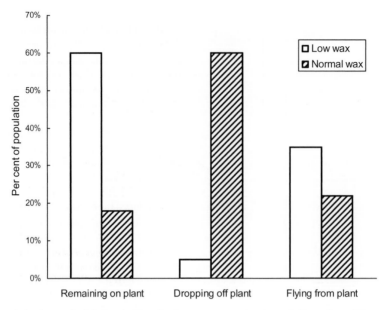

Figure 4.6 Different behaviours of adult female *Hippodamia convergens* on two isogenic lines of pea (*Pisum sativum*), one with a leaf surface with reduced wax cover ('low wax') and the other covered with a normal wax bloom ('normal wax') (redrawn from Eigenbrode et al. 1998).

escape was by dropping to the ground. The effect of plant surfaces on searching activity is well documented. Reduction in surface waxbloom in isogenic lines of pea (*Pisum sativum*) improved the effectiveness of movement and prey capture by adult *Hip. convergens* significantly at low prey densities (Eigenbrode et al. 1998) (Fig. 4.6). Indeed, the waxbloom on most pea varieties is slippery enough to cause many *Hip. convergens* and *C. septempunctata* to fall. Experiments by Kareiva and Sahakian (1990) showed percentage falls for the two coccinellids to be, respectively, 32 and 47%, while the comparable figures on 'leafless' pea varieties were only 9 and 26%. The latter varieties are resistant to powdery mildew *Erisyphe polygoni* because the leaf area has been largely replaced by a profusion of green stipules and tendrils; particularly the tendrils provide a firm hold for ladybirds. As a result, predation is much improved, and leafless peas show an apparent resistance to *A. pisum* which they do not possess in the absence of ladybirds. Similarly, predation of the lepidopteran *Plutella xylostella* larvae by *Hip. convergens* adult females was significantly greater on the less waxy 'glossy' than on normal waxbloom isogenic plants, again because of better adhesion on the former

(Eigenbrode & Kabalo 1999). Glandular trichomes of tobacco leaves significantly decreased the search speed of *Hip. convergens* larvae (Belcher & Thurston 1982).

4.3.2.3 Microclimate

The effects of microclimate are obviously associated with the density and structure of host plants. It is thus hard to discriminate between the effect of both, and the importance of microclimatic effects is therefore frequently underestimated.

The impact of microclimate on the distribution of coccinellid adults and larvae in cereals was demonstrated by Honěk (1979, 1982a). The **density** of cereal crops may vary widely and this influences plant, air and soil surface temperatures as well as humidity. On bright June days, the temperature difference between the coolest and warmest sites within a cereal crop may be as high as 17°C. Adult *C. septempunctata* and *C. quinquepunctata* preferred sparse, well-insolated areas with a warm microclimate, while *P. quatuordecimpunctata* was less choosy and tolerated dense areas with a cooler microclimate. Adult preferences at different

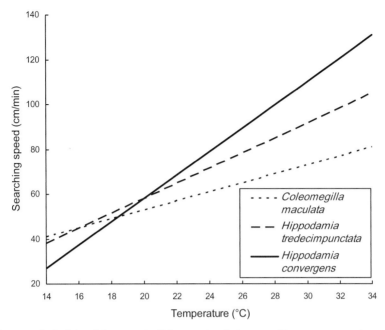

Figure 4.7 Searching speed of adults of three coccinellid species in relation to ambient temperature in stands of spring cereals in central USA (redrawn from Elliott et al. 2000).

places were correlated with larval density and apparently also with larval preferences (Honěk 1983). Preferences for a particular microclimate may be reflected in the daily movements of coccinellids on plants. During the cool morning hours, adult *C. septempunctata* may bask on the upper parts of the plants or on bare ground (Honěk 1985a). Within-plant movement may be facilitated by **innate circadian activity** of the species, as in *Col. maculata* (Benton & Crump 1981) or *C. septempunctata* (Zotov 1983, Nakamuta 1987; 5.4.1.1).

The causes of differences in vertical distribution of particular coccinellids on host plants are often preferences for **specific microclimate**. In crops of maize and barley, Ewert & Chiang (1966) found that *Hip. convergens* and *Hip. tredecimpunctata* preferred the upper parts of the plants, while *Col. maculata* preferred the lower parts. These differences could be attributed to the vertical gradient of decreasing humidity towards the top of the crop. The preferences of the coccinellids for plant stratum were correlated with their resistance to desiccation. *Coccinella septempunctata* prefers the top section of plants regardless of their height, probably because this provides the opportunity of **thermoregu-**

latory basking. A similar vertical distribution of *C. septempunctata* was observed on tall hop plants (J. Zelený, unpublished) and on about 10-times shorter dense cereal stands (A. Honěk, unpublished).

Recently, microclimate has attracted little attention but behavioural studies support earlier conclusions. Thus *C. septempunctata*, which prefer a warm microclimate, lay eggs at warm places which they may often find on the surface of unnatural objects including metal cans, wood and concrete structures (Sakuratani & Nakamura 1997). For pupation, the larvae select sites exposed to solar radiation, and pupate in positions that optimize body warming (Sakuratani et al. 1991). In warm-adapted *Hip. convergens*, *Hip. tredecimpunctata* and *Col. maculata*, the proportion of time spent in flight activity, searching for aphids and handling time to eat them were positively correlated with temperature (Elliott et al. 2000; Fig. 4.7).

4.3.2.4 Landscape

Recently much attention has been paid to the **effects of surroundings**, spatial distribution and composition of plant stands as well as to the effects of orographic diversity of landscape, on the coccinellid community at a particular habitat. The effect of alien plant species was first shown as an influence of other plants mixed into the stand of host plant of the prey. Coccinellids frequently benefit from diversification created by **planting different crop** species **together**. The abundance of *Col. maculata* was increased by growing beans (*Phaseolus vulgaris*) or squash (*Cucurbita maxima*) with maize providing a source of pollen and aphids (Wetzler & Risch 1984, Andow & Risch 1985). A positive **effect of weeds** in a crop stand on reducing phytophagous pests and/or increasing coccinellid density has frequently been demonstrated (Horn 1981). As an example of this, the abundance of *Col. maculata* was higher in maize crops containing the weed hophornbeam copperleaf (*Acalypha ostryaefolia*) than in weed-free stands (Cottrell & Yeargan 1998b) because the ladybird preferred to oviposit on the weed (Cottrell & Yeargan 1999).

Coccinellid communities on the prey host plant are also influenced by plants growing at the **edge of a crop**. The adults leave crops to evaluate the quality of surrounding habitats, a movement termed 'resource mapping'. This opens access to spatially separated resources and may cause an 'edge effect', i.e. the accumulation of individuals at the edge of the crop (Ries et al. 2004). **Adjacent uncultivated** land may also provide refuge and food for ladybirds when prey is scarce on the agricultural crops and to bridge gaps in synchronization with the pest species serving as prey. Parts of a field may also have a similar function (van den Bosch et al. 1959). These results led to the idea of '**strip cutting**' of alfalfa in California in the 1950s (Schlinger & Dietrick 1960). This practice allowed the movement of adult *C. transversalis* from harvested to unharvested strips during the 24 hours following a harvest. The coccinellids colonized these strips near the edge; therefore densities decreased with increasing distance from the borders with the harvested strips (Hossain et al. 2002). Such a decline in density of coccinellids also occurs on non-crop plants at the edge of the crop. Wormwood (*Artemisia vulgaris*), tansy (*Tanacetum vulgare*) and stinging nettle (*Urtica dioica*) planted as bordering vegetation, significantly increased the density of adults and larvae of *C. septempunctata*, *A. bipunctata* and *P. quatuordecimpunctata* on lettuce (*Lactuca sativa*) plants compared to the control lettuce field not surrounded by weedy margins (Sengonca et al. 2002). In northern Italy, coccinellids were also more abundant in crops with weedy field edges

bordered by hedgerows of young trees than in crops bordered by hedgerows consisting of old trees (Burgio et al. 2006).

The increasing interest in nature conservation has made it clear that **fragmentation** of the habitat, the size of fragments and their distance one from the other can affect coccinellid abundance. In a pioneer study, Kareiva (1987) showed that, in a stand of Canada golden-rod (*Solidago canadensis*) populated by *Uroleucon nigrotuberculatum* aphids, that fragmentation of the stand into 1×1 m plots led to more frequent local explosions of aphid than in unfragmented area. The population dynamics of *U. nigrotuberculatum* in the fragmented plots became less stable than in the unfragmented area because of differential predation by *C. septempunctata*. Experimentally established aphid colonies were eliminated (eaten by ladybirds) twice as frequently in the unfragmented area (67% of colonies destroyed) than in the fragmented area (only 30% of colonies destroyed). In contrast, Grez et al. (2004) showed that dividing an area of alfalfa into several plots separated by bare ground, resulted in coccinellid abundance being only slightly influenced by plot size and distance between them. The optimum for coccinellid abundance was fragmentation into small patches a short distance apart. The **effect was transitory** and influenced *Hip. variegata* and *Hip. convergens* more than *Eriopis connexa*. The same authors later showed that the effect of the loss of habitat area (84% of the alfalfa crop removed from the 30×30 m experimental plots) and extent of its fragmentation (experimental plots divided into 1, 4 or 16 fragments) did not affect overall coccinellid abundance but that particular species, *Hyperaspis sphaeridioides* and *A. bipunctata* were most abundant in plots where 84% of alfalfa stand had been removed. The difference in population density persisted for long time only for *A. bipunctata* (Grez et al. 2008). Coccinellids avoided bare ground, remained longer in alfalfa plots than on bare ground and were more prone to move between crop fragments if these were close to each other. Coccinellids tended to stay longer in non-fragmented than in fragmented plots, although aphid abundance was similar throughout the experimental area (Grez et al. 2005). In experiments by Banks (1999), strips of broccoli were interspersed with sections sown to weeds representing 25, 50 and 75% of the total strip area. The strips containing particular percentages of weed cover were sown each in three replicates with weedy area divided into 2, 4 or 8 fragments. The abundance of *C. septempunctata* was not affected by the proportion of total strip area sown by weeds but was significantly but not consistently influenced by the extent of its fragmentation.

Finally, the global **character of the landscape**, which includes not only fragmentation and composition of the vegetation cover but also the agriculture practices used, has its effects on coccinellid abundance. Several examples have accumulated in the last years. Bianchi et al. (2007) showed that the long-term decrease in the abundance of *C. septempunctata* in the Czech Republic may have been caused by **changes in agricultural practices**, including reduced use of fertilizers and changes to the crop species grown. These changes occurred after agricultural production ceased to be subsidized. *C. septempunctata* only partially coped with these changes in spite of the plasticity of its life cycle based on 'phenotypic polymorphism', i.e. the element of its plasticity that enables it to occupy a wide ecological niche and extensive geographical area (Hodek & Michaud 2008). In the USA, Gardiner et al. (2009) showed that domestic *Hip. convergens* was replaced by the invasive *C. septempunctata*, *Har. axyridis*, *Hip. variegata* and *P. quatuordecimpunctata*, but less so in grassland dominated landscapes with low structural diversity and low amount of forests than in landscapes with other characteristics. Intensively managed agricultural areas therefore represent **landscape-scale refuges** for native coccinellid diversity (Gardiner et al. 2009). The effect of landscape varies with geographic area and more information on the impact on coccinellids is needed from the tropics. In a study in Swaziland, Magagula & Samways (2001) showed that the highest coccinellid abundance and diversity was in orchards, smaller in riparian areas and least in dry savannah. Prey availability was the most important determinant of coccinellid presence; humidity, temperature and ground insolation had no effect.

Switching between habitats is important for the maintenance of coccinellid populations, which in this way can exploit **seasonal changes** of quality in different habitats. A number of case studies have documented coccinellids switching between stands of wild herbs, trees, shrubs and crops. In a typical case, which represents the life cycle of coccinellids in temperate zones, adults leave the overwintering sites to feed on wild herbs, shrubs or trees populated with prey early in the season, move between them, then fly to crops which have been infested early by aphids; they occasionally change between crops where they reproduce. Adults of the last generation of the year finally

again move to tree, shrub and wild herb hosts before hibernation. These movements have been amply documented for coccinellids in Europe (Banks 1955, Lusis 1961, Hodek et al. 1962, 1966, Iperti 1965, Honěk 1982b,1989, Hemptinne & Naisse 1988) and central Asia (Yakhontov 1966, Aleksidze 1970, Savoiskaya 1970) and hold also for North America where the distribution of coccinellid species follows the sequence of aphid populations as they become available in different crops (French et al. 2001). The factor determining these seasonal migrations between habitats is the search for food. In the spring, *Col. maculata* **migrate from hibernacula** to first arrive in wheat, but later on in June–August they move to maize because of the changing availability of aphids (Nault & Kennedy 2000). Similar food-motivated changes of habitat were observed in subtropical and tropical areas (Wiesmann 1955, Brown 1969, Iperti et al. 1970, Laudého et al. 1970).

Movement between habitats is by **flight** (6.3.1.3). Most displacements are probably accomplished by trivial flights, which occur near the ground. Boiteau et al. (1999) studied the altitudinal distribution of flights using window traps placed 0.8–14.3 m above the ground. Maximum catches, consisting mainly of *Hip. convergens*, *C. septempunctata* and *C. trifasciata*, were at 0.8 m. The flight of the other 18 species caught in this study was more evenly spread with respect to altitude. Sarospataki & Marko (1995) also established abundant trivial flight activity in a short period at the end of June of *C. septempunctata* at 0–2 m above ground level while migrants captured above the vegetation at 25–27 m height were very few.

4.3.3 Dominance, diversity and niche differentiation

The structure of coccinellid communities in different habitats shares common features that enable a few generalizations to be made. The spatial and trophic capacity of any habitat is limited, and the length of time that suitable conditions persist is rather short. Thus only a limited number of individuals and species may successfully survive and breed at a given place. This determines the general similarity in the numbers of species and their order of dominance in coccinellid communities in different parts of the world.

At any habitat, the coccinellid community will contain a few **dominant species**. Usually two to four species will represent more than 90% of the individuals. The other coccinellids, although often several times greater in number of species than the dominant ones, yet represent only a small fraction of the total number of individuals present. Comparative studies of diversity and dominance were made in maize crops and apple orchards in Hungary (Loevei 1981, Radwan & Loevei 1982, 1983, Loevei et al. 1991). Larval, pupal and adult communities in the maize crops grown under three different regimes consisted respectively of 7, 6 and 11 species, and the proportion of the most abundant species in the total population of each development stage ranged between 0.42–0.93, 0.31–0.47 and 0.41–0.68, respectively. The results for the apple orchards were similar regardless of whether they were treated with insecticides or not.

A frequent aim of habitat studies is to define the **niches** of individual coccinellid species and empirically compare the differences between these niches. The concept of niche with regard to coccinellids was reviewed by Snyder (2009). The species may be separated one from the other by using different habitats (**spatial niche**), using the same habitat at different time (**temporal niche**) or using a specific prey (**prey choice niches**). The knowledge necessary to recognize different kinds of niche separation in the open is to determine the relative abundance of a species which means its proportion in the sample of the coccinellid community of a habitat, and identifying the habitat by a number of characteristics each of which could influence coccinellid abundance. Honěk & Rejmánek (1982) classified 45 samples of adults captured in stands of 14 crop, wild herb and tree species (habitats). Five groups of communities were distinguished by differences in the dominance of particular species. Three groups consisted of communities of **herbaceous stands** inhabited by aphid populations of low or moderate density, mainly field crops. Communities of each of these three groups were dominated by one of the common species: *P. quatuordecimpunctata*, *C. septempunctata* or *Hip. variegata*. The other two groups consisted of communities inhabiting **trees or wild herbs** with dense aphid populations. These differed each other in the relative proportions of *A. bipunctata* and *C. septempunctata*. Using this data retrieved from habitat studies then enabled the calculation of **niche breadth for each species** and the overlap with the niches of other species (Honěk 1985b). Nedvěd (1999) developed this analysis further, and classified the preferences of particular species as reflected by their presence

on host plant species which were each infested by a single species of aphid. He distinguished eight groups of coccinellid species which we list here with their main representatives: (i) tree canopy species (*Myrrha octodecimguttata*, *Har. quadripunctata*), (ii) steppe species also inhabiting field localities (*C. quinquepunctata*, *Scymnus frontalis*), (iii) tree-inhabiting species not restricted to feeding on aphids (*Calvia quatuordecimguttata*), (iv) generalist species characteristic of shrubs and trees (*A. bipunctata*), (v) common field species (*C. septempunctata*), (vi) species restricted to spruce (*A. conglomerata*), (vii) species restricted to oak (*Scymnus auritus*, *Stethorus punctillum*) and (viii) species restricted to pine (*Anatis ocellata*). Coderre et al. (1987) measured niche differences for oviposition preference, between four species of aphidophagous predators, including *Col. maculata* and *Hip. tredecimpunctata*. Oviposition preferences were clearly separable on each of the three principal axes which may be interpreted as (i) aphid abundance, (ii) timing of oviposition and vertical distribution of egg clusters on the host plant, and (iii) distance from the field margin.

4.4   COCCINELLID COMMUNITIES OF PARTICULAR HABITATS

As was pointed out in 4.1, the identification of coccinellid habitats according to units created by humans may be unjustified. By dividing the landscape in this way we may overlook differences in environmental quality that are unimportant from the human perspective but important for coccinellids. An apparently uniform stand thus may represent a mosaic of conditions perceived by coccinellids as different habitats. Nevertheless, faunas of different stands as distinguished by humans continue to be studied and here we summarize some recent results. The stands are classified into three groups: crops, wild herbaceous plants and trees. For each group the results are presented separately for the Palearctic (Europe) and Nearctic (USA and Canada) regions.

Stands of **crops** host ladybird communities made up of only few species. This is true for Europe (Table 4.1) where coccinellid communities were dominated (dominant species are those that make more than 5% of the total coccinellid population) by *C. septempunctata*. Less important were *C. quinquepunctata* and *P. quatuordecimpunctata*. The crop stands have not yet been dominated by the invasive *Har. axyridis* (Ameixa 2010).

Table 4.1 Relative abundance (%) of coccinellid species on crops, Europe.

	A	B	C	D
Adalia bipunctata	—	4.7	—	0.3
Adalia decempunctata	—	0.6	—	—
Anisosticta novemdecimpunctata	—	—	—	0.1
Aphidecta obliterata	—	—	—	0.1
Chilocorus bipustulatus	—	—	—	0.1
Chilocorus renipustulatus	—	—	—	0.1
Coccidula rufa	10.7	—	0.4	2.2
Coccinella hieroglyphica	+	—	—	—
Coccinella quinquepunctata	2.8	—	9.1	6.8
Coccinella septempunctata	86.0	89.2	90.5	80.1
Coccinula quatuordecimpustulata	0.1	—	—	0.2
Hippodamia tredecimpunctata	—	1.4	—	0.6
Hippodamia variegata	+	1.4	—	—
Myzia oblongoguttata	—	—	—	0.1
Propylea quatuordecimpunctata	0.4	2.5	—	9.0
Scymnus frontalis	—	—	—	0.1
Scymnus haemorrhoidalis	+	—	—	0.2
Scymnus suturalis	—	—	—	0.1
Scymnus sp.	—	0.3	—	—

+ species recorded in <0.1% frequency.
A Finland, cereals, N = 5824 (Clayhills & Markkula 1974);
B Serbia, cereals, N = 361 (Tomanovic et al. 2008);
C Finland, potato. N = 504 (Clayhills & Markkula 1974);
D Finland, forage leguminosae, N = 1752 (Clayhills & Markkula 1974).

The communities of crops in the Nearctic are more diversified. Six species of coccinellids may become dominant in the communities of cereals (Table 4.2) and nine species in the communities of other crops (Table 4.3). This is because in communities of crops the dominant positions may be taken not only by native species typical for the herbaceous stratum, *Col. maculata*, *Cycloneda munda*, *Hip. convergens*, *Hip. parenthesis*, *Hip. sinuata* and *Hip. tredecimpunctata*, but also by species introduced from the Palearctic, *C. septempunctata* and *P. quatuordecimpunctata*. *Harmonia axyridis*, a species which in Europe is typical for communities of tree stands, became dominant in maize (Lucas et al. 2007, Musser & Shelton 2003), medicinal crops (Lucas et al. 2007) and soybean (Lucas et al. 2007, Schmidt et al. 2008). The coccinellid communities of crops in the Nearctic are thus originally more diversified but apparently less resistant to invasion of alien species than the communities of Europe.

The coccinellid communities inhabiting stands of **wild herbs** in Europe (Table 4.4) are more diversified than the communities inhabiting crops. This is because the coccinellid fauna of wild herbs consists both of species typical for the herbaceous stratum (*C. septempunctata*, *Hip. variegata*, *P. quatuordecimpunctata*) and of species typical for trees (*A. bipunctata*).

Trees host still more diversified communities of coccinellids. Recent studies in Europe (Table 4.5) have concerned communities of deciduous trees in the western and northern parts of the continent (Clayhills & Markkula 1974, Leather et al. 1999). Interestingly, the community of a **Mediterranean** citrus orchard was similar to that of deciduous trees (Kavallieratos et al. 2004). In contrast, a study of communities of **coniferous** trees (Selyemova et al. 2007) pointed out the distinct difference from broad-leaved trees. Several studies of North American apple orchards (Table 4.6) showed a dominance of alien species *C. septempunctata*, *Har. axyridis* and *P. quatuordecimpunctata* over native species (Brown & Miller 1998, Brown 2004, Lucas et al. 2007). On wild trees (Table 4.7) the dominance of alien species in the coccinellid communities was less prominent but *Har. axyridis* dominated in one study of Oregon (LaMana & Miller 1996).

Unlike European and North American communities, those of other parts of the world have been less studied (Table 4.8). Data from the East Palearctic (Yu 1999) and Africa (Woin et al. 2000, Magagula & Samways 2001) indicate that communities of **tropical** and **subtropical** areas may be more diverse than those of temperate zones. This is, of course, expected considering the greater species richness in the tropics.

The above data illustrate the limitations of current work on coccinellid communities. First, the examples of these communities have nearly all been retrieved from papers published in high-impact journals. Although this may guarantee quality, the other side of the coin is that the data are always published to illustrate something other than coccinellid community structure. Because of the difficulties in publishing raw community data in 'high quality' journals, the majority probably ends up in what is called the 'grey literature', i.e. local publications, conference proceedings or university theses. How much better would it be if this information was collected and subjected to meta-analysis! The second drawback is that there is enormous variation in the quality of species determination. If the data on aphidophagous species were limited to the subfamily coccinellinae, different studies

Table 4.2 Relative abundance (%) of coccinellid species on small grain cereals and maize, North America.

	A	B	C*	D	E	F	G	H	I
Adalia bipunctata	—	—	—	—	0.5	—	0.2	0.4	—
Coccinella septempunctata	7.5	6.5	1	2.0	3.9	2.5	2.1	5.5	3.3
Coccinella transversoguttata	—	1.3	6	—	—	—	1.3	—	—
Coleomegilla maculata	5.0	7.7	2	45.3	35.1	63.1	28.6	49.1	65.9
Cycloneda munda	—	1.1	5	2.0	—	28.0	0.6	—	1.6
Harmonia axyridis	—	—	—	—	54.8	—	—	32.5	—
Hippodamia convergens	59.4	45.6	1	32.4	—	5.1	41.9	—	27.6
Hippodamia glacialis	—	—	—	—	—	—	—	—	—
Hippodamia parenthesis	3.2	10.0	3	6.2	—	0.2	0.7	—	—
Hippodamia sinuata	—	—	5	—	—	—	—	—	—
Hippodamia sp.	—	—	—	—	5.7	—	—	—	—
Hippodamia tredecimpunctata	24.9	27.9	5	12.1	—	1.1	24.7	—	1.6
Olla v-nigrum	—	—	5	—	—	—	—	—	—
Propylea quatuordecimpunctata	—	—	—	—	0.4	—	—	12.6	—
Scymnus spp.	—	—	4	—	—	—	—	—	—

+ species recorded in <0.1% frequency.
*Abundances ranked starting from the most (1) to the least frequent (6) species.
A USA, South Dakota, spring wheat, N = 4883 (Elliott & Kieckhefer 2000); B USA, South Dakota, small grain cereals, N = 5835 (Elliott et al. 1996); C USA, Midwest, wheat, N = not indicated (Obrycki et al. 2000); D USA, Nebraska, N = not indicated (Wright & DeVries 2000); E USA, New York, maize, N = not indicated (Musser & Shelton 2003); F USA, Iowa, maize, sticky trap, N = 1221 (Bruck & Lewis 1998); G USA, South Dakota, maize, N = 20753 (Elliott et al. 1996); H Canada, Quebec, maize, N = 8450 (Lucas et al. 2007); I USA, Nebraska, maize, N = not indicated (Wright & DeVries 2000).

Table 4.3 Relative abundance (%) of coccinellid species on stands of medicinal crops, forage leguminosae, sorghum and soybean, North America.

	A	B	C*	D	E	F	G	H	I	J
Adalia bipunctata	—	—	5	—	0.5	—	—	—	—	—
Anatis quindecimpunctata	—	—	—	—	—	—	—	—	—	0.3
Brachiacantha ursina	—	—	—	—	—	—	—	—	0.5	0.3
Coccinella californica	—	—	—	—	1.8	—	—	—	—	—
Coccinella septempunctata	20.1	5.7	1	17.2	27.4	0.2	0.2	5.2	8.2	5.5
Coccinella transversoguttata	—	1.2	7	—	—	—	—	—	—	—
Coccinella trifasciata	—	—	—	—	48.2	—	—	—	—	—
Coccinella undecimpunctata	—	—	—	—	3.6	—	—	—	—	—
Coleomegilla maculata	51.1	8.6	1	22.6	—	3.5	1.0	23.9	8.3	2.2
Cycloneda munda	—	0.6	3	1.2	—	—	—	0.4	—	2.5
Cycloneda polita	—	—	—	—	1.4	—	—	—	—	—
Harmonia axyridis	9.0	—	—	—	3.7	—	—	—	21.0	78.1
Hippodamia convergens	14.9	48.9	1	54.5	9.4	50.3	63.5	65.4	—	9.9
Hippodamia glacialis	—	—	6	—	—	—	—	—	—	—
Hippodamia parenthesis	—	16.5	2	4.0	—	—	0.1	0.8	1.0	1.0
Hippodamia sinuata	—	—	5	—	7.4	45.0	33.9	—	—	—
Hippodamia tredecimpunctata	—	18.4	5	0.6	—	—	—	4.7	—	0.1
Hippodamia variegata	—	—	—	—	—	—	—	—	1.0	—
Mulsantina picta	—	—	—	—	0.4	—	—	—	—	—
Olla v-nigrum	—	—	5	—	—	+	+	—	—	—
Propylea quatuordecimpunctata	4.8	—	—	—	—	—	—	—	60.0	—
Scymnus loewii	—	—	—	—	—	—	1.2	—	—	—
Scymnus spp.	—	—	4	—	—	0.2	—	—	—	—
Unidentified	—	—	—	—	—	0.2	—	—	—	—

+ species recorded in <0.1% frequency.
*Abundances ranked starting from the most (1) to the least frequent (7) species.
A Canada, Quebec, medicinal crops, N = 305 (Lucas et al. 2007); B USA, South Dakota, alfalfa, N = 7974 (Elliott et al. 1996); C USA, Midwest, alfalfa, N = not indicated (Obrycki et al. 2000); D USA, Nebraska, alfalfa, N = not indicated (Wright & DeVries 2000); E USA, Oregon, clover, alfalfa and peppermint, N = 780 (LaMana & Miller 1996); F USA, Texas, sorghum, N = 2872 (Michels et al. 1996); G USA, Texas, sorghum, N = 29,354 (Michels & Matis 2008); H USA, Nebraska, sorghum, N = not indicated (Wright & DeVries 2000); I Canada, Quebec, soybean, N = 334 (Lucas et al. 2007); J USA, Iowa, soybean, N = 1739 (Schmidt et al. 2008).

Table 4.4 Relative abundance (%) of coccinellid species on wild herbs, Europe.

	A	B	C
Adalia bipunctata	8.3	0.7	50.9
Chilocorus bipustulatus	0.5	—	—
Coccidula rufa	—	+	—
Coccinella septempunctata	4.9	12.0	49.1
Hippodamia variegata	21.6	48.1	—
Oenopia conglobata	2.8	0.3	—
Platynaspis luteorubra	—	+	—
Propylea quatuordecimpunctata	32.8	12.2	—
Psyllobora vigintiduopunctata	6.3	2.1	—
Scymnus apetzi	—	7.1	—
Scymnus auritus	—	1.4	—
Scymnus frontalis	—	6.3	—
Scymnus interruptus	—	0.1	—
Scymnus rubromaculatus	—	8.6	—
Scymnus subvillosus	—	0.4	—
Scymnus sp.	17.3	—	—
Stethorus pusillus	5.1	0.6	—

+ species recorded in <0.1% frequency.
A Italy, mixed herbs adjacent to old hedgerows, $N = 656$ (Burgio et al. 2006);
B Italy, mixed herbs adjacent to young hedgerows, $N = 2112$ (Burgio et al. 2006);
C United Kingdom, grasses and herbs, $N = 112$ (Leather et al. 1999).

Table 4.5 Relative abundance (%) of coccinellid species on trees, Europe.

	A	B	C	D	E
Adalia bipunctata	16.4	36.4	1.1	0.4	0.4
Adalia conglomerata	—	—	—	—	16.3
Adalia decempunctata	0.8	—	—	—	—
Anatis ocellata	—	—	—	—	7.7
Aphidecta obliterata	—	—	—	—	50.3
Calvia quatuordecimguttata	—	—	14.7	0.1	0.6
Ceratomegilla notata	—	—	—	—	2.8
Ceratomegilla undecimnotata	0.8	—	—	—	—
Coccinella quinquepunctata	—	—	—	11.0	+
Coccinella septempunctata	10.8	42.4	84.2	63.0	17.4
Coccidula rufa	—	—	—	0.3	—
Exochomus quadripustulatus	—	—	—	—	+
Halyzia sedecimguttata	—	—	—	—	0.6
Hippodamia variegata	—	—	—	—	0.1
Myzia oblongoguttata	—	—	—	—	0.6
Propylea quatuordecimpunctata	3.0	3.0	—	17.8	1.0
Psyllobora vigintiduopunctata	—	18.2	—	—	0.2
Scymnus abietis	—	—	—	—	1.6
Scymnus apetzi	51.4	—	—	—	—
Scymnus rubromaculatus	6.5	—	—	—	—
Scymnus subvillosus	10.2	—	—	—	—
Tytthaspis sedecimpunctata	—	—	—	—	+

+ species recorded in <0.1% frequency.
A Greece, citrus, $N = 675$ (Kavallieratos et al. 2004); B United Kingdom, deciduous trees, $N = 33$ (Leather et al. 1999); C Finland, apple, $N = 95$ (Clayhills & Markkula 1974); D Finland, black and red currant, $N = 73$ (Clayhills & Markkula 1974); E Slovakia, spruce, $N = 3636$ (Selyemova et al. 2007).

Table 4.6 Relative abundance (%) of coccinellid species in apple orchards, North America.

	A	B	C	D
Adalia bipunctata	3.2	—	—	+
Anatis labiculata	2.2	—	—	—
Brachiacantha ursina	—	—	—	22.0
Chilocorus stigma	—	—	—	+
Coccinella septempunctata	87.8	9.5	26.4	4.5
Coleomegilla maculata	1.7	14.0	2.0	30.0
Cycloneda munda	3.7	—	—	—
Harmonia axyridis	—	76.5	65.5	12.0
Hippodamia parenthesis	—	—	—	+
Hyperaspis binotata	—	—	—	+
Hyperaspis undulata	—	—	—	+
Nephus flavifrons	—	—	—	+
Olla v-nigrum	1.2	—	—	—
Propylea quatuordecimpunctata	—	—	—	15.0
Psyllobora vigintimaculata	—	—	—	+
Scymnus sp.	—	—	6.1	—

+ species recorded in <0.1% frequency.
A USA, West Virginia, 1989–93, N = 306 (Brown & Miller 1998); B USA, West Virginia, 1995–96, N = 276 (Brown & Miller 1998); C USA, West Virginia, N = 148 (Brown 2004); D Canada, Quebec, N = 691 (Lucas et al. 2007).

Table 4.7 Relative abundance (%) of coccinellid species on wild tree stands, North America.

	A	B	C
Adalia bipunctata	13.3	0.3	—
Anatis quindecimpunctata	—	0.3	—
Calvia quatuordecimguttata	0.6	—	—
Chilocorus sp.	0.1	0.3	7.1
Coccinella californica	0.2	—	—
Coccinella septempunctata	5.0	—	—
Coccinella trifasciata	0.8	1.1	—
Coleomegilla maculata	—	0.3	—
Cycloneda polita	3.9	—	—
Exochomus quadripustulatus	3.7	—	—
Harmonia axyridis	69.6	2.4	4.8
Hippodamia convergens	0.6	—	—
Hippodamia sinuata	1.0	—	—
Mulsantina picta	1.2	—	—
Mulsantina sp.	—	—	14.3
Myzia subvittata	0.1	—	—
Propylea quatuordecimpunctata	—	32.1	—
Psyllobora vigintimaculata	—	63.3	73.8

A USA, Oregon, mixed trees and shrubs, N = 2984 (LaMana & Miller 1996); B USA, Maine, yellow traps, deciduous forest, N = 1510 (Finlayson et al. 2008); C USA, Maine, yellow traps, coniferous forests, N = 126 (Finlayson et al. 2008).

Table 4.8 Studies of coccinellid communities out of Europe and North America.

Area	Host plant	No. of coccinellid species in the community	Reference
Cameroon	rice	13	Woin et al. (2000)
China	*Pinus armandii*	18	Yu (1999)
Swaziland	garden– savanna	23 (incl. 3 Epilachninae)	Magagula & Samways (2001)

would be more or less comparable in quality. However, when other subfamilies are included into sampling plans, the accurate determination of the collected material to species is far from common. Third, as stated above, more quantitative sampling of subtropical and tropical communities is necessary to fill the gaps ('white spots') in the distribution map of coccinellid communities.

4.5   CONCLUSION

Here we review knowledge of the distribution of coccinellid species among habitats. This includes the delimitation and characterization of what the habitat is, a description of its coccinellid community and a determination of the factors that attract particular species and hold the community together. We believe that this knowledge in itself is not only fundamental for coccinellid studies but also encourages the proposition of hypotheses. The review part of this chapter carries the risk that gaps which still exist in our knowledge may be hidden, in particular where the delimitation of a habitat has been erroneous, the time required for the development of coccinellid communities may have been underestimated, real causes determining species presence may have been overlooked, and, to us the most dangerous possibility, illusory causes may have been suggested and erroneous proofs of their effects have been done. In our view the new developments of discipline replicated results and strengthened hypotheses provided in Honěk and Hodek (1996). Our rather long experience of coccinellid habitat studies, however, may have made our evaluation of facts and causes too rigid. Our perspective is perhaps too much determined by the geographic conditions in which we have worked. We hope that future researchers, working under different conditions, may bring new ideas and make significant progress in the topic of this chapter.

ACKNOWLEDGEMENTS

The work was supported by grant # 522/08/1300 of the Grant Agency of the Czech Republic.

REFERENCES

Agarwala, B. K. and P. Bardhanroy. 1999. Numerical response of ladybird beetles (Col., Coccinellidae) to aphid prey (Hom., Aphididae) in a field bean in north-east India. *J. Appl. Entomol.* 123: 401–405.

Aleksidze, G. N. 1970. *Adalia bipunctata. Zashchita Rastenij ot Vreditelej i Boleznei* 1970, 12. (In Russian.)

Ameixa, O. M. C. C. 2010. Aphids in changing world. In P. Kindmann, A. F. G. Dixon and J. P. Michaud (eds). *Aphid biodiversity under Environmental Change.*, pp. 21–40. Springer, Dordrecht.

Andow, D. A. and S. J. Risch. 1985. Predation in diversified agroecosystems: relations between a coccinellid predator *Coleomegilla maculata* and its food. *J. Appl. Ecol.* 22: 357–372.

Arefin, V. S. and L. A. Ivliev. 1988. Spatial distribution and determination of population density of cocinellids in soybean agrocenoses in Primorie. *In Rol nasekomykh v biocenozakh Dalnego Vostoka.* Dalnevostokhnyj Otdel Akademii Nauk SSSR Vladivostok. pp. 4–12. (In Russian.)

Banks, C. J. 1955. An ecological study of Coccinellidae associated with *Aphis fabae* Scop. on *Vicia faba. Bull. Entomol. Res.* 46: 561–587.

Banks, J. E. 1999. Differential response of two agroecosystem predators, *Pterostichus melanarius* (Coleoptera: Carabidae) and *Coccinella septempunctata* (Coleoptera: Coccinellidae), to habitat composition and fragmentation-scale manipulations. *Can. Entomol.* 131: 645–657.

Bechinski, E. J. and L. P. Pedigo. 1982. Evaluation of methods for sampling predatory arthropods in soybeans. *Environ. Entomol.* 11: 756–761.

Belcher, D. W. and R. Thurston. 1982. Inhibition of movement of larvae of the convergent lady beetle by leaf trichomes of tobacco. *Environ. Entomol.* 11: 91–94.

Ben Saad, A. A. and G. W. Bishop. 1976. Effect of artificial honeydews on insect communities in potato fields. *Environ. Entomol.* 5: 453–457.

Benton, A. H. and A. J. Crump. 1981. Observations on the spring and summer behavior of the 12-spotted ladybird beetle, *Coleomegilla maculata* (De Geer) (Coleoptera: Coccinellidae). *J. N. Y. Entomol. Soc.* 89: 102–108.

Bianchi, F. J. J. A., A. Honěk and W. van der Werf. 2007. Changes in agricultural land use can explain population decline in a ladybeetle species in the Czech Republic: evidence from a process-based spatially explicit model. *Landscape Ecol.* 22: 1541–1554.

Boiteau, G., Y. Bousquet and W. P. L. Osborn. 1999. Vertical and temporal distribution of Coccinellidae (Coleoptera) in flight over an agricultural landscape. *Can. Entomol.* 131: 269–277.

Boldyrev, M. I., W. H. A. Wilde and B. C. Smith. 1969. Predaceous coccinellid oviposition responses to *Juniperus* wood. *Can. Entomol.* 101: 1199–1206.

van den Bosch, R., E. I. Schlinger, E. J. Dietrick and I. M. Hall. 1959. The role of imported parasites in the biological control of the spotted alfalfa aphid in southern California in 1967 (Coccinellid activity). *J. Econ. Entomol.* 52: 142–154.

Bosque-Perez, N. A., J. B. Johnson, D. J. Schotzko and L. Unger. 2002. Species diversity, abundance, and phenology of aphid natural enemies on spring wheats resistant and susceptible to Russian wheat aphids. *BioControl* 47: 667–684.

Brown, H. D. 1969. *The Predacious Coccinellidae Associated with the Wheat Aphid*, Schizaphis graminum (Rond.) *in the Orange Free State.* Unpubl. PhD thesis, University of Stellenbosch, South Africa.

Brown, M. W. 2004. Role of aphid predator guild in controlling spiraea aphid populations on apple in West Virginia, USA. *Biol. Control* 29: 189–198.

Brown, M. W. and S. S. Miller. 1998. Coccinellidae (Coleoptera) in apple orchards of eastern West Virginia and the impact of invasion by *Harmonia axyridis. Entomol. News* 109: 143–151.

Brown, M. W. and J. J. Schmitt. 2001. Seasonal and diurnal dynamics of beneficial insect populations in apple orchards under different management intensity. *Environ. Entomol.* 30: 415–424.

Bruck, D. J. and L. C. Lewis. 1998. Influence of adjacent cornfield habitat, trap location, and trap height on capture numbers of predators and a parasitoid of the European corn borer (Lepidoptera: Pyralidae) in central Iowa. *Environ. Entomol.* 27: 1557–1562.

Burgio, G., R. Ferrari, L. Boriani, M. Pozzati and J. van Lenteren. 2006. The role of ecological infrastructures on Coccinellidae (Coleoptera) and other predators in weedy field margins within northern Italy agroecosystems. *Bull. Insectology* 59: 59–67.

Carter, N. and A. F. G. Dixon. 1984. Honeydew: an arrestant stimulus for coccinellids. *Ecol. Entomol.* 9: 383–387.

Carter, N., A. F. G. Dixon and R. Rabbinge. 1982. *Cereal Aphid Populations: Biology, Simulation and Prediction.* PUDOC, Wageningen. 73 pp.

Clayhills, T. and M. Markkula. 1974. The abundance of coccinellids on cultivated plants. *Ann. Entomol. Fenn.* 40: 49–55.

Coderre, D., L. Provencher and J. L. Tourneur. 1987. Oviposition and niche partitioning in aphidophagous insects on maize. *Can. Entomol.* 119: 195–203.

Colunga-Garcia, M. and S. H. Gage. 1998. Arrival, establishment, and habitat use of the multicolored Asian lady beetle (Coleoptera: Coccinellidae) in a Michigan landscape. *Environ. Entomol.* 27: 1574–1580.

Cosper, R. D., M. J. Gaylor and J. C. Williams. 1983. Intraplant distribution of three insect predators on cotton, and seasonal effects of their distribution on vacuum sampler efficiency. *Environ. Entomol.* 12: 1568–1571.

Costamagna, A. C. and D. A. Landis. 2007. Quantifying predation on soybean aphid through direct field observations. *Biol. Control* 42: 16–24.

Cottrell, T. E. and K. V. Yeargan. 1998a. Effect of pollen on *Coleomegilla maculata* (Coleoptera: Coccinellidae) population density, predation, and cannibalism in sweet corn. *Environ. Entomol.* 27: 1402–1410.

Cottrell, T. E. and K. V. Yeargan. 1998b. Influence of a native weed, *Acalypha ostryaefolia* (Euphorbiaceae), on *Coleomegilla maculata* (Coleoptera: Coccinellidae) population density, predation, and cannibalism in sweet corn. *Environ. Entomol.* 27: 1375–1385.

Cottrell, T. E. and K. V. Yeargan. 1999. Factors influencing dispersal of larval *Coleomegilla maculata* from the weed *Acalypha ostryaefolia* to sweet corn. *Entomol. Exp. Appl.* 90: 313–322.

Dhillon, M. K. and H. C. Sharma. 2009. Effect of *Bacillus thuringiensis* -endotoxins Cry1Ab and Cry1Ac on the coccinellid beetle, *Cheilomenes sexmaculatus* (Coleoptera, Coccinellidae) under direct and indirect exposure conditions. *Biocont. Sci. Technol.* 19: 407–420.

Dixon, A. F. G. and J. L. Hemptinne. 2001. Body size distribution in predatory ladybird beetles reflects that of their prey. *Ecology* 82: 1847–1856.

Eigenbrode, S. D. and N. N. Kabalo. 1999. Effects of *Brassica oleraca* waxblooms on predation and attachement by *Hippodamia convergens*. *Entomol. Exp. Appl.* 91: 125–130.

Eigenbrode, S. D., C. White, M. Rohde and J. C. Simon. 1998. Behavior and effectiveness of adult *Hippodamia convergens* (Coleoptera: Coccinellidae) as a predator of *Acyrthosiphon pisum* (Homoptera: Aphididae) on a wax mutant of *Pisum sativum*. *Environ. Entomol.* 27: 902–909.

Ellington, J., K. Kiser, G. Ferguson and M. Cardenas. 1984. A comparison of sweepnet, absolute, and insectvac sampling methods in cotton ecosystems. *J. Econ. Entomol.* 77: 599–605.

Elliott, N. C. and R. W. Kieckhefer. 1990. Dynamics of aphidophagous coccinellid assemblages in small grain fields in eastern South Dakota. *Environ. Entomol.* 19: 1320–1329.

Elliott, N. C. and R. W. Kieckhefer. 2000. Response by coccinellids to spatial variation in cereal aphid density. *Popul. Ecol.* 42: 81–90.

Elliott, N., R. Kieckhefer and W. Kauffman. 1996. Effects of an invading coccinellid on native coccinellids in an agricultural landscape. *Oecologia* 105: 537–544.

Elliott, N. C. and G. J. Michels. 1997. Estimating aphidophagous coccinellid populations in alfalfa. *Biol. Control* 8: 43–51.

Elliott, N. C., R. W. Kieckhefer and W. C. Kauffman. 1991. Estimating adult coccinellid populations in wheat fields by removal, sweepnet, and visual count sampling. *Can. Entomol.* 123: 13–22.

Elliott, N. C., R. W. Kieckhefer and D. A. Beck. 2000. Adult coccinellid activity and predation on aphids in spring cereals. *Biol. Control* 17: 218–226.

Evans, E. W. 2003. Searching and reproductive behaviour of female aphidophagous ladybirds (Coleoptera: Coccinellidae): a review. *Eur. J. Entomol.* 100: 1–10.

Evans, E. W. 2004. Habitat displacement of North American ladybirds by an introduced species. *Ecology* 85: 637–647.

Evans, E. W. and D. R. Richards. 1997. Managing the dispersal of ladybird beetles (Col.: Coccinellidae): use of artificial honeydew to manipulate spatial distributions. *Entomophaga* 42: 93–102.

Evans, E. W. and N. N. Youssef. 1992. Numerical responses of aphid predators to varying prey density among Utah alfalfa fields. *J. Kans. Entomol. Soc.* 65: 30–38.

Ewert, M. A. and H. C. Chiang. 1966. Effect of some environmental factors on the distribution of three species of Coccinellidae in their microhabitat. *In* I. Hodek (ed.). *Ecology of Aphidophagous Insects*. Academia, Prague and Dr. W. Junk, The Hague. pp. 195–219.

Ferran, A., M. O. Cruz de Boelpaepe, H. Schanderl and M. M. Larroque. 1984. Les aptitudes trophiques et reproductrices des femelles de *Semiadalia undecimnotata* (Col.: Coccinellidae). *Entomophaga* 29: 151–170.

Finlayson, C. J., K. M. Landry and A. V. Alyokhin. 2008. Abundance of native and non-native lady beetles (Coleoptera: Coccinellidae) in different habitats in Maine. *Ann. Entomol. Soc. Am.* 101: 1078–1087.

Francis, F., G. Lognay, J. P. Wathelet and E. Haubruge. 2001. Effects of allelochemicals from first (Brassicaceae) and second (*Myzus persicae* and *Brevicoryne brassicae*) trophic levels on *Adalia bipunctata*. *J. Chem. Ecol.* 27: 243–256.

Francke, D. L., J. P. Harmon, C. T. Harvey and A. R. Ives. 2008. Pea aphid dropping behavior diminishes foraging efficiency of a predatory beetle. *Entomol. Exp. Appl.* 127: 118–124.

Frazer, B. D. and N. Gilbert. 1976. Coccinellids and aphids: a quantitative study of the impact of adult lady-birds (Coleoptera: Coccinellidae) preying on field populations of pea aphids (Homoptera: Aphididae). *J. Entomol. Soc. B. C.* 73: 33–56.

Frazer, B. D. and D. A. Raworth. 1985. Sampling for adult coccinellids and their numerical response to strawberry aphids (Coleoptera: Coccinellidae: Homoptera: Aphididae). *Can. Entomol.* 117: 153–161.

Frechette, B., A. F. G. Dixon, C. Alauzet, N. Boughenou and J. L. Hemptinne. 2006. Should aphidophagous ladybirds be reluctant to lay eggs in the presence of unsuitable prey? *Entomol. Exp. Appl.* 118: 121–127.

French, B. W., N. C. Elliott, S. D. Kindler and D. C. Arnold 2001. Seasonal occurrence of aphids and natural enemies in wheat and associated crops. *Southwest. Entomol.* 26: 49–61.

Gagné, W. C. and J. L. Martin. 1968. The insect ecology of red pine plantations in central Ontario. *Can. Entomol.* 100: 835–864.

Gardiner, M. M., D. A. Landis, C. Gratton et al. 2009. Landscape composition influences patterns of native and exotic lady beetle abundance. *Divers. Distrib.* 15: 554–564.

Ghanim, A. E. B., B. Freier and T. Wetzel. 1984. Zur Nahrungsafnahme und Eiablage von *Coccinella septempunctata* L. bei unterschiedlichem Angebot von Aphiden der Arten *Macrosiphum avenae* (Fabr.) und *Rhopalosiphum padi* (L.). *Arch. Phytopathol. Pflanzenschutz* 20: 117–125.

Giles, K. L., R. D. Madden, R. Stockland, M. E. Payton and J. W. Dillwith 2002. Host plants affect predator fitness via the nutritional value of herbivore prey: investigation of a plant–aphid–ladybeetle system. *BioControl* 47: 1–21.

Giles, K. L., R. Stockland, J. L. Madden, M. E. Payton and J. W. Dillwith. 2001. Preimaginal survival and development of *Coleomegilla maculata* and *Hippodamia convergens* (Coleoptera: Coccinellidae) reared on *Acyrthosiphon pisum*: effects of host plants. *Environ. Entomol.* 30: 964–971.

Girling, R. D. and M. Hassall. 2008. Behavioural responses of the seven-spot ladybird *Coccinella septempunctata* to plant headspace chemicals collected from four crop Brassicas and *Arabidopsis thaliana*, infested with *Myzus persicae*. *Agric. Forest Entomol.* 10: 297–306.

Grez, A. A., T. Zaviezo, S. Diaz, B. Camousseigt and G. Cortes. 2008. Effects of habitat loss and fragmentation on the abundance and species richness of aphidophagous beetles and aphids in experimental alfalfa landscapes. *Eur. J. Entomol.* 105: 411–420.

Grez, A. A., T. Zaviezo and M. Rios. 2005. Ladybird (Coleoptera: Coccinellidae) dispersal in experimental fragmented alfalfa landscapes. *Eur. J. Entomol.* 102: 209–216.

Grez, A., T. Zaviezo, L. Tischendorf and L. Fahrig. 2004. A transient, positive effect of habitat fragmentation on insect population densities. *Oecologia* 141: 444–451.

Gumos, H. and J. Wisniewski. 1960. Intensity of appearing of Coccinellidae in pine woods. *Pol. Pismo Entomol. (B)* 3/4: 217–223.

Hagen, K. S., E. F. Sawall and R. L. Tassan. 1971. The use of food sprays to increase effectiveness of entomophagous insects. *Proc. Tall Timbers Conf. Ecol. and Anim. Control Habit. Manage., Tallahassee, February 1970*: 59–80.

Hagler, J. R. and V. P. Jones. 2010. A protein-based approach to mark arthropods for mark-capture type research. *Entomol. Exp. Appl.* 135: 177-192.

Hagler, J. R. and S. E. Naranjo. 2004. A multiple ELISA system for simultaneously monitoring intercrop movement and feeding activity of mass-released insect predators. *Int. J. Pest Manage.* 50: 199-207.

Hamilton, R. A., E. B. Dogan, G. B. Schaalje and G. M. Booth. 1999. Olfactory response of the lady beetle *Hippodamia convergens* (Coleoptera: Coccinellidae) to prey related odors, including a scanning electron microscopy study of the antennal sensila. *Environ. Entomol.* 28: 812–822.

Harmon, J. P., E. Stephens and J. Losey. 2007. The decline of native coccinellids (Coleoptera: Coccinellidae) in the United States and Canada. *J. Insect Conserv.* 11: 85–94.

Hatano, E., G. Kunert, J. P. Michaud and W. W. Weisser. 2009. Chemical cues mediating aphid location by natural enemies. *Eur. J. Entomol.* 105: 797–806.

Hemptinne, J. L. and A. F. G. Dixon. 1991. Why ladybirds have generally been so ineffective in biological control? *In* L. Polgár, R. J. Chambers, A. F. G. Dixon and I. Hodek (eds). *Behaviour and Impact of Aphidophaga*. SPB Academic Publishing, The Hague. pp. 149–157.

Hemptinne, J. L. and J. Naisse. 1988. Life cycle strategy of *Adalia bipunctata* (L.) (Col., Coccinellidae) in temperate country. *In* E. Niemczyk and A. F. G. Dixon (eds). *Ecology and Effectiveness of Aphidophaga*. SPB Academic Publishers, The Hague. pp. 71–77.

Hemptinne, J. L., J. Naisse and S. Os. 1988. Glimps [sic] of the life history of *Propylea quatuordecimpunctata* (L.) (Coleoptera: Coccinellidae). *Meded. Fac. Landbouw. Rijksuniv. Gent* 53: 1175–1182.

Hemptinne, J. L., M. Doumbia and A. F. G. Dixon. 2000a. Assessment of plant quality by ladybirds: role of aphid and plant phenology. *J. Insect Behav.* 13: 353–359.

Hemptinne, J. L., M. Gaudin, A. F. G. Dixon and G. Lognay. 2000b. Social feeding in ladybird beetles: adaptive significance and mechanism. *Chemoecology* 10: 149–152.

Hodek, I. and A. Honěk. 1996. *Ecology of Coccinellidae.* Kluwer Academic Publishers, Dordrecht. 464 pp.

Hodek, I. and J. P. Michaud. 2008. Why is *Coccinella septempunctata* so successful? (A point-of-view). *Eur. J. Entomol.* 105: 1–12.

Hodek, I., P. Stary and P. Stys. 1962. The natural enemy complex of *Aphis fabae* and its effectiveness in control. *Proc. 11 Int. Congr. Entomol, Vienna, 1962* 2: 747–749.

Hodek, I. J. Holman, P. Stary, P. Stys and J. Zeleny. 1966. *Natural Enemies of* Aphis fabae *in the CSSR.* Academia, Prague. 144 pp.

Honěk, A. 1978a. The losses of *Coccinella septempunctata* L. populations during the first cutting of forage leguminosae. *Sb. UVTIZ Ochr. Rostl.* 14: 233–236.

Honěk, A. 1978b. Trophic regulation of postdiapause ovariole maturation in *Coccinella septempunctata* (Col.: Coccinellidae). *Entomophaga* 23: 213–216.

Honěk, A. 1979. Plant density and occurrence of *Coccinella septempunctata* and *Propylea quatuordecimpunctata* (Coleoptera, Coccinellidae) in cereals. *Acta Entomol. Bohemoslov.* 76, 308–312.

Honěk, A. 1980. Population density of aphids at the time of settling and ovariole maturation in *Coccinella septempunctata* (Col., Coccinellidae). *Entomophaga* 25: 427–430.

Honěk, A. 1981. Aphidophagous Coccinellidae (Coleoptera) and Chrysopidae (Neuroptera) on three weeds: factors determining the composition of populations. *Acta Entomol. Bohemoslov.* 78: 303–310.

Honěk, A. 1982a. Factors which determine the composition of field communities of adult aphidophagous Coccinellidae (Coleoptera). *Z. Angew. Entomol.* 94: 157–168.

Honěk, A. 1982b. The distribution of overwintered *Coccinella septempunctata* L. (Col., Coccinellidae) adults in agricultural crops. *Z. Angew. Entomol.* 94: 311–319.

Honěk, A. 1983. Factors affecting the distribution of larvae of aphid predators (Col., Coccinellidae and Dipt., Syrphidae) in cereal stands. *Z. Angew. Entomol.* 95: 336–343.

Honěk, A. 1985a. Activity and predation of *Coccinella septempunctata* adults in the field (Col., Coccinellidae). *Z. Angew. Entomol.* 100: 399–409.

Honěk, A. 1985b. Habitat preferences of aphidophagous coccinellids (Coleoptera). *Entomophaga* 30: 253–264.

Honěk, A. 1989. Overwintering and annual changes of abundance of *Coccinella septempunctata* in Czechoslovakia (Coleoptera, Coccinellidae). *Acta Entomol. Bohemoslov.* 86: 179–192.

Honěk, A. and F. Kocourek. 1986. The flight of aphid predators to a light trap: possible interpretations. *In* I. Hodek (ed.). *Ecology of Aphidophaga.* Academia, Prague. pp. 333–338.

Honek, A. and M. Rejmánek 1982. The communities of adult aphidophagous Coccinellidae (Coleoptera): a multivariate analysis. *Acta Oecol. Oecol. Appl.* 3: 95–104.

Honěk, A., Z. Martinkova and S. Pekar. 2005. Temporal stability of morph frequency in central European populations of *Adalia bipunctata* and *A. decempunctata* (Coleoptera: Coccinellidae). *Eur. J. Entomol.* 102: 437–442.

Honek, A., A. F. G. Dixon and Z. Martinkova. 2007. Body size, reproductive allocation and maximum reproductive rate of two species of aphidophagous Coccinellidae exploiting the same resource. *Entomol. Exp. Appl.* 1: 1–11.

Hoogendoorn, M. and G. E. Heimpel. 2004. Competitive interactions between an exotic and a native ladybeetle: a field cage study. *Entomol. Exp. Appl.* 111: 19–28.

Horn, D. J. 1981. Effect of weedy backgrounds on colonization of collards by green peach aphid, *Myzus persicae,* and its major predators. *Environ. Entomol.* 10: 285–289.

Hossain, Z., G. M. Gurr, S. D. Wratten and A. Raman. 2002. Habitat manipulation in lucerne *Medicago sativa*: arthropod population dynamics in harvested and 'refuge' crop strips. *J. Appl. Ecol.* 39: 445–454.

Iperti, G. 1965. Contribution á l'étude de la specificité chez les principales coccinelles aphidiphages des Alpes-Maritimes at des Basses-Alpes. *Entomophaga* 10: 159–178.

Iperti, G. 1966. Comportement naturel des Coccinelles aphidiphages du Sud-Est de la France. Leur type de spécificité, leur action prédatrice sur *Aphis fabae* L. *Entomophaga* 11: 203–210.

Iperti, G., Y. Laudého, J. Brun and E. Choppin de Janvry. 1970. Les entomophages de *Parlatoria blanchardi* Targ. dans les palmeraies de l'Adrar mauritanien. III. Introduction acclimation et efficacité d'un nouveau prédateur Coccinellidae: *Chilocorus bipustulatus* L. (Souche d'Iran). *Ann. Zool. Ecol. Anim.* 2: 617–638.

Iperti, G., L. Lapchin, A. Ferran, J. M. Rabasse and J. P. Lyon. 1988. Sequential sampling of adult *Coccinella septempunctata* L. in wheat fields. *Can. Entomol.* 120: 773–778.

Ives, P. M. 1981a. Estimation of coccinellid numbers and movement in the field. *Can. Entomol.* 113: 981–997.

Ives, P. M. 1981b. Feeding and egg production of two species of coccinellids in the laboratory. *Can. Entomol.* 113: 999–1005.

Ivliev, L. A., V. N. Kuznetsov and E. G. Matis. 1975. Ecology and faunistics of coccinellids (Coleoptera, Coccinellidae) of the extreme North-East of the USSR. *Trudy Biol.-pochv. Inst. Vostoch. Centra Akad. Nauk. SSSR* 27: 5–20.

Jamal, E. and G. C. Brown. 2001. Orientation of *Hippodamia convergens* (Coleoptera: Coccinellidae) larvae to volatile chemicals associated with *Myzus nicotianae* (Homoptera: Aphididae). *Environ. Entomol.* 30: 1012–1016.

Kalushkov, P. 1998. Ten aphid species (Sternorrhyncha: Aphididae) as prey for *Adalia bipunctata* (Coleoptera: Coccinellidae). *Eur. J. Entomol.* 95: 343–349.

Kareiva, P. 1987. Habitat fragmentation and the stability of predator–prey interactions. *Nature* 326: 388–390.

Kareiva, P. and R. Sahakian. 1990. Tritrophic effects of a simple architectural mutation in pea plants. *Nature* 345: 433–434.

Kavallieratos, N. G., G. J. Stathas, C. G. Athanassiou and G. T. Papadoulis. 2002. *Dittrichia viscosa* and *Rubus ulmifolius* as reservoirs of aphid parasitoids (Hymenoptera: Braconidae: Aphidiinae) and the role of certain coccinellid species. *Phytoparasitica* 30: 231–242.

Kavallieratos, N. G., C. G Athanassiou, Z. Tomanovic, G. D Papadopoulos and B. J. Vayias. 2004. Seasonal abundance and effect of predators (Coleoptera, Coccinellidae) and parasitoids (Hymenoptera: Braconidae, Aphidiinae) on *Myzus persicae* (Hemiptera, Aphidoidea) densities on tobacco: a two-year study from Central Greece. *Biologia* 59: 613–619.

Kawauchi, S. 1981. The number of oviposition, hatchability and the term of oviposition of *Propylea japonica* Thunberg (Coleoptera, Coccinellidae) under different food condition. *Kontyu* 49: 183–191.

Klausnitzer, B. 1968. Zur Biologie von *Myrrha octodecimguttata* (L.) (Col. Coccinellidae). *Entomol. Nachr.* 12: 102–104.

Klewer, N., Z. Ruzicka and S. Schulz. 2007. (Z)-Pentacos-12-ene, an oviposition-deterring pheromone of *Cheilomenes sulphurea*. *J. Chem. Ecol.* 33: 2167–2170.

Koch, R. L. and W. D. Hutchinson. 2003. Phenology and blacklight trapping of the multicolored Asian lady beetle (Coleoptera: Coccinellidae) in a Minnesota agricultural landscape. *J. Entomol. Sci.* 38: 477–480.

Koch, R. L., E. C. Burkness, S. J. Wold Burkness and W. D. Hutchinson. 2004. Phytophagous preferences of the multicolored Asian lady beetle (Coleoptera: Coccinellidae) for autumn-ripening fruit. *J. Econ. Entomol.* 97: 539–544.

Koji, S. and K. Nakamura. 2002. Population dynamics of a thistle-feeding lady beetle *Epilachna niponica* (Coccinellidae: Epilachninae) in Kanazawa, Japan. 1. Adult demographic traits and population stability. *Popul. Ecol.* 44: 103–112.

Kokubu, H. 1986. *Migration rates, in situ reproduction, and flight characteristics of aphidophagous insects (Chrysopidae, Coccinellidae, and Syrphidae) in cornfields.* Unpubl. PhD thesis, University of Basel, Switzerland.

Kuznetsov, V. N. 1975. Zoogeographical analysis of coccinellid fauna (Coleoptera, Coccinellidae) in the Primorye Territory. *Trudy Biol.-pochv. Inst. Vostoch. Centra Akad. Nauk. SSSR* 27: 153–163. (In Russian.)

Labrie, G., D. Coderre and E. Lucas. 2008. Overwintering strategy of multicolored Asian lady beetle (Coleoptera: Coccinellidae): cold-free space as a factor of invasive success. *Ann. Entomol. Soc. Am.* 101: 860–866.

LaMana, M. L. and J. C. Miller. 1996. Field observations on *Harmonia axyridis* Pallas (Coleoptera: Coccinellidae) in Oregon. *Biol. Control.* 6: 232–237.

Lapchin, L, A. Ferran, G. Iperti, J. M. Rabasse and J. P. Lyon. 1987. Coccinellids (Coleoptera: Coccinellidae) and syrphids (Diptera: Syrphidae) as predators of aphids in cereal crops: a comparison of sampling methods. *Can. Entomol.* 119: 815–822.

Laubertie, E., X. Martini, C. Cadena et al. 2006. The immediate source of the oviposition-deterring pheromone produced by larvae of *Adalia bipunctata* (L.) (Coleoptera, Coccinellidae). *J. Insect Behav.* 19: 231–240.

Laudého, Y., E. C. de Janvry, G. Iperti and J. Brun. 1970. Intervention bio-écologique contre la cochenille blanche du palmier-dattier (*Parlatoria blanchardi* Targ.)(Coccoidea-Diaspididae) en Ardar Mauritanien. *Fruits* 25: 147–160.

Leather, S. R., R. C. A. Cooke, M. D. E. Fellowes and R. Rombe. 1999. Distribution and abundance of ladybirds (Coleoptera: Coccinellidae) in non-crop habitats. *Eur. J. Entomol.* 96: 23–27.

Losey, J. E. and R. F. Denno. 1998. The escape response of pea aphids to foliar-foraging predators: factors affecting dropping behaviour. *Ecol. Entomol.* 23: 53–56.

Losey, J. E. and R. F. Denno. 1999. Factors facilitating synergistic predation: the central role of synchrony. *Ecol. Appl.* 9: 378–386.

Loevei, G. L. 1981. Coccinellid community in an apple orchard bordering a deciduous forest. *Acta Phytopath. Acad. Sci. Hung.* 16: 143–150.

Loevei, G. L., M. Sarospataki and Z. A. Radwan. 1991. Structure of ladybird (Coleoptera: Coccinellidae) assemblages in apple: changes through developmental stages. *Environ. Entomol.* 20: 1301–1308.

Lucas, E., C. Vincent, G. Labrie et al. 2007. The multicolored asian lady beetle *Harmonia axyridis* (Coleoptera: Coccinellidae) in Quebec agroecosystems ten years after its arrival. *Eur. J. Entomol.* 104: 737–743.

Lundgren, J. G. 2009. Nutritional aspects of non-prey foods in the life histories of predaceous Coccinellidae. *Biol. Control* 51: 294–305.

Lundgren, J. G. and R. N. Wiedenmann. 2004. Nutritional suitability of corn pollen for the predator *Coleomegilla maculata* (Coleoptera: Coccinellidae). *J. Insect Physiol.* 50: 567–575.

Lundgren, J. G., A. A. Razzak and R. N. Wiedenmann. 2004. Population responses and food consumption by predators *Coleomegilla maculata* and *Harmonia axyridis* (Coleoptera: Coccinellidae) during anthesis in an Illinois cornfield. *Environ. Entomol.* 33: 958–963.

Lusis, YaYa. 1961. On the biological meaning of colour polymorphism of lady-beetle *Adalia bipunctata* L. *Latv. Entomol.* 4: 2–29.

Magagula, C. N. and M. J. Samways. 2001. Maintenance of ladybeetle diversity across a heterogeneous African agricultural/savanna land mosaic. *Biod. Conserv.* 10: 209–222.

Majerus, M. E. N. 1994. *Ladybirds.* Harper Collins, London. 367 pp.

Michaud, J. P. 2004. Natural mortality of Asian citrus psyllid (Homoptera: Psyllidae) in central Florida. *Biol. Control.* 29: 260–269.

Michels, G. J. and R. W. Behle. 1992. Evaluation of sampling methods for lady beetles (Coleoptera: coccinellidae) in grain sorghum. *J. Econ. Entomol.* 85: 2251–2257.

Michels, G. J., N. C. Elliott, R. L. Romero and T. D. Johnson. 1996. Sampling aphidophagous Coccinellidae in grain sorghum. *Southwest. Entomol.* 21: 237–246.

Michels, G. J., N. C. Elliott, R. L. Romero and W. B. French. 1997. Estimating populations of aphidophagous Coccinellidae (Coleoptera) in winter wheat. *Environ. Entomol.* 26: 4–11.

Michels, G. J. and J. H. Matis. 2008. Corn leaf aphid, Rhopalosiphum maidis (Hemiptera: Aphididae), is a key to greenbug, *Schizaphis graminum* (Hemiptera: Aphididae), biological control in grain sorghum, *Sorghum bicolor. Eur. J. Entomol.* 105: 513-520.

Musser, F. R. and A. F. Shelton. 2003. Factors altering the temporal and within-plant distribution of coccinellids in corn and their impact on potential intra-guild predation. *Environ. Entomol.* 32: 575–583.

Nakamuta, K. 1987. Diel rhythmicity of prey-searching and its predominance over starvation in the lady beetle, *Coccinella septempunctata bruckii. Physiol. Entomol.* 12: 91–98.

Narayandas, G. K. and A. V. Alyokhin. 2006. Interplant movement of potato aphid (Homoptera: Aphididae) in response to environmental stimuli. *Environ. Entomol.* 35: 733–739.

Nault, B. A. and G. G. Kennedy. 2000. Seasonal changes in habitat preference by *Coleomegilla maculata*: implications for Colorado potato beetle management in potato. *Biol. Control.* 17: 164–173.

Nedvěd, O. 1999. Host complexes of predaceous ladybeetles (Col., Coccinellidae). *J. Appl. Entomol.* 123: 73–76.

Nelson, E. H. and J. A. Rosenheim. 2006. Encounters between aphids and their predators: the relative frequencies of disturbance and consumption. *Entomol. Exp. Appl.* 118: 211–219.

Neuenschwander, P., Hagen, K. S. and R. F. Smith. 1975. Predation of aphids in California's alfalfa fields. *Hilgardia* 43: 53–78.

Nichols, P. R. and W. W. Neel. 1977. The use of food wheast as a supplemental food for *Coleomegilla maculata* (DeGeer) (Coleoptera: Coccinellidae) in the field. *Southwest. Entomol.* 2: 102–105.

Ninkovic, V. and J. Pettersson. 2003. Searching behaviour of the seven-spotted ladybird, *Coccinella septempunctata*: effects of plant–plant odour interaction. *Oikos* 100: 65–70.

Ninkovic, V., S. AlAbbasi and J. Pettersson. 2001. The influence of aphid-induced plant volatiles on ladybird searching behavior. *Biol. Control.* 21: 191–195.

Obrycki, J. J., N. C. Elliott and K. L. Giles. 2000. Coccinellid introductions: potential for and evaluation of nontarget effects. *In* P. A. Follett and J. J. Duan (eds). *Nontarget Effects of Biological Control.* Kluwer Academic Publishers, Boston. pp 127-145.

Obrycki, J. J., J. D. Harwood, T. J. Kring and R. J. O' Neil. 2009. Aphidophagy by Coccinellidae: application of biological control in agroecosystems. *Biol. Control.* 51: 244–254.

Ofuya, T. I. 1991. Aspects of the ecology of predation in two coccinellid species on the cowpea aphid in Nigeria. *In* L.

Polgár, R. J. Chambers, A. F. G. Dixon and I. Hodek (eds). *Behaviour and Impact of Aphidophaga.* SPB Academic Publishing, The Hague. pp. 213–220.

Oztemiz, S., M. Karacaoglu and F. Yarpuzlu. 2008. Natural enemies of *Ceroplastes* species (Homoptera: Coccidae), their efficiency and population movement in citrus orchards in the eastern Mediterranean region of Turkey. *J. Entomol. Res. Soc.* 10: 35–46.

Parajulee, M. N. and J. E. Slosser. 2003. Potential of yellow sticky traps for lady beetle survey in cotton. *J. Econ. Entomol.* 96: 239–254.

Patt, J. M., G. C. Hamilton and J. H. Lashomb. 1997. Impact of strip-insectary intercropping with flowers on conservation biological control of the Colorado potato beetle. *Adv. Hort. Sci.* 11: 175–181.

Pettersson, J., V. Ninkovic, R. Glinwood et al. 2008. Chemical stimuli supporting foraging behaviour of *Coccinella septempunctata* L.. Coleoptera: Coccinellidae): volatiles and allelobiosis. *Appl. Entomol. Zool.* 43: 315–321.

Poutsma, J., A. J. M. Loomans, B. Aukema and T. Heijerman. 2008. Predicting the potential geographical distribution of the harlequin ladybird, *Harmonia axyridis*, using the CLIMEX model. *BioControl* 53: 103–125.

Pruszynski, S. and J. J. Lipa. 1970. Observations on life cycle and food specialization of *Adalia bipunctata* (L.) (Coleoptera, Coccinellidae). *Prace Nauk. Inst. Ochr. Roślin* 12: 99–116.

Radcliffe, E. B., R. W. Weires, R. E. Stucker and D. K. Barnes. 1976. Influence of cultivars and pesticides on pea aphid, spotted alfalfa aphid, and associated arthropod taxa in a Minnesota alfalfa ecosystem. *Environ. Entomol.* 5: 1195–1207.

Radwan, Z. and G. L. Loevei. 1982. Distribution and bionomics of ladybird beetles (Col., Coccinellidae) living in an apple orchard near Budapest, Hungary. *Z. Angew. Entomol.* 94: 169–175.

Radwan, Z. and G. L. Loevei. 1983. Structure and seasonal dynamics of larval, pupal, and adult coccinellid (Col., Coccinellidae) assemblages in two types of maize fields in Hungary. *Z. Angew. Entomol.* 95: 396–408.

Rana, J. S., A. F. G. Dixon and V. Jarosik. 2002. Costs and benefits of prey specialization in a generalist insect predator. *J. Anim. Ecol.* 71: 15–22.

Riddick, E. W., G. Dively and P. Barbosa. 2000. Season-long abundance of generalist predators in transgenic versus nontransgenic potato fields. *J. Entomol. Sci.* 35: 349–359.

Ries, L., R. J. Fletcher, J. Battin and T. D. Sisk. 2004. Ecological responses to habitat edges: mechanisms, models, and variability explained. *Annu. Rev. Ecol. Syst.* 35: 491–522.

Rosen, D. and U. Gerson. 1965. Field studies of *Chilocorus bipustulatus* (L.) on citrus in Israel. *Ann. Epiph.* 16: 71–76.

Ruzicka, Z. 2006. Oviposition-deterring effects of conspecific and heterospecific larval tracks of *Cheilomenes sexmaculata* (Coleoptera: Coccinellidae). *Eur. J. Entomol.* 103: 757–763.

Sakuratani, Y. 1977. Population fluctuations and spatial distributions of natural enemies of aphids in corn fields. *Jpn J. Ecol.* 27: 291–300.

Sakuratani, Y. and Y. Nakamura. 1997. Oviposition strategies of *Coccinella septempunctata* (Col.: Coccinellidae). *Entomophaga* 42: 33–40.

Sakuratani, Y., K. Ikeuchi and T. Ioka. 1991. Seasonal changes in angle of pupation of *Coccinella septempunctata bruckii* in relation to solar altitude. *In* L. Polgár, R. J. Chambers, A. F. G. Dixon and I. Hodek (eds). *Behaviour and Impact of Aphidophaga*. SPB Academic Publishing, The Hague. pp. 259–264.

Sakuratani, Y., Y. Sugihara, M. Isida, S. Kuwahara and T. Sugimoto. 1983. Aggregative response of adults of *Coccinella septempunctata brucki* Mulsant (Coleoptera: Coccinellidae) to aphid population density. *Mem. Fac. Agric. Kinki Univ.* 16: 49–54.

Samways, M. J., R. Osborn, H. Hastings and V. Hattingh. 1999. Global climate change and accuracy of prediction of species' geographical ranges: establishment success of introduced ladybirds (Coccinellidae, *Chilocorus* spp.) worldwide. *J. Biogeog.* 26: 795–812.

Sarmento, R. A., M. Venzon, A. Pallini, E. E. Oliveira and A. Janssen. 2007. Use of odours by *Cycloneda sanguinea* to assess patch quality. *Entomol. Exp. Appl.* 124: 313–318.

Sarospataki, M. and V. Marko. 1995. Flight activity of *Coccinella septempunctata* (Coleoptera: Coccinellidae) at different strata of a forest in relation to migration to hibernation sites. *Eur. J. Entomol.* 92: 415–419.

de Sassi, C., C. B. Müller and J. Kraus. 2006. Fungal plant endosymbiosis alter life history and reproductive success of aphid predators. *Proc. R. Soc. Lond. (B)* 273: 1301–1306.

Savoiskaya, G. I. 1970. Introduction and acclimatisation of some coccinellids in the Alma-Ata reserve. *Trudy Alma-Atin. Gos. Zapov.* 9: 138–162.

Schlinger, E. I. and E. J. Dietrick. 1960. Biological control of insect pests aided by stripfarming alfalfa in experimental program. *Calif. Agric.* 14: 8–9.

Schmidt, N. P., M. E. O' Neal and P. M. Dixon. 2008. Aphidophagous predators in Iowa soybean, a community comparison across multiple years and sampling methods. *Ann. Entomol. Soc. Am.* 101: 341–350.

Seko, T., K. Yamashita and K. Miura. 2008. Residence period of a flightless strain of the ladybird beetle *Harmonia axyridis* Pallas (Coleoptera: Coccinellidae) in open fields. *Biol. Control* 47: 194–198.

Selyemova, D., P. Zach, D. Nemethova et al. 2007. Assemblage structure and altitudinal distribution of lady beetles (Coleoptera, Coccinellidae) in the mountain spruce forests of Polana Mountains, the West Carpathians. *Biologia* 62: 610–616.

Sengonca, C., J. Kranz and P. Blaeser. 2002. Attractiveness of three weed species to polyphagous predators and their influence on aphid populations in adjacent lettuce cultivations. *J. Pest Sci.* 75: 161–165.

Shah, M. A. 1983. A stimulant in *Berberis vulgaris* inducing oviposition in coccinellids. *Entomol. Exp. Appl.* 33: 119–120.

Skirvin, D. J., J. N. Perry and R. Harrington. 1997. The effect of climate change on an aphid–coccinellid interaction. *Global Change Biol.* 3: 1–11.

Sloggett, J. J. and M. E. N. Majerus. 2000. Habitat preferences and diet in the predatory Coccinellidae (Coleoptera): an evolutionary perspective. *Biol. J. Linn. Soc.* 70: 63–88.

Snyder, W. E. 2009. Coccinellids in diverse communities: Which niche fits? *Biol. Control* 51: 323–335.

Southwood, T. R. E. and P. A. Henderson. 2000. *Ecological Methods*. 3rd edn. Blackwell Science, Oxford, UK. 575 pp.

Spellman, B., M. W. Brown and C. R. Mathews. 2006. Effect of floral and extrafloral resources on predation of *Aphis spiraecola* by *Harmonia axyridis* on apple. *BioControl* 51: 715–724.

Stephens, E. J. and J. E. Losey. 2004. Comparison of sticky cards, visual and sweep sampling of coccinellid populations in alfalfa. *Environ. Entomol.* 33: 535–539.

Storck-Weyhermueller, S. 1988. Einfluß natürlicher Feinde auf die Populationsdynamik der Getreideblattläuse im Winterweizen Mittelhessens (Homoptera: Aphididae). *Entomol. Gener.* 13: 189–206.

Tamaki, G., B. Annis and M. Weiss. 1981. Response of natural enemies to the green peach aphid in different plant cultures. *Environ. Entomol.* 10: 375–378.

Thofelt, L. 1975. Studies on leaf temperature recorded by direct measurement and by thermography. *Acta Univ. Upsal.* 12: 3–143.

Timms, J. E. L. and S. R. Leather. 2007. Ladybird egg cluster size: relationships between species, oviposition substrate and cannibalism. *Bull. Entomol. Res.* 97: 613–618.

Tomanovic, Z., N. G. Kavallieratos, P. Stary et al. 2008. Cereal aphids (Hemiptera: Aphidoidea) in Serbia: seasonal dynamics and natural enemies. *Eur. J. Entomol.* 105: 495–501.

Torres, J. B. and J. R. Ruberson. 2005. Canopy- and ground-dwelling predatory arthropods in commercial Bt and non-Bt cotton fields: patterns and mechanisms. *Environ. Entomol.* 34: 1242–1256.

Turchin, P. and P. Kareiva. 1989. Aggregation in *Aphis varians*: an effective strategy for reducing predation risk. *Ecology* 70: 1008–1016.

Udayagiri, S., C. E. Mason and J. D. Pesek. 1997. *Coleomegilla maculata*, *Coccinella septempunctata* (Coleoptera: Coccinellidae), *Chrysoperla carnea* (Neuroptera: Chrysopidae), and *Macrocentrus grandii* (Hymenoptera: Braconidae) trapped on colored sticky traps in corn habitats. *Environ. Entomol.* 26: 983–988.

Van der Werf, W., E. W. Evans and J. Powell. 2000. Measuring and modelling the dispersal of *Coccinella septempunctata* (Coleoptera: Coccinellidae) in alfalfa fields. *Eur. J. Entomol.* 97: 487–493.

Van Emden, H. F. 2010. Cooking up biological control of aphids with mixed ingredients. *International Symposium*

Ecology of Aphidophaga 11. Provincia di Perugia, Perugia, p. 39.

Wetzler, R. A. and S. J. Risch 1984. Experimental studies of beetle diffusion in simple and complex habitats. *J. Anim. Ecol.* 53: 1–19.

Wheeler, A. G. and E. R. Hoebke. 2008. Rise and fall of an immigrant lady beetle: is *Coccinella undecimpunctata* L: (Coleoptera: Coccinellidae) still present in North America? *Proc. Entomol. Soc. Wash.* 110: 817–823.

Wiesmann, R. 1955. Untersuchungen an den Prädatoren der Baumwollschadinsekten in Aegypten im Jahre 1951/52. *Acta Trop.* 12: 222–239.

Woin, N., C. Volkmar and T. Wetzel. 2000. Seasonal activity and diversity of ladybirds (Coleoptera: Coccinellidae) as ecological bioindicators in paddy fields. *Mitt. Deut. Ges. Allg. Angew. Entomol.* 12: 203-206

Wratten, S. D. 1973. The effectiveness of the coccinellid beetle, *Adalia bipunctata* (L.), as a predator of the lime aphid, *Eucallipterus tiliae* L. *J. Anim. Ecol.* 42: 785–802.

Wright, E. J. and J. E. Laing. 1980. Numerical response of coccinellids to aphids in corn in southern Ontario. *Can. Entomol.* 112: 977–988.

Wright, R. J. and T. A. DeVries. 2000. Species composition and relative abundance of Coccinellidae (Coleoptera) in south central Nebraska field crops. *J. Kans. Entomol. Soc.* 73: 103–111.

Yakhontov, V. V. 1966. Diapause in Coccinellidae of Central Asia. *In* I. Hodek (ed). *Ecology of Aphidophagous Insects.* Academia, Prague and Dr. W. Junk, The Hague. pp. 107–108.

Yasuda, H., T. Takagi and K. Kogi. 2000. Effects of conspecific and heterospecific larval tracks on the oviposition behaviour of the predatory ladybird, *Harmonia axyridis* (Coleoptera: Coccinellidae). *Eur. J. Entomol.* 97: 551–553.

Yu, G. Y. 1999. Lady beetles (Coleoptera: Coccinellidae) from *Pinus armandii* Franchet infested with *Pineus* sp. *Entomotaxonomia.* 21: 281–287.

Zhu, J., A. A. Cosse, J. J. Obrycki, K. S. Boo and T. C. Baker. 1999. Olfactory reactions of the twelve-spotted lady beetle, *Coleomegilla maculata* and the green lacewing, *Chrysopa carnea* to semiochemicals released from their prey and host plant: Electroantennogram and behavioral responses. *J. Chem. Ecol.* 25: 1163–1177.

Zotov, V. A. 1983. Exogenous and endogenous components of the diel rhythm of activity in *Coccinella septempunctata* (Coleoptera, Coccinellidae). *Zool. Zh.* 62: 1654–1661.

Chapter 5

FOOD RELATIONSHIPS

Ivo Hodek[1] and Edward W. Evans[2]

[1] Institute of Entomology, Academy of Sciences, CZ 370 05, České Budějovice, Czech Republic
[2] Department of Biology, Utah State University, Logan, UT 84322 USA

Ecology and Behaviour of the Ladybird Beetles (Coccinellidae), First Edition. Edited by I. Hodek, H.F. van Emden, A. Honěk.
© 2012 Blackwell Publishing Ltd. Published 2012 by Blackwell Publishing Ltd.

5.1 INTRODUCTION

Studies of food relations in coccinellids were among the earliest fields of research on this family, because of the evident interest in ladybirds as natural enemies of aphid, coccid and mite pests. The history of studies of coccinellid food relations can be divided into three periods. In the earliest period, studies focussed on what the ladybirds eat and lists of the prey species eaten were compiled. In the middle phase, hypotheses and models were constructed that were based on only fragmentary experiments, usually executed in extremely artificial settings. This remained characteristic well into the end decades of the last century. In agreement with Burk (1988), the need for empirical hypothesis testing in future decades was stressed by Hodek (1996).

Fortunately a new generation of researchers has undertaken more comprehensive causal studies in several branches of coccinellid food research and important progress has been achieved within a relatively short period. Here, in the introduction, we briefly call attention to some of the studies that are discussed in this chapter.

In 5.2, Sloggett's (2008b) prey size–density hypothesis, based on concrete observations, opens a new dimension of prey specificity: size relation between the prey and predator instead of (or together with) physiological suitability and/or chemical composition of the prey. Body size may be more important than prey chemistry as a universal factor underlying dietary specialization.

New light has been thrown on nutritional complementarity by Evans and co-authors (Kalaskar & Evans 2001, Evans et al. 2004, Evans & Gunther 2005) and Michaud & Jyoti (2008), and differences in the larval versus adult food specificity was stressed by Michaud (2005). Some problems of food ecology are being solved by detailed field experiments, often using the addition of artificial food and sometimes the mark–recapture techniques (Evans & Toler 2007, Honěk et al. 2008a, b, Sloggett 2008a, b, Sloggett et al. 2008; 5.2.7).

Important recent advances have also been made concerning patterns and processes of food consumption (5.3). Close study has revealed new insights, for example, as to how prey limitation early in life affects ladybird performance later on (Dmitriew & Rowe 2007), and how the biochemical properties of prey versus predator may influence a ladybird's ability to balance its nutritional needs (Specty et al. 2003, Kagata & Katayama 2006).

The importance of behavioural studies and their relative shortage in relation to insect predators has recently been stressed (Mills & Kean 2010). However, in these studies (5.4) decisive, qualitative progress has been made. Sensory responses, which were previously questioned, have not only been repetitively documented by many teams in several countries, but also the chemical composition of some attracting volatiles has been revealed; thus the volatiles produced by herbivore-injured plants have been found to attract enemies to those herbivores. The adaptive phenomenon of oviposition deterring substances, left on the substrate by crawling larvae, has been discovered (Ruzicka 1997) and then recorded in a number of coccinellid species, and one chemical responsible has already been identified (Klewer et al. 2007).

Under the pressure of the typically discontinuous occurrence of aphid prey, the evolution of coccinellid feeding behaviour evidently has not led to development of **gustatory discrimination** in generalists, at least in aphidophagous coccinellids. It has been recorded repeatedly that the ladybirds do not avoid less suitable or even toxic food. They even do not prefer better food when given a choice (e.g. Nielsen et al. 2002, Nedvěd & Salvucci 2008, Snyder & Clevenger 2004; 5.2.1.1, 5.2.5, 5.2.6.1).

We compile the findings in this chapter in the optimistic expectation that they represent just the beginning of a new fruitful period for studies of coccinellid food and feeding.

5.2 FOOD SPECIFICITY

5.2.1 Food range

Although a tendency to feed on the same or similar prey can be observed in taxonomic groups of coccinellids, one may find prey **specialists** even **within** individual **tribes**. Thus, for example, in the generally aphidophagous tribe Coccinellini there are also non-predaceous species, such as the phytophagous *Bulaea lichatschovi* (Capra 1947, Dyadechko 1954, Savoiskaya 1966, 1970a) and the pollinivorous and mycophagous *Tytthaspis sedecimpunctata* (Dauguet 1949, Turian 1969, Ricci 1982, Ricci et al. 1983; 5.2.6). Within this tribe there are further species which specialize on immature stages of Coleoptera, such as *Aiolocaria* spp.

(Iwata 1932, 1965, Savoiskaya 1970a, 1983, Kuznetsov 1993), *Calvia quindecimguttata* (Kanervo 1940) and *Coccinella hieroglyphica* (Hippa et al. 1984; Table 5.1) which feed on pre-imaginal stages of Chrysomelidae, or *Neocalvia* spp. which prey on larvae of mycophagous coccinellids of the tribe Psylloborini (Camargo 1937). Thus, not only entomophagy, but also phytophagy and mycophagy are represented within the tribe Coccinellini. Phytophagy is typical for the subfamily Epilachninae, mycophagy in the tribe Psylloborini (Sutherland & Parrella 2009) and feeding on mites in Stethorini (Biddinger et al. 2009). The

classification of coccinellids is still partially artificial (Giorgi et al. 2009; Chapter 1) and thus discussing the food specialization of a whole tribe cannot but have its limitations (Hodek 1996).

Predaceous coccinellids have a wide range of **accepted food**. Apart from feeding on sternorrhynchan Hemiptera and phytophagous mites, they often prey also on Thysanoptera and young instars of holometabolan insects (Evans 2009) and even food of plant origin as well as fungi. Taxonomic differences in the rate of coccinellid development are related to the predominant type of food consumed (Table 5.2).

Table 5.1 Larval mortality and pupal weight of *Coccinella hieroglyphica* fed on aphids or chrysomelids (Hippa et al. 1984).

Prey	Larval mortality %	n	Fresh weight of ♀ pupae (mg) mean	S.E.	n
*Myzus persicae**	20.9	43	9.78	0.13	12
*Symydobius oblongus**	22.9	70	9.89	0.12	25
Galerucella sagittariae[†]	—	—	—	—	—
eggs	67.7	31	10.19	0.72	5
larvae	10.8	74	12.47	0.22	17

*Aphidoidea.
[†]Chrysomelidae.

Table 5.2 Rate of natural increase (r_m) of acariphagous, aphidophagous and coccidophagous coccinellid species (after Roy et al. 2003).

Species	Prey relationship	Temp (°C)	r_m (d^{-1})
Parastethorus nogripes	Acariphagous	25.0	0.152
Stethorus madecassus	Acariphagous	25.0	0.155
Stethorus punctum picipes	Acariphagous	24.0	0.121
Stethorus pusillus	Acariphagous	24.0	0.100
Coccinella septempunctata	Aphidophagous	26.0	0.190
Coleomegilla maculata lengi	Aphidophagous	25.0	0.110
Olla v-nigrum	Aphidophagous	25.7	0.160
Propylea quatuordecimpunctata	Aphidophagous	26.0	0.140
Diomus hennesseyi	Coccidophagous*	25.0	0.103
Exochomus flaviventris	Coccidophagous*	25.0	0.050
Hyperaspis notata	Coccidophagous*	25.0	0.081
Hyperaspis raynevali	Coccidophagous*	25.0	0.081
Hyperaspis senegalensis hottentotta	Coccidophagous*	26.0	0.070
Rodolia iceryae	Coccidophagous*	27.0	0.064

*Coccidophagous ladybirds have characteristically slower development: 5–10 % per day.

The great variety of food consumed led to the assumption that food specificity in coccinellids concerns only major taxonomic groupings: e.g. aphidophagous ladybirds only eat aphids. Observed **acceptance** has been **mistaken for suitability** of prey, even by experienced workers. For example Balduf (1935) concluded, on the basis of a survey of accepted food, that no special groups of aphids are selected by ladybirds of the tribes Hippodamiini and Coccinellini. Kanervo (1940) treated the six species of aphid prey as a single complex because coccinellids did not show any great preference for individual species of aphids. Such an assumption has long been perpetuated even though contradictory evidence has been reported. This is notably the case in the incorrect characterization of prey of *Adalia bipunctata*. Although habitat preference is important, overgeneralizations may often be erroneous. The conclusion that all aphids living on shrubs or trees represent suitable prey for *A. bipunctata*, while aphids from other habitats are unfavourable prey for this ladybird (Pruszynski & Lipa 1971), has already been refuted by earlier data on *Aphis sambuci* (Hodek 1956, Blackman 1965) as well as by more recent data discussed in this chapter.

Early studies of predator–prey relationships were mostly limited to compiling lists of observed feeding in the field or concurrent presence of predator and prey on the same plant (Hodek 1996 for references). The reliability of such lists was questioned quite early by Thompson (1951): 'The various species of ladybirds do not actually feed . . . the host insects with which they are associated in the records'. Thompson warned further that: 'The gradual accumulation of such records in the literature finally gives a picture which may be completely inaccurate in so far as the real behaviour and food habits of the species are concerned.'

However, when experimental analysis of the food relationships of particular species of coccinellids is missing, the lists of **observed associated occurrence** of predator and prey can be useful. They provide preliminary orientation on the relations of ladybird species to natural prey and habitats. Lists of **accepted** prey have been published for the eight most common aphidophagous coccinellid species from Far East Russia (Kuznetsov 1975, 1993) and for 36 species of aphidophagous Coccinellidae in India (Agarwala & Ghosh 1988). Agarwala et al. (1987) discriminated between the 'accepted' and 'common' aphid prey of four coccinellid species from northeastern India. The data is based on examinations of gut contents and field observations.

Majerus (1994) gives a useful list of principal and secondary foods of British coccinellids (although without defining these two categories). Another helpful food list of Klausnitzer and Klausnitzer (1997) for central European Coccinellidae does not differentiate the suitabilities of different prey, but just gives examples of prey consumed by ladybirds based on the authors' observations. Species of psyllids and aleyrodids observed to be preyed on by coccinellids have recently been listed, while an other compilation in the same paper lists essential coccid prey (Hodek & Honěk 2009).

After the invasion of the soybean pest *Aphis glycines* to the USA, Wu et al. (2004) compiled a list of its natural enemies (including seven coccinellid species) in China and southern Korea. *Propylea japonica* (62%) and *Har. axyridis* (10%) were earlier reported as dominant ladybeetles on *A. glycines* in China (Wang & Ba 1998).

5.2.1.1 Methods for detection of food range

The **natural food range** of coccinellids can be ascertained in several ways (Weber & Lundgren 2009). The classical method is the microscopic detection and identification of **prey remnants** from the guts or excreta (e.g. Agarwala et al. 1987, Triltsch 1999, Ricci & Ponti 2005, Ricci et al. 2005, Davidson & Evans 2010) which can be compared with the whole specimens of insects from the same community.

In **serological assays** (also Chapter 10.7) the specific proteins of the prey are identified by their reaction with the serum of a sensitized mammal (usually a rabbit). In the 1980s, the most popular assays were ELISAs (e.g. Crook & Sunderland 1984, Sunderland et al. 1987, Hagley & Allen 1990). A double antibody method that can be scored by eye (Stuart & Greenstone 1990, Greenstone & Trowell 1994) is still used (Santos et al. 2009).

The interpretation of **molecular gut-content** data depends on the rate of decay in their detectability, expressed as DNA half-life. The predators with longer period of prey-detectability might be wrongly considered as more important, when compared with predators that have short detectability. **DNA half-life** depends on types of ingestion and digestion. When one egg of *Leptinotarsa decemlineata* was used as prey, DNA half-life was 7.0 hours in *Col. maculata* and 50.9 hours in the spined soldier bug, *Podisus maculiventris* (Green-stone et al. 2007). Two serological methods have recently been used to prove that eggs of a noctuid are prey of *Hip. variegata* (Mansfield et al. 2008). The **polymerase chain reaction** technique (PCR) effectively amplifies DNA prey residues in the gut contents of arthropod predators by use of specific primers. Species can be identified based on gene amplification fragments which are observed as bands in the stained agarose gel following electrophoresis. The use of data obtained by PCR-based methods requires careful evaluation of all possible factors that may affect interpretation. *Hip. variegata* had the shortest median detection success 17 hours at 20°C, while it was 36 hours in a nabid bug and 50 hours in a spider. The rate of detection decreases with increasing temperature in *Hip. variegata*, but not in the spider, although sex and weight of *Hip. variegata* did not influence detection of prey DNA (Hosseini et al. 2008). PCR can only be used to detect prey DNA in the gut if the predator was captured shortly before testing. In the case of *Rhopalosiphum padi* as prey and *A. bipunctata* as predator, temperature (21 or 14°C) had no significant effect on *R. padi* detection (McMillan et al. 2007). For a novel use of **chromatographic analysis** see Sloggett et al. (2009b; also Chapter 10).

5.2.2   Nutritional suitability of food

In spite of their considerable polyphagy as regards **accepted food**, many coccinellid species are specialists as far as **nutritionally suitable** food is concerned. Predator–prey relationships should be viewed in this

sense to rightly assess the possible impact of a coccinellid on any given pest.

Different criteria of food suitability have been used in food research in coccinellids. As stressed above and in 5.2.1, **acceptance** has been an invalid criterion of food suitability. **Preference** for certain prey seemed to provide useful evidence: Strauss' (1982) linear index of prey selection was used in the analysis of prey preference in *Stethorus punctum* (Houck 1986). However, experiments with *Megoura viciae* (Blackman 1967a) and *Aphis sambuci* (Hodek 1956, Nedvěd & Salvucci 2008) have shown that ladybirds feed on toxic aphids even if provided with a better alternative.

The **presence of larvae** of the predator in association with a prey species is considered to be good evidence for evaluating food specificity in the field, although it may be misleading in a habitat with several potential prey species. Mills (1981) warned that, in early and late season, the adults can be observed in association with prey types and habitats that are not used during reproduction, and the presence of larvae served as criterion for his list of essential prey for *A. bipunctata*. A great advantage of this list (Mills 1981) is the inclusion of the type of **habitat** and **host plant**. Most listed prey would be essential prey. *A. bipunctata*, however, develops on the listed *Aphis sambuci* with a significantly lower larval survival and weight of emerging adults, compared with good essential prey (Blackman 1965, 1967b; Table 5.3). *Aphis sambuci* is thus a rather unsuitable prey for *A. bipunctata*. In terms of the fecundity of *A. bipunctata* females, *Drepanosiphum platanoidis* on sycamore and *Chaitophorus capreae* on sallow were the least suitable prey (Mills 1981). In rare cases the predator–prey relationship has been safely

Table 5.3 Larval development and oviposition of *Adalia bipunctata* on different aphids (Blackman 1965, 1967b).

Aphid species	Larval development (days)	Larval mortality (%)	Weight of adults at emergence (mg)	Fecundity (total eggs laid)	Fertility (% of viable eggs)
Myzus persicae	10.4	17.8	11.8	676.2	89.4
Neomyzus circumflexus	9.5	16.7	11.9	—	—
Acyrthosiphon pisum	10.8	13.9	12.6	—	—
Microlophium carnosum	10.6	9.1	12.4	—	—
Aphis fabae	13.0	27.6	7.9	249.6	55.9
Aphis sambuci	13.4	25.0	8.0	—	—
*Brevicoryne brassicae**	21; 23	(66.7)	5.1; 6.1	—	—

*Only 2 out of 6 larvae completed development.

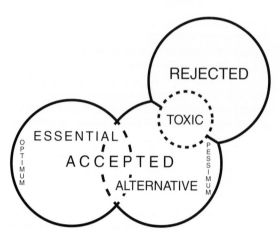

Figure 5.1 Types of physiological suitability of coccinellid foods (modified from Hodek 1996).

identified by another indirect method: that of systematic and **long term observation**. In the course of 5 years, Eastop and Pope (1966, 1969) found 99% of specimens of *Scymnus auritus* on oaks infested with *Phylloxera glabra*.

In most cases only **experimental evidence**, when main life history parameters of predators are measured on a prey species, is an entirely safe way to assess physiological suitability. This became apparent when it was found that a toxic aphid, *Aphis sambuci*, was eaten by *C. septempunctata*, but exerted detrimental effects (Hodek 1956). It was then suggested that physiological parameters, such as growth, reproduction and mortality, should be used as the decisive criterion. It was proposed to discriminate between two principal kinds of food: **essential** (Fig. 5.1; 5.2.11) that ensure oviposition and the completion of larval development, and **alternative** that serve only as a source of energy and thus prolong survival in comparison with starvation (Hodek 1962, 1967, 1973, 1993). This approach has been widely accepted (e.g. Mills 1981, Evans et al. 1999, Giorgi et al. 2009, Berkvens et al. 2010).

The experimental evaluation of the suitability of prey does have some disadvantages, however. If the prey is offered to the predator in a small space, the escape and defence behaviour of prey is limited and predator's searching behaviour may be modified. By such experiments a prey species can be ascertained as essential, although in the field it does not coincide with

the predator in space and time (Iperti 1965, 1966, Brun & Iperti 1978). It appears that alternative feeding on various foods is advantageous for predatory ladybirds (5.2.5). The listing of 'unnatural' essential prey (in 5.2.11) can be useful information for culturing ladybirds. However, it is important not to overestimate the records on nutritive quality of the prey (5.2.6, 5.2.11). Complex predator–prey relations have recently been discussed in a comprehensive and novel way (Sloggett 2008a; 5.2.3).

Ideally, to characterize 'natural' essential prey, observational and experimental approaches should be combined. The relationships of 11 strains of five coccinellid species (*C. septempunctata, Hip. tredecimpunctata, Hip. variegata, P. quatuordecimpunctata* and *Cer. undecimnotata*) with two species of essential aphid prey (*Diuraphis noxia* and *Schizaphis graminum*) were evaluated by Michels and Flanders (1992). The comparison of the main life-history parameters indicated important differences in prey suitability, between both coccinellid species and strains of the same species. Thus, for example, in the Moldavian strain of *P. quatuordecimpunctata* the number of eggs laid with *D. noxia* as food was only a fraction of that with *S. graminum*. The Ukrainian strain of the same coccinellid species, however, had twice the fecundity when fed on *D. noxia* than on *S. graminum*. Michaud (2000) compared the performance of seven coccinellid species fed on *Aphis spiraecola* and *Toxoptera citricidus* and Ozder and Saglam (2003) evaluated *A. bipunctata* on four and *C. septempunctata* on five aphid species (Table 5.4).

In Europe, the effect of a range of aphid species on life history parameters has been compared in *A. bipunctata* (Kalushkov 1998), *Anatis ocellata* and *Calvia quatuordecimguttata* (Kalushkov & Hodek 2001), *C. septempunctata* (Kalushkov & Hodek 2004) and *P. quatuordecimpunctata* (Kalushkov & Hodek 2005). In all four papers the laboratory experiments were complemented by observation of predator–prey relations in the field. In India, Omkar and his colleagues made comparative studies for four species: *C. septempunctata* (Omkar & Srivastava 2003), *C. leonina transversalis* (Omkar & James 2004), *Menochilus sexmaculatus* (Omkar & Bind 2004), and *P. dissecta* (Pervez & Omkar 2004). They compared the performance of ladybirds when reared on seven aphid species as prey. In these four experiments the sequence of suitability of all aphids used appeared surprisingly regular when measured by survival of larvae, survival of adults and fecundity

Table 5.4 Total development time and survival of two coccinellid species as affected by aphid prey (recorded at $25 \pm 1°C$, 16L:8D in Tekirdag, Turkey) (modified after Ozder & Saglam 2003).

	Aphid species				
	S. avenae	**R. padi**	**M. cerasi**	**H. pruni**	**M. dirhodum**
	Adalia bipunctata				
Total development (days)	20.7 ± 1.03 (19–22) c	16.7 ± 0.76 (16–18) a	17.9 ± 0.80 (17–20) b	19.8 ± 1.14 (18–21) c	—
Survival (%)	67 ab	78 ab	82 a	50 b	—
	Coccinella septempunctata				
Total development (days)	18.9 ± 1.07 (16–21) a	19.0 ± 1.15 (18–21) a	17.5 ± 0.84 (16–18) a	20.8 ± 1.60 (19–23) b	18.6 ± 0.89 (18–20) a
Survival (%)	84 a	70 ab	60 ab	37 b	63 a

Within rows, means followed by a common letter do not differ significantly ($P < 0.05$).
Values are means ± SE; the range between minimum and maximum values is given in parentheses ($n = 30$).
Sitobion avenae, Rhopalosiphum padi, Myzus cerasi, Hyalopterus pruni, Metopolophium dirhodum.

(Tables 5.5, 5.6). In other studies the effect of prey on the parameters was more variable (see next paragraph). When Omkar and co-authors experimented with *Aphis nerii*, it was always the least suitable prey. *Aphis craccivora* and *A. gossypii* were optimal for three ladybird species, but not for *C. septempunctata*. *Lipaphis pseudobrassicae* was optimal prey for this species, while this aphid was the second worst prey for *M. sexmaculatus*. For *C. leonina transversalis* and *P. dissecta*, *L. pseudobrassicae* was the third optimal prey. Earlier Indian papers also recorded *L. pseudobrassicae* as a very good prey for *C. septempunctata*, but it was unsuitable for the reproduction of *Micraspis discolor* (Atwal & Sethi 1963, Agarwala et al. 1987).

More variable effects of aphids on coccinellids' parameters have been recorded by other workers. A high percentage of *Har. axyridis* larvae survived to adult (70%) when reared on *Aphis spiraecola* (Table 5.7), but the females did not lay eggs on this diet (Michaud 2000). Similarly the fecundity of *C. septempunctata* was optimum on *Acyrthosiphon pisum* and *Sitobion avenae*, but the male longevity was lower on these two aphids than on *A. fabae* and *A. craccivora* – although the fresh weight of emerged males, reared on the former two aphids, was among the highest (Kalushkov & Hodek 2004).

For the sake of easier orientation, differences in the nutritive suitability of foods will be discussed in particular subchapters. Alternative food will be discussed in the sections on low quality prey (5.2.6), non-insect food (5.2.9), and non-hemipteran prey (5.2.7). There are overlapping cases; the categories of food should never be used dogmatically and inflexibly.

5.2.3 Prey size–density hypothesis

In almost the entirety of Section 5.2 the discussion of food specificity is based on the nutritional and chemical qualities of prey, i.e. mostly on the nature of allelochemicals derived from host plants. The research has stressed these aspects for the past several decades, and the contents of the chapter reflect this.

However, prey specificity has recently begun to be viewed from another angle: the relation between the **body size** of aphidophagous ladybirds and aphid prey is considered as more important (or at least similarly important) than chemistry. The turning point may be represented by the concept of Sloggett (2008b): 'body size trade-offs are the likely most important universal factor underlying dietary specialization in aphidophagous coccinellids'. As our short explanation here will necessarily remain a simplification, the reader is advised to study directly the two recent papers by Sloggett (2008a, b), which contain many stimulating thoughts.

Sloggett's (2008a) **prey size–density hypothesis** has been formulated in response to a related hypothesis

Table 5.5 Life-history parameters of *Coccinella septempunctata* as affected by aphid prey (modified after Omkar and Srivastava 2003).

Aphid species	L. erysimi	M. persicae	A. craccivora	A. gossypii	U. compositae	A. nerii
Development duration (days)	13.9 ± 0.1 a	15.2 ± 0.1 b	16.8 ± 0.1 c	17.9 ± 0.1 d	21.0 ± 0.1 e	22.9 ± 0.1 b
Weight (mg) Male	38.2 ± 0.3 a	33.9 ± 0.1 b	29.9 ± 0.2 c	27.1 ± 0.2 d	23.8 ± 0.1 e	20.8 ± 0.2 b
Female	40.6 ± 0.2 a	37.7 ± 0.2 b	33.4 ± 0.1 c	31.3 ± 0.2 d	25.9 ± 0.1 e	22.3 ± 0.4 b
Immature survival (%)	73.5 ± 0.9 a	66.6 ± 0.8 b	64.2 ± 2.2 bc	61.2 ± 3.1 cd	52.1 ± 2.4 e	43.9 ± 1.3 f
Adult emergence (%)	90.1 ± 1.4 a	88.5 ± 1.2 ab	85.0 ± 1.5 ab	81.1 ± 2.1 ab	80.2 ± 2.9 bc	71.7 ± 2.8 d
Consumption (aphids/female/day)	72.1 ± 1.4 a	65.3 ± 1.5 b	59.2 ± 1.3 c	53.8 ± 1.6 cd	34.8 ± 1.5 e	26.3 ± 2.1 f
Female longevity (days)	85.7 ± 1.5	80.0 ± 1.0	76.6 ± 1.4	73.7 ± 1.8	62.5 ± 1.3	53.5 ± 1.0
Relative growth rate (RGR)	1.52 ± 0.02	1.27 ± 0.03	0.99 ± 0.01	0.86 ± 0.02	0.57 ± 0.01	0.49 ± 0.02
Oviposition period (days)	69.8 ± 1.3 a	57.9 ± 1.0 b	50.7 ± 1.1 c	44.8 ± 1.6 d	30.2 ± 1.1 e	16.4 ± 0.6 f
Fecundity (eggs/female)	1764 ± 8.46 a	1198.5 ± 0.1 b	1060.7 ± 25.8 c	739.2 ± 32.0 d	488.1 ± 16.4 c	203.2 ± 11.8 d
Egg viability (%)	87.9 ± 1.1 a	79.9 ± 0.6 b	74.7 ± 0.73 c	71.12 ± 0.9 d	58.6 ± 1.3 e	48.7 ± 2.1 f

Lipaphis erysimi, *Myzus persicae*, *Aphis craccivora*, *Aphis gossypii*, *Uroleucon compositae*, *Aphis nerii*.
Mean ± SE followed by the same letters in the same row are not significantly different at $P < 0.001$.
$n = 10$, $25 \pm 2°C$

Table 5.6 Life-history parameters of *Propylea dissecta* as affected by aphid prey (modified after Pervez and Omkar 2004).

Aphid species	A. gossypii	A. craccivora	L. erysimi	U. compositae	B. brassicae	R. maidis	M. persicae
Immature survival (%)	77.1 ± 1.1 a	74.1 ± 1.3 ab	71.8 ± 1.5 b	67.7 ± 1.9 bc	67.2 ± 1.9 c	65.3 ± 1.3 cd	63.0 ± 1.9 d
Adult emergence (%)	93.2 ± 0.8 a	92.0 ± 0.9 ab	88.9 ± 1.5 bc	85.3 ± 1.0 c	84.9 ± 1.0 d	82.8 ± 1.5 e	81.7 ± 1.8 e
Female longevity (days)	62.4 ± 1.9 a	58.4 ± 0.9 ab	53.0 ± 2.6 bc	51.9 ± 1.4 c	51.0 ± 1.4 cd	50.9 ± 2.3 d	49.4 ± 2.3 d
Oviposition period (days)	50.3 ± 2.0 a	44.8 ± 0.9 b	34.0 ± 1.6 c	29.0 ± 1.3 c	27.0 ± 1.5 cd	24.2 ± 1.3 d	18.0 ± 1.4 e
Fecundity (eggs/female)	856.0 ± 30.0 a	750.0 ± 36.7 b	506.0 ± 24.1 c	456.8 ± 21.5 d	414.0 ± 17.7 de	374.6 ± 16.8 e	212.0 ± 18.2 f
Egg viability (%)	96.4 ± 0.3 a	95.4 ± 0.4 a	87.0 ± 1.4 b	83.9 ± 1.7 bc	81.2 ± 1.8 c	76.1 ± 1.7 cd	72.5 ± 2.8 d

Aphis gossypii, *Aphis craccivora*, *Lipaphis erysimi*, *Uroleucon compositae*, *Brevicoryne brassicae*, *Rhopalosiphum maidis*, *Myzus persicae*.
$n = 10$; mean ± SE in the same row not followed by the same letter are significantly different at $P < 0.001$.

Table 5.7 Survival rates for larvae of seven coccinellid species fed on *Aphis spiraecola* and *Toxoptera citricida* (modified after Michaud 2000).

	Larvae surviving to adult (%)		
	A. spiraecola	*T. citricida*	*P*
Coccinella septempunctata	5.0	0.0	ns*
Coelophora inaequalis	0.0	0.0	—
Coleomegilla maculata fuscilabris	0.0	45.0	0.01[†]
Cycloneda sanguinea	60.0	100.0	0.001*
Harmonia axyridis	70.0	95.0	ns*
Hippodamia convergens	68.0	0.0	0.001[†]
Olla v-nigrum	0.0	0.0	—

Larvae (*n* = 20) were fed from eclosion on either *A. spiraecola* or *T. citricida*.
*Significant differences based on one-way ANOVA; ns indicates no significant difference.
[†]Significant differences based on Mann-Whitney *U* test; ns indicate no significant difference.

(Dixon 2007, see also Dixon & Stewart 1991, Stewart et al. 1991, Dixon & Hemptinne 2001). From calculations of minimum mass of prey needed for egg laying and the surface searched within a time unit, both of which scale with body weight, Dixon (2007) maintained that 'there appears to be no association between the size of aphidophagous predators and that of the species or the age structure of the aphid colonies they exploit'.

In Dixon's hypothesis (Fig. 5.2a) smaller ladybirds reproduce at lower total aphid densities (thus earlier in the season). The idea that the body of aphidophagous ladybirds and the size of their aphid prey are unrelated has become embedded in the literature, based on the absence of correlation between the weight of seven species of aphidophagous ladybirds of different sizes and the weight of their prey (Dixon & Stewart 1991, Stewart et al. 1991). Furthermore, comparisons between the largest and smallest species showed no difference (Stewart et al. 1991). Comparisons were based on species-specific prey lists from the literature where typical prey was not differentiated from rarely eaten prey. Furthermore, nymphs of early instars are much smaller than adult aphids, and large and small ladybirds evidently exploit them differently. Particularly in the case of large aphid species the relative size affects capture efficiency. This was described by Klingauf (1967) in coccinellid larvae and observed by Sloggett (2008b), when e.g. adults of large *Cinara* conifer aphids were easily caught by adults of the large

Anatis ocellata, but kicked away the small *Myrrha octodecimguttata*.

A further size-related dilemma commented upon by Sloggett (2008b) is that of the relation between body size of aphidophagous ladybirds and aphid density and size. While Dixon's (2007) model considers only prey density (Fig. 5.2a), Sloggett (2008b) argues that the relationship is more complex and comprises **both prey size and prey density** (Fig. 5.2b, c). Dixon's (2007) model holds true when the prey is a small aphid species: large and small ladybirds have the same chance of catching the prey. This is not the case with large aphid species. A small ladybird can only catch young instars (Fig. 5.2b, c). Thus it exploits only a proportion of aphids and needs a higher prey density to catch the same number of aphids (Fig. 5.2b). The largest aphids caught by large ladybirds represent most energy. Small ladybirds are many times less efficient and are likely to need a higher prey density for reproduction (Fig. 5.2c). Simply put, **small ladybirds** can exploit lower densities of small aphid species because they need less food. By contrast, **large ladybirds** can exploit lower densities of large aphid species because they are more efficient at capture. Thus there is a relationship between predator size and prey size, but it is not the rigid relationship as tested by Stewart et al. (1991). In such a situation, with large aphid species, small ladybirds will not reproduce earlier, but later (or not at all) (Sloggett 2008b).

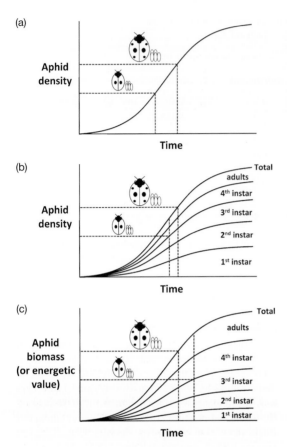

(a)

Aphid
density

Time

(b)

Aphid
density

Total
adults
4th instar
3rd instar
2nd instar
1st instar

Time

(c)

Aphid
biomass
(or energetic
value)

Total
adults
4th instar
3rd instar
2nd instar
1st instar

Time

Figure 5.2 Relationship between prey abundance and ladybird body size. In each case the horizontal dashed lines represent the prey threshold necessary for reproduction by the large (upper line) and small (lower line) ladybirds. The vertical dashed lines show the total aphid abundance and the time at which reproduction is possible for each of the species. (a) The established view of Dixon (2007): here there is no relationship between ladybird body size and capture efficiency as might occur with small aphid species. Small ladybirds require a lower density of aphids for reproduction on account of their lower energetic requirements (see text for details) and can thus reproduce earlier in the development of an aphid population. (b) Total aphid density required by a small ladybird is increased if the small ladybird, on account of its size, is unable to catch larger aphid instars (in this case fourth instars and adults); this might occur with large aphid species. Aphid density is broken down into the proportions of the different instars occurring, and does not take into account the differences in size and energetic content of the different aphid instars but only their numerical abundance. (c) A similar, but more realistic graph than (b) taking into account the different energetic values of different instars. Because the small aphid instars which can be caught by a small ladybird are less energy rich than the average aphid, this acts to make the total abundance of aphids necessary for reproduction higher still, leading to a situation where a small ladybird reproduces later than a large one (from Sloggett 2008b).

Sloggett (2008b) quotes several pieces of evidence, supporting his prey size–density hypothesis. These include: (i) Banks (1955) recorded on nettles (*Urtica dioica*) infested by the larger aphid *Microlophium carnosum*, earlier arrival of large numbers of *C. septem-punctata* and only later appearance of *A. bipunctata* and *P. quatuordecimpunctata*. (ii) Evans (2004) reported that *C. septempunctata* can exploit the large pea aphids (*A. pisum*) on alfalfa at lower densities than can the smaller native ladybirds. The topic was revisited in Honěk et al. (2008a, b), when aphids were rather small and ladybirds thus followed Dixon's (2007) argument.

5.2.4   Euryphagous and stenophagous species / Generalist and specialist species

In most of 5.2 the discussion is focussed on specific prey types. In this section the focus is on the characteristics of the coccinellid species, in particular the contrast between those that can successfully develop and reproduce on only a narrow range of foods, that is specialist or stenophagous species, and those with a rather wide range of essential prey / foods – generalist or euryphagous (polyphagous) species. The latter species may be more easily manipulated when used in biological control (Chapter 11) as they can be more easily mass-produced on artificial diets (5.2.10) for periodic colonization, or their numbers augmented in the field by providing alternative foods.

In relation to prey, the terms generalists versus specialists have naturally remained rather vague, because, as Sloggett (2008b) rightly points out 'dietary breadth is a continuum, rather than a dichotomy'. Thus we have hesitated (Hodek & Michaud 2008) over which of these terms to use for *A. bipunctata* to indicate its difference from *C. septempunctata*: we considered

Table 5.8 Dichotomization of the British aphidophagous Coccinellini into dietary generalists and specialists. For discussion on conifer specialists see Sloggett and Majerus (2000) (modified after Sloggett 2008b).

Generalists	Specialists (habitat)
Adalia bipunctata	Adalia decempunctata (trees and shrubs)
Calvia quatuordecimguttata	Anatis ocellata (conifers)
Coccinella magnifica	Anisosticta novemdecimpunctata (reed beds)
Coccinella quinquepunctata	Aphidecta obliterata (conifers)
Coccinella septempunctata	Harmonia quadripunctata (conifers)
Coccinella undecimpunctata	Hippodamia tredecimpunctata (reed beds, marsh)
Hippodamia variegata	Myrrha octodecimguttata (conifers)
Propylea quatuordecimpunctata	Myzia oblongoguttata (conifers)

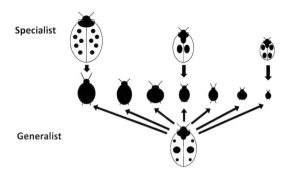

Figure 5.3 Body sizes of ladybirds which are dietary specialists and generalists in relation to aphid size. Specialists will closely match their prey in size whereas generalists, which feed on prey of a diversity of sizes, will be medium sized. As a result, specialists are expected to display a wider diversity of sizes than generalists (from Sloggett 2008b).

C. septempunctata to be a habitat generalist, but more specialized to prey (24 essential prey in Hodek's 1996 list). *A. bipunctata* seems to us more habitat specialized, with a clear preference for shrubs and trees, but less specialized towards prey (40 essential prey in Hodek's 1996 list). Both *A. bipunctata* and *C. septempunctata* appeared in the same group of generalists in a recent paper (Sloggett 2008b), where 16 British aphidophagous Coccinellini are divided into eight generalists and eight specialists (Table 5.8).

As underlined by Sloggett (2008b) evolution of specialization in predators operates by fitness trade-offs between types of prey. Adaptations enabling high fitness on one type cause a decrease in fitness with other prey types. Such a trade-off holds also for generalist versus specialist predators. Specialists are very successful on only a few types of prey, but perform poorly on many others, while for generalists there is less variation in performance on a wider range of prey. More specialised aphidophagous ladybirds are less mobile due to their greater tolerance of lower aphid densities (Sloggett et al. 2008).

Interest has always been concentrated on trade-offs related to prey nutritive suitability, as affected by allelochemicals derived from prey host plants. However, no unequivocal demonstration of such a chemical trade-off has been made for aphidophagous ladybirds (Sloggett 2008a) (see also discussion at the end of 5.2.5). Sloggett (2008b) therefore suggests that body size may play such a role (5.2.3). A specialist is likely to have the right size to perform well on the special prey: it is big enough to catch the old aphid instars (and adults), but not so big as to need too high a prey density for reproduction, i.e. its size will fit the prey. A generalist will, by contrast, tend to be of medium size and therefore prey on a wide range of aphid sizes (Fig. 5.3, Sloggett 2008b).

Sloggett (2008b) tested his **'body size–dietary breadth' hypothesis** on 16 native British aphidophagous Coccinellini (Table 5.8, Sloggett 2008a). He compared sizes between eight specialists and eight generalists (Fig. 5.4, Sloggett 2008a). The size range of specialists is greater and both the largest (*Anatis ocellata*, *Myzia oblongoguttata*) and the smallest two species (*Anisosticta novemdecimpunctata*, *Aphidecta obliterata*) are specialists. In conclusion the author argues that 'body size is a trade-off of likely considerable importance in determining prey specialization due to its relationship with both prey size and density'. He is right to invite a re-evaluation of the hypotheses on body size, such as those claiming that there is no relationship between predator body size and prey size (5.2.3).

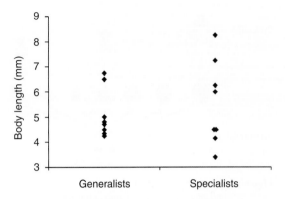

Figure 5.4 The distribution of body sizes in dietary generalists and specialists belonging to the native British aphidophagous Coccinellini (from Sloggett 2008b).

5.2.4.1   Generalists

Coleomegilla maculata

A number of papers have indicated that *Col. maculata* is a widely polyphagous species. Early data on its polyphagy were reported by Putman (1957). The rate of larval development was not decreased by feeding on the **mite** *Tetranychus urticae* compared with feeding on aphids; all larvae completed development, adult survival was very high and oviposition was not inhibited. Another mite, *Panonychus ulmi* also appeared to be suitable food for larvae of *Col. maculata*, and the adults fed much more extensively on crawlers of *Pulvinaria vitis* than did the other coccinellids in the study. Three *Coccinella* spp., *Cycloneda munda* and *A. bipunctata* were less euryphagous.

Studies of this type are also used to imply broad dietary breadth within individual groups of prey, particularly across aphid species. Some such data have been supplied for *C. septempunctata* and *A. bipunctata* by Blackman (1967a, b) and for *Har. axyridis* and *C. s. brucki* by Japanese authors (Okamoto 1961, 1966, Hukusima & Sakurai 1964, Takeda et al. 1964, Hukusima & Kamei 1970). Other examples are also discussed in 5.2.2.

Coleomegilla maculata preys so intensively on **eggs** of some pest insects that it has been studied in the control of the lepidopterans *Heliothis virescens* (Ables et al. 1978) and *Ostrinia nubilalis* (Risch et al. 1982) as well as the Colorado potato beetle *Leptinotarsa decemlineata*

(Hazzard & Ferro 1991, Hilbeck et al. 1997, Nault & Kennedy 2000, Mallampalli et al. 2005). Wiebe and Obrycki (2004) reported *Col. maculata* feeding on eggs of another chrysomelid, *Galerucella pusilla*, but the larvae did not thrive well on this prey. The functional response of *Col. maculata* to eggs of *L. decemlineata* was modified by the presence of aphids, but not maize pollen.

An important aspect of euryphagy of *Col. maculata* is its **pollinivory** (5.2.9 for recent data) and an apparent preference for **powdered food**. The preference for lower strata of corn plants discussed in Chapter 4 is apparently related to the pollinivory of this species (Ewert & Chiang 1966). Laboratory tests have proved that pollen of several plants is essential food for *Col. maculata*. The pollens of *Betula populifolia* and *Carpinus caroliniana* were found to be superior to pollen of *Zea mays*, *Cannabis sativa*, *Juglans cinerea* and *Typha latifolia*, but all of these pollens enabled the completion of larval development in *Col. maculata*. By contrast, the pollen of coniferous trees, and also of *Quercus rubra* and *Ambrosia artemisiifolia* was unsuitable (Smith 1960, 1961). When Atallah and Newsom (1966) used stamens of cotton as a source of pollen, *Col. maculata* larvae failed to develop and died within 4 days. Pollinivory of coccinellids, including *Col. maculata*, has been reviewed by Lundgren (2009).

Although *Col. maculata* can develop satisfactorily on mites and coccids (Putman 1957), live aphids (Smith 1965b), and on holometabolan eggs and larvae, its specific relationship to powdered food was shown by preference for aphids in such a form. Only 30% of the larvae completed development if fed with live *Acyrthosiphon pisum*, and the resulting adults weighed only 8.5 mg; the same prey in a dry powdered state, however, gave a much higher survival (90%) and larger adults (12.0 mg) (Smith 1965b). Dry aphids appeared marginally more suitable than pollen (Smith 1965b). In another study, however, *Col. maculata* adults had a doubled pre-oviposition period and a halved fecundity when fed on corn (*Zea mays*) pollen in comparison with live aphids *A. pisum* (Hodek et al. 1978).

A limit to the polyphagy of *Col. maculata* was shown when both adults and larvae refused nymphal leafhoppers of several species, which were, however, eaten by larvae of *C. novemnotata* and adults of *C. trifasciata* (Yadava & Shaw 1968). The nutritional suitability of this prey was not determined.

Harmonia axyridis

Already before the invasion of *Har. axyridis* to the Nearctic region several Japanese authors had reported its polyphagy in relation to aphids (Okamoto 1961, 1966, Hukusima & Sakurai 1964, Takeda et al. 1964, Hukusima & Kamei 1970) and particularly to a substitutive food, represented by a lyophilized powder of drone honey bee larvae (Niijima et al. 1986; 5.2.7). Very high (100%, the same as in *C. septempunctata*) successful completion of larval development was recorded on *Aphis fabae* on sugar beet in California (Ehler et al. 1997). Also after the arrival to Europe, *Har. axyridis* appears to have remained a generalist predator. In Belgium, for example, three aphid species have recently been reported as its prey with *Microlophium carnosum* on *Urtica* preferred before *Acyrthosiphon pisum* on *Pisum* and *Sitobion avenae* on *Triticum* (Alhmedi et al. 2008). *Har. axyridis* was also reported as the most abundant predator of the hemlock wooly adelgid *Adelges tsugae* on *Tsuga* species in southeastern United States (Wallace & Hain 2000). For feeding on psyllids see Hodek and Honěk (2009).

Adalia bipunctata

Wide polyphagy was documented by Mills (1979, 1981, Hodek 1993). Mills reported 28 aphid species as the prey of adults; with seven of them not accepted by the larvae (Table 5.9). An experimental analysis of predation in the field has shown a great variation in the suitability of prey aphids. In terms of the number of coccinellid eggs produced, *Eucallipterus tiliae* on lime appeared the most beneficial, with *Euceraphis punctipennis* on birch as the second best, while *Drepanosiphum platanoidis* on sycamore and *Chaitophorus capreae* on sallow were the least beneficial. Mills concluded that the suitability of an aphid species as prey was most influenced by capture efficiency. This amounted to 58% for the most suitable prey, *E. tiliae*, but only to 21% for the least suitable *D. platanoidis*. The nutritive value of the aphids for *A. bipunctata* was also considered to play a role in prey selection in the field (Mills 1979). Hodek (1996) listed 40 species as essential prey of *A. bipunctata* and this list has not increased much (5.2.11).

Rhopalosiphum padi was less suitable than *Aphis pomi* for egg laying in overwintered females of *A. bipunctata*

Table 5.9 Aphids on which *Adalia bipunctata* was observed to feed by Mills (1979).

Host plant		Aphid
Alnus glutinosa	—	Pterocallis alni
Fagus sylvatica	—	Phyllaphis fagi
Betula alba	—	Euceraphis punctipennis
Prunus spinosa	—	Brachycaudus helichrysi
Sarothamnus scoparius	A	Aphis sp.
Prunus cerasus	—	Myzus cerasi
Rumex obtusifolius	A	Aphis sp.
Sambucus nigra	—	Aphis sambuci
Viburnum opulus	A	Aphis sp.
Crataegus monogyna	A	Dysaphis sp.
Corylus avellana	—	Myzocallis coryli
Larix decidua	—	Adelgid sp.
Tilia europaea	—	Eucallipterus tiliae
Acer platanoides	A	Periphyllus testudinaceus
Urtica dioica	—	Microlophium carnosum
Quercus robur	—	Tuberculatus annulatus
Vinca major	A	Aulacorthum solani
Phragmites communis	A	Hyalopterus pruni
Rosa canina	—	Macrosiphum rosae
Salix caprea	—	Chaitophorus capreae
Pinus sylvestris	—	Adelgid sp.
Castanea sativa	—	Myzocallis castanicola
Acer pseudoplatanus	—	Drepanosiphum platanoidis
Cirsium arvense	—	Uroleucon cirsii
Quercus cerris	—	Myzocallis boerneri
Juglans regia	—	Chromaphis juglandicola
Salix fragilis	—	Cavariella sp.
Salix fragilis	—	Tuberolachnus salignus

A, Adults but not larvae of *A. bipunctata* were observed feeding on these prey species.

(Semyanov 1970). Overwintered females had a much greater fecundity than the first generation on the problematic prey *H. pruni*. The host plants of the prey for the two coccinellid generations were not specified. Also the successful rearing of *A. bipunctata* on an artificial diet (Kariluoto 1980) or on *Ephestia kuehniella* eggs (de Clercq et al. 2005) is an indication of polyphagy. In a more recent experiment, *A. bipunctata* larvae developed with highest mortality on *H. pruni* (50%); *Sitobion avenae* was less detrimental (33%), while *R. padi* (22%) and *Myzus cerasi* (18%) might be considered essential larval prey (Ozder & Saglam 2003).

Hippodamia *spp.*

The relative euryphagy of these species is very favourable for natural biological control on sorghum. On the Texas High Plains, *Rhopalosiphum maidis* provides an early season food source for native coccinellids, mostly *Hip. convergens* (57.0%) and *Hip. sinuata* (36.6%), that lay eggs there. The 'captive' larval population then feeds on the later arriving *Schizaphis graminum*. For the prevention of damaging levels of *S. graminum*, the presence of *R. maidis* is therefore a key factor (Michels & Matis 2008).

Other generalists

Oenopia (= Synharmonia) conglobata (Kanervo 1940, 1946) and **Coccinella hieroglyphica** (Hippa et al. 1984, Sloggett & Majerus 1994) show tendencies towards polyphagy; both aphids and pre-imaginal stages of chrysomelids were found to be essential prey for these species. Kanervo (1940, 1946) claimed that **Calvia quindecimguttata** was a specialized feeder on chrysomelids and that aphids were accepted merely as alternative prey. **Calvia quatuordecimguttata** can develop well both on psyllids and aphids, while psyllids (particularly *Psylla mali*) appear to be their preferred food, enabling faster larval development, greater weight of pupae and higher fecundity (Semyanov 1980). Feeding and reproduction on all stages of psyllids, mainly *Psylla jucunda* on *Acacia*, was recorded in **Harmonia conformis** in Australia over a period of 2 years (Hales 1979). Indications of polyphagy in **Propylea quatuordecimpunctata** are discussed in relation to non-insect food in 5.2.9.

Two species of **Coelophora** both endemic to New Caledonia, exhibit specificity to different prey groups, namely aphids and coccids. Larvae of the aphidophagous *C. mulsanti*, which feed on the more mobile prey, are more active than larvae of the coccidophagous *C. quadrivittata*. Thus even between congenerics the typical differences in behaviour-related prey mobility are maintained. While the coccidophagous species appears to be rather stenophagous (only one prey is known, *Coccus viridis*), the aphidophagous *C. mulsanti* was euryphagous, being reared on several aphids (*Rhopalosiphum padi*, *Hyperomyzus lactucae*, *Aphis gossypii*); it is also reported to feed on cicadellid larvae in the field (Sallée & Chazeau 1985).

5.2.4.2   Specialists

The thermophilous coccinellid **Clitostethus arcuatus** (tribus Scymnini) appears to be a specialized predator of aleyrodids (Hodek & Honěk 2009), ovipositing e.g. on *Aleurodes proletella* (Bathon & Pietrzik 1986). A study of prey specificity of **Rodolia cardinalis** before its release to Galapagos confirmed its stenophagy: it feeds specifically on Margarodidae and did not complete development on scale insects from the families Pseudococcidae, Eriococcidae and Coccidae (Causton et al. 2004).

The stenophagy of the genus **Coccinella** was already indicated by the findings of Putman (1957) and Smith (1965a; 5.2.10). Of three *Coccinella* spp., two (*C. novemnotata* and *C. trifasciata*) could not complete their larval development on dry powdered aphids. *C. transversoguttata* succeeded at least on one of the three aphids tested (on dry *Acyrthosiphon pisum*) (Smith 1961). Whereas Hodek (1996) listed 24 species of essential prey for *C. septempunctata*, i.e. much less than the 40 species for *A. bipunctata*, both species are often considered as generalists (e.g. Sloggett 2008a, b). Reduced values of *C. septempunctata* life-history parameters on low quality foods are discussed in 5.2.6. The strange report that the larvae of **Coccinella undecimpunctata** can complete their development when fed solely on fresh manure (Hawkes & Marriner 1927) has been rightly questioned by Benham & Muggleton (1970) with the plausible explanation that cannibalism was overlooked. The attraction to manure is evidently related to halophily of *C. undecimpunctata* (Hodek et al. 1978 as *Cer. undecimnotata* (sic)). Harpaz (1958) in Israel achieved complete larval development of *C. undecimpunctata* on *Aphis pomi*.

Adelgidae and Lachnidae have been reported as favoured prey of **Aphidecta obliterata** (Majerus & Kearns 1989, Klausnitzer & Klausnitzer 1997).

Non-abundant nutrient concept

A novel view of the relation between prey and a ladybird was suggested by Cohen and Brummett (1997). They criticized the '**rule of sameness**' (House 1974) which maintains that nutritional requirements of insects are in principle (i.e. excluding consideration of allelochemical content) 'similar regardless of systematic position or feeding habit of the species'. Despite caveats expressed by House (1974), his rule has been interpreted broadly to imply that nutritional differences are negligible between various prey species. Cohen and Brummett (1997) point out that this liberal interpretation has resulted in a neglect of important differences in the nutritional composition of various prey. Inspired by Liebig's 'law of the minimum', Cohen and Brummett measured the body content of **methionine** (the least abundant amino acid) in three predators and three prey. Thus *E. kuehniella* eggs contain 11.2 ng/mg, while *Bemisia* larvae only have 0.2 ng/mg of methionine. The adult ladybird *Serangium parcesetosum* contains only 6.1 ng/mg, while *Chrysopa* or *Geocoris* have 10.0 ng/mg of methionine. *S. parcesetosum*, with its lower carcass methionine content but also smaller size, shorter handling time and shorter life span, is more suited to be a successful, long-term predator of *Bemisia* than are either *Chrysopa* or *Geocoris*.

This principle does not only apply to methionine. Other substances, such as other essential amino acids as well as sterols, polyunsaturated fatty acids or vitamins, may be limiting. Cohen and Brummett (1997) stress that 'explanation of fitness of predators to their prey is more likely to be found in considering nutrients that are non-abundant rather than energy'.

5.2.5 Mixed and combined diet

The role of **mixing** of different dietary resources has been studied quite frequently in phytophagous insects and has been found to be positive in most cases (for references see Unsicker et al. 2008). The favourable effect of mixed diets, more often documented in non-coccinellid predators (recently e.g. by Harwood et al. 2009) has been explained by several theories: either by the necessity to get a more complete and balanced range of the nutrients needed (Pulliam 1975, Raubenheimer & Simpson 1999), or to dilute allelochemicals present in individual host plants and therefore to decrease their toxic effect (Freeland & Janzen 1974, Behmer et al. 2002).

In predaceous ladybirds this approach has long been neglected. Feeding on several kinds of food has been interpreted as a rather emergency feature, compelled by shortage of the 'right' food. Frequently this is true and we may then rightly speak about alternative, or substitution food. In particular cases, in species such as *Tytthaspis sedecimpunctata* (Ricci 1982; 1986a; Ricci et al. 1983), *Rhyzobius litura* (Ricci 1986b), *Illeis galbula* (Anderson 1982) and perhaps

also *Propylea* spp. (Turian 1971, Hukusima & Itoh 1976) and other polyphagous species, when even taxonomically distant organisms serve as food (5.2.7, 5.2.9), mixed feeding appears favourable, and is a usual phenomenon.

The coccinellids may select a favourable balance of important nutrients from pollen and conidia of mildew, or pollen and aphids etc., as is assumed by the model of **self-selection** of optimal diets (Waldbauer & Friedman 1991). Some workers have already attempted to check by experiments the suitability of **food combinations** for coccinellids. The simultaneous presence of two prey has been found to increase the predation rate of *Har. axyridis* (Lucas et al. 2004; Fig. 5.5). A positive effect of coccinellid feeding on curculionid larvae in

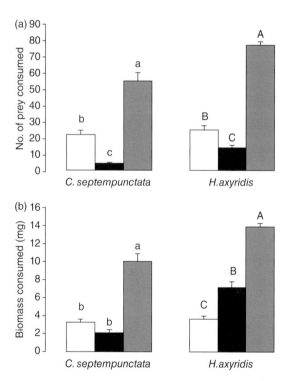

Figure 5.5 Overall predator voracity (mean ± SE) of *Coccinella septempunctata* and *Harmonia axyridis* in 24 hours, when offered 100 *Aphis pomi* (white bars), 30 larvae of *Choristoneura rosaceana* (black bars) or both prey species (grey bars). (a) Total number of prey predated. (b) Total biomass consumed. Different letters indicate a significant ($P < 0.05$) difference among the different treatments for the same coccinellid species (modified from Lucas et al. 2004).

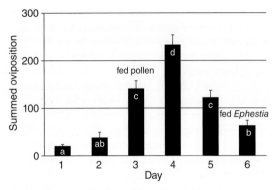

Figure 5.6 Total egg production (mean ± SEM) by *Hippodamia convergens* females (*n* = 113) over 135 days as a function of food provisioning showing the peak in oviposition every fourth day following provision with animal protein. Females paired with males were held on a 6-day feeding cycle (eggs of *Ephestia kuehniella* once every 6 days followed by pollen 3 days later) with continuous access to sunflower petioles). Means bearing the same letter were not significantly different (Tukey test, α > 0.05) (modified from Michaud & Qureshi 2006).

fields of alfalfa, when essential aphid prey is available only at low levels, was studied by Evans and collaborators (5.2.7). The importance of animal protein from *E. kuehniella* for egg laying was documented in *Hip. convergens* (Michaud & Qureshi 2006; Fig. 5.6). Related findings by Snyder et al. (2000) are discussed in 5.2.8.2.

In some experiments, however, no benefit from mixing was observed. Four species of ladybirds, reared on the mixture of *Myzus persicae* and eggs of *Leptinotarsa decemlineata*, had lower survival and development rate than the controls reared on aphids alone (Snyder & Clevenger 2004). They did not avoid eating the chrysomelid eggs, although there was a surplus of aphids. Similarly *C. septempunctata* did not benefit from a mixed diet of three aphid species, *Sitobion avenae*, *Metopolophium dirhodum* and *Rhopalosiphum padi*. The value of this mixture was intermediate between the higher quality single species diets of one of the first two species and the lower quality *R. padi*. The cause of quality differences, either nutrient content (Cohen & Brummett 1997) or presence of allelochemicals (Hauge et al. 1998), remained unknown. It is most interesting that in the studies cited above the authors found that the same ranking of the three aphid prey

for quality also applied for spiders and carabids. Nielsen et al. (2002) used a mix of very good essential prey (the aphid *Metopolophium dirhodum*) and two slightly lower quality essential prey (*Myzus persicae* and *Sitobion avenae*) for *C. septempunctata*. Again no benefit of mixing was recorded although the mixed diet contained entirely essential aphid prey, whether the aphids had been raised on the same or different host plants. Inclusion in the mix of a toxic aphid (*A. sambuci*) considerably decreased the larval fitness of *C. septempunctata*. This result again demonstrates the inability of coccinellids to avoid toxic prey. A similar finding for the same predator and prey (Nedvěd & Salvucci 2008) is discussed in 5.2.6.1.

Males of *Har. axyridis* showed a consistent feeding preference for *Myzus persicae* versus *A. fabae*, while the females did not show any preference. Mixing the aphids in varying ratios of *M. persicae* to *A. fabae* did not affect oviposition rate (Soares et al. 2004). Similarly Phoofolo et al. (2007) did not find any difference between a one-species diet and a mixed diet of *Schizaphis graminum* and *Rhopalosiphum padi* for larvae of *Hip. convergens*.

One-species prey, but a **mixture of different developmental stages**, may appear more suitable for the predator than a homogeneous prey consisting of only one stage. Transfer of females of *Chil. nigritus* after 10 days feeding on a heterogeneous mixture of all stages of the diaspid *Abgrallaspis cyanophyli* to homogeneous prey resulted in an important decrease in oviposition during the next days (Ponsonby & Copland 2007b; Table 5.10). It has also been suggested for other insects that treating a prey population as a homogenous entity can lead to erroneous conclusions (Rudolf 2008).

5.2.5.1   Complementation across stages

By changing the diet between larvae and adults of the highly polyphagous *Col. maculata*, Michaud and Jyoti (2008) revealed what they called 'dietary complementation across life stages'. Coccinellid larvae appear to process some types of food more efficiently than adults. Eggs of the pyralid *Ephestia kuehniella* were the best diet for the larvae and resulted in the largest adults, with highest fecundity (182 eggs) and shortest pre-oviposition period when the adults were then fed on aphids (*Schizaphis graminum*). This combination thus led to higher fecundity than continuous feeding on aphids (139 eggs) or continuous feeding on

Table 5.10 Decrease in total egg laying after a change in prey population structure in *Chilocorus nigritus* fed on *Abgrallaspis cyanophyli* at 26°C ($n = 5$) (modified after Ponsonby & Copland 2007b).

Prey stage	No. of eggs laid (±SD)
Treatment 1 control	
10 days on all stages	165.0 (57.6)
10 days on all stages	174.2 (26.4)
Treatment 2	
10 days on all stages	141.2 (33.4)
10 days on first-instar nymphs	91.0 (37.6)
Treatment 3	
10 days on all stages	138.6 (48.1)
10 days on second-instar nymphs	69.6 (41.3)
Treatment 4	
10 days on all stages	135.8 (37.8)
10 days on adult females	71.4 (57.1)

Table 5.11 Mean values for the reproductive performance of female *Coleomegilla maculata* revealing interactions between larval and adult diets. Larvae were fed one of three diets and the adults obtained in each treatment further subdivided into two groups, each got one of two adult diets (after Michaud & Jyoti 2008).

	Adult diet	
Larval diet	Greenbug*	Flour moth eggs[†]
Pre-reproductive period (days)		
Greenbug	14.9	14.9
Flour moth eggs	10.9	16.4
Pollen	15.8	14.3
Reproductive period (days)		
Greenbug	30.1	20.1
Flour moth eggs	22.2	23.1
Pollen	28.5	23.7
Fecundity (number of eggs/female)		
Greenbug	138.7	161.3
Flour moth eggs	182.4	146.3
Pollen	125.8	106.5
Fertility (percentage of egg hatch)		
Greenbug	42.4	65.8
Flour moth eggs	48.8	68.1
Pollen	49.1	46.7
Fertility (number of larvae/female)		
Greenbug	58.0	108.1
Flour moth eggs	129.4	101.1
Pollen	66.8	50.3

*Schizaphis graminum.
[†]*Ephestia kuehniella.*

lepidopteran eggs (146 eggs) (Table 5.11). The authors conclude that both foods are essential, but nutritional demands for larval growth and development versus adult dispersal and reproduction may differ. *E. kuehniella* eggs were also combined with *Acyrthosiphon pisum* in similar experiments on *P. japonica* (Hamasaki & Matsui 2006; Table 5.12).

5.2.5.2   Prey switching

Can switching of prey have a significant effect on the predator? *Chilocorus nigritus* larvae can switch from feeding on *Coccus hesperidum* (Coccidae) to *Abgrallaspis cyanophyli* (Diaspididae) and vice versa with only minor detrimental effects. Subsequent generations proved that both these prey were essential (Ponsonby & Copland 2007a).

5.2.5.3   Prey specialization through selection

The performance gradually increased over six *A. bipunctata* generations when they were reared and selected on *Aphis fabae*, a rather poor prey (5.2.6.3); e.g. the mortality decreased from 58 to 0%. The achieved specialization to *A. fabae* resulted in poorer performance on the previously (and generally) highly suitable prey, *Acyrthosiphon pisum*. The authors assume that this ability to adapt enables *A. bipunctata* to 'switch'

from the more suitable tree-dwelling aphids, such as e.g. lime aphid *Eucallipterus tiliae*, to field aphids occurring later in the season (Rana et al. 2002). While the results of such apparent selection are very important, the general validity of this interpretation is contentious (Sloggett 2008b). Furthermore, tree aphids are poor prey, while some field aphids (such as *A. pisum*) are excellent prey. Also some aphids return from fields to primary hosts (trees, shrubs) in late season.

5.2.6   Lower quality prey (mostly aphids)

5.2.6.1   Toxic prey

The reasons why particular prey species are harmful to particular coccinellids have not yet been fully explained.

Table 5.12 Reproductive traits of *Propylea japonica* reared on *Acyrthosiphon pisum* or *Ephestia kuehniella* eggs (modified after Hamasaki and Matsui 2006).

Larval diet	A. pisum	E. kuehniella eggs	A. pisum	E. kuehniella eggs	A. pisum	E. kuehniella eggs
Adult diet	Pre-oviposition period (d)		Eggs laid per 20 d*		Egg hatchability (%)[†]	
A. pisum	3.9 ± 1.4 (15)	3.7 ± 1.5 (19)	184.4 ± 106.8 (15)	210.9 ± 95.9 (19)	60.8 ± 9.2 (15)	63.7 ± 20.0 (19)
E. kuehniella eggs	6.8 ± 3.6 (13)	5.3 ± 1.4 (15)	58.0 ± 33.8 (13)	76.5 ± 52.1 (15)	77.5 ± 15.5 (13)	66.5 ± 21.0 (15)

Mean ± SD. Values in parentheses indicate the number of samples.
*Total number of eggs laid per female for 20 d after emergence.
[†]Percentage of hatched larvae.

That death is caused by starvation, resulting from a low feeding rate on unsuitable food, may be discounted because in experiments comparably reduced feeding on other, essential food did not cause a substantial rise in mortality (e.g. Hodek 1957a). There remain two other possibilities. Either the unsuitable aphids contain some special **substances (allelochemicals derived from the plants) poisonous** to coccinellids, or such aphids are deficient in nutritive value. Most studies have focussed on the former possibility.

Aphis sambuci

Jöhnssen (1930) observed a marked increase in feeding by *C. septempunctata*, when he switched it from *Aphis sambuci* to *A. hederae*. In two experiments, Hodek (1956) also found a rather low intake of *A. sambuci* by *C. septempunctata* and showed that this aphid is inadequate food for both larvae and adults of *C. septempunctata*; the larvae could not complete their development and died within 25–26 days. In the first experiment (started 24 June) two-thirds of the larvae died in the fourth instar after a higher feeding rate, while in the second experiment (started 22 July) all but one died in the third instar after a much lower feeding rate (Fig. 5.7). This difference may have been caused by differing chemical compositions in different phenological phases of the host plant and consequently also in the aphids.

All **freshly emerged** *C. septempunctata* adults died on average after 17.5 days when fed only on *A. sambuci*. The control adults, fed on *Aphis fabae* (= *A.cirsiiacanthoidis*) or on *Uroleucon aeneus* had at most a mortality of 16.6%. Compared to the control, the survival of

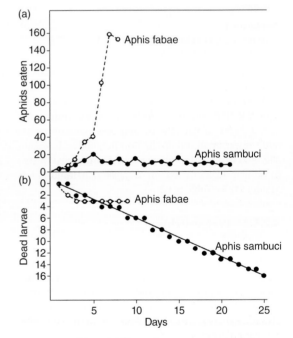

Figure 5.7 Effect of *Aphis sambuci* in comparison with *Aphis fabae* on the larval development of *Coccinella septempunctata*. (a) Daily feeding rate of larvae. (b) Mortality of larvae (from Hodek 1956).

overwintered adults on *A. sambuci* was not decreased, but egg laying was inhibited (Hodek 1957b) and ovaries did not mature. Not until suitable food had been provided for 9 days, did the adults begin to oviposit. As the aphid host plant *Sambucus nigra* often

Table 5.13 Larvae of *Coccinella septempunctata* fed on reduced daily rates of *Aphis fabae* (Hodek 1957a).

	Duration of development (days)						
	Larval instars						
Daily feeding rate in 4th instar (number of aphids)	1 + 2 + 3	4	Prepupa	Pupa	Total	Mortality (%)	n
surplus (aver. consumed 70)	5.2	2.6	1.0	4.0	12.8	8	13
30	4.8	3.7	0.8	4.2	13.5	0	18
10	4.9	5.9	0.9	4.2	15.9	10	20

Table 5.14 Survival to adulthood, larval times, total developmental times and adult weights of *Coccinella septempunctata* larvae on different aphid diets (modified after Nielsen et al. 2002).

Aphid diets	n	No. survived*	Larval time (days)[†]	Total developmental time (days)[†]	Adult weight (mg)[‡]
Experiment 1					
Metopolophium dirhodum	30	26 a	11.27 ± 0.20 a	15.27 ± 0.22 a	25.54 ± 0.84 a
Myzus persicae	30	20 ab	12.95 ± 0.17 b	16.70 ± 0.15 b	18.88 ± 0.67 b
Aphis sambuci	30	0 c	—	—	—
M. dirhodum and *M. persicae*	30	18 b	11.22 ± 0.21 a	15.06 ± 0.12 a	25.44 ± 0.81 a
M. dirhodum and *A. sambuci*	30	2 c	13.50 ± 0.50 ab	17.00 ± 0.00 b	14.58 ± 1.78 b
M. persicae and *A. sambuci*	30	0 c	—	—	—
M. dirhodum, *M. persicae* and *A. sambuci*	30	5 c	17.80 ± 0.38 b	21.60 ± 0.51 c	16.14 ± 1.21 b
Experiment 2					
M. dirhodum	50	27 a	11.26 ± 0.17 a	15.33 ± 0.19 a	27.57 ± 0.95 a
Sitobion avenae	50	35 a	12.09 ± 0.26 b	16.11 ± 0.27 b	28.09 ± 0.73 a
M. dirhodum and *S. avenae*	50	34 a	11.41 ± 0.20 c	15.18 ± 0.17 a	28.63 ± 0.90 a

Mean ± SE; within columns and experiments the same letters indicate no significant difference between treatments.
*χ^2 test; numbers in this column are sample sizes for next three columns.
[†]Multiple comparisons after Kruskal–Wallis tests.
[‡]Fisher LSD test.

grows at or near *C. septempunctata* hibernation sites, the adults may be found in colonies of *A. sambuci* in the spring, but they do not lay eggs there.

The possibility that the insufficient feeding rate on *A. sambuci* (Fig. 5.7) might cause the mortality of *C. septempunctata* larvae by starvation was excluded by rearing the most voracious fourth instar (Hodek 1957a) on substantially reduced food rations of the suitable food – *A. fabae* (Table 5.13). The reduction of feeding rate to one-seventh prolonged the fourth instar, but there was no increase in mortality. Development was completed with only about one-third of the total larval food intake, with a surplus of aphids. If only some larval instars were fed with *A. sambuci* (Hodek

1957a), most (68–76%) larvae pupated and larval development was prolonged. In a later study in England *A. sambuci*, although a very poor food, enabled half the specimens of *C. septempunctata* to develop to extremely small adults (Blackman 1965, 1967b). A negative effect of *A. sambuci* on *C. septempunctata* was also documented by Nielsen et al. (2002; Table 5.14), and on *C. magnifica* by Sloggett et al. (2002).

The larvae of *A. bipunctata* completed development on *A. sambuci*, although at a slower rate than on *A. fabae* (Hodek 1957a). This was confirmed by Blackman (1965, 1967b). In central Europe this species breeds mostly on shrubs and trees, i.e. in the same habitat as *A. sambuci*. The larvae of *Scymnus subvillosus* seem

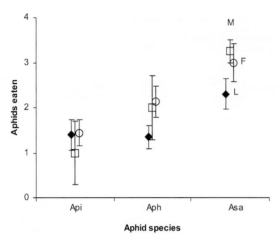

Figure 5.8 Consumption rates (mean ± SE) per 4 hours by *Coccinella septempunctata* larvae (L, solid diamonds), males (M, open squares), and females (F, open circles) on three aphid species. Api, *Acyrthosiphon pisum*; Aph, *Aphis cirsiiacanthoidis*; Asa, *Aphis sambuci* (from Nedvěd & Salvucci 2008).

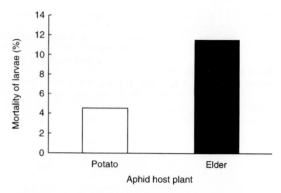

Figure 5.9 Mortality of *Harmonia axyridis* larvae 7 days after rearing began, when they were fed with *Aulacorthum magnoliae* reared on potato (*n* = 243) or elder (*n* = 243) (from Fukunaga & Akimoto 2007).

to be adapted to feeding on *A. sambuci*; Klausnitzer (1992) was able to rear late instar larvae to adults on this diet.

For *Aphis sambuci* we may assume passage of the **glycoside sambunigrin**, contained in the host plant *Sambucus nigra*, from the plant into the aphid. This glycoside may be split enzymatically into hydrocyanic acid in the body of the coccinellid, for enzymes which split glycosides have been identified in the body of *Coccinella* spp. (Kuznetsov 1948).

Contrary to expectation, *Aphis sambuci* was preferred in a preliminary, 4-hour choice experiment to *Aphis cirsiiacanthoidis* and *Acyrthosiphon pisum* (Nedvěd & Salvucci 2008; Fig. 5.8). Starved fourth instar or adult *C. septempunctata* consumed twice as many *A. sambuci* as of the two other aphid species. These preference tests again show how misleading observations of prey acceptance can be for evaluating prey suitability. Nedvěd & Salvucci (2008) recorded *C. septempucntata* fourth instars dying when fed on *A. sambuci*, thus confirming the earlier findings of Hodek (1956), Blackman (1965) and Nielsen et al. (2002).

Aulacorthum magnoliae

Similar or identical allelochemical substances are also the likely lethal agent in *A. magnoliae*. This aphid, also

from an elder (*Sambucus racemosa* ssp. *sieboldiana*), was found to be toxic to *C. septempunctata brucki* and *Har. axyridis* (Okamoto 1966). This was confirmed by Fukunaga and Akimoto (2007; Fig. 5.9) who found high larval mortality of *Har. axyridis* that preyed on *A. magnoliae* fed on *S. r. sieboldiana*. The authors consider that the aphid absorbs toxic substances or their precursors from *Sambucus*.

Aphis craccivora

Aphis craccivora from a number of different host plants was recorded as unsuitable food for *Har. axyridis* already in early studies (Okamoto 1966, 1978, Hukusima & Kamei 1970 in Hodek 1996) and again more recently by Ueno (2003; Table 5.15), who however found lower toxicity than were reported in the earlier studies. Hukusima & Kamei (1970) found that *Har. axyridis* larvae did not develop on this prey, and adults also died within 4–6 days if the aphid was collected from *Robinia pseudoacacia* or six other host plants. However, *A. craccivora* was a favourable food when fed on *Vicia sativa* or *Vigna unguiculata* subsp. *cylindrica*. In contrast to *Har. axyridis*, both larvae and adults of *P. japonica* were resistant to the detrimental effect of *A. craccivora* from *R. pseudoacacia*. In Okamoto's (1966) experiments, *Har. axyridis* larvae also died when *A. craccivora* was reared on *Vicia faba* instead of *R. pseudoacacia*.

Hukusima & Kamei (1970) suggested **allelochemicals** as the explanation for the probably toxic effect of

Table 5.15 Larval period and pupal mass of *Harmonia axyridis* reared on *Acyrthosiphon pisum*, *Aphis craccivora*, or artificial diet (after Ueno 2003).

	Food		
	A. pisum	**A. craccivora**	**Artificial diet**
	Larval period (days)		
Male	13.8 ± 0.5 (44)	18.8 ± 0.7 (32)	19.7 ± 0.6 (32)
Female	13.1 ± 0.3 (45)	20.6 ± 0.8 (22)	21.1 ± 0.6 (41)
	Pupal mass (mg)		
Male	27.2 ± 0.5 (44)	19.6 ± 0.6 (32)	20.9 ± 0.9 (32)
Female	28.5 ± 0.7 (45)	22.0 ± 0.9 (22)	23.1 ± 0.7 (41)

Mean ± SE. Sample sizes are given in parentheses.
Both aphids were reared on bean plant seedlings in glass tubes.

A. craccivora on *Har. axyridis*. The authors reported that aphids from *R. pseudoacacia* contained two extra amino acids (glycine and arginine) in contrast to aphids from other host plants, though these are not normally considered to be toxic substances. In later experiments, also in Japan, the lethal effect of *A. craccivora* fed on *R. pseudoacacia* was ascribed to the **amines canavanine** and **ethanolamine** isolated both from the host plant and the aphid by Obatake and Suzuki (1985).

Cer. undecimnotata can neither complete larval development nor oviposit when fed on *Aphis craccivora* (from *Vicia faba*), though on the same host plant *A. craccivora* is an essential food for *C. septempunctata* (Hodek 1960). Azam and Ali (1970) reported *A. craccivora* as lethal for *C. septempunctata* when the aphid was fed on *Glycinidia*. The effect of host plant chemistry on the unsuitability of prey might explain an earlier observation that the larvae of *C. septempunctata brucki* and *P. japonica* die when fed on *A. craccivora* in spring, but not in summer (Takeda et al. 1964). Different host plants were probably involved. **Different concentrations of allelochemical substances** in the same plant but in different seasons of the year might be another plausible explanation, indicated by findings from another aphidophagous insect, *Chrysopa perla*. When the lacewings were fed on *Aphis nerii* from *Nerium oleander* in June, none survived, but the survival was much higher (78%) in winter (Canard 1977).

Megoura viciae

While **gradual poisoning** is one of the possible explanations of the detrimental effect of some unsuitable aphids (as *A. sambuci* or *A. craccivora*), in some instances

an **acute toxic effect** has been proven or at least strongly indicated.

Larvae of *A. decempunctata* attacked and ate *Aphis fabae* and *M. viciae*, but after about 2 minutes the larvae released the prey and vomited and 50% of larvae provided with *M. viciae* died (Dixon 1958). The toxicity of *M. viciae* to *A. bipunctata* was studied in detail by Blackman (1965, 1967b) who found this aphid to be lethal to all larval stages and to adults. Neither larvae nor adults avoid attacking this aphid when mixed with suitable prey. Even when *M. viciae* and the non-toxic *Acyrthosiphon pisum* were provided in the ratio of 1:9, all the larvae failed to reach the pupal stage. The first instar larvae, when fed on this prey, perished more rapidly than those starved. The fourth instars accepted *M. viciae* readily, but after about 4 min. of feeding they suddenly rejected their prey and vomited. When fed on this prey, all adults died within a week of emergence. Frechette et al. (2006; Table 5.16) have more recently confirmed the toxicity of *M. viciae* for *A. bipunctata*. Even the larvae of the polyphagous *Har. axyridis* cannot develop when fed exclusively on *M. viciae*. They succeeded only when at least the first instar larvae were fed on *Aphis gossypii* that evidently is essential prey for this coccinellid (Tsaganou et al. 2004). *M. viciae* is toxic to *Exochomus quadripustulatus*: the larvae died within 2 days (Radwan & Lövei 1983), and the aphid is very unsuitable for *Cycloneda sanguinea* (Isikber & Copland 2002; Table 5.17).

The rapid response of *A. bipunctata* and *A. decempunctata* to *M. viciae* indicates a toxic effect. A toxic substance suspected (Dixon 1958) was, however, not found in chemical analysis of the aphid (Dixon et al. 1965). In contrast to the results from the

Table 5.16 Number of eggs laid by *Adalia bipunctata* females that oviposited after 8 and 24 hours, and the percentage of mortality after 24 hours in the presence of three aphid species, *Acyrthosiphon pisum*, *Aphis fabae* and *Megoura viciae* (modified after Frechette et al. 2006).

| | Previously fed on | Treatment | *n* | No. of eggs | | Mortality (%) |
				8 h	24 h	
Experiment 1	*A. pisum*	*A. pisum*	17	13.00 ± 1.94 (*n* = 9)	14.40 ± 1.33 (*n* = 15)	0
	A. pisum	*A. fabae*	17	11.40 ± 1.69 (*n* = 10)	13.92 ± 1.76 (*n* = 13)	0
	A. pisum	*M. viciae*	20	7.63 ± 1.80 (*n* = 8)	9.63 ± 1.38 (*n* = 16)	20
Experiment 2	*A. fabae*	*A. pisum*	18	13.10 ± 1.80 (*n* = 10)	15.79 ± 2.17 (*n* = 14)	0
	A. fabae	*A. fabae*	18	12.88 ± 1.76 (*n* = 8)	16.58 ± 2.56 (*n* = 12)	0

Mean ± SE; negative effect of *M. viciae* was shown by both parameters, egg laying and mortality.

Table 5.17 Developmental time and survival rate of *Cycloneda sanguinea* on four aphid species (modified after Isikber & Copland 2002).

	A. gossypii	*A. fabae*	*M. persicae*	*M. viciae*	P value
Overall development*	14.6 ± 0.2 b (12)	15.2 ± 0.2 b (13)	14.5 ± 0.3 b (10)	16.7 ± 0.8 a (7)	P < 0.01
Overall survival[†]	0.86 a (14/12)	0.81 a (16/13)	0.77 a (13/10)	0.39 b (18/7)	P = 0.012

Aphis gossypii, *Aphis fabae*, *Myzus persicae*, *Megoura viciae*
Values within rows with the same letter are not significantly different (LSD at 1% level). One way ANOVA was applied for data analysis.
*Overall development (larva + prepupa + pupa). Figures in brackets show the number of individuals as replicates.
[†]Overall survival (larva + prepupa + pupa). Figures in brackets show the numbers of individuals at the start and at the end of development, respectively.

aforementioned species, more than 60% of *C. septempunctata* larvae successfully completed development on *M. viciae*, although a slightly negative effect was shown on the length of development and weight of emerged adults (Blackman 1965, 1967b; Table 5.18).

Aphis nerii

Aphis nerii infests host plants in the families Asclepiadaceae (milkweeds) and Apocyanaceae (oleander). Oleanders are toxic due to high levels of the **cardiac glycoside cardenolides**, particularly **oleandrin** and **neriin**. The cardenolides are ingested by the aphid, sequestered and excreted in the honeydew (Rothschild et al. 1970, Malcolm 1990). Malcolm identified 25

cardenolides in *Nerium oleander* and 17 of them in the aphids; 20 were detected in the honeydew. *A. nerii* on *N. oleander* is poisonous to most coccinellids that have been tested, including *C. septempunctata*, *Cer. undecimnotata*, *P. quatuordecimpunctata* and *A. bipunctata* (Iperti 1966), and *Harmonia dimidiata*, *C. septempunctata brucki* and *C. leonina* (Tao & Chiu 1971). An exception is *Hip. variegata*, which developed normally on this prey (Iperti 1966). Bristow (1991) observed that honeydew produced by leaf-feeding *A. nerii* was less palatable to ants (*Linepithema humile*) than honeydew from floral colonies. She also offered aphids from these two plant parts to *Hip. convergens* adults, and these consumed significantly more leaf than floral aphids. The ladybirds often failed to consume the aphid entirely and

Table 5.18 Development of *Coccinella septempunctata* larvae on different aphids (Blackman 1965, 1967b).

Aphid species	Development (days)	Mortality (%)	Weight of adult at emergence (mg)
Myzus persicae	13.0	12.5	36.4
Aphis fabae	13.6	9.1	36.3
Acyrthosiphon pisum	13.3	18.6	37.2
Megoura viciae	14.8	13.4	33.5
Brevicoryne brassicae	16.1	26.1	30.9
*Aphis sambuci**	19.5	50.0	18.4

*Only 6 out of 12 larvae completed development.

spent 1–10 minutes cleaning their mouthparts after each attack. The parallel response of the ants and coccinellids may indicate a lower content of cardenolides in the floral tips.

Rather surprising, therefore, are the reports that larvae of three coccinellid species completed their development with quite a high survival on *A. nerii*, reared in this experiment, however, on *Calotropis procera*. Survival was as follows: *C. septempunctata*, 43.9% (Omkar & Srivastava 2003); *C. leonina transversalis*, 37.8% (Omkar & James 2004); *Menochilus sexmaculatus*, 32.3% (Omkar & Bind 2004).

Whereas the aphids from oleander are poisonous for coccinellids, the scale *Aspidiotus nerii* from oleander is a more suitable prey for the coccinellid *Chilocorus infernalis* than *Asterolecanium* sp. on giant bamboo *Dendrocalamus giganteus* (Hattingh & Samways 1991). It is not clear whether the scales avoid taking in glycosides, or whether they are or are not sequestered by the coccinellid.

Macrosiphum albifrons

Macrosiphum albifrons, the lupin aphid, feeding on *Lupinus mutabilis* plants with a high content of **quinolizidine alkaloids** (>0.1% alkaloids in fresh matter), was toxic to *C. septempunctata* larvae and adults (Gruppe & Roemer 1988). The aphids themselves preferred plants with high alkaloid content. Adults of *A. bipunctata* survived longer (max. 15 days) than *C. septempunctata* (10 days) when fed on aphids from high alkaloid plants. Other *Lupinus* spp. cultivars with a high content of the alkaloid lupanin were 100% lethal to the larvae of *C. septempunctata*. The only exception was *L. luteus*, although its total alkaloid content is similar. This host plant contains 70% **spartein** and only 30% **lupanin** (Emrich 1991).

Toxoptera citricidus

Toxoptera citricidus caused the death of larvae in all coccinellid species studied in Taiwan: *Har. dimidiata*, *Har. axyridis*, *C. septempunctata brucki*, *C. leonina*, *Menochilus sexmaculatus* and *Synonycha grandis* (Tao & Chiu 1971). The lethal effect was also recorded in Venezuela on the larvae of *Cycloneda sanguinea*, while *Uroleucon ambrosiae* served as an essential prey (Morales & Burandt 1985). In contrast, Michaud (2000) found that *C. sanguinea* from Florida citrus groves completed larval development on *T. citricidus* as did *Har. axyridis*, with respectively 100 and 95% survival. The adults of *C. sanguinea* were also found to oviposit when they were fed on *T. citricidus* and their voracity was so high that they had the potential to control the aphid. Larvae of four other coccinellid species, *C. septempunctata*, *Coelophora inaequalis*, *Olla v-nigrum* and *Hip. convergens*, did not complete development on *T. citricidus*; however, the females of these four species produced viable eggs on the same diet (Michaud 2000). This is one of the observations that call for more precision in the definition of essential food, along with other cases in which prey suitability for larvae and adults differs (5.2.11). Similar complex interrelations were recorded by Michaud (2000) for *Aphis spiraecola* on citrus.

Aphis jacobaeae

Feeding of *C. septempunctata* on *A. jacobaeae* containing **pyrolizidine alkaloids** apparently differs from the intoxication of coccinellids by other toxic aphids mentioned above. *A. jacobaeae* feeds on several species of *Senecio* where a large range of pyrolizidine alkaloids are present in the form of N-oxides. A high content of **senecionine, seneciphylline, jacobine** and erucifoline was found in *S. jacobaea*, and similarly in *A. jacobaeae* and its predator *C. septempunctata* (Witte et al. 1990). The aphid contained 0.8–3.5 mg of pyrolizidine alkaloids per 1 g fresh weight and the adult coccinellid 0.3–4.9 mg. The **precoccinellines** (alkaloids which the coccinellids produce) amounted to a mean of 10.5 mg/g fresh weight (Chapter 9).

Both aphids and coccinellids store their pyrrolizidine alkaloids as tertiary alkaloids, while other insects (*Tyria*, *Arctia*) store them as N-oxides (Witte et al. 1990). These authors proposed that the high pyrrolizidine alkaloid content may protect ladybirds against birds. This probably was the first reported case of coccinellids obtaining defence chemicals from their prey, but it was not measured to what extent the alkaloid content affected development and oviposition of the coccinellids.

The **hydroxamic acid DIMBOA** (2,4-dihydroxy-7-methoxy-1,4-benzoxazin-3-one) present in wheat extracts is associated with resistance of wheat to aphids. The effect of this secondary metabolite on larvae of the coccinellid *Eriopis connexa* was examined by feeding them on aphids (*Rhopalosiphum padi*) that had fed for 24 hours on wheat cultivars with differing DIMBOA content (Martos et al. 1992). The allelochemical was less deleterious than the toxic compounds discussed above. The greatest negative effect was paradoxically achieved by intermediate levels of DIMBOA (Table 5.19). Aphids feeding on wheat seedlings with an intermediate DIMBOA level accumulated most DIMBOA in their bodies because aphids on the high DIMBOA varieties could detect the compound and this deterred their feeding. Thus high DIMBOA concentrations, which would increase the resistance to

aphids, would actually reduce the effect of the secondary metabolite on the predator (Martos et al. 1992).

Some authors have attempted to explain the unsuitability of certain foods in terms of **nutrient deficiency**. Atwal & Sethi (1963) suggested that the higher protein content of *Lipaphis pseudobrassicae* makes this aphid more favourable than *Aphis gossypii* for *C. septempunctata*. Hariri (1966b) similarly supposed that *Acyrthosiphon pisum* is more nutritious for *A. bipunctata* than is *Aphis fabae*. The low suitability of *A. fabae* for larvae of *A. bipunctata* is explained by Blackman (1967b) as due to two factors: difficulty in ingesting the food and low nutritive value. The two may be related; nutritive value may be low because some essential nutrients are left behind in the non-ingested carcass.

General remarks on toxic prey

As described above, there are several examples of aphids which, like many other herbivores, protect themselves by toxins sequestered from their host plants. Pasteels (2007) calls this method 'chemical piracy' to stress its difference from the synthesis of toxins *de novo* in animal bodies (for specific coccinellid alkaloids see Chapter 9).

The toxicity of certain prey operates selectively. For example, *Megoura viciae* is toxic to *Adalia* spp. and *Exochomus quadripustulatus*, but most *C. septempunctata* larvae develop on this prey. *Aphis sambuci* is toxic for *C. septempunctata*, but much less so to *A. bipunctata*. The differences are often related to habitat specificity: *Coccinella septempunctata* prefer habitats with low plants, which is where *M. viciae* occurs, while *Adalia* spp. and

Table 5.19 Effect of DIMBOA levels in cultivars of wheat on development of *Eriopis connexa*. Aphids (*Rhopalosiphum padi*) were allowed to feed on the plants for 24 hours before they were given as prey to coccinellid larvae (Martos et al. 1992).

Content of DIMBOA* in wheat plants (µg/g fresh weight)	n	Development duration (days)		Survival (%)	
		larvae[†]	pupae[†]	larvae	pupae
140	10	10.4 a	4.0 a	98	100
270	10	10.7 b	4.0 a	93	100
440	10	9.1 c	3.8 a	100	100

*DIMBOA = 2,4-dihydroxy-7-methoxy-1,4-benzoxazin-3-one is a secondary metabolite found in cereal extracts; its level indicates resistance to aphid.
[†]Means within a column followed by the same letter are not significantly different (P = 0.05).

E. quadripustulatus, which prefer shrubs and trees, come into contact with *A. sambuci* more frequently, but not with *M. viciae*.

In the relationship between aphids (and prey in general) and coccinellids we are witnessing a similar process of evolution of specificity that has been intensively studied in phytophages and plants. In fact, we may witness here a further level of the evolutionary process within tritrophic relations. Moreover, in the case of the tritrophic chain *Senecio jacobaea–Aphis jacobaeae–C. septempunctata*, we may see additional evolutionary elements: *C. septempunctata* not only seems unaffected by pyrrolizidine alkaloids in aphids, but keeps the chemicals in their body and probably uses them as a defensive chemical, together with their own alkaloids. It is a pity that this interesting model has not been studied in more detail since 1990.

In general, much research is needed on the adaptive mechanisms in individual species of coccinellids to toxic prey. Are the toxic compounds detoxified or excreted in a non-sequestered state or are they stored in a transformed inert form? *Har. axyridis* appears to detoxify the alkaloids of intraguild coccinellid prey (Sloggett & Davis 2010). An important message emerges for further studies: because the prey insect mostly gets toxins from the host plant, not only the plant species should be recorded, but also other aspects, the season, plant variety or cultivar etc., because the concentration of allelochemicals will vary.

5.2.6.2   Rejected prey

While certain accepted prey do not facilitate development (5.2.6.3) or can even be toxic (5.2.6.1), other prey are rejected and not eaten. The placement of unsuitable prey into these three groups remains rather arbitrary. However, in some prey species considered to be rejected prey due to unknown factors (e.g., *Icerya purchasi* or *Brevicoryne brassicae*), toxic substances have been found in later analyses. These relations have most often been described for aphids and aphidophagous coccinellids, but also occur between other prey and predators (Fig. 5.1).

Delphiniobium junackianum feeds on *Aconitum* which contains the poisonous compound **aconitin**. This allelochemical may be the reason why coccinellids reject this prey, although Hawkes (1920) suspected that the intense colour of the aphid was probably a warning. In other examples, unpalatability is connected with a **waxy surface**, as in *Brevicoryne*

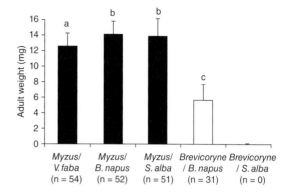

Figure 5.10 Effect of the aphid host plant (*Sinapis alba*, *Brassica napus* and *Vicia faba*) on adult weight of *Adalia bipunctata* fed with *Myzus persicae* or *Brevicoryne brassicae*. n, number of individuals reaching the adult stage. Different letters indicate significant differences at P = 0.001 (from Francis et al. 2000).

brassicae or, to a lesser extent, in *Hyalopterus pruni*. Telenga and Bogunova (1936) observed that the coccinellid *Har. axyridis* refused *B. brassicae* in the field, and in the laboratory the females ceased oviposition when transferred from *H. pruni* to *B. brassicae*. George (1957) also noticed that *B. brassicae* was avoided by Coccinellidae.

The effect of **glucosinolates**, the main **allelochemicals of Brassicaceae**, on *A. bipunctata* was studied in a tritrophic context (other tritrophic studies in 5.2.12.). Two aphid species, the generalist *Myzus persicae* and the brassica specialist *B. brassicae* were fed on two brassicas: white mustard, *Sinapis alba*, with a high glucosinolate content and oil seed rape, *Brassica napus*, where the level of glucosinolates is six times lower. Four types of prey were fed to *A. bipunctata* and survival, adult weight, developmental rate and reproduction were recorded (Francis et al. 2000; Figs. 5.10, 5.11). *Myzus persicae* from *S. alba* gave low larval mortality of *A. bipunctata* but negatively affected oviposition. *Brevicoryne brassicae* from *S. alba* was clearly a toxic food, causing prolonged development of early instars; and no individual completed development.

This difference was explained by gas chromatographic analysis: significant amounts of glucosinolates were detected in *M. persicae*, but no degradation products. However, these highly toxic degradation products, **isothiocyanates**, were identified in *B. brassicae*. Moreover, myrosinase, which catalyses

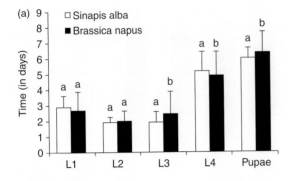

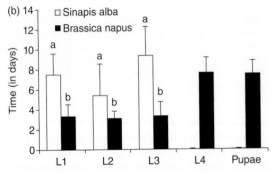

Figure 5.11 Effect of the aphid host plant (*Sinapis alba*, *Brassica napus*) on larval and pupal development of *Adalia bipunctata* fed with *Myzus persicae* (a) and *Brevicoryne brassicae* (b). Different letters indicate significant differences at P = 0.001 for pairwise comparisons (from Francis et al. 2000).

glucosinolate degradation, was detected in *B. brassicae* but not in *M. persicae* (Francis et al. 2000). When fed exclusively on *B. brassicae*, *Har. axyridis* larvae also did not complete development (Tsaganou et al. 2004). Aphid diet containing glucosinolates affects first instar *A. bipunctata* more seriously than *C. septempunctata*. As little as 0.2% of sinigrin in the diet leads to 100% mortality in the former species; by contrast, first instars of *C. septempunctata* survived when fed with *B. brassicae* reared on *Brassica nigra* or artificial diet containing up to 1% sinigrin. However, development rate was decreased (Pratt et al. 2008; Table 5.20).

We see here a similar relation between allelochemical tolerance and habitat as with *Aphis sambuci*. Again the common occurrence of the coccinellid species in the preferred habitat with the toxic aphid appears to lead to **evolution of adaptive resistance** to the alellochemical. *Coccinella septempunctata* comes in contact with brassicas and therefore their glucosinolates more often than *A. bipunctata*, which prefers tree and shrub habitats.

Hyalopterus pruni was described as unsuitable for *A. bipunctata* by Hawkes (1920). It was rejected by larvae of *A. decempunctata* immediately on piercing the body wall. In subsequent attacks this aphid was rejected as soon as the larva touched it with its palps (Dixon 1958). By contrast, Hodek (1959) found this aphid to be essential larval prey for *C. septempunctata* in spite of the waxy covering. Ozder & Saglam (2003) recorded 63% larval mortality in *C. septempunctata* and

Table 5.20 Survival of *Adalia bipunctata* or *Coccinella septempunctata* first instar larvae to second instar (a) fed *Brevicoryne brassicae* or *Myzus persicae* reared on *Brassica nigra*, (b) fed *Brevicoryne brassicae* reared on artificial diets containing a range of concentrations of sinigrin (modified after Pratt et al. 2008).

(a) Survival to second instar (%)	Aphid species	
	Myzus persicae	*Brevicoryne brassicae*
Adalia bipunctata	90	0
Coccinella septempunctata	90	90

(b) Survival to second instar (%)	Sinigrin content in aphid artificial diet					
	0%	0.2%	0.4%	0.6%	0.8%	1.0%
Adalia bipunctata	90	0	0	0	0	0
Coccinella septempunctata	100	90	90	80	70	80

n = 9–10.

50% in *A. bipunctata* when *H. pruni* was the prey, while in *Parexochomus nigromaculatus* the larval mortality was only 26% (Atlıhan & Kaydan 2002, Atlıhan & Ozgokce 2002). A marshland species *Anisosticta bitriangularis* preferred starvation to feeding on an unusual prey **Schizolachnus piniradiatae** (Gagné & Martin 1968).

An interesting study was made about 80–90 years ago on the famous **Rodolia cardinalis**. This coccidophagous ladybird refused to feed on its normal host **Icerya purchasi** when the scale fed on *Spartium* or *Genista*. It was thought that such plants with sparse leaves failed to provide shade (Savastano 1918) or that the smell of the plants repelled the ladybirds (Balachowsky 1930). However, even if the coccids which had fed on *Spartium* or *Genista* were offered to the ladybirds without host plants they were still rejected (Poutiers 1930). These unsuitable host plants contain respectively the yellow pigment **genistein** and the alkaloid **spartein**. It has therefore been hypothesized (Hodek 1956) that substances imbibed by *I. purchasi* from the plants make them unpalatable for *Rodolia*, similar to the toxic effects of alellochemicals to other coccinellids. Thus the unsuitability of *I. purchasi* from *Spartium* or *Genista* was considered another case of acquired toxicity (Hodek 1973; 5.2.6.1).

In a later study, the survival and development time of *R. cardinalis* and *Chilocorus bipustulatus* was measured when they were reared on four prey. These were three scale insects (a margarodid *I. purchasi*, a diaspidid *Lepidosaphes ulmi* and a pseudococcid *Planococcus citri*) and *Aphis craccivora*. The prey were reared on the alkaloid-bearing legumes *Erythrina corallodendron* and *Spartium junceum* and on non-toxic plants as a control. Survival of both ladybirds was significantly reduced and the development time of *R. cardinalis* increased by preying on insects from the toxic plants (Mendel et al. 1992). However, this was not the case with *Cryptolaemus montrouzieri*. This negative effect of *E. corallodendron* and *S. junceum* rendering *I. purchasi* toxic, prevented colonization of these plants by this ladybird (Mendel & Blumberg 1991).

Aiolocaria hexaspilota, a specialized **predator of chrysomelids**, has been reported as feeding both as larvae and adults on the pre-imaginal stages of several chrysomelids (Iwata 1932, 1965, Savoiskaya 1970a, 1983, Kuznetsov 1975), but rejected another chrys-

omelid, *Agelastica coerulea* from *Alnus japonica* (Iwata 1965) and two central-Asiatic chrysomelids, (vernacular names in Savoiskaya 1970a). In Kazakhstan, its essential prey is the chrysomelid *Melasoma populi*. In later experiments *A. hexaspilota* accepted larvae and pupae of *Galeruca interrupta armeniaca* rather reluctantly (Savoiskaya 1983).

5.2.6.3 'Problematic' prey

Coccinellids accept some food which is not adequate and worsens life-history parameters, although it is not toxic.

This can occur when coccinellids, specialized to certain taxonomic groups of prey, are fed **prey from other groups**. Although adults and larvae of *A. bipunctata* were occasionally found feeding on tetranychid mites (Robinson 1951) and the gut of *A. bipunctata* and of three *Coccinella* spp. contained remains of the mite *Panonychus ulmi* (Putman 1964), the coccinellids could not develop on this prey (Robinson 1951, Putman 1957). Conversely, the adults of the acarophagous *Stethorus pusillus* did not oviposit when fed on aphids, and the larvae could not complete their development (Putman 1955). Also **other prey of the generally appropriate group** may be inadequate. This acarophagous coccinellid refuses the mite *Bryobia praetiosa*. If *Stethorus gilvifrons* preys on *Bryobia rubrioculus*, the adults do not oviposit and the larvae die in the second or third instar (Dosse 1967).

Aphis fabae poses quite a problem because of its taxonomy. Its host-plant-adapted biotypes have been redefined as different species, and thus the authors quoted below probably worked with a variety of species. When the host plant is not reported, one can only guess at the exact aphid used. This may explain why the suitability of *A. fabae* differs between the studies. An aphid used by Hodek (1956, 1957a, b) is what is now probably *A. cirsiiacanthoidis*, while Ehler et al. (1997) defined their prey as the '*A. fabae* complex on sugar beet'.

As early as 1954, Dyadechko found *A. fabae* somewhat less suitable than *Toxoptera graminum* for *Coccinula quatuordecimpustulata*. Nine out of ten species of coccinellids including the polyphagous *Col. maculata* tested by Smith (1965b) failed to complete their larval development when fed on *Aphis fabae*. The 10th species, *Hip. tredecimpunctata*, reached the adult stage, but

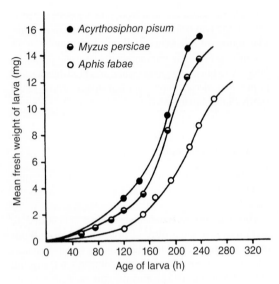

Figure 5.12 Development of *Adalia bipunctata* larvae on different aphid foods (from Blackman 1967a).

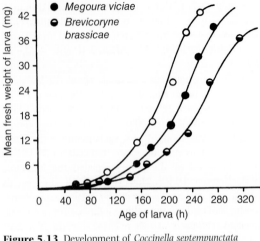

Figure 5.13 Development of *Coccinella septempunctata* larvae on different aphid foods (from Blackman 1967b).

the beetles were smaller and their development was significantly slower than when fed on *Acyrthosiphon pisum* or *Rhopalosiphum maidis*.

In a thorough study, Blackman (1965, 1967b) tested the suitability of several aphid foods for *A. bipunctata* and *C. septempunctata* (Tables 5.3 and 5.18, Figs. 5.12 and 5.13). Although *A. fabae* and *A. sambuci* are natural prey for *A. bipunctata* in the field, they were relatively unsuitable food for this coccinellid, as was shown by the longer larval development and the greater larval mortality. *A. bipunctata* also had a lower fecundity and fertility when fed on *A. fabae*. Hariri (1966a, b) reported similar results. When the larvae were fed on this aphid the resulting adults weighed less, they contained less fat and glycogen and their fecundity was halved. Iperti (1966) ascertained an adverse effect of *A. fabae* on vitellogenesis in *A. bipunctata*.

In contrast to *A. bipunctata*, larvae of *C. septempunctata* are able to develop on *A. fabae* successfully. Hodek, Blackman and Ehler recorded almost 100% successful completion of larval development by *C. septempunctata* on *A. fabae*: Hodek (1956, 1957a), 92–100%; Blackman (1965, 1967b), 91%; Ehler et al. (1997), 100%. *Aphis fabae* is toxic to the larvae of *Exochomus quadripustulatus* (Radwan & Lövei 1983), but rather suitable for *Cer. undecimnotata* (under the

wrong name *Hip. undecimnotata*) (Papachristos & Milonas 2008).

Atwal and Sethi (1963) found that the larvae and resulting pupae of *C. septempunctata* were heavier when fed on *Lipaphis pseudobrassicae* than on **Sitobion avenae** and **Aphis gossypii**. When studying food specificity in *Harmonia axyridis*, Hukusima and Kamei (1970) found three aphids (**Aphis pomi**, **Brevicoryne brassicae** and **Hyalopterus pruni**) out of nine to be less suitable for the larvae, prolonging development by about one-third, compared with diets of six more suitable species, including *Acyrthosiphon pisum* and *Myzus persicae*. *Aphis pomi* was the least suitable of four essential aphid prey for *Exochomus quadripustulatus*: the mortality of larvae was 19% (vs. 7–11%) and the adult fresh weight 8 mg (vs. 12–14 mg) (Radwan & Lövei 1983). Olszak (1988) found *A. pomi* highly detrimental for larvae of *C. septempunctata* (100% mortality), *A. bipunctata* (94%) and *P. quatuordecimpunctata* (73%).

Chilocorus bipustulatus is unable to reproduce when feeding on the olive scale **Saissetia oleae** (Huffaker & Doutt 1965). *Chilocorus stigma* has a markedly lower fecundity and survival and a prolonged pre-oviposition period if fed on the mussel scale *Lepidosaphes beckii* (Muma 1955), whereas the scale *Chrysomphalus aonidum* is a suitable prey.

Table 5.21 Effect of four different host plants of the mealybug *Phenacoccus manihoti* on life-history parameters in *Exochomus flaviventris* (modified after Le Rü & Mitsipa 2000).

Glycosides	Cassava (*Manihot esculenta*)			*Talinum*
	Incoza var.	Zanaga var.	hybrid[1]	
	n = 11 high content	*n* = 10 low content	*n* = 11 —	*n* = 11 lacks glycosides
Immature mortality (%)	24.4	15.9	23.1	14.8
Oviposition time (days)	28.9	47.4	25.8	38.6
Total fecundity (eggs/female)	265.1	384.5	199.5	269.0

[1]Hybrid of *M. esculenta* and *M. glaziovii*.

The adults of the coccidophagous *Cryptognatha nodiceps* also feed on the **aleurodid** *Aleurodicus cocois*, but the females do not lay eggs (Lopez et al. 2004).

Prey feeding on resistant plants

Aphis fabae feeding on a partially resistant cultivar of *Vicia faba* (that reduced aphid numbers by 63%) was not a favourable prey for *C. septempunctata*: fecundity, fertility and oviposition period were decreased while the pre-oviposition period was prolonged compared to feeding on prey from the susceptible cultivar. The combined action of the partial resistance of the crop and the slightly handicapped predator was more effective in lessening aphid numbers than either effect alone (Shannag & Obeidat 2008).

Similar observations were made on cassava. A host plant effect on the polyphagous African ladybird *Exochomus flaviventris* was recorded when it fed on the cassava mealybug **Phenacoccus manihoti**, when reared on varieties of cassava (*Manihot esculenta*) and a weed (*Talinum triangulare*) that sometimes hosted large populations of the mealybug. Cassava varieties contained different levels of **cyanogenic glycosides** that were associated with their resistance to mealybugs. The mealybugs that fed on the low-resistance variety 'Zanaga' appeared to be the most suitable for *E. flaviventris* (Table 5.21). The worst effects on the duration and survival in larval development of ladybirds were caused by mealybugs from host plants with intermediate levels of glycosides and surprisingly not those with the highest (Le Rü & Mitsipa 2000). Similar greatest negative effects of intermediate allelochemical content on coccinellids were observed with

DIMBOA in wheat, where the aphid prey found higher DIMBOA levels deterrent (5.2.6.1).

5.2.7 Prey other than aphids/coccids

5.2.7.1 Developmental stages of Holometabola

The rearing of specific prey for aphidophagous or coccidophagous coccinellids requires a large amount of space, human resources and energy, and rearing coccinellids on artificial diets (5.2.10) has met with limited success. Therefore some authors have tried to use substitute insect prey that are easier to produce, mostly the **eggs of lepidopteran** stored product pests. Substitute prey often appear to be suitable. A recent review is by Evans (2009).

When *Cryptolaemus montrouzieri* larvae are reared on the eggs of *Sitotroga cerealella* neither survival during development nor the weight of emerging adults are reduced (Pilipjuk et al. 1982). The euryphagous *Har. axyridis* was reared more successfully on eggs of the pyralid moth *Ephestia kuehniella* than on *Acyrthosiphon pisum*. The pyralid eggs were UV-killed (Schanderl et al. 1988) (Table 5.22) or frozen (Berkvens et al. 2008). *E. kuehniella* eggs were, however, unsuitable for *Cer. undecimnotata* (Iperti & Trepanier-Blais 1972, Schanderl et al. 1988, Table 5.22) and *P. quatuordecimpunctata* (Bazzocchi et al. 2004), although this diet was reported as suitable for the oligophagous *C. septempunctata* (Sundby 1966) and *A. bipunctata* (de Clercq et al. 2005).

Upon being switched from a diet of aphids to an aqueous solution of sucrose, females of *C. tranversalis* ceased oviposition after 3 days and laid almost no eggs

Table 5.22 Eggs of *Ephestia kuehniella* as food for *Harmonia axyridis* and *Ceratomegilla undecimnotata* (Schanderl et al. 1988).

Coccinellid species	Generation*	Food[†]	Pre-imag. development		Adult weight (mg)	Reproduct. (%)	Oviposition rate (eggs/day)
			mortality (%)	duration (days)			
Harmonia axyridis	control	aphid	11.4	14.8 ± 0.1	29.6 ± 0.9	80.0	22.7 ± 4.1
	1	eggs	2.6	14.1 ± 0.1	26.8 ± 0.9	85.0	38.6 ± 4.8
	2	eggs	11.7	16.0 ± 0.1	33.0 ± 1.1	80.0	40.9 ± 4.6
	3	eggs	4.7	15.9 ± 0.1	28.7 ± 1.5	72.2	34.2 ± 4.7
Ceratomegilla undecimnotata	control	aphid	6.0	12.3 ± 0.1	29.1 ± 0.7	100.0	20.4 ± 2.8
	1	eggs	16.8	15.1 ± 0.2	23.4 ± 0.8	55.0	†8.6 ± 3.7
	2	eggs	45.2	17.3 ± 0.2	23.6 ± 1.0	60.0	11.1 ± 4.0
	3	eggs	51.0	22.3 ± 0.7	19.7 ± 1.0	55.0	13.8 ± 3.4

*Both larvae and adults of all generations were fed on eggs of *E. kuehniella*, killed by x-rays.
[†]An essential aphid prey, *Acyrthosiphon pisum*, was used as control food.

on a diet of second instars of the moth *Helicoverpa armigera*. However, females laid eggs in small numbers (on average 2.7 eggs per day) when provided with a combination of both these alternative diets (Evans 2000; 5.2.5). Natural feeding by *Hip. variegata* on the eggs of *Helicoverpa armigera* in Australian cotton fields was demonstrated by ELISA (5.2.1; Mansfield et al. 2008).

Col. maculata preyed on eggs and larvae of three lepidopterans in the laboratory: *Plutella xylostella* was preyed upon more than *Pieris rapae* and *Trichoplusia ni*. *Col. maculata* adults preyed more on eggs, while the fourth instars preferred larvae (Roger et al. 2000; also 5.2.4.1). Roger et al. (2001) observed that *Col. maculata* larvae selected eggs of the noctuid *T. ni* according to their age: one day old eggs were preferred over three day old eggs, regardless of whether the eggs were parasitized by *Trichogramma evanescence* or not. In maize fields, *Col. maculata* is recorded as a very important predator of eggs of the noctuid *Helicoverpa zea*, contributing to about 45% of the observed predation (Pfannenstiel & Yeargan 2002).

It has even been recorded that the adults and third instars of *Har. axyridis* feed on the eggs and first instars of the monarch butterfly, *Danaus plexippus* (Koch et al. 2003).

The rearing of *Har. axyridis* for several years on eggs of *E. kuehniella* substantially decreased coccinellid **intensive searching behaviour** for aphid prey (5.4.1.2) compared to coccinellids continuously reared

on aphids (Ettifouri & Ferran 1993). Any possible change in behaviour patterns, produced by mass-rearing coccinellids on a substitute diet should always be looked for, as it might hamper their efficiency in the biological control of natural prey.

Developmental stages of **Coleoptera** and other insect orders are also eaten by ladybirds. In the laboratory *Har. axyridis* and *C. septempunctata* consumed larvae of the **dipteran** cecidomyiid *Contarinia nasturtii*. *Har. axyridis* showed higher voracity than *C. septempunctata* and did not prefer *Myzus persicae* over larvae of *C. nasturtii*, but not on infested broccoli plants (Corlay et al. 2007).

Coleomegilla maculata fed on larvae of the **chrysomelids** *Xanthogaleruca luteola* (Weber & Holman 1976) and *Galerucella nymphaeae* (Schlacher & Cronin 2007). In spite of the reports of *Col. maculata* feeding on the eggs of *Leptinotarsa decemlineata* (5.2.4.1), these are not essential food for some ladybirds: larvae of four coccinellid species were not able to complete larval development on a diet solely of eggs of this chrysomelid (Snyder & Clevenger 2004). *Cleobora mellyi* and *Har. conformis* adults, however, mated and oviposited when supplied with eggs of another chrysomelid, *Chrysophtharta bimaculata* (Elliott & de Little 1980).

A series of studies on the curculionid *Hypera postica*, that is only an alternative prey for ladybirds in alfalfa fields, opens a new horizon in the field of coccinellid diets, stressing the importance of complementary food

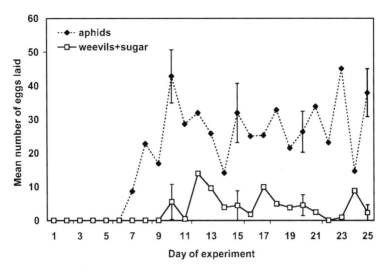

Figure 5.14 Effect of 25 days initial feeding by *Harmonia axyridis* females, on alternative foods (*Hypera postica* larvae, sugar water) versus aphids, on egg laying when pea aphids were provided (*n* = 6–8), the bars represent SE (from Evans & Gunther 2005).

(5.2.5). *H. postica* larvae feed on alfalfa together with aphids. In the laboratory only 5% of fourth instar *C. septempunctata* developed to adult when provided with live *H. postica* larvae. The proportion of successful development increased to 70% if the ladybirds received dead larvae. Observations revealed that *C. septempunctata* larvae were often deterred by the defensive wriggling of live *H. postica* larvae (Kalaskar & Evans 2001). Other predator–prey relations may quite often be affected by similar behavioural obstacles. Ladybirds are probably successful in preying on *H. postica* larvae when the latter moult.

The importance of alternative prey can be best assessed when it is an addition to a low amount of essential prey. The very low oviposition rate of *C. septempunctata* females with a limited supply of aphids is increased by addition of alfalfa weevil (*H. postica*) larvae, although this prey alone does not facilitate *C. septempunctata* egg laying. Consumption of coleopteran larvae apparently serves for self-maintenance of ladybird females; however, combined with essential nutrients from aphids it enables greater egg production (Evans et al. 2004). Also for the polyphagous *Har. axyridis*, larvae of *H. postica* are only alternative prey, as the females lay no eggs on this diet alone and larvae do not develop successfully (Kalaskar & Evans 2001).

In the absence of aphids, this food facilitates survival and bodily maintenance of females, and speeds up oviposition after a switch to essential aphid food (Evans & Gunther 2005; Fig. 5.14). In alfalfa fields in Utah (USA), different coccinellid species respond differently to alternative prey: the native species *Hip. convergens* and *Hip. quinquesignata quinquesignata* respond by aggregating only when the essential prey *Acyrthosiphon pisum* is at high density. By contrast, the invasive species *C. septempunctata* also aggregates in response to high abundances of the alternative prey, *H. postica* larvae (Evans & Toler 2007; Fig. 5.15). The authors assume that this numerical response of *C. septempunctata* may partly account for the displacement of native ladybirds from alfalfa when aphid numbers decline.

In an experiment, both *C. septempunctata* adults and third instars preferred preying on **parasitized aphids**. *Aphis fabae* parasitized by *Lysiphlebus fabarum* were killed by adult ladybirds 3.5 times more often than non-parasitized aphids, and by ladybird larvae twice as often (Meyhöfer & Klug 2002). *Coccinella undecimpunctata* showed no preference for healthy *Myzus persicae* over those parasitized by *Aphidius colemani* (Bilu & Coll 2009). Fourth instars of *C. septempunctata* and *Hip. convergens* preyed upon fully formed mummies of

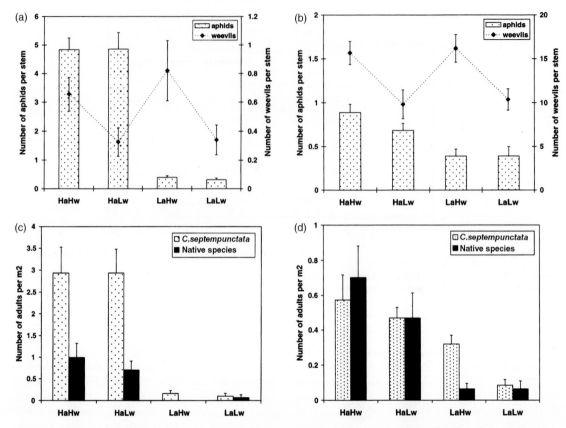

Figure 5.15 Role of alternative prey (larvae of *Hypera postica*) and essential prey (*Acyrthosiphon pisum*) shown in a field experiment where the populations of both types of prey on alfalfa plots were manipulated to produce four combinations: Ha and La, high and low numbers of aphids; Hw and Lw, high and low numbers of weevils. (a) and (b) mean number (± SE) per stem of aphids (columns) and weevils (points and lines) at the outset (24 May) and conclusion of experiment (31 May), respectively. (c) and (d) mean number (± SE) per m² of *Coccinella septempunctata* and several native ladybird species on 24 May and 31 May, respectively (from Evans & Toler 2007).

Schizaphis graminum, but a pure diet of mummies was unsuitable for the complete development to adults. A mixture of mummies and live aphids resulted in delayed development and smaller adults (Royer et al. 2008; Table 5.23).

Rodolia cardinalis avoided *Icerya purchasi* when the prey was **parasitized** by *Cryptochaetum iceryae*. The beetles would starve if only scales containing pupae of this parasitic fly were available (Quezada & de Bach 1973). *Delphastus pusillus*, a predator of all stages of whiteflies, exhibited a marked tendency to avoid fourth instar *Bemisia tabaci* containing third instar and pupal aphelinid endoparasitoids (Hoelmer et al. 1994).

Intraguild (IG) predation on heterospecific ladybird eggs and cannibalistic feeding on conspecific eggs is dealt with in 5.2.8 and Chapter 7. However, some recent findings are also discussed here. Eggs of *Col. maculata*, *Hip. convergens* and *Olla v-nigrum* as food (and also conspecific eggs) enable complete larval development of *Har. axyridis*. In contrast, the larvae of the former three species, native to North America, cannot develop to adult on eggs of *Har. axyridis* (Cottrell 2004, 2007; Fig. 5.16). Thus *Har. axyridis* appears to be a kind of generalist predator of ladybird eggs, but with certain limits: eggs of *A. bipunctata* are reported to be quite toxic to *Har. axyridis* (Sato & Dixon 2004,

Table 5.23 Consumption of *Schizaphis graminum* mummies parasitized by *Lysiphlebus testaceipes* by larvae of *Coccinella septempunctata* and *Hippodamia convergens* in a 24-hour no-choice test (after Royer et al. 2008).

		Mummies consumed (*n*)	
Stage	Mummies provided (*n*)	Coccinella septempunctata	Hippodamia convergens
1st instars (*n* = 5)	10	0	0
2nd instars (*n* = 5)	20	19*	17*
3rd instars (*n* = 5)	80	80	80
4th instars (*n* = 5)	150	150	150
Adults (*n* = 5)	150	150	150

*Feeding was attempted but mummies were only partially <50% consumed.

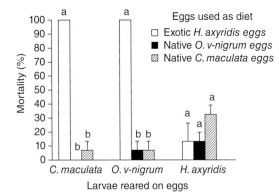

Figure 5.16 Mortality of larvae of *Coleomegilla maculata, Olla v-nigrum* or *Harmonia axyridis* reared from the first instar only on eggs of *Coleomegilla maculata, Olla v-nigrum* or *Harmonia axyridis*. Means separated by Tukey's HSD test and significant differences (P < 0.05) within groups of bars are indicated by different letters above individual bars (modified from Cottrell 2004).

Table 5.24 Suitability of different diets for growth and development in *Harmonia axyridis* and *Adalia bipunctata* (after Ware et al. 2009).

Parameter	Harmonia axyridis	Adalia bipunctata
Survival to adulthood	(A = C = D) > B	(A = C) > (B = D)
Larval development time	A > C > D > B	A > C > (B = D)
Maximum pronotal width	A > (C = D) > B	A > (C = B) > D

A, unlimited aphids; B, limited aphids; C, limited aphids plus conspecific eggs; D, limited aphids plus non-conspecific eggs. In this table '>' indicates a more suitable diet and '=' indicates two diets of equal suitability. Suitability hierarchies are worked out on the basis of the significance of differences observed between diets.

but see Ware et al. 2009, also Sloggett et al. 2009a, Sloggett & Davis 2010). The latter study investigates the metabolic pathways of coccinellid alkaloids in the predator. The eggs of other species serve as suitable food also mutually for *Col. maculata, Hip. convergens* and *O. v-nigrum*, but not for *Cycloneda munda* (Cottrell 2007). The eggs of *Har. axyridis* were found harmful also for *A. bipunctata* in an experiment with mixed diets (Ware et al. 2009; Table 5.24).

In two species, *Har. axyridis* and *A. bipunctata*, laying single **trophic eggs**, which were eaten by the female immediately after oviposition, was reported in the absence of aphid prey (Santi & Maini 2007; 5.2.8.1). The laying of infertile eggs for the nutrition of sibling neonate larvae in the same clutch represents another adaptation, as they do not serve as food for the females (Perry & Roitberg 2005, 2006; 5.2.8.1).

For some insect females, the **spermatophore** is an important source of proteins. Thus the idea that this might also apply in ladybirds is quite plausible, because the females have been observed to eat

Table 5.25 Results of rearing aphidophagous coccinellids on drone honeybee powder[a] (Niijima et al. 1986).

Species	Larval development	Adult longevity	Fecundity	Generations reared
Adalia bipunctata	+	+	−	−
Anatis halonis	+	++	±	−
Calvia muiri	++	++	+	11
Coccinula crotchi	++	++	+	3
Coccinella explanata	+	++	+	2
C. septemp. brucki	+	++	+	4
Coelophora biplagiata	++	++	++	2
Harmonia axyridis	++	++	++	16
H. yedoensis	++	++	++	8
Hippodamia convergens	+	+	−	−
H. tredecimpunctata	++	+	+	2
Menochilus sexmaculatus	++	++	++	25
Propylea japonica	++	++	++	6
Scymnus hilaris	−	++	−	−
S. otohime	−	+	−	−

[a]++, good; +, fair; ±, occasionally successful; −, unsuccessful.

spermatophores after copulation (Obata & Hidaka 1987). Omkar & Mishra (2005) assumed that *C. septempunctata* compensate costs for multiple mating by feeding on spermatophores. This assumption has recently been rejected by Perry & Rowe (2008). Neither spermatophore consumption, nor multiple matings (three and five times) influenced female longevity.

A powder from lyophilized **larvae of drone honey bees** proved a satisfactory substitute food for several aphidophagous coccinellid species. At first *Har. axyridis* was reared on non-modified larvae or pupae of males from *Apis mellifera* (Okada 1970). 'Drone powder' was used successfully with *Har. axyridis* (and *Har. yedoensis*), and later tried on other aphidophagous and coccidophagous coccinellids. The achievements of Okada's team were reviewed (Niijima et al. 1986) and the suitability of the drone powder for 15 coccinellid species compared (Table 5.25). For two oligophagous species of the genus *Coccinella* the drone powder was less suitable. *Adalia bipunctata* and *Hip. convergens* did not oviposit at all. A similarly negative result was received with another oligophagous species *Cer. undecimnotata* (Ferran et al. 1981).

Great stability is an advantage of the drone powder. It keeps its original quality at +5°C for several months and even at room temperature its nutritional value is the same after 3 months (Niijima et al. 1986). Drone powder appeared to be of about the same suitability as lyophilized aphids for *Har. axyridis*, but for the weight

of the emerged adults it was much better than the powdered *Acyrthosiphon pisum*. Diets developed for *Chilocorus* spp. were also based on honey-bee brood. The most satisfactory diets contained royal jelly and other supplements (Hattingh & Samways 1993).

5.2.7.2   Non-aphid hemipterans

These have also been reported as coccinellid prey. Among 13 natural enemies species, *Har. axyridis* and *P. japonica* were predators of the **aleyrodid** *Bemisia tabaci* in cotton fields of northern China in 2003 through 2005 (Lin et al. 2008). In the laboratory, *C. septempunctata* adults preferred *Thrips tabaci* over the aleyrodid *Trialeurodes vaporariorum* on tomato. The aphid *Macrosiphum euphorbiae* was, however, preferred over both non-aphid prey (Deligeorgidis et al. 2005). **Psyllids**, which are also near relatives of aphids, both being Sternorrhyncha, unsurprisingly can be essential prey for ladybirds that are not specialists on psyllids. The invasive Asian citrus psyllid *Diaphorina citri* was found to be essential prey for four coccinellid species (*Olla v-nigrum*, *Exochomus childreni*, *Curinus coeruleus* and *Har. axyridis*); however, *Cycloneda sanguinea* females ceased egg laying after the second day on this prey. The performance of *C. sanguinea* larvae on *D. citri* was much lower than the other four coccinellid species (Michaud & Olsen 2004).

Adelgids, the nearest relatives of aphids, have not often been studied as coccinellid prey, but *Har. axyridis* was reported to be the most abundant predator (81%) of the hemlock woolly adelgid (*Adelges tsugae*) on two *Tsuga* species. *Har. axyridis* was moderately well synchronized with the adelgid life cycle; this may indicate that *A. tsugae* could be a suitable prey for this coccinellid (Wallace & Hain 2000), but it was reported that the larvae do not always complete development on this adelgid (Butin et al. 2004). Prey other than Hemiptera have also been recently reviewed by Evans (2009).

5.2.8   Cannibalism

As predators, ladybirds at times engage in cannibalism. Eggs, pupae and especially newly moulted individuals may serve as intraspecific prey. Cannibalism complicates mass production of ladybirds as biological control agents. It requires either isolation of vulnerable individuals, or the creation of sufficient physical complexity in rearing to reduce encounters between hungry, active individuals and vulnerable, inactive conspecifics (e.g. Shands et al. (1970) filled rearing cages with wood shavings). Cannibalism by ladybirds also often occurs in natural settings, raising questions about the adaptive significance and population consequences (Cushing 1992, Martini et al. 2009). Potential costs include risk of injury in attacking conspecifics, loss of inclusive fitness from consuming relatives, and transmission of disease from infected victims (Dixon 2000). However, the widespread nature of cannibalism among ladybirds suggests that there may often also be considerable benefits (Osawa 1992c). Two distinct kinds of cannibalism occur (Fox 1975). In **sibling cannibalism**, newly hatched first instars eat unhatched eggs in the same egg cluster. In **non-sibling cannibalism**, unrelated individuals (as larvae or adults) cannibalize eggs, larvae, pupae, or newly emerged adults. Aphidophagous species of ladybirds appear especially prone to cannibalism; coccidophagous coccinellids may encounter each other less frequently and be less aggressive. Also they are more likely to complete development before populations of their prey collapse (Dixon 2000).

Mills (1982b) and Osawa (1989, 1993, 2002) drew attention to egg cannibalism in natural populations, by determining the relative frequency of sibling and non-sibling cannibalism over the growing season.

Sibling egg cannibalism may be much less frequent than non-sibling cannibalism of eggs (e.g. Mills 1982b), and it appears to occur without any influence of the availability of prey or the seasonal timing of oviposition (Osawa 1989). In contrast, the incidence of non-sibling cannibalism of eggs has been observed to increase over the growing season in studies of *Har. axyridis* on a peach tree (Osawa 1989) and artichoke (Osawa 1992a). (For feeding on heterospecific coccinellid eggs see 5.2.7.1 and Chapter 7.)

5.2.8.1   Sibling egg cannibalism

Often hatching asynchronously, young ladybird larvae remain at the egg cluster for approximately 8–36 hours before searching for food (e.g. Banks 1956, Dixon 1959, Brown 1972, Nakamura et al. 2006). Although early hatching larvae do not attack each other on the egg cluster (Brown 1972, Hodek 1996), they cannibalize late hatching and infertile eggs before dispersing (e.g. Kawai 1978, Osawa 1992c). In the laboratory, most (50–60%) eggs of *Cycloneda sanguinea*, *Har. axyridis* and *Olla v-nigrum*, hatched within 10 minutes. Of the remaining eggs, some hatched later but most were cannibalized (Michaud & Grant 2004; Fig. 5.17). Larvae remained at clusters until all unhatched eggs had been consumed and dispersal of

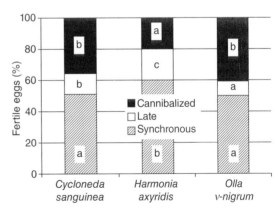

Figure 5.17 Percentages of fertile eggs of three coccinellid species that hatched synchronously (within 10 minutes of first hatch), late (>10 minutes after first hatch) or were cannibalized by siblings. Bars bearing the same letter were not significantly different among species within a category (test of proportions, P > 0.05) (from Michaud & Grant 2004).

larvae was delayed by several hours (see also Kawai 1978, Nakamura et al. 2006).

Sibling egg cannibalism by ladybirds may have considerable adaptive significance. Dispersing first instars risk starvation before finding and overcoming their first prey (Banks 1954, Dixon 1959, Kawai 1976, Wratten 1976). Initial consumption of sibling eggs may supply critical energy and nutrients, speeding development and increasing survival (Banks 1954, 1956, Kawai 1978, Roy et al. 2007). Osawa (1992c) recorded increasing larval survival to the second instar in *Har. axyridis* on citrus trees, at both low and high aphid density, due to having fed on up to three sibling eggs.

The two sexes may differ in how much they gain from sibling egg cannibalism. In a laboratory experiment, Osawa (2002) found that sibling consumption of a single egg (with aphid consumption thereafter) especially benefited males of *Har. axyridis*. Sibling egg cannibalism promoted both larger body size and weight (an advantage for mating success) and faster development time. Omkar et al. (2007) reported similar results for *P. dissecta* and *C. leonina transversalis*. Michaud and Grant (2004), however, reasoned that females are more resource-sensitive than males, and therefore would benefit most from sibling egg cannibalism. In support, they found that females of *Har. axyridis* and *O. v-nigrum* (but not of *C. sanguinea*) weighed more as adults, after engaging in sibling egg cannibalism and maturing on frozen eggs of *Ephestia kuehniella*. These authors noted that eggs of *E. kuehniella* were superior to *Aphis spiraecola* used by Osawa (2002). Thus, either males or females may benefit most from sibling egg cannibalism, depending on whether the diet subsequently is of low or high quality.

Sibling cannibalism may benefit females most when eggs are infected by bacteria that kill male embryos (Hurst & Majerus 1993, Majerus & Hurst 1997, Nakamura et al. 2006). Male-killing bacteria (including *Rickettsia, Wolbachia, Spiroplasma, Flavobacterium*, and gamma-proteobacteria) attack many species of ladybirds (Majerus 2006; Chapter 8). Because vertical transmission occurs only via egg cytoplasm, bacterial fitness is enhanced by killing males. Female-biased cannibalism (better termed scavenging) of male eggs killed by bacteria can further strengthen a strongly female-skewed, population sex ratio (e.g. Nomura & Niijima 1997). Nakamura et al. (2006), for example, report that the per capita consumption (sibling

cannibalism) of unhatched eggs by female first instars of *Har. axyridis* was increased 4–14 times in egg clusters of a Japanese population highly infected with *Spiroplasma*. Cannibalism itself did not result in horizontal transmission of male-killing bacteria when uninfected larvae (fourth instar) of *Menochilus sexmaculatus* fed on infected conspecific eggs (or larvae and pupae) (Nomura & Niijima 1997).

Osawa (1992c) focussed on the critical issue of survival of first instars to moulting, and considered the inclusive fitness of three potential beneficiaries of sibling egg cannibalism: the cannibal, the victim, and the mother of the victim. Inclusive fitnesses were determined for cases in which the cannibal (*Har. axyridis*) consumed up to three sibling eggs. Survival to the second instar was measured in field releases of first instars on citrus trees with low or high densities of aphids (*Aphis spiraecola*). Osawa concluded that full sibling cannibalism may be adaptive for both cannibal and victim when aphid density is low, but not for the victim and only sometimes for the cannibal when prey density is high. The inclusive fitness of the mother did not vary with the degree of sibling cannibalism of her eggs when aphid density was low, but decreased with increasing cannibalism when prey density was high. Thus it appears that the evolution of sibling egg cannibalism is driven especially by the fitness gains of both cannibals and victims when prey (other than the victims) are scarce. This condition of prey scarcity may often hold for aphidophagous ladybirds that lay their eggs when aphids are abundant, with subsequent decline in aphid numbers as the larvae develop (e.g. Hemptinne et al. 1992).

Osawa (2003) further investigated the possibility of the mother influencing sibling egg cannibalism and concluded that the female may have little ability to manage sibling egg cannibalism in an adaptive fashion. Perry and Roitberg (2005) also explored whether female as well as offspring fitness may be promoted by laying infertile or late-hatching eggs. These authors placed reproductively active females of *Har. axyridis* in alternating low and high resource (prey) environments for 24 hours. Females laid 56% more infertile eggs in the low resource environment (Perry & Roitberg 2005; Fig. 5.18). The authors concluded that this was not a simple, direct result of resource limitation; the females were in good condition when introduced to low prey conditions, and were well able to tolerate such low food conditions for the following 24 hours. Instead, females

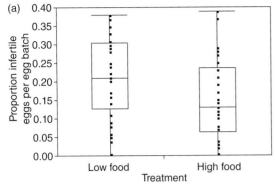

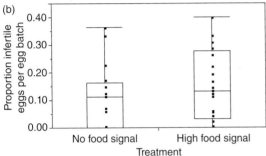

Figure 5.18 Proportion of trophic eggs produced when females of *Harmonia axyridis* were given information that there was low or high food availability through: (a) the internal cue of prey encounter and consumption or (b) the external cue of prey scent. Horizontal lines represent means, and whiskers indicate the range of observations. The proportion of trophic eggs was smaller under the low food treatment in the internal cue but not the external cue experiment (from Perry & Roitberg 2005).

may have withheld sperm to produce infertile eggs as an adaptive, maternal strategy to provide eggs as food for offspring when resource conditions were poor (Alexander 1974).

The conclusions of Osawa (1992c, 2003) and Perry and Roitberg (2005) differ strikingly as to whether (and how) sibling egg cannibalism may be controlled, at least in part, as an adaptive strategy of the ovipositing female (see also Osawa & Ohashi 2008). Another means of adaptive manipulation of sibling egg cannibalism was suggested by Michaud and Grant (2004), who found that eggs laid singly were less likely to be cannibalized by siblings than were eggs laid near each other. Roy et al. (2007) investigated whether hatching

asynchrony (and therefore potential for sibling egg cannibalism) varied with dispersion pattern amongst eggs of *A. bipunctata* laid together, but found no difference between eggs laid linearly or as a cluster.

Trophic egg laying is another form of egg consumption by kin that is clearer in its benefits for ovipositing females. Thus Santi and Maini (2007) reported that when prey became scarce, starving (yet still gravid) females of *Har. axyridis* and *A. bipunctata* (but interestingly, not of *P. quatuordecimpunctata* and *Hip. variegata*) laid a single, inviable egg which they immediately consumed. If the experimenter removed the eggs, the female searched the local area intensively (seemingly searching for the egg). Laying of trophic eggs was repeated only when the egg was removed. When given prey, the same females laid eggs that hatched (60–70%).

5.2.8.2 Non-sibling egg cannibalism

Consumption of eggs

Dispersed ladybird larvae and adults often eat conspecific (and sometimes heterospecific) eggs, larvae and pupae (e.g. Wright & Laing 1982, Takahashi 1989, Yasuda & Shinya 1997). Eggs are highly vulnerable (5.4.1.3) when laid near or in aphid colonies; clusters of eggs are encountered frequently by foraging larvae and adults (e.g. Banks 1955, Osawa 1989). Species-specific oviposition site preferences can result in enhanced cannibalism over intraguild predation (e.g. Schellhorn & Andow 1999a). Mills (1982b) recorded cannibalism of *A. bipunctata* eggs on lime trees infested with *Eucallipterus tiliae*: adults (22%), first (19%), second (33%), and third instars (26%). Cannibalism was density dependent. The threat of egg cannibalism may underlie the tendency of aphidophagous ladybirds to oviposit at some distance from, rather than in an aphid colony (Osawa 1989, 1992b, 5.4.1.3). Dixon (2000) notes that eggs of aphidophagous ladybirds mature and hatch relatively quickly; he suggests that this too may reflect selection as imposed by cannibalism.

Non-sibling cannibalism may occur most when food is scarce (e.g. Takahashi 1989) and may be rare under less stressful conditions (e.g. Triltsch 1997). This has been confirmed in the laboratory (Agarwala & Dixon 1992, Cottrell 2005). In the field, egg cannibalism by *Col. maculata* (which is able to develop and reproduce on maize pollen alone; 5.2.9) occurred less frequently during anthesis in control plots (with abundant pollen) than in experimental plots of detassled maize, i.e. without pollen (Cottrell & Yeargan 1998). Similarly among field populations of *Col. maculata*, *A. bipunctata* and *Hip. convergens* foraging on maize, Schellhorn & Andow (1999b) found that larval and pupal cannibalism occurred only when formerly abundant aphids disappeared. This study illustrates well why cannibalism is considered a 'life-boat strategy' (e.g. Cushing 1992).

Non-sibling cannibalism may be adaptive, both in providing critical energy and nutrients to the cannibal (especially when other food is scarce), and in eliminating a cannibal's potential competitors (Fox 1975, Polis 1981, Agarwala & Dixon 1992). Larvae of the coccidophagous ladybird *Exochomus flavipes* generally developed faster when fed on conspecific eggs than when fed on their usual prey, *Dactylopius opuntiae* (Geyer 1947). Similarly, *Col. maculata* larvae gained more weight by preying on conspecific eggs than on aphids (Gagné et al. 2002). Agarwala and Dixon (1992) found that conspecific eggs of *A. bipunctata* were superior to aphids as food for fourth instars; consumption of a greater biomass of aphids was required to support a given rate of growth. Furthermore, the first instars preferred conspecific eggs over aphids, and preferred aphids painted with egg extract over eggs painted with aphid extract. First instars were apparently attracted to eggs by chemical cues. Omkar et al. (2006b) obtained similar results when *C. tranversalis*, *P. dissecta*, and *Coelophora saucia* were presented with conspecific eggs and the essential prey, *Aphis gossypii*.

In some earlier reports conspecific eggs or other insect eggs were found inferior to aphids as food for coccinellid larvae. Larvae of *C. septempunctata* developed more slowly and attained smaller sizes on an exclusive diet of eggs than on aphids (Koide 1962), and even the first and second instars of *C. septempunctata brucki* moulted only to the next stage (Takahashi 1987). Larvae of *Col. maculata* took longer to develop when reared on conspecific eggs than on eggs of the fall webworm, *Hyphantria cunea* (Warren & Tadić 1967).

Egg cannibalism appears more common than intraguild egg predation in part because chemical defenses deter other ladybird species from attacking eggs (e.g. Agarwala & Dixon 1992, Hemptinne et al. 2000a, b). For example, adults of *Menochilus sexmaculatus* and *C. leonina transversalis* ate conspecific eggs much more readily than each other's eggs (Agarwala et al. 1998). Similarly, fourth instar and adult *Har. axyridis* consumed conspecific eggs more than eggs of *A. bipunctata*. Such feeding behaviour with a clear tendency towards cannibalism (i.e. stronger intraspecific than interspecific effect), may ultimately limit the adverse effects on the native species (*A. bipunctata*) of this exotic intraguild top predator (Burgio et al. 2002).

Consumption of larvae

The expression of cannibalistic larval behaviour towards other larvae, as influenced by prey availability, varies among genetic lines of *Har. axyridis* (Wagner et al. 1999) and also among species (e.g. Yasuda et al. 2001, Pervez et al. 2006). Pervez et al. (2006) reported stronger cannibalistic behaviour of larvae of *P. dissecta* than of *C. leonina transversalis*. The importance of any size disparity between attacking and attacked larvae was illustrated by Yasuda et al. (2001), who found that *C. septempunctata brucki* and *Har. axyridis* larvae generally escaped when attacked by conspecifics of the same age, but did so less frequently when attacked by an older larva. Omkar et al. (2006a) found similar results for escapes of *C. leonina transversalis* larvae from would-be cannibals, but interestingly also found that *P. dissecta* larvae generally escaped more often when attacked by older conspecific larvae than by larvae of the same age. Such studies on insects confined in limited space prevent emigration in response to declining prey density, a response that may be important in natural populations (Schellhorn & Andow 1999b, Sato et al. 2003). Osawa (1992a), for example, found that pupae of *Har. axyridis* fell victim to cannibalism less frequently when they pupated away from, rather than near an aphid colony.

Conspecific larvae often, but not always, may be relatively unsuitable (yet marginally adequate) for each other as food. Yasuda and Ohnuma (1999) found that fourth instar *C. septempunctata* gained significantly less weight on a diet of conspecific immobilized (i.e. the legs amputated) fourth instars than on *Aphis gossypii*. In contrast, fourth instars of the more polyphagous *Har. axyridis* (which were toxic as prey for *C. septempunctata*) developed just as well on a diet of conspecific fourth instars (or those of *C. septempunctata*) as on a diet of aphids. Interestingly, a diet of dead, conspecific larvae alone (i.e. without aphids in the diet) was unsuitable over the long term for *Har. axyridis* reared from hatching (Snyder et al. 2000). Survivorship and weight as adults were reduced and development time increased, when larvae of *Cycloneda sanguinea*, *Olla v-nigrum* and *Har. axyridis* were reared from hatching on a diet of frozen conspecific larvae versus a diet of frozen eggs of *Ephestia kuehniella* and bee pollen. Among all three species, the incidence of cannibalism increased with reduced provision of *E. kuehniella* eggs and with increased larval density (Michaud 2003). Pervez et al. (2006) compared conspecific eggs or first instar larvae with aphids (*Aphis gossypii*) as diets for larval *P. dissecta* and *C. leonina transversalis*. For both species, survivorship to the pupal stage and adult weight were reduced by feeding on conspecifics, especially on larvae. Larval performance was overall somewhat better on a diet of conspecific eggs than of larvae (both were inferior to aphid diet), perhaps because the less developed eggs are easier to digest. Nomura and Niijima (1997) also reported that fourth instars of *M. sexmaculatus* suffered from malnutrition and showed high mortality when reared on conspecific first and second instars and pupae, but not when reared on conspecific eggs.

Cannibalism on conspecific larvae is nonetheless important for ladybird development and survival, especially under low prey availability (Schellhorn & Andow 1999b, Wagner et al. 1999, Snyder et al. 2000). Wagner et al. (1999) reared larvae of *Har. axyridis* to the third instar under low or high aphid diets (*Myzus persicae*). When reared under low aphid availability, those third instars that cannibalized a single first instar developed significantly faster thereafter than controls, even though all individuals were provided with aphids *ad libitum* from the third instar onwards. Thus, cannibalism offset the adverse effects of earlier low aphid availability on larval growth rate that persisted even after aphids became highly available. Snyder et al. (2000) provided another example of cannibalism counteracting negative effects of a poor diet. Survival and rate of development increased in larvae of *Har. axyridis* feeding on aphids of relatively low nutritional quality (*Aphis fabae* and *Aphis nerii*), when the larvae cannibalized dead conspecific larvae every other day. The effect was intensified when the cannibalized larvae had fed on high quality food (pupae of *Apis mellifera*).

Kin recognition

A cost of cannibalism can lie in consuming kin, with subsequent reduction in inclusive fitness. But in at least some cases, ladybirds appear able to recognize and avoid cannibalizing their own kin. Agarwala and Dixon (1993) found that females of *A. bipunctata*, and their offspring (second instars), consumed fewer of their own eggs than eggs of other females when confined with equal numbers of both kinds of eggs. However, males ate as many of their own eggs as eggs sired by other males (perhaps because males invest much less energy in eggs and cannot safely distinguish them). Joseph et al. (1999) found that third instar *Har. axyridis* placed together were much less likely to cannibalize siblings than unrelated individuals. Michaud (2003) also found that larvae of *Har. axyridis*, but not of *Cycloneda sanguinea* and *Olla v-nigrum*, were less likely to cannibalize sibling than non-sibling larvae. Pervez et al. (2005) reported the same for third instars of *P. dissecta* and *C. leonina transversalis* but not for fourth instars, perhaps because of their greater voracity.

Population consequences

Cannibalism has intriguing potential consequences for population dynamics (e.g. Cushing 1992). Mills (1982b) and Osawa (1993) drew attention to the considerable potential of non-sibling egg cannibalism in regulating population size in ladybirds, perhaps often at levels below those necessary for effective biological control of target pests. As a strong intraspecific interaction limiting population size, cannibalism furthermore can reduce adverse interspecific effects on populations of other ladybird species, for example by **top intraguild predators** such as *Har. axyridis* (e.g. Burgio et al. 2002).

Cannibalism can limit population size through **disease transmission** as well as through victim mortality. Saito and Bjornson (2006) reported 100% horizontal transmission when first instars of *Hip. convergens* consumed conspecific eggs that were infected with an unidentified microsporidian (as noted above, however, Nomura and Niijima (1997) found that male-killing bacteria were not transmitted horizontally through egg cannibalism).

Egg cannibalism is also recorded in **phytophagous** ladybirds. Nakamura et al. (2004) assume that egg cannibalism causes the marked population cycle that occurs each generation in a lowland (but not a highland) population of *Epilachna vigintioctopunctata* in Indonesia.

5.2.9   Non-insect food (pollen, nectar, spores of fungi)

Among food of plant origin, **pollen** and **nectar** (both from flowers and from extrafloral nectaries), form an important food for even explicitly carnivorous coccinellids (see also 5.2.13, 5.2.14). This plant food allows the coccinellids to survive with a reduced mortality when insect food is scarce intermittently, or at the end of the season, when they accumulate reserves for diapause (5.2.5, Chapter 6).

Pollinivory was recorded very early in several coccinellid species (for refs see Hodek 1996 and Lundgren 2009). Pollinivory was found by dissections of the gut of *P. quatuordecimpunctata* (Hemptinne et al. 1988) and *A. bipunctata* (Hemptinne & Desprets 1986, Hemptinne & Naisse 1987) particularly in early spring. Polinivory of *A. bipunctata* has recently been studied by de Clercq and his team (de Clercq et al. 2005, Jalali

et al. 2009a). Experimental feeding with pollen showed that for most entomophagous ladybirds, it represents alternative food which alone does not allow the oocytes to mature but may contribute to an earlier start of oviposition after aphids appear.

Adults of the introduced *Chilocorus kuwanae* were observed feeding on the nectar and pollen of two *Euonymus* spp. in North Carolina. The presence of this pollen in the gut of this species was documented through dissection (Nalepa et al. 1992). Non-aphid food was also documented in central Italy by gut analysis in *C. septempunctata* adults after their dispersal from matured crops (Ricci et al. 2005). The ladybirds fed on pollen of Asteraceae and Apiaceae, as well as on fungal spores of *Alternaria* spp. and *Cladosporium* spp. (Triltsch 1999, Ricci et al. 2005; Fig. 5.19).

Two years analysis of gut contents of *Cer. notata* brought the first details on the diet of this rare species. Both larvae and adults contained aphids, but in their

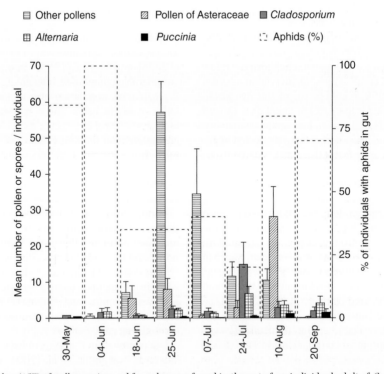

Figure 5.19 Number (±SE) of pollen grains and fungal spores found in the gut of an individual adult of *Coccinella septempunctata*, compared to the percentage of coccinellid adults with aphids in their gut (dotted bars) on different dates in 2002 in central Italy (from Ricci et al. 2005).

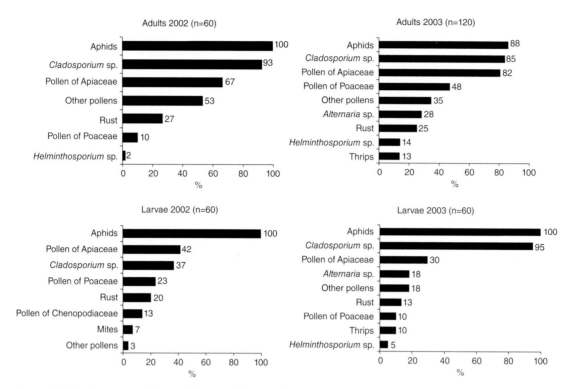

Figure 5.20 Food remains in the gut of *Ceratomegilla notata* adults and larvae in the northern Italian Alps (from Ricci & Ponti 2005).

absence the ladybirds consumed thrips. Also the incidence of pollen of Apiaceae and Poaceae, and fungal spores, mainly of *Cladosporium* spp., was very high (Ricci & Ponti 2005; Fig. 5.20).

Feeding on pollen can enable **reserves** to be accumulated to provide for long-term starvation during **dormancy** (Hagen 1962). An increase of reserves in the fat body, after feeding on maize pollen in the laboratory, was recorded in adult *C. septempunctata* collected during migration to hibernation sites (Ceryngier et al. 2004).

Pollens from different plants have different compositions (e.g. the pollen of *Pinus* contains much less protein than other pollens used in the study) and are not equally adequate as food for coccinellids (Smith 1960, 1961). Sunflower pollen proved fatal to both larvae and adults of *Col. maculata*; it adhered to the insect cuticle due to its surface structure (Michaud & Grant 2005). It was reported that pollen expressing

the insecticidal Cry 3Bb1 protein for control of corn root worm had no measurable negative effect on the development to pupation of *Col. maculata* larvae nor on adult survival and reproduction (Duan et al. 2002).

For some coccinellid species, **pollen** may represent an **essential food**, and apparently the only essential food, so that they are in fact phytophagous. Particularly the high-altitude alpine coccinellid species, such as ***Coccinella reitteri*** or ***Ceratomegilla barovskii kiritschenkoi***, are adapted to feeding on pollen, often on edelweiss (*Leontopodium alpinum*), because their habitats often lack aphids (Savoiskaya 1970b). ***Bulaea lichatschovi*** is often mentioned as an example of pollinivory (e.g. Savoiskaya 1983), but detailed rearing experiments have not yet been undertaken to elucidate the physiological value of pollen in this species (5.2.13).

More is known about another pollen feeder, the polyphagous ***Coleomegilla maculata*** (5.2.4.1). It was found that *Col. maculata* may complete its full development on pollen of a number of plants (*Zea mays, Betula populifolia, Cannabis sativa, Carpinus caroliniana*) equally successfully as on aphids (Smith 1960). The species was reported to prefer pollen to aphids (Ewert & Chiang 1966), with its observed preference for the middle strata of crop plants considered as an indication of a predominant feeding on pollen. Preference for pollen was also recorded in maize in Kentucky where predation on *Helicoverpa zea* and cannibalism on conspecific eggs decreased in plots with abundant pollen (Cottrell & Yeargan 1998). **Maize pollen** was experimentally shown to be **essential food** for *Col. maculata* (Hodek et al. 1978). When fed exclusively with maize pollen (*Zea mays*), however, the adults have a doubled pre-oviposition period and a halved fecundity, compared with preying on aphids (Hodek et al. 1978; Fig. 5.21). It thus appears that pollinivory represents just one aspect of the wide polyphagy of this species, shown also by its acarophagy (5.2.4) and feeding on the eggs of Lepidoptera and Coleoptera (5.2.7).

In an Illinois cornfield, the density of *Col. maculata* eggs increased during anthesis and it was strongly correlated with the rate of pollen shed. Only a minority of *Col. maculata* larvae and adults preyed on aphids, although they were also abundant during anthesis. In contrast, the majority of larvae and adults of *Har. axyridis* preyed on aphids. This was also shown by dissection of guts of larvae and adults of both species (Lundgren and Wiedenmann 2004; Fig. 5.22). The organic content of corn pollen was strongly correlated with the survival of young adults of *Col.*

maculata, suggesting that some critical micronutrient was present at suboptimal levels in some of the pollen. The efficiency with which *Col. maculata* larvae convert pollen into biomass increased as the larvae aged. The authors assume that there is a physiological change in late instar larvae which allows them to produce more biomass from the same amount of pollen. This is not the case with aphid prey (Lundgren et al. 2004).

Michaud and Grant (2005) compared the nutritive value of lepidopteran eggs with several pollens (maize, sorghum, bee-collected) for development of *Col. maculata*. When water was available, larval survival on pollen was 100%, as it was on frozen eggs of *E. kuehniella*. However, in a simulated drought treatment, larval survival on pollen was significantly reduced. The *Ephestia* egg diet facilitated a shorter development time and resulted in heavier adults. Thus not only aphids (Hodek et al. 1978), but also *Ephestia* eggs, are more suitable for *Col. maculata* then pollen (Michaud & Grant 2005). Also, for *Har. axyridis*, bee-collected pollen is a less suitable food than frozen eggs of *E. kuehniella*. A diet of pollen alone allowed 35–48% larvae to reach adulthood, but the development was 31–49% longer and the adults were 37–68% lighter. When provided exclusively with pollen both as larvae and adults, only

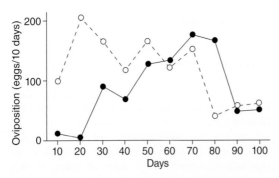

Figure 5.21 Effect of food (closed circles, pollen of *Zea mays*; open circles, *Acyrthosiphon pisum*) on the oviposition rate of *Coleomegilla maculata lengi* (*n* = 20) (modified from Hodek et al. 1978).

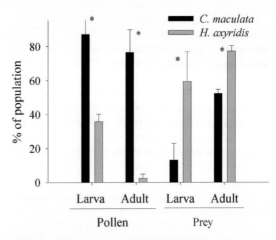

Figure 5.22 Percentage of *Harmonia axyridis* and *Coleomegilla maculata* adults that had pollen or prey in their digestive tracts. Only the individuals collected during anthesis were used. Error bars represent SEMs, and bars with asterisk above are significantly different (*t*-test, $\alpha = 0.05$) (from Lundgren et al. 2004).

about 40% of females were able to produce a small number of viable eggs (Berkvens et al. 2008).

Extrafloral nectaries were reported early as a source of substitution food by Watson and Thompson (1933) for *Leis conformis* (the plant was *Crotalaria striata*) and later, for example, by Putman (1955) for *Stethorus pusillus* (on young leaves of peach trees), by Ibrahim (1955b) for *C. undecimpunctata menetriesi* (nectaries on the midrib of cotton leaves) and by Anderson (1982) for *Apolinus lividigaster* (glands on the leaves of the euphorbiacean *Glochidion fernandii*). The importance of feeding on the extrafloral nectaries of cotton plants was documented by using nectariless varieties. The abundance of *Hip. convergens* was significantly reduced in the nectariless plots compared to plots with nectaried cotton (Schuster et al. 1976). In contrast, predation of *Har. axyridis* on *Aphis spiraecola* was reduced in the presence of extrafloral nectar (Spellman et al. 2006).

A long-standing question concerning the food relations of **Tytthaspis sedecimpunctata** was solved by Ricci (1982) in a detailed morpho-ethological study. The mandibles of the larvae in *T. sedecimpunctata* and *Coccinella nigrovittata* are equipped with 20–22 spine-like processes which form a kind of comb or rake. This structure enables the larvae to exploit their food sources: **pollen from *Lolium*** perenne and *L. multiflorum*, and spores of *Alternaria* sp. The larvae detach pollen and spores from their support by an upward movement of the head (Ricci 1982; Fig. 5.23, 5.4.3). The examination of adult gut contents after the mowing of meadows showed the pollen from Poaceae (most common was *L. perenne*), which is the preferred food, and **conidia and spores of fungi** (most often *Alternaria* sp. and *Cladosporium* sp.) (Fig. 5.24). Mites and thrips were also present (Ricci et al. 1983).

The **alternation** in feeding on grains of **pollen** and **spores** of fungi documented so precisely in *T. sedecimpunctata* by Ricci (1982, Ricci et al. 1983) is probably not exceptional. Other records indicate the same habit in *Illeis galbula* (conidia and hyphae of *Oidium*, pollen of *Ligustrum* spp. and *Acacia* spp., Anderson 1982) and in *Rhyzobius litura* (aphids, conidia of *Oidium*, *Alternaria* and *Cladosporium*, spores of *Puccinia*, pollen of Poaceae and *Mercurialis annua* (Ricci 1986b). Larvae of *R. litura* develop equally well on conidia of *Oidium monilioides* and on the aphid *Rhopalosiphum padi* (5.2.5).

Hukusima and Itoh (1976) attempted to rear four coccinellid species on pollen and powdery mildew.

(a)

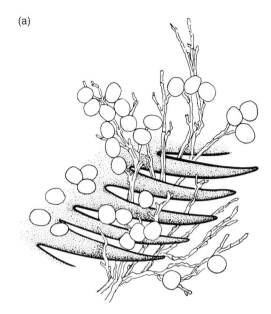

(b)

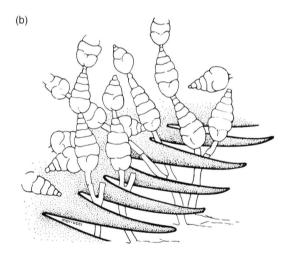

Figure 5.23 The use of the comb-like structure on the mandible of *Tytthaspis sedecimpunctata* for collecting pollen of *Lolium perenne* (a) and spores of *Alternaria* sp. (b) (from Ricci 1982).

They failed with *C. septempunctata brucki* and *Menochilus sexmaculatus*. Meagre success was achieved in *P. japonica*; only 7% of individuals reached the adult stage on powdery mildew, while 10 and 16% reached adulthood on maize and rye pollen, respectively. In the notorious generalist *Har. axyridis*, the results were even

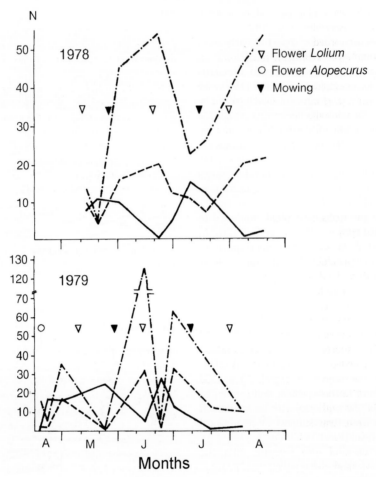

Figure 5.24 Number of grains of pollen and conidia of fungi in the gut contents of *Tytthaspis sedecimpunctata* adults in relation to flowering of *Lolium perenne* and *Alopecurus pratensis* and the hay harvest. Unbroken line, pollen; dot and dashed line, conidia of *Cladosporium* sp.; broken line, conidia of *Alternaria*; open triangles, flowering of grasses; closed triangles, hay harvest (from Ricci et al. 1983).

lower, 0, 10 and 7%, respectively. The authors quoted numerous early observations by Ninomiya of *C. s. brucki* feeding on powdery mildew on oak.

It should be noted that we still have few records of the food habits of **Propylea** *quatuordecimpunctata* and *P. japonica* and thus the polyphagy of this genus cannot be ruled out, although both species are usually considered to be specialized aphid feeders (e.g. Klausnitzer & Klausnitzer 1986 and Olszak 1988). However, Turian (1971) had already added *P. quatuordecimpunctata* to the list of micromycetophagous coccinellids, based on three observations of adults feeding on mildews

(*Microsphaera alphitoides* on *Quercus*, *Sphaerotheca cast-agnei* on *Taraxacum*). In spite of the results above by Hukusima and Itoh (1976), pollen and mildew can only be considered a good alternative diet for *Propylea* spp.

Some coccinellids have been observed **consuming** small amounts of **leaf material**. The behaviour by *C. septempunctata* of scraping plant surfaces was considered unique by Legrand and Barbosa (2003), but observations on *C. septempunctata* feeding in spring on young leaves had already been mentioned by Hodek (1973). Feeding on leaf tissue by larvae of *Col.*

maculata and *Har. axyridis* has been recently reported (Moser et al. 2008, Moser & Obrycki 2009).

Koch et al. (2004) recorded that *H. axyridis* adults fed in autumn on damaged **ripening fruit**, such as apples, grapes, pumpkins and raspberries. Feeding by *A. bipunctata* adults on cherries and plums has also been observed recently (I. Hodek, unpublished). Coccinellids have even occasionally been reported **biting** fairly strongly **into human skin** (Klausnitzer 1989, Majerus & Kearns 1989), probably when drinking sweat.

5.2.10   Substitute diets and food supplements (sprays)

Aphids and coccids are available from the field for only part of the year, and can be demanding to rear in sufficient numbers to support large laboratory populations of ladybirds. Considerable effort therefore has been spent in developing substitute foods for ladybirds (see also 5.2.7). Dried or frozen aphids have been fed to ladybirds with mixed success. Using **dried aphids**, Smith (1965b) succeeded in rearing *Anatis mali*, *A. bipunctata*, *Hip. tredecimpunctata* and *Col. maculata*, but not *Coccinella* spp. In contrast, *C. septempunctata* was reared successfully with **quick-frozen aphids** (Shands et al. 1966), as was *Hip. convergens* (Hagen 1962).

Given the challenges of developing mixtures of solely chemically defined substances (i.e. **holidic diets**), most effort has been devoted to developing diets that include, in addition to such mixtures, natural substances (e.g. honey, yeast, royal jelly of bees, and vertebrate tissues) either in limited amount (**meridic diets**) or in sufficient amounts to provide most dietary requirements (**oligidic diets**; Racioppi et al. 1981). Smirnoff (1958) reported good results in rearing larvae and adults of multiple species of aphidophagous, coccidophagous, acarophagous, and mycophagous coccinellids, when the ladybirds were fed with diets including royal jelly, agar, cane sugar, honey, alfalfa flour, yeast and water, plus small quantities of their natural foods. Similarly, Chumakova (1962) reared *Cryptolaemus montrouzieri* on **small amounts of natural prey** mixed with a variety of substances. These successes could not be repeated by others, however (Tanaka & Maeta 1965, G. Iperti, unpublished).

To date, most success in developing a laboratory diet has been achieved for the polyphagous and pollinivorous **Coleomegilla maculata** (Hodek et al. 1978). Szumkowski (1952, 1961b) used liver, meat or fish mixed with vitamin C, honey or sugar, and found finally that a mixture of fresh yeast with glucose or sucrose solution was superior to the liver diet in promoting larval development. Szumkowski's diet of pig's liver with vitamins supported egg production of *Col. maculata* (Warren & Tadić 1967). However, these diets failed to support egg-laying and larval development of *Cycloneda sanguinea* and *Hip. convergens* (Szumkowski 1961a). Smith (1965b) succeeded in rearing *Col. maculata* larvae on a diet based on brewer's yeast and sucrose. When the yeast was replaced by liver, this diet also supported oviposition by adults. Other diets (including a banana diet) varied in their abilities to support oviposition in 13 species of coccinellids (Smith 1965a).

Based on analyses of aphids and calf liver, Atallah and Newsom (1966) formulated 16 diets for *Col. maculata*, six of which supported larval development and oviposition. The most successful diets included a high proportion of liver, and carotenoids and sterols extracted from cotton leaves. Addition of vitamin E stimulated copulation (see Hodek 1996, pp. 191–2). This diet was not satisfactory for *C. novemnotata*, *Cycloneda* spp., *Hip. convergens*, and *Olla v-nigrum*. Modifications, including adding carrot lipid, whole egg, beef liver and honey, made the diet suitable for rearing larvae of *A. bipunctata*. Although the resultant adults failed to oviposit when given the artificial diet, they did so when given aphids (Kariluoto et al. 1976). Adult emergence on this diet was promoted by adding antifungal agents (sorbic acid and methyl-p-hydroxybenzoate), not only with *A. bipunctata* but also with *C. septempunctata*, *C. transversoguttata*, *Hip. tredecimpunctata* and *P. quatuordecimpunctata* (Kariluoto 1978, 1980). Females of *A. bipunctata* (after a delay, compared to feeding on aphids), and *C. septempunctata* (to a lesser degree), produced eggs on this improved diet.

Another polyphagous species, **Harmonia axyridis**, has over the years proved challenging to rear on artificial diet. Limited success with a mixture including agar, cellulose powder, yeast, saccharose, and amino acids was achieved by Hukusima and Takeda (1975). A powder from lyophilized larvae of honey-bee drones has proved a satisfactory (essential) food for *Har. axyridis*, *Har. yedoensis*, *Coelophora*

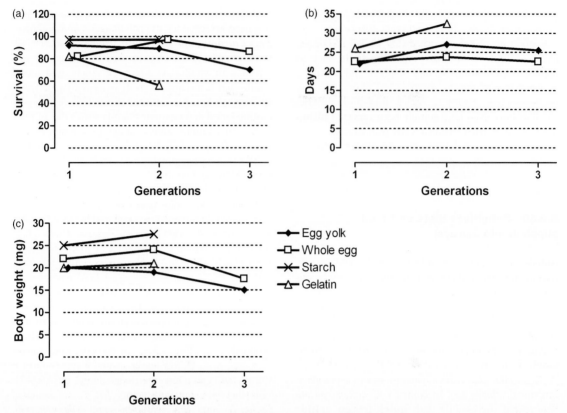

Figure 5.25 Survival rates (a), mean development times (b) and mean adult body weight (c) of *Harmonia axyridis* when maintained for successive generations on four artificial diets (from Dong et al. 2001).

biplagiata, Menochilus sexmaculatus, and P. japonica (e.g. Niijima et al. 1986; for more on 'drone powder' diet, see 5.2.7.1) Substitutive diets potentially producing multiple generations of *Har. axyridis* included chicken liver, brewer's yeast, sugars, salts, and vitamins, to which was added either whole egg or starch (most satisfactory), or egg yolk or gelatin (less satisfactory) (Dong et al. 2001; Fig. 5.25). Sighinolfi et al. (2008) used pork liver mixed with oils, sucrose, glycerine, yeast extract and vitamins, to rear *Har. axyridis* from egg to reproductive maturity, but immature development was delayed and adult emergence rate, weight, and fecundity reduced, compared to rearing on *Ephestia kuehniella* eggs.

Oviposition by **Hippodamia convergens** occurred on diets of banana, fresh liver, and casein + yeast + sugar (Smith 1965a). Racioppi et al. (1981) succeeded in rearing larvae of this species on a liver-based, oligidic

diet, but the resultant adults did not oviposit unless also provided with aphids. With modifications of liver-based diets, Hussein and Hagen (1991) succeeded in rearing larvae of *Hip. convergens* (small resultant adults) and Bain et al. (1984) managed to rear the Australian coccinellid *Cleobora mellyi*.

Building on the work of Szumkowski (1952) and Smirnoff (1958), Chinese scientists developed diets for supporting both larval growth and oviposition of the oligophagous **Coccinella septempunctata**. These liver-based diets included royal jelly, soybean oil or corn oil, and a juvenoid in olive oil (e.g. Chen et al. 1980, 1984, 1989, Chen & Qin 1982, Qin 1988).

A fruitful avenue for the development of artificial diets for coccinellids is the identification of key nutrient requirements (Silva et al. 2009, Pilorget et al. 2010). Atallah and Killebrew (1967), for example, identified specific amino acid requirements for *Col. maculata*, i.e. amino acids that must be ingested or derived from essential precursors in the food as opposed to those that the ladybird can synthesize itself. Svoboda & Robbins (1979) documented differences in saturated versus unsaturated sterols of the phytophagous *Epilachna varivestis* and the aphidophagous *C. septempunctata*. These differences reflect the sterols obtained from a plant versus insect diet, and can guide efforts to include sterols in artificial diets. Sighinolfi et al. (2008) characterized the amino acid and fatty acid content in *Har. axyridis* prepupae and adults and of their food, when the predators were reared on either a pork-based artificial diet (see above) or eggs of *E. kuehniella*. The analyses enabled Sighinolfi and colleagues to identify nutritional deficiencies in the artificial diet, and suggest possible improvements. A fuller discussion of artificial diets is provided by Hodek (1996) and recently by Bonte *et al.* (2010).

Sprays of food substitutes to crops to enhance numbers and activities of entomophagous insects (Hagen et al. 1971, Hagen 1986) have been explored. **Sugar** (usually sucrose, dissolved in water), when sprayed onto plant foliage as a substitute for floral nectar and honeydew, promotes local ladybird aggregation (e.g. Ewert & Chiang 1966, Schiefelbein & Chang 1966, Carlson & Chiang 1973, Mensah & Madden 1994, Evans & Richards 1997, van der Werf et al. 2000). Feeding on sugar (compared with only water) in the period of prey absence improved the reproduction of *Stethorus japonicus* in the subsequent period of ample prey: the pre-oviposition period was shortened from 2.25 to 1.44 days and only 203.7 *Tetranychus urticae* eggs were needed for oviposition to start, versus 368 eggs on water (Kishimoto & Adachi 2010).

Even stronger results are obtained from sprays that **mix protein supplements** (typically yeast-based) **with** the **sugar** (e.g. Hagen et al. 1971, 1976, Ben Saad & Bishop 1976b, Nichols & Neel 1977, Evans & Swallow 1993, Mensah 1997, 2002a, b, Mensah & Singleton 2003). The aggregating effect of food sprays appears to derive from arresting the ladybirds so that they linger longer in treated areas; cage experiments indicate that ladybird adults are not attracted by applications of protein supplement and sugar to alfalfa and cotton (Hagen et al. 1971, 1976, Nichols & Neel 1977). Ladybird adults (*Hippodamia* spp., *C. transversoguttata*, and *Scymnus postpinctus*) have been reported to be attracted, however, to potato plants sprayed with honey, molasses or tryptophan (Ben Saad & Bishop 1976a). The effects of a single food spray often dissipate within a week or so (e.g. Evans & Swallow 1993), but repeated applications can result in elevated ladybird densities over much of the growing season (e.g. Mensah 2002b, Mensah & Singleton 2003).

The goal of using food sprays is to build up large numbers of ladybirds and other natural enemies in crops early in the season while pest numbers are still low (e.g. Hagen et al. 1971, Ehler et al. 1997, Mensah & Singleton 2003). Other uses include concentrating the predators in protected areas prior to the application of an insecticide elsewhere (e.g. Evans & Richards 1997). While the results to date are intriguing, the mechanisms and benefits associated with responses of ladybirds and other natural enemies to field applications of food sprays are still little understood, and the implementation of this potential component of programmes in conservation biological control remains in its early stages (Wade et al. 2008).

5.2.11 Essential foods

The 'historical' and functional reasons for the introduction of the terms 'essential' and 'alternative' food/prey are explained in 5.2.2. The terms were defined more than 40 years ago (Hodek 1962); essential food supports successful immature development and adult reproduction, while alternative food merely maintains survival (Table 5.26, Fig. 5.1). Such a simple dichotomised definition is useful by focussing on the principal difference between the food classes and has been accepted by the scientific community (e.g. Mills 1981, Majerus 1994, Evans et al. 1999, Ware et al. 2008, Giorgi et al. 2009, Berkvens et al. 2010).

Naturally, there are **transitions** between both types. Thus essential foods show varying degrees of favourability, enabling different levels of developmental rate, fecundity and survival. Alternative foods may range from highly toxic (5.2.6.1) to quite suitable (5.2.6.3) for enabling survival in periods of scarcity of essential food by providing a supply of energy to compensate for metabolic losses or even to accumulate reserves for dormancy.

It may be considered a shortcoming that the two terms cannot be defined by numerical criteria. Decisions within the 'grey zone' of intermediate cases can only be subjective. One may hesitate as to whether 65% completion of larval development indicates somewhat

Table 5.26 List of essential foods.

Adalia bipunctata	Adelges laricis (L)	Mills 1981
	Acyrthosiphon pisum	Blackman 1965, 1967b, Hariri 1966a, b, Fye 1981
	A. pisum (dry powdered)	Smith 1965b
	Aphis fabae (L)	Mills 1981
	Aphis hederae (L)	Mills 1981
	Aphis pomi	Iperti 1965
	Aphis sambuci (L)	Mills 1981
	Aulacorthum solani (L)	Mills 1981
	Betulaphis quadrituberculata (L)	Mills 1981
	Brachycaudus helichrysi (L)	Mills 1981
	Brachycaudus persicae	Iperti 1965
	Brachycaudus tragopogonis	Iperti 1965
	Cavariella spp. (L)	Mills 1981
	Chaitophorus capreae (L)	Mills 1981
	Chaitophorus versicolor (L)	Mills 1981
	Chromaphis juglandicola (L)	Mills 1981
	Drepanosiphum platanoidis (L)	Mills 1981
	Eucallipterus tiliae	Wratten 1973
	E. tiliae (L)	Mills 1981
	Euceraphis punctipennis (L)	Mills 1981
	Hyalopterus pruni	Semyanov 1970
	H. pruni (L)	Mills 1981
	Macrosiphum rosae	Brun & Iperti 1978
	M. rosae (L)	Mills 1981
	Microlophium carnosum (L)	Mills 1981
	Microlophium carnosum	Blackman 1965, 1967b, Hariri 1966a, b
	Myzocallis boerneri (L)	Mills 1981
	Myzocallis carpini (L)	Mills 1981
	Myzocallis castanicola (L)	Mills 1981
	Myzocallis coryli (L)	Mills 1981
	Myzus cerasi	Iperti 1965, Ozder & Saglam 2003
	M. cerasi (L)	Mills 1981
	Myzus persicae	Blackman 1965, 1967b, Kariluoto 1980, Fye 1981
	Neomyzus circumflexus	Blackman 1965, 1967b
	Periphyllus lyropictus (L)	Mills 1981
	Phorodon humuli (L)	Mills 1981
	Phyllaphis fagi (L)	Mills 1981
	Pineus pini (L)	Mills 1981
	Pterocallis alni (L)	Mills 1981
	Rhopalosiphum maidis (dry) (slightly slower development)	Smith 1965b
	Rhopalosiphum padi (lower ovip.)	Semyanov 1970, Ozder & Saglam 2003
	Schizaphis graminum	Fye 1981
	Sitobion avenae	Ozder & Saglam 2003
	Tuberculatus annulatus (L)	Mills 1981
	Tuberolachnus salignus (L)	Mills 1981
	Uroleucon cirsii (L)	Mills 1981

Table 5.26 (*Continued*)

Adalia	*Aphis pomi*	Iperti 1965
decempunctata	*Brachycaudus persicae*	Iperti 1965
	Chaitophorus capreae (L)	Mills 1981
	Cinara palestinensis	Bodenheimer & Neumark 1955
	Drepanosiphum platanoidis (L)	Mills 1981
	Eucallipterus tiliae (L)	Mills 1981
	Euceraphis punctipennis (L)	Mills 1981
	Matsucoccus josephi	Bodenheimer & Neumark 1955[a]
	Myzocallis boerneri (L)	Mills 1981
	Myzocallis coryli (L)	Mills 1981
	Phyllaphis fagi (L)	Mills 1981
	Rhopalosiphum maidis	Iperti 1965
	Thelaxes dryophila (L)	Mills 1981
	Tuberculatus annulatus (L)	Mills 1981
Aiolocaria hexaspilota	*Chrysomela populi* (pre-imag.stages)	Savoiskaya 1983 (p. 155)
Anatis mali	*Acyrthosiphon pisum*	Smith 1965a
	Mindarus abietinus	Berthiaume et al. 2000
	Rhopalosiphum maidis	Smith 1965a
Anatis ocellata	*Euceraphis punctipennis* (L)	Mills 1981
	Myzus persicae (less suitable)	Kesten 1969
	Pineus pini (L)	Mills 1981
	Rhopalosiphum padi (less suitable)	Kesten 1969
	Schizolachnus pineti	Kesten 1969
	S. pineti (L)	Mills 1981
Aphidecta obliterata	*Adelges cooleyi*	Wylie 1958
	Adelges nusslini	Wylie 1958
	Elatobium abietinum (L)	Mills 1981, Day et al. 2006
	Pineus pini (L)	Mills 1981
	Rhopalosiphum padi	Oliver et al. 2006
Axinoscymnus cardilobus	*Bemisia tabaci*	Huang et al. 2006, 2008
Azya orbigera	*Coccus viridis*	Liere & Perfecto 2008
Brumoides suturalis	*Ferrisia virgata* (better for dev.)	Gautam 1990
	Planococcus minor (better for oviposition)	Gautam 1990
Calvia muiri	drone powder	Niijima 1979
Calvia quatuordecimguttata	*Acyrthosiphon pisum*	Ruzicka 2006
	Aphis pomi (L)	Mills 1981
	Eucallipterus tiliae (L)	Mills 1981
	Euceraphis punctipennis (L)	Mills 1981
	Apis mellifera – drone powder	Niijima 1979
	Psylla mali	Semyanov 1980
	Schizaphis graminum	Fye 1981
Calvia quindecimguttata	*Melasoma aenea* (pre-imag. stages)	Kanervo 1946
Ceratomegilla undecimnotata	*Aphis fabae*	Hodek 1960, Iperti 1965, Brun & Iperti 1978
	Acyrthosiphon pisum	Ruzicka 2006
	Myzus persicae	Brun & Iperti 1978
Cheilomenes lunata	*Aphis craccivora*	Ofuya & Akingbohungbe 1988
Cheilomenes propinqua vicina	*Aphis craccivora*	Mandour et al. 2006
Chilocorus bipustulatus	*Aspidiotus nerii*	Uygun & Elekcioglu 1998
	Chrysomphalus aonidum	Yinon 1969
Chilocorus circumdatus	*Aonidiella orientalis*	Elder & Bell 1998

(*Continued*)

Table 5.26 (*Continued*)

Chilocorus kuwanae	*Chionaspis salicis*	Kuznetsov & Pantyuchov 1988
	Chionaspis alnus	Kuznetsov & Pantyuchov 1988
	Chrysomphalus bifasciculatus	Tanaka 1981
	Unaspis euonymi	Ricci et al. 2006
	Unaspis yanonensis	Nohara 1962a, Tanaka 1981
Chilocorus malasiae	*Aonidiella orientalis*	Elder & Bell 1998
Chilocorus nigritus	*Abgrallaspis cyanophylli*	Ponsonby & Copland 1996, 1998, 2000, 2007b
	Acutaspis umbonifera	Ponsonby & Copland 2007a
	Aonidiella aurantii	Samways 1986
	A. aurantii (preferred by adults)	Samways & Tate 1986, Samways & Wilson 1988
	Aspidiotus nerii	Erichsen et al. 1991
	A. nerii (preferred by larvae)	Samways & Wilson 1988
	Coccus hesperidum	Ponsonby & Copland 2007a
	Pinnaspis buxi	Ponsonby & Copland 2007a
	Saissetia coffeae	Ponsonby & Copland 2007a
Chilocorus renipustulatus	*Chionaspis salicis* (L)	Mills 1981
Chilocorus rubidus	*Eulecanium caraganae* (eggs for larvae, eggs and larvae for adults)	Pantyukhov 1968
Chilocorus stigma (*bivulnerus*)	*Chrysomphalus aonidum*	Muma 1955
Clitostethus arcuatus	*Aleurodes proletella*	Bathon & Pietrzik 1986
	Aleurotuba jelinekii (L)	Mills 1981
	Siphoninus phillyreae	Bellows et al. 1992
Clitostethus oculatus	*Bemisia tabaci*	Liu et al. 1997, Liu & Stansly 1999, Ren et al. 2002
Coccinella hieroglyphica	*Galerucella sagittariae* (larvae)	Hippa et al. 1984
	Myzus persicae (less suitable than *G. sagittariae*)	Hippa et al. 1984
Coccinella magnifica	*Acyrthosiphon pisum*	Sloggett et al. 2002
	Aphis fabae	Sloggett et al. 2002
	Microlophium carnosum	Sloggett et al. 2002
Coccinella septempunctata	*Acyrthosiphon pisum*	Blackman 1965, 1967b, Schanderl et al. 1988, Obrycki and Orr 1990
	Aphis craccivora	Hodek 1960, Iperti 1965
	Aphis fabae	Hodek 1956, Blackman 1965, 1967a, Iperti 1965, Brun & Iperti 1978
	A. fabae (L)	Mills 1981
	Aphis glycines	Costamagna et al. 2008
	Aphis gossypii	Iperti 1965, Zhang 1992
	A. gossypii (G)	Agarwala et al. 1987
	Aphis jacobaeae (L)	Mills 1981
	Aphis nerii (on *Calotropis procera*)(G)	Agarwala et al. 1987
	Aphis urticata	Iperti 1965
	Brevicoryne brassicae (L)	Mills 1981
	Diuraphis noxia	Michels & Flanders 1992, Formusoh & Wilde 1993
	Hyalopterus pruni	Hodek 1960
	Hyperomyzus lactucae (L)	Mills 1981
	Lipaphis pseudobrassicae	Atwal & Sethi 1963, Sethi & Atwal 1964

Table 5.26 (*Continued*)

	L. pseudobrassicae (G)	Agarwala et al. 1987
	Longiunguis donacis	Iperti 1965
	Macrosiphoniella artemisiae	Iperti 1965
	Megoura viciae (slightly less suitable)	Blackman 1965, 1967b
	Metopolophium dirhodum (L)	Mills 1981
	Microlophium carnosum (L)	Mills 1981
	Myzus cerasi	Ozder & Saglam 2003
	Myzus persicae	Blackman 1965, 1967b, Brun & Iperti 1978, Kariluoto 1980
	M. persicae (L)	Mills 1981
	Myzus persicae nicotianae	Katsarou et al. 2005
	Rhopalosiphum maidis	Obrycki & Orr 1990
	Rhopalosiphum padi	Ozder & Saglam 2003
	Schizaphis graminum	Fye 1981, Michels & Behle 1991, Formusoh & Wilde 1993
	Sitobion avenae (L)	Mills 1981, Ozder & Saglam 2003
	Sitobion avenae	Ghanim et al. 1984
	Uroleucon aeneus	Hodek 1960
	Uroleucon cirsii (L)	Mills 1981
Coccinella septempunctata brucki	*C. s. brucki* (eggs) (development prolonged)	Koide 1962
	Myzus malisuctus	Hukusima & Sakurai 1964
	Myzus persicae	Hukusima & Sakurai 1964
	Neophyllaphis podocarpi	Maeta 1965
	Rhopalosiphum padi	Okamoto 1966
	Sitobion avenae	Hukusima & Sakurai 1964
	Vesiculaphis caricis	Takeda et al. 1964
Coccinella transversalis (*C. leonina transversalis*)	*Aphis craccivora*	Debaraj & Singh 1990, Agarwala & Yasuda 2001a
	A. craccivora (G)	Agarwala et al. 1987
	Aphis gossypii	Pervez et al. 2006
	Lipaphis pseudobrassicae (G)	Agarwala et al. 1987
Coccinella transversoguttata	*Myzus persicae*	Kariluoto 1980
	Phorodon humuli	Campbell & Cone 1999
	Schizaphis graminum	Fye 1981
Coccinella transversoguttata richardsoni	*Acyrthosiphon pisum* (dry)	Smith 1965b
Coccinella undecimpunctata	*Aleyrodes proletella*	Moura et al. 2006
	Aphis fabae	Moura et al. 2006, Soares & Serpa 2007
	Aphis pomi	Harpaz 1958
	Laingia psammae (L)	Mills 1981
	Metopolophium dirhodum (L)	Mills 1981
	Myzus persicae	Cabral et al. 2006
	Sitobion avenae (L)	Mills 1981
Coccinella undecimpunctata menetriesi	*Aphis gossypii*	Ibrahim 1955a, b
	Aphis laburni	Ibrahim 1955a, b
	Aphis nerii	Ibrahim 1955a, b
	Aphis punicae	Ibrahim 1955a, b
	Hyalopterus pruni	Ibrahim 1955a, b
	Lipaphis pseudobrassicae	Ibrahim 1955a, b
	Macrosiphoniella sanborni	Ibrahim 1955a, b
	Myzus persicae	Ibrahim 1955a, b

(*Continued*)

Table 5.26 (*Continued*)

Coelophora biplagiata	*Aphis gossypii*	Yu et al. 2005
Coelophora mulsanti	*Rhopalosiphum padi*	Sallée & Chazeau 1985
Coelophora quadrivittata	*Coccus viridis*	Chazeau 1981
Coelophora saucia	*Myzus persicae*	Omkar & Pathak 2006
Coleomegilla maculata	*Acyrthosiphon pisum* (also dry)	Smith 1965b
(ssp. *lengi* incl.)	*Aphis glycines*	Mignault et al. 2006
	C. maculata (eggs)	Warren & Tadić 1967
	Ephestia kuehniella (eggs)	Michaud & Grant 2005
	Hyphantria cunea (eggs)	Warren & Tadić 1967
	Leptinotarsa decemlineata (eggs) (less suitable than aphids or pollen)	Ferro & Hazzard 1991, Hazzard & Ferro 1991
	Myzus persicae	Ferro & Hazzard 1991, Hazzard & Ferro 1991
	Rhopalosiphum maidis (also dry) (slightly less suitable)	Smith 1965b
	corn pollen	Ferro & Hazzard 1991, Hazzard & Ferro 1991
	pollen (corn, sorghum)	Michaud & Grant 2005
Cryptognatha nodiceps	*Aspidiotus destructor*	Lopez et al. 2004
Cryptolaemus montrouzieri	*Maconellicoccus hirsutus*	Persad & Khan 2002
	Planococcus citri	Garcia & O'Neil 2000
	Pulvinaria psidii (eggs)	Mani & Krishnamoorthy 1990
Curinus coeruleus	*Diaphorina citri*	Michaud & Olsen 2004
	Heteropsylla cubana	Chazeau et al. 1992, da Silva et al. 1992
Cycloneda ancoralis	*Aphis gossypii*	Elliott et al. 1994
	Aphis helianthi	Elliott et al. 1994
	Diuraphis noxia	Elliott et al. 1994
	Lipaphis pseudobrassicae	Elliott et al. 1994
Cycloneda limbifer	*Aphis craccivora*	Zeleny 1969
Cycloneda munda	*Acyrthosiphon pisum*	Smith 1965b
Cycloneda sanguinea	*Aphis fabae*	Isikber & Copland 2002
	Aphis gossypii	Isikber & Copland 2001, 2002
	Myzus persicae	Isikber & Copland 2002
Delphastus catalinae	*Bemisia tabaci*	Pickett et al. 1999
		Simmons & Legaspi 2004, Zang & Liu 2007
Delphastus pusillus	*Bemisia tabaci*	Hoelmer et al. 1993, 1994, Guershon & Gerling 1999
Diomus austrinus	*Phenacoccus madeirensis*	Chong et al. 2005
	Planococcus citri	Chong et al. 2005
Exochomus childreni	*Diaphorina citri*	Michaud & Olsen 2004
Exochomus flavipes	*Dactylopius opuntiae*	Geyer 1947
	Matsucoccus josephi	Bodenheimer & Neumark 1955
Exochomus quadripustulatus	*Acyrthosiphon pisum*	Radwan & Lövei 1983
	Chionaspis salicis (L)	Mills 1981
	Dysaphis devecta	Radwan & Lövei 1983
	Dysaphis plantaginea	Radwan & Lövei 1983
	Planococcus citri	Katsoyannos & Laudeho 1977
	Pseudochermes fraxini (L)	Mills 1981
	Saissetia oleae	Katsoyannos & Laudeho 1977

Table 5.26 (*Continued*)

Harmonia axyridis	*Acyrthosiphon pisum*	Hukusima & Kamei 1970, Fye 1981, Schanderl et al. 1988
	Adelges tsugae	Wallace & Hain 2000, Flowers et al. 2005
	Aphis craccivora	Mogi 1969
	Aphis fabae	Soares et al. 2005
	Aphis glycines	Mignault et al. 2006, Costamagna et al. 2008
	Aphis pomi	Hukusima & Kamei 1970
	Apis mellifera-drone powder	Niijima 1979
	Brevicoryne brassicae (less suitable)	Hukusima & Kamei 1970
	Capitophorus elaeagni	Osawa 1992
	Diaphorina citri	Michaud & Olsen 2004
	Ephestia kuehniella (eggs)	Schanderl et al. 1988
	Hyalopterus pruni (less suitable)	Hukusima & Kamei 1970
	Hyperomyzus carduellinus	Hukusima & Kamei 1970
	Megoura viciae japonica	Hukusima & Kamei 1970
	Myzus persicae	Hukusima & Kamei 1970, Schanderl et al. 1985, Soares et al. 2005
	Nasonovia lactucae	Hukusima & Ohwaki 1972
	Neophyllaphis podocarpi	Maeta 1965
	Periphyllus californensis	Hukusima & Kamei 1970
	Rhopalosiphum padi	Okamoto 1966
	Schizaphis graminum	Fye 1981
	Sitobion ibarae	Hukusima & Kamei 1970
Harmonia conformis	*Acyrthosiphon pisum*	Fye 1981
	Aphis punicae	Moursi & Kamal 1946
	Macrosiphum rosae	Maelzer 1978
	Psylla jucunda	Hales 1979
	Schizaphis graminum	Fye 1981
Harmonia dimidiata	*Acyrthosiphon pisum*	Semyanov 1999, Ruzicka 2006
	Myzus persicae	Fye 1981, Semyanov 1999
	Eriosoma lanigerum	Chakrabarti et al. 1988
	Rhopalosiphum maidis	Semyanov 1999
	Sitotroga (eggs)	Semyanov 1999
Harmonia sedecimnotata	*Myzus persicae*	Semyanov 2000
Hippodamia convergens	*Diuraphis noxia*	Formusoh & Wilde 1993
	Phorodon humuli	Campbell & Cone 1999
	Rhopalosiphum padi	Phoofolo et al. 2007
	Schizaphis graminum	Michels & Behle 1991, Formusoh & Wilde 1993
		Phoofolo et al. 2007
	Therioaphis maculata	Nielson & Currie 1960
Hippodamia parenthesis	*Acyrthosiphon pisum* (dry)	Smith 1965b
	A. pisum	Orr & Obrycki 1990
	Schizaphis graminum (less suitable than *A. pisum*)	Orr & Obrycki 1990
Hippodamia quinquesignata	*Acyrthosiphon pisum*	Kaddou 1960
Hippodamia sinuata	*Rhopalosiphum maidis*	Michels & Behle 1991
	Schizaphis graminum	Michels & Behle 1991
Hippodamia tredecimpunctata	*Acyrthosiphon pisum* (dry powdered)	Smith 1965b
	Diuraphis noxia (less suitable)	Michels & Flanders 1992
	Rhopalosiphum maidis (dry powdered) (slower development)	Smith 1965b
	Schizaphis graminum (more suitable)	Michels & Flanders 1992

(*Continued*)

Table 5.26 (*Continued*)

Hippodamia variegata	*Acyrthosiphon pisum*	Obrycki & Orr 1990
	Aphis craccivora	Iperti 1965
	Aphis fabae (on beans)	Brun & Iperti 1978
	Aphis glycines	Costamagna et al. 2008
	Aphis gossypii	El Habi et al. 2000
	Aphis nerii	Iperti 1965
	Aphis pomi (L)	Brun & Iperti 1978
	Diuraphis noxia (low fecundity)	Michels & Flanders 1992
	Dysaphis crataegi	Kontodimas & Stathas 2005
	Macrosiphoniella artemisiae	Iperti 1965
	Myzus persicae	Iperti 1965
	Rhopalosiphum maidis	Obrycki & Orr 1990
	Schizaphis graminum	Michels & Bateman 1986
	Schizaphis graminum (low larv. surv.)	Michels & Flanders 1992
Hyperaspis desertorum	*Orthezia urticae* (L)	Savoiskaya 1983 (p. 152)
Hyperaspis lateralis	*Pseudococcus aurilanatus*	McKenzie 1932
	Pseudococcus sequoiae (eggs, young larvae preferred prey)	McKenzie 1932
Hyperaspis notata	*Ferrisia virgata*	Staubli Dreyer et al. 1997b
	Phenacoccus madeirensis	Staubli Dreyer et al. 1997b
	Phenacoccus manihoti	Staubli Dreyer et al. 1997a
Hyperaspis pantherina	*Orthezia insignis*	Booth et al. 1995
Hyperaspis raynevali	*Phenacoccus herreni*	Kiyindou & Fabres 1987
Hyperaspis senegalensis hottentotta	*Phenacoccus manihoti*	Fabres & Kiyindou 1985
Lioadalia flavomaculata	*Schizaphis graminum*	Michels & Bateman 1986
Lindorus lophantae	*Aspidiotus nerii*	Honda & Luck 1995, Cividanes & Gutierrez 1996, Stathas 2000
	Phoenicococcus marlatti	Gomez 1999
Macroilleis hauseri	*Podosphaera leucotricha* (mildew)	Liu 1950
Megalocaria dilatata	*Pseudoregma bambucicola*	Puttarudriah & Channabasavanna 1952
Menochilus sexmaculatus	*Aphis craccivora*	Srikanth & Lakkundi 1990, Agarwala et al. 2001, Agarwala & Yasuda 2000, 2001a Omkar et al. 2005
	Aphis craccivora (G)	Agarwala et al. 1987
	Aphis gossypii	Agarwala & Yasuda 2000
	Aphis nerii (G)	Agarwala et al. 1987
	Aphis spiraecola	Agarwala & Yasuda 2000
	Aphis spiraecola (G)	Agarwala et al. 1987
	Lipaphis pseudobrassicae (G)	Agarwala et al. 1987
	Nasonovia lactucae	Hukusima & Komada 1972
	Schizaphis graminum	Fye 1981
	Sitobion ibarae	Hukusima & Komada 1974
Micraspis discolor	*Aphis craccivora* (G)	Agarwala et al. 1987
	Aphis spiraecola (G)	Agarwala et al. 1987
Myrrha octodecimguttata	*Pineus pini* (L)	Mills 1981
Nephus bilucernarius	*Dysmicoccus* spp.	Gonzáles-Hernández et al. 1999
Nephus bisignatus	*Planococcus citri*	Kontodimas et al. 2005
Nephus includens	*Planococcus citri*	Kontodimas et al. 2005

Table 5.26 (*Continued*)

Oenopia conglobata	*Aphis pomi*	Iperti 1965
	Brachycaudus persicae	Iperti 1965
	Cinara palestinensis	Bodenheimer & Neumark 1955
	Galerucella lineola (pre-imag.)	Kanervo 1946
	Matsucoccus josephi	Bodenheimer & Neumark 1955[a]
	Rhopalosiphum maidis	Iperti 1965
	Schizaphis graminum	Fye 1981
Oenopia conglobata contaminata	*Agonoscena pistaciae*	Mehrnejad & Jalali 2004
Olla v-nigrum	*Diaphorina citri*	Michaud & Olsen 2004
	Heteropsylla cubana	Chazeau et al. 1991
	Psylla uncatoides	Chazeau et al. 1991
	Trialeurodes vaporariorum	Chazeau et al. 1991
Paranaemia vittigera	*Diuraphis noxia*	Robinson 1992
Parastethorus nigripes	*Tetranychus urticae*	Bailey & Caon 1986
Pharoscymnus numidicus	*Parlatoria blanchardi*	Kehat 1968
	Prodenia litura (eggs)	Kehat 1968
Propylea dissecta	*Aphis craccivora*	Omkar & Mishra 2005
	Aphis gossypii	Omkar & Mishra 2005, Pervez et al. 2006
	Lipaphis pseudobrassicae	Omkar & Mishra 2005
	Rhopalosiphum maidis	Omkar & Mishra 2005
	Uroleucon compositae	Omkar & Mishra 2005
Propylea japonica	*Aphis gossypii*	Zhang 1992
	Ephestia kühniella (eggs, lower ovip.)	Hamasaki & Matsui 2006
	Nasonovia lactucae	Hukusima & Komada 1972
	Nilaparvata lugens	Bai et al. 2006
	Sitobion akabiae	Kawauchi 1979
	Sitobion ibarae	Hukusima & Komada 1972
	Uroleucon formosanum	Hukusima & Komada 1972
Propylea quatuordecimpunctata	*Acyrthosiphon pisum*	Obrycki & Orr 1990
	Aphis fabae (L)	Mills 1981
	Aphis glycines	Mignault et al. 2006
	Brachycaudus helichrysi (L)	Mills 1981
	Diuraphis noxia (low fecundity)	Michels & Flanders 1992
	Eucallipterus tiliae (L)	Mills 1981
	Metapolophium dirhodum (L)	Mills 1981
	Pterocallis alni (L)	Mills 1981
	Rhopalosiphum maidis	Brun & Iperti 1978, Obrycki & Orr 1990
	Schizaphis graminum	Fye 1981, Michels & Flanders 1992
	Sitobion avenae (L)	Mills 1981
	Uroleucon cirsii (L)	Mills 1981
	Uroleucon jaceae (L)	Mills 1981
Rhyzobius ventralis	*Eriococcus coriaceus*	Richards 1985
Rhyzobius litura	*Sitobion avenae* (L)	Mills 1981
	Uroleucon cirsii (L)	Mills 1981
	Uroleucon jaceae (L)	Mills 1981
Rodatus major	*Monophlebulus pilosior* (eggs)	Richards 1985
Sasajiscymnus tsugae	*Adelges tsugae*	Cheah & McClure 1998, Flowers et al. 2005
Scymnus auritus	*Phylloxera glabra*	Eastop & Pope 1966, 1969
Scymnus coccivora	*Maconellicoccus hirsutus*	Persad & Khan 2002
Scymnus frontalis	*Diuraphis noxia*	Naranjo et al. 1990, Farid et al. 1997
Scymnus levaillanti	*Aphis gossypii*	Isikber & Copland 2001
Scymnus marinus	*Saissetia oleae* (eggs)	Ba M'hamed & Chemseddine 2001

(*Continued*)

Table 5.26 (*Continued*)

Scymnus posticalis	*Aphis gossypii*	Agarwala & Yasuda 2001b
Scymnus sinuanodulus	*Adelges tsugae*	Lu & Montgomery 2001
Scymnus subvillosus	*Aphis sambuci*	Klausnitzer 1992
Serangium	*Bemisia tabaci*	Yigit 1992
parcesetosum	*Dialeurodes citri*	Yigit et al. 2003
	Coccus hesperidum	Yigit et al. 2003
Stethorus bifidus	*Tetranychus urticae*	Charles et al. 1985
Stethorus gilvifrons	*Panonychus ulmi*	Dosse 1967
	Tetranychus cinnabarinus (eggs)	Dosse 1967
	Tetranychus urticae (eggs)	Dosse 1967
Stethorus japonicus	*Tetranychus urticae*	Mori et al. 2005
	Tetranychus mcdanieli	Roy et al. 2003
	Tetranychus urticae	Rott & Ponsonby 2000
Stethorus punctum	*Panonychus ulmi*	Houck 1986
	Tetranychus urticae	Houck 1986, 1991
Stethorus pusillus	*Panonychus ulmi*	Putman 1955
	Tetranychus bimaculatus	Putman 1955
Stethorus tridens	*Tetranychus evansi*	Fiaboe et al. 2007
Synonycha grandis	*Paraoregma alexandri*	Shantibala et al. 1992
	Pseudoregma bambucicola	Puttarudriah & Channabasavanna 1952, Puttarudriah & Maheswariah 1966

(G) based on examination of gut content – 'common' food (Agarwala et al. 1987).
(L) based on observation of larvae of the coccinellid occurring together with the recorded prey (Mills 1981).
[a]Savoiskaya (1983, p. 142) doubts whether *M. josephi* is really an essential prey of *A. decempunctata* and *O. conglobata*. She is right mentioning that generally coccids are very unfavourable prey of Coccinellini.

less suitable essential prey or quite a good alternative prey.

An important refinement of the definitions was made by Michaud (2005), when he called attention to special cases where the relation to prey is not identical in predator larvae and adults. Michaud (2000) found, for example, that females of *Hip. convergens* had a good fecundity on *Toxoptera citricidus*, that this appears essential prey for females, while the larvae could not complete development on this prey. Michaud (2005) gave other examples of such **'non-symmetric' relations to essential food** and proposed to add the term **'complete food'** for prey that enables both larval development and female oviposition. Consequently we can recognize, for example, 'larval essential prey' of a coccinellid species, if it is not also recorded as essential for adults. To arrive at a numerical criterion for evaluating prey quality, Michaud (2005) suggests rearing ladybirds on conspecific eggs as a reference diet. However, Sloggett and Lorenz (2008) recorded nutritional differences of conspecific eggs in different species.

5.2.12 Tritrophic studies

The effect of the plant on a carnivorous insect via the herbivore is discussed in several sections of 5.2. The effect of plant **allelochemical substances** (secondary chemicals of Fraenkel 1959, 1969) was most probably the earliest topic studied in this respect. The observations on the toxic effect of allelochemicals from the elder *Sambucus* spp., acting through elder aphids (*Aphis sambuci* and *Aulacorthum magnoliae*) on *C. septempunctata* (Hodek 1956, Blackman 1965, Okamoto 1966) represented the very onset of studies of the wide field of prey specificity in coccinellids. Prey acquiring their toxicity from plants (5.2.6.1) are also included in this chapter.

Plants with a high **nutritional value** may have a positive effect, also transferred through the prey, on ladybirds. Compared with *Acyrthosiphon pisum* reared on *Vicia faba*, *A. pisum* reared on alfalfa stored significantly more fatty acids (particularly six times more myristic acid) which resulted in a 1.17-fold increase in available calories. Feeding on this more

Table 5.27 Survival of *Coccinella septempunctata* larvae, and ratio of female size at 24°C resulting from increasing daily levels of *Acythosiphon pisum* reared on either *M. sativa* or *V. faba* at 22°C (after Giles et al. 2002).

Variable	A. pisum (mg/day)							
	2.2	**2.2**	**4.3**	**4.3**	**8.2**	**8.2**	**16.4**	**16.4**
	*M.s.**	*V.f.†*	*M.s.*	*V.f.*	*M.s.*	*V.f.*	*M.s.*	*V.f.*
Larva	0.333	0.050‡	0.817	0.617	0.897	0.864	0.930	0.867
Female size§	0.167	0	0.158	0.174	0.540	0.325	0.780	0.432
n**	60	60	60	60	58	59	57	60

*Pea aphids reared on *Medicago sativa*.
†Pea aphids reared on *Vicia faba*.
‡Paired underlined values represent significant differences ($p < 0.05$) for $2 \times 2\ \chi^2$ tests between host-plants at each mg level.
§Elliptical body area was measured to estimate adult size (Obrycki et al. 1998).
**Number of *C. septempunctata* larvae at beginning of experiment.

nutritive prey increased larval survival, decreased development time and resulted in larger adults of *C. septempunctata* (Giles et al. 2002; Table 5.27). One of the tritrophic systems even occurs in a land-water ecotone: water lily–*Galerucella nymphaeae*–*Col. maculata* (Schlacher & Cronin 2007).

Another interesting tritrophic system consists of the willow, a leaf beetle (*Plagiodera versicolora*) and the predatory coccinellid *Aiolocaria hexaspilota* (see also 5.4.1.2 for this tritrophic relation). Cutting off willow trees leads to an **increase** in their **leaf nitrogen**. The relative growth rate of the chrysomelid increased when feeding on the nitrogen rich leaves, but nitrogen level did not increase in the body of the leaf beetle. Thus it is not clear why the relative growth rate also increased in the coccinellid larvae preying on the chrysomelid larvae feeding on N-rich leaves (Kagata & Ohgushi 2007; Table 5.28). Tritrophic studies based on pest-resistant plants are discussed in 5.2.6.3.

Elevated air CO₂ concentration significantly increased tannin and gossypol content and decreased N content in cotton. The survival of *Aphis gossypii* increased with higher CO_2 concentrations while their fatty acid content decreased. In contrast, the survival and fecundity of *P. japonica* was not significantly affected, but its development time was longer. As global CO_2 levels rise the authors speculated that *A. gossypii* may become a more serious pest due to its increased survival and the slower development of the ladybird (Gao et al. 2008).

Table 5.28 Qualitative traits of leaves, chrysomelids (*Plagiodera versicolora*) and ladybirds (*Aiolocaria hexaspilota*) on cut and uncut willows (modified after Kagata & Ohgushi 2007).

Willows	Cut	Uncut (control)
Leaves of host plant		
% Water (FM)	63.82 ± 0.51	62.12 ± 0.62
% Carbon (DM)	44.90 ± 0.75	45.23 ± 0.59
% Nitrogen (DM)	2.64 ± 0.11	2.31 ± 0.06
Chrysomelid prey		
Larval mass (mg FM)	5.20 ± 0.33	3.71 ± 0.19
% Water (FM)	76.29 ± 0.57	76.52 ± 0.42
% Carbon (DM)	48.73 ± 1.52	48.65 ± 1.31
% Nitrogen (DM)	8.43 ± 0.20	8.48 ± 0.16
Ladybird larva		
Larval mass (mg FM)	64.70 ± 4.10	50.81 ± 4.54
% Water (FM)	78.80 ± 0.51	78.93 ± 0.58
% Carbon (DM)	45.75 ± 1.96	42.47 ± 1.23
% Nitrogen (DM)	10.68 ± 0.50	9.23 ± 0.28

FM, fresh mass; DM, dry mass; Means ± SE.

Physiological condition (senescence) of the bean plant modified the incidence of adult diapause in coccinellids via the aphid prey (Rolley et al. 1974; Table 6.8, in Chapter 6).

A particular kind of tritrophic phenomenon is the so called 'crying for help' of **plants injured through herbivory**, i.e. the production of attractants for natural enemies of the herbivores (detected for the first time in predatory acari). Such attraction has already been observed in several coccinellid species, attracted for example by methylsalicylate produced at a higher level due to aphid infestation of plants (Zhu & Park 2005, Karban 2007; several cases are discussed in 5.4.1.2).

5.2.13   Food of phytophagous Coccinellidae

Phytophagous ladybirds of the subfamily Epilachninae feed especially (but not exclusively) on host plants of the families Solanaceae and Cucurbitaceae (Schaefer 1983). Some species have been studied intensively in relation to their feeding on economically important host plants. These include the Mexican bean beetle, *Epilachna varivestis*, a pest with a worldwide distribution (e.g. Fujiyama et al. 1998, Abe et al. 2000, Shirai & Yara 2001), and *H. vigintioctomaculata* (Fujiyama & Katakura 2002). Interestingly, a pest of cucurbits, *Henosepilachna elaterii* (=*Epilachna chrysomelina*) has been reported as also feeding on aphids (El Khidir 1969).

Closely related, **co-occurring species of Henosepilachna (=Epilachna)** differ in their abilities to feed on particular host plant species. In Japan, *H. vigintioctomaculata* preferred potato plants, and failed to develop on eggplant, whereas *H. vigintioctopunctata* fed and developed on both potatoes and eggplant (Iwao 1954). Similarly, among three closely related species of the *H. vigintioctomaculata* complex, the wild herb *Solanum japonense* was most suitable for development of *H. pustulosa*, less so for *H. niponica*, and least for *H. yasutomii* (Fujiyama & Katakura 2002).

The usual food of *H. pustulosa* is thistle (*Cirsium* spp.), but Iwao and Machida (1961) found that a varying proportion of individuals also accept potato plants, with a positive correlation between a female's acceptance of potato and her offspring's acceptance of this host plant. As a result of either pre-imaginal conditioning or negative selection (with high mortality on potatoes), a greater percentage of individuals that fed as larvae on potato versus thistles (74% vs. 31%) subsequently accepted potato as adults. Furthermore, individuals could also be conditioned during the first days of adult life to prefer either host plant thereafter (Iwao & Machida 1961). By contrast, Shirai and Morimoto (1999) found no difference between populations associated with potato and wild thistle in their abilities

to feed on potato. Koji et al. (2004) compared feeding responses of two Japanese *H. niponica* populations, separated by 150 km and associated with different thistle species, to several geographically restricted populations of thistles. Although the two (Asiu and Yuwaku) populations were similar in many of their responses to the host plants (i.e. overall they shared a conserved hierarchy of feeding preferences and growth performance), they differed strikingly in their response to a host plant used naturally as a secondary food source by only the Yuwaku population. Asiu adults thus avoided *Cirsium kagamontanum*, and their offspring failed to develop on this host plant. Yuwaku adults, in contrast, preferred this thistle, and approximately 10% of their offspring developed on this host. Similarly, Fujiyama et al. (2005) found that adult preference and larval performance on an unusual host plant, the deciduous tree *Pterostyrax hispidus*, was higher for a population of *Henosepilachna yasutomii* that occurs on this tree than for other populations of this ladybird species naturally associated with other host plants (see also Kikuta et al. 2010, Kuwajima et al. 2010).

The **genetic basis** that may underlie such patterns of host use in *Epilachna* spp. has received considerable attention. The major host plants of *E. vigintioctopunctata* in Southeast Asia are solanaceous species (Shirai & Katakura 1999). Shirai and Katakura (2000) found natural populations of this ladybird using the legume *Centrosema pubescens* in Malaysia and Indonesia. When these legume-feeding strains were maintained for a number of generations on *Solanum nigrum*, however, they rapidly lost their ability to use the legume (Shirai & Katakura 2000; Fig 5.26). Interestingly, the authors were unable to select conversely for a shift in host use by rearing *Solanum*-feeding ladybird strains on the legume. Ueno et al. (1999) studied siblings of *H. vigintioctomaculata* collected from *Solanum tuberosum* on Honshu or from the novel host *Schizopepon bryoniaefolius* on Hokkaido. Since Ueno et al. (1999) did not find negative genetic correlations in established and novel host plant use, they suggested that the coccinellid expansion onto *S. bryoniaefolius* on Honshu is prevented by ecological and behavioural factors, rather than genetic constraints. A subsequent study (Ueno et al. 2001), using beetles associated with *S. bryoniaefolius* on Hokkaido, found lower heritabilities for growth performance when the beetles were reared on *S. bryoniaefolius* rather than on *S.*

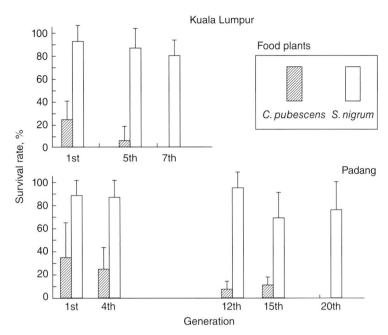

Figure 5.26 Survival rates (emergence rates from pupae) in successive laboratory-reared generations of *Henosepilachna vigintioctopunctata*, for two populations (top, from Kuala Lumpur; bottom, Padang in Indonesia) found feeding in the field on the legume *Centrosema pubescens*. Laboratory colonies were maintained on *Solanum nigrum*, with some individuals at each generation transferred as eggs to *C. pubescens* (from Shirai & Katakura 2000).

tuberosum, suggesting local adaptation in host plant use.

Intraspecific variation among plant genotypes of a given host species has been studied as well, for its influence on ladybird feeding and performance. In laboratory experiments, Fujiyama and Katakura (1997) examined responses of a population of *H. pustulosa* from Hokkaido to different clones of two host plants, thistle (*Cirsium kamtschaticum*) and blue cohosh (*Caulophyllum robustum*). Feeding rates of adult beetles differed significantly among clones for both of the host species, as did larval performance (measured by eclosion rate, duration of development and resultant adult size, when reared on a particular plant clone). When two thistle clones were transplanted to a common garden, the difference in their suitability for larvae narrowed, thus demonstrating also an environmental influence (Fujiyama & Katakura 2001). As further evidence of major environmental influence, eclosion rates increased significantly when larvae fed on shoots exposed to increasing light intensity, but adult female feeding preference was not affected. This suggested that females might respond primarily to host genotype, even when environmental conditions affect the suitability of that genotype for their offspring (Fujiyama & Katakura 2001). Examination of egg laying among clones by females in the field, however, affirmed the importance in host selection of environmental factors (clone size, soil moisture, and exposure to sunlight). As the growing season proceeded, females changed the distribution of eggs among clones from aggregated to more nearly random (Fujiyama et al. 2003).

Chemical and physical attributes of plants influencing host selection by phytophagous ladybirds have been explored in detail, especially among pest species (e.g. Napal et al. 2010). From analyses of potato leaf extracts, Endo et al. (2004) found that **methyl linolenate** acts synergistically with sugars (glucose and fructose) to stimulate feeding behaviour in *E. vigintioctomaculata*. **Luteolin 7-0-glucoside** in *Physalis alkekengi* stimulates *H. vigintioctopunctata* to feed on this host plant (Hori et al. 2005). Plant defences have been examined especially in plant resistance studies for the Mexican bean beetle, *E. varivestis* (Hammond & Coope 1999). For example, larval feeding on isolines of soybean with **trichomes** (versus a glabrous variety) resulted in increased mortality and reduced pupal mass (Gannon & Bach 1996). The 'Davis' soybean cultivar releases volatiles that attract females of *E. varivestis*, but this 'death-trap' cultivar is an unsuitable host: females that feed on it lay few eggs, and their offspring die by the third instar (Burden & Norris 1994). The importance of environmental factors on resistance expression was highlighted by Jenkins et al. (1997), who reported that resistance among varieties of soybean strengthened with decreasing soil moisture, since Mexican bean beetle larvae took longer to develop and suffered greater mortality. Induced resistance to the Mexican bean beetle was found when larvae fed on soybean plants that had been subjected to previous beetle feeding damage, or had been treated with jasmonic acid (Iverson et al. 2001). Underwood et al. (2000) found substantial variation among soybean genotypes in their constitutive and induced antifeedant resistance to *E. varivestis*, but the levels of these two forms of resistance were uncorrelated among varieties.

Non-pest as well as pest populations of *Epilachna* can reach high numbers and defoliate their host plants, with consequences for the number of beetles in future generations (e.g. Koji & Nakamura 2002). Food relationships therefore play a central role in the population dynamics of herbivorous ladybirds, especially among introduced populations that are subject to reduced pressure from natural enemies and that reach especially high abundance (Ohgushi & Sawada 1998). At such high abundance, there is strong intraspecific competition among larvae as they defoliate their host plants, leading to reduced overwintering survival with decreasing adult size (Ohgushi & Sawada 1995, Ohgushi 1996). Food relationships can also underlie major changes in population dynamics associated with the shift of *Epilachna* species from wild hosts to potatoes. Shirai and Morimoto (1997, 1999) documented key life history changes, including larger adult body size, higher fecundity, faster larval development and reduced life span, associated with such host shifts.

In contrast to the Epilachninae, which feed on vegetative parts of plants, the isolated **genus *Bulaea*** displays a more specialized form of phytophagy. Capra

(1947) reported that *B. lichatschovi* and its close relatives were 'predominantly if not exclusively' pollinivorous both as larvae and adults, with a preference for Chenopodiaceae. Others (Dyadechko 1954, Bielawski 1959, Savoiskaya 1966,1970b) also consider *B. lichatschovi* as phytophagous. Savoiskaya (1966, 1970b) reported this species as pollinivorous on a variety of plants (*Euphorbia*, *Artemisia*, *Eurotia*, *Atriplex*, *Nitraria*, *Tamarix* and *Clematis*) in Kazakhstan. Savoiskaya (1970) further reported that, in central Asia, this species fed on pollen (of *Tamarix*, *Euphorbia*, *Artemisia*, *Eurotia ceratoides* and *Atriplex*), nectar (*Nitraria*, *Clematis*) and leaves of sugar beet and young apple trees. Goidanich (1943) reported pollinivory in another ladybird genus, *Micraspis*. (See also 5.2.9.)

5.2.14   Food of mycophagous Coccinellidae

Although Coccinellidae are taxonomically related to mostly mycophagous coleopteran families, ladybird mycophagy (together with phytophagy, aphidophagy, etc.) appears to be derived from an ancestral coccidophagous feeding behaviour (Giorgi et al. 2009). Sutherland and Parrella (2009) rightly divide mycophagy of coccinellids into facultative (5.2.9) and obligatory. The obligate mycophagous species of the genera ***Psyllobora* (=*Thea*), *Vibidia*, *Illeis*** and ***Halyzia*** belong to the tribe Halyziini, now Psylloborini (Vandenberg 2002). They feed on powdery mildew (Erisyphaceae). Species of these genera have often been mistakenly considered as carnivorous (see Hodek 1996). Their mycophagy is reflected in the distinctive shapes of their mandibles (Kovář 1996). Davidson (1921) checked in the laboratory that these fungi are their essential food: both larvae and adults of *Psyllobora vigintimaculata* died when offered arthropod food such as aphids, coccids and spider mites.

Strouhal (1926a, b) discussed the history of ecological and morphological differentiation among members of the tribe Halyziini and its consequences for taxonomy. Strouhal also compared adult and larval mouthparts among species in detail, and reviewed observations on mycophagy. Development of the mycophagous ***Illeis koebelei*** from egg to adult on drone powder (5.2.9) was possible, but the adults were small and did not oviposit (Niijima 1979). Fig. 5.27 shows the food relations of this species (Takeuchi et al. 2000).

Tytthaspis sedecimpunctata (tribe Coccinellini) has usually been reported as aphidophagous

Powdery mildew (host plant)	Apr May Jun Jul Aug Sep Oct Nov
Sphaerotheca pannosa	——— A
(*Rosa multiflora*)	— · — · – A,L,P
Oidium sp.	——————— A,E,L,P
(*Pyracantha coccinea*)	— · A,E,L,P
Microsphaera pulchra var. *pulchra*	———————— A,E,L,P
(*Benthamidia florida*)	— · — · — · A,E,L,P
Sphaerotheca fusca	— A
(*Cosmos* sp.)	
Podosphaera tridactyla var. *tridactyla*	A,E,L,P —
(*Prunus sp.*)	— · — · — · — · A,E,L,P
Phyllactinia moricola	A,E,L,P ———————
(*Morus australis*)	A,E,L,P — · — · — · — · — · — · —
Sphaerotheca cucurbitae	A,E,L,P ————————
(*Trichosanthes kirilowii* var. *japonica*)	A,E,L,P — · — · — · —

Figure 5.27 Seasonal occurrence of *Illeis koebelei* on different powdery mildews. Horizontal bars, observation period. A, adult; E, egg; L, larva; P, pupa. Full line, 1997; broken line, 1998 (modified from Takeuchi et al. 2000).

(Dyadechko 1954, Semyanov 1965, Klausnitzer 1966). Based on observations and preliminary experiments, Turian (1969) published the first precise information on its food, demonstrating that this species feeds on Erysiphaceae. Adults of *T. sedecimpunctata* seemingly show no specificity in ingesting various species of powdery mildew. Turian (1969) also observed *Psyllobora* (=*Thea*) *vigintiduopunctata* larvae and adults feeding on various Erysiphaceae, and proposed the term 'micromycetophagy' to describe feeding on lower fungi. For additional studies of ladybirds feeding on mildew (along with pollen and aphids) see 5.2.5.

5.3 QUANTITATIVE ASPECTS OF FOOD RELATIONS

The number or biomass of prey consumed by larval and adult ladybirds varies widely depending on specific circumstances (e.g. species of prey, rearing temperature). To complete their larval development, individuals may consume from 100 to well over 1000 aphids of various species; most food intake (typically 60–80%) occurs during the fourth instar (see Hodek 1996, and for coccinellids preying on scale insects part 2.3 in Hodek & Honěk 2009). Adult females, especially when ovipositing, feed more than males.

Methodological differences probably account for some of the variation in reported consumption rates of ladybirds. Baseline consumption may be estimated by comparing prey mortality in cages with and without predators (e.g. Hodek 1956), and by distinguishing predation mortality from other causes (e.g. Kaddou 1960). In some cases, the weight of prey consumed has been recorded (e.g. Blackman 1967, Ives 1981b, Ferran et al. 1984a, b); however, care must then be taken to consider changes in prey weight by dehydration. Isotope labelling (Ferran et al. 1981) and indirect measurement by faecal production (Honěk 1986) are useful additional methods. It is difficult to relate consumption rate to faecal production rate quantitatively without detailed information on digestion. The weight of faeces produced by *C. septempunctata* individuals following collection, however, provides an informative index of consumption rates of ladybirds in the field (Honěk 1986). Consumption rates can also be estimated from field observations (e.g. Latham & Mills 2009), or by assessing the hunger level of field-collected individuals. 'Hunger curves' were drawn from aphid consumption rates of *C. trifasciata* and *C. californica* after females or males were starved for 0–50 hours, and then field hunger levels were estimated by measuring their rates of aphid consumption in the laboratory (Frazer & Gilbert 1976, Frazer & Gill 1981; Fig. 5.28).

New molecular techniques, such as those using species-specific DNA sequences, can identify prey eaten and give quantitative estimates (e.g. Hoogendoorn & Heimpel 2001). Harwood and Obrycki (2005) have reviewed the diverse methods used with aphidophagous predators, which also include gut dissection, stable isotopes, protein analyses (by electrophoresis and use of antibodies) and chromatography to detect prey pigments (5.1; Chapter 10).

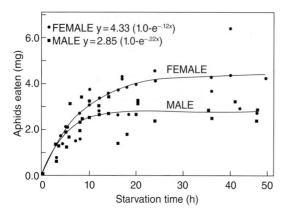

Figure 5.28 Hunger curves of male and female *Coccinella californica* (from Frazer & Gill 1981).

5.3.1 Effect of physical factors on consumption

5.3.1.1 Daily consumption rate

At temperatures favourable for growth and survival, the consumption rates of both larvae and adults generally increase with increasing temperature, and are correlated with the increase in developmental rate (Hodek 1996). Adults of *Delphastus catalinae* consumed increasing numbers per day of eggs and nymphs of the sweet potato whitefly *Bemisia tabaci* as ambient temperature increased from 14 to 35°C (Simmons & Legaspi 2004). Similar results were reported for larvae and adults of *Hip. convergens* attacking eggs of the cotton bollworm *Helicoverpa zea* (Parajulee et al. 2006) and for larvae of *Hip. convergens* and *C. septempunctata* consuming adult *Myzus persicae nicotianae* from tobacco (Katsarou et al. 2005). The number of prey (first and second instars of *Leptinotarsa decemlineata*) killed per day by larvae and adults of *Col. maculata* increased linearly with temperature (10–30°C) (Giroux et al. 1995).

At excessively high temperatures, ladybird foraging, development and survival are adversely affected (e.g. Alikhan & Yousef 1986, Huang et al. 2008, Taghizadeh et al. 2008). Consumption rates of fourth instars and adults of *Har. axyridis* were reduced at high temperature over the range of 10–30°C, especially for the darker morph *nigra*, and the thermal optimum (near 25°C) was 3.7°C lower for adults (but not

larvae) of this morph (Soares et al. 2002, Fig. 5.29). Overall, the larger, darker (melanic) adults of the *nigra* morph appeared better adapted for cold regions, where this morph prevails in frequency over the *aulica* morph.

At temperatures below the development threshold, coccinellids cease to consume prey (e.g. at 7.5°C for *Col. maculata*, Giroux et al. 1995). Ricci et al. (2006) studied the foraging behaviour of adult *Chil. kuwanae* preying on overwintering females of the scale insect *Unaspis euonymi* on *Euonymus japonicus*. The ambient temperature was gradually reduced from 15 to 2°C (exposing the beetles to each, successively lower, temperature for 10 days), before the final increase to 15°C. With the lowering of temperature, the adults consumed scales at a gradually reduced rate and at 4 and 2°C they consumed almost no prey but continued, at reduced rates, to lift scale covers without consuming the scales; such scales then died.

5.3.1.2   Total food consumption

Hodek (1996, pp. 204–205) concluded from a review of the literature that, in general, the total food consumption during the entire larval development tends to remain relatively constant over a wide range of **constant temperatures**. Coccinellids appear similar to other insects in this respect (see Rubner 1908, cited in Allee et al. 1949). Katsarou et al. (2005) reported highest larval prey consumption both in *Hip. convergens* and *C. septempunctata* at 23°C versus at 14, 17 and 20°C. In *A. bipunctata* total larval food intake (mg wet weight) declined from 65 mg at 15°C to 52 mg at 25°C when larvae received an excess of *Sitobion avenae*. When the larvae were provided with a limited number of aphids, however, the intake was highest at intermediate temperature (28 mg at 20°C) (Schuder et al. 2004).

The consumption patterns of coccinellids have mostly been studied at constant temperatures. Hodek (1957a), however, found that total larval food consumption was doubled when larvae of *C. septempunctata* developed at **naturally fluctuating** summer **temperatures**. Sundby (1966) found no difference in the amount of food consumed by larvae of *C. septempunctata* under more limited alternation between only 16 and 21°C (the control was constant 16 or 21°C), but such alternating conditions nonetheless resulted in heavier pupae (R.A. Sundby unpublished,

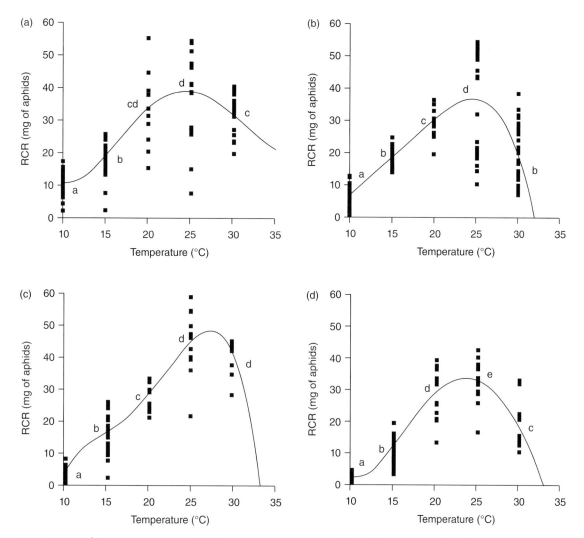

Figure 5.29 Relative consumption rate (RCR), at varying temperatures, of larval *Harmonia axyridis aulica* (A) and *nigra* (B), and adult *aulica* (C) and *nigra* (D) (after Soares et al. 2002).

in Ellingsen 1969). Temperatures fluctuating between 8 and 28°C resulted in an increase (10.6% more than at constant 18°C) in total food consumption by larvae of *A. bipunctata* (Ellingsen 1969). An important increase in food intake at alternating temperatures was also found in females of *Chil. nigritus* (Ponsonby & Copland 2000; Table 5.29).

Although little studied, **humidity** probably affects food consumption by ladybirds. Coccinellids may com-

pensate for high evaporative water loss by consuming more prey with a high **water content**; Hodek et al. (1965) found indications of increased consumption of aphids in *C. septempuncata* at low relative humidities in the air. Foods with a low water content (e.g. pollen) may become unsuitable for ladybirds when free water is not available and humidity is low. There was no correlation, however, between initial water content of different foods and survivorship of larvae of *Col. maculata*,

Table 5.29 Significantly higher food intake at alternating temperatures: effect of temperature on number and weight of pre-ovipositing adult female *Abgrallaspis cyanophylli* eaten in a 24-hour period by adult *Chilocorus nigritus* 2 to 6 weeks after eclosion (modified after Ponsonby & Copland 2000).

Temperature (°C)	Predator	Mean number eaten (±SD)	Estimated weight eaten (mg)
15	Female	1.94 (2.89) a	0.394
	Male	1.28 (1.45) a	0.259
20	Female	3.94 (1.83) b	0.801
	Male	4.00 (1.45) b	0.812
26	Female	8.83 (4.11) c	1.793
	Male	5.17 (2.28) bd	1.049
30	Female	7.17 (3.31) c	1.455
	Male	6.94 (3.44) cd	1.409
14/30 (12h/12h)	Female	12.33 (4.92) e	2.504
	Male	7.22 (2.49) c	1.466

Means with the same letter within the column are not significantly different (LSD at 5% level = 1.9616). *n* = 18 males and 18 females at each temperature level.

when reared under simulated drought conditions (Michaud & Grant 2005).

5.3.2 Effect of prey density on consumption: functional response

Pioneering ecologists distinguished key aspects of predation as it may influence prey population dynamics. Solomon (1949, 1964) defined the concept of a predator's **functional response** as the rate of prey consumption (i.e. number of prey consumed per unit time) as a function of prey density. Solomon also defined the **numerical response**, i.e. the change in numbers of predators occurring (through immigration or aggregation and reproduction) in response to a prey density. A crucially important development was Holling's (1959a, b, 1965) proposal of three basic types of functional response which reflect that the number of prey consumed by a predator as prey density increases may increase linearly (type I), in a decelerating fashion (type II) or in a sigmoidal fashion (type III).

In recent years, functional responses have been viewed broadly as the rate of prey consumption of an individual predator (Abrams & Ginzburg 2000) and often studied in coccinellids. The functional response can then be modelled as it varies not only with prey density, but also with other interacting factors (which may influence predator foraging) such as predator density (including in combination with dependence on

the prey density to yield a ratio; e.g. Arditi & Ginzburg 1989, Mills & Lacan 2004, Schenk et al. 2005), densities of alternate prey (e.g. Tschanz et al. 2007), densities of the predator's natural enemies (e.g. Krivan & Sirot 2004), foraging substrate and abiotic conditions (e.g. temperature; Jalali et al. 2010, Khan & Khan 2010).

Most commonly ladybirds are studied foraging singly in simple laboratory arenas, or less commonly on caged host plants in the field, to estimate the number of prey consumed in response to experimentally varied densities of available prey. From these data, the functional response is then characterized to type.

The most common form for ladybirds is the type II functional response (Table 5.30). But simple extrapolation of laboratory estimates to field settings is questionable (e.g. O'Neil 1989, 1997). For example, *Col. maculata* larvae ate Colorado potato beetle eggs on caged potato plants in the field at only about half the rate as they did on excised potato leaves in the laboratory; in both settings, however, a type II response was observed (Munyaneza & Obrycki 1997).

In some instances, ladybirds have shown type III responses (e.g. Sarmento et al. 2007) that may arise from predator learning (e.g. as in **prey-switching**; 5.2.5), and such a response has the potential to stabilize predator–prey interactions (Hassell 1978). In a laboratory experiment, Murdoch and Marks (1973) tested for prey-switching in *C. septempunctata* when presented with varying ratios of two prey

Table 5.30 Examples of experimental studies of functional responses, illustrating both the settings and foraging substrates, and the prey attacked by the predators. Results are characterized by basic response type, as concluded by the authors. 'Petri dish' or 'Cage' indicates an absence of plant material.

Coccinellid species	Prey species	Setting	Response type	Reference
Adalia bipunctata	*Myzus persicae*	Petri dish	II	Jalali et al. (2010)
Aphidecta obliterata (adults)	*Elatobium abietinum*	Petri dish (lab)	II	Day et al. (2006)
Aphidecta obliterata (4th instars and adults)	*Elatobium abietinum*	Spruce section	II	Timms et al. (2008)
Coccinella septempunctata (adults)	*Macrosiphum euphorbiae*	Rose leaves (lab)	II	Deligeorgidis et al. (2005)
Coccinella undecimpunctata (4th instars and adults)	*Myzus persicae*	Cage (lab)	II	Cabral et al. (2009)
Coleomegilla maculata (4th instars)	*Leptinotarsa decemlineata* eggs	Petri dish (lab) Potato plants (greenhouse, field)	II	Munyaneza and Obrycki (1997)
Delphastus pusillus (female)	*Bemisia tabaci* nymphs	Cotton leaves (lab)	II	Guershon and Gerling (1999)
Eriopis connexa (female)	*Macrosiphum euphorbiae*	Cage (lab)	III	Sarmento et al. (2007a)
	Tetranychus evansi	Cage (lab)	II	
Harmonia axyridis (larvae and adults)	*Aphis gossypii*	Cucumber leaves (lab)	II	Lee and Kang (2004)
Harmonia axyridis (larvae and adults)	*Danaus plexippus* eggs and larvae	Petri dish (lab) larvae eating eggs and larvae, adults eating eggs	II I	Koch et al. (2003)
Harmonia conformis (adults and larvae)	*Eriosoma lanigerum*	Petri dish (lab)	II	Asante (1995)
Harmonia dimidiata (females)	*Cervaphis quercus*	Cage (lab)	II	Agarwala et al. (2009)
Orcus australasiae (adults)	*Eriosoma lanigerum*	Petri dish (lab)	II	Asante (1995)
Propylea dissecta (larvae and adults)	*Aphis gossypii*	Leaves of *Lagenaria vulgaris* (lab)	II	Omkar and Pervez (2004)
Propylea quatuordecimpunctata (adults)	*Diuraphis noxia*	Petri dish (lab)	II	Messina and Hanks (1998)
		Indian rice grass (lab)	II	
		Crested wheatgrass (lab)	III	
Scymnus creperus (larvae and adults)	*Aphis gossypii*	Cotton leaves (lab)	II	Wells et al. (2001)
Stethorus tridens (females)	*Tetranychus evansi*	*Solanum* leaves	II	Britto et al. (2009)

(*Acyrthosiphon pisum* and *Aphis fabae*) which occur in different microhabitats. That these authors failed to find evidence for prey-switching re-inforced the consensus that the functional responses of ladybirds are most likely to be type II.

The behavioral mechanisms underlying a given functional response are sometimes examined explicitly, but more often are simply inferred by extracting estimates from a standard model applied to experimental data. Type II responses are typically fitted (most often following polynomial regression) using Holling's (1966) **disc equation**, or Rogers' (1972) variant for the disc equation when consumed prey are not replaced (e.g. see Parajulee et al. 2006). Attack coefficients and handling time are then estimated as assumed constants across the range of prey densities studied. However, the mechanistic basis of a type II (or III) functional response is often far from clear (e.g. Holling 1966, Mills 1982a, van Rijn et al. 2005), despite widespread use of such assumed constants. Models incorporating these assumptions therefore may serve better for simulating population dynamics than for understanding the foraging behaviour of predators that in fact underlies these dynamics.

Foraging ladybirds in fact may **spend less time handling and processing** individual prey as prey density increases, either because the predator processes prey more quickly or because it consumes less of each prey. Particularly in the latter case, this can increase the 'killing power' of the predator. The handling time of adults of the mite-feeding *Stethorus bifidus*, for example, decreases at increased densities of *Tetranychus linearius* as the predator extracts less of the killed prey contents (Peterson et al. 2000; Fig. 5.30). Because this promotes an increase in the proportion of the prey killed as prey density increases, the predator may contribute to the regulation of prey numbers.

The nature of the host plant as a **substrate for foraging** may interact with prey density to influence a ladybird's functional response (Guershon & Gerling 1999, Siddiqui et al. 1999; 5.4.1.1). Larvae of *Hip. convergens* foraging for *Acyrthosiphon pisum* on normal or reduced-wax bloom pea varieties had different type II responses as the instantaneous search rate was higher on reduced-wax bloom plants (Rutledge & Eigenbrode 2003). Most probably the predator was better able to attach to plants with reduced-wax bloom, and its search rate was elevated. Handling time of prey,

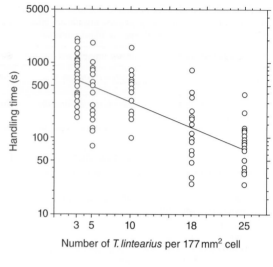

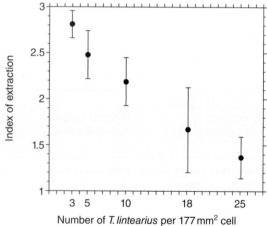

Figure 5.30 The handling time (top) and (bottom) the index of extraction (reflecting the proportion of body contents removed from an individual prey) of *Stethorus bifidus* adults foraging at different densities of the mite *Tetranychus linearius* (from Peterson et al. 2000).

however, did not differ between pea plants with reduced wax and normal wax bloom.

The foraging substrate can even shift the basic nature of the functional response. Messina and Hanks (1998) found that, in attacking the Russian wheat aphid *Diuraphis noxia*, *P. quatuordecimpunctata* followed a type II response on Indian ricegrass (*Oryzopsis hymenoides*), but type III on crested wheatgrass (*Agropyron desertorum*). As prey density increased on the wheatgrass, a decreasing proportion of the aphids apparently occurred in locations (e.g. within rolled leaves) where they escaped the predators.

Through their effects on prey quality, **host plants** may additionally influence the functional responses of predators. In Petri dish pairings of larvae of *C. septempunctata* with varying densities of aphid nymphs of *Lipaphis pseudobrassicae*, Kumar et al. (1999) recorded differences in the rate with which percentage prey consumption decreased with an increase in prey density for prey reared on *Brassica campestris*, *B. oleracea* or *Rhaphanus sativus*. For each of these three prey, however, the predator's functional response was type II.

The **activity of other predators** may also influence a predator's functional response, as formalized in the concept of **mutual interference** (Hassell 1978). The searching efficiency of *C. septempunctata* females decreased linearly with an increase in density of conspecifics (Siddiqui et al. 1999). A similar effect was recorded for *P. dissecta* (Omkar & Pervez 2004).

The availability of **co-occurring prey** may further influence a predator's functional response through **prey selection**. Typically, the predator consumes fewer alternative prey when an essential prey is also available (5.2.7). Hazzard and Ferro (1991) found in the laboratory, and Mallampalli et al. (2005) in field cages, that *Col. maculata* consumed fewer Colorado potato beetle eggs (its alternative prey; Snyder & Clevenger 2004) when essential prey, *Myzus persicae* was present. *Coleomegilla maculata* and *Har. axyridis* attacked fewer eggs of *Ostrinia nubilalis* when *Rhopalosiphum maidis* were present (Musser & Shelton 2003).

Additional examples are provided by Lucas et al. (2004) and Koch et al. (2005), as discussed further by Evans (2008).

Prey activity may influence the relative amounts of co-occurring prey consumed. *Har. axyridis* selected *Tetranychus urticae* and not the cicadellid *Hyaliodes vitripennis* that could defend itself, but selected the latter as the better prey when it had been immobilized (Provost et al. 2006).

5.3.3 Effects of consumption on growth and reproduction

5.3.3.1 Larval development

Predaceous ladybirds often have to cope with prey scarcity, including its intermittent absence. In particular, aphidophagous species have a pronounced ability to adjust to this. Larval development can be completed even when rates of prey consumption are very low; larvae of *C. septempunctata* allowed to consume only 33–40% of the number of aphids consumed when provided in excess still completed their development (Hodek 1957, Sundby 1966). Nonetheless, **reduced rates of prey consumption** result in slower development and greater mortality (both of larvae and adults) (Phoofolo et al. 2008; Table 5.31) as well as lower weights and smaller sizes of pupae and adults (for earlier papers see Hodek 1996). The number of ovarioles may also vary among adult ladybird females as a function of larval food consumption (Rhamhalinghan 1985, Dixon & Guo 1993).

Variable **prey quality** similarly affects larval growth and development (e.g. Smith 1965a, b, Obrycki & Orr

Table 5.31 Effect of food deprivation in the fourth instars on the completion of development to adults in three ladybird species (modified after Phoofolo et al. 2008).

Food deprivation period (days)	*Coleomegilla maculata*		*Harmonia axyridis*		*Hippodamia convergens*	
	n	% female	*n*	% female	*n*	% female
0	20	50.0	15	71.4	10	50.0
1	19	42.1	16	50.0	12	50.0
2	20	55.0	16	62.5	12	50.0
3	11	18.2	10	10.0	13	30.8

Number of consecutive days of the fourth instar during which lady beetle larvae were deprived of food until they either pupated or died.

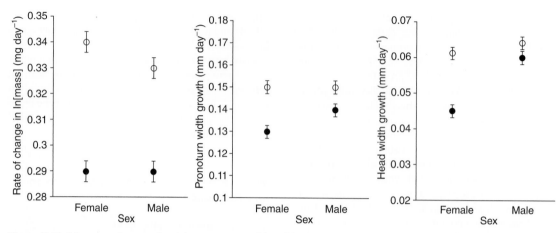

Figure 5.31 Mean growth rates of weight, pronotum width and head width of *Harmonia axyridis* larvae provisioned with high food availability, following an initial six-day period (upon egg hatch) of low (open circles) or high (filled circles) food availability. Growth rates were based on measurements made at the end of the initial period and at eclosion (from Dmitriew & Rowe 2007).

1990, Giles et al. 2002, Omkar & James 2004; 5.2). The quality and quantity of prey consumed may have interactive effects. Phoofolo et al. (2007) reared larvae of *Hip. convergens* on *ad libitum* versus limited (4 mg per day) quantities of *Schizaphis graminum* and *Rhopalosiphum padi*, both aphids from winter wheat. The larvae grew at the same rates on *ad libitum* diets of the two aphids, but the development of fourth instars was slowed significantly more on limited diets of *R. padi* than of *S. graminum*.

Different species of ladybirds differ in their developmental responses to increasing availability of a given prey species (e.g., Phoofolo et al. 2008, 2009). When provided variable quantities of *Acyrthosiphon pisum*, larvae of *Anatis mali* were more flexible than larvae of *Col. maculata* in **adjusting their rates of development** and their adult weight. Thus, with reduced prey consumption, larvae of *A. mali*, but not of *Col. maculata*, were able to convert a greater percentage of food consumed into body weight (Smith 1965b).

Smith's (1965a) study illustrates the considerable flexibility in developmental responses that ladybirds show to varying prey quantity and quality. Furthermore, as shown by a split-brood full-sib experiment with families of *Har. axyridis* larvae that were provided with aphids or an artificial diet, such phenotypic plasticity in developmental time, adult size, and other characters, can have a strong underlying genetic basis (Grill et al. 1997), and thus quite plausibly can be

shaped by natural selection. Dmitriew and Rowe (2007; Fig. 5.31) investigated the lifetime consequences for *Har. axyridis*, and the **compensatory ability** of older larvae, in response to temporary food shortage early in development. First and second instars received abundant prey (*A. pisum*) but only on alternate days. As third instars, they were fed continuously. Then following diet restriction, the larvae grew more slowly and achieved smaller weights and sizes than did continuously fed larvae. But when ample food was restored, the diet-restricted individuals fully compensated for lost growth, and achieved the same weight and size at adult eclosion as the control. Compensation arose both through extended time until pupation and accelerated larval growth rates. Thus, temporary food shortage and compensatory larval responses thereafter resulted in a small lifetime fitness cost in terms of a prolonged larval period, greater larval mortality and more rapid death when food was removed 100 days after experimental subjects reached adulthood. Longevity, lifetime female fecundity and male mating success of adults that had been subjected to diet restriction as larvae were unaffected.

5.3.3.2   Adult performance

The foods consumed by adult lady beetles can vary greatly in quality (5.2) as reflected both in how much they promote longevity (e.g. Omkar & James 2004), and in how well they support **reproduction** (e.g. Omkar & James 2004, Michaud & Qureshi 2005). Effects of the foods depend on the physical conditions (e.g. temperature or humidity; e.g. Michaud & Grant 2004) and on their consumption as either sole diet or as part of a mixed diet (e.g. Evans et al. 1999, Soares et al. 2004; 5.2.5). Furthermore, the suitability of a particular food for reproduction versus larval development may be quite different (Michaud 2005), and its suitability for reproduction may vary, depending on the diet experienced previously during the larval stages (Michaud & Jyoti 2008).

Egg laying depends on the quantity of consumed food of a given type (review of earlier evidence in Hodek 1996). Ives (1981a, b) found that females of *C. trifasciata* and *C. californica* did not produce eggs at very low rates of aphid consumption. Above a threshold (presumably set by the need for self-maintenance), egg production increased linearly with increasing aphid consumption (see also Ibrahim 1955 and Ferran et al. 1984b). Decreased egg laying and increased incidence of oosorption was studied in relation to temporary food restriction in *Harmonia axyridis* (Osawa 2005, see also Kajita & Evans 2009; Fig. 5.32).

When self-maintenance needs in *C. septempunctata* were met by alternative prey (weevil larvae), the females produced more eggs with lower aphid consumption than in the absence of the alternative prey (Evans et al. 2004). Such a complementary pattern of egg production at very low levels of essential food may enable mobile, widely searching ladybird females to oviposit quickly upon discovery of local patches with favourable prey conditions for their offspring (Evans 2003).

5.3.4 Conversion and utilization of consumed food

Larvae of *Har. axyridis* compensated for earlier food shortages by accelerating their growth rates when given the same amount of food as well fed controls throughout larval development (Dmitriew & Rowe 2007; 5.3.3.1). Thus it appears that the food conversion rate, or ECI (efficiency of conversion of ingested material (Waldbauer 1968); i.e. percentage of consumed prey biomass converted into predator biomass) varied with food conditions, so as to reduce the adverse effects of food shortage (see also Smith 1965). Food conversion rates of *A. bipunctata* larvae consuming

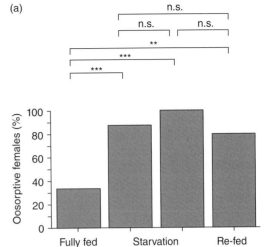

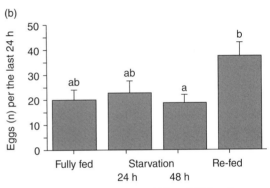

Figure 5.32 Effect of temporary (24- and 48-hour) starvation and resumed 24-hour feeding (after 24-hour starvation) on (a) incidence of oosorptive individuals, and (b) egg laying; for (a), statistical differences at $P = 0.01$, $P = 0.001$, $P = 0.0001$ and $P > 0.05$ in a χ^2-test on the number of individuals indicated by horizontal bars with signs above; for (b) vertical lines indicate SE, and different letters indicate statistical differences at $P = 0.05$ using Scheffe's range test (modified from Osawa 2005).

Sitobion avenae increased to 34–40% from 22–29% (wet weight) when the predators received a reduced number of aphids, only 25% of what larvae ate when they were offered aphids *ad libitum* (Schuder et al. 2004). Similar increased rates of ECI at low levels of food consumption were found in aphidophagous mirids and chrysopids (Glen 1973, Zheng et al. 1993). Aphidophages thus apparently feed with less than full efficiency of conversion at high prey levels, although the fitness benefits of so doing are not yet clear (Dmitriew & Rowe 2007).

Food conversion rates of ladybirds can also vary for a given prey depending on the age of the predator, the morph of the prey, and the host plant on which the prey feeds. Fourth instars of *Scymnus levaillanti* and *Cycloneda sanguinea* that fed on *Aphis gossypii* had lower ECIs (approximately 50% and 20%, respectively, of dry weight) at several temperatures than did earlier instars. This may reflect the metabolic costs of preparation for pupation (Isikber & Copland 2001). Wipperfuerth et al. (1987) reported higher egg production by *Hip. convergens* feeding on apterous virginoparous nymphs of *Myzus persicae* than on the same weight of alatiform gynoparous ones. Shannag and Obeidat (2006) found a higher ECI (19% dry weight basis) throughout the larval period of *C. septempunctata* when feeding on *Aphis fabae* from susceptible *Vicia faba*, when compared with an ECI of 14% on aphids from a resistant cultivar.

ECIs differ between ladybird species in their **nature of feeding** and **type of prey**. The higher ECIs of larvae of *S. levaillanti* than of *C. sanguinea* when both species fed on cotton aphids reflects greater pre-oral digestion in *S. levaillanti* (Isikber & Copland 2001; see also Cohen 1989 and 5.4.3). Specty et al. (2003) reported a much higher ECI of eggs of *E. kuehniella* (with high protein and lipid content) than of pea aphids (with high carbohydrate but similar water content) by fourth instar *Har. axyridis*. Mills (1982a) hypothesized that better assimilation of nutrients in coccidophagous versus aphidophagous ladybirds might contribute to the success of coccidophages in biological control. Based on a literature survey of conversion efficiencies that did not differ between aphidophages and coccidophages (varying between 0.10 and 0.30) Dixon (2000, table on p. 206) opposed Mills' (1982b) hypothesis.

In general, ECIs depend both on the predator's ability to **digest** and **assimilate consumed food** (e.g. Mills 1982b, Bilde & Toft 1999, Bilde et al. 2000, Jalali et al. 2009a, b, Lundgren 2009, Lundgren & Weber 2010) and on the **allocation** of assimilated nutrients towards maintenance, growth (biomass accumulation) and activities such as foraging and reproduction (e.g. O'Neil & Wiedenmann 1987, Nakashima & Hirose 1999, Dixon & Agarwala 2002, Agarwala et al. 2008, Kajita et al. 2009). When maintained on a diet of *Hypera postica* larvae compared with a diet of pea aphids, adult females of *Har. axyridis* not only had lower rates of consumption, but also lower rates of assimilation of consumed prey, and they allocated more of the assimilated nutrients and energy to searching (Evans & Gunther 2005). Both rate of consumption and assimilation efficiency (as measured by the weight ratio of faeces produced to weight of aphids consumed) decreased (along with egg production) in females of *Menochilus sexmaculatus*, *C. transversalis* and *Har. axyridis* (Dixon & Agarwala 2002). The improved ability of ladybirds to grow and reproduce given a selection of different prey (Rana et al. 2002) is likely to include increased assimilation as well as consumption of prey.

The predator's utilization of consumed food will also reflect its ability to balance its nutritional needs against the nutritional properties of its prey. Specty et al. (2003) compared the body composition of prey and adults of the polyphagous *Har. axyridis* when reared on eggs of *E. kuehniella* with feeding on nymphs and adults of pea aphids. The biochemical profiles of the ladybird adults reflected those of the prey: adults reared on eggs had higher protein and lipid content than adults reared on aphids, but the difference in body composition was less marked between the two groups of predators than between the two types of prey.

In an interesting test of how the biochemical composition of prey and predator may influence consumption patterns of ladybirds, Kagata and Katayama (2006) examined nutritional aspects of intraguild predation by *Har. axyridis* and *C. septempunctata* (Chapter 7). The authors addressed the hypothesis that intraguild predation is an adaptive response to **nitrogen limitation in the diet** (Denno & Fagan 2003). These two ladybird species prey on each other in a strongly asymmetric fashion, with *Har. axyridis* acting as the intraguild predator far more frequently than *C. septempunctata* (e.g. Yasuda et al. 2001, 2004). Kagata and Katayama tested predictions that *Har. axyridis* has higher nitrogen content than *C. septempunctata* when both feed on aphids, and that *Har. axyridis* has lower nitrogen-use efficiency than *C. septempunctata*. Both considerations might favour greater intraguild predation by *Har. axyridis*, such that the predator could capitalize by consuming the relatively nitrogen-rich tissues of the intraguild prey (versus nitrogen-poor tissues of aphids); however, neither prediction was supported. Hence, it does not appear that nitrogen shortage promotes the strong tendency of *Har. axyridis* to engage in intraguild predation.

5.3.5 Aggregative numerical response

5.3.5.1 Temporal and spatial patterns

The collective outcome of individual ladybirds searching for prey can be evaluated by examining the degree to which the predators concentrate in areas of high prey density. This population-level phenomenon is the aggregative component of the predators' numerical response to varying prey density (Solomon 1949, 1964). After finding a prey, an individual ladybird increases the thoroughness of searching. Its movement becomes slower, often with turning after only a short distance. This change from extensive to intensive searching (5.4.1.2) has generally been assumed to be the basis (or at least one of the important mechanisms) of the aggregative numerical response.

Having highly mobile adults which range widely across landscapes, ladybirds are well known to aggregate (and reproduce) in response to **local prey density**. Thus, large numbers of adults often occur near colonies of prey such as aphids (e.g. Banks 1956), mites (e.g. Hull et al. 1976) and chrysomelid beetles (e.g. Matsura 1976). Numbers of aphidophagous ladybirds (and their offspring) generally rise and fall in parallel with changes in aphid population size, although the temporal matching is often less than perfect (e.g. Radcliffe et al. 1976, Wright & Laing 1980, Frazer & Raworth 1985, Hemptinne et al. 1992, Agarwala & Bardhanroy 1999, Osawa 2000, Rana 2006, Kajita & Evans 2010).

Prey density varies widely over time and space at varying scales. Ladybirds frequently aggregate most strongly in areas **within a habitat** (e.g., within an agricultural field) where prey are most abundant (Frazer et al. 1981, Sakuratani et al. 1983, Cappuccino 1988, Turchin & Kareiva 1989, Obata & Johki 1990, Giles et al. 1994, Elliott & Kieckhefer 2000, Osawa 2000, Evans & Toler 2007). The degree of predator–prey **spatial matching** within a habitat can vary widely over the growing season. Park and Obrycki (2004) mapped spatial correlations over the season between the local abundance of *Rhopalosiphum maidis* and adults and larvae of *Har. axyridis*, *Col. maculata* and *C. septempunctata*). Coccinellids aggregated at sites of local aphid abundance during the peak population period (Park & Obrycki 2004; Fig 5.33).

Ladybirds also aggregate at locally high concentrations of their prey at the **landscape scale**, as for example shown in differences among crop fields

(Honěk 1982, Evans & Youssef 1992, Elliott et al. 2002a, b, Brown 2004, Evans 2004). Ives et al. (1993) found that individual adults of *C. septempunctata* and *Hip. variegata* responded very weakly to aphid density on individual stems of fireweed (*Chamerion angustifolium*), but responded from moderately to very strongly to aphid density among populations of fireweed stems. Similarly, Schellhorn and Andow (2005) found that, whereas adults of *A. bipunctata* and *Hip. tredecimpunctata* responded to variation in aphid abundance on individual corn plants within 10 m × 10 m plots, *Col. maculata* adults were sensitive to spatial variation in aphid abundance at a larger scale, and responded to differences in aphid numbers between 10 m × 10 m plots. A fourth species (*Hip. convergens*) responded to variation in aphid abundance at both spatial scales. The authors noted that such differing responses among the four ladybird species were complementary, and together might lead to greater aphid suppression.

5.3.5.2 Modeling of aggregative responses

Kareiva and Odell (1987) parameterized quantitative models of ladybird movements within habitats by using field observations of foraging individuals of *C. septempunctata* (e.g. the rate with which the beetles reversed their direction, taken as a function of hunger level, when moving along linear arrays of host plants). From these models of individual beetle behaviour, the authors demonstrated how observed aggregations of the predator population can develop at local patches of high prey density. In addressing a larger spatial scale, Krivan (2008) tested alternate models of emigration and immigration behaviour to account for the observed increase in numbers of adults of *C. septempunctata* in different alfalfa fields with increasing aphid density, as reported by Honěk (1982). The model best accounting for the field data was one in which the emigration rate was assumed to increase with decreasing local prey density (as discussed above), and the immigration rate was assumed to be independent of local prey density.

5.3.5.3 Factors other than focal prey density

The presence of other vegetation in the vicinity of host plants can influence the numerical response of

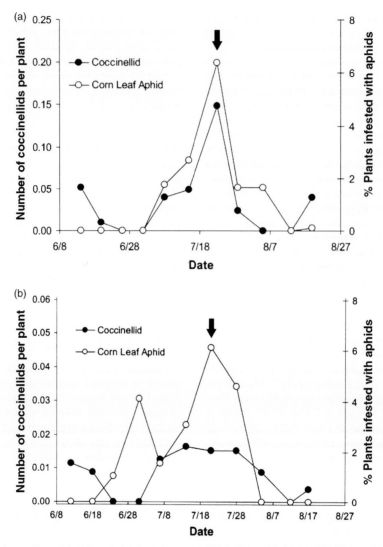

Figure 5.33 Population dynamics of corn leaf aphids (*Rhopalosiphum maidis*) and ladybirds (*Harmonia axyridis, Coleomegilla maculata, Coccinella septempunctata*) in June, July and August. Arrow, the week of peak corn anthesis. (a) large-scale study (970 samples in a 8-ha cornfield). (b) small-scale study (196 samples in a 50 m × 50 m cornfield) (from Park & Obrycki 2004).

ladybirds. Within alfalfa fields, Harmon et al. (2000) found that adults of *Col. maculata* aggregated in patches with high densities of dandelions (yielding pollen for this pollinivorous species; 5.2.9). These patches supported only low densities of pea aphids, apparently because of the local build-up in ladybird numbers. Similarly, Sengonca et al. (2002) found that *C. septempunctata, A. bipunctata* and *P. quatuordecimpunctata*

occurred in greater numbers in plots of lettuce that included weeds (wormwood, tansy or stinging nettle), and aphid (especially *M. persicae*) populations were correspondingly reduced. The authors suggested that the weeds attracted predators by **plant volatiles** as well as providing shelter and additional prey.

The potential importance of **additional prey** for the aggregative response of ladybirds was established by placing potted nettles with *Microlophium carnosum*, adjacent to grass plots infested by *Rhopalosiphum padi* (Müller & Godfray 1997). Coccinellids accumulated in large numbers in the highly infested grass and rapidly drove numbers of *R. padi* to low levels. The predators, many now swollen with eggs, then shifted to the nearby potted nettles where they laid their eggs, and where they and their offspring fed on *Microlophium carnosum*.

At least some aphidophagous ladybirds may also be responsive to the availability of **non-aphid prey** within a habitat. This possibility has been most explored in alfalfa (e.g. Evans & Youssef 1992, Giles et al. 1994), but probably occurs in a variety of settings. The introduced *C. septempunctata* has become the most common aphidophagous ladybird in alfalfa fields of the inter-mountain west of North America. In contrast to native coccinellids, this generalist predator colonizes and persists in alfalfa even when aphid densities are low, and shows a positive numerical response to an alternative non-aphid prey, the abundant larvae of the alfalfa weevil (Evans & Toler 2007). Significant numbers of *C. septempunctata* early in the growing season may now be a key factor in restricting pea aphid population growth, with a consequence that native coccinellids have recently largely abandoned alfalfa fields in the absence of sufficient numbers of aphids to retain them (Evans 2004).

A numerical response of coccinellids to one prey can lead to a subsequent numerical response later in the season to **another prey in the same habitat** or nearby. The presence of the non-damaging *Rhopalosiphum maidis* on grain sorghum early in the season elicits an aggregative numerical response by *Hippodamia* spp., followed by a strong reproductive response by colonizing adults. This leads to heavy ladybird predation on *Sitobion graminum* that appear on sorghum later in the season (Michels & Burd 2007). Rand and Louda (2006) recorded the dispersal of adult coccinellids from prey on agricultural crops into adjacent grassland where they heavily attacked an aphid species (*Bipersona* sp.) on a native thistle.

The **importance of alternate prey** for aggregative numerical responses of ladybirds may be one reason that this response is very much influenced by **vegetation diversity** within fields and by increasing patchiness of non-cultivated land surrounding a field. This aspect has been explored in a variety of agricultural settings (Elliott et al. 1998, 2002a). For example, Elliott et al. (2002b) found that factors associated with

the landscape matrix in which individual fields of alfalfa occur (e.g. percentage of the surrounding landscape that is wetlands, woodlands or crops) can sometimes overshadow the direct numerical response of ladybirds (*Hip. convergens*, *Hip. parenthesis* and *C. septempunctata*) to the aphid populations within these fields.

5.4 FOOD-RELATED BEHAVIOUR

5.4.1 Foraging behaviour

In studies of ladybird behaviour, a number of paradoxical features are manifest. Ladybirds, particularly adults, move in a seemingly chaotic way which appears maladaptive. For example, they sometimes walk over suitable aphids without attacking them. Their behaviour appears so 'aimless' in contrast to that of parasitoids that Murdie (1971) once called them 'blundering idiots'. In fact many aspects of ladybird foraging behaviour are rather sophisticated, e.g. the oviposition deterrent cues (5.4.1.3) and the attraction of ladybirds to volatiles from injured plants (5.4.1.2).

Armed with the axiom of optimality, i.e. that females should produce an optimal number of progeny, we might be astonished at the sight of huge numbers of ladybirds at the seashore that are victims of 'non-economic' overproduction of offspring (5.4.1.5). However, is it truly a waste? The production of very numerous offspring under diverse environmental conditions potentially ensure wide genetic variance and thus increase the probability of overwintering survival and also successful dispersal to new areas.

The reductionist approach suggests limiting the number of species studied and attempting to generalize for the whole family, or at least for one of its food guilds. However, as stressed by Sloggett (2005), coccinellids exhibit considerable diversity in their ecology and ethology. More specifically Harmon et al. (1998), in their analysis of the role of vision in foraging in four species, say the same: 'while the historical trend has been to generalize behavioural traits found in one species to all coccinellids, our study demonstrates the wide variation in the use of foraging cues'. Such interspecific comparisons can lead to understanding why individual species may co-exist in one ecosystem or why they become invasive.

The field of predator foraging behaviour is an area of behavioural ecology marked by two isolated methodological approaches: while empirical data continue to be collected and analyzed, optimality models are constructed by computing fitness components. As long ago as 1988 it was proposed that: 'The next twenty years in insect behavioural ecology will be dominated by the **empirical testing of hypotheses**.' (Burk 1988); this empirical testing still needs to be carried out today. Although there are stimulating concepts and models, we need data as 'specific quantitative tests of particular hypotheses and focussed on individuals and their phenotypic variation. With such data, we will be in a position to decide whether acceptable models have been devised' (Burk 1988). For the research of spatial relations in predator–prey interactions, Inchausti and Ballesteros (2008) invite the 'experimental and field biologists to investigate carefully, how the tendency to switch patches relates to foraging success' of predators. Roitberg (1993) warned that testable hypotheses cannot be built on oversimplified assumptions. Currently, a wide gap seemingly remains between the two above-mentioned approaches as far as coccinellids are concerned, while some studies on insect parasitoids have achieved a much higher standard of research (Wajnberg et al. 2008).

Observations must be performed (and subsequently verified) under experimental conditions as similar as possible to those in nature. Although this prerequisite seems obvious, many experiments have been undertaken under unnatural conditions, which are likely to have introduced **artefacts**. Danks (1983) and Okuyama (2008) provide another important warning: not to work with only 'central measures' such as averages of behaviour parameters. **Variation** is typically treated as a nuisance in many behavioural studies. However, this variation can provide important ecological perspectives: behavioural variation, because it is very large, can cause a substantial change in community dynamics (Okuyama 2008). It is important to test **multiple working hypotheses** and alternatives to optimal models for predator foraging (Ward 1992, 1993, Nonacs & Dill 1993). Lederberg's (1992) conviction that 'the most revolutionary discoveries have arisen out of observation that did not fit prevailing scientific doctrine, and required a **re-examination of the fundamental concepts**' is of great importance. One of the research lines showing hopeful progress in this much needed direction is the prey size–density study (Sloggett 2008b; 5.2.3).

General information on optimal foraging theory and related topics is provided by e.g. Krebs & McCleery

(1984), Bell (1985, 1990), Waage & Greathead (1986), Ward (1992, 1993), Nonacs & Dill (1993), Godfray (1994), Jervis (2005), Inchausti and Ballesteros (2008) and Raubenheimer et al. (2009). Foraging behaviour of coccinellid larvae has been reviewed by Ferran and Dixon (1993) and of ovipositing females by Seagraves (2009).

The location of prey by ladybirds may be divided into three steps (Vinson 1977, Hodek 1993): (i) finding a habitat, (ii) finding the prey within the habitat and (iii) accepting the prey. Both the prey and the host plant, particularly when damaged by herbivores, emit attractive volatiles (5.4.1.2). In this chapter we prefer not to use the classification of semiochemicals into pheromones, allomones, kairomones and synomones (see e.g. Hatano et al. 2008), as the terms are often used ambiguously or arbitrarily.

We broadly follow a sequence of prey-related behavioural steps. The first aim of adults is to find food (5.4.1.1 and 5.4.1.2) and mate, and then find a suitable place for oviposition (5.4.1.3). Finding a suitable prey-bearing **habitat**, while ranging across the landscape, remains a rather mysterious process: how, for example, does a ladybird discover a crop field with initially scarce, small aphid colonies? One explantion might be that ladybirds orient to prey using plumes of volatile semiochemicals (5.4.1.5).

Rather more evidence is available on phase (ii). Foraging on the **host plants** is guided partly indirectly, by taxes and the structure and surface of host plants (5.4.1.1). Studies on this have evolved from an earlier assumption of fully random foraging to the idea that foraging is partly guided by visual and olfactory cues (5.4.1.2). A very important moment is the first **encounter** with prey, an encounter that strikingly changes behaviour from extensive to intensive searching (5.4.1.2).

Well-fed and mated females look for **oviposition sites** (5.4.1.3). Prey suitable for larvae is usually also used by females, although not always (5.2).

First instar larvae can increase their vitality by feeding on sibling eggs (5.2.8), which can be either unfertilized or otherwise non-viable or even trophic eggs (5.2.7). After dispersal from egg clutches, larval behaviour is similar to that of adults, though they are probably equipped with less complex sensory capabilities (5.4.1.2, 5.4.1.4). Aphidophagous coccinellids, due to the inherently intermittent occurrence of aphids, often face prey shortage and resort to cannibalism (5.2.8) and/or feeding on alternative foods (5.2.9).

Table 5.32 Time budget of adult *Coccinella septempunctata* behaviour in the experiments (after Minoretti & Weisser 2000).

Behaviour	Aphid-free plant (*n* = 13)	Aphid-infested plant	
		10-aphid colony (*n* = 10)	30-aphid colony (*n* = 9)
Searching	67.0 ± 4.9	63.7 ± 6.8	40.3 ± 6.1
Grooming	4.0 ± 1.7	9.9 ± 3.1	14.6 ± 3.6
Resting	12.3 ± 5.0	8.9 ± 3.9	8.6 ± 4.3
Handling prey	0	16.9 ± 2.7	24.0 ± 3.4
Feeding on plant tissue	5.4 ± 2.5	1.0 ± 0.8	0.5 ± 0.5
Feeding on nectar	5.4 ± 3.8	2.2 ± 1.3	11.1 ± 4.9

Values are percentages of the total time spent on a plant (mean percentage ± SE).

5.4.1.1 Indirect factors in foraging

Indirect factors in foraging include taxes, plant structure, plant surface, diurnal periodicity. Coccinellids spend most of their time searching for prey (Table 5.32). On a plant, the encounter of predator and prey is made more probable by certain regular behaviours such as taxes, and features of the structure and surface of plants.

Taxes

Encounters are made more likely, at least between aphidophagous and acarophagous species and their prey, by similar tactic responses. Coccinellids, like mites and aphids, exhibit positive phototaxis and negative geotaxis (Fleschner 1950, Dixon 1959, Kaddou 1960, Baensch 1964, 1966, Kesten 1969). However, satiated larvae of *Menochilus sexmaculatus* were observed to be negatively phototactic (Ng 1986).

Plant structure

To study the movements of predator **larvae** on the stem and twigs of plants, Baensch (1964, 1966) used a 50 cm high model tree. When moving upwards on the stem, coccinellids stop at bifurcations, but soon continue walking upwards to follow the main axis.

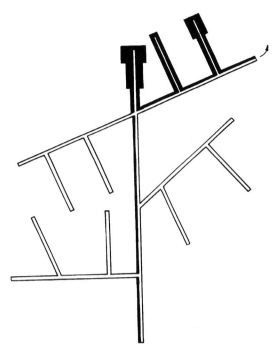

Figure 5.34 Search pattern of one *Adalia bipunctata* adult on a model tree until flying away (arrow) after 23 minutes. The thickness of the line indicates the searching frequency of the predator (modified from Baensch 1964).

After an unsuccessful search on the top the larvae return down the stem and search upwards at bifurcations. If unsuccessful on a twig, they return to the main stem and move to the top again. So, on the model tree, the larvae were more or less 'trapped' between the apex and the tops of branches (Fig. 5.34). If the **larvae** eventually reach the ground, they do not move upwards until they have covered a certain sideways distance. The **adults** soon fly away after an unsuccessful search. Kareiva (1990) compared the effect of plant structure on the movement of *C. septempunctata* adults. There was strong aggregation on goldenrod (*Solidago* sp.), weak aggregation on beans (*Vicia* sp.) and no aggregation on pea (Fig. 5.35). The importance of plant morphology was shown in a comparison of 'normal' peas and a 'leafless' variety. *C. septempunctata* beetles fall off less frequently from the latter plants and moved faster. One day after release they were aggregated on aphid-infested leafless plants, but still randomly distributed on 'normal' peas (Kareiva 1990).

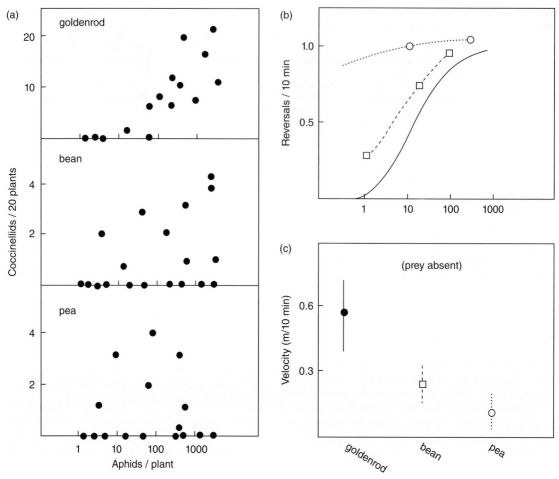

Figure 5.35 Left, aggregations of *Coccinella septempunctata* beetles two days after their release in fields of goldenrod (*Solidago* sp.), bean (*Vicia* sp.) and pea (*Pisum* sp.). A positive association between the number of beetles per 20 plants (sampled in 1 m² plots) and the density of aphids on the plants indicates aggregation; slopes can be used as a measure of the rate of aggregation. (a) Movement of *Coccinella septempunctata*, starved for 12 hours, on rows of three plant species; closed circle and unbroken line, goldenrod; open squares, bean; open circles, pea. (b) Number of reversals per 10 minutes. (c) Velocity (m/10 min) (from Kareiva 1990).

The different structure of two host plants of *Uroleucon* aphids affected the searching efficiency of *C. septempunctata* fourth instars. They spent most of the time on the shoots of *Carduus crispus*, yet only detected 55% of first instar aphids; in contrast, 90% of first instar aphids were detected by larvae on *Centaurea jacea*, even though the larvae searched on shoots only one third of the time (Table 5.33). The explanation was that first instar aphids were hidden on the structurally complex stem of *C. crispus*, which are covered with small leaves, spines and hairs, whereas they were exposed on *C. jacea* (Stadler 1991).

Foraging efficiency may be increased by the tendency of coccinellids to crawl along an edge or a raised surface. Thus **on leaves**, **veins** influence the direction of movement, and it is near veins that colonies of aphids most often occur (Banks 1957, Dixon 1959, Baensch 1964, 1966, Marks 1977, Shah 1982). There are, of course, modifications of this general scheme. While prominent veins of sycamore and lime leaves were successfully searched by *A. bipunctata* (Dixon 1970, Wratten 1973), the larvae of *C. septempunctata* searched mainly the **edge of leaves** and the **stem** on pea and bean plants, where the veins of leaves are not

Table 5.33 Proportion of *Uroleucon* spp. colonies encountered (A) and aphids eaten (B) by 4th instar larvae of *Coccinella septempunctata* on five host plants (Stadler 1991).

Host plant ——— Aphid	*Centaurea jacea* ——— *Uroleucon (Uromelan) jaceae*	*Centaurea scabiosa* ——— *Uroleucon jaceae* ssp.*henrichi*	*Cirsium arvense* ——— *Uroleucon (Uromelan) cirsii*	*Carduus crispus* ——— *Uroleucon (Uromelan) aeneus*	*Cichorium intybus* ——— *Uroleucon (Uromelan) cichorii*
First instar (A)	90	80	60	55	100
Adult aphid (A)	100	100	70	80	100
First instar (B)	90	90	80	70	90
Adult aphid (B)	20	15	15	25	10

prominent. Their searching was thus not very successful on pea and bean, as the aphids occurred mostly around the veins (Carter et al. 1984). Similar observations were made on wheat leaves: when the last instar larva of *C. septempunctata* searches for *Sitobion avenae* it walks along the leaf edge, so that the central area is not searched and aphids there are not found (Ferran & Deconchat 1992). The aphid *Rhopalosiphum padi* may often be found in the space between the stem and the ear of wheat which cannot be entered by third and fourth instar larvae of *C. septempunctata* (Ferran & Dixon 1993).

Plant surface

Banks (1957) observed how important the plant surface is for the foraging of coccinellid larvae (Table 5.34). **Glandular trichomes** on tobacco plants impeded foraging of larvae in *Col. maculata* (Elsey 1974) and *Hip. convergens* (Belcher & Thurston 1982). In a study of *A. bipunctata* larvae, no feeding was observed on tomato and tobacco plants covered with long, dense, upright trichomes and glandular hairs, or on bush beans with hook-shaped hairs. Leaves of another six plants without these characteristics were suitable (Shah 1982). Foraging for nymphs of *Heteropsylla cubana* by *Curinus coeruleus* adults was significantly affected by the leaf surface of different species of *Leucaena* host plant (Da Silva et al. 1992).

Crop plants selected for herbivore-resistant pubescence are also unsuitable for entomophagous insects (but see 5.4.1.3, egg protection). On the richly glandular pubescent leaves of aphid-resistant potato, newly hatched larvae of *C. transversoguttata*, *C. septempunctata*, *Col. maculata* and *Hip. convergens* accumulated trichome exudates on their appendages and bodies and

Table 5.34 Mean rates of movements of first instar larvae of *Propylea quatuordecimpunctata* on various surfaces (Banks 1957).

	Paper	Clean bean leaf	Honey-dewed bean	Hairy potato leaf
mm/min	151	154	104	54
m/1–1.5 day		225–300	150–200	75–100

moved only c. 5 mm from the egg within 48 hours. By contrast, on the normal potato plants, ladybird larvae walked over 100 mm (Obrycki & Tauber 1984). The movement of 2 day old larvae of *P. quatuordecimpunctata* was impeded on the hairy leaves of egg-plant (speed 6.3–8.8 cm/minute), while the glabrous leaves of pepper (22.6 cm/minute) and particularly the lower surface of maize leaves (30.0 cm/min) were suitable. The larvae usually followed the veins of the maize leaves (Quilici & Iperti 1986).

For the acarophagous larvae of *Stethorus pusillus*, the surface of plant leaves is also important. Leaves of tomato and pepper plants facilitate the highest consumption of *Tetranychus urticae* by ladybird larvae (0.68 and 0.55 prey eaten/10 minutes), respectively, while aubergine (0.26 prey/10 minutes) was intermediate and cucumber plants (0.19 prey/10 minutes) the least favourable; the responsible factor was not specified (Rott & Ponsonby 2000; Table 5.35). **Surface waxes** may also significantly affect the suitability of host plants for coccinellids. They captured fewer aphids on the leaves of *Brassica oleracea* than on other comparable plant species, because of the thick, slippery wax layer (Shah 1982). Although *A. pisum* densities

Table 5.35 Effect of plant surface on behaviour of *Stethorus pusillus* observed on leaf discs with *Tetranychus urticae* as prey (modified after Rott & Ponsonby 2000).

	Distance (mm) moved in 10 min	Time moving (%)	Prey encountered	Prey eaten
Tomato	23.4 ± 2.73 a	11 ± 1.0 a	1.44 ± 0.11 a	0.68 ± 0.05 a
Pepper	24.7 ± 2.50 a	8 ± 0.8 a	0.85 ± 0.07 b	0.55 ± 0.04 a
Aubergine	10.9 ± 0.76 b	4 ± 0.4 b	0.30 ± 0.04 c	0.26 ± 0.04 b
Cucumber	15.7 ± 1.20 b	5 ± 0.4 b	0.24 ± 0.04 c	0.19 ± 0.03 b

Mean ± SE within a treatment followed by the same letter are not significantly different at $P = 0.05$ (Tukey's Test).

were significantly lower on peas with reduced surface wax, numbers of predatory coccinellids did not differ consistently between reduced-wax and normal-wax peas. This failed to support a hypothesis that higher predator numbers on reduced-wax peas contribute to lower aphid density there (White & Eigenbrode 2000). On normal-wax peas *Hip. convergens* larvae exhibited higher rates of falling and lower search rate than on reduced-wax peas (Rutledge & Eigenbrode 2003). A strong reduction of surface wax bloom in mutated peas appeared to eliminate the avoidance response by fourth instar *Hip. convergens* larvae to leaves exposed to conspecifics of the same instar (Rutledge et al. 2008; 5.4.1.3). Predation efficacy of *C. septempunctata* on *Acyrthosiphon pisum* was found to be affected differentially on three types of pea plants that differed in surface waxes and content of allelochemicals (Legrand & Barbosa 2003).

Larvae of individual ladybird species vary in their ability to adhere to plant surfaces. When comparing the **attachment ability** of larvae of five species to pea plants with crystalline epicuticular waxes, Eigenbrode et al. (2008) recorded high ability for *A. bipunctata* larvae. These authors also measured the attachment force on clean glass and found much higher values in the primarily arboreal species *Har. axyridis* (11.4 mN) and *A. bipunctata* (6.3 mN) than in the eurytopic, primarily field species *C. septempunctata*, *C. transversoguttata* and *Hip. convergens* (2.6–2.8 mN). A morphological study of the attachment system in *Cryptolaemus montrouzieri* was combined with measurements of attachment forces generated by beetles on the plant surface (Gorb et al. 2008).

Diurnal periodicity of foraging

C. septempunctata brucki adults locate their prey visually from a very short distance, but only in light

(5.4.1.2). In complete darkness they can only capture prey after contact (Nakamuta 1984, 1985). This may be why, under 16L:8D, their searching activity is highest between 09.00 and 16.00 hours (with lights on at 04.00) (Nakamuta 1987). The periodicity of *C. septempunctata* is similar to that of *Har. axyridis* (Miura & Nishimura 1980) and, perhaps, also to that of other species which are mostly found foraging in the field from 09.00 to 17.00 (*Hip. convergens*, *C. transversoguttata* and *Col. maculata*; Mack & Smilowitz 1978). Just over half (53%, $n = 36$) of unspecified ladybird larvae foraged from 15.00 to 21.00 on sugar beet in August (Meyhöfer 2001). Larvae and both sexes of adult *Stethorus punctum* cease their feeding activity at dusk and resume it at dawn (Hull et al. 1977). Under constant light the rhythm persists (i.e. it is endogenous), but the endogenous period is shorter than 24 hours, as in other diurnal animals (Nakamuta 1987). *Col. maculata* lay eggs mostly in the early afternoon (Fig. 5.36; Staley & Yeargan 2005).

When *C. septempunctata brucki* adults are starved, their activity increases for several hours and then it decreases with continuing **starvation**, so that it is highest on the first day (Nakamuta 1987). Similarly, the activity of starving larvae of *P. quatuordecimpunctata* increases during the first day, remains high until the middle of the second day and then decreases (Quilici & Iperti 1986).

5.4.1.2   The role of senses in foraging

Recent findings suggest that, at least for coccinellid adults and in case of short distance perception, Thompson (1951) was right when he maintained that Coccinellidae have 'sense organs and there is no doubt that they can perceive their hosts at a distance. Certainly there seems no reason to think . . . that their behaviour can be described in any sense as random action'.

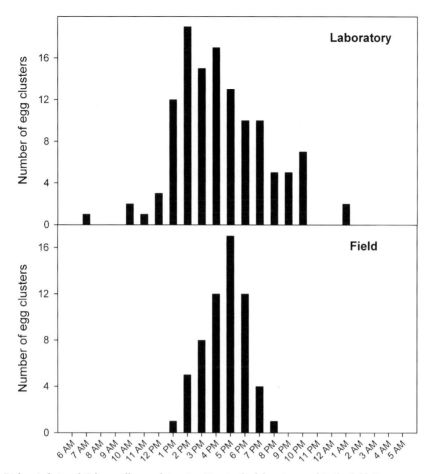

Figure 5.36 Diel periodicity of *Coleomegilla maculata* oviposition in the laboratory and in the field. Bars represent number of egg clusters laid during the hour that ended at the time shown on the *x*-axis (from Staley & Yeargan 2005).

There is a contrast between this statement and Murdie's (1971) assumption that coccinellids are just 'blundering idiots'. Many other specialists have also believed in such **'non-sensory'** behaviour of ladybirds (see Hodek 1996 for early references, and Liu & Stansly 1999 for the case of the psyllophagous *Delphastus catalinae* and *Clitostethus oculatus*). It was thought for a long time that prey are not detected by a coccinellid until actual contact occurs and reports were that prey may not be detected even when only a few millimeters away (Banks 1957), and even when the wind carries the smell of the prey to the coccinellid (Baensch 1964, 1966, experiments on larvae). Some individuals indeed do not react to prey until they touch it and often it is observed both in the laboratory and in the field that

'coccinellids walk right over aphids without eating them' (McAllister & Roitberg 1987). This may be because the coccinellid is at that moment in another phase of its behaviour, i.e. a satiated adult, a male searching for a mate, or a female looking for a place to oviposit (Hodek 1993).

One reason why ladybirds' crawling was frequently believed to be **random**, was the observation that they **revisit** places already searched while neglecting other areas. For example, a first instar larva of *P. quatuordecimpunctata*, moving for 3 hours on a clump of 14 bean stems, spent 52% of the time searching on leaves visited more than once and only 12% on leaves visited only once, the remaining time being spent on stems or stationary (Banks 1957). Marks (1977) suggested there might be adaptive value in behaviour that leads the coccinellid to re-cross its former track. After disturbance, many aphids such as *Acyrthosiphon pisum* or *Aphis fabae* withdraw their stylets and move away from where they were feeding. These aphids can then be captured by coccinellid larvae the next time they traverse their original search path.

Effect of encounter

After feeding, the predator increases the thoroughness of subsequent searching by slower, winding movements in the immediate vicinity, often turning frequently after only short distances (Banks 1957, for later reports see Hodek 1996). This behavioural change is generally considered to enhance the efficiency of searching for aggregated resource items (Curio 1976). Thus, for example, the larvae of *C. septempunctata* which have found aphid colonies remain close to them for long periods (e.g. 18 hours; Banks 1957; Fig. 5.37). In fact, this behaviour is recorded also in non-entomophagous insects (for examples see Bell 1985, Nakamuta 1985b).

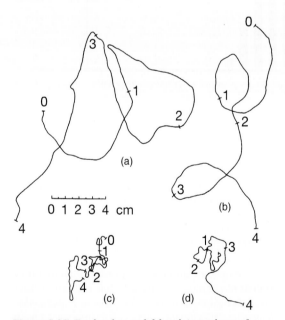

Figure 5.37 Tracks of an unfed fourth instar larva of *Adalia bipunctata* on paper, before (a, b) and after (c, d) feeding on an aphid (the four successive intervals of 15 seconds are marked by numbers along the tracks) (from Banks 1957).

Following the early reports by Fleschner (1950) and Banks (1957), this phenomenon has been variously termed as 'area-concentrated' (Curio 1976, Nakamuta 1985a, b), 'success motivated' (Vinson 1977) or 'area-restricted' searching (Carter & Dixon 1982). It seems best to speak about a '**switchover from extensive to intensive search behaviour**'. However, the analysis of the mechanism and its variability did not make much progress until experiments by Nakamuta and Ferran in the 1980s and 1990s (see below), and our understanding is still far from complete.

Switching to intensive search seems to be **learned**; it is not shown by newly hatched adults and only develops after some experience of prey capture (Ettifouri & Ferran 1992). Nakamuta (1985a, b) discriminated in experiments between contact with an aphid (*Myzus persicae*), biting an aphid and consumption of an aphid; he suggested that **contact with a prey**, rather than its consumption, is the cue that elicits the switchover. This may be an adaptation to situations where the contacted aphid escapes capture (5.4.2). Ferran and Dixon (1993) suggested, however, that the **ingestion** is the cue eliciting the switch. The **duration** of intensive search in *C. septempunctata* adults depends on the quality of the signal (Nakamuta 1985a, b; Table 5.36). Consumption of an aphid produced the longest response. The parameters of the intensive search in fourth instar *C. septempunctata* are affected by the duration of fasting before the encounter with prey (Carter & Dixon 1982). In *C. septempunctata* larvae, the change to intensive search may also be induced by the presence of honeydew (Carter & Dixon 1984). Intraspecific variability in searching behaviour, neglected in the optimal foraging approach (5.4.1; Bell 1990), has been reported in *Cer. undecimnotata*. The larvae show a highly variable response to prey capture, 'adopting intensive search immediately, later or never' (Ferran & Dixon 1993).

In two coccidophagous species, *Chil. bipustulatus* and *Chil. kuwanae*, the same change from extensive to intensive search was observed on encountering prey (Podoler & Henen 1986; Table 5.37). The phenomenon has also been observed in acarophagous coccinellids (Houck 1991; Table 5.38).

Earlier observations on sensorial perception

The 'early' observations (before and in the 1980s) on prey perception are mostly indirect, and a combination of olfactory and visual cues has mostly just been assumed. Not all early workers rejected the possibility of sensory orientation by coccinellids towards their prey. Fleschner (1950) reported distant perception of prey by *Stethorus punctum picipes*, but only from about 0.5 mm. In *Har. axyridis*, the perception of olfactory (together with visual) cues was reported in adults, when they were attracted to gauze and polyethylene bags contaning aphid-infested leaves (Obata 1986, 1997).

Two early observations on two **large coccinellids** preying on coleopteran and lepidopteran larvae indicated prey-orientation behaviour. Older instars of *Aiolocaria hexaspilota*, a predator of chrysomelid larvae,

actively pursue their prey (Savoiskaya 1970a). Similar behaviour was reported for the adults of *Anatis ocellata* preying on larvae of the tortricid *Choristoneura pinus*. Foraging beetles stopped within a distance of 13–19 mm from the prey before moving forward and quickly snatching a caterpillar in their mandibles (Allen et al. 1970).

Two other early reports suggest that coccinellids are guided by **olfactory cues**. Coccinellid adults were observed to follow the **trails of** *Formica polyctena* **ants** (Bhatkar 1982) that tolerated them, and in this way arrive at aphid colonies. In the laboratory, the coccinellids followed such a trail 150 mm long, previously used by 20 ants and apparently chemically marked by them.

Table 5.36 Duration of area-concentrated search of 24-hour starved ladybird beetles, *Coccinella septempunctata brucki*, exposed to different feeding stimuli (Nakamuta 1985a, b).

Types of feeding stimuli	Duration (s)
No stimulation	2.1 ± 0.7a
Contact with an aphid	5.3 ± 2.0b
Biting an aphid	12.9 ± 6.3c
Consumption of an aphid	19.8 ± 9.6d
Contact with an agar block	5.4 ± 2.1b
Consumption of an agar block coated with aphid body fluid	16.7 ± 7.3e

Values are means ± SD of 10 different individuals. Means followed by different letters are significantly different at 0.05 level by Cochran-Cox's *t*-test.

Table 5.38 Proportion of time *Stethorus punctum* females and 3rd instar larvae spent searching, feeding and resting, when they encountered an abundance of all instars of *Tetranychus urticae* (Houck 1991).

	Proportion of time (%) spent		
	searching	feeding	resting
Satiated females	45.1	14.4	40.5
Starved females*	24.4	43.3	32.3
Satiated larvae	78.4	21.6	0
Starved larvae*	47.1	52.9	0

*The increased feeding time was due to a more complete removal of prey contents.

Table 5.37 Response (within 10 seconds) of *Chilocorus bipustulatus* to the contact with prey, a diaspidid scale (modified after Podoler & Henen 1986).

Predator stage	Prey instar	Mean speed (cm/sec)	Mean change in angle (degrees/cm)	Mean number of turns
Adult	Control	0.80a	26.00a	6.80a
	1st	0.37b	116.57b	7.93a
	2nd	0.43b	70.43c	7.97a
	3rd	0.48b	61.43c	7.43a
Larva	Control	0.52a	34.43a	3.00a
	1st	0.30b	80.28b	5.60b
	2nd	0.35b	59.71c	5.47b
	3rd	0.36b	52.43ac	5.43b

Within columns and species groups, numbers followed by different letters are significantly different at P < 0.05 (Duncan's Multiple Range Test).

Marks (1977) observed that fourth instar *C. septempunctata* searched much longer (for 215 seconds) the first time than in repeat searches (5–40 seconds) and supposed that the larva leaves a **chemical marker** when it dabs the surface with the anal disc. As the larvae never ignored plants previously visited by other larvae, the author assumed that there was an individual-specific means of marking a surface. Moving larvae held their maxillary palps (bearing fine distal setae) close to the surface, apparently to perceive the trail. Marking of the surface by larvae was later confirmed in studies on oviposition deterrence of adult females (Ruzicka 1997; 5.4.1.3). Furthermore, after 30 years, Marks' finding on the ability of ladybird larvae to perceive trails of conspecific larvae, and avoid searching the marked surfaces, finally appears to have been confirmed (Reynolds 2007, Rutledge et al. 2008; 5.4.1.3). Thus not only adult females, but **also larvae** seem to **perceive sensory cues** left by larvae.

Heidari and Copland (1992) reported that the fourth instar *Cryptolaemus montrouzieri* perceive prey only by physical contact, while adults were able to detect their mealybug (Pseudococcidae) prey by **sensory stimuli**. The distance at which the adults stop and then 'jump' towards the prey was reported to be 14mm, slightly greater than the **perception distance** experimentally established for two subspecies of *C. septempunctata*. Nakamuta (1984b) and Nakamuta and Saito (1985) found that adults of *C. s. brucki* orientated themselves towards aphid prey from a distance of about 7mm in light, while they were unable to locate it in darkness. Thus, these results suggest that adult coccinellids may perceive the prey also visually at close proximity. In adults of *C. s. septempunctata*, **visual** perception of aphids was reported to be up to 10mm, while for **olfactory** perception by fourth instars this distance was 7mm (Stubbs 1980). The experiments with larvae used an aphid crushed onto filter paper: an intact aphid might elicit a different, perhaps less intense olfactory stimulus. Nakamuta's results (1984a) further suggest that the aphid body fluid is the final cue to prey recognition. Many **later findings** also support **sensory orientation** (see Vision, olfaction). The reason why visual and olfactory perception was overlooked may be that the distances involved are very small (Stubbs 1980).

Honeydew, sucrose

With **honeydew**, acting either as an arrestant for *C. septempunctata* larvae (Carter & Dixon 1984) or an oviposition stimulant (Evans & Dixon 1986), olfaction is clearly involved. Honeydew has a similar effect to that of prey. Larvae do not discriminate between different amounts of honeydew, although *C. septempunctata* larvae do discriminate between differences in the density of aphid honeydew droplets. They climbed more often over a surface contaminated with honeydew of the ant-tended *Aphis craccivora* (but here deprived of ants) than over a similar surface with the less dense honeydew droplets of the non-ant-tended *Acyrthosiphon pisum* (Ide et al. 2007).

The role of prey traces (honeydew, exuviae) in eliciting responses from coccinellids was later documented using a Y-tube olfactometer. More *Har. axyridis* beetles preferred the odour of leaves of buckthorn naturally infested with *Aphis glycines* to that of artificially infested or uninfested buckthorn (Bahlai et al. 2008; Table 5.39). Similar observations with prey traces have been made on coccidophages. Wax and honeydew of mealybugs (*Phenacoccus manihoti*, *Planococcus citri*) arrested *Exochomus flaviventris* and *Diomus* sp., which are predators of *P. manihoti*, the cassava mealybug (van den Meiracker et al. 1990).

The **sex pheromones of coccids** attract coccidophagous ladybirds. *Rhyzobius* sp. responded to sex pheromones of two matsucoccids, *Matsucoccus feytaudi* and *M. matsumurae* (Branco et al. 2006).

The **arrestant effect** of honeydew has been used **in augmentative biological control**, in which honeydew being replaced by sucrose. **Sucrose** dissolved in water (150g/l) applied to alfalfa arrested *C. septempunctata* and *C. transversoguttata richardsoni:*, within 24–48 hours ladybird densities increased 2–13 times in the centre of the treated plots, whereas in the surrounding fields densities of ladybirds decreased to less than two-thirds of their former density. 'Artificial honeydew' might be used not only to bring ladybirds to fields with still only few aphids, but also – by spraying an adjacent crop – push the coccinellids away from fields scheduled for insecticide treatment (Evans & Richards 1997). In a study of dispersal using mark–release–recapture, the residence time of beetles was 20–30% longer in sugar-sprayed plots and the density of unmarked beetles rose by a factor of 10–20 in sugar-sprayed plots during the first 4–6 hours following early morning spraying. Such a **fast aggregation** seems difficult to explain by random arrival. Because sugar is not volatile, the authors hypothesized that later arriving ladybirds were attracted by earlier arrivals (Van der Werf et al. 2000; Chapter 9.4).

Table 5.39 Preferences of collected adult *Harmonia axyridis* recorded in a Y-type tube olfactometer for cues derived from a hedgerow ecosystem (modified after Bahlai et al. 2008).

Comparison				% Choice	
T1	T2	*n*	% Response	T1	T2
R. cathartica*	Blank	30	75.0	70.0	30.0[‡]
R. cathartica	*M. domestica* odor[†]	31	77.5	32.3	67.7[‡]
R. cathartica	Artificially aphid infested *R. cathartica* odor	32	80.0	53.1	46.9
R. cathartica	Naturally aphid infested *R. cathartica* odor	38	95.0	31.6	68.4[‡]

**Rhamnus cathartica* (buckthorn), winter host of *Aphis glycines*, frequent in agricultural hedgerows.
[†]*Malus domestica*, apple.
[‡]Significant, nonrandom responses.

Sensory receptors

The spectral sensitivity and structure of the compound eyes of *C. septempunctata* adults have been studied (Agee et al. 1990, Lin & Wu 1992, Lin et al. 1992) and photoreceptors for three electromagnetic wavelength ranges have been documented: UV, green and a third between UV and green. Storch (1976) found visual perception to be of little importance to fourth instar *C. transversoguttata*; however, the **prolegs** were important in the detection of prey as they bear sensillae which detect prey at a distance. Kesten (1969) thought that the **maxillary palps** are perhaps the most important sensory organs through which both adults and larvae of *Anatis ocellata* recognize prey. Yan et al. (1982) assumed that chemoreception is situated in the sensillae on the terminal segment of the **labial palp** of *C. septempunctata* adults. According to Nakamuta (1985), prey contact by maxillary palps or maxillae elicits capture behaviour. Da Silva et al. (1992) observed that *Curinus coeruleus* adults seemed unaware of their prey (*Heteropsylla cubana*) until contact was made, usually with the maxillary palps. The number of sensory receptors on the maxillary palps was compared across coccinellid adults with different food specializations. The aphidophagous polyphage *C. septempunctata* has 1800 chemoreceptors on each palp, the aphidophagous oligophage *Cer. undecimnotata* has 1100 receptors and a coccidophagous *Chilocorus* sp. has 830 receptors. In all three cases **olfactory** receptors are more numerous than gustatory ones. The **phytophagous** *Epilachna* 'chrysomelina', however,

has more gustatory than olfactory receptors, in total only 480 (Barbier et al. 1996). The importance of the maxillary palps for discriminating between clean surfaces and those with deterrent larval tracks (5.4.1.3) was proved by maxillary palp amputation in experiments with *Cycloneda limbifer* and *Cer. undecimnotata* (Ruzicka 2003). Sensilla on antennae of *Har. axyridis* were described in detail by Chi et al. (2009).

Vision

A strong positive response to yellow (chosen from seven colours) sticky panels was exhibited by *C. septempunctata* adults, while *Hip. parenthesis* did not show visual orientation to any particular colour (Maredia et al. 1992). Lorenzetti et al. (1997) reported a significantly greater abundance of coccinellids on stressed and therefore yellow maize plants, while chrysopids preferred greener control plants. Laboratory experiments also examined the role of vision in close-proximity foraging behaviour in three of four species, but **colour vision** in only two species (Harmon et al. 1998). It is not clear how pollinivory in *Col. maculata* is related to the finding in this study that this species appears not to use colour or any other type of visual cue. When responding to red or green morphs of *Acyrthosiphon pisum* on green or red backgrounds, *C. septempunctata* could distinguish between red and green morphs and used colour contrast with the background in foraging. *Har. axyridis*, however, consumed more red than green morphs regardless of background. *Hip. convergens* did not respond to colour, but its foraging was decreased in the dark (Harmon et al. 1998).

In an arena, *Har. axyridis* adults, particularly females, strongly preferred yellow paper pillars over green ones. This has recently been confirmed by Adedipe and Park (2010). No female visited the green pillars (Mondor & Warren 2000). These authors also found a clear **effect of conditioning** once the females had located the food on top of the green pillars over a 3-week period; they then spent more time searching green than yellow pillars in a subsequent trial. These interesting findings, however, do not seem to prove that *Har. axyridis* adults 'effectively locate prey over large distances' as interpreted by Rutledge et al. (2004). These important results relate to the question posed recently by Pasteels (2007): 'To what degree are ladybird responses to various cues innate or learned?' Also coccinellid larvae can learn: rejection of lower quality prey (parasitized old versus non-parasitized young lepidopteran eggs) by fourth instars of *Col. maculata* gradually increased at subsequent encounters (Boivin et al. 2010). This learned behaviour was partly forgotten after 48 hours.

In visual bioassays in a tube arena, *Har. axyridis* significantly chose **silhouettes** over blank spaces (Bahlai et al. 2008). As with a study on *Chil. nigritus* (Hattingh & Samways 1995), the authors interpret the finding as indicating long-range visual orientation for location of prey habitats. Another tenable explanation might be that the response to silhouettes was related to the hypsotactic orientation to hibernation sites (Chapter 6.3.1).

Olfaction

A full understanding of the olfactory response of coccinellids to their prey, particularly Sternorrhyncha, has to be based on complex studies of the chemoecological properties, not only of the phytophagous prey but also of their host plants. Chemical cues from both damaged and intact plants, and sex pheromones and alarm pheromones of the prey have to be included in research. Recent general reviews may be found in Pickett and Glinwood (2007), Hatano et al. (2008) and Pettersson et al. (2008) (also Chapter 9). In the response to combined signals there may be a high degree of **mutual synergism**, e.g. to prey plus prey waste (Bahlai et al. 2008; Table 5.39) or inhibition

(e.g. inhibition of response to (E)-β-farnesene by the alarm pheromone inhibitor (–)-β-caryophyllene; Al Abassi et al. 2000). Thus it is not surprising that there are inconsistencies across experimental results, particularly when individual olfactory cues are studied in isolation.

Olfactory responses were detected many times in coccinellids in the late 1990s and 2000s. when **olfactometers and electroantennograms (EAG)** began to be used. However, already in the 1960s, Colburn and Asquith (1970) had found that *Stethorus punctum* adults were attracted to their prey by smell in an olfactory cage. Sengonca and Liu (1994) demonstrated in an eight-arm olfactometer that adults of *C. septempunctata* were attracted to odours of two aphid species, while the odour of a non-prey insect, *Epilachna varivestis*, was not attractive. In contrast, in a Y-tube olfactometer and a choice arena, Schaller and Nentwig (2000) did not find a positive response of *C. septempunctata* to its essential prey *Acyrthosiphon pisum* or its honeydew, while they found positive orientation to plant stimuli and of males to females. A strong odour of aphids (1000 *Myzus persicae*) stimulated female *Har. sedecimnotata*, that had been deprived of aphid prey for 30 days, to lay eggs rapidly (Semyanov & Vaghina 2001). The females of *Cycloneda sanguinea* reacted more intensively to a superior prey, *Macrosiphum euphorbiae*, than to an inferior prey, *Tetranychus evansi*, while they responded negatively to plants hosting another coccinellid *Eriopis connexa* (Sarmento et al. 2007).

In several experiments coccinellids have been found to respond to the cues emanating from **intact plants without prey**. In *Col. maculata*, the greatest EAG responses were to a plant cue, the maize volatiles alpha-terpineol and 2-phenylethanol, and to the aphid sex pheromone (4aS, 7S, 7aR)-nepetalactone (Zhu et al. 1999). The response of *Col. maculata* to maize volatiles may be related to the feeding of this species on maize pollen (J.J. Sloggett, unpublished). *Coccinella septempunctata* also responded positively in a four-arm olfactometer to the single plant volatile (Z)-jasmone. However, the response of *C. septempunctata* to a combination of cues from barley and two weeds, *Cirsium arvense* and *Elytrigia repens* was higher than to barley alone, both in the field and in an olfactometer (Birkett et al. 2000). In contrast, neither *Myzus persicae* alone, nor cabbage alone (also when dammaged mechanically) were attractive for larvae and adults of *Har.*

Table 5.40 Response of aphid predators choosing tea shoot-aphid complexes (*Toxoptera aurantii*) or blank control in a Y-tube olfactometer (modified after Han & Chen 2002).

Number of aphids/100		2	4	6	8	12	20	28	36
Number of aphid-damaged shoots		4	10	15	20	30	50	70	90
Number of natural enemies choosing odor or control									
Chrysopa sinica	Odor	10	11	12	13	13	15*	17**	19**
	Control	10	9	8	7	7	5	3	1
Coccinella septempunctata	Odor	10	12	13	14	15*	17**	18**	19**
	Control	10	8	7	6	5	3	2	1

χ^2 test is used to determine the difference between number of natural enemies choosing odor and number of choosing control.
* and ** indicate statistical significance at $P < 0.05$ and $P < 0.01$, respectively.

axyridis. Only cabbages with 60 aphids were attractive (Yoon et al. 2010).

A. bipunctata adults responded to a **combined stimulus of host plant** (*Vicia faba*) **and aphid** (*Aphis fabae*) and not to the control odour source which was the plant alone (Raymond et al. 2000). *Coccinella septempunctata* adults responded to several sources of semiochemicals related to tea aphids, *Toxoptera aurantii* (rinses of aphid cuticle, honeydew and volatiles emitted from aphid-damaged tea shoots). An increase in dose of any of the above cues decreased the EAG responses of *C. septempunctata* (Han & Chen 2002; Table 5.40).

Larvae and adults of *A. bipunctata* were attracted by crushed *A. pisum* and *M. persicae*, but not *Brevicoryne brassicae*. (E)-β-farnesene (EBF), the **aphid alarm pheromone**, was proved by bioassay with the pure compound to be the effective semiochemical involved in the attraction. Plant leaves alone (*V. faba*, *Brassica napus*, *Sinapis alba*) or leaves with intact aphids did not attract ladybirds. The volatile sesquiterpene EBF attracted both male and female *Har. axyridis* and elicited EAG responses (Verheggen et al. 2007). **Volatilization** of EBF was **quantified** for 60 minutes: after the initial burst it declined exponentially, but detectable amounts were still present after 30 minutes (Schwartzberg et al. 2008). The absence of attraction by *B. brassicae* has been assumed to be due to **inhibition** of the aphid **alarm pheromone** by isothiocyanates that are emitted from this aphid (Francis et al. 2004). Similar inhibition of the attractiveness of EBF, this time

by (–)-β-caryophyllene, was demonstrated by Al Abassi et al. (2000). Such inhibition might explain why *B. brassicae* siphuncle droplets (containing EBF) were not attractive to *Har. axyridis* in contrast to those from *A. pisum* (Mondor & Roitberg 2000).

The phenomenon that **plants damaged** by herbivores produce **volatiles attracting their natural enemies**, originally discovered in phytophagous and predaceous acari (Dicke et al. 1990) as well as in caterpillars and their parasitoids (Turlings et al. 1990), has also been studied in ladybirds (Ninkovic et al. 2001). Significantly more methylsalicylate was released from aphid-infested than from uninfested soybean plants. In addition, (D)-limonene and (E,E)-α-farnesene were released. Studies coupling gas chromatography to EAG with volatile extracts from soybean plants showed that **methylsalicylate** elicited significant electrophysiological responses in *C. septempunctata*. In field tests, traps baited with methylsalicylate were highly attractive to *C. septempunctata* adults, but not to *Har. axyridis* (Zhu & Park 2005). Attraction by volatiles from a damaged host plant (sagebrush, *Artemisia tridentata*) has also been demonstrated for two further coccinellid species, *Hip. convergens* and *Brumoides septentrionis* (Karban 2007). *Aiolocaria hexaspilota*, a predator of chrysomelid larvae, also responds to volatiles (E-β-ocimene is the most abundant) from willows damaged by the larvae but not the adults of the chrysomelid (Yoneya et al. 2009). The production of volatiles depends on the severity of plant damage. Thus *Rhopalosiphum maidis*, which uses intercellular stylet penetration, does not damage plant cells of maize and this is probably why emissions of volatiles were not recorded by Turlings et al. (1998).

Volatiles can be **destroyed** by air pollutants. Common herbivore-induced plant volatiles, such as inducible terpenes and green-leaf volatiles, are completely degraded by exposure to moderately enhanced atmospheric ozone (O_3) levels (60 and 120 nl/l). However, orientation of natural enemies was not disrupted in assays of either of two tritrophic systems involving parasitoids and acarophagous acari, and a similar situation in coccinellids may be assumed. Other herbivore induced volatiles, such as a nitrile and methyl-salicylate, were not reduced by elevated O_3 and might thus replace the above less stable volatiles as attractants (Pinto et al. 2007).

General remarks on sensory orientation

After a long period in which sensory orientation of ladybirds was considered unimportant (with exceptions such as Fleschner 1950, Allen et al. 1970, Marks 1977, Obata 1986 and in the case of large ladybirds), there now seems to be a general consensus in favour of **non-contact sensory orientation**, both visual and olfactory, to conspecifics and prey, and also olfactory orientation to volatiles from prey-damaged plants (Zhu & Park 2005, Karban 2007) or even intact plants (Zhu et al. 1999, Birkett et al. 2000). An important novel observation appears to open new directions of research: this is that the **learned response** to colours recorded by Mondor & Warren 2000) may not be limited just to colours. The olfactory responses to volatiles indicate that they involve complex synergistic relations between the cues emanating from host plants and prey, but also by antagonistic relations such as between allelochemicals from plants and aphid alarm pheromone.

Agrawal (2011) considers jasmonic acid a master regulator of plant defensive responses.

An irony of research history lies with Marks' (1977) two findings and interpretations. These were not believed at the time by experienced authors, but rediscovered much later in the deposition of a chemical marker by the larval anal disc (Laubertie et al. 2006) and **perception** of the cue **by the maxillary palps** in adults (Ruzicka 2003) and larvae (Reynolds 2007, Rutledge et al. 2008).

Attempts to understand perception and processing of olfactory signals from food (and conspecifics) by coccinellids are still in their infancy. **In the field**, the volatile semiochemicals are mixed and diluted, creating a **dynamic olfactory environment**. To interpret the complex natural situation, researchers should not content themselves with only studying coccinellid literature. Studies on this topic in parasitoids are more advanced; general reviews, such as e.g. Vet and Dicke (1992), De Bruyne and Baker (2008), van Emden et al. (2008), Lei and Vickers (2008) and Riffell et al. (2008) can provide important insights for coccinellid researchers.

The notorious temporary absence of aphid prey and the frequent need to feed on alternative prey evidently worked against the evolution of the **gustatory senses** in aphidophagous and other predaceous ladybirds. This is indicated both by their inability to discriminate between suitable and toxic prey (5.2.6.1) and the low number of gustatory sensors on the palps compared to much larger number in the Epilachninae (p. 223).

5.4.1.3   Finding an oviposition site

Females foraging for prey search mostly in parallel with their search for suitable oviposition sites. Thus much of the evidence dealing with sensory orientation (5.4.1.2) is equally important when considering oviposition. The most important cue would seem to be the **amount and quality of prey** available; thus ensuring that the hatching larvae do not have to travel too far before finding aphids. Honěk (1980) ascertained the aphid density necessary for the induction of egg laying by *C. septempunctata* on several crops. However, this is not the only important factor; here there is 'a **trade-off** between two opposing risk factors: potential progeny starvation if eggs are laid **too far** from the food supply, and their potential loss to cannibalism or predation if they are laid in **too close** proximity' (Michaud & Jyoti 2007). This paper reports how *Hip. convergens* females solve that dilemma on sorghum plants infested by *Rhopalosiphum maidis* in Kansas fields. The egg clusters are typically placed on the undersides of lower, non-infested leaves and stems, some 30–50 cm distant from the aphid colonies that are higher up on the plants, in the whorl. Osawa also (1989, 2003) stressed that *Har. axyridis* females oviposit a metre distant from the nearest aphid colonies on *Prunus persica* trees. This behaviour apparently evolved under the pressure of intense predation on ladybirds near aphid colonies. Eggs of *C. magnifica*

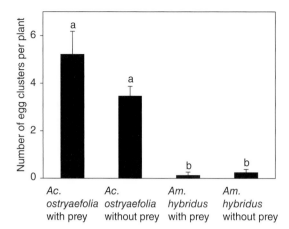

Figure 5.38 Number (±SE) of *Coleomegilla maculata* egg clusters oviposited over an 8-day period on *Acalypha ostryaefolia* and *Amaranthus hybridus* plants with and without prey (200 *Heliothis zea* eggs) added every other day. Means sharing the same letter are not significantly different (P > 0.05, Fisher protected LSD test) (from Griffin & Yeargan 2002).

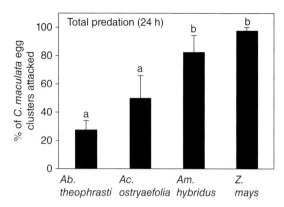

Figure 5.39 Percentage (±SE) of *Coleomegilla maculata* egg clusters attacked by predators on selected plants over a 24 hour period. Plants: *Abutilon theophrasti, Acalypha ostryaefolia, Amaranthus hybridus, Zea mays*. Means sharing the same letter are not significantly different (P > 0.05, Fisher protected LSD test) (from Griffin & Yeargan 2002).

were, however, laid as close as <10 cm from aphids (Sloggett & Majerus 2000).

Another **egg protection behaviour** reported in two tritrophic systems is the occurrence of a large number of eggs of coccinellids on **highly pubescent plants**. Higher egg predation on plants with less or no pubescence has probably selected for this oviposition preference. Although the presence of trichomes on leaves of aphid-resistant potato hinders the movement of coccinellid larvae (*C. septempunctata, C. leonina transversalis, Col. maculata, Hip. convergens*), particularly of first instars, such potato plants often harbour many eggs. Highly pubescent plants had the highest percentage of coccinellid eggs, while plants with the lowest densities of trichomes had the highest percentage of beetles. The difference was not caused by a difference in the abundance of aphids. It is possible that the eggs were more readily removed by **cannibalism on the low-pubescence plants** (Obrycki & Tauber 1984, 1985). *Coleomegilla maculata* females also prefer to lay eggs on plants with glandular trichomes of velvet leaf (*Abutilon theophrasti*) or *Acalypha ostryaefolia*, although they spent more time on smooth pigweed plants (*Amaranthus hybridus*) (Griffin & Yeargan 2002, Staley & Yeargan 2005, Seagraves & Yeargan 2006;

Figs 5.38, 5.39). Almost 10 times more *Col. maculata* eggs were found on the pubescent tomato companion crop than on less pubescent maize and the eggs on tomato had 2.6–5.9 times higher survival.

Aggregation

Localized aphid densities are usually variable within fields and the predators' aggregation to patches of high aphid density has often been observed. The aggregation results during intensive search from slowed movement adopted after finding prey (5.4.1.2), and has been termed the 'aggregated numerical response' (5.3.5). This spatial dimension in predator–prey interactions has been emphasized by Kareiva and Odell (1987) and Kareiva (1990) (5.4.1.1 for details).

Age of prey colony

The adaptive significance of reduced attractiveness to predators of very old colonies seems obvious. But it is not easy to define the cues used by the ladybirds to determine colony age. On 200 *Pittosporum tobira* trees in the open, Johki et al. (1988) observed that adults of several coccinellid species were less attracted to aphid colonies which were so old that the trees were sticky with honeydew. The number of coccinellid adults increased with **increasing prey density** only until

Table 5.41 Abundance of coccinellids (mean number per tree) in relation to aphid infestation (*Toxoptera odinae, Aphis spiraecola*) on trees of *Pittosporum tobira* (Johki et al. 1988).

	Infestation grade*				
	0	**I**	**II**	**III**	**IV**
Harmonia axyridis	0	0.87	1.41	1.60	0.32
Coccinella septempunctata	0	0.25	0.27	0.36	0.32
Menochilus sexmaculatus	0	0.31	0.64	0.57	0.26
Propylea japonica	0	0.21	0.29	0.42	0.21
Scymnus posticalis	0	0.28	0.25	0.47	0.05
Number of trees	8	61	59	53	19

*I–III, each colony consists of a fundatrix and few young nymphs (I), of 10–20 nymphs (II), of 21–50 nymphs (III); IV,colonies extend from leaf to stem (honeydew makes the tree sticky). Kyoto, Japan, 20 May.

the colonies contained 50 nymphs (Table 5.41). It is a pity that the number of eggs laid is not given. The negative effect of the excessively high aphid infestation was strong for the small *Scymnus posticalis* and negligible for *C. septempunctata*.

In the laboratory, a significantly lower number of *Chilocorus nigritus* adults were found on *Cucurbita* fruits with high infestations of *Aspidiotus nerii* (>60 scales per cm²) than on less heavily infested fruits (24 scales per cm²) (Erichsen et al. 1991). Based also on earlier findings (Samways 1986) the authors assumed that the coccids become less suitable prey at such a high density. This is one of rare reports on this topic in coccidophagous ladybirds, where the decision on oviposition site is less critical due to the coccid population being much more stable in time than aphids.

Females of *C. septempunctata* laid significantly more eggs in a 4-hour period on previously uninfested 3-week old bean plants infested with immature aphids than on previously infested 2-month old plants with adult aphids (Hemptinne et al. 1993). The authors related the decrease in oviposition with the combination of older plants and older aphids (the 'old combination' of the paper) to the presence of adult aphids indicating a greater age for the aphid colony. However, the presence of adult aphids as an indication of 'risky' old colonies was not accepted by Hodek (1996, p. 214) and later the hypothesis was withdrawn (Dixon 2000, p. 128): 'aphidophagous ladybirds do not appear to use cues associated with the age structure of

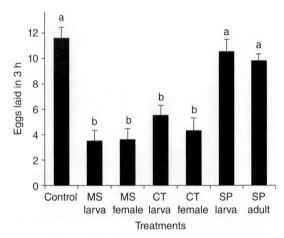

Figure 5.40 The effect of mutual sizes on deterrence. Number (±SE) of eggs laid in 3 hours (vertical axis) by *Menochilus sexmaculatus* (MS) females in interaction with larvae or adults of conspecific or heterospecific predators (horizontal axis). Same letters above bars indicate no difference between the treatments (Scheffé multiple range test, P > 0.05). CT, *Coccinella leonina transversalis*; SP, *Scymnus pyrocheilus* – small size (modified from Agarwala et al. 2003a, b).

the aphid colony or phenological age of the plant, but respond positively to high aphid abundance'. Majerus (1994, p. 315) reported coccinellids ovipositing at old aphid colonies on nettles. In the above observations by Johki et al. (1988) and Hemptinne et al. (1993), the negative cue was probably the **excessive presence of honeydew** that can impede movement of coccinellids, particularly of small species. An adequate level of honeydew, however, does act as an arrestant (Carter & Dixon 1984; 5.4.1.2).

Oviposition deterrence

The observation that *A. bipunctata* females did not oviposit in patches where they encountered conspecific larvae led at first to the erroneous assumption that it is **physical contact** that produces the inhibition (Hemptinne et al. 1992), although Marks (1977) had already suggested much earlier that coccinellid larvae sign the surface with chemical markers from their anal disc. Action of volatile semiochemicals was dismissed: 'Larval odour had no significant effect on ladybird oviposition' (Hemptinne et al. 1992, p. 241). More recently, Agarwala et al. (2003b; Fig. 5.40) again

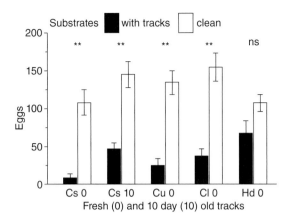

Figure 5.41 Number (±SE) of eggs laid by 10 *Menochilus sexmaculatus* females on substrates with and without larval tracks in choice tests with tracks of first instar larvae of: Cs, conspecifics; Cu, *Ceratomegilla undecimnotata*; Cl, *Cycloneda limbifer*; Hd, *Harmonia dimidiata*. (Wilcoxon paired-sample test: **, P < 0.01; ns, not significant) (modified from Ruzicka 2006).

experimented with **direct encounters**. *Menochilus sexmaculatus* females reduced their oviposition when exposed to immobilized conspecifics or *C. leonina transversalis* individuals, but not larvae or adults of *Scymnus pyrocheilus*; this small ladybird apparently does not represent a great risk of being an intraguild predator (Chapter 7).

The above interpretations pre-date an important discovery: crawling larvae of aphidophagous predators deposit **oviposition-deterring semiochemicals** (ODS) that **inhibit conspecific females from ovipositing**. This discovery was made both with chrysopids (Ruzicka 1994, 1996) and coccinellids (Ruzicka 1997, Doumbia et al. 1998). The lasting deterrence, based on persistent semiochemicals left in the trail, is obviously more efficient than direct encounters. The probability of inhibition is also spatially increased. The distribution of offspring among numerous prey patches certainly reduces intraspecific competition and cannibalism (Fig. 5.41).

Although, after only about 10 years of research, robust comparisons are not yet possible, the deterrence of conspecific larval tracks seems **stronger with smaller coccinellid species** (Ruzicka 2001). While also operating in *C. septempunctata*, this deterrent mechanism is weaker than in other species, e.g. *Cer. undecimnotata* (Ruzicka 2001; Table 5.42). The ODS left

by coccinellid larvae also vary among species in their degree of **environmental persistence**. Whereas the residues of *Cycloneda limbifer* lasted >30 days (Ruzicka 2002), 10 days in *A. bipunctata* (Hemptinne et al. 2001), 5 days in *Cer. undecimnotata* and only 1 day in *C. septempunctata* (Ruzicka 2002). In the predator of adelgids and the green spruce aphid, *Aphidecta obliterata*, the inhibition effect of larval track density was studied simultaneously with the stimulatory effect of prey density (Oliver et al. 2006; Fig. 5.42). Oviposition deterrence depends on the physiological condition of the ovipositing female. Old females of *A. bipunctata* were less reluctant than young females to lay eggs in patches with ODS. Oviposition deterrence similarly diminishes due to continuous exposure for 15 days to an ODS contaminated paper (Frechette et al. 2004; Fig. 5.43). Intraspecific oviposition deterrence was also recorded in *Har. axyridis* females of a flightless strain on cucumbers infested with *A. gossypii*: the females moved to the control half of the greenhouse free from conspecific larvae (Gil et al. 2004).

The overall less intensive response of *C. septempunctata* females is shown by their low sensitivity to **tracks of heterospecific larvae**, for example, of *Cycloneda limbifer* (Ruzicka 2001). This may indicate that the large *C. septempunctata* is less sensitive to the risk of intraguild predation compared to smaller native coccinellids. Intraspecific oviposition deterrence was confirmed in the aggressive intraguild predator *Har. axyridis* (Chapter 7), but this species, not surprisingly, exhibited no significant response to larval tracks of *C. septempunctata* (Yasuda et al. 2000). The authors plausibly assumed that this low oviposition-deterring response reflected a low risk of predation on *Har. axyridis* eggs by *C. septempunctata*, as they observed in the field that *C. septempunctata* avoided eating *Har. axyridis* eggs.

Michaud & Jyoti (2007) studied the intensity of heterospecific oviposition deterrence in relation to **niche overlap**. On sorghum plants *Hip. convergens* responded intensively to conspecific oviposition-deterring cues. Heterospecific oviposition deterrence of *Hip. convergens* females to *Col. maculata* larval trails was weaker, evidently due to a substantial niche overlap between the two species (Figs. 5.44, 5.45). Magro et al. (2007) arrived at a contrasting conclusion on the effect of niche overlap when they recorded a strong oviposition-deterring response to heterospecific tracks in two *Adalia* species, while *C. septempunctata* females were not deterred by the tracks of both *Adalia* species. Magro

Table 5.42 Egg laying by females of *Ceratomegilla undecimnotata, Harmonia dimidiata, Cycloneda limbifer, Coccinella septempunctata* on clean substrates with tracks of first instar larvae in choice test (modified after Ruzicka 2001).

	Coccinellid larvae tested							
	C. limbifer		*C. undecimnotata*		*C. septempunctata*		*H. dimidiata*	
Females tested	clean	tracks	clean	tracks	clean	tracks	clean	tracks
C. limbifer								
eggs	200 (21)**	33 (13)	276 (20)**	38 (12)	174 (39) ns	151 (32)	230 (35)**	91 (33)
% eggs	86	14	88	12	56	43	73	27
C. undecimnotata								
eggs	149 (25) ns	59 (18)	180 (27)**	49 (20)	150 (21) ns	85 (17)	212 (24)**	16 (14)
% eggs	70	30	80	20	63	37	93	7
C. septempunctata								
eggs	196 (27) ns	191 (23)	167 (15) ns	205 (33)	367 (30)*	208 (32)	236 (23) ns	185 (39)
% eggs	50	50	47	53	64	36	59	41
H. dimidiata								
eggs	151 (26) ns	154 (21)	137 (22) ns	93 (26)	128 (22) ns	147 (19)	173 (22) ns	109 (27)
% eggs	49	51	61	39	46	54	63	37

Mean number per replicate (SE in brackets) and mean percentage. Ten females of each species were tested in 10 replicates. Numbers of eggs on substrates were compared with Wilcoxon paired sample test, *, $P < 0.05$; **, $P < 0.01$; ns, not significantly different ($P \geq 0.05$).

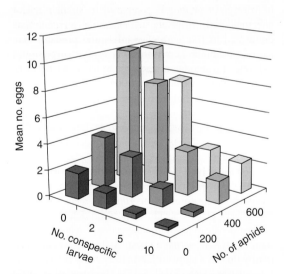

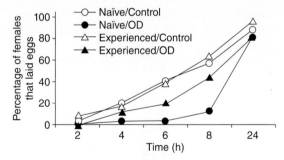

Figure 5.43 The percentage of experienced (triangles) and naïve (circles) females of *Adalia bipunctata* that laid eggs in oviposition-deterrence (OD) (black lines) and control treatments (grey lines) after 2, 4, 6, 8, and 24 hours (modified from Frechette et al. 2004).

Figure 5.42 Mean number of eggs laid by *Aphidecta obliterata* females over 24 hours on filter paper substrates differentially contaminated with conspecific larval tracks made by different numbers of larvae across a range of prey densities (from Oliver et al. 2006).

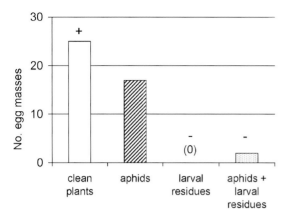

Figure 5.44 Number of egg masses laid by 38 *Hippodamia convergens* females on four types of three-leaf sorghum plants when tested individually in experimental arenas for 24–48 hours. Plant treatments: no additional stimulus (clean plants); a colony of *Schizaphis graminum*; residues of conspecific larvae (larval residues); a colony of *S. graminum* and conspecific larval residues. Plus and minus plant types, respectively, receiving significantly more or fewer egg masses than would be expected by chance if no plant type were preferred over any other (χ^2 goodness-of-fit test, P < 0.05) (from Michaud & Jyoti 2007).

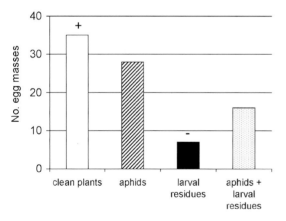

Figure 5.45 Numbers of egg masses laid by 56 *Hippodamia convergens* females on four types of three-leaf sorghum plants when tested individually in experimental arenas for 24–48 hours. Plant treatments: no additional stimulus (clean plants); a colony of *Schizaphis graminum* (aphids); residues of *Coleomegilla maculata* larvae (larval residues); a colony of *S. graminum* and *C. maculata* larval residues (aphids + larval residues). For other explanations see Fig. 5.44 (from Michaud & Jyoti 2007).

et al. (2007) assume that the three studied species do have overlapping habitats but, in fact, *A. decempunctata* and *C. septempunctata* only exceptionally share the same habitat (Majerus 1994, p. 142, Honěk 1985; also Table 5.12 and Fig. 5.16 in Hodek & Honěk 1996).

With *Menochilus sexmaculatus*, the behavioural function of the first **chemically defined ODS** was finally proved in coccinellids. A cuticular alkene, (Z)-pentacos-12-ene, was found in the chloroform extracts of first instars. It was then synthesized and its oviposition-deterring activity **proved by bioassay** (Klewer et al. 2007; Table 5.43). Several semiochemical substances were also found in *A. bipunctata* (Hemptinne et al. 2001), *A. decempunctata* and *C. septempunctata* (Magro et al. 2007), but their deterrent activity was not checked by bioassays.

Data on the **movement of females** (Ruzicka & Zemek 2003) **seemingly conflict** with Ruzicka's earlier findings (1997, 2001, 2002). However, the parameters measured here were different: not the number of eggs laid, but the **time spent on the substrates** and the distance walked on them. Females of *Cycloneda limbifer* stayed longer on a surface with conspecific tracks than on a clean surface. In contrast, they stayed less time where there were heterospecific tracks left by *Cer. undecimnotata* (Ruzicka & Zemek 2003). *Cer. undecimnotata* also spent longer on clean substrates than on those with conspecific tracks; that is, it behaved in a way similar to the oviposition-deterrence response. When number of eggs is recorded (Ruzicka 2003), *C. limbifer* shows the usual oviposition-deterrence response.

Ruzicka and Zemek (2008; Fig. 5.46) have recently documented with an automatic video tracking system that **larvae** (fourth instars of *C. limbifer*) also respond to the tracks of conspecific first instars. Inhibition of larval origin by conspecific larvae was found in *Hip. convergens* (Rutledge et al. 2008). These larval responses to larval tracks recall the old findings by Marks (1977).

Faeces of coccinellids appear to have a **similar function** as the larval tracks. Females of *Har. axyridis* and *P. japonica* reduce feeding and oviposition when exposed to conspecific larval and adult faeces or a faecal water extract. This mechanism might prevent repeated foraging on the same areas. *P. japonica* also responded to the faeces of *Har. axyridis* with the response being even stronger than that to conspecific faeces. As *P. japonica* is a potential IG prey while *Har.*

Table 5.43 Number of eggs laid by *Menochilus sexmaculatus* females on filter paper treated with a chloroform extract of conspecific first-instar larvae (*) and with different amounts of (Z)-pentacos-12-ene (+) versus solvent controls (after Klewer et al. 2007).

Concentration µg/cm² or larval equivalent/cm² (*)	Treatment	Control	Wilcoxon test
0.25*	46 ± 6.7	81.5 ± 4.7	P = 0.002
1.25(+)	121.5 ± 12.4	207.5 ± 10.9	P = 0.004
0.125(+)	105.0 ± 16.3	166.7 ± 11.0	P = 0.006
0.0125(+)	148.7 ± 10.4	135.9 ± 10.4	P = 0.625

Mean number of eggs laid in dual-choice tests with 10 replicates of 10 females. Altogether 100 individuals were tested in 10 replicates in one test. Wilcoxon paired sample test.

(a)

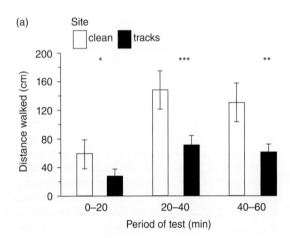

(b)

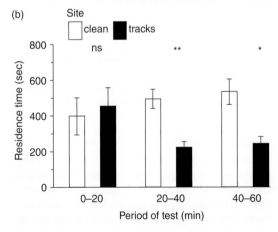

Figure 5.46 Effects of conspecific tracks of first instar *Cycloneda limbifer* on (a) the distance walked and (b) residence time of fourth instars (modified from Ruzicka & Zemek 2008).

Table 5.44 Effect of water extract of larval and adult feces of conspecifics and heterospecifics on oviposition by *Propylea japonica* and *Harmonia axyridis*; cues of the small species do not affect oviposition of *H. axyridis* (after Agarwala et al. 2003a).

Treatment by water extract of	Eggs laid after 24 hr by female (mean ± SE)	
	P. japonica	*H. axyridis*
P. japonica feces		
Adults	7.8 ± 0.95 (a)	40.5 ± 2.58 (a)
Larvae	7.3 ± 0.53 (a)	37.5 ± 2.31 (a)
H. axyridis feces		
Adults	3.7 ± 0.81 (b)	23.3 ± 2.57 (b)
Larvae	3.1 ± 0.67 (b)	21.8 ± 1.73 (b)
Water only (control)	14.8 ± 1.14 (b)	38.9 ± 2.00 (a)

Different letters in a column indicate significant difference at P < 0.05; Scheffé's multiple range test (*n* = 10).

axyridis a strong IG predator, the response may help *P. japonica* to avoid dangerous encounters (Agarwala et al. 2003a; Table 5.44).

Coccinellid tracks are even avoided by **parasitoids**. *Aphidius ervi* avoided the trails of *C. septempunctata* adults or fourth instars (Nakashima & Senoo 2003). The signal lost its function rather quickly, after about 18–24 hours. An OD response to **eggs**, but not larval tracks, was found in a **phytophagous** coccinellid *Henosepilachna niponica*, a specialized herbivore on *Cirsium kagamontanum* (Ohgushi & Sawada 1985).

The discovery of OD caused by conspecific larval tracks (Ruzicka 1994, 1997) makes the scenario likely that an aphid patch becomes – after a certain delay of time – unsuitable for ovipositing females due to a **critical density of larval tracks** from an earlier oviposition.

In a hypothetical concept Kindlmann and Dixon (1993) expressed a general assumption that 'the coccinellids should lay a few eggs early in the development of an aphid colony'. The authors rightly mention that their model is relevant only for isolated aphid patches, escape from which might br dangerous for the larvae because of long distance to alternative patches, (i.e. 'when no other aphid colonies are available', in Kindlmann and Dixon 1993, p. 448). In most situations, however, the larvae can reach another food resource; even just emerged first instar larvae may walk without prey for 1.0–1.5 days (Banks 1957). Ruzicka's (1994) discovery of oviposition deterrence was incorporated into the concept that closing of the period suitable for oviposition ends due to 'the adults' response to the tracks left by conspecific larvae' (Dixon 2000, p. 106).

Coccinellid species often have specific preferences for certain microhabitats or **vegetation strata** and this affects their choice of oviposition sites. Thus *Col. maculata*, which prefers more shaded and more humid environments and feeds on maize pollen that has fallen onto the lower leaf axils (Ewert & Chiang 1966, Foott 1973), deposits all its eggs on lower leaves of the maize, while *C. septempunctata* and *Hip. tredecimpunctata* lay about one-third of their eggs on upper leaves (Coderre & Tourneur 1986). The location of the egg batches may also be affected by **temperature preferences**; thus in early spring *C. septempunctata* prefer the warm lower parts of wheat plants (Honěk 1979, 1983). Sometimes the eggs are laid in unexpected places. Females sometimes avoid the growing parts of the plants, which are usually the most aphid-infested, and instead oviposit in places unsuitable for the hatching larvae, e.g. on the **soil or stones** (Ferran et al. 1989). In the mild winter of central Honshu, pupae and/or eggs of *C. s. brucki* were found in directly insolated but unusual microhabitats such as metal cans or paper and wooden material on sites exposed to solar radiation (Ohashi et al. 2005).

5.4.1.4 Foraging of first instars

It has been suggested that egg size is constrained by the need to assure energetic reserves for the neonate larvae (Stewart et al. 1991). Osawa (2003) questions this premise with the alternative suggestion that newly hatched larvae get energy by eating conspecific eggs. Sloggett and Lorenz (2008) suggested that **egg nutrient content** rather than egg size is the most important means of parental provision of energetic reserves. They compared three ladybird species (*A. bipunctata*, *A. decempunctata*, *Anisosticta novemdecimpunctata*) that exhibited a decrease in egg mass as food specialisation increased. As the more specialised ladybirds can persist at lower aphid densities than generalists (Sloggett & Majerus 2000, Sloggett 2008b), it would appear that the more specialized ladybirds in this comparison are not provisioned adequately for foraging at low prey density. However, analysis of the eggs showed that the smaller eggs of *A. novemdecimpunctata* contained relatively more **lipid and glycogen**, and are thus actually the richer in energy (Sloggett & Lorenz 2008). A part of the lipid remains in the larva as an energy source.

Before moving on to the following section on the stresses of neonate larvae, we draw the reader's attention to sections 5.2.8 and 8.4.5.2.

Newly hatched first instars risk heavy mortality when searching for prey when it is at a very low prey density. The first instar larvae of aphidophagous ladybirds spend some time on the empty eggshells (Banks, 1957, reported 12–24 hours for *P. quatuordecimpunctata*) and eventually indulge in cannibalism. Then they can maintain their active search for prey for 25–35 hours before becoming inactive due to exhaustion. Therefore, they need to find food within 1.0–1.5 days (Table 5.34, p. 217). Ng (1988) reported similar evidence, quoting a survival range of 15–32 hours for several species.

Sibling cannibalism on eggs of the same clutch usually prolongs survival (5.2.8) and the searching capacity of newly hatched larvae. The consumption of just a single egg nearly doubled the survival of the first instars (Banks 1954, species not given). Significantly increased survival of first instar larvae after feeding on sibling eggs that hatch late, are unfertilized or killed by microorganisms (Chapter 8.4), was also later recorded in other species (see Hodek 1996). Hurst and Majerus (1993) reported not only an increased survival (by 75%), but also an increased speed of movement in neonate larvae of *A. bipunctata* after eating an egg. First instar *Cheilomenes lunata* moulted to the second instar after eating only two eggs (Brown 1972). This was achieved in *A. bipunctata* (Banks 1956), and in *Har. axyridis* (Osawa 1992b) after eating three eggs.

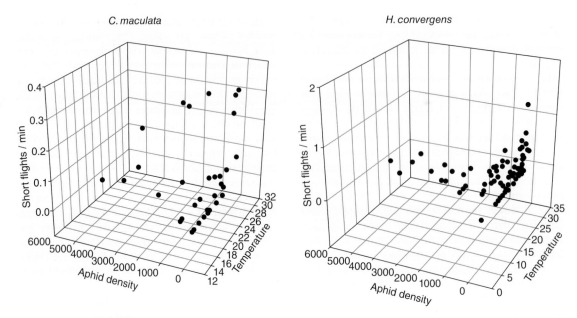

Figure 5.47 The frequency of short (<2 m) flights (number of flights/min.) by *Coleomegilla maculata* and *Hippodamia convergens* in relation to ambient temperature (°C) and cereal aphid density. Note differences in scales of the vertical axes (modified from Elliott et al. 2000).

5.4.1.5   Movement among habitats/patches in the landscape

After small scale foraging (e.g. at the level of the individual plant) discussed above, we now discuss **larger scale foraging**. Highly mobile coccinellid generalists range widely and feed and reproduce in many habitats. When a prey patch becomes unsuitable for oviposition, the females search for other favourable patches; their offspring are thus widely scattered. Dispersion is aided by the deterrent effect on oviposition of larval tracks (5.4.1.3), preventing clumping of pre-imaginal stages in the richest patches of prey and reducing intraspecific competition.

Interspecific differences in vagility were noted by Elliott et al. (1998, 2000; Fig. 5.47), who observed that the more vagile *C. septempunctata* and *Hip. convergens* were more affected by broader scale variation in landscape structure than the less vagile *Col. maculata* and *Hip. parenthesis* (see also Brewer & Elliott 2004). Evans (2004) reported that, in alfalfa fields, *C. septempunctata* females are more abundant when *Acyrthosiphon pisum* populations are low than are native ladybird species. Snyder et al. (2004) suggested

that such an early arrival of *C. septempunctata* to developing aphid colonies, and their earlier oviposition there, gave larvae a size advantage over the later-developing offspring of native species.

Habitat features (Chapter 4) can reduce the success of ladybirds in locating incipient outbreaks of prey. Kareiva (1987), for example, found that **habitat fragmentation** impaired the searching ability of *C. septempunctata* adults foraging in experimental arrays of goldenrod, with isolated aphid populations. The ability of ladybirds to discover localized prey populations reflects effective resource tracking, brought about by constant movement of individuals between host plants as shown for *Har. axyridis* (Osawa 2000). Building on Kareiva's (1987) results, With et al. (2002) found that adults of *Har. axyridis* tracked aphid population dynamics more effectively than did adults of *Col. maculata* within and across experimentally fragmented landscapes with 10–80% cover of red clover planted in clumped or fragmented fashion. Grez et al. (2005, 2008) assessed and distinguished between the effects of habitat loss and habitat fragmentation on population levels of ladybirds and aphids in differently fragmented plots of alfalfa.

The **use of alternative habitats** by *C. septempunctata* was followed in detail over two seasons in Michigan, USA, by Maredia et al. (1992). A diverse agroecosystem, composed of a mosaic of habitats, supported *C. septempunctata* from late April till late August, with the coccinellid feeding on various aphid species on annual and perennial crops (wheat, maize, soybean and alfalfa), poplar and weeds. The ability of ladybirds to track prey populations through space and time can lead to rapid large-scale changes in resource use by a species, as in the colonization of orchards by *Olla v-nigrum* following the invasion of the citrus psyllid in Florida (Michaud 2001). In northern Italy, Burgio et al. (2006) observed seasonal movements of *Hip. variegata, C. septempunctata* and *P. quatuordecimpunctata* among arable crops, fallow, vegetable crops and weedy field margins.

Immigration/emigration

A heterogeneous landscape presents a mosaic of potential patches within which highly mobile coccinellid adults can select those that are suitable. The **initial discovery** by predators of local prey populations of low density is intriguing. When such responses are rapid and strong, ladybirds may be effective in preventing large-scale outbreaks of their prey (Frazer et al. 1981). Field experiments reveal that small, highly localized populations of mites and aphids can be quickly discovered and reduced in size by ladybirds (Frazer et al. 1981, Congdon et al. 1993, Evans 2004). Evidence is still lacking with respect to the sensory cues employed by **flying** coccinellid **adults** to **locate profitable patches**. There are records of adult olfactory responses to aphids at short distances, but these cues have not yet been checked as long-distance attractants (5.4.1.2). Visual cues (e.g. host discolouration from aphid damage) might act over longer distances (e.g. Lorenzetti et al. 1997). Generally, long-distance searching behaviour of ladybirds needs further intensive investigation.

Many workers have suggested that immigration rates of ladybirds are independent of local prey density (Ives et al. 1993, Osawa 2000, van der Werf et al. 2000, Cardinale et al. 2006). In the study by Cardinale et al. (2006), immigration rates of adult *C. septempunctata, Col. maculata* and *Har. axyridis* were independent of aphid density in adjacent $1 \, m^2$ plots. Flying adults arrive in patches at random and then make decisions to stay or leave based on the presence or abundance of

prey, as well as signals left by conspecific larvae (5.4.1.3). Frazer (1988) estimated that as many as 50% of ladybirds emigrate from a patch every day only to be replaced by immigrants. Thus the abundance of adult coccinellids in a patch reflects the net balance between rates of immigration and emigration. Empirical evidence indicates that there is greater **emigration** from patches of lower quality than from patches of higher quality (Kareiva 1990). When placed on individual potted plants in the laboratory, adults of *C. septempunctata* varied in how long they remained before leaving, depending on the number of aphids present (Minoretti & Weisser 2000). In a diversified garden, Osawa (2000) found that marked adults of *Har. axyridis* tended to stay longer at sites with greater numbers of aphids. The rate of emigration from an alfalfa field was particularly high in one of three mark–recapture experiments in which the resident aphid density was especially low (van der Werf et al. 2000). On a smaller spatial scale, Cardinale et al. (2006) used visual censuses to measure emigration rates of adults from patches of alfalfa with one of three aphid densities. Emigration rates of *C. septempunctata* and *Col. maculata* (but not *Har. axyridis*) increased with decreasing aphid density among patches. In a **model**, based on empirical data on dispersal of aphidophagous coccinellids, the best fit for three scenarios was the one including unconditional immigration rates (Krivan 2008; 5.3.5). Detailed studies on the emigration of ladybirds are warranted and should focus on the levels of adult satiation, as suggested by Evans (2003).

Foraging over landscape in swarms

In spite of behavioural traits predicted by the optimal foraging theory (5.4.1), in order to keep the abundance of progeny within the limits of optimal fitness, coccinellids can exhibit population explosions resulting in travelling swarms of young 'hungry' beetles. The 'swarming' beetles may ultimately get into lakes or sea, are then washed up and form aggregations on beaches (Klausnitzer 1989, Isard et al. 2001, Denemark & Losey 2010). Although the ladybird 'swarming' is not yet fully understood, it has been recorded many times and described in reviews and books (e.g. Hagen 1962, Hodek 1973, p. 82, Hodek et al. 1993, Majerus 1994, p. 186, Majerus & Majerus 1996, Nalepa et al. 1998, Turnock & Wise 2004). The species often described as behaving in this manner is *C. septempunctata*. In Surrey,

UK, Majerus (1994) recorded densities of over 1000 pupae of *C. septempunctata* per square metre in a nettle patch. There were 126 individuals on a single nettle stem. For *Hip. convergens*, Dickson et al. (1955) estimated 54,000 individuals per 0.4 ha of alfalfa. In some years, when coccinellids emerge from their pupae, aphids are scarce, having already been consumed by them as larvae or by other aphidophaga. The deteriorating physiological state of host plants, emigration of aphids and their infestation by fungi also play a role in the disappearance of aphids.

This phenomenon has rarely been reported in other than aphidophagous coccinellids. The tendency towards the 'risky' overproduction of progeny, particularly in *C. septempunctata*, has not been out-selected, although it seems to work against the constraints and models of an 'optimal' strategy. The idea that overpopulation is a 'suboptimal behaviour' leading to 'low numbers the following year' (Kindlmann & Dixon 1993) is not in concert with observations. There is no evidence that fewer adults enter dormancy in years of population explosion.

Maximum reproductive potential is realized in spite of the risk that a great proportion of the offspring is lost. This might be a result of the evolutionary trade-off between such loss and the fact that individual parents thereby increase the probability of random survival of their offspring. Evidently enough individuals originating from a population explosion survive to reproduce the next spring and perpetuate this life-history trait. Perhaps this feature is one reason why *C. septempunctata* has proved to be an exceptionally successful competitor during the rapid expansion of its distribution area in the Nearctic region (Krafsur et al. 2005, Snyder & Evans 2006, Frank & McCoy 2007, Harmon et al. 2007) from the Atlantic to the Pacific coast.

5.4.1.6   Ladybird foraging and ants

Many species of ants attend honeydew-producing Sternorrhyncha (see also Chapter 8.1.4). Early observations and experiments showing that attendant ants are hostile to enemies (including Coccinellidae) of Sternorrhyncha are summarized by Nixon (1951) and particularly by Way (1963). Nixon supposed that the ants do not protect the honeydew producers actively as a source of food, but merely incidentally. Way (1963) suggested that the ants are **aggressive towards intruders** on their food sources, but that intruders are tolerated away from the food source. From a study of a

colony of *Aphis fabae* on *Cirsium arvense* tended by *Myrmica ruginodis*, Jiggins et al. (1993) also concluded that the ants vigorously **defend the aphid colony** against coccinellids that are in the colony or nearby. *Lasius niger* workers are not hostile to coccinellid adults that they meet away from attended aphids (Bhatkar 1982). Thus, the protection given by ants to Sternorrhyncha is related to their value as a food source.

Defence provided by attendant **ants** can **nullify the effectiveness** of aphid predators (Powell & Silverman 2010). In laboratory experiments, *Aphis gossypii* was tended by the red imported fire ant, *Solenopsis invicta*, which was killing both adults and larvae of *Hip. convergens*, and also larval chrysopids and syrphids (Vinson & Scarborough 1989). A project of inundative biological control of aphids (*Brachycaudus persicae*, *B. prunicola*, *Hyalopterus pruni*, *Myzus persicae*) in peach orchards by *Adalia bipunctata* larvae was impaired by ants (*Lasius niger*, *Formica rufibarbis*) until the latter were excluded from the trees by glue bands (Kreiter & Iperti 1986). Ants may kill both adult and larval coccinellids (Sloggett et al. 1999). Even the wax-producing larvae of *Cryptognatha nodiceps*, *Pseudoazya trinitatis* and *Zagloba aeneipennis* were eaten by several species of ant attending the coccid *Aspidiotus destructor* in the New Hebrides (Cochereau 1969). However, the specific wax with ultraviolet reflectiveness (unique in Coccinellidae) does protect pupae of *Apolinus lividigaster* (Richards 1980). *Scymnus* larvae with experimentally removed wax cover were more vulnerable to the attacks by carabids and ants, while the wax cover did not influence cannibalism or interspecific coccinellid predation (Voelkl & Vohland 1996).

Larvae of *C. septempunctata* had a higher predation rate on *Aphis craccivora* than on *Acyrthosiphon pisum* when neither was guarded by ants. *A. pisum* is a non-tended aphid, while *A. craccivora* is facultatively ant-tended. The difference in foraging efficiency of ladybirds on these two aphids can be explained by aphid behaviour (Table 5.45). In the non-protected *A. pisum* dropping off from plants has evolved as a response to alarm pheromone. This predator-avoidance behaviour decreases the efficiency of foraging coccinellid larvae. When *A. craccivora* was guarded by ants, the predation by *C. septempunctata* larvae was much lower than in absence of the ants (Suzuki & Ide 2008). An increase in coccinellid populations (mainly *A. decempunctata* and *Scymnus interruptus*) was observed in citrus trees infested by *Aphis spiraecola* (Piñol et al. 2009).

Table 5.45 Behaviour of *Coccinella septempunctata* larvae and two species of aphids on *Vicia faba* as affected by ants (*Lasius japonicus*) (modified after Suzuki & Ide 2008).

| Parameters of behaviour | In the absence of ants | | | | In the presence of ants | | |
| | *Aphis craccivora* | | *Acyrthosiphon pisum* | | *Aphis craccivora* | | |
	n	Mean	*n*	Mean	*n*	Mean	*P**
Aphids attacked (*n* per hr)	20	14.09 a	20	14.97 a	100	5.70 b	<0.0001
Predation success rate	20	0.98 a	20	0.74 b	81	0.81 ab	0.0001
Aphids eaten (*n* per hr)	20	13.74 a	20	10.98 b	100	4.22 c	<0.0001
Aphids escaped by dropping (*n* per hr)	20	9.36 a	20	59.05 b	100	6.50 a	<0.0001
Aphids escaped by walking (*n* per hr)	20	12.86 a	20	2.26 b	100	4.18 c	<0.0001
Resident time on plant (s)	20	7942.80 a	20	3500.60 b	100	1438.20 c	<0.0001

*Kruskal-Wallis test.
Values followed by different letters indicate significant differences (Mann-Whitney U-test with Bonferroni adjustments for multiple tests, P < 0.0167), (values of SE omitted).

The predation rate of the myrmecophilous coccinellid *Azya orbigera*, an important predator of the green coffee scale *Coccus viridis*, is not decreased in the presence of the mutualistic ant *Azteca instabilis*. Furthermore, the ant showed aggressive behaviour toward *A. orbigera*'s parasitoids and its presence effectively decreased the parasitisation (Liere & Perfecto 2008). Similar behaviour was observed in *Platynaspis luteorubra* (Voelkl 1995).

It is generally assumed that the Argentine ant, *Linepithema humile*, tends honeydew-excreting scale insects and its presence impacts negatively on natural enemies. This was analysed by exclusion experiments in Californian vineyards infested by the mealybugs *Pseudococcus maritimus* and *Pseudococcus viburni*. Argentine ants, however, increased the density of the ladybird *Cryptolaemus montrouzieri* (Table 5.46). The ants also increased the population density of mealybugs, but not by disrupting the activity of natural enemies. The effect was rather due to the removal of honeydew; when ants did not remove the honeydew, mealybug crawlers became trapped in it. The larvae of *C. montrouzieri*, which resemble mealybugs by also being covered with wax, successfully mimic mealybugs to avoid detection by ants (Daane et al. 2007).

Ant protection of trophobionts may be ineffective against specialized predators (including coccinellids) that circumvent attack by trophobiont-tending ants. The profiles of three aphid predators' cuticular hydrocarbons (CHCs) are strikingly convergent to those of

Table 5.46 Density per vine per sample of Argentine ants, obscure mealybugs (MB), and mealybug destroyer on vines in ant-exclusion and ant-tended treatments in two vineyards in the Californian Central Coast wine grape region (modified after Daane et al. 2007).

Vineyard	Insect	Ant-tended	Ant-excluded
A	Argentine ant*	33.8 ± 7.4	0
	Obscure MB[†]	303.2 ± 45.8	11.8 ± 4.1
	MB destroyer[‡]	0.31 ± 0.09	0.07 ± 0.04
B	Argentine ant	38.6 ± 7.0	0.4 ± 0.3
	Obscure MB	129.2 ± 20.5	59.2 ± 11.7
	MB destroyer	0.29 ± 0.09	0.09 ± 0.05

**Linepithema humile*.
[†]*Pseudococcus viburni*.
[‡]*Cryptolaemus montrouzieri*.

ant-attended aphids, while the CHC profiles of ants are different. Chemically mimicking the CHCs of prey allows the predators to avoid detection both by the aphids and by tending ants (Lohman et al. 2006). Larvae of *Platynaspis luteorubra* also mimic the CHCs of prey aphids (cited by Majerus et al. 2007).

When tending the soybean aphid *Aphis glycines*, the ant *Monomorium minimum* was observed harassing or killing *Har. axyridis* and an anthocorid. Ant-attendance resulted in reduced predation and increased aphid numbers up to 10-fold. Coccinellid adults were attacked by ants immediately, while the larvae were not attacked until direct physical contact had occurred (Herbert & Horn 2008). Small colonies of *Aphis craccivora* were less preyed upon by *C. septempunctata* when the host plant had extrafloral nectaries that attracted the aphid-attending ants (Katayama & Suzuki 2010).

5.4.2   Prey capture

When an encounter betwen a coccinellid and a prey occurs, the coccinellid may fail to capture and feed on the latter. This aspect of predator–prey relationships may be of paramount importance for the impact of the predator (Varley & Gradwell 1970), but has not yet attracted enough attention.

Although **aphids** are generally considered to be completely 'helpless' (Imms 1947), their **defensive and escape behaviour** may actually be quite efficient. Several authors have studied such responses of aphids towards two *Adalia* spp., both larvae and adults of *A. decempunctata* (Dixon 1958, 1959), and larvae of *A. bipunctata* (Klingauf 1967, Wratten 1976). Stadler (1991) compared such behaviour within the genus *Uroleucon* (Fig. 5.48). The **aphids** show various **defence** responses, e.g. kicking movements, move-

ments of the body, pulling free the appendage seized by the coccinellid and 'waxing' the coccinellid with an oily liquid appearing on the tip of the siphunculi. Moreover, the aphids may **escape** by walking away or dropping off the plant (Prasad & Snyder 2010). The intensity of the aphid response varies with the stage of the predator; thus *A. decempunctata* adults induce escape rather than defence in *Microlophium carnosum* (Dixon 1959). The success or failure of a predator to **capture the prey** depends on their relative sizes. Older larvae of *A. bipunctata* (9 mm long) succeed in capturing 90–100% of first instar and about 60–70% of adult *Myzus persicae* or *Neomyzus circumflexus*, but only 0–50% of the much larger *Acyrthosiphon pisum* (Klingauf 1967). Aphids with long appendages are more difficult for *Hip. quinquesignata* larvae to capture than are more compact aphids with short appendages (Kaddou 1960).

The suitability of an aphid species as prey for *A. bipunctata* was strongly determined by capture efficiency (Mills 1979). The latter amounted to 58% for the most suitable prey, *Eucallipterus tiliae*, but only to 21% for the least suitable *Drepanosiphum platanoidis*. The level of predation by *A. bipunctata* larvae on three birch aphid species was also dependent upon aphid behaviour (Hajek & Dahlsten 1987). The long-legged *Euceraphis betulae*, the most successful escapee, was highly mobile and frequently walked

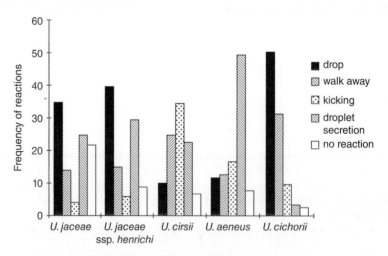

Figure 5.48 Relative frequency of defence and escape reactions of different *Uroleucon* species to attacks by *Coccinella septempunctata* larvae. Reactions are arranged according to the defence value for the aphids (drop > walk away > kicking > droplet secretion > no reaction) (from Stadler 1991).

Table 5.47 Results of encounters between three species of birch aphids, *Betulaphis brevipilosa*, *Callipterinella calliptera*, *Euceraphis betulae* and larvae of *Adalia bipunctata* (Hajek & Dahlsten 1987).

	Aphid species	Total encounters	% captures	% active escapes	% passive escapes	% aphid winning over coccinellid
		L1 *Adalia bipunctata*				
L1	*B. brevipilosa*	33	60.6	0.0	39.4	0.0
L4	*B. brevipilosa*	56	35.7	8.9	55.4	0.0
L1	*C. calliptera*	51	39.2	41.2	19.6	0.0
L4	*C. calliptera*	214*	8.4	71.5	20.1	0.0
L1	*E. betulae*	216*	6.9	75.5	0.0	17.6
L4	*E. betulae*	260*	0.0	71.5	0.0	28.5[†]
		L4 *Adalia bipunctata*				
L1	*B. brevipilosa*	65	30.8	1.5	67.7	0.0
L4	*B. brevipilosa*	38	52.6	0.0	47.4	0.0
L1	*C. calliptera*	33	60.6	9.1	30.3	0.0
L4	*C. calliptera*	66*	30.3	60.6	9.1	0.0
L1	*E. betulae*	104*	19.2	78.8	1.0	1.0
L4	*E. betulae*	375*	1.9	82.6	0.0	15.5

*The numbers of encounters were increased for these mobile aphids.
[†]A leg or rostrum of L4 *E. betulae* was often grasped by L1 *A. bipunctata*. The coccinellid did not hold the prey long before the aphid simply pulled away with no evidence of impairment.

away from the predator. In contrast *Betulaphis brevipilosa*, a flat and sessile species, was the least successful active escapee, but often passively escaped detection (Table 5.47).

Wratten (1976) stressed the importance of the direction of the predator's approach to the prey, *Eucallipterus tiliae*. The larvae of *A. bipunctata* achieved the highest contact rate when they approached fourth instars from the rear, suggesting that aphid vision is important in predator avoidance. Earlier, it was supposed that the escape response of aphids resulted from their **visual ability**. According to Klingauf (1967), *A. pisum* perceives the predator at a distance of 4–10 mm, while some predators perceive the prey only by contact. This time advantage for the prey can be used for escape. However, the ability to perceive the prey at a short distance (7–10 mm) has now been found in several aphidophagous coccinellid species (5.4.1.2) and consequently the above explanation of successful escape by aphids has partly lost its plausibility. An aphid may gain time when a small coccinellid larva at first shows a fright response and only attacks the prey after a delay (Baensch 1964). The intensity of the fright response depends on the size of the prey and is suppressed

by learning once several prey have been captured (Baensch 1964, Klingauf 1967).

Another efficient escape/defence mechanism is the secretion of **alarm pheromone** (5.4.1.2) by attacked aphids (early numerous references in Hodek 1996). Most authors worked with the aphid *A. pisum*, which easily drops off plants when disturbed. *A. bipunctata* adults forage more vigorously at higher temperatures and this producei greater vibrations and a high rate of aphid dropping behaviour. Pea aphids react to vibrations which enhance their responsiveness to later alarm pheromone stimuli (Clegg & Barlow 1982, Brodsky & Barlow 1986).

Of the aphid species tested for an alarm pheromone response, *Schizaphis graminum* has been the most sensitive. In field tests on sorghum plants, *S. graminum* adults and larvae were dislodged five and four times more often, respectively, than they were consumed by *C. septempunctata* adults (McConnell & Kring 1990; Table 5.48). The ratio between disturbed and consumed aphids is species-specific. In *A. pisum*, this ratio is 30:1, while in *Aphis gossypii* it is only 1:14. The level of disturbance is also coccinellid-specific. Thus *A. bipunctata* causes little disturbance and so the

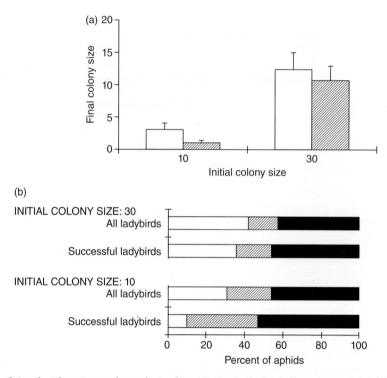

Figure 5.49 Fate of *Acyrthosiphon pisum* on bean plants after visit of a single *Coccinella septempunctata* adult, as affected by initial size of the aphid colony (*n* = 9–11). (a) Number of aphids remaining on the plants at the end of the experiments; open bars, all replicates; hatched bars, only replicates in which ladybirds killed at least one aphid. (b) Aphids eaten (hatched); emigrated from the host plant, i.e. dislodged (black); remaining on the plant (open) (from Minoretti & Weisser 2000).

Table 5.48 Number of aphids consumed and dislodged by *Coccinella septempunctata* in laboratory and field tests (mean ± SEM) (McConnell & Kring 1990).

Aphid stage	Consumed	Dislodged
Laboratory		
Adults	6.7 ± 0.4	12.0 ± 1.0
Nymphs	26.9 ± 2.2	34.8 ± 2.9
Field		
Adults	6.7 ± 0.4	33.7 ± 2.6
Nymphs	8.9 ± 0.6	35.6 ± 2.6

percentage of prey consumed is high (28:72 and 19:81) whereas in *C. septempunctata* the proportion of disturbed aphids dominates (94:6 for *A. pisum* and 82:18 for *Schizaphis graminum*) (Nelson & Rosenheim 2006). The ratio of aphids disturbed to aphids consumed also depends on the size of their colonies (Minoretti & Weisser 2000; Fig. 5.49).

Predation by *Har. axyridis* on *A. pisum* was reduced by approximately 40% by aphids dropping (Francke et al. 2008). The ratio between aphid dropping and aphid consumption is correlated with the predator's stage of development and its rate of movement. *Har. axyridis* daubed with aphid pheromone droplets cause a greater proportion of aphids to drop off a plant (Mondor & Roitberg 2004).

Dispersal of dropping pea aphids, caused by adult coccinellids (*C. californica*), may have a negative effect on crops by spreading plant viruses (Roitberg & Myers 1978, Roitberg et al. 1979).

When the host plants of aphids are infected by **endophytic fungi** the production of winged aphids, as a response to predator (*A. bipunctata* larvae) threat, is decreased (Zuest et al. 2008).

Some aphids, e.g. *Ceratovacuna* and *Pseudoregma* spp., produce sterile **soldiers** that **attack aphid predators** (Aoki et al. 1981, Arakaki 1992b, c; also Chapter 8.1.5). *Sasajiscymnus kurohime* is a specialized predator of such aphids, successfully attacking the sugar cane woolly aphid, *Ceratovacuna lanigera*, in spite of its soldier-like first instars that defend the colony. This is because the coccinellid larvae can move among the woolly aphids without eliciting an escape response from them. Only 7.5–12.9% of the aphids secreted alarm pheromone when attacked by larvae of *S. kurohime*, while the figure was 82.5% for attack by adults. It is assumed that the coccinellid larva is protected by its waxy covering, similar to that of the aphid (Arakaki 1992b, c). However, chemical mimicry may be more likely (J.J. Sloggett, unpublished), similar to the case described by Daane et al. (2007; 5.4.1.6).

In addition to defensive movements the aphids use also **chemo-mechanical defences**. Early on, Palmer (1914) reported that large aphids of the genus *Macrosiphum* smear the mouthparts of *Hip. convergens* with a glue excreted from the siphunculi and thus cause starvation of the coccinellid. This defence seems successful only at low temperatures; at higher temperatures the coccinellids can ingest the glue. A similar defence mechanism was reported in *Hyalopterus pruni* against *A. bipunctata* (Hawkes 1920). The monophagous aphid, *Aulacorthum nipponicum*, feeding exclusively on *Paederia foetida*, is seldom attacked by coccinellids. Biting into the aphid is immediately followed by dropping the prey, salivating and by a persistent grooming of mouthparts. The droplets on the tip of the siphunculi contain a potent **deterrent**, an iridoid glycoside **paederoside**, originating from the host plant (Nishida & Fukami 1989).

5.4.3 Food intake

The epilachnine **phyllophagous** coccinellids scrape the parenchyma off the leaves and suck it out, imbibing only fluid. The sucked-out parenchyma remains on the leaf, leaving a characteristic feeding pattern (Butt 1951, Klausnitzer 1965). The **pollinivorous** and **mycophagous** larva of *Tytthaspis sedecimpunctata* uses a comb-like mandible to collect pollen and spores (Ricci 1982; Fig. 5.42, 5.2.9).

The food of **aphidophagous** coccinellids mostly consists not only of the sucked body fluid of aphids, but also frequently solid parts of the aphid body, such as legs, antennae, etc. (Butt 1951, Triltsch 1999). The type of feeding depends on the comparative size of prey and predator. When the prey is too large, it is merely sucked out and the skin with appendages is discarded (Butt 1951, Harpaz 1958, Hagen 1962, Kesten 1969). This type of feeding behaviour is normal in younger larvae (first and second instar). Older larvae and adults may consume smaller prey completely (Butt 1951, Hagen 1962, Kesten 1969, Triltsch 1999).

Most coccinellids, particularly when larvae, show **extra-intestinal digestion** (extra-oral digestion of Richards & Goletsos 1991); they periodically regurgitate fluid from the gut into the chewed prey and suck back the pre-digested food. Some coccinellid larvae even perform extra-intestinal digestion through their **sucking mandibles**, as do the larvae of *Dytiscus*, *Chrysopa*, etc. (Wigglesworth 1953). This type of pre-digestion has been observed in larvae of *Stethorus* spp. (Blattny & Osvald 1949, Fleschner 1950, Putman 1955, Hagen 1962) and in the unusually shaped larvae of *Platynaspis luteorubra*. These larvae have repeatedly been observed sucking aphids hollow merely via one of the aphid's legs; during this process the body of the aphid is alternately deflated and inflated (I. Hodek unpublished). Ricci (1979) gives a detailed description of extra-intestinal digestion in *P. luteorubra*.

Feeding can easily be observed, especially with pale-coloured aphids, for the internal and surrounding tissues of the part seized by the larva's mandibles become darker during the injection process and return to their natural colour a few seconds later during reimbibition (for details see Hodek 1996). The extra-intestinal (extra-oral) feeding of the larvae of *Apolinus lividigaster* (Richards & Goletsos 1991) is similar to that of *P. luteorubra*. The adults of *A. lividigaster* have bifid mandibles and chew their prey, while the sickle-shaped unidentate mandibles of the larvae pierce and then hold prey during the repeated sucking. In the larvae of *Scymnus* s. str., *Scymnus subvillosus* and perhaps also *Nephus includens* it appears that feeding is similar (Ricci 1979).

A special type of feeding behaviour was observed by Richards (1980, 1985). *Rhyzobius ventralis* larvae sometimes attack the vulnerable females of the coccid *Eriococcus coriaceus* through a large opening in the latter's sac, through which the tip of its abdomen protrudes and crawlers escape. The coccinellid larva usually makes a hole in the lower part of the sac, curls itself around it and remains there feeding until the prey is empty. Larvae and adults of the large Australian coccidophagous species *Rodatus major* (originally described as *Rhizobius*) are specialized on the eggs of *Monophlebulus pilosior* (Margarodidae), and the fourth instar feeds on the eggs as they are laid 'lying with its mouth close to the margarodid's genital opening' (Richards 1985). A similar behaviour has been recorded in first instar *Sasajiscymnus kurohime*, attacking the larvae of the 'soldier' producing aphid *Ceratovacuna lanigera* at the moment when they are born (Arakaki 1992a; 5.4.2).

5.5   CONCLUSIONS

The analysis of coccinellid food relations is not only very interesting and important from a theoretical point of view, but also for its practical consequences – for **predicting** ladybird **activity** in the field and for **improving** their **impact** in the contexts of conservation and augmentative biological control.

It is worth **comparing** the progress achieved in **individual** food-related **fields**. A very great qualitative step forwards has been made in **ethological research** (5.4) and thus partially matches the earlier greater progress with parasitoids. Fifteen years ago, it was still an enigma as to whether ladybirds find their prey by random movements – in spite of a few observations of non-random behaviour. In addition to indirect clues on matters such as taxes or characteristics of host plants (5.4.1.1), the only mechanism definitely known in foraging was the **role of encounter** that switches ladybird movement to a concentrated search (5.4.1.2). In contrast, there is now a wealth of evidence on cues that guide coccinellid behaviour. At least four types should be mentioned: **volatiles of plants** and/or **of prey**, aphid alarm pheromones (all in 5.4.1.2) and oviposition-deterrent chemicals in **larval tracks** (5.4.1.3). Study of ladybird orientation to volatiles has not progressed beyond laboratory tests, with field assays remaining to be completed. In searching for an oviposition site, there is a trade-off for coccinellids between two factors: whether to lay eggs near to prey, so that the first instars find prey early, or whether to reduce attacks by conspecific or heterospecific ladybirds on offspring by ovipositing further away (5.4.1.3).

Although the research discussed in 5.2 and 5.3 has not stagnated in the course of the last 15 years, not so many breakthroughs have been made in these fields.

Since about 50 years ago, when **prey specificity** in coccinellids was discovered and the principal definition of essential v. alternative food established (5.2.2, 5.2.11), further evidence has accumulated, and new **allelochemicals** recorded (5.2.6.1) – on which the trophic specificity is mostly based. However, recently prey specificity has begun to be discussed also from a novel angle: **size relations** appear to be more important in predator–prey relations, particularly in evolutionary terms, than the chemical composition of the prey (5.2.3). Support for this view is based on records that ladybirds do not refuse unsuitable/toxic prey and it is not avoided even when when offered in a mixture with suitable prey. This is perhaps due to a less developed **gustatory sense** of coccinellids. However, high (or total) larval mortality and inhibition of egg laying (5.2.2) on non-essential prey/food certainly plays an important role for a coccinellid species in the predator–prey relationship (and eventually also in the impact on the prey).

Thus the **adequacy of prey** can only be safely estalbihsed by experimentally recorded levels of life-history parameters, such as successful completion of larval development and high reproductive performance. While earlier research has worked with single prey/ foods, some recent studies have revealed the importance of **mixed food** (5.2.5). Although modern serological and molecular methods were developed for the detection of prey consumed by predators (5.2.1; also Chapter 10), the classical microscopical analysis of food remnants from the gut can also yield precise evidence.

While the food resources of phytophagous (but also coccidophagous) ladybirds are rather stable, **specialized adaptive characteristics** have evolved **in aphidophagous** species under the pressure exerted by the inherently ephemeral occurrence of their prey. This constraint has led to the evolution of a number of adaptive life-history traits, such as, for example, fast development (Chapter 3), high adult mobility, high speed of larval movement and reversible oosorption (Chapter 6).

(See also the general comments in 5.1 and 5.4.1 and partial conclusions in 5.4.1.2.)

ACKNOWLEDGEMENTS

The authors are most thankful to John J. Sloggett for very helpful comments.

IH's contribution to this chapter was supported by the project of the Institute of Entomology, Academy of Sciences of the Czech Republic No. AVOZ50070508.

REFERENCES

Abe, Y., T. Nakamura and H. Inoue. 2000. Exploitation of Fabaceae plants by the Mexican bean beetle *Epilachna varivestis* (Coleoptera: Coccinellidae) in Japan. *Appl. Entomol. Zool.* 35: 81–85.

Ables, J. R., S. L. Jones and D. W. McCommas Jr. 1978. Response of selected predator species to different densities of *Aphis gossypii* and *Heliothis virescens* eggs. *Environ. Entomol.* 7: 402–404.

Abrams, P. A. and L. R. Ginzburg. 2000. The nature of predation: prey dependent, ratio dependent or neither? *Trends Ecol. Evol.* 15: 337–341.

Adedipe, F. and Y.-L. Park. 2010. Visual and olfactory preference of *Harmonia axyridis* (Coleoptera: Coccinellidae) adults to various companion plants. *J. Asia Pac. Entomol.* 13: 319–323.

Agarwala, B. K. and P. Bardhanroy. 1999. Numerical response of ladybird beetles (Col., Coccinellidae) to aphid prey (Hom., Aphididae) in a field bean in north-east India. *J. Appl. Entomol.* 123: 401–405.

Agarwala, B. K. and A. F. G. Dixon. 1992. Laboratory study of cannibalism and interspecific predation in ladybirds. *Ecol. Entomol.* 17: 303–309.

Agarwala, B. K. and A. F. G. Dixon. 1993. Why do ladybirds lay eggs in clusters? *Funct. Ecol.* 7: 541–548.

Agarwala, B. K. and A. K. Ghosh. 1988. Prey records of aphidophagous Coccinellidae in India. A review and bibliography. *Trop. Pest Manag.* 34: 1–14.

Agarwala, B. K. and H. Yasuda. 2000. Competitive ability of ladybird predators of aphids: a review of *Cheilomenes sexmaculata* (Fabr.) (Coleoptera: Coccinellidae) with a worldwide checklist of preys. *J. Aphidol.* 14: 1–20.

Agarwala, B. K. and H. Yasuda. 2001a. Larval interactions in aphidophagous predators: effectiveness of wax cover as defence shield of *Scymnus* larvae against predation from syrphids. *Entomol. Exp. Appl.* 100: 101–107.

Agarwala, B. K. and H. Yasuda. 2001b. Overlapping oviposition and chemical defense of eggs in two co-occurring species of ladybird predators of aphids. *J. Ethol.* 19: 47–53.

Agarwala, B. K., S. Bhattacharya and P. Bardhanroy. 1998. Who eats whose eggs? Intra- versus inter-specific interactions in starving ladybird beetles predaceous on aphids. *Ethol. Ecol. Evol.* 10: 361–368.

Agarwala, B. K., S. Das and A. K. Bhaumik. 1987. Natural food range and feeding habits of aphidophagous insects in north east India. *J. Aphidol.* 1: 18–22.

Agarwala, B. K., H. Yasuda and Y. Kajita. 2003a. Effect of conspecific and heterospecific feces on foraging and oviposi-

tion of two predatory ladybirds: role of fecal cues in predator avoidance. *J. Chem. Ecol.* 29: 357–376.

Agarwala, B. K., H. Yasuda and S. Sato. 2008. Life history response of a predatory ladybird, *Harmonia axyridis* (Pallas) (Coleoptera: Coccinellidae), to food stress. *Appl. Entomol. Zool.* 43: 183–189.

Agarwala, B. K., P. Bardhanroy, H. Yasuda and T. Takizawa. 2001. Prey consumption and oviposition of the aphidophagous predator *Menochilus sexmaculatus* (Coleoptera: Coccinellidae) in relation to prey density and adult size. *Environ. Entomol.* 30: 1182–1187.

Agarwala, B. K., P. Bardhanroy, H. Yasuda and T. Takizawa. 2003b. Effects of conspecific and heterospecific competitors on feeding and oviposition of a predatory ladybird: a laboratory study. *Entomol. Exp. Appl.* 106: 219–226.

Agarwala, B. K., T. K. Singh, R. K. Lokeshwari and M. Sharmila. 2009. Functional response and reproductive attributes of the aphidophagous ladybird beetle, *Harmonia dimidiata* (F.) in oak trees of sericultural importance. *J. Asia Pac. Entomol.* 12: 179–182.

Agee, H. R., E. R. Mitchell and R. V. Flanders. 1990. Spectral sensitivity of the compound eye of *Coccinella septempunctata* (Coleoptera: Coccinellidae). *Ann. Entomol. Soc. Am.* 83: 817–819.

Agrawal, A. A. 2011. Current trends in the evolutionary ecology of plant defence. *Func. Ecol.* 25: 420–432.

Al Abassi, S., M. A. Birkett, J. Pettersson et al. 2000. Response of the seven-spot ladybird to an aphid alarm pheromone and an alarm pheromone inhibitor is mediated by paired olfactory cells. *J. Chem. Ecol.* 26: 1765–1771.

Alexander, R. D. 1974. The evolution of social behavior. *Annu. Rev. Ecol. Syst.* 5: 325–383.

Alhmedi, A., E. Haubruge and F. Francis. 2008. Role of prey–host plant associations on *Harmonia axyridis* and *Episyrphus balteatus* reproduction and predatory efficiency. *Entomol. Exp. Appl.* 128: 49–56.

Alikhan, M. A. and M. Yousef. 1986. Temperature and food requirements of *Chilomenes sexmaculata* (Coleoptera: Coccinellidae). *Environ. Entomol.* 15: 800–802.

Allee, W. C., A. E. Emerson, K. P. Schmidt, O. Park and T. Park. 1949. *The Principles of Animal Ecology.* Saunders, Philadelphia. 837 pp.

Allen, D. C., F. B. Knight and J. L. Foltz. 1970. Invertebrate predators of the jack-pine budworm, *Choristoneura pinus*, in Michigan. *Ann. Entomol. Soc. Am.* 63: 59–64.

Anderson, J. M. E. 1982. Seasonal habitat utilization and food of the ladybirds *Scymnodes lividigaster* (Mulsant) and *Leptothea galbula* (Mulsant) (Coleoptera: Coccinellidae). *Aust. J. Zool.* 30: 59–70.

Aoki, S., S. Akimoto and S. Yamane. 1981. Observations on *Pseudoregma alexanderi* (Homoptera, Pemphigidae), an aphid species producing pseudoscorpion-like soldiers on bamboos. *Kontyû* 49: 355–366.

Arakaki, N. 1992a. Feeding behaviour of the ladybird *Pseudoscymnus kurohime*. *J. Ethol.* 10: 7–13.

Arakaki, N. 1992b. Feeding by soldiers of the bamboo aphid *Pseudoregma koshunensis* (Homoptera: Aphididae). *J. Ethol.* 10: 147–148.

Arakaki, N. 1992c. Shortened longevity of soldiers of the bamboo aphid *Pseudoregma koshunensis* (Takahashi) (Homoptera: Aphididae) due to attack behaviour. *J. Ethol.* 10: 149–151.

Arditi, R. and L. R. Ginzburg. 1989. Coupling in predator–prey dynamics: ratio dependence. *J. Theor. Biol.* 139: 311–326.

Asante, S. K. 1995. Functional responses of the European earwig and 2 species of coccinellids to densities of *Eriosoma lanigerum* (Hausmann) (Hemiptera, Aphididae). *J. Aust. Entomol. Soc.* 34: 105–109.

Atallah, Y. H. and R. Killebrew. 1967. Ecological and nutritional studies on *Coleomegilla maculata* (Coleoptera: Coccinellidae). Amino acid requirements of the adult determined by the use of C^{14}-labeled acetate. *Ann. Entomol. Soc. Am.* 60: 186–188.

Atallah, Y. H. and L. D. Newsom. 1966. Ecological and nutritional studies on *Coleomegilla maculata* DeGeer (Coleoptera: Coccinellidae). I. The development of an artificial diet and a laboratory rearing technique. *J. Econ. Entomol.* 59: 1173–1179.

Atlhan, R. and M. B. Kaydan. 2002. Development, survival and reproduction of three coccinellids feeding on *Hyalopterus pruni* (Geoffr.) (Homoptera: Aphididae). *Turk. J. Agric. For.* 26: 119–124.

Atlhan, R. and M. S. Ozgokce. 2002. Development, fecundity and prey consumption of *Exochomus nigromaculatus* feeding on *Hyalopterus pruni*. *Phytoparasitica* 30: 443–450.

Atwal, A. S. and S. L. Sethi. 1963. Biochemical basis for the food preference of a predator beetle. *Curr. Sci.* 32: 511–512.

Azam, K. M. and M. H. Ali. 1970. A study of factors affecting the dissemination of the predatory beetle, *Coccinella septempunctata* L. *Final. Techn. Rep.* (F6-IN-249, A7-ENT-40) Dept. Entomol. Andhra Pradesh Agr. Univ., Rajendranagar, Hyderabad, India.

Baensch, R. 1964. Vergleichende Untersuchungen zur Biologie und zum Beutefangverhalten aphidovorer Coccinelliden, Chrysopiden und Syrphiden. *Zool. Jb. Syst.* 91: 271–340.

Baensch, R. 1966. On prey-seeking behaviour of aphidophagous insects. *In* I. Hodek (ed.). *Ecology of Aphidophagous Insects*. Academia, Prague and W. Junk, The Hague. pp. 123–128.

Ba M'Hamed, T. and M. Chemseddine. 2001. Assessment of temperature effects on the development and fecundity of *Pullus mediterraneus* (Col., Coccinellidae) and consumption of *Saissetia oleae* eggs (Hom., Coccoida). *J. Appl. Ent.* 125: 527–531.

Bahlai, C. A., J. A. Welsman, E. C. MacLeod et al. 2008. Role of visual and olfactory cues from agricultural hedgerows in the orientation behavior of multicolored Asian lady beetle

(Coleoptera: Coccinellidae). *Environ. Entomol.* 37: 973–979.

Bai, Y. Y., M. X. Jiang, J. A. Cheng and D. Wang. 2006. Effects of Cry1Ab toxin on *Propylea japonica* (Thunberg) (Coleoptera: Coccinellidae) through its prey, *Nilaparvata lugens* Stal (Homoptera: Delphacidae), feeding on transgenic *Bt* rice. *Environ. Entomol.* 35: 1130–1136.

Bailey, P. and G. Caon. 1986. Predation of twospotted mite, *Tetranychus urticae* Koch (Acarina: Tetranychidae) by *Haplothrips victoriensis* Bagnall (Thysanoptera: Phlaeothripidae) and *Stethorus nigripes* Kapur (Coleoptera: Coccinellidae) on seed lucerne crops in South Australia. *Aust. J. Zool.* 34: 515–525.

Bain, J., P. Singh, M. D. Ashby and R. J. van Boven. 1984. Laboratory rearing of the predatory coccinellid *Cleobora mellyi* (Col.: Coccinellidae) for biological control of *Paropsis charybdis* (Col.: Chrysomelidae) in New Zealand. *Entomophaga* 29: 237–244.

Balachowsky, A. 1930. L'extension de la cochenille australienne (*Icerya purchasi* Mask.) en France et de son prédateur *Novius cardinalis* Muls. *Annls Epiphyt.* 16: 1–24.

Balduf, W. V. 1935. *The Bionomics of Entomophagous Coleoptera*. (13. Coccinellidae lady-beetles). John S. Swift, Chicago, New York. 220 pp.

Banks, C. J. 1954. The searching behaviour of coccinellid larvae. *Br. J. Anim. Behav.* 2: 37–38.

Banks, C. J. 1955. An ecological study of Coccinellidae (Col.) associated with *Aphis fabae* Scop. on *Vicia faba*. *Bull. Entomol. Res.* 46: 561–587.

Banks, C. J. 1956. Observations on the behaviour and mortality in Coccinellidae before dispersal from the egg shells. *Proc. R. Entomol. Soc. Lond. (A)* 31: 56–60.

Banks C. J. 1957. The behaviour of individual Coccinellid larvae on plants. *Br. J. Anim Behav.* 5: 12–24.

Barbier, R., J. Le Lannic and J. Brun. 1996. Récepteurs sensoriels des palpes maxillaires de coccinellidae adultes aphidiphages, coccidiphages et phytophages. *Bull. Soc. Zool. Fr.* 121: 255–268.

Bathon, H. and J. Pietrzik. 1986. Zur Nahrungsaufnahme des Bogen-Marienkäfers, *Clitostethus arcuatus* (Rossi) (Col., Coccinellidae), einem Vertilger der Kohlmottenlaus, *Aleurodes proletella* Linné (Hom., Aleurodidae). *J. Appl. Entomol.* 102: 321–326.

Bazzocchi, G. G., A. Lanzoni, G. Accinelli and G. Burgio. 2004. Overwintering, phenology and fecundity of *Harmonia axyridis* in comparison with native coccinellid species in Italy. *BioControl* 49: 245–260.

Behmer, S. T., S. J. Simpson and D. Raubenheimer. 2002. Herbivore foraging in chemically heterogeneous environments: nutrients and secondary metabolites. *Ecology* 83: 2489–2501.

Belcher, D. W. and R. Thurston. 1982. Inhibition of movement of larvae of the convergent lady beetle by leaf trichomes of tobacco. *Environ. Entomol.* 11: 91–94.

Bell, W. J. 1985. Sources of information controlling motor patterns in arthropod local search orientation. *J. Insect Physiol.* 31: 837–847.

Bell, W. J. 1990. Searching behavior patterns in insects. *Annu. Rev. Entomol.* 35: 447–467.

Bellows, T. S., T. D. Paine and D. Gerling. 1992. Development, survival, longevity, and fecundity of *Clitostethus arcuatus* (Coleoptera: Coccinellidae) on *Siphoninus phillyreae* (Homoptera: Aleyrodidae) in the laboratory. *Environ. Entomol.* 21: 659–663.

Ben Saad, A. A. and G. W. Bishop. 1976a. Attraction of insects to potato plants through use of artificial honeydews and aphid juice. *Entomophaga* 21: 49–57.

Ben Saad, A. A. and G. W. Bishop. 1976b. Effect of artificial honeydews on insect communities in potato fields. *Environ. Entomol.* 5: 453–457.

Benham, B. R. and J. Muggleton. 1970. Studies on the ecology of *Coccinella undecimpunctata* Linn. (Col. Coccinellidae). *Entomologist* 103: 153–170.

Berkvens, N., J. Bonte, D. Berkvens et al. 2008. Pollen as an alternative food for *Harmonia axyridis*. *BioControl* 53: 201–210.

Berkvens, N., C. Landuyt, K. Deforce et al. 2010. Alternative foods for the multicoloured Asian ladybeetle *Harmonia axyridis* (Coleoptera, Coccinellidae). *Eur. J. Entomol* 107: 189–195.

Berthiaume, R., C. Herbert and C. Cloutier. 2000. Predation on *Mindarus abietinus* infesting balsam fir grown as Christmas trees: the impact of coccinellid larval predation with emphasis on *Anatis mali*. *BioControl* 45: 425–438.

Bhatkar, A. P. 1982. Orientation and defense of ladybeetles (Col., Coccinellidae), following ant trail in search of aphids. *Folia Entomol. Mexic.* 53: 75–85.

Biddinger, D. J., D. C. Weber and L. A. Hull. 2009. Coccinellidae as predators of mites: Stethorini in biological control. *Biol. Contr.* 51: 268–283.

Bielawski, R. 1959. *Biedronki-Coccinellidae. Klucze do oznaczania owadów Polski*. Part 19, Tom 76. Panstw. wyd. nauk. Warszawa. 92 pp. (In Polish.)

Bilde, T. and S. Toft. 1999. Prey consumption and fecundity of the carabid beetle *Calathus melanocephalus* on diets of three cereal aphids: high consumption rates of low-quality prey. *Pedobiologia* 43: 422–429.

Bilde, T., J. A. Axelsen and S. Toft. 2000. The value of Collembola from agricultural soils as food for a generalist predator. *J. Appl. Ecol.* 37: 672–683.

Bilu, E. and M. Coll. 2009. Parasitized aphids are inferior prey for a coccinellid predator: implications for intraguild predation. *Environ. Entomol.* 38: 153–158.

Birkett, M. A., C. A. M. Campbell, K. Chamberlain et al. 2000. New roles for cis-jasmone as an insect semiochemical and in plant defense. *Proc. Nat. Acad. Sci. USA* 97: 9329–9334.

Blackman, R. L. 1965. Studies on specificity in Coccinellidae. *Ann. Appl. Biol.* 56: 336–338.

Blackman, R. L. 1967a. Selection of aphid prey by *Adalia bipunctata* L. and *Coccinella 7-punctata* L. *Ann. Appl. Biol.* 59: 331–338.

Blackman, R. L. 1967b. The effects of different aphid foods on *Adalia bipunctata* L. and *Coccinella 7-punctata* L. *Ann. Appl. Biol.* 59: 207–219.

Blattny, C. and C. V. Osvald. 1949. Contribution à la connaisance des signes génériques et de la biologie du *Stethorus punctillum* Weise. *Věst. Čs. spol. zool.* 13: 30–40. (In Czech with English summary.)

Bodenheimer, F. S. and S. Neumark. 1955. *The Israel Pine Matsucoccus (Matsucoccus josephi n. sp.)*. Kiryath Sepher, Jerusalem. 122 pp.

Boivin, G., C. Roger, D. Coderre and E. Wajnberg. 2010. Learning affects prey selection in larvae of a generalist coccinellid predator. *Entomol. Exp. Appl.* 135: 48–55.

Bonte M., M. A. Samih, P. De Clercq. 2010. Development and reproduction of *Adalia bipunctata* on factitious and artificial foods. *BioControl* 55: 485–491.

Booth, R. G., A. E. Cross, S. V. Fowler and R. H. Shaw. 1995. The biology and taxonomy of *Hyperaspis pantherina* (Coleoptera: Coccinellidae) and the classical biological control of its prey, *Orthezia insignis* (Homoptera: Ortheziidae). *Bull. Entomol. Res.* 85: 307–314.

Branco, M., J. C. Franco, E. Dunkelblum et al. 2006a. A common mode of attraction of larvae and adults of insect predators to the sex pheromone of their prey (Hemiptera: Matsucoccidae). *Bull. Entomol. Res.* 96: 179–185.

Branco, M., M. Lettere, J. C. Franco, A. Binazzi and H. Jactel. 2006b. Kairomonal response of predators to three pine bast scale sex pheromones. *J. Chem. Ecol.* 32: 1577–1586.

Brewer, M. J. and N. C. Elliott. 2004. Biological control of cereal aphids in North America and mediating effects of host plant and habitat manipulation. *Annu. Rev. Entomol.* 49: 219–242.

Bristow, C. M. 1991. Are ant–aphid associations a tritrophic interaction? Oleander aphids and Argentine ants. *Oecologia* 87: 514–521.

Britto, E. P. J., M. G. C. Gondim, J. B. Torres et al. 2009. Predation and reproductive output of the ladybird beetle *Stethorus tridens* preying on tomato red spider mite *Tetranychus evansi*. *BioControl* 54: 363–368.

Brodsky, L. M. and C. A. Barlow. 1986. Escape responses of the pea aphid, *Acyrthosiphon pisum* (Harris) (Homoptera: Aphididae): influence of predator type and temperature. *Can. J. Zool.* 64: 937–939.

Brown, H. D. 1972. The behaviour of newly hatched coccinellid larvae (Coleoptera: Coccinellidae). *J. Entomol. Soc. south. Africa* 35: 149–157.

Brown, M. W. 2004. Role of aphid predator guild in controlling spirea aphid populations on apple in West Virginia, USA. *Biol. Contr.* 29: 189–198.

Brun, J. and G. Iperti. 1978. Influence de l'alimentation sur la fécondité des Coccinelles aphidiphages. *Ann. Zool.-Ecol. Anim.* 10: 449–452.

de Bruyne, M. and T. C. Baker. 2008. Odor detection in insects: volatile codes. *J. Chem. Ecol.* 34: 882–897.

Burden, B. J. and D. M. Norris. 1994. Ovarian failure induced in *Epilachna varivestis* by a 'death-trap' 'Davis' variety of *Glycine max*. *Entomol. Exp. Appl.* 73: 183–186.

Burgio, G., F. Santi and S. Maini. 2002. On intra-guild predation and cannibalism in *Harmonia axyridis* (Pallas) and *Adalia bipunctata* L. (Coleoptera: Coccinellidae). *Biol. Contr.* 24: 110–116.

Burgio, G., R. Ferrari, L. Boriani, M. Pozzati and J. Van Lenteren. 2006. The role of ecological infrastructures on Coccinellidae (Coleoptera) and other predators in weedy field margins within northern Italy agroecosystems. *Bull. Insectol.* 59: 59–67.

Burk, T. 1988. Insect behavioral ecology: Some future paths. *Annu. Rev. Entomol.* 33: 319–335.

Butin, E. E., N. P. Havill, J. S. Elkinton and M. E. Montgomery. 2004. Feeding preference of three lady beetle predators of the hemlock woolly adelgid (Homoptera: Adelgidae). *J. Econ. Entomol.* 97: 1635–1641.

Butt, F. H. 1951. Feeding habitats and mechanism of the Mexican bean beetle. *Cornell Univ. Agr. Exp. Sta., Mem.* 306, 32 pp.

Cabral, S., A. O. Soares, R. Moura and P. Garcia. 2006. Suitability of *Aphis fabae*, *Myzus persicae* (Homoptera: Aphididae) and *Aleyrodes proletella* (Homoptera: Aleyrodidae) as prey for *Coccinella undecimpunctata* (Coleoptera: Coccinellidae). *Biol. Contr.* 39: 434–440.

Cabral, S., A. O. Soares and P. Garcia. 2009. Predation by *Coccinella undecimpunctata* L. (Coleoptera: Coccinellidae) on *Myzus persicae* Sulzer (Homoptera: Aphididae): Effect of prey density. *Biol. Contr.* 50: 25–29.

Camargo, F. 1937. Notas taxonomicas e biologicas sobre alguns Coccinellideos do genera *Psyllobora* Chevrolat (Col. Coccinellidae). *Revta Entomol. Rio de J.* 7: 362–377.

Campbell, C. A. M. and W. W. Cone. 1999. Consumption of damson-hop aphids (*Phorodon humuli*) by larvae of *Coccinella transversoguttata* and *Hippodamia convergens* (Coleoptera: Coccinellidae). *Biocontr. Sci. Technol.* 9: 75–78.

Canard, M. 1977. Diminution du taux de survie du prédateur *Chrysopa perla* (L.) (Neuroptera, Chrysopidae) en relation avec le comportement du puceron *Aphis nerii* B. de F. (Homoptera, Aphididae). *In* J. Médioni and E. Boesiger (eds). *Mécanismes Ethologiques de l'Evolution*. Masson, Paris. pp. 49–51.

Cappuccino, N. 1988. Spatial patterns of goldenrod aphids and the response of enemies to patch density. *Oecologia* 76: 607–610.

Capra, F. 1947. Note sui Coccinellidi (Col.) III. La larva ed il regime pollinivoro di *Bulaea lichatschovi* Hummel. *Mem. Soc. Entomol. Ital.*, Suppl. 26: 80–86.

Cardinale, B. J., J. J. Weis, A. E. Forbes, K. J. Tilmon and I. R. Ives. 2006. Biodiversity as both a cause and consequence of resource availability: a study of reciprocal causality in a predator–prey system. *J. Anim. Ecol.* 75: 497–505.

Carlson, R. E. and H. C. Chiang. 1973. Reduction of an *Ostrinia nubilalis* population by predatory insects attracted by sucrose sprays. *Entomophaga* 18: 204–211.

Carter, M. C. and A. F. G. Dixon. 1982. Habitat quality and the foraging behaviour of coccinellid larvae. *J. Anim. Ecol.* 51: 865–878.

Carter, M. C. and A. F. G. Dixon. 1984. Honeydew: an arrestant stimulus for coccinellids. *Ecol. Entomol.* 9: 383–387.

Carter, M. C., D. Sutherland and A. F. G. Dixon. 1984. Plant structure and the searching efficiency of coccinellid larvae. *Oecologia* 63: 394–397.

Causton, C. E., M. P. Lincango and T. G. A. Poulsom. 2004. Feeding range studies of *Rodolia cardinalis* (Mulsant), a candidate biological control agent of *Icerya purchasi* Maskell in the Galapagos islands. *Biol. Contr.* 29: 315–325.

Ceryngier, P., J. Havelka and I. Hodek. 2004. Mating and activity of gonads in pre-dormant and dormant ladybirds (Coleoptera: Coccinellidae). *Invert. Repr. Dev.* 45: 127–135.

Chakrabarti, S., D. Ghosh and N. Debnath. 1988. Developmental rate and larval voracity in *Harmonia* (*Leis*) *dimidiata* (Col., Coccinellidae), a predator of *Eriosoma lanigerum* (Hom., Aphididae) in western Himalaya. *Acta Entomol. Bohemoslov.* 85: 335–339.

Charles, J. G., E. Collyer and V. White. 1985. Integrated control of *Tetranychus urticae* with *Phytoseiulus persimilis* and *Stethorus bifidus* in commercial raspberry gardens. *N. Zealand J. Exp. Agr.* 13: 385–393.

Chazeau, J. 1981. Données sur la biologie de *Coelophora quadrivittata* (Col.: Coccinellidae), prédateur de *Coccus viridis* (Hom.: Coccidae) en Nouvelle-Calédonie. *Entomophaga* 26: 301–312.

Chazeau, J., E. Bouye and L. B. Delarbogne. 1991. Development and life table of *Olla v-nigrum* (Col., Coccinellidae), a natural enemy of *Heteropsylla cubana* (Hom., Psyllidae) introduced in New Caledonia. *Entomophaga* 36: 275–285.

Chazeau, J., I. Capart and L. Bonnet de Larbogne. 1992. Biological control of the Leucaena psyllid in New Caledonia: the introduction of *Curinus coeruleus* (Coccinellidae). *Leucaena Res. Reports* 13: 59–61.

Cheah, C. A. S. J. and M. S. McClure. 1998. Life history and development of *Pseudoscymnus tsugae* (Coleoptera: Coccinellidae), a new predator of the hemlock wooly adelgid (Homoptera: Adelgidae). *Environ. Entomol.* 27: 1531–1536.

Chen, Z.-H. and J. Qin. 1982. The nutritional role of water content in the artificial diet of *Coccinella septempunctata*. *Acta Entomol. Sinica* 25: 141–146. (In Chinese with English summary.)

Chen, Z.-H., E.-Y. Chen and F.-S. Yan. 1980. Effects of diets on the feeding and reproduction of *Coccinella septempunctata* L. *Acta Entomol. Sinica* 23: 141–148. (In Chinese with English summary.)

Chen, Z.-H., J.-D. Qin and C.-L. Shen. 1989. Effects of altering composition of artificial diets on the larval growth and development of *Coccinella septempunctata*. *Acta Entomol. Sinica* 32: 385–392.

Chen, Z.-H., J.-D. Qin, X.-M. Fan and X.-L. Li. 1984. Effects of adding lipids and juvenoid into the artificial diet on feeding and reproduction of *Coccinella septempunctata* L. *Acta Entomol. Sinica* 27: 136–146. (In Chinese with English summary.)

Chi, D. F., G. L. Wang, J. W. Liu, Q. Y. Wu and Y. P. Zhu. 2009. Antennal morphology and sensilla of Asian multicolored ladybird beetle, *Harmonia axyridis* Pallas (Coleoptera: Coccinellidae). *Entomol. News* 120: 139–152.

Chong, J. H., R. D. Oetting and L. S. Osborne. 2005. Development of *Diomus austrinus* Gordon (Coleoptera: Coccinellidae) on two mealybug prey species at five constant temperatures. *Biol. Contr.* 33: 39–48.

Chumakova, B. M. 1962. Experiments in rearing of predatory beetle *Cryptolaemus montrouzieri* Muls. on an artificial diet. *Zesz. Probl. Postep Nauk Roln.* 35: 195–200.

Cividanes, F. J. and A. P. Gutierrez. 1996. Modeling the age-specific per capita growth and reproduction of *Rhizobius lophanthae* (Col.: Coccinellidae). *Entomophaga* 41: 257–266.

Clegg, J. M. and C. A. Barlow. 1982. Escape behaviour of the pea aphid *Acyrthosiphon pisum* (Harris) in response to alarm pheromone and vibration. *Can. J. Zool.* 60: 2245–2252.

Cochereau, P. 1969. Controle biologique d'*Aspidiotus destructor* Signoret (Homoptera, Diaspinae) dans l'Ile Vaté (Nouvelles Hébrides) au moyen de *Rhizobius pulchellus* Montrouzier (Coleoptera, Coccinellidae). *Cahiers O.R.S.T.O.M.* 8: 57–100.

Coderre, D. and J. C. Tourneur. 1986. Vertical distribution of aphids and aphidophagous insects on maize. *In* I. Hodek (ed). *Ecology of Aphidophaga*. Academia, Prague and W. Junk, Dordrecht. pp. 291–296.

Cohen, A. C. 1989. Ingestion efficiency and protein consumption by a heteropteran predator. *Ann. Entomol. Soc. Am.* 82: 495–499.

Cohen, A. C. and D. L. Brummett. 1997. The non-abundant nutrient (NAN) concept as a determinant of predator–prey fitness. *Entomophaga* 42: 85–91.

Colburn, R. and D. Asquith. 1970. A cage used to study the finding of a host by the ladybird beetle, *Stethorus punctum*. *J. Econ. Entomol.* 63: 1376–1377.

Congdon, B. D., C. H. Shanks and A. L. Antonelli. 1993. Population interaction between *Stethorus punctum picipes* (Coleoptera, Coccinellidae) and *Tetranychus urticae* (Acari, Tetranychidae) in red raspberries at low predator and prey densities. *Environ. Entomol.* 22: 1302–1307.

Corlay, F., G. Boivin and G. Belair. 2007. Efficiency of natural enemies against the swede midge *Contarinia nasturtii* (Diptera: Cecidomyiidae), a new invasive species in North America. *Biol. Contr.* 43: 195–201.

Costamagna, A. C., D. A. Landis and M. J. Brewer. 2008. The role of natural enemy guilds in *Aphis glycines* suppression. *Biol. Contr.* 45: 368–379.

Cottrell, T. E. 2004. Suitability of exotic and native lady beetle eggs (Coleoptera: Coccinellidae) for development of lady beetle larvae. *Biol. Contr.* 31: 362–371.

Cottrell, T. E. 2005. Predation and cannibalism of lady beetle eggs by adult lady beetles. *Biol. Contr.* 34: 159–164.

Cottrell, T. E. 2007. Predation by adult and larval lady beetles (Coleoptera: Coccinellidae) on initial contact with lady beetle eggs. *Environ. Entomol.* 36: 390–401.

Cottrell, T. E. and K. V. Yeargan. 1998. Effect of pollen on *Coleomegilla maculata* (Coleoptera: Coccinellidae) population density, predation and cannibalism in sweet corn. *Environ. Entomol.* 27: 1402–1410.

Crook, N. E. and K. D. Sunderland. 1984. Detection of aphid remains in predatory insects and spiders by ELISA. *Ann. Appl. Biol.* 105: 413–422.

Curio, E. 1976. *The Ethology of Predation*. Springer, Berlin. 250 pp.

Cushing, J. M. 1992. A size-structured model for cannibalism. *Theor. Popul. Biol.* 42: 347–361.

Daane, K. M., K. R. Sime, J. Fallon and M. L. Cooper. 2007. Impacts of Argentine ants on mealybugs and their natural enemies in California's coastal vineyards. *Ecol. Entomol.* 32: 583–596.

Danks, H. V. 1983. Extreme individuals in natural populations. *Bull. Entomol. Soc. Am.* 29(1): 41–46.

Dauguet, P. 1949. *Les Coccinellini de France*. Paris. 46 pp.

Davidson, L. N. and E. W. Evans. 2010. Frass analysis of diets of aphidophagous lady beetles (Coleoptera: Coccinellidae) in Utah alfalfa fields. *Environ. Entomol.* 39: 576–582.

Davidson, W. M. 1921. Observations on *Psyllobora taedata* LeConte, a coccinellid attacking mildews. *Entomol. News* 32: 83–89.

Day, K. R., M. Docherty, S. R. Leather and N. A. C. Kidd. 2006. The role of generalist insect predators and pathogens in suppressing green spruce aphid populations through direct mortality and mediation of aphid dropping behavior. *Biol. Contr.* 38: 233–246.

Debaraj, Y. and T. K. Singh. 1990. Biology of an aphidophagous coccinellid predator, *Coccinella transversalis* Fab. *J. Biol. Control* 4: 93–95.

De Clercq, P., M. Bonte, K. Van Speybroeck, K. Bolckmans and K. Deforce. 2005. Development and reproduction of *Adalia bipunctata* (Coleoptera: Coccinellidae) on eggs of *Ephestia kuehniella* (Lepidoptera: Phycitidae) and pollen. *Pest Manage. Sci.* 61: 1129–1132.

Deligeorgidis, P. N., C. G. Ipsilandis, G. Kaltsoudas and G. Sidiropoulos. 2005. An index model on predatory effect of female adults of *Coccinella septempunctata* L. on *Macrosiphum euphorbiae* Thomas. *J. Appl. Entomol.* 129: 1–5.

Deligeorgidis, P. N., C. G. Ipsilandis, M. Vaiopoulou, G. Kaltsoudas and G. Sidiropoulos. 2005. Predatory effect of *Coccinella septempunctata* on *Thrips tabaci* and *Trialeurodes vaporariorum. J. Appl. Entomol.* 129: 246–249.

Denmark, E. and J. E. Losey. 2010. Causes and consequences of ladybug washups in the Finger Lakes Region of New York State (Coleoptera: Coccinellidae). *Entomol. Am.* 116: 78–88.

Denno, R. F. and W. F. Fagan. 2003. Might nitrogen limitation promote omnivory among carnivorous arthropods? *Ecology* 84: 2522–2531.

Dicke, M., M. W. Sabelis, J. Takabayashi, J. Bruin and M. A. Posthumus. 1990. Plant strategies of manipulating predator–prey interactions through allelochemicals: prospects for application in pest control. *J. Chem. Ecol.* 16: 3091–3118.

Dickson, R. C., E. F. Laird and C. R. Pesho. 1955. The spotted alfalfa aphid (*Therioaphis maculata*).(Predator relationship: Hemiptera, Coccinellidae, Syrphidae, Neuroptera). *Hilgardia* 24: 93–118.

Dixon, A. F. G. 1958. The escape response shown by certain aphids to the presence of the coccinelid *Adalia decempunctata* (L.). *Trans. R. Entomol. Soc. Lond.* 110: 319–334.

Dixon, A. F. G. 1959. An experimental study of the searching behaviour of the predatory coccinellid beetle *Adalia decempunctata* (L.). *J. Anim. Ecol.* 28: 259–281.

Dixon, A. F. G. 1970. Factors limiting the effectiveness of the coccinellid beetle, as a predator of the sycamore aphid, *Drepanosiphum platanoides* (Schr.). *J. Anim. Ecol.* 39: 739–751.

Dixon, A. F. G. 2000. *Insect Predator–Prey Dynamics. Ladybird Beetles and Biological Control.* Cambridge University Press, Cambridge, UK. 257 pp.

Dixon, A. F. G. 2007. Body size and resource partitioning in ladybirds. *Popul. Ecol.* 49: 45–50.

Dixon, A. F. G. and B. K. Agarwala. 2002. Triangular fecundity function and ageing in ladybird beetles. *Ecol. Entomol.* 27: 433–440.

Dixon, A. F. G. and Y. Guo. 1993. Egg and cluster size in ladybird beetles (Coleoptera: Coccinellidae): The direct and indirect effects of aphid abundance. *Eur. J. Entomol.* 90: 457–463.

Dixon, A. F. G. and J. L. Hemptinne. 2001. Body size distribution in predatory ladybird beetles reflects that of their prey. *Ecology* 82: 1847–1856.

Dixon, A. F. G., M. Martin-Smith and G. Subramanian. 1965. Constituents of *Megoura viciae* Buckton. *J. Chem. Soc.* 1965: 1562–1564.

Dixon, A. F. G. and L. A. Stewart. 1991. Size and foraging in ladybird beetles. *In* L. Polgar, R. J. Chambers, A. F. G. Dixon

and I. Hodek (eds). *Behaviour and Impact of Aphidophaga.* SPB Acad. Publ., The Hague. pp. 123–132.

Dmitriew, C. and L. Rowe. 2007. Effects of early resource limitation and compensatory growth on lifetime fitness in the ladybird beetle (*Harmonia axyridis*). *J. Evol. Biol.* 20: 1298–1310.

Dong, H., J. J. Ellington and M. D. Remmenga. 2001. An artificial diet for the lady beetle *Harmonia axyridis* Pallas (Coleoptera: Coccinellidae). *Southw. Entomol.* 26: 205–213.

Dosse, G. 1967. Schadmilben des Libanons und ihre Prädatoren. *Z. Angew. Entomol.* 59: 16–48.

Doumbia, M., J. L. Hemptinne and A. F. G. Dixon. 1998. Assessment of patch quality by ladybirds: role of larval tracks. *Oecologia* 113: 197–202.

Duan, J. J., G. Head, M. J. McKee et al. 2002. Evaluation of dietary effects of transgenic corn pollen expressing Cry3Bb1 protein on a non-target ladybird beetle, *Coleomegilla maculata. Entomol. Exp. Appl.* 104: 271–280.

Dyadechko, N. P. 1954. *Coccinellids of the Ukrainian SSR.* Kiev. 156 pp. (In Russian.)

Eastop, V. F. and R. D. Pope. 1966. Notes on the ecology and phenology of some British Coccinellidae. *Entomologist* 99: 287–289.

Eastop, V. F. and R. D. Pope. 1969. Notes on the biology of some British Coccinellidae. *Entomologist* 102: 162–164.

Ehler, L. E., R. F. Long, M. G. Kensey and S. K. Kelley. 1997. Potential for augmentative biological control of black bean aphid in California sugarbeet. *Entomophaga* 42: 241–256.

Eigenbrode, S. D., J. E. Andreas, M. G. Cripps et al. 2008. Induced chemical defenses in invasive plants: a case study with *Cynoglossum officinale* L. *Biol. Invasions* 10: 1373–1379.

El Habi, M., A. Sekkat, L. El Jadd and A. Boumezzough. 2000. Biology of *Hippodamia variegata* Goeze (Col., Coccinellidae) and its suitability against *Aphis gossypii* Glov (Hom., Aphididae) on cucumber under greenhouse conditions. *J. Appl. Entomol* 124: 365–374.

El Khidir, E. 1969. A contribution to the biology of *Epilachna chrysomelina* F., the melon lady-bird beetle in the Sudan (Col., Coccinellidae). *Sudan Agric. J.* 4: 32–37.

Elder, R. J. and K. L. Bell. 1998. Establishment of *Chilocorus* spp. (Coleoptera: Coccinellidae) in a *Carica papaya* L. orchard infested by *Aonidiella orientalis* (Newstead) (Hemiptera: Diaspididae). *Aust. J. Entomol.* 37: 362–365.

Elliot, H. J. and D. W. De Little. 1980. Laboratory studies on predation of *Chrysopharta bimaculata* (Olivier) (Col.: Chrysomelidae) eggs by the coccinellids *Cleobora mellyi* Mulsant and *Harmonia conformis* (Boisduval). *Gen. Appl. Entomol.* 12: 33–36.

Elliott, N. C. and R. W. Kieckhefer. 2000. Response by coccinellids to spatial variation in cereal aphid density. *Popul. Ecol.* 42: 81–90.

Elliott, N. C., R. W. Kieckhefer and D. A. Beck. 2000. Adult coccinellid activity and predation on aphids in spring cereals. *Biol. Contr.* 17: 218–226.

Elliott, N. C., R. W. Kieckhefer and D. A. Beck. 2002a. Effect of aphids and the surrounding landscape on the abundance of Coccinellidae in cornfields. *Biol. Contr.* 24: 214–220.

Elliott, N. C., B. W. French, G. J. Michels and D. K. Reed. 1994. Influence of four aphid prey species on development, survival, and adult size of *Cycloneda ancoralis*. *Southwest. Entomol.* 19: 57–61.

Elliott, N. C., R. W. Kieckhefer, J.-H. Lee and B. W. French. 1998. Influence of within-field and landscape factors on aphid predator populations in wheat. *Landscape Ecol.* 14: 239–252.

Elliott, N. C., R. W. Kieckhefer, G. J. Michels Jr. and K. L. Giles. 2002b. Predator abundance in alfalfa fields in relation to aphids, within-field vegetation, and landscape matrix. *Environ. Entomol.* 31: 253–260.

Ellingsen, J.-L. 1969. Effect of constant and varying temperature on development, feeding, and survival of *Adalia bipunctata* L. (Col., Coccinellidae). *Norsk Entomol. Tidsskr.* 16: 121–125.

Elsey, K. D. 1974. Influence of plant host on searching speed of two predators. *Entomophaga* 19: 3–6.

van Emden, H. F., A. P. Storeck, S. Douloumapaka et al. 2008. Plant chemistry and aphid parasitoids (Hymenoptera: Braconidae): Imprinting and memory. *Eur. J. Entomol.* 105: 477–483.

Emrich, B. H. 1991. Erworbene Toxizität bei der Lupinenblattlaus *Macrosiphum albifrons* und ihr Einfluß auf die aphidophagen Prädatoren *Coccinella septempunctata*, *Episyrphus balteatus* und *Chrysoperla carnea*. *J. Plant Diseases Prot.* 98: 398–404.

Endo, N., M. Abe, T. Sekine and K. Matsuda. 2004. Feeding stimulants of Solanaceae-feeding lady beetle, *Epilachna vigintioctomaculata* (Coleoptera: Coccinellidae) from potato leaves. *Appl. Entomol. Zool.* 39: 411–416.

Erichsen, C., M. J. Samways and V. Hattingh. 1991. Reaction of the ladybird *Chilocorus nigritus* (F.) (Col., Coccinellidae) to a doomed food resource. *J. Appl. Entomol.* 112: 493–498.

Ettifouri, M. and A. Ferran. 1992. Influence d'une alimentation prealable et du jeune sur l'apparition de la recherche intensive des proies chez *Semiadalia undecimnotata*. *Entomol. Exp. Appl.* 65: 101–111.

Ettifouri, M. and A. Ferran. 1993. Influence of larval rearing diet on the intensive searching behaviour of *Harmonia axyridis* (Col.: Coccinellidae) larvae. *Entomophaga* 38: 51–59.

Evans, E. W. 2000. Egg production in response to combined alternative foods by the predator *Coccinella transversalis*. *Entomol. Exp. Appl.* 94: 141–147.

Evans, E. W. 2003. Searching and reproductive behaviour of female aphidophagous ladybirds (Coleoptera: Coccinellidae): a review. *Eur. J. Entomol.* 100: 1–10.

Evans, E. W. 2004. Habitat displacement of North American ladybirds by an introduced species. *Ecology* 85: 637–647.

Evans, E. W. 2008. Multitrophic interactions among plants, aphids, alternate prey and shared natural enemies: a review. *Eur. J. Entomol.* 105: 369–380.

Evans, E. W. 2009. Lady beetles as predators of insects other than Hemiptera. *Biol. Contr.* 51: 255–267.

Evans, E. W. and A. F. G. Dixon. 1986. Cues for oviposition by ladybird beetles (Coccinellidae): response to aphids. *J. Anim. Ecol.* 55: 1027–1034.

Evans, E. W. and D. I. Gunther. 2005. The link between food and reproduction in aphidophagous predators: a case study with *Harmonia axyridis* (Coleoptera: Coccinellidae). *Eur. J. Entomol.* 102: 423–430.

Evans, E. W. and D. R. Richards. 1997. Managing the dispersal of ladybird beetles (Col.: Coccinellidae): use of artificial honeydew to manipulate spatial distributions. *Entomophaga* 42: 93–102.

Evans, E. W. and J. G. Swallow. 1993. Numerical responses of natural enemies to artificial honeydew in Utah alfalfa. *Environ. Entomol.* 22: 1392–1401.

Evans, E. W. and T. R. Toler. 2007. Aggregation of polyphagous predators in response to multiple prey: ladybirds (Coleoptera: Coccinellidae) foraging in alfalfa. *Popul. Ecol.* 49: 29–36.

Evans, E. W. and N. N. Youssef. 1992. Numerical responses of aphid predators to varying prey density among Utah alfalfa fields. *J. Kans. Entomol. Soc.* 65: 30–38.

Evans, E. W., D. R. Richards and A. Kalaskar. 2004. Using food for different purposes: female responses to prey in the predator *Coccinella septempunctata* L. (Coleoptera: Coccinellidae). *Ecol. Entomol.* 29: 27–34.

Evans, E. W., A. T. Stevenson and D. R. Richards. 1999. Essential versus alternative foods of insect predators: benefits of a mixed diet. *Oecologia* 121: 107–112.

Ewert, M. A. and H. C. Chiang. 1966a. Dispersal of three species of coccinellids in corn fields. *Can. Entomol.* 98: 999–1003.

Ewert, M. A. and H. C. Chiang. 1966b. Effects of some environmental factors on the distribution of three species of Coccinellidae in their microhabitat. *In* I. Hodek (ed.). *Ecology of Aphidophagous Insects.* Academia, Prague and W. Junk, The Hague. pp. 195–219.

Fabres, G. and A. Kiyindou. 1985. Comparaison du potentiel biotique de deux coccinelles (*Exochomus flaviventris* et *Hyperaspis senegalensis hottentotta*, Col. Coccinellidae) predatrices de *Phenacoccus manihoti* (Hom. Pseudococcidae) au Congo. *Acta Œcol.* 6: 339–348.

Farid, A., J. B. Johnson and S. S. Quisenberry. 1997. Compatibility of a coccinellid predator with a Russian wheat aphid resistant wheat. *J. Kansas Entomol. Soc.* 70, 114–119.

Ferran, A. and M. Deconchat. 1992. Exploration of wheat leaves by *Coccinella septempunctata* L. (Coleoptera, Coccinellidae) larvae. *J. Insect Behav.* 5: 147–159.

Ferran, A. and A. F. G. Dixon. 1993. Foraging behaviour of ladybird larvae (Coleoptera: Coccinellidae). *Eur. J. Entomol.* 90: 383–402.

Ferran, A., L. A. Buscarlet and M. M. Larroque. 1981. Utilization de HT^{18}O pour mesurer la consommation alimentaire chez les larves agées de *Semiadalia 11notata* (Col.: Coccinellidae). *Entomophaga* 26: 71–77.

Ferran, A., M. O. Cruz de Boelpaepe, H. Schanderl and M. M. Larroque. 1984a. Les aptitudes trophiques et reproductrices des femelles de *Semiadalia undecimnotata* (Col.: Coccinellidae). *Entomophaga* 29: 157–170.

Ferran, A., M. O. Cruz de Boelpaepe, L. A. Buscarlet, M. M. Larroque and H. Schanderl. 1984b. Les relations trophiques entre les larves de la coccinelle *Semiadalia undecimnotata* Schn. et le puceron *Myzus persicae* Sulz.: generalization à d'autres couples 'proieprédateur' et influence des conditions d'élevage de l'auxiliaire. *Acta Oecol.* 5: 85–97.

Ferran, A., P. Gubanti, G. Iperti, A. Migeon and J. Onillon. 1989. La répartition spatiale des différents stades de *Coccinella septempunctata* dans un champ de blé: variation au cours de la saison. *Entomol. Exper. Appl.* 53: 229–236.

Fiaboe, K. K. M., M. G. C. Gondim, G. J. De Moraes, C. K. P. O. Ogol and M. Knapp. 2007. Bionomics of the acarophagous ladybird beetle *Stethorus tridens* fed *Tetranychus evansi*. *J. Appl. Entomol.* 131: 355–361.

Fleschner, C. A. 1950. Studies on the searching capacity of the larvae of three predators of the Citrus red mite (*Paratetranychus citri*) (*Stethorus picipes, Conwentzia hageni, Chrysopa californica*). *Hilgardia* 20: 233–265.

Flowers, R. W., S. M. Salom and L. T. Kok. 2005. Competitive interactions among two specialist predators and a generalist predator of hemlock wooly adelgid, *Adelges tsugae* (Homoptera: Adelgidae), in the laboratory. *Environ. Entomol.* 34: 664–675.

Foott, W. H. 1973. Observations on Coccinellidae in corn fields in Essex county, Ontario. *Proc. Entomol. Soc. Ontario* 104: 16–21.

Formusoh, E. S. and G. E. Wilde. 1993. Preference and development of two species of predatory coccinellids on the Russian wheat aphid and greenbug biotype E (Homoptera: Aphididae). *J. Agr. Entomol.* 10: 65–70.

Fox, L. R. 1975. Cannibalism in natural populations. *Annu. Rev. Ecol. Syst.* 6: 87–106.

Fraenkel, G. 1959. The raison d'être of secondary plant substances. *Science* 129: 1466–1470.

Fraenkel, G. 1969. Evaluation of our thoughts on secondary plant substances. *Entomol. Exp. Appl.* 12: 473–486.

Francis, F., E. Haubruge and C. Gaspar. 2000. Influence of host plants on specialist/generalist aphids and on the

development of *Adalia bipunctata* (Coleoptera: Coccinellidae). *Eur. J. Entomol.* 97: 481–485.

Francis, F., G. Lognay and E. Haubruge. 2004. Olfactory responses to aphid and host plant volatile releases: (E)-β-farnesene an effective kairomone for the predator *Adalia bipunctata*. *J. Chem. Ecol.* 30: 741–755.

Francis, F., G. Lognay, J. P. Wathelet and E. Haubruge. 2001. Effects of allelochemicals from first (Brassicaceae) and second (*Myzus persicae* and *Brevicoryne brassicae*) trophic levels on *Adalia bipunctata*. *J. Chem. Ecol.* 27: 243–256.

Francke, D. L., J. P. Harmon, C. T. Harvey and A. R. Ives. 2008. Pea aphid dropping behavior diminishes foraging efficiency of a predatory ladybeetle. *Entomol. Exp. Appl.* 127: 118–124.

Frank, J. H. and E. D. McCoy. 2007. The risk of classical biological control in Florida. *Biol. Contr.* 41: 151–174.

Frazer, B. D. 1988. Coccinellidae. *In* A. K. Minks and P. Harrewijn (eds). *Aphids: Their Biology, Natural Enemies and Control*. Volume 2B. Elsevier, Amsterdam. pp. 231–247.

Frazer, B. D. and N. Gilbert. 1976. Coccinellids and aphids: a quantitative sudy of the impact of adult lady-birds (Coleoptera: Coccinellidae) preying on field populations of pea aphids (Homoptera: Aphididae). *J. Entomol. Soc. Br. Columbia* 73: 33–56.

Frazer, B. D. and B. Gill. 1981. Hunger, movement, and predation of *Coccinella californica* on pea aphids in the laboratory and in the field. *Can. Entomol.* 113: 1025–1033.

Frazer, B. D. and D. A. Raworth. 1985. Sampling for adult coccinellids and their numerical response to strawberry aphids (Coleoptera: Coccinellidae; Homoptera: Aphididae). *Can. Entomol.* 117: 153–161.

Frazer, B. D., N. Gilbert, V. Nealis and D. A. Raworth. 1981. Control of aphid density by a complex of predators. *Can. Entomol.* 113: 1035–1041.

Frechette, B., A. F. G. Dixon, C. Alauzet and J. L. Hemptinne. 2004. Age and experience influence patch assessment for oviposition by an insect predator. *Ecol. Entomol.* 29: 578–583.

Frechette, B., A. F. G. Dixon, C. Alauzet, N. Boughenou and J. L. Hemptinne. 2006. Should aphidophagous ladybirds be reluctant to lay eggs in the presence of unsuitable prey? *Entomol. Exp. Appl.* 118: 121–127.

Freeland, W. J. and D. H. Janzen. 1974. Strategies in herbivory by mammals: the role of plant secondary compounds. *Am. Nat.* 108: 269–289.

Fujiyama, N. and H. Katakura. 1997. Individual variation in two host plants of the ladybird beetle, *Epilachna pustulosa* (Coleoptera: Coccinellidae). *Ecol. Res.* 12: 257–264.

Fujiyama, N. and H. Katakura. 2001. Variable correspondence of female host preference and larval performance in a phytophagous ladybird beetle *Epilachna pustulosa* (Coleoptera: Coccinellidae). *Ecol. Res.* 16: 405–414.

Fujiyama, N. and H. Katakura. 2002. Host plant suitability of *Solanum japonense* (Solanaceae) as an alternative larval

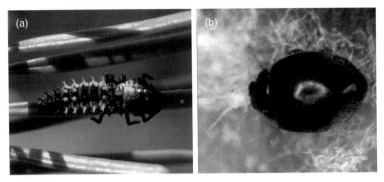

Plate 7.5 Coccinellids as intraguild predators. (a) Intraguild predation by a fourth instar larva of *Harmonia axyridis* preying upon a pupa of *Adalia bipunctata*. Picture by Serge Laplante; (b) Intraguild predation by an adult of *Stethorus punctillum* preying upon a first instar of the mullein bug *Campylomma verbasci*. Picture by Olivier Aubry (Laboratoire de Lutte biologique, UQAM, 2008).

Plate 7.6 Coccinellids as intraguild prey. (a) Intraguild predation by an adult *Podisus maculiventris* on a larva of *Coleomegilla maculata lengi* in a soybean field. Picture by Florent Renaud, 2007; (b) Intraguild predation by a *Thanatus* sp. spider on an adult of *Harmonia axyridis* in an apple orchard. Picture by Jennifer de Almeida, 2008 (Both pictures from Laboratoire de Lutte biologique, UQAM).

Ecology and Behaviour of the Ladybird Beetles (Coccinellidae), First Edition. Edited by I. Hodek, H.F. van Emden, A. Honěk.
© 2012 Blackwell Publishing Ltd. Published 2012 by Blackwell Publishing Ltd.

Plate 8.8 The underside of the elytron of an *Adalia bipunctata* infected with *Coccipolipus hippodamiae* (six large adult female mites and their eggs are visible) (photo courtesy of Emma Rhule).

Plate 8.6 *Cowperia indica* (photo courtesy of J. Poorani).

Plate 8.9 Thalli of *Hesperomyces coccinelloides* on the elytra of *Stethorus pusillus* (photo courtesy of Johan Bogaert).

Plate 8.7 A pupa of *Nothoserphus mirabilis* attached to the ventral side of its killed host, a larva of *Menochilus sexmaculatus* (photo courtesy of J. Poorani).

Plate 8.10 A clutch of *Adalia bipunctata* eggs and young larvae in which only half the offspring have hatched, due to the action of a male-killing *Rickettsia* (Photo: Remy Ware).

Plate 10.1 Exclusion cages mounted in an alfalfa field to determine impact of coccinellid species feeding on pea aphids, *Acyrthosiphon pisum* (photo: Edward Evans).

Plate 10.2 An exclusion cage designed for assessing impact of coccinellids and other natural enemies on aphids and psyllids infesting expanding grapefruit terminals. The cage is constructed from a clear plastic 2-litre soda bottle with mesh sleeves attached to either end with silicone and fastened to the branch with a length of wire secured to the inside of the bottle with packing tape. The sleeve mesh can be selected to permit the passage of particular insects (in this case, parasitoids), while excluding larger ones (coccinellids). The basal sleeve is attached tightly around the branch with a zip tie; the terminal sleeve is sufficiently long to permit unimpeded growth of the shoot for 2–3 weeks and can be untied to permit periodic access and counting of insects (photo: J.P. Michaud).

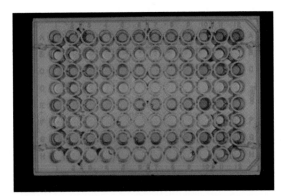

Plate 10.5 A 96-well microtitre plate following enzyme-linked immunosorbent assay. The amount of antigen–antibody binding is signified by the intensity of the reaction; absorbance is typically monitored by spectrophotometer to infer qualitative (and occasionally semi-quantitative) assessments of predation (photo: James Harwood).

Plate 11.1 An assassin bug (Reduviidae) preying on an adult *Hippodamia convergens* (J.P. Michaud).

Plate 11.5 Close-up of an adult *Hippodamia convergens* drinking extra-floral nectar from the petiole of a sunflower plant *Helianthus annuus* (J.P. Michaud).

Plate 11.2 The vedalia beetle *Rodolia cardinalis* with eggs and neonate larva on a mature cottony cushion scale (Jack Kelly Clark, courtesy UC Statewide IPM Program).

Plate 11.3 Aggregation of *Harmonia axyridis* attempting to enter a house under a door (courtesy of Marlin Rice).

Plate 11.6 A 'beetle bank' comprising a strip of perennial grasses forming dense tussocks to serve as overwintering habitat for coccinellids and other beneficial insects (Otago Regional Council, New Zealand).

food for three closely related *Epilachna* ladybird beetles (Coleoptera: Coccinellidae). *Appl. Entomol. Zool.* 37: 551–557.

Fujiyama, N., H. Katakura and Y. Shirai. 1998. Report of the Mexican bean beetle, *Epilachna varivestis* (Coleoptera: Coccinellidae) in Japan. *Appl. Entomol. Zool.* 33: 327–331.

Fujiyama, N., T. Koizumi and H. Katakura. 2003. Conspecific thistle plant selection by a herbivorous ladybird beetle, *Epilachna pustulosa*. *Entomol. Exp. Appl.* 108: 33–42.

Fujiyama, N., K. Matsumoto, N. Kobayashi, Y. Ohta and H. Katakura. 2005. Differentiation in the ability to utilize *Pterostyrax hispida* (Ebenales: Styracaceae) among four local populations of the phytophagous ladybird beetle *Henosepilachna yasutomii* (Coleoptera: Coccinellidae). *Popul. Ecol.* 47: 91–98.

Fukunaga, Y. and S.-I. Akimoto. 2007. Toxicity of the aphid *Aulacorthum magnoliae* to the predator *Harmonia axyridis* (Coleoptera: Coccinellidae) and genetic variance in the assimilation of the toxic aphids in *H. axyridis* larvae. *Entomol. Sci.* 10: 45–53.

Gagné, I., D. Coderre and Y. Mauffette. 2002. Egg cannibalism by *Coleomegilla maculata lengi* neonates: preference even in the presence of essential prey. *Ecol. Entomol.* 27: 285–291.

Fye, R. E. 1981. Rearing and release of coccinellids for potential control of pear psylla. *Adv. Agric. Technol.*, US Dept. Agric. West. Ser. 20: 1–9.

Gagné, W. C. and J. L. Martin. 1968. The insect ecology of red pine plantations in central Ontario. V. The Coccinellidae (Coleoptera). *Can. Entomol.* 100: 835–846.

Gannon, A. J. and C. E. Bach. 1996. Effects of soybean trichome density on Mexican Bean Beetle (Coleoptera: Coccinellidae) development and feeding preference. *Environ. Entomol.* 25: 1077–1082.

Gao, F., S.-R. Zhu, Y. C. Sun et al. 2008. Interactive effects of elevated CO_2 and cotton cultivar on tritrophic interaction of *Gossypium hirsutum*, *Aphis gossypii* and *Propylea japonica*. *Environ. Entomol.* 37: 29–37.

Garcia, J. F. and R. J. O'Neil. 2000. Effect of *Coleus* size and variegation on attack rates, searching strategy, and selected life history characteristics of *Cryptolaemus montrouzieri* (Coleoptera: Coccinellidae). *Biol. Contr.* 18: 225–234.

Gautam, R. D. 1990. Mass-multiplication technique of coccinellid predator, ladybird beetle (*Brumoides suturalis*). *Indian J. Agr. Sci.* 60: 747–750.

George, K. S. 1957. Preliminary investigations on the biology and ecology of the parasites and predators of *Brevicoryne brassicae* (L.). *Bull. Entomol. Res.* 48: 619–629.

Geyer, J. W. C. 1947. A study of the biology and ecology of *Exochomus flavipes* Thunb. (Coccinellidae, Coleoptera). Part 1. 2. *J. Entomol. Soc. South. Africa* 9: 219–234; 10: 64–109.

Ghanim, A. E.-B., B. Freier and T. Wetzel. 1984. Zur Nahrungsaufnahme und Eiablage von *Coccinella septempunctata* L. bei unterschiedlichem Angebot von Aphiden der Arten *Macrosiphum avenae* (Fabr.) und *Rhopalosiphum padi* (L.). *Arch. Phytopathol. PflSch.* 20: 117–125.

Gil, L., A. Ferran, J. Gambier et al. 2004. Dispersion of flightless adults of the Asian lady beetle, *Harmonia axyridis*, in greenhouses containing cucumbers infested with the aphid *Aphis gossypii*: effect of the presence of conspecific larvae. *Entomol. Exp. Appl.* 112: 1–6.

Giles, K. L., J. J. Obrycki and T. A. DeGooyer. 1994. Prevalence of predators associated with *Acyrthosiphum pisum* (Homoptera: Aphididae) and *Hypera postica* Gyllenhal (Coleoptera: Curculionidae) during growth of the first crop of alfalfa. *Biol. Contr.* 4: 170–177.

Giles, K. L., R. D. Madden, R. Stockland, M. E. Payton and J. W. Dillwith. 2002. Host plants affect predator fitness via the nutritional value of herbivore prey: Investigation of a plant–aphid–ladybeetle system. *BioControl* 47: 1–21.

Giorgi, J. A., N. J. Vandenberg, J. V. McHugh et al. 2009. The evolution of food preferences in Coccinellidae. *Biol. Contr.* 51: 215–231.

Giroux, S., R. M. Duchesne and D. Coderre. 1995. Predation of *Leptinotarsa decemlineata* (Coleoptera, Chrysomelidae) by *Coleomegilla maculata* (Coleoptera, Coccinellidae) – comparative effectiveness of predator developmental stages and effect of temperature. *Environ. Entomol.* 24: 748–754.

Glen, D. M. 1973. The food requirements of *Blepharidopterus angulatus* (Heteroptera: Miridae) as a predator of the lime aphid, *Eucallipterus tiliae*. *Entomol. Exp. Appl.* 16: 255–267.

Godfray, H. C. J. 1994. *Parasitoids: Behavioral and Evolutionary Ecology*. Princeton University Press, Princeton, New Jersey. 488 pp.

Goidanich, A. 1943. Due Coccinelle igrofile e pollinivore sul riso (*Hippodamia 13-punctata* L., *Anisosticta 19-punctata* L.) *Risicoltura* 33: 145–156, 169–177.

Gomez Vives, S. 1999. Biology of *Lindorus lophantae* Blaisdell (Col.: Coccinellidae), a candidate biocontrol agent of *Phoenicococcus marlatti* (Hom.: Phoenicococcidae) in the date-palm grove of Elche, Spain. *In* M. Canard and Beyssat-Arnaouty (eds). *Proceedings of the 1st Regional Symposium for Applied Biological Control in Mediterranean Countries*, Cairo, Egypt, 1998. Toulouse, France.

Gonzáles-Hernández, H., N. J. Reimer and M. W. Johnson. 1999. Survey of the natural enemies of *Dysmicoccus* mealybugs on pineapple in Hawai. *BioControl* 44: 47–58.

Gorb, E., D. Voigt, S. D. Eigenbrode and S. Gorb. 2008. Attachment force of the beetle *Cryptolaemus montrouzieri* (Coleoptera, Coccinellidae) on leaflet surfaces of mutants of the pea *Pisum sativum* (Fabaceae) with regular and reduced wax coverage. *Arthropod–Plant Interact.* 2: 247–259.

Greenstone, M. H. and S. C. Trowell. 1994. Arthropod preda-tion: a simplified immunodot format for predator gut analy-sis. *Ann. Entomol. Soc. Am.* 87: 214–217.

Greenstone, M. H., D. L. Rowley, D. C. Weber, M. E. Payton and D. J. Hawthorne. 2007. Feeding mode and prey detectability half-lives in molecular gut-content analysis: an example with two predators of the Colorado potato beetle. *Bull. Entomol. Res.* 97: 201–209.

Grez, A. A., T. Zaviezo and M. Rios. 2005. Ladybird (Coleop-tera: Coccinellidae) dispersal in fragmented alfalfa land-scapes. *Eur. J. Entomol.* 102: 209–216.

Grez, A. A., T. Zaviezo, S. Diaz, B. Comousseight and G. Cortes. 2008. Effects of habitat loss and fragmentation on the abundance and species richness of aphidophagous beetles and aphids in experimental alfalfa landscapes. *Eur. J. Entomol.* 105: 411–420.

Griffin, M. L. and K. V. Yeargan. 2002. Factors potentially affecting oviposition site selection by the lady beetle *Cole-omegilla maculata* (Coleoptera: Coccinellidae). *Environ. Entomol.* 31: 112–119.

Grill, C. P., A. J. Moore and E. D. Brodie III. 1997. The genetics of phenotypic plasticity in a colonizing population of the ladybird beetle, *Harmonia axyridis. Heredity* 78: 261–269.

Gruppe, A. and P. Roemer. 1988. The lupin aphid (*Macro-siphum albifrons* Essing, 1911. (Hom., Aphididae) in West Germany: its occurrence, host plants and natural enemies. *Z. Angew. Entomol.* 106: 135–143.

Guershon, M. and D. Gerling. 1999. Predatory behavior of *Delphastus pusillus* in relation to the phenotypic plasticity of *Bemisia tabaci* nymphs. *Entomol. Exp. Appl.* 92: 239–248.

Hagen, K. S. 1962. Biology and ecology of predaceous Coc-cinellidae. *Annu. Rev. Entomol.* 7: 289–326.

Hagen, K. S. 1986. Ecosystem analysis: plant cultivars (HPR), entomophagous species and food supplements. *In* D. J. Boethel and R. D. Eikenbary (eds). *Interactions of Plant Resistance and Parasitoids and Predators of Insects.* Wiley, New York. pp. 151–197.

Hagen, K. S., P. Greany, E. F. Sawall Jr. and R. L. Tassan. 1976. Tryptophan in artificial honeydews as a source of an attractant for adult *Chrysopa carnea. Environ. Entomol.* 5: 458–468.

Hagen, K. S., E. F. Sawall Jr. and R. L. Tassan. 1971. The use of food sprays to increase effectiveness of entomophagous insects. *Proc. Tall Timbers Conf. Ecol. Anim. Control Habitat Manage.* 2: 59–81.

Hagley, E. A. C. and W. R. Allen. 1990. The green apple aphid, *Aphis pomi* DeGeer (Homoptera: Aphididae), as prey of poly-phagous arthropod predators in Ontario. *Can. Entomol.* 122: 1221–1228.

Hajek, A. E. and D. L. Dahlsten. 1987. Behavioral interactions between three birch aphid species and *Adalia bipunctata* larvae. *Entomol. Exper. Appl.* 45: 81–87.

Hales, D. 1979. Population dynamics of *Harmonia conformis* (Boisd.) (Coleoptera: Coccinellidae) on acacia. *Gen. Appl. Entomol.* 11: 3–8.

Hamasaki, K. and M. Matsui. 2006. Development and repro-duction of an aphidophagous coccinellid, *Propylea japonica* (Thunberg) (Coleoptera: Coccinellidae), reared on an alter-native diet, *Ephestia kuehniella* Zeller (Lepidoptera: Pyrali-dae) eggs. *Appl. Entomol. Zool.* 41: 233–237.

Hammond, R. B. and R. L. Coope. 1999. Antibiosis of released soybean germplasm to Mexican bean beetle (Coleoptera: Coccinellidae). *J. Entomol. Sci.* 34: 183–190.

Han, B. and Z. M. Chen. 2002. Behavioral and electrophysi-ological responses of natural enemies to synomones from tea shoots and kairomones from tea aphids, *Toxoptera aurantii. J. Chem. Ecol.* 28: 2203–2219.

Hariri, G. 1966a. Laboratory studies on the reproduction of *Adalia bipunctata* (L.) (Coleoptera: Coccinellidae). *Entomol. Exp. Appl.* 9: 200–204.

Hariri, G. 1966b. Changes in metabolic reserves of three species of aphidophagous Coccinellidae (Coleoptera) during metamorphosis. *Entomol. Exp. Appl.* 9: 349–358.

Harmon, J. P., A. R. Ives, J. E. Losey, A. C. Olson and K. S. Rauwald. 2000. *Coleomegilla maculata* (Coleoptera: Coc-cinellidae) predation on pea aphids promoted by proximity to dandelions. *Oecologia* 125: 543–548.

Harmon, J. P., J. E. Losey and A. R. Ives. 1998. The role of vision and color in the close proximity foraging behav-ior of four coccinellid species. *Oecologia* 115: 287–292.

Harmon, J. P., E. Stephens and J. Losey. 2007. The decline of native coccinellids (Coleoptera: Coccinellidae) in the United States and Canada. *J. Insect Conserv.* 11: 85–94.

Harpaz, I. 1958. Bionomics of the 11-spotted ladybird beetle, *Coccinella undecimpunctata* L., in a subtropical climate. *Proc. 10th Int. Congr. Entomol.* Montreal 1956, 2: 657–659.

Harwood, J. D. and J. J. Obrycki. 2005. Quantifying aphid pre-dation rates of generalist predators in the field. *Eur. J. Entomol.* 102: 335–350.

Harwood, J. D., S. W. Phillips, J. Lello et al. 2009. Invertebrate biodiversity affects predator fitness and hence potential to control pests in crops. *Biol. Contr.* 51: 499–506.

Hassell, M. P. 1978. *The Dynamics of Arthropod Predator–Prey Systems.* Monographs in Population Biology 13. Princeton Univ. Press, Princeton. 237 pp.

Hatano, E., G. Kunert, J. P. Michaud and W. W. Weisser. 2008. Chemical cues mediating aphid location by natural enemies. *Eur. J. Entomol.* 105: 797–806.

Hattingh, V. and M. J. Samways. 1991. A forced change in prey type during field introductions of coccidophagous bio-control agents *Chilocorus* species (Coleoptera: Coccinelli-dae): is it an important consideration in achieving establishment? *In* L. Polgár, R. J. Chambers, A. F. G. Dixon and I. Hodek (eds). *Behaviour and Impact of Aphidophaga.* SPB Acad. Publ., The Hague. pp. 143–148.

Hattingh, V. and M. J. Samways. 1993. Evaluation of artificial diets and two species of natural prey as laboratory food for *Chilocorus* spp. *Entomol. Exp. Appl.* 69: 13–20.

Hattingh, V. and M. J. Samways. 1995. Visual and olfactory location of biotopes, prey patches, and individual prey by the ladybeetle *Chilocorus nigritus*. *Entomol. Exp. Appl.* 75: 87–98.

Hauge, M. S., F. H. Nielsen and S. Toft. 1998. The influence of three cereal aphid species and mixed diet on larval survival, development and adult weight of *Coccinella septempunctata*. *Entomol. Exp. Appl.* 89: 319–323.

Hawkes, O. A. M. 1920. Observations on the life-history, biology, and genetics of the lady-bird, *Adalia bipunctata* (Mulsant). *Proc. Zool. Soc. Lond.* 33: 475–490.

Hawkes, O. A. M. and T. F. Marriner. 1927. A preliminary account of the life-history of *Coccinella 11–punctata* (L.). *Trans. R. Entomol. Soc. Lond.* 75: 47–52.

Hazzard, R. V. and D. N. Ferro. 1991. Feeding responses of adult *Coleomegilla maculata* (Coleoptera: Coccinellidae) to eggs of the Colorado potato beetle (Coleoptera: Chrysomelidae) and green peach aphids (Homoptera: Aphididae). *Environ. Entomol.* 20: 644–651.

Heidari, M. and M. J. W. Copland. 1992. Host finding by *Cryptolaemus montrouzieri* (Col., Coccinellidae) a predator of mealybugs (Hom., Pseudococcidae). *Entomophaga* 37: 621–625.

Hemptinne, J.-L. and A. Desprets. 1986. Pollen as a spring food for *Adalia bipunctata*. *In* I. Hodek (ed.). *Ecology of Aphidophaga*. Academia, Prague and W. Junk, Dordrecht. pp. 29–35.

Hemptinne, J.-L. and J. Naisse. 1987. Ecophysiology of the reproductive activity of *Adalia bipunctata* L. (Col., Coccinellidae). *Med. Fac. Landbouww. Rijksuniv. Gent* 52: 225–233.

Hemptinne, J.-L., A. F. G. Dixon and J. Coffin. 1992. Attack strategy of ladybird beetles (Coccinellidae): factors shaping their numerical response. *Oecologia* 90: 238–245.

Hemptinne, J.-L., A. F. G. Dixon and C. Gauthier. 2000a. Nutritive cost of intraguild predation on eggs of *Coccinella septempunctata* and *Adalia bipunctata* (Coleoptera: Coccinellidae). *Eur. J. Entomol.* 97: 559–562.

Hemptinne, J.-L., J. Naisse and S. Os. 1988. Glimps of the life history of *Propylea quatuordecimpunctata* (L.) (Coleoptera: Coccinellidae). *Med. Fac. Landbouww. Rijksuniv. Gent* 53: 1175–1182.

Hemptinne, J.-L., A. F. G. Dixon, J. L. Doucet and J. E. Petersen. 1993. Optimal foraging by hoverflies (Diptera: Syrphidae) and ladybirds (Coleoptera: Coccinellidae): mechanisms. *Eur. J. Entomol.* 90: 451–455.

Hemptinne, J.-L., G. Lognay, M. Doumbia and A. F. G. Dixon. 2001. Chemical nature and persistence of the oviposition deterring pheromone in the tracks of the larvae of the two spot ladybird, *Adalia bipunctata* (Coleoptera: Coccinellidae). *Chemoecology* 11: 43–47.

Hemptinne, J.-L., G. Lognay, C. Gauthier and A. F. G. Dixon. 2000b. Role of surface chemical signals in egg cannibalism and intraguild predation in ladybirds (Coleoptera: Coccinellidae). *Chemoecology* 10: 123–128.

Herbert, J. J. and D. J. Horn. 2008. Effect of ant attendance by *Monomorium minimum* (Buckley) (Hymenoptera: Formicidae) on predation and parasitism of the soybean aphid *Glycines macumura* (Hemiptera: Aphididae). *Envir. Entomol.* 37: 1258–1263.

Hilbeck, A., C. Eckel and G. G. Kennedy. 1997. Predation on Colorado potato beetle eggs by generalist predators in research and commercial potato plantings. *Biol. Control* 8: 191–196.

Hippa, H., S. Koponen and R. Roine. 1984. Larval growth of *Coccinella hieroglyphica* (Col., Coccinellidae) fed on aphids and preimaginal stages of *Galerucella sagittariae* (Col., Chrysomelidae). *Rep. Kevo Subarctic Res. Stat.* 19: 67–70.

Hodek I. 1956. The influence of *Aphis sambuci* L. as prey of the ladybird beetle *Coccinella septempunctata* L. *Věst. Čs. Spol. Zool.* 20: 62–74. (In Czech, Engl. summ.).

Hodek, I. 1957a. The influence of *Aphis sambuci* L. as food for *Coccinella 7-punctata* L. II. *Čas. Čs. Spol. Entomol.* 54: 10–17. (In Czech, Engl. summ.).

Hodek, I. 1957b. The larval food consumption of *Coccinella 7-punctata* L. *Zool. Listy (Folia Zoologica)* 6: 3–11. (In Czech with English summary.)

Hodek, I. 1959. Ecology of aphidophagous Coccinellidae. *In Proc. Int. Conf. Insect Path. Biol. Control*, Academia, Prague, 1958: 543–547.

Hodek, I. 1960. The influence of various aphid species as food for two lady-birds *Coccinella 7-punctata* L. and *Adalia bipunctata* L. *In The Ontogeny of Insects.* (Proc. Symp. Praha 1959). Academia, Prague. pp. 314–316.

Hodek, I. 1962. Essential and alternative food in insects. *Proc. 11th Int. Congr. Entomol.* Vienna 1960, 2: 696–697.

Hodek, I. 1967. Bionomics and ecology of predaceous Coccinellidae. *Annu. Rev. Entomol.* 12: 79–104.

Hodek, I. 1973. *Biology of Coccinellidae.* Academia, Prague and W. Junk, The Hague. 260 pp.

Hodek. I. 1993. Habitat and food specificity in aphidophagous predators. *Biocont. Sci. Tech.* 3: 91–100.

Hodek, I. 1996. Food relationships. *In* I. Hodek and A. Honěk (eds). *Ecology of Coccinellidae.* Kluwer, Dordrecht. pp. 143–238.

Hodek, I. and A. Honěk, eds. 1996. *Ecology of Coccinellidae.* Kluwer, Dordrecht. 464 pp.

Hodek, I. and A. Honěk. 2009. Scale insects, mealybugs, whiteflies and psyllids (Hemiptera, Sternorrhyncha) as prey of ladybirds. *Biol. Contr.* 51: 232–243.

Hodek, I. and J. P. Michaud. 2008. Why is *Coccinella septempunctata* so successful? (A point-of-view). *Eur. J. Entomol.* 105: 1–12.

Hodek, I., G. Iperti and M. Hodková. 1993. Long-distance flights in Coccinellidae (Coleoptera). *Eur. J. Entomol.* 90: 403–414.

Hodek, I., Z. Ruzicka and M. Hodková. 1978. Pollinivorie et aphidiphagie chez *Coleomegilla maculata*. *Ann. Zool.-Ecol. Anim.* 10: 453–459.

Hodek, I., K. Novák, V. Skuhravý and J. Holman. 1965. The predation of *Coccinella septempunctata* L. on *Aphis fabae* Scop. on sugar beet. *Acta Entomol. Bohemoslov.* 62: 241–253.

Hoelmer, K. A., L. S. Osborne and R. K. Yokomi. 1993. Reproduction and feeding behaviour of *Delphastus pusillus* (Coleoptera: Coccinellidae), a predator of *Bemisia tabaci* (Homoptera: Aleyrodidae). *J. Econ. Entomol.* 86: 322–329.

Hoelmer, K. A., L. S. Osborne and R. K. Yokomi. 1994. Interactions of the whitefly predator *Delphastus pusillus* (Coleoptera: Coccinellidae) with parasitized sweetpotato whitefly (Homoptera: Aleyrodidae). *Environ. Entomol.* 23: 136–139.

Holling C. S. 1959a. Some characteristics of simple types of predation and parasitism. *Can. Entomol.* 91: 385–398.

Holling, C. S. 1959b. The components of predation as revealed by a study of small mammal predation of the European sawfly. *Can. Entomol.* 91: 293–320.

Holling, C. S. 1965. The functional response of predators to prey density and its role in mimicry and population regulation. *Mem. Entomol. Soc. Can.* 45: 3–60.

Holling, C. S. 1966. Functional response of invertebrate predators to prey density. *Mem. Entomol. Soc. Can.* 48: 1–87.

Honda, J. Y. and R. F. Luck. 1995. Scale morphology effects on feeding behaviour and biological control potential of *Rhyzobius lophantae* (Coleoptera: Coccinellidae). *Ann. Entomol. Soc. Am.* 88: 441–450.

Honěk, A. 1979. Plant density and occurrence of *Coccinella septempunctata* and *Propylea quatuordecimpunctata* (Coleoptera, Coccinellidae) in cereals. *Acta Entomol. Bohemoslov.* 76: 308–312.

Honěk, A. 1980. Population density of aphids at the time of settling and ovariole maturation in *Coccinella septempunctata* (Col.: Coccinellidae). *Entomophaga* 25: 427–430.

Honěk, A. 1982. Factors which determine the composition of field communities of adult aphidophagous Coccinellidae (Coleoptera). *Z. Angew. Entomol.* 94: 157–168.

Honěk, A. 1983. Factors affecting the distribution of larvae of aphid predators (Col., Coccinellidae and Dipt., Syrphidae) in cereal stands. *Z. Angew. Entomol.* 95: 336–343.

Honěk, A. 1985. Habitat preferences of aphidophagous coccinellids (Coleoptera). *Entomophaga* 30: 253–264.

Honěk, A. 1986. Production of faeces in natural populations of aphidophagous coccinellids (Col.) and estimation of predation rates. *J. Appl. Entomol.* 102: 467–476.

Honěk, A., A. F. G. Dixon and Z. Martinkova. 2008a. Body size, reproductive allocation, and maximum reproductive rate of two species of aphidophagous Coccinellidae exploiting the same resource. *Entomol. Exp. Appl.* 127: 1–9.

Honěk, A., A. F. G. Dixon and Z. Martinkova. 2008b. Body size and the temporal sequence in the reproductive activity of two species of aphidophagous coccinellids exploiting the same resource. *Eur. J. Entomol.* 105: 421–425.

Hoogendoorn, M. and G. E. Heimpel. 2001. PCR-based gut content analysis of insect predators: using ribosomal ITS-1 fragments from prey to estimate predation frequency. *Mol. Ecol.* 10: 2059–2068.

Hori, M., Y. Araki, W. Sugeno, Y. Usui and K. Matsuda. 2005. Luteolin 7-O-glucoside in hozuki leaves, *Physalis alkekengi*, is involved in feeding stimulation in *Epilachna vigintioctopunctata*. *Jpn. J. Appl. Entomol. Zool.* 49: 251–254.

Hosseini, R., O. Schmidt and M. A. Keller. 2008. Factors affecting detectability of prey DNA in the gut contents of invertebrate predators: a polymerase chain reaction-based method. *Entomol. Exp. Appl.* 126: 194–202.

Houck, M. A. 1986. Prey preference in *Stethorus punctum* (Coleoptera: Coccinellidae). *Environ. Entomol.* 15: 967–970.

Houck, M. A. 1991. Time and resource partitioning in *Stethorus punctum* (Coleoptera: Coccinellidae). *Environ. Entomol.* 20: 494–497.

House, H. L. 1974. Digestion. In M. Rockstein (ed.). *The Physiology of Insects*. Academic Press, New York. pp. 63–120.

Huang, Z., S.-X. Ren and P. D. Musa. 2008. Effects of temperature on development, survival, longevity, and fecundity of the *Bemisia tabaci* Gennadius (Homoptera: Aleyrodidae) predator, *Axinoscymnus cardilobus* (Coleoptera: Coccinellidae). *Biol. Contr.* 46: 209–215.

Huang, Z., S.-X. Ren and S.-L. Yao. 2006. Life history of *Axinoscymnus cardilobus* (Col., Coccinellidae), a predator of *Bemisia tabaci* (Hom., Aleyrodidae). *J. Appl. Entomol.* 130: 437–441.

Huffaker, C. B. and R. L. Doutt. 1965. Establishment of the coccinellid, *Chilocorus bipustulatus* Linneus, in California olive groves. *Pan-Pacif. Entomol.* 44: 61–63.

Hukusima, S. and K. Itoh. 1976. Pollen and fungus as food for some coccinellid beetles. *Res. Bull. Fac. Agr. Gifu Univ.* 39: 31–38.

Hukusima, S. and M. Kamei. 1970. Effects of various species of aphids as food on development, fecundity and longevity of *Harmonia axyridis* Pallas (Coleoptera: Coccinellidae). *Res. Bull. Fac. Agric. Gifu Univ.* 29: 53–66.

Hukusima, S. and N. Komada. 1972. Effects of several species of aphids as food on development and nutrition of *Propylaea japonica* Thunberg (Coleoptera: Coccinellidae). *Proc. Kansai Plant Prot. Soc.* 14: 7–13. (In Japanese with English summary.)

Hukusima, S. and S. Kouyama. 1974. Life histories and food habits of *Menochilus sexmaculatus* Fabricius (Coleoptera: Coccinellidae). *Res. Bull. Fac. Agric. Gifu Univ.* 36: 19–29.

Hukusima, S. and T. Ohwaki. 1972. Further notes on feeding biology of *Harmonia axyridis* Pallas (Coleoptera: Coccinellidae). *Res. Bull. Fac. Agric. Gifu Univ.* 33: 75–82.

Hukusima, S. and H. Sakurai. 1964. Aphid consumption by adult *Coccinella septempunctata bruckii* Mulsant in relation to temperature (Coleoptera: Coccinellidae). *Annu. Rev. Plant Protection North Japan* 15: 126–128. (In Japanese.)

Hukusima, S. and S. Takeda. 1975. Artificial diets for larvae of *Harmonia axyridis* Pallas (Coleoptera: Coccinellidae), an insect predator of aphids and scale insects. *Res. Bull. Fac. Agr. Gifu Univ.* 38: 49–53.

Hull, L. A., D. Asquith and P. D. Mowery. 1976. Distribution of *Stethorus punctum* in relation to densities of the European red mite. *Environ. Entomol.* 5: 337–342.

Hull, L. A., D. Asquith and P. D. Mowery. 1977a. The functional responses of *Stethorus punctum* to densities of the European red mite. *Environ. Entomol.* 6: 85–90.

Hull, L. A., D. Asquith and P. D. Mowery. 1977b. The mite searching ability of *Stethorus punctum* within an apple orchard. *Environ. Entomol.* 6: 684–688.

Hurst, G. D. D. and M. E. N. Majerus. 1993. Why do maternally inherited microorganisms kill males? *Heredity* 71: 81–95.

Hussein, M. Y. and K. S. Hagen. 1991. Rearing of *Hippodamia convergens* on artificial diet of chicken liver, yeast and sucrose. *Entomol. Exp. Appl.* 59: 197–199.

Ibrahim, M. M. 1955a. Studies on *Coccinella undecimpunctata aegyptiaca* Reiche. 1. Preliminary notes and morphology of the early stages. *Bull. Soc. Entomol. Egypte* 39: 251–274.

Ibrahim, M. M. 1955b. Studies on *Coccinella undecimpunctata aegyptiaca* Reiche. 2. Biology and life-history. *Bull. Soc. Entomol. Egypte* 39: 395–423.

Ide, T., N. Suzuki and N. Katayama. 2007. The use of honeydew in foraging for aphids by larvae of the ladybird beetle, *Coccinella septempunctata* L. (Coleoptera: Coccinellidae). *Ecol. Entomol.* 32: 455–460.

Imms, A. D. 1947. *Insect Natural History*. London.

Inchausti, P. and S. Ballesteros. 2008. Intuition, functional responses and the formulation of predator–prey models when there is a large disparity in the spatial domains of the interacting species. *J. Anim. Ecol.* 77: 891–897.

Iperti, G. 1965. Contribution a l'étude de la spécificité chez les principales Coccinelles aphidiphages des Alpes-Maritimes et des Basses-Alpes. *Entomophaga* 10: 159–178.

Iperti, G. 1966. Comportement naturel des Coccinelles aphidiphages du Sud-Est de la France. Leur type de spécificité, leur action prédatrice sur *Aphis fabae* L. *Entomophaga* 11: 203–210.

Iperti, G. and N. Trepanier-Blais. 1972. Valeur alimentaire des oeufs d'*Anagasta kuehniella* Z. (Lep., Pyralidae) pour une coccinelle aphidiphage: *Adonia undecimnotata* Schn. (Col., Coccinellidae). *Entomophaga* 17: 437–441.

Isard, S. A., D. A. R. Kristovich, S. H. Gage, C. J. Jones and N. F. Laird. 2001. Atmospheric motion systems that influence the redistribution and accumulation of insects on the beaches of the Great Lakes in North America. *Aerobiologia* 17: 275–291.

Isikber, A. A. and M. J. W. Copland. 2001. Food consumption and utilization by larvae of two coccinellid predators,

Scymnus levaillanti and *Cycloneda sanguinea*, on cotton aphid, *Aphis gossypii*. *BioControl* 46: 455–467.

Isikber, A. A. and M. J. W. Copland. 2002. Effects of various aphid foods on *Cycloneda sanguinea*. *Entomol. Exp. Appl.* 102: 93–97.

Iverson, A. L., R. B. Hammond and L. R. Iverson. 2001. Induction of soybean resistance to the Mexican Bean Beetle (Coleoptera: Coccinellidae). *J. Kans. Entomol. Soc.* 74: 185–191.

Ives, A. R., P. Kareiva and R. Perry. 1993. Response of a predator to variation in prey density at three hierarchical scales: lady beetles feeding on aphids. *Ecology* 74: 1929–1938.

Ives, P. M. 1981a. Estimation of coccinellid numbers and movement in the field. *Can. Entomol.* 113: 981–997.

Ives, P. M. 1981b. Feeding and egg production of two species of coccinellids in the laboratory. *Can. Entomol.* 113: 999–1005.

Iwao, S. 1954. On the distributions of *Epilachna sparsa orientalis* Dieki and *E. vigintioctomaculata* Motschulsky and the boundary of their geographical distributions (I). *Oyo-Kontyû* 9: 135–141. (In Japanese with English summary.)

Iwao, S. and A. Machida. 1961. Further experiments on the host-plant preference in a phytophagous ladybeetle, *Epilachna pustulosa* Kono. *Insect Ecol.* 9: 9–16. (In Japanese with English summary.)

Iwata, K. 1932. On the biology of two large lady-birds in Japan. *Trans. Kansai Entomol. Soc.* 3: 13–26.

Iwata, K. 1965. Supplement on the biology of two large ladybirds in Japan. *Acta Coleopter.* 2: 57–68.

Jalali, M. A., L. Tirry and P. De Clercq. 2009a. Effects of food and temperature on development, fecundity and life-table parameters of *Adalia bipunctata* (Coleoptera: Coccinellidae). *J. Appl. Entomol.* 113: 615–625.

Jalali, M. A., L. Tirry and P. De Clercq. 2009b. Food consumption and immature growth of *Adalia bipunctata* (Coleoptera: Coccinellidae) on a natural prey and a factitious food. *Eur. J. Entomol.* 106: 193–198.

Jalali, M. A., L. Tirry and P. De Clercq. 2010. Effect of temperature on the functional response of *Adalia bipunctata* to *Myzus persicae*. *BioControl* 55: 261–269.

Jenkins, E. B., R. B. Hammond, S. K. St. Martin and R. L. Cooper. 1997. Effect of soil moisture and soybean growth stage on resistance to Mexican bean beetle (Coleoptera: Coccinellidae). *J. Econ. Entomol.* 90: 697–703.

Jervis, M. A., ed. 2005. *Insects as Natural Enemies: A Practical Perspective*. Kluwer, Dordrecht, The Netherlands. 764 pp.

Jiggins, C., M. Majerus and U. Gough. 1993. Ant defence of colonies of *Aphis fabae* Scopoli (Hemiptera: Aphididae), against predation by ladybirds. *Br. J. Ent. Nat. Hist.* 6: 129–137.

Johki, Y., S. Obata and M. Matsui. 1988. Distribution and behaviour of five species of aphidophagous ladybirds (Coleoptera) around aphid colonies. *In* E. Niemczyk and A. F. G.

256 I. Hodek and E. W. Evans

Dixon (eds). *Ecology and Effectiveness of Aphidophaga*. SPB Acad. Publ., The Hague. pp. 35–38.

Jöhnssen, A. 1930. Beiträge zur Entwicklungs- und Ernährungsbiologie einheimischer Coccinelliden unter besonderer Berücksichtigung von *Coccinella septempunctata* L. *Z. Angew. Entomol.* 16: 87–158.

Joseph, S. B., W. E. Snyder and A. J. Moore. 1999. Cannibalizing *Harmonia axyridis* (Coleoptera: Coccinellidae) larvae use endogenous cues to avoid eating relatives. *J. Evol. Biol.* 12: 792–797.

Kaddou, I. K. 1960. The feeding behavior of *Hippodamia quinquesignata* (Kirby) larvae. *Univ. Calif. Publ. Entomol.* 16: 181–230.

Kagata, H. and N. Katayama. 2006. Does nitrogen limitation promote intraguild predation in an aphidophagous ladybird? *Entomol. Exp. Appl.* 119: 239–246.

Kagata, H. and T. Ohgushi. 2007. Carbon–nitrogen stoichiometry in the tritrophic food chain: willow, leaf beetle, and predatory ladybird beetle. *Ecol. Res.* 22: 671–677.

Kajita, Y. and E. W. Evans. 2009. Ovarian dynamics and oosorption in two species of predatory lady beetles (Coleoptera: Coccinellidae). *Physiol. Entomol.* 34: 185–194.

Kajita, Y. and E. W. Evans. 2010. Alfalfa fields promote high reproductive rate of an invasive predatory lady beetle. *Biol. Invasions* 12: 2293–2302.

Kajita, Y., E. W. Evans and H. Yasuda. 2009. Reproductive responses of invasive and native predatory lady beetles (Coleoptera: Coccinellidae) to varying prey availability. *Environ. Entomol.* 38: 1283–1292.

Kalaskar, A. and E. W. Evans. 2001. Larval responses of aphidophagous lady beetles (Coleoptera: Coccinellidae) to weevil larvae versus aphids as prey. *Ann. Entomol. Soc. Am.* 94: 76–81.

Kalushkov, P. 1998. Ten aphid species (Sternorrhyncha: Aphididae) as prey for *Adalia bipunctata* (Coleoptera: Coccinellidae). *Eur. J. Entomol.* 95: 343–349.

Kalushkov, P. and I. Hodek. 2001. New essential aphid prey for *Anatis ocellata* and *Calvia quatuordecimguttata* (Coleoptera: Coccinellidae). *Biocontr. Sci. Technol.* 11: 35–39.

Kalushkov, P. and I. Hodek. 2004. The effects of thirteen species of aphids on some life history parameters of the ladybird *Coccinella septempunctata*. *BioControl* 49: 21–32.

Kalushkov, P. and I. Hodek. 2005. The effects of six species of aphids on some life history parameters of the ladybird *Propylea quatuordecimpunctata* (Coleoptera: Coccinellidae). *Eur. J. Entomol* 102: 449–452.

Kanervo, V. 1940. Beobachtungen und Versuche zur Ermittlung der Nahrung einiger Coccinelliden. *Ann. Entomol. Fenn.* 6: 89–110.

Kanervo, V. 1946. Studien über die natürlichen Feinde des Erlenblattkäfers, *Melasoma aenea* L. (Col., Chrysomelidae). *Ann. Zool. Soc. Vanamo* 12: 206. (In Finish with German summary.)

Karban, R. 2007. Damage to sagebrush attracts predators but this does not reduce herbivory. *Entomol. Exp. Appl.* 125: 71–80.

Kareiva, P. 1987. Habitat fragmentation and the stability of predator–prey interactions. *Nature* 326: 388–390.

Kareiva, P. 1990. The spatial dimension in pest–enemy interactions. In M. Mackauer, L. E. Ehler and J. Roland (eds). *Critical Issues in Biological Control*. Intercept Andover, Hants. pp. 213–227.

Kareiva, P. and G. Odell. 1987. Swarms of predators exhibit 'preytaxis' if individual predators use area-restricted search. *Am. Nat.* 130: 233–270.

Kariluoto, K. T. 1978. Optimum levels of sorbic acid and methyl-p-hydroxybenzoate in an artificial diet for *Adalia bipunctata* L. (Coleoptera, Coccinellidae) larvae. *Ann. Entomol. Fenn.* 44: 94–97.

Kariluoto, K. T. 1980. Survival and fecundity of *Adalia bipunctata* (Coleoptera, Coccinellidae) and some other predatory insect species on an artificial diet and a natural prey. *Ann. Entomol. Fenn.* 46: 101–106.

Kariluoto, K. T., E. Junnikkala and M. Markkula. 1976. Attempts at rearing *Adalia bipunctata* L. (Col., Coccinellidae) on different artificial diets. *Ann. Entomol. Fenn.* 92: 91–97.

Katayama, N. and N. Suzuki. 2010. Extrafloral nectaries indirectly protect small aphid colonies via ant-mediated interactions. *Appl. Entomol. Zool.* 45: 505–511.

Katsarou, I., J. T. Margaritopoulos, J. A. Tsitsipis, D. Ch. Perdikis and K. D. Zarpas. 2005. Effect of temperature on development, growth and feeding of *Coccinella septempunctata* and *Hippodamia convergens* reared on the tobacco aphid, *Myzus persicae nicotianae*. *BioControl* 50: 565–588.

Katsoyannos, P. and Y. Laudeho. 1977. Contribution a la mise au point de l' elevage d' *Exochomus quadripustulatus* L. (Col., Coccinellidae). *Biologia Gallo-Hellenica* 6: 251–258.

Kawai, A. 1976. Analysis of the aggregation behavior in the larvae of *Harmonia axyridis* Pallas (Coleoptera: Coccinellidae) to prey colony. *Res. Popul. Ecol.* 18: 123–134.

Kawai, A. 1978. Sibling cannibalism in the first instar larvae of *Harmonia axyridis* Pallas (Coleoptera: Coccinellidae). *Kontyû* 46: 14–19.

Kawauchi, S. 1979. Effects of prey density on the rate of prey consumption, development and survival of *Propylea japonica* Thunberg (Coleoptera: Coccinellidae). *Kontyû* 47: 204–212.

Kehat, M. 1968. The feeding behaviour of *Pharoscymnus numidicus* (Coccinellidae), predator of the date palm scale *Parlatoria blanchardi*. *Entomol. Exp. Appl.* 11: 30–42.

Kesten, U. 1969. Zur Morphologie und Biologie von *Anatis ocellata* (L.) (Coleoptera, Coccinellidae). *Z. Ang. Entomol.* 63: 412–455.

Khan, M. R. and M. R. Khan. 2010. The relationship between temperature and the functional response of *Coccinella sep-*

tempunctata (L.) (Coleoptera: Coccinellidae). *Pakistan J. Zool.* 42: 461–466.

Kikuta, S., N. Fujiyama and H. Katakura. 2010. Local variation in the thistle species *Cirsium grayanum* affects its utilization by the herbivorous ladybird beetle *Henosepilachna niponica*. *Entomol. Exp. Appl.* 136: 262–270.

Kindlmann, P. and A. F. G. Dixon. 1993. Optimal foraging in ladybird beetles (Coleoptera: Coccinellidae) and its consequences for their use in biological control. *Eur. J. Entomol.* 90: 443–450.

Kishimoto, H. and I. Adachi. 2010. Effects of sucrose on survival and oviposition of the predacious insects *Stethorus japonicus* (Coleoptera: Coccinellidae), *Oligota kashmirica benefica* (Coleoptera: Staphylinidae), and *Scolothrips takahashii* (Thysanoptera: Thripidae). *Appl. Entomol. Zool.* 45: 621–626.

Kiyindou, A. and G. Fabres. 1987. Etude de la capacite d'accroisement chez *Hyperaspis raynevali*, predateur introduit au Congo pour la regulation des populations de *Phenacoccus manihoti* (Hom.: Pseudococcidae). *Entomophaga* 32: 181–189.

Klausnitzer, B. 1965. Zur Biologie der *Epilachna argus* Geoffr. (Col. Coccinellidae). *Entomol. Nachr.* 9: 87–89.

Klausnitzer, B. 1966. Übersicht über die Nahrung der einheimischen Coccinellidae (Col.). *Entomol. Ber.* 1966: 91–102.

Klausnitzer, B. 1989. Marienkäferansammlungen am Ostseestrand. *Entomol. Nachr. Ber.* 33: 189–194.

Klausnitzer, B. 1992. Coccinelliden als Prädatoren der Holunderblattlaus (*Aphis sambuci* L.) im Wärmefrühjahr 1992. *Entomol. Nachr. Ber.* 36: 185–190.

Klausnitzer, B. and H. Klausnitzer. 1986. *Marienkäfer (Coccinellidae)*. (3rd ed.). A. Ziemsen Verlag, Wittenberg-Lutherstadt. 104 pp.

Klausnitzer, B. and H. Klausnitzer. 1997. *Marienkäfer – Coccinellidae* (4th ed.). Westarp Wissenschaften, Magdeburg. 175 pp.

Klewer, N., Z. Ruzicka and S. Schulz. 2007. (Z)-pentacos-12-ene, an oviposition-deterring pheromone of *Cheilomenes sexmaculata*. *J. Chem. Ecol.* 33: 2167–2170.

Klingauf, F. 1967. Abwehr- und Meideraktionen von Blattläusen (Aphididae) bei Bedrohung durch Räubern und Parasiten. *Z. Angew. Entomol.* 60: 269–317.

Koch, R. L., E. C. Burkness, S. J. W. Burkness and W. D. Hutchison. 2004. Phytophagous preferences of the multicolored Asian lady beetle (Coleoptera: Coccinellidae) for autumn-ripening fruit. *J. Econ. Entomol.* 97: 539–544.

Koch, R. L., W. D. Hutchison, R. C. Venette and G. E. Heimpel. 2003. Susceptibility of immature monarch butterfly, *Danaus plexippus* (Lepidoptera: Nymphalidae: Danainae), to predation by *Harmonia axyridis* (Coleoptera: Coccinellidae). *Biol. Contr.* 28: 265–270.

Koch, R. L., R. C. Venette and W. D. Hutchison. 2005. Influence of alternate prey on predation of monarch butterfly (Lepidoptera: Nymphalidae) larvae by the multicolored Asian lady beetle (Coleoptera: Coccinellidae). *Environ. Entomol.* 34: 410–416.

Koide, T. 1962. Observations on the feeding habit of the larva of *Coccinella septempunctata bruckii* Mulsant. The feeding behaviour and number of prey fed under different temperatures. *Kontyû* 30: 236–241. (In Japanese with English summary.)

Koji, S. and K. Nakamura. 2002. Population dynamics of a thistle-feeding lady beetle *Epilachna niponica* (Coccinellidae: Epilachninae) in Kanazawa, Japan. 1. Adult demographic traits and population stability. *Popul. Ecol.* 44: 103–112.

Koji, S., K. Nakamura and M. Yamashita. 2004. Adaptive change and conservatism in host specificity in two local populations of the thistle-feeding ladybird beetle *Epilachna niponica*. *Entomol. Exp. Appl.* 112: 145–153.

Kontodimas, D. C. and G. J. Stathas. 2005. Phenology, fecundity and life table parameters of the predator *Hippodamia variegata* reared on *Dysaphis crataegi*. *BioControl* 50: 223–233.

Kontodimas, D. C., P. A. Eliopoulos, G. J. Stathas and L. P. Economou. 2004. Comparative temperature-dependent development of *Nephus includens* (Kirsch) and *Nephus bisignatus* (Boheman) (Coleoptera: Coccinellidae) preying on *Planococcus citri* (Risso) (Homoptera: Pseudococcidae): Evaluation of a linear and various nonlinear models using specific criteria. *Envir. Entomol.* 33: 1–11.

Kovář, I. 1996. Morphology and anatomy. *In* I. Hodek and A. Honěk (eds). *Ecology of Coccinellidae*. Kluwer, Dordrecht. pp. 1–18.

Krafsur, E. S., J. J. Obrycki and J. D. Harwood. 2005. Comparative genetic studies of native and introduced Coccinellidae in North America. *Eur. J. Entomol.* 102: 469–474.

Krebs, J. R. and R. H. McCleery. 1984. Optimization in behavioural ecology. *In* J. R. Krebs and N. B. Davies (eds). *Behavioural Ecology: An Evolutionary Approach* (2nd edn). Blackwell Sci. Publ., Oxford. pp. 91–121.

Kreiter, S. and G. Iperti. 1986. Effectiveness of *Adalia bipunctata* against aphids in a peach orchard with special reference to ant/aphid relationships. *In* I. Hodek (ed.). *Ecology of Aphidophaga*. Academia, Prague and W. Junk, Dordrecht. pp. 537–543.

Krivan, V. 2008. Dispersal dynamics: distribution of lady beetles (Coleoptera: Coccinellidae). *Eur. J. Entomol.* 105: 405–409.

Krivan, V. and E. Sirot. 2004. Do short-term behavioural responses of consumers in tri-trophic food chains persist at the population time-scale? *Evol. Ecol. Res.* 6: 1063–1081.

Kumar, A., N. Kumar, A. Siddiqui and C. P. M. Tripathi. 1999. Prey–predator relationship between *Lipaphis erysimi* Kalt. (Hom., Aphididae) and *Coccinella septempunctata* L. (Col., Coccinellidae). II. Effect of host plants on the functional response of the predator. *J. Appl. Entomol.* 123: 591–596.

Kuwajima, M., N. Kobayashi, T. Katoh and H. Katakura. 2010. Detection of ecological hybrid inviability in a pair of sympatric phytophagous ladybird beetles (*Henosepilachna* spp.). *Entomol. Exp. Appl.* 134: 280–286.

Kuznetsov, N. Ya. 1948. *Principles of Insect Physiology.* Vol. I. Izd. Akad. Nauk SSSR. 380 pp. (In Russian.)

Kuznetsov, V. N. 1975. Fauna and ecology of coccinellids (Coleoptera, Coccinellidae) in Primorye region. *Tr. Biol.-pochv. Inst.* 28: 3–24. (In Russian.)

Kuznetsov, V. N. 1993. *Coccinellids (Coleoptera, Coccinellidae) of the Russian Far East.* Dalnauka, Vladivostok. 334 pp. (In Russian.)

Kuznetsov, V. N. and G. A. Pantyuchov. 1988. Ecology of *Chilocorus kuwanae* Silv. (Coleoptera, Coccinellidae) and its acclimatization in Adjar ASSR. *In The Role of Insects in the Biocoenoses of Far East.* DVO AN SSSR, Vladivostok. pp. 48–54. (In Russian.)

Latham, D. R. and N. J. Mills. 2009. Quantifying insect predation: a comparison of three methods for estimating daily per capita consumption of two aphidophagous predators. *Environ. Entomol.* 38: 1117–1125.

Laubertie, E., X. Martini, C. Cadena et al. 2006. The immediate source of the oviposition-deterring pheromone produced by larvae of *Adalia bipunctata* (L.) (Coleoptera, Coccinellidae) J. *Insect Behav.* 19: 231–240.

Lederberg, J. 1992. The interface of science and medicine. *Mount Sinai J. Med.* 59: 380–383.

Lee, J. H. and T. J. Kang. 2004. Functional response of *Harmonia axyridis* (Pallas) (Coleoptera: Coccinellidae) to *Aphis gossypii* Glover (Homoptera: Aphididae) in the laboratory. *Biol. Contr.* 31: 306–310.

Legrand, A. and P. Barbosa. 2003. Plant morphological complexity impacts foraging efficiency of adult *Coccinella septempunctata* L. (Coleoptera: Coccinellidae). *Environ. Entomol.* 32: 1219–1226.

Le Rü, B. and A. Mitsipa. 2000. Influence of the host plant of the cassava mealybug *Phenacoccus manihoti* on life-history parameters of the predator *Exochomus flaviventris*. *Entomol. Exp. Appl.* 95: 209–212.

Lei, H. and N. Vickers. 2008. Central processing of natural odor mixtures in insects. *J. Chem. Ecol.* 34: 915–927.

Liere, H. and I. Perfecto. 2008. Cheating on a mutualism: Indirect benefits of ant attendance to a coccidophagous coccinellid. *Environ. Entomol.* 37: 143–149.

Lin, J.-T. and C.-Y. Wu. 1992. A comparative study on the color vision of four coleopteran insects. *Bull. Inst. Zool., Acad. Sinica* 31: 81–88.

Lin, J.-T., C.-Y. Wu and Y.-T. Chang. 1992. Semifused rhabdom of the ladybird beetle *Coccinella septempunctata* Linnaeus (Coleoptera: Coccinellidae). *Bull. Inst. Zool., Acad. Sinica* 31: 261–269.

Lin, K., K. Wu, Y. Zhang and Y. Guo. 2008. Naturally occurring populations of *Bemisia tabaci*, biotype B and associated natural enemies in agro-ecosystems in northern China. *Biocontr. Sci. Techn.* 18: 169–182.

Liu, C. L. 1950. Contribution to the knowledge of Chinese Coccinellidae. X. Occurrence of *Perilitus coccinellae* (Schrank), a parasite of adult Coccinellidae in North China (Hymenoptera, Braconidae). *Entomol. News* 61: 207–208.

Liu, T. and P. A. Stansly. 1999. Searching and feeding behavior of *Nephaspis oculatus* and *Delphastus catalinae* (Coleoptera: Coccinellidae), predators of *Bemisia argentifolii* (Homoptera: Aleyrodidae). *Environ. Entomol.* 28: 901–906.

Liu, T.-X., P. A. Stansly, K. A. Hoelmer and L. S. Osborne. 1997. Life history of *Nephaspis oculatus* (Coleoptera: Coccinellidae), a predator of *Bemisia argentifolii* (Homoptera: Aleyrodidae). *Ann. Entomol. Soc. Am.* 90: 776–782.

Lohman, D. J., Q. Liao and N. E. Pierce. 2006. Convergence of chemical mimicry in a guild of aphid predators. *Ecol. Entomol.* 31: 41–51.

Lopez, V. F., M. T. K. Kairo and J. A. Irish. 2004. Biology and prey range of *Cryptognatha nodiceps* (Coleoptera: Coccinellidae), a potential biological control agent for the coconut scale, *Aspidiotus destructor* (Hemiptera: Diaspididae). *Biocontr. Sci. Techn.* 14: 475–485.

Lorenzetti, F., J. T. Arnason, B. J. R. Philogene and R. I. Hamilton. 1997. Evidence for spatial niche partitioning in predaceous aphidophaga: use of plant colour as a cue. *Entomophaga* 42: 49–56.

Lu, W. H. and M. E. Montgomery. 2001. Oviposition, development, and feeding of *Scymnus* (*Neopullus*) *sinuanodulus* (Coleoptera: Coccinellidae): A predator of *Adelges tsugae* (Homoptera: Adelgidae). *Ann. Entomol. Soc. Am.* 94: 64–70.

Lucas, E., S. Demougeot, C. Vincent and D. Coderre. 2004. Predation upon the oblique-banded leafroller, *Choristoneura rosaceana* (Lepidoptera: Tortricidae), by two aphidophagous coccinellids (Coleoptera: Coccinellidae) in the presence and absence of aphids. *Eur. J. Entomol.* 101: 37–41.

Lundgren, J. G. 2009. Nutritional aspects of non-prey foods in the life histories of predaceous Coccinellidae. *Biol. Contr.* 51: 294–305.

Lundgren, J. G. and D. C. Weber. 2010. Changes in digestive rate of a predatory beetle over its larval stage: Implications for dietary breadth. *J. Insect Physiol.* 56: 431–437.

Lundgren, J. G. and R. N. Wiedenmann. 2004. Nutritional suitability of corn pollen for the predator *Coleomegilla maculata* (Coleoptera: Coccinellidae). *J. Insect Physiol.* 50: 567–575.

Lundgren, J. G., A. A. Razzak and R. N. Wiedenmann. 2004. Population responses and food consumption by predators *Coleomegilla maculata* and *Harmonia axyridis* (Coleoptera: Coccinellidae) during anthesis in an Illinois cornfield. *Environ. Entomol.* 33: 958–963.

Mack, T. P. and Z. Smilowitz. 1978. Diurnal, seasonal, and relative abundance of *Myzus persicae* (Sulzer) predators. *J.N.Y. Entomol. Soc.* 86: 305.

Maelzer, D. A. 1978. The growth and voracity of larvae of *Leis conformis* (Boisd.) (Coleoptera: Coccinellidae) fed on the rose aphid *Macrosiphum rosae* (L.) (Homoptera: Aphididae) in the laboratory. *Aust. J. Zool.* 26: 293–304.

Maeta, Y. 1965. Some observations of the habits od two predacious coccinellid beetles, *Harmonia axyridis* Pallas and *Coccinella septempunctata bruckii* Mulsant. *Tohoku Konchu Kenkyu* 1: 83–94. (In Japanese.)

Magro, A., J. N. Téné, N. Bastin, A. F. G. Dixon and J.-L. Hemptinne. 2007. Assessment of patch quality by ladybirds: relative response to conspecific and heterospecific larval tracks a consequence of habitat similarity? *Chemoecology* 17: 37–45.

Majerus, M. E. N. 1994. *Ladybirds.* Harper Collins, London. 367 pp.

Majerus M. E. N. 2006. The impact of male-killing bacteria on the evolution of aphidophagous coccinellids. *Eur. J. Entomol.* 103: 1–7.

Majerus, M. E. N. and G. D. D. Hurst. 1997. Ladybirds as a model system for the study of male-killing symbionts. *Entomophaga* 42: 13–20.

Majerus, M. E. N. and P. Kearns. 1989. *Ladybirds.* Richmond Publishing Co., London. 103 pp.

Majerus, M. E. N. and T. M. O. Majerus. 1996. Ladybird population explosions. *Brit. J. Entomol. Nat. Hist.* 9: 65–76.

Majerus, M. E. N., J. J. Sloggett, J.-F. Godeau and J.-L. Hemptinne. 2007. Interactions between ants and aphidophagous and coccidophagous ladybirds. *Popul. Ecol.* 49: 15–27.

Malcolm, S. B. 1990. Chemical defence in chewing and sucking insect herbivores: plant-derived cardenolides in the monarch butterfly and oleander aphid. *Chemoecology* 1: 12–21.

Mallampalli, N., F. Gould and P. Barbosa. 2005. Predation of Colorado potato beetle eggs by a polyphagous ladybeetle in the presence of alternate prey: potential impact on resistance evolution. *Entomol. Exp. Appl.* 114: 47–54.

Mandour, N. S., N. A.-S. El-Basha and T.-X. Liu. 2006. Functional response of the ladybird, *Cydonia vicina nilotica* to cowpea aphid, *Aphis craccivora* in the laboratory. *Insect Sci.* 13: 49–54.

Mani, M. and A. Krishnamoorthy. 1990. Evaluation of the exotic predator *Cryptolaemus montrouzieri* Muls. (Coccinellidae, Coleoptera) in the suppression of green shield scale, *Chloropulvinaria psidii* (Maskell) (Coccidae, Hemiptera) on guava. *Entomon* 15: 45–48.

Mansfield, S., J. R. Hagler and M. E. A. Whitehouse. 2008. A comparative study on the efficacy of a pest-specific and prey-marking enzyme-linked immunosorbent assay for detection of predation. *Entomol. Exp. Appl.* 127:199–206.

Maredia, K. M., S. H. Gage, D. A. Landis and J. M. Scriber. 1992. Habitat use patterns by the seven-spotted lady beetle (Coleoptera: Coccinellidae) in a diverse agricultural landscape. *Biol. Contr.* 2: 159–165.

Marks, R. J. 1977. Laboratory studies of plant searching behaviour by *Coccinella septempunctata* L. larvae. *Bull. Entomol. Res.* 67: 235–341.

Martini, X., P. Haccou, I. Olivieri and J.-L. Hemptinne. 2009. Evolution of cannibalism and female's response to oviposition-deterring pheromone in aphidophagous predators. *J. Anim. Ecol.* 78: 964–972.

Martos, A., A. Givovich and H. M. Niemeyer. 1992. Effect of DIMBOA, an aphid resistance factor in wheat, on the aphid predator *Eriopis connexa* Germar (Coleoptera: Coccinellidae). *J. Chem. Ecol.* 18: 469–479.

Matsura, T. 1976. Ecological studies of a coccinellid, *Aiolocaria hexaspilota* Hope. I. Interaction between field populations of *A. hexaspilota* and its prey, the walnut leaf beetle (*Gastrolina depressa* Baly). *Jap. J. Ecol.* 26: 147–156. (In Japanese with English summary.)

McAllister, M. K. and B. D. Roitberg. 1987. Adaptive suicidal behaviour in pea aphids. *Nature* 328: 797–799.

McConnell, J. A. and T. J. Kring. 1990. Predation and dislodgment of *Schizaphis graminum* (Homoptera: Aphididae), by adult *Coccinella septempunctata* (Coleoptera: Coccinellidae). *Environ. Entomol.* 19: 1798–1802.

McKenzie, H. L. 1932. The biology and feeding habits of *Hyperaspis lateralis* Mulsant (Coleoptera–Coccinellidae). *Univ. Calif. Publ.* 6: 9–20.

McMillan, S., A. K. Kuusk, A. Cassel-Lundhagen and B. Ekbom. 2007. The influence of time and temperature on molecular gut content analysis: *Adalia bipunctata* fed with *Rhopalosiphum padi*. *Insect Sci.* 14: 353–358.

Mehrnejad, M. R. and M. A. Jalali. 2004. Life history parameters of the coccinellid beetle, *Oenopia conglobata contaminata*, an important predator of the common pistachio psylla, *Agonoscena pistaciae* (Hemiptera: Psylloidea). *Biocontr. Sci. Techn.* 14: 701–711.

van den Meiracker, R. A. F., W. N. O. Hammond and J. J. M. van Alphen. 1990. The role of kairomones in prey finding by *Diomus* sp. and *Exochomus* sp., two coccinellid predators of the cassava mealybug, *Phenacoccus manihoti*. *Entomol. Exp. Appl.* 56: 209–217.

Mendel, Z. and D. Blumberg. 1991. Colonization trials with *Cryptochaetum iceryae* and *Rodolia cardinalis* for improved biological control of *Icerya purchasi* in Israel. *Biol. Contr.* 1: 68–74.

Mendel, Z., D. Blumberg, A. Zehavi and M. Weisenberg. 1992. Some polyphagous Homoptera gain protection from their natural enemies by feeding on the toxic plants *Spartium junceum* and *Erythrina corallodendrum* (Leguminosae). *Chemoecology* 3: 118–124.

Mensah, R. K. 1997. Local density responses of predatory insects of *Helicoverpa* spp. to a newly developed food 'Envirofeast' in commercial cotton in Australia. *Int. J. Pest Manage.* 43: 221–225.

Mensah, R. K. 2002a. Development of an integrated pest management programme for cotton. Part 1: Establishing and utilizing natural enemies. *Int. J. Pest Manage.* 48: 87–94.

Mensah, R. K. 2002b. Development of an integrated pest management programme for cotton. Part 2: Integration of a lucerne/cotton interplant system, food supplement sprays with biological and synthetic insecticides. *Int. J. Pest Manage.* 48: 95–105.

Mensah, R. K. and J. L. Madden. 1994. Conservation of two predator species for biological control of *Chrysophtharta bimaculata* (Col., Chrysomelidae) in Tasmanian forests. *Entomophaga* 39: 71–83.

Mensah, R. K. and A. Singleton. 2003. Optimum timing and placement of a supplementary food spray Envirofeast® for the establishment of predatory insects of *Helicoverpa* spp. in cotton systems in Australia. *Int. J. Pest Manage.* 49: 163–168.

Messina, F. J. and J. B. Hanks. 1998. Host plant alters the shape of the functional response of an aphid predator (Coleoptera: Coccinellidae). *Environ. Entomol.* 27: 1196–1202.

Meyhöfer, R. 2001. Intraguild predation by aphidophagous predators on parasitized aphids: the use of multiple video cameras. *Entomol. Exp. Appl.* 100: 77–87.

Meyhöfer, R. and T. Klug. 2002. Intraguild predation on the aphid parasitoid *Lysiphlebus fabarum* (Marshall) (Hymenoptera: Aphidiidae): mortality risks and behavioral decisions made under the threats of predation. *Biol. Contr.* 25: 239–248.

Michaud, J. P. 2000. Development and reproduction of ladybeetles (Coleoptera: Coccinellidae) on the citrus aphids *Aphis spiraecola* Patch and *Toxoptera citricida* (Kirkaldy) (Homoptera: Aphididae). *Biol. Contr.* 18: 287–297.

Michaud, J. P. 2001. Numerical response of *Olla v-nigrum* (Coleoptera: Coccinellidae) to infestations of Asian citrus psyllid (Hemiptera: Psyllidae) in Florida. *Florida Entomol.* 84: 608–612.

Michaud, J. P. 2003. A comparative study of larval cannibalism in three species of ladybird. *Ecol. Entomol.* 28: 92–101.

Michaud, J. P. 2005. On the assessment of prey suitability in aphidophagous Coccinellidae. *Eur. J. Entomol.* 102: 385–390.

Michaud, J. P. and A. K. Grant. 2004. Adaptive significance of sibling egg cannibalism in Coccinellidae: comparative evidence from three species. *Ann. Entomol. Soc. Am.* 97: 710–719.

Michaud, J. P. and A. K. Grant. 2005. Suitability of pollen sources for the development and reproduction of *Coleomegilla maculata* (Coleoptera: Coccinellidae) under simulated drought conditions. *Biol. Contr.* 32: 363–370.

Michaud, J. P. and J. L. Jyoti. 2007. Repellency of conspecific and heterospecific larval residues to *Hippodamia convergens* (Coleoptera: Coccinellidae) ovipositing on sorghum plants. *Eur. J. Entomol.* 104: 399–405.

Michaud, J. P. and J. L. Jyoti. 2008. Dietary complementation across life stages in the polyphagous lady beetle *Coleomegilla maculata*. *Entomol. Exp. Appl.* 126: 40–45.

Michaud, J. P. and L. E. Olsen. 2004. Suitability of Asian citrus psyllid, *Diaphorina citri*, as prey for ladybeetles. *BioControl* 49: 417–431.

Michaud, J. P. and J. A. Qureshi. 2005. Induction of reproductive diapause in *Hippodamia convergens* (Coleoptera: Coccinellidae) hinges on prey quality and availability. *Eur. J. Entomol.* 102: 483–487.

Michaud, J. P. and J. A. Qureshi. 2006. Reproductive diapause in *Hippodamia convergens* (Coleoptera: Coccinellidae) and its life history consequences. *Biol. Contr.* 39: 193–200.

Michels, G. J. and R. V. Flanders. 1992. Larval development, aphid consumption and oviposition for five imported coccinellids at constant temperature on Russian wheat aphids and greenbugs. *Southw. Entomol.* 17: 233–243.

Michels, G. J. and A. C. Bateman. 1986. Larval biology of two imported predators of the greenbug, *Hippodamia variegata* Goeze and *Adalia flavomaculata* DeGeer, under constant temperatures. *Southw. Entomol.* 11: 23–30.

Michels, G. J. and R. W. Behle. 1991a. Effects of two prey species on the development of *Hippodamia sinuata* (Coleoptera: Coccinellidae) larvae at constant temperatures. *J. Econ. Entomol.* 84: 1480–1484.

Michels, G. J. and R. W. Behle. 1991b. A comparison of *Coccinella septempunctata* and *Hippodamia convergens* larval development on greenbugs at constant temperatures. *Southw. Entomol.* 16: 73–80.

Michels, G. J., Jr. and J. D. Burd. 2007. IPM Case Studies: Sorghum. *In* H. F. van Emden and R. Harrington (eds). *Aphids as Crop Pests.* CAB International, Wallingford, UK. pp. 627–638.

Michels, G. J., Jr. and J. H. Matis. 2008. Corn leaf aphid, *Rhopalosiphum maidis* (Hemiptera: Aphididae), as a key to greenbug, *Schizaphis graminum* (Hemiptera: Aphididae), biological control in grain sorghum, *Sorghum bicolor*. *Environ. Entomol.* 105: 513–520.

Mignault, M. P., M. Roy and J. Brodeur. 2006. Soybean aphid predators in Quebec and the suitability of *Aphis glycines* as prey for three Coccinellidae. *BioControl* 51: 89–106.

Mills, N. J. 1979. *Adalia bipunctata (L.) as a generalist predator of aphids*. PhD thesis, University East Anglia, Norwich.

Mills, N. J. 1981. Essential and alternative foods for some British Coccinellidae (Coleoptera). *Entomol. Gaz.* 32: 197–202.

Mills, N. J. 1982a. Satiation and the functional response: a test of a new model. *Ecol. Entomol.* 7: 305–315.

Mills, N. J. 1982b. Voracity, cannibalism, and coccinellid predation. *Ann. Appl. Biol.* 101: 144–148.

Mills, N. J. and J. M. Kean. 2010. Behavioral studies, molecular approaches, and modelling: Methodological contributions to biological control success. *Biol. Contr.* 52: 255–262.

Mills, N. J. and I. Lacan. 2004. Ratio dependence in the functional response of insect parasitoids: evidence from *Trichogramma minutum* foraging for eggs in small host patches. *Ecol. Entomol.* 29: 208–216.

Minoretti, N. and W. W. Weisser. 2000. The impact of individual ladybirds (*Coccinella septempunctata*, Coleoptera:

Coccinellidae) on aphid colonies. *Eur. J. Entomol.* 97: 475–479.

Miura, T. and S. Nishimura. 1980. The larval period and predaceous activity of an aphidophagous coccinellid, *Harmonia axyridis* Pallas. *Bull. Fac. Agric. Shimane Univ.* 14: 144–148.

Mogi, M. 1969. Predation response of the larvae of *Harmonia axyridis* Pallas (Coccinellidae) to the different prey density. *Jap. J. Appl. Entomol. Zool.* 13: 9–16.

Mondor, E. B. and B. D. Roitberg. 2000. Has the attraction of predatory coccinellids to cornicle droplets constrained aphid alarm signalling behavior? *J. Insect Behav.* 13: 321–329.

Mondor, E. B. and B. D. Roitberg. 2004. Inclusive fitness benefits of scent-marking predators. *Proc. R. Soc. London B* 271: S341–S343.

Mondor, E. B. and J. L. Warren. 2000. Unconditioned and conditioned responses to colour in the predatory coccinellid, *Harmonia axyridis* (Coleoptera: Coccinellidae). *Eur. J. Entomol.* 97: 463–467.

Morales, J. and C. L. Burandt Jr. 1985. Interactions between *Cycloneda sanguinea* and the brown citrus aphid: adult feeding and larval mortality. *Environ. Entomol.* 14: 520–522.

Mori, K., M. Nozawa, K. Arai and T. Gotoh. 2005. Life-history traits of the acarophagous lady beetle, *Stethorus japonicus* at three constant temperatures. *BioControl* 50: 35–51.

Moser, S. E. and J. J. Obrycki. 2009. Non-target effects of neonicotinoid seed treatments; mortality of coccinellid larvae related to zoophytophagy. *Biol. Contr.* 51: 487–492.

Moser, S. E., J. D. Harwood and J. J. Obrycki. 2008. Larval feeding on *Bt* hybrid and non-*Bt* corn seedlings by *Harmonia axyridis* (Coleoptera: Coccinellidae) and *Coleomegilla maculata* (Coleoptera: Coccinellidae). *Environ. Entomol.* 37: 525–533.

Moura, R., P. Garcia, S. Cabral and A. O. Soares. 2006. Does pirimicarb affect the voracity of the euryphagous predator, *Coccinella undecimpunctata* L. (Coleoptera: Coccinellidae)? *Biol. Contr.* 38: 363–368.

Moursi, A. A. and M. Kamal. 1946. Notes on the biology and feeding habits of the introduced beneficial insect *Leis conformis* Boisd. (Coccinell.) *Bull. Soc. Fouad I. Entomol.* 30: 63–74.

Müller, C. B. and H. C. J. Godfray. 1997. Apparent competition between two aphid species. *J. Anim. Ecol.* 66: 57–64.

Muma, M. H. 1955. Some ecological studies on the twice stabbed lady beetle, *Chilocorus stigma* (Say). *Ann. Entomol. Soc. Am.* 48: 493–498.

Munyaneza, J. and J. J. Obrycki. 1997. Functional response of *Coleomegilla maculata* (Coleoptera: Coccinellidae) to Colorado potato beetle eggs (Coleoptera: Chrysomelidae). *Biol. Contr.* 8: 215–224.

Murdie, G. 1971. Simulation on the effects of predators/parasite models on prey/host spatial distribution. *In* G. P. Patil, E. C. Pielou, W. E. Waters (eds). *Statistical Ecology 1.*

Pennsylvania State University Press, Harrisburg. pp. 215–223.

Murdoch, W. W. and J. R. Marks. 1973. Predation by coccinellid beetles: experiments on switching. *Ecology* 54: 159–167.

Musser, F. R. and A. M. Shelton. 2003. Predation of *Ostrinia nubilalis* (Lepidoptera: Crambidae) eggs in sweet corn by generalist predators and the impact of alternative foods. *Environ. Entomol.* 32: 1131–1138.

Nakamura, K., N. Hasan, L. Abbas, H. C. J. Godfray and M. B. Bonsall. 2004. Generation cycles in Indonesian lady beetle populations may occur as a result of cannibalism. *Proc. R. Soc. Lond. B* 271: S501–S504.

Nakamura, K., K. Miura, P. De Jong and H. Ueno. 2006. Comparison of the incidence of sibling cannibalism between male-killing *Spiroplasma* infected and uninfected clutches of a predatory ladybird beetle, *Harmonia axyridis* (Coleoptera: Coccinellidae). *Eur. J. Entomol.* 103: 323–326.

Nakamuta, K. 1984a. Aphid body fluid stimulates feeding of a predatory ladybeetle, *Coccinella septempunctata* L. (Coleoptera: Coccinellidae). *Appl. Entomol. Zool.* 19: 123–125.

Nakamuta, K. 1984b. Visual orientation of a ladybeetle, *Coccinella septempunctata* L., (Coleoptera: Coccinellidae), towards its prey. *Appl. Entomol. Zool.* 19: 82–86.

Nakamuta, K. 1985a. Area-concentrated search in adult *Coccinella septempunctata* L. (Coleoptera: Coccinellidae): Releasing stimuli and decision of giving-up time. *Jpn. J. Appl. Entomol. Zool.* 29: 55–60. (In Japanese with English summary.)

Nakamuta, K. 1985b. Mechanism of the switchover from extensive to area-concentrated search behaviour of the ladybird beetle, *Coccinella septempunctata bruckii*. *J. Insect Physiol.* 31: 849–856.

Nakamuta, K. 1987. Diel rhythmicity of prey-searching activity and its predominance over starvation in the lady beetle, *Coccinella septempunctata bruckii*. *Physiol. Entomol.* 1: 91–98.

Nakamuta, K. and T. Saito. 1985. Recognition of aphid prey by the lady beetle, *Coccinella septempunctata brucki* Mulsant (Coleoptera: Coccinellidae). *Appl. Entomol. Zool.* 20: 479–483.

Nakashima, Y. and Y. Hirose. 1999. Effects of prey availability on longevity, prey consumption, and egg production in the insect predators *Orius sauteri* and *O. tantillus* (Hemiptera: Anthocoridae). *Ann. Entomol. Soc. Am.* 92: 537–541.

Nakashima, Y. and N. Senoo. 2003. Avoidance of ladybird trails by an aphid parasitoid *Aphidius ervi*: active period and effects of prior oviposition experience. *Entomol. Exp. Appl.* 109: 163–166.

Nalepa, C. A., S. B. Bambara and A. M. Burroughs. 1992. Pollen and nectar feeding by *Chilocorus kuwanae* (Silvestri) (Coleoptera: Coccinellidae). *Proc. Entomol. Soc. Wash.* 94: 596–597.

Nalepa, C. A., K. R. Ahlstrom, B. A. Nault and J. L. Williams. 1998. Mass appearance of ladybeetles (Coleoptera: Coccinellidae) on North Carolina beaches. *Entomol. News* 109: 277–281.

Napal, G. N. D., M. T. Defago, G. R. Valladares and S. M. Palacios. 2010. Response of *Epilachna paenulata* to two flavonoids, Pinocembrin and Quercetin, in a comparative study. *J. Chem. Ecol.* 36: 898–904.

Naranjo, S. E., R. L. Gibson and D. D. Walgenbach. 1990. Development, survival, and reproduction of *Scymnus frontalis* (Coleoptera: Coccinellidae), an imported predator of Russian wheat aphid, at four fluctuating temperatures. *Ann. Entomol. Soc. Am.* 83: 527–531.

Nault, B. A. and G. G. Kennedy. 2000. Seasonal changes in habitat preference by *Coleomegilla maculata*: implications for Colorado potato beetle management in potato. *Biol. Contr.* 17: 164–173.

Nedvěd, O. and S. Salvucci. 2008. Ladybird *Coccinella septempunctata* (Coleoptera: Coccinellidae) prefers toxic prey in laboratory choice experiment. *Eur. J. Entomol.* 105: 431–436.

Nelson, E. and J. A. Rosenheim. 2006. Encounters between aphids and their predators: the relative frequencies of disturbance and consumption. *Entomol. Exp. Appl.* 118: 211–218.

Ng, S. M. 1986a. Effects of sibling egg cannibalism on the first instar larvae of four species of aphidophagous Coccinellidae. *In* I. Hodek (ed.). *Ecology of Aphidophaga*. Academia, Prague and W. Junk, Dordrecht, The Netherlands. pp. 69–75.

Ng, S. M. 1986b. The geotactic and phototactic responses of four species of aphidophagous coccinellid larvae. *In* I. Hodek (ed.). *Ecology of Aphidophaga*. Academia, Prague and W. Junk, Dordrecht, The Netherlands. pp. 57–68.

Ng, S. M. 1988. Observations on the foraging behaviour of starved aphidophagous coccinellid larvae (Coleoptera: Coccinellidae). *In* E. Niemczyk and A. F. G. Dixon (eds). *Ecology and Effectiveness of Aphidophaga*. SPB Acad. Publ., The Hague. pp. 29–33.

Nichols, P. R. and W. W. Neel. 1977. The use of Food Wheast® as a supplemental food for *Coleomegilla maculata* (DeGeer) (Coleoptera: Coccinellidae) in the field. *Southw. Entomol.* 2: 102–105.

Nielsen, F. H., M. S. Hauge and S. Toft. 2002. The influence of mixed aphid diets on larval performance of *Coccinella septempunctata* (Col., Coccinellidae). *J. Appl. Entomol.* 126: 194–197.

Nielson, M. W. and W. E. Currie. 1960. Biology of the convergent lady beetle when fed a spotted alfalfa aphid diet. *J. Econ. Entomol.* 53: 257–259.

Niijima, K. 1979. Further attempts to rear coccinellids on drone powder with field observation. *Bull. Fac. Agric. Tamagawa Univ.* 19: 7–12.

Niijima, K., M. Matsuka and I. Okada. 1986. Artificial diets for an aphidophagous coccinellid, *Harmonia axyridis*, and

its nutrition. (Minireview). *In* I. Hodek (ed.). *Ecology of Aphidophaga*. Academia, Prague and W. Junk, Dordrecht, The Netherlands. pp. 37–50.

Ninkovic, V., S. AlAbassi and J. Pettersson. 2001. The influence of aphid-induced plant volatiles on ladybird beetle searching behavior. *Biol. Contr.* 21: 191–195.

Nishida, R. and H. Fukami. 1989. Host plant iridoid-based chemical defense of an aphid, *Acyrthosiphon nipponicus*, against ladybird beetles. *J. Chem. Ecol.* 15: 1837–1845.

Nixon, G. E. 1951. *The Association of Ants with Aphids and Coccids*. Commonwealth Inst. Entomol., London. 36 pp.

Nomura, M. and K. Niijima. 1997. Effect of contagion and cannibalism on the abnormal sex ratio in *Menochilus sexmaculatus* (Fabricius)(Coleoptera: Coccinellidae). *Appl. Entomol. Zool.* 32: 501–504.

Nonacs, P. and L. M. Dill. 1993. Is satisficing an alternative to optimal foraging theory? *Oikos* 67: 371–375.

O'Neil, R. J. 1989. Comparison between laboratory and field measurements of the functional response of *Podisus maculiventris* (Heteroptera: Pentatomidae). *J. Kans. Entomol. Soc.* 62: 148–155.

O'Neil, R. J. 1997. Functional response and search strategy of *Podisus maculiventris* (Heteroptera: Pentatomidae) attacking Colorado potato beetle (Coleoptera: Chrysomelidae). *Environ. Entomol.* 26: 1183–1190.

O'Neil, R. J. and R. N. Wiedenmann. 1987. Adaptations of arthropod predators to agricultural systems. *Florida Entomol.* 70: 40–48.

Obata, S. 1986. Mechanisms of prey finding in the aphidophagous ladybird beetle, *Harmonia axyridis* (Col.: Coccinellidae). *Entomophaga* 31: 303–311.

Obata, S. 1997. The influence of aphids on the behaviour of adults of the ladybird beetle, *Harmonia axyridis* (Col.: Coccinellidae). *Entomophaga* 42: 103–106.

Obata, S. and T. Hidaka. 1987. Ejection and ingestion of the spermatophore by the female beetle, *Harmonia axyridis* Pallas (Coleoptera: Coccinellidae). *Can. Entomol.* 119: 603–604.

Obata, S. and Y. Johki. 1990. Distribution and behaviour of adult ladybird, *Harmonia axyridis* Pallas (Coleoptera, Coccinellidae), around aphid colonies. *Jpn. J. Entomol.* 58: 839–845.

Obatake, H. and H. Suzuki. 1985. On the isolation and identification of canavanine and ethanolamine contained in the young leaves of black locus, *Robinia pseudoacacia*, lethal for the lady beetle, *Harmonia axyridis*. *Techn. Bull. Fac. Agr., Kagawa Univ.* 36: 107–115.

Obrycki, J. J. and C. J. Orr. 1990. Suitability of three prey species for Nearctic populations of *Coccinella septempunctata*, *Hippodamia variegata*, and *Propylea quatuordecimpunctata* (Coleoptera: Coccinellidae). *J. Econ. Entomol.* 83: 1292–1297.

Obrycki, J. J. and M. J. Tauber. 1984. Natural enemy activity on glandular pubescent potato plants in the greenhouse: an unreliable predictor of effects in the field. *Environ. Entomol.* 13: 679–683.

Obrycki, J. J. and M. J. Tauber. 1985. Seasonal occurrence and relative abundance of aphid predators and parasitoids on pubescent potato plants. *Can. Entomol.* 117: 1231–1237.

Obrycki, J. J., K. L. Giles and A. M. Ormord. 1998. Interactions between an introduced and indigenous coccinellid species at different prey densities. *Oecologia* 117: 279–285.

Ofuya, T. I. and A. E. Akingbohungbe. 1988. Functional and numerical responses of *Cheilomenes lunata* (Fabricius) (Coleoptera: Coccinellidae) feeding on the cowpea aphid, *Aphis craccivora* Koch (Homoptera: Aphididae). *Insect Sci. Applic.* 9: 543–546.

Ohashi, K., Y. Sakuratani, N. Osawa, S. Yano and A. Takafuji. 2005. Thermal microhabitat use by the ladybird beetle, *Coccinella septempunctata* (Col., Coccinellidae), and its life cycle consequences. *Environ. Entomol.* 34: 432–439.

Ohgushi, T. 1996. Consequences of adult size for survival and reproductive performance in a herbivorous ladybird beetle. *Ecol. Entomol.* 21: 47–55.

Ohgushi, T. and H. Sawada. 1985. Population equilibrium with respect to available food resource and its behavioural basis in a herbivorous lady beetle, *Henosepilachna niponica*. *J. Anim. Ecol.* 54: 781–796.

Ohgushi, T. and H. Sawada. 1995. Demographic attributes of an introduced herbivorous lady beetle. *Res. Popul. Ecol.* 37: 29–36.

Ohgushi, T. and H. Sawada. 1998. What changed the demography of an introduced population of an herbivorous lady beetle? *J. Anim. Ecol.* 67: 679–688.

Okada, I. 1970. A new method of artificial rearing of coccinellid, *Harmonia axyridis* Pallas. *Heredity, Tokyo* 24: 32–35. (In Japanese.)

Okamoto, H. 1961. Comparison of ecological characters of the predatory ladybird *Coccinella septempunctata bruckii* fed on the apple grain aphids, *Rhopalosiphum prunifoliae* and the cabbage aphids, *Brevicoryne brassicae*. *Jap. J. Appl. Entomol. Zool.* 5: 277–278. (In Japanese.)

Okamoto, H. 1966. Three problems of prey specificity of aphidophagous coccinellids. *In* I. Hodek (ed.) *Ecology of Aphidophagous Insects.* Academia, Prague and W. Junk, The Hague. pp. 45–46.

Okamoto, H. 1978. Laboratory studies on food ecology of aphidophagous lady beetles (Coleoptera: Coccinellidae). *Mem. Fac. Agric. Kagawa Univ.* 32: 1–94. (In Japanese.)

Okuyama, T. 2008. Individual behavioral variation in predator–prey models. *Ecol. Res.* 23: 665–671.

Oliver, T. H., J. E. L. Timms, A. Taylor and S. R. Leather. 2006. Oviposition responses to patch quality in the larch ladybird *Aphidecta obliterata* (Coleoptera: Coccinellidae): effects of aphid density, and con- and heterospecific tracks. *Bull. Entomol. Res.* 96: 25–34.

Olszak, R. W. 1988. Voracity and development of three species of Coccinellidae, preying upon different species of aphids.

In E. Niemczyk and A. F. G. Dixon (eds). *Ecology and Effectiveness of Aphidophaga.* SPB Acad. Publ., The Hague. pp. 47–53.

Omkar and R. B. Bind. 2004. Prey quality dependent growth, development and reproduction of a biocontrol agent, *Cheilomenes sexmaculata* (Fabricius) (Coleoptera: Coccinellidae). *Biocontr. Sci. Techn.* 14: 665–673.

Omkar and B. E. James. 2004. Influence of prey species on immature survival, development, predation and reproduction of *Coccinella transversalis* Fabricius (Col., Coccinellidae). *J. Appl. Entomol.* 128: 150–157.

Omkar and G. Mishra. 2005. Preference–performance of a generalist predatory ladybird: a laboratory study. *Biol. Contr.* 34: 187–195.

Omkar and S. Pathak. 2006. Effects of different photoperiods and wavelengths of light on the life-history traits of an aphidophagous ladybird, *Coelophora saucia* (Mulsant). *J. Appl. Entomol.* 130: 45–50.

Omkar and A. Pervez. 2004. Functional and numerical responses of *Propylea dissecta* (Col., Coccinellidae). *J. Appl. Entomol.* 128: 140–146.

Omkar and S. Srivastava. 2003. Influence of six aphid prey species on development and reproduction of a ladybird beetle, *Coccinella septempunctata*. *BioControl* 48: 379–393.

Omkar, A. Pervez and A. K. Gupta. 2006a. Attack, escape and predation rates of two aphidophagous ladybirds during conspecific and heterospecific interactions. *Biocontr. Sci. Techn.* 16: 295–305.

Omkar, A. Pervez and A. K. Gupta. 2006b. Why do neonates of aphidophagous ladybird beetles preferentially consume conspecific eggs in presence of aphids? *Biocontr. Sci. Techn.* 16: 233–243.

Omkar, A. Pervez and A. K. Gupta. 2007. Sibling cannibalism in aphidophagous ladybirds: its impact on sex-dependent development and body weight. *J. Appl. Entomol.* 131: 81–84.

Omkar, G. Mishra, S. Srivastava, A. K. Gupta and S. K. Singh. 2005. Reproductive performance of four aphidophagous ladybirds on cowpea aphid, *Aphis craccivora* Koch. *J. Appl. Entomol.* 129: 217–220.

Orr, C. J. and J. J. Obrycki. 1990. Thermal and dietary requirements for development of *Hippodamia parenthesis* (Coleoptera: Coccinellidae). *Environ. Entomol.* 19: 1523–1527.

Osawa, N. 1989. Sibling and non-sibling cannibalism by larvae of a lady beetle *Harmonia axyridis* Pallas (Coleoptera, Coccinellidae) in the field. *Res. Popul. Ecol.* 31: 153–160.

Osawa, N. 1992a. A life table of the ladybird beetle *Harmonia axyridis* Pallas (Coleoptera, Coccinellidae) in relation to the aphid abundance. *Jpn. J. Entomol.* 60: 575–579.

Osawa, N. 1992b. Effect of pupation site on pupal cannibalism and parasitism in the ladybird beetle *Harmonia axyridis* Pallas (Coleoptera, Coccinellidae). *Jpn. J. Entomol.* 60: 131–135.

Osawa, N. 1992c. Sibling cannibalism in the ladybird beetle *Harmonia axyridis*: fitness consequences for mother and offspring. *Res. Popul. Ecol.* 34: 45–55.

Osawa, N. 1993. Population field studies of the aphidophagous ladybird beetle *Harmonia axyridis* (Coleoptera: Coccinellidae): life tables and key factor analysis. *Res. Popul. Ecol.* 35: 335–348.

Osawa, N. 2000. Population field studies on the aphidophagous ladybird beetle *Harmonia axyridis* (Coleoptera: Coccinellidae): resource tracking and population characteristics. *Popul. Ecol.* 42: 115–127.

Osawa, N. 2002. Sex-dependent effects of sibling cannibalism on life history traits of the ladybird beetle *Harmonia axyridis* (Coleoptera: Coccinellidae). *Biol. J. Linn. Soc.* 76: 349–360.

Osawa, N. 2003. The influence of female oviposition strategy on sibling cannibalism in the ladybird beetle *Harmonia axyridis* (Coleoptera: Coccinellidae). *Eur. J. Entomol.* 100: 43–48.

Osawa, N. 2005. The effect of prey availability on ovarian development and oosorption in the ladybird beetle *Harmonia axyridis* (Coleoptera: Coccinellidae). *Eur. J. Entomol.* 102: 503–511.

Osawa N. and K. Ohashi. 2008. Sympatric coexistence of sibling species *Harmonia yedoensis* and *H. axyridis* (Coleoptera: Coccinellidae) and the roles of maternal investment through egg and sibling cannibalism. *Eur. J. Entomol.* 105: 445–454.

Ozder, N. and O. Saglam. 2003. Effects of aphid prey on larval development and mortality of *Adalia bipunctata* and *Coccinella septempunctata* (Coleoptera: Coccinellidae). *Biocontr. Sci. Techn.* 13: 449–453.

Palmer, M. A. 1914. Some notes on life history of lady beetles. *Ann. Entomol. Soc. Am.* 7: 213–238.

Pantyukhov, G. A. 1968. A study of ecology and physiology of the predatory beetle *Chilocorus rubidus* Hope (Coleoptera, Coccinellidae). *Zool. Zh.* 47: 376–386. (In Russian with English summary.)

Papachristos, D. P. and P. G. Milonas. 2008. Adverse effects of soil applied insecticides on the predatory coccinellid *Hippodamia undecimnotata* (Coleoptera: Coccinellidae). *Biol. Contr.* 47: 77–81.

Parajulee, M. N., R. B. Shrestha, J. F. Leser, D. B. Wester and C. A. Bianco. 2006. Evaluation of the functional response of selected arthropod predators on bollworm eggs in the laboratory and effect of temperature on their predation efficiency. *Environ. Entomol.* 35: 379–386.

Park, Y. L. and J. J. Obrycki. 2004. Spatio-temporal distribution of corn leaf aphids (Homoptera: Aphididae) and lady beetles (Coleoptera: Coccinellidae) in Iowa cornfields. *Biol. Contr.* 31: 210–217.

Pasteels, J. M. 2007. Chemical defence, offence and alliance in ants–aphids–ladybirds relationships. *Popul. Ecol.* 49: 5–14.

Perry, J. C. and B. D. Roitberg. 2005. Ladybird mothers mitigate offspring starvation risk by laying trophic eggs. *Behav. Ecol. Sociobiol.* 58: 578–586.

Perry, J. C. and B. D. Roitberg. 2006. Trophic egg laying: hypothesis and tests. *Oikos* 112: 706–714.

Perry, J. C. and L. Rowe. 2008. Neither mating rate nor spermatophore feeding influences longevity in a ladybird beetle. *Ethology* 114: 504–511.

Persad, A. and A. Khan. 2002. Comparison of life table parameters for *Maconellicoccus hirsutus*, *Anagyrus camali*, *Cryptolaemus montrouzieri* and *Scymnus coccivora*. *BioControl* 47: 137–149.

Pervez, A. and Omkar. 2004. Prey dependent life attributes of an aphidophagous ladybird beetle, *Propylea dissecta* (Coccinellidae). *Biocontr. Sci. Techn.* 14: 385–396.

Pervez, A., A. K. Gupta and Omkar. 2005. Kin recognition and avoidance of kin cannibalism by the larvae of co-occurring ladybirds: a laboratory study. *Eur. J. Entomol.* 102: 513–518.

Pervez, A., A. K. Gupta and Omkar. 2006. Larval cannibalism in aphidophagous ladybirds: Influencing factors, benefits and costs. *Biol. Contr.* 38: 307–313.

Peterson, P. G., P. G. McGregor and B. P. Springett. 2000. Density dependent prey feeding time of *Stethorus bifidus* (Coleoptera: Coccinellidae) on *Tetranychus lintearius* (Acari: Tetranychidae). *N. Z. J. Zool.* 27: 41–44.

Pettersson, J., V. Ninkovic, R. Glinwood et al. 2008. Chemical stimuli supporting foraging behaviour of *Coccinella septempunctata* L. (Coleoptera: Coccinellidae): volatiles and allelobiosis. *Appl. Entomol. Zool.* 43: 315–321.

Pfannenstiel, R. S. and K. V. Yeargan. 2002. Identification and diel activity patterns of predators attacking *Helicoverpa zea* (Lepidoptera: Noctuidae) eggs in soybean and sweet corn. *Environ. Entomol.* 31: 232–241.

Phoofolo, M. W., N. C. Elliott and K. L. Giles. 2009. Analysis of growth and development in the final instar of three species of predatory coccinellids under varying prey availability. *Entomol. Exp. Appl.* 131: 264–277.

Phoofolo, M. W., K. L. Giles and N. C. Elliott. 2007. Quantitative evaluation of suitability of the greenbug, *Schizaphis graminum*, and the bird cherry-oat aphid, *Rhopalosiphum padi*, as prey for *Hippodamia convergens* (Coleoptera: Coccinellidae). *Biol. Contr.* 41: 25–32.

Phoofolo, M. W., K. L. Giles and N. C. Elliott. 2008. Larval life history responses to food deprivation in three species of predatory lady beetles (Coleoptera: Coccinellidae). *Environ. Entomol.* 37: 315–322.

Pickett, C. H., K. A. Casanave, S. E. Schoenig and K. M. Heinz. 1999. Rearing *Delphastus catalinae* (Coleoptera: Coccinellidae): Practical experience and a modeling analysis. *Can. Entomol.* 131: 115–129.

Pickett, J. A. and R. T. Glinwood. 2007. Chemical ecology. *In* H. F. van Emden and R. Harrington (eds). *Aphids as Crop Pests*. CABI, UK. pp. 235–260.

Pilipjuk, V. I., L. N. Bugaeva and E. V. Baklanova. 1982. On the possibility of breeding of the predatory beetle *Cryptolaemus montrouzieri* Muls. (Coleoptera, Coccinellidae) on the eggs of *Sitotroga cerealella* Ol. *Entomol. Obozr.* 1: 50–52. (In Russian.)

Pilorget, L., J. Buckner and J. G. Lundgren. 2010. Sterol limitation in a pollen-fed omnivorous lady beetle (Coleoptera: Coccinellidae). *J. Insect Physiol.* 56: 81–87.

Piñol, J., X. Espalader, N. Canellas and N. Perez. 2009. Effects of the concurrent exclusion of ants and earwigs on aphid abundance in an organic citrus grove. *BioControl* 54: 515–527.

Pinto, D. M., J. D. Blande, R. Nykaenen et al. 2007. Ozone degrades common herbivore-induced plant volatiles: does this affect herbivore prey location by predators and parasitoids? *J. Chem. Ecol.* 33: 683–694.

Podoler, H. and J. Henen. 1986. Foraging behavior of two species of the genus *Chilocorus* (Coccinellidae: Coleoptera): a comparative study. *Phytoparasitica* 14: 11–23.

Polis, G. A. 1981. The evolution and dynamics of intraspecific predation. *Annu. Rev. Ecol. Syst.* 12: 225–251.

Ponsonby, D. J. and M. J. W. Copland. 1996. Effect of temperature on development and immature survival in the scale insect predator, *Chilocorus nigritus* (F.) (Coleoptera: Coccinellidae). *Biocontr. Sci. Techn.* 6: 101–109.

Ponsonby, D. J. and M. J. W. Copland. 1998. Environmental influences on fecundity, egg viability and egg cannibalism in the scale insect predator, *Chilocorus nigritus. BioControl* 43: 39–52.

Ponsonby, D. J. and M. J. W. Copland. 2000. Maximum feeding potential of larvae and adults of the scale insect predator, *Chilocorus nigritus* with a new method of estimating food intake. *BioControl* 45: 295–310.

Ponsonby, D. J. and M. J. W. Copland. 2007a. Aspects of prey relations in the coccidophagous ladybird *Chilocorus nigritus* relevant to its use as a biological control agent of scale insects in temperate glasshouses. *BioControl* 52: 629–640.

Ponsonby, D. J. and M. J. W. Copland. 2007b. Influence of host density and population structure on egg production in the coccidophagous ladybird, *Chilocorus nigritus* F. (Coleoptera: Coccinellidae). *Agr. Forest Entomol.* 9: 287–296.

Poutiers, R. 1930. Sur le comportement du *Novius cardinalis* vis-à-vis de certain alcaloides. *C. R. Séanc. Soc. Biol.* 103: 1023–1025.

Powell, B. E. and J. Silverman. 2010. Impact of *Linepithema humile* and *Tapinoma sessile* (Hymenoptera: Formicidae) on three natural enemies of *Aphis gossypii* (Hemiptera: Aphididae). *Biol. Contr.* 54: 285–291.

Prasad, R. P. and W. E. Snyder. 2010. A non-trophic interaction chain links predators in different spatial niches. *Oecologia* 162: 747–753.

Pratt, C., T. W. Pope, G. Powell and J. T. Rossiter. 2008. Accumulation of glucosinolates by the cabbage aphid *Brevicoryne brassicae* as a defense against two coccinellid species. *J. Chem. Ecol.* 34: 323–329.

Provost, C., E. Lucas, D. Coderre and G. Chouinard. 2006. Prey selection by the lady beetle *Harmonia axyridis*: the influence of prey mobility and prey species. *J. Insect Behav.* 19: 265–277.

Pruszynski, S. and J. J. Lipa. 1971. The occurrence of predatory Coccinellidae on alfalfa crops. *Ekol. Pol.* 19: 365–386.

Pulliam, H. R. 1975. Diet optimization with nutrient constraints. *Am. Nat.* 109: 765–768.

Putman, W. L. 1955. Bionomics of *Stethorus punctillum* Weise in Ontario. *Can. Entomol.* 87: 9–33.

Putman, W. L. 1957. Laboratory studies on the food of some coccinellids (Coleoptera) found in Ontario peach orchards. *Can. Entomol.* 89: 527–579.

Putman, W. L. 1964. Occurrence and food of some coccinellids (Coleoptera) in Ontario peach orchards. *Can. Entomol.* 96: 1149–1155.

Puttarudriah, M. and G. P. Channabasavanna. 1952. Two giant ladybird beetles predaceous on aphids in Mysore. *Indian J. Entomol.* 14: 1.

Puttarudriah, D. M. and B. M. Maheswariah. 1966. Biology, fecundity and rate of feeding of *Synonycha grandis* Thunberg. *Proc. 2nd All-Ind. Congress*, Varanasi 1962, pt. 2.

Qin, J. 1988. Effect on the performance of *Coccinella septempunctata* of altering the components in a simplified diet devoid of insect material. *In* E. Niemczyk and A. F. G. Dixon (eds). *Ecology and Effectiveness of Aphidophaga*. SPB Acad. Publ., The Hague. pp. 55–59.

Quezada, J. R. and P. DeBach. 1973. Bioecological and population studies of the cottony-cushion scale, *Icerya purchasi* Mask., and its natural enemies, *Rodolia cardinalis* Mul. and *Cryptochaetum iceryae* Will., in southern California. *Hilgardia* 41: 631–688.

Quilici, S. and G. Iperti. 1986. The influence of host–plant on the searching ability of first instar larvae of *Propylea quatuordecimpunctata*. *In* I. Hodek (ed.). *Ecology of Aphidophaga*. Academia, Prague and W. Junk, Dordrecht, The Netherlands. pp. 99–106.

Racioppi, J. V., R. L. Burton and R. Eikenbary. 1981. The effects of various oligidic synthetic diets on the growth of *Hippodamia convergens*. *Entomol. Exp. Appl.* 30: 68–72.

Radcliffe, E. B., R. W. Weires, R. E. Stucker and D. K. Barnes. 1976. Influence of cultivars and pesticides on pea aphid, spotted alfalfa aphid, and associated arthropod taxa in a Minnesota alfalfa ecosystem. *Environ. Entomol.* 5: 1195–1207.

Radwan, Z. and G. L. Lövei. 1983. Aphids as prey for the coccinellid *Exochomus quadripustulatus*. *Entomol. Exp. Appl.* 34: 283–286.

Rana, J. S. 2006. Response of *Coccinella septempunctata* and *Menochilus sexmaculatus* (Coleoptera: Coccinellidae) to their aphid prey, *Lipaphis erysimi* (Hemiptera: Aphididae) in rapeseed-mustard. *Eur. J. Entomol.* 103: 81–84.

Rana, J. S., A. F. G. Dixon and V. Jarosik. 2002. Costs and benefits of prey specialization in a generalist insect predator. *J. Anim. Ecol.* 71: 15–22.

Rand, T. A. and S. M. Louda. 2006. Spillover of agriculturally subsidized predators as a potential threat to native insect herbivores in fragmented landscapes. *Conserv. Biol.* 20: 1720–1729.

Raubenheimer, D. and S. J. Simpson. 1999. Integrating nutrition: a geometrical approach. *Entomol. Exp. Appl.* 91: 67–82.

Raubenheimer, D., S. J. Simpson and D. Mayntz. 2009. Nutrition, ecology and nutritional ecology: toward an integrated framework. *Func. Ecol.* 23: 4–16.

Raymond, B., A. C. Darby and A. E. Douglas. 2000. The olfactory responses of coccinellids to aphids on plants. *Entomol. Exp. Appl.* 95: 113–117.

Ren, S.-X., P. A. Stansly and T.-X. Liu. 2002. Life history of the whitefly predator *Nephaspis oculatus* (Coleoptera: Coccinellidae) at six constant temperatures. *Biol. Contr.* 23: 262–268.

Reynolds, A. M. 2007. Avoidance of conspecific odour trails results in scale-free movement patterns and the execution of an optimal searching strategy. *Europhysics Lett.* 79.

Rhamhalinghan, M. 1985. Intraspecific variations in ovariole number/ovary in *Coccinella septempunctata* L. (Coleoptera: Coccinellidae). *Ind. Zool.* 9: 91–97.

Ricci, C. 1979. L'apparato boccale pungente succhiante della larva di *Platynaspis luteorubra* Goeze (Col., Coccinellidae). *Boll. Lab. Entomol. Agr. 'F. Silvestri'* 36: 179–198. (In Italian with English summary.)

Ricci, C. 1982. Constitution and function of the mandibles in *Tytthaspis sedecimpunctata* (L.) and *Tytthaspis trilineata* (Weise) larvae. *Frustula Entomol.* 3: 205–212. (In Italian with English summary).

Ricci, C. 1986a. Food strategy of *Tytthaspis sedecimpunctata* in different habitats. *In* I. Hodek (ed.). *Ecology of Aphidophaga*. Academia, Prague and W. Junk, Dordrecht, The Netherlands. pp. 311–316.

Ricci, C. 1986b. Seasonal food preferences and behaviour of *Rhyzobius litura*. *In* I. Hodek (ed.). *Ecology of Aphidophaga*. Academia, Prague and W. Junk, Dordrecht, The Netherlands. pp. 119–123.

Ricci, C. and L. Ponti. 2005. Seasonal food of *Ceratomegilla notata* (Coleoptera: Coccinellidae) in mountain environments of northern Italian Alps. *Eur. J. Entomol.* 102: 527–530.

Ricci, C., G. Fiori and S. Colazza. 1983. Regime alimentare dell'adulto di *Tytthaspis sedecimpunctata* (L.) (Coleoptera Coccinellidae) in ambiente a influenza antropica primaria: prato polifita. *Atti XIII Congr. Naz. Ital. Entomol.*, Sestriere-Torino, 691–698. (In Italian with English summary.)

Ricci, C., L. Ponti and A. Pires. 2005. Migratory flight and pre-diapause feeding of *Coccinella septempunctata* (Coleoptera) adults in agricultural and mountain ecosystems of Central Italy. *Eur. J. Entomol.* 102: 531–538.

Ricci, C., A. Primavera and V. Negri. 2006. Effects of low temperatures on *Chilocorus kuwanae* (Coleoptera: Coccinellidae) trophic activity. *Eur. J. Entomol.* 103: 547–551.

Richards, A. M. 1980. Defensive adaptations and behaviour in *Scymnodes lividigaster* (Coleoptera: Coccinellidae). *J. Zool. Lond.* 192: 157–168.

Richards, A. M. 1981. *Rhyzobius ventralis* (Erichson) and *R. forestieri* (Mulsant) (Coleoptera: Coccinellidae), their biology and value for scale insect control. *Bull. Entomol. Res.* 71: 33–46.

Richards, A. M. 1985. Biology and defensive adaptations in *Rodatus major* (Coleoptera: Coccinellidae) and its prey, *Monophlebulus pilosior* (Hemiptera: Margarodidae). *J. Zool. Lond.* 205: 287–295.

Richards, A. M. and C. Goletsos. 1991. Feeding behaviour in Australian aphidophagous Coccinellidae. *In* L. Polgar, R. J. Chambers, A. F. G. Dixon and I. Hodek (eds). *Behaviour and Impact of Aphidophaga*. SPB Acad. Publ., The Hague. pp. 227–234.

Riffell, J. A., L. Abrell and J. G. Hildebrand. 2008. Physical processes and real-time chemical measurement of the insect olfactory environment. *J. Chem. Ecol.* 34: 837–853.

Risch, S. J., R. Wrubel and D. Andow. 1982. Foraging by a predaceous beetle, *Coleomegilla maculata* (Coleoptera: Coccinellidae), in a polyculture: effects of plant density and diversity. *Environ. Entomol.* 11: 949–950.

Robinson, A. G. 1951. Annotated list of predators of Tetranychid mites in Manitoba. *Rep. Entomol. Soc. Ont.* 82: 33–37.

Robinson, J. 1992. Predators and parasitoids of Russian wheat aphid in Central Mexico. *Southw. Entomol.* 17: 185–186.

Roger, C., D. Coderre and G. Boivin. 2000. Differential prey utilization by the generalist predator *Coleomegilla maculata lengi* according to prey size and species. *Entomol. Exp. Appl.* 94: 3–13.

Roger, C., D. Coderre, C. Vigneault and G. Boivin. 2001. Prey discrimination by a generalist coccinellid predator: effect of prey age or parasitism? *Ecol. Entomol.* 26: 163–172.

Rogers, D. J. 1972. Random search and insect population models. *J. Anim. Ecol.* 41: 369–383.

Roitberg, B. D. 1993. Towards a general theory of host acceptance by aphid parasitoids. *Eur. J. Entomol* 90: 369–376.

Roitberg, B. D. and J. H. Myers. 1978. Effect of adult coccinellidae on the spread of a plant virus by an aphid. *J. Appl. Ecol.* 15: 775–779.

Roitberg, B. D., J. H. Myers and B. D. Frazer. 1979. The influence of predators on the movement of apterous pea aphids between plants. *J. Anim. Ecol.* 48: 111–122.

Rolley, F., I. Hodek and G. Iperti. 1974. Influence de la nourriture aphidienne (selon l'âge de la plante–hôte à partir de laquelle les pucerons se multiplient) sur l'induction de la dormance chez *Semiadalia undecimnotata* Schn. (Coleoptera: Coccinellidae). *Ann. Zool.-Écol. Anim.* 6: 53–60.

Rothschild, M., J. von Euw and T. Reichstein. 1970. Cardiac glycosides in the oleander aphid *Aphis nerii*. *J. Insect Physiol.* 16: 1191–1195.

Rott, A. S. and D. J. Ponsonby. 2000. The effects of temperature, relative humidity and host plant on the behaviour of *Stethorus punctillum* as a predator of the two-spotted spider mite, *Tetranychus urticae*. *BioControl* 45: 155–164.

Roy, H. E., H. Rudge, L. Goldrick and D. Hawkins. 2007. Eat or be eaten: prevalence and impact of egg cannibalism on two-spot ladybirds, *Adalia bipunctata*. *Entomol. Exp. Appl.* 125: 33–38.

Roy, M., J. Brodeur and C. Cloutier. 2003. Effect of temperature on intrinsic rates of natural increase (r_m) of a coccinellid and its spider mite prey. *BioControl* 48: 57–72.

Royer, T. A., K. L. Giles, M. M. Lebusa and M. E. Payton. 2008. Preference and suitability of greenbug, *Schizaphis graminum* (Hemiptera: Aphididae) mummies parasitized by *Lysiphlebus testaceipes* (Hymenoptera: Aphidiidae) as food for *Coccinella septempunctata* and *Hippodamia convergens* (Coleoptera: Coccinellidae). *Biol. Contr.* 47: 82–88.

Rudolf, V. H. W. 2008. Consequences of size structure in the prey for predator–prey dynamics: the composite functional response. *J. Anim. Ecol.* 77: 520–528.

Rutledge, C. E. and S. D. Eigenbrode. 2003. Epicuticular wax on pea plants decreases instantaneous search rate of *Hippodamia convergens* larvae and reduces attachment to leaf surfaces. *Can. Entomol.* 135: 93–101.

Rutledge, C. E., S. D. Eigenbrode and H. Ding. 2008. A plant surface mutation mediates predator interference among ladybird larvae. *Ecol. Entomol.* 33: 464–472.

Rutledge, C. E., R. J. O'Neil, T. B. Fox and D. A. Landis. 2004. Soybean aphid predators and their use in integrated pest management. *Ann. Entomol. Soc. Am.* 97: 240–248.

Ruzicka, Z. 1994. Oviposition-deterring pheromone in *Chrysopa oculata* (Neuroptera: Chrysopidae). *Eur. J. Entomol.* 91: 361–370.

Ruzicka, Z. 1996. Oviposition-deterring pheromone in Chrysopidae (Neuroptera): Intra- and interspecific effects. *Eur. J. Entomol.* 93: 161–166.

Ruzicka, Z. 1997. Recognition of oviposition-deterring allomones by aphidophagous predators (Neuroptera: Chrysopidae, Coleoptera: Coccinellidae). *Eur. J. Entomol.* 94: 431–434.

Ruzicka, Z. 2001. Oviposition responses of aphidophagous coccinellids to tracks of ladybird (Coleoptera: Coccinellidae) and lacewing (Neuroptera: Chrysopidae) larvae. *Eur. J. Entomol.* 98: 183–188.

Ruzicka, Z. 2002. Persistence of deterrent larval tracks in *Coccinella septempunctata*, *Cycloneda limbifer* and *Semiadalia undecimnotata* (Coleoptera: Coccinellidae). *Eur. J. Entomol.* 99: 471–475.

Ruzicka, Z. 2003. Perception of oviposition-deterring larval tracks in aphidophagous coccinellids *Cycloneda limbifer* and *Ceratomegilla undecimnotata* (Coleoptera: Coccinellidae). *Eur. J. Entomol.* 100: 345–350.

Ruzicka, Z. 2006. Oviposition-deterring effects of conspecific and heterospecific larval tracks on *Cheilomenes sexmaculata* (Coleoptera: Coccinellidae). *Eur. J. Entomol.* 103: 757–763.

Ruzicka, Z. and I. Hodek. 1978. Observations prèliminaires sur l'halophilie chez *Coccinella undecimnotata* L. *Ann. Zool.-Ecol. Anim.* 10: 367–371.

Ruzicka, Z. and R. Zemek. 2003. Effect of conspecific and heterospecific larval tracks on mobility and searching patterns of *Cycloneda limbifer* (Coleoptera: Coccinellidae) females. *Arquipélago (Life Mar. Sci) Suppl* 5: 85–93.

Ruzicka, Z. and R. Zemek. 2008. Deterrent effects of larval tracks on conspecific larvae in *Cycloneda limbifer*. *BioControl* 53: 763–771.

Saito, T. and S. Bjornson. 2006. Horizontal transmission of a microsporidium from the convergent lady beetle, *Hippodamia convergens* Guerin-Meneville (Coleoptera: Coccinellidae), to three coccinellid species of Nova Scotia. *Biol. Control* 39: 427–433.

Sakuratani, Y., Y., Sugiura, M. Isida, S. Kuwahara and T. Sugimoto. 1983. Aggregative response of adults of *Coccinella septempunctata bruckii* Mulsant (Coleoptera: Coccinellidae) to aphid population density. *Mem. Fac. Agric. Kinki Univ.* 16: 49–54.

Sallée, B. and J. Chazeau. 1985. Cycle de développement, table de vie, et taux intrinsèque d'accroissement en conditions contrôlées de *Coelophora mulsanti* (Montrouzier), Coccinellidae aphidiphage de Nouvelle-Calédonie (Coleoptera). *Annls Soc. Entomol. France* 21: 407–412.

Samways, M. J. 1986. Combined effect of natural enemies (Hymenoptera: Aphelinidae and Coleoptera: Coccinellidae) with different niche breadths in reducing high populations of red scale, *Aonidiella aurantii* (Maskell) – (Hemiptera: Diaspididae). *Bull. Entomol. Res.* 76: 671–683.

Samways, M. J. and B. A. Tate. 1986. Mass-rearing of the scale predator *Chilocorus nigritus* (F.) (Coccinellidae). *Citrus Subtrop. Fruits J.* 1986: 9–14.

Samways, M. J. and S. J. Wilson. 1988. Aspects of the feeding behaviour of *Chilocorus nigritus* (F.) (Col., Coccinellidae) relative to its effectiveness as a biocontrol agents. *J. Appl. Entomol.* 106: 177–182.

Santi, F. and S. Maini. 2007. Ladybirds mothers eating their eggs: is it cannibalism? *Bull. Insectol.* 60: 89–91.

Santos, S. A. P., J. A. Pereira, M. C. Rodrigues et al. 2009. Identification of predator–prey relationships between coccinellids and *Saissetia oleae* (Hemiptera: Coccidae), in olive groves, using an enzyme-linked immunosorbent assay. *J. Pest Sci.* 82: 101–108.

Sarmento, R. A., A. Pallini, M. Venzon et al. 2007a. Functional response of the predator *Eriopis connexa* (Coleoptera:

Coccinellidae) to different prey types. *Brazilian Arch. Biol. Tech.* 50: 121–126.

Sarmento, R. A., M. Venzon, A. Pallini, E. E. Oliveira and A. Janssen. 2007b. Use of odours by *Cycloneda sanguinea* to assess patch quality. *Entomol. Exp. Appl.* 124: 313–318.

Sato, S. and A. F. G. Dixon. 2004. Effect of intraguild predation on the survival and development of three species of aphidophagous ladybirds: consequences for invasive species. *Agr. Forest Entomol.* 6: 21–24.

Sato, S., A. F. G. Dixon and H. Yasuda. 2003. Effect of emigration on cannibalism and intraguild predation in aphidophagous ladybirds. *Ecol. Entomol.* 28: 628–633.

Savastano, L. 1918. Talune notizie sul *Novius* e l'*Icerya* riguardanti l'arboricoltore. *Boll. R. Staz. Sper. Agrum. Fruttic.* 32: 1–2.

Savoiskaya, G. I. 1966. The significance of Coccinellidae in the biological control of apple-tree aphids in the Alma-Ata fruit-growing region. *In* I. Hodek (ed.). *Ecology of Aphidophagous Insects.* Academia, Prague and W. Junk, The Hague. pp. 317–319.

Savoiskaya, G. I. 1970a. Coccinellids of the Alma-Ata reserve. *Trudy Alma-Atin. Gos. Zapov.* 9: 163–187. (In Russian.)

Savoiskaya, G. I. 1970b. Introduction and acclimatisation of some coccinellids in the Alma-Ata reserve. *Trudy Alma-Atin. Gos. Zapov.* 9: 138–162. (In Russian.)

Savoiskaya, G. I. 1983. *Kokcinellidy.* Izdatelstvo Nauka Kazachskoi SSR, Alma-Ata. 246 pp. (In Russian.)

Schaefer, P. W. 1983. Natural enemies and host plants of species in the Epilachninae (Coleoptera: Coccinellidae). A world list. *Agric. Exp. Stn., Univ. Delaware Bull.* 445: 42.

Schaller, M. and W. Nentwig. 2000. Olfactory orientation of the seven-spot ladybird beetle, *Coccinella septempunctata* (Coleoptera: Coccinellidae): attraction of adults to plants and conspecific females. *Eur. J. Entomol.* 97: 155–159.

Schanderl, H., A. Ferran and V. Garcia. 1988. L'élevage de deux coccinelles *Harmonia axyridis* et *Semiadalia undecimnotata* à l'aide d'oeufs d'*Anagasta kuehniella* tués aux rayons ultraviolets. *Entomol. Exp. Appl.* 49: 235–244.

Schanderl, H., A. Ferran and M.-M. Larroque. 1985. Les besoins trophiques et thermiques des larves de la coccinelle *Harmonia axyridis* Pallas. *Agronomie* 5: 417–421.

Schellhorn, N. A. and D. A. Andow. 1999a. Cannibalism and interspecific predation: role of oviposition behavior. *Ecol. Appl.* 9: 418–428.

Schellhorn, N. A. and D. A. Andow. 1999b. Mortality of coccinellid (Coleoptera: Coccinellidae) larvae and pupae when prey become scarce. *Environ. Entomol.* 28: 1092–1100.

Schellhorn, N. A. and D. A. Andow. 2005. Response of coccinellids to their aphid prey at different spatial scales. *Popul. Ecol.* 47: 71–76.

Schenk, D., L. F. Bersier and S. Bacher. 2005. An experimental test of the nature of predation: neither prey- nor ratio-dependent. *J. Anim. Ecol.* 74: 86–91.

Schiefelbein, J. W. and H. C. Chiang. 1966. Effects of spray of sucrose solution in a corn field on the populations of predatory insects and their prey. *Entomophaga* 11: 333–339.

Schlacher, T. A. and G. Cronin. 2007. A trophic cascade in a macrophyte-based food web at the land-water ecotone. *Ecol. Res.* 22: 749–755.

Schuder, I., M. Hommes and O. Larink. 2004. The influence of temperature and food supply on the development of *Adalia bipunctata* (Coleoptera: Coccinellidae). *Eur. J. Entomol.* 101: 379–384.

Schuster, M. F., M. J. Lukefahr and F. G. Maxwell. 1976. Impact of nectariless cotton on plant bugs and natural enemies. *J. Econ. Entomol.* 69: 400–402.

Schwartzberg, E. G., G. Kunert, C. Stephan et al. 2008. Real-time analysis of alarm pheromone emission by the pea aphid (*Acyrthosiphon pisum*) under predation. *J. Chem. Ecol.* 34: 76–81.

Seagraves, M. P. 2009. Lady beetle oviposition behavior in response to the trophic environment. *Biol. Contr.* 51: 313–322.

Seagraves, M. P. and K. V. Yeargan. 2006. Selection and evaluation of a companion plant to indirectly augment densities of *Coleomegilla maculata* (Coleoptera: Coccinellidae) in sweet corn. *Environ. Entomol.* 35: 1334–1341.

Semyanov, V. P. 1965. Fauna, biology and usefulness of coccinellids (Coleoptera: Coccinellidae) in Belorus SSR. *Zap. Leningr. SelKhoz. Inst.* 95: 106–120. (In Russian.)

Semyanov, V. P. 1970. Biological properties of *Adalia bipunctata* L. (Coleoptera: Coccinellidae) in conditions of Leningrad region. *Zashch. Rast. Vredit. Bolez.* 127: 105–112. (In Russian.)

Semyanov, V. P. 1980. Biology of *Calvia quatuordecimguttata* L. (Coleoptera, Coccinellidae). *Entomol. Obozr.* 59: 757–763. (In Russian.)

Semyanov, V. P. 1999. Biology of coccinellids (Coleoptera) from Southeast Asia. I. *Leis dimidiata* (Fabr.) *Entomol. Obozr.* 78: 537–544. (In Russian with English summary.)

Semyanov, V. P. 2000. The biology of the coccinellids (Coleoptera) from Southeast Asia. II. *Harmonia sedecimnotata* F.). *Entomol. Obozr.* 79: 3–9. (In Russian with English summary).

Semyanov, V. P. and N. P. Vaghina. 2001. The odor of aphids as a signal for termination of the trophic diapause in the lady beetle *Harmonia sedecimnotata* (Fabr.) (Coleoptera, Coccinellidae). *Proc. Zool. Inst. Russ. Acad. Sci.* 289: 161–166.

Sengonca, C. and B. Liu. 1994. Responses of the different instar predator, *Coccinella septempunctata* L. (Coleoptera: Coccinellidae), to the kairomones produced by the prey and non-prey insects as well as the predator itself. *Z. PflKrankh. PflSchutz.* 101: 173–177.

Sengonca, C., J. Kranz and P. Blaeser. 2002. Attractiveness of three weed species to polyphagous predators and their influence on aphid populations in adjacent lettuce cultivations. *J. Pest Sci.* 75: 161–165.

Sethi, S. L. and A. S. Atwal. 1964. Influence of temperature and humidity on the development of different stages of *Coccinella septempunctata* (Coleoptera, Coccinellidae). *Ind. J. Agric. Sci.* 34: 166–171.

Shah, M. A. 1982. The influence of plant surfaces on the searching behaviour of Coccinellid larvae. *Entomol. Exp. Appl.* 31: 377–380.

Shands, W. A., R. L. Holmes and G. W. Simpson. 1970. Improved laboratory production of eggs of *Coccinella septempunctata*. *J. Econ. Entomol.* 63: 315–317.

Shands, W. A., M. K. Shands and G. W. Simpson. 1966. Technique for massproducing *Coccinella septempunctata*. *J. Econ. Entomol.* 59: 102–103.

Shannag, H. K. and W. M. Obeidat. 2006. Voracity and conversion efficiency by larvae of *Coccinella septempunctata* L. (Coleoptera: Coccinellidae) on *Aphis fabae* Scop. (Homoptera: Aphididae) reared on two faba bean cultivars with different levels of resistance. *Appl. Entomol. Zool.* 41: 521–527.

Shannag, H. K. and W. M. Obeidat. 2008. Interaction between plant resistance and predation of *Aphis fabae* (Homoptera: Aphididae) by *Coccinella septempunctata* (Coleoptera: Coccinellidae). *Ann. Appl. Biol.* 152: 331–337.

Shantibala, K., L. S. Singh and T. K. Singh. 1992. Development and larval voracity of a coccinellid predator, *Synonycha grandis* (Thunberg) on bamboo aphid, *Paraoregma alexandri* (Takahashi). *Manipur J. Agric. Sci.* 1: 55–57.

Shirai, Y. and H. Katakura. 1999. Host plants of the phytophagous ladybird beetle, *Epilachna vigintioctopunctata* (Coleoptera: Coccinellidae) in Southeast Asia and Japan. *Appl. Entomol. Zool.* 34: 75–83.

Shirai, Y. and H. Katakura. 2000. Adaptation to a new host plant, *Centrosema pubescens* (Fabales: Leguminosae), by the phytophagous ladybird beetle, *Epilachna vigintioctopunctata*. Coleoptera: Coccinellidae), in tropical Asia. *Popul. Ecol.* 42: 129–134.

Shirai, Y. and N. Morimoto. 1997. Life history traits of pest and non-pest populations in the phytophagous ladybird beetle, *Epilachna niponica* (Coleoptera, Coccinellidae). *Res. Popul. Ecol.* 39: 163–171.

Shirai, Y. and N. Morimoto. 1999. A host shift from wild blue cohosh to cultivated potato by the phytophagous ladybird beetle, *Epilachna yasutomii* (Coleoptera, Coccinellidae). *Res. Popul. Ecol.* 41: 161–167.

Shirai, Y. and K. Yara. 2001. Potential distribution area of the Mexican bean beetle, *Epilachna varivestis* (Coleoptera: Coccinellidae) in Japan, estimated from its high-temperature tolerance. *Appl. Entomol. Zool.* 36: 409–417.

Siddiqui, A., A. Kumar, N. Kumar and C. P. M. Tripathi. 1999. Prey–predator relationship between *Lipaphis erysimi* Kalt. (Homoptera: Aphididae) and *Coccinella septempunctata* Linn. (Coleoptera: Coccinellidae): III. Effect of host plants on the searching strategy, mutual interference and killing power of the predator. *Biol. Agric. Hort.* 17: 11–17.

Sighinolfi, L., G. Febvay, M. L. Dindo et al. 2008. Biological and biochemical characteristics for quality control of *Harmonia axyridis* (Pallas) (Coleoptera, Coccinellidae) reared on a liver-based diet. *Arch. Insect Biochem. Physiol.* 68: 26–39.

da Silva, P. G., K. S. Hagen and A. P. Gutierrez. 1992. Functional response of *Curinus coeruleus* (Col.: Coccinellidae) to *Heteropsylla cubana* (Hom.: Psyllidae) on artificial and natural substrates. *Entomophaga* 37: 555–564.

Silva, R. B., J. C. Zanuncio, J. E. Serrao et al. 2009. Suitability of different artificial diets for development and survival of stages of the predaceous ladybird beetle *Eriopis connexa*. *Phytoparasitica* 37: 115–123.

Simmons, A. M. and J. C. Legaspi. 2004. Survival and predation of *Delphastus catalinae* (Coleoptera: Coccinellidae), a predator of whiteflies (Homoptera: Aleyrodidae), after exposure to a range of constant temperatures. *Environ. Entomol.* 33: 839–843.

Sloggett, J. J. 2005. Are we studying too few taxa? Insights from aphidophagous ladybird beetles (Coleoptera: Coccinellidae). *Eur. J. Entomol.* 102: 391–398.

Sloggett, J. J. 2008a. Habitat and dietary specificity in aphidophagous ladybirds (Coleoptera: Coccinellidae): Explaining specialization. *Proc. Neth. Entomol. Soc. Meet.* 19: 95–113.

Sloggett, J. J. 2008b. Weighty matters: Body size, diet and specialization in aphidophagous ladybird beetles (Coleoptera: Coccinellidae). *Eur. J. Entomol.* 105: 381–389.

Sloggett, J. J. and A. J. Davis. 2010. Eating chemically defended prey: alkaloid metabolism in an invasive ladybird predator of other ladybirds (Coleoptera: Coccinellidae). *J. Exper. Biol.* 213: 237–241.

Sloggett, J. J. and M. W. Lorenz. 2008. Egg composition and reproductive investment in aphidophagous ladybird beetles (Coccinellidae: Coccinellini): egg development and interspecific variation. *Physiol. Entomol.* 33: 200–208.

Sloggett, J. J. and M. E. N. Majerus. 1994. *The Cambridge Ladybird Survey* 17: 13–14.

Sloggett, J. J. and M. E. N. Majerus. 2000. Habitat preferences and diet in the predatory Coccinellidae (Coleoptera): an evolutionary perspective. *Biol. J. Linn. Soc.* 70: 63–88.

Sloggett, J. J. and M. E. N. Majerus. 2003. Adaptations of *Coccinella magnifica*, a myrmecophilous coccinellid to aggression by wood ants (*Formica rufa* group). II. Larval behaviour, and ladybird oviposition location. *Eur. J. Entomol.* 100: 337–344.

Sloggett, J. J., K. F. Haynes and J. J. Obrycki. 2009a. Hidden costs to an invasive intraguild predator from chemically defended native prey. *Oikos* 118: 1396–1404.

Sloggett, J. J., J. J. Obrycki and K. F. Haynes. 2009b. Identification and quantification of predation: novel use of gaz chromatography-mass spectrometric analysis of prey alkaloid markers. *Func. Ecol.* 23: 416–426.

Sloggett, J. J., A. Manica, M. J. Day and M. E. N. Majerus. 1999. Predation of ladybirds (Coleoptera, Coccinellidae) on ants

Formica rufa L. (Hymenoptera, Formicidae). *Entomol. Gaz.* 50: 217–221.

Sloggett, J. J., W. Völkl, W. Schulze, J. H. G. Schulenburg and M. E. N. Majerus. 2002. The ant-associations and diet of the ladybird *Coccinella magnifica* (Coleoptera: Coccinellidae). *Eur. J. Entomol.* 99: 565–569.

Sloggett, J. J., I. Zeilstra and J. J. Obrycki. 2008. Patch residence by aphidophagous ladybird beetles: Do specialists stay longer? *Biol. Contr.* 47: 199–206.

Smirnoff, W. A. 1958. An artificial diet for rearing coccinellid beetles. *Can. Entomol.* 90: 563–565.

Smith, B. C. 1960. A technique for rearing Coccinellid beetles on dry foods, and influence of various pollens on the development of *Coleomegilla maculata lengi* Timb. (Coleoptera: Coccinellidae). *Can. J. Zool.* 38: 1047–1049.

Smith, B. C. 1961. Results of rearing some coccinellid (Coleoptera: Coccinellidae) larvae on various pollens. *Proc. Entomol. Soc. Ont.* 91: 270–271.

Smith, B. C. 1965a. Differences in *Anatis mali* Auct. and *Coleomegilla maculata lengi* Timberlake to changes in the quality and quantity of the larval food. *Can. Entomol.* 97: 1159–1166.

Smith, B. C. 1965b. Growth and development of coccinellid larvae on dry foods (Coleoptera, Coccinellidae). *Can. Entomol.* 97: 760–768.

Snyder, W. E. and G. M. Clevenger. 2004. Negative dietary effects of Colorado potato beetle eggs for the larvae of native and introduced ladybird beetles. *Biol. Contr.* 31: 353–361.

Snyder, W. E. and E. W. Evans. 2006. Ecological effects of invasive arthropod generalist predators. *Annu. Rev. Ecol. Evol. Syst.* 37: 95–122.

Snyder, W. E., S. N. Ballard, S. Yang et al. 2004. Complementary biocontrol of aphids by the ladybird beetle *Harmonia axyridis* and the parasitoid *Aphelinus asychis* on greenhouse roses. *Biol. Contr.* 30: 229–235.

Snyder, W. E., S. B. Joseph, R. F. Preziosi and A. J. Moore. 2000. Nutritional benefits of cannibalism for the lady beetle *Harmonia axyridis* (Coleoptera: Coccinellidae) when prey quality is poor. *Environ. Entomol.* 29: 1173–1179.

Soares, A. O. and A. Serpa. 2007. Interference competition between ladybird beetle adults (Coleoptera: Coccinellidae): effects on growth and reproductive capacity. *Popul. Ecol.* 49: 37–43.

Soares, A. O., D. Coderre and H. Schanderl. 2002. Effect of temperature and intraspecific allometry on predation by two phenotypes of *Harmonia axyridis* Pallas (Coleoptera: Coccinellidae). *Environ. Entomol.* 32: 939–944.

Soares, A. O., D. Coderre and H. Schanderl. 2004. Dietary self-selection behaviour by the adults of the aphidophagous ladybeetle *Harmonia axyridis* (Coleoptera: Coccinellidae). *J. Anim. Ecol.* 73: 478–486.

Soares, A. O., D. Coderre and H. Schanderl. 2005. Influence of prey quality on the fitness of two phenotypes of *Harmonia axyridis* adults. *Entomol. Exp. Appl.* 114: 227–232.

Solomon, M. E. 1949. The natural control of animal populations. *J. Anim. Ecol.* 18: 1–35.

Solomon, M. E. 1964. Analysis of processes involved in the natural control of insects. *Adv. Ecol. Res.* 2: 1–58.

Specty, O., G. Febvay, S. Grenier et al. 2003. Nutritional plasticity of the predatory ladybeetle *Harmonia axyridis* (Coleoptera: Coccinellidae): comparison between natural and substitution prey. *Arch. Insect Biochem. Physiol.* 52: 81–91.

Spellman, B., M. W. Brown and C. R. Mathews. 2006. Effect of floral and extrafloral resources on predation of *Aphis spiraecola* by *Harmonia axyridis* on apple. *BioControl* 51: 715–724.

Srikanth, J. and N. H. Lakkundi. 1990. Seasonal population fluctuations of cowpea aphid *Aphis craccivora* Koch and its predatory coccinellids. *Insect Sci. Applic.* 11: 21–26.

Stadler, B. 1991. Predation success of *Coccinella septempunctata* when attacking different *Uroleucon* species. *In* L. Polgár, R. J. Chambers, A. F. G. Dixon and I. Hodek (eds). *Behaviour and Impact of Aphidophaga.* SPB Acad. Publ., The Hague. pp. 265–271.

Staley, A. C. and K. V. Yeargan. 2005. Oviposition behavior of *Coleomegilla maculata* (Coleoptera: Coccinellidae): Diel periodicity and choice of host plants. *Environ. Entomol.* 34: 440–445.

Stathas, G. J. 2000. The effect of temperature on the development of the predator *Rhyzobius lophanthae* and its phenology in Greece. *BioControl* 45: 439–451.

Staubli Dreyer, B., P. Neuenschwander, J. Baumgartner and S. Dorn. 1997b. Trophic influences on survival, development and reproduction of *Hyperaspis notata* (Col., Coccinellidae). *J. Appl. Entomol.* 121: 249–256.

Staubli Dreyer, B., P. Neuenschwander, B. Bouyjou, J. Baumgartner and S. Dorn. 1997a. The influence of temperature on the life table of *Hyperaspis notata*. *Entomol. Exp. Appl.* 84: 85–92.

Stewart, L. A., J.-L. Hemptinne and A. F. G. Dixon. 1991. Reproductive tactics of ladybird beetles: relationships between egg size, ovariole number and developmental time. *Funct. Ecol.* 5: 380–385.

Storch, R. H. 1976. Prey detection by fourth stage *Coccinella transversoguttata* larvae (Coleoptera: Coccinellidae). *Anim. Behav.* 24: 690–693.

Strauss, R. E. 1982. Influence of replicated sub-samples and sub-sample heterogeneity on the linear index of food selection. *Trans. Am. Fisheries Soc.* 111: 517–522.

Strouhal, H. 1926a. Die Larven der Palaearktischen Coccinellini und Psylloborini (Coleopt.). *Arch. Naturgesch.* 92(A)3: 1–63.

Strouhal, H. 1926b. Pilzfressende Coccinelliden (Tribus Psylloborini) (Col.). *Z. Wiss. Insekt-Biol.* 21: 131–143.

Stuart, M. K. and M. H. Greenstone. 1990. Beyond ELISA: a rapid, sensitive, specific immunodot assay for identification of predator stomach contents. *Ann. Entomol. Soc. Am.* 83: 1101–1107.

Stubbs, M. 1980. Another look at prey detection by coccinellids. *Ecol. Entomol.* 5: 179–182.

Sundby, R. A. 1966. A comparative study of the efficiency of three predatory insects, *Coccinella septempunctata* L. (Coleoptera: Coccinellidae), *Chrysopa carnea* St. (Neuroptera: Chrysopidae) and *Syrphus ribesii* L. (Diptera: Syrphidae) at two different temperatures. *Entomophaga* 11: 395–404.

Sunderland, K. D., N. E. Crook, D. L. Stacey and B. J. Fuller. 1987. A study of feeding by polyphagous predators on cereal aphids using ELISA and gut dissection. *J. Appl. Ecol.* 24: 907–933.

Sutherland, A. M. and M. P. Parrella. 2009. Mycophagy in Coccinellidae: review and synthesis. *Biol. Contr.* 51: 284–293.

Suzuki, N. and T. Ide. 2008. The foraging behaviors of larvae of the ladybird beetle, *Coccinella septempunctata* (Coleoptera: Coccinellidae) towards ant-tended and non-ant-tended aphids. *Ecol. Res.* 23: 371–378.

Svoboda, J. A. and W. E. Robbins. 1979. Comparison of sterols from a phytophagous and predacious species of the family Coccinellidae. *Experientia* 35: 186–187.

Szumkowski, W. 1952. Observations of Coccinellidae. II. Experimental rearing of *Coleomegilla* on a non-insect diet. *Proc. 9th Int. Congr. Entomol. Amsterdam* 1951: 781–785.

Szumkowski, W. 1961a. Aparicion de un coccinelido predato nuevo para Venezuela. *Agronomía Trop. (Venezuela)* 11: 33–37.

Szumkowski, W. 1961b. Dietas sin insectos vivos para la cria de *Coleomegilla maculata* Deg. (Coccinellidae, Coleoptera). *Agronomía Trop. (Venezuela)* 10: 149–154.

Taghizadeh, R., Y. Fathipour and K. Kamali. 2008. Influence of temperature on life-table parameters of *Stethorus gilvifrons* (Mulsant) (Coleoptera: Coccinellidae) fed on *Tetranychus urticae* Koch. *J. Appl. Entomol.* 132: 638–645.

Takahashi, K. 1987. Cannibalism by the larvae of *Coccinella septempunctata bruckii* Mulsant (Coleoptera: Coccinellidae) in mass-rearing experiments. *Jap. J. Appl. Entomol. Zool.* 31: 201–205. (In Japanese with English summary.)

Takahashi, K. 1989. Intra- and inter-specific predations by lady beetles in spring alfalfa fields. *Jpn. J. Entomol.* 57: 199–203.

Takeda, S., S. Hukusima and H. Yamada. 1964. Seasonal abundance in coccinellid beetles. *Res. Bull. Fac. Agric. Gifu Univ.* 19: 55–63. (In Japanese with English summary.)

Takeuchi, M., Y. Sasaki, C. Sato et al. 2000. Seasonal host utilization of mycophagous ladybird *Illeis koebelei* (Coccinellidae: Coleoptera). *Jpn. J. Appl. Entomol. Zool.* 44: 89–94.

Tanaka, M. 1981. Biological control of arrowhead scale, *Unaspis yanonensis* (Kuwana), in the implementations of IPM programs of citrus orchards in Japan. *Proc. Int. Soc. Citriculture* 2: 636–640.

Tanaka, M. and Y. Maeta. 1965. Rearing of some predacious coccinellid beetles by artificial diets. *Bull. Hort. Res. Stn. (D)* 63: 17–35. (In Japanese with English summary.)

Tao, C. C. and S. C. Chiu. 1971. Biological control of citrus, vegetables and tobacco aphids. *Spec. Publ. Taiwan Agr. Res. Inst.* 10: 1–110.

Telenga, N. A. and M. V. Bogunova. 1936. The most important predators of coccids and aphids in the Ussuri region of Far East and their utilisation. *Zashch. Rast.*, 1939(10): 75–87. (In Russian.)

Thompson, W. R. 1951. The specificity of host relations in predaceous insects. *Can. Entomol.* 83: 262–269.

Timms, J. E., T. H. Oliver, N. A. Straw and S. R. Leather. 2008. The effect of host plant on the coccinellid functional response: Is the conifer specialist *Aphidecta obliterata* (L.) (Coleoptera: Coccinellidae) better adapted to spruce than the generalist *Adalia bipunctata* (L.) (Coleoptera: Coccinellidae)? *Biol. Contr.* 47: 273–281.

Triltsch, H. 1997. Contents in field sampled adults of *Coccinella septempunctata* (Col.: Coccinellidae). *Entomophaga* 42: 125–131.

Triltsch, H. 1999. Food remains in the guts of *Coccinella septempunctata* (Coleoptera: Coccinellidae) adults and larvae. *Eur. J. Entomol.* 96: 355–364.

Tsaganou, F. C., C. J. Hodgson, C. G. Athanassiou, N. G. Kavallieratos and Z. Tomanovic. 2004. Effect of *Aphis gossypii* Glover, *Brevicoryne brassicae* (L.), *and Megoura viciae* Buckton (Hemiptera: Aphidoidea) on the development of the predator *Harmonia axyridis* (Pallas) (Coleoptera: Coccinellidae). *Biol. Contr.* 31: 138–144.

Tschanz B., L. F. Bersier and S. Bacher. 2007. Functional responses: a question of alternative prey and predator density. *Ecology* 88: 1300–1308.

Turchin, P. and P. Kareiva. 1989. Aggregation in *Aphis varians*: an effective strategy for reducing predation risk. *Ecology* 70: 1008–1016.

Turian, G. 1969. Coccinelles micromycétophages (Col.). *Mit. Schweiz. Entomol. Ges.* 42: 52–57.

Turian, G. 1971. *Thea 22-punctata* et autres Coccinelles micromycétophages. Nature du pigment élytral jaune. *Bull. Soc. Entomol. Suisse* 44: 277–280.

Turlings, T. C. J., J. H. Tumlinson and W. J. Lewis. 1990. Exploitation of herbivore-induced plant odors by host seeking parasitic wasps. *Science* 250: 1251–1253.

Turlings, T. C. J., M. Bernasconi, R. Bertossa et al. 1998. The induction of volatile emissions in maize by three herbivore species with different feeding habits: possible consequences for their natural enemies. *Biol. Contr.* 11: 122–129.

Turnock, W. J. and I. L. Wise. 2004. Density and survival of lady beetles (Coccinellidae) in overwintering sites in Manitoba. *Can. Field-Nat.* 118: 309–317.

Ueno, H. 2003. Genetic variation in larval period and pupal mass in an aphidophagous ladybird beetle (*Harmonia axyridis*) reared in different environments. *Entomol. Exp. Appl.* 106: 211–218.

Ueno, H., N. Fujiyama, K. Irie, Y. Sato and H. Katakura. 1999. Genetic basis for established and novel host plant use in a herbivorous ladybird beetle, *Epilachna vigintioctomaculata*. *Entomol. Exp. Appl.* 91: 245–250.

Ueno, H., Y. Hasegawa, N. Fujiyama and H. Katakura. 2001. Comparison of genetic variation in growth performance on normal and novel host plants in a local population of a herbivorous ladybird beetle, *Epilachna vigintioctomaculata*. *Heredity* 87: 1–7.

Underwood, N., W. Morris, K. Gross and J. R. Lockwood. 2000. Induced resistance to Mexican bean beetles in soybean: variation among genotypes and lack of correlation with constitutive resistance. *Oecologia* 122: 83–89.

Unsicker, S. B., A. Oswald, G. Köhler and W. W. Weisser. 2008. Complementarity effects through dietary mixing enhance the performance of a generalist insect herbivore. *Oecologia* 156: 313–324.

Uygun, N. and N. Z. Elekcioglu. 1998. Effect of three diaspididae prey species on development and fecundity of the ladybeetle *Chilocorus bipustulatus* in the laboratory. *BioControl* 43: 153–162.

Van der Werf, W., E. W. Evans and J. Powell. 2000. Measuring and modelling the dispersal of *Coccinella septempunctata* (Coleoptera: Coccinellidae) in alfalfa fields. *Eur. J. Entomol.* 97: 487–493.

Van Rijn, P. C. J., F. M. Bakker, W. A. D. van der Hoeven and M. W. Sabelis. 2005. Is arthropod predation exclusively satiation-driven? *Oikos* 109: 101–116.

Vandenberg, N. J. 2002. Family 93. Coccinellidae Latreille 1807. *In* R. H. Arnett Jr., M. C. Thomas, P. E. Skelley and J. H. Frank (eds). *American Beetles. Polyphaga: Scarabaeoidea through Curculionoidea*, Vol. 2. CRC Press, Boca Raton, FL. pp. 371–389.

Varley, G. C. and G. R. Gradwell. 1970. Recent advances in insect population dynamics. *Annu. Rev. Entomol.* 15: 1–24.

Verheggen, F. J., Q. Fagel, S. Heuskin et al. 2007. Electrophysiological and behavioral responses of the multicolored Asian lady beetle, *Harmonia axyridis* Pallas, to sesquiterpene semiochemicals. *J. Chem. Ecol.* 33: 2148–2155.

Vet, L. E. M. and M. Dicke. 1992. Ecology of infochemical use by natural enemies in a tritrophic context. *Annu. Rev. Entomol.* 37: 141–172.

Vinson, S. B. 1977. Behavioral chemicals in the augmentation of natural enemies. *In* R. L. Ridgway and S. B. Vinson (eds). *Biological Control by Augmentation of Natural Enemies.* Plenum Press, New York. pp. 237–279.

Vinson, S. B. and T. A. Scarborough. 1989. Impact of the imported fire ant on laboratory populations of cotton aphid (*Aphis gossypii*) predators. *Florida Entomol.* 72: 107–111.

Voelkl, W. 1995. Behavioral and morphological adaptations of the coccinellid, *Platynaspis luteorubra* for exploiting ant-attended resources (Coleoptera, Coccinellidae). *J. Insect Behav.* 8: 653–670.

Voelkl, W. and K. Vohland. 1996. Wax covers in larvae of two *Scymnus* species: Do they enhance coccinellid larval survival? *Oecologia* 107: 498–503.

Waage, J. and D. Greathead (eds). 1986. *Insect Parasitoids.* Academic Press, London. 389 pp.

Wade, M. R., M. P. Zalucki, S. D. Wratten and K. A. Robinson. 2008. Conservation biological control of arthropods using artificial food sprays: current status and future challenges. *Biol. Contr.* 45: 185–199.

Wagner, J. D., M. D. Glover, J. B. Moseley and A. J. Moore. 1999. Heritability and fitness consequences of cannibalism in *Harmonia axyridis*. *Evol. Ecol. Res.* 1: 375–388.

Wajnberg, E., C. Bernstein and J. van Alphen. 2008. *Behavioral Ecology of Insect Parasitoids: From Theoretical Approaches to Field Applications.* Blackwell, Oxford. 445 pp.

Waldbauer, G. P. 1968. The consumption and utilization of food by insects. *Adv. Insect Physiol.* 5: 229–288.

Waldbauer, G. P. and S. Friedman. 1991. Self-selection of optimal diets by insects. *Annu. Rev. Entomol.* 36: 43–63.

Wallace, M. S. and F. P. Hain. 2000. Field surveys and evaluation of native and established predators of the hemlock woolly adelgid (Homoptera: Adelgidae) in the southeastern United States. *Environ. Entomol.* 29: 638–644.

Wang, Y. Z. and F. Ba. 1998. Study on optimum control of the soybean aphid. *Acta Phyt. Sinica* 25: 152–155.

Ward, D. 1992. The role of satisficing in foraging theory. *Oikos* 63: 312–317.

Ward, D. 1993. Foraging theory, like other fields of science, needs multiple working hypotheses. *Oikos* 67: 376–378.

Ware, R. L., B. Yguel and M. E. N. Majerus. 2008. Effects of larval diet on female reproductive output of the European coccinellid *Adalia bipunctata* and the invasive species *Harmonia axyridis* (Coleoptera: Coccinellidae). *Eur. J. Entomol.* 105: 437–443.

Ware, R., B. Yguel and M. Majerus. 2009. Effects of competition, cannibalism and intra-guild predation on larval development of the European coccinellid *Adalia bipunctata* and the invasive species *Harmonia axyridis*. *Ecol. Entomol.* 34: 12–19.

Warren, L. O. and M. Tadić. 1967. Biological observations on *Coleomegilla maculata* and its role as a predator of the fall webworm. *J. Econ. Entomol.* 60: 1492–1496.

Watson, J. R. and W. L. Thompson. 1933. Food habits of *Leis conformis* Boisd. (Chinese lady-beetle). *Florida Entomol.* 17: 27–29.

Way, M. J. 1963. Mutualism between ants and honeydew-producing Homoptera. *Annu. Rev. Entomol.* 8: 307–344.

Weber, D. C. and J. G. Lundgren. 2009. Assessing the trophic ecology of the Coccinellidae: their roles as predators and as prey. *Biol. Contr.* 51: 199–214.

Weber, R. G. and M. V. Holman. 1976. *Coleomegilla maculata* (DeGeer), (Coleoptera: Coccinellidae), a new coleopteran predator of the elm leaf beetle, *Pyrrhalta luteola* (Müller), (Coleoptera: Chrysomelidae). *J. Kansas Entomol. Soc.* 49: 160.

Wells, M. L., R. M. McPherson, J. R. Ruberson and G. A. Herzog. 2001. Coccinellids in cotton: population response to pesticide application and feeding response to cotton aphids (Homoptera: Aphididae). *Environ. Entomol.* 30: 785–793.

White, C. and S. D. Eigenbrode. 2000. Effects of surface wax variation in *Pisum sativum* on herbivorous and entomophagous insects in the field. *Environ. Entomol.* 29: 773–780.

Wiebe, A. P. and J. J. Obrycki. 2004. Quantitative assessment of predation of eggs and larvae of *Galerucella pusilla* in Iowa. *Biol. Contr.* 31: 16–28.

Wigglesworth, V. B. 1953. *The Principles of Insect Physiology.* 5th ed. with add. Methuen, London. 546 pp.

Wipperfuerth, T., K. S. Hagen and T. E. Mittler. 1987. Egg production by the coccinellid *Hippodamia convergens* fed on two morphs of the green peach aphid, *Myzus persicae*. *Entomol. Exp. Appl.* 44: 195–198.

With, K. A., D. M. Pavuk, J. L. Worchuck, R. K. Oates and J. L. Fisher. 2002. Threshold effects of landscape structure on biological control in agroecosystems. *Ecol. Appl.* 12: 52–65.

Witte, L., A. Ehmke and Th. Hartmann. 1990. Interspecific flow of pyrrolizidine alkaloids. *Naturwissenschaften* 77: 540–543.

Wratten, S. D. 1973. The effectiveness of the Coccinellid beetle, *Adalia bipunctata* (L.), as a predator of the lime aphid, *Eucallipterus tiliae* L. *J. Anim. Ecol.* 42: 785–802.

Wratten, S. D. 1976. Searching by *Adalia bipunctata* (L.) (Coleoptera: Coccinellidae) and escape behaviour of its aphid and cicadellid prey on lime (*Tilia x vulgaris* Hayne). *Ecol. Entomol.* 1: 139–142.

Wright, E. J. and J. E. Laing. 1980. Numerical response of coccinellids to aphids in corn in southern Ontario. *Can. Entomol.* 112: 977–988.

Wright, E. J. and J. E. Laing. 1982. Stage-specific mortality of *Coleomegilla maculata lengi* Timberlake on corn in southern Ontario. *Environ. Entomol.* 11: 32–37.

Wu, Z., D. Schenk-Hamlin, W. Zhan, D. W. Ragsdale and G. E. Heimpel. 2004. The soybean aphid in China: A historical review. *Ann. Entomol. Soc. Am.* 97: 209–218.

Wylie, H. G. (1958) Observations on *Aphidecta obliterata* (L.) (Coleoptera: Coccinellidae), a predator of conifer-infesting Aphidoidea. *Can. Entomol.* 90: 518–522.

Yadava, C. and F. R. Shaw. 1968. The preference of certain coccinellids for pea aphids, leafhoppers and alfalfa larvae. *J. Econ. Entomol.* 61: 1104–1105.

Yan, F.-S., J. Qin and X.-F. Xiang. 1982. The fine structure of the chemoreceptors on the labial palps of *Coccinella septempunctata*. *Acta Entomol. Sinica* 25: 135–140. (In Chinese with English summary.)

Yasuda, H. and N. Ohnuma. 1999. Effect of cannibalism and predation on the larval performance of two ladybird beetles. *Entomol. Exp. Appl.* 93: 63–67.

Yasuda, H. and K. Shinya. 1997. Cannibalism and interspecific predation in two predatory ladybirds in relation to prey abundance in the field. *Entomophaga* 42: 153–163.

Yasuda, H., T. Takagi and K. Kogi. 2000. Effects of conspecific and heterospecific larval tracks on the oviposition behaviour of the predatory ladybird, *Harmonia axyridis* (Coleoptera: Coccinellidae). *Eur. J. Entomol.* 97: 551–553.

Yasuda, H., E. W. Evans, Y. Kajita, K. Urakawa and T. Takizawa. 2004. Asymmetric larval interactions between introduced and indigenous ladybirds in North America. *Oecologia* 141: 722–731.

Yasuda, H., T. Kikuchi, P. Kindlmann and S. Sato. 2001. Relationships between attack and escape rates, cannibalism, and intraguild predation in larvae of two predatory ladybirds. *J. Insect Behav.* 14: 373–383.

Yigit, A. 1992. Method for culturing *Serangium parcesetosum* Sicard on *Bemisia tabaci* Genn. (Hom.: Aleyrodidae). *J. Plant Dis. Prot.* 99: 525–527.

Yigit, A., R. Canhilal and U. Ekmekci. 2003. Seasonal population fluctuations of *Serangium parcesetosum* (Coleoptera: Coccinellidae), a predator of citrus whitefly, *Dialeurodes citri* (Homoptera: Aleyrodidae) in Turkey's eastern Mediterranean citrus groves. *Environ. Entomol.* 32: 1105–1114.

Yinon, U. 1969. Food consumption of the armored scale ladybeetle, *Chilocorus bipustulatus* (Coleoptera, Coccinellidae). *Entomol. Exp. Appl.* 12: 139–146.

Yoneya, K., S. Kugimiya and J. Takabayashi. 2009. Can herbivore-induced plant volatiles inform predatory insect about the most suitable stage of its prey? *Physiol. Entomol.* 34: 379–386.

Yoon, C., D.-K. Seo, J.-O. Yang, S.-H. Kang and G.-H. Kim. 2010. Attraction of the predator *Harmonia axyridis* (Coleoptera: Coccinellidae), to its prey, *Myzus persicae* (Hemiptera: Aphididae), feeding on Chinese cabbage. *J. Asia Pac. Entomol.* 13: 255–260.

Yu, J.-H., H. Chi and B.-H. Chen. 2005. Life table and predation of *Lemnia biplagiata* (Coleoptera: Coccinellidae) fed on *Aphis gossypii* (Homoptera: Aphididae) with a proof on relationship among gross reproduction rate, net reproduction rate, and preadult survivorship. *Ann. Entomol. Soc. Am.* 98: 475–482.

Zang, L. S. and T. X. Liu. 2007. Intraguild interactions between an oligophagous predator, *Delphastus catalinae* (Coleoptera: Coccinellidae), and a parasitoid, *Encarsia sophia* (Hymenoptera: Aphelinidae), of *Bemisia tabaci* (Homoptera: Aleyrodidae). *Biol. Contr.* 41: 142–150.

Zeleny, J. 1969. A biological and toxicological study of *Cycloneda limbifer* Casey (Coleoptera, Coccinellidae). *Acta Entomol. Bohemoslov.* 66: 333–344.

Zhang, Z. Q. 1992. The natural enemies of *Aphis gossypii* Glover (Hom., Aphididae) in China. *J. Appl. Entomol.* 114: 251–262.

Zheng, Y., K. S. Hagen, K. M. Daane and T. E. Mittler. 1993. Influence of larval dietary supply on the food consumption, food utilization efficiency, growth and development of the lacewing *Chrysoperla carnea*. *Entomol. Exp. Appl.* 67: 1–7.

Zhu, J. and K.-C. Park. 2005. Methyl salicylate, a soybean aphid-induced plant volatile attractive to the predator *Coccinella septempunctata*. *J. Chem. Ecol.* 31: 1733–1746.

Zhu, J., A. A. Cosse, J. J. Obrycki, K. S. Boo and T. C. Baker. 1999. Olfactory reactions of the twelve-spotted beetle, *Coleomegilla maculata* and the green lacewing, *Chrysoperla*

carnea to semiochemicals released from their prey and host plant: electroantennogram and behavioral responses. *J. Chem. Ecol.* 25: 1163–1177.

Zuest, T., S. A. Haerri and C. B. Müller. 2008. Endophytic fungi decrease available resources for the aphid *Rhopalosiphum padi* and impair their ability to induce defences against predators. *Ecol. Entomol.* 33: 80–85.

Chapter 6

DIAPAUSE/ DORMANCY

Ivo Hodek

Institute of Entomology, Academy of Sciences, CZ 370 05, České Budějovice, Czech Republic

6.1 INTRODUCTION: MECHANISMS AND DEFINITIONS

Some generalities need to be explained to make the reading of this chapter easier. The adaptive functions of diapause are: (i) to **synchronize** the development of active stages with favourable conditions and (ii) to enhance the **survival** potential during unfavourable periods. The modern definitions were coined by Lees (1955). He defined **quiescence** as direct inhibition of development, caused by the direct effect of ambient conditions (low temperature, lack of humidity), which can be terminated immediately by favourable conditions. **Diapause** is caused by conditions which do not directly prevent development, but which are merely signals of seasonal changes (cues, seasonal tokens; a typical signal is daylength). Thus diapause can begin before the onset of an unfavourable environmental change. For insects Danks (1987) also used the term **dormancy**, previously employed only for vertebrates and plants, to cover all states of suppressed development (i.e. both diapause and quiescence). Additional terms were coined to emphasize specific types of diapause (for details and relevant references see Hodek 1996, Saunders 2002).

A system of terms can never encompass the striking complexity of diapause mechanisms. Moreover, the same individual has the potential for alternative ways of arrest or resumption of development; this is clearly seen in *Chilocorus bipustulatus* (6.2.13). Therefore most authors use the term 'diapause' in a broad sense for any **adaptive** arrest of development which is accompanied by behavioural, structural and biochemical changes in insects, i.e. other than quiescence. The **variation in the ecological mechanisms** which govern the induction, maintenance and termination of diapause have to be a focus of our studies. For discussion of the period sensitive to diapause induction see 6.2.3.

Similar to the antithetic definitions for quiescence and diapause, there is a contrasting definition of the two basic types of diapause. If the potential for diapause is not realized in each generation in multivoltine (i.e. producing several generations per year) species or populations, diapause is termed '**facultative**'. A good example is *Har. axyridis* (Roy & Wajnberg 2008; 6.2.9). Its onset is regulated by appropriate environmental cues, e.g. photoperiod – the **photoperiodic response**. The **critical photoperiod** is the daylength that induces 50% response, i.e. in half of the sample diapause is induced, in the other half it is not. It should be stressed that **potentially multivoltine species** may show a univoltine life cycle in regions where the period of suitable conditions is unfavourably short. By contrast, '**obligatory**' diapause is entered by virtually every individual in each generation of the so-called obligatory univoltines, **regardless of the environment** (Lees 1955).

The original definition of obligatory diapause has become outdated by experiments with changing photoperiods. Changing conditions (e.g. exposure to short days followed by long days) for insects entering 'obligatory' diapause under, for example, a constant photoperiod in the laboratory, may be a prerequisite for their development. Zaslavskii (1970) demonstrated such phenomenon in *Chil. bipustulatus* (6.2.13). Changing photoperiodic regimes have not yet become a regular experimental procedure; we can expect that many examples of diapause currently regarded as obligatory may in future be considered facultative. A study of populations from northern Europe indicates this possibility for *C. septempunctata* (Semyanov 1978a; 6.2.1.3). It is therefore advisable not to restrict oneself to the use of constant photoperiods when studying the regulation of diapause.

6.1.1 Hibernation and aestivation

For quite a long time it is mostly hibernation diapause (induced principally by short days and low temperature) that has been studied, probably because in earlier studies diapause was thought of only as an adaptation to surviving sub-zero temperatures. Later, aestivation diapause was recognized, where the inducing environmental cues are long daylength and high temperature. While hibernation diapause can be completed spontaneously without any change of conditions, aestivation diapause is often terminated by a decrease in daylength and temperature (Masaki 1980). *Coccinella septempunctata brucki* shows this type of diapause in central Honshu, Japan (6.2.2.1). Typically, aestivation diapause serves to bridge the periods of drought in subtropical regions. Until recently it was wrongly assumed that diapause is superfluous in the tropics, but many cases have been quoted by Denlinger (1986), such as that of *Stenotarsus rotundus*. Some ethological and ecological traits of diapausing

S. rotundus are similar to those of coccinellids, but also changes in moisture operate in diapause regulation (Hodek 2003 and references therein).

6.1.2 Termination/completion of diapause

Ideas concerning the end of diapause have developed substantially. The classical concept assumed that the prerequisite for the termination of diapause was an exposure to 'chilling', i.e. low temperature above zero. The earlier variation of this concept contained the assumption that the inhibition of development was 'broken' by exposure to cold, whereas in the second variation Andrewartha (1952) stressed that at low temperature, often in the range from 5 to 10°C, a gradual **diapause development** ('physiogenesis') must be completed before normal development (morphogenesis) can be resumed. Because Andrewartha's assumption that 'chilling' is a prerequisite for the completion of a 'healthy' physiogenesis was later overgeneralized, the possibility that the programmed course of events might be modified by environmental signals such as an increase in temperature, was often neglected. In fact, there are **multiple pathways to diapause completion**; at least two should be considered here. Apart from **horotelic** processes of diapause development, diapause can be completed by **tachytelic** processes where the insects are **activated** by some environmental stimuli (Hodek 1983). According to Henderson et al. (1953), horotelic means 'evolving at the standard rate', while tachytelic means 'evolving at a rate faster than the standard rate'. Such a distinction has also been made by Danks (1987, 2001) when he speaks (1987, p. 397) about internal (genetically programmed) and external (environmentally determined) mechanisms.

Under natural conditions, the prerequisites for both diapause development and activation are usually fulfilled by seasonal changes. For example, in autumn and winter the development of hibernation diapause can proceed during the period of short daylength and low temperature. Vernal activation coincides with an increase in photoperiod, light intensity and temperature, the appearance of essential food (Chapter 5.2.11) and sometimes the arrival of rains. Chilling is not a prerequisite for the completion of hibernation diapause in many species (reviewed by Hodek & Hodková 1988). **Low temperatures** are important; they (i)

conserve metabolic reserves, (ii) prevent a premature resumption of post-diapause morphogenesis and thus synchronize the life cycle and (iii) can provide a contrast to later increases in temperature, so that they are a component of the activating stimulus.

6.1.3 Phases of dormancy

In temperate zones diapause usually covers only the first periods of hibernation or aestivo-hibernation. In mid-winter or even in autumn, the potential for the resumption of development is already recovered. This **post-diapause** phase is, in fact, **a mere quiescence** because development is inhibited only environmentally (mostly by low temperature) and may be resumed in the laboratory. As stressed above by Andrewartha's term 'physiogenesis', diapause is a **dynamic state**. Insects undergo a series of changes in the course of diapause and only some of these are well known. There are conspicuous **ethological** changes, often a migratory phase in pre-diapause and end of dormancy (undertaken by coccinellids), changes in photo- and geotaxis, etc. **Physiological changes** include: (i) diapause intensification soon after its onset; (ii) decrease in diapause intensity due to the progress of horotelic processes of diapause development; (iii) consumption of food reserves accumulated during pre-diapause; (iv) dynamics of cryoprotectants (trehalose, polyols, antifreeze proteins) and ice nucleators, regulated both endogenously and by environmental changes (6.4.4); (v) a change in temperature prerequisites. The final phase of diapause and an early phase of post-diapause morphogenesis overlap.

Photoperiodic activation occurs in many insects (see e.g. Tauber et al. 1986, Danks 1987, Saunders 2002). As well as a change in temperature and/or photophase, both disturbance and injury can **terminate diapause** or at least accelerate its termination (Hodek et al. 1977).

6.1.4 Endocrinological aspects of adult diapause

Presumably exogenous stimuli influence the **neurosecretory cells of the brain** which are the prime movers of the neuroendocrine system. In the ecophysiological analysis of adult diapause in coccinellids, it is helpful to have a general idea of the underlying endocrinological pathways: (i) Environmental signals are perceived by **receptors** (photoreceptors are located in the brain or in compound eyes). (ii) Evaluation and storage of the information involves a **photoperiodic clock**; the exact nature of these processes is not yet fully understood. A recent survey is given in Saunders (2002). (iii) The stored information is transmitted to neuroendocrine organs; in adults these are neurosecretory cells of the brain, the **corpora cardiaca** (CC) and **corpora allata** (CA). The neurosecretory material is released from the CC where the axons of the neurosecretory cells terminate. The CC also contain glandular cells which produce their own hormones. The CA produce juvenile hormones (JHs), whereas the ovaries and possibly other tissues (testes, oenocytes, epidermis) produce ecdysteroids in adults. (iv) The neuroendocrine system controls the expression of diapause at the level of **target tissues** (ovaries, fat body, etc.). The most conspicuous feature of adult diapause is the **suppression of reproductive functions** (maturation of ovaries and probably male accessory glands and mating activity; 6.4.1.3). Metabolic reserves (glycogen, lipids) and cryoprotectants (polyols, trehalose, hysteresis proteins) accumulate and the metabolic rate is reduced. Specific diapause proteins are synthesized instead of vitellogenins (6.4.1.2, 6.4.3). For a survey of hormonal control of diapause (including the associated changes in gene expression) see Denlinger et al. (2005).

6.2 ECOPHYSIOLOGICAL REGULATION OF DIAPAUSE IN COCCINELLIDS

In the early studies, hypotheses were based on observations. While Dobrzhanskii (1922a, b; 6.2.1) (better known under his later spelling Th. Dobrzhansky) underrated environmental cues (at least in *C. septempunctata*), other authors in the years 1922–48 acknowledged the involvement in diapause regulation of temperature, humidity and shortage of prey (for details see Hodek 1996).

6.2.1 *Coccinella septempunctata*

Regulation of diapause in this species is complicated, at least in the European populations. In two early studies of populations from the region of Kiev, **Ukraine** Dobrzhanskii (1922a, b) described a **bivoltine** developmental cycle for *C. septempunctata* and assumed a genetically fixed alternation of a generation with unbroken development and a generation entering diapause (details on p. 287 in Hodek 1996). Since then experimental research has indicated a very wide plasticity in *C. septempunctata* (Obrycki & Tauber 1981, Phoofolo & Obrycki 2000, Hodek & Michaud 2008) enabling heterogeneous activity in different fractions of populations (6.2.1.6). The two short studies by Dobrzhanskii (1922a, b) are important, because for the first time imaginal diapause in Coccinellidae was recognized together with its adaptive significance for survival, i.e. enabling the species to withstand a long period of shortage of aphids.

6.2.1.1   Central Europe

Jöhnssen (1930) reported a univoltine cycle for *C. septempunctata* in **Germany**. However, he admitted the possibility of a second generation occurring under favourable conditions, though it would later die out in the egg or larval stage.

In **Bohemia** (50° N, western Czech Republic), the **population in the autumn** was found to consist of **two fractions.** Although in some years aggregations of both sexes of dormant *C. septempunctata* may be found in their hibernation quarters from early August onwards, one can also find actively feeding coccinellids on vegetation with aphids (often on different weeds, such as *Carduus* spp. and Daucaceae) for the whole of September and in early October (Hodek 1962). This agrees with observations by Telenga (1948), who also used to find *C. septempunctata* in the Ukraine both partly dormant and partly feeding at that same season.

The **physiological condition** of these two fractions from Bohemia was determined by dissection immediately after sampling, and after rearing (Hodek 1962; Table 6.1). Whereas the alimentary canal in the sampled **dormant beetles** was empty of food and there were no traces of vitellinization in the ovaries, the digestive tract was full of food in more than half of the sampled **active adults** and 13–20% of females possessed one or more vitellinized oocytes or even eggs. The difference between the dormant and active parts of the population became more striking when beetles from **both fractions** were **reared** for 3 weeks under long-days, at 19–22.5°C and with plentiful essential aphid food. The ovarioles of about 85–90% of dormant

Table 6.1 Difference between two population fractions of autumnal adults of *Coccinella septempunctata* (Hodek 1962).

| Date sampled[*,†] | Condition | n | Dissection of sampled or reared adults | Digestive tract | | Fat body | | | Ovarioles with | |
				empty	full of food	+	++	+++	germaria only	at least one vitellinized oocyte
18.9.[*]	active	30	sampled	4	26	3	8	19	26	4
27.9.[†]	active	20	sampled	9	11	5	3	12	16	4
18.9.[*]	active	18	reared[‡]	—	—	—	—	—	2	16
8.8., 6.9.[*]	dormant	17	sampled	17	0	2	5	10	17	0
8.8.[*]	dormant	29	reared[‡]	—	—	—	—	—	26	3
	dormant	25	reared[‡]	—	—	—	—	—	21	4

[*]Collected in N. Bohemia, near Louny.
[†]Collected in S.E. Slovakia, near Kral. Chlumec.
[‡]3 weeks, long days, 19–22.5°C, surplus of essential aphids.

females remained without any vitellinization, while about 90% of the females collected on plants possessed vitellinized oocytes after this period of rearing (Hodek 1962; Table 6.1). Dissections in **summer** (mid-July) of females collected **outdoors** in central Bohemia (50° N) a fortnight after adult emergence indicated a strong **tendency to univoltinism**: 84–93% of the females entered diapause.

Preliminary experiments

The incidence of diapause was high (65–80%) when small groups of coccinellids were **reared indoors** under conditions approaching those in the field (except for extremes) and with the natural photoperiod of late June and early July. In addition to these experimental small groups dissected, two simultaneous massive cultures of 200 adults each were reared to check the reproduction. A very low oviposition was observed in spite of plenty of essential aphid food and heating with a lamp during cool days. Most adults aggregated in corners of the cages from mid-July, leaving only 5–10 beetles moving about, even though the aggregations were disturbed every day. Only one or two egg batches were obtained per day, and in late October about 80% of females were still alive. Adults from the second generation, reared from the first egg batches, emerged from pupae from mid-August onwards; they remained very active and feeding for about three weeks, but then aggregated after four weeks (Hodek 1962).

The above experiments demonstrated that *C. septempunctata* adults feed before they enter diapause (in contrast to Yakhontov's 1962 assumption for *Cer. undecimnotata*) and that the onset of diapause is not prevented by surplus essential food (in contrast to Hagen's 1962 report on *Hip. convergens*). The data showed the plasticity of the *C. septempunctata* life cycle and a weak effect of daylength on the prevention of diapause, since in some replicates the beetles entered diapause in spite of having emerged from pupae under the longest possible daylength. The slight difference between the outdoor and indoor results suggested the possibility that temperature extremes play some part in inducing diapause.

Attempts at diapause prevention in Slovakia (48° N)

A premature development of the first generation (the adults emerged from pupae as early as mid-May 1968) was employed to assess the **potential for bivoltinism** in the population from a warm region in southwestern Slovakia (Zohor near Bratislava). Almost natural conditions were used: extremes were excluded by the transfer of beetles on cold days to the greenhouse, or into the laboratory at a constant 25°C, into shadow during strong sunshine and indoors for the night. The proportion of diapausing females (i.e. ovarioles only consisting of germaria without any trace of vitellinization) ranged in the replicates between 66 and 90% (average = 79%, n = 184) (Hodek 1973). In spite of particularly favourable conditions (a warm spring, a warm region) the tendency to a univoltine cycle prevailed in a high percentage of *C. septempunctata* females. The last beetles were dissected on the summer solstice, 21 June: enterring diapause could therefore not be affected by the shortening of long days.

Laboratory diapause prevention and selection

A series of experiments were carried out under controlled photoperiod and temperature conditions (Hodek & Cerkasov 1960, 1961, Hodek 1962, for details see Hodek 1996). Cultures were started in five successive seasons (1956–60) with beetles collected from the field after hibernation. The eggs originated from non-diapausing females and thus there was **selection against the tendency to diapause** and a gradual decrease in diapause incidence was achieved. In spite of favourable conditions (i.e. long-day conditions of 16L:8D or 18L:6D, a constant 25°C and an excess of essential aphids), the **incidence of diapause** remained rather high in the first generations, and usually **fluctuated between 60 and 90%**. These fluctuations were apparently related to varying conditions: e.g. the number and origin of specimens starting the culture, and the age of the females founding the progeny. In the subsequent generations a progressive decrease was usually obtained within the first three

generations, as in 1956 or in 1958. A steady excess of essential prey, cleanliness and a population density not exceeding 25 pairs in a cage of about 8l is needed to achieve a very low incidence or even absence of diapause in later generations. Due to long laboratory breeding under constant conditions and inbreeding, the vitality of the beetles decreased, so that in the sixth generation the oviposition period and longevity had decreased to 8 and 45 days respectively (Hodek & Cerkasov 1960, 1961, Hodek 1962). A decrease in diapause incidence down the sequence of generations indicates selection against a tendency for obligatory diapause under long days. Such results are not exceptional in the literature on diapause (Hodek & Honěk 1970). It is assumed that multiple genes control the tendencies for diapause (Tauber et al. 1986; Danks 1987).

Induction of diapause in selected lines

When the tendency to the 'obligatory' entry into diapause was selected out, the way was open to study the effect of **environmental factors** on diapause induction (Hodek & Cerkasov 1960, 1961). This was done gradually, with coccinellids obtained from different generations and cultures (Table 6.2). Up to the time that the larvae were transferred to the experimental conditions, they had been reared under the usual

Table 6.2 The incidence of *Coccinella septempunctata* females entering diapause under laboratory conditions (Hodek & Cerkasov 1961).*

Culture/ generation	Age of larvae at experiment onset a (days)	Temperature (°C) (at dark/light phase)	Photophase (h/24 h)	Age at dissection (days)	*n*	Diapausing females (%)
I/7	3	17–18 (D) / 20–22 (L)	12	42–51	15	94
I/7	pupae	17–18 (D) / 20–22 (L)	12	56–62	15	87
I/9	5–8	17–18 (D) / 20–21 (L)	12	24–41	30	87
III/4,5	2–3	18 ± 0.5	12	26–43	20	85
III/7	6–7	18.5 ± 0.7	12	20–37	38	50
I/9	5–7	22 ± 0.5	8	33–35	10	70
II/4	4–5	22 ± 1.0	12	24–34	42	60
I/9	3–7	25 ± 0.5	12	30–33	40	33
I/6	eggs	24–25 (D) / 27–28 (L)	8	24–26	20	10
III/4, 5	2–3	18 ± 0.5	19	26–43	24	13
III/7	6–7	18.5 ± 0.7	19	20–37	46	4

*Before the transfer to experimental conditions the insects were reared at 25°C, 16 or 18 hours photophase; both before and during the experiment an excess of essential food was supplied.

breeding conditions described above. Although the age of the larvae differed in individual experiments, the results are consistent, and separated out the effect of two environmental variables, photoperiod and temperature. The **importance of photoperiod** was **dominant**: diapause was prevented by long-day conditions (19L:5D) even at low temperatures of 18 or 18.5°C in, respectively, 87 and 96% of females. The response of diapause to short-day conditions was considerably **modified by temperature** (Table 6.2). At lower temperatures around 18°C the incidence of diapause reached 85 to 94% (with one unexplained exception of only 50%). An increase in temperature to 22 and 25°C led to a marked decrease in the incidence of diapause, and at temperatures fluctuating between 24–25°C (night) and 27–28°C (day) it amounted to just 10%. No significant differences were found in the incidence of diapause between the lengths of short photophase of 12L or 8L, or between the exposure of younger or older larvae or of pupae to the experimental conditions. The critical photoperiod was not ascertained; >16L:8D prevents diapause.

The central European population of *C. septempunctata* is evidently heterogeneous, but with a prevalence of univoltines (Hodek 1962). As this species is reported as univoltine in northern Europe (6.2.1.3) and obligatory (Dobrzhanskii 1922a, b) or facultatively bivoltine in the Ukraine (Dyadechko 1954), it seems probable that central Europe lies in a **transition zone** (in the sense of Bodenheimer & Vermes 1957) **between** the distribution areas of **uni- and multivoltine populations** (Table 6.3; see also 6.2.1.6 and 6.2.16).

The above studies of diapause regulation in *C. septempunctata* from central Europe did not include experiments with variable photoperiods. However, it has been assumed (Hodek 1973) that obligatory entry into diapause may result when the sequence of long day after short day conditions is not experienced under effective temperatures. Such a sequence may be a prerequisite for maturation and oviposition. In the studies described above, a gradual negative selection for this requirement appears to have been executed. Such an experimental response to the increase in daylength after a period of short days was then reported in populations from northern Europe (Semyanov 1978b; Zaslavsky & Semyanov 1983; 6.2.1.3) and could also operate in central Europe.

Diapause development/photoperiodic activation

The course of diapause development was followed first by **preliminary checks.** While in the August samples only 10–20% (n = 72) of females were activated by transfer to long days and a surplus of suitable aphid prey, in November it was 78–100% (n = 40) (Hodek 1962). To analyze approximately the role of temperature in diapause development, the August samples were also exposed for 3 or 6 weeks to 0, 5 or 12°C before transfer to the above re-activating conditions. The 6 week exposure resulted in 27–56% more reproducing females than the 3 week exposure, and there was no significant difference between the effect of 5 versus 12°C. The temperature of 0°C was the least effective (Hodek 1970; also p. 298 in Hodek 1996).

Table 6.3 Hypothetical condition of *Coccinella septempunctata* populations in early autumn (Hodek 1962).

Activity	Ovaria	Voltinism	Type of diapause	Generation	Origin
Dormant at hibernation quarters	Without vitellinization	univoltine	'obligatory'	1st	(early batches)
Active (feeding) on plants	Without vitellinization	univoltine	'obligatory'	1st	(later batches or slower development)
	(primarily or retrogressed reproductive	multivoltine	facultative	1st or 2nd	(later batches or slower development)
		multivoltine	facultative	0	(still surviving overwintered adults)
				1st	(early batches)

Table 6.4 Response to photoperiod in *Coccinella septempunctata* females transferred from a dormancy site to laboratory in October (Hodek & Ruzicka 1979).*

| Photoperiod | n | Normal oviposition | | | | | Transient oviposition | | | No oviposition |
| | | after short pre-OP | | after long pre-OP | | | | | | |
		%	duration of pre-OP (days)	%	duration of pre-OP (days)	%	duration of pre-OP (days)	duration of OP (days)		%
18L:6D	31	74.1	13.6 (9–26)	6.5	38, 52	0[†]				19.4
12L:12D	32	3	13	25	67.5 (56–72)	9.5	13, 15, 18			62.5

*The experiment was discontinued the 73rd day; the total duration of oviposition (OP) and post-oviposition was not followed.
[†]Till the 26th day of oviposition, when the females were transferred to short day.

These preliminary experiments established (i) that in **August**, soon after arrival at dormancy sites, the females are refractory to photoperiodic activation; (ii) the decrease in diapause intensity with time, both in the open between August and **November** and in the laboratory; (iii) the temperature of 12°C as suitable for diapause development.

The progress of diapause development was later investigated in more detail by **comparative transfers** to laboratory short- and long-day regimes of October and May samples from hibernation sites. In **October**, when diapause development was not yet completed, the photoperiodic response still controlled the reproductive activity of females from Bohemia, and many more females reproduced under long day (81%) than under short day conditions (36%). This experiment revealed the polyphenic composition of the population: at a short daylength only 3% of females oviposited normally, the other females either showed a much delayed oviposition (mean pre-oviposition period of 68 days) or transient oviposition lasting less than 3 weeks (Table 6.4). Diapause was completed in late winter/early spring; then the maturation of ovaries was prevented only by low ambient temperature. By **May** some maturation occurred due to the increase in temperature in the field and the pre-oviposition period after transfer was therefore very short at both photoperiods (Hodek & Ruzicka 1979).

In general, the above two experimental results from Bohemia are consistent with those from the Paris region (Bonnemaison 1964) and from southeastern France (Hodek et al. 1977; 6.2.1.2) and with findings

on other insects (Hodek 1983): (i) diapause development proceeds with time, i.e. the **intensity of diapause** gradually diminishes in the course of late summer, autumn and early winter, and is **completed** in almost the entire population **by December**; (ii) processes of diapause development that end diapause can proceed at 12, 15 and even 25°C, and thus low temperatures near 0°C are not a prerequisite for diapause completion; (iii) the photoperiodic response disappears with the termination of diapause.

6.2.1.2   Western Europe: France

Diapause induction

Although the climate around **Paris** (Ile-de-France, c. 49°N) is different from that of central Europe, the results from a study of diapause in *C. septempunctata* in this region (Bonnemaison 1964) are **similar to** those from **Bohemia** (6.2.1.1). A large proportion (85–95%) of the first generation entered diapause despite favourable conditions (long day of 16L:8D or 18L:6D, 22°C, excess of aphids), under which selection proceeded from 85% diapausing individuals by steps to 40, 20, 5 and 5% in the first five generations to a culture consisting solely of non-diapausing coccinellids by the sixth generation. In the second generation the incidence of diapause was increased by short days or shortage of food. Bonnemaison accepted the hypothesis of a mixture of uni- and multivoltine phenotypes and assumed that immigrants are brought by air movements and human transport. He did not think the development of a second generation in the Paris area possible.

Diapause in males

As explained in 6.4.1.3, diapause in coccinellids is not connected with inactivity of the tissues of the testicular follicles. Because of difficulties of method, diapause regulation has only rarely been studied in coccinellid males (Hodek & Landa 1971, Ceryngier et al. 1992, 2004) and accessory glands were not included in the analysis. From knowledge of other insects, one might expect diapause to result in inhibition of the accessory glands and the copulatory aptitude of males. When Bonnemaison (1964) reared emerged males of the fourth selected generation of *C. septempunctata* under 12L:12D and 14 or 18°C for 15 or 25 days, he could not find these symptoms of diapause. After transfer to favourable conditions where females were being reared (16L:8D, 20°C), the males copulated after 2–7 days, and the females laid apparently (not specified) viable eggs after 3–23 days.

Diapause development

Near Paris, Bonnemaison (1964) found that diapause lasted 3–6 months in *C. septempunctata* females which had emerged from pupae in early August and were kept at 20–22°C and under a natural photoperiod. In contrast to the Czech results, he failed to activate the diapausing females (probably collected too early between late July and early September) by an exposure to 5 or 8°C for 5, 9 or 13 weeks and by subsequent rearing at 20°C and 16L:8D for 15, 11 or 7 weeks. His findings on the onset of previtellogenesis in females in the field from late September onwards indicate, very similarly to the results from Bohemia (6.2.1.1) some kind of lifting of the diapause inhibition in autumn. In contrast to central Europe, however, the apparently **higher temperature** of the Paris region enabled the **first stage of maturation** to proceed **in the field** and to be recorded by the dissection of field samples, whereas in central Europe the potential for maturation could only be revealed by laboratory rearing (Table 6.1).

As pointed out above, the French population does not differ from the general picture. Diapause was completed **in December** when the **photoperiodic response** was **lost** in almost all adults. This was indicated by almost identical incidence of reproduction

and duration of the pre-oviposition period at 18L:6D and 12L:12D photoperiod regimes.

In **southeastern France** (Basses Alpes, c. 44°N), *C. septempunctata* has been reported as univoltine or partly bivoltine (Iperti 1966a). Later, two observations on diapause development were made on this population: (i) a comparison of the coccinellids from a hill (about 600 m) in the plain with those from a high mountain (Cousson, 1512 m), both close to the town of Digne, and (ii) the activating effect of injury. There is a strong indication that the intensity of diapause in early December was greater at the **high altitude** (medians of oviposition delay were 27 days at 12L:12D and 17 days at 18L:6D) than at the **lower altitude** (13 and 10 days, respectively; Hodek et al. 1977). The reason for this difference may be the high incidence of 'obligatory' univoltines in the high altitude dormancy sites, which are usually occupied earlier and by larger individuals (Honěk 1989; 6.3.2.3).

Half of the sample from Digne were **injured** by cutting off the second pair of wings. This treatment shortened the mean pre-oviposition period in the beetles from both altitudes (Hodek et al. 1977). Wounding has been reported in several insect species as a stimulus that decreases diapause intensity (Hodek 1983). It may act directly, through metabolic changes, or indirectly via sensory pathways affecting the neuroendocrine system. A similarly activating effect of parasitization on diapause was observed in another study (Ceryngier et al. 2004).

6.2.1.3 Northern Europe

From **northern Europe,** most authors have reported a univoltine cycle for *C. septempunctata* (Banks 1954 from **England**; Sundby 1968 from **Norway**; Semyanov 1978a from **northern Russia**).

However, in a sample from Helsinki (**Finland**), surprisingly all beetles reproduced in the first generation when reared under constant long day conditions, i.e. they responded like potential multivoltines (Hämäläinen & Markkula 1972). Such a response, never found in central or western Europe, is unexpected so far north and is different from the populations from **northern Russia** (Semyanov 1978a).

Semyanov activated beetles, sampled at the beginning of dormancy, by exposing them to 18L:6D at 25°C for at least 30 days and he considered the reproducing individuals as multivoltines. In fact, activation by photoperiod was involved (similar to that of the Czech *C. septempunctata*; 6.2.1.1). Activation was achieved in 33% in the population from the Khibiny Mountains and 55% in the St. Petersburg population. This may be compared with 81% activated females in the October sample in the population from Bohemia.

Zaslavsky and Semyanov (1983) reported that the 'obligatory' univoltines need a short-day sensitization before exposure to long days in order to reproduce. Semyanov (1978a) executed such experiments with a population (from the Khibiny Mountains on the Kola peninsula) having the **critical photoperiod** (term explained in 6.1) of 17L:7D at 25°C. He used 10L:14D at 20°C or 14L:10D at 25°C for the 'short-day sensitization' and then 20L:4D at 20°C or 18L:6D at 25°C as the long day regime, and achieved reproduction in all individuals. The principal treatises on diapause (Tauber et al. 1986; Danks 1987; Saunders 2002) refer to a short day/long day requirement for diapause prevention. The surprising results mentioned earlier from Finland with the *C. septempunctata* population from the Helsinki region (Hämäläinen & Markkula 1972) could have been produced by a similar sequence of conditions, and the authors may have omitted the information that young adults (or larvae) were kept under short days before exposure to long days.

In experiments on populations from the Pskov and Novgorod regions of Russia, diapause of *C. septempunctata* was intensified by the combined action of short days and the absence of aphid prey (Zaslavsky & Vaghina 1996).

6.2.1.4   Mediterranean region

On the coastal plain of **Israel**, Bodenheimer (1943) recorded reproduction in late September after aestivation diapause that was not induced by a lack of aphids. There *C. septempunctata* develops one complete and one partial generation in spring which is repeated in autumn. Hibernation of *C. septempunctata* is apparently a quiescence rather than diapause in this region.

In **northern Greece** near **Thessaloniki** (about 41°N), the coccinellids aggregate in autumn on mountain tops, although the absence of a photoperiodic response recorded in the sample collected in mid-November (Hodek et al. 1989) would enable them to reproduce also in short days. It cannot be excluded that the populations from Thessaloniki belong to a Mediterranean biotype of *C. septempunctata*, less dependent on

photoperiod in the regulation of its life cycle. Preliminary results obtained with the population from southern Spain also indicate this tendency (Hodek & Okuda 1993).

Although no similar aggregation of *C. septempunctata* adults was found in autumn in the **Athens** region, coccinellids resting individually on dry plants or small conifers were sampled on the slopes of hills. The beetles were apparently on their way to hibernation sites. This autumnal behaviour was taken as an indication that winter dormancy also occurs in central Greece, at least in a proportion of the population (Hodek et al. 1989).

This assumption was confirmed by two detailed field and laboratory studies on populations from **central Greece** (Katsoyannos et al. 1997a, b), although some aspects, e.g. natural voltinism and the probability of a second reproduction period in autumn, were not completely resolved. Katsoyannos's (1997b) statement that *C. septempunctata* is a multivoltine species was based on laboratory rearings; in shaded outdoor cages in the Institute's yard, four generations were achieved (Fig. 6.1a, b). While there was evidently a natural photoperiod, the other conditions (continuous feeding on essential prey and protection from direct insolation) were far from what happens in the field. Eggs, larvae and pupae were abundant in the plain only between April and June, and absent in winter. In June, most adults migrated to **mountains**, where they were recorded the whole year except in May, but with immature ovaries and empty guts (Katsoyannos et al. 1997b). That aestivation diapause continues in the open as hibernation quiescence was shown by reproduction after a short pre-oviposition period in late autumn and winter samples transferred from dormancy sites to the laboratory (Fig. 6.2). Field observations therefore indicate a univoltine life cycle in central Greece, but the tendency to obligate univoltinism is probably present in only a low proportion of the population. This is shown by the high incidence of oviposition in the first generation in the vial cultures (Fig. 6.1b). Beetles in outdoor cages spent the winter in dormancy and resumed egg laying in spring (Katsoyannos et al. 1997a, Fig. 6.1a, b).

Iperti (1966a) recognizes *C. septempunctata* as univoltine or partly bivoltine in the **French Riviera**. The first generation only oviposits exceptionally before diapause.

In the mountainous region of Sweida, in **southern Syria** (1400–1500 m) *C. septempunctata* are actively

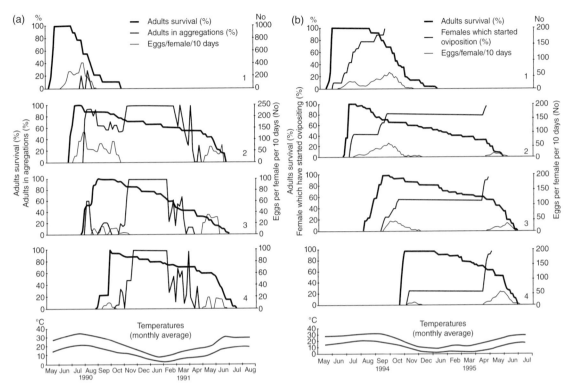

Figure 6.1 Subsequent cultures (1–4) of *Coccinella septempunctata* adults reared outdoors under temperatures (monthly averages of minimum and maximum) indicated below; Kifissia, Athens (from Katsoyannos et al. 1997a). (a) Ladybirds were reared in outdoor cages in 1990–91; (b) Outdoor rearings in vials 1994–95.

preying on *Eriosoma lanigerum* in April and May. The authors recorded aestivation quiescence from late June to early September and no reproduction before the winter diapause (Almatni et al. 2002). They agree with Bodenheimer's (1943, 1957) interpretation, although it is in fact different (see above).

In the region of **Ankara** (**Turkey**), Bodenheimer (1943) did not record much difference from the central-European pattern of aestivo-hibernation terminating in late May and June. Although Bodenheimer (1943) found active adults of *C. septempunctata* in late March and in September, he maintained that there is only one annual generation in the Ankara region. From similar observations in central Europe, the possibility of a partial second generation has been based on dissections and experiments (6.2.1.1; Table 6.1).

Long-term phenological observations also in Italy and Spain would clarify the cycle of *C. septempunctata* in this region. Oportunistic traits may be involved such

as those described e.g. in coccinellids from eastern Australia (6.2.10, 6.2.11).

6.2.1.5 Nearctic region

The establishment of *C. septempunctata* in the Nearctic region was first recorded in New Jersey during the years 1973–74, and the species was reported as almost entirely univoltine, with only occasional females producing a second generation (Angalet et al. 1979). Obrycki and Tauber (1981) also found a univoltine cycle in New York State. In a comparison between four populations, Phoofolo and Obrycki (2000) recorded, however, that 47 and 70% of ladybirds from Iowa and Delaware (USA), respectively, laid their first eggs within the first two weeks. These figures agree with those obtained over 5 years in the first generations of Czech populations (6.2.1.1).

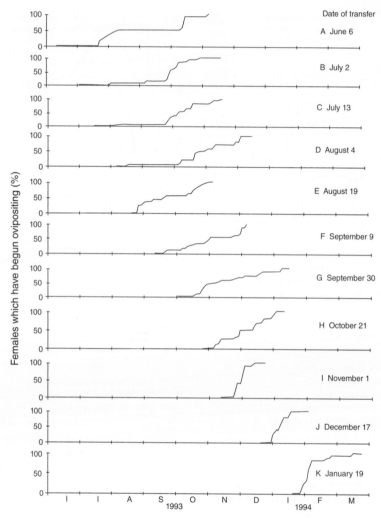

Figure 6.2 Pre-oviposition periods of *Coccinella septempunctata* females collected in aestivo-hibernation from Mt. Kitheron (1993–94) and transferred at different dates (a–k) to 25°C, 16L:8D and presence of aphids in the laboratory (from Katsoyannos et al. 1997b).

6.2.1.6   Potential multivoltines in univoltine populations

It seems remarkable that more than 85 years after the classic Dobrzhanskii (1922) papers, we cannot yet fully answer the question: how is the voltinism of *C. septempunctata* regulated? (See later that this is not different in other species, such as *Hip. convergens*, 6.2.7.) While in temperate climates of Europe and USA we see the

scenario of a rare and partial development of a second generation, a similar situation often obtains in the Mediterranean climate, with the exception of the reports from the coastal plain of Israel (Bodenheimer 1943) and an isolated observation from southern Spain (Hodek & Okuda 1993). A bivoltine life cycle with two reproduction periods in spring and autumn, isolated by two dormancies, has also been found in central Japan in *C. s. brucki*, but there it is based on reverse photoperiodic responses (6.2.2.1). The low level of understanding of diapause regulation in *C. septempunctata* is caused by the great plasticity of this species, evolved due to a constant instability of the occurrence of aphid prey (Chapter 5) and evidently maintained by the vagility that prevents the formation of isolated gene pools in individual regions.

To simplify, we may say that researchers bring to the laboratory samples of a bimodal population, composed of 'obligatory' univoltines and potential multivoltines. The proportion of these two fractions of the population varies between localities and most probably also between years. We cannot be sure that our 'favourable' experimental conditions enable the maximum expression of the potential multivoltinism. Some of the conditions used (temperature, prey, population density, relative humidity) may decrease the incidence of reproducing females. Thus, we can only state what proportion of our sample produces a second generation under our experimental conditions. We do not, however, know the maximum proportion of potential multivoltines in a sample of a given population. Anyway, the potential for multivoltinism is only rarely realized in the field.

Phoofolo and Obrycki (2000) carried out a detailed comparison of multivoltine tendencies in four geographically distant populations. Perhaps a two-stage response (Danks 1987) underlies the adult development of *C. septempunctata*. This phenomenon may be similar to the northern population of *Chil. bipustulatus* (Zaslavskii 1970). Therefore the possibility of an 'artificial' multivoltine cycle may be revealed in the laboratory under a short day/long day photoperiodic programme.

6.2.2 *Coccinella septempunctata brucki*

6.2.2.1 Central Japan (central Honshu)

The life cycle of *C. s. brucki* in central Japan is quite the reverse of that of *C. s. septempunctata* as described from Europe and North America. Regulation of diapause was first studied in the regions of Nagoya and Tokyo in central Honshu. The conditions of the Nagoya plain are suitable for reproduction and development of larvae in spring (from mid-April to late June) and again in autumn (in September and October). These periods of active life alternate with periods of developmental arrest, a deep aestivation (summer diapause) and a

weak hibernation which is a mere quiescence ((Sakurai et al. 1981a; 6.1). In spite of the natural photoperiod in winter, hibernating adults were quickly activated by a simple increase in temperature to 25°C (Sakurai et al. 1981a, 1982, 1983, 1986, 1987a, b, Okuda & Chinzei 1988).

Diapause induction/prevention

Preliminary experiments indicated that **short days** and low temperature **prevent diapause** in beetles from the Nagoya region (Hirano et al. 1982). A similar but much weaker response was reported in the Tokyo population (Niijima & Kawashita 1982). When the Nagoya coccinellids were reared under short days (11L:13D or 10L:14D) and a low temperature (18°C), the beetles showed a high respiration rate and females oviposited. High temperature (25°C) weakened the diapause-preventing effect of short days: the females oviposited for only 10 days and their respiration rate was low (Sakurai et al. 1987a).

Short days prevented diapause, but not in the whole sample of the Nagoya population. The experimental photoperiod and temperature, which roughly simulated autumn conditions in that region (13.5L:10.5D, 18°C), stimulated reproduction in 15 females (62%) within 36 days, while only one female (5%) oviposited within the 50 days of rearing under control conditions of long days (18L:6D) and high temperature (25°C). Another sample responded similarly to a still shorter photophase (12L:12D) combined with high temperature (25°C): 79% (n = 29) females oviposited within 30 days while in the control (18L:6D, 25°C) only 36% females oviposited (Okuda & Hodek 1983, Hodek et al. 1984).

Populations of *C. s. septempunctata* are very sensitive to shortage or lower quality of food and respond by increasing the incidence of diapause (6.2.1.1 and Chapter 5). This phenomenon was also observed in *C. s. brucki* (Niijima & Kawashita 1982, Kawauchi 1985) and might have partially affected the above-mentioned variation in incidence of ovipositing females. Anyway, the same tendency was recorded in all quoted experiments: In central Japan, *C. s. brucki* is a short-day insect and the induction of diapause of the populations from central Honshu is regulated by the reverse photoperiodic response of that found in *C. s. septempunctata* (which is a long-day insect; 6.2.1).

The bivoltine cycle of *C. s. brucki* in central Honshu may be modified if the pupae (and/or eggs) develop in directly insolated artificial microhabitats, such as metal cans or paper and wooden material on slopes exposed to solar radiation. While most beetles of the population hibernate in quiescence, reproductive adults can therefore occur at the same time (Ohashi et al. 2005).

Diapause development and termination

A short-day photoperiodic response also operates in the development and termination of diapause. Adults collected in September in aestivation sites in the Nagoya region were **activated** remarkably well **by autumn-like conditions** (13.5L:10.5D, 18°C) with 36% of the females ovipositing within just a fortnight and 77% within 80 days ($n = 13$). Under long days and high temperature (18L:6D, 25°C), only 18 and 29% of control females oviposited within 14 and 80 days, respectively ($n = 17$) (Hodek et al. 1984).

In contrast to hibernation, aestivation diapause is often terminated by environmental factors (Masaki 1980), i.e. by tachytelic processes (*sensu* Hodek 1981, 1983; 6.1). The aestivation of *C. s. brucki* is thus terminated in the usual way. This laboratory finding needs to be verified by observation in the field; the laboratory results may reflect only one of the possible multiple ways of diapause termination and another pathway may be operating in the natural situation. In general, hibernation diapause can be terminated by photoperiodic activation in the laboratory, but usually this is not the case in nature where diapause ends spontaneously around the winter solstice (Hodek 1971b; Tauber & Tauber 1976).

Life cycle in central Honshu

In the Nagoya plain, the progeny of hibernated females can develop from mid-April to late June and is thus exposed to an increase in daylength from 14 to 16 hours and in average temperature from 15 to 25°C. These conditions ought to induce aestivation diapause in a large proportion of adults. We recorded a high incidence (64 and 95%) of diapause under a constant, but longer daylength of 18 hours. The resumption of reproductive activity in September coincides with shortening daylengths from 14 to 13 hours and with a decrease in mean temperature from 25 to 20°C. Our laboratory conditions, which simulated the autumn in Nagoya, stimulated the resumption of oviposition. Thus *C. s. brucki* from central Japan shows environ-

mental regulation of the life cycle which is the reverse of that of *C. s. septempunctata*. All results indicate that the photoperiodic and temperature responses are directed towards the induction of aestivation diapause.

The reversal of the photoperiodic response might be the consequence of **divergence in allopatric populations** during the process of geographic differentiation which has reached the subspecies stage (Mayr 1970). If the distribution of *C. s. brucki* is limited to Japan, the subspecies probably represents a peripheral isolate (Mayr 1970) whose **first arrivals in the south** of the Japanese archipelago may have been multivoltine individuals predominantly lacking the photoperiodic response. The short-day type of photoperiodic response was then acquired under selective pressure of the climate with hot summers and mild winters. Of course, this speculation relates to just one of several possible evolutionary pathways. An analysis of the problem should begin with the study of photoperiodic responses of *C. s. brucki* (or eventually other subspecies) populations not only from Japan but also from adjacent regions, particularly from southern Korea and Ryukyu.

6.2.2.2 Sapporo, Hokkaido (Japan)

In a population from Sapporo, long-day photoperiod combined with high temperature averted diapause (Okuda & Hodek 1994). This was evidenced both by the relatively fast activation of the entire autumnal sample and by the incidence of reproductive females in the F1 progeny. In long days of 16L:8D and 25°C, all females collected in early September started oviposition within 32 days. Short days of 12:12D combined with 20°C, however, inhibited the reproductive activity of most females from the sample ($n = 34$), so that only three started ovipositing after 10 weeks. In the F1 generation, diapause was averted in 37 and 63% of the females when a long-day photoperiod of 16L:8D was combined with 25 or 30°C respectively. In short days of 12L:12D and 20°C, the beetles of the F1 generation aggregated after a short period of feeding and did not begin to oviposit, at least until they were 68 days old, when the experiment was discontinued (Okuda & Hodek 1994).

In contrast to the populations from central Honshu, these results strongly indicate that the **Sapporo population** of *C. s. brucki* has the **long-day photoperiodic response**. In this respect *C. s. brucki* from Hokkaido is similar to *C. s. septempunctata* from **central Europe** (Hodek & Cerkasov 1961, Hodek & Ruzicka 1979) and some other European populations (Bonnemaison 1964, Hämäläinen & Markkula 1972, Hodek et al. 1977, Semyanov 1978b). This similarity in photoperiodic response is appropriate to ensure a similar life cycle, i.e. to induce winter diapause in two climatically similar regions with long harsh winters.

6.2.2.3 Northern Honshu (Japan)

An intriguing question still remains to be solved: what happens in northern Honshu in the transition zone between the two above populations. A group of Japanese researchers has recently tackled this problem by sampling and dissecting *C. s. brucki* samples in Hokkaido and several regions of Honshu (Ohashi et al. 2003). The absence of summer diapause in Hokkaido and its presence in the plains of central Honshu was corroborated.

At **higher altitudes of central Honshu**, and in **northern Honshu** that is an intermediate area between regions of short-day and long-day populations, the frequency of diapause expression varied greatly among samples from different localities and years. Moreover, diapausing and non-diapausing adults co-existed in 31% of summer samples. The authors rightly concluded that there is genetic variation in diapause tendency within the local populations. They suggest that diapause is prevented there if the average daily mean air temperature in July is lower than 21.5°C (Ohashi et al. 2003).

6.2.3 *Coccinella novemnotata*

The **bivoltine** *C. novemnotata* (McMullen 1967a, b) undergoes diapause twice during the annual cycle. The adults of the **spring generation** pass the hot, dry summer months in diapause and lay eggs in the early autumn. The adults of the **autumn generation** pass the winter in diapause and reproduce in the early spring. In the spring, the teneral adults are subjected to a photophase which increases from 17.5 to 18 hours and in the autumn decreases from about 14 to 13 hours (Fig. 6.3). Temperature and the amount of prey are also involved: Particularly low temperature contributes to the induction of diapause in the autumn generation. The intermediate photophase of 16 hours consistently determines non-diapause development in

90–100% of coccinellids, even when feeding rates are reduced to one-quarter and the temperature ranges between 15.5 and 32°C.

One of the studies (McMullen 1967b) analyses the **stage sensitive to diapause induction** (Fig. 6.4). The conclusion that it is the young adult aged 1–7 days that is sensitive is, however, rather questionable because the percentage of diapause was estimated by the dissection of females when only 14 days old, i.e. only 4–7 days after transfer. In such a short time the processes controlling the maturation or regression of ovaries could not have been completed at 21°C. In order to exclude what seems indeed very probable, i.e. that adults aged more than 7 days are also responsive, it would be necessary to dissect females much later after the transfer. These experiments did not exclude the possibility that, as in other insects diapausing as adults (Hodek 1971a), the pre-imaginal stages are also sensitive to the cues controlling diapause induction. This was recorded in *Chil. bipustulatus* (Tadmor & Applebaum 1971; Table 6.12; 6.2.13). In *C. novemnotata* the sensitivity of pre-imaginal stages is indicated

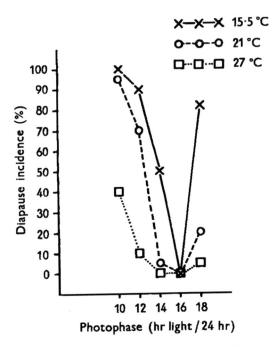

Figure 6.3 Modification of photoperiodic response by temperature in *Coccinella novemnotata* (from McMullen 1967b).

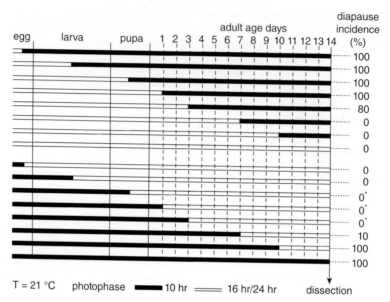

Figure 6.4 Effect of change in photoperiod on diapause incidence in *Coccinella novemnotata*. Females were dissected at the age of 14 days. Under the treatments marked by * ovogenesis in non-diapause individuals was retarded, compared with controls (from McMullen 1967b).

by retarded ovogenesis in females (marked with 'a' on Fig 6.4) which were transferred to a non-diapause photoperiod as teneral or young adults.

The **sensitivity** of earlier stages may be **masked** when later stages are influenced by the **reverse stimuli**; the effect may thus be reversed. The sensitivity of the pre-imaginal stages may be distinguished when the adults are kept under neutral conditions. In fact, we do not search for a specific sensitive stage, but rather for the stage of development in which we can still succeed in preventing oviposition completely (i.e. to reverse the former opposite effect) by a transfer from non-diapause to diapause-inducing conditions. A suitable criterion for the sensitivity of larvae may be the length of the pre-oviposition period in females kept as larvae for different periods of time under diapause conditions (Hodek 1971a).

6.2.4   *Adalia bipunctata*

Due to the low developmental threshold (Chapter 3.6) for post-diapause and non-diapause development

A. bipunctata can start mating and ovipositing early in spring (Hodek 1960, Obrycki et al. 1983, Hemptinne & Naisse 1987) and have a **multivoltine** life cycle. Four generations are reported from the region of Bologna, northern Italy (Bazzocchi et al. 2004). Short daylengths induce diapause, while long daylengths prevent diapause in all individuals. The **critical photoperiod** for diapause induction in the Ithaca, New York, USA (about 42°N) population lies between 13L:11D and 14L:10D at 23°C (Obrycki et al. 1983; Table 6.5). In two samples from southeastern France, Iperti and Prudent (1986) found oviposition in 54 and 85% of females at 13L:11D and 25°C, while at 12L:12D no female reproduced during a period of 4 weeks. Thus, the critical photoperiod of both populations, collected in different continents, appears to be very similar.

Diapause completion does not require exposure to low temperatures. At 23°C, duration of diapause is quantitatively related to photoperiod. The progress of **diapause development** is indicated by the photoperiodic response diminishing substantially in late December and being almost absent in March (Obrycki et al. 1983; Table 6.6).

Table 6.5 Induction and duration of diapause in *Adalia bipunctata* under a series of constant photoperiods ($23 \pm 1°C$) (Obrycki et al. 1983).

Photoperiod	Preoviposition period (x ± SD) (days)	Diapause (%)	Ovipositing females (*n*)
16L:8D	8.4 ± 2.2	0	18
15L:9D	8.9 ± 2.3	0	11
14L:14D*	10.4 ± 2.4 (A)	—	7
	26.7 ± 6.4 (B)	46	6
13L:11D	61.8 ± 16.2	100	13
12L:12D	98.5 ± 8.2	100	11
10L:14D	113.3 ± 18.1	100	11

*Group A females oviposited in about the same time as observed at 16L:8D and 15L:9D, no diapause; group B females took substantially longer to oviposit, weak diapause.

Table 6.6 Pre-oviposition period in *Adalia bipunctata* after transfer from outdoors to various photoperiods (Obrycki et al. 1983).

Sample date	Photoperiod				
	16L:8D	14L:10D	12L:12D	10L:14D	Natural
22 Oct.	14.6 ± 1.6 (9/9)	12.3 ± 3.3 (8/10)	75.0 ± 8.4 (3/9)	144.5 ± 34.6 (2/10)	116.0 ± 37 (4/10)
22 Dec.	A -	10.8 ± 1.8 (10/11)	20.8 ± 7.7 (7/13)	18.0 ± 5.3 (3/10)	21.5 ± 4.9 (2/10)
22 Dec.	B -		117 (1/6)	97.0 ± 10.8 (3/7)	61.3 ± 30.3 (3/8)
7 Mar.	—	7.0 ± 1.7 (6/8)	10.8 ± 3.1 (8/12)	12.0 ± 4.3 (8/11)	12.6 ± 2.7 (7/8)
21 May	—	2.8 ± 1.3 (7/8)	2.5 ± 1.0 (6/6)	2.4 ± 1.0 (7/9)	3.0 ± 1.2 (4/4)

Numbers in parentheses indicate number of ovipositing females per total number of females in each condition. Temperature under constant photoperiods, $23 \pm 1°C$; temperature under natural daylength, $23 \pm 2°C$. For the definition of groups A and B see Table 6.5.

Adalia bipunctata has recently **invaded Japan**. Larvae and pupae are found in April and May, while during summer (July, August) the inactive adults are dormant in groups of up to 10 in rolled dry leaves. In autumn some actively foraging and preying adults were observed, but no larvae. In addition to the sites recorded during aestivation, the bark of maple is used for overwintering (Sakuratani et al. 2000).

6.2.5 *Propylea quatuordecimpunctata* and *P. dissecta*

Obrycki et al. (1993) studied the photoperiodic induction of diapause in three populations of *P. quatuordecimpunctata* from widely separated geographical regions (Quebec 45°N, Turkey 40°N, southern France

44°N). Little difference was found in their **critical photoperiod** for diapause induction. The authors explained this similarity by the previous 2–4 generations reared under long days. However, the critical photoperiod depends mostly on the latitude, and this was similar in all three regions. At long days and 26°C Phoofolo and Obrycki (2000) recorded in the above three populations, respectively, 83, 56 and 80% of females ovipositing within 2 weeks of emergence, which is a higher proportion of 'multivoltines' than in *C. septempunctata* (6.2.1.5).

In **northern India**, active adults of *P. dissecta* are found from February to April and again from August to October (Omkar & Pervez 2000). These activity periods are a little different from those reported for *C. septempunctata* for the same region: December to March and July to September (Omkar & Pervez 2000).

6.2.6   *Hippodamia tredecimpunctata*

Populations of *Hip. tredecimpunctata* from Maine (USA) are bivoltine. At a constant temperature of 21°C, diapause can be prevented in all females by long days of 16L:8D. The photoperiod of 14L:10D (near the critical threshold) enabled reduced (i.e.halved) oviposition in 70% of females. At 12L:12D still one third of females laid eggs for 14 to 69 days. The results were slightly modified by the photoperiod during larval rearing, indicating that also the larvae are sensitive to factors inducing adult diapause (Storch & Vaundell 1972).

Below, the data for **three long distance migrants** are discussed. Fifteen years ago, it was surprising (Hodek 1996) how meagre was the experimental evidence on the regulation of diapause in the three best-known long-distance migrants, the Nearctic *Hip. convergens*, the Palaearctic *Cer. undecimnotata* and *Har. axyridis* from Far East Asia. There were only studies in southeastern France and central Europe on *Cer. undecimnotata* (6.2.8). This situation has much improved in the last decade.

6.2.7   *Hippodamia convergens*

In **northern California,** Hagen (1962) assumed that there were three types of dormancy in *Hip. convergens*. The majority of the population has a univoltine cycle. This is the original pattern, as before irrigation was introduced, the species was dependent on aphids on prairie grasses in the spring. The irrigated crops enable introduced aphids to maintain themselves during the summer and autumn. Some *Hippodamia* spp. (including *Hip. convergens*) have reacted to this later abundance of aphids by reproduction in the summer and multivoltinism. The **multivoltine** adults of *Hip. convergens* enter hibernation in the autumn. Their diapause is induced mainly by **photoperiod and temperature** (Hagen 1962). For the Californian populations, the photoperiod of 14L:10D and 25°C was reported as preventing diapause (Davis & Kirkland 1982).

Hippodamia convergens often joins the other *Hippodamia* spp. in the valley aggregations. Most **univoltine** *Hip. convergens* adults show a **facultative diapause** that appears to be largely **nutritionally induced**. In the laboratory, however, there were occasions when 10–20% of the beetles entered diapause, although conditions were optimal and the beetles were supplied

with an excess of essential aphid food (Chapter 5.2.11, for the term). Hagen (1962, p. 305 therein) supposed that strains of *Hip. convergens* exist which possess an obligatory diapause (see the discussion on *C. septempunctata*, 6.2.1.6).

In the **Great Plains region** of central USA, various cases of nutritional regulation of reproductive diapause were analyzed in females of *Hip. convergens* (Michaud & Qureshi 2005, 2006). The **importance of drinking sap** on sunflowers in the summer months in West Kansas was examined in this arid region. Sunflower petioles and pollen as well as lepidopteran eggs were provided to the beetles collected in early June. While these females did not oviposit in the absence of protein food, feeding on eggs of *Ephestia kuehniella* followed by pollen enabled 66% ($n = 171$) of the females to lay viable eggs at a low rate of 6.6 eggs/day. The females, transferred on 14 August to essential aphid food (*Schizaphis graminum*), laid six times more eggs.

These experiments stressed the **adaptive role** of the life cycle in *Hip. convergens* in that it enables survival during arid summer conditions when there is a shortage of the essential food, aphids. In the absence of protein-rich food, the first generation can enter diapause. Another tactic could be to wait in a **state of lowered metabolism** (but less lowered than in diapause) for the re-appearance of essential aphid food, relying meanwhile on alternate foods. Then a switch to intensive egg laying can be quick, as was shown by a short oviposition delay of only 4 or 6–9 days on essential prey (Michaud & Qureshi 2005, 2006). For *Hip. convergens*, we still need to know more about the combined action of food and photoperiod.

In the upper coastal plain of **South Carolina**, diapausing adults of *Hip. convergens* were recorded feeding on eggs and larvae of the moth *Heliothis zea* in spite of a short day photoperiod of 12L:12D, when they were transferred during December/January to temperatures >15.5°C (Roach & Thomas 1991). This may indicate that **diapause development** was already **completed**.

6.2.8   *Ceratomegilla (=Semiadalia)* *undecimnotata*

Detailed studies in central Greece (Katsoyannos et al. 1997a, b, 2005) have widened our knowledge on diapause of *Cer. undecimnotata*. In one season, five generations were reared in outdoor cages with a surplus of

Table 6.7 Duration of preoviposition period of females of *Ceratomegilla undecimnotata* collected during their aestivo-hibernation from the summit of Mount Kitheron (1409 m) and transferred to the laboratory (25°C, 16L:8D, and presence of aphids) (Katsoyannos et al. 2005).

Collection date	Females collected	Duration of preoviposition (days)				
		Median	Mean	sd	Minimum	Maximum
04-Jul-93	16	92	94	4	89	101
21-Jul-93	15	68	72	9	68	103
04-Aug-93	22	64	64	4	58	68
19-Aug-93	20	42	44	6	40	60
09-Sep-93	17	21	22	7	11	34
28-Sep-93	18	24	30	16	15	66
21-Oct-93	19	29	30	6	21	48
01-Nov-93	18	30	31	3	28	42
18-Nov-93	19	28	29	10	13	46
17-Dec-93	20	17	22	10	17	49
19-Jan-94	16	16	20	7	12	37
30-Mar-94	24	14	15	5	9	29

aphids. It was demonstrated that **diapause is facultative** in a part of population: about 30% of females remained immature in the first three generations. Thus the population appears to be **heterogeneous** as regards the **induction** of diapause (similar to *C. septempunctata*, 6.2.1.4 and 6.2.1.6, and *Hip. convergens*, 6.2.7). The results from these cultures corresponds to that from the dissection of samples **from the plain**, where about 40–50% non-reproductive females were recorded from mid-June. At that time most females (70–100% in different years) collected **on mountain summits** were immature (Katsoyannos et al. 2005). Regularly transferring samples from the mountain tops to long days of 16L:8D and prey surplus at 25°C led to activation. Females laid eggs after a gradually shortened pre-oviposition period, that was long in summer (92 and 64 days in July and August, respectively) and decreased to about 20 days in September. This clearly showed the **progress of diapause development**, i.e. a decrease in diapause intensity (Table 6.7). Katsoyannos et al. (2005) plausibly assumed that since late autumn the ladybirds are quiescent rather than diapausing. The transfer experiment indicates that *Cer. undecimnotata* is a long-day insect. The authors believed (apparently wrongly) that diapause of *Cer. undecimnotata* in Greece is induced by long days and high temperatures, because of the above-mentioned 30% of immature females in cages in summer in spite of a surplus of aphids.

The evidence obtained from *Cer. undecimnotata* in France (Iperti & Hodek 1974, Hodek & Iperti 1983) also shows it to be a long-day insect while supporting the possibility that **lack of food** also plays some role in diapause induction. In spite of the important studies made in Greece, the regulation of diapause induction in the field is not yet clear. While under experimentally improved conditions several generations can be reared within a year, the natural life cycle seems univoltine and the relative role of the factors (photoperiod and food) inducing diapause outdoors is not clear (see later for Czech populations).

Similarly to *Hip. convergens* (Hagen 1962; Michaud & Qureshi 2005, 2006) *Cer. undecimnotata* shows indications of a **nutritional induction** of diapause. When the young adults of *Cer. undecimnotata* were reared at 20°C for only 2 days on *Myzus persicae* and then for 5 days on a 1:1 mixture of honey and agar, they completed the regression of ovaries and accumulated large reserves in the fat body (Iperti & Hodek 1974). The survival of these adults, induced to diapause by alternative food, was quite long at 5–8°C (the average, the median and the maximum longevity were respectively 124, 136 and 198 days). The other 11 combinations of aphid and carbohydrate food yielded similar results. Diapausing adults collected in the hibernation sites in December lived about twice as long at 5–8°C, with median values between 230 and 260 days (Iperti & Hodek 1974). Although these results

Table 6.8 Effect of the physiological condition of the host plant of aphids (*Aphis fabae*), used as food for larvae of *Ceratomegilla undecimnotata*, on the incidence of diapause in the adult coccinellids (Rolley et al. 1974).

Host-plants of prey-aphids fed to larvae	Incidence of reproducing females*		Frequency of quantity of fat reserves (%)				
	n	%	n	no or low	medium	high	Sex
Young, reared in the lab	—	—	21	75.5	10	14.5	females
	21	95	24	33.5	37.5	29	males
Senescent, collected in the field	—	—	27	11	0	89	females
	27	44	17	6	6	88	males

The coccinellid adults (emerged from pupae in mid-July) were fed with aphids reared on young plants from the laboratory. Both larvae and adults were reared in screened outdoor cages. The mean of daily minimum temperatures was 14°C and of maximum temperatures 28°C, whilst the absolute extremes were 10 and 31°C.
*Females with vitellinized oocytes or eggs.

might indicate that there are nutritional components in diapause induction in *Cer. undecimnotata*, the adults have been observed to migrate to dormancy sites when aphids are still abundant in the valleys where they breed (G. Iperti, unpublished). Thus the situation appears different from that reported by Hagen (1962) for *Hip. convergens*.

Also the physiological **age of the host plant** of aphids plays an important role. This was tested by rearing *Cer. undecimnotata* larvae on two food regimes: aphids from young versus old bean plants. The incidence of diapause was increased by feeding **larvae on aphids from old plants** (Rolley et al. 1974; Table 6.8). All adults reared as larvae in the two food regimes were fed after emergence from pupae with aphids reared on young plants. Both larvae and adults were reared in screened outdoor cages. The mean daily minimum and maximum temperature was 14 and 28°C, respectively.

An **inhibition** of reproduction **by short daylength** (12L:12D) in *Cer. undecimnotata* at 25°C is weaker than in some other coccinellids such as *C. septempunctata*. Although the difference in the duration of the pre-oviposition period between the two photoperiods (12L:12D versus 18L:6D) was great (60 versus 8 days, respectively), the short days merely delayed, and did not prevent reproduction, except 10% of females, and the resulting total fecundity was almost twice as high (824 eggs) than at long days (455 eggs) (Table 6.9; Fig. 6.5). Also the oviposition rate was substantially higher

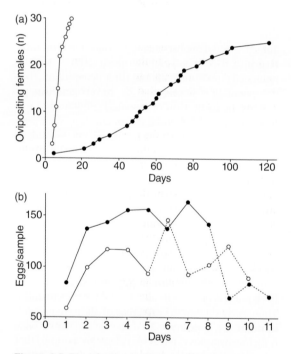

Figure 6.5 Reproduction in *Ceratomegilla undecimnotata* at 25°C, 12L:12D (solid circles) and 18L:6D (open circles) (from Hodek & Iperti 1983). (a) Duration of pre-oviposition period; (b) Oviposition rate, oviposition by individuals aligned for a uniform start in both photoperiods.

Table 6.9 Comparison of reproductive activity of *Ceratomegilla undecimnotata* reared at 25 ± 1°C and 12L:12D or 18L:6D (Hodek & Iperti 1983).

Photophase	n		Duration (days) pre-oviposition	oviposition	post-oviposition	Fecundity (eggs/female)
12L	25	aver.	60.3	38.5	2.7	823.6
		med.	61	39	3	789
		range	5–121	7–78	1–9	35–2277
18L	30	aver.	7.7	25.5	2.9	454.4
		med.	7.5	23	3	414
		range	4–14	4–71	1–7	65–1284

at a short daylength (Hodek & Iperti 1983). It can be speculated that, in this mostly univoltine population, the **precocious photoperiodic activation** by long daylengths (i.e. by tachytelic processes) was not adequate for the 'normal course of events', which involve the horotelic termination of diapause under short daylength (for definitions see 6.1.2). It is still too early to distinguish definitively the relative importance of individual factors (food, photoperiod, temperature and others) in diapause induction in *Cer. undecimnotata*, but it seems that photoperiod is a major cue also in this species.

In *Cer. undecimnotata* the photoperiodic response is lost during **diapause development**, so that after diapause, in spring, adults reproduce without respect to daylength (Hodek & Ruzicka 1979). The development of diapause (monitored by the decrease in the duration of pre-oviposition) follows the same course in south-eastern France and in central Europe (Iperti & Hodek 1974). In early diapause (before mid-August) the French samples had two patterns of pre-oviposition duration. The long durations were around 25 days. The other pattern in summer was short pre-oviposition periods (around 10 days – the duration normally found in December–January), apparently occurring in the females which had just arrived in the hibernation sites; their diapause was not yet fixed. This phenomenon of **diapause fixation** (6.1.3) has already been reported from several other insect species (Hodek 1983; Danks 1987). Adult diapause of *Cer. undecimnotata* females could be prematurely but efficiently terminated (and oviposition started) in September by application of 6 of 18 juvenoids (artificial analogues of juvenile hormone) tested (Hodek 1973; 6.1.4).

6.2.9 *Harmonia axyridis*

Telenga and Bogunova (1936) reported from **East Siberia** that only those adults of *Har. axyridis* which emerge from the pupa before mid-August can oviposit in the same season. Ulyanova (1956) confirmed the **multivoltine** character of this coccinellid after it was imported to the warm climate of **central Asia** (Tashkent, Uzbekistan), where the beetles terminated hibernation in mid-February. When the first generation then emerged from pupae in April and the second in June, the females were reproductive. In the next two generations that emerged in July and August, most females were diapausing. In neither of these two papers were the diapause inducing factors discussed.

In **central Japan** *Har. axyridis* has a **bivoltine cycle** interrupted twice, by aestivation and hibernation. **In contrast to C. s. brucki**, which has a similar life cycle in that region, but is a short-day insect **aestivating in diapause** (6.2.2.1), a study suggests that *Har. axyridis* **hibernates in diapause** while its **aestivation** may be a **mere quiescence** (Sakurai et al. 1988). This assumption of direct inhibition by high temperature is supported by observation of mating and oviposition in August. In adults of the second generation, appearing in mid-October, the ovaries remained undeveloped and the corpora allata small, and thus no vitellogenins were found by electrophoresis in the haemolymph of females. Mating and oviposition were not observed before April. Low temperature (18°C) and short daylength (10L:14D) was reported to stimulate the respiration rate, which was suppressed by contrasting conditions (25°C, 16L:8D). These data would not support the above assumption that the aestivation is a

Table 6.10 Results of dissections (% of females) of *Harmonia axyridis* collected at hibernation sites. F, full; E, empty; s, small; m, medium; l, large; Im, immature; M, mature; R, regressed; transfer, date of transfer to the hibernation site (Iperti & Bertand 2001, modified).

Locality	Midgut		Fat body			Spermatheca		Ovaries		
Month	F	E	s	m	l	E	F	Im	M	R
Alpilles 230 m a.s.l. n = 1690 transfer: 16 Dec.										
Dec.	50	50	20	40	40	50	50	90	0	10
Jan.	0	100	0	37	63	50	50	100	0	0
Febr.	0	100	50	10	40	100	0	100	0	0
March	0	100	40	0	60	50	50	100	0	0
April	30	60	90	10	0	25	75	100	0	0
Roquebrune 280 m a.s.l. n = 1500 transfer: 9 Dec.										
Dec.	30	70	10	10	80	90	10	90	0	10
Jan.	0	100	0	10	90	70	30	100	0	0
Febr.	0	100	0	70	30	100	0	100	0	0
March	0	100	10	30	60	90	10	100	0	0
April	0	100	90	10	0	10	90	70	0	30
Tourniol 1180 m a.s.l. n = 1788 transfer: 10 Jan.										
Jan.	25	75	50	50	0	25	75	100	0	0
Febr.	0	100	33	33	34	66	34	100	0	0
March	0	100	0	40	60	50	50	70	30	0
April	0	100	0	100	0	20	80	60	40	0
May	20	80	100	0	0	10	90	40	60	0
Courbons 1200 m a.s.l. n = 1818 transfer: 23 Dec.										
Dec.	20	80	10	20	70	78	22	100	0	0
Febr.	0	100	0	30	70	100	0	100	0	0
March	0	100	0	60	40	70	30	100	0	0
April	0	100	100	0	0	40	60	80	20	0
May	0	100	90	10	0	23	77	56	44	0

mere quiescence. Both the early studies by Telenga and Bogunova (1936) and Ulyanova (1956) and the more recent studies described below indicate that *Har. axyridis* is a long-day insect entering hibernation diapause in response to short days. Therefore it is strange that the autumnal conditions inducing diapause merely double the duration of the pre-oviposition period (Sakurai et al. 1988).

In **southeastern France**, the imported population of *Har. axyridis* (probably from China) hibernated successfully with <10% mortality (at some sites it was only around 2%) on five sites varying in altitude from 40 to 1200 m. The beetles were **fed before overwintering on eggs** of the moth *E. kuehniella* at outdoor conditions of decreasing temperature of <22°C and a photophase (<14 hours 40 minutes) that induced diapause. Ovaries did not mature before April, but the spermathecae contained sperm in March in 30–50% of females. A much higher proportion of females (75–90%) contained

sperm in their spermathecae later, before the flight from hibernation sites (Table 6.10). The intensity of diapause, indicated by the pre-oviposition period after transfer to 22°C and 16L:8D, gradually decreased during overwintering. The delay of oviposition in March was only 7–8 days (Iperti & Bertand 2001; Table 6.11). In **northeast Canada** no winter survival of *Har. axyridis* was observed outside buildings (Labrie et al. 2008) because they do not withstand temperatures around −20°C (Koch et al. 2004).

Bazzocchi et al. (2004) reported four generations per year in the region of Bologna (**northern Italy**) for *Har. axyridis* of commercial origin. They did not record aestivation diapause (reported by Sakurai et al. 1988, for central Japan) probably due to the continuous availability of aphid prey, the screening from sunlight and favourable air humidity caused by surrounding vegetation. In the commercial culture, *Har. axyridis* was fed on frozen eggs of the moth *E. kuehniella*.

Table 6.11 Preoviposition period (days) in *Harmonia axyridis* females collected at hibernation sites and reared at $22 \pm 1°C$ and 16L:8D on abundance of eggs of *Ephestia kuehniella* (Iperti & Bertand 2001, modified).

Dates\Sites	Antibes	Roquebrune	Alpilles	Courbons	Tourniol
November	11 (7–19)	—	—	—	—
December	16 (12–20)	—	—	—	—
January	14 (10–24)	12 (7–17)	14 (11–18	—	—
February	9 (7–12)	11 (10–13)	12 (9–14)	11 (8–14)	11 (8–13)
early March	—	9 (6–10)	12 (11–17)	13 (12–14)	13 (10–18)
late March	7 (4–10)	7 (6–9)	8 (6–12)	—	—
April	—	—	—	7 (4–12)	9 (7–11)
May	—	—	—	—	7 (5–13)

In experiments executed in the region of Ghent, **Belgium** (Berkvens et al. 2008), diapause was induced by short days of 12L:12D at 23°C, similar to the experiments in France (Ongagna & Iperti 1994). The diapause lasted 1–3 months when the ladybirds were fed on *Acyrthosiphon pisum*, but was longer when frozen *E. kuehniella* eggs were the food. A laboratory population of commercial origin was reared for c. 50 generations at long days of 16L:8D; its response to diapause induction was weaker than in populations founded by beetles collected in the field (Berkvens et al. 2008).

6.2.10 *Coccinella leonina (=repanda)*

The facultative diapause of *C. leonina*, a common and widely distributed aphidophagous coccinellid of **eastern Australia**, is reported to be induced mainly by non-aphid diet, i.e. pollen and sugar. When fed with aphids, almost all females reproduce at temperatures of 20, 22, 28 and 32°C without respect to the photoperiods 10L:14D, 12L:12D or 14L:10D. There is some incidence of diapause at the highest temperature, but there is also a certain tendency to diapause at the lowest temperature (Anderson et al. 1986). The authors conclude that this result, together with field data, may indicate the capacity to enter both summer and winter diapause. They characterize the life cycle strategy of *C. leonina* as opportunistic, enabling the use of fluctuating food supplies.

6.2.11 *Apolinus (=Scymnodes) lividigaster* and *Illeis (=Leptothea) galbula*

Around Sydney, **Australia**, the aphidophagous *A. lividigaster* and mycophagous *I. galbula*, both multivoltine,

show a very **plastic reproductive strategy** adapted to the unpredictable conditions there. Apart from the periods of dormancy (aestivation and hibernation in *A. lividigaster* and only hibernation in *I. galbula*), **diapausing** adults also **occurred concurrently with ovipositing** females during the periods when reproduction normally occurs (Anderson 1981). In another paper, Anderson et al. (1986) supposed that the diapause of *I. galbula* was controlled by photoperiod.

6.2.12 *Harmonia sedecimnotata*

An experimental culture of this Asian ladybird was founded from individuals sampled near the town of Guangzou, southeastern China. The species does not show a photoperiodically induced diapause. In reproducing 30 day old adults fed on aphids (*M. persicae*), **trophic dormancy** was induced by feeding them on only 10% sucrose solution. In the course of 4 weeks the ovaries were resorbed and only small germaria were found by dissection. The beetles aggregated similarly to diapausing individuals. When feeding with aphids was resumed, the females recommenced egg laying, even with a very low ration of prey. The authors assumed that the beetles were activated by a food signal and not metabolically, and thus considered this dormancy a diapause (Zaslavsky et al. 1998). More probably a quiescence was concerned, as both induction and termination of the dormancy were directly produced by absence or presence of suitable food (6.1).

In a culture with only **intermittent presence** of aphid prey, monthly oviposition was doubled (207 eggs, $n = 39$) compared to a culture with a continuous surplus of aphids (107 eggs, $n = 48$). Furthermore, the longevity of females was increased by this manipulation (Semyanov & Vaghina 2003).

Adults in trophic dormancy can be used for **storage** at 12°C with the low mortalities (n = 420) of 9.6, 25.5, 41.8 and 55.1% in the first, second, third and fourth month respectively. At 25°C, 25% of beetles (n = 1145) died in the first month (Semyanov 2000). Similar trophic dormancy was induced in *Cer. undecimnotata* (Iperti & Hodek 1974; 6.2.8).

6.2.13 *Chilocorus* spp.

Chilocorus rubidus, a species from **eastern Siberia** is strictly **univoltine**. The dormancy lasts from late August/mid-September to late April, while diapause (which is marked by very low oxygen consumption) ends as early as late December to early January. The rest of the dormancy is spent in mere quiescence during which oviposition can be started by transferring the beetles to a higher temperature. Pantyukhov (1968a) found that neither low temperature, nor any particular photoperiod, was necessary for the **termination of diapause** and that the passage of 3.5–4 months was sufficient under field conditions or at temperatures of 25, 20 or 5.8°C.

In contrast to *Chil. rubidus*, diapause can be prevented in the majority of **Chil. renipustulatus** individuals in both populations studied. Thus only 12% of beetles from Maikop (**southern Russia**, 45°N, 40°E) entered diapause under long days, whereas in the population from **St. Petersburg**, 60°N, the figure was 38%. The critical photoperiod (6.1) is insensitive to temperature (within the range 20–25°C) and is 2 hours longer in the beetles from St. Petersburg than in the Maikop strain (Pantyukhov 1968b).

Chilocorus bipustulatus from **St. Petersburg** (60°N), a northern population, reproduced when a three week exposure to **short days** (9L:15D) was **followed by long days**; they do not lay eggs if the larvae, pupae and adults are reared continuously in long days (20L:4D) (Zaslavskii 1970). The short day treatment is equally effective when given to the larvae. The alternative way to achieve oviposition is an exposure to +7°C for about a month followed by long days. If the beetles are constantly left in long days, diapause ends spontaneously after 2–4 months. However, after about 1.5–2 months of egg laying in continuous long days, the beetles gradually cease ovipositing and have to be acti-

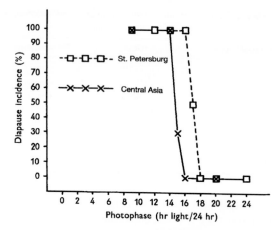

Figure 6.6 Effect of photoperiod on diapause incidence in strains of *Chilocorus bipustulatus* (from Zaslavskii 1970).

vated from this secondary diapause by a fresh experience of about 25 days in short days.

The **critical photoperiod** for the beetles from St. Petersburg is 2 hours longer (17L:7D) than for the central Asian population (Fig. 6.6). It does not change with temperature within the range of 20–27°C. The results of Zaslavskii (1970) are of extreme importance, as they were an early confirmation that a number of alternative pathways can lead to the triggering of the neuroendocrine system regulating oviposition.

In *Chilocorus bipustulatus* from **Central Asia** (40–41°N), **diapause** is **induced** by a photophase shorter than 15 hours and at 20°C 100% beetles enter diapause. At 24°C, however, diapause is completely prevented in all females in spite of the short day (9 or 11 hours). Spontaneous **termination of diapause** does not occur when the beetles are maintained at diapause-inducing conditions. An increase in temperature or a prolongation of photophase beyond the above-mentioned thresholds terminates diapause very quickly: oviposition then takes place within a few days. It is just as easy to terminate diapause 7–10 days after its induction as it is after 3 months. By contrast, transfer of the ovipositing beetles from long days to diapause-inducing conditions gradually suppresses oviposition, while a return to long days restores it. Alternatively, after an exposure to +8°C for 30–40 days, the beetles mature even in short days. Thus, sensitivity to photoperiod could be nullified by an exposure to cold (Zaslavskii & Bogdanova 1965).

Table 6.12 Effect of photoperiod and temperature on oviposition in *Chilocorus bipustulatus* in Israel (Tadmor & Applebaum 1971).

Photophase (h/24h) and temperature (°C)				Ovipositung females (%)	Mean pre-oviposition period (days)	Number of replicates
larva and pupa		adult				
24	28°	10	8°/20°*	46	21	11
24	28°	10	18°	60	17	10
10	18°	10	18°	1	21	11
24	28°	10	22°	100	10	7
24	28°	14	18°	93	14	14

*8°C during scotophase, 20°C during photophase.

The populations of **Chilocorus bipustulatus** in **Israel** (32–33° N), in the eastern Mediterranean area, differ from those of central Asia in their diapause threshold both for photoperiod and temperature (Tadmor & Applebaum 1971). A photophase of 14 hours prevents diapause, whereas a shorter photophase (10 hours) tends to induce it (Table 6.12). Thus, consistent with the findings of Danilevskii (1965), the **critical photoperiod** is about 2 hours shorter in populations from Israel than from central Asia. The effect of the short photophase is modified by temperature: at 22°C it is completely nullified so that all females lay eggs. A cumulative **sensitivity to diapause induction** is exhibited both in the pre-imaginal stages and in the adults.

There is contradiction between earlier and later reports from the field conditions about the summer occurrence and biocontrol efficiency of *Chil. bipustulatus* in Israel. Early reports concluded that *Chil. bipustulatus* was common in spring and extremely scarce in summer, and was therefore an inefficient predator of scale insects (Hecht 1936; Bodenheimer 1957). The high mortality in summer was supposed to have been caused by a dry hot wind from the desert. Later findings show, however, that *Chil. bipustulatus* is rather abundant in the summer (the population may even peak in early summer) and plays an important role in retarding the build-up of scales in this period (Nadel & Biron 1964; Avidov & Rosen 1965; Rosen & Gerson 1965; Kehat 1968; Ben-Dov & Rosen 1969). This has led to speculation (Hodek 1967) that the difference between the earlier and later reports has perhaps resulted from the improved environmental conditions due to irrigation, as the same difference also exists between older (Bodenheimer 1957) and more recent

reports (Plaut 1965) on the abundance of *Stethorus pusillus*. In citrus groves, M. Kehat, S. Greenberg and D. Gordon (unpublished) found there is a considerable decline in numbers of *Chil. bipustulatus* females with well-developed ovaries during July, and also from October to December. When the non-reproductive females were transferred to the laboratory (28°C) and provided with coccids, their ovaries matured in both these seasons. The failure to induce diapause at 16 combinations of light conditions was apparently due to a high temperature of 28°C (Tadmor & Applebaum 1971). An even higher temperature (35°C) additionally decreases oviposition and survival.

In the **central Asian** populations of **Chilocorus geminus** (Tashkent, Uzbekistan, 40–41° N), induction and termination of diapause is very similar to that in *Chil. bipustulatus*. The only difference is a higher temperature threshold: as much as 26°C is needed to prevent diapause in short days (Zaslavskii & Bogdanova 1965).

6.2.14 *Stethorus punctum picipes* and *S. japonicus*

The **South Californian populations** of *S. p. picipes*, a predator on spider mites, have a weak facultative diapause which can be quickly terminated by an increase in temperature and/or photoperiod, and induced again by the reverse changes (McMurtry et al. 1974); it may be a mere quiescence. Long days (16L:8D) at 21–22°C stimulated almost all females to oviposit, but about half the females also oviposited in short days (10L:14D). An increase in temperature by about 5 to 26.7°C enabled oviposition by almost all females. In the mild conditions near the ocean the females also reproduce in winter, when they can find prey. All developmental stages were found in mid-winter on mite-infested plants, but on mite-free oaks the females had small or shrivelled ovarioles (19 of 21 females).

In the Ibaraki region (central Japan), diapause of **Stethorus japonicus** is induced at 18°C under photoperiods <13L:11D, but at 14L:10D and 16L:8D it is prevented in 60% of females. Active adults appear on trees from June, and the photoperiod 13L:11D occurs in the region in mid- to late September. Thus the 5–7 generations per year, estimated by Mori et al. (2005), appears unrealistically high.

6.2.15   *Scymnus (Neopullus) sinuanodulus*

The first generation (the offspring of the diapausing one) of this predator on adelgids does not oviposit until the following spring (Lu & Montgomery 2001). Although the authors deny the existence of diapause, it evidently is entered. Oviposition is achieved in autumn after an exposure to low temperature (5–10°C) for 1.5 months.

6.2.16   Ecophysiological regulation of diapause in coccinellids

The life history of coccinellids depends on regional climatic conditions (Hagen 1962; Fig. 6.7). It is not easy to 'distil' a general view on the regulation of diapause and voltinism from the very diverse evidence presented in this chapter. The evidence is very varied and defies meaningful generalizations for several reasons. (i) The reported evidence still provides just a small proportion of knowledge that would be needed; even for the frequently studied species the **analysis is incomplete**. Unfortunately, research activity in this field has received little attention in the last 10–15 years, particularly when compared with other areas of coccinellid study. (ii) It is difficult to make comparisons between species, because the direction of **research** has varied **arbitrarily**, as biased by conditions and the different inclinations of the researchers. Thus, for example, we lack detailed studies on the effect of photoperiod on *Hip. convergens*, while for both subspecies of *C. septempunctata* such studies are available. The varied experimental approach is probably the reason why particular

regulatory environmental factors have become attached to individual species. Complex relationships cannot be well understood, when the results of studies are **biased by individual experimental plans**. (iii) **Diversity** of responses is **inherent to diapause** because of its **adaptive role** (6.1); thus diapause mechanisms inevitably resist an easy generalisation. (iv) In the case of aphidophagous coccinellids, the selection pressure of the **unpredictable** intermittent availability of **aphid prey** naturally leads to a certain convergence, in the direction to a **greater plasticity** of diapause mechanisms within this guild. Plastic responses to anthropogenic environmental changes have been observed both in the field (irrigation for *Hip. convergens* in California or in the coccidophagous *Chil. bipustulatus* in Israel as well) and in experiments (surplus of aphid prey, and perhaps the shading of outdoor cages for *Cer. undecimnotata* in Greece).

Despite the above constraints, some species have been shown to respond to a **complex of environmental cues**, though these factors may be differentially operational in different environmental situations or life-cycle phases. Thus, for example, *Cer. undecimnotata* responds to at least both the photoperiod and food. We may assume that, superimposed on the **basic photoperiodic response** (based on the annual astronomically precise repetition of daylength), are the less rigid reactions to less predictable environmental changes in food availability and quality, and other factors such as temperature, humidity and population density. The archetypal **primary nutritive factor** seems to be 'prepared to enter the game' under unpredicted events of prey abundance – thanks to phenotypic plasticity.

We might envisage a scenario which **combines plasticity with resilience**. One aspect of the life-cycle strategies is the 'safety' ('insurance') factor, the univoltine trait, permanently perpetuated in the gene pool and maintained (i.e. not selected out) in spite of any momentary favourable conditions that are unreliable in the long run. However, the polygenes 'watch' for changes in the environment. If there is a promising improvement they 'open the gate' for multivoltine development, more or less appropriate to the kind of improvement. The system appears resilient by **maintaining** the **univoltine trait** quite intensively.

This scenario is adequate for *C. septempunctata* and species or populations with a similar life history living

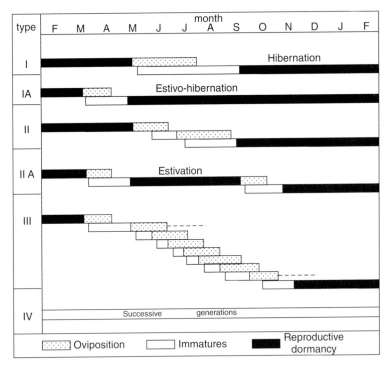

Figure 6.7 Types of voltinism among coccinellids in the northern hemisphere (from Hagen 1962).

in temperate regions/climate. In coccinellids with different prey or from different climatic areas, the regulation of voltinism can differ, as e.g. in tropical areas or in coccidophaga.

6.3 BEHAVIOUR PATTERNS RELATED TO DIAPAUSE

Coccinellids induced to diapause change their behaviour. Some species make lengthy **migratory flights** and form large **aggregations** in a dormant state, in which they may remain for 9 months. This behaviour is most pronounced in *Hip. convergens* (6.3.2.7), *Cer. undecimnotata* (6.3.2.1) and *Har. axyridis* (6.3.2.2). In general, these adaptations of migration and aggregation have developed **most often in aphidophagous** coccinellids of the tribes Coccinellini, Hippodamiini and Anisostictini. Ample descriptive literature exists on such behaviour of coccinellids, while studies on causal relationships are rarer.

What is **required** is a series of **comparative experimental studies** on migration and aggregation behaviour both in the laboratory and field. Because dormancy behaviour is not the same in different species of coccinellids, there is a danger of misleading generalizations based on only a few well-documented cases. Therefore specific variations in individual species will be discussed after a general description of the typical behavioural phases.

6.3.1 Phases of dormancy behaviour

6.3.1.1 Pre-diapause

Most coccinellid adults accumulate energy reserves for hibernation (6.4.1.1, 6.4.2) before migrating to hibernation sites (e.g. *C. septempunctata*, 6.3.2.3), while others continue to feed after the main phase of migration (e.g. *Hip. convergens*, 6.3.2.7).

In *C. septempunctata*, diapause is determined in late larval and early imaginal life (6.2.1). The adults in which the 'points' have been set in the direction of diapause feed voraciously and the ingested food is used for building up **large metabolic reserves** in the fat body. These reserves are mainly fat and glycogen, and serve as a source of energy during the long time without food that follows (6.3.2.1, 6.3.2.3). It can be assumed that it is this accumulation of sufficient reserves that represents the stimulus for the beginning of diapause.

In *Hip. convergens* it is presumed that diapause is induced mainly by an absence of aphid food (Hagen 1962; 6.2.7). When aphids are lacking, the coccinellids feed on **alternative plant food** (nectar of flowers or pollen; Chapter 5.2.9) and accumulate reserves from this food. They migrate into the mountain forests, without having built up any great amount of fat. There they feed further, deposit sufficient fat for aestivo-hibernation and move to the final overwintering sites (Hagen 1962).

The assumption that *Cer. undecimnotata* adults do not feed before dormancy and manage with the reserves accumulated by the larva (Yakhontov 1962) has subsequently been rejected (6.2.8).

6.3.1.2 Migration

There is some controversy regarding the use of the term 'migration'. Hagen (1962) argues that only the flight toward aggregational sites can be considered as migration since it is **directional** and under partial control of the beetle. The disbanding of aggregations is a simple **dispersal** flight, as it is not directional. According to Johnson's (1969) concept, both flights are migration, as the direction of both displacements is greatly affected by wind and only at the concluding phase of migration to dormancy sites do the beetles control their direction in relation to visual or other cues. Although this disagreement seems only semantic, it actually goes to the heart of the problem: do coccinellids fly to hibernation sites by a directional flight, or are they brought there passively by wind currents? There is some circumstantial evidence for both viewpoints.

The **aggregations** for dormancy are often formed **at prominent features** of the landscape. For those coccinellids that make lengthy migratory flights (e.g. *Cer. undecimnotata* or *Har. axyridis*), these features may be summits of hills, large rocks or high buildings. They may, however, also be a forest edge, a terrain wave, a shrub, a tree, or a post in a flat landscape for

coccinellids that aggregate within the breeding area but usually in a different habitat or microhabitat (e.g. *C. septempunctata*). These sites are **the same year after year** if the relief remains the same or change if the relief is changed. The creation or removal of triangulation posts, fences and huts provides unintentional experiments which show that the aggregation sites of *Cer. undecimnotata* can change (Hodek 1996).

It seems improbable that coccinellids would always be transported passively to the same places, or that wind currents would be so drastically changed, for example by the erection of a post of 15 cm or less in diameter, that the coccinellids are carried elsewhere. Also direct observations of beetles landing on the top of a hill (in Raná, northern Bohemia) showed that *Cer. undecimnotata* landed actively, e.g. on a shifting person who at that moment was the most prominent object (Hodek 1996). Any air currents existing were too weak to be noticed by the observer. Some authors have even observed that coccinellid adults move against the prevailing wind, but in the direction of mountain peaks (Mani 1962).

Obata (1986) observed that *Har. axyridis* approaching the hibernation site **changed direction in flight**; she regarded their landing as an active response (6.3.2.2). Therefore an analogy has been sought between the aggregation of coccinellids at prominent objects and the habit of flying around hill tops shown by ants and other insects (Alcock 1987) or the flight to feeding sites on trees by *Melolontha*. Such **hypsotactic orientation** has also been widely accepted in ladybirds (Hodek 1960; Hagen 1962; Nalepa et al. 2000, 2005).

Coccinellids have also been observed to make use of **air-currents**, especially in mountain valleys. Savoiskaya (1966) reported that, in the Zailiiskii Alatau mountains of Kazakhstan, the coccinellids fly up the valleys to their hibernation places with the help of a steady breeze blowing up in the daytime. Also *Hip. convergens* seems to be transported to the mountains and back to the valley by air currents (Hagen 1962, 1966; Fig. 6.8).

Individual species differ in migration behaviour. Hagen (1962) considered *Col. maculata* a **'climatotactic' aggregator**. In many species behavioural responses of both types most probably apply (Hodek 1967), but one type may dominate. Johnson (1969) judges that orientation of a migrant for some perceived but far-distant hibernation site is not a tenable concept, but a 1.5 km distant mountain was experimentally

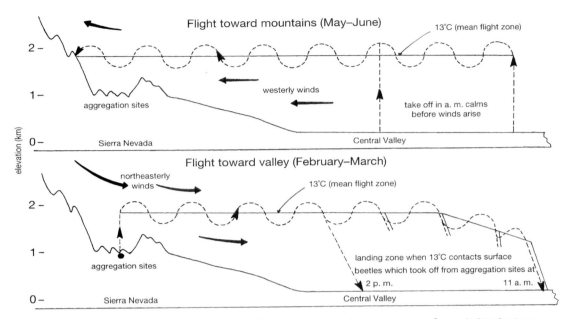

Figure 6.8 Suspected temperature-controlled flight oscillations which *Hippodamia convergens* undergoes in its migratory flights (from Hagen 1962).

established to be a visual cue for migrating *Cer. undec-imnotata* (Iperti & Buscarlet 1972, 1986). In the long distance migrations which Hagen (1962, 1966) hypothesized for *Hip. convergens* (6.3.2.7; Fig. 6.8) it is only towards the end of these long migrations that the hypsotactic response may be shown and that a visually oriented directional flight can guide the coccinellids to a prominent feature. Hypsotaxis may also apply to dormancy aggregations of *Hip. convergens* on mountain tops (e.g. Stewart et al. 1967, 6.3.2.7).

6.3.1.3   Flight and methods for its study

The **tendency to migratory flight** has been studied and **quantified** in several ways. Solbreck (1974) and Rankin and Rankin (1980) used the method of **tethered flight,** respectively for *Col. maculata* and *Hip. convergens*, although details were not identical (see below). Rankin and Rankin (1980) used a low speed fan to **stimulate** the coccinellids to fly. The recorded absolute values for the flight durations are thus different. While in *Hip. convergens* flights >30 minutes were considered to be migratory (6.3.2.7), less than 10% of *Col. maculata* adults flew >10 minutes and this only occurred

during the most suitable period in mid-May, and only aproximately 20% flew for about 1 minute.

The changes in flight behaviour in the second phase of dormancy (December–May) were studied by Solbreck (1974) in *Col. maculata*, also using the tethered flight technique but without additional stimuli other than loss of tarsal contact. A 30 second duration of flight was used as the best criterion for comparing flight tendency of beetles sampled at different dates or exposed to various temperatures and photoperiods before the flight tests. An important 'maturation' of flight behaviour, i.e. an increase in the percentage of individuals flying for 30 seconds or more, was found between mid-April and mid-May, i.e. in the season of dispersal from hibernation sites. An increase in temperature strongly increased the rate of the flight maturation process above a threshold somewhere between 15 and 19°C. Short photoperiods significantly delayed the process in December, but not in late March or even later. Diapause development was evidently not yet completed in December.

The much poorer flight performance of beetles of post-diapause *Hip. convergens* **parasitized** by the braconid *Dinocampus coccinellae* (Chapter 8.3.2.1) was recorded with tethered beetles in a **recording flight mill** (Ruzicka 1984, Ruzicka & Hagen 1985, 1986). The durations recorded for non-parasitized adults, collected before dispersal from hibernation sites, corresponded with Rankin and Rankin's (1980) data for migratory flight, i.e. 45–50 minutes. The speed of flight recorded on the mill averaged 60–70 m/minute (Ruzicka 1984).

Another method for quantifying flight behaviour is to record the incidence of **spontaneous take-offs**. This approach was used with *Myrrha octodecimguttata* (Pulliainen 1964) and *C. septempunctata* (Zaslavsky & Semyanov 1983, Okuda et al. 1986, Okuda & Hodek 1989, Honěk 1990, Nedvěd et al. 2001). This method is **only qualitative**, it simply divides the sample into 'fliers' and 'non-fliers', but does not measure the duration of the flight. Thus we cannot estimate whether the flight is migratory or just trivial. Recording take-off is suitable for comparing samples from different habitats, of different ages or phases of the life cycle. The experimental conditions must of course be identical. The importance of such conditions is shown by two studies on *C. septempunctata*. Very high **take-off frequencies** were recorded for *C. septempunctata* by Honěk (1990) when at 30°C the beetles were **stimulated to fly** by a mild air-current applied for 10 seconds, at the start and in the middle of a 3 minute test period. Under such circumstances, 60% take-offs were recorded in December in the sample from the hibernation site. When the beetles of the same species and region were left to **take-off spontaneously**, without any stimulation, at 25°C and less intense illumination (Okuda & Hodek 1983, Okuda et al. 1986, Okuda & Hodek 1989), take-off frequencies were much lower and more appropriate for the relevant phases of the life cycle. Thus from late July to mid-August, after arrival at their hibernation sites, only 10–20% took off. In spring, before the dispersal from hibernation sites, the incidence of take-off ranged from 30–70%, while after dispersal to the fields it was 20–50%.

Another interesting method for monitoring the tendency to migratory flight (Khrolinsky 1964) was used with *C. septempunctata* (Semyanov 1978a, Zaslavsky & Semyanov 1983, 1986). Within 20 minutes of a short narcosis with ether, the beetles show their urge to fly by opening their elytra and spreading their wings. In outdoor beetles this response coincided well with the periods of migration from and to hibernation sites. A certain doubt about the reliability of the 'wing spreading response' as a criterion of migratory

flight might arise from its occurrence in all young beetles, irrespective of diapause-inducing or preventing photoperiods.

The discovery of **regeneration of flight muscles** in *C. septempunctata* (Okuda et al. 1986, Okuda & Hodek 1989; 6.3.2.3, 6.4.1.4; Fig. 6.12) in October in the hibernation site (seven months before dispersal from there) conflicts with the general assumption that ladybirds do not prepare for dispersal until as late as when they end dormancy in spring under the influence of increasing temperatures. Later more detailed 2-year studies on dormant beetles of the Czech *C. septempunctata* population confirmed the unexpected increases in **respiration** in the 2 years: rising from about 450 to 750 μl/g/h O_2 from late August to mid-September 1994 and from 750 to 1050 μl/g/h O_2 from mid-September to 10 October 1995. In 1995 it was also found that the volume of **flight muscles** gradually increased from 1.8 μl in early September to 2.15 μl on 23 October (the last record) (Nedvěd et al. 2001). Thus changes in both the volume of flight muscles and the respiration rate indicate a very **early tendency to re-activation**, about two months before diapause is completed in mid-winter (Hodek & Ruzicka 1979). Such an apparently non-adaptive event remains an enigma as the greater muscles could consume more reserves during any later periods of temporary increase in temperature. We might hypothesize that the unexpected findings on respiration and flight muscles (Okuda et al. 1986, Okuda & Hodek 1989, Nedvěd et al. 2001) are related to the bivoltine cycles of some Mediterranean populations with separated hibernation and aestivation periods (6.2.1.4). After a hot and dry summer a regrowth of vegetation and aphid populations, thanks to higher humidity, may enable reproduction. However, no detailed study reports such an intercalated reproductive period in the field. There are just indications of such a possibility from the coastal plain of Israel (Bodenheimer 1943) and from one locality on the French Riviera (Iperti 1966a).

Iperti (1986) studied the **migration flight**, particularly the **take-off**, in a complex manner by relating the migration behaviour to measured temperature and wind speed and by marking the insects. The effect of environmental conditions on take-off can only be understood if the individual factors are considered as a complex by the use of Richardson's index of air-turbulence (Iperti 1986), which indicates the relation between the relative influence of the mechanical and thermal energy of air. **Temperature and wind speed** were recorded at the same two heights, 0.25 m and 1.25 m, in a study of *Cer. undecimnotata* (Iperti et al. 1983, 1988, Iperti 1986). A decrease in Richardson's index was positively correlated with a decrease in the incidence of females taking-off for a migratory flight. **Marking with iridium** (the stable isotope 191), an element normally absent from the body of coccinellids, made it possible to show that the released *Cer. undecimnotata* adults directed their flight from a place on the plain (at 395 m) to a 732 m hill 1.5 km away. In samples of beetles from the hill the isotope 191 was modified in the lab into 192 by irradiation and then detected by gamma-spectroscopy (Iperti & Buscarlet 1972, 1986, Iperti & Rolley 1973).

Data on coccinellids trapped during four seasons (1992–95) in **window traps** mounted on a 15 m high tower in New Brunswick, Canada, give an interesting picture of height and **seasonal distribution of flights** for several abundant species (Table 6.13). The median height at which *C. septempunctata* (*n* = 349), *Hip. convergens* (*n* = 279) and *A. bipunctata* (*n* = 551) were trapped was 3.8 m, 0.8 m and 5.3 m, respectively (Boiteau et al. 1999). *Hippodamia convergens* was not caught in flight till as late as August, while *A. bipunctata* was already flying in spring; maximum catches were recorded for *A. bipunctata* in July and for *C. septempunctata* about a month later (Boiteau et al. 1999; Table 6.14). Both the heights and dates of these catches may indicate trivial rather than long-distance flights.

Observations in Hungary showed the different character of pre- and post-hibernation movements in *C. septempunctata* in terms of the **height of flight**. In a year when meteorological conditions were favourable for flight, negligible numbers of beetles were caught in spring by Malaise traps at 12.5–14.5 m and 25–27 m, but high numbers were trapped in late summer (Sarospataki & Marko 1995). It is conceivable that the long distance migratory flight (in contrast to gradual dispersal) operates at a greater height than the dispersion flight after hibernation. Beetles caught at shrub level apparently represent individuals which hibernate near to their breeding sites.

In the course of 2000–2002, *Har. axyridis* was the most abundant species (*n* = 325) caught in **blacklight traps** (Koch & Hutchison 2003). The second most abundant was *Hip. tredecimpunctata*. The other species were very rare: *C. septempunctata* (2 individuals), *Col. maculata* (1), *Cycloneda munda* (1). Non-specified coccinellids were reported to be significantly more attracted (*n* = 30 samples) to blacklight fluorescent lamps (mean number of individuals = 2.46) than to black light blue (1.56) or to cool light (0.36). Blacklight and 'blacklight–blue' lamps have a major peak in the ultraviolet region at around 365 nm. The black light had other peaks in the visible range around 430 and 540 nm (Nabli et al. 1999).

In relation to the **adaptive aspects of dormancy behaviour,** Honěk (1989) stresses the fact that it is often warm sites that are chosen for hibernation (usually the south, southwest, or west aspect) and assumes that an appropriate temperature/humidity relationship and sufficient aeration may **prevent the spread of** diseases (particularly **mycosis**) during winter. In addition to the hypsotactic visual orientation, he suspects that *C. septempunctata* also perceive the overall temperature of the site on warm days. The preference for north-facing situations on trees, reported from England for *Chil. bipustulatus* (Majerus & Kearns 1989), is a rare exception.

Semyanov (1965b) states that the tendency to hibernate on at least slightly elevated ground may have a survival advantage in lowland regions when extensive parts of the **plains** are **flooded**. Many other authors have speculated on the adaptive significance of hibernation in the mountains and as aggregations (summarized in Hodek 1960). It is quite possible that mass hibernation facilitates **contact** with beetles of the **opposite gender** which will be important for less abundant species and leads to crossing between beetles that have developed under different climatic conditions. In regions with warm climates, hibernation in high mountains may **retard** the beginning of **spring activity** of the beetles till aphids in the plain have multiplied sufficiently. A similar function may be attributed to hibernation among the litter in plains, under mosses and in other humid microhabitats where cooling is caused by evaporation.

6.3.1.4 Aggregations

Aggregating is a specific behavioural feature connected with dormancy in many Coccinellidae. Coccinellids are led to form aggregations when brought passively (wind currents) and actively (hypsotaxis; 6.3.1.2) repeatedly to the same dormancy locality and habitat (e.g. a hill or a line of high trees). After arriving at the site, the beetles are led to specific portions of habitats by their responses to physical factors (**hydrotaxis, thermotaxis**), and **negative phototaxis, geotaxis** and

Table 6.13 Weekly vertical flight frequency distribution for three coccinellid species (week 1, 17–23 May), New Brunswick, Canada (Boiteau et al. 1999, modified).

	Week																			
	1	2	3	4	5	6	7	8	9	10	11	12	13	14	15	16	17	18	19	20
Adalia bipunctata																				
Height 0.8–6.8 m	--8--		5	13	--18--		58	40	28	27	18	--14--			--10--				0	0
Height 8.8–14.3 m	--16--		7	9	--16--		26	17	8	10	5	--4--			--5--				0	0
Coccinella septempunctata																				
Height 0.8–6.8 m	10		--9--		--10--		--29--		16	38	39			--18--				–	–	
Height 8.8–14.3 m	5		--4--		--8--		--4--		4	6	17			--2--				–	–	
Hippodamia convergens																				
Height 0.8–6.8 m	0	0		0	0		--8--				53	77	0	74		--12--				
Height 8.8–14.3 m	0	0		0	0		--0--				2	2	0	1		--0--				

Data are from 1992 for *Adalia bipunctata* and *Coccinella septempunctata* and from 1995 for *Hippodamia convergens*.

Table 6.14 Seasonal distribution of coccinellid species trapped between 1992 and 1995 with annual catch greater than 40 (week 1, 17–23 May), New Brunswick, Canada (Boiteau et al. 1999, modified).

Species	No. of individuals per year				Height of 50% capture (m)	Week									
	1992	1993	1994	1995		1-2	3-4	5-6	7-8	9-10	11-12	13-14	15-16	17-18	19-20
Hyperaspis bigeminata	7	7	10	47	5.3	0	46	18	0	0	3	1	2	1	0
Anatis mali	27	1	9	40	5.3	1	36	9	3	17	7	3	0	0	1
Coccinella septempunctata	244	32	20	53	3.8	20	48	40	43	82	91	12	3	6	2
Adalia bipunctata	362	69	51	32	5.3	25	64	116	182	109	43	12	10	4	6
Hippodamia convergens	5	15	30	229	0.8	0	2	2	1	6	134	76	10	21	27

Less abundant species (i.e. ≤40) are as follows: *Propylea quatuordecimpunctata, Hippodamia tredecimpunctata, Myzia pullata randalli, Coccinella hieroglyphica kirbyi, Hyperaspis undulata, Brachyacantha ursina, Hippodamia parenthesis, Anisosticta bitriangularis, Chilocorus stigma, Mulsantina hudsonica, Scymnus brullei, Coccinella trifasciata perplexa, Calvia quatuordecimguttata, Psyllobora vigintimaculata* and *Harmonia axyridis*.

thigmotaxis cause them to hide in microhabitats such as the space under a stone, the crevices in a rock, in a grass tussock, or in tree bark. For example, the beetles hibernating in litter are apparently guided by negative phototaxis and positive geotaxis (possibly also positive hydrotaxis) to hide on or near the ground after a period of sitting on the plants. During this **period of 'waiting'** their behavioural patterns gradually change from those which led the beetles to the hibernation habitat during migration. Hydrotactic responses prevail in climatotactic species such as *Hip. convergens* and *Col. maculata* (Hagen 1962; 6.3.1.2). In aestivating *C. novemnotata* (McMullen 1967a) and hibernating *Cer. undecimnotata*, ventilated crevices are preferred (Hodek 1960, Yakhontov 1962, Iperti 1966a) as they apparently **reduce the danger of mycosis** (Iperti 1964, 1966b) (Chapter 8.4.3). Honěk et al. (2007) first studied for hibernation aggregation in **Hip. variegata**.

Pulliainen (1963, 1964) studied the responses of **Myrrha octodecimguttata** collected in bark crevices in the final period of dormancy and re-activated by a higher temperature in the laboratory. The beetles showed a strong hydro-negative reaction, which only changed after a prolonged desiccation. They reacted indifferently to long-wave light, and photo-negatively to short-wave light. After desiccation, they became strongly photo-negative to long-wave light also. The optimum relative humidity was about 30–40%. Novák (1966) found that the geotactic reaction was slightly positive in dormant *C. septempunctata* collected when 'waiting' on trees before they hid in the litter. In humid environments, the response changed to be slightly geo-negative.

In the laboratory, the formation of clusters of *Hip. convergens* adults increased with a decrease in temperature from 35°C to 15°C. Temperature also affected the response to light: the beetles became photo-negative when the temperature dropped below 5°C (Copp 1983). These experiments were performed from January to March, i.e. relatively late in dormancy (August–April) when diapause had most probably been terminated (6.1).

The direct stimuli which result in aggregation may – apart from thigmotaxis – also include **chemotaxis**, i.e. **attraction by semiochemicals** (Al Abassi et al. 1998; Chapter 9.7). The specific odour of coccinellids is penetratingly noticeable to humans, and speculation has been published on the role of pheromones in the formation of aggregations. However, the evidence is still limited. Savoiskaya (1966) assumed an aggregative function for odour, based on her observations on the genus *Adalia*. It may be that a pheromone attracts later migrants to those beetles which have arrived earlier. Yakhontov (1962), Majerus and Kearns (1989) and Majerus (1994) supposed that dead beetles remaining from the previous hibernation had an attraction, but this has been refuted by G. Iperti (unpublished). Indeed, Novák (1965) has proved experimentally that dead beetles are more likely to act as a repellent than as an attractant.

Copp (1983) reports the ability of **Hip. convergens** adults to locate clusters of living conspecifics by **olfactory cues** in special test chamber in which visual cues and direct contact were eliminated (Table 6.15). The volatile sesquiterpene (-)-beta-caryophylene, suspected to act as an aggregation semiochemical, attracted both genders of **Har. axyridis** in a bioassay, while only males responded in a four-arm olfactometer (Verheggen et al. 2007).

In coccinellids, which are distasteful prey with aposematic colouration, gregariousness may have

Table 6.15 Distribution of adults of *Hippodamia convergens* in the arena above a chamber that contains a beetle cluster in one quadrant (Copp 1983).

Quadrant of arena	Contents in the quadrant below	Without CO$_2$ absorbent		With CO$_2$ absorbent	
		total no.of beetles	mean ± SD	total no.of beetles	mean ± SD
1	beetle cluster	162	7.7 ± 1.9	53	6.6 ± 1.4
2	blank	15	0.7 ± 0.9	11	1.4 ± 1.6
3	blank	20	0.9 ± 1.4	5	0.6 ± 0.5
4	blank	20	0.9 ± 1.1	11	1.4 ± 1.2

Ten adults were tested in one replicate; they were counted after 48 hours.

Table 6.16 The mean numbers (*n*) per m² in autumn litter samples and spring emergence samples, the proportion of living beetles in the autumn samples, and the overwintering survival (density spring emergents/density in autumn), for *Hippodamia tredecimpunctata* (H13), *Coccinella septempunctata* (C7), *Coccinella transversoguttata* (CT), *Hippodamia convergens* (HC), and *Hippodamia parenthesis* (HP), and all coccinellids in the beach-ridge forest at the Delta Marsh Field Station, Manitoba, Canada (Turnock & Wise 2004).

Species		1992 Autumn	1993 Spring	1993 Autumn	1994 Spring	1995 Spring
H13	n/m²	47.3	17.5	65.6	9.1	27.9
	Survival	0.95	0.37	0.97	0.14	—
	n	355	54	492	28	86
C7	n/m²	31.3	0.65	45.3	2.3	1.9
	Survival	0.72	0.02	0.73	0.05	—
	n	235	2	340	7	6
CT	n/m²	6.9	0.32	8.5	0	0
	Survival	0.37	0.05	0.16	0	—
	n	52	1	64	0	0
HC	n/m²	101	0	2.1	0	0
	Survival	0.38	0	0.5	0	—
	n	8	0	16	0	0
HP	n/m²	0.4	0	0.13	0	0
	Survival	0.67	0	1	0	—
	n	3	0	1	0	0
Total	n/m²	87.1	18.5	122.4	11.7	30.5
	Survival	0.81	0.21	0.81	0.1	—
	n	653	57	913	36*	94*

*Includes one specimen of *Anisosticta bitriangularis* in 1994 and two specimens of *Calvia quatuordecimguttata* in 1995.

evolved as a means of **defence against predation**. Above a certain minimum group size, group members have a lower rate of death from predation than solitary individuals (Sillén-Tullberg & Leimar 1988).

Both **monospecific** and **heterospecific aggregations** of coccinellids may be formed. As a general rule, only monospecific clusters are formed whenever several species hibernate in the same habitat (McMullen 1967a), though the presence of a few adults of other species has been reported. Only twice have mixed clusters of *Cer. undecimnotata* with *A. bipunctata* been observed. It is more usual to find heterospecific aggregations of *C. septempunctata* with *Cer. undecimnotata* in grass tussocks on the tops of hills (Hodek 1960), especially where rocks are missing. Pulliainen (1966) has reported that *Scymnus suturalis* and *Aphidecta obliterata* were admixed to aggregations of *Myrrha octodecimguttata*. Kuznetsov (1977) found multispecies aggrega-

tions most often when *Har. axyridis* and sometimes also *Oenopia* (= *Synharmonia*) *conglobata* were together with *Aiolocaria hexaspilota* (= *mirabilis*), as well as other insects. Heterospecific aggregations of overwintering coccinellids (Table 6.16) were found in leaf litter in late October in a beach-ridge forest on the southern shore of Lake Manitoba (Manitoba, Canada) (Turnock & Wise 2004).

Some of the mass aggregations, e.g. those washed up on beaches (Lee 1980a) are, however, not directly related to diapause, but originally produced by hunger and thirst when the emergence of large numbers of ladybirds coincides with the disappearance of aphids in the landscape (Chapter 5.4.1.5).

Transmission between beetles of the ectoparasitic **fungus** *Hesperomyces virescens* between beetles occurred during the hibernation of *Har. axyridis* in aggregations. It is both the social contact in multi-layered piles in aggregations and mating that enables such transmission in the winter (Nalepa & Weir 2007; Chapter 8.4.3).

6.3.1.5 Emergence from dormancy sites

At the end of dormancy, the beetles gradually change their behaviour. The inactivity in the hiding places changes to **slight mobility**, sometimes already long before dispersal. This final stage is conspicuous and has often been observed in coccinellids hibernating in litter at forest edges. Here coccinellids stay for a prolonged period on the trees (especially on young pines) in the dormancy site **before dispersal** (e.g. Bielawski 1961; 6.3.2.3). During this closing period of dormancy, the coccinellids have often been observed **mating**. Mating was verified by dissection of the spermathecae before dispersal from the hibernation sites of *Cer. undecimnotata* and *C. septempunctata* (Hodek & Landa 1971, Hodek & Ceryngier 2000, Ceryngier et al. 2004) and of *Col. maculata* (Solbreck 1974). About half of *C. septempunctata* females already had sperm in their spermathecae in September when they were in early diapause in their hibernation site (Hodek & Ceryngier 2000, Ceryngier et al. 2004; 6.4.1.2).

Dispersal from dormancy sites continues **over several weeks** (Hodek 1960, Bielawski 1961, Savoiskaya 1965, McMullen 1967a). The onset and progress of dispersal is regulated by the ambient temperature, as the second phase of dormancy is simple quiescence (6.1). Usually a prolonged **increase of average temperature** over 10°C is required to induce the beetles to emerge (Hodek 1960). In the laboratory, the distance walked by diapausing *C. septempunctata* was recorded; the best correlation for mobility was with a previous exposure of 10–15 days to the increased temperatures (Ruzicka & Kindlmann 1991).

Different species, even if they have hibernated at the same site, **do not disperse simultaneously**. In Bohemia (western Czech Republic) *A. bipunctata* is the earliest species, followed by *C. quinquepunctata*, *P. quatuordecimpunctata* and *Tytthaspis sedecimpunctata*, then by *C. septempunctata* and *Hip. variegata*, while *Cer. undecimnotata* leaves very late (Hodek 1960). A similar, but not identical picture was reported by Bielawski (1961) at a hibernation site under young pines in Poland (Warszawa–Bielany). *Adalia bipunctata*, *A. decempunctata* and *C. quinquepunctata* leave about a month earlier than *C. septempunctata* and *P. quatuordec-*

impunctata. Whereas these two observations agree that *A. bipunctata* is an early and *C. septempunctata* a late disperser, Banks (1955) reported the opposite observation from England: *C. septempunctata* appeared on nettles one month earlier than *A. bipunctata* and *P. quatuordecimpunctata*. However, for *C. septempunctata* the nettles were probably the first habitat visited after dormancy, while *A. bipunctata* and *P. quatuordecimpunctata* (due to their habitat preference; Chapter 4) had previously spent some time in another habitat of shrubs or trees.

The flight of coccinellids from hibernation sites to breeding habitats is considered a **dispersal flight** (Hodek et al. 1993; 6.3.1.3). It is not a long-distance flight, but usually a **step-wise process**. Beetles of several species have been observed to fly from the dormancy site only to adjacent or nearby fields at first (Hodek 1960, Okuda & Hodek 1983). If suitable aphids (Chapter 5.2.2) are there in sufficient quantity, the coccinellids will only gradually disperse further afield. In the dispersing beetles, the previous great depletion of reserves during dormancy would most probably cause hunger; thus the dispersal flight may gradually change into the **normal, trivial (appetitive) flight** (Hodek et al. 1993).

6.3.2 Behaviour of individual species

For a detailed survey on behaviour of ladybirds related to phases of dormancy and the difference between climatotactic and hypsotactic aggregators, see Hagen's (1962) classic review. A comprehensive survey of varied hibernation habitats of ladybirds in Britain is on pp. 158–159 in Majerus's (1994) book.

6.3.2.1 *Ceratomegilla (=Semiadalia)* *undecimnotata*

The migratory and aggregation behaviour of this species is similar in the three regions of central Asia (Dobzhansky 1925, Radzievskaya 1939, Yakhontov 1966), the Czech and Slovak Republics at the northern limit of the distribution area (Hodek 1960, 1967) and southeastern France (Iperti 1966b). The dormancy sites are situated **at prominent features on hills or mountains** (large rocks, heaps of stones, shrubs or other plants, or artificial structures, such as posts, triangulation points, but only very rarely buildings or their ruins). It is assumed that such sites are chosen as the result of **hypsotactic responses** (6.3.1.2). *Ceratomegilla undecimnotata* prefer rock cracks, especially those exposed to wind, where their mortality is reduced. In the absence of such cracks, the beetles hibernate at the base of bushes, but then there is a great danger of mycosis. The petrography of the site is not crucial, hibernation quarters can be found both on limestone and on igneous rocks (e.g. basalt). The **altitude** of the site above sea level varies: in central Asia it is much higher (2000 m) than in the Czech Republic and Slovakia (400–900 m) or in France (usually above 700 m). The majority of dormancy sites in these regions are situated on a **southwest aspect**. According to Iperti (1966b) the attraction of this aspect may be the afternoon insolation of these sites (6.2.8).

The coccinellids arrive at the hibernation places over a period of several weeks on warm calm days and leave similarly. At the onset and end of hibernation they creep around close to the shelter during the day, forming numerous small clusters during the night or colder spells. For the cold part of the period of dormancy they aggregate in a few large clusters. Their spring **emergence** depends on the **ambient temperature**. In the Louny-hills (northern Bohemia) emergence occurs after a period of mean temperature at 12–14.5°C (Hodek 1960). Yakhontov (1962) assumed that a certain degree of gonad maturation is the signal for emergence. In Bohemia, however, females leaving the hibernation sites had ovaries at different stages of maturation. The females leaving the hibernation sites early lack the stage of the first oocyte. If they disperse later, this stage has already been attained at the dormancy site. In males, the tissue of the testicular follicles is already fully active 2–4 weeks before emergence and the males fertilize the females at the dormancy site (Hodek & Landa 1971, Hodek & Ceryngier 2000, Ceryngier et al. 2004; 6.4.1.3). Year after year, *Cer. undecimnotata* uses the same dormancy sites if the silhouette of the hill has not changed.

6.3.2.2 Hypsotactic species *Harmonia axyridis, Har. conformis* and *Aiolocaria hexaspilota* (=*mirabilis*)

Harmonia axyridis has dormancy behaviour similar to that of *Cer. undecimnotata*. After a short report from east Siberia (Telenga & Bogunova 1936), the first detailed study was undertaken in **Japan**, near Kyoto.

All 14 hibernation sites visited by Obata (1986) were situated near **prominent features** of a whitish or light colour. When the beetles were offered black, red, green, yellow and white boards near one of the aggregation sites below a large rock at the top of Mt. Shiroyama, they landed mostly on the white boards. The white board was more attractive when the sun shone brightly. Obata (1986) and Obata et al. (1986) suppose that the whitish or light colour may be an indication of a dry site (but see Nalepa et al. 2005 below). Obata's observation that *Har. axyridis* adults changed flight direction towards the white board from some distance is important; thus their landing appears to be directional and visually oriented (Obata 1986).

After its successful introduction to the USA and Canada, and the subsequent increase in population, *Har. axyridis* has become a **nuisance by entering buildings** in search of dormancy shelter (Koch & Galvan 2008). Sometimes the beetles **also** use **partly open structures**. Since 1993 repeated winter aggregations of thousands of *Har. axyridis* have been observed in a concrete observation tower (20 m high) built on the crest of a ridge (at 326 m) in south-central Pennsylvania. During very cold winters (e.g. 1994) a majority of the beetles died (Schaefer 2003; see 6.4.4).

In habitations, *Har. axyridis* are totally unwelcome. They irritate people just by their mass presence, but also provoke allergic reactions. This problem has initiated an increased interest in the ethological aspects of their hibernation with the practical aim of preventing beetles entering buildings. A series of studies in the USA focussed on the **orientation** of *Har. axyridis* to their dormancy sites. Nalepa et al. (2000) **contested distant orientation** of arriving beetles **to** the **conspecifics** present in these sites. They found little evidence for volatile aggregation pheromones, but did not exclude contact chemoreception with those conspecifics that had arrived earlier. Thus the suggestion that pheromones are important for the orientation of the beetles remains open (9.7).

Important progress in our understanding of the behaviour of *Har. axyridis*, when searching for dormancy sites, has come from a field study in which black stripes on white backgrounds but with different degrees of contrast were offered to **alighting ladybirds**. The beetles significantly **preferred** the targets with **contrast** (Nalepa et al. 2005). This showed that *Har. axyridis* is **not attracted to white colours per se**.

This pre-dormancy flight occurred when the temperature rose above 21°C with peaks between 14.00

and 16.00 hours, shifting to earlier hours as the season advanced (Nalepa et al. 2005). In Monticello, Florida, *Har. axyridis* adults arrived at the hibernation sites as late as mid-November to early December, due to the extended summer season (Riddick et al. 2000). This indicates a multivoltine life cycle (6.2.9).

To deter the invasion of homes by *Har. axyridis*, menthol or camphor were used in Maryland and Florida to treat the crevices through which the beetles enter houses. Camphor was more efficient than menthol, but both substances evaporated rather quickly (Riddick et al. 2000).

The dormancy behaviour in eastern Australia of the introduced species ***Harmonia conformis*** has also been described as **hypsotactic** (Hales et al. 1986). Both in summer and winter the aggregating coccinellids were found on mountain tops (800–1000 m) in crevices of radio/TV towers and telegraph/beacon poles, till 22°N (in Queensland). On one occasion a 1 m × 1 m × 0.5 m box was completely filled with dormant *Har. conformis*.

Aiolocaria hexaspilota, a univoltine predator of chrysomelids which lives in the Ussuri region of **east Siberia**, also uses **cracks** in **rocks on bare hills** as dormancy sites (Kuznetsov 1977). From late August to early September the beetles fly first from the forests to their edges and land on well-insolated trees. Only later, after the first spells of frost, when the average temperature falls to around 12°C, do the beetles migrate to well-insolated slopes of rocky hills or also to buildings. Exceptionally, the earliest beetles may arrive in late August but usually the migration occurs from late September to early October (Table 6.17). On the hills, the

Table 6.17 Arrival of adults of *Aiolocaria hexaspilota* to their dormancy sites on a rocky hill in the Ussuri region of east Siberia (monitored area 0.25 m², time of counting 20 min) (Kuznetsov 1977).

Date 1970	Number of beetles counted at given hours					
	8	10	12	14	16	18
28 Sept.	0	8	32	25	21	1
29 Sept.	0	19	38	69	29	5
30 Sept.	0	21	15	87	20	6
2 Oct.	0	17	27	80	19	13
3 Oct.	0	7	18	41	8	4
6 Oct.	0	2	24	31	2	0
9 Oct.	0	0	1	3	0	0

aggregation sites are situated at any altitude, on the southern, southwestern or western slopes and the site is determined by the presence of rocks. Clusters from 5–20 to several hundred individuals occur in the crevices which may be as deep as 2 m. In such crevices the temperature is much higher than ambient: even if the temperature falls to as low as −32°C, it does not decrease below −6°C at 50 cm deep in a crevice. In the spring the adults mate and disperse in May at an average temperature of 10–12°C.

6.3.2.3 *Coccinella septempunctata* and other species dormant in the litter

Attention should be drawn to the fact that old references to hibernation of *C. septempunctata*, which describe mass assemblages between rocks on mountain summits (for earlier references see Hodek 1960) may refer to *Cer. undecimnotata*. The seven-spotted form of *Cer. undecimnotata* is often confused with *C. septempunctata*, even by professional entomologists. G. Iperti (unpublished) has verified that such confusion occurred in Fabre's (1879) data from Mont Ventoux and St. Amand (France). The same confusion has probably occurred elsewhere.

Dormancy behaviour of *C. septempunctata*, in contrast to that of *Cer. undecimnotata*, is heterogeneous even within the same geographical area. The only safe generalization is that *C. septempunctata* always hibernates **on the ground**: under stones, in litter, in holes in the soil surface, near the base of plants in grass tussocks, but never in the cracks of tree bark or walls – in contrast to *A. bipunctata* (6.3.2.5). The species mostly forms only small- or medium-sized aggregations, not exceeding tens of beetles. In lowlands the usual dormancy sites of *C. septempunctata* are situated at forest edges, in clearings or in windbreaks. If no such site is close by, *C. septempunctata* hibernates near isolated shrubs or other plants, or on slight unevennesses of the terrain. No special differences seem to occur between individual countries (Hodek 1996).

The flight of *C. septempunctata* to dormancy sites is often step-wise, particularly when the beetles have not completed feeding and continue to forage on the way (Nedvěd et al. 2001). Such pre-hibernation roving of *C. septempunctata* adults was monitored in a detailed way over 5 years (1998–2002) in **central Italy** (Ricci et al. 2005). After the field crops matured and the aphids there had disappeared, the beetles moved to wild plants in increasingly elevated habitats: to about

1250 m in early July and later up to 1800 m. Their summer feeding on several species of aphids, pollen and spores of fungi was observed and quantified by gut analysis. In the period between mid-June and mid-July, less than 40% of adults had aphids in their gut. This percentage increased again later. Also in central Italy, individual matings may be observed during summer, which provide sperm for storage by the females during winter – similar to findings in central Europe (6.4.1.2, 6.4.1.3). In August some adults stop feeding, enter diapause and aggregate under stones on mountain tops or in plants, e.g. among the leaves of *Verbascum thapsus* (Ricci et al. 2005).

After their arrival at the dormancy site, *C. septempunctata* remain clustered on the plants close to their subsequent hiding places (during August and September in **Bohemia**) and only hide gradually as the air becomes cooler. Very often they can be found at the ends of twigs of young pines (Hodek 1960, Bielawski 1961, Semyanov 1965a,b, Klausnitzer 1967) or between the ear spathes of maize plants left after harvest (Hemptinne 1988). In a similar way they stay on the vegetation near the hibernation site before their dispersal in spring.

The largest aggregations of dormant beetles have been found in mountains or on hills (Hodek et al. 1977, 1989, Honěk 1989, Hodek & Okuda 1993, Honěk et al. 2007). The populations from **montane** and lowland dormancy **sites** differ. **Intensity of diapause** (as measured by duration of oviposition delay; 6.1) was much **greater** in individuals from 1500 m (peak of Mt. Cousson) than in the beetles sampled from a nearby hill of about 600 m (both sites in the region of Digne, France) (Hodek et al. 1977; 6.2.1.2). The coccinellids from about 1400–1500 m (peaks of the Giant Mountains = Krkonose, northern Bohemia) were much larger than those in lowland hibernation sites at forest edges, at about 300–400 m (Honěk 1989). *Coccinella septempunctata* has a plastic life cycle (6.2.1; Tables 6.1 and 6.3) which is again reflected by the above differences. Presumably the beetles entering diapause earliest in the season are those individuals with a strong tendency to a univoltine life cycle. Honěk (1989) and Honěk et al. (2007) assume that larger individuals which develop earlier in the breeding season have a better ability to complete long range flight. Thus the larger individuals with deeper diapause occupy the dormancy sites in mountains.

Winter **survival** of *C. septempunctata* was recorded in artificial hibernacula from early November 1993 to early April 1995 in Bohemia. Hill top populations from the Giant Mountains (Krkonose, 1480 m) and Rana (400 m) were compared with those from the lowland (300–360 m). One particular site at 1100 m was not on a hill top but in a meadow. An interesting difference was found between the two altitudes. A high proportion in the meadow of small individuals, less capable of long flight, indicated poor nutrition of larvae (and/or pre-hibernation adults) and presence of local ladybirds. In this sample medium sized individuals had only 1.7% survival, while in the high altitude sample the survival was 83.3% in medium sized males and 86.7% in medium sized females (Zhou et al. 1995, Honěk 1997). As the cause of mortality was not studied, it cannot be excluded that the high altitude hill top sample was less prone to fungal infection (Chapter 8.4.3) and this contributed to higher survival.

Of the common coccinellids, **C. quinquepunctata** has a hibernation behaviour similar to that of *C. septempunctata*. However, it is found far more often than the latter on young pine trees at forest edges (Hodek 1960, Bielawski 1961, Semyanov 1965a, b, Klausnitzer 1967). It is reported from shingle banks (Majerus 1994), and can also be found among small stones on hills (e.g. the Louny-hills in north Bohemia) used as hibernation quarters by *Cer. undecimnotata* and *C. septempunctata* (Hodek 1960 and unpublished). *Coccinella quinquepunctata* and *P. quatuordecimpunctata* were the most abundant species hidden in pine cones during hibernation (Ruzicka & Vostrel 1985) on the same hills.

Some **other species** are usually present in similar hibernation sites on the plain (the litter or upper soil layer at forest edges) (Hodek 1960, Bielawski 1961, Semyanov 1965a, b, Klausnitzer 1967, Novák & Grenarová 1967). Such species are *Coccinula quatuordecimpustulata*, *P. quatuordecimpunctata*, *Hip. variegata* (in dry places), other *Hippodamia* spp. (if moist habitats are nearby), the phytophagous *Subcoccinella vigintiquatuorpunctata* and the mycophagous *Psyllobora* (= *Thea*) *vigintiduopunctata*, especially in rather moist places. *Tytthaspis sedecimpunctata* formed the largest aggregation (over 10,000 individuals) ever found in Britain (Majerus & Kearns 1989). Evans (1936) found a dormancy aggregation of **Tytthaspis sedecimpunctata** on a wooden post in Berkshire, England in December. Legay & De Reggi (1962) found aggregations (50–150 individuals per 100 m²) of the same species (together with 1% of **Psyllobora vigintiduopunctata**) near Lyon, France under litter at the base of trees on a small rise overtopping the surrounding fields by only 20 m. Whereas in central Europe *T. sedecimpunctata* is usually found in dry places, in central Italy its usual hibernation sites are the wild vegetation of ditches and embankments. In central Italy this habitat is also used for hibernation by **Rhyzobius litura** which, however, remains there also during the breeding period; at least 90% of the apterous population do so (Ricci 1986).

The Australian species **C. leonina** was found in similar dormancy sites as *C. septempunctata* both in summer and winter: in soil cracks, near the roots of herbaceous plants, under pieces of cloth, and once at the top of a hill close to its feeding habitats (Hales et al. 1986; also 6.2.10). In addition to the above species, Savoiskaya (1983, p.174) reported from Kazakhstan that *Vibidia duodecimguttata*, *Bulaea lichatschovi*, *Exochomus flavipes* and *Parexochomus melanocephalus*, and from Siberia that *C. hieroglyphica* and *Hip. septemmaculata*, also hibernated in litter or in the upper soil layers.

Besides the above species for which litter in the widest sense typically provides shelter, *Calvia quatuordecimguttata* and *Exochomus quadripustulatus* sometimes also occur in litter. These two species are equally abundant in bark crevices. Litter also serves as the dormancy site for *Stethorus pusillus* (Putman 1955, Berker 1958, Savoiskaya 1983) and *S. punctum punctum* (Felland & Hull 1996), although some authors report *Stethorus* spp. hibernating in the bark (Moter 1959).

Hibernation behaviour of the abundant **P. quatuordecimpunctata** was not well known until the detailed study on populations from an agricultural but extensively wooded area of the Province of Hainaut in western Belgium (Hemptinne 1988). Hemptinne compared litter samples from the forest edge and 20 m into the forest, and found that *P. quatuordecimpunctata* preferred the inner site for hibernation, with 1.11 beetles/m^2 at the edge compared with 1.41 beetles/m^2 inside the wood. By contrast, *C. septempunctata* preferred the edge (0.98 beetles/m^2) to the interior of the wood (0.78 beetles/m^2), and an orientation facing south or west. The compass orientation was not important for *P. quatuordecimpunctata*. The well-known preference for higher elevations was again shown by *C. septempunctata* but was less obvious with *P. quatuordecimpunctata*. In Poland *P. quatuordecimpunctata* hibernates more often in pine than in oak litter (Bielawski 1961). For England, Majerus (1994) reports a diversity of hibernation sites near to the ground, and small groups of 2–3 *P. quatuordecimpunctata* are found.

6.3.2.4 *Coleomegilla maculata*

Moist places are preferred by *Col. maculata* (Hodson 1937, Solbreck 1974, Benton & Crump 1979, Jean et al. 1990, Roach & Thomas 1991). The vertical distribution of active adults on the plants and their susceptibility to desiccation also indicate a preference for higher humidity (eg. Ewert & Chiang 1966; Chapter 4). The migration to dormancy sites occurs at low level and the aggregations are found at or near the edges of open fields at the base of dominant trees (Parker et al. 1977, Benton & Crump 1979). Three dormancy sites, described by the latter authors in New York State, were at the base of willow trees and a fourth was at the base of a large poplar. In one case the aggregation was located near a small pond. Also a large aggregation near Montréal (D. Coderre, unpublished) was situated below a large willow tree near a small reservoir. In Vermont, USA, the attractive trees were mostly maples (Parker et al. 1977) and pecans in the upper coastal plain of South Carolina (Roach & Thomas 1991). All these sites were repeatedly visited by the beetles year after year. In *Col. maculata* (similarly to *Hip. convergens*) the orientation to the dormancy sites is mainly by **climatotaxis**, and it is most probable that one of the major factors is humidity. During the influx, the beetles in New York alighted on undergrowth vegetation, including wild raspberry (*Rubus occidentalis*) and sensitive fern (*Onoclea sensibilis*). With drops in temperature, the beetles moved down to the ground. According to Roach and Thomas (1991) the hibernation sites are characterized by a deep layer of largely decomposed organic matter. In the mid-Atlantic states of the USA, *Col. maculata* overwintered in the greatest numbers in leaf litter in windbreaks that were adjacent to maize fields. This type of overwintering site was preferred to forest edges adjacent to maize (Nault & Kennedy 2000).

While the migration to hibernation sites lasts only 1–2 weeks, the spring dispersal is much longer (Solbreck 1974, Benton & Crump 1981). Adults that had overwintered were observed to feed on pollen at or near the dormancy site, on *Populus*, dandelions or cowslip (*Caltha palustris*) (see also 6.3.1.3).

6.3.2.5 *Adalia bipunctata*

This coccinellid is known in early dormancy for frequent invasions of buildings, where it appears in cracks in walls, in lofts, or behind windows and even in rooms (e.g. Hawkes 1920, Semyanov 1970). However, recently it has been replaced by *Har. axyridis* in the USA and western Europe (6.3.2.2). It can be assumed that the coccinellids are brought to buildings by their hypsotactic behaviour. Such behaviour is also indicated by a small aggregation of *A. bipunctata* (20–30 individuals) found close to a large aggregation of *Cer. undecimnotata* on a hill top (I. Hodek, unpublished).

However, a fair proportion of *A. bipunctata* apparently does not leave the orchard, park or forest habitats, and hibernate together with *Chil. bipustulatus* and *Stethorus pusillus*, either in crevices in tree bark or even in paper bands around trees (Speyer 1934).

Mass hibernation by *A. bipunctata* in the thick bark of old Tyan-shan spruces (*Picea schrenkiana*) is described by Savoiskaya (1983) from Kazakhstan. Every autumn, the beetles migrate in masses to the same trees in mountain valleys, flying up the valley; a rate of 50–55 beetles/min was observed in late September. It was noticed that they turned back sharply just behind the tree (evidently pushed there by air turbulence), swarmed in the wind shelter and dropped rapidly onto the trunk. Up to 6000 beetles were found on one tree. Similar sites on *Picea* were also used by *A. bipunctata* (= *fasciatopunctata*) and by *Oenopia conglobata*. The latter species was most abundant up to 1500 m, while *Adalia* species preferred higher altitudes. Winter aggregations of *O. conglobata* and *Halyzia tschitscherini* were found in the bark crevices of poplars in the valleys of montane rivers in Kazakhstan (Savoiskaya 1983). Aggregations of up to 200 *A. bipunctata* adults were observed by Smee (1922) on elm trees (*Ulmus*) in England (as many as 1000 beetles on one tree). Smee was the first to succeed in attracting coccinellids to cages containing tree bark, i.e. artificial hibernacula. In Belgium, *A. bipunctata* often overwinters in poplar plantations with a preference for the trees on the southern edge. Most of the aggregations (mean size, 2–6 individuals; range, 1–36) are situated in bark crevices on the south-west sides of the trees (Hemptinne 1985, Hemptinne & Naisse 1988).

Adalia bipunctata, being a multivoltine species, enters diapause relatively late (in early October in Germany; Speyer 1934) and emerges very early in spring (in March in Belgium; Hemptinne 1985). *Adalia bipunctata* is often found together with *Chil. bipustulatus* (Speyer 1934, Bielawski 1961, Savoiskaya 1965).

6.3.2.6   *Myrrha octodecimguttata* and other forest species which hibernate in bark crevices

Bark crevices of Scots pine and other conifers are reported as hibernation sites for *Har. quadripunctata*, *M. octodecimguttata* and *Aphidecta obliterata* (Majerus 1994). According to this author, the dormancy sites of **Myzia oblongoguttata** and **Anatis ocellata** are unknown in England. Not much is known about the hibernation of other coccinellid species of coniferous forests. According to recent observations, they do not seem to leave their breeding habitat.

Myrrha octodecimguttata, which lives and breeds in the crowns of pine trees in Germany (Klausnitzer 1968), hibernates in bark crevices of old pines; this has been established also for Poland by Bielawski (1961), for Finland by Pulliainen (1963, 1964, 1966) and for England by Majerus (1994). Pulliainen observed that the coccinellids preferred to hibernate in the lowest 10 cm of the trunk (93% of beetles) and the south and east sides of the tree (62%). On these sides the tendency to aggregate was most pronounced; the aggregations averaged 3–8 individuals (maximum 14).

Bark crevices of old pines are used as dormancy sites by many other forest species (but 6.3.2.5) including: **Scymnus suturalis** (Delucchi 1954, Pulliainen 1966, while Bielawski 1961 found this species more frequently in the bark crevices of deciduous trees), **Scymnus nigrinus** (Bielawski 1961), and **Aphidecta obliterata** (Pulliainen 1966, Parry 1980). **Exochomus quadripustulatus** uses bark crevices and litter with about equal frequency for its dormancy site (Bielawski 1961, Klausnitzer 1967). Other conifers than pines may also be visited, e.g. cedar by *Hip. tredecimpunctata* (Thomas 1932). Some coccinellids prefer the bark crevices of deciduous trees, e.g. *Nephus quadrimaculatus* prefers chestnuts (Bielawski 1961), and *Oenopia conglobata* and *Har. quadripunctata* chestnut and poplar (Bielawski 1961; I. Hodek, unpublished).

6.3.2.7   *Hippodamia convergens*

The progeny of diapausing beetles emerge from pupae in early May. If there are not enough aphids nearby, they migrate long distances towards the mountains (in northern California to the Sierra Nevada). Hagen (1962, 1966) proposed a hypothesis to explain the mechanism of migration (Fig. 6.8; 6.2.7). Migration is initiated by take-off during the calm of morning: the subsequent vertical upward primary flight is apparently assisted by convection currents. This goes on as long as adequate warm temperature permits the beetles to fly. At the lower temperature threshold of about 11–13°C flight is inhibited and the beetles fall about 300 m into a warmer layer of air (13–18°C) where flight is resumed. The beetles which are thought to undergo these vertical oscillations are simultaneously blown sideways. Each day during the late morning in May and June, westerly **winds carry the beetles** toward the mountains (Fig. 6.8). The migration is terminated when the air currents contact the mountains (Hagen 1962).

If aphids are found near the landing place, the beetles produce eggs there. Usually, however, they feed on a non-insect diet (e.g. pollen) and build up reserves. After a week or so the coccinellids exhibit **secondary, directional flight** near the ground, flying up and down mountain creeks in search for **summer aggregation** sites. There the adults remain rather inactive until October, when they become mobile again, presumably disturbed by the first rains. The **tertiary flight** during warm periods usually leads them to lower parts of the creek where new, larger **overwintering aggregations** are formed (Hagen 1962).

Hippodamia convergens adults are guided in their search for the aggregation sites by a series of factors. In contrast to *Cer. undecimnotata*, they do not always respond to prominent objects. Sometimes assemblages of *Hip. convergens* are found at mountain tops (Douglass 1930, Throne 1935, Sherman 1938, Stewart et al. 1967), but according to Hagen's (1962) hypothesis they reach summits in a semi-passive way. **Moisture and light** are considered to be the most important factors influencing **selection of aggregation sites**. At first the coccinellids assemble on bushes and trees in large clearings near creeks. After several days they move close to the creeks, and settle down along the banks, often in spots exposed to afternoon sunlight. Thus the aggregations are formed in similar places year after year, but also new clearings are occupied. The beetles exhibit a marked preference for litter with 20% moisture (Hodson 1937). Hagen (1962) assumed that *Hip. convergens* **drinks water** in order to maintain a constant water content and that a physiological requirement for free water may be the basic characteristic of **climatotactic aggregators** *(Hip. convergens, Col. maculata)* in contrast to hypsotactic species such as *Cer. undecimnotata* (6.3.2.1).

On the eastern edge of the Rocky Mountains, massive aggregations of *Hip. convergens* (mixed with the curculionid *Hypera postica*) were found below stones about 1 m above the edge of the water of an irrigation reservoir in mid-March. Most ladybirds dispersed before the end of March (Simpson & Welborn 1975). This finding might also indicate the preference for moist habitats in dormant *Hip. convergens*.

Large winter colonies (from one, about 40 million beetles have been collected) are formed by many non-contiguous small aggregations under leaves, at the bases of bushes, and on tree trunks. At higher altitudes, snow often covers *Hip. convergens* aggregations for about three months. The majority, however, hibernate near the snow line in the Sierra Nevada, and during most of the winter they are not covered with snow (Hagen 1962).

The beetles aggregate in canyons and are thus in shade. When they become heated to above 14°C in the early spring, they take off vertically and are eventually caught up by winds (blowing in the opposite direction to that in summer) which carry them over ridges that lie to the west between the aggregation sites and the plains. The fall in air temperature in the evening brings the beetles to the ground, terminating the flight. Not all coccinellids, however, get so far. The beetles that overwinter on the highest peaks of the Sierra are found after dispersal to have reached only the mountain valleys just below these peaks (Hagen 1962).

Hagen's findings were complemented by **laboratory experiments** using tethered flight (Rankin & Rankin 1980). Individual coccinellids were suspended from their pronotum by a toothpick attached with melted wax and were **stimulated to fly** by a low-speed fan; the duration of flight was measured. **Long tethered flight** (>30 minutes) was considered a good **indication of migratory behaviour** (6.3.1.3). *Hippodamia convergens* was found to be a typical migrant: the long flights are post-teneral and pre-reproductive in females, and are associated with adult diapause. Starvation greatly enhances migratory behaviour and, under optimal feeding conditions, short daylength has the same effect (Rankin & Rankin 1980). Optimal feeding for 7 days induces maturation of ovaries and greatly reduces the tendency to long tethered flight.

To monitor the **tendency to long flights in the field**, active beetles were collected in aphid-infested fields from March to June and diapausing beetles were sampled on the top of a mountain from July to December (Fig. 6.9). The adults, newly arrived at the fields in

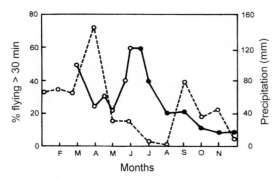

Figure 6.9 Flight activity of *Hippodamia convergens* and monthly precipitation in 1977 (dashed line) (from Rankin & Rankin 1980). The adults were collected in fields around Austin (open circles, solid line) and on top of Mt Locke, Davis Mts, W. Texas (closed circles).

March, had a relatively high proportion (50%) of individuals with a migratory tendency, which decreased to about 20–30% in the next two months. By late May and June, about 60% of coccinellids again displayed long tethered flight. The beetles behaved similarly when sampled in the field (before migration) and at the mountain top, early after their arrival at the dormancy site. The incidence of migratory tendency then dropped to about 20% in August and to less than 10% in the last months of the year (Fig. 6.9; Rankin & Rankin 1980).

6.3.2.8   *Hippodamia quinquesignata*

The location of hibernation sites by *Hip. quinquesignata*, as described by Edwards (1957) and Harper and Lilly (1982), seems also to be **hypsotactic** as in *Cer. undecimnotata*. At least all aggregations described were found in the mountains at altitudes ranging between 1677 m and 3354 m in the USA (Edwards 1957) and 1250–2744 m in Alberta, Canada (Harper & Lilly 1982). *Hippodamia quinquesignata* uses the same aggregation sites each year. They are usually located on upper, exposed slopes (often west-facing) with sparse vegetation and covered with rocks. The beetles are found under rocks and debris, in crevices in rocks, logs or pine cones, or at the base of junipers. Sometimes *Hip. caseyi* is admixed in small numbers.

In spring, mating occurs before dispersal. In California, however, *Hip. quinquesignata* does not migrate long distances and spends its **facultative** summer **diapause**, as well as its winter diapause, at the edges of

fields. Thus the adults can easily **respond to** later increases in **aphid numbers** by producing summer and autumn generations (Hagen 1962, Neuenschwander et al. 1975).

6.3.2.9   Hibernation of mycophagous and phytophagous species

Hibernation in aggregations is most common in entomophagous coccinellids but also two mycophagous species have been reported hibernating in aggregations in litter. In mid-April Ruscinsky (1933) found four large aggregations of **Vibidia duodecimguttata** among fallen leaves at the base of trees on top of a hill in erstwhile Rumania (Bessarabia). Each aggregation covered about 1 m^2 and consisted of some 2500 individuals. **Psyllobora (= Thea) vigintiduopunctata** was often found in moist litter at forest edges visited by *C. septempunctata* and related species (Hodek 1960, Bielawski 1961) (6.3.2.3).

Aggregations have also been observed in **phytophagous coccinellids** in the subfamily Epilachninae. In Africa, Poulton (1936) observed aestivation aggregations of immobile **Epilachna dregei**, once in Bechuanaland and once in northern Uganda on prominent features (a hill, a termite-hill) in mid-July, i.e. in the middle of the dry season. In February, Kapur (1954) found four aggregations of several thousand individuals of **Aphidentula bisquadripunctata** (=*Epilachna*) at an altitude of about 400 m in India (Chota Nagpur, Bihar) at the base of grass c. 0.7 m high, in the vicinity of an almost dry brook. The coccinellids were in diapause (empty guts, a large fat body, unripe ovaries). In both *Epilachna* spp. observed, hypsotactic aggregation may have been involved. Ghabn (1951) mentions winter migrations of **E. 'chrysomelina'** from the fields into the surrounding desert in Egypt without, however, describing the hibernation sites.

6.4   ANATOMICAL AND PHYSIOLOGICAL CHANGES RELATED TO DORMANCY

6.4.1   Anatomical state

6.4.1.1   Fat body and digestive tract

A greatly enlarged fat body and voided digestive tract are the conspicuous features of diapausing coccinellids of both sexes. The characteristics of midguts in *C. novemnotata* are typical for all coccinellid species: the midguts of dormant beetles are reduced to whitish, opaque, thick-walled tubes containing a brown fluid (McMullen 1967a).

In *C. s. brucki*, an important difference between feeding and dormant adults was found in the **proteolytic activity** and **ultrastructure of the midgut** (Morikawa et al. 1989, Sakurai et al. 1991). Trypsin activity was very low in summer and completely inhibited in January. The columnar cells of the midgut epithelium of active beetles contained abundant rough endoplasmic reticulum (RER) and the **mitochondria** showed distinct cristae. In dormant coccinellids the mitochondrial cristae were obscure and RER was rarely found. These differences are evidently due to feeding versus absence of feeding; they are identical in the aestival diapause and winter quiescence (6.2.2).

6.4.1.2   Ovary, spermatheca

The **developmental progress** of ovaries was quantified in *Har. axyridis* on a six-stage scale (Obata 1988a, b, Osawa 2005): (1) no oocytes; (2) one oocyte; (3) two whitish oocytes; (4) two oocytes, with the basal oocyte filled with whitish yellow yolk; (5) two oocytes, with the basal oocyte with yellow yolk; (6) the basal oocyte grown to the maximum size. Stages 5 and 6 are considered as matured ovarioles (Osawa 2005). The next reproductive state is represented by the eggs present in the calyx of the ovary. Stages 3 and 4 were considered as one stage in a five-stage scale for *C. septempunctata* (Okuda et al. 1986, Okuda & Hodek 1989).

As coccinellids, particularly the aphidophagous species, often encounter a scarcity or absence of suitable prey (Chapter 5.3.3), oosorptive ovaries may also be found. **Oosorption** occurs during the intermediate developmental stage of ovarioles (Osawa 2005). The survival of females is maintained by egg resorption because of the possibility of future oviposition when aphid resources again become available, often after long- or short-distance movement (6.3.1.2). Oosorption also takes place when diapause is induced in reproducing females, i.e. 'secondarily'. Egg resorption in the late reproductive season and due to bad condition of host plants has been recorded in several studies on the *Henosepilachna* complex. When the conditions improved, oviposition was resumed (Ohgushi 1996 and the references therein). Osawa (2005) gives two main characteristics of an oosorptive oocyte: (i) dark

yellow or orange colour, (ii) modified shape (see Osawa 2005, p. 504, figure 2 therein).

In a European study of **gonads during dormancy** in four coccinellid species, Osawa's stages 1–3 (and the stage during which resorption occurs) were recorded. However, the transparency of the oocytes was emphasized rather than their number, i.e. (1) no oocyte, (2) transparent oocytes, (3) early phase of vitellinisation (Ceryngier et al. 2004). From the four univoltine species sampled at dormancy sites (*C. septempunctata*, *C. quinquepunctata*, *C. magnifica* and *Cer. undecimnotata*), only the last-named species completely lacked any indication of reproductive activity in early dormancy in September. All females had completely inactive ovaries (stage 1) and none of them had sperm in their spermathecae (Figs 6.10, 6.11; Ceryngier et al. 2004). This finding is in agreement with older data for *Cer. undecimnotata* (Hodek & Landa 1971; Table 6.18). While all *C. septempunctata* females had inactive ovaries similarly to *Cer. undecimnotata*, 10% of the ovaries of *C. quinquepunctata* were in stage 2 and one *C. magnifica* female had resorbed oocytes. In all *Coccinella* species, some females had sperm in their spermathecae; the incidence of mated females was highest (30%) in *C. septempunctata* (Fig. 6.11; Ceryngier et al. 2004). More than 10 years earlier (Ceryngier et al. 1992) a higher percentage (50%) was recorded.

Parasitization of *C. septempunctata* by the braconid *Dinocampus coccinellae* (8.3) surprisingly increased not only the incidence of mated females (to 50% in September), but also the state of maturity of the ovaries (Ceryngier et al. 2004). Two females that contained second instar larvae of *D. coccinellae* had vitellinized oocytes, while another two parasitized by younger stages had inactive ovaries. This observation still needs a detailed experimental study to verify whether parasitization can really cause a kind of activation similar to that caused by injury (in a French population of *C. septempunctata*; Hodek et al. 1977; 6.2.1.2).

For England, Majerus (1994) reported that *C. septempunctata* only rarely mate before hibernation. In North Carolina, the females of *Har. axyridis*, collected from buildings shortly after they arrived there, mostly in October, varied greatly in the proportion that had mated: in 1993 the mean was 11.8% (range 0–25%; n = 246) and in 1994 it was 41.4% (range 0–70%; n = 461) (Nalepa et al. 1996).

In general, the **receptivity of female** coccinellids for mating is evidently not dependent on the physiological state of their ovaries. As mentioned above, in

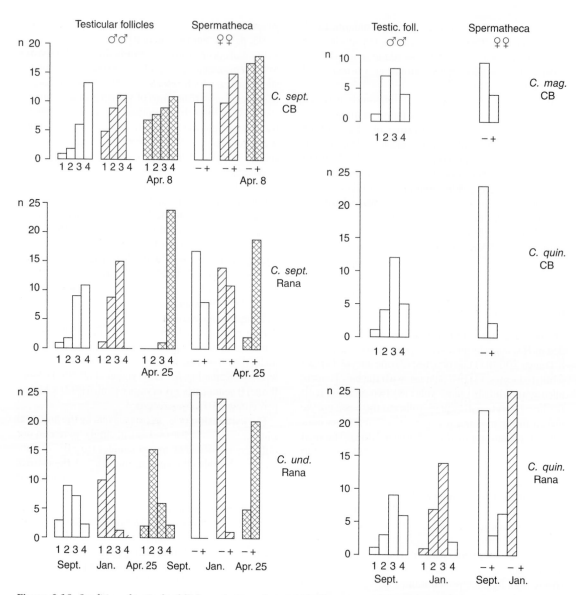

Figure 6.10 Condition of testicular follicles and spermatheca. *Coccinella septempunctata, Ceratomegilla undecimnotata, Coccinella magnifica* and *Coccinella quinquepunctata*. 1, lowest activity; 4, highest activity; −, empty; +, filled with sperm; České Budějovice (CB, southern Bohemia; Rana, northern Bohemia) (from Hodek & Ceryngier 2000).

C. septempunctata 30–50% of females entering diapause have the spermathecae full of sperm (Ceryngier et al. 1992, 2004) and the first females of *Cer. undecimnotata* contain sperm in their spermathecae 3 weeks before the ovaries mature (Hodek & Landa 1971; Table 6.18).

Also in old studies (e.g. on *P. quatuordecimpunctata*, Hariri 1966; *C. novemnotata*, McMullen 1967a) the ovaries of diapausing females are described as consisting of mere germaria.

In *A. bipunctata*, however, continual activity of the ovaries was reported from Belgium when the

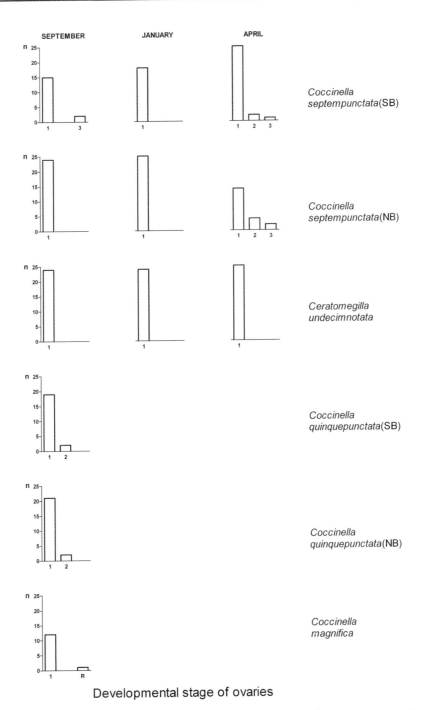

Developmental stage of ovaries

Figure 6.11 Condition of ovaries in coccinellid females. 1, no oocytes in ovarioles; 2, first, transparent oocytes; 3, small vitellinized oocytes; R, vitellinized oocytes partly resorbed. Only non-parasitized coccinellids are included. Origin of samples: southern Bohemia (SB), northern Bohemia (NB) (from Ceryngier et al. 2004).

Table 6.18 Gonads in dormant *Ceratomegilla undecimnotata* collected in Louny hills (N. Bohemia) (Hodek & Landa 1971).

Date of sampling	Ovaria		Spermatheca		Spermatocytes in testicular follicles*	
	no egg chambers	one oocyte	no sperm	with sperm	absent	present
16.12.57	20	—	20	—	20	—
11.02.58	20	—	20	—	20	—
08.04.58	30	—	30	—	20	—
23.04.58	20	—	18	2	11	9
29.04.58	20	—	12	8	7	13
05.05.58	20	—	4	16	—	20
13.05.58	10	10	—	20	—	20
19.05.58	7	13	—	20	—	20

*Vesiculae seminales always full of sperm.

differentiated and growing oocytes were resorbed and not vitellinized for most of the season: vitellinized and chorionated oocytes were only found in May. This report seems somewhat to contradict the 3–4 generation cycle reported for Belgium (Hemptinne & Naisse 1987). Usually a bivoltine life cycle has been recorded for *A. bipunctata* in cold temperate climates (e.g. Obrycki et al. 1983, Majerus 1994, Klausnitzer & Klausnitzer 1997).

In central European populations of *C. septempunctata*, which are **heterogeneous as to voltinism** (6.2.1.1), dissections of samples from summer and early autumn show both processes, differentiation and resorption. In July, about 40% of females show the start of vitellogenesis, but in the next month the oocytes begin to be resorbed, earlier in the beetles from dormancy sites and slightly later in the beetles still remaining in breeding sites (Fig. 6.12; Okuda et al. 1986). A similar result was obtained in the laboratory at $25 \pm 2°C$ and short daylength (12L:12D): a maximum of aproximately 40% females had vitellogenic oocytes on the third day and resorption was terminated on the ninth day. Under long daylength (18L:6D) the process of resorption was delayed; it was only terminated after about 3 weeks (Okuda et al. 1986).

As late as in mid- and late **September**, a fraction of the central European populations of *C. septempunctata* may still be found actively feeding on remaining aphids. Most of these coccinellids have their guts full of digested aphids and some of them (**10–20%**) also have **vitellinized oocytes** (Hodek 1962; Table 6.1). This finding was confirmed in a recent study on populations from

southern Bohemia (Ceryngier et al. 2004). It has been assumed that this heterogeneity in the maturation of ovaries found in the Czech populations is related to the partial tendency to multivoltinism (Hodek 1962; Table 6.3; 6.2.1.6, 6.2.16).

In the **bivoltine** *C. s. brucki* in central Japan the females have matured oocytes in spring and in autumn. They spend both aestivation and hibernation with previtellogenic ovaries. However, oogenesis was reported to advance gradually during hibernation (in a relatively mild winter), while it was completely suppressed in the course of aestivation (Sakurai et al. 1983, 1986). These observations indicate that, from the physiological point of view, only the aestivation is diapause (6.2.2.1).

In *Chil. renipustulatus*, the majority of females diapause with ovaries in which development is blocked at an early stage; in about 7% of the females development is stopped even earlier and follicles are completely missing (Pantyukhov 1965, 1968a). The females with no follicles have a substantially greater longevity (by 1–2 months in the laboratory) than those where some development has occurred. The situation is probably similar in *Chil. bipustulatus* and *Chil. geminus*. Zaslavskii and Bogdanova (1965) report that in some diapausing females of these two species, the development of the ovarioles is inhibited at the stage of one small follicle with some yolk deposited.

It is at the **end of dormancy** in spring, often still at dormancy sites while no food is taken, that previtellogenesis and the formation of the first oocyte occur. In *Cer. undecimnotata* (Hodek & Landa 1971; Ceryngier

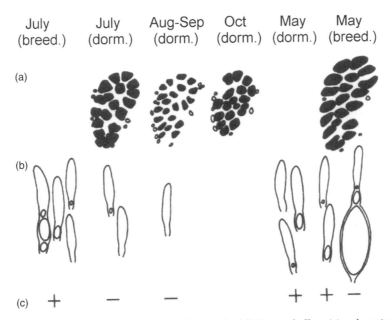

July
(breed.)   July
(dorm.)   Aug-Sep
(dorm.)   Oct
(dorm.)   May
(dorm.)   May
(breed.)

(a)

(b)

(c)   +   −   −   +   +   −

Figure 6.12 Schematic representation of changes in dorso-longitudinal flight muscle fibres (a) and ovaries (b) in relation to flight tendency (c) in *Coccinella septempunctata* adults. Breeding (breed.) and dormancy (dorm.) sites; +, flight tendency high; −, flight tendency low or absent (from Okuda et al. 1986).

et al. 2004; Table 6.18; Fig. 6.11) this happens much later, around mid-May, than in *C. septempunctata* (Okuda et al. 1986, Ceryngier et al. 2004; Fig. 6.12).

In three studied species, *C. septempunctata* and *Cer. undecimnotata* (Hodek & Landa 1971, Ceryngier et al. 2004) and *A. bipunctata* (Hemptinne & Naisse 1987), the spermathecae of at least part of the sampled females contain **spermatozoids before** the **dispersal** from the hibernation sites. *Adalia bipunctata* have empty spermathecae from the onset of dormancy until spring (Hemptinne & Naisse 1987).

The females of *S. pusillus* often live for 2 years and enter a **second dormancy**. The mature ovaries are largely resorbed and the metabolic reserves are re-accumulated in the enlarged fat body. This resorption of ovaries even occurs during the first hibernation of those females which have emerged from pupae before late July and have already oviposited in that season (Putman 1955). This latter situation may be expected to apply in every multivoltine or at least partly bivoltine life cycle, as occurs in *C. septempunctata* (Hodek 1962). It is also unlikely that a second hibernation is an

exceptional event; it has been proved to occur in *P. quatuordecimpunctata* by Hariri (1966) in England, in *C. septempunctata* by Sundby (1968) in Norway, and in two species (*Har. axyridis* and *Aiolocaria hexaspilota*) introduced to central Asia from the Far East (Savoiskaya 1970a, b).

6.4.1.3   Male gonads

Contrary to what is often assumed, the activity of the tissue of coccinellid testicular follicles is not arrested when diapause is induced. Testes already mature in the pupa, and the males are ready to mate shortly after adult emergence. While the males of *C. septempunctata* acquire all other characteristics of the diapause syndrome, such as accumulation of reserves and behavioural characteristics, i.e. migration and change of taxes, they are nevertheless able to **fertilize females** just **prior** to entering **diapause**. On 22 September 1998, the males of *C. septempunctata* still had highly active testicular follicles (at stage 3.5 on average, when estimated at 4 stages), in spite of being dormant in the tussocks on the hill top (Fig. 6.10; Hodek & Ceryngier 2000). This is not the same for all coccinellid species; see *Cer. undecimnotata* in Fig. 6.10 and *A. bipunctata* (Hemptinne & Naisse 1987).

In the next 4 months of the cold period, up to 22 January, the activity of the testicular follicles in most males of *C. septempunctata* had on average regressed to stage 2.2, due to the low temperature prevailing over that period. There were loose spermatocytes and spermatids in the centre of the testicular follicles, with spermatogonies (apical cells) at the distal top and spermatodesms (sperm bundles) at the proximal end. Spermatogenesis ceases at the stage of the division of spermatogonies or of young spermatocytes. In January the average state of activity of the testicular follicles was again much lower in *Cer. undecimnotata* than in *C. septempunctata* (Fig. 6.10).

The degree of testicular activity differed between autumn samples of four species (Fig. 6.10). The least active were the testes of *Cer. undecimnotata* (mean degree of activity: 2.4) and the next were those of *C. magnifica* (2.8), *C. quinquepunctata* (3.2) and *C. septempunctata* (3.6).

In spring, the increase in ambient temperature (to about 12°C; Hodek 1960, 1973) initiates the opposite process. In April the **activity** of the **testicular follicles** is **resumed**; this occurs earlier in *C. septempunctata* than in *Cer. undecimnotata* (Fig. 6.10). In the plain near České Budějovice (South Bohemia), the progress in activity is more varied than in the hills in northern Bohemia. Spermatogenesis begins: the spermatogonies begin to divide and groups of growing spermatocytes enveloped by cyst cells are formed which quickly fill the whole follicle. In some years the tissue of the testicular follicles is fully active and spermiogenesis is complete in all males by mid-April (Hodek & Ceryngier 2000, Ceryngier et al. 2004). Such a situation was already reported earlier with photographs (Hodek & Landa 1971).

However, the males do not have to wait for the new sperm to mature before mating, as they have kept **reserves of sperm in** their **seminal vesicles**. Also, if the females disperse early, about one half of them could use sperm from their **spermathecae** for fertilizing their eggs. Mating at the dormancy sites may be advantageous in central Europe for *Cer. undecimnotata* that occurs there on the northernmost boundary of its distribution area and is rather rare, so that the probability of genders meeting away from the hibernation

sites is much lower than in *C. septempunctata* (Hodek & Ceryngier 2000).

A biochemical study on the **Japanese subspecies C. septempunctata brucki** reported interesting results. In central Japan, this subspecies has a summer diapause and a winter quiescence (Okuda & Hodek 1983; 6.2.2). In the laboratory at 25°C and long days (i.e. conditions inducing summer diapause) Okuda (2000) found that **DNA synthesis in** the **testicular follicles** decreased within 30 days of adult life to less than half the levels recorded in pre-diapause (adult age 1 day) and post-diapause (age 120 days). Thus DNA synthesis **never ceased** completely and spermatocytes, although reduced in number, were present in diapausing males. In fact this is not particularly surprising, because the activity of the tissue of the testicular follicles is affected by the ambient temperature (Hodek & Landa 1971). The metabolic rate of Okuda's beetles at 25°C remained high since they did not experience the marked decrease in temperature of the Czech outdoor beetles. In the mild winter of central Japan temperature is rather high also in the field (Okuda 2000).

In Belgian populations of the multivoltine *A. bipunctata*, the degeneration of differentiated spermatogonia begins in September. Spermatogenesis ceases in January and February and resumes fully in May (Hemptinne & Naisse 1987).

Mating activity just **prior to dispersing** from aggregations has also been reported in California for *Hip. quinquesignata punctulata*, *Hip. sinuata* and *Hip. parenthesis* (Hagen 1962). Well-developed testes in *Hip. convergens* in dormancy sites have been reported by Stewart et al. (1967).

Similar to the findings on four Czech coccinellid populations (Ceryngier et al. 2004), Zaslavskii and Bogdanova (1965) could not find any effect of diapause-inducing conditions on the state of the testes in *Chil. bipustulatus* and *Chil. geminus*. However, gametogenesis evidently ceases very early in the life of both these species, since it was very difficult to find mitotic divisions in active or diapausing males older than 7–10 days.

The presence of spermatozoa in the spermathecae of some **C. septempunctata** females in autumn (Hodek & Cerkasov 1961) and particularly the fertilization of females by males kept under diapause conditions, that was recorded in *C. septempunctata* by Bonnemaison (1964) and in *Stethorus pusillus* by Putman (1955), would suggest that the activity of the **male accessory**

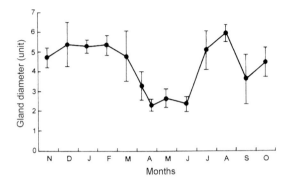

Figure 6.13 Seasonal changes in the size of male accessory glands in *Coccinella septempunctata brucki*. (*n* = 10; one unit = 25 μm) (from Sakurai et al. 1982).

glands might also be independent of diapause. However, in *C. s. brucki* the glands were much smaller in diapausing aestivating males than in quiescent hibernating ones in which they were developed (Sakurai et al. 1982; Fig. 6.13).

It has been shown experimentally with *Cer. undecimnotata* that the activity of the **testicular follicles** depends on the ambient temperature: under favourable temperature it may be continued or resumed, when alternative food (e.g. honey solution) is provided. At a temperature of around 12°C, the males can embark on a normal and gradual decrease in gonad activity (Hodek & Landa 1971).

In *C. septempunctata*, photoperiod did not exert any influence on the activity of the testicular follicles that was increased by exposure to 20°C, even in starved beetles. Activity of the follicles was positively affected by feeding the males on honey solution and still more by them preying on aphids, while it was lower in males parasitized by larvae of *Dinocampus coccinellae* (Ceryngier et al. 1992). This enhancing effect of feeding is in agreement with earlier findings on males of *Cer. undecimnotata* (Hodek & Landa 1971).

6.4.1.4 Flight muscles

In late summer *C. septempunctata* adults migrate to dormancy sites to return in spring to breeding sites (see also 6.3.1.2 and 6.3.1.5). Flight muscle fibres are well developed immediately after arrival at the aestivo-hibernation sites, but are gradually resorbed to about

80% of their maximum diameter by August and September. While this resorption in **early dormancy** was expected, the regeneration recorded already in October, 6 months before the dispersal flight, was rather surprising (Fig. 6.12). These histological findings (Okuda et al. 1986) were corroborated several years later by measuring changes in the volume of the flight muscles (Nedvěd et al. 2001). Changes in oxygen consumption and take-off also indicated a tendency to flight resumption a long time before the spring dispersal (Nedvěd et al. 2001). For a discussion of this unexpected phenomenon, see 6.3.1.3. The decrease in flight tendency recorded **after dispersal** to the breeding sites was not followed by flight muscle autolysis (Okuda et al. 1986). Evidently the flight muscles remain well developed as they are needed for trivial flight for foraging. Coccinellids foraging for their aphid prey, that quickly changes in abundance, often undertakes flights of relatively great distance.

In *C. s. brucki* from central Honshu, Japan, autolysis of flight muscles occurs **during aestivation**, while the flight muscles of hibernating adults are not autolyzed, and in this population aestivation and hibernation are separated by a period of reproduction (Okuda 1981, Sakurai et al. 1982, 1986). **Regeneration** of the flight muscles **after aestivation** is thus necessary for dispersal in this population of *C. s. brucki*. In central Europe, *C. s. septempunctata* undergoes a long uninterrupted aestivo-hibernation. It is a pity that neither the condition of flight muscles, nor flight tendency have been studied in warmer European regions, where two (or more) generations of *C. s. septempunctata* could develop.

Resorption of flight muscles during aestivation of *C. s. brucki* decreases the metabolic rate which, in spite of their diapause condition, would be high because of the high ambient temperature in summer. During hibernation, metabolism is lowered by the low ambient temperature.

6.4.2 Metabolic changes related to diapause

6.4.2.1 Lipids

Diapausing insects typically show an increased fat content (Lees 1955, Hodek & Cerkasov 1963, 1965, Tauber et al. 1986, Danks 1987, 2001). Their survival during dormancy largely depends on the amount of metabolic reserves they can accumulate before diapause.

In many coccinellids the deposition of reserves in the fat body is quite extensive, and a **hypertrophied fat body** has always been reported in diapausing coccinellids. Numerical data based on analysis of the amount of fat accumulated are, however, less common. For comparison of early findings in six coccinellid species, see the table on p. 247 in Hodek (1996).

In the laboratory, slightly less fat is accumulated in **C. septempunctata** than in the field (Hodek & Cerkasov 1961). Females, sampled in the hibernation sites in early October, when an important proportion of the reserves had already been utilized, still had 0.205 mg of lipids per mg total weight. Females, in which diapause was induced in the laboratory at 12L:12D and 18°C had 20% less lipids (0.164 mg).

Zhou et al. (1995) reported, also for the Czech Republic, a slightly higher content of fat in *C. septempunctata* females, sampled in mid-September in **mountains** (1420 m a.s.l.: 9.4 mg per individual) than in those from a hill (400 m a.s.l.: 8.9 mg per individual). During hibernation there was a lower decrease in fat content at the higher altitude (to 6.6 mg per individual, i.e. by about 30%) than at lower sites (to 4.9 mg per individual, i.e. a decrease of about 50%).

To monitor **changes during dormancy**, it is advisable to express lipids and glycogen in **absolute values** of weight per specimen. Then changes in individual components of body weight are independent, unlike the situation if, for example, lipids are expressed as percentage of dry or fresh weight. Whereas in reality the fat content in **Cer. undecimnotata** males substantially decreased from late August to late October from 6.4 to 3.6 mg per beetle, i.e. by 44%, in the relative expression, in contrast, the initial 41% of fat per dry weight only falls to 38% (Hodek & Cerkasov 1963). While the latter expression is an important drawback, some authors use it and thus their data have to be reported here in this way. Weights that follow are all given as absolute weight per individual unless stated otherwise.

In *Cer. undecimnotata* 6–7 mg of fat per female were recovered **in late August**, i.e. about 40% weight (Hodek & Cerkasov 1963). As the first sample was taken 1 month after the arrival of the beetles in the hibernation sites, it can be supposed that the initial fat content was substantially higher. *Ceratomegilla undecimnotata* has the **highest** relative **fat content** of all

insect species which were analyzed before this study. In three coccinellid species, Hariri (1966) found 5–9 mg fat per beetle in the much heavier **C. septempunctata**, 1–3 mg in **A. bipunctata** and 1–2 mg in **P. quatuordecimpunctata**. Coccidophagous **Chil. rubidus** only has 26.5% dry weight as fat at the beginning of hibernation (Pantyukhov 1968b) and **Chil. renipustulatus** 25% (Pantyukhov 1968a) or 20% (Pantyukhov 1965).

Two Australian coccinellids, the aphidophagous **Apolinus lividigaster** and the mycophagous **Illeis galbula**, doubled their fat content before reproductive diapause to about 20–30%. Fat deposition was associated with consumption of alternative food such as pollen (Anderson 1981).

While Jean et al. (1990) recorded only 1.9 mg of fat for **Col. maculata** (comparable to the values above) in November, Labrie et al. (2008) reported a surprisingly high content of lipids in this species, also in Quebec and also in November: 73.2% of dry weight in females and 61.4% in males. The values of the latter authors for **Har. axyridis** were in the usual range of 35.5% in females and 26.4% in males.

Rate of decrease

The decrease in the substantial fat reserves during hibernation is drastic: for example, during the 8 months of dormancy of *Cer. undecimnotata* the absolute amount of **fat decreased to a** mere **quarter** (Hodek & Cerkasov 1963).

This decrease is much smaller in some Heteroptera and Coleoptera (Hodek 1996, p. 248) which hibernate in soil, isolated from temperature changes. The proportions of the initial fat content consumed throughout the whole hibernation period in *C. septempunctata*, *A. bipunctata* and *P. quatuordecimpunctata* were, for males and females, respectively, as follows: 49 and 61, 75 and 87, and 71 and 60% (Hariri 1966). *Chil. rubidus* catabolized 32% of the initial amount of fat during dormancy (Pantyukhov 1968b).

The rate of decrease in fat reserves in **Cer. undecimnotata** is related to the **impact of ambient temperature**. The rapid decrease in fat content commences in late summer and early autumn. In September 1957 it was 0.082 mg of lipids per day, and in October 1958 0.022 mg per day. During the cold period of November to March the decrease of fat was 0.009 mg per day. In April and early May the utilization of fat reserves again increased, and varied in the three springs studied due to different temperatures: the decrease of fat in *Cer. undecimnotata* males amounted to 0.018 mg, 0.045 mg and 0.027 mg per day in 1957, 1958 and 1959, respectively (Hodek & Cerkasov 1963). Findings by Hariri (1966) and (Pantyukhov 1968a, b) agree with the above.

Also in the small ***Aphidecta obliterata*** fat reserves decreased in early diapause from 0.75 to 0.5 mg per beetle, then remained unchanged during hibernation and decreased again in spring from April to May to 0.3 mg per female (Parry 1980). Even taking into account that adults of *A. obliterata* are small, the fat content is still rather low. In *A. bipunctata* adults, the initial fat content of 3.3 mg per beetle (54% dry weight in August and September) was reduced to 1.2 mg (30% dry weight in April). The average daily consumption was 0.01 mg fat per day (Mills 1981). This rate of decrease was very similar to 0.009 mg fat per day calculated for *Cer. undecimnotata* above (Hodek & Cerkasov 1963).

A laboratory experiment established the effect of six temperatures on the decrease in lipid reserves and survival in diapausing adults of ***Col. maculata lengi*** (Jean et al. 1990). At the start of the experiment in November, lipid reserves averaged 1.9 mg per insect and they diminished rapidly as the temperature increased. The rate of lipid decrease was significantly higher at 20, 10, 4 (and strangely also at −10°C) than at −0.5 and −4°C. The relationship between temperature and daily lipid consumption was stable between −0.5 and −4°C and exponential between 0 and 20°C (Jean et al. 1990).

Just before emerging from hibernation, *Cer. undecimnotata* males contained 17–19% and females 18–22% dry weight as fat (Hodek & Cerkasov 1963), *A. bipunctata* 14%, *P. quatuordecimpunctata* 19% and *C. septempunctata* 23% dry weight (Hariri 1966).

Sexual differences

On most occasions in the study by Hodek and Cerkasov (1963), *Cer. undecimnotata* females had a higher absolute fat content (by 0.1–1.2 mg) than males. Taking the amount of fat in males as 100%, then that in females ranged between 100 and 160%. In the three species studied by Hariri (1966), females contained 1.4 to 1.7 times more fat than males at the start of hibernation.

In addition to recording fat content, Pantyukhov (1968b) also observed changes in the iodine number during hibernation of *Chil. rubidus*. Although the **'iodine number'** before the onset of frosts in September and October amounted to 88–89, in January and February this index of the proportion of unsaturated fatty acids in fat rose to 93–94. As the unsaturated acids have a lower freezing point than the saturated ones, their rise would increase cold resistance. In *Chil. renipustulatus* (Pantyukhov 1965) similar increases in the iodine number from autumn (59–60) to winter (62–64) was observed.

6.4.2.2 Glycogen

Like lipids, glycogen reserves are accumulated during pre-diapause. However, glycogen is unstable in comparison with the more stable fat; thus both absolute and relative glycogen content greatly fluctuated in the course of dormancy in *Cer. undecimnotata* (Hodek & Cerkasov 1963).

The amount of glycogen in dormant coccinellids was roughly 10 times less than the amount of fat (for a comparison among four species see Hodek 1996, p. 251). In late August *Cer. undecimnotata* adults had about 0.4 mg of glycogen per beetle which represented about **2–3% of dry weight** (Hodek & Cerkasov 1963; Hodek 1996, table 7.03 therein). Values ascertained by Hariri (1966) for *A. bipunctata* in mid-October are one-tenth of those for *Cer. undecimnotata* – 0.04 mg of glycogen per beetle. In contrast to the above aphidophagous coccinellids, the coccidophagous ladybirds of the genus **Chilocorus** are reported to contain much more glycogen (**10–12% of dry weight**) at the beginning of diapause (Pantyukhov 1965, 1968a, b).

Rate of decrease

In all coccinellid species studied, glycogen content decreased during dormancy. In *Cer. undecimnotata* the rate of decrease (as for fat content) depended on temperature. In late summer and early autumn the decrease was **very steep** so that the **daily decrease** averaged 2.5 µg in males and 2.8 µg in females. During the 8 months of dormancy in *Cer. undecimnotata*, glycogen decreased in males to about 22% and in females to 25% of the initial amount (Hodek & Cerkasov 1963).

Before emergence from the dormancy site, *Cer. undecimnotata* contained 50–85 μg and 90–105 μg of glycogen in males and females respectively, which equalled 0.7–1.3% of dry weight. In *A. bipunctata* it was 5–8 μg in males and 6–9 μg in females and equalled about 0.2% of dry weight (Hodek & Cerkasov 1963).

Sexual differences

Markedly higher absolute glycogen content was found in females than in males of *Cer. undecimnotata*: if the content in males is taken as 100%, then the amount of glycogen in females represented 100–200% (Hodek & Cerkasov 1963). At the start of hibernation, *C. septempunctata* and *A. bipunctata* females again contained more glycogen than did the males, whereas both genders of *P. quatuordecimpunctata* contained almost equal amounts (Hariri 1966).

Since glycerol and other polyols are responsible for frost resistance in some insects, both fat and glycogen must be considered not only as a source of energy, but also as precursors for polyols (6.4.4).

6.4.2.3   Water

Water content in monthly samples of *Hip. convergens* taken from Sierra Nevada aggregations remained remarkably constant (Hagen 1962). Also in *Cer. undecimnotata* there was no general tendency for a decrease or increase in **absolute water content** during the 8 months of dormancy (Hodek & Cerkasov 1963). The water content usually amounted to **10–12 mg** per male and to **11–14 mg** per female. The continual increase in the **relative water content** from about 50–55% to about 60–63% which has been observed during dormancy is simply caused by the fall in dry weight, particularly of fat. The same applies for the increase to 64 and 68% from January to April in *Chil. renipustulatus* and *Chil. rubidus* (Pantyukhov 1965, 1968a, b).

Beetles may increase their body water content by **drinking** or by the production of **metabolic water** when splitting fat. The loss of body water by transpiration is affected by air humidity. Hagen (1962) assumes, for *Hip. convergens*, that the water balance is maintained by imbibing water. If the beetles from aggregations are kept in a refrigerator for a month or so in the absence of litter, a distinct water loss results. When these beetles are then exposed to water, they drink avidly. In field samples of *Cer. undecimnotata*,

increases in water content usually coincided with rainfall and high humidity of the air, while decreases coincided with drought (Hodek & Cerkasov 1963).

6.4.2.4   Metabolic rate

Consistent with one of the adaptive roles of diapause (6.1), its onset is invariably associated with a striking drop in the metabolic level (Tauber et al. 1986, Danks 1987), e.g. in the chrysomelid *Leptinotarsa decemlineata* a drop to 15–20% of the normal respiratory rate was observed in dormant beetles (De Wilde 1969).

A **decrease in oxygen consumption** has also been demonstrated in diapausing coccinellids. When measuring the respiratory rate at 18°C in diapausing *Hip. convergens*, Stewart et al. (1967) obtained the value of 12 μl O_2/beetle/h immediately after the arrival at the aestivation sites in the Pinnacle mountains (Arkansas, USA) in late June. From July to February, oxygen consumption fluctuated between 5 and 9 μl, while in reproducing beetles in the lowlands it was 29 μl in March. During diapause, the level of metabolism was thus reduced to about 30 to 15%.

In diapausing *Col. maculata lengi* from Vermont (USA) the metabolic rate was low (0.18 μl O_2/mg body weight/15 min), but was about twice higher in summer samples (0.34 μl O_2/mg /15 min; Parker et al. 1977).

In two diapausing ladybirds, *Hip. convergens* and *Col. maculata*, Lee (1980b) compared the **effects of cold acclimation** at 6°C and **warm acclimation** at 20°C on the oxygen consumption within a range of five temperatures from 0 to 20°C. Within this temperature range the respiration rate in cold acclimated *Hip. convergens* increased from 0.1 to 0.75 μl O_2/mg/h and in warm acclimated beetles the values at all five temperatures were slightly but significantly lower by 0.05–0.2 μl O_2/mg/h. In *Col. maculata* this effect was reversed: the five oxygen consumption levels were lower in cold acclimated beetles. Lee (1989a) attributes these reversed responses to different acclimation temperatures associated with the different type of hibernation sites of the two species: *Hip. convergens* is more exposed to ambient temperatures (6.3.2.7), while *Col. maculata* hibernates in leaf litter under a snow cover (6.3.2.4). In a later experiment, *Hip. convergens* was collected in January and held for 6 weeks in the dark and at 4°C, and the metabolic rate was measured after exposure to 20°C and 12L:12D; within 14 days the oxygen consumption decreased to 40% of the initial value, from 1.13 to about $0.5\,\mu l\,O_2/mg/h$ (Bennett & Lee 1989).

Pantyukhov (1968a, b) similarly recorded a much lower oxygen consumption in diapausing *Chil. rubidus* ($0.1\,\mu l\,O_2/mg/h$) than in active adults ($0.8\,\mu l\,O_2/mg/h$) (Pantyukhov 1968a), and similar difference between diapausing *Chil. renipustulatus* ($0.52\,\mu l\,O_2/mg/h$) and active adults ($1.25\,\mu l\,O_2/mg/h$) (Pantyukhov 1968b). Oxygen consumption by diapausing *Chil. bipustulatus* females reared in the laboratory was only a little more than half that of non-diapausing individuals (Tadmor & Applebaum 1971).

In Japan, Sakurai (1969) ascertained a decrease in respiration rate from $2–3\,\mu l\,O_2/mg/h$ in active beetles to $0.4–1.3\,\mu l\,O_2$ in *C. s. brucki* during aestivation diapause (measured at 30°C). Hibernating *Cer. undecimnotata* have a respiration rate around $0.8\,\mu l\,O_2/mg/h$ (measured at 25°C) but, if re-activated in the laboratory, the beetles show a doubled oxygen consumption, so that in ovipositing females it amounts to 1.6 (1.1–2.0) $\mu l\,O_2/mg/h$ (Hodek & Cerkasov 1958, Hodek 1970). In populations of *A. obliterata* from eastern Scotland the lowest level of respiration ($1.43\,\mu l\,O_2/mg/h$, measured at 20°C) was recorded in diapausing adults in mid-February, while the metabolic rate of post-hibernating beetles in early May was almost three times higher at $3.63\,\mu l\,O_2/mg/h$ (Parry 1980).

6.4.3 *Corpora allata* and regulation of vitellogenesis

As in other insects, reproduction and adult diapause of coccinellids are regulated by the neuroendocrine system (see also 6.1.4). While evidence for the transfer of environmental signals to the neurosecretory cells of the *pars intercerebralis* of the brain and for the secretion and function of the brain hormone is still missing in coccinellids, there are some data on the subsequent pathways of the neuroendocrine regulation, i.e. on the action of the **juvenile hormone** (JH) secreted by the *corpora allata* (CA). In reproducing males, JH stimulates the development of the accessory glands and, in the maturation of the ovaries, JH is necessary for the synthesis of vitellogenins which are deposited in the developing oocytes as **yolk proteins** (vitellins). Although the **active inhibition of the CA** found in a heteropteran, *Pyrrhocoris apterus* (Hodková 1976, Hodková et al. 2001), may be a general feature in adult diapause of insects, we still

lack similar evidence for coccinellids. Most data are concerned only with the so-called **'passive inhibition' due to the inactivity of the CA** or the very low titre of JH during adult diapause. All data have come from only two coccinellid species: *C. septempunctata* and *Cer. undecimnotata*.

The **volume of the CA** is the usual measure of their activity, but this has been questioned. In the bivoltine *C. s. brucki*, the size of the CA in hibernating beetles is similar to that in active beetles, but more than twice as large as during aestivation (Sakurai et al. 1981b, 1983). This has been interpreted as one of the important indications that aestivating adults of this subspecies enter diapause, while hibernation is a mere quiescence. Also during the hibernation diapause of the central European population of *C. s. septempunctata*, the onset of diapause was accompanied by a decrease in the size of the CA. In spring, the CA increased in size again with the resumption of activity and with the progress of ovarian maturation (Okuda 1984, Okuda et al. 1986). In the same region the same tendency was observed in *Cer. undecimnotata* (Okuda 1984, Okuda & Hodek 1989).

One of the ways in which the function of JH in the reproductive activity of females can be demonstrated is the monitoring of **electrophoretic patterns** of **haemolymph proteins**. Topical application of a JH-analogue (methoprene) to aestivating (i.e. diapausing) females of *C. s. brucki* caused the disappearance of bands specific for diapause, while a band appeared which was presumed to be vitellogenin. That band also appeared in late May and in mid-October, i.e. during, respectively, the spring and autumnal reproduction periods of young adults. It remains to be established why a distinct band of vitellogenin was not found after aestivation in late September though it was very distinct after hibernation in mid-April (Sakurai et al. 1987b).

Okuda and Chinzei (1988) studied the synthesis of **vitellogenin**, the yolk-protein precursor in *C. s. brucki*. While in pre-diapause the vitellogenin level was negligible and not detectable during aestivation diapause, it did occur after diapause, but then only after feeding.

Electrophoretic methods were also used to investigate the synthesis of vitellogenin in individual organs in *C. septempunctata*. Synthesis was observed in the **fat body** and to a lesser extent in the **ovaries** of mature females. Vitellogenin synthesis was not detectable in the brain or the thoracic muscles of mature females, or in the fat body of males (Zhai et al. 1984). When the coccinellids were reared on a substitute diet (raw pig liver, honey and sugar; Chapter 5.2.10) very low vitellogenin synthesis occurred; this was consistent with the observation that the oocytes in most females developed only to the pre-vitellogenic stage (Zhai et al. 1984). The synthesis of vitellogenin in the fat body of non-reproductive females fed on a substitute diet was induced by a JH-analogue (Zhai & Zhang 1984, Zhang & Zhai 1985).

Quantitative changes of vitellogenin synthesis in the fat body and the ovary were studied by **radioimmunoassay** in *C. septempunctata*. The fat body secreted more than 90% of the vitellogenin newly synthesized there, whereas the ovaries retained most of the vitellogenin they synthesized. The vitellogenin synthesized in the ovaries of active vitellogenic females was about 20% of that produced in the fat body (Zhai et al. 1985).

The correlation between the activity of the CA (i.e. synthesis of JH) and the development of oocytes was documented also in *C. septempunctata*, with both a **bioassay** using the wax moth, *Galleria mellonella* (Fu & Chen 1984) and a short-term radiochemical assay (Guan & Chen 1986).

6.4.4   Cold-hardiness

Studies focussing on the effect of low temperatures on coccinellids are of practical importance when a species has been introduced to a different climate for biological control, and the potential for survival in the new region needs to be estimated. This was done for *Har. axyridis* after invasion to the Nearctic region (see below).

The level of cold-hardiness achieved at the onset of diapause as a result of the **diapause syndrome** is usually increased through **cold acclimation** under the influence of decreasing temperatures in winter. Paradoxically, this process may continue after the winter solstice, after the end of diapause (Hodková & Hodek 2004). As the temperature falls, **supercooling** occurs which is followed by the freezing of the tissues accompanied by a rapid rise in internal temperature. The external temperature at which this occurs is defined as the **supercooling point** (SCP). For the general theory of cold-hardiness and definitions of terms see Salt (1961, 1964), Lee and Denlinger (1991), Leather et al. (1993) and Hodková and Hodek (2004).

The results obtained for coccinellids are similar to those for most insects. Resistance to temperatures below zero is rather high in the middle of dormancy, after cold acclimation resulting from the gradual decrease in temperature. Early or late frosts, however, can be damaging. The degree of cold-hardiness varies naturally in different species and is related to their type of hibernation. **Subnivean** species which hibernate in litter, and are usually covered by snow, are certainly more sensitive to freezing than the more exposed **supranivean** species which hibernate in crevices of bark or rocks.

Early evidence for cold-hardiness in coccinellids came from the studies by Pantyukhov on two *Chilocorus* species. The common *Chil. bipustulatus* has low cold-hardiness. Most individuals have their supercooling point at −8 to −9°C, the limit for survival being −10 to −12°C (Pantyukhov 1965). A prolonged decrease in temperature down to −5 to −6°C at the soil surface in the hibernation sites causes rather high mortality. Considerable mortality is to be expected where there is a thin snow cover or where the spring temperature fluctuates. A considerable increase of cold-hardiness in *Chil. rubidus* was recorded between September and January (Pantyukhov 1968a). In spring the cold-hardiness decreased again. Similar changes have been noticed in measurements of the supercooling points (Table 6.19). The lowest temperature that could still be survived by a considerable number of beetles (40%) in January was −13.5°C for 2 days. *Chil. rubidus* hibernates in the litter and, if this is additionally covered by a sufficient layer of snow, this subnivean beetle is not exposed to extreme low temperatures. In the Petersburg region, Pantyukhov (1968b) measured the minimum **temperature** on the ground **under** the **snow**. In the winter of 1964/65 this was −2.5°C, and −7°C the next winter; the survival of the beetles was 67–82% and 52–70%, respectively. By contrast, in Alma-Ata, Kazakhstan, where the minimum temperature (without snow) was −10°C, only 12% of the beetles survived. In the Far East region of Russia the survival was 21% (Pantyukhov 1968b).

In the **freezing-intolerant** *Aphidecta obliterata* (overwintering beneath bark, the supercooling points of adults without visible **gut contents** were much lower (in late autumn from −28.0 to −30.2°C) than those of the more abundant adults with guts containing food material (from −13.7 to −13.0°C) (Parry 1980). The difference is evidently caused by gut contents forming nucleating agents for freezing.

Table 6.19 Supercooling point and freezing point in *Chilocorus rubidus* (Pantyukhov 1968a).

Months	*n*	Supercooling point (°C)			Freezing point (°C)		
		aver.	min.	max.	aver.	min.	max.
Sep.	41	−8.2	−2.7	−14.2	−2.6	−0.4	−5.2
Dec.	57	−13.2	−4.9	−17.4	−3.2	−0.7	−6.8
Jan.	50	−12.9	−4.3	−17.2	−3.2	−0.8	−6.7
Feb.	40	−13.5	−5.6	−17.8	−3.4	−1.1	−6.9
April	32	−3.4	−0.2	−5.5	−0.8	−0.1	−2.7

Overwintering with the digestive tract containing food material is unusual in coccinellids, and may be associated with the mild climate of eastern Scotland where the study was carried out. The highest mean supercooling point in spring was −7.5°C. *Aphidecta obliterata* contains no polyols.

Although warm winds may melt the snow in southern Alberta, Canada, *Hip. quinquesignata* can be considered as a **subnivean hibernator**. The supercooling points in late autumn and winter averaged between −18.5 and −22.2°C, with maximum supercooling capacity around −27°C. In spring, before dispersal, the supercooling point was higher, with means between −12 and −13°C (Harper & Lilly 1982). In outdoor beetles, the supercooling point correlated well with the content of glycerol, but the beetles reared in the laboratory and containing no glycerol or sorbitol, still supercooled to −18.5°C; supercooling is evidently also due to other factors.

Adalia bipunctata showed a much higher resistance to low moisture and larger extremes of temperature than the coccinellid species that hibernate in litter (Novák & Grenarová 1967). When placed under a roof or into the crown of *Picea* trees, *A. bipunctata* suffered only 13 to 40% mortality, while *C. septempunctata*, *C. quinquepunctata* and *Exochomus quadripustulatus* died out completely or by more than 90% within 4.5 months (mid-October to early March). There were no differences, however, in the winter survival in the grass environment among the four species tested. The higher tolerance of *A. bipunctata* is obviously connected with its usual hibernation site in the drier microhabitat of bark crevices where it is not protected from ambient extreme conditions.

The low survival of the litter-hibernator *E. quadripustulatus* may be considered surprising as very low values of the supercooling point (SCP) were ascertained in hibernating individuals (Parry 1986, Nedvěd 1993). Parry emphasized the difference in the supercooling point in late October between *A. obliterata* (hibernating in bark crevices (SCP, −14.3°C) and *E. quadripustulatus* (SCP, −19.4°C); this difference was the reverse of expectation. The samples had, however, different sources: *A. obliterata* adults were not collected in bark crevices but from the needles of Douglas fir. This means that *A. obliterata* was still feeding and would have contained food material in the gut (see above in this chapter), while *E. quadripustulatus* was inactive in the litter.

A detailed comparison has been undertaken between *Hip. convergens* and *Col. maculata lengi*, which are both freezing-intolerant insects. The diapausing adults collected from hibernation sites were stored for one month in screened field cages covered with snow. In February, the SCPs were similar for both species: −18°C for *Col. m. lengi* and −15°C for *Hip. convergens*. However, during an exposure of five days to 25°C, the SCP for *Col. m. lengi* adults increased to −9°C, while it remained low and constant for *Hip. convergens* (Lee 1980b). The difference could again be related to the condition of the hibernation sites. In Lee's **supranivean** hibernacula, diapausing *Hip. convergens* are in nature exposed to the direct influence of ambient temperature and insolation (which may have raised the body temperature) and have evolved an adaptation keeping cold-hardiness resistant to temperature impact. In contrast, *Col. m. lengi* hibernates in **subnivean** hibernacula and thus lacks such adaptation (Lee 1980b).

In central Europe a comparison was made between a **subnivean** hibernator *C. septempunctata* (mostly in grass tussocks), and *Cer. undecimnotata*, hibernating predominantly in **supranivean** sites (rock crevices). When the SCP is measured in samples of beetles taken from the field in the coldest period, the values for

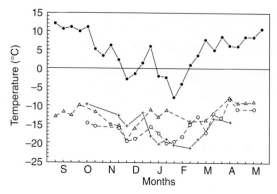

Figure 6.14 Seasonal changes in supercooling point in dormant coccinellid adults (below) and temperature means (above). *Coccinella septempunctata* collected in natural insulated hibernacula (grass tussocks, triangles); *C. septempunctata* exposed in outdoor insectary (crosses); *Ceratomegilla undecimnotata* exposed in outdoor insectary (open circles) (from Nedvěd 1993).

Table 6.20 Overwintering survival in caged coccinellid adults (Bushland, Texas, USA) (Michels et al. 1997, modified).

Species	Percentage survival by year			
	1990	**1991**	**1992**	**1993**
Snow cover (cm)	45.7	36.0	57.2	17.8
Cycloneda ancoralis	—	—	12.9	7.8
Eriopis connexa	—	—	31.1	2.7
Hippodamia tredecimpunctata	58.6	—	27.4	—
Hippodamia variegata	29.4	20.4	17.9	4.3
Oenopia conglobata	—	—	14.1	10.9
Propylea quatuordecimpunctata	33.9	25.6	16.5	—
Scymnus frontalis	2.4	5.1	3.9	—

C. septempunctata were much higher than for *Cer. undecimnotata* (Fig. 6.14). The lower SCP for *Cer. undecimnotata* was evidently induced by the exposure to ambient temperature. When the adults of *C. septempunctata* were transferred from their insulated hibernaculum to a field-cage (where they were exposed to ambient temperature), their SCP decreased to the level of *Cer. undecimnotata* (Nedvěd 1993).

The importance of snow cover for the overwintering survival was also reported in the Texas High Plains for seven imported coccinellid species (Michels et al. 1997). The beetles were caged on native grasses during four winter periods. Survival ranged from 59% for *Hip. tredecimpunctata* to only 3% for *Scymnus frontalis* (Table 6.20). Higher survival was recorded in colder years with snow cover. An absence of snow cover combined with cold temperature, particularly in April and May, was probably detrimental in 1993, the last year.

The SCP is not always a good indicator of cold-hardiness. In *Hip. convergens* adults exposed at the end of diapause to 20°C for 20 days (Bennett & Lee 1989), the SCP increased only slightly from −16 to −12.9°C, while the survival at −5°C for 2 hours decreased strongly from almost 100% to only 50%. The adaptive maintenance of supercooling discussed in an earlier paper by Lee (1980b) may thus have a lower value for

survival than expected. With *Cer. undecimnotata*, however, the survival of a 24 hour exposure to low temperature is a good match to changes in the SCP (Nedvěd 1993). Also in *Har. axyridis*, the SCP was found to be a good indicator of cold-hardiness for summer and winter samples from Georgia and Minnesota, USA. It decreased by about 12°C from August to December (Koch et al. 2004).

Watanabe (2002) studied *Har. axyridis* in a **region with mild winters**, Tsukuba, Japan (36°N). The seasonal change of SCP and the 50% lower lethal temperature were both at their minimum in winter between December and February. The seasonal change in the polyol myo-inositol was well and negatively correlated with the changes in the SCP and lower lethal temperature (Fig. 6.15). Dynamics of SCP during hibernation indoor and outdoor was compared in *Har. axyridis* adults (Berkvens et al. 2010; Fig. 6.16).

While in central Japan the conditions are suitable for winter survival of *Har. axyridis*, there are harder winters in **cold regions of North America**. Although the SCP of *Har. axyridis* was there lower in winter (−23°C) (it was only −18°C in Japan), this evidently was not enough for survival as the minimum air temperatures were below −23°C. Therefore survival of extreme winter conditions depends on finding overwintering sites that provide adequate insulation (Carrillo et al. 2004). Koch et al. (2004) recorded 100% mortality after 24 hours at −20°C. Outside buildings, Labrie et al. (2008) did not observe any survival in

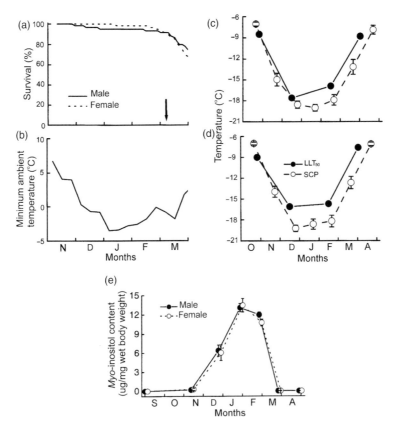

Figure 6.15 Seasonal changes of survival rate in *Harmonia axyridis* adults (a) and 10-day-minimum temperatures in an artificial hibernaculum (b). Arrow, date of first observed mating. Seasonal changes of supercooling point (SCP, dashed line with open circles) and 50 % lower lethal temperature (LLT$_{50}$, solid line with closed circles) in male (c) and female (d) collected outdoors. (e) Seasonal changes of *myo*-inositol content in adults (all from Watanabe 2002).

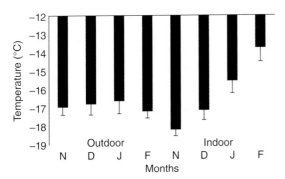

Figure 6.16 Supercooling point (mean ± SE) of outdoor and indoor overwintering *Harmonia axyridis* adults (from Berkvens et al. 2010).

the winter 2003/2004 in northeast Canada, when the temperature ranged between 5 and −25°C. Adults hibernating on hill tops near Montreal (71–73°W, 45°N) were dead at the end of winter (G. Labrie, unpublished)

Size and melanism can both affect winter survival, as shown by Osawa (2001) for *Har. axyridis*. In the cold winter of 1987/88 (with a great temperature range) a size-related mortality probably occurred in melanic females, with melanic females in spring 1988 being on average larger than in spring 1987. Small melanic females may have had lower survival because of lower reserves, with the reserves exhausted more easily in melanics because of their greater heat absorption during spells of warm sunny weather (Osawa 2001). Ohgushi (1986) showed that it was sex- and/or size-dependent mortality that resulted in larger adults and a higher proportion of females after overwintering in the herbivorous *Henosepilachna niponica*.

Some studies have aimed at finding the **storage temperature** for the developmental stages of ladybirds. Survival of eggs, larvae and pupae of *C. undecimpunctata* was quite high (65–85%) after 7-day storage at 6.0°C. Survival of adults was longer after 5–10 days of feeding on aphids at 28°C with a 14L:10D photoperiod (50%), than when the adults were exposed to cold immediately after emergence (30–40%) (Abdel-Salam & Abdel-Baky 2000). No attempt was made to induce diapause, which would have increased the survival (Iperti & Hodek 1974; 6.2.8). Survival of the second and third instars of *Col. maculata lengi* at 4 and 8°C was close to 100% for the first two weeks, but decreased drastically to 60% after 3 weeks and to 0% after 5 weeks. Development was resumed without negative effects when the 2-week larvae were returned to 24°C (Gagne & Coderre 2001).

6.5   CONCLUSIONS AND LACUNAE IN KNOWLEDGE

Diapause is an important component of ladybirds' life history. Knowledge of its timing (particularly of its end) is undispensable in improving biological control. Although the once popular method – using ladybirds from hibernation aggregations in gardens or crop fields – has almost been abandoned because of the subsequent fast dispersal of the released ladybirds, ecological methods of biological control and IPM depend on understanding **when and why** the **ladybirds arrive in crops**. This is particularly important when the crops have to be treated beforehand, e.g. by the application of alternative foods or other attractants (see Chapters 5 and 11).

Similarly, knowledge about **voltinism and its regulation** is needed for rational support of ladybird activity in the field or for rearing. Without insight into the role of photoperiod in diapause induction, one may make unrealistic high estimates of voltinism when they are based only on the effect of temperature on development.

Apart from its applied aspects, diapause of coccinellids is an interesting and useful **model for research**

in several fields. The important critical remark by Sloggett (2005), that our research is focused on **only a few coccinellid species,** is highly applicable to diapause studies. Only lately, when *Har. axyridis* invaded both America and Europe, has this species begun to compete for research interest with *C. septempunctata*. Surprisingly diapause studies have not focussed on *A. bipunctata*, popular in other studies. Not only have **most species** been **ignored**, but even in monospecific studies, numerous for *C. septempunctata*, intraspecific variation has rarely been considered. Comparative studies on **diapause intensity** or **overwintering survival**, such as by Hodek et al. (1977) and Zhou et al. (1995), have remained rare and should be extended.

Also, **geographic variation in diapause** expression (which has often attracted interest in other insect groups) has been a rare topic even with common coccinellid species, such as *C. septempunctata*. Yet there are clearly regions where local authors could make interesting findings (Mediterranean Europe, India) because of the wide temperature and photoperiod ranges there. 'One-off' visits by foreign workers can produce only fragmentary results.

Still only partly solved is the problem why migrating coccinellids often **fly to prominent structures** and repeatedly visit the same hibernation sites. What is the contribution of passive transport by air currents versus active flight together with hypsotactic orientation? Similarly open is the role of various taxes versus pheromones in the formation of massive **aggregations** (also Chapter 9).

Another problem, belonging partly to the sphere of population genetics, is the **inheritance of** tendencies to the **expression of univoltinism and multivoltinism** (6.2.16). Another task for population genetics is how the seemingly deleterious (at least from the view of optimal foraging theory (OFT)) trait to **overproduction of offspring** in some species is perpetuated. Is an increased variance a satisfactory benefit in the trade-off between the 'OFT economy' and the overproduction of offspring?

One of important adaptations, essential for the aphidophagous coccinellids, but analyzed rather exceptionally (Osawa 2005), is the **postponement of egg laying** and reversible **resorption of eggs**.

It is a pity that the frequency of studies focussed on diapause decreased in the last two decades. Hopefully, the above-mentioned topics show that such research is rewarding.

ACKNOWLEDGEMENTS

The work was supported by the project of the Institute of Entomology, Academy of Sciences of the Czech Republic No. AVOZ50070508.

REFERENCES

Abdel-Salam, A. H. and N. F. Abdel-Baky. 2000. Possible storage of *Coccinella undecimpunctata* (Col., Coccinellidae) under low temperature and its effect on some biological characteristics. *J. Appl. Entomol.* 124: 169–176.

Al Abassi, S., M. A. Birkett, J. Pettersson, J. A. Pickett and C. M. Woodcock. 1998. Ladybird beetle odour identified and found to be responsible for attraction between adults. *Cell. Mol. Life Sci.* 54: 876–879.

Alcock, J. 1987. Leks and hilltopping in insects. *J. Nat. Hist.* 21: 319–328.

Almatni, W., M. Z. Mahmalji and H. Al Rouz. 2002. A preliminary study of some natural enemies of woolly apple aphid *Eriosoma lanigerum* (Hausmann) (Homoptera: Aphididae) in Sweida, Syria. *Damascus Univ. J. Agric. Sci.* 18: 117–129.

Anderson, J. M. E. 1981. Seasonal field analyses of fat content, live weight, dry weight and water content of the aphidophagous *Scymnus lividigaster* (Mulsant) and mycophagous *Leptothea galbula* (Mulsant) (Coleoptera: Coccinellidae). *Austr. J. Zool.* 29: 679–689.

Anderson, J. M. E., D. F. Hales and K. A. van Brunschot. 1986. Parasitisation of coccinellids in Australia. In: Hodek I (ed.): *Ecology of Aphidophaga.* Academia, Prague & W. Junk, Dordrecht. pp. 519–524.

Andrewartha, H. G. 1952. Diapause in relation to the ecology of insects. *Biol. Rev.* 27: 50–107.

Angalet, G. W., J. M. Tropp and A. N. Eggert. 1979. *Coccinella septempunctata* in the United States: recolonizations and notes on its ecology. *Environ. Entomol.* 8: 896–901.

Avidov, Z. and D. Rosen. 1965. Studies towards an integrated control programme of citrus pests in Israel. *12th Int. Congr. Entomol.,* London 1964. pp. 572–573.

Banks, C. J. 1954. Random and non-random distributions of Coccinellidae. *J. Soc. Br. Entomol.* 4: 211–215.

Banks, C. J. 1955. An ecological study of Coccinellidae associated with *Aphis fabae* Scop. on *Vicia faba. Bull. Entomol. Res.* 46: 561–587.

Bazzocchi, G. G., A. Lanzoni, G. Accinelli and G. Burgio. 2004. Overwintering, phenology and fecundity of *Harmonia axyridis* in comparison with native coccinellid species in Italy. *BioControl* 49: 245–260.

Ben-Dov, Y. and D. Rosen. 1969. Efficacy of natural enemies of the California red scale on citrus in Israel. *J. Econ. Entomol.* 62: 1057–1060.

Bennett, L. E. and R. E Lee Jr. 1989. Simulated winter to summer transition in diapausing adults of the lady beetle

(*Hippodamia convergens*): supercooling point is not indicative of cold-hardiness. *Physiol. Entomol.* 14: 361–367.

Benton, A. H. and A. J. Crump. 1979. Observations on aggregation and overwintering in the coccinellid beetle *Coleomegilla maculata* (DeGeer). *J. N. Y. Entomol. Soc.* 87: 154–159.

Benton, A. H. and A. J. Crump 1981. Observations on the spring and summer behavior of the 12-spotted ladybird beetle, *Coleomegilla maculata* (DeGeer) (Coleoptera: Coccinellidae). *J. N. Y. Entomol. Soc.* 89: 102–108.

Berker, J. 1958. Die natürlichen Feinde der Tetranychiden (*Stethorus punctillum*). *Z. Angew. Entomol.* 43: 115–172.

Berkvens, N., J. S. Bale, D. Berkvens, L. Tirry and P. De Clercq. 2010. Cold tolerance of the harlequin ladybird *Harmonia axyridis* in Europe. *J. Insect Physiol.* 56, 438–444.

Berkvens, N., J. Bonte, D. Berkvens, L. Tirry and P. De Clercq. 2008. Influence of diet and photoperiod on development and reproduction of European populations of *Harmonia axyridis* (Pallas) (Coleoptera: Coccinellidae). *BioControl* 53: 211–221.

Bielawski, R. 1961. Die in einem Krautpflanzenverein und in einer Kieferschonung in Warszawa/Bielany auftretenden Coccinellidae (Coleoptera). *Fragm. Faun. (Warsaw)* 8: 485–525.

Bodenheimer, F. S. 1943. Studies on the life-history and ecology of Coccinellidae. I. The life-history of *Coccinella septempunctata* L. in four different zoogeographical regions. *Bull. Soc. Fouad 1er Entomol.* 27: 1–28.

Bodenheimer, F. S. and P. M. Vermes. 1957. The genetical aspects of diapause in the transitory zone. *Stud. Biol., Jerusalem* 1: 106–109.

Boiteau, G., Y. Bousquet and W. P. L. Osborn. 1999. Vertical and temporal distribution of Coccinellidae (Coleoptera) in flight over an agricultural landscape. *Can. Entomol.* 131: 269–277.

Bonnemaison, L. 1964. Observations écologiques sur la Coccinelle à 7 points (*Coccinella septempunctata* L.) dans la région parisienne (Col.). *Bull. Soc. Entomol. Fr.* 69: 64–83.

Carrillo, M.A., R. L. Koch, R. C. Venette, C. A. Cannon and W. D. Hutchison. 2004. Response of the multicolored Asian lady beetle (Coleoptera: Coccinellidae) to low temperatures: implications for winter survival. *Am. Entomol.* 50: 157–158.

Ceryngier, P., P. Kindlmann, J. Havelka, et al. 1992. Effect of food, parasitization, photoperiod and temperature on gonads and sexual activity of males of *Coccinella septempunctata* (Coleoptera, Coccinellidae) in autumn. *Acta Entomol. Bohemoslav.* 89: 97–106.

Ceryngier, P., J. Havelka and I. Hodek. 2004. Mating and activity of gonads in pre-dormant and dormant ladybirds (Coleoptera: Coccinellidae). *Invert. Repr. Dev.* 45: 127–135.

Copp, N. H. 1983. Temperature-dependent behaviours and cluster formation by aggregating ladybird beetles. *Anim. Behav.* 31: 424–430.

Danks, H. V. 1987. *Insect Dormancy: An Ecological Perspective.* Biological Survey of Canada (Terrestrial arthropods), National Museum of Natural Sciences, Ottawa. 439 pp.

Danks, H. V. 2001. The nature of dormancy responses in insects. *Acta Soc. Zool. Bohemoslov.* 65: 169–179.

Danilevskii, A. S. 1965. *Photoperiodism and Seasonal Development of Insects.* 1st English edition. Oliver and Boyd. Edinburgh and London.

Davis, J. R. and R. L. Kirkland. 1982. Physiological and environmental factors related to the dispersal flight of the convergent lady beetle, *Hippodamia convergens* (Guerin-Meneville). *J. Kansas Entomol. Soc.* 55: 187–196.

Delucchi, V. 1954. *Pullus impexus* (Muls.) (Coleoptera, Coccinellidae), a predator of *Adelges piceae* (Ratz.) (Hemiptera, Adelgidae), with notes on its parasites. *Bull. Entomol. Res.* 45: 243–278.

Denlinger, D. L. 1986. Dormancy in tropical insects. *Annu. Rev. Entomol.* 31: 239–264.

Denlinger, D. L., G. D. Yocum and J. P. Rinehart. 2005. Hormonal control of diapause. *In* L. I. Gilbert, K. Iatrou and S. S. Gill (eds). *Comprehensive Molecular Insect Science.* Elsevier, Amsterdam. pp. 615–650.

Dobrzhanskii, F. G. 1922a. Imaginal diapause in Coccinellidae. Mass aggregations and migrations in Coccinellidae. *Izv. Otd. Prikl. Entomol.* 2: 103–124.

Dobrzhanskii, F. G. 1922b. Imaginal diapause in Coccinellidae. Mass aggregations and migrations in Coccinellidae. *Izv. Otd. Prikl. Entomol.* 2: 229–234.

Dobzhansky, T. 1925. Über das Massennauftreten einiger Coccinelliden im Gebirge Turkestans. *Z. Wiss. InsektBiol.* 20: 249–256.

Douglass, J. R. 1930. Hibernation of the convergent ladybeetle *Hippodamia convergens* Guer., on a mountain peak in New Mexico. *J. Econ. Entomol.* 23: 288.

Dyadechko, N. P. 1954. *Coccinellids of the Ukrainian SSR.* Kiev. 156 pp. (In Russian.)

Edwards, J. G. 1957. Entomology above the timberline: II. The attraction of ladybird beetles to mountain tops. *Coleop. Bull.* 11: 41–46.

Evans, A. C. 1936. A note on the hibernation of *Micraspis sedecimpunctata* L. (var. *12-punctata* L.) (Coccinell.) at Rothamsted Experimental Station. *Proc. R. Entomol. Soc. Lond.* 11: 116–119.

Ewert, M. A. and H. C. Chiang. 1966. Effects of some environmental factors on the distribution of three species of Coccinellidae in their microhabitat. *In* I. Hodek (ed.). *Ecology of Aphidophagous Insects.* Academia, Prague and W. Junk, The Hague. pp. 195–219.

Fabre, J.-H. 1879. *Souvenirs entomologiques, I. Etudes sur l'instinct et les moeurs des insectes,* 5th edn. Paris.

Felland, C. M. and L. A. Hull. 1996. Overwintering of *Stethorus punctum punctum* (Coleoptera: Coccinellidae) in apple orchard ground cover. *Environ. Entomol.* 25: 972–976.

Fu, Y. and Z. Chen. 1984. The concentration of juvenile hormone in female adults of *Coccinella septempunctata* during ovarian development. *Acta Entomol. Sinica* 27: 268–274.

Gagne, I. and D. Coderre. 2001. Cold storage of *Coleomegilla maculata* larvae. *Biocont. Sci. Tech.* 11: 361–369.

Ghabn, A. 1951. Studies on the biology and control of *Epilachna chrysomelina* L. in Egypt. *Bull. Soc. Fouad 1er Entomol.* 35: 77–106.

Guan, X. and E. Chen. 1986. Corpus allatum activity in the female *Coccinella septempunctata* L. adults. *Acta Entomol. Sinica* 29: 10–14.

Hagen, K. S. 1962. Biology and ecology of predaceous Coccinellidae. *Annu. Rev. Entomol.* 7: 289–326.

Hagen, K. S. 1966. Suspected migratory flight behaviour of *Hippodamia convergens. In* I. Hodek (ed.). *Ecology of Aphidophagous Insects.* Academia, Prague and W. Junk, The Hague. pp. 135–136.

Hales, D. F., J. M. E. Anderson and K. A. van Brunschot. 1986. Aggregation in Australian ladybirds. In: Hodek I (ed.) *Ecology of Aphidophaga.* Academia, Prague and W. Junk, Dordrecht. pp. 205–210.

Hämäläinen, M. and M. Markkula. 1972. Possibility of producing *Coccinella septempunctata* (L.) (Col., Coccinellidae) without a diapause. *Ann. Entomol. Fenn.* 38: 193–194.

Hariri, G. 1966. Studies on the physiology of hibernating Coccinellidae (Coleoptera): changes in the metabolic reserves and gonads. *Proc. R. Entomol. Soc. Lond. (A)* 41: 133–144.

Harper, A. M. and C. E. Lilly. 1982. Aggregations and winter survival in southern Alberta of *Hippodamia quinquesignata* (Coleoptera, Coccinellidae), a predator of the pea aphid (Homoptera: Aphididae). *Can. Entomol.* 114: 303–309.

Hawkes, O. A. M. 1920. Observations on the life-history, biology, and genetics of the lady-bird beetle, *Adalia bipunctata* (Mulsant). *Proc. Zool. Soc. Lond.* 33: 475–490.

Hecht, O. 1936. Studies on the biology of *Chilocorus bipustulatus* (Coccinell.) an enemy of the red scale *Chrysomphalus aurantii. Bull. Soc. Fouad I. Entomol.* 20: 299–326.

Hemptinne, J.-L. 1985. Les sites d'hivernation de la coccinelle *Adalia bipunctata* (L.) (Col., Coccinellidae) en Belgique. *Acta Ecol. Ecol. Appl.* 6: 3–13.

Hemptinne, J.-L. 1988. Ecological requirements for hibernating *Propylea quatuordecimpunctata* (L.) and *Coccinella septempunctata* L. (Col.: Coccinellidae). *Entomophaga* 33: 505–515.

Hemptinne, J.-L. and J. Naisse. 1987. Ecophysiology of the reproductive activity of *Adalia bipunctata* L. (Col., Coccinellidae). *Meded. Fac. Landbouw. Rijksuniv. Gent* 52: 225–233.

Hemptinne, J.-L. and J. Naisse. 1988. Life cycle strategy of *Adalia bipunctata* (L.) (Col., Coccinellidae) in temperate country. *In* E. Niemczyk and A. F. G. Dixon (eds). *Ecology and Effectiveness of Aphidophaga.* SPB Academic Publishing, The Hague. pp. 71–77.

Henderson, I. F., J. H. Henderson and J. H. Kenneth. 1953. *A Dictionary of Scientific Terms,* 5th edn. D. van Nostrand, New York.

Hirano, T., H. Sakurai and S. Takeda. 1982. Some factors inducing aestivation of *Coccinella septempunctata*. *Proc. Kansai Plant Prot. Soc.* 24: 61. (In Japanese.)

Hodek, I. 1960. Hibernation-bionomics in Coccinellidae. *Cas. Cs. Spol. Entomol.* 57: 1–20. (In Czech with English summary.)

Hodek, I. 1962. Experimental influencing of the imaginal diapause in *Coccinella septempunctata* L. (Col., Coccinellidae), 2nd part. *Cas. Cs. Spol. Entomol.* 59: 297–313.

Hodek, I. 1967. Bionomics and ecology of predaceous Coccinellidae. *Annu. Rev. Entomol.* 12: 79–104.

Hodek, I. 1970. Termination of diapause in two coccinellids (Coleoptera). *Acta Entomol. Bohemoslov.* 67: 218–222.

Hodek, I. 1971a. Sensitivity of larvae to photoperiod controlling the adult diapause of two insects. *J. Insect Physiol.* 17: 205–216.

Hodek, I. 1971b. Termination of adult diapause in *Pyrrhocoris apterus* (Heteroptera: Pyrrhocoridae) in the field. *Entomol. Exp. Appl.* 14: 212–222.

Hodek, I. 1973. *Biology of Coccinellidae*. Academia, Prague and W.Junk, The Hague. 260 pp.

Hodek, I. 1981. La role des signaux de l'environnement et des processus endogènes dans la régulation de reproduction par la diapause imaginale. *Bull. Soc. Zool. Fr.* 106: 317–325.

Hodek, I. 1983. Role of environmental factors and endogenous mechanisms in the seasonality of reproduction in insects diapausing as adults. *In* V. K. Brown and I. Hodek (eds). *Diapause and Life Cycle Strategies in Insects*. W. Junk, The Hague. pp. 9–33.

Hodek I. 1996. Dormancy. *In* I. Hodek and A. Honěk (eds). *Ecology of Coccinellidae*. Kluwer, Dordrecht. pp. 239–318.

Hodek, I. 2003. Role of water and moisture in diapause development (A review). *Eur. J. Entomol.* 100: 223–232.

Hodek, I. and J. Cerkasov. 1958. A study of the imaginal hibernation of *Semiadalia undecimnotata* Schneid. (Coccinellidae, Col.) in the open, I. *Vest. Cs. Spol. Zool.* 22: 180–192. (In Czech with English summary.)

Hodek, I. and J. Cerkasov. 1960. Prevention and artificial induction of the imaginal diapause in *Coccinella 7-punctata* L. *Nature* 187: 345.

Hodek, I. and J. Cerkasov. 1961. Prevention and artificial induction of imaginal diapause in *Coccinella septempunctata* L. (Col., Coccinellidae). *Entomol. Exp. Appl.* 4: 179–190.

Hodek, I. and J. Cerkasov. 1963. Imaginal dormancy in *Semiadalia undecimnotata* Schneid. (Coccinellidae, Col.) II. Changes in water, fat and glycogen content. *Vest. Cs. Spol. Zool.* 27: 298–318.

Hodek, I. and J. Cerkasov. 1965. Biochemical changes in *Semiadalia undecimnotata* (Schneider) adults during diapause. *Nature* 205: 925–926.

Hodek, I. and P. Ceryngier. 2000. Sexual activity in Coccinellidae (Coleoptera): a review. *Eur. J. Entomol.* 97: 449–456.

Hodek, I. and M. Hodková. 1988. Multiple role of temperature during insect diapause: a review. *Entomol. Exp. Appl.* 49: 153–165.

Hodek, I. and A. Honěk. 1970. Incidence of diapause in *Aelia acuminata* (L.) populations from southwest Slovakia (Heteroptera). *Vest. Cs. Spol. Zool.* 34: 170–183.

Hodek, I. and G. Iperti. 1983. Sensitivity to photoperiod in relation to diapause in *Semiadalia undecimnotata* females. *Entomol. Exp. Appl.* 34: 9–12.

Hodek, I. and V. Landa. 1971. Anatomical and histological changes during dormancy in two Coccinellidae. *Entomophaga* 16: 239–251.

Hodek, I. and J. P. Michaud. 2008. Why is *Coccinella septempunctata* so successful? (A point of view). *Eur. J. Entomol.* 105: 1–12.

Hodek, I. and T. Okuda. 1993. A weak tendency to 'obligatory' diapause in *Coccinella septempunctata* from southern Spain. *Entomophaga* 38: 139–142.

Hodek, I. and Z. Ruzicka. 1979. Photoperiodic response in relation to diapause in *Coccinella septempunctata* (Coleoptera). *Acta Entomol. Bohemoslov.* 76: 209–218.

Hodek, I., M. Hodková and V. P. Semyanov. 1989. Physiological state of *Coccinella septempunctata* adults from northern Greece sampled in mid-hibernation. *Acta Entomol. Bohemoslov.* 86: 241–251.

Hodek, I., G. Iperti and M. Hodková. 1993. Long distance flights in Coccinellidae (a review). In: *Behavioural Ecology of Aphidophaga* (Proc. 5 Int. Symp. Ecol. Aphidophaga, Antibes, Sept. 1993). *Eur. J. Entomol.* 90: 403–414.

Hodek, I., G. Iperti and F. Rolley. 1977. Activation of hibernating *Coccinella septempunctata* (Coleoptera) and *Perilitus coccinellae* (Hymenoptera) and photoperiodic response after diapause. *Entomol. Exp. Appl.* 21: 275–286.

Hodek, I., T. Okuda and M. Hodková. 1984. Reverse photoperiodic responses in two subspecies of *Coccinella septempunctata* L. (Col., Coccinellidae). *Zool. Jb. Syst.* 111: 439–448.

Hodková, M. 1976. Nervous inhibition of corpora allata by photoperiod in *Pyrrhocoris apterus*. *Nature* 263: 521–523.

Hodková, M. and I. Hodek. 2004. Photoperiod, diapause and cold-hardiness (mini-review). *Eur. J. Entomol.* 101: 445–458.

Hodková, M., T. Okuda and R. M. Wagner. 2001. Regulation of corpora allata in females of *Pyrrhocoris apterus* (Heteroptera) (a mini-review). *In Vitro Cell Dev. Biol. Anim.* 37: 560–563.

Hodson, A. C. 1937. Some aspects of the role of water in insect hibernation. *Ecol. Monogr.* 7: 271–315.

Honěk, A. 1989. Overwintering and annual changes of abundance of *Coccinella septempunctata* in Czechoslovakia (Coleoptera, Coccinellidae). *Acta Entomol. Bohemoslov.* 86: 179–192.

Honěk, A. 1990. Seasonal changes in flight activity of *Coccinella septempunctata* L. (Coleoptera, Coccinellidae). *Acta Entomol. Bohemoslov.* 87: 336–341.

Honĕk, A. 1997. Factors determining winter survival in *Coccinella septempunctata* (Col., Coccinellidae). *Entomophaga* 42: 119–124.

Honĕk, A., Z. Martinkova and S. Pekar. 2007. Aggregation characteristics of three species of Coccinellidae (Coleoptera) at hibernation sites. *Eur. J. Entomol.* 104: 51–56.

Iperti, G. 1964. Les parasites des coccinelles aphidiphages dans les Alpes-Maritimes et les Basses-Alpes. *Entomophaga.* 9: 153–180.

Iperti, G. 1966a. Migration of *Adonia undecimnotata* in southeastern France. *In* I. Hodek (ed.). *Ecology of Aphidophagous Insects.* Academia, Prague and W. Junk, The Hague. pp. 137–138.

Iperti, G. 1966b. Protection of coccinellids against mycosis. *In* I. Hodek (ed.). *Ecology of Aphidophagous Insects.* Academia, Prague and W. Junk, The Hague. pp. 189–190.

Iperti, G. 1966c. Comportement naturel des Coccinelles aphidiphages du Sud-Est de la France. Leur type de spécificité, leur action prédatrice sur *Aphis fabae* L. *Entomophaga.* 11: 203–210.

Iperti, G. 1986. Ecobiologie des coccinelles aphidiphages: les migrations. *Colloques de l'INRA* 36: 107–120.

Iperti, G. and E. Bertand. 2001. Hibernation of *Harmonia axyridis* (Coleoptera: Coccinellidae) in south-eastern France. *Acta Soc. Zool. Bohemoslov.* 65: 207–210.

Iperti, G. and L.-A. Buscarlet. 1972. Contribution a l'étude d'une migration d'*Adonia 11-notata* Schn. (Coleoptera, Coccinellidae) par marquage avec l'iridium 191 stable. *Ann. Zool. Ecol. Anim.* 4: 249–254.

Iperti, G. and L. A. Buscarlet. 1986. Seasonal migration of the ladybird *Semiadalia undecimnotata*. *In* I. Hodek (ed.). *Ecology of Aphidophaga.* Academia, Prague and W. Junk, Dordrecht. pp. 199–204.

Iperti, G. and I. Hodek. 1974. Induction alimentaire de la dormance imaginale chez *Semiadalia undecimnotata* Schn. (Coleoptera: Coccinellidae) pour aider a la conservation des coccinelles elevees au laboratoire avant une utilisation ultérieure. *Ann. Zool.-Ecol. Anim.* 6: 41–51.

Iperti, G. and P. Prudent. 1986. Effect of photoperiod on egg-laying in *Adalia bipunctata*. *In* I. Hodek (ed.). *Ecology of Aphidophaga.* Academia, Prague and W. Junk, Dordrecht. pp. 245–246.

Iperti, G. and F. Rolley. 1973. Étude de l'acquisition de l'état de dormance chez une coccinelle aphidiphage et migrante, *Adonia 11-notata* Schn. (Col. Coccinellidae) à l'aide d'une technique de marquage (avec l'iridium 191 stable). *Ann. Zool. Ecol. Anim.* 5: 255–259.

Iperti, G., J. Brun and J. P. Bordet. 1988. Model for forecasting the migratory take-off behaviour of the aphidophagous coccinellid: *Semiadalia undecimnotata* Schn. *In* E. Niemczyk and A. F. G. Dixon (eds). *Ecology and Effectiveness of Aphidophaga.* SPB Academic Publishing, The Hague. pp. 129–134.

Iperti, G., J. Brun and C. Samie. 1983. Influence des facteurs climatiques, et plus particulièrement des mouvements atmosphériques turbulents de l'air sur l'envol migratoire d'une coccinelle aphidiphage, *Semiadalia undecimnotata*. *OEPP Bull.* 13: 235–240.

Jean, C., D. Coderre and J-.C. Tourneur. 1990. Effects of temperature and substrate on survival and lipid consumption of hibernating *Coleomegilla maculata lengi* (Coleoptera: Coccinellidae). *Environ. Entomol.* 19: 1657–1662.

Johnson, C. G. 1969. *Migration and Dispersal of Insects by Flight.* Methuen, London, 763 pp.

Jöhnssen, A. 1930. Beiträge zur Entwicklungs- und Ernährungsbiologie einheimischer Coccinelliden unter besonderer Berücksichtigung von *Coccinella septempunctata* L. *Z. Angew. Entomol.* 16: 87–158.

Kapur, A. P. 1954. Mass assemblage of the Coccinellid beetle *Epilachna bisquadripunctata* (Gylenhal) in Chota Nagpur. *Curr. Sci.* 23: 230–231.

Katsoyannos, P., D. C. Kontodimas and G. J. Stathas. 1997b. Summer diapause and winter quiescence of *Coccinella septempunctata* (Col. Coccinellidae) in central Greece. *Entomophaga* 42: 483–491.

Katsoyannos, P., D. C. Kontodimas and G. Stathas. 2005. Summer diapause and winter quiescence of *Hippodamia (Semiadalia) undecimnotata* (Coleoptera: Coccinellidae) in central Greece. *Eur. J. Entomol.* 102: 453–457.

Katsoyannos, P., G. J. Stathas and D. C. Kontodimas 1997a. Phenology of *Coccinella septempunctata* (Col.: Coccinellidae) in central Greece. *Entomophaga* 42: 435–444.

Kawauchi, S. 1985. Effects of photoperiod on the induction of diapause, the live weight of emerging adult and the duration of development of three species of aphidophagous coccinellids (Coleoptera, Coccinellidae). *Kontyu* 53: 536–546.

Kehat, M. 1968. The phenology of *Pharoscymnus* spp. and *Chilocorus bipustulatus* L. (Coccinellidae) in date palm plantations in Israel. *Ann. Epiphyties.* 19: 605–614.

Khrolinsky, L. G. 1964. *Curculionids of the genus* Apion *of the Chernovitsy region.* Summary of a thesis. Plant Protection Institute, Leningrad. pp. 22

Klausnitzer, B. 1967. Zur Kenntnis der Beziehungen der Coccinellidae zu Kiefernwäldern *(Pinus silvestris* L.). *Acta Entomol. Bohemoslov.* 64: 62–68.

Klausnitzer, B. 1968. Zur Biologie von *Myrrha octodecimguttata* (L.). (Col. Coccinellidae). *Entomol. Nachr.* 12: 102–104.

Klausnitzer, B. and H. Klausnitzer. 1997. *Marienkaefer: Coccinellidae*, 4th edn. Westarp, Magdeburg, 175 pp.

Koch, R. L., M. A. Carrillo, R. C. Venette, C. A. Cannon and W. D. Hutchison. 2004. Cold hardiness of the multicolored Asian lady beetle (Coleoptera: Coccinellidae). *Environ. Entomol.* 33: 815–822.

Koch, R. L. and T. L. Galvan. 2008. Bad side of a good beetle: the North American experience with *Harmonia axyridis*. *BioControl* 53: 23–35.

Koch, R. L. and W. D. Hutchison 2003. Phenology and black-light trapping of the multicolored Asian lady beetle (Coleoptera: Coccinellidae) in a Minnesota agricultural landscape. *J. Entomol. Sci.* 38: 477–480.

Kuznetsov, V. N. 1977. The biology of the ladybug *Aiolocaria mirabilis* Motsch. (Coleoptera, Coccinellidae) in Primorye, USSR. *Trudy Biol.-pochv. Inst. Akad. Nauk., Vladivostok* 44: 108–117.

Labrie, G., D. Coderre and E. Lucas. 2008. Overwintering strategy of multicolored Asian lady beetle (Coleoptera: Coccinellidae): Cold-free space as a factor of invasive success. *Ann. Entomol. Soc. Am.* 101: 860–866.

Leather, S. R., K. F. A. Walters and J. S. Bale. 1993. *The Ecology of Insect Overwintering*. Cambridge University Press, Cambridge. 255 pp.

Lee R. E. Jr. 1980a. Aggregation of lady beetles on the shores of lakes (Coleoptera: Coccinellidae). *Am. Midl. Nat.* 104: 295–304.

Lee, R. E., Jr. 1980b. Physiological adaptations of Coccinellidae to supranivean and subnivean hibernacula. *J. Insect Physiol.* 36: 135–138.

Lee R. E., Jr. and D. L. Denlinger. (eds). 1991. *Insects at Low Temperature*. Chapman and Hall, London. 513 pp.

Lees, A. D. 1955. *The Physiology of Diapause in Arthropods*. Cambridge University Press, Cambridge. 151 pp.

Legay, J.-M. and M. de Reggi 1962. Sur un lieu d'hibernation de la Coccinelle *Tytthaspis sedecimpunctata* dans la région Lyonnaise. *Bull. Mens. Soc. Linn. Lyon.* 9: 267–269.

Lu, W. and M. E. Montgomery. 2001. Oviposition, development and feeding of *Scymnus (Neopullus) sinuanodulus* (Coleoptera: Coccinellidae): a predator of *Adelges tsugae* (Homoptera: Adelgidae). *Ann. Entomol. Soc. Am.* 94: 64–70.

Majerus, M. E. N. 1994. *Ladybirds*. Harper Collins, London. 367 pp.

Majerus, M. E. N. and P. Kearns. 1989. *Ladybirds*. Richmond, London. 103 pp.

Mani, M. S. 1962. *Introduction to High Altitude Entomology*. Insect life above the timber-line in the north-west Himalaya. Methuen, London. 302 pp.

Masaki, S. 1980. Summer diapause. *Annu. Rev. Entomol.* 25: 1–25.

Mayr, E. 1970. *Populations, Species and Evolution*. Harvard University Press, Cambridge, MA. 453 pp.

McMullen, R. D. 1967a. A field study of diapause in *Coccinella novemnotata* (Coleoptera: Coccinellidae). *Can. Entomol.* 99: 42–49.

McMullen, R. D. 1967b. The effects of photoperiod, temperature and food supply on rate of development and diapause in *Coccinella novemnotata*. *Can. Entomol.* 99: 578–586.

McMurtry, J.A., G. T. Scriven and R. S. Malone. 1974. Factors affecting oviposition of *Stethorus picipes* (Coleoptera: Coccinellidae), with special reference to photoperiod. *Environ. Entomol.* 3: 123–127.

Michaud, J. P. and J. A. Qureshi 2005. Induction of reproductive diapause in *Hippodamia convergens* (Coleoptera: Coccinellidae) hinges on prey quality and availability. *Eur. J. Entomol.* 102: 483–487.

Michaud, J. P. and J. A. Qureshi. 2006. Reproductive diapause in *Hippodamia convergens* (Coleoptera: Coccinellidae) and its life history consequences. *Biol. Control* 39: 193–200.

Michels, G. J., R. V. Flanders and J. B. Bible. 1997. Overwintering survival of seven imported coccinellids in the Texas High Plains. *Southwest Entomol.* 22: 157–166.

Mills, N. J. 1981. The mortality and fat content of *Adalia bipunctata* during hibernation. *Entomol. Exp. Appl.* 30: 265–268.

Mori, K., M. Nozawa, K. Arai and T. Gotoh. 2005. Life history traits of the acarophagous ladybeetle, *Stethorus japonicus* at three constant temperatures. *BioControl* 50: 35–51.

Morikawa, K., H. Sakurai, T. Miyosi and S. Takeda. 1989. Changes of ultrastructure and protease activity in midgut related to diapause of the lady-bird beetle, *Coccinella septempunctata bruckii*. *Res. Bull. Fac. Agric. Gifu Univ.* 54: 71–79. (In Japanese).

Moter, G. 1959. *Untersuchungen zur Biologie von* Stethorus punctillum *Weise*. Unpubl. PhD thesis, University of Cologne, Germany.

Nabli, H., W. C. Bailey and S. Necibi. 1999. Beneficial insect attraction to light traps with different wavelengths. *Biol. Control* 16: 185–188.

Nadel, D. and S. Biron. 1964. Laboratory studies and controlled mass rearing of *Chilocorus bipustulatus* L. a scale predator in Israel. *Riv. Parassit.* 25: 165–206.

Nalepa, C. A., K. A. Kidd and K. R. Ahlstrom. 1996. Biology of *Harmonia axyridis* (Coleoptera: Coccinellidae) in winter aggregations. *Ann. Entomol. Soc. Am.* 89: 681–685.

Nalepa, C. A., G. G. Kennedy and C. Brownie. 2005. Role of visual contrast in the alighting behavior of *Harmonia axyridis* (Coleoptera: Coccinellidae) at overwintering sites. *Environ. Entomol.* 34: 425–431.

Nalepa, C. A., K. A. Kidd and D. I. Hopkins. 2000. The multicolored Asian lady beetle (Coleoptera: Coccinellidae): orientation to aggregation sites. *J. Entomol. Sci.* 35: 150–157.

Nalepa, C. A. and A. Weir. 2007. Infection of *Harmonia axyridis* (Coleoptera: Coccinellidae) by *Hesperomyces virescens* (Ascomycetes: Laboulbeniales): role of mating status and aggregation behavior. *J. Invert. Path.* 94: 196–203.

Nault, B. A. and G. G. Kennedy. 2000. Seasonal changes in habitat preference by *Coleomegilla maculata*: implications for Colorado potato beetle management in potato. *Biol. Control* 17: 164–173.

Nedvěd, O. 1993. Comparison of cold hardiness in two ladybird beetles (Coleoptera: Coccinellidae) with contrasting hibernation behaviour. *Eur. J. Entomol.* 90: 465–470.

Nedvěd, O., P. Ceryngier, M. Hodková and I. Hodek. 2001. Flight potential and oxygen uptake during early dormancy in *Coccinella septempunctata*. *Entomol. Exp. Appl.* 99: 371–380.

Neuenschwander, P., K. S. Hagen and R. F. Smith. 1975. Predation of aphids in California's alfalfa fields. *Hilgardia.* 43: 53–78.

Niijima, K. and T. Kawashita. 1982. Studies on the ovarian development in *Coccinella septempunctata bruckii* Mulsant. *Bull. Fac. Agric. Tamagawa Univ.* 22: 7–13.

Novák, B. 1965. Beitrag zu den Labor-Untersuchungen der Aggregation von Coccinelliden. *1. Konf. Schaedl. Hackfruechte, Prague, 1965.* Institute for Plant Protection, Prague. pp. 85–89.

Novák, B. 1966. Die Bewirkung der Georeaktionen von *Coccinella septempunctata* L. durch die Feuchtigkeit. *Sb. Praci Prir. Fak. Palack. Univ. Olomouc* 22: 147–151. (In Czech.)

Novák, B. and K. Grenarová. 1967. Coccinelliden an der Grenze des Feld- und Waldbiotops – Hibernationsversuche mit den Imagines fuehrender Arten. *3. Konf. Schaedl. Hackfruechte, Prague, 1967.* Institute for Plant Protection, Prague. pp. 49–59. (In Czech).

Obata, S. 1986. Determination of hibernation site in the ladybird beetle, *Harmonia axyridis* Pallas (Coleoptera, Coccinellidae). *Kontyu.* 54: 218–223.

Obata, S. 1988a. Mating behaviour and sperm transfer in the ladybird beetle, *Harmonia axyridis* Pallas (Coleoptera: Coccinellidae). *In* E. Niemczyk and A. F. G. Dixon (eds). *Ecology and Effectiveness of Aphidophaga.* SPB Academic Publishing, The Hague. pp. 39–42.

Obata, S. 1988b. Mating refusal and its significance in female of the ladybird beetle, *Harmonia axyridis. J. Physiol. Entomol.* 13: 193–199.

Obata, S., Y. Johki and T. Hidaka. 1986. Location of hibernation sites in the ladybird beetle, *Harmonia axyridis. In* I. Hodek (ed.). *Ecology of Aphidophaga.* Academia, Prague and W. Junk, Dordrecht. pp. 193–198.

Obrycki, J. J. and M. J. Tauber. 1981. Phenology of three coccinellid species: thermal requirements for development. *Ann. Entomol. Soc. Am.* 74: 31–36.

Obrycki, J. J., D. B. Orr, C. J. Orr, M. Wallendorf and R. F. Flanders. 1993. Comparative developmental and reproductive biology of 3 populations of *Propylea quatuordecimpunctata* (Coleoptera, Coccinellidae). *Biol. Control.* 3: 27–33.

Obrycki, J. J., M. J. Tauber, C. A. Tauber and B. Gollands. 1983. Environmental control of the seasonal life cycle of *Adalia bipunctata* (Coleoptera: Coccinellidae). *Environ. Entomol.* 12: 416–421.

Ohashi, K., S. E. Kawauchi and Y. Sakuratani. 2003. Geographic and annual variation of summer-diapause expression in the ladybird beetle, *Coccinella septempunctata* (Coleoptera: Coccinellidae), in Japan. *Appl. Entomol. Zool.* 38: 187–196.

Ohashi, K., Y. Sakuratani, N. Osawa, S. Yano and A. Takafuji. 2005. Thermal microhabitat use by the ladybird beetle, *Coccinella septempunctata* (Col., Coccinellidae), and its life cycle consequences. *Environ. Entomol.* 34: 432–439.

Ohgushi, T. 1986. Population dynamics of an herbivorous lady beetle, *Henosepilachna niponica*, in a seasonal environment. *J. Anim. Ecol.* 55: 861–879.

Ohgushi, T. 1996. A reproductive trade off in an herbivorous ladybeetle: egg resorption and female survival. *Oecologia* 106: 345–351.

Okuda, T. 1981. *Studies on diapause in* Coccinella septempunctata bruckii *Mulsant, with special reference to histological changes during aestivation and hibernation*. MSc thesis, Gifu University, Gifu, Japan. 54 pp.

Okuda, T. 1984. Anatomical and ecophysiological study of *Coccinella septempunctata* L. and *Semiadalia undecimnotata* Schneid. in relation to diapause. Unpubl. PhD thesis, Czechoslovak Academy of Sciences, Prague, Czechoslovakia. 130 pp.

Okuda, T. 2000. DNA synthesis by testicular follicles in a ladybeetle, *Coccinella septempunctata brucki* (Coleoptera: Coccinellidae) in relation to aestivation diapause. *Invert. Repr. Dev.* 38: 71–79.

Okuda, T. and Y. Chinzei. 1988. Vitellogenesis in a lady-beetle, *Coccinella septempunctata* in relation to the aestivation-diapause. *J. Insect Physiol.* 34: 393–401.

Okuda, T. and I. Hodek. 1983. Response to constant photoperiods in *Coccinella septempunctata brucki* populations from central Japan (Coleoptera, Coccinellidae). *Acta Entomol. Bohemoslov.* 80: 74–75.

Okuda, T. and I. Hodek. 1989. Flight tendency of two coccinellids, *Semiadalia undecimnotata* and *Coccinella septempunctata*, in relation to diapause. *In* M. Tonner, T. Soldan and B. Bennettova (eds). *Regulation of Insect Reproduction IV.* Academia, Prague. pp. 385–400.

Okuda, T. and I. Hodek. 1994. Diapause and photoperiodic response in *Coccinella septempunctata brucki* Mulsant (Coleoptera: Coccinellidae) in Hokkaido, *Jpn J. Appl. Entomol. Zool.* 29: 549–554.

Okuda, T., M. Hodková and I. Hodek. 1986. Flight tendency in *Coccinella septempunctata* in relation to changes in flight muscles, ovaries and corpus allatum. *In* I. Hodek (ed.). *Ecology of Aphidophaga.* Academia, Prague and W. Junk, Dordrecht. pp. 217–223.

Omkar and A. Pervez. 2000. Biodiversity of predaceous coccinellids (Coleoptera: Coccinellidae) in India: a review. *J. Aphidol.* 14: 41–66.

Ongagna, P. and G. Iperti. 1994. Influence de la temperature et de la photoperiode chez *Harmonia axyridis* Pall. (Col., Coccinellidae): obtention d'adultes rapidement feconds ou en dormance. *J. Appl. Entomol.* 117: 314–317.

Osawa, N. 2001. The effect of hibernation on the seasonal variations in adult body size and sex ratio of the polymorphic ladybird beetle *Harmonia axyridis*: the role of thermal melanism. *Acta Soc. Zool. Bohemoslov.* 65: 269–278.

Osawa, N. 2005. The effect of prey availability on ovarian development and oosorption in the ladybird beetle *Harmonia axyridis* (Coleoptera: Coccinellidae). *Eur. J. Entomol.* 102: 503–511.

Pantyukhov, G. A. 1965. Influence of temperature and relative humidity on development of *Chilocorus renipustulatus* Scriba (Col. Coccinellidae). *Trudy Zool. Inst. Akad. Nauk., Leningrad* 36, 70–85. (In Russian).

Pantyukhov, G. A. 1968a. A study of ecology and physiology of the predatory beetle *Chilocorus rubidus* Hope (Coleoptera, Coccinellidae). *Zool. Zh.* 47: 376–386. (in Russian).

Pantyukhov, G. A. 1968b. On photoperiodic reaction of *Chilocorus renipustulatus* Scriba (Coleoptera, Coccinellidae). *Entomol. Obozr.* 47: 376–385. (In Russian).

Parker, B. L., M. E. Whalon and M. Warshaw. 1977. Respiration and parasitism in *Coleomegilla maculata lengi* (Coleoptera: Coccinellidae). *Ann. Entomol. Soc. Am.* 70: 984–987.

Parry, W. H. 1980. Overwintering of *Aphidecta obliterata* (L.) (Coleoptera: Coccinellidae) in north east Scotland. *Acta Ecol. Ecol. Appl.* 1: 307–316.

Parry, W. H. 1986. The overwintering strategy of *Aphidecta obliterata* in Scottish coniferous forests. *In* I. Hodek (ed.). *Ecology of Aphidophaga.* Academia, Prague and W. Junk, Dordrecht. pp. 179–184.

Phoofolo, M. W. and J. J. Obrycki. 2000. Demographic analysis of reproduction in Nearctic and Palearctic populations of *Coccinella septempunctata* and *Propylea quatuordecimpunctata.* *BioControl* 45: 25–43.

Plaut, H. N. 1965. On the phenology and control value of *Stethorus punctillum* Weise as a predator of *Tetranychus cinnabarinus* Boisd. in Israel. *Entomophaga.* 10: 133–137.

Poulton, E. B. 1936. Assembles of coccinellid beetles observed in N. Uganda (1927) by Prof. Hale Carpenter and in Bechualand (1935) by Dr W.A. Lamborn. *Proc. R. Entomol. Soc. Lond. (A)* 11: 99–100.

Pulliainen, E. 1963. Preliminary notes on the humidity reactions of *Myrrha 18-guttata* L. (Col., Coccinellidae). *Ann. Entomol. Fenn.* 29: 240–246.

Pulliainen, E. 1964. Studies on the humidity and light orientation and the flying activity of *Myrrha 18-guttata* L. (Col., Coccinellidae). *Ann. Entomol. Fenn.* 30: 117–141.

Pulliainen, E. 1966. On the hibernation sites of *Myrrha octodecimguttata* L. (Col., Coccinellidae) on the butts of the pine (*Pinus silvestris* L.). *Ann. Entomol. Fenn.* 32: 99–104.

Putman, W. L. 1955. Bionomics of *Stethorus punctillum* Weise in Ontario. *Can. Entomol.* 87: 9–33.

Radzievskaya, S. B. 1939. Concerning the hibernation of the ladybird and the struggle with the aphids. *Vop. Ekol. Biotsen.* 4, 268–275. (In Russian).

Rankin, M. A. and S. Rankin. 1980. Some factors affecting presumed migratory flight activity of the convergent ladybeetle, *Hippodamia convergens* (Coccinellidae: Coleoptera). *Biol. Bull.* 158: 356–369.

Ricci, C. 1986. Habitat distribution and migration to hibernation sites of *Tytthaspis sedecimpunctata* and *Rhyzobius litura* in central Italy. *In* I. Hodek (ed.). *Ecology of Aphidophaga.* Academia, Prague and W. Junk, Dordrecht. pp. 211–216.

Ricci, C., L. Ponti and A. Pires. 2005. Migratory flight and pre-diapause feeding of *Coccinella septempunctata* (Coleoptera) adults in agricultural and mountain ecosystems of Central Italy. *Eur. J. Entomol.* 102: 531–538.

Riddick, E. W., J. R. Aldrich, A. De Milo and J. C. Davis. 2000. Potential for modifying the behavior of the multicolored Asian lady beetle (Coleoptera: Coccinellidae) with plant-derived natural products. *Ann. Entomol. Soc. Am.* 93: 1314–1321.

Roach, S. H. and W. M. Thomas. 1991. Overwintering and spring emergence of three coccinellid species in the coastal plain of South Carolina. *Environ. Entomol.* 20: 540–544.

Rolley, F., I. Hodek and G. Iperti. 1974. Influence de la nourriture aphidienne (selon l'age de la plante-hote a partir de laquelle les pucerons se multiplient) sur l'induction de la dormance chez *Semiadalia undecimnotata* Schn. (Coleoptera: Coccinellidae). *Ann. Zool.-Ecol. Anim.* 6: 53–60.

Rosen, D. and U. Gerson. 1965. Field studies of *Chilocorus bipustulatus* (L.) on citrus in Israel. *Ann. Épiphyties.* 16: 71–76.

Roy, H. E. and E. Wajnberg (eds). 2008. *From Biological Control to Invasion: the Ladybird Harmonia axyridis as a Model Species.* Springer, IOBC. 287 pp.; *BioControl* 53: 1–287.

Ruscinsky, A. 1933. Über Anhäufungen von *Vibidia duodecimguttata* Poda in Wäldern von Palanea (Bessarabien). *Bul. Muz. Natn. Ist. Nat. Chisinau* 5: 162–163.

Ruzicka, Z. 1984. Two simple recording flight mills for the behavioural study of insect. *Acta Entomol. Bohemoslov.* 81: 429–433.

Ruzicka, Z. and K. S. Hagen. 1985. Impact of parasitism on migratory flight performance in females of *Hippodamia convergens* (Coleoptera). *Acta Entomol. Bohemoslov.* 82: 401–406.

Ruzicka, Z. and K. S. Hagen. 1986. Influence of *Perilitus coccinellae* on the flight performance of overwintered *Hippodamia convergens*. *In* I. Hodek (ed.). *Ecology of Aphidophaga.* Academia, Prague and W. Junk, Dordrecht. pp. 229–232.

Ruzicka, Z. and P. Kindlmann. 1991. The effect of changes in temperature on the locomotory activity of hibernating *Coccinella septempunctata*. *In* L. Polgar, R. J. Chambers, A. F. G. Dixon and I. Hodek (eds) *Behaviour and Impact of Aphidophaga.* SPB Academic Publishing, The Hague. pp. 249–254.

Ruzicka, Z. and J. Vostrel. 1985. Hibernation of *Coccinella quinquepunctata* and *Propylea quatuordecimpunctata* (Coleoptera, Coccinellidae) in pinecones. *Vest. Cs. Spol. Zool.* 49: 281–284.

Sakurai, H. 1969. Respiration and glycogen contents in the adult life of the *Coccinella septempunctata* Mulsant and *Epilachna vigintioctopunctata* Fabricius (Coleoptera: Coccinellidae). *Appl. Entomol. Zool.* 4: 55–57.

Sakurai, H., Y. Mori and S. Takeda. 1981a. Studies on the diapause of *Coccinella septempunctata brucki* Mulsant. I. Physiological changes of adults related to aestivation and hibernation. *Res. Bull. Fac. Agric. Gifu Univ.* 45: 9–15. (In Japanese.)

Sakurai, H., K. Goto, Y. Mori and S. Takeda. 1981b. Studies on the diapause of *Coccinella septempunctata bruckii* Mulsant. II. Role of corpus allatum related with diapause. *Res. Bull. Fac. Agric. Gifu Univ.* 45: 17–23. (In Japanese.)

Sakurai, H., S. Takeda and T. Kawai. 1988. Diapause regulation in the lady beetle, *Harmonia axyridis*. In: Niemczyk E. and Dixon A. F. G. (eds): *Ecology and Effectiveness of Aphidophaga*. SPB Acad. Publ., The Hague. pp. 67–70.

Sakurai, H., T. Okuda and S. Takeda. 1982. Studies on the diapause of *Coccinella septempunctata brucki* Mulsant. *Res. Bull. Fac. Agric. Gifu Univ.* 46: 29–40. (In Japanese.)

Sakurai, H., K. Goto and S. Takeda. 1983. Emergence of the ladybird beetle, *Coccinella septempunctata bruckii* Mulsant in the field. *Res. Bull. Fac. Agric. Gifu Univ.* 48: 37–45. (In Japanese.)

Sakurai, H., T. Hirano and S. Takeda. 1986. Physiological distinction between aestivation and hibernation in the lady beetle, *Coccinella septempunctata bruckii* (Coleoptera: Coccinellidae). *Appl. Entomol. Zool.* 21: 424–429.

Sakurai, H., T. Hirano, K. Kodama and S. Takeda. 1987a. Conditions governing diapause induction in the lady beetle, *Coccinella septempunctata brucki* (Coleoptera: Coccinellidae). *Appl. Entomol. Zool.* 22: 133–138.

Sakurai, H., T. Hirano and S. Takeda. 1987b. Change in electrophoretic pattern of haemolymph protein in diapause regulation of the lady beetle, *Coccinella septempunctata bruckii* (Coleoptera: Coccinellidae). *Appl. Entomol. Zool.* 22: 286–291.

Sakurai, H., S. Takeda and K. Morikawa. 1991. Digestive function during dormancy of the ladybird beetle *Coccinella septempunctata bruckii*. In L. Polgar, R. J. Chambers, A. F. G. Dixon and I. Hodek (eds): *Behaviour and Impact of Aphidophaga*. SPB Academic Publishing, The Hague. pp. 255–258.

Sakuratani, Y., Y. Matsumoto, M. Oka et al. 2000. Life history of *Adalia bipunctata* (Coleoptera: Coccinellidae) in Japan. *Eur. J. Entomol.* 97: 555–558.

Salt, R. W. 1961. Principles of insect cold-hardiness. *Annu. Rev. Entomol.* 6: 55–74.

Salt, R. W. 1964. Trends and needs in the study of insect cold-hardiness. *Can. Entomol.* 96: 400–405.

Sarospataki, M. and V. Marko. 1995. Flight activity of *Coccinella septempunctata* at different strata of a forest in relation to migration to hibernation sites. *Eur. J. Entomol.* 92: 415–419.

Saunders, D. S. 2002. *Insect Clocks.* Elsevier, Amsterdam. 560 pp.

Savoiskaya, G. I. 1965. Biology and perspectives of utilisation of coccinellids in the control of aphids in south-eastern Kazakhstan orchards. *Trudy Inst. Zashch. Rast., Alma-Ata* 9: 128–156. (In Russian.)

Savoiskaya, G. I. 1966. Hibernation and migration of coccinellids in south-eastern Kazakhstan. *In* I. Hodek (ed.). *Ecology of Aphidophagous Insects.* Academia, Prague and W. Junk, The Hague. pp. 139–142.

Savoiskaya, G. I. 1970a. Introduction and acclimatisation of some coccinellids in the Alma-Ata reserve. *Trudy Alma-Atin. Gos. Zapov.* 9: 138–162.

Savoiskaya, G. I. 1970b. Coccinellids of the Alma-Ata reserve. *Trudy Alma-Atin. Gos. Zapov.* 9: 163–187. (In Russian.)

Savoiskaya, G. I. 1983. *Coccinellids.* Izdatelstvo Nauka Kazakhskoi SSR, Alma-Ata. 246 pp. (In Russian.)

Schaefer, P. W. 2003. Winter aggregation of *Harmonia axyridis* (Coleoptera: Coccinellidae) in a concrete observation tower. *Entomol. News* 114: 23–28.

Semyanov, V. P. 1965a. Fauna, biology and usefulness of coccinellids (Coleoptera: Coccinellidae) in Belorus SSR. *Zap. Leningr. SelKhoz. Inst.* 95: 106–120. (In Russian.)

Semyanov, V. P. 1965b. Fauna and distribution in habitats of coccinellids (Coleoptera: Coccinellidae) in the Leningrad Area. *Entomol. Obozr.* 44: 315–323. (In Russian.)

Semyanov, V. P. 1970. Biological properties of *Adalia bipunctata* L. (Coleoptera: Coccinellidae) in conditions of Leningrad region. *Zash. Rast. Vred. Bolez.* 127: 105–112. (In Russian.)

Semyanov V. P. 1978a. Migration and migration state in *Coccinella septempunctata* L. *In Morfologia, Systematika i Evolyutsia Zhivotnykh.* Zool. Inst., Akademia Nauk SSSR, Leningrad. pp. 75–76. (In Russian.)

Semyanov, V. P. 1978b. Structure of populations and pecularities of the photoperiodic reaction of *Coccinella septempunctata* L. (Col., Coccinellidae). *Trudy Zool. Inst. Akad. Nauk, Leningrad.* 69: 110–113. (In Russian.)

Semyanov, V. P. 2000. Biology of coccinellids (Coleoptera, Coccinellidae) from Southeast Asia: II. *Harmonia sedecimnotata* (F.). *Entomol. Obozr.* 79: 3–9. (In Russian.)

Semyanov, V. P. and N. P. Vaghina. 2003. Effect of the trophic diapause on fecundity and longevity of a tropical coccinellid, *Harmonia sedecimnotata* (Fabr.) (Coleoptera: Coccinellidae). *Entomol. Obozr.* 82: 3–5. (In Russian.)

Sherman, F. 1938. Massing of convergent ladybeetle at summits of mountains in southeastern United States. *J. Econ. Entomol.* 31: 320–322.

Sillén-Tullberg, B. and O. Leimar. 1988. The evolution of gregariousness in distasteful insects as a defense against predators. *Am. Nat.* 132: 723–734.

Simpson, R. G. and C. E. Welborn. 1975. Aggregations of alfalfa weevils, *Hypera postica*, convergent lady beetles, *Hippodamia convergens*, and other insects. *Environ. Entomol.* 4: 193–194.

Sloggett, J.J. 2005. Are we studying too few taxa? Insights from aphidophagous ladybird beetles (Coleoptera : Coccinellidae). *Eur. J. Entomol.* 102: 391–398.

Smee, C. 1922. British ladybird beetles. Their control of aphids. *Fruit Grower* 53: 675–676, 717–718, 759.

Solbreck, C. 1974. Maturation of post-hibernation flight behaviour in the coccinellid *Coleomegilla maculata* (DeGeer). *Oecologia* 17: 265–275.

Speyer, W. 1934. Die an der Niederelbe in Obstbaumfanggürteln überwinternden Insekten. III. Mitteilung. Coleoptera: Coccinellidae. *Z. PflKrankh. PflPath. PflSchutz* 44: 321–330.

Stewart, J. W., W. H. Whitcomb and K. O. Bell. 1967. Estivation studies of the convergent lady beetle in Arkansas. *J. Econ. Entomol.* 60: 1730–1735.

Storch, R. H. and W. L. Vaundell. 1972. The effect of photoperiod on diapause induction and inhibition in *Hippodamia tredecimpunctata* (Coleoptera: Coccinellidae). *Can. Entomol.* 104: 285–288.

Sundby, R. A. 1968. Some factors influencing the reproduction and longevity of *Coccinella septempunctata* Linnaeus (Coleoptera: Coccinellidae). *Entomophaga* 13: 197–202.

Tadmor, U. and S. W. Applebaum. 1971. Adult diapause in the predaceous coccinellid, *Chilocorus bipustulatus*: photoperiodic induction. *J. Insect Physiol.* 17: 1211–1215.

Tauber, M. J. and C. A. Tauber. 1976. Insect seasonality: diapause maintenance, termination, and post-diapause development. *Annu. Rev. Entomol.* 21: 81–107.

Tauber, M. J., C. A. Tauber and S. Masaki. 1986. *Seasonal Adaptation of Insects*. Oxford University Press, Oxford. 411 pp.

Telenga, N. A. 1948. *Biological Methods of Insect Pest Control (Predaceous Coccinellids and their Utilisation in the USSR)*. Izdatelstvo Akademii Nauk SSSR, Kiev. 120 pp. (In Russian.)

Telenga, N. A. and Bogunova, M. V. 1936. The most important predators of coccids and aphids in the Ussuri region of Far East and their utilisation. *Zash. Rast.* 1939 (10): 75–87. (In Russian.)

Thomas, W.A. 1932. Hibernation of the 13-spotted ladybeetle. *J. Econ. Entomol.* 25: 136.

Throne, A. H. 1935. An unusual occurrence of the convergent ladybeetle. *Ecology* 16: 125.

Turnock, W. J. and I. L. Wise. 2004. Density and survival of lady beetles (Coccinellidae) in overwintering sites in Manitoba. *Can. Field-Nat.* 118: 309–317.

Ulyanova, L.S. 1956. On the possibility of acclimatisation of the Far-East coccinellid *Harmonia axyridis* Pall. in the conditions of Uzbekistan. *Trudy Inst. Zool. Parazit., Tashkent.* 6: 111–119. (In Russian.)

Verheggen, F. J., Q. Fagel, S. Heuskin et al. 2007. Electrophysiological and behavioral responses of the multicolored Asian lady beetle, *Harmonia axyridis* Pallas, to sesquiterpene semiochemicals. *J. Chem. Ecol.* 33: 2148–2155.

Watanabe, M. 2002. Cold tolerance and *myo*-inositol accumulation in overwintering adults of a lady beetle, *Harmonia axyridis* (Coleoptera: Coccinellidae). *Eur. J. Entomol.* 99: 5–9.

de Wilde, J. 1969. Diapause and seasonal synchronization in the adult Colorado beetle. *Symp. Soc. Exp. Biol.* 23: 263–284.

Yakhontov, V. V. 1962. Seasonal migrations of lady-birds *Brumus octosignatus* Gebl. and *Semiadalia undecimnotata* Schneid. in Central Asia. *Proc. 11th Int. Congr. Entomol., Vienna, 1960.* 1: 21–23.

Yakhontov, V. V. 1966. Diapause in Coccinellidae of central Asia. *In* I. Hodek (ed.). *Ecology of Aphidophagous Insects.* Academia, Prague and W. Junk, The Hague. pp. 107–108.

Zaslavskii, V. A. 1970. Geographical races of *Chilocorus bipustulatus* L. (Coleoptera, Coccinellidae). I. Two types of photoperiodical reaction controlling the imaginal diapause in the northern race. *Zool. Zh.* 64: 1354–1365. (In Russian.)

Zaslavskii, V. A. and T. P. Bogdanova. 1965. Properties of imaginal diapause in two *Chilocorus* species (Coleoptera, Coccinellidae). *Trudy Zool. Inst. Akad. Nauk, Leningrad* 36: 89–95. (In Russian.)

Zaslavsky, V. A. and V. P. Semyanov. 1983. The migratory state in the ladybeetle (*Coccinella septempunctata*). *Zool. Zh.* 62: 878–891. (In Russian.)

Zaslavsky, V. A. and V. P. Semyanov. 1986. Migratory behaviour of coccinellid beetles. *In* I. Hodek (ed.). *Ecology of Aphidophaga.* Academia, Prague and W. Junk, Dordrecht. pp. 225–227.

Zaslavsky, V. A. and N. P. Vaghina. 1996. Joint and separate effects of photoperiodic and alimentary induction of diapause in *Coccinella septempunctata* (Coleoptera, Coccinellidae). *Zool. Zh.* 75: 1474–1481. (In Russian.)

Zaslavsky, V. A., V. P. Semyanov and N. P. Vaghina. 1998. Food as a cue factor controlling adult diapause in the lady beetle, *Harmonia sedecimnotata* (Coleoptera: Coccinellidae). *Zool. Zh.* 77: 1383–1388. (In Russian.)

Zhai, Q. and J. Zhang. 1984. Studies on the vitellogenesis of *Coccinella semtpempunctata*: synthesis of vitellogenin by fat body in vitro. *Acta Entomol. Sinica* 27: 361–367.

Zhai, Q., J. H. Postlethwait and J. W. Bodley. 1984. Vitellogenin synthesis in the lady beetle *Coccinella septempunctata*. *Insect Biochem.* 14: 299–305.

Zhai, Q., X. Guan, E. Chen and J. Zhang. 1985. On the vitello-genesis of *Coccinella septempunctata*: synthesis and secretion of vitellogenin by the fat body and the ovary. *Acta Entomol. Sinica* 28: 362–368.

Zhang, J. and Q. Zhai. 1985. On the vitellogenesis of *Coccinella septempunctata*: regulation of vitellogenin synthesis by juvenile hormone analogue. *Acta Entomol. Sinica* 28: 121–128.

Zhou, X., A. Honěk, W. Powell and N. Carter. 1995. Variations in body length, weight, fat content and survival in *Coccinella septempunctata* at different hibernation sites. *Entomol. Exp. Appl.* 75: 99–107.

Chapter 7

INTRAGUILD INTERACTIONS

Éric Lucas

Département des Sciences Biologiques, Université du Québec à Montréal, C.P. 8888
Succ. Centre-ville, Montréal, Québec, Canada, H3C 3P8

Ecology and Behaviour of the Ladybird Beetles (Coccinellidae), First Edition. Edited by I. Hodek, H.F. van Emden, A. Honěk.
© 2012 Blackwell Publishing Ltd. Published 2012 by Blackwell Publishing Ltd.

7.1 SCOPE

This chapter deals with **interactions between coccinellid** species, and **between coccinellids and other intraguild organisms**. Intraspecific predation (cannibalism; 5.2.8), interactions with the extraguild prey (Chapters 5 and 11), interactions with non-guild natural enemies (Chapter 8) and intraspecific competition will not be discussed here.

7.2 COCCINELLIDS AS GUILD MEMBERS

A **guild** is a group of species in a community that **share similar resources** (food or space) regardless of differences in tactics of resource acquisition and in taxonomic position (Polis et al. 1989). More than **4000** coccinellid **species** belong to guilds according to their **essential food**, ensuring the completion of larval development and oviposition (Hodek 1996; 5.2.1 and 5.2.11; Fig. 5.1). However, most predaceous coccinellid species also exploit **alternative food** sources. *Har. axyridis* for example consumes not only aphids and coccids, but also non-hemipteran prey (McLure 1986, Lucas et al. 1997b, 1998b, 2002, 2004a;. 5.2.7) and material of plant origin such as pollen and nectar (Kovach 2004, Lucas et al. 2007a; 5.2.9). This species is also an intraguild predator (Alhmedi et al. 2010; 7.8).

The guild of a coccinellid species may change with season (Triltsch 1997). In Eastern Canada, most temperate aphidophagous species are **pollinivorous** during spring (when animal prey is scarce) and later become **zoophagous**. Several species, such as *Har. axyridis*, become **frugivorous** in autumn (Kovach 2004, Lucas et al. 2007a). *Col. maculata* is a **zoophytophagous** species completing its life cycle on plant or animal material (5.2.9).

Apart from ladybirds (e.g. Evans 1991, Loevei et al. 1991, Lucas et al. 2007b), the guilds include also **non-coccinellid predators**, **parasitoids**, and **pathogens** (e.g. Obrycki & Tauber 1985, Coderre & Tourneur 1986, Brown 2004).

7.3 COCCINELLIDS AS NEUTRALISTS

Coccinellids avoid interacting with other intraguild members by **evolutionary** or **ecological mechanisms**. Partition of the niche allows **coexistence** of species exploiting the same resource.

7.3.1 Temporal guild partition

Temporal guild partition has often been reported (Lövei & Radwan 1988, Musser & Shelton 2003, Brown 2004, Dixon 2007). The **activity** of predators exploiting *Adelges tsugae* varies **diurnally**, some being more active at night (Flowers et al. 2007). Also Interactions between *A. bipunctata* and *Har. axyridis* in Japan may be avoided by **desynchronisation** of their occurrences (Toda & Sakuratani 2006).

7.3.2 Spatial guild partition

Adults and larvae of co-occurring species commonly exploit **different plant parts/heights** (Coderre & Tourneur 1986, Coderre et al. 1987, Chang 1996, Schellhorn & Andow 1999a, Lucas et al. 2002, Musser & Shelton 2003, Hoogendoorn & Heimpel 2004, Evans & Toler 2007, Flowers et al. 2007). They may also avoid interactions by changing their **within-plant distribution** (Hoogendoorn & Heimpel 2004). **Plant and habitat characteristics** may have an impact on foraging efficiency or influence contact between intraguild competitors (Lucas & Brodeur 1999, Lucas et al. 2004c, Janssen et al. 2007). *Coccinella transversoguttata* and *Scymnus lacustris* dominate in younger red pine stands, whereas *Mulsantina picta* and *Anatis mali* do so in older stands (Gagne & Martin 1968). Coccinellids were more abundant on yellow (nutrient-stressed) maize plants than on control (greener) plants, while lacewings were more numerous on control (Lorenzetti et al. 1997). Spatial guild partition occurs also at a **larger scale**, when different species exploit **different crops** in a landscape (Colunga-Garcia et al. 1997).

7.3.3 Thermal guild partition

Coccinellidae, Syrphidae and Chrysopidae differ in their **lower developmental threshold** and **speed of development** (Honěk & Kocourek 1988, 1990), and in their **resource exploitation efficiency** (Dixon et al. 2005). In northern regions, different **tolerance to cold** leads to different **overwintering strategies** and reduces interaction opportunities. In Eastern Canada, *Har. axyridis* does not survive outside and overwinters in buildings, while the indigenous *Col. maculata* overwinters outside (Labrie et al. 2008).

7.3.4   Body size guild partition

The **body size** of ladybirds also determines a partition in resource exploitation as the size imposes **geometrical** and **physiological constraints** in term of resource density needed (Dixon & Hemptinne 2001, Dixon 2007, Sloggett 2008). Smaller species would thus exploit lower aphid population densities more successfully than larger species (but see Evans 2004).

7.4   COCCINELLIDS AS INTERACTING ORGANISMS

In spite of guild partition, coccinellids are repeatedly involved in **intraguild interactions** at different periods of their life cycle (Fig. 7.1). These interactions are promoted by the fact that (i) many other insects share the **same guilds with many coccinellid species**; (ii) the extraguild prey of most predaceous coccinellids have an **aggregated distribution in time and space** leading to natural enemies aggregation; and (iii) coccinellid eggs and early instars occur **close to** their **food source** (Lucas 2005). For example, the large **temporal overlap** of *P. japonica*, *Har. axyridis*, and *C. septempunctata brucki* exploiting *Aphis gossypii* on *Hibiscus* trees in Japan (Kajita et al. 2006) enhances interaction opportunities.

All types of **classical ecological interactions** (mutualism, commensalism, competition, and predation) might occur within the guild. For any of these interactions, many **direct effects** (from direct physical interactions) and **indirect effects** (through intermediary species) (Wootton 1994, Abrams et al. 1996) may impact the focal species, its intraguild competitors, its principal (extraguild) food source and its higher-order (extraguild) natural enemies (Fig. 7.1). These interactions may affect the density (**density-mediated effects**) of the species considered, or any morphological, physiological or behavioural traits (**traits-mediated effects**).

History of the protagonists is one of key factors in intraguild interactions (Lucas 2005) that are frequently sudden and **ephemeral events** occurring at specific times and stages. The size, morphology, physiology, behaviour, vigour and autonomy, as well as the competitive, defensive and predatory aptitudes of an individual **change drastically throughout its life cycle**. All traits affecting the ladybird's **stage** at the time of a potential interaction (such as **time of colonization**, **life cycle length**, **voltinism**, etc.) influence the probability of its occurrence, its type, and its outcome. For instance, phoretic mites will use specific stages of ladybirds (7.5). Also, a third ladybird instar may be an intraguild prey of a chrysopid larva, whereas in the fourth instar it could be the predator of a similar size chrysopid larva (Fig. 7.2).

7.5   COCCINELLIDS AS INTRAGUILD COMMENSALISTS AND MUTUALISTS

Direct or **indirect commensalism** occurs whenever an organism derives a **benefit from** a second organism while the latter is not affected (Wootton 1994). **Phoresy** is an example: an organism is transported (dispersed) by an unaffected host (Holte et al. 2001). The coccinellid acts as a **vector** of hemisarcoptid mites when the non-feeding hypopal stage attaches itself to the elytra of the coccinellid adults (O'Connor & Houck 1989). Several predatory phoretic mites belong to the same guild as their coccinellid carrier (e.g. Houck & O'Connor 1991, Hurst et al. 1997, Holte et al. 2001).

Coccinellids may similarly act as **vectors of intraguild pathogens** (Pell et al. 1997, Roy et al. 1998, 2001, Thomas et al. 2006). This interaction is considered commensalism when the pathogens have negligible direct effects on ladybirds. If the coccinellids are infested by the pathogens, the interaction is detrimental (7.9.2).

Coccinellids may **indirectly benefit other guild members** through their action on the shared resource (Losey & Denno 1998, 1999, Aquilino et al. 2005), an interaction known as '**predator facilitation**' (Charnov 1976). For instance, the foliar-foraging *C. septempunctata* generates a strong dropping response in *Acyrthosiphon pisum* increasing its availability for the ground-foraging carabid *Harpalus pennsylvanicus* (Losey & Denno 1998, 1999; Fig. 7.3).

Direct or **indirect mutualism** is **mutually beneficial** for two different populations/species (Wootton 1994, Abrams et al. 1996) that share the same resource. Hypopodes of the astigmatid mite *Hemisarcoptes*

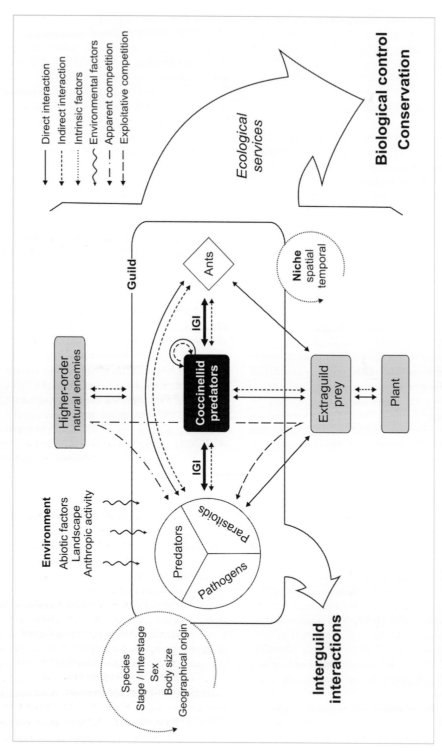

Figure 7.1 Schematic representation of intraguild interactions (IGI) involving predatory ladybirds. Any type of interaction may generate indirect effects on other protagonists within and outside the guild.

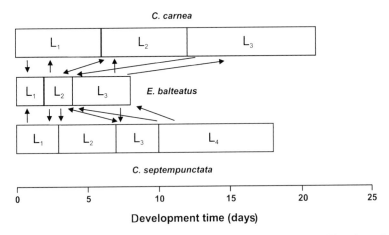

Figure 7.2 Linkage between developmental time of larval instars and occurrence of intraguild predation between *Episyrphus balteatus*, *Chrysoperla carnea* and *Coccinella septempunctata* in a perfectly synchronized community. Arrows point toward intraguild prey. (After Hindayana et al. 2001, with permission).

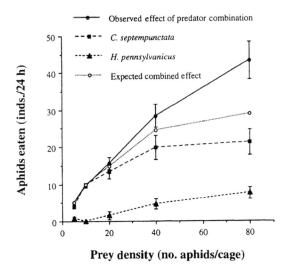

Figure 7.3 Coccinelids as commensalists. Effect of aphid prey density and predator treatment on the consumption of prey aphids. Predator treatments: foliar-foraging predator only *(Coccinella septempunctata)*; ground-foraging predator only *(Harpalus pennsylvanicus)*; both predators in combination (C. *septempunctata* plus *H. pennsylvanicus*). The expected effect is based on the sum of the foliar and ground predators acting in isolation. (After Losey & Denno 1998, with permission).

cooremani benefit from *Chil. cacti* by phoretic dispersal and resource acquisition. The ladybird may benefit by acquiring resources from the hypopode and thus the interaction was assumed to be potentially mutualistic (Holte et al. 2001). The boundary between parasitism, mutualism and commensalism remains to be clarified.

7.6 COCCINELLIDS AS COMPETITORS

7.6.1 Exploitative competition

Exploitative competition is an **indirect interaction** between individuals of two competing species: one species reduces the abundance of a shared resource and consequently affects the second species (Wootton 1994, Abrams et al. 1996). For example, a coccinellid larva or adult affects another intraguild member by **consuming** or **disturbing** their shared prey. Smaller coccinellids with lower food requirements have a competitive advantage over larger species at low aphid densities (Obrycki et al. 1998b).

The recruitment of coccinellid predators is described by the **aggregative numerical response**, the

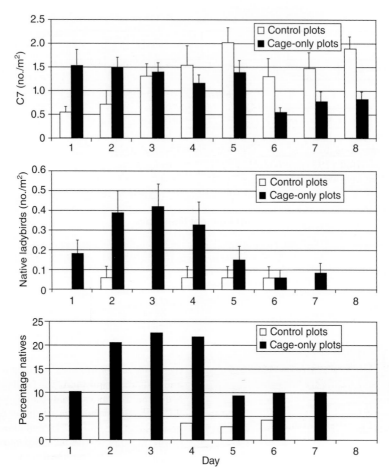

Figure 7.4 Coccinellids as exploitative competitors. Numbers of adults of *Coccinella septempunctata* (C7) and native ladybirds observed in visual censuses on successive days (*x*-axis) of experimental plots that had either been caged previously (cage-only plots) or left uncaged (control plots) in lucerne (top and middle panels; means + 1 SE), and percentage of all individuals (of C7 or natives) observed that belonged to native species (bottom panel). Day 1 is the day after cage removal. (After Evans 2004, with permission).

threshold of which is species specific (5.3.5). Evans (2004) showed that the arrival of the Palearctic *C. septempunctata* in Utah (USA) lucerne fields caused a decline in pea aphid populations and then a decline in native coccinellids (1992–2001) (Fig. 7.4). An artificial increase in aphid densities stimulated the return of native coccinellids to the focal crop. Evans proposed that *C. septempunctata* prevented aphid outbreaks, reducing the level at which lucerne crops retain native coccinellids.

Whitefly egg consumption in multi-specific treatments (one *Delphastus catalinae* adult and one *Col. maculata* adult) was significantly lower than the individual consumptions of the two coccinellid species, even though no behavioural interference was observed (Lucas et al. 2004c).

Exploitative competition may also take the form of **induced defensive responses and dispersion of a shared prey** following ladybird attacks (Dixon 1958, Nelson & Rosenheim 2006). These responses can affect **subsequent predation efficiency** by intraguild competitors (conspecifics or heterospecifics). Aphid dropping induced by *Har. axyridis* reduced its predation efficiency by approximately 40% (Francke et al. 2008) and potentially also that of other foliar-dwelling guild members. On apple saplings, *C. septempunctata* caused significantly greater dispersion of mite colonies than did *Har. axyridis* (Lucas et al. 2002).

7.6.2   Apparent competition

This indirect interaction arises when **two prey** species share a **common natural enemy** and when an increase in one prey leads to an increase in the shared natural enemy so as to cause a decline in the other prey species (Holt 1977, Wootton 1994). The invasion of *Har. axyridis* provided the potential for such an interaction. *Harmonia axyridis* is a low-quality host for the generalist parasitoid *Dinocampus coccinellae* that does not seem to discriminate between coccinellid species; the abundance of this new host may thus constitute **an egg sink**, leading to a decrease in *D. coccinellae* populations and consequently to a **reduction of the parasitism on the native coccinellids** (Hoogendoorn & Heimpel 2002, Firlej et al. 2005, 2006, Koyama & Majerus 2007; also 8.3.2.1).

7.7   COCCINELLIDS AS VECTORS OF MALE-KILLING BACTERIA

Coccinellids are suspected of transmitting **male-killing bacteria, such as *Wolbachia***, although clear empirical data are lacking (Hurst et al. 2003, Tinsley & Majerus 2007; see also 8.4.5.2). This interaction may have **detrimental or beneficial** effects (Majerus et al. 1998, Engelstadter & Hurst 2007) on the other coccinellids and on infected competitors, but is positive for the bacteria. These bacteria are transmitted **vertically** from mother to daughter, but **horizontal transmission is strongly suspected** in coccinellids (Hurst et al. 2003). Horizontal transfer is more likely to occur by direct transfer of infective material (e.g. predation) between closely related species and organisms in close confinement (Tinsley & Majerus 2007).

7.8   COCCINELLIDS AS INTRAGUILD PREDATORS

7.8.1   General rules of intraguild predation involving coccinellid predators

Intraguild predation (IGP), defined as **predation on a competitor** (Polis et al. 1989, Polis & Holt 1992),

includes (i) **effective IGP**: prey is killed and consumed (IGP *sensu stricto*); (ii) **interspecific killing**: prey is killed but not consumed; and (iii) **IGP risk**: prey is at risk of being killed and/or consumed (Lucas 2005). IGP usually involves an **intraguild predator**, an **intraguild prey** and a shared resource (**extraguild prey**). IGP is a widespread force **structuring animal assemblages** (Arim & Marquet 2004). IGP is also important in **biological control** systems (Rosenheim et al. 1993, 1995; 7.11.2, Chapter 11).

IGP by or on predaceous coccinellids is quite common (Dixon 2000, Lucas 2005, Gagnon 2010) due to **aggregations** of the shared prey (mainly gregarious Sternorrhyncha) (Dixon 1985, 1987), **richness and abundance** of competitors (Frazer 1988, Drea & Gordon 1990), and **drastic changes in body size** and **mobility** during the **life cycle**. IGP involving coccinellids is common in the field even at high extraguild prey densities (Gardiner & Landis 2007, Gagnon 2010, but see Hemptinne & Dixon 2005).

IGP is **described** by its (i) **intensity (probability of occurrence)**, (ii) **direction** (mutual or unidirectional), and (iii) **symmetry** (dominance or not of a species). IGP was mostly demonstrated in microcosm where the intraguild prey cannot escape so that the **real IGP intensity** in the field remained **uncertain** until recently (e.g. Hindayana et al. 2001, Kindlmann & Houdkova 2006). Using molecular tools, Gagnon (2010) detected coccinellid intraguild prey in 52.9% of the coccinellid intraguild predators in Quebec soybean crops. Direct observations on cotton revealed that 6.6% of **neonate lacewings died from IGP** in a period equivalent to about **1.4% of their total life cycle** (Rosenheim et al. 1999). Such results indicate **IGP** as a **major source of mortality** for lacewing neonate larvae and also for young instars of coccinellids.

Studies tend to demonstrate that the intensity of IGP involving aphidophagous coccinellids is possibly higher than in other systems such as those involving aleyrodophagous mirids (Lucas & Alomar 2002a, b) due to the active searching of coccinellids and to their low mobility compared to mirids.

Five general rules on IGP involving coccinellids can be established **at species level**:

1 **IGP intensity decreases as extraguild prey density increases** (Sengonca & Frings 1985, Lucas et al. 1998a, Schellhorn & Andow 1999b, Hindayana et al. 2001, Yasuda et al. 2004, Gagnon 2010). For exceptions see Lucas et al. (1998a), Phoofolo

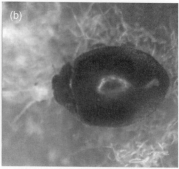

Figure 7.5 Coccinellids as intraguild predators. (a) Intraguild predation by a fourth instar larva of *Harmonia axyridis* preying upon a pupa of *Adalia bipunctata*. Picture by Serge Laplante; (b) Intraguild predation by an adult of *Stethorus punctillum* preying upon a first instar of the mullein bug *Campylomma verbasci*. Picture by Olivier Aubry (Laboratoire de Lutte biologique, UQAM, 2008). (See colour plate.)

and Obrycki (1998), and Snyder et al. (2004b). IGP is frequently motivated by **nutritive needs**, which decrease in the presence of alternative food. Furthermore, an increase in extraguild prey density can generate a **dilution effect** protecting furtive intraguild prey (Lucas & Brodeur 2001).

2 **IGP intensity increases as intraguild prey density increases** (Noia et al. 2008), thus increasing the **probability of encounters**. These first two rules may theoretically be compensatory depending on the numerical response of intraguild prey to extraguild prey density (but see Lucas & Rosenheim 2011).

3 **IGP is unidirectional** (the predator and prey status of each protagonist remain constant; Polis et al. 1989) **and directed toward a specialist** organism (predator, parasitoid or pathogen) (Lucas et al. 1998a, Yasuda & Ohnuma 1999, Hindayana et al. 2001).

4 **IGP is mutual** (each species may prey on the other one during its life cycle; Polis et al. 1989) **when two non-specialized species are involved** (Lucas et al. 1998a, Felix & Soares 2004, but see Mallampalli et al. 2002 and de Clercq et al. 2003 for exceptions). Due to drastic changes in size, mobility, vigour and defensive capacity during the life cycle, each species may at specific times be the intraguild prey or the intraguild predator (Lucas 2005).

5 **Mutual IGP is asymmetrical** (i.e. one species acts as the predator significantly more often than does the other; Lucas 2005) **in favour of the larger species** that are generally more successful in confrontations (Lucas et al. 1998a, Obrycki et al. 1998a, Hindayana et al. 2001), but exceptions may occur (Sengonca & Frings 1985, Lucas et al. 1998a, Snyder et al. 2004b).

7.8.2 Occurrence of intraguild predation by coccinellids

Most coccinellid species may prey upon intraguild competitors (Fig. 7.5). **Older larval instars** (third and fourth) and **adults** are more often **intraguild predators** than younger stages. Because of **an important (at least 5-fold) increase in body size** during immature development, an encounter between small, young larval stages (or eggs), and larger, more powerful older larvae (or adults) often results in antagonistic interactions. In contrast, an encounter between **similar sized** ladybird and/or lacewing larvae usually generates a **low intensity of IGP** even at low extraguild prey densities (Chang 1996, Lucas et al. 1998a, Obrycki et al. 1998a). Older larvae interact with each other more strongly than young larvae (Yasuda et al. 2004).

The **significant benefits** of IGP for coccinellids include: (i) **elimination of a potential predator**, (ii) **elimination of a competitor**, (iii) **consumption of a protein-rich meal** (Polis et al. 1989), and (iv) **acquisition of toxins from intraguild prey** (Hautier et al. 2008). Four hypotheses have been built (Lucas 2005):

1 **Protective IGP hypothesis**: the predator attacks the IG prey to **protect itself before a period of**

high vulnerability (Fig. 7.7). The consumption of the prey is facultative.

2 **Competitive IGP hypothesis**: the predator **eliminates a competitor** and the consumption of the IG prey is facultative. Neither protective nor competitive IGP have yet been reported in coccinellids.

3 **Nutritional IGP hypothesis**: IGP occurs during **extraguild food shortage** or when the **alternative food** is **scarce** or has a lower nutritive value than the IG prey (e.g. Lucas et al. 2009). In temperate areas, pollen is used as a food source at the beginning of the season, but does not allow maturation of ovaries in most predatory ladybird species (5.2.9); at this time, any predation event would be highly beneficial. Also, *Har. axyridis* is among the last insects to be found in autumn in maize fields of Eastern Canada (as larvae and pupae); third and fourth instars are highly aggressive at this time, and IGP/cannibalism is extremely common (Lucas, unpublished). For *Har. axyridis*, supplementing a limited aphid diet with *A. bipunctata* eggs provides **nutritional benefits during extraguild prey shortages** (Ware et al. 2009). Such ability to effectively exploit coccinellid intraguild prey as food (**specialization**) may be related to a **lower efficiency in extraguild prey exploitation** (Sato et al. 2008). Based on the fact that **nitrogen content** is generally higher in predatory insects than in herbivores, Kagata and Katayama (2006) tested and rejected the hypothesis that nitrogen shortages promote IGP between aphidophagous coccinellids.

4 **Opportunistic IGP hypothesis**: the predator **selects the prey according to its size**, regardless of its guild. The lacewing *Chrysoperla rufilabris* seems to show **no preference** for intra- or extraguild prey, and IGP occurrence is mainly determined by the **encounter rate** (Lucas et al. 1998a). Many coccinellids species are polyphagous and prey selection is greatly influenced by **capture efficiency**, which is directly related to **predator and prey relative sizes** (Dixon 1959, Klingauf 1967; 5.2.3) and to the **prey defensive abilities** (Provost et al. 2006).

Preying upon a competitor which is also a predator can be **risky**: the intraguild predator **may be injured** or may itself **become the prey** (Polis et al. 1989, Dixon 2000). However, such a situation seems rare in the field since coccinellids usually avoid confrontation with prey of similar size (Lucas et al. 1998a). Ladybirds may also **be contaminated by generalist entomopathogens** infesting their intraguild prey. Also, IGP may increase a predator exposure to pesticides (Provost et al. 2003). Finally, many coccinellids are protected by **toxic alkaloids** (7.9.1, 5.2.6.1, 9.2).

All in all, the **cost/benefit ratio of IGP** by coccinellid predators would depend whether the **intraguild prey are** predators, parasitoids, or pathogens.

7.8.3   Intraguild predation on intraguild coccinellids (by coccinellids)

Most extraguild prey are exploited by an assemblage of ladybird species (Chazeau 1985, Frazer 1988, Drea & Gordon 1990). Potential for antagonistic interactions and IGP between ladybird predators is therefore high (e.g. Takahashi 1989, Schellhorn & Andow 1999a, b, Yasuda & Ohnuma 1999, Yasuda et al. 2001, 2004, Michaud & Grant 2003, Felix & Soares 2004, Snyder et al. 2004a, Kajita et al. 2006, Majerus et al. 2006, Hodek & Michaud 2008).

Since morphology and development are relatively similar among ladybirds, **IGP between coccinellids** follows **six general rules**:

1 **IGP intensity is usually high among predaceous coccinellid species** due to their **great voracity, their similar foraging strategies**, and to a **contagious distribution of their resource** (Hodek 1996).

2 **IGP is mutual.** A ladybird is an **intraguild predator at old larval/adult stages** or **intraguild prey at young larval/egg stages**.

3 **The larger individual usually preys upon the smaller**.

4 **Polyphagous species are less affected by alkaloids** of their intraguild prey than specialists (Yasuda & Ohnuma 1999).

5 **Eggs, pupae, and moulting individuals are highly susceptible to IGP** (exceptions are linked to chemical protection; 7.9.1).

6 **Adults are never successfully attacked**.

Important **differences among species** modulate the above rules. The **attack rates** (the number of attacks divided by the number of individuals contacted) of *Har. axyridis* toward *C. septempunctata* exceeded 50%, while the **attack rates** of *C. septempunctata* toward *Har.*

axyridis was less than 20% (Yasuda et al. 2001). The average **escape rates** (escapes divided by attacks) of *Har. axyridis* following *C. septempunctata* attacks was higher than the escape rates of *C. septempunctata* following *Har. axyridis* attacks. In the pair *Har. axyridis* and *C. undecimpunctata*, a smaller advantage in body weight was required for *Har. axyridis* to become an intraguild predator of *C. undecimpunctata* than vice versa (Felix & Soares 2004). Snyder et al. (2004) found that the **exotic** species *C. septempunctata* and *Har. axyridis* had a significant advantage in IGP confrontations over the **native** *Hip. convergens* and *C. transversoguttata*. Especially *Har. axyridis* attacked more successfully and escaped more frequently when attacked. In contrast to native species, **cannibalism** was a **greater threat for Har. axyridis** than IGP (5.2.8).

As intraguild predators, coccinellids frequently have **sub-lethal effects** on intraguild prey. The presence of *Har. axyridis* **slowed** *A. bipunctata* **larval development** at high extraguild prey density, whereas IGP by *Har. axyridis* or by *C. septempunctata* occurred at low extraguild prey density (Kajita et al. 2000). In a cage experiment, *Har. axyridis* caused a **decrease in** *C. undecimpunctata* **fecundity**, even when the extraguild resource was abundant (Soares & Serpa 2007). In field cages, *Har. axyridis* larvae (intraguild predator) had no effect on survival or weight gain of *Col. maculata* larvae (intraguild prey) (Hoogendoorn & Heimpel 2004), but the intraguild prey **modified its distribution**, possibly to avoid interactions with the predator. Adult *Col. maculata*, *C. septempunctata* or *Har. axyridis* did not modify their **distribution** when in presence of adults of the other species (Lucas et al. 2002).

Finally, **key factors** influencing IGP among coccinellids are (i) **time of colonisation**, **voltinism**, and **speed of development** which determine the intraguild prey/intraguild predator status of the protagonists at a specific time, and (ii) **oviposition, moulting and pupation site**, since the susceptibility of **non-mobile stages/periods** is strongly linked to their position (e.g. distance from the shared resource) (Lucas et al. 2000, Lucas 2005).

7.8.4 Intraguild predation on intraguild non-coccinellid predators

Significant mortality from IGP by coccinellid predators has been demonstrated on lacewings (e.g. Michaud

& Grant 2003, Gardiner & Landis 2007), syrphids (Hindayana et al. 2001), and cecidomyiids (e.g. Lucas et al. 1998a, Gardiner & Landis 2007).

7.8.4.1 Intraguild neuropterans

Green/brown lacewings and ladybirds often **co-occur spatially and temporally** in the fields (Phoofolo & Obrycki 1998). **Mutual IGP** has been reported both in the laboratory and in the field (Sengonca & Frings 1985, Lucas et al. 1998a, Phoofolo & Obrycki 1998). For example, **lacewing eggs**, despite the presence of the pedicel and a tough chorion, were easily eaten by all stages of *Col. maculata* (Lucas 1998). However, when **reared** on lacewing eggs, coccinellids did not complete their pre-imaginal development (*C. septempunctata*) or developed into smaller adults (*Col. maculata* and *Har. axyridis*) (Phoofolo & Obrycki 1998). **Lacewing pupae** were not successfully attacked by *Col. maculata* (Lucas et al. 1998a).

In the presence of a *C. septempunctata* first instar, first instars of *Chrysoperla plorabunda* sheltered themselves in axils after feeding (Chang 1996). *Chrysopa perla* and *Chrysopa oculata* **avoided oviposition** on substrates with **tracks** of *Har. (= Leis) dimidiata* first instars and *Col. maculata* third instars, respectively. Tracks of first instars of some other coccinellid species had no impact on *C. perla* and *C. oculata* oviposition (Ruzicka 2001b, Chauhan & Weber 2008).

7.8.4.2 Intraguild dipterans

Vermiform dipteran larvae seem **highly susceptible** to IGP by coccinellids (Lucas et al. 1998a, Gardiner & Landis 2007). *Harmonia axyridis* caused 40% direct mortality of the cecidomyiid *Aphidoletes aphidimyza* larvae after 2 hours in a microcosm experiment (Voynaud 2008). The aphidophagous chamaemyiid *Leucopis* spp. larvae are occasionally victims of IGP by ladybird and lacewing larvae (Sluss & Foote 1973). Both cecidomyiid and chamaemyiid larvae are **furtive predators** that exploit aphid colonies without generating aphid **defensive response** (Lucas et al. 1998a, Lucas & Brodeur 2001, Frechette et al. 2008). They benefit thus from **a dilution effect** (the probability of being attacked by a predator decreases as colony size increases; Edmunds 1974), in large aphid colonies which reduces their susceptibility to IGP (Lucas & Brodeur 2001).

Coccinellids have an **indirect effect** by reducing aphid density through predation and aphid dispersion (Voynaud 2008). In addition to having less food available, midges suffer from a reduced **dilution effect** that increases IGP risks (Lucas & Brodeur 2001).

Leaf **trichomes** provide **refuges** for *A. aphidimyza* eggs against IGP by *Col. maculata* and play a role in **oviposition site selection** (Lucas & Brodeur 1999). Also, *A. aphidimyza* laid fewer eggs on plants previously exposed to second instar *C. septempunctata* (Ruzicka & Havelka 1998; but see Lucas & Brodeur 1999).

When attacked by coccinellids (Hindayana et al. 2001), older syrphid larvae defend themselves by **oral secretions (slime)** or rarely by **counter-attacks**. *Episyrphus balteatus* **pupae** were **never attacked** by *C. septempunctata* adults or larvae (Hindayana et al. 2001). In Petri dishes, *E. balteatus* laid fewer eggs in presence of *Har. axyridis* larval tracks (Almohamad et al. 2010).

7.8.4.3   Intraguild hemipterans

The **great mobility** of mirids should severely lower interactions with intraguild members as well as the possibility of relevant observations. **Asymmetrical IGP** in favour of the coccinellid was observed between *Har. axyridis* and the mirid *Hyaliodes vitripennis*. A **low intensity** of IGP was also noted when compared to coccinellid–coccinellid interactions. The presence of less mobile extraguild prey (phytophagous mites), to which the attacks of mirids were directed, further reduced IGP intensity (Provost et al. 2005, 2006).

Small acarophagous ladybirds also attack young stages of acarophagous bugs, such as *Stethorus pusillus* preying on first instar mirid *Campylomma verbasci* (Fig. 7.5). IGP by coccinellids on eggs, young larvae, or moulting individuals of other intraguild heteropterans is highly probable but has not yet been documented.

7.8.4.4   Other intraguild predators

Predatory mites of the phytoseiid *Amblyseius andersoni* are frequently found in **acarodomatia** (tufts of hair or invaginations on the leaf surface of several plants). These structures protect the mites against IGP by *C. septempunctata* and *Hip. variegata* adults, but not against second instar lacewings (*Chrysoperla rufilabris*) (Norton et al. 2001).

IGP by ladybirds on eggs of the **derodontid** beetle *Laricobius nigrinus* has also been reported (Flowers et al. 2005). IGP by coccinellids could also affect **other predators** (Cantharidae, Dermaptera, spiders, etc.).

7.8.5   Intraguild predation on intraguild parasitoids

Coccinellids can affect **intraguild parasitoids:** (i) by **direct predation** of parasitized extraguild prey, or (ii) by **non-lethal impact** on adult parasitoids. Ladybirds often **avoid eating parasitized extraguild prey**, or **prefer unparasitized prey** (5.2.7.1).

The coccidophagous *Rodolia* sp. was reluctant to consume parasitized coccids (Quezada & DeBach 1974). *Cryptolaemus montrouzieri* **feeds on** citrus mealybugs **parasitized** by *Anagyrus pseudococci* but not on individuals older than 4 days (Mutsu et al. 2008).

The aleyrophagous *Serangium parcesetosum* larvae and adults tended to **avoid** *Bemisia tabaci* pupae parasitized by *Eretmocerus mundus* (Al-Zyoud 2007). *Delphastus catalinae* adults **did not discriminate** whiteflies parasitized by *Encarsia sophia*, while second instars **preferred unparasitized** whiteflies (Zang & Liu 2007).

In the aphidophagous *Col. maculata*, larvae did not discriminate between healthy *Trichoplusia ni* eggs and those parasitized by *Trichogramma evanescens* (Roger et al. 2001) and *C. septempunctata* did not discriminate *Lysiphlebus testaceipes* mummies from unparasitized *Schizaphis graminum* (Royer et al. 2008). Despite intense (98–100%) predation on *L. testaceipes* mummies, *Hip. convergens* exhibited a **partial preference for unparasitized** hosts (Colfer & Rosenheim 2001; but see Ferguson & Stiling 1996). *Coccinella undecimpunctata* larvae had no preference for parasitized or healthy aphids even though **parasitized aphids had inferior nutritive values** (Bilu & Coll 2009). Coccinellids **destroyed more than 95%** of the immature parasitoids of the psyllid, *Diaphorina citri*, in Florida citrus groves (Michaud 2004).

The larvae of *C. undecimpunctata* disturbed *A. colemani* **adults** and **reduced parasitization** (Bilu & Coll 2007). Adults of the parasitoid *Lysiphlebus fabarum*, **avoided** plants with coccinellids (Raymond et al. 2000). Nakashima et al. (2004, 2006) also demonstrated that three aphid parasitoid species **avoided leaves previously visited by adult ladybirds**.

Figure 7.6 Coccinellids as intraguild prey. (a) Intraguild predation by an adult *Podisus maculiventris* on a larva of *Coleomegilla maculata lengi* in a soybean field. Picture by Florent Renaud, 2007; (b) Intraguild predation by a *Thanatus* sp. spider on an adult of *Harmonia axyridis* in an apple orchard. Picture by Jennifer de Almeida, 2008 (Both pictures from Laboratoire de Lutte biologique, UQAM). (See colour plate.)

7.8.6 Intraguild predation on intraguild pathogens

Coccinellids interact with intraguild **entomopathogens** when eating infected prey and the outcome depends whether **the pathogen** is **specific** to the extraguild prey or **able to infect coccinellids** as well. For instance, aphids (or aphid cadavers) infected by aphidopathogenic fungus *Pandora neoaphidis* are preyed upon by *Har. axyridis* and *C. septempunctata* adults and larvae (Roy et al. 1998, 2003, 2008a, Roy & Pell 2000). Despite an overall **preference** by both predators for uninfected aphids, *C. septempunctata* is **more selective** than *Har. axyridis*, the Japanese population of *Har. axyridis* is more selective than the UK population, and satiated individuals are more selective than starved ones (8.4).

7.8.7 Coccinellids as top predators

Among coccinellids, *Har. axyridis*, is considered a top predator, dominating both the aphidophagous and coccidophagous guilds (Dixon 2000, Majerus et al. 2006, Pervez & Omkar 2006). A **top (or apex) predator** is largely **free from predation pressure** and is regulated more by **bottom-up** than **top-down** forces

(Gittleman & Gompper 2005). It should have a **stronger impact** on **intermediate predators** (the intraguild prey) than on the **shared prey** (the extraguild prey) (Dixon 2007). Indeed, numerous (mainly laboratory) studies confirm that *Har. axyridis* **dominates confrontations** with most coccinellid species (Pell et al. 2008, Alhmedi et al. 2010). This is explained by a **large body size**, **strong larval spines**, **chemical protection**, **rapid larval development**, great **nutritional plasticity** and high **aggressiveness** (Labrie et al. 2006, Majerus et al. 2006, Pervez & Omkar 2006, Sato et al. 2008). The predator could also have a better capacity to process ingested defence chemicals (Kajita et al. 2010). Labrie et al. (2006) reported the development of a **fifth larval instar** in laboratory conditions.

An alternative assumption is that *Har. axyridis* is **one species among other** large generalist aphidophagous coccinellids (Soares et al. 2008):

1 Generalist aphidophagous species are a threat to *Har. axyridis* younger/smaller instars (Felix & Soares 2004, Burgio et al. 2008). Furthermore, *Har. axyridis* **is sometimes dominated** in intraguild interactions, for example, by the pentatomid *Podisus maculiventris* (de Clercq et al. 2003), by spiders (Yasuda & Kimura 2001, Fig. 7.6b) or by the

ladybird *Anatis ocellata* (Ware & Majerus 2008).
Thus it is **not** always **free from significant
predation**;

2 Its stronger impact on the intermediate predator
than on the shared prey remains to be demonstrated
(Lucas et al. 2007a);

3 Also other **eurytopic** and/or **euryphagous** species,
such as *C. septempunctata* (Hodek & Michaud 2008)
and *P. quatuordecimpunctata* (Lucas et al. 2007b)
have achieved success in invasion.

According to still another view, **IGP does simply
not determine predator dominance**. IGP between
ladybird species has no significant impact on their
abundance, since **defensive traits evolved to avoid
IGP** (Kindlmann & Houdkova 2006, Dixon 2007).
Cannibalism may be the **key factor regulating** coccinellid abundance.

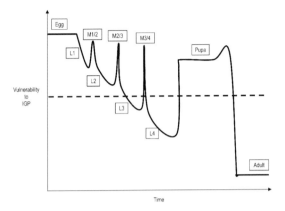

Figure 7.7 Susceptibility of a coccinellid to intraguild
predation during its life cycle. The horizontal line refers to a
threshold of vulnerability linked to a potential intraguild
predator. M1/2, Moult from first to second larval instar; L1,
first larval instar. (After Lucas 2005).

7.9 COCCINELLIDS AS INTRAGUILD PREY

Any predatory ladybird risks being attacked by
intraguild generalists (predators, parasitoids, or
pathogens) (Figs. 7.1 and 7.6). This vulnerability to
IGP is **species- and stage-dependent**: some species
are more protected by **size, unpalatability, behaviour** or other **protective devices** (Dixon 2000, de
Clercq et al. 2003, Lucas 2005, Ware & Majerus
2008).

Even though all developmental stages are **potential
intraguild prey, non-mobile stages** (eggs and
pupae), **young instars** and **molting individuals** are
much more vulnerable (Sengonca & Frings 1985,
Lucas et al. 1997a, 1998a, 2000, Obrycki et al. 1998a,
Hindayana et al. 2001; Fig. 7.7).

Finally, apart from **direct attacks** (e.g. death or
injury), coccinellid intraguild prey may also suffer from
IGP risks (e.g. **retarded development)** and respond
by defensive mechanisms (7.9.1).

7.9.1 Defensive mechanisms of coccinellids against intraguild predation

Defensive mechanisms protect individuals during
susceptible periods, some developed specifically
against intraguild threats, while others against several

types of enemies. Protection is achieved by **successive
lines of defence** (Lucas 2005).

7.9.1.1 Defence of all stages

Many coccinellid species are protected by **toxic compounds** either **synthesized *de novo***, or **sequestered**
from their food (**chemical piracy**) (Pasteels 2007,
Kajita et al. 2010; 9.3): the **trade-off between the
cost (production, toxicity) and the benefits**
(defence) is not clear (Pasteels 2007). **Biparental
endowment** of chemical defences into eggs is possible: in *Epilachna paenulata*, males transfer alkaloids to
females at mating (Camarano et al. 2009).

The toxic compounds may be encountered **passively** by the attacker, or may be **actively released**
(reflex bleeding; 8.2.1.2 and 9.2.1) in response to
attacks (Cuenot 1896, Pasteels et al. 1973, Eisner et al.
1986, Holloway et al. 1991, de Jong et al. 1991, Attygalle et al. 1993). As a rule, these compounds seem to
have a **limited effect against cannibalism** but a **significant one against IGP** (Agarwala et al. 1998,
Hemptinne et al. 2000b). Reflex bleeding of larvae
could have a cost on the weight of resulting adults
(Sato et al. 2009).

Another type of defence is exploiting a **protected
habitat** (such as aphid galls or **ant-attended aphid
colonies**; 7.10) that provides an **intraguild enemy-
free space**.

7.9.1.2   Defence of eggs

Defence of eggs has been extensively studied (e.g. Agarwala & Dixon 1993, Yasuda & Shinya 1997, Hemptinne & Dixon 2000, Cottrell 2005). The first line of defence is the **selection of an enemy-free space** for laying eggs (5.4.1.3). **This behaviour** reflects the trade-off between **egg susceptibility** to intraguild predators and cannibals, and **starvation risk** in **neonates** (Schellhorn & Andow 1999a, Magro et al. 2007, Seagraves 2009).

Females may **avoid oviposition sites** where intraguild predators are present (e.g. Iablokoff-Khnzorian 1982). Adult females are able to detect **oviposition-deterring semiochemicals** (ODS) from **conspecific larval tracks** (e.g. Ruzicka 1997, 2001a, 2003, Doumbia et al. 1998, Hemptinne et al. 2000b, Frechette et al. 2003; 5.4.1.3). For **heterospecific ODS**, several studies failed to detect any response (Doumbia et al. 1998, Yasuda et al. 2000, Oliver et al. 2006), while others demonstrated a significant heterospecific effect (Ruzicka 1997, 2001a, 2006, Michaud & Jyoti 2007) or inconsistent results (Ruzicka 2001b, Ruzicka & Zemek 2003).

Faeces of *Har. axyridis* (intraguild predator) represented cues that **reduced feeding and oviposition** in *P. japonica* (intraguild prey) (Agarwala et al. 2003). However, *Har. axyridis* behaviour was not affected by *P. quatuordecimpunctata* faeces.

In the presence of the intraguild predator *Chrysoperla carnea*, *Serangium parcesetosum* laid more eggs between the plant veins (Al-Zyoud et al. 2005).

The presence of **immobilized individuals of the coccinellid intraguild predator** *C. leonina transversalis* (but not the smaller *Scymnus pyrocheilus*) **reduced oviposition** in *Menochilus sexmaculatus* (Agarwala et al. 2003).

A second line of defence of ladybird eggs is the **chemical protection** (9.6) that varies according to the intraguild prey and the intraguild predator species (Sato & Dixon 2004, Flowers et al. 2005, Rieder et al. 2008, Ware et al. 2008). The chemicals involved may be partially or totally **repellent**, and may involve **nutritive costs** (Agarwala & Dixon 1992, Hemptinne et al. 2000a, Cottrell 2007, Kajita et al. 2010). For example, chemical protection efficiently protects *Calvia quatuordecimguttata* against *Har. axyridis* (Ware et al. 2008), *Har. quadripunctata* (Dyson 1996), *A. bipunctata*, *P. quatuordecimpunctata*, *Calvia decemguttata* and *A. decempunctata* (Vanhove 1998). Eating **heterospecific**

eggs may **slow** the predator's **development**, **decrease larval weight gain** and **adult size**, and even **cause death** (Phoofolo & Obrycki 1998, Hemptinne et al. 2000a, b, Cottrell 2004, Rieder et al. 2008). Egg chemical defence could be linked to **co-evolution of defences against sympatric** ladybird species (Agarwala & Yasuda 2001, Rieder et al. 2008).

Despite chemical defence, eggs remain highly **susceptible to cannibalism** (e.g. Agarwala et al. 1998, Omkar et al. 2004, Rieder et al. 2008, Sato & Dixon 2004). **Larvae prefer** (and are less affected by) consuming **conspecific eggs** (but see Cottrell 2005). In situations of aphid scarcity, egg cannibalism may help starving larvae to survive (Agarwala & Yasuda 2001). **Cannibalizing eggs may protect larvae against IGP**, either by **additional toxin acquisition**, or by **acceleration of their development**, thus reducing the time they are susceptible to IGP (Gagne et al. 2002, Michaud & Grant 2004, Omkar et al. 2007).

Another defensive trait of ladybird eggs is linked to the fact that they are laid in **clusters** and are thus less vulnerable to predation than when laid singly (due to a **dilution effect**) (Agarwala & Yasuda 2001). Furthermore, the **cluster may concentrate the toxic compounds** of eggs at a single point, increasing efficiency (Agarwala & Yasuda 2001, Omkar et al. 2004).

7.9.1.3   Defence of larvae

The susceptibility of young instars to IGP is linked to the **oviposition site** often **selected** by the female close to resources (5.4.1.3). Older larvae may **actively avoid** sites occupied by potential predators (Hoogendoorn & Heimpel 2004).

Morphological characteristics, such as abdominal **spines** or **wax**, may provide defence for coccinellid larvae against IGP (Voelkl & Vohland 1996, Ware & Majerus 2008). Such traits are well developed in some species and absent in others (Pope 1979, Michaud & Grant 2003, Ware & Majerus 2008).

Large body size remains one of the most effective defences of coccinellid larvae against IGP (Lucas et al. 1997a, 1998a, Felix & Soares 2004, Ware & Majerus 2008). Thus any mechanism able to provide a **larger size** at the time of the intraguild confrontation (e.g. an earlier time of colonization) will **increase the potential intraguild prey survival probability**. Also **diet** can significantly affect the growth of the larvae. The **slow-growth–high-mortality (SGHM) hypothesis**, which predicts that prolonged development results in greater exposure to natural enemies and in a subsequent increase in mortality (Clancy & Price 1987), applies to IGP involving coccinellids. Furthermore, one can implicitly propose a second hypothesis: as a result of prolonged development, the body size of an individual at any particular time is expected to be smaller and therefore its susceptibility to IGP should be higher (Fig. 7.2).

Following an encounter, coccinellid larvae frequently **escape** by different means (Hough-Goldstein et al. 1996, Lucas et al. 1997a, 1998a, de Clercq et al. 2003; 5.4.2). The susceptibility of *Col. maculata lengi* larvae to IGP by chrysopid third instar (*C. rufilabris*) is **age specific** and is influenced by the coccinellid **behavioural defensive strategies** (Lucas et al. 1997a): young larvae exhibited escape reactions (**dropping, fleeing and retreating**) but did not survive once caught by the lacewings, while older larvae used **wriggling** or **biting** representing 7.5% to 11% of all successful defences (Lucas et al. 1997a). A **drastic difference in dropping** from the plant was observed between *A. bipunctata* and *C. septempunctata* when facing *Har. axyridis* larvae: 43.3% of the *C. septempunctata* first instars dropped and 54.5% were killed by IGP whereas <2% of the *A. bipunctata* first instars dropped. The mortality of the latter from IGP was therefore 95% (Sato et al. 2005). Since IGP is influenced by extraguild prey density, **emigration when food is scarce** has a potential defensive value. Emigration tendency varies among species: 80% of *C. septempunctata brucki* larvae emigrated prior to aphid population extinction, whereas less than 20% of *Har. axyridis* and *P. japonica* did so (Sato et al. 2003).

Protection of coccinellid larvae by **alkaloids** differs according to intraguild predators and prey (Yasuda & Ohnuma 1999). For example, both *Cycloneda sanguinea* and *Har. axyridis* larvae preferred to feed on dead *C. sanguinea* larvae than on dead *Har. axyridis* larvae. Moreover, *Har. axyridis* larvae that fed on *C. sanguinea* larvae reached the adult stages, while no *C. sanguinea* larvae completed development on *Har. axyridis* larvae (Michaud 2002).

7.9.1.4 Defence of moulting individuals and pupae

Site selection for pupation and ecdysis influences ladybird **susceptibility to IGP**. In *Col. maculata lengi*, vulnerability of pupae and newly moulted larvae to IGP by *Chrysopa rufilabris* larvae depended on **site selection**, leaves with aphid colonies being the most risky sites (Lucas et al. 2000; also Osawa 1992, Michaud & Jyoti 2007; 5.4.1.3). Larval ecdysis generally occurs (60%) near the aphid colonies exploited by the larvae (Lucas et al. 2000); by doing so, the larvae stay close to the resources and reduce the risk of encountering an intraguild predator during displacement. Since the moulting process lasts less than 20 min (except the later sclerotization of the cuticle), **IGP risk** is relatively low.

In contrast, 90% of the larvae **left the plant to pupate**; this reflects predation risks on the plant. In *Col. maculata*, pupation lasts about 20% of preimaginal developmental time (Warren & Tadić 1967). Remaining on the plant would thus expose pupae to natural enemies for a longer period than moulting larvae. Furthermore, the proximity of a food source is less important since the emerging adults are alate and can **disperse**. Finally, when searching for a pupation site, IGP risks are lower for fourth instar larvae than it would be in younger instars.

Microhabitat selection for ecdysis and pupation thus reflects a **trade-off between the advantages of remaining close to resources and the costs of being exposed to IGP** (Lucas et al. 2000).

Pupae may also reduce IGP risk through **flipping behaviour, 'gin traps'** (sclerotized teeth along junctions of movable abdominal segments; e.g. Eisner & Eisner 1992), and **warning colouration** (Majerus 1994). **Gregarious pupation** (of 2–5 pupae) has been reported for some species (*Har. axyridis, C. septempunctata*) in the field (G. Labrie, unpublished) and may also provide defence against IGP through a passive **dilution effect**. Specific defences such as **hair cover** is a protection against intraguild ants (Voelkl 1995). In some species, these hairs secrete **defensive droplets** (Attygalle et al. 1993). **Anachoresis**, the selection of shelters (holes, crevices etc.) (Edmunds 1974), also protects pupae against predation (Richards 1985).

7.9.1.5 Defence of adults

Adults are better defended than any other stage (Sengonca & Frings 1985, Lucas et al. 1998a, Mallampalli et al. 2002, de Clercq et al. 2003), but may be preyed upon by intraguild predators such as chrysopids (Lucas et al. 1998a), pentatomids (de Clercq et al. 2003; Fig. 7.6a), reduviids (E. Lucas, unpublished) or spiders (Fig. 7.6b).

Adults may **avoid sites** with intraguild natural enemies (7.9). Adults may escape by **fleeing, dropping** or **flying away** (Hough-Goldstein et al. 1996, de Clercq et al. 2003). The shape (rounded, compact) of coccinellid adults and body sclerotization should contribute to the low susceptibility of adults to IGP (de Clercq et al. 2003).

Unpalatability of adults linked to the presence of **greater concentrations of toxic alkaloids** than in larvae may also lower the susceptibility of adults to IGP, even though this is species-specific (Mallampalli et al. 2002, de Clercq et al. 2003).

The **aposematic colouration** (bright and contrasting patterns) of adults is considered a **warning signal** (for toxicity) and an effective defence against visual predators (8.2). This aposematic pattern is common to many sympatric species, which may **reinforce the strength of the warning message** (**Mullerian mimicry**) (Holloway et al. 1991, Dolenska et al. 2009). However, no study has demonstrated any impact of aposematism against IGP.

7.9.2 Intraguild predation on coccinellids

Intraguild predators of ladybirds belong to a variety of groups (especially chrysopids, Hemiptera, syrphids and coccinellids). **Coccinellids** represent possibly the most dangerous intraspecific (cannibalism) and interspecific (IGP) threat for other coccinellids (7.8.3). Intraguild predation may also come from generalist parasitoids or pathogens.

7.9.2.1 Intraguild predation by intraguild predators

Chrysopid larvae frequently attack all ladybird stages (Sengonca & Frings 1985, Lucas et al. 1998a) and are usually reported to be **superior in confrontations** with similar-size coccinellid larvae (Sengonca & Frings 1985, Lucas et al. 1997a, 1998a, Phoofolo & Obrycki 1998, Michaud & Grant 2003). This superiority may be linked to their more **elongated mouthparts** allowing chrysopids to keep coccinellids at a safe distance (Lucas et al. 1997a, 1998a, Michaud & Grant 2003). *Coccinella septempunctata* females detected the presence

of **ODS** and laid fewer eggs on sites previously exposed to second lacewing instars (Ruzicka 1997).

Syrphid larvae preyed upon coccinellid larvae (Hindayana et al. 2001): in small arena, 1.1% of *C. septempunctata* larvae were killed and eaten by *Episyrphus balteatus* larvae and 7.4% were killed but **not consumed**. Coccinellids **avoided** pea plants when *E. balteatus* eggs or larvae were present (Alhmedi et al. 2007a, b).

Pentatomids also preyed on intraguild coccinellids, the interaction being **asymmetrical** in favour of the bugs (Mallampalli et al. 2002, de Clercq et al. 2003; Fig. 7.6a). By contrast, the **anthocorid** bug *Orius laevigatus* was not able to feed on *A. bipunctata* or *Har. axyridis* eggs (Santi & Maini 2006). However, **molecular gut-content analysis** showed that 2.5% of immature *Orius insidiosus* had consumed *Har. axyridis* material in the field (Harwood et al. 2009). IGP by **mirids** on coccinellids has not been reported and may be rare since mirids select slow-moving easy and safe prey (Kullenberg 1944, Frechette et al. 2007). It is not known whether coccinellid eggs can be attacked by mirids.

Spiders are important ladybird predators (Ceryngier & Hodek 1996, Yasuda & Kimura 2001; Fig. 7.6b). Most spiders are generalists and frequently share a common prey with generalist coccinellids. The **crab spider** *Misumenops tricuspidatus* attacked larvae of aphidophagous coccinellids (Yasuda & Kimura 2001).

IGP by the **derodontid** beetle *Laricobius nigrinus* on eggs of the ladybird *Sasajiscymnus tsugae* has also been reported, both species preying on the hemlock woolly adelgid (Flowers et al. 2005).

7.9.2.2 Intraguild predation by intraguild parasitoids

IGP on coccinellids can result from parasitism by **generalist parasitoids** also attacking extraguild prey. Babendreier et al. (2003) investigated the impact on natural enemies of **mass releases** of *Trichogramma brassicae* against the European corn borer *Ostrinia nubilalis*. No parasitoid emerged from *C. septempunctata* and *A. bipunctata* eggs; however, **young larval instars** of *Trichogramma* were recorded in *A. bipunctata* eggs and **egg mortality increased**. Since several coccinellids have been reported as predators of *O. nubilalis* eggs (Musser & Shelton 2003), the interaction is considered as IGP (or **intraguild parasitism**).

7.9.2.3   Intraguild predation by intraguild pathogens

IGP on coccinellids can result from **contamination by a generalist pathogen** infesting both the extraguild prey and the coccinellid hosts. Contamination by the fungus *Beauveria bassiana* has been observed in **hibernation sites** (Iperti 1964, 1966, Cottrell & Shapiro-Ilan 2003). In **biological control programmes**, contamination by generalist pathogens of both coccinellids and the target pests has been extensively demonstrated with **fungi** (Hodek 1973, Todorova et al. 1996, 2000, Cottrell & Shapiro-Ilan 2003, 2008), **bacteria/by-products of bacteria** (e.g. Giroux et al. 1994) and **nematodes** (Mracek & Ruzicka 1990, Lemire et al. 1996, Shapiro-Ilan & Cottrell 2005). Impacts are highly variable according to the **strain/species** considered (Krieg et al. 1984, Lucas et al. 2004b, Cottrell & Shapiro-Ilan 2008).

Some coccinellid species **avoid** preying upon **sporulating cadavers** of extraguild prey, **reducing contamination risks** (Roy et al. 2008b).

7.10   COCCINELLIDS INTERACTING WITH INTRAGUILD ANTS

Intraguild interactions between **ants** and ladybirds mainly result from **aphid-attending by ants**, but also may include **ant predation by coccinellids** (Majerus et al. 2007; see also 5.4.1.6 and 8.2.4).

Ants attend **honeydew** producing Hemiptera; 40% of aphid species are obligatorily tended by ants (Kunkel & Kloft 1985). Since they feed on honeydew and sometimes on the hemipterans producing it, ants belong to the **same guilds** as aphidophagous or coccidophagous coccinellids. Ants **protecting honeydew** against predators and parasitoids (e.g. Way 1963, Bradley 1973, Addicott 1979, Ceryngier & Hodek 1996) are frequently **keystone species** (producing disproportionately large effects on the abundance of interacting species in a community). Consequently the **natural enemy guild structure** is **drastically changed** by the presence of ants (e.g. Paine 1969, Sloggett & Majerus 2000, Eubanks & Styrsky 2006, Guenard 2007, Majerus et al. 2007).

Interactions between coccinellids and ants depend on: **species and density of** (i) **honeydew-producers**, (ii) **ants**, and (iii) **coccinellids** (Sloggett & Majerus

2000, Harmon & Andow 2007, Majerus et al. 2007). The **aggressiveness** of the ants and the **ants–hemipteran relative density** determine the **protection intensity** or **bodyguard effect** (Harmon & Andow 2007). The **numerical response** of ants and natural enemies will be related to the attractiveness of the resource (chemicals produced and insect density).

The aptitude to exploit ant-tended aphid colonies reflects the degree of **myrmecophily** of ladybirds. Per definition, myrmecophilous species would take more benefits when exploiting colonies tended by ants than unattended ones, which would not be the case for non-myrmecophilous species.

7.10.1   Non-myrmecophilous coccinellid species

Ants mostly show **aggressiveness toward non-myrmecophilous coccinellids** (e.g. Way 1963, Tedders et al. 1990, Eubanks 2001, Herbert & Horn 2008). In extreme cases, ants may even kill all coccinellid developmental stages (Voelkl 1995, Ceryngier & Hodek 1996, Kaplan & Eubanks 2002, Eubanks & Styrsky 2006, Oliver et al. 2008).

Ladybird larvae appear to be **more susceptible** to ant attacks **than adults**, and Sternorrhyncha are usually exploited only by adult ladybirds (Reimer et al. 1993, Guenard 2007). Ant-attended sternorrhynchans may provide **enemy-free space for potential intraguild prey** of non-myrmecophilous coccinellids, such as **myrmecophilous coccinellids** (7.10.2), **furtive predators** (Guenard 2007) or **parasitoids** (Voelkl 1992, Fischer et al. 2001).

Mutualism between ants and honeydew-producers may include alarm signals for the attention of ants. The treehopper *Publilia concava* produces an **acoustic vibration alarm signal** when in contact with *Har. axyridis*: this signal increases both ant activity and the probability of the ants detecting the ladybird (Morales et al. 2008).

Consequently, non-myrmecophilous coccinellids frequently leave ant-attended aphid colonies, and may also **avoid laying eggs** in these patches (Oliver et al. 2008).

According to Sloggett and Majerus (2000), there is a **great variability in the degree of association** between non-myrmecophilous coccinellids and ants, co-occurrence being linked to an extreme food specialization and/or to a scarcity of non-attended colonies.

7.10.2 Myrmecophilous coccinellid species

Myrmecophilous ladybirds possess **morphological, behavioural or chemical traits** that efficiently protect them from attacks by ants (Pontin 1960, Majerus 1989, Voelkl 1995, 1997, Voelkl & Vohland 1996, Sloggett et al. 1998, Majerus et al. 2007). Some species are also able to **follow ant trails** (Godeau et al. 2003). These species are usually **restricted to ant-attended systems** during at least a part of their life cycle. The **benefits** of being able to exploit ant-attended prey include: (i) **avoiding competition with non-myrmecophilous competitors** and (ii) **avoiding predation/IGP/parasitism in an enemy-free space** (Guenard 2007).

For example, the myrmecophilous ladybird *Azya orbigera* is a predator of the green coffee scale, *Coccus viridis*. When these coccids are attended by the ant *Azteca instabilis*, **the ants attack the parasitoids** of *A. orbigera*, **reducing the parasitization rate** on the ladybird (Liere & Perfecto 2008).

The myrmecophilous *Scymnus posticalis* is never attacked by ants and is found within aphid colonies together with another ladybird, *Phymatosternus lewisii* (Kaneko 2007a). Furthermore, *S. posticalis* is an intraguild predator of the aphid parasitoid *Lysiphlebus japonicus* within ant-attended aphid colonies (Kaneko 2007b). This means that ants can **protect** myrmecophilous coccinellids (Voelkl 1995) and other intraguild prey such as furtive predators **against non-myrmecophilous** intraguild predators and/or intraguild parasitoids (Guenard 2007), but do not protect them against other myrmecophilous species.

7.11 APPLIED ASPECTS OF INTRAGUILD INTERACTIONS

7.11.1 Conservation

Intraguild interactions may affect **coccinellid co-existence** and consequently **coccinellid diversity** conservation. This is illustrated by **biological invasions** that occurred in North America, notably with *C. septempunctata*, *Har. axyridis*, and *P. quatuordecimpunctata*, but also *Hip. variegata* (Day et al. 1994, Coderre

et al. 1995, Brown & Miller 1998, Brown 2003, Turnock et al. 2003, Alyokhin & Sewell 2004, Lucas et al. 2007a, b). These **invasions** have greatly **modified** the structure and dynamics of the **coccinellid assemblages** (e.g. Evans 1991, Elliott et al. 1996, Horn 1996, Brown & Miller 1998, Michaud 2002, Turnock et al. 2003, Lucas et al. 2007b; but see Brown 2003). In Eastern Canada, for example, the **exotic species** *P. quatuordecimpunctata*, *C. septempunctata*, and *Har. axyridis* are **dominant** in most agricultural ecosystems that have aphids (Lucas et al. 2007b). By the end of the 1970s, the dominant species in Quebec maize fields was *Hip. tredecimpunctata tibialis*. Following the invasion of exotic ladybirds this species **disappeared completely** from maize ecosystem (Coderre & Tourneur 1986, Lucas et al. 2007b), but it is **still present in wild environments** (S. Laplante, unpublished).

In northeastern USA, **successive invasions** by the same exotic species caused a significant decline in *Hip. tredecimpunctata* and *C. transversoguttata* **abundances**, but increased coccinellid **diversity** (Alyokhin & Sewell 2004). A positive correlation between the densities of the three invaders was possibly a result of **biotic facilitation**. Similarly, populations of *Cycloneda sanguinea*, the dominant aphidophagous coccinellid in Florida citrus ecosystems, decreased in this system following *Har. axyridis* establishment (Michaud 2002). **Interactions** can also occur **between invasive species:** in the midwestern USA, *Har. axyridis* replaced *C. septempunctata*, allowing **the return** of several **native** coccinellid species to American orchards (previously excluded by *C. septempunctata*) (Horn 1996, Brown 1999, 2003).

If biological invasions are **detrimental to ladybird specialists**, they probably result in **local displacement of euryphagous species** (Evans 2004, Mills 2006). Evans (2004) verified the **shifting habitat hypothesis** in Utah alfalfa fields: native coccinellids abandoned the crop when an invading species (*C. septempunctata*) kept aphid populations at low densities, but returned to that crop when aphid densities were artificially increased (see Fig. 7.4).

In order **to co-exist, the intraguild prey should be more efficient in resource exploitation than the intraguild predator** (Holt & Polis 1997, Mylius et al. 2001, Arim & Marquet 2004, Borer et al. 2007). However, the predictions of most **models** are not valid in the field, due to the presence of more than two natural enemy species, of alternative resources, of other interactions such as cannibalism, of a temporal sequence of predators, of refuges, of anti-predatory mechanisms, or others (Okuyama & Ruyle 2003, Kindlmann & Houdkova 2006, Amarasekare 2007, Borer et al. 2007, Holt & Huxel 2007, Janssen et al. 2007, Rudolf 2007).

Obrycki et al. (1998b) proposed that **smaller coccinellid species** would have a **competitive advantage** over **larger** ones **at low aphid densities** because of their lower food requirements. **At higher aphid densities**, large species might have an advantage by IGP-interference. This may promote **co-existence** between species.

Which mechanisms cause changes in coccinellid assemblages? It is very difficult to clearly **establish the link** between **field studies** demonstrating modifications in terms of composition, abundance, and dynamics, and **laboratory studies** on interaction processes. The reality is complex, with changes involving **direct interactions** (e.g. IGP), **indirect interactions** (e.g. exploitative competition) and **interguild effects** (7.12).

Finally, **intraguild interactions may also prevent biological invasion**. It has been proposed that the failure of *A. bipunctata* to invade Japan (Sakuratani et al. 2000) could be due to **heavy mortality from IGP** by *Har. axyridis* and *C. septempunctata* (Kajita et al. 2006; but see Toda & Sakuratani 2006).

7.11.2 Biological control

Intraguild interactions may affect **biological control** of pests. Coccinellids act as **natural control agents** of herbivorous pests and several species are **used commercially** as **biocontrol agents**. Intraguild interactions may affect pest control **synergistically**, **additively**, or **antagonistically** and influence biocontrol mainly by: (i) **intraguild predation**, (ii) **predator facilitation** and (iii) **ant interference**.

7.11.2.1 Intraguild predation and biocontrol

In a **meta-analysis of literature**, the effects on herbivorous pest control depended on the type of IGP involved (Rosenheim & Harmon 2006).

Observations recorded **opposite effects of IGP by/ on coccinellids on biological control**. IGP may **disrupt** biological control by ladybirds. Adding the spider *M. tricuspidatus* (IGP of coccinellid larvae) generated a lower level of aphid control than by ladybirds alone (Yasuda & Kimura 2001). On the other hand, invasion of West Virginia (USA) apple orchards by *Har. axyridis* drastically modified the coccinellid community, but **improved** *Aphis spiraecola* **natural control** (Brown & Miller 1998). Snyder et al. (2004a), Aquilino et al. (2005) and Bilu and Coll (2007) suggest a **complementarity** of ladybirds and other natural enemies. Similarly, Weisser (2003) reported **an additive effect of coccinellids and parasitoids** against the pea aphid.

In an aphidophagous guild, IGP by *Har. axyridis* on *Aphidoletes aphidimyza* and *Chrysoperla carnea* did not affect the control of soybean aphids, *Aphis glycines*, either in the laboratory or in field cages (Costamagna et al. 2007, Gardiner & Landis 2007). The authors explained this by the fact that *Har. axyridis* and *C. septempunctata* had such a **strong impact on aphids** that **IGP did not disrupt biological control**.

In Florida citrus groves *Har. axyridis*, *Olla v-nigrum*, *Cycloneda sanguinea*, and *Exochomus childreni* were responsible for more than 95% mortality of immature stages of the psyllid parasitoid, *Tamarixia radiata* (Michaud 2004). Removing the ladybirds improved *Diaphorina citri* maturation success by 120-fold; IGP did not decrease biocontrol.

Despite high IGP in greenhouse cages by the coccinellid *Delphastus pusillus* on two parasitoids (*Encarsia formosa* and *Encarsia pergandiella*), **no disruption** of whitefly control occurred (Heinz & Nelson 1996).

When considering IGP between coccinellids and pathogens, the negative impact of the predator (consumption of infested pests) may be reduced by the **transportation of infective material** by the coccinellid (such as conidia) and **subsequent contamination** of uninfected hosts (Thomas et al. 2006).

7.11.2.2 Facilitation and biocontrol

Predator facilitation may occur when the activity of one predator increases the susceptibility of a shared prey to another predator (Losey & Denno 1998, 1999). In the complex of *Col. maculata*, *Har. axyridis*, and *Nabis* sp., the proportion of pea aphids consumed was increased by 0.14 when **enemy richness** increased from one to three, due to predator facilitation and potentially to a **decrease in intraspecific competition** (Aquilino et al. 2005). Nevertheless, predator facilitation may be rare in the field, since **three key elements are required** for this interaction:

(i) **synchrony** of the predatory species, (ii) **predator-induced escape behaviour** of the prey resulting in **habitat switching** and encounters with **new predators**, and (iii) **minimal negative interaction** between the predatory species (Losey & Denno 1999).

7.11.2.3 Ants and biocontrol

Ants exert **disruptive impact** on the biological control of honeydew-producing pests by coccinellids (Ceryngier & Hodek 1996, Kaplan & Eubanks 2002, Herbert & Horn 2008). For instance, the presence of the ant *Pheidole megacephala* prevented effective control of *Coccus viridis* by coccidophagous coccinellids, mainly *Azya luteipes* and *Curinus coeruleus* in Hawaiin coffee trees (Reimer et al. 1993).

The mutualism between ants and honeydew-producers leads to an increase of both taxa but also to a greater **suppression of other herbivorous species** such as caterpillars and phytophagous bugs, and may thus have an **overall beneficial effect** (Eubanks & Styrsky 2006).

7.11.2.4 Intraguild interactions and biocontrol approaches

Intraguild interactions involving ladybirds may have more or less significant impacts depending on the **type of biocontrol** implemented. In **classical biological control**, the main effect of intraguild interaction would be **biotic interference** between local and released agents (Goeden & Louda 1976, Stiling 1993). Introduced coccinellid species rarely lowered the pest control levels. For example, the **introductions** of *Har. axyridis*, *C. septempunctata* or *P. quatuordecimpunctata* in North America did not reduce the natural control of agricultural pests, despite strong effects on native ladybird species (Gardiner & Landis 2007, Lucas et al. 2007a), but see the case of *Aphis spiraecola* above. **As local species**, ladybirds can also interact with the released agent through competition and IGP. Releases of *Aphidius colemani* improved the control of soybean aphids, despite IGP by *Har. axyridis* and *Chrysoperla carnea* (Chacón et al. 2008).

Contrasting effects of IGP have been reported in **augmentative biocontrol**. IGP may influence **inundative biological control** when released agents (i) attack local intraguild members, (ii) are attacked by local intraguild members, (iii) are attacked by other released agents. Heavy predation upon released agents

(and **biocontrol disruption**) has been observed in some experiments (that do not involve ladybirds) (e.g. Rosenheim et al. 1993, 1999). By contrast, despite intense IGP, aphid **control was greatest** in treatments using both *Hip. convergens* and the parasitoid *L. testaceipes*, probably due to a **partial preference** of the coccinellid for unparasitized aphids (Colfer & Rosenheim 2001).

In a **conservation biocontrol** approach, most programmes tend to diversify the environment, in order to increase natural enemy **biodiversity/abundance** and/or to **anticipate their colonization** of focal crops (Thomas et al. 1992, Landis et al. 2000). A more complex environment may, for example, provide more **refuges** for intraguild prey and reduce **encounter frequencies** (Fincke & Denno 2002). Increasing the richness of natural enemies may also enable **predator facilitation**.

The **impact of management** on crop **colonization** by natural enemies is of critical importance. For example, **colonization sequence** may determine the **intraguild predator/prey status** of each guild member and consequently could lead to avoidance mechanisms. In rape, pea and wheat crops, the **aphidophagous guild** in the margin strips of stinging nettle differed from that of the crops (Alhmedi et al. 2007a, c): while *Har. axyridis* was more common in the **field margins** than in the crops, *C. septempunctata* showed the reverse pattern.

7.12 INTERGUILD EFFECTS

Because of a wide range of food, predaceous coccinellids **belong to different guilds**. Ladybirds respond **numerically** (Evans & Toler 2007; 5.3.5) or **functionally** (Lucas et al. 2004a; 5.3.2) to the simultaneous presence of several prey species. Intraguild interactions affecting coccinellid populations would thus affect the different guilds concerned. For example the honeydew produced by the aphid *Rhopalosiphum maidis* on maize plants in the laboratory drastically increased the longevity and parasitism performance of the parasitoid *Trichogramma ostriniae*. Since this parasitoid is released in **augmentative control** programmes against the European corn borer, any impact on aphid populations by aphidophagous predators (such as coccinellids) would decrease honeydew production and consequently reduce the **parasitoid efficiency** against the target pest (Fuchsberg et al.

2007). The situation is even more complicated since several predatory coccinellids belong both to the aphidophagous guild and to the guild that exploit corn borers (Musser & Shelton 2003). Additionally, *Trichogramma* spp. can theoretically parasitize aphidophagous insect eggs (Babendreier et al. 2003, Mansfield & Mills 2004).

7.13   CONCLUSION

In the previous books on ladybirds, no chapter was specifically dedicated to **intraguild interactions** (Hodek 1973, Hodek & Honěk 1996). Studies on interactions involving ladybirds focussed mainly on **vertical interactions** (such as predaceous coccinellids versus prey). In the last 20 years, **horizontal (intraguild) interactions** have been studied increasingly and it is now clear that these interactions dramatically influence the **composition, structure and dynamics** of guilds and consequently of entomological **communities**. Intraguild interactions have generated **ecological and evolutionary responses** by ladybirds such as **defensive responses**, etc.

Despite this recent interest, intraguild interactions remain difficult to study and many questions are pending. At the **methodological** level, intraguild interactions (specifically IGP) are **difficult to detect and quantify** in the field. **Traditional methods** in the laboratory, in field cages or in the field provide a rough estimate of the interactions. The **intensity** of the interaction (Hindayana et al. 2001), as well as its **direction** (Frechette et al. 2007), can change according to complexity and size of the experimental arena. Alternative, but very time-consuming methods include **gut dissection** (Triltsch 1997; 5.2.1) and **direct observations** (e.g. Rosenheim et al. 1999; Chapter 10) that provide **precise and realistic information** on the interaction and permit **impact quantification**. Fortunately, **powerful new tools** (such as molecular methods) have been developed (Chapter 10).

At the **conceptual** level, studies considering the complexity of the **whole system** are needed. It is crucial that future studies consider not only the intraguild organisms, but also other **co-occurring species**, other **potential prey**, and **higher-order natural enemies**. Since most coccinellid species studied are **generalist predators** (i.e. belong to multiple guilds) future studies should thus also consider **interguild effects**. Doing this may lead to a different interpretation of the impact of an interaction on **species coexistence** or on **biological control efficiency** (Eubanks & Styrsky 2006, Straub & Snyder 2006).

Future studies should also consider the **multiple co-occurring types of interactions**, e.g. IGP and defences and their **indirect effects**.

There is crucial need to consider intraguild interactions at **larger temporal** (covering several generations) and **spatial (landscape)** scales, and to consider **individual, population, community** and **ecosystem** levels. The real impact of an intraguild interaction remains very difficult to assess at a large spatial scale. Alternatively, it is speculative to link an **ecological phenomenon evaluated at a large scale** with a specific **intraguild interaction**. For example, it is extremely difficult to establish if a species is excluding another one via IGP at the landscape scale. Furthermore, most studies on intraguild interactions are carried out over **short temporal scales**.

Finally, even if **biotic** (e.g. biological invasion), **abiotic** (e.g. global climatic change) or **anthropic** factors (e.g. biological control programmes) have been studied by focussing on vertical interactions, their impact on intraguild interactions remains poorly understood.

Studies on intraguild interactions are **concentrated** on just a few systems and some guilds are poorly studied; the present chapter is consequently **biased toward aphidophagous guilds**. The literature available is also **biased toward laboratory results** and considers only a **few species**. Sloggett (2005) showed that, from 1995 to 2004, 76% of the available literature on intraguild relations of aphidophagous ladybirds was **concentrated on five species**: *C. septempunctata* 33%, *Har. axyridis* 19%, *Col. maculata* 12%, *A. bipunctata* 7% and *Hip. convergens* 5%. Thus there is a great need for studies on **less-studied** species or within **less-studied guilds**, and especially **in the field**.

An interesting result from Sloggett's analysis is that **before 1974, no study on intraguild interactions was available**, whereas 108 studies out of a total of 623 (>17%) were reported in the 1995–2004 period (Sloggett 2005).

To conclude, it can be claimed that intraguild interactions now **retain the attention of the scientific community** and generate a huge amount of literature. Intraguild interactions are fascinating and yet remain mainly *terra incognita*.

REFERENCES

Abrams, P. A., B. A. Menge, G. G. Mittlebach, D. A. Spiller and P. Yodzis. 1996. The role of indirect effects in food webs. *In* G. A. Polis and K. O. Winemiller (eds). *Food Webs: Integration of Patterns and Dynamics.* Chapman and Hall, New York. pp. 371–399.

Addicott, J. F. 1979. A multispecies aphid–ant association: density dependence and species–specific effects. *Can. J. Zool.* 57: 558–569.

Agarwala, B. K. and A. F. G. Dixon. 1992. Laboratory study of cannibalism and interspecific predation in ladybirds. *Ecol. Entomol.* 17: 303–309.

Agarwala, B. K. and A. F. G. Dixon. 1993. Kin recognition: egg and larval cannibalism in *Adalia bipunctata* (Coleoptera: Coccinellidae). *Eur. J. Entomol.* 90: 45–50.

Agarwala, B. K. and H. Yasuda. 2001. Overlaping oviposition and chemical defense of eggs in two co-occurring species of ladybird predators of aphids. *J. Ethol.* 19: 47–53.

Agarwala, B. K., S. Bhattacharya and P. Bardhanroy. 1998. Who eats whose eggs? Intra- versus inter-specific interactions in starving ladybird beetles predaceous on aphids. *Ethol. Ecol. Evol.* 10: 361–368.

Agarwala, B. K., H. Yasuda and Y. Kajita. 2003. Effect of conspecific and heterospecific feces on foraging and oviposition of two predatory ladybirds: role of fecal cues in predator avoidance. *J. Chem. Ecol.* 29: 357–376.

Alhmedi, A., F. Francis, B. Bodson and E. Haubruge. 2007a. Évaluation de la diversité des pucerons et de leurs ennemis naturels en grandes cultures à proximité de parcelles d'orties. *Notes fauniques Gembloux* 60(4): 147–152.

Alhmedi, A., F. Francis, B. Bodson and E. Haubruge. 2007b. Intraguild interactions of aphidophagous predators in fields: effect of *Coccinella septempunctata* and *Episyrphus balteatus* occurrence on aphid infested plants. *Comm. Agric. Appl. Biol. Sci.* 72(3) 381–390.

Alhmedi, A., E, Haubruge, B. Bodson and F. Francis. 2007c. Aphidophagous guilds on nettle (*Urtica dioica*) strips close to fields of green pea, rape and wheat. *Insect Sci.* 14: 419–424.

Alhmedi, A., E. Haubruge and F. Francis. 2010. Intraguild interactions implicating invasive species: *Harmonia axyridis* as a model species. *Biotechnol. Agron. Soc. Environ.* 14(1): 187–201.

Almohamad, R., F. J. Verheggen, F. Francis and E. Haubruge. 2010. Intraguild interactions between the predatory hoverfly *Episyrphus balteatus* (Diptera: Syrphidae) and the Asian ladybird, *Harmonia axyridis* (Coleoptera: Coccinellidae): effect of larval tracks. *Eur. J. Entomol.* 107: 41–45.

Alyokhin, A. and G. Sewell. 2004. Changes in lady beetle community following the establishment of three alien species. *Biol. Invasions* 6: 463–471.

Al-Zyoud, F. A. 2007. Prey species preferences of the predator *Serangium parcesetosum* Sicard (Col., Coccinellidae) and its interaction with another natural enemy. *Pak. J. Biol. Sci.* 10: 2159–2165.

Al-Zyoud, F. A., N. Tort and C. Segonca. 2005. Influence of leaf portion and plant species on the egg-laying behaviour of the predatory ladybird *Serangium parcesetosum* Sicard (Col., Coccinellidae) in the presence of a natural enemy. *J. Pest Sci.* 78: 167–174.

Amarasekare, P. 2007. Trade-offs, temporal variation, and species coexistence in communities with intraguild predation. *Ecology* 88: 2720–2728.

Aquilino, K. M., B. J. Cardinale and A. R. Ives. 2005. Reciprocal effects of host plant and natural enemy diversity on herbivore suppression: an empirical study of a model tritrophic system. *Oikos* 108: 275–282.

Arim, M. and P. A. Marquet. 2004. Intraguild predation: a widespread interaction related to species biology. *Ecol. Letters* 7: 557–564.

Attygalle, A. B., K. D. McCormick, C. L. Blankespoor, T. Eisner and J. Meinwald. 1993. Azamacrolides: a family of alkaloids from the pupal defensive secretion of a ladybird beetle (*Epilachna varivestis*). *Proc. Nat. Acad. Sci. USA* 90: 5204–5208.

Babendreier, D., M. Rostas, M. C. J. Hofte, S. Kuske and F. Bigler. 2003. Effects of mass releases of *Trichogramma brassicae* on predatory insects in maize. *Entomol. Exp. Appl.* 108: 115–124.

Bilu, E. and M. Coll. 2007. The importance of intraguild interactions to the combined effect of a parasitoid and a predator on aphid population suppression. *BioControl* 52: 753–763.

Bilu, E. and Coll, M. 2009. Parasitized aphids are inferior prey for a coccinellid predator: implications for intraguild predation. *Environ. Entomol.* 38: 153–158.

Borer, E. T, C. J. Briggs and R. D. Holt. 2007. Predators, parasitoids, and pathogens: a cross-cutting examination of intraguild predation theory. *Ecology* 88: 2681–2688.

Bradley, G. A. 1973. Effect of *Formica obscuripes* (Hymenoptera: Formicidae) on the predator–prey relationship between *Hyperaspis congressis* (Coleoptera: Coccinellidae) and *Toumeyella numismaticum* (Homoptera: Coccidae). *Can. Entomol.* 105: 1113–1118.

Brown, M. W. 1999. Temporal changes in the aphid predator guild in eastern North America. *Integrated Plant Protection in Orchards, IOBC/WPRS Bull.* 22(7): 7–11.

Brown, M. W. 2003. Intraguild responses of aphid predators on apple to the invasion of an exotic species, *Harmonia axyridis. BioControl* 48: 141–153.

Brown, M. W. 2004. Role of aphid predator guild in controlling spirea aphid populations on apple in West Virginia, USA. *Biol. Control* 29: 189–198.

Brown, M. W. and S. S. Miller. 1998. Coccinellidae (Coleoptera) in apple orchards of eastern West Virginia and the impact of invasion by *Harmonia axyridis*. *Entomol. News* 109: 143–151.

Burgio, G., A. Lanzoni, G. Accinelli and S. Maini. 2008. Estimation of mortality by entomophages on exotic *Harmonia axyridis* versus native *Adalia bipunctata* in semi-field conditions in northern Italy. *In* H. E. Roy and E. Wajnberg (eds). *From Biological Control to Invasion: the Ladybird* Harmonia axyridis *as a Model Species.* Springer, Netherlands. pp. 277–287.

Camarano, S., A. Gonzalez and C. Rossini. 2009. Biparental endowment of endogenous defensive alkaloids in Epilachna paenulata. *J. Chem. Ecol.* 35: 1–7.

Ceryngier, P. and I. Hodek. 1996. Enemies of Coccinellidae. *In* I. Hodek and A. Honěk (eds). *Ecology of Coccinellidae.* Kluwer Academic Publishers, Boston. pp. 319–350.

Chacón, J. M., D. A. Landis and G. E. Heimpel. 2008. Potential for biotic interference of a classical biological control agent of the soybean aphid. *Biol. Control* 46: 216–225.

Chang, G. C. 1996. Comparison of single versus multiple species of generalist predators for biological control. *Environ. Entomol.* 25: 207–622.

Charnov, E. L. 1976. Optimal foraging: the marginal value theorem. *Theor. Popul. Biology* 9: 126–136.

Chauhan, K. R. and D. C. Weber. 2008. Lady beetle (Coleoptera: Coccinellidae) tracks deter oviposition by the goldeneyed lacewing. *Chrysopa oculata. Biocontrol Sci. Technol.* 18: 727–731.

Chazeau, J. 1985. Predaceous insects. *In* W. Helle and M. W. Sabelis (eds). *World Crop Pests, Spider Mites. Their Biology, Natural Enemies and Control. Vol. 1B.* Elsevier, New York. pp. 211–246.

Clancy, K. M. and P. W. Price. 1987. Rapid herbivore growth enhances enemy attack: sublethal plant defenses remain a paradox. *Ecology* 68: 736–738.

de Clercq, P., I. Peters, G. Vergauwe and O. Thas. 2003. Interaction between *Podisus maculiventris* and *Harmonia axyridis*, two predators used in augmentative biological control in greenhouse crops. *BioControl* 48: 39–55.

Coderre, D. and J. C. Tourneur. 1986. Vertical distribution of aphids and aphidophagous insects on maize. In: Hodek I (ed) *Ecology of Aphidophaga*, Academia, Prague. pp. 291–296.

Coderre, D., E. Lucas and I. Gagné. 1995. The occurrence of *Harmonia axyridis* Pallas (Coleoptera: Coccinellidae) in Canada. *Can. Entomol.* 127: 609–611.

Coderre, D., L. Provencher and J. C. Tourneur. 1987. Oviposition and niche partitioning in aphidophagous insects on maize. *Can. Entomol.* 119: 195–203.

Colfer, R. G. and J. A. Rosenheim. 2001. Predation on immature parasitoids and its impact on aphid suppression. *Oecologia* 126: 292–304.

Colunga-Garcia, M., S. H. Gage and D. A. Landis. 1997. The response of an assemblage of Coccinellidae (Coleoptera) to a diverse agricultural landscape. *Environ. Entomol.* 26: 797–804.

Costamagna, A. C., D. A. Landis and C. D. Difonzo. 2007. Suppression of soybean aphid by generalist predators results in a trophic cascade in soybeans. *Ecol. Appl.* 17: 441–451.

Cottrell, T. E. 2004. Suitability of exotic and native lady beetle eggs (Coleoptera: Coccinellidae) for development of lady beetle larvae. *Biol. Control* 31: 362–371.

Cottrell, T. E. 2005. Predation and cannibalism of lady beetle eggs by adult lady beetles. *Biol. Control* 34: 159–164.

Cottrell, T. E. 2007. Predation by adult and larval lady beetles (Coleoptera: Coccinellidae) on initial contact with lady beetle eggs. *Environ. Entomol.* 36: 390–401.

Cottrell, T. E. and D. I. Shapiro-Ilan. 2003. Susceptibility of a native and an exotic lady beetle (Coleoptera: Coccinellidae) to *Beauveria bassiana. J. Invert. Pathol.* 2: 137–144.

Cottrell, T. E. and D. I. Shapiro-Ilan. 2008. Susceptibility of endemic and exotic North American ladybirds to endemic fungal entomopathogens. *Eur. J. Entomol.* 105: 455–460.

Cuenot, L. 1896. Sur la saignee reflexe et les moyens défense de quelques insectes. *Arch. Zool. Exp. Gen.* 4: 655–680.

Day, W. H., D. R. Prokrym, D. R. Ellis and R. J. Chianese. 1994. The known distribution of the predator *Propylea quatuordecimpunctata* (Coleoptera, Coccinellidae) in the United States, and thoughts on the origin of this species and five other exotic lady beetles in Eastern North America. *Entomol. News* 105: 244–256.

Dixon, A. F. G. 1958. The escape responses shown by certain aphids to the presence of the coccinellid *Adalia decempunctata* (L.). *Trans. R. Entomol. Soc. Lond.* 110: 319–334.

Dixon, A. F. G. 1959. An experimental study of the searching behaviour of the predatory coccinellid beetle *Adalia decempunctata* (L.). *J. Anim. Ecol.* 28: 259–281.

Dixon, A. F. G. 1985. *Aphid Ecology.* Blackie and Son Ltd., New York. 157 pp.

Dixon, A. F. G. 1987. The way of life of aphids: host specificity, speciation and distribution. *In* A. K. Minks and P. Hanewin (eds), *Word Crop Pests Aphids: Their Biology, Natural Enemies and Control. Vol.2A.* Elsevier, Amsterdam. pp. 197–207.

Dixon, A. F. G. 2000. *Insect Predator–Prey Dynamic: Ladybird Beetles and Biological Control.* Cambridge University Press, Cambridge, UK. 257 pp.

Dixon, A. F. G. 2007. Body size and resource partitioning in ladybirds. *Popul. Ecol.* 49: 45–50.

Dixon, A. F. G. and J. L. Hemptinne. 2001. Body size distribution in predatory ladybird beetles reflects that of their prey. *Ecology* 82: 1847–1856.

Dixon, A. F. G., V. Jarosik and A. Honěk. 2005. Thermal requirements for development and resource partitioning in aphidophagous guilds. *Eur. J. Entomol.* 102: 407–411.

Dolenska, M., O. Nedvěd, P. Vesely, M. Tesarova and R. Fuchs. 2009. What constitutes optical warning signals of ladybirds (Coleoptera: Coccinellidae) towards bird predators: colour, pattern or general look? *Biol. J. Linn. Soc.* 98: 234–242.

Doumbia, M., J. L. Hemptinne and A. F. G. Dixon. 1998. Assessment of patch quality by ladybirds: role of larval tracks. *Oecologia* 113: 197–202.

Drea, J. J. and R. D. Gordon. 1990. Coccinellidae. *In* D. Rosen (ed). *World Crop Pests, Armored Scale Insects. Vol. 4B*. Elsevier Science Publishers, New York. pp. 19–40.

Dyson, E. 1996. An investigation of *Calvia quatuordecimpunctata*: egg cannibalism, egg predation and evidence for a male-killer. Undergraduate dissertation, University of Cambridge.

Edmunds, M. 1974. *Defence in Animals*. Longman Inc., New York. 358 pp.

Eisner, T. and M. Eisner. 1992. Operation and defensive role of 'gin traps' in a coccinellid pupa (*Cycloneda sanguinea*). *Psyche* 99: 265–273.

Eisner, T., M. Goetz, D. Aneshansley et al. 1986. Defensive alkaloid in blood of Mexican bean beetle (*Epilachna varivestis*). *Experientia* 42: 204–207.

Elliott, N., R. Kieckhefer and W. Kauffman. 1996. Effects of an invading coccinellid on native coccinellids in an agricultural landscape. *Oecologia* 105: 537–544.

Engelstadter, J. and G. D. D. Hurst. 2007. The impact of male-killing bacteria on host evolutionary processes. *Genetics* 175: 245–254.

Eubanks, M. D. 2001. Estimates of the direct and indirect effects of red imported fire ants on biological control in field crops. *Biol. Control* 21: 35–43.

Eubanks, M. D. and J. D. Styrsky. 2006. Ant–hemipteran mutualism: keystone interactions that alter food web dynamics and influence plant fitness. *In* J. Brodeur and G. Boivin (eds). *Progress in Biological Control: Trophic and Guild Interactions in Biological Control*. Springer, Netherland. pp. 171–190.

Evans, E. W. 1991. Intra versus interspecific interactions of ladybeetles (Coleoptera: Coccinellidae) attacking aphids. *Oecologia* 87: 401–408.

Evans, E. W. 2004. Habitat displacement of North American ladybirds by an introduced species. *Ecology* 85: 637–665.

Evans, E. W. and T. R. Toler. 2007. Aggregation of polyphagous predators in response to multiple prey: ladybirds (Coleoptera: Coccinellidae) foraging in alfalfa. *Popul. Ecol.* 49: 29–36.

Felix, S. and A. O. Soares. 2004. Intraguild predation between the aphidophagous ladybird beetles *Harmonia axyridis* and *Coccinella undecimpunctata* (Coleoptera: Coccinellidae): the role of bodyweight. *Eur. J. Entomol.* 101: 237–242.

Ferguson, K. I. and P. Stiling. 1996. Non-additive effects of multiple natural enemies on aphid populations. *Oecologia* 108: 375–379.

Fincke, D. L. and R. F. Denno 2002. Intraguild predation diminished in complex-structured vegetation: implication for prey suppression. *Ecology* 83: 643–652.

Firlej, A., G. Boivin, E. Lucas and D. Coderre. 2005. First report of parasitism of *Harmonia axyridis* parasitism by *Dinocampus coccinellae* Schrank in Canada. *Biol. Invasions* 7: 553–556.

Firlej, A., E. Lucas, D. Coderre and G. Boivin. 2006. Teratocytes growth pattern reflects host suitability in a host–parasitoid assemblage. *Physiol. Entomol.* 32: 181–187.

Fischer, S., J. Samietz, F. L. Wäckers and S. Dorn. 2001. Interaction of vibrational and visual cues in parasitoid host location. *J. Comp. Physiol. (A)* 187: 785–791.

Flowers, R. W., S. M. Salom and L. T. Kok 2005. Competitive interactions among two specialist predators and a generalist predator of hemlock woolly adelgid, *Adelges tsugae* (HemipteraL Adelgidae), in the laboratory. *Environ. Entomol.* 34: 664–675.

Flowers, R. W., S. M. Salomb, L. T. Kokc and D. E. Mullins 2007. Behavior and daily activity patterns of specialist and generalist predators of the hemlock woolly adelgid, *Adelges tsugae. J. Insect Sci.* 7(44): 1–20.

Francke, D. L., J. P. Harmon, C. T. Harvey and A. R. Ives. 2008. Pea aphid dropping behavior diminishes foraging efficiency of a predatory ladybeetle. *Entomol. Exp. Appl.* 127: 118–124.

Frazer, B. D. 1988. Coccinellidae. *In* A. K. Minks, P. Harrewijn and W. Helle (eds). *World Crop Pests, Aphids. Vol. 2B*. Elsevier, New York. pp. 231–247.

Frechette, B., C. Alauzet and J.-L. Hemptinne. 2003. Oviposition behaviour of the two spots ladybird beetle *Adalia bipunctata* (L.) (Coleoptera: Coccinellidae) on plants with conspecific larval tracks. In: Soares AO, Ventura MA, Garcia V and Hemptinne J-L (eds). *Proc. 8th Int. Symp. Ecol. Aphidophaga. Arquipélago, Life Mar. Sci. (Suppl.)* 5: 73–77.

Frechette, B., S. Rojo, O. Alomar and E. Lucas. 2007. Intraguild predation among mirids and syrphids. Who is the prey and who is the predator? *BioControl* 52: 175–191.

Frechette, B., Larouche, F. and E. Lucas. 2008. *Leucopis annulipes* larvae (Diptera: Chamaemyiidae) use a furtive predation strategy within aphid colonies. *Eur. J. Entomol.* 105: 399–403.

Fuchsberg, J. R., T. H. Yong, J. E. Losey, M. E. Carter and M. P. Hoffmann. 2007. Evaluation of corn leaf aphid (*Rhopalosiphum maidis*; Homoptera : Aphididae) honeydew as a food source for the egg parasitoid *Trichogramma ostriniae* (Hymenoptera : Trichogrammatidae). *Biol. Control* 40: 230–236.

Gagne, I., D. Coderre and Y. Mauffette. 2002. Egg cannibalism by *Coleomegilla maculata lengi* neonates: preference even in the presence of essential prey. *Ecol. Entomol.* 27: 285–291.

Gagne, W. C. and J. L. Martin. 1968. The insect ecology of red pine plantations in central Ontario. V. The Coccinellidae (Coleoptera). *Can. Entomol.* 100: 835–846.

Gagnon, A. E. 2010. Prédation intraguilde chez les coccinellidae : impact sur la lutte biologique au puceron du soya. Unpubl. PhD dissertation, Université Laval, Québec, Canada.

Gardiner, M. M. and D. A. Landis. 2007. Impact of intraguild predation by adult *Harmonia axyridis* (Coleoptera: Coccinellidae) on *Aphis glycines* (Hemiptera: Aphididae) biological control in cage studies. *Biol. Control* 40: 386–395.

Giroux, S., J. C. Cote, C. Vincent, P. Martel and D. Coderre. 1994. Bacteriological insecticide M-One effects on predation efficiency and mortality of adult *Coleomegilla maculata lengi* (Coleoptera: Coccinellidae). *J. Econ. Entomol.* 87: 39–43.

Gittleman, J. L. and M. E. Gompper. 2005. Plight of predators: the importance of carnivores for understanding patterns of conservation and biodiversity and extinction risks. In Barbosa P and Castellanos I (eds) *Ecology of Predator–Prey interactions*. Oxford University Press, USA, pp. 370–388.

Godeau, J. F., J. L. Hemptinne and J. C. Verhaeghe. 2003. Ant trail: a highway for *Coccinella magnifica* Redtenbacher (Coleoptera: Coccinellidae). *In* A. O. Soares, M. A. Ventura, V. Garcia and J.-L. Hemptinne (eds). *Proc. 8th Int. Symp. Ecol. Aphidophaga. Arquipélago, Life Mar. Sci. (Suppl.)* 5: 79–83.

Goeden, R. D. and S. M. Louda. 1976. Biotic interference with insects imported for weed control. *Annu. Rev. Entomol.* 21: 325–342.

Guenard, B. 2007. Mutualisme fourmis pucerons et guilde aphidiphage associée : le cas de la prédation furtive. Unpubl. MSc dissertation, Université du Québec à Montréal, Québec, Canada.

Harmon, J. P. and D. A. Andow 2007. Behavioral mechanisms underlying ants' density-dependent deterrence of aphid-eating predators. *Oikos* 116: 1030–1036.

Harwood, J. D., H. J. S. Yoo, M. H. Greenstone, D. L. Rowley and R. J. O'Neil. 2009. Differential impact of adults and nymphs of a generalist predator on an exotic invasive pest demonstrated by molecular gut-content analysis. *Biol. Invasions* 11: 895–903.

Hautier, L., J.-C Grégoire, J. De Schauwers et al. 2008. Intraguild predation by *Harmonia axyridis* on coccinellids revealed by exogenous alkaloid sequestration. *Chemoecology* 18: 191–196.

Heinz, K. M. and J. M. Nelson 1996. Interspecific interactions among natural enemies of *Bemesia* in an inundative biological control program. *Biol. Control* 6: 384–393.

Hemptinne, J. L. and A. F. G. Dixon. 2000. Defence, oviposition and sex: semiochemical parsimony in two species of ladybird beetles (Coleoptera: Coccinellidae)? A short review. *Eur. J. Entomol.* 97: 443–447.

Hemptinne, J. L. and A. F. G. Dixon 2005. Intraguild predation in aphidophagous guilds: does it exist? *Proc. Int. Symposium Biol. Control Aphids Coccids. Tsuruoka, Japan.* (http://72.14.205.104/search?q=cache:ROmacSkS6_gJ:www.net.sfsi.co.jp/shoko-travel/symposium/symPDF/S5/Hemptinne.pdf+hemptinne+dixon+IGPandhl=frandct=clnkandcd=1andgl=ca.)

Hemptinne, J. L., A. F. G. Dixon and C. Gauthier 2000a. Nutritive cost of intraguild predation on eggs of *Coccinella septempunctata* and *Adalia bipunctata* (Coleoptera: Coccinellidae). *Eur. J. Entomol.* 97: 559–562.

Hemptinne, J. L., G. Lognay, C. Gauthier and A. F. G. Dixon 2000b. Role of surface chemical signals in egg cannibalism and intraguild predation in ladybirds (Coleoptera: Coccinellidae). *Chemoecology* 10: 123–128.

Herbert, J. J. and D. J. Horn. 2008. Effect of ant attendance by *Monomorium minimum* (Buckley) (Hymenoptera: Formicidae) on predation and parasitism of the soybean aphid *Aphis glycines* Matsumura (Hemiptera: Aphididae). *Environ. Entomol.* 37: 1258–1263.

Hindayana, D., R. Meyhofer, D. Scholz and H. M. Poehling 2001. Intraguild predation among the hoverfly *Episyrphus balteatus* de Geer (Diptera: Syrphidae) and other aphidophagous predators. *Biol. Control* 20: 236–246.

Hodek, I. 1973. *Biology of Coccinellidae*. Academia Press, Prague. 292 pp.

Hodek, I. 1996. Food relationships. *In* I. Hodek and A. Honěk (eds). *Ecology of Coccinellidae*. Kluwer Academic Publishers, Boston, USA. pp. 143–238.

Hodek, I. and A. Honěk. 1996. *Ecology of Coccinellidae*. Kluwer Academic Publishers, Netherlands. 480 pp.

Hodek, I. and J. P. Michaud. 2008. Why is *Coccinella septempunctata* so successful? (A point-of-view). *Eur. J. Entomol.* 105: 1–12.

Holloway, G. J., P. W. de Jong, P. M. Brakefield and H. de Vos. 1991. Chemical defence in ladybird beetles (Coccinellidae). I. Distribution of coccinelline and individual variation in defence in 7-spot ladybirds (*Coccinella septempunctata*). *Chemoecology* 2: 7–14.

Holt, R. D. 1977. Predation, apparent competition and the structure of prey communities. *Theor. Popul. Ecol.* 12: 197–229.

Holt, R. D. and G. R. Huxel. 2007. Alternative prey and the dynamics of intraguild predation: theoretical prespectives. *Ecology* 88: 2706–2712.

Holt, R. D. and G. A. Polis. 1997. A theoretical framework for intraguild predation. *Am. Nat.* 149: 745–764.

Holte, A. E., M. A. Houck and N. L. Collie. 2001. Potential role of parasitism in the evolution of mutualism in astigmatid mites: *Hemisarcoptes cooremani* as a Model. *Exp. Appl. Acarol.* 25: 97–107.

Honěk, A. and F. Kocourek. 1988. Thermal requirements for development of aphidophagous Coccinellidae (Coleoptera), Chrysopidae, Hemerobiidae (Neuroptera), and Syrphidae (Diptera): some general trends. *Oecologia* 76: 455–460.

Honěk, A. and F. Kocourek. 1990. Temperature and development time in insects: a general relationship between thermal constants. *Zool. Jahrb. Abt. Syst. Öekol. Geogr. Tiere* 117: 401–439.

Hoogendoorn, M. and G. E. Heimpel. 2002. Indirect interactions between and introduced and a native ladybird beetle species mediated by a shared parasitoid. *Biol. Control* 25: 224–230.

Hoogendoorn, M. and G. E. Heimpel. 2004. Competitive interactions between an exotic and a native ladybeetle: a field cage study. *Entomol. Exp. Appl.* 111: 19–28.

Horn, D. J. 1996. Impact of introduced Coccinellidae on native species in nontarget ecosystems. (Abstract.) Symposium of the IOBC working group 'Ecology of Aphidophaga'. University of Agriculture, Gembloux (unpaginated).

Houck, M. A. and B. M. O'Connor. 1991. Phoresy in the acariform acari. *Annu. Rev. Entomol.* 36: 611–636.

Hough-Goldstein, J., J. Cox and A. Armstrong. 1996. *Podisus maculiventris* (Hemiptera: Pentatomidae) predation on ladybird beetles (Coleoptera: Coccinellidae). *Fla. Entomol.* 79: 64–68.

Hurst, G. D. D., M. E. N. Majerus and A. Fain. 1997. Coccinellidae (Coleoptera) as vectors of mites. *Eur. J. Entomol.* 94: 317–319.

Hurst, G. D. D., J. H. Graf von der Schulenburg, T. M. O. Majerus et al. 2003. Invasion of one insect species, *Adalia bipunctata*, by two different male-killing bacteria. *Insect Mol. Biol.* 8: 133–139.

Iablokoff-Khnzorian, S. M. 1982. *Les coccinelles; Coléoptères–Coccinellidae*. Société Nouvelle des Éditions Boubée, Paris. 568 pp.

Iperti, G. 1964. Les parasites des coccinelles aphidiphages dans les Alpes-Maritimes et les Basses-Alpes. *Entomophaga* 9: 153–180.

Iperti, G. 1966. Comportement naturel des coccinelles aphidiphages du sud-est de la France: leur type de spécificité, leur action prédatrice sur *Aphis fabae* L. *Entomophaga* 11: 203–210.

Janssen, A., M. W. Sabelis, S. Magalhães, M. Montserrat and T. van der Hammen. 2007. Habitat structure affects intraguild predation. *Ecology* 88: 2713–2719.

de Jong, P. W., G. J. Holloway, P. M. Brakefield and H. de Vos. 1991. Chemical defence in ladybird beetles (Coccinellidae). II. Amount of reflex fluid, the alkaloid adaline and individual variation in defence in 2-spot ladybirds (*Adalia bipunctata*). *Chemoecology* 2: 15–19.

Kagata, H. and N. Katayama. 2006. Does nitrogen limitation promote intraguild predation in an aphidophagous ladybird? *Entomol. Exp. Appl.* 119: 239–246.

Kajita, Y., J. J. Obrycki, J. J. Sloggett and K. F. Haynes. 2010. Intraspecific alkaloid variation in ladybird eggs and its effects on con- and hetero-specific intraguild predators. *Oecologia* 163: 313–322.

Kajita, Y., F. Takano, H. Yasuda and B. K. Agarwala. 2000. Effects of indigenous ladybird species (Coleoptera: Coccinellidae) on the survival of an exotic species in relation to prey abundance. *Appl. Entomol. Zool.* 35: 473–479.

Kajita, Y., F. Takano, H. Yasuda and E. W. Evans. 2006. Interactions between introduced and native predatory ladybirds (Coleoptera, Coccinellidae): factors influencing the success of species introductions. *Ecol. Entomol.* 31: 58–67.

Kaneko, S. 2007a. Larvae of two ladybirds, *Phymatosternus lewisii* and *Scymnus posticalis* (Coleoptera: Coccinellidae),

exploiting colonies of the brown citrus aphid *Toxoptera citricidus* (Homoptera: Aphididae) attended by the ant *Pristomyrmex pungens* (Hymenoptera: Formicidae). *Appl. Entomol. Zool.* 2: 181–187.

Kaneko, S. 2007b. Predator and parasitoid attacking ant-attended aphids: effects of predator presence and attending ant species on emerging parasitoid numbers. *Ecol. Res.* 22: 451–458.

Kaplan, I. and M. D. Eubanks. 2002. Disruption of cotton aphid (Homoptera: Aphididae) – natural enemy dynamics by red imported fire ants (Hymenoptera: Formicidae). *Environ. Entomol.* 31: 1175–1183.

Kindlmann, P. and K. Houdkova. 2006. Intraguild predation: fiction or reality? *Popul. Ecol.* 48: 317–322.

Klingauf, F. 1967. Abwehr- und Meidereaktionen von Blattläusen (Aphididae) bein Bedrohung durch Räuber und parasiten. *Z. Angew. Entomol.* 60: 269–317.

Kovach, J. 2004. Impact of the multicolored Asian lady beetle as a pest of fruit and people. *Am. Entomol.* 50: 165–167.

Koyama, S. and M. E. N. Majerus. 2007. Interactions between the parasitoid wasp *Dinocampus coccinellae* and two species of coccinellid from Japan and Britain. *In* H. E. Roy and E. Wajnberg (eds). *From Biological Control to Invasion: The Ladybird* Harmonia axyridis *as a Model Species*. Springer, Netherlands. pp. 253–264.

Krieg, A., A. Huger, G. Langenbruch and W. Schnetter. 1984. New results on *Bacillus thuringiensis* var. *tenebrionis* with special regard to its effects on the Colorado potato beetle (*Leptinotarsa decemlineata*). *Anz. Schaedlkd. Pflanzenschutz Umweltschutz* 57: 145–150.

Kullenberg, B. 1944. Studien über die Biologie der Capsiden. *Zool. Bidrag Från Uppsala (Suppl.)* 23: 1–522.

Kunkel, H. and W. J. Kloft. 1985. Die Honigtauerzeuger des Waldes. *In* W. J. Kloft and H. Kunkel (eds). *Waldtracht und Waldhonig in der Imkerei*. Ehrenwirth, Munich. pp. 48–264.

Labrie, G., E. Lucas and D. Coderre. 2006. Can developmental and behavioral characteristics of the multicolored Asian lady beetle *Harmonia axyridis* explain its invasive success? *Biol. Invasions* 8: 743–754.

Labrie, G., D. Coderre and E. Lucas. 2008. Overwintering strategy of the multicolored Asian lady beetle (Coleoptera: Coccinellidae): cold-free space as a factor of invasive success. *Ann. Entomol. Soc. Am.* 101: 860–866.

Landis, D. A., S. D. Wratten and G. M. Gurr. 2000. Habitat management to conserve natural enemies of arthropod pests in agriculture. *Annu. Rev. Entomol.* 45: 175–201.

Lemire, S., D. Coderre, C. Vincent and G. Bélair. 1996. Lethal and sublethal effects of the entomogenous nematode, *Steinernema carpocapsae*, on the coccinellid *Harmonia axyridis*. *Nematropica* 26: 284–285.

Liere, H. and I. Perfecto. 2008. Cheating on a mutualism: indirect benefits of ant attendance to a coccidophagous coccinellid. *Ecol. Entomol.* 37: 143–149.

Lorenzetti, F., J. T. Arnason, B. J. R. Philogène and R. I. Hamilton. 1997. Différentiation spatiale de niche chez des aphidiphages prédateurs: la couleur de la plante comme critère de sélection. *BioControl* 42: 49–56.

Losey, J. E. and R. F. Denno. 1998. Positive predator–predator interactions: enhanced predation rates and synergistic suppression of aphid populations. *Ecology* 79: 2143–2152.

Losey, J. E. and R. F. Denno. 1999. Factors facilitating synergistic predation: the central role of synchrony. *Ecol. Appl.* 9: 378–386.

Lövei, G. L. and Z. A. Radwan. 1988. Seasonal dynamics and microhabitat distribution of coccinellid developmental stages in apple orchard. *In* E. Niemezyk and A. F. G. Dixon (eds). *Ecology and Effectiveness of Aphidophaga*. SPB Academic Publishing, The Hague. pp. 275–277.

Loevei, G. L., M. Sárospataki and Z. A. Radwan. 1991. Structure of ladybird (Coleoptera: Coccinellidae) assemblage in apple: changes through developmental stages. *Environ. Entomol.* 20: 1301–1308.

Lucas, E. 1998. How do ladybirds (*Coleomegilla maculata lengi*, Coleoptera: Coccinellidae) feed on green lacewing eggs (*Chrysoperla rufilabris*, Neuroptera: Chrysopidae). *Can. Entomol.* 130: 547–548.

Lucas, E. 2005. Intraguild predation among aphidophagaous predators. *Eur. J. Entomol.* 102: 351–364.

Lucas, E. and O. Alomar. 2002a. Impact of *Macrolophus caliginosus* presence on damage production by *Dicyphus tamaninii* (Heteroptera: Miridae) on tomato fruits. *J. Econ. Entomol.* 95: 1123–1126.

Lucas, E. and O. Alomar. 2002b. Impact of the presence of *Dicyphus tamaninii* Wagner (Heteroptera: Miridae) on whitefly (Homotera: Aleyrodidae) predation by *Macrolophus caliginosus* (Wagner) (Heteroptera: Miridae). *Biol. Control.* 25: 123–128.

Lucas, E. and J. Brodeur. 1999. Oviposition site selection by *Aphidoletes aphidimyza* Rondani (Diptera: Cecidomyiidae). *Environ. Entomol.* 28: 622–627.

Lucas, E. and J. Brodeur. 2001. A fox in a sheep-clothing: dilution effect for a furtive predator living inside prey aggregation. *Ecology* 82: 3246–3250.

Lucas, E. and J.A. Rosenheim. 2011. Influence of extraguild prey density on intraguild predation by heteropteran predators: A review of the evidence and a case study. *Biol. Control.* 59: 61–67.

Lucas, E., D. Coderre and J. Brodeur. 1997a. Instar-specific defense of *Coleomegilla maculata lengi* (Coccinellidae): influence on attack success of the intraguild predator *Chrysoperla rufilabris* (Chrysopidae). *Entomophaga* 42: 3–12.

Lucas, E., D. Coderre and C. Vincent 1997b. Voracity and feeding preferences of *Coccinella septempunctata* and *Harmonia axyridis* (Coleoptera: Cocinellidae) on *Tetranychus urticae* and *Aphis citricola*. *Entomol. Exp. Appl.* 85: 151–159.

Lucas, E., D. Coderre and J. Brodeur. 1998a. Intraguild predation among three aphid predators: Characterization and influence of extra-guild prey density. *Ecology* 79: 1084–1092.

Lucas, E., S. Lapalme and D. Coderre. 1998b. Voracité comparative de trois coccinelles aphidiphages sur le tétranyque rouge du pommier (Acarina: Tetranychidae). *Phytoprotection* 78: 117–123.

Lucas, E., D. Coderre and J. Brodeur. 2000. Selection of molting and pupating site by *Coleomegilla maculata lengi* Timberlake (Coleoptera: Coccinellidae): avoidance of intraguild predation? *Environ. Entomol.* 29: 454–459.

Lucas, E., I. Gagne and D. Coderre. 2002. Impact of the arrival of *Harmonia axyridis* on adults *Coccinella septempunctata* and *Coleomegilla maculata* (Coleoptera : Coccinellidae). *Eur. J. Entomol.* 99: 457–463.

Lucas, E., B. Frechette and O. Alomar. 2009. Resource quality, resource availability, and intraguild predation among omnivorous mirids. *Biocontrol Sci. Technol.* 19: 555–572.

Lucas, E., S. Demougeot, C. Vincent and D. Coderre. 2004a. Predation upon the oblique-banded leafroller, *Choristoneura rosaceana* (Lepidoptera: Tortricidae), by two aphidophagous coccinellids (Coleoptera: Coccinellidae) in the presence and absence of aphids. *Eur. J. Entomol.* 101: 37–41.

Lucas, E., S. Giroux, S. Demougeot, R. M. Duchesne and D. Coderre 2004b. Compatibility of a natural enemy, *Coleomegilla maculata lengi* (Coleoptera: Coccinellidae) and four insecticides used against the Colorado Potato Beetle (Coleoptera: Chrysomelidae). *J. Appl. Entomol.* 128: 233–239.

Lucas, E., C. Labrecque and D. Coderre. 2004c. *Delphastus catalinae* and *Coleomegilla maculata lengi* (Coleoptera: Coccinellidae) as biological control agents of the greenhouse whitefly, *Trialeurodes vaporariorum* (Homoptera: Aleyrodidae). *Pest Manag. Sci.* 60: 1073–1078.

Lucas, E., G. Labrie, C. Vincent and J. Kovach. 2007a. The multicoloured Asian ladybeetle: beneficial or nuisance organism? *In* C. Vincent, M. Goettel and G. Lazarovits (eds.). *Biological Control: A Global Perspective*. CABI International, UK. pp. 38–52.

Lucas, E., C. Vincent, G. Labrie et al. 2007b. The multicolored Asian ladybeetle *Harmonia axyridis* in Quebec agroecosystems ten years after its arrival? *Eur. J. Entomol.* 104: 737–743.

Magro, A., J. N. Tene, N. Bastin, A. F. G. Dixon and J. L. Hemptinne. 2007. Assessment of patch quality by ladybirds: relative response to conspecific and heterospecific larval tracks a consequence of habitat similarity? *Chemoecology* 17: 37–45.

Majerus, M. E. N. 1989. *Coccinella magnifica* (Redtenbacher): A myrmecophilous ladybird. *Br. J. Entomol. Nat. Hist.* 2: 97–106.

Majerus, M. E. N. 1994. *Ladybirds. New Naturalist Series 81*. Harper Collins, London. 368 pp.

Majerus, T. M. O., M. E. N. Majerus, B. Knowles et al. 1998. Extreme variation in the prevalence of inherited male-killing microorganisms between three populations of

Harmonia axyridis (Coleoptera: Coccinellidae). *Heredity* 81: 683–691.

Majerus, M. E. N., J. J. Slogett, J. F. Godeau and J. L. Hemptinne. 2007. Interactions between ants and aphidophagous and coccidophagous ladybirds. *Popul. Ecol.* 49: 15–27.

Majerus, M. E. N., V. Strawson and H. Roy. 2006. The potential impacts of the arrival of the harlequin ladybird, *Harmonia axyridis* (Pallas) (Coleoptera: Coccinellidae), in Britain. *Ecol. Entomol.* 31: 207–215.

Mallampalli, N., I. Castellanos and P. Barbosa. 2002. Evidence for intraguild predation by *Podisus maculiventris* on a lady-beetle, *Coleomegilla maculata*: implications for biological control of Colorado potato beetle, *Leptinotarsa decemlineata*. *BioControl* 47: 387–398.

Mansfield, S. and N. J. Mills. 2004. A comparison of method-ologies for the assessment of host preference of the gregari-ous egg parasitoid *Trichogramma platneri*. *Biol. Control* 29: 332–340.

McLure, M. S. 1986. Role of predators in regulation of endemic populations of *Matsucoccus matsumurae* (Homop-tera: Margarodidae) in Japan. *Environ. Entomol.* 15: 976–983.

Michaud, J. P. 2002. Invasion of the Florida citrus ecosystem by *Harmonia axyridis* (Coleoptera: Coccinellidae) and asym-metric competition with a native species, *Cycloneda san-guinea*. *Environ. Entomol.* 31: 827–835.

Michaud, J. P. 2004. Natural mortality of Asian psyllid (Homoptera: Psyllidae) in central Florida. *Biol. Control* 29: 260–269.

Michaud, J. P. and A. K. Grant. 2003. Intraguild predation among ladybeetles and a green lacewing: do the larval spines of *Curinus coeruleus* (Coleoptera: Coccinellidae) serve as a defensive function? *Bull. Entomol. Res.* 93: 499–505.

Michaud, J. P. and A. K. Grant. 2004. Adaptive significance of sibling egg cannibalism in Coccinellidae: comparative evidence from three species. *Ann. Entomol. Soc. Am.* 97: 710–719.

Michaud, J. P. and J. L. Jyoti. 2007. Repellency of conspecific and heterospecific larval residues to *Hippodamia convergens* (Coleoptera: Coccinellidae) ovipositing on sorghum plants. *Eur. J. Entomol.* 104: 399–405.

Mills, N. 2006. Interspecific competition among natural enemies and single versus multiple introductions in biologi-cal control. *In* J. Brodeur and G. Boivin (eds) *Progress in Biological Control: Trophic and Guild Interactions in Biological Control.* Springer, The Netherlands. pp. 191–220.

Morales, M. A., J. L. Barone and C. S. Henry. 2008. Acoustic alarm signalling facilitates predator protection of treehop-pers by mutualist ant bodyguards. *Proc. R. Soc. Lond. (Biol)* 275: 1935–1941.

Mracek, Z. and Z. Ruzicka. 1990. Infectivity and development of *Steinernema sp.* strain Hylobius (Nematoda, Stein-ernematidae) in aphids and aphidophagous coccinellids. *J. Appl. Entomol.* 110: 92–95.

Musser, F. R. and A. M. Shelton. 2003. Predation of *Ostrinia nubilalis* (Lepidoptera: Crambidae) eggs in sweet corn by generalist predators and the impact of alternative foods. *Environ. Entomol.* 32: 1131–1138.

Mutsu, M., N. Kilinçer, S. Ulgenturk and M. B. Kaydan. 2008. Feeding behavior of *Cryptolaemus montrouzieri* on mealy-bugs parasitized by *Anagyrus pseudococci*. *Phytoparasitica* 36: 360–367.

Mylius, S. D., K. Klumpers, A. M. de Roos and L. Persson. 2001. Impact of intraguild predation and stage structure on simple communities along a productivity gradient. *Am. Nat.* 158: 259–276.

Nakashima, Y., M. A. Birkett, B. J. Pye, J. A. Pickett and W. Powell. 2004. The role of semiochemicals in the avoidance of the seven-spot ladybird, *Coccinella septempunctata*, by the aphid parasitoid, *Aphidius ervi*. *J. Chem. Ecol.* 30: 1103–1116.

Nakashima, Y., M. A. Birkett, B. J. Pye and W. Powell. 2006. Chemically mediated intraguild predator avoidance by aphid parasitoids: interspecific variability in sensitivity to semiochemical trails of ladybird predators. *J. Chem. Ecol.* 32: 1989–1998.

Nelson, E. H. and J. A. Rosenheim. 2006. Encounters between aphids and their predators: the relative frequencies of dis-turbance and consumption. *Entomol. Exp. Appl.* 118: 211–219.

Noia, M., I. Borges and A. O. Soares. 2008. Intraguild preda-tion between the aphidophagous ladybird beetles *Harmonia axyridis* and *Coccinella undecimpunctata* (Coleoptera: Coc-cinellidae): the role of intra and extraguild prey densities. *Biol. Control* 46: 140–146

Norton, A. P., G. English-Loeb and E. Belden. 2001. Host plant manipulation of natural enemies: leaf domatia protect ben-eficial mites from insect predators. *Oecologia* 126: 535–542.

Obrycki, J. J. and M. J. Tauber. 1985. Seasonal occurrence and relative abundance of aphid predators and parasitoids on pubescent potato plants. *Can. Entomol.* 117: 1231–1237.

Obrycki, J. J., K. L. Giles and A. M. Ormord. 1998a. Experimen-tal assessment of interactions between larval *Coleomegilla maculata* and *Coccinella septempunctata* (Coleoptera: Coc-cinellidae) in field cages. *Environ. Entomol.* 27: 1280–1288.

Obrycki, J. J., K. L. Giles and A. M. Ormord. 1998b. Interac-tions between an introduced and indigenous coccinellid species at different prey densities. *Oecologia* 117: 279–285.

O'Connor, B. M. and M. A. Houck. 1989. Two new genera of Hemisarcoptidae (Acari: Astigmata) from the Huron Mountains of Northern Michigan. *Gt Lakes Entomol.* 22: 1–10.

Okuyama, T. and R. L. Ruyle. 2003. Analysis of adaptive for-aging in an intraguild predation system. *Web Ecol.* 4: 1–6.

Oliver, T. H., J. E. L. Timms, A. Taylor and S. R. Leather. 2006. Oviposition responses to patch quality in the larch ladybird

Aphidecta obliterata (Coleoptera: Coccinellidae): effects of aphid density, and con- and heterospecific tracks. *Bull. Entomol. Res.* 96: 25–34.

Oliver, T. H., I. Jones, J. M. Cook and S. R. Leather. 2008. Avoidance responses of an aphidophagous ladybird *Adalia bipunctata*, to aphid-tending ants. *Ecol. Entomol.* 33: 523–528.

Omkar, G. M., A. Pervez and A. K. Gupta. 2004. Role of surface chemicals in egg cannibalism and intraguild predation by neonates of two aphidophagous ladybirds, *Propylea dissecta* and *Coccinella transversalis*. *J. Appl. Entomol.* 128: 691–695.

Omkar, G. M., A. Pervez, A. K. Gupta. 2007. Sibling cannibalism in aphidophagous ladybirds: its impact on sex-dependent development and body weight. *J. Appl. Entomol.* 131: 81–84.

Osawa, N. 1992. Effect of pupation site on pupal cannibalism and parasitism in the ladybird beetle *Harmonia axyridis* Pallas (Coleoptera, Coccinellidae). *Jap. J. Entomol.* 60: 131–135.

Paine, R. T. 1969. A note on trophic complexity and community stability. *Am. Nat.* 103: 91–93.

Pasteels, J. M. 2007. Chemical defence, offence and alliance in ants–aphids–ladybirds relationships. *Popul. Ecol.* 49: 5–14.

Pasteels, J. M., C. Deroe, B. Tursch et al. 1973. Distribution et activités des alcaloïdes défensifs des Coccinellidae. *J. Insect Physiol.* 19: 1771–1784.

Pell, J. K., R. Pluke, S. J. Clark, M. G. Kenward and P. G. Alderson. 1997. Interactions between two aphid natural enemies, the entomopathogenic fungus *Erynia neoaphidis* Remaudière and Hennebert (Zygomycetes: Entomophthorales) and the predatory beetle *Coccinella septempunctata* L. (Coleoptera: Coccinellidae). *J. Invert. Pathol.* 69: 261–268.

Pell, J. K., J. Baverstock, H. E. Roy, R. L. Ware and M. E. N. Majerus. 2008. Intraguild predation involving *Harmonia axyridis* : a review of current knowledge and future perspectives. *In* H. E. Roy and E. Wajnberg (eds). *From Biological Control to Invasion: The Ladybird* Harmonia axyridis *as a Model Species*. Springer, Netherlands. pp. 147–168.

Pervez, A. and Omkar. 2006. Ecology and biological control application of multicolored asian ladybird, *Harmonia axyridis*: a review. *Biocontrol Sci. Tech.* 16: 111–128.

Phoofolo, M. W. and J. J. Obrycki. 1998. Potential for intraguild predation and competition among predatory Coccinellidae and Chrysopidae. *Entomol. Exp. Appl.* 89: 47–55.

Polis, G. A. and R. D. Holt. 1992. Intraguild predation: the dynamics of complex trophic interactions. *Trends Ecol. Evol.* 7: 151–154.

Polis, G. A., C. A. Myers and R. D. Holt. 1989. The ecology and evolution of intraguild predation: Potential competitors that eat each other. *Annu. Rev. Ecol. Syst.* 20: 297–330.

Pontin, A. J. 1960. Some records of predators and parasites adapted to attack aphids attended by ants. *Entomol. Mon. Mag. Lond.* 95: 154–155.

Pope, R. D. 1979. Wax production by coccinellid larvae (Coleoptera). *Syst. Entomol.* 4: 171–196.

Provost, C., D. Coderre, E. Lucas, G. Chouinard and N. Bostanian. 2003. Impact d'une dose sublétale de lambda-cyhalothrin sur les prédateurs intraguildes d'acariens phytophages en vergers de pommiers. *Phytoprotection* 84: 105–113.

Provost, C., D. Coderre, E. Lucas, G. Chouinard and N. Bostanian. 2005. Impact of Intraguild predation and lambda-cyhalothrin on predation efficiency of three acarophagous predators. *Pest Manag. Sci.* 61: 532–538.

Provost, C., E. Lucas, D. Coderre and G. Chouinard. 2006. Prey selection by the lady beetle *Harmonia axyridis*: the influence of prey mobility and prey species. *J. Insect Behav.* 19: 265–277.

Quezada, J. R. and P. DeBach. 1974. Bioecological and population studies of the cottony cushion scale, *Icerya purchasi* Mask. and its natural enemies, *Rodolia cardinalis* Mulsant and *Chryptochaetum iceryae* Will., in Southern California. *Hilgardia* 41: 631–688.

Raymond, B., A. C. Darby and A. E. Douglas. 2000. Intraguild predators and the spatial distribution of a parasitoid. *Oecologia* 124: 367–372.

Rieder, J. P., T. A. S. Newbold, S. Sato, H. Yasuda and E. W. Evans. 2008. Intra-guild predation and variation in egg defence between sympatric and allopatric populations of two species of ladybird beetles. *Ecol. Entomol.* 33: 53–58.

Reimer, N. J., M. L. Cope and G. Yasuda. 1993. Interference of *Pheidole megacephala* (Hymenoptera: Formicidae) with biological control of *Coccus viridis* (Homoptera: Coccidae) in coffee. *Environ. Entomol.* 22: 483–488.

Richards, A. M. 1985. Biology and defensive adaptations in *Rodatus major* (Coleoptera: Coccinellidae) and its prey, *Monophlebulus pilosior* (Hemiptera: Margarodidae). *J. Zool.* 205: 287–295.

Roger, C., D. Coderre, C. Vigneault and G. Boivin. 2001. Prey discrimination by a generalist coccinellid predator: effect of prey age or parasitism? *Ecol. Entomol.* 26: 163–172.

Rosenheim, J. A. and J. P. Harmon. 2006. The influence of intraguild predation on the suppression of a shared prey population: an empirical reassessment. *In* J. Brodeur and G. Boivin (eds). *Progress in Biological Control: Trophic and Guild Interactions in Biological Control*. Springer, Netherland. pp. 1–20.

Rosenheim, J. A., R. Wilhoit and C. A. Armer. 1993. Influence of intraguild predation among generalist insect predators on the suppression of an herbivore population. *Oecologia* 96: 439–449.

Rosenheim, J. A., H. K. Kaya, L. E. Ehler, J. J. Marois and B. A. Jaffee. 1995. Intraguild predation among biological control agents: theory and evidence. *Biol. Control* 5: 303–335.

Rosenheim, J. A., D. D. Limburg and R. G. Colfer. 1999. Impact of generalist predators on a biological control agent, *Chrysoperla carnea*: direct observations. *Ecol. Appl.* 9: 409–417.

Roy, H. E. and J. K. Pell. 2000. Interactions between entomopathogenic fungi and other natural enemies: implications for biological control. *Biocontrol Sci. Tech.* 10: 737–752.

Roy, H. E., J. K. Pella, S. J. Clark and P. G. Aldersonb. 1998. Implications of predator foraging on aphid pathogen dynamics. *J. Invert. Pathol.* 71: 236–247.

Roy, H. E., J. K. Pell and P. G. Alderson. 2001. Targeted dispersal of the aphid pathogenic fungus *Erynia neoaphidis* by the aphid predator *Coccinella septempunctata*. *Biocontrol Sci. Tech.* 11: 99–110.

Roy, H. E., P. G. Alderson and J. K. Pella. 2003. Effect of spatial heterogeneity on the role of *Coccinella septempunctata* as an intra-guild predator of the aphid pathogen *Pandora neoaphidis*. *J. Invert. Pathol.* 82: 85–95.

Roy, H. E., J. Baverstock, R. L. Ware et al. 2008a. Intraguild predation of the aphid pathogenic fungus *Pandora neoaphidis* by the invasive coccinellid *Harmonia axyridis*. *Ecol. Entomol.* 33: 175–182.

Roy, H. E., J. Brown, P. Rothery, R. L. Ware and M. E. N. Majerus 2008b. Interactions between the fungal pathogen *Beauvaria bassiana* and three species of coccinellid: *Harmonia axyridis*, *Coccinella septempunctata* and *Adalia bipunctata*. In H. E. Roy and E. Wajnberg (eds). *From Biological Control to Invasion: the Ladybird Harmonia axyridis as a Model Species*. Springer, The Netherlands. pp. 265–276.

Royer, T. A., K. L. Giles, M. M. Lebusa and M. E. Payton. 2008. Preference and suitability of greenbug, *Schizaphis graminum* (Hemiptera: Aphididae) mummies parasitized by *Lysiphlebus testaceipes* (Hymenoptera: Aphidiidae) as food for *Coccinella septempunctata* and *Hippodamia convergens* (Coleoptera: Coccinellidae). *Biol. Control* 47: 82–88.

Rudolf, V. H. 2007. The interaction of cannibalism and omnivory: consequences for community dynamics. *Ecology* 88: 2697–2705.

Ruzicka, Z. 1997. Recognition of oviposition-deterring allomones by aphidophagous predators (Neuroptera: Chrysopidae, Coleoptera: Coccinellidae). *Eur. J. Entomol.* 94: 431–434.

Ruzicka, Z. 2001a. Oviposition responses of aphidophagous coccinellids to tracks of ladybird (Coleoptera: Coccinellidae) and lacewing (Neuroptera: Chrysopidae) larvae. *Eur. J. Entomol.* 98: 183–188.

Ruzicka, Z. 2001b. Responses of chrysopids (Neuroptera) to larval tracks of aphidophagous coccinellids (Coleoptera). *Eur. J. Entomol.* 98: 283–285.

Ruzicka, Z. 2003. Perception of oviposition-deterring larval tracks in aphidophagous coccinellids *Cycloneda limbifer* and *Ceratomegilla undecimnotata* (Coleoptera: Coccinellidae). *Eur. J. Entomol.* 100: 345–350.

Ruzicka, Z. 2006. Oviposition-deterring effects of conspecific and heterospecific larval tracks on *Cheilomenes sexmaculata* (Coleoptera: Coccinellidae). *Eur. J. Entomol.* 103: 757–763.

Ruzicka, Z. and J. Havelka. 1998. Effects of oviposition-deterring pheromone and allomones on *Aphidoletes aphidimyza* (Diptera: Cecidomyiidae). *Eur. J. Entomol.* 95: 211–216.

Ruzicka, Z. and R. Zemek. 2003. Effects of conspecific and heterospecific larval tracks on mobility and searching patterns of *Cycloneda limbifer* Say (Coleoptera: Coccinellidae) females. In A. O. Soares, M. A. Ventura, V. Garcia and J-.L. Hemptinne (eds). *Proc. 8th Int. Symp. Ecol. Aphidophaga. Arquipélago, Life Mar. Sci. (Suppl.)* 5: 85–93.

Sakuratani, Y., Y. Matsumoto, M. Oka et al. 2000. Life history of *Adalia bipunctata* (Coleoptera: Coccinellidae) in Japan. *Eur. J. Entomol.* 97: 555–558.

Santi, F. and S. Maini. 2006. Predation upon *Adalia bipunctata* and *Harmonia axyridis* eggs by *Chrysoperla carnea* larvae and *Orius laevigatus* adults. *Bull. Insectol.* 59: 53–58.

Sato, S. and A. F. G. Dixon. 2004. Effect of intraguild predation on the survival and development of three species of aphidophagous ladybirds: consequences for invasive species. *Agric. Forest Entomol.* 6: 21–24.

Sato, S., A. F. G. Dixon and H. Yasuda. 2003. Effect of emigration on cannibalism and intraguild predation in aphidophagous ladybirds. *Ecol. Entomol.* 28: 628–633.

Sato, S., H. Yasuda and E. W. Evans. 2005. Dropping behaviour of larvae of aphidophagous ladybirds and its effects on incidence of intraguild predation: interactions between the intraguild prey, *Adalia bipunctata* (L.) and *Coccinella septempunctata* (L.), and the intraguild predator, *Harmonia axyridis* Pallas. *Ecol. Entomol.* 30: 220–224.

Sato, S., R. Jimbo, H. Yasuda and A. F. G. Dixon. 2008. Cost of being an intraguild predator in predatory ladybirds. *Appl. Entomol. Zool.* 43: 143–147.

Sato, S., K. Kushibuchi and H. Yasuda. 2009. Effect of reflex bleeding of a predatory ladybird beetle, Harmonia axyridis (Pallas) (Coleoptera: Coccinellidae), as a means of avoiding intraguild predation and its cost. *Appl. Entomol. Zool.* 44: 203–206.

Schellhorn, N. A. and D. A. Andow. 1999a. Cannibalism and interspecific predation: role of oviposition behavior. *Ecol. Appl.* 9: 418–428.

Schellhorn, N. A. and D. A. Andow. 1999b. Mortality of coccinellid (Coleoptera: Coccinellidae) larva and pupae when prey become scarce. *Popul. Ecol.* 28: 1092–1100.

Seagraves, M. P. 2009. Lady beetle oviposition behavior in response to the trophic environment. *Biol. Control* 51: 313–322.

Sengonca, C. and B. Frings. 1985. Interference and competitive behaviour of the aphid predators, *Chrysoperla carnea* and *Coccinella septempunctata* in the laboratory. *Entomophaga* 30: 245–251.

Shapiro-Ilan, D. I. and T. E. Cottrell 2005. Susceptibility of lady beetles (Coleoptera: Coccinellidae) to entomopathogenic nematodes. *J. Invert. Pathol.* 2: 150–156.

Sloggett, J. J. 2005. Are we studying too few taxa? Insights from aphidophagous ladybird beetles (Coleoptera: Coccinellidae). *Eur. J. Entomol.* 102: 391–398.

Sloggett, J. J. 2008. Weighty matters: body size, diet and specialization in aphidophagous ladybird beetles (Coleoptera: Coccinellidae). *Eur. J. Entomol.* 105: 381–389.

Sloggett, J. J. and M. E. N. Majerus 2000. Habitat preferences and diet in the predatory Coccinellidae (Coleoptera): an evolutionary perspective. *Biol. J. Linn. Soc.* 70: 63–88.

Sloggett, J. J., R. A. Wood and M. E. N. Majerus. 1998. Adaptations of *Coccinella magnifica* Redtenbacher, a myrmecophilous coccinellid, to aggression by wood ants (*Formica rufa* group). I. Adult behavioral adaptation, its ecological context and evolution. *J. Insect Behav.* 11: 889–904.

Sluss, T. P. and B. A. Foote. 1973. Biology and immature stages of *Leucopis verticalis* (Diptera: Chamaemyiidae). *Can. Entomol.* 103: 1427–1434.

Snyder, W. E., S. N. Ballard, S. Yang et al. 2004a. Complementary biocontrol of aphids by the ladybird beetle *Harmonia axyridis* and the parasitoids *Aphelinus asychis* on greenhouse roses. *Biol. Control.* 30: 229–235.

Snyder, W. E., G. M. Clevenger and S. D. Eigendrode. 2004b. Intraguild predation and successful invasion by introduced ladybird species. *Oecologia* 140: 559–565.

Soares, A. O. and A. Serpa. 2007. Interference competition between ladybird beetle adults (Coleoptera: Coccinellidae): effects on growth and reproductive capacity. *Popul. Ecol.* 49: 37–43.

Soares, A. O., I. Borges, P. A. V. Borges, G. Labrie and E. Lucas. 2008. *Harmonia axyridis*: what will stop the invader? *In* H. E. Roy and E. Wajnberg (eds). *From Biological Control to Invasion: The Ladybird* Harmonia axyridis *as a Model Species.* Springer, The Netherlands. pp. 127–146.

Stiling, P. 1993. Why do natural enemies fail in classical biological control programs? *Am. Entomol.* 39: 31–37.

Straub, C. S. and W. E. Snyder. 2006. Experimental approaches to understanding the relationship between predator biodiversity and biological control. *In* J. Brodeur and G. Boivin (eds) *Progress in Biological Control: Trophic and Guild Interactions in Biological Control.* Springer, The Netherlands. pp. 221–240.

Takahashi, K. 1989. Intra- and inter-specific predations by lady beetles in spring alfalfa fields. *Jap. J. Entomol.* 57: 199–203.

Tedders, W. L., C. C. Reilly, B. W. Wood, R. K. Morrison and C. S. Lofgren. 1990. Behavior of *Solenopsis invicta* (Hymenoptera: Formicidae) in pecan orchards. *Environ. Entomol.* 19: 44–53.

Thomas, M. B., S. D. Wratten and N. W. Sotherton 1992. Creation of 'island' habitats in farmland to manipulate populations of beneficial arthropods: predator densities and species composition. *J. Appl. Ecol.* 29: 524–531.

Thomas, M. B., S. P. Arthurs and E. L. Watson. 2006. Trophic and guild interactions and the influence of multiple species on disease. *In* J. Brodeur and G. Boivin (eds).

Progress in Biological Control: Trophic and Guild Interactions in Biological Control. Springer, The Netherlands: pp. 101–122.

Tinsley, M. C. and M. E. N. Majerus 2007. Small steps or giant leaps for male-killers? Phylogenetic constraints to male-killer host shifts. *BMC Evol. Biol.* 7: 238.

Toda, Y. and Y. Sakuratani. 2006. Expansion of geographical distribution of an exotic ladybird beetle, *Adalia bipunctata* (Coleoptera: Coccinellidae), and its interspecific relationships with native ladybird beetles in Japan. *Ecol. Res.* 21: 292–300.

Todorova, S. I., D. Coderre and J. C. Cote. 2000. Pathogenicity of *Beauvaria bassiana* isolates toward *Leptinotarsa decemlineata* (Coleoptera: Chrysomelidae), *Myzus persicae* (Homoptera: Aphididae) and their predator *Coleomegilla maculata lengi* (Coleoptera: Coccinellidae). *Phytoprotection* 81: 15–22.

Todorova, S. I., J. C. Cote and D. Coderre. 1996. Evaluation of the effects of two *Beauveria bassiana* (Balsamo) Vuillemin strains on the development of *Coleomegilla maculata lengi* Timberlake (Col., Coccinellidae). *J. Appl. Entomol.* 120: 159–163.

Triltsch, H. 1997. Gut contents in field sampled adults of *Coccinella septempunctata* (Col.: Coccinellidae). *Entomophaga* 42: 125–131.

Turnock, W. J., I. L. Wise and F. O. Matheson 2003. Abundance of some native coccinellines (Coleoptera: Coccinellidae) before and after the appearance of *Coccinella semptempunctata. Can. Entomol.* 135: 391–404.

Vanhove, F. 1998. Impact de la défense des oeufs sur la structure des communautés de Coccinellidae. Unpubl. Master dissertation, University of Gembloux, Begium.

Voelkl, W. 1992. Aphids or their parasitoids: who actually benefits from ant-attendance? *J. Anim. Ecol.* 61: 273–281.

Voelkl, W. 1995. Behavioral and morphological adaptations of the coccinellid, *Platynaspis luteorubra* for exploiting ant-attended resources (Coleoptera: Coccinellidae). *J. Insect Behav.* 8: 653–670.

Voelkl, W. 1997. Interactions between ants and aphid parasitoids: patterns and consequences for resource utilization. *Ecol. Stud.* 130: 225–240.

Voelkl, W. and K. Vohland. 1996. Wax cover in larvae of *Scymnus* species: do they enhance coccinellid larval survival? *Oecologia* 107: 498–503.

Voynaud, L. 2008. Prédation intraguilde entre prédateurs actif et furtif au sein d'une guilde aphidiphage. Unpubl. MSc dissertation, Université du Québec à Montréal, Québec, Canada.

Ware, R. L. and M. E. N. Majerus. 2008. Intraguild predation of immature stages of British and Japanese coccinellids by the invasive ladybird *Harmonia axyridis. In* H. E. Roy and E. Wajnberg (eds). *From Biological Control to Invasion: The Ladybird* Harmonia axyridis *as a Model Species.* Springer, The Netherlands. pp. 169–188.

Ware, R. L., B. Ygel and M. E. N. Majerus. 2009. Effects of competition, cannibalism and intra-guild predation on larval development of the European coccinellid *Adalia bipunctata* and the invasive species *Harmonia axyridis*. *Ecol. Entomol.* 34: 12–19.

Ware, R. L., F. Ramon-Portugal, A. Magro et al. 2008. Chemical protection of *Calvia quatuordecimpunctata* eggs against intraguild predation by the invasive ladybird *Harmonia axyridis*. *In* H. E. Roy and E. Wajnberg (eds). *From Biological Control to Invasion: The Ladybird* Harmonia axyridis *as a Model Species.* Springer, The Netherlands. pp. 189–200.

Warren, L. O. and M. Tadić. 1967. Biological observations on *Coleomegilla maculata* and its role as a predator of the fall webworm. *J. Econ. Entomol.* 60: 1492–1496.

Way, M. J. 1963. Mutualism between ants and honeydew-producing homoptera. *Annu. Rev. Entomol.* 8: 307–344.

Weisser, W. W. 2003. Additive effects of pea aphid natural enemies despite intraguild predation. *In* A. O. Soares, M. A. Ventura, V. Garcia and J-.L. Hemptinne (eds). *Proc. 8th Int. Symp. Ecol. Aphidophaga. Arquipélago, Life Mar. Sci. (Suppl.)* 5: 11–15.

Wootton, J. T. 1994. The nature and consequences of indirect effects in ecological communities. *Annu. Rev. Ecol. Syst.* 25: 443–466.

Yasuda, H. and T. Kimura. 2001. Interspecific interactions in a tritrophic arthropod system: effects of a spider on the survival of larvae of three predatory ladybirds in relation to aphids. *Entomol. Exp. Appl.* 98: 17–25.

Yasuda, H. and N. Ohnuma. 1999. Effect of cannibalism and predation on the larval performance of two ladybird beetles. *Entomol. Exp. Appl.* 93: 63–67.

Yasuda, H. and K. Shinya. 1997. Cannibalism and inter-specific predation in two predatory ladybirds in relation to prey abundance in the field. *Entomophaga* 42: 153–163.

Yasuda, H., T. Takagi and K. Kogi. 2000. Effects of conspecific and heterospecific larval tracks on the oviposition behavior of the predatory ladybird, *Harmonia axyridis* (Coleoptera: Coccinellidae). *Eur. J. Entomol.* 97: 551–553.

Yasuda, H., T. Kikuchi, P. Kindlmann and S. Sato. 2001. Relationships between attack and escape rates, cannibalism, and predation in larvae of two predatory ladybirds. *J. Insect Behav.* 14: 373–384.

Yasuda, H., E. W. Evans, Y. Kajita, K. Urakawa and T. Takizawa. 2004. Asymmetric larval interactions between introduced and indigenous ladybirds in North America. *Oecologia* 141: 722–731.

Zang, L. S. and T. X. Liu. 2007. Intraguild interactions between an oligophagous predator, *Delphastus catalinae* (Coleoptera: Coccinellidae), and a parasitoid, *Encarsia sophia* (Hymenoptera: Aphelinidae), of *Bemesia tabaci* (Homoptera: Aleyrodidae). *Biol. Control* 41: 142–150.

Chapter 8

NATURAL ENEMIES OF LADYBIRD BEETLES

Piotr Ceryngier,[1] Helen E. Roy[2] and Remy L. Poland[3]

[1] Centre for Ecological Research, Polish Academy of Sciences, Dziekanow Lesny, 05-092 Lomianki, Poland
[2] NERC Centre for Ecology & Hydrology, Crowmarsh Gifford, Oxfordshire, OX10 8BB, UK
[3] Clifton College, 32 College Road, Clifton, Bristol, BS8 3JH, UK

Ecology and Behaviour of the Ladybird Beetles (Coccinellidae), First Edition. Edited by I. Hodek, H.F. van Emden, A. Honěk.
© 2012 Blackwell Publishing Ltd. Published 2012 by Blackwell Publishing Ltd.

8.1 INTRODUCTION

The term **natural enemy** is often used to denote an organism that draws nutrition from another organism, such as a predator (or herbivore), parasitoid, parasite or pathogen (e.g. DeBach & Rosen 1991, Thacker 2002). In this review, we consider natural enemies in a broader sense and define them after Flint and Dreistadt (1998) as 'organisms that kill, decrease the reproductive potential, or otherwise reduce the numbers of another organism'. Such a definition would include some organisms that interact with the prey organism in ways other than exploitation, e.g. through competition or self-defence. Most interactions between ladybirds and their competitors, including intra-guild predation, are discussed elsewhere (Chapter 7). Here, we consider the relations of predatory Coccinellidae with Hemiptera-tending ants, and devote a short section to mites phoretic on Coccinellidae. Although this phoretic relationship has not been shown to be detrimental to ladybirds, the shared food resources of mites and ladybirds (both prey on the same hemipterans) make them competitors. Another unusual interaction that may harm ladybirds is that with some of their prey, i.e. social aphids which produce aggressive soldiers defending their colonies against predators. This is also briefly discussed here. All remaining organisms considered as natural enemies of ladybirds in this review behave in an exploitative manner.

8.2 PREDATION AND RELATED PHENOMENA

8.2.1 Anti-predator defences

Ladybirds display a range of general defence reactions, from escape (by flying away, running away or dropping to the ground) to immobilization (so-called thanatosis) and also other more specific anti-predator adaptations.

8.2.1.1 Aposematic colouration and other visual signals

Adults of many members of Coccinellidae are conspicuously coloured, often with contrasting red-and-black or yellow-and-black patterns on their elytra. Such patterns usually serve as aposematic (warning) colouration (Moore et al. 1990, Joron 2003). Larvae and pupae may also be aposematically coloured with dark

and bright areas on their surface (e.g. Richards 1985, Holloway et al. 1991). The function of warning colouration is to advertise to potential predators that the bearer is unpalatable, toxic or nutritionally unprofitable (Joron 2003, Blount et al. 2009). Indeed, ladybirds are often distasteful and toxic to vertebrate and invertebrate predators (Daloze et al. 1995). A recent experimental study by Dolenska et al. (2009) suggests that not only warning colours but also any spotted pattern and general ladybird appearance (oval and convex body shape) may be signals of prey unprofitability for some optically oriented vertebrate predators, such as the great tit (*Parus major*).

8.2.1.2 Reflex bleeding

The bitter taste and toxic properties of many ladybird beetles can be attributed to a variety of **defensive alkaloids** (Daloze et al. 1995). About 50 different alkaloids have been identified in coccinellids (Laurent et al. 2005). The alkaloids are produced in ladybird haemolymph and are distributed by the haemolymph throughout the insect's body. When disturbed, ladybirds secrete droplets of haemolymph through tibiofemoral articulations. The released fluid is called 'reflex blood' and the corresponding defence mechanism – 'reflex bleeding'. Reflex bleeding is commonly used by adults of many ladybird species and by larvae and/or pupae of some species, but in the latter cases, the fluid is usually released from pores in the dorsal body surface (Holloway et al. 1991, Daloze et al. 1995). The pupae of some Epilachninae are known to exude the droplets of defensive fluid by specialized, glandular hairs (Schroeder et al. 1998). In adult ladybirds, reflex bleeding is often associated with thanatosis (Daloze et al. 1995, Ceryngier & Hodek 1996).

Various ladybird species have been shown to exert differing degrees of response and harmful effects on predators which is attributed to differences in their alkaloid composition. Marples et al. (1989) reported that blue tit (*Cyanistes caeruleus*) nestlings suffer severe mortality when supplied with food containing homogenized *Coccinella septempunctata* beetles, but no toxic effect was observed when homogenized *Adalia bipunctata* were added to the nestlings' food. *Adalia decempunctata* also had no apparent toxic effect on blue tit nestlings (Marples 1993). The high toxicity of *C. septempunctata* was associated with severe pathological changes in the blue tits' livers (Marples et al. 1989), and this was probably caused by coccinelline, the main

C. septempunctata alkaloid. In contrast to coccinelline, the major alkaloid compound of *Adalia* species, adaline, appears rather benign to young tits. These striking differences led the authors to the conclusion that light-coloured (typical) *A. bipunctata* and *A. decempunctata* are Batesian mimics of their toxic relative, *C. septempunctata*. Melanic *Adalia* individuals are, according to this hypothesis, the mimics of another model species, *Exochomus quadripustulatus*.

Elytra colouration in *Harmonia axyridis* (at least in its light form *succinea*) seems to represent a true aposematic signal. Bezzerides et al. (2007) found that the proportion of light areas on light-and-black patterned elytra of this form was positively correlated with the concentration of the alkaloid harmonine in ladybird bodies.

Reflex blood also contains **methylalkylpyrazines**, which are involved in ladybird defence. In contrast to alkaloids, pyrazines are volatile, so they can be olfactorily detected by predators. They are responsible for the odour that, in addition to aposematic colouration, acts as a signal in highlighting the unprofitability of a ladybird as food. It was found that pyrazines tend to be absent from cryptically coloured Coccinellidae, but are often present in aposematic species (Moore et al. 1990, Daloze et al. 1995).

Reflex bleeding may also act as a **mechanical defence** against some invertebrate predators. The fluid coagulates quickly on exposure to air and may stick to a predator's legs, antennae and mouthparts (Eisner et al. 1986).

8.2.1.3 Morphological anti-predator adaptations

Adult Coccinellidae, like most beetles, are relatively well protected against many predators by their exoskeleton and elytra. Ladybird pupae also have relatively hard cuticles and, in many species, are additionally protected by the final larval skin. Larvae, and sometimes pupae, may be defended by spiny projections or wax covers (Pope 1979, Richards 1980, Majerus et al. 2007). Pupae of many Coleoptera, including Coccinellidae, have deep intersegmental clefts with heavily sclerotized margins between some of the abdominal tergites. These devices, called 'gin traps', act as jaws when a disturbed pupa rapidly raises and drops its body. Gin traps of *Cycloneda sanguinea* pupae have been found an effective defence against ant attacks (Eisner & Eisner 1992).

8.2.2 Vertebrate predators

Despite being distasteful and toxic, ladybird beetles have been reported to be eaten by various vertebrate and invertebrate predators. Predation by vertebrates concerns virtually all the main groups: fish (e.g. Gomiero et al. 2008), amphibians (e.g. Cicek & Mermer 2007), reptiles (e.g. Pal et al. 2007), birds (e.g. Mizer 1970) and mammals (e.g. Chapman et al. 1955).

Predation of Coccinellidae by **birds** has been analyzed in detail. The contribution of Coccinellidae to the diet of birds was highlighted by Mizer (1970) in Ukraine through analysis of food remnants in almost 7000 stomachs of birds belonging to 234 species. Mizer (1970) detected ladybirds in the diet of 20% of bird species, but only in 2% of the stomachs (individuals) examined (Fig. 8.1). Ladybirds found in bird stomachs belonged to 23 species, of which two, *C. septempunctata* and *Hippodamia variegata*, accounted for more than a half of the total number (Fig. 8.2).

Mean numbers of ladybirds per bird stomach are given in Fig. 8.3, for those bird species which were represented in Mizer's data by at least 20 individuals. For almost all bird species considered, there was less than half a ladybird per stomach. The only exception was the house martin (*Delichon urbica*) with the average number of 0.75 ladybirds per stomach, which is consistent with the findings of other researchers, that birds feeding on the wing such as swifts, swallows and martins often ingest ladybirds (Muggleton 1978, Majerus & Majerus 1997a).

From Fig. 8.3, it is apparent that the tree sparrow (*Passer montanus*) is the next most frequent bird predator of Coccinellidae. This species is partly granivorous in its adult life, but feeds its nestlings mostly with insects. In the surroundings of Bratislava (Slovakia), and several localities in Poland, coccinellid larvae, pupae and adults were found to constitute one of the major fractions of the nestling diet of tree sparrows and house sparrows (*Passer domesticus*) (Wieloch 1975, Kristin 1986, Kristin et al. 1995).

8.2.3 Invertebrate predators

Spiders are frequently reported as preying on ladybirds, especially **web-building spiders** (e.g. Nentwig 1983, Laing 1988, Richardson & Hanks 2009, Sloggett 2010). However, the experimental study by Nentwig

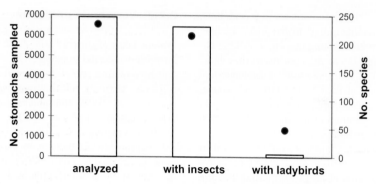

Figure 8.1 Analysis of stomach contents of Ukrainian birds (data extracted from Mizer 1970). Bars, numbers of stomachs; dots, numbers of bird species.

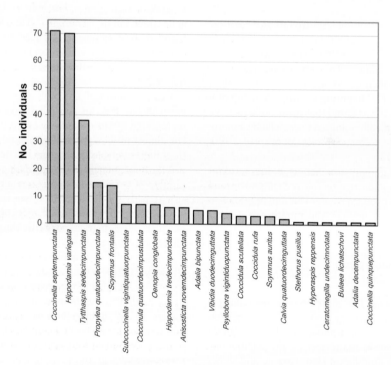

Figure 8.2 Species composition of Coccinellidae found in stomachs of Ukrainian birds (data extracted from Mizer 1970).

(1983) suggests that ladybird beetles, even if caught in a spider web, may be not eaten.

In contrast to web-builders, **actively hunting spiders** can capture their prey in a more targeted manner. It was found that these predators relatively rarely attack coleopterans because of their thick and hard cuticle (Nentwig 1986, Nyffeler 1999). In the

case of Coccinellidae, defensive alkaloids may act as additional protection (Eisner et al. 1986; Camarano et al. 2006).

Insects of various orders (e.g. Hemiptera, Diptera, Coleoptera, Neuroptera) have been reported to prey on Coccinellidae, and many of them are intraguild predators (Chapter 7). Insects not belonging to the same

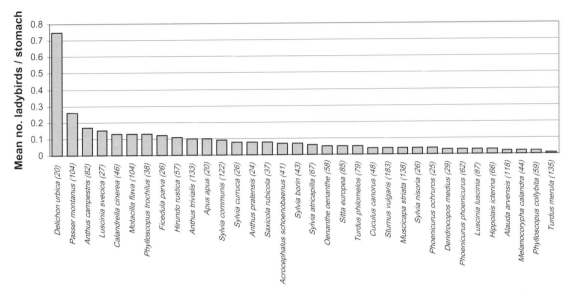

Figure 8.3 Mean numbers of ladybirds in the stomachs of individual bird species (data extracted from Mizer 1970). The number of stomachs analysed is given in brackets.

guild as their predatory ladybird prey include gomphid dragonflies (Odonata: Gomphidae) (Conrad 2005), robber flies (Diptera: Asilidae) (Ghahari et al. 2007) or vespid wasps (Hymenoptera: Vespidae) (Gambino 1992).

Predatory insects known to hunt **phytophagous ladybirds** include beetles (entomophagous Coccinellidae, Carabidae, Cantharidae), true bugs (Anthocoridae, Nabidae, Reduviidae, Pentatomidae, Lygaeidae), neuropterans (Chrysopidae, Myrmeleontidae), butterflies (larvae of some Noctuidae), earwigs (Forficulidae) and ants (Formicidae) (Howard & Landis 1936, Ohgushi 1986). Ants are also known as occasional predators of predatory coccinellids (Sloggett et al. 1999, Majerus et al. 2007), but the majority of interactions between these two groups of insects are of a competitive nature.

8.2.4 Hemiptera-tending ants

The mutualistic relationship in which ants tend honeydew-producing Hemiptera is a very common and widespread phenomenon (Styrsky & Eubanks 2007). The great majority of ants taking part in such associations belong to phylogenetically advanced sub-families Dolichoderinae, Formicinae and Myrmicinae. The hemipterans most frequently attended are aphids (Aphididae), soft scales (Coccidae) and mealybugs (Pseudococcidae) (Delabie 2001). Ants benefit from this association mainly by feeding on honeydew, a predictable and renewable source of carbohydrates and other compounds. The most important gains of hemipterans are sanitation of their colonies and the protection from natural enemies (Delabie 2001, Majerus et al. 2007, Styrsky & Eubanks 2007).

By protecting honeydew producers, ants come into antagonistic interactions with potential enemies of the former, including ladybirds (Majerus et al. 2007). The level of ant aggression towards enemies of the tended hemipterans increases with increasing proximity to the tended colonies (Hanks & Sadof 1990, Sloggett et al. 1998, Dejean et al. 2002, Sloggett & Majerus 2003, Majerus et al. 2007). As a consequence of a high ant aggression, ladybird adults and especially their soft-bodied larvae are sometimes killed when feeding on ant-tended resources (Cochereau 1969, Majerus 1989, Dejean et al. 2002). More often, adults are chased from the colony, while larvae may be picked up and carried away or dropped off the plant (Hanks & Sadof 1990, Majerus et al. 2007). The number of coccinellids is usually lower in the presence of ants than in their absence (Bradley 1973, Reimer et al. 1993, Takizawa & Yasuda 2006).

Many ladybird species can reduce the effects of ant aggression using various kinds of behavioural, physical and chemical defences. Mechanisms of ladybird defence against ants were recently reviewed by Majerus et al. (2007) and include running or flying away,

dropping to the ground, taking a 'clamp down' posture, reflex bleeding, mechanical protection by hard exoskeleton or structures such as spines or wax filaments.

More specific adaptations allowing avoidance of ant aggression can be found in those ladybird species that specialize in feeding on ant-tended hemipterans (Table 8.1). Some of these **myrmecophilous ladybirds**

Table 8.1 Ladybirds adapted to feed on prey tended by ants.

Ladybird	Ant	Tended prey group	Region	Adaptations to ant aggression	References
COCCINELLINAE					
Coccinellini					
Coccinella magnifica	*Formica* spp.	aphids	Palaearctic	possibly chemical and/or behavioural protection	2, 11, 12, 13
SCYMNINAE					
Scymnini					
Scymnus interruptus	*Lasius niger*	aphids	Germany	smeary, adhesive wax cover of the larva repels ant attack	15
Scymnus nigrinus	*Formica polyctena*	aphids	Germany	smeary, adhesive wax cover of the larva repels ant attack	15
Scymnus (Pullus) posticalis	*Lasius niger, Pristomyrmex pungens*	aphids	Japan	larvae ignored by ants (wax cover may function as chemical camouflage)	3, 4
Hyperaspidini					
Hyperaspis conviva	*Formica obscuripes*	coccid	Manitoba (Canada)	larvae rarely attacked, wax cover and reflex bleeding – sufficient protection	1
Hyperaspis reppensis	*Tapinoma nigerrimum*	fulgorid (Auchenorrhyncha)	Italy	larvae not attacked, adults – 'cowering behaviour' when attacked	10
Brachiacanthadini					
Brachiacantha quadripunctata	*Lasius umbratus*	coccids	Massachusetts (USA)	larvae covered with waxy tufts resemble giant coccids	16
Brachiacantha ursina	*Lasius claviger*	aphids	Washington DC (USA)	larvae with waxy cover not attacked by ants	9

Table 8.1 (*Continued*)

Ladybird	Ant	Tended prey group	Region	Adaptations to ant aggression	References
Platynaspidini					
Phymatosternus lewisii	*Pristomyrmex pungens*	aphids	Japan	coccid–like larvae not attacked by ants	4
Platynaspis luteorubra	*Lasius niger, Myrmica rugulosa, Tetramorium caespitum*	aphids	Europe	larvae – morphological adaptations (flattened shape, long setae, short legs) and presumably chemical camouflage pupae – dense hair cover adults – 'cowering behaviour'	6, 14
ORTALIINAE					
Ortaliini					
Apolinus lividigaster	*Crematogaster* sp., *Paratrechina* sp.	aphids	eastern Australia	very long body projections and wax cover in larva and pupa, defensive behaviour of larva, waxy smear around pupa	7
COCCIDULINAE					
Coccidulini					
Rodatus major	*Iridomyrmex* sp.	margarodid (Coccoidea) eggs	eastern Australia	larval feeding inside margarodid ovisac, wax cover and defensive behaviour of the larva, wax shroud resembling margarodid ovisac produced by prepupa	8
Azyini					
Azya orbigera	*Azteca instabilis*	coccid	Mexico	sticky, waxy filaments of the larva prevent effective ant attack	5

References: 1, Bradley (1973); 2, Godeau et al. (2009); 3, Kaneko (2002); 4, Kaneko (2007); 5, Liere & Perfecto (2008); 6, Majerus et al. (2007); 7, Richards (1980); 8, Richards (1985); 9, Schwarz (1890); 10, Silvestri (1903); 11, Sloggett et al. (1998); 12, Sloggett et al. (2002); 13, Sloggett & Majerus (2003); 14, Volkl (1995); 15, Volkl & Vohland (1996); 16, Wheeler (1911).

(or their particular life stages) are ignored by the ants and, hence, have access to prey resources effectively defended from other predators. To avoid ant aggression, myrmecophilous ladybirds may use chemical camouflage and/or may morphologically mimic their ant-tended prey. Wax filaments produced by the larvae of many ladybird species make them very similar to some of their prey, e.g. coccids or mealybugs. According to Pope (1979), protection against ants is one of the main functions of the larval wax.

8.2.5   Social aphids with a soldier caste

Some aphids in the subfamilies Hormaphidinae (mainly of the tribe Cerataphidini) and Pemphiginae are known to produce morphologically distinct 'soldiers' that defend reproducing individuals in the colony from natural enemies (Stern & Foster 1996). Soldier-producing species are usually those that form long-lived galls or large, exposed colonies (Aoki et al. 2001). In most species, soldiers are a sterile caste constituting first or second instar nymphs which do not moult to older instars. They often differ from normal nymphs by having a more elongate body, a shorter proboscis and enlarged, sclerotized legs (Ito 1989, Stern & Foster 1996, Ijichi et al. 2005, Kurosu et al. 2008). To fight with potential enemies, soldiers of certain aphids use their stylets, through which some of them inject paralyzing venom (Kutsukake et al. 2004). In other aphids, heads of the soldiers are equipped with a pair of frontal horns which are used to pierce their victims. Characteristics of soldiers found in different aphid groups are given in Table 8.2.

A single soldier aphid can rarely cause fatal damage to a ladybird. Shibao (1998) found that individual soldiers of the bamboo aphid, *Pseudoregma bambucicola*, can destroy (with their frontal horns) a small victim such as a syrphid egg or hatchling, but are unable to kill even a first instar larva of predatory ladybird, *Synonycha grandis*. Therefore, soldiers attack in unison; up to three soldiers have been observed to attack a newly hatched *S. grandis* larva, and up to 16 soldiers with an older larva. Very large (20–25 mm in length) fourth instar larvae of *S. grandis*, even if attacked by many soldiers, are usually thrown down to the ground rather than killed. Nevertheless, experiments performed by Shibao (1998) showed that the mortalities of first and third–fourth instar larvae of *S. grandis* were positively correlated with the density of *P. bambucicola* soldiers.

The developmental stage of a ladybird that is most vulnerable to attack from soldier aphids is the egg. Some ladybirds feeding on soldier-producing aphids have developed adaptations to protect their eggs from soldiers. Females of *Sasajiscymnus kurohime*, for example, cover their eggs with a layer of faeces-like secretion, which supposedly serves as a protection against the soldiers of *Ceratovacuna lanigera* and *P. bambucicola* (Arakaki 1988, 1992, Joshi & Viraktamath 2004). Other ladybird species, *S. grandis* and *Megalocaria dilatata*, lay their eggs a safe distance from aphid colonies (Arakaki 1992, Joshi & Viraktamath 2004). Larvae and adults of the latter two species have been observed to reflex bleed in response to soldier attacks (Joshi & Viraktamath 2004). Interestingly, larvae of *S. kurohime* are not attacked by the soldiers of *C. lanigera* (Arakaki 1992).

Table 8.2 Characteristics of aphid soldiers and colonies they defend.

Aphid group	Colony type	Developmental stage of soldiers	Soldier weapon	References
Cerataphidini on primary host plants (*Styrax* spp.)	in galls	2nd instar	stylets	4, 7, 9, 10, 11, 14
Cerataphidini on secondary host plants (monocotyledons)	exposed	1st instar	frontal horns	2, 5, 7, 8, 12, 13, 14
Pemphiginae	in galls or exposed	usually 1st instar	stylets	1, 3, 6, 7, 14

References: 1, Aoki et al. (2001); 2, Aoki et al. (2007); 3, Aoki & Kurosu (1986); 4, Aoki & Kurosu (1989); 5, Arakaki (1988); 6, Ijichi et al. (2005); 7, Ito (1989); 8, Joshi & Viraktamath (2004); 9, Kurosu & Aoki (2003); 10, Kurosu et al. (2008); 11, Kutsukake et al. (2004); 12, Shibao (1998); 13, Stern (1998); 14, Stern & Foster (1996).

8.3 PARASITOIDS

Ladybirds are attacked by a wide array of insect parasitoids. Some are specific for limited ranges of taxa within the Coccinellidae, while others are broad polyphages, which can parasitize members of various insect families and orders. Furthermore, there are also many secondary parasitoids or hyperparasitoids associated with the primary parasitoids. Summarized data on primary and secondary ladybird parasitoids and their host and distribution records are presented in Tables 8.3 and 8.4, but these lists are certainly far from

comprehensive. Many taxa are probably still omitted, some should perhaps be removed as misidentifications, and others transferred to another group of parasitoids of ladybirds. In particular it is often not easy to determine whether a particular parasitoid is primary or secondary, or even tertiary, especially as in most cases we only know its taxonomic position.

Hereafter, we first describe the characteristics of parasitoids reported from Coccinellidae, as based on this imperfect set of data in Tables 8.3 and 8.4 and the other tables referred to in Table 8.3. Then we review the available information on selected parasitoid taxa.

Table 8.3 Primary and presumed primary parasitoids of Coccinellidae.

Parasitoid types: I, imaginal; P, pupal; L–P, larval–pupal (oviposition into host larva, emergence from host pupa); L, larval; E, egg parasitoid.

Parasitoid taxon	Reported ladybird host taxa	Parasitoid type	Distribution records	References*
GENERA SPECIFIC FOR COCCINELLIDAE				
Hymenoptera: Braconidae				
Dinocampus D. coccinellae	see Table 8.5	I	cosmopolitan	see Table 8.5
Hymenoptera: Chalcididae				
Uga several species, see Table 8.8	Epilachninae	L–P	Asia, Africa, Australia, Indonesia	see Table 8.8
Hymenoptera: Encyrtidae				
Cowperia several species, see Table 8.9	see Table 8.9	P	Europe, Asia, Indonesia	see Table 8.9
Homalotylus many species, see Table 8.10	see Table 8.11	L, L–P	cosmopolitan	see Table 8.10
Hymenoptera: Proctotrupidae				
Nothoserphus several species, see Table 8.12	see Table 8.12	L	Europe, Asia, Indonesia	see Table 8.12
Hymenoptera: Pteromalidae				
Metastenus M. concinnus	*Cryptognatha signata, Cryptolaemus montrouzieri, Scymnus impexus, Scymnus apetzi, Scymnus* sp.	P	Europe, Asia, Argentina	23

(Continued)

Table 8.3 (Continued)

Parasitoid taxon	Reported ladybird host taxa	Parasitoid type	Distribution records	References*
M. townsendi	Azya orbigera, Cryptognatha nodiceps, Cryptognatha simillima, Hyperaspis lateralis, Hyperaspis sp., Microweisea sp., Nephus guttulatus, Pentilia insidiosa, Pseudoazya trinitatis, Scymnus otohime, Scymnus sp.	P	Carribean, Mexico, USA, Japan	23

SPECIES SPECIFIC FOR COCCINELLIDAE

Diptera: Phoridae

Phalacrotophora several species, see Table 8.13	Coccinellinae, Chilocorinae	P	see Table 8.13	see Table 8.13

Diptera: Tachinidae

Euthelyconychia epilachnae (syn. Paradexodes epilachnae)	Epilachna varivestis, Epilachna defecta, Henosepilachna vigintisexpunctata	L, P	Mexico	11, 15, 28
Pseudebenia epilachnae	Epilachna quadricollis	L	South Korea	33

Hymenoptera: Braconidae

Centistes scymni	Scymnus (Pullus) impexus	I	Switzerland, Germany	9
Centistes subsulcatus	Propylea quatuordecimpunctata	I	Belgium	13

Hymenoptera: Encyrtidae

Anagyrus australiensis	Diomus pumilio	P	Australia	23
Ooencyrtus azul	Chilocorus nigripes	P	Kenya	23
Ooencyrtus bedfordi	Chilocorus cacti, Chilocorus discoideus, Chilocorus distigma	P	South Africa, Uganda	23
Ooencyrtus camerounensis	Epilachna eckloni, Chnootriba similis	E	Cameroon, Ethiopia, Senegal, South Africa	4, 23
Ooencyrtus epilachnae	Epilachna dregei	E	South Africa, Uganda	23
Ooencyrtus puparum	Platynaspis sp.	P	Senegal	23
Ooencyrtus sinis	Exochomus flavipes, Exochomus flaviventris	P	South Africa	23
Prochiloneurus nigriflagellum	Orcus australasiae	L	Queensland (Australia)	23

Hymenoptera: Eulophidae

Aprostocetus neglectus	Chilocorini, Scymnini, ?Coccinellini	L–P	Europe, Asia, Nearctic	23
Chrysocharis johnsoni	Henosepilachna vigintioctopunctata	?	India	23
Chrysonotomyia appannai	Henosepilachna vigintioctopunctata	E	India	23
Oomyzus mashhoodi	undet. Coccinellidae	L	India	16
Oomyzus scaposus	Coccinellini, Chilocorini, Scymnini	L–P	cosmopolitan	23
Pediobius foveolatus	Epilachninae	L, L–P, P	Africa, Asia, Australia, Pacific islands, Nearctic (introduced)	23
Pediobius nishidai	Epilachna mexicana	L	Costa Rica	23

Table 8.3 (*Continued*)

Parasitoid taxon	Reported ladybird host taxa	Parasitoid type	Distribution records	References*
Quadrastichus ovulorum	Epilachninae	E	Africa, India, Sri Lanka, Melanesia	23
Sigmoepilachna indica†	*Epilachna* sp.	E	India	23
Tetrastichus decrescens	*Henosepilachna vigintioctomaculata*	?	China	23
Tetrastichus epilachnae	Epilachninae, Coccinellini, Chilocorini, Scymnini	L–P	Europe, Asia, Morocco	23
Hymenoptera: Eupelmidae				
Eupelmus vermai	*Epilachna* sp.	L	India	23
Hymenoptera: Pteromalidae				
Inkaka quadridentata	*Cryptolaemus montrouzieri*	?	Australia, New Zealand	23
Merismoclea rojasi	*Coccidophilus citricola*	?	Argentina, Chile	23
Mesopolobus secundus	undet. Scymninae, *Hyperaspis senegalensis*	L	Uganda, Kenya	8, 9
Ophelosia bifasciata	undet. Coccinellidae	L	Australia	5
Oricoruna orientalis	*Rodolia fumida*	L	India	23
Oricoruna sp.	*Exochomus* sp., *Rodolia occidentalis*	?	Nigeria	23

TAXA REPORTED FROM VARIOUS HOSTS INCLUDING COCCINELLIDAE

Diptera: Phoridae				
Megaselia spp.	*Chilocorus distigma*	P	East Africa	10
	Chilocorus quadrimaculatus	P	Kenya	10
	Epilachna varivestis	L, P	USA	10, 15
Diptera: Sarcophagidae				
Boettcheria latisterna (syn. *Sarcophaga latisterna*)	*Epilachna varivestis*	I	Ohio (USA)	11, 15, 28
Helicobia rapax (syn. *Sarcophaga helicis*)	*Epilachna varivestis*	L	Alabama (USA)	11, 15, 28
Ravinia errabunda (syn. *Sarcophaga reinhardii*)	*Epilachna varivestis*	L	Mexico	11, 15, 28
Diptera: Tachinidae				
Chetogena claripennis	*Epilachna borealis, Epilachna varivestis*	L, P	North America	11, 15, 24, 28
Lydinolydella metallica	*Epilachna eusema, Epilachna marginella, Epilachna* sp.	L	South America	3, 11, 28
Lypha slossonae	*Epilachna varivestis*	?	North America	11, 24, 28
Medina spp. incl. *M. separata, M. collaris* and *M. melania*‡	Epilachninae, Coccinellinae, Chilocorinae	I	Palaearctic	2, 6, 29
Myiopharus doryphorae (syn. *Doryphorophaga doryphorae*)	*Epilachna varivestis*	I	North America	15, 24
Policheta unicolor	*Subcoccinella vigintiquatuorpunctata*	?	France	14

(Continued)

Table 8.3 (*Continued*)

Parasitoid taxon	Reported ladybird host taxa	Parasitoid type	Distribution records	References*
Strongygaster triangulifera	*Coccinella trifasciata, Coleomegilla maculata, Harmonia axyridis, Epilachna varivestis*	I	North America	21, 24
Hymenoptera: Encyrtidae				
Anagyrus sp.	*Telsimia* sp.	L	Australia	23
Cerchysiella sp.	*Chilocorus bipustulatus*	L	Israel	36
Eupoecilopoda perpunctata	*Scymnus* sp.	P	Iran	19
Isodromus niger	*Chilocorus similis*	?	SE Europe, Asia, USA	23
Hymenoptera: Eulophidae				
Baryscapus thanasimi	*Chilocorus stigma*	L	USA	23
Omphale sp.§	*Henosepilachna vigintioctopunctata*	E	India	26
Oomyzus sempronius	*Chilocorus bipustulatus*	L	Europe, Egypt, Turkey	23
Parachrysocharis sp.	*Hippodamia variegata*	?	India	23
Pnigalio agraules	*Chilocorus bipustulatus*	?	Palaearctic	23
Tetrastichus cydoniae	*Cheilomenes propinqua vicina, Chnootriba similis*	P	western Africa	23, 31
Tetrastichus orissaensis	*Epilachna* sp, *Subcoccinella vigintiquatuorpunctata*	?	India, Hungary, Italy, former Yugoslavia	23
Hymenoptera: Eupelmidae				
Anastatus spp.	*Chilocorus bipustulatus, Hyperaspis* sp.	?	Israel, Nigeria	23, 36
Eupelmus urozonus	*Chilocorus bipustulatus*	?	?	23
Eupelmus sp.	*Chilocorus bipustulatus*	?	Israel	36
Hymenoptera: Pteromalidae				
Austroterobia sp.	*Rodolia iceryae*	?	Kenya	23
Mesopolobus sp.	*Chnootriba similis*	P	Ethiopia	4
Pseudocatolaccus sp.	*Chilocorus bipustulatus, Cryptolaemus montrouzieri*	?	Russia	23
Trichomalopsis acuminata	*Propylea quatuordecimpunctata*	P	Hungary	27
Hymenoptera: Trichogrammatidae				
Trichogramma sp.	*Epilachna* sp.	E	Indonesia	23

*See Table 8.4 for list of references

†*Sigmoepilachna indica* is the only described species of the recently erected genus *Sigmoepilachna* and its only known host is *Epilachna* sp. In our opinion, however, placing *Sigmoepilachna* among the genera specific for Coccinellidae would be premature, taking into account the highly insufficient data on this genus.

‡Most older records on the parasitization of ladybirds by *Medina* spp. erroneously refer to *Medina luctuosa*.

§This record of *Omphale* sp. as an egg parasitoid of *Henosepilachna vigintioctopunctata* may refer to *Chrysonotomyia appannai* (syn. *Omphale epilachni*) considering that both were reported from the same region (southern India) and host species.

Table 8.4 Hyperparasitoids and presumed hyperparasitoids of Coccinellidae.

Parasitoid taxon	Primary host	Secondary (coccinellid) host	Distribution records	References
Hymenoptera: Ceraphronidae				
Aphanogmus sp.	?	*Chilocorus bipustulatus*	Israel	36
Hymenoptera: Chalcididae				
Brachymeria carinatifrons	*Euthelyconychia epilachnae, Lydinolydella metallica*	*Epilachna mexicana, Epilachna varivestis, Epilachna* sp.	Texas (USA), Mexico, Venezuela, Brazil	3, 23
Conura porteri	*Dinocampus coccinellae, Perilitus stuardoi*	?	Chile	23
Conura paranensis	*Dinocampus coccinellae*	*Cycloneda sanguinea*	Argentina	23
Conura petioliventris	*Dinocampus coccinellae*	*Hippodamia convergens*	California (USA)	23
Conura sp.	?	*Adalia deficiens*	South America	23
	Lydinolydella metallica	*Epilachna eusema, Epilachna* sp.	Argentina	3, 23
Hymenoptera: Encyrtidae				
Cheiloneurus carinatus	*Homalotylus* sp.	?	Africa	23
Cheiloneurus cyanonotus	*Homalotylus* spp., Tetrastichinae	Epilachninae, Chilocorinae, Coccinellinae	Africa	23
Cheiloneurus liorhipnusi	?	*Chnootriba similis*	Kenya, Senegal	23
Cheiloneurus orbitalis	*Homalotylus* sp.	undet. Coccinellidae	South Africa	7
Coccidoctonus trinidadensis	*?Homalotylus* sp.	*Cryptolaemus* sp.	Central America	11, 23
Homalotyloidea dahlbomii	*Homalotylus* spp.	*Chilocorus bipustulatus, ?Rhyzobius litura*	Europe, Canary Islands, Israel	23, 35
Ooencyrtus distatus	*Homalotylus* sp.	*Scymnus morelleti*	South Africa	23
Ooencyrtus polyphagus	?	*Chnootriba similis, Henosepilachna elaterii, Exochomus flavipes, Platynaspis* sp.	Cameroon, Mali, Senegal	23
Prochiloneurus aegyptiacus	*Homalotylus* spp.	*Chilocorus bipustulatus, Exochomus flavipes, Hyperaspis aestimabilis*	Africa, Asia, southern Europe	23
	Metastenus sp.	*Exochomus* sp., *Hyperaspis* sp.	Israel	36

(Continued)

Table 8.4 (*Continued*)

Parasitoid taxon	Primary host	Secondary (coccinellid) host	Distribution records	References
Hymenoptera: Eulophidae				
Aprostocetus esurus	?	*Chilocorus similis*	USA	20
Elasmus sp.	?	*Henosepilachna vigintioctopunctata*	India	35
Pediobius amaurocoelus	?	*Chnootriba similis*	Ghana	17
Pediobius sp.	Tetrastichinae	*Henosepilachna vigintioctopunctata*	India	35
	?	*Exochomus* sp.	Ghana, Nigeria, Zambia	23
Tetrastichinae	*Pediobius foveolatus*	*Henosepilachna vigintioctopunctata*	India	35
Hymenoptera: Eupelmidae				
Eupelmus sp.	?	*Henosepilachna vigintioctopunctata*	India	35
Hymenoptera: Eurytomidae				
Aximopsis sp.	*Pediobius foveolatus*	*Henosepilachna vigintioctopunctata*	India	23, 35
Hymenoptera: Ichneumonidae				
Gelis agilis (syn. *Gelis instabilis*)	*Dinocampus coccinellae*	*Coccinella septempunctata*	Poland	6
Gelis melanocephalus	?	*Coccinella septempunctata*	England	12, 32
Gelis sp.	*Dinocampus coccinellae*	*Coleomegilla maculata*	Ontario (Canada)	6
Syrphoctonus tarsatorius	?	*Coccinella septempunctata*	?	12, 32
Phygadeuon subfuscus	*Euthelyconychia epilachnae*	*Epilachna varivestis*	North America	11
Hymenoptera: Megaspilidae				
Dendrocerus spp. (syn. *Lygocerus, Atritomellus*), incl. *D. ergensis*	*Homalotylus* spp.	*Chilocorus bipustulatus, Coccinella septempunctata, Pharoscymnus numidicus, Pharoscymnus ovoideus, Scymnus* sp.	Japan, Israel, France, Italy, Spain, North Africa	1, 6, 11, 34, 36
Hymenoptera: Pteromalidae				
Catolaccus spp.	?	*Henosepilachna vigintioctopunctata*	China	23
	Oomyzus scaposus	*Coccinella septempunctata*	Poland	25
Dibrachys microgastri	*Dinocampus coccinellae*	*Coleomegilla maculata, Coccinella septempunctata*	Illinois (USA), England	23, 37
Ophelosia crawfordi	*Homalotylus* sp.	?	?	11

Table 8.4 (*Continued*)

Parasitoid taxon	Primary host	Secondary (coccinellid) host	Distribution records	References
Pachyneuron albutius (syn. *P. syrphi*)	?	*Chilocorus bipustulatus, Chilocorus renipustulatus, Exochomus quadripustulatus, Coccinella septempunctata*	Former USSR	22, 23
Pachyneuron altiscuta	?	*Harmonia axyridis*	North Carolina (USA)	30
Pachyneuron chilocori	*Homalotylus* sp.	*Chilocorus bipustulatus*	Italy, Israel	23, 36
Pachyneuron muscarum (syn. *P. concolor, P. siculum*)	*Homalotylus* sp.	*Chilocorus bipustulatus, Coccinella septempunctata*	Europe, Israel	18, 23, 36
Pachyneuron solitarium	*Homalotylus* sp.	*Anatis ocellata, Coccinella septempunctata, Myzia oblongoguttata*	Hungary, Poland, West Siberia	6
Pachyneuron spp.	*Homalotylus* spp., Tetrastichinae	*Calvia quatuordecimguttata, Cheilomenes lunata, Chilocorus inornatus, Exochomus flavipes, Harmonia axyridis, Hippodamia tredecimpunctata, Hippodamia variegata, Hyperaspis senegalensis, Menochilus sexmaculatus, Nephus kiesenwetteri, Nephus ornatus, Nephus soudanensis, Scymnus quadrillum, Subcoccinella vigintiquatuorpunctata*	Asia, Africa, Europe, North America	6, 23
Trichomalopsis dubia	*Dinocampus coccinellae*	*Coleomegilla maculata*	North America	23
	Pediobius foveolatus	*Epilachna varivestis*	North America	23
Trichomalopsis (syn. *Eupteromalus*) sp.	*Dinocampus coccinellae*	*Coccinella undecimpunctata, Coleomegilla maculata*	Egypt, Ontario (Canada)	6
	Oomyzus scaposus	*Coccinella septempunctata*	Poland	25
Hymenoptera: Signiphoridae				
Chartocerus subaeneus	?	*Nephus* sp.	?	23

References to Tables 8.3 and 8.4: 1, Alekseev & Radchenko (2001); 2, Belshaw (1993); 3, Berry & Parker (1949); 4, Beyene et al. (2007); 5, Boucek (1988); 6, Ceryngier & Hodek (1996); 7, Compere (1938); 8, Crawford (1912); 9, Delucchi (1954); 10, Disney et al. (1994); 11, Domenichini (1957); 12, Elliott & Morley (1907); 13, Hemptinne (1988); 14, Hodek (1973); 15, Howard & Landis (1936); 16, Husain & Khan (1986); 17, Kerrich (1973); 18, Klausnitzer (1967); 19, Lotfalizadeh (2010); 20, Marlatt (1903); 21, Nalepa & Kidd (2002); 22, Nikol'skaya (1934); 23, Noyes (2011); 24, O'Hara (2009); 25, Pankanin–Franczyk & Ceryngier (1999); 26, Patnaik & Mohapatra (2004); 27, Radwan & Lovei (1982); 28, Richerson (1970); 29, Richter (1971); 30, Riddick et al. (2009); 31, Risbec (1951); 32, Schaefer & Semyanov (1992); 33, Shima et al. (2010); 34, Smirnoff (1957); 35, Usman & Thontadarya (1957); 36, Yinon (1969); 37, R. Comont (unpublished).

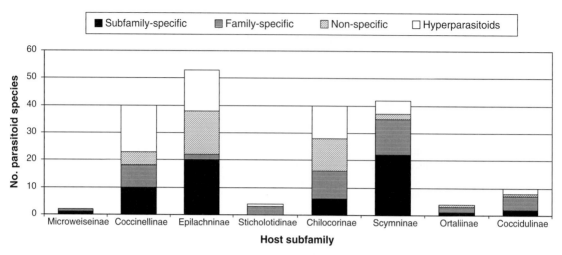

Figure 8.4 Number and host specificity of parasitoid species reported from members of different subfamilies of Coccinellidae. Subfamily-specific species are reported only from hosts belonging to a particular ladybird subfamily; Family-specific species are reported from more than one ladybird subfamily but not from other hosts; Non-specific species are reported from ladybird hosts as well as hosts belonging to other insect families or orders; Hyperparasitoids are species developing in primary ladybird parasitoids.

In allocating the ladybird genera to subfamilies and tribes we have used the system proposed by Nedvĕd and Kovář (Chapter 1).

8.3.1   General characteristics of parasitoids of ladybirds

8.3.1.1   Host specificity of parasitoids recorded from Coccinellidae

Of almost 160 parasitoids listed here, about 40 have been classified as **hyperparasitoids**. The remaining 117 taxa, representing three dipteran and eight hymenopteran families, are considered as **primary parasitoids**. The most numerous group among these consists of 46 species belonging to the six genera believed to be entirely specific to Coccinellidae. This number would rise by a further 26 species if we count all described species in these six genera, including those whose hosts have not yet been discovered.

The remaining parasitoid species include a group of about 39 species presumably specific for Coccinellidae, as well as about 32 species of parasitoids not specific for ladybirds but recorded from them. It is possible that some taxa in these two groups are, in fact, hyperparasitoids.

8.3.1.2   Parasitoids of different subfamilies of Coccinellidae

Ladybird hosts of the listed parasitoids belong to eight subfamilies (Fig. 8.4). The majority of records concern four of them, Coccinellinae, Epilachninae, Chilocorinae and Scymninae, with Epilachninae being associated with the highest number (53) of parasitoid species. Interestingly, most parasitoids reported from the Coccinellinae, Epilachninae and Chilocorinae are polyphages and hyperparasitoids, while those of Scymninae, as well as of the remaining four subfamilies (Microweiseinae, Sticholotidinae, Ortaliinae and Coccidulinae), are mainly host subfamily- and family-specific taxa.

8.3.1.3   Parasitism in different developmental stages of Coccinellidae

All developmental stages of ladybirds are subject to attack from parasitoids, though parasitization of eggs and adult beetles is relatively rare and the former is restricted to phytophagous ladybirds of the subfamily Epilachninae. It has been supposed that predatory ladybirds are devoid of **egg parasitoids** due to the cannibalistic habit of the newly hatched larvae. Successful parasitoid development in the eggs prior to being cannibalized is unlikely because egg parasitoids usually emerge from parasitized eggs after the non-parasitized eggs have hatched (Klausnitzer 1969). However, newly hatched larvae of some epilachnine ladybirds are also known to cannibalize unhatched eggs in the batch (Nakamura 1976, Nakamura & Ohgushi 1981).

Parasitization of adult ladybirds, although only a few parasitoid species are involved, has received disproportionately more attention from researchers than parasitization of other stages. Such over-representation is largely due to one species, the euphorine braconid, *Dinocampus coccinellae*, which, since the first systematic studies by Ogloblin (1924) and Balduf (1926), has continuously attracted the attention of successive generations of entomologists.

The predominant group of parasitoids of ladybirds is formed by the **parasitoids attacking larvae and pupae**. However, only very few species in this group, such as the flies *Phalacrotophora fasciata* and *P. berolinensis*, or the wasps *Oomyzus scaposus* and *Pediobius foveolatus*, have been studied in any detail. The biology and ecology of the latter species has been relatively well recognized because of its economic importance as a biological control agent of phytophagous ladybirds.

8.3.2   Review of the more important taxa

Most of the research on ladybird parasitoids concerns a few species found in the Palaearctic region (especially in its European part) and, to a lesser degree, in the Nearctic region. Apart from host and distribution records, we know nearly nothing about the majority of species parasitizing Coccinellidae in the remaining parts of the world.

The intention of this review is not only to discuss the relatively well-known European and North American species, but also to draw the attention of the reader to certain less known but, in our opinion, not less important parasitoids. We applied two criteria in selecting parasitoid taxa for more detailed presentation: (i) the degree of their specificity for ladybirds and (ii) their importance as a mortality factor of Coccinellidae from a more global perspective. Thus, we review the literature for ladybird-specific genera (*Dinocampus*, *Uga*, *Cowperia*, *Homalotylus*, *Nothoserphus* and *Metastenus*) and those species groups (*Phalacrotophora* spp.) or single species (*O. scaposus*, *P. foveolatus*) which play a

significant role in limiting ladybird populations in many parts of the world.

8.3.2.1   *Dinocampus* Foerster (Hymenoptera: Braconidae, Euphorinae)

The only known species certainly belonging to the genus *Dinocampus* is **D. coccinellae** (Schrank) (synonyms: *Dinocampus terminatus* (Nees), *Perilitus americanus* Riley, *P. coccinellae* (Schrank), *P. terminatus* (Nees)). *Dinocampus nipponicus* recently described from Japan (Belokobylskij 2000) was, soon after its description, transferred to the genus *Centistina* (Belokobylskij 2001). The current taxonomic status of *Perilitus stuardoi*, the other euphorine wasp which might belong to *Dinocampus*, is not known to us. *Perilitus stuardoi* was reported only from Chile as a parasitoid of *Adalia deficiens*, *A. bipunctata*, *Eriopis connexa* and *Cryptolaemus montrouzieri* (Smith 1953) and more recently of *Coleomegilla quadrifasciata* (Gonzales 2006).

Dinocampus coccinellae nowadays has a cosmopolitan **distribution** covering all continents except Antarctica, and many islands (Table 8.5). Although it has most frequently been reported from the temperate zone of the Holarctic, there are also records from the Arctic (Itilleq in Greenland, van Achterberg 2006) as well as from the tropics (e.g. Bangalore in India, Ghorpade 1977; and Planaltina in Brazil, Santos & Pinto 1981) and the temperate regions of the Southern Hemisphere (e.g. south-eastern Australia, Anderson et al. 1986; New Zealand and Tasmania, Alma 1980). The natural geographic range of *D. coccinellae* is difficult to reconstruct. According to Balduf (1926), the wasp might either have been introduced to North America from Europe or may be native to both North America and Eurasia. It is thought that *D. coccinellae* arrived in some parts of its present distribution together with ladybirds released for biological control purposes. It was probably transferred to the Hawaiian islands with the introduced *Olla v-nigrum* (Timberlake 1918), and to New Zealand with another biocontrol agent, *Coccinella undecimpunctata* (Gourlay 1930).

Mode of reproduction

Dinocampus coccinellae usually reproduces by **thelytokous parthenogenesis** (Ceryngier & Hodek 1996). However, as males can sporadically be produced (Muesebeck 1936, Hudon 1959, Wright 1978, Geoghegan et al. 1998a, Shaw et al. 1999, O. Nedvěd,

Table 8.5 Host records of *Dinocampus coccinellae* (from Ceryngier & Hodek 1996 supplemented and updated).

Host	Region	Reference
COCCINELLINAE		
Adalia bipunctata	Europe	4
	USA	4
Adalia decempunctata	Great Britain	4
Anatis ocellata	Great Britain	4
Anisosticta sibirica	Russian Far East	4
Calvia muiri	Japan	4
Calvia quatuordecimguttata	Europe	4
	Russian Far East	4
Ceratomegilla undecimnotata	Europe	4
Cheilomenes sulphurea	South Africa	4
Cheilomenes propinqua	South Africa	4
Cleobora mellyi	Tasmania	1
Coccinella californica	USA	4
Coccinella hieroglyphica	Great Britain	4
Coccinella leonina	Asia	4
	Australia	4
	Hawaii	4
Coccinella magnifica	Europe	4
	India	4
Coccinella novemnotata	North America	4
Coccinella quinquepunctata	Europe	4
	West Siberia	4
Coccinella septempunctata	Europe	4
	Asia	4
	USA	4
Coccinella septempunctata algerica	Tunisia	2
Coccinella septempunctata brucki	Japan	4
Coccinella trifasciata	West Siberia	4
	North America	4
Coccinella undecimpunctata	Europe	4
	Egypt	4
	New Zealand	4
	USA	4
Coccinula quatuordecimpustulata	Europe	4
	West Siberia	4
Coccinula sinensis	Russian Far East	4
Coelophora (=Lemnia) biplagiata	Taiwan	4
Coelophora inaequalis	Australia	4
	Hawaii	4
Coleomegilla maculata	North America	4
	Brazil	14
Cycloneda munda	USA	4
Cycloneda sanguinea	Brazil	13, 14
Eriopis connexa	Brazil	14
Halyzia sedecimguttata	Great Britain	4
Harmonia antipoda	New Zealand	1
Harmonia axyridis	Asia	4
	Japan	4
	North America	6, 8
	Europe	3, 7

Table 8.5 (*Continued*)

Host	Region	Reference
Harmonia conformis	Australia	4
Harmonia dimidiata	Taiwan	5
Harmonia octomaculata	Asia	4
Harmonia quadripunctata	Europe	4
	West Siberia	4
Hippodamia (Adonia) arctica	Russian Far East	4
Hippodamia convergens	USA	4
Hippodamia parenthesis	USA	4
Hippodamia tredecimpunctata	Europe	4
	Asiatic part of Russia	4
	USA	4
Hippodamia (Adonia) variegata	Europe	4
	Asia	4
	Chile	12
Illeis cincta	India	4
Lioadalia flavomaculata	South Africa	4
Macronaemia hauseri	China	4
Menochilus sexmaculatus	India	4
	Taiwan	4
	Japan	4
Micraspis discolor	Taiwan	4
	Vietnam	9
Myrrha octodecimguttata	Great Britain	15
Myzia oblongoguttata	Great Britain	4
Oenopia conglobata	Europe	4
	Russian Far East	4
Olla v–nigrum	USA	4
	Hawaii	4
	Brazil	14
Propylea japonica	Japan	4
	Russian Far East	4
Propylea quatuordecimpunctata	Europe	4
	Asiatic part of Russia	4
Psyllobora vigintiduopunctata	Great Britain	17
Tytthaspis sedecimpunctata	Great Britain	11
CHILOCORINAE		
Exochomus troberti concavus	South Africa	4
Exochomus quadripustulatus	Great Britain	10
Priscibrumus (=Exochomus) lituratus	Himalaya	4
Priscibrumus uropygialis	Himalaya	4
COLEOPTERA: CURCULIONIDAE		
Sitona discoideus	New Zealand	16

References: 1, Alma (1980); 2, Ben Halima–Kamel (2010); 3, Berkvens et al. (2010); 4, Ceryngier & Hodek (1996); 5, Chou (1981); 6, Firlej et al. (2005); 7, Koyama & Majerus (2008); 8, LaMana & Miller (1996); 9, Long & Belokobylskij (2003); 10, Mabbott (2006); 11, Majerus (1997); 12, Rebolledo et al. (2009); 13, Santos & Pinto (1981); 14, Silva et al. (2009); 15, Sloggett & Majerus (2000); 16, Wightman (1986); 17, R. Comont (unpublished).

unpublished), at least some of its strains are actually deuterotokous.

Several *D. coccinellae* **males** (one reared in Ontario, Canada, from *Col. maculata* and four in Scotland from *C. septempunctata*), when placed with females, displayed **courtship behaviour**, involving wing vibration (Wright 1978; Geoghegan et al. 1998a; Shaw et al. 1999) and walking in tight circles (Wright 1978). Moreover, each of those males tried to mount females but the mounting attempts of all but one male were rejected. The exception was the Canadian male, which mated with four females for about 20 minutes per female. Of 74 ladybirds offered to those four mated females, only 24 were parasitized and from 11 of them adult wasps developed, all of them being females (Wright 1978).

Life cycle

From host location to oviposition

Dinocampus coccinellae mostly oviposits into adult ladybird beetles but, especially when adult hosts are scarce, larvae and pupae can also be parasitized (Smith 1960, Maeta 1969, Filatova 1974).

Like some other euphorines, e.g. *Perilitus rutilus* or *Microctonus* spp. (Jackson 1928, Barratt & Johnstone 2001), *D. coccinellae* most frequently parasitizes its hosts when they are mobile (Bryden & Bishop 1945, Walker 1961, Richerson & DeLoach 1972, Orr et al. 1992). **Pre-oviposition and oviposition behaviour** of the parasitoid may be categorized as a sequence of the following activities: (i) pursuit and investigation of the host without extending the ovipositor, (ii) ovipositional stance with the ovipositor extended ventrally and forwards between the legs, and (iii) ovipositional attack, i.e. a thrust into the host with the ovipositor, which, if successful, is followed by egg deposition into the host's body (Richerson & DeLoach 1972, Geoghegan et al. 1998b). The existing evidence indicates that these activities are stimulated both visually (by the movement, colour, and size of potential hosts) and olfactorily (by host derived substances). The importance of **visual stimulation** was shown in tests using artificial hosts such as metal, wooden, plastic or paper models (Walker 1961, Richerson & DeLoach 1972). If such models were moving and were red or black, they caused wasp responses including pursuits, ovipositional stances and ovipositional attacks. Additionally, when the models were smeared with a ladybird's

defensive secretion, numbers of the first two responses, but not of ovipositional attacks, increased. This may indicate that **odour** is an important stimulus for host recognition but not for host acceptance (Richerson & DeLoach 1972). More recently, Al Abassi et al. (2001) found that *D. coccinellae* is strongly attracted by some ladybird-produced substances (the alkaloids precoccinelline and myrrhine) but is not attracted by others (hippodamine, pyrazines).

Movement of the host not only facilitates its location by the parasitoid, but also makes **oviposition** easier. When a ladybird is walking, its elytra are slightly raised and the head is extended forward, which gives the female parasitoid better access to the soft membranes between the posterior abdominal segments or between the head and thorax, the host areas into which *D. coccinellae* most often oviposits (Balduf 1926, Iperti 1964, Sluss 1968, Richerson & DeLoach 1972). If a potential host is immobile, the wasp will, before ovipositing, stimulate such a host to walk by palpating it with the antennae, encircling it and probing with the ovipositor (Balduf 1926, Walker 1961, Richerson & DeLoach 1972). Immobilization of a host in response to examination by *D. coccinellae* is a frequently displayed **defensive behaviour against parasitization**. Other behaviours involved in ladybird defence are rapid escape, attacks on the parasitoid with mandibles, and attempts to kick the parasitoid ovipositor with hind legs (Firlej et al. 2010). *Dinocampus coccinellae* oviposition has to be very rapid because of host mobility and defensive reactions, and takes only a fraction of a second (Balduf 1926).

Development

The ripe ovarian **egg of *D. coccinellae*** is elongate and measures about 250 μm by 30 μm. After deposition in a host, it starts to grow and assume a more oval shape. Its average dimensions prior to hatching are 1010 μm by 570 μm (Sluss 1968). The hatched **larva** passes through **three instars** (Ogloblin 1924, Kadono-Okuda et al. 1995, Okuda & Kadono-Okuda 1995). As in many other parasitic wasps (Quicke 1997), the first instar larva, but not later instars, is equipped with grasping mandibles. It is believed that these serve to destroy other parasitoids within the host (Ogloblin 1924). Indeed, although *D. coccinellae* is a **solitary parasitoid**, more than one egg and/or first instar larva, sometimes as many as a few dozen, can be found in a single host (Ogloblin 1913, Balduf

1926, Maeta 1969), but only one larva survives to the second instar (Balduf 1926, Sluss 1968). Also the rate of **superparasitism** (proportion of parasitized hosts harbouring more than one parasitoid individual) in field samples may be quite high. Maeta (1969) found that 33–49% of parasitized *Coccinella septempunctata brucki* in Kurume (Fukuoka Prefecture, Japan) were superparasitized and Ceryngier (2000) recorded a superparasitism rate in *C. septempunctata* in Poland ranging between 22 and 64%. The frequent superparasitization of hosts by *D. coccinellae* may indicate a weak ability of this species to discriminate between unparasitized and already parasitized hosts. This was confirmed in laboratory tests by Okuda & Ceryngier (2000).

At moderate laboratory temperatures (20–26°C), the egg–larva **development of non-diapausing D. coccinellae** lasts 2–3 weeks (Sluss 1968, Obrycki & Tauber 1978, Obrycki et al. 1985, Obrycki 1989, Kadono-Okuda et al. 1995, Firlej et al. 2007), of which, 5–10 days are occupied with embryonic development (Balduf 1926, Sluss 1968, Kadono-Okuda et al. 1995). Development time in less suitable hosts, or those parasitized as juveniles, is longer than in suitable adult hosts (Obrycki et al. 1985, Obrycki 1989, Firlej 2007).

When the host stage parasitized is a larva or pupa, parasitoid development is arrested at the first instar larva until the eclosion of the host to adult (Kadono-Okuda et al. 1995). In dormant adult ladybirds, *D. coccinellae* usually diapauses as a first instar larva, and occasionally as an egg (Balduf 1926, Wright & Laing 1982). The **diapause** of *D. coccinellae* and that of its host ladybird are interrelated processes; in diapausing hosts, the parasitoid will not develop beyond the first larval instar (Kadono-Okuda et al. 1995) but causes a decrease in the duration of its host's diapause (Ceryngier et al. 2004).

In the early phase of *D. coccinellae* development, a stimulatory effect of the parasitoid on the maturation of the female host's gonads may be observed. Later, however, vitellin accumulated in the host oocytes is reabsorbed (Kadono-Okuda et al. 1995) and **ovarian maturation is inhibited** (Balduf 1926, Walker 1961, Maeta 1969, Wright & Laing 1978). The gonads of male hosts also seem to be affected by development of *D. coccinellae*, since an **inhibition of spermatogenesis** activity was found in the testes of parasitized *C. septempunctata* males (Ceryngier et al. 1992, 2004).

The **mature D. coccinellae larva** emerges from the host ladybird through the membrane between the fifth and sixth or between the sixth and seventh abdominal tergites, and then spins its **cocoon** between the legs of the host and pupates (Balduf 1926, Sluss 1968, Maeta 1969). About 30 minutes before the emergence of the larva, the ladybird becomes immobile and usually stays in this condition until its death, which most often happens within a few days. The legs of the immobilized beetle can only clasp the cocoon and so it is assumed that their extensor muscles are paralyzed (Balduf 1926, Bryden & Bishop 1945, Walker 1961).

The **duration of the pupal stage** of *D. coccinellae* at moderate temperatures (22–26°C) is about 7–10 days (Obrycki 1989, Obrycki et al. 1985, Firlej et al. 2007). After completing its pupal development, the wasp leaves the cocoon by biting through it at the cephalic end (Balduf 1926).

Larval nutrition

During most of its development, the *D. coccinellae* larva does not feed directly on the host tissues. As observed by Sluss (1968), only first instar larvae may directly consume the host's fat body. Later on, larval nutrition is mediated by **teratocytes** – cells derived from the serosa of the parasitoid egg. When the egg hatches, its serosal membrane dissociates into individual teratocytes which are released into the host haemocoel. Although in some hymenopteran parasitoids teratocytes may have functions other than parasitoid nutrition (e.g. host immunosupression, host growth regulation) (Dahlman 1990), it was found by Ogloblin (1924) that in *D. coccinellae* the major function of teratocytes is to provide food for the developing parasitoid larva. More recently, Okuda & Kadono-Okuda (1995) characterized a specific polypeptide synthesized by the teratocytes from amino acids absorbed from the host haemolymph. They detected this polypeptide in the guts of *D. coccinellae* larvae, proving that it was ingested by them. In the course of parasitism, teratocytes grow due to synthesis and accumulation of this specific polypeptide (Okuda & Kadono-Okuda 1995; Kadono-Okuda et al. 1998), and their number decreases as a result of larval feeding (Sluss 1968, Kadono-Okuda et al. 1995). The average initial number of teratocytes in a suitable host, such as adult *C. septempunctata brucki*, or adult *Col. maculata*, is more than 500 (Kadono-Okuda et al. 1995, Firlej et al. 2007). However, that number may be lower in less suitable hosts, e.g. larval *Har. axyridis* (Firlej et al. 2007).

The number of teratocytes gradually decreases with development of the parasitoid, until their complete or almost complete depletion when the larva emerges (Sluss 1968, Kadono-Okuda et al. 1995, Firlej et al. 2007). Meanwhile, the linear size of teratocytes that are not ingested increases more than 10-fold during the course of development from about 46–47 µm in diameter just after egg hatch to about 500 µm in diameter at emergence of the larva (Sluss 1968, Kadono-Okuda et al. 1995). In a less suitable host (*Har. axyridis* parasitized as a larva) the final size of teratocytes may be even larger (up to 900 µm in diameter), perhaps because of their excessive polyploidization and the prolonged development time of the parasitoid (Firlej et al. 2007).

During developmental arrest of the parasitoid larva in a juvenile or diapausing host, teratocyte development is also arrested; the diameter of such 'diapausing' teratocytes was found to be 45–50 µm in diapausing *Hippodamia convergens* adults (Sluss 1968) and 100–150 µm in diapausing *C. septempunctata brucki* adults (Kadono-Okuda et al. 1995).

As a consequence of the indirect feeding of *D. coccinellae* larva, most of the host organs remain intact during the development of the parasitoid larva. The only organs found to be strongly affected are the gonads, the development of which is inhibited, and the fat body that degenerates and considerably decreases in size (Sluss 1968). Due to the relatively little damage, the **ladybirds usually survive the emergence of the parasitoid larva**, although they are considerably weakened and paralyzed. Most of them die within the next few days, but some may recover (Timberlake 1916, Bryden & Bishop 1945, Walker 1961, Anderson et al. 1986, Triltsch 1996). Triltsch (1996) found that some laboratory parasitized *C. septempunctata* females not only survived parasitization but also started laying eggs about 12 days after the emergence of the parasitoid larva.

Host range and host suitability

D. coccinellae is generally considered a parasitoid of ladybirds of the subfamily **Coccinellinae**. It can successfully parasitize a wide array of species belonging to this subfamily, but some evidence indicates that representatives of other subfamilies or even families may also serve as occasional hosts (Table 8.5). Apart from

the field records listed in Table 8.5, there are also cases of successful laboratory parasitization by *D. coccinellae* of **non-Coccinellinae hosts.** Richerson & DeLoach (1972) bred a wasp from *Brachiacantha ursina* (Scymninae) and O. Nedvěd (unpublished) from *E. quadripustulatus* (Chilocorinae). However, laboratory tests (Klausnitzer 1969, Ghorpade 1979) showed that ladybirds of subfamilies other than Coccinellinae are not accepted by *D. coccinellae*. This suggests that, either more than one parasitoid species is involved, or different strains of *D. coccinellae* have different ranges of accepted and suitable hosts.

Suitability of ladybird species for development of *D. coccinellae* has usually been quantified as the **rate of successful parasitism** (proportion of hosts attacked to those allowing the emergence of adult parasitoids). Its reported values vary greatly, even within the same species and development stage of host (Table 8.6). Such variability may be a result of different methods used by different authors to determine the rates, but it may also reflect real differences in suitability, related, for example, to the physiological state of the hosts, their colour morphs (Berkvens et al. 2010), or the origin of parasitoid (Koyama & Majerus 2008) and host populations (Orr et al. 1992; Koyama & Majerus 2008).

Juvenile stages (larvae and pupae) of ladybirds are usually less suitable for *D. coccinellae* than adults, and younger larval instars are less suitable than older ones (David & Wilde 1973, Obrycki et al. 1985, Geoghegan et al. 1998b). However, Firlej et al. (2007, 2010) and Berkvens et al. (2010) found fourth instar larvae of *Har. axyridis* to be more suitable than the adults.

Host preferences and parasitization rates

The level of parasitization of ladybirds by *D. coccinellae* may fluctuate considerably depending on the locality, season and host. Sometimes, parasitization rates may reach 70% or more (Geoghegan et al. 1997, Ceryngier 2000), but usually much lower values are reported (for a review see Hodek 1973, pp. 203–204).

Ladybirds are often more heavily affected when aggregating at overwintering sites than when they are active (Iperti 1964, Parker et al. 1977, Anderson et al. 1986). Parasitization rates of such aggregating ladybirds may differ remarkably between different overwintering sites in the same region. In the mountainous region of southwestern Poland, Ceryngier (2000) found parasitization ranging from about 15 to 25% in *C. septempunctata* overwintering on mountain tops,

Table 8.6 Rates of successful parasitism in *Dinocampus coccinellae* of various geographic origins in relation to species, developmental stage and geographic origin of the hosts.

Host species	*D. coccinellae /* host* origin	Host stage parasitized	Rate of successful parasitism	Reference
Coccinella novemnotata	Missouri, USA	A	0.96	Richerson & DeLoach (1972)
Coleomegilla maculata	Missouri, USA	A	0.96	Richerson & DeLoach (1972)
	New Jersey, USA	A	0.80	Cartwright et al. (1982)
	Minnesota, USA	A	0.75	Hoogendoorn & Heimpel (2002)
	Iowa and Georgia, USA	A	0.61	Orr et al. (1992)
	Quebec, Canada	A	0.49	Firlej et al. (2007)
	New York, USA	A	0.40	Obrycki et al. (1985)
	Iowa, USA	A	0.34	Obrycki (1989)
	Quebec, Canada	A	0.18	Firlej et al. (2005)
	New York, USA	P	0.28	Obrycki et al. (1985)
	Quebec, Canada	L4	0.58	Firlej et al. (2007)
	New York, USA	L4	0.26	Obrycki et al. (1985)
	New York, USA	L3	0.19	Obrycki et al. (1985)
	New York, USA	L2	0.08	Obrycki et al. (1985)
	New York, USA	L1	0.00	Obrycki et al. (1985)
Hippodamia convergens	Missouri, USA	A	0.92	Richerson & DeLoach (1972)
	Iowa, USA	A	0.30	Obrycki (1989)
	Kansas, USA	L4	0.34	David & Wilde (1973)
	Kansas, USA	L3	0.13	David & Wilde (1973)
	Kansas, USA	L2	0.23	David & Wilde (1973)
	Kansas, USA	L1	0.10	David & Wilde (1973)
Coccinella septempunctata	UK	A	0.78	Sloggett et al. (2004)
	UK	A	0.64	Koyama & Majerus (2008)
	New Jersey, USA	A	0.58	Cartwright et al. (1982)
	Honshu, Japan / UK	A	0.57	Koyama & Majerus (2008)
	Iowa and Georgia, USA	A	0.49	Orr et al. (1992)
	Iowa, USA	A	0.32	Obrycki (1989)
C. septempunctata brucki	Honshu, Japan	A	0.55	Koyama & Majerus (2008)
	UK / Honshu, Japan	A	0.49	Koyama & Majerus (2008)
Hippodamia parenthesis	Missouri, USA	A	0.72	Richerson & DeLoach (1972)
Cycloneda munda	Iowa, USA	A	0.57	Obrycki (1989)
	Missouri, USA	A	0.12	Richerson & DeLoach (1972)
Coccinella undecimpunctata	New Jersey, USA	A	0.42	Cartwright et al. (1982)
Olla sp.	New Jersey, USA	A	0.39	Cartwright et al. (1982)
Harmonia axyridis	UK / Honshu, Japan	A	0.26	Koyama & Majerus (2008)
	Honshu, Japan	A	0.25	Koyama & Majerus (2008)
	Honshu, Japan / UK	A	0.17	Koyama & Majerus (2008)
	Belgium	A[†]	0.17	Berkvens et al. (2010)
	Minnesota, USA	A	0.10	Hoogendoorn & Heimpel (2002)
	Belgium	A[‡]	0.02	Berkvens et al. (2010)
	Belgium	A[§]	0.00	Berkvens et al.(2010)
	Quebec, Canada	A	0.00	Firlej et al. (2005)
	Quebec, Canada	A	0.00	Firlej et al. (2007)
	UK	A	0.00	Koyama & Majerus (2008)
	Belgium	L4[§]	0.15	Berkvens et al. (2010)
	Belgium	L4[‡]	0.07	Berkvens et al. (2010)
	Quebec, Canada	L4	0.12	Firlej et al. (2007)
	Belgium	L3[§]	0.10	Berkvens et al. (2010)

(Continued)

Table 8.6 *(Continued)*

Host species	*D. coccinellae* / host* origin	Host stage parasitized	Rate of successful parasitism	Reference
Hippodamia variegata	Iowa and Georgia, USA / Canada	A	0.17	Orr et al. (1992)
	Iowa and Georgia, USA / France	A	0.09	Orr et al. (1992)
	Iowa, USA	A	0.00	Obrycki (1989)
Adalia bipunctata	Missouri, USA	A	0.12	Richerson & DeLoach (1972)
	New Jersey, USA	A	0.00	Cartwright et al. (1982)
Brachiacantha ursina	Missouri, USA	A	0.04	Richerson & DeLoach (1972)
Propylea quatuordecimpunctata	Iowa, USA	A	0.02	Obrycki (1989)
	Iowa and Georgia, USA / Canada	A	0.01	Orr et al. (1992)
	Iowa and Georgia, USA / Turkey	A	0.00	Orr et al. (1992)
Coccinella magnifica	UK	A	0.00	Sloggett et al. (2004)

Host stages: A, adult, P, pupa, L1–L4, larval instars.
*Host origin is given only if different from *D. coccinellae* origin.
[†]Melanic hosts from long–term laboratory population (61–82 generations in the laboratory).
[‡]Melanic hosts from recently established laboratory population (3–6 generations in the laboratory).
[§]Non–melanic hosts from recently established laboratory population (3–6 generations in the laboratory).

while it was approximately three times higher at a hibernaculum at the foot of the mountain.

Due to parasitoid preferences, differential parasitization rates in relation to the species, sex, age and developmental stage of the hosts have been recorded. Majerus (1997), while comparing parasitization of adults of various British ladybirds, found mean parasitization rates of about 20% for *C. undecimpunctata*, *Har. quadripunctata* and *C. septempunctata*, the mean rate of 9.7% for *Coccinella quinquepunctata* and rates below 5% for nine other species (Table 8.7).

It was noted by several authors (Maeta 1969, Parker et al. 1977, Cartwright et al. 1982) that female ladybirds are parasitized to a higher degree than males, and Davis et al. (2006) confirmed experimentally the preference of British *D. coccinellae* for ovipositing into female rather than male *C. septempunctata* hosts. Moreover, British *D. coccinellae*, having a choice between young (newly eclosed) and old (overwintered) *C. septempunctata* adults, oviposited preferentially into young ones. Such behaviour was considered adaptive because young hosts are considered more suitable than old hosts due to their longer expected lifespan (Majerus et al. 2000a). *Dinocampus coccinellae* also

shows a clear preference for adult over juvenile hosts (Geoghegan et al. 1998b) and for older over younger larval hosts (Obrycki et al. 1985). British strains of the parasitoid appeared reluctant to oviposit in *C. septempunctata* larvae and especially pupae (Geoghegan 1998b), while in North American, Japanese and central European *D. coccinellae*, such a tendency has not been found (Maeta 1969, David & Wilde 1973, Obrycki et al. 1985, Berkvens et al. 2010, P. Ceryngier, unpublished). The reluctance of British *D. coccinellae* to parasitize pre-imaginal hosts is possibly not so strong when immatures of the invasive *Har. axyridis* serve as hosts. Ware et al. (2010) reported emergence of *D. coccinellae* from nine *Har. axyridis* adults collected in the wild as pupae, while in earlier samples of several thousand immatures of *C. septempunctata* no *D. coccinellae* were recovered (Geoghegan et al. 1998b).

It can be assumed that host preferences of a parasitoid are usually adaptive, i.e. positively related with host suitability. An interesting case of an almost certainly maladaptive relationship between host preference and suitability has been found for the system comprising Canadian *D. coccinellae* and invasive *Har.*

Table 8.7 Parasitization of British ladybirds by *Dinocampus coccinellae* measured as the rate of emergence of parasitoid larvae from the host beetles (from Majerus 1997).

Ladybird species	Total of samples	Number parasitized	Mean parasitization rate (%)
Coccinella undecimpunctata	262	58	22.1
Harmonia quadripunctata	284	55	19.4
Coccinella septempunctata	4222	734	17.4
Coccinella quinquepunctata	113	11	9.7
Calvia quatuordecimguttata	98	4	4.1
Hippodamia variegata	113	3	2.7
Myzia oblongoguttata	108	2	1.9
Propylea quatuordecimpunctata	562	10	1.8
Anatis ocellata	244	4	1.6
Tytthaspis sedecimpunctata	10537	148	1.4
Halyzia sedecimguttata	490	4	0.8
Coccinella magnifica	279	1	0.4
Coccinella hieroglyphica	234	1	0.4
Adalia bipunctata	4077	0	0
Adalia decempunctata	185	0	0
Anisosticta novemdecimpunctata	1381	0	0
Aphidecta obliterata	254	0	0
Myrrha octodecimguttata	57	0	0
Psyllobora vigintiduopunctata	143	0	0

axyridis. Although *Har. axyridis* adults are marginal hosts of a very low suitability for *D. coccinellae* (Burling et al. 2010; Table 8.6), in choice tests they were parasitized no less frequently than highly suitable *Col. maculata* adults. Furthermore, the wasps preferred to oviposit into *Har. axyridis* adults than much more suitable larvae of this species (Firlej et al. 2010).

8.3.2.2   *Uga* Girault (Hymenoptera: Chalcididae)

Synonym: *Neotainania* Husain & Agarwal.

Information on these wasps is scant. They form a highly specialized group of ladybird parasitoids and are distributed from Africa through southern Asia to Australia. Of seven described species, six have been proven to be parasitoids of larvae and pupae of phytophagous ladybirds of the subfamily **Epilachninae** (Table 8.8).

Probably the most widely distributed and most common of *Uga* species is *U. menoni* (Fig. 8.5). Kerrich (1960) reports that, in some seasons, this species can almost entirely eliminate *Epilachna* spp. in Orissa (southern India). In the Kyonggido area of Korea, *U. menoni* was found to be a **solitary larval–pupal**

parasitoid parasitizing older larvae (third and fourth instars) and pupae of *Henosepilachna vigintioctopunctata*. The wasp was present in the field from June to September and the highest parasitization rate was recorded in July (Lee et al. 1988).

8.3.2.3   *Cowperia* Girault (Hymenoptera: Encyrtidae)

Synonym: *Aminellus* Masi

Three of five described species of the genus *Cowperia* have been found to be **larval and pupal parasitoids** of ladybirds predaceous on mealybugs (Pseudococcidae) (Noyes & Hayat 1984) and whiteflies (Aleyrodidae) (Clausen 1934). Host associations of the remaining two species are not yet known (Table 8.9). *Cowperia* adults have characteristic, stout body form (Fig. 8.6).

More detailed studies on these parasitoids and their effects on host populations are lacking or unknown. Clausen (1934) mentions that, in the locality of Kaban Djahne in Sumatra, larvae and pupae of *Scymnus smithianus* were heavily parasitized by a species of *Aminellus* (= *Cowperia*) subsequently described by Kerrich (1963) as *A. sumatraensis*.

Table 8.8 Host and distribution records of *Uga* species (compiled from Noyes 2011 and Swaine & Ironside 1983 where indicated).

Uga species	Host records	Distribution records
colliscutellum	*Henosepilachna vigintioctopunctata, Henosepilachna guttatopustulata**	Queensland (Australia)
coriacea	*Epilachna canina, Epilachna* sp.	South Africa, Uganda
digitata	—	China
hemicarinata	*Henosepilachna vigintioctomaculata*	China
javanica	*Henosepilachna vigintioctopunctata*	Java (Indonesia)
menoni	*Epilachna* sp., *Henosepilachna ocellata, Henosepilachna vigintioctomaculata, Henosepilachna vigintioctopunctata*	India, Korea, Taiwan
sinensis	*Epilachna* sp.	China

*Host reported by Swaine & Ironside (1983).

8.3.2.4 *Homalotylus* Mayr (Hymenoptera: Encyrtidae)

Synonyms: *Anisotylus* Timberlake, *Echthroplectis* Foerster, *Hemaenasioidea* Girault, *Lepidaphycus* Blanchard, *Mendozaniella* Brethes, *Neoaenasioidea* Agarwal, *Nobrimus* Thomson.

This cosmopolitan genus is strictly associated with Coccinellidae (Timberlake 1919, Noyes & Hayat 1984, Trjapitzin & Triapitsyn 2003). Of 63 known species (Noyes 2011), 30 have been shown to develop in ladybirds although it is predicted that the remaining 17 also use ladybirds as a host (Table 8.10). Although there are also many reports of *Homalotylus* parasitizing aphids and coccids, they almost certainly resulted from erroneous host assignment due to mass rearing of the hemipterans together with their ladybird predators.

The majority of data on the biology and ecology of *Homalotylus* have been reported for *H. flaminius*, although most of them probably refer to *H. eytelweinii*. These two widely distributed species were historically regarded as one species (Timberlake 1919), and, for decades, both were usually reported by non-taxonomists as *H. flaminius*. However, most specialists now agree

Figure 8.5 *Uga menoni* in side view. Note strongly raised scutellum – a very characteristic feature of the genus *Uga* (from Kerrich 1960, Arthur Smith del., with permission).

that *H. eytelweinii* is a distinct species (Hoffer 1963, Graham 1969, Trjapitzin & Triapitsyn 2003, Noyes 2011). According to Klausnitzer (1969) *H. eytelweinii* is a parasitoid of ladybirds in the tribes Coccinellini, Psylloborini (= Halyziini) and Chilocorini, and *H. flaminius* parasitizes ladybirds in the Scymnini.

Life cycle

Homalotylus species are **endoparasitoids of ladybird larvae and pupae**. The species that develop in small hosts, e.g. of the tribe Scymnini, are usually **solitary** (Klausnitzer 1969, Vakhidov 1975, Lotfalizadeh & Ebrahimi 2001, Fallahzadeh et al. 2006), while those associated with larger ones are **gregarious** (Iperti 1964, Klausnitzer 1969, Kulman 1971).

Table 8.9 Host and distribution records of *Cowperia* species.

Cowperia species	Ladybird host records	Distribution records	References
areolata	*Cryptolaemus montrouzieri, Scymnus apetzi*	southern Europe, Tadzhikistan, Turkey, Georgia, Armenia	Noyes (2011)
indica	*Cryptolaemus montrouzieri, Jauravia* sp., *Nephus* sp., *Scymnus* sp.	India, Sri Lanka	Noyes (2011)
punctata	—	Singapore, China	Noyes (2011)
subnigra	—	China	Li & Chai (2008)
sumatraensis	*Scymnus smithianus*	Sumatra	Clausen (1934)
Cowperia sp.	*Cryptogonus kapuri*	India	Poorani (2008)

Figure 8.6 *Cowperia indica* (photo courtesy of J. Poorani). (See colour plate.)

Female wasps lay eggs into larvae of young instars (Kato 1968, Filatova 1974, Kuznetsov 1987), often when they are attached to the substrate at ecdysis (Iperti 1964). Larvae of the gregarious *Homalotylus* species avoid direct contact with their siblings and at the end of development they occupy separate

chambers with thin walls made of remnants of dry host tissue. In such cases, each emerging adult makes its own opening and emerges from the host (Iperti 1964, Filatova 1974). Some *Homalotylus* species (e.g. *H. eytelweinii, H. shuvakhinae*) tend to complete their development in mummified older larvae of ladybirds (Kato 1968, Yinon 1969, Semyanov 1986, Trjapitzin & Triapitsyn 2003), but others (e.g. *H. platynaspidis, H. nigricornis*) complete their development in the pupae (Vakhidov 1975, Myartseva 1981, Volkl 1995).

Homalotylus, at least certain species of the *flaminius* group, **overwinter** inside the mummified host as a prepupa, a fully grown larva (Iperti 1964, Filatova 1974), or as a pupa (Smirnoff 1957, Kuznetsov 1987). The reported **duration of development** in non-overwintering wasps ranges, depending on the region, season and host species, from only 7–9 days to 45 days (Telenga 1948, Rubtsov 1954, Smirnoff 1957, Iperti 1964, Kato 1968). The adults are sexually mature at emergence (Filatova 1974) and feed on aphid and coccid honeydew (Rubtsov 1954, Filatova 1974).

Host specificity

Different *Homalotylus* species show varying degrees of host specificity (Table 8.11). Many species are known to be associated with hosts belonging to only one tribe of Coccinellidae, but some (*H. eytelweinii, H. africanus, H. terminalis, H. quaylei*) can parasitize ladybirds of various subfamilies and tribes. Although most data on *Homalotylus* refers to a few species parasitizing ladybirds of the subfamilies Coccinellinae and Chilocorinae (*H. eytelweinii, H. terminalis*), the majority of known host associations in *Homalotylus* concern two tribes of the subfamily Scymninae: the Scymnini and Hyperaspidini (Table 8.11).

Table 8.10 Described species of *Homalotylus* with their distribution records and references reporting parasitization of Coccinellidae (compiled from Noyes 2011 and other sources where indicated). Lack of reference means that a given *Homalotylus* species has not so far been reported to parasitize ladybirds.

Homalotylus species	Distribution	Reference reporting parasitization of Coccinellidae
affinis	USA: California	Noyes (2011)
africanus	central and southern Africa	Noyes (2011)
agarwali	India	—
albiclavatus	India, Iran	Shafee & Fatma (1984), Fallahzadeh & Japoshvili (2010)
albifrons	Japan	Noyes (2011)
albitarsus	USA: Maryland	Timberlake (1919)
aligarhensis	India	—
balchanensis	Turkmenistan	—
brevicauda	Mexico	Timberlake (1919)
cockerelli	Mexico, USA: Texas	Noyes (2011)
ephippium	Europe, Iran, Russia: Yakutia	Noyes (2011)
eytelweinii	Europe, Asia, Africa, Central and South America	Noyes (2011)
ferrierei	India	Shafee & Fatma (1984)
flaminius	cosmopolitan	Noyes (2011)
formosus	India	—
hemipterinus	Europe, Asia, Africa, Indonesia, Malaysia, Fiji, Central and South America	Noyes (2011)
himalayensis	China	—
hybridus	Slovakia	—
hyperaspicola	Japan	Noyes (2011)
hyperaspidis	North and Central America	Noyes (2011)
indicus	India	Noyes (2011)
latipes	Paraguay	—
longicaudus	China	—
longipedicellus	India	Shafee & Fatma (1984)
mexicanus	Mexico	Trjapitzin et al. (1999)
mirabilis	South America	Noyes (2011)
mundus	Philippines, Taiwan	—
nigricornis	Europe, Asia, Canary Islands	Noyes (2011)
oculatus	Philippines	Noyes (2011)
pallentipes	USA: Arizona, Missouri	—
platynaspidis	Europe, western Asia	Noyes (2011)
punctifrons	USA: Florida	—
quaylei	Europe, Africa, Asia, South America	Noyes (2011)
rubricatus	Russian Far East	—
scutellaris	China	—
scymnivorus	India, Japan, Mongolia	Noyes (2011)
shuvakhinae	Mexico	Noyes (2011)
similis	USA	Noyes (2011)
sinensis	China, Iran	Noyes (2011)
singularis	Czech Republic	—
terminalis	North America, Antilles, Uruguay	Noyes (2011)
trisubalbus	China	—
turkmenicus	Turkmenistan, India, Iran	Noyes (2011)
vicinus	Africa, Madagascar, Iran	Noyes (2011), Fallahzadeh & Japoshvili (2010)
yunnanensis	China	—
zhaoi	China	—

After submitting of this Chapter, 17 new species of *Homalotylus* were described from Costa Rica, of which one, *H. hypnos* Noyes, has a known coccinellid host (*Hyperaspis* sp.) (Noyes 2011). These 17 species are not listed here.

Table 8.11 Reported associations of *Homalotylus* species with the subfamilies and tribes of their ladybird hosts (compiled from host records quoted by Noyes 2011 and other sources where indicated).

Homalotylus species	Coccinellinae		Epilachninae	Sticholotidinae	Chilocorinae	Scymninae			Ortaliinae	Coccidulinae	
	Coccinellini	Halyziini	Epilachnini	Sticholotidini	Chilocorini	Scymnini	Hyperaspidini	Platynaspidini	Noviini	Coccidulini	Azyini
affinis	–	–	–	–	–	–	+	–	–	–	–
africanus	–	–	–	–	+	+	+	–	–	+	–
albiclavatus	–	–	–	–	–	+	–	–	–	–	–
albifrons	–	–	–	–	–	–	–	–	–	–	–
brevicauda	–	–	–	–	–	+†	–	–	–	–	–
cockerelli	–	–	–	–	–	–	+	–	–	–	–
ephippium	–	–	–	–	+	–	–	–	–	–	–
eytelweinii + flaminius*	+	+	+	+	+	+	+	+	–	–	–
hemipterinus	+	–	–	–	+	–	–	–	+	–	–
hyperaspicola	–	–	–	–	–	–	+	–	–	–	–
hyperaspidis	–	–	–	–	–	–	+	–	–	–	–
indicus	–	–	–	–	–	+	–	–	–	–	–
mirabilis	+	–	–	–	–	+	–	–	–	–	–
nigricornis	–	–	–	–	–	+	+	–	–	–	–
oculatus	–	–	–	–	–	+	–	–	–	–	–
platynaspidis	–	–	–	–	–	+	–	+	–	–	–
quaylei	–	–	–	+	–	+	+	–	–	–	–
scymnivorus	–	–	–	–	–	+	–	–	–	–	–
shuvakhinae	–	–	–	–	–	–	–	–	–	–	+
similis	–	–	–	–	–	+	+	–	–	–	–
sinensis	+	–	–	–	+	+	–	–	–	–	–
terminalis	+	+	–	–	+	+	–	–	–	–	–
turkmenicus	+	–	–	–	+‡	+§	+	–	–	–	–
vicinus	–	–	–	–	–	+	+	–	–	–	–
Number of species	6	2	1	2	7	16	11	2	1	1	1

*Data for *H. eytelweinii* and *H. flaminius* are combined because these species have not been differentiated by many authors.
†Timberlake (1919).
‡Fallahzadeh & Japoshvii (2010).
§Lotfalizadeh (2010).

Parasitization rates

In regions with temperate and cold climates, the Old World *Homalotylus* of the *flaminius* group (presumably *H. eytelweinii*) usually show low rates of parasitization. In the Russian Far East, Western Siberia, Turkmenistan and Uzbekistan, it has only been recorded sporadically from several aphidophagous ladybirds (Filatova 1974, Vakhidov 1975, Semyanov 1986, Kuznetsov 1987). In northeastern and central parts of Europe, *Homalotylus* is again not common, although it may sometimes parasitize ladybirds at 20–30% (Klausnitzer 1967, Semyanov 1986, Pankanin-Franczyk & Ceryngier 1999). In the Mediterranean and Black Sea basins, parasitization often reaches moderate or high levels. Stathas et al. (2008) recorded up to 50% parasitization of *Chilocorus bipustulatus* on sour orange trees in southern Greece. In southeastern France, Iperti (1964) noted rates of parasitization exceeding 80% for *C. septempunctata* and 50% for *Propylea quatuordecimpunctata*. Both Stathas et al. (2008) and Iperti (1964) found that parasitization rates tend to increase as the host population develops. According to Smirnoff (1957), 95% of *Chil. bipustulatus* individuals developing on date palms on the Atlantic coast of Morocco were destroyed by the parasitoid. In equatorial Africa (Republic of the Congo), Fabres (1981) found parasitization rates of *Exochomus flaviventris* ranging between 7 and 10% when feeding on the cassava mealybug.

In Germany, Volkl (1995) recorded that parasitization of myrmecophilous *Platynaspis luteorubra* by *H. platynaspidis* may even be higher than 50%. However, the parasitization rate largely depended on the absence of ants (*Lasius niger*) tending aphids exploited by *P. luteorubra* larvae. In ant-tended aphid colonies, especially in simply structured habitats such as plant stems, parasitization rates were significantly lower. Although ants were not found to be aggressive towards *H. platynaspidis*, the searching efficiency of the parasitoid might be highly reduced by continual encounters with ants.

8.3.2.5 *Nothoserphus* Brues (Hymenoptera: Proctotrupidae)

Synonyms: *Thomsonina* Hellen, *Watanabeia* Masner.

The genus *Nothoserphus* seems to be geographically confined to **Palaearctic and Oriental regions** (Johnson 1992). Of 11 species described so far, five have been found to be **solitary parasitoids of ladybird larvae**, while the host records for the remaining

six species are lacking (Table 8.12). The genus is divided into **three species-groups** – the *boops-*, *afissae-* and *mirabilis*-groups – and this corresponds with the earlier taxonomic allocation of the *Nothoserphus* species to three distinct genera: *Thomsonina*, *Watanabeia* and *Nothoserphus* (Townes & Townes 1981, Lin 1987). According to the known host records (Table 8.12), species of the *boops*-group are parasitoids of the Scymninae, those of the *afissae*-group parasitize Epilachninae and those of the *mirabilis*-group parasitize Coccinellinae.

Nothoserphus females lay eggs in host larvae. *Nothoserphus afissae* prefers to oviposit in the second and third larval instars of *Henosepilachna vigintioctomaculata*, which are also much more suitable for the development of the parasitoid than either the younger or older larvae (Kovalenko 2002). If the host larvae are parasitized at the first instar, they suffer 100% mortality before the parasitoid larvae has completed development. Oviposition into fourth instar larvae leads to 60% mortality of hosts, and, subsequently no successful development of *N. afissae*. According to Semyanov (1998), females of *N. mirabilis* preferentially parasitize third and fourth instar larvae of *Menochilus sexmaculatus* and do not react to the presence of pupae and first instar larvae.

As a consequence of parasitoid feeding, the host dies at the prepupal stage. The fully grown parasitoid larva emerges from the dead host through the membrane between the two last abdominal sternites and pupates in characteristic posture (Fig. 8.7) on the ventral side of the host's body (Semyanov 1998, Kovalenko 2002).

Kovalenko (2002) and Kovalenko & Kuznetsov (2005) found that *N. afissae* may be an **effective control agent of the herbivorous *H. vigintioctomaculata*** on potato fields in the Russian Far East. During larval development of the beetle population, which takes place between late June and late August, three generations of the parasitoid are produced. **Parasitization rates** gradually increase and may even reach 100% in August. In Japan, however, only 1.5% of the same ladybird species (*H. vigintioctomaculata*) feeding on the same host plant (potatoes) was parasitized (Nakamura 1987). Similarly, low parasitization rates were recorded in relation to two Japanese species feeding on thistles (*Cirsium*) – *Henosepilachna niponica* (Ohgushi 1986) and *H. pustulosa* (Nakamura & Ohgushi 1981).

In a single sample taken from maize field in southeastern China, 19% of larvae of the predatory

Table 8.12 Host and distribution records of *Nothoserphus* species.

Nothoserphus species	Host records	Distribution records	References
***boops*–group (*Thomsonina*)**			
boops	*Scymnus nigrinus*	Scandinavia, Czech Republic	7, 8, 13
fuscipes	—	Taiwan	4
partitus	—	Taiwan	4
scymni	*Scymnus dorcatomoides, Scymnus* sp.	Japan	7
townesi	—	Taiwan	4
***afissae*–group (*Watanabeia*)**			
aequalis	—	Nepal, Taiwan	4
afissae	*Epilachna admirabilis, Epilachna varivestis*, Henosepilachna niponica, Henosepilachna pustulosa, Henosepilachna vigintioctomaculata*	Japan, Korea, Russian Far East	1, 2, 3, 5, 6, 7
debilis	–	Nepal, Taiwan	4
epilachnae	*Epilachna admirabilis, Henosepilachna vigintioctopunctata*	Java, China	7
***mirabilis*–group (*Nothoserphus*)**			
admirabilis	—	Taiwan	4
mirabilis	*Coccinella leonina transversalis, Illeis bielawskii, Illeis cincta, Illeis* sp., *Menochilus sexmaculatus, Synona obscura*, undetermined coccinellid	south–eastern China, India, Java, Nepal, Taiwan	4, 9, 10, 11, 12

*Record from the area in Japan invaded by *E. varivestis* (outside its native range).
References: 1, Fujiyama et al. (1998); 2, Kovalenko (2002); 3, Lee et al. (1988); 4, Lin (1987); 5, Nakamura & Ohgushi (1981); 6, Ohgushi (1986); 7, Pschorn–Walcher (1958); 8, Pschorn–Walcher (1971); 9, Poorani (2008); 10, J. Poorani, unpublished; 11, Poorani et al. (2008); 12, Semyanov (1998); 13, Zeman & Vanek (1999).

Figure 8.7 A pupa of *Nothoserphus mirabilis* attached to the ventral side of its killed host, a larva of *Menochilus sexmaculatus* (photo courtesy of J. Poorani). (See colour plate.)

M. sexmaculatus were parasitized by *N. mirabilis* (Semyanov 1998).

8.3.2.6 *Metastenus* Walker (Hymenoptera: Pteromalidae)

Synonyms: *Scymnophagus* Ashmead, *Tripolycystus* Dodd.

Five species of the genus *Metastenus* have been described so far, and two of them, *M. concinnus* and *M. townsendi*, appear to be **pupal parasitoids** of ladybirds belonging to the subfamilies Scymninae (tribes Scymnini, Hyperaspidini, Pentiliini and Cryptognathini), Coccidulinae (Azyini) and Microweiseinae (Microweiseini) (Table 8.3). Hosts of the remaining species, i.e. the Australian *M. sulcatus*, the Hungarian

M. caliginosus and the Indian *M. indicus*, are not known (Noyes 2011).

Metastenus concinnus is generally reported from the Palaearctic, with the exception of a single record from Argentina, while the **distribution** of *M. townsendi* includes the Nearctic and Neotropical regions and Japan (Noyes 2011). An undetermined species of *Metastenus* has also been reported as a parasitoid of *Exochomus* and *Hyperaspis* species from several localities in tropical Africa (Neuenschwander et al. 1987). These records possibly relate to *Mesopolobus secundus* – a species reported from Uganda and Kenya (Table 8.3), which resembles *M. townsendi* (Crawford 1912, Delucchi 1954).

Life cycle

Most information on the biology and life cycle of *Metastenus* can be found in Delucchi's (1954) characterization (as *Scymnophagus mesnili*), of *Metastenus concinnus*. Delucchi (1954) reports parasitization of *Scymnus* (*Pullus*) *impexus* by this species in Switzerland and Germany. Female parasitoids were found ovipositing into host pupae in late spring (May and June). The parasitoids either developed directly, emerging as adults between late June and mid-July, or entered larval diapause inside the dead host pupa, emerging the next spring. It is not known which hosts the second generation wasps (the progeny of non-diapausing individuals) developed in.

The pupae of *S. impexus* parasitized by *M. concinnus* are easily distinguishable because they change colour from reddish-brown to light brown. A single parasitized pupa usually contains two parasitoid individuals, which lie face-to-face in their pupal stage (Delucchi 1954).

The recorded impact of *M. concinnus* on *S. impexus* is not very high. Delucchi (1954) mentions **parasitization rates** in three large samples (each of at least 1000 pupae) that range between 0.8 and 12.6%.

M. concinnus probably reproduces by **thelytokous parthenogenesis**, since males of this species are unknown (Delucchi 1954). In *M. townsendi*, however, both sexes are known (Peck 1963, Tachikawa 1972).

8.3.2.7 *Phalacrotophora* Enderlein (Diptera: Phoridae)

The worldwide-distributed genus *Phalacrotophora* comprises more than 50 described species (Brown 2007).

Some of them are known as **gregarious endoparasitoids of ladybird pupae** (Table 8.13).

Host ranges

For most of the *Phalacrotophora* species parasitizing ladybirds, records of host species are very scarce (Table 8.13) and thus tell us little about the real host spectra of the flies. The two exceptions are the European *P. fasciata* and *P. berolinensis*. Both have been reported relatively frequently from a variety of ladybirds in the subfamilies Coccinellinae and Chilocorinae (see Disney et al. 1994 and Ceryngier & Hodek 1996). In contrast, another European species, *P. beuki*, was recognized as a monophagous parasitoid of *Anatis ocellata* (Durska et al. 2003).

Life cycle

A sexually mature *Phalacrotophora* female will locate and select a ladybird prepupa, and then start to attract males. As the females in this genus possess complex glands on the abdomen which probably serve for the production of pheromones, Disney et al. (1994) assume that the males are attracted pheromonally. Disney et al. (1994) observed that one or more males may arrive to mate with the female, most often on or near the selected ladybird prepupa. A pair usually spends several minutes *in copula* (Disney et al. 1994).

After mating, the male leaves the female. The female still attends the prepupa and starts to parasitize it, usually at the beginning of its pupation (Disney et al. 1994, Ceryngier & Hodek 1996), although in rare cases it is the prepupa that is parasitized (Filatova 1974, Disney & Chazeau 1990, Disney et al. 1994). It has been shown that females of *P. fasciata* and/or *P. berolinensis* can assess the age of host prepupae and select older ones (Hurst et al. 1998). This is considered adaptive because the female loses less time waiting for a host ecdysis before ovipositing and consequently increases her oviposition rate.

Females of *P. fasciata* and *P. berolinensis* were observed **laying eggs** either on the surface of the pupa or internally. Usually, the ventral thoracic region of the pupa, and less frequently its dorsal surface, was chosen as the areas for oviposition (Disney et al. 1994). Females of *P. quadrimaculata* were seen to attach their eggs to the sides of the host pupa or, to a lesser extent, to its ventral surface (Disney & Chazeau 1990).

Table 8.13 *Phalacrotophora* species reported as ladybird parasitoids.

Phalacrotophora species	Distribution	Reported hosts	References
berolinensis	Europe	many Coccinellinae and Chilocorinae	Disney et al. (1994), Ceryngier & Hodek (1996)
beuki	Europe	*Anatis ocellata*	Durska et al. (2003)
decimaculata	China	unidentified ladybird	Liu (2001)
delageae	Europe	*Adalia bipunctata* *Adalia decempunctata* *Calvia quatuordecimguttata* *Coccinella septempunctata* *Harmonia axyridis* *Sospita vigintiguttata*	Disney & Beuk (1997), Triltsch (1999), Durska et al. (2003), P. Ceryngier, E. Durska & R. H. L. Disney (unpublished), S. Harding, R. H. L. Disney & R. L. Poland (unpublished)
fasciata	Europe, Siberia, Russian Far East, China, Japan	many Coccinellinae and Chilocorinae	Disney et al. (1994), Ceryngier & Hodek (1996), Miura (2010), Lengyel (2011)
indiana	India	*Harmonia eucharis*	Colyer (1961)
nedae	South America	*Neda marginalis* *Neocalvia anastomozans* *Cycloneda sanguinea*	Disney et al. (1994), Aguiar–Menezes et al. (2008)
philaxyridis	Japan (and probably part of continental Asia)	*Harmonia axyridis*	Disney (1997)
quadrimaculata	Taiwan, Sulawesi, New Caledonia, China	*Olla v–nigrum*	Disney & Chazeau (1990), Liu (2001)
Phalacrotophora sp.	Australia	*Harmonia conformis* *Cleobora mellyi*	New (1975), Anderson et al. (1986)
Phalacrotophora sp.	India	*Harmonia expalliata*	Disney et al. (1994)

The eggs hatch within 24 h, and the larvae then enter the host (Wylie 1958, Disney & Chazeau 1990). **Larval development** in *Phalacrotophora* proceeds rapidly. According to Lichtenstein (1920, cited by Disney et al. 1994) and Disney and Chazeau (1990), it may be as short as 2–3 days, but longer times are more often reported. Wylie (1958) indicates 5 days, Filatova (1974) 7–9 days and Kuznetsov (1987) 8–12 days.

Fully grown larvae leave the host through an irregular hole, ventrally between the head and thorax, drop to the ground and **pupate** on its surface or in the upper soil layer (Wylie 1958, Disney & Chazeau 1990). The flies either emerge as adults after 15–25 days or overwinter and emerge the next season (Disney et al. 1994, Ceryngier & Hodek 1996). *Phalacrotophora* **overwinter as fully formed adults within the puparia** (Disney 1994, Durska et al. 2003).

The number of *Phalacrotophora* individuals developing in a single host is related to host size. Generally, larger host species bear a higher number of parasitoid larvae than smaller ones. In pupae of *An. ocellata*,

7–10 larvae were usually reported (Filatova 1974), the average number given for *C. septempunctata* is similar (Semyanov 1986, Disney et al. 1994), but only around two develop from *A. bipunctata* (Disney 1979, Disney et al. 1994).

Many authors (Richards 1926, Colyer 1952, 1954, Delucchi 1953, Disney et al. 1994) have reported that *Phalacrotophora* **feeds on the haemolymph of ladybird** pupae or, less frequently, that of larvae. The fly cuts into either the ventral (Colyer 1952) or dorsal (Disney et al. 1994) surface of the pupa and then imbibes fluid from the wound. This behaviour may allow the fly to obtain appropriate protein-rich food and/or facilitate penetration of the pupa by the hatching larvae (Disney et al. 1994). The small wound and loss of haemolymph, as a consequence of the fly's feeding, do not prevent the development of unparasitized ladybird pupae or the development of flies in parasitized pupae (Disney et al. 1994). Apart from host feeding, adults probably feed on the sap of injured trees. Lengyel (2009) reports high numbers of

Phalacrotophora spp. (mostly *P. beuki*) gathering at the sap exuding from wounds on elm and, less frequently, poplar and maple trees.

Another interesting behaviour of adult *Phalacrotophora* is the **swarming** of the flies around the bases of tree trunks. Early observations of this phenomenon (Colyer 1952, 1954) mostly refer to *P. berolinensis*, but recently Irwin (2006) found a swarm of another species, *P. delageae* in England. It turned out that these swarms, both of *P. berolinensis* and *P. delageae*, exclusively, or almost exclusively, consist of females. Individual flies may temporarily separate from the swarm to settle on the tree trunk with their wings outstretched and the abdomen convulsively expanding and contracting (Schmitz 1929, cited by Colyer 1952), or they may take up a posture in which the wings are vibrating and the membranous patch at the base of the fifth tergite is displayed (Irwin 2006). The function of the swarming and associated 'display postures' is unknown. It could be to attract males (Irwin 2006), but males have not been reported to respond to such female behaviour.

Parasitization rates

Parasitization of ladybird pupae by *Phalacrotophora* may differ greatly depending on the year, season, host species and region, sometimes reaching high values. For example, Disney et al. (1994) and Hurst et al. (1998) recorded combined parasitization rates of *C. septempunctata* by *P. fasciata* and *P. berolinensis* that exceeded 80%.

Distinct differences in parasitization rates of particular host species have been recorded, even for the most polyphagous species *P. fasciata*. In the St. Petersburg region, Lipa and Semyanov (1967) found that parasitization of *Myzia oblongoguttata* might reach 25%, while in other species it was less than 5%. In West Siberia, Filatova (1974) recorded up to 45% parasitization of *C. septempunctata*, *Anatis ocellata* and *Hip. tredecimpunctata*, up to 25–30% parasitization of *Har. axyridis*, *A. bipunctata* and *E. quadripustulatus*, and below 15% parasitization of other species. In Britain, Disney et al. (1994) found, that of the three arboreal species, *A. decempunctata*, *A. bipunctata* and *Calvia quatuordecimguttata*, the most affected by the flies was the least abundant *C. quatuordecimguttata*.

Data on the parasitization rates of ladybird pupae by *Phalacrotophora* species other than *P. fasciata* are scant. Wylie (1958) reports more than 50% parasitization of

Aphidecta obliterata by *P. berolinensis* on conifers in the Vosges Mountains (eastern France). During three consecutive study seasons in central Poland, Durska et al. (2003) recorded mean parasitization rates of *An. ocellata* by *P. beuki* ranging between 36 and 41%. Parasitization of *Har. axyridis* by *Phalacrotophora* sp. (possibly *P. philaxyridis*) was found to reach up to 17.7% in Kyoto (central Japan) (Osawa 1992a), and to fluctuate between 0.4 and 6.7% during May–October in the Chuncheon area of Korea (Park et al. 1996). In New Caledonia, Disney & Chazeau (1990) noted *P. quadrimaculata* parasitizing 15–79% (39% on average) of an introduced ladybird species, *Olla v-nigrum*.

Multiparasitism

The parasitization of a single host by two species, *P. fasciata* and *P. berolinensis*, has been recorded several times. The host species involved were *A. bipunctata* (Disney 1979, Disney et al. 1994), *M. oblongoguttata* and *C. magnifica* (Ceryngier & Hodek 1996).

Host defence

The usual response of ladybird prepupae and pupae to physical disturbances, including those caused by predators and parasitoids, is 'flicking', i.e. rapid and repeated raising and dropping of their anterior end. Disney et al. (1994) found that this behaviour is relatively ineffective as a defence against *Phalacrotophora* species. Ovipositing *Phalacrotophora* species were only deterred in 17 out of 61 observations by flicking of *A. bipunctata* prepupae and pupae. Disney & Chazeau (1990) reported that flicking by *O. v-nigrum* did not seem to deter *P. quadrimaculata* from oviposition.

The common defensive behaviour of ladybirds against enemies, reflex bleeding, may also be shown by the prepupae defending themselves from *Phalacrotophora*. Disney et al. (1994) found *C. septempunctata* prepupae to reflex bleed in response to ovipositor insertion by the flies. In one instance of five prepupae observed, fluid secreted from the dorsal surface contaminated the legs of the fly, and this deterred it from ovipositing.

8.3.2.8   *Oomyzus scaposus* (Thomson) (Hymenoptera: Eulophidae, Tetrastichinae)

Synonyms: *Tetrastichus coccinellae* Kurdjumov, *T. melanis* Burks, *T. sexmaculatus* Chandy Kurian, *Syntomosphyrus taprobanes* Waterston.

Oomyzus scaposus is probably the most widely distributed and most common species of ladybird larval and pupal parasitoids belonging to the eulophid subfamily Tetrastichinae. Several other species of lesser importance can be found within the genera *Aprostocetus*, *Baryscapus*, *Oomyzus* and *Tetrastichus* (Table 8.3).

According to the **distribution** records quoted by Noyes (2011), *O. scaposus* occurs in the Nearctic, Palaearctic, Oriental and Australasian regions. The reported localities extend from high latitudes in the Northern Hemisphere, even beyond the Arctic Circle (the Murmansk region of Russia), through the tropics (Colombia, southern India, Sri Lanka, Indonesia), to the temperate areas in the Southern Hemisphere (New Zealand). The only big land masses where this species has not been recorded are South America (except Colombia) and sub-Saharan Africa.

The **host range** of *O. scaposus* reported by Noyes (2011) is also quite wide. It mostly includes ladybird species of the tribes Coccinellini (subfamily Coccinellinae) and Chilocorini (Chilocorinae), but there are also single reports of infection of *Scymnus* sp. (Scymninae) and *Chrysopa* spp. (Neuroptera: Chrysopidae).

Tetrastichinae is an extremely difficult group taxonomically (LaSalle 1993) and hence many records concerning *O. scaposus* and related species should be treated with caution.

Life cycle

Females of *O. scaposus* prefer to **lay eggs in third and fourth instar ladybird larvae**, although they have also been reported ovipositing in pupae (Iperti 1964, Klausnitzer 1969), as well as younger larvae (Filatova 1974). According to Semyanov (1986), the eggs are usually inserted between the thorax and abdomen of the host, sometimes between its pleura, and sporadically into the head capsule. The process of oviposition lasts from one to 2.5 minutes. After removing her ovipositor, the **female feeds on the fluid from the wound** (Ogloblin 1913, Semyanov 1986). The female may oviposit up to three times into the same host (Semyanov 1986).

A parasitized ladybird larva usually develops to the pupal stage and then dies and becomes darker. Adult wasps emerge from one, or sometimes more, small openings bitten through the cuticle in the dorsal side of the pupa (Filatova 1974). If young larvae (first and second instar) are parasitized, wasps may emerge before the host pupates (Filatova 1974).

Many *O. scaposus* individuals can develop successfully within a single host; up to 47 were recorded by Semyanov (1986) emerging from individual pupae of *C. septempunctata*. The wasps are sexually mature at emergence and mate within a few minutes of emerging (Iperti 1964, Filatova 1974).

The **development** of *O. scaposus* has been reported to last 20–32 days in southeastern France (Iperti 1964), about 20 days in the Poltava region (Ukraine) (Ogloblin 1913) and 15–26 days in the Russian Far East (Kuznetsov 1987). Filatova (1974) found that wasps developed in 15–18 days at a mean daily temperature of 23°C.

Due to relatively fast development, *O. scaposus* can produce **several generations per year** under favourable conditions (Telenga 1948, Iperti 1964). Parasitoids developing later in the season enter **diapause** to overwinter as prepupae inside their dead hosts (Telenga 1948, Iperti 1964, Filatova 1974).

Parasitization rates

Oomyzus scaposus is sometimes reported as an important mortality factor of ladybirds, especially of multivoltine coccidophagous *Chilocorus* spp. in warmer parts of the Palaearctic region. The level of parasitization of successive generations of these ladybirds tends to rapidly increase, so that, late in the season, high parasitization rates can be observed (Rubtsov 1954, Murashevskaya 1969, Stathas 2001, Stathas et al. 2008).

Aphidophagous ladybirds are more dispersive than species of *Chilocorus* and such time-dependent trends of parasitization rates by *O. scaposus* are more difficult to demonstrate. Nevertheless, Iperti (1964) found that, in southeastern France, parasitization of *C. septempunctata* developing early in the season (April–May) was 0–1%, while later (June–July) it was almost 20%. The rates of parasitization by *O. scaposus* reported for aphidophagous Coccinellidae are generally very variable and may fluctuate considerably from year to year (Dean 1983, Semyanov 1986, Pankanin-Franczyk & Ceryngier 1999).

8.3.2.9 *Pediobius foveolatus* (Crawford) (Hymenoptera: Eulophidae, Entedoninae)

Generic synonym: *Pleurotropis* Foerster.

Species synonyms: *Pediobius epilachnae* (Rohwer), *P. mediopunctata* (Waterston), *P. simiolus* (Takahashi), *Mestocharis lividus* Girault.

Pedobius foveolatus is a **gregarious larval endoparasitoid of phytophagous ladybirds** of the subfamily Epilachninae (Lall 1961, Kerrich 1973). The report on predaceous *C. septempunctata* and *Menochilus sexmaculatus* as hosts of this wasp (Bhatkar & Subba Rao 1976) certainly relates to a different parasitoid species (Bledsoe et al. 1983).

Pedobius foveolatus is widely **distributed** in Afrotropical, Oriental, southern Palaearctic and Australasian regions (Kerrich 1973). Additionally, it was **introduced to the Nearctic** region (USA) to control *Epilachna varivestis*. However, establishment of the parasitoid in that region is impeded in winter. Therefore, laboratory-bred wasps are released annually (Stevens et al. 1975a, 1975b, Schaefer et al. 1983).

Life cycle

Pedobius foveolatus reproduces arrhenotokously (Stevens et al. 1975b). The adults mate soon after emergence from their dead (mummified) hosts (Lal 1946, Lall 1961, Stevens et al. 1975b). The process of mating usually lasts 15–30 seconds (Lal 1946, Lall 1961).

Female *P. foveolatus* oviposit into host larvae, preferentially of older instars (Appanna 1948, Lall 1961, Stevens et al. 1975b), and sometimes into freshly moulted pupae (Appanna 1948). Lall (1961) found that young *Epilachna* larvae (first and second instar) are not suitable for *P. foveolatus* development because, in such hosts, parasitoid larvae die before pupation. Appanna (1948) observed first instar larvae of *Henosepilachna vigintioctopunctata* to be stung by *P. foveolatus*, but without oviposition. Nevertheless, the attack caused the darkening of the larvae that is typical of parasitization.

According to Appanna's (1948) detailed observations of *P. foveolatus* **oviposition behaviour**, the female stings a host larva several times. The first two stings are brief and are not accompanied by oviposition. They cause sudden quick movements of the host; this behaviour is not displayed during later stings, which are associated with oviposition. It is possible that the initial stings inject paralyzing venom into the host's body, so the wasp can begin egg laying, a process lasting 3–4 minutes. An individual female may oviposit three times into the same host at intervals of about 15–20 minutes. Eggs are laid just below the larval skin through the soft intersegmental membranes, usually on the dorsal surface of the larva.

At 25°C, **egg incubation** lasts about 2 days. The hatched **larvae pass through three instars** and pupate inside the host mummy about one week after hatching. The **pupal stage** lasts 4–5 days. **Emergence of adults** takes place approximately 13.5 days after egg deposition (Bledsoe et al. 1983).

Parasitization by *P. foveolatus* leads to the death of the host larva or pupa and its **mummification**. A few days after parasitization, the host larva stops feeding, dies, darkens and hardens (Lal 1946, Lall 1961, Appanna 1948, Stevens et al. 1975b). Adult *P. foveolatus* emerge from the host mummy through one or, more rarely, two, holes on the dorsal side (Appanna 1948). The reported number of individuals emerging from a single field-collected host usually ranges from one or a few to 20–30 (Lal 1946, Appanna 1948, Lall 1961, Barrows & Hooker 1981). As the mean clutch size of *P. foveolatus* in an unparasitized larva of *E. varivestis* was found in laboratory tests to be about 13 (range 5–22) (Hooker & Barrows 1992), the higher values within the reported range might refer to cases of **superparasitism**. It has been shown that females of *P. foveolatus* can superparasitize, in spite of their ability to discriminate between previously parasitized and unparasitized hosts (Shepard & Gale 1977, Hooker & Barrows 1992). **Host discrimination** in this species is expressed by much less frequent ovipositions into parasitized than unparasitized hosts (Shepard & Gale 1977), and by reduction of the clutch size laid in the former (Hooker & Barrows 1992).

Parasitization rates

Pedobius foveolatus has often been reported to limit considerably the populations of epilachnine ladybirds. On potato fields around Bangalore (India), Appanna (1948) recorded average parasitization of *H. vigintioctopunctata* to exceed 40%; it was about 30% in February and March, but reached much higher values (60–77%) later in the season. Parasitization of the same host species and in the same region, reported by Venkatesha (2006) on a medicinal plant, *Withania somnifera*, was similarly high (range 52–70%). In Sumatra, parasitization rates of pupae of *Epilachna* sp. on bitter melon (*Momordica charantia*) were around 25% between March and September, and 60% between October and December (Abbas & Nakamura 1985).

Pedobius foveolatus *as a biocontrol agent*

Pedobius foveolatus can parasitize many phytophagous ladybirds, including those which are serious pests of cultivated plants, and so, attempts have been made to use this parasitoid as a biological control agent. The best-known example involves releases of the wasp in the eastern USA against the notorious pest of soy and other beans, the Mexican bean beetle (*E. varivestis*). It was found that *P. foveolatus* of Indian and Japanese origin can successfully parasitize *E. varivestis* (Angalet et al. 1968, Schaefer et al. 1983), as well as some other American Epilachninae (Romero-Napoles et al. 1987). However, neither Indian nor Japanese wasps can over-winter in the areas where they were released (Schaefer et al. 1983). Schaefer et al. (1983) supposed that it was not climatic constraints, but the lack of a suitable winter host that prevents establishment of *P. foveolatus* in North America. They hypothesized that in its native range, *P. foveolatus* spends the winter months as dia-pausing larvae inside overwintering larvae of certain Epilachninae, e.g. *Epilachna admirabilis*. Since all eastern North American Epilachninae overwinter as adults, there is no possibility for the parasitoid to survive the winter in the field.

Due to the **inability of *P. foveolatus* to overwin-ter in North America**, control of the Mexican bean beetle by this species can only be achieved through the releases of laboratory-reared wasps. Preliminary results of such releases in soybean fields and vegetable gardens appeared promising (Stevens et al. 1975a, Barrows & Hooker 1981) and so the wasps became commercially produced for gardeners and farmers (Schaefer et al. 1983).

Less known, but no less successful, was the applica-tion of *P. foveolatus* against *Henosepilachna vigintisex-punctata* on Saipan (Mariana Islands). This ladybird was accidentally introduced to the island in 1948 and became a serious pest of solanaceous crops. Releasing *P. foveolatus* in 1985 rapidly suppressed the beetle pop-ulation. An island-wide survey performed in 1989 revealed that about 80% of *H. vigintisexpunctata* larvae were parasitized (Chiu & Moore 1993).

8.4   PARASITES AND PATHOGENS

8.4.1   Acarina

8.4.1.1   Phoretic mites

Mites found on coccinellids may be divided into those that are parasitic, and those that simply use the coc-cinellid as a means of transport between hosts.

This latter, phoretic group includes species in the families Hemisarcoptidae, Winterschmidtiidae and Acaridae of the order Astigmata, which prey on coccids and other hemipterans. The hypopus stage of the mite does not feed, but attaches itself to the outer surface of an arthropod to be vectored to new host plants and prey colonies. Mites of the genus *Hemisarcoptes* (Hemisarcoptidae) are important predators of dias-pidid scale insects (Gerson et al. 1990, Izraylevich & Gerson 1993, Ji et al. 1994), and are generally vec-tored between prey colonies by members of the genus *Chilocorus*, since these species also prey on coccids (Houck & O'Connor 1991). Four new species of hemisarcoptid mite were discovered in surveys of the mite fauna of coccinellids collected in southern England, Holland and Belgium in the early 1990s (Fain et al. 1995, 1997). It is likely that many new species of mites that are phoretic on coccinellids await discovery.

Although these mites are not actually parasitic on coccinellids and have no known detrimental impact, their phoresy is clearly important in biological control. Species of *Chilocorus* may be selected as biological control agents against coccids, and their efficiency in control programs may be increased by ensuring that released individuals carry hypopi of mites that will also attack the coccids.

A variety of truly parasitic mites have also been recorded from ladybirds. These include some species, such as *Leptus ignotus*, that parasitize a wide variety of arthropods (Hurst et al. 1997a), and mites of the genus *Coccipolipus* that specialize on coccinellids.

8.4.1.2   *Coccipolipus* Husband (Prostigmata: Podapolipidae)

The mite genus *Coccipolipus* was erected by Husband in 1972 and comprises 14 species known to be parasitic upon coccinellids in the subfamilies Coccinellinae, Epilachninae and Chilocorinae (Husband 1984b, Ceryngier & Hodek 1996; Table 8.14). Although many of these species are tropical, the best researched is the widely distributed *Coccipolipus hippodamiae*, which has been recorded from the United States, Russia, central and eastern Europe, and the Demo-cratic Republic of the Congo (Table 8.14). Detailed work on this species in Europe has shown that it is the causative agent of a **sexually transmitted disease** (Hurst et al. 1995, Webberley et al. 2004, Webberley et al. 2006a).

Table 8.14 Described species of the genus *Coccipolipus* and their known host and distribution records (adapted and updated from Ceryngier & Hodek 1996).

Coccipolipus species	Hosts and distribution	References
africanae	*Epilachna* spp. (DR Congo, Rwanda)	Husband (1984a)
arturi	*Henosepilachna vigintioctopunctata* (Sumatra)	Haitlinger (1998)
benoiti	*Henosepilachna elaterii* (South Africa)	Husband (1981)
bifasciatae	*Henosepilachna bifasciata* (Zimbabwe)	Husband (1984a)
cacti	*Chilocorus cacti* (Mexico)	Husband (1989)
camerouni	*Epilachna nigrolimbata* (Cameroun)	Husband (1984a)
chilocori	*Chilocorus* spp. (DR Congo, Kenya)	Husband (1981)
cooremani	*Cheilomenes* spp. (DR Congo, Rwanda)	Husband (1983)
epilachnae	*Epilachna* spp. (El Salvador, Guatemala, Honduras, Mexico, USA)	Smiley (1974); Schroder & Schroder (1989)
hippodamiae	*Hippodamia convergens* (USA)	McDaniel & Morrill (1969); Husband
	Adalia bipunctata (USA, Russia, Poland, France, Italy, Germany, Austria, Czech Republic, Sweden, Ukraine, Georgia Rep.)	(1981); Eidelberg (1994); Majerus (1994); Hurst et al. (1995); Zakharov & Eidelberg (1997);
	Adalia decempunctata (Poland, Hungary, Germany, Russia)	Webberley et al. (2004, 2006b); Rhule & Majerus (2008); Rhule
	Oenopia conglobata (Poland)	et al. (2010); Riddick (2010);
	Calvia quatuordecimguttata (Poland)	J. J. Sloggett, unpublished
	Harmonia quadripunctata (France)	
	Harmonia axyridis (Poland, USA)	
	Coccinella magnifica (UK)	
	Exochomus troberti concavus (DR Congo)	
	Exochomus fulvimanus (DR Congo)	
	Coccinella septempunctata (Ukraine)	
macfarlanei	*Cycloneda sanguinea* (Trinidad, El Salvador)	Husband (1972, 1981, 1984b,
	Coccinella leonina transversalis (Australia, New Hebrides)	1989); Smiley (1974); Eidelberg (1994); Hajiqanbar et al. (2007);
	Coccinella transversoguttata (USA)	R. W. Husband & P. Ceryngier,
	Coccinella septempunctata (Iran, Ukraine, Poland)	unpublished
	Coccinella undecimpunctata (Iran)	
micraspisi	*Micraspis* spp. (DR Congo)	Husband (1983, 1984a)
	Declivitata spp. (DR Congo, Rwanda)	
oconnori	*Chilocorus stigma* (USA)	Husband (1989)
solanophilae	*Epilachna karisimbica* (DR Congo)	Cooreman (1952)

Life cycle

All stages of *C. hippodamiae* live on the underside of the elytra of coccinellids (Husband 1981, Majerus 1994; Fig. 8.8). The mouthparts of adult females become embedded into the host's elytra, or occasionally into the dorsal surface of the abdomen, allowing them to feed on host haemolymph. It is thought that female mites are fertilized in their final larval instar, since colonies of egg-laying females have resulted from the artificial transfer of just one larva to a previously uninfected host (E.L. Rhule, unpublished.). Eggs hatch into motile larvae, and these migrate between hosts (Knell & Webberley 2004). Webberley and Hurst (2002) report that **transmission** almost always occurs through sexual contact, although, on rare occasions, larvae may be transferred through close physical contact during host overwintering. Once on a novel host, larvae embed their mouthparts and metamorphose into adults. Thereafter, adult mites are entirely sedentary. Establishment and subsequent maintenance of the mite within a host coccinellid

Figure 8.8 The underside of the elytron of an *Adalia bipunctata* infected with *Coccipolipus hippodamiae* (six large adult female mites and their eggs are visible) (photo courtesy of Emma Rhule). (See colour plate)

population is dependent on two key factors: high levels of promiscuity to permit horizontal transmission between host individuals, and overlapping generations to facilitate vertical transmission down generations (Majerus 1994).

Prevalence

Webberley et al. (2006a) report an increase in the prevalence of *C. hippodamiae* in *A. bipunctata* populations through the year, with up to 90% of some populations of the beetle being infected by the latter part of the breeding season. This is thought to be a direct consequence of mating rate and the extent that matings occur between individuals of successive generations.

Variation in the prevalence of the mite on different host species has also been attributed to the frequency of mating of the hosts (Webberley et al. 2004). Highest prevalence is seen in *A. bipunctata*, which is more promiscuous than the less commonly infected *A. decempunctata* and *Oenopia conglobata*. Lowest prevalence was recorded in *Calvia quatuordecimguttata*. This species requires an overwintering diapause before breeding, leading to a lack of consistent sexual activity between generations and hence a barrier to mite vertical transmission. It is suggested that, in some parts of Europe, *C. hippodamiae* is lacking from *A. bipunctata* populations (Britain, coastal areas of northwest continental Europe) or is scarce (Scandinavia), due to limited intergenerational mating of the ladybird (Majerus 1994,

Webberley et al. 2006b). In some years in northwest Europe, the old generation dies before the new generation is reproductively mature.

The only coccinellid from which *Coccipolipus* mites have been recorded in Britain is *C. magnifica* (J.J. Sloggett, unpublished). The presence of mites on this species is probably due to the extended longevity of *C. magnifica*, promoted by its symbiotic relationship with ants (Sloggett et al. 1998). This means that some adults always survive until the first of the next generation have eclosed, allowing some intergenerational mating each year, and transmission of the mite down host generations (J.J. Sloggett, unpublished).

Effects on hosts

Infection of coccinellids by *C. hippodamiae* has strong negative effects, whereby female hosts become infertile. This has been particularly well studied in *A. bipunctata*, in which complete sterility was induced within approximately three weeks of infection (Hurst et al. 1995, Webberley et al. 2004). Hurst et al. (1995) speculate that sterility results from mite-infection interfering with the production of the chorion of the egg, since eggs laid by infected females were observed to shrivel and desiccate within 24 hours of being laid. Webberley and Hurst (2002) also demonstrated that infected *A. bipunctata*, especially males, were less likely to survive overwintering.

The interactions between other *Coccipolipus* species and their hosts have been less studied, although it is thought that patterns of transmission are likely to be similar as for *C. hippodamiae*, with similar effects on host fitness. Indeed, Schroder (1982) reports that *C. epilachnae* reduces fecundity of *E. varivestis* by two-thirds, and increases mortality by 40%.

Coccipolipus *in biological control*

Coccipolipus epilachnae has been introduced from Central America into the USA to control *E. varivestis*, which is a pest of soybean, although there is some disagreement in the literature over whether it actually suppresses host populations (Hochmuth et al. 1987, Schroder 1982, Cantwell et al. 1985).

Work is currently being undertaken to assess the potential of using *C. hippodamiae* to control the invasive coccinellid *Har. axyridis* in Britain and Europe (Rhule & Majerus 2008, Rhule et al. 2010). The fact that *C. hippodamiae* has been recovered from its congener, *Har.*

quadripunctata (Rhule & Majerus 2008), led the authors to suppose that *Har. axyridis* may also be a suitable host and, if similar effects on female fertility are found in *Har. axyridis* as were documented in *A. bipunctata*, infection with this mite may represent a promising avenue for controlling invasive *Har. axyridis* populations. Rhule et al. (2010) were able to establish successful mite colonies, consisting of reproducing adult females and their eggs, through artificial transfer onto *Har. axyridis*. Thereafter, infection was transmitted horizontally through copulation, and infected females were found to become completely sterile within 3 weeks of infection. This work demonstrates the potential for *C. hippodamiae* to become established on *Har. axyridis* as a novel host, and to significantly reduce host fitness. Further work is required to ascertain whether the mite would actually regulate host population numbers in the long term. Recent sampling of *Har. axyridis* in Torun, Poland, has revealed natural infection by *C. hippodamiae* (Rhule et al. 2010), so it seems that, even in the absence of human intervention, this mite may naturally propagate through *Har. axyridis* populations, at least in parts of continental Europe.

8.4.2   Nematodes

Nemathelminthes attacking Coccinellidae belong almost exclusively to the Nematoda, with only a single report of a primitive **Nematomorpha** attacking *Coccinella leonina transversalis* (Anderson et al. 1986).

Field and laboratory studies have demonstrated that ladybirds may be susceptible to entomopathogenic nematodes belonging to several families, such as Steinernematidae (Mracek & Ruzicka 1990, Abdel-Moniem & Gesraha 2001, Shapiro-Ilan & Cottrell 2005), Heterorhabditidae (Abdel-Moniem & Gesraha 2001, Shapiro-Ilan & Cottrell 2005), Allantonematidae (Iperti 1964, Hariri 1965, Narsi Reddy & Narayan Rao 1984) or Mermithidae (Delucchi 1953, Iperti 1964, Rhamhalinghan 1986a). However, only the members of the latter two families have been reported to parasitize ladybirds in the wild.

8.4.2.1   Allantonematidae (Tylenchida)

Within the Allantonematidae, *Parasitilenchus coccinellinae* has been reported from *P. quatuordecimpunctata*, *A. bipunctata*, *Oenopia conglobata*, *Hip. variegata* and

Ceratomegilla undecimnotata in Europe (Iperti 1964, Iperti & van Waerebeke 1968), and *Menochilus sexmaculatus* and *Illeis indica* in India (Narsi Reddy & Narayan Rao 1984).

The **level of parasitization** of coccinellids by *P. coccinellinae* varies with host species. In France, Iperti & van Waerebeke (1968) found that multivoltine species were most heavily parasitized, especially *P. quatuordecimpunctata* (up to 70%), with *O. conglobata* less so (20%), and *A. bipunctata* and *Hip. variegata* the least heavily parasitized. The univoltine species *Cer. undecimnotata* was only found infected occasionally. *Menochilus sexmaculatus* in India was found parasitized at a rate of 22%, while only 2–3% of *I. indica* were attacked (Narsi Reddy & Narayan Rao 1984).

Over 100 adult female *P. coccinellinae* may infest a single host, with up to 10,000 larvae and young adults (Iperti & van Waerebeke 1968, Narsi Reddy & Narayan Rao 1984). The method of transmission from one host to the next is unknown, but it may be sexually transmitted. Certainly, *P. coccinellinae* is common in the reproductive organs of its host. Ceryngier & Hodek (1996) also suggest that transmission may occur through host tracheae or through soft parts of the cuticle between sclerites. They also speculate that transmission may occur in coccinellid overwintering aggregations, where damp conditions may facilitate nematode propagation. Ceryngier & Hodek (1996) provide details of the **life cycle** of *P. coccinellinae*. Within 4 days of penetrating the body cavity, the ovaries of the fertilized female nematode become enlarged and soon the uterus becomes filled with free larvae and developing eggs. The first larval moult occurs inside the mother, while the remaining two occur within the host body cavity. The third moult either gives rise to adult males, or a final instar of larval females with ovarial primordia of 5–9 cells. Larval females are fertilized by the males at this stage and will not undergo their fourth and final moult until penetrating a new host.

Parasitilenchus coccinellinae is not usually fatal to its host, but retards maturation of the ovaries, consumes host resources and is known to have a general debilitating effect (Ceryngier & Hodek 1996).

There is one report of allantonematid nematodes of the genus *Howardula* infesting *A. bipunctata* larvae in England (Hariri 1965). These nematodes did not seem to affect the gonads of their host as other allantonematids do, but did result in a reduction in size of host fat bodies.

8.4.2.2 Mermithidae (Mermithida)

The immature stages of several members of the Mermithidae are known to be **solitary endoparasites** of adult ladybirds, although Delucchi (1953) recorded parasitization of the larvae of *Aphidecta obliterata*. Identification of the worms to species level is difficult because only the juvenile stages are found in coccinellids. Iperti (1964) found nematodes of the genus *Mermis* in four aphidophagous coccinellids in southeast France. From 2.5 to 4.2% of overwintering *Hip. variegata* were infected; aestivating populations of *Cer. undecimnotata* were infected at a prevalence of 1.2%; while active *P. quatuordecimpunctata* and *C. septempunctata* were only occasionally infected (Iperti 1964).

Rhamhalinghan (1986 a,b,c, 1987 a,b,c, 1988) carried out detailed studies of the infection of *C. septempunctata* by nematodes of the genus *Hexamermis* (= *Coccinellimermis*) in India. These nematodes reduce the weight, respiratory rate and the size of the fat body of their hosts. Host ovary growth and development is retarded and the worm can cause physical damage to vital organs such as the ovaries, tracheae, alimentary canal, Malpighian tubules, nervous system and heart. Parasitized females undergo a marked change in behaviour: they do not mate, eat fewer aphids and become hyperactive. Infection ultimately results in **paralysis and death of the host** within about 17 days, after which the worm exits.

8.4.3 Fungal pathogens

8.4.3.1 Hypocreales (Ascomycota)

Fungal pathogens of the order Hypocreales are widely regarded as important natural enemies of coccinellids, but their role in regulating coccinellid populations is poorly understood (Majerus 1994, Ceryngier & Hodek 1996, Roy & Cottrell 2008). Most research on the interactions between pathogenic hypocrealean fungi and coccinellids has focused on the impacts of fungal-based biorational pesticides on non-target insects such as coccinellids (James et al. 1995, Roy & Pell 2000, Riddick et al. 2009). There is little information on the interactions of coccinellids with naturally occurring fungal pathogens (Ceryngier 2000, Roy & Cottrell 2008, Riddick et al. 2009, Roy et al. 2009, Steenberg & Harding 2009, 2010), even though fungi, such as

Beauveria bassiana, are reported as major mortality factors of coccinellids particularly during overwintering (Iperti 1966, Ceryngier & Hodek 1996, Barron & Wilson 1998, Ormond et al. 2006).

There are several hypocrealean fungi that have been found infecting ladybirds: *B. bassiana* (James et al. 1995, Ceryngier 2000, Cottrell & Shapiro-Ilan 2003, Roy et al. 2008, Steenberg & Harding 2009), *Metarhizium anisopliae* (Ginsberg et al. 2002), *Isaria farinosa* (syn. *Paecilomyces farinosus*) (Ceryngier & Hodek 1996, Ceryngier 2000, Steenberg & Harding 2009), *I. fumosorosea* (syn. *P. fumosoroseus*) (Ceryngier & Hodek 1996) and *Lecanicillium* (syn. *Verticillium*) *lecanii* (Ceryngier 2000, Steenberg & Harding 2009).

Beauveria

The most well studied genus of hypocrealean fungi infecting coccinellids is *Beauveria* (Roy & Cottrell 2008), and the phylogeny and corresponding taxonomy of this single genus is undergoing major review (Rehner & Buckley 2005, Rehner et al. 2006). It is now thought appropriate to consider *B. bassiana* in the broadest sense as *B. bassiana* sensu lato because recognition and identification of *B. bassiana* as a distinct species has not been possible (Rehner & Buckley 2005). Indeed, *B. bassiana* s.l. appears to exist as non-monophyletic morphospecies and currently it is difficult to resolve separate species with certainty (Rehner & Buckley 2005, Ormond et al. 2010). It is important to keep in mind that the species reported as *B. bassiana* is likely to be one of many from within the species complex *B. bassiana* s.l.

Life cycle

The general life cycles of hypocrealean entomopathogens are remarkably similar despite their taxonomic diversity (Roy et al. 2006). They produce infective spores (conidia) that attach, germinate and penetrate directly through the host cuticle. So, unlike viral and bacterial pathogens, there is no requirement for ingestion of fungal spores. Once within the host, they proliferate as protoplasts, blastospores and hyphal bodies utilizing the host as a nutritional resource. Ultimately, the host is killed, and the fungus produces infective conidia for further transmission, or resting structures, such as sexual or asexual resting spores, chlamydospores or mummified hosts, for survival in the absence of new hosts or under adverse environmental conditions. The life cycles of hypocrealean fungi are **hemibiotrophic**, i.e. they switch from a parasitic, biotrophic phase in the haemocoel to a saprophytic phase colonizing the body after death.

Fungal activity is strongly influenced by abiotic and biotic conditions; high humidity (in excess of 95%) is required for conidium germination, infection, and sporulation, and the speed of kill is influenced by temperature (Vega et al. 2009). Fungal species exhibit a spectrum of adaptations that reflect the need to overcome environmental limitations and the host's defences (Roy et al. 2006, Ormond et al. 2010).

Prevalence of infection

Kuznetsov (1997) reviewed research on natural enemies attacking coccinellids in the Primorsky Territory of Russia (Far East Siberia) and summarized that the population dynamics of coccinellids in that territory are not significantly influenced by entomopathogens. *Henosepilachna vigintioctomaculata*, a serious pest of potato in the Primorsky Territory, was found infected by *Beauveria* species at frequencies of about 5.5–7%. Other coccinellids found in this territory infected by *Beauveria* species included *C. septempunctata*, *Har. axyridis*, *Calvia quatuordecimguttata* and *Hip. tredecimpunctata*.

Cottrell and Shapiro-Ilan (2003) took field-collected adults of *Olla v-nigrum* and *Har. axyridis* to the laboratory during autumn months, and found that 33% of the total mortality in the September sample and 81% in the October sample of *O. v-nigrum* could be attributed to infection by *B. bassiana*. No infection, however, was found in analogous samples of *Har. axyridis*.

Sublethal effects

Sublethal effects of hypocrealean pathogens on host insects are varied (Roy et al. 2006). There has been limited research on the premortality effects of these pathogens on coccinellids. Poprawski et al. (1998) found that moribund *Serangium parcesetosum* larvae infected with *B. bassiana* were less voracious than uninfected larvae, but they did not detect sublethal effects of either *B. bassiana* or *I. fumosorosea* on development of this predator. Roy et al. (2008) measured mortality of *C. septempunctata*, *Har. axyridis* (populations from Japan and Britain) and *A. bipunctata* and fecundity of the two latter species when exposed to *B. bassiana*. Mortality of both *C. septempunctata* and *A. bipunctata*

was higher relative to the Japanese and British populations of *Har. axyridis* but an impact of *B. bassiana* on *Har. axyridis* (in Britain) was detected via reduced fecundity at all *B. bassiana* doses tested (10^5, 10^7 and 10^9 conidia/ml).

8.4.3.2 *Hesperomyces* spp. (Ascomycota: Laboulbeniales, Laboulbeniaceae)

Laboulbeniales are **obligate ectoparasites** that infect many insect and non-insect hosts, but especially Coleoptera (Weir & Hammond 1997). They occur from the tropics to the temperate and polar regions on both terrestrial and aquatic hosts (Weir 2002, Harwood et al. 2006). Coccinellids are infected by several species of the genus *Hesperomyces* (Table 8.15; Fig. 8.9).

Most Laboulbeniales do not penetrate the insect cuticle, but *Hesperomyces virescens* (the most commonly reported laboulbenialean on coccinellids) exhibits rhizoidal penetration into the host body by production of a circular appressorium, which attaches and penetrates the host cuticle (Weir & Beakes 1996).

In general, laboulbenialean fungi do not directly cause mortality of their hosts (Weir & Beakes 1996), but a few **negative fitness effects** have been documented including reduced longevity of *Chil. bipustulatus* in Israel (Kamburov et al. 1967). Heavy infections can supposedly impede flight, mating, foraging and feeding but this requires further investigation (Nalepa & Weir 2007).

Hesperomyces virescens is often on the ventroposterior of males and the dorsoposterior of females; a sexual dimorphism that reflects the major **transmission** mechanism which is thought to be direct contact during mating (Weir & Beakes 1996, Welch et al. 2001, Riddick & Schaefer 2005). However, the distribution of *H. virescens* thalli on aggregating beetles are not explained by sexual transmission (Riddick & Schaefer 2005, Riddick 2006, Nalepa & Weir 2007); fungal thalli on overwintering *Har. axyridis* are located on the anterior part of the body which accords with direct contact through aggregation (Nalepa & Weir 2007). A similar pattern was reported for overwintering *A. bipunctata* which had fungal thalli distributed at the margins and front angles of the elytra (Weir & Beakes 1996). Nalepa & Weir (2007) conclude that transmission of this fungus is through contact with conspecifics: sexual contact is of primary importance in the mating season but aggregation in winter also plays a significant role.

Table 8.15 Host and distribution records of *Hesperomyces* species parasitizing Coccinellidae.

Hesperomyces species	Host	Distribution	Reference
chilomenis	*Cheilomenes lunata*	Kenya	Thaxter (1931)
coccinelloides	*Diomus seminulus*	Brazil	Rossi & Bergonzo (2008)
	Diomus sp.	Ecuador	Castro & Rossi (2008)
	Scymnus tardus	Panama	Thaxter (1931)
	Scymnus sp.	Spain	Santamaría (1995)
	Stethorus pusillus	Belgium	De Kesel (2011)
	undet. Coccinellidae	USA	Benjamin (1989) (cited in Castro & Rossi 2008)
	undet. Scymninae	Grenada	Thaxter (1931)
	undet. Scymninae	Jamaica	Thaxter (1931)
	undet. Scymninae	Philippines	Thaxter (1931)
	undet. Scymninae	Borneo	Thaxter (1931)
hyperaspidis	*Hyperaspis* sp.	Trinidad	Thaxter (1931)
virescens	*Adalia bipunctata*	England	Weir & Beakes (1996)
		Austria	Christian (2001)
		France	Webberley et al. (2006b)
		Germany	Webberley et al. (2006b)
		Italy	Webberley et al. (2006b)
		Sweden	Webberley et al. (2006b)
		The Netherlands	Webberley et al. (2006b)
	Adalia decempunctata –	Italy	Castaldo et al. (2004)
	Adalia sp.	Belgium	A. De Kesel, unpublished
	Brachiacantha quadripunctata	USA	Harwood et al. (2006a)
	Chilocorus bipustulatus	Israel	Kamburov et al. (1967)
	Chilocorus renipustulatus	England	Hubble (2011)
	Chilocorus stigma	USA	Thaxter (1931)
	Coccinella septempunctata	USA	Harwood et al. (2006b)
	Coccinula crotchi	Japan	M. E. N. Majerus & R. L. Ware, unpublished
	Coccinula quatuordecimpustulata	Greece	Castaldo et al. (2004)
	Coccinula sinensis	Japan	M. E. N. Majerus & R. L. Ware, unpublished
	Cycloneda munda	USA	Harwood et al. (2006b)
	Cycloneda sanguinea	? England*	Tavares (1979)
	Eriopis connexa	Argentina	Thaxter (1931)
	Harmonia axyridis	USA	Garces & Williams (2004)
		Germany	Steenberg & Harding (2010), K. Twardowska, unpublished
		Belgium	De Kesel (2011)
		The Netherlands	De Kesel (2011)
	Hippodamia convergens	USA	Thaxter (1931)
	Olla v–nigrum	Fiji	Weir & Beakes (1996)
		USA	Roy & Cottrell (2008)
	Propylea quatuordecimpunctata	Spain	Santamaría (2003)
	Psyllobora sp.	France	Tavares (1985)
	Psyllobora vigintiduopunctata	Spain	Santamaría (2003)
		Belgium	De Kesel (2011)
	Psyllobora vigintimaculata	USA	Harwood et al. (2006)
	Tytthaspis sedecimpunctata	Greece	Castaldo et al. (2004)

*The reported locality (Rustington, England) is either erroneous or concerns laboratory culture of the host native to Americas.

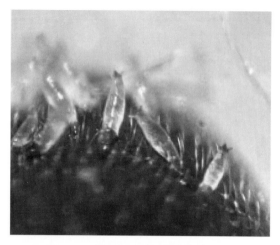

Figure 8.9 Thalli of *Hesperomyces coccinelloides* on the elytra of *Stethorus pusillus* (photo courtesy of Johan Bogaert). (See colour plate.)

Prevalence of infection

Harwood et al. (2006a) examined the prevalence of *H. virescens* on coccinellids in Kentucky, USA, and found that 82.3% of *Har. axyridis* were infected compared with only 4.7, 4.2 and 2.5% of *Psyllobora vigintimaculata*, *Brachiacantha quadripunctata* and *Cycloneda munda*, respectively. This ectoparasitic fungus was not recovered from *Col. maculata* or *Hyperaspis signata*. In Georgia, USA, the prevalence of natural *H. virescens* infection of *O. v-nigrum* is similar as of *Har. axyridis* in Kentucky (T. Cottrell, unpublished data reported in Roy & Cottrell 2008).

The prevalence of infection of *H. virescens* on *A. bipunctata* in London was higher at the centre of the city than at the periphery and, indeed, infection was rare or absent outside the city (Welch et al. 2001). This variation in prevalence over such a short distance (less than 25 km) is unusual, but indicates an association between urbanization and disease dynamics. It is thought that the prevalence of the fungus is linked to elevated urban temperatures which increase the probability of generations of the host overlapping and consequent interbreeding between cohorts.

8.4.3.3   Nosematidae (Microsporidia)

Until recently, microsporidia have been grouped with the protozoa but recent molecular evidence suggests that they should be placed within the fungal kingdom

(e.g. Lee et al. 2008). They are intracellular pathogens with fascinating but complex life cycles.

Life cycle

Each microsporidial spore comprises a long, coiled tube called the polar filament. Once a spore is in the host, the polar filament is extruded and acts as a needle injecting the spore contents into the host cell to begin infection. Once inside the appropriate host cell, the intracellular stage (schizont) proliferates asexually and then undergoes an autogamous sexual process producing sporonts. In the genus *Nosema*, each sporont gives rise to one spore (Ceryngier & Hodek 1996). **Transmission** between hosts is dependent on the release of spores into the environment and ingestion by a suitable host. Microsporidian species are often highly specific not just to host species but to specific tissues within the host such as the fat body, midgut wall or the reproductive tissues. Infections by most microsporidia result in chronic diseases and often the host can appear asymptomatic, although body size, longevity and fecundity are often reduced and development time increased (Hajek 2004). These pathogens are often vertically transmitted from mother to egg, a strategy that enables microsporidia to persist in low density or scattered host populations. Some microsporidia are vectored by parasitoids to other insects. There are only a few studies examining the interactions between nosematid microsporidia and coccinellid hosts. There are undoubtedly many more examples yet to be revealed.

Nosematid species infecting Coccinellidae

Brooks et al. (1980) found two microsporidia, subsequently described as *Nosema epilachnae* and *Nosema varivestis* (Brooks et al. 1985), to infect nearly 100% of laboratory colonies of *E. varivestris*. Unusually for a microsporidial infection, the pathogens decimated these colonies. Further investigations revealed both pathogens in field populations of *E. varivestis* in North and South Carolina, USA, with up to 28% infection rates by the more virulent *N. epilachnae*, and a few per cent rates by the less virulent *N. varivestis*. The two *Nosema* species were mechanically transmitted between infected and uninfected hosts by the parasitoid *P. foveolatus*. This parasitoid is highly susceptible to both *N. epilachnae* and *N. varivestis* and so represents a shared host (Brooks et al. 1980).

Another *Nosema* species, *N. henosepilachnae*, has been found infecting various tissues of *Henosepilachna elaterii* in Senegal. This microsporidiosis is not directly lethal, but reduces fecundity in adult hosts (Toguebaye & Marchand 1984).

There have been three *Nosema* species identified from predatory coccinellids. *Nosema tracheophila* was found in the tracheal epithelium and connective tissues of *C. septempunctata* (Cali & Briggs 1967) and *Nosema coccinellae* in the midgut epithelium, Malpighian tubules, gonads, nervous and muscle tissues of *C. septempunctata*, *Hip. tredecimpunctata* and *Myrrha octodecimguttata* (Lipa 1968a). *Nosema hippodamiae* was identified from the midguts and fat bodies of *Hip. convergens* in 1959 (Lipa & Steinhaus 1959) and, just after a decade later, microsporidial spores resembling *N. hippodamiae* were observed in the fat body, muscle, gut, Malpighian tubules and testes of 50% of the *Hip. convergens* (Sluss 1968). Interestingly, Sluss (1968) noted that this microsporidium prevented development of *Hip. convergens* parasitoid, *Dinocampus coccinellae*.

Recently, an unidentified nosematid pathogen of *Hip. convergens* was reported by Bjornson and co-authors (Saito & Bjornson 2006, 2008, Joudrey & Bjornson 2007, Bjornson 2008). After ultrastructure and molecular examination, it was described as a new species, *Tubulinosema hippodamiae* (Bjornson et al. 2011). However, it is currently unknown as to whether *T. hippodamiae* actually represents a new species or is one of the coccinellid pathogens already described under the genus *Nosema*. To check this, molecular characterization of type specimens of *N. hippodamiae*, *N. coccinellae* and *N. tracheophila* should be made (Bjornson et al. 2011).

Bjornson (2008) noted rather low prevalence of infection of *Hip. convergens* with *T. hippodamiae*; only 1% of individuals collected in the winter from the field for commercial retail were infected. Infection of *Hip. convergens* by this species has been shown to increase development time and reduce longevity and female fecundity (Joudrey & Bjornson 2007). Saito & Bjornson (2006) demonstrated the efficacy of horizontal transmission of the microsporidium from *Hip. convergens* to larval *C. septempunctata*, *C. trifasciata perplexa* and *Har. axyridis* when these larvae consumed infected eggs in the laboratory. Indeed, the microsporidium was transmitted with 100% efficiency when first instar larvae were fed infected eggs, and, in all cases, larval development was significantly longer for microsporidia-infected individuals than for uninfected individuals, but the microsporidium had no effect on larval mortality (Saito & Bjornson 2008). Therefore, the practice

of redistributing *Hip. convergens* from overwintering locations to agricultural ecosystems could facilitate the dispersal of microsporidia to other coccinellids. Laboratory studies demonstrated that invasive non-native species of coccinellid, such as *Har. axyridis*, were less susceptible to infection by *T. hippodamiae* than native species (Saito & Bjornson 2006, 2008); infection was as heavy in *C. trifasciata perplexa* (a native coccinellid) as it was in *Hip. convergens* (original host), but lighter in the introduced species *C. septempunctata* and *Har. axyridis*.

8.4.4   Protozoan pathogens

8.4.4.1   Septate eugregarines (Apicomplexa: Eugregarinida: Septatorina)

Eugregarines are large unicellular organisms (sometimes more than 0.5 mm in length) that inhabit alimentary canals, coelomic spaces or reproductive vesicles of many invertebrates (Rueckert & Leander 2008).

Several species of the eugregarine suborder Septatorina have been reported to inhabit intestines of ladybirds (Table 8.16). Most of them have been placed in the genus *Gregarina* of the family Gregarinidae. However, recent studies by Clopton (2009) showed polyphyly of both Gregarinidae and *Gregarina*, and thus both are likely to be split in the near future.

Life cycle

The direct **transmission** of gregarines usually takes place by the host orally ingesting oocysts. Eight sporozoites emerge from each oocyst, and, using their apical complexes, attach to or invade the epithelial cells of the host. The sporozoites begin to feed and develop into larger trophozoites that subsequently detach from the epithelial cells and live freely in the lumen of the host's gut. Two mature trophozoites pair up in a process called syzygy and develop into gamonts. A gametocyst wall forms around each pair of gamonts and the gamonts begin to divide to produce hundreds of gametes through a process called gametogony. Gamete pairs fuse, form zygotes and become surrounded by an oocyst wall. The process of sporogony, involving meiotic and mitotic divisions within the oocyst, yields eight spindle-shaped sporozoites. Hundreds of oocysts are formed within each gametocyst and these are released in host faeces (Watson 1915, Leander et al. 2006, Rueckert & Leander 2008).

Table 8.16 Eugregarines reported from ladybird hosts (the numbers in brackets apply to references listed below the table).

Eugregarine	Host / region
Brustiophoridae	
Brustiospora indicola	*Stethorus* sp. / India (8)
Gregarinidae	
Anisolobus indicus	*Coccinella septempunctata* / India (5)
Gregarina barbarara	*Adalia bipunctata* / New York, USA (20), Germany (4)
	Coccinella sp. / New York, USA (22)
	Coccinella trifasciata / California, USA (11)
	Exochomus quadripustulatus / Silesia, Poland (3)*
	Hippodamia convergens / California, USA (11)
	Hippodamia sinuata / California, USA (11)
	Tytthaspis sedecimpunctata / Silesia, Poland (3)*
Gregarina californica	*Coccinella californica* / California, USA (11)
Gregarina chilocori	*Chilocorus rubidus* / Japan (16)
Gregarina coccinellae	*Coccinella quinquepunctata* / St Petersburg region, Russia (12)
	Coccinella septempunctata / Poland (10)
	Exochomus quadripustulatus / Poland (13)
	Harmonia quadripunctata / Poland (13)
	Hippodamia tredecimpunctata / Poland (10)
	Myrrha octodecimguttata / St Petersburg region, Russia (11), Poland (13)
Gregarina dasguptai	*Coccinella septempunctata* / India (14)
Gregarina fragilis	*Coccinella* sp. / Illinois, USA (22)
	Coccinella trifasciata / USA, California (11)
Gregarina hyashii	*Coccinella leonina transversalis* / India (18)
Gregarina katherina	*Aiolocaria hexaspilota* / Japan (21)
	Ceratomegilla undecimnotata / Slovakia (15)
	Coccinella californica / California, USA (11)
	Coccinella novemnotata / New York, USA (22)
	Coccinella septempunctata / Silesia, Poland (3)†, Slovakia (15)
	Coccinella septempunctata brucki / Japan (21)
	Coccinella trifasciata / California, USA (11)
	Coccinula quatuordecimpustulata / Silesia, Poland (3)†
	Pharoscymnus anchorago / Mauretania (9)
Gregarina ruszkowskii	*Adalia bipunctata* / Poland (13)
	Coccinella quinquepunctata / Poland (10)
	Coccinella septempunctata / Poland (10)
	Exochomus quadripustulatus / Poland (13)
Gregarina straeleni	*Epilachna* spp. / DR Congo (20)
Undetermined eugregarines	*Adalia bipunctata* / Nova Scotia, Canada (17)‡
	Adalia decempunctata / France (7)
	Coccinella septempunctata / France (7)
	Henosepilachna pustulosa/ Japan (6)
	Hippodamia convergens / California, USA (1)‡
	Hippodamia variegata / France (7)
	Oenopia conglobata / France (7)
	Propylea quatuordecimpunctata / France (7)
	Scymnus apetzi / France (19)
	Scymnus (Pullus) impexus / Switzerland and Germany (2)

*According to (10), this record concerns *G. coccinellae*.
†According to (10), this record concerns *G. ruszkowskii*.
‡Possibly three different gregarine species.
References: 1, Bjornson (2008); 2, Delucchi (1954); 3, Foerster (1938); 4, Geus (1969); 5, Haldar et al. (1988); 6, Hoshide (1980); 7, Iperti (1964); 8, Kundu & Haldar (1981); 9, Laudeho et al. (1969); 10, Lipa (1967); 11, Lipa (1968b); 12, Lipa & Semyanov (1967); 13, Lipa et al. (1975); 14, Mandal et al. (1986); 15, Matis & Valigurova (2000); 16, Obata (1953), cited in Laudeho et al. 1969; 17, Saito & Bjornson (2008); 18, Sengupta & Haldar (1996); 19, Sezer (1969), cited in Laudeho et al. 1969; 20, Theodorides & Jolivet (1959), cited in Laudeho et al. 1969; 21, Tsugawa (1951); 22, Watson (1915); 23, Watson (1916).

Prevalence

Gregarine diseases seem to be more common in regions with warm climates than in colder ones. The prevalence of *Gregarina katherina* in coccinellids occupying palm groves in Mauretania was high: the proportion of infected adult *Pharoscymnus anchorago* fluctuated between 50 and 100% during the year, and two introduced coccinellids (*Chil. bipustulatus* and *Chil. stigma*) were infected to a similarly high degree. No eugregarines, however, were found in *Chil. distigma* (Laudeho et al. 1969). Iperti (1964) reported that five aphidophagous coccinellids in southeastern France harboured eugregarines, and one of them, *P. quatuordecimpunctata*, at a prevalence of about 10%. Lipa and Semyanov (1967) observed only a few per cent of *C. quinquepunctata* and *Myrrha octodecimguttata* infected by *G. coccinellae* in the St Petersburg region of Russia. Similarly low prevalences of *G. coccinellae* and *G. ruszkowskii* were recorded in four ladybird species in Poland (Lipa et al. 1975). Bjornson (2008) has examined the prevalence of eugregarines in commercially available *Hip. convergens* that were collected from their overwintering sites in California. Although only 0.2% of the beeetles were found to be infected, the eugregarines involved probably belonged to three different species, and none of them were similar in size to *Gregarina barbara*, the only eugregarine reported from *Hip. convergens*.

Effects on hosts

Eugregarines developing in Coccinellidae are considered to be weak pathogens that destroy intestinal cells in coccinellid larvae and adults and derive nourishment from their digestive tracts (Ceryngier & Hodek 1996). However, Laudeho et al. (1969) reported that gregarine infection could cause reduced fecundity and longevity of *P. anchorago* and, in heavily infected individuals, the gametocysts could even cause death by blocking the intestinal tract.

8.4.5   Bacteria

8.4.5.1   General pathogenic bacteria

With the exception of one group of bacteria with a particularly interesting life cycle, rather little is known about the bacterial diseases of coccinellids. Those with experience of culturing coccinellids in the laboratory may have observed mortality in stocks that are not hygienically maintained, with larvae showing symptoms of enteric disease characteristic of bacterial or viral infections of the gut. However, there are only a few empirical studies of bacterial pathogens in coccinellids, and little detail on effects in natural populations. In the quest to find biological methods of controlling some phytophagous Epilachninae, which act as crop pests, a number of authors have found evidence of susceptibility to bacteria. A Chinese strain of *Bacillus thuringiensis* was found to be harmful to larvae, but not adults, of *Henosepilachna vigintioctomaculata* (Ping et al. 2008); and Pena et al. (2006) reported insecticidal activity of a Mexican strain of *B. thuringiensis* against *E. varivestis*. Otsu et al. (2003) found that a chitinase secreting strain of the bacterium *Alcaligenes paradoxus* inhibited feeding and oviposition by *H. vigintioctopunctata* adults, but had no effects on longevity. Rather less is known of the effects of such bacteria on entomophagous species, although Giroux et al. (1994) found no lethal effects of the commercial M-one strain of *B. thuringiensis* on *Col. maculata lengi*.

8.4.5.2   Male-killing bacteria

Most studies of bacterial infections of Coccinellidae concern the male-killing bacteria (Majerus & Hurst 1997). The first observation of a female-biased sex ratio in a coccinellid was reported by Lusis (1947), who noted that some female *A. bipunctata* produced only or predominantly female progeny, and that this trait was maternally inherited. 'Abnormal sex ratios' were also later observed in *Har. axyridis* (Matsuka et al. 1975) and *Menochilus sexmaculatus* (Niijima & Nakajima 1981). Hurst et al. (1992) reported similar findings from a British sample of *A. bipunctata*, and postulated the causative agent of the 'sex ratio' trait to be a cytoplasmically inherited bacterium which kills males, since normal sex ratios were obtained after oral administration of tetracycline to affected females. The male-killing bacterium was later identified as a *Rickettsia* (Werren et al. 1994).

Diversity of male-killing bacteria in the Coccinellidae

Since this first discovery in *A. bipunctata*, male-killing bacteria have now been identified in 13 other coccinellid host species, and are suspected to occur in five others (Table 8.17). Bacteria of five different groups have been identified as male-killers of coccinellids (*Rickettsia*, *Wolbachia*, *Spiroplasma*, Flavobacteria and γ-proteobacteria), suggesting that male-killing has evolved independently many times. Given the still small

Table 8.17 Male–killing bacteria in the Coccinellidae (adapted from Majerus 2006).

Host species	Countries	Agent	Prevalence	Evidence*	References
Adalia bipunctata	England, Scotland, Holland, Denmark, Luxembourg, Belgium, France, Germany, Poland, Russia, Kyrgyzstan, Norway	*Rickettsia*	0.01–0.23	f-bsr, lhr, mi, as, hs, m, sDNA, PCR	Hurst et al. (1992, 1999a); Werren et al. (1994); Majerus et al. (2000b); Schulenburg (2000); Zakharov & Shaikevich (2001); Tinsley (2003)
Adalia bipunctata	Germany, Russia, Sweden	*Spiroplasma* (Group VI)	0.03–0.53	f-bsr, lhr, mi, as, hs, m, sDNA, PCR	Hurst et al. (1999a); Majerus et al. (2000b); Tinsley (2003)
Adalia bipunctata	Russia, Sweden	Two strains of *Wolbachia*	0.02–0.1	f-bsr, lhr, mi, as, hs, m, sDNA, PCR	Hurst et al. (1999c); Majerus et al. (2000b)
Adalia decempunctata	Germany, England	*Rickettsia*	0.09	f-bsr, lhr, mi, as, hs, m, sDNA, PCR	Schulenburg et al. (2001); Majerus (2003)
Anisosticta novemdecimpunctata	England	*Spiroplasma* (Group VI)	0.31	f-bsr, lhr, mi, as, sDNA, PCR	Tinsley & Majerus (2006)
Calvia quatuordecimguttata	England, Canada	Unknown	<0.05	f-bsr, lhr, mi	Majerus (2003)
Coccidula rufa	Germany	*Rickettsia*	0.59	f-bsr, PCR	Weinert et al. (2007)
Coccidula rufa	Germany	*Wolbachia*	0.78	f-bsr, PCR	Weinert et al. (2007)
Coccinella septempunctata	England Germany	Unknown	<0.01	f-bsr, lhr	J. J. Sloggett, unpublished
Coccinella septempunctata brucki	Japan	Unknown	0.13	f-bsr, lhr, mi	M. E. N. Majerus, unpublished
Coccinella undecimpunctata	Egypt	*Wolbachia*	0.5	f-bsr, lhr, mi, sDNA, PCR	Elnagdy (2008)
Coccinula crotchi	Japan	Flavobacterium	0.23	f-bsr, lhr, mi, as, sDNA, PCR	M. E. N. Majerus, unpublished
Coccinula sinensis	Japan	Flavobacterium	0.23	f-bsr, lhr, mi, as, PCR	Majerus & Majerus (2000)
Coleomegilla maculata	USA	Flavobacterium	0.23	f-bsr, lhr, mi, as, sDNA, PCR	Hurst et al. (1996, 1997b)
Harmonia axyridis	Japan, Russia, South Korea	*Spiroplasma* (Group VI)	0.02–0.86	f-bsr, lhr, mi, as, hs, m, sDNA, PCR	Matsuka et al. (1975); Majerus et al. (1998, 1999); Majerus (2001, 2003)

Table 8.17 (*Continued*)

Host species	Countries	Agent	Prevalence	Evidence*	References
Harmonia quadripunctata	France	Flavobacterium	0.11	f-bsr, lhr, mi, as, sDNA, PCR	M. E. N. Majerus & M. C. Tinsley, unpublished
Hippodamia quinquesignata	USA	Unknown	Unknown	f-bsr, lhr, mi	Shull (1948)
Hippodamia variegata	Turkey, England	Flavobacterium	0.07–0.13	f-bsr, lhr, mi, as, sDNA, PCR	Hurst et al. (1999b)
Menochilus sexmaculatus	Japan	γ–proteobacterium	0.13	f-bsr, lhr, mi, as, PCR	Majerus (2001, 2003)
Mulsantina picta	USA	Unknown	Unknown	f-bsr, mi	J. J. Sloggett, unpublished
Propylea japonica	Japan	*Rickettsia*	0.07–0.26	f-bsr, lhr, mi, as, PCR	Majerus (2001, 2003)
Rhyzobius litura	Germany	*Rickettsia*	0.84	f-bsr, PCR	Weinert et al. (2007)
Rhyzobius litura	Germany	*Wolbachia*	0.89	f-bsr, PCR	Weinert et al. (2007)

*Evidence given as: f-bsr, female biased sex ratio; lhr, low egg hatch rate in infected lines; mi, maternal inheritance; as, antibiotic sensitive; hs, heat sensitive; m, microscopy; sDNA, DNA sequencing; PCR, detection of symbiont using symbiont–specific PCR reaction.

number of coccinellid species in which male-killing has been sought and identified, it is unlikely that the full taxonomic diversity of male-killers in coccinellids has yet been revealed (Weinert et al. 2007).

Evolutionary rationale of male-killing in the Coccinellidae

For cytoplasmic symbionts, the existence within a male is an evolutionary 'dead-end', since there is no opportunity for vertical transmission to the next generation. Instead, such symbionts may increase their fitness indirectly by killing male hosts and in doing so preferentially favour female hosts, which carry clonally identical copies of themselves. In coccinellids, this is achieved by causing death early in male embryonic development, and thereby 'reallocating' resources that would have been used by males to their female siblings (Hurst 1991, Majerus 2003). **Resource reallocation** in coccinellids comes in two forms. First, infected females suffer less from competition than they would when growing up with male siblings. In addition, upon hatching, they are able to consume the dead embryos of their brothers, which provide a substantial nutritional advantage.

Majerus and Hurst (1997) identified a number of behavioural and ecological **properties of aphidophagous coccinellids** that make this group particularly prone to male-killer infection. First, aphidophagous coccinellids lay their eggs in clutches, meaning that the likelihood of fitness compensation accruing to hosts harbouring the same male-killer is high. Secondly, populations of their aphid prey are highly unstable in space and time. And finally, coccinellids are highly cannibalistic and indulge in sibling egg consumption (Majerus & Majerus 1997b). **Cannibalistic behaviour** (5.2.8) is of particular relevance in male-killer infected clutches, in which half of the eggs will fail to develop as they are male. Daughters of infected females gain a significant survival advantage by feeding upon the unused soma and dead embryos of their brothers since, on average, each female larva has one dead egg to feed on before dispersal from the clutch (Fig. 8.10), giving significant increase in larval survival time.

Evolutionary dynamics of male-killing in the Coccinellidae

In considering the dynamics of early male-killers, Hurst (1991) and Hurst et al. (1997c) showed that

Figure 8.10 A clutch of *Adalia bipunctata* eggs and young larvae in which only half the offspring have hatched, due to the action of a male-killing *Rickettsia* (Photo: Remy Poland). (See colour plate.)

three parameters affect the spread of a male-killer in a host population. These are: the level of fitness compensation accrued to infected females (to offset the cost of male losses), the vertical transmission efficiency of the male-killer, and any direct cost that infection imposes on infected females.

In coccinellids **fitness compensation** results chiefly from resource reallocation to sisters through consumption of the dead males, the relative level of which will depend largely on the availability of prey. However, the vertical transmission efficiency and direct costs of male-killers vary as a result of interactions between different male-killers and their various hosts. In most species, **vertical transmission** is not 100%. Values obtained from laboratory cultures vary from 72% for a *Rickettsia* from a Russian population of *A. bipunctata* (Majerus et al. 2000) to 99.98% for a *Spiroplasma* from a Japanese population of *Har. axyridis* (Majerus et al. 1999).

The **cost to females of bearing a male-killing bacterium**, in addition to the loss of male progeny, has been assessed in five species of coccinellid (Matsuka et al. 1975, Hurst et al. 1994, Hurst et al. 1999b, Majerus 2001). Each case has shown some negative fitness effects, such as decreased oviposition rates, lower overall fecundity, higher infertility levels or shorter adult life-span.

Adalia bipunctata is a particularly interesting species in which to study male-killing, as it has been shown to be host to four different male-killers (a *Rickettsia*, two

Wolbachias and a *Spiroplasma*), with all four occurring together within a single sample from Moscow (Majerus et al. 2000). Weinert et al. (2007) provide evidence for the co-infection of German *Rhyzobius litura* and *Coccidula rufa* populations with both *Rickettsia* and *Wolbachia*. Theoretical models of the evolution of male-killing suggest that two or more male-killers cannot coexist in a single host population at equilibrium, except in the presence of male-killer suppressors (Randerson et al. 2000).

A recent study has demonstrated that the vertical transmission of a γ-proteobacterium to *Menochilus sexmaculatus* is influenced by the male; the prevalence of the male-killing trait varies depending on the male a female has mated with (Majerus & Majerus 2010). Further analysis demonstrated that a single dominant allele (rescue gene) functions to rescue male progeny of infected females from the male-killing effects of this microbe (Majerus & Majerus 2010). Furthermore, presence of the rescue gene in either parent does not significantly affect the inheritance of the symbiont.

Evolutionary implications of male-killing in the Coccinellidae

Male-killer infection can have significant evolutionary implications on host populations, resulting both from the presence of the male-killer itself, and the female-biased population sex ratios that it produces. Since both male-killing bacteria and host mitochondria are maternally inherited, an invading male-killer will be in linkage disequilibrium with the mitotype of the first host that it invades. Johnstone & Hurst (1996) described how the invasion of a male-killer can cause mitotype selective sweeps in the host population. This will be the case when the population is host to a single male-killer. For populations co-infected with more than one male-killer, mitochondrial polymorphism would be expected. Indeed, Jiggins & Tinsley (2005) have linked an ancient mitochondrial polymorphism in *A. bipunctata* to the presence of multiple *Rickettsia* strains. The association between male-killers and host mitotypes should be remembered by those working on coccinellid population genetics and phylogenetics, since mtDNA sequences are likely to show significant deviations from neutrality (Chapter 2).

The maternal inheritance of male-killing bacteria means that they are in direct conflict with the rest of the host genome. Since most nuclear genes are biparentally inherited, they will be under selection to resist

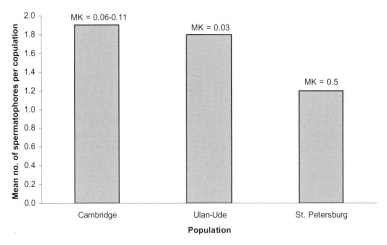

Figure 8.11 Number of spermatophores produced per copulation by male *Adalia bipunctata* from three populations (Cambridge, UK; Ulan Ude, Russia; and St. Petersburg, Russia). Male-killer prevalences (MK) of each population are shown above. Adapted with permission from Majerus (2003).

the mortality imposed on them by the pathological action of the male-killer. This intra-genomic conflict may lead to the evolution of suppressor systems acting against the male-killer. Majerus (2003, 2006) suggested that suppression may be achieved by killing the male-killer itself, reducing its vertical transmission efficiency or blocking its action. Unpublished data reported in Majerus (2003) indicated the presence of a male-killer rescue gene in *M. sexmaculatus* infected with a γ–proteobacterium.

It may be expected that, in populations that have a female-biased operational sex ratio due to the presence of a male-killer, the strength of selection for males to compete and females to choose will be reduced. In very strongly female-biased populations, in which males rather than females become the limiting sex, full sex role reversal may result. Reduction or reversal of sexual selection may be manifest in a number of ways. These effects have been most comprehensively studied in the *Wolbachia*-infected butterfly *Acraea encedon* (Jiggins et al. 1999), and limited published data is available for coccinellids. However, some interesting effects on female mating preferences have been documented in some species. Osawa & Nishida (1992) reported varying levels of female choice in Japanese populations of *Har. axyridis*, and it has been suggested that this is correlated with the presence of male-killing *Spiroplasma* (Majerus 2006).

A female-bias in the operational sex ratio may select for males to invest less in each mating than they would when the sexes are closer to parity. This has been demonstrated in *A. bipunctata*, in which males insert variable numbers of spermatophores during copulation. Majerus (2003) found that males from a strongly female-biased population were observed to pass significantly fewer spermatophores per copulation than those from less female-biased populations (Fig. 8.11).

Finally, female-biased sex ratios can affect the way in which sexually transmitted diseases spread through the host population. Coccinellids are highly promiscuous and so the rate of spread of a sexually transmitted disease will be increased in populations exhibiting a sex ratio bias. An association between prevalence of the sexually transmitted mite *C. hippodamiae* and male-killer prevalence has been documented in Scandinavian *A. bipunctata* populations (Tinsley 2003). Further investigation into the association between male-killer infection and sexually transmitted diseases is warranted.

8.5 IMPACT OF NATURAL ENEMIES ON LADYBIRD POPULATIONS

In the last part of the chapter we discuss the role of natural enemies in influencing numerical changes in the populations of phytophagous and predatory ladybirds.

8.5.1 Impact on phytophagous Coccinellidae

There are several field studies investigating the causes of **mortality of juvenile epilachnine** ladybirds by means of life table construction and key factor analysis. Some of these studies were performed in the tropical regions of Sumatra (Indonesia) (latitude 1° S) (Abbas & Nakamura 1985, Nakamura et al. 1988, Inoue et al. 1993) and Ethiopia (7° N) (Beyene et al. 2007), and others from the temperate Japanese islands Honshu (35–36° N) (Nakamura 1976, 1987, Nakamura & Ohgushi 1981, Shirai 1987, 1988) and Hokkaido (43° N) (Kimura & Katakura 1986). The most striking difference between the mortality factors affecting tropical and temperate epilachnines concerns parasitoids. Tropical ladybirds suffered high mortality from egg parasitoids (Tetrastichinae in Sumatra and *Ooencyrtus camerounensis* in Ethiopia) but egg parasitoids were entirely absent from temperate regions (Fig. 8.12a). Similarly, larval and pupal parasitoids, although present in tropical and temperate zones, tended to exert a much stronger effect on host populations in the tropics than in temperate regions (Fig. 8.12 b, c).

Quantification of the impact of predators is more difficult than that of parasitoids. Two phenomena, inter- and intraspecific predation (cannibalism), are usually indistinguishable and categories such as 'disappearance' or 'unknown causes' are frequently found among mortality factors in the life tables. In Fig. 8.12d–f, the maximum possible rates of predation on tropical and temperate Epilachninae are juxtaposed. In calculating these rates, we combined rates of intra- and interspecific predation, and assumed that every individual that 'disappeared' had been consumed. Although differences between tropical and temperate populations were not very clear, maximum predation rates for eggs and larvae of temperate ladybirds tended to be quite high (Fig. 8.12d and e). Indeed, the role of predators, especially an earwig, *Anechura harmandi*, in reducing the numbers of eggs and larvae of *Henosepilachna* spp. is often emphasized in Japanese studies (Nakamura & Ohgushi 1981, Nakamura 1983, Kimura & Katakura 1986, Ohgushi 1986, Shirai 1987).

Depletion of food resources due to feeding of epilachnine larvae is often reported in the tropics (Abbas & Nakamura 1985, Nakamura et al. 1988, Inoue et al. 1993), but also in temperate regions (Nakamura 1976, 1983, 1987, Kimura & Katakura 1986), indicating inability of natural enemies to regulate herbivore populations in a density-dependent manner. However, in temperate latitudes of the Russian Far East (44° N), Kovalenko and Kuznetsov (2005) found good synchronization between population dynamics of *H. vigintioctomaculata* and its parasitoid, *N. afissae*, hinting at a density-dependent regulatory mechanism.

A parasitoid, *P. foveolatus*, used in classical and augmentative biological control programmes, has been found to provide an effective suppression of the populations of certain noxious Epilachninae (8.3.2.9).

Although reported **mortality of adult phytophagous Coccinellidae** during dormancy may exceed 95% (Nakamura 1983, Inoue et al. 1993), little is known about the factors responsible for this mortality.

8.5.2 Impact on predatory Coccinellidae

Japanese studies on the mortality of *Aiolocaria hexaspilota* (Matsura 1976) and *Har. axyridis* (Osawa 1992b, 1993), with the field data structured in the form of life tables, suggest negligible role of parasitoids and interspecific predators as mortality factors of these ladybirds. The most important impact on the numbers of **immature stages** of both species was cannibalism of eggs and larvae, and prey shortage during larval feeding. Of the three mentioned papers, only one (Osawa 1993) reported any effect of parasitoids. According to this paper, parasitism of *Har. axyridis* pupae by *Phalacrotophora* sp. in two study years accounted for 4.9% and 18.6% mortality of pupae.

On maize crops in Ontario (Canada), egg predation (without distinguishing between interspecific predation and cannibalism) was recognized as an important mortality factor of *Col. maculata*. By conducting daily observations of selected egg clusters, Wright & Laing (1982) found the predation rate of eggs to be 45% in one year and 49% in another.

The majority of available data do not allow reliable estimation of the regulatory effects of enemies on the populations of predatory ladybirds. However, high

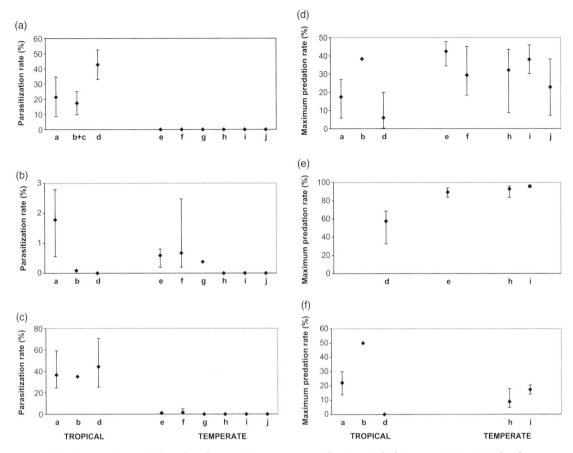

Figure 8.12 Impact of parasitoids and predators on immature stages of various Epilachninae species in tropical and temperate regions (square, mean value; bar, recorded range).

(a) Parasitization of eggs; (b) proportion of hatched larvae dying from parasitoids during larval development; (c) proportion of pupae dying from parasitoids; (d) maximum predation rates of eggs; (e) maximum predation rates of larvae; (f) maximum predation rates of pupae.

Maximum predation rate equals pooled rates of recorded predation (incl. cannibalism) and 'disappearance'.

a, *Henosepilachna septima* on Sumatra (latitude 1°S) (after Abbas & Nakamura 1985).

b and c, *Henosepilachna vigintioctopunctata* on Sumatra (1°S) (after Nakamura et al. 1988 (b) and Inoue et al. 1993 (c)).

d, *Chnootriba similis* in Ethiopia (7°N) (after Beyene et al. 2007).

e, *Henosepilachna pustulosa* on Honshu (35°N) (after Nakamura & Ohgushi 1981).

f, *Henosepilachna vigintioctopunctata* on Honshu (35°N) (after Nakamura 1976).

g, *Henosepilachna vigintioctomaculata* on Honshu (36°N) (after Nakamura 1987).

h, *Henosepilachna niponica* on Honshu (36°N) (after Shirai 1987).

i, *Henosepilachna yasutomii* on Honshu (36°N) (after Shirai 1988).

j, *Henosepilachna pustulosa* on Hokkaido (43°N) (after Kimura & Katakura 1986).

parasitization rates of larvae and pupae, recorded in many studies, especially by *Homalotylus* spp. (8.3.2.4), *Phalacrotophora* spp. (8.3.2.7) and *O. scaposus* (8.3.2.8), indicate that, in certain circumstances, larval and pupal parasitoids may be important mortality factors.

Several generalisations can be made concerning parasitization of immature entomophagous Coccinellidae by hymenopterans (*Homalotylus* spp., *O. scaposus*): (i) relatively sedentary coccidophagous species tend to be parasitized to a higher degree than their more dispersive aphidophagous relatives, (ii) parasitization rates are usually higher in warmer regions than in colder ones, (iii) parasitization rates tend to increase with the progress of host population development and (iv) successive host generations tend to be more and more affected by the parasitoids. Thus, in warm regions of Morocco and southeastern Europe, larvae and pupae of late generations of coccidophagous *Chilocorus* spp. can be parasitized at rates of up to 80–95% (Rubtsov 1954, Smirnoff 1957, Murashevskaya 1969, Stathas et al. 2008).

Parasitization of predatory species by flies of the genus *Phalacrotophora* may also reach high values (around 80%), although rates of parasitization are very variable (Disney & Chazeau 1990, Disney et al. 1994, Hurst et al. 1998).

Two biotic factors are often ranked among the most important causes of mortality in **adult predatory ladybirds**: the braconid parasitoid *D. coccinellae*, and entomopathogenic fungi particularly of the genus *Beauveria*.

Parasitization rates by *D. coccinellae* are sometimes quite high, especially in overwintering ladybird populations. However, the rates of host mortality caused by parasitoid development are clearly lower than the parasitization rates. For example, parasitization rates of *C. septempunctata* recorded in early dormancy in two overwintering sites in Poland were 15 and 74%, while the rates of parasitoid emergence were only 4 and 24%, respectively (Ceryngier 2000).

Fungal pathogens are the organisms most frequently reported as responsible for high winter mortality of Coccinellidae (Iperti 1966, Lipa et al. 1975, Olszak 1986, Ceryngier 2000, Ormond et al. 2006). Evaluation of their effects on naturally overwintering ladybirds is also complex. The susceptibility of an insect to the pathogen largely depends on the condition of the host and its associated immunity. It was found that various entomopathogenic fungi may opportunistically infect moribund ladybirds, while healthy ones are resistant to their attack (Ceryngier & Hodek 1996, Ceryngier 2000).

8.5.3   Concluding note

Some data suggest that natural enemies may exert stronger and more persistent effects on the populations of phytophagous than entomophagous Coccinellidae, and, of the latter, coccidophagous species may be more affected than aphidophagous ones. This may be related to differences in the prevalent lifestyles within these groups. Riddick et al. (2009) suppose that phytophagous species, which are typically sedentary, are more easily located by natural enemies, especially parasitoids, than are more mobile predatory species. However, regardless of which group of Coccinellidae is considered, there is little evidence for top-down regulation of their populations. Further studies are needed to clarify the role that natural enemies play in population dynamics of both phytophagous and predatory ladybirds.

REFERENCES

Abbas, I. and K. Nakamura. 1985. Adult population parameters and life tables of an epilachnine beetle (Coleoptera: Coccinellidae) feeding on bitter cucumber in Sumatra. *Res. Popul. Ecol.* 27: 313–324.

Abdel-Moniem, A. S. H. and M. A. Gesraha. 2001. Evaluation of certain entomopathogenic nematodes against the melon ladybird, *Epilachna chrysomelina* F. (Coleoptera: Coccinellidae). *Arch. Phytopathol. Plant Prot.* 34: 327–336.

van Achterberg, C. 2006. The Braconidae (Hymenoptera) of Greenland. *Zool. Med. Leiden.* 80: 13–62.

Aguiar-Menezes, E. L., E. Durska, A. L. S. Resende and A. T. Lixa. 2008. Phoridae (Diptera) as primary parasitoid of ladybird pupae (Coleoptera: Coccinellidae). *In XXII Congresso Brasileiro de Entomologia*, 24–29 Aug. 2008, Umberlandia-MG (Resumos).

Al Abassi, S., M. A. Birkett, J. Pettersson et al. 2001. Response of the ladybird parasitoid *Dinocampus coccinellae* to toxic alkaloids from the seven-spot ladybird, *Coccinella septempunctata. J. Chem. Ecol.* 27: 33–43.

Alekseev, V. N. and T. D. Radchenko 2001. Ceraphronoid wasps (Hymenoptera, Ceraphronoidea) of the fauna of the Ukraine. Communication 1. *Vest. Zool.* 35: 3–16.

Alma, P. J. 1980. Observations on some coccinellids in New Zealand and their significance to the biological control of *Paropsis charybdis* (Coleoptera: Chrysomelidae). *N. Z. Entomol.* 7: 164–165.

Anderson, J. M. E., D. F. Hales and K. A. van Brunschot. 1986. Parasitisation of coccinellids in Australia. *In* I. Hodek (ed.). *Ecology of Aphidophaga*. Academia, Prague and Dr W. Junk, Dordrecht. pp. 519–524.

Angalet, G. W., L. W. Coles and J. A. Stewart. 1968. Two potential parasites of the Mexican bean beetle from India. *J. Econ. Entomol.* 61: 1073–1075.

Aoki, S. and U. Kurosu. 1986. Soldiers of a European gall aphid, *Pemphigus spyrotecae* (Homoptera: Aphidoidea): why do they molt? *J. Ethol.* 4: 97–104.

Aoki, S. and U. Kurosu. 1989. Soldiers of *Astegopteryx styraci* (Homoptera, Aphidoidea) clean their gall. *Jpn. J. Entomol.* 57: 407–416.

Aoki, S., U. Kurosu and C. D. von Dohlen. 2001. Colony defense by wingpadded nymphs in *Grylloprociphilus imbricator* (Hemiptera: Aphididae). *Fla. Entomologist.* 84: 431–434.

Aoki, S., U. Kurosu and W. Sirikajornjaru. 2007. A new soldier-producing aphid species, *Pseudoregma baenzigeri*, sp. nov, from Northern Thailand. *J. Insect Sci.* 7: 38–48.

Appanna, M. 1948. The larval parasite, *Pleurotropis foveolatus*, C., of the potato beetle, *Epilachna 28-punctata. Curr. Sci.* 5: 154–155.

Arakaki, N. 1988. Egg protection with feces in the ladybeetle, *Pseudoscymnus kurohime* (Miyatake) (Coleoptera: Coccinellidae). *Appl. Entomol. Zool.* 23: 495–497.

Arakaki, N. 1992. Predators of the sugar cane woolly aphid, *Ceratovacuna lanigera* (Homoptera: Aphididae) in Okinawa and predator avoidance of defensive attack by the aphid. *Appl. Entomol. Zool.* 27: 159–161.

Balduf, W. V. 1926. The bionomics of *Dinocampus coccinellae* Schrank. *Ann. Entomol. Soc. Am.* 19: 465–498.

Barratt, B. I. P. and P. D. Johnstone. 2001. Factors affecting parasitism by *Microctonus aethiopoides* (Hymenoptera: Braconidae) and parasitoid development in natural and novel host species. *Bull. Entomol. Res.* 91: 245–253.

Barron, A. and K. Wilson. 1998. Overwintering survival in the seven spot ladybird, *Coccinella septempunctata* (Coleoptera: Coccinellidae). *Eur. J. Entomol.* 95: 639–642.

Barrows, E. M. and M. E. Hooker 1981. Parasitization of the Mexican bean beetle by *Pediobius foveolatus* in urban vegetable gardens. *Environ. Entomol.* 10: 782–786.

Belokobylskij, S. A. 2000. New species of the subfamily Euphorinae (Hymenoptera, Braconidae) from East Palaearctic. Part I. *Far Eastern Entomology.* 87: 1–28.

Belokobylskij, S. A. 2001. First record of the genus *Centistina* Enderlein, 1912 from the Palaearctic Region (Hymenoptera: Braconidae. Euphorinae). *Zoosystematica Rossica.* 10: 166.

Belshaw, R. 1993. Tachinid flies, Diptera: Tachinidae. *Handbooks for the Identification of British Insects* 10 (4a (i)), Royal Entomol. Soc., London. 170 pp.

Ben Halima-Kamel, M. 2010. Les ennemis naturels de *Coccinella algerica* Kovář dans la region du Sahel de Tunisie. *Entomologie Faunistique – Faunistic Entomology.* 62: 97–101.

Berkvens, N., J. Moens, D. Berkvens et al. 2010. *Dinocampus coccinellae* as a parasitoid of the invasive ladybird *Harmonia axyridis* in Europe. *Biol. Contr.* 53: 92–99.

Berry, P. A. and H. L. Parker. 1949. Investigations on a South American *Epilachna* sp. and the importation of its parasite *Lydinolydella metallica* Tns. into the United States (Coleoptera, Coccinellidae; Diptera, Larvaevoridae). *Proc. Entomol. Soc. Wash.* 51: 93–103.

Beyene, Y., T. Hofsvang and F. Azerefegne. 2007. Population dynamics of tef epilachna (*Chnootriba similis* Thunberg) (Coleoptera, Coccinellidae) in Ethiopia. *Crop Prot.* 26: 1634–1643.

Bezzerides, A. L., K. J. McGraw, R. S. Parker and J. Husseini. 2007. Elytra color as a signal of chemical defense in the Asian ladybird beetle *Harmonia axyridis. Behav. Ecol. Sociobiol.* 61: 1401–1408.

Bhatkar, A. P. and B. R. Subba Rao. 1976. Biology of *Pediobius foveolatus* (Hymenoptera: Eulophidae), a parasite of predatory ladybeetles in India. *Entomol. Germ.* 3: 242–247.

Bjornson, S. 2008. Natural enemies of the convergent lady beetle, *Hippodamia convergens* Guerin-Meneville: their inadvertent importation and potential significance for augmentative biological control. *Biol. Contr.* 44: 305–311.

Bjornson, S., J. Le, T. Saito and H. Wang. 2011. Ultrastructure and molecular characterization of a microsporidium, *Tubulinosema hippodamiae*, from the convergent lady beetle, *Hippodamia convergens* Guerin-Meneville. *J. Invert. Pathol.* 106: 280–288.

Bledsoe, L. W., R. V. Flanders and C. R. Edwards. 1983. Morphology and development of the immature stages of *Pediobius foveolatus* (Hymenoptera: Eulophidae). *Ann. Entomol. Soc. Am.* 76: 953–957.

Blount, J. D., M. P. Speed, G. D. Ruxton and P. A. Stephens. 2009. Warning displays may function as honest signals of toxicity. *Proc. R. Soc. Lond. (B)* 276: 871–877.

Boucek, Z. 1988. *Australian Chalcidoidea (Hymenoptera). A Biosystematic Revision of Genera of Fourteen Families, with a Reclassification of Species.* C.A.B. International, Wallingford, 832 pp.

Bradley, G. A. 1973. Effect of *Formica obscuripes* (Hymenoptera: Formicidae) on the predator–prey relationship between *Hyperaspis congressis* (Coleoptera: Coccinellidae) and *Toumeyella numismaticum* (Homoptera: Coccidae). *Can. Entomol.* 105: 1113–1118.

Brooks, W. M., D. B. Montross, R. K. Sprenkel and G. Carner. 1980. Microsporidiosis of coleopterous pests of soybeans. *J. Invert. Pathol.* 35: 93–95.

Brooks, W. M., E. I. Hazard and J. Becnel. 1985. Two new species of *Nosema* (Microsporidia: Nosematidae) from the Mexican bean beetle *Epilachna varivestis* (Coleoptera: Coccinellidae). *J. Protozool.* 32: 525–535.

Brown B. 2007. Kinds of Phoridae. http://www.discoverlife.org/mp/20o?act=x_checklistandguide=Phoridae (accessed 25 February 2008).

Bryden, J. W. and M. W. H. Bishop 1945. *Perilitus coccinellae* (Schrank) (Hym: Braconidae) in Cambridgeshire. *Entomol. Mon. Mag.* 81: 51–52.

Burling, J. M., R. L. Ware and L. L. Handley 2010. Is *Harmonia axyridis* a marginal host for the parasitoid wasp *Dinocampus coccinellae?* Morphological and molecular genetic analyses. *IOBC/wprs Bull.* 58: 23–25.

Cali, A. and J. D. Briggs. 1967. The biology and life history of *Nosema tracheophila* sp. n. (Protozoa: Cnidospora: Microsporidea) found in *Coccinella septempunctata* Linnaeus (Coleoptera: Coccinellidae). *J. Invert. Pathol.* 9: 515–522.

Camarano, S., A. Gonzalez and C. Rossini. 2006. Chemical defense of the ladybird beetle *Epilachna paenulata*. *Chemoecology.* 16: 179–184.

Cantwell, G. E., W. W. Cantelo and M. A. Cantwell. 1985. Effect of a parasitic mite, *Coccipolipus epilachnae*, on fecundity, food consumption and longevity of the Mexican bean beetle. *J. Entomol. Sci.* 20: 199–203.

Cartwright, B., R. D. Eikenbary and G. W. Angalet. 1982. Parasitism by *Perilitus coccinellae* (Hym.: Braconidae) of indigenous coccinellid hosts and the introduced *Coccinella septempunctata* (Col.: Coccinellidae), with notes on winter mortality. *Entomophaga.* 27: 237–243.

Castaldo D., W. Rossi and F. Sabatini. 2004. Contribution to the knowledge of the Laboulbeniales from Greece. *Plant Biosystems.* 138: 261–269.

Castro, A. C. P. and W. Rossi. 2008. New records of Laboulbeniales (Fungi, Ascomycota) from Ecuador. *Biodiversity of South America, I. Memoirs on Biodiversity.* 1: 11–18.

Ceryngier, P. 2000. Overwintering of *Coccinella septempunctata* (Coleoptera: Coccinellidae) at different altitudes in the Karkonosze Mts, SW Poland. *Eur. J. Entomol.* 97: 323–328.

Ceryngier, P., J. Havelka and I. Hodek. 2004. Mating and activity of gonads in pre–dormant and dormant ladybirds (Coleoptera: Coccinellidae). *Invert. Reprod. Develop.* 45: 127–135.

Ceryngier, P. and I. Hodek. 1996. Enemies of Coccinellidae. *In* I. Hodek and A. Honěk (eds). *Ecology of Coccinellidae.* Kluwer Acad. Publ., Dordrecht. pp. 319–350.

Ceryngier, P., P. Kindlmann, J. Havelka et al. 1992. Effect of food, parasitization, photoperiod and temperature on gonads and sexual activity of males of *Coccinella septempunctata* (Coleoptera, Coccinellidae) in autumn. *Acta Entomol. Bohemoslov.* 89: 97–106.

Chapman, J. A., J. I. Romer and J. Stark. 1955. Ladybird and army cutworm adults as food for grizzly bears in Montana. *Ecology.* 36: 156–158.

Cicek, K. and A. Mermer. 2007. Food composition of the marsh frog, *Rana ridibunda* Pallas, 1771, in Thrace. *Turk. J. Zool.* 31: 83–90.

Chiu, C. H. and A. Moore. 1993. Biological control of the Philippine ladybeetle, *Epilachna philippinensis* (Hymenoptera: Eulophidae), on Saipan. *Micronesica* suppl. 4: 79–80.

Chou, L. 1981. A preliminary list of Braconidae (Hymenoptera) of Taiwan. *J. Agric. Res. China.* 30: 71–88.

Christian, E. 2001. The coccinellid parasite *Hesperomyces virescens* and further species of the order Laboulbeniales (Ascomycotina) new to Austria. *Ann. Naturhist. Mus. Wien.* 103 B: 599–603.

Clausen, C. P. 1934. The natural enemies of Aleyrodidae in tropical Asia. *Philippine J. Sci.* 53: 253–265.

Clopton, R. E. 2009. Phylogenetic relationships, evolution, and systematic revision of the septate gregarines (Apicomplexa: Eugregarinorida: Septatorina). *Comp. Parasitol.* 76: 167–190.

Cochereau, P. 1969. Controle biologique d'*Aspidiotus destructor* Signoret (Homoptera, Diaspinae) dans l'Ile Vate (Nouvelles Hebrides) au moyen de *Rhizobius pulchellus* Montrouzier (Coleoptera, Coccinellidae). *Cah. ORSTOM, ser. Biol.* 8: 57–100.

Colyer, C. N. 1952. Notes on the life-histories of the British species of *Phalacrotophora* Enderlein (Dipt., Phoridae). *Entomol. Mon. Mag.* 88: 135–139.

Colyer, C. N. 1954. Further notes on the life-histories of the British species of *Phalacrotophora* Enderlein (Dipt., Phoridae). *Entomol. Mon. Mag.* 90: 208–210.

Colyer, C. N. 1961. A new species of *Phalacrotophora* (Diptera, Phoridae) from India. *Broteria.* 30: 93–97.

Compere, H. 1938. A report on some miscellaneous African Encyrtidae in the British Museum. *Bull. Entomol. Res.* 29: 315–337.

Conrad, A. 2005. *Adalia bipunctata* als Beute von *Gomphus flavipes* (Coleoptera: Coccinellidae; Odonata: Gomphidae). *Libellula.* 24: 237–239.

Cooreman, J. 1952. Acariens Podapolipidae du Congo Belge. *Bull. Inst. Roy. Soc. Nat. Belg.* 28: 1–10.

Cottrell, T. E. and D. I. Shapiro-Ilan 2003. Susceptibility of a native and an exotic lady beetle (Coleoptera: Coccinellidae) to *Beauveria bassiana*. *J. Invert. Pathol.* 84: 137–144.

Crawford, J. C. 1912. Descriptions of new Hymenoptera, no. 5. *Proc. U. S. Nat. Mus.* 43: 163–188.

Dahlman, D. L. 1990. Evaluation of teratocyte functions: an overview. *Arch. Insect Bioch. Physiol.* 13: 159–166.

Daloze, D., J-.C. Braekman and J. M. Pasteels. 1995. Ladybird defence alkaloids: structural, chemotaxonomic and biosynthetic aspects (Col.: Coccinellidae). *Chemoecology.* 5/6: 173–183.

David, M. H. and G. Wilde. 1973. Susceptibility of the convergent lady beetle to parasitism by *Perilitus coccinellae* (Schrank) (Hymenoptera: Braconidae). *J. Kansas Entomol. Soc.* 46: 359–362.

Davis, D. S., S. L. Stewart, A. Manica and M. E. N. Majerus. 2006. Adaptive preferential selection of female coccinellid hosts by the parasitoid wasp *Dinocampus coccinellae* (Hymenoptera: Braconidae). *Eur. J. Entomol.* 103: 41–45.

Dean, G. J. 1983. Survival of some aphid (Hemiptera: Aphididae) predators with special reference to their parasites in England. *Bull. Entomol. Res.* 73: 469–480.

DeBach, P. and D. Rosen. 1991. *Biological Control by Natural Enemies*, 2nd edn. Cambridge University Press, Cambridge. 440 pp.

Dejean, A., J. Orivel and M. Gibernau. 2002. Specialized predation on plataspid heteropterans in a coccinellid beetle: adaptive behavior and responses of prey attended or not by ants. *Behav. Ecol.* 13: 154–159.

De Kesel, A. 2011. *Hesperomyces* (Laboulbeniales) and coccinellid hosts. *Sterbeeckia*. 30: 32–37.

Delabie, J. H. C. 2001. Trophobiosis between Formicidae and Hemiptera (Sternorrhyncha and Auchenorrhyncha): an overview. *Neotrop. Entomol.* 30: 501–516.

Delucchi, V. 1953. *Aphidecta obliterata* L. (Coleoptera, Coccinellidae) als Rauber von *Dreyfusia* (*Adelges*) *piceae* Ratz. *Pflanzenschutz Berichte.* 11: 73–83.

Delucchi, V. 1954. *Pullus impexus* (Muls.) (Coleoptera, Coccinellidae), a predator of *Adelges piceae* (Ratz.) (Hemiptera, Adelgidae), with notes on its parasites. *Bull. Entomol. Res.* 45: 243–278.

Disney, R. H. L. 1979. Natural history notes on some British Phoridae (Diptera) with comments on a changing picture. *Entomol. Gaz.* 30: 141–150.

Disney, R. H. L. 1994. *Scuttle Flies: The Phoridae.* Chapman and Hall, London. pp. xii, 467.

Disney, R. H. L. 1997. A new species of Phoridae (Diptera) that parasitises a widespread Asian ladybird beetle (Coleoptera: Coccinellidae). *Entomologist.* 116: 163–168.

Disney, R. H. L. and P. L.Th. Beuk. 1997. European *Phalacrotophora* (Diptera: Phoridae). *Entomol. Gaz.* 48: 185–192.

Disney, R. H. L. and J. Chazeau. 1990. The recognition and biology of *Phalacrotophora quadrimaculata* (Diptera: Phoridae) parasitising *Olla v-nigrum* (Coleoptera: Coccinellidae). *Ann. Parasitol. Hum. Comp.* 65: 98–100.

Disney, R. H. L., M. E. N. Majerus and M. J. Walpole. 1994. Phoridae (Diptera) parasitising Coccinellidae (Coleoptera). *Entomologist.* 113: 28–42.

Dolenska, M., O. Nedvěd, P. Vesely, M. Tesarova and R. Fuchs. 2009. What constitutes optical warning signals of ladybirds (Coleoptera: Coccinellidae) towards bird predators: colour, pattern or general look? *Biol. J. Linn. Soc.* 98: 234–242.

Domenichini, G. 1957. Contributo alla conoscenza dei parassiti e iperparassiti dei Coleoptera Coccinellidae. *Boll. Zool. Agr. Bachic.*, Milano. 22: 215–246.

Durska, E., P. Ceryngier and R. H. L. Disney. 2003. *Phalacrotophora beuki* (Diptera: Phoridae), a parasitoid of ladybird pupae (Coleoptera: Coccinellidae). *Eur. J. Entomol.* 100: 627–630.

Eidelberg, M. M. 1994. Mites of the family Podapolipidae (Heterostigmata, Tarsonemina) of Ukraine and adjacent areas with description of a new species. *Vestn. Zool.* 1: 37–42. (In Russian with English summary.)

Eisner, T. and M. Eisner. 1992. Operation and defensive role of 'gin traps' in a coccinellid pupa (*Cycloneda sanguinea*). *Psyche.* 99: 265–273.

Eisner, T., M. Goetz, D. Aneshansley, G. Ferstandig-Arnold and J. Meinwald. 1986. Defensive alkaloid in blood of Mexican bean beetle (*Epilachna varivestis*). *Experientia.* 42: 204–207.

Elliott, E. and C. Morley. 1907. On the hymenopterous parasites of Coleoptera. *Trans. Entomol. Soc. Lond.* 1907: 7–75.

Elnagdy, S. M. 2008. *Comparative analysis of bacterial male-killing in Coccinellidae from different climatic regions.* Unpubl. PhD thesis, University of Cambridge, UK.

Fabres, G. 1981. Entomophagous insects associated with the cassava mealybug in Congo. *Trop. Pest Manage.* 27: 145–146.

Fain, A., G. D. D. Hurst, J. C. Tweddle et al. 1995. Description and observations of two new species of Hemisarcoptidae from deutonymphs phoretic on Coccinellidae (Coleoptera) in Britain. *Int J. Acarol.* 21: 1–8.

Fain, A., G. D. D. Hurst, C. Fassotte et al. 1997. New observations on the mites of the family Hemisarcoptidae (Acari: Astigmata) phoretic on Coccinellidae (Coleoptera). *Entomologie.* 67: 89–94.

Fallahzadeh, M. and G. Japoshvili. 2010. Checklist of Iranian Encyrtids (Hymenoptera: Chalcidoidea) with descriptions of new species. *J. Insect Sci.* 10:68–93.

Fallahzadeh, M., M. Shojaei, H. Ostovan and K. Kamali. 2006. The first report of two parasitoid wasps on the larvae of *Nephus bipunctatus* (Col.: Coccinellidae) from Iran. *J. Entomol. Soc. Iran* 26: 95–96. (In Persian with English summary.)

Filatova, I. T. 1974. The parasites of Coccinellidae (Coleoptera) in West Siberia. *In* N. G. Kolomyietz (ed.), *The Fauna and Ecology of Insects from Siberia.* Publ. House 'Nauka', Siberian Branch, Novosibirsk. pp. 173–185. (In Russian.)

Firlej, A., G. Boivin, E. Lucas and D. Coderre. 2005. First report of *Harmonia axyridis* Pallas being attacked by *Dinocampus coccinellae* Schrank in Canada. *Biol. Inv.* 7: 553–556.

Firlej, A., E. Lucas, D. Coderre and G. Boivin. 2007. Teratocytes growth pattern reflects host suitability in a host-parasitoid assemblage. *Physiol. Entomol.* 32: 181–187.

Firlej, A., E. Lucas, D. Coderre and G. Boivin. 2010. Impact of host behavioral defenses on parasitization efficacy of a larval and adult parasitoid. *BioControl.* 55: 339–348.

Flint, M. L. and S. H. Dreistadt. 1998. *Natural Enemies Handbook: The Illustrated Guide to Biological Pest Control.* Univ. Calif. Div. Agric. Nat. Res., Berkeley. 154 pp.

Foerster, H. 1938. Gregarinen in schlesischen Insekten. *Z. Parasitenk.* 10: 157–209.

Fujiyama, N., H. Katakura and Y. Shirai. 1998. Report of the Mexican bean beetle, *Epilachna varivestis* (Coleoptera: Coccinellidae) in Japan. *Appl. Entomol. Zool.* 33: 327–331.

Gambino, P. 1992. Yellowjacket (*Vespula pensylvanica*) predation at Hawaii Volcanoes and Haleakala National Parks: identity of prey items. *Proc. Haw. Entomol. Soc.* 31: 157–164.

Garces, S. and Williams, R. 2004. First Record of *Hesperomyces virescens* Thaxter (Laboulbeniales: Ascomycetes) on *Harmonia axyridis* (Pallas) (Coleoptera: Coccinellidae). *J. Kansas Entomol. Soc.* 77: 156–158.

Geoghegan, I. E., W. P. Thomas and M. E. N. Majerus. 1997. Notes on the coccinellid parasitoid *Dinocampus coccinellae* (Schrank) (Hymenoptera: Braconidae) in Scotland. *Entomologist.* 116: 179–184.

Geoghegan, I. E., T. M. O. Majerus and M. E. N. Majerus. 1998a. A record of a rare male of the parthenogenetic parasitoid *Dinocampus coccinellae* (Schrank) (Hym.: Braconidae). *Entomol Rec. J. Var.* 110: 171–172.

Geoghegan, I. E., T. M. O. Majerus and M. E. N. Majerus. 1998b. Differential parasitisation of adult and pre-imaginal *Coccinella septempunctata* (Coleoptera: Coccinellidae) by *Dinocampus coccinellae* (Hymenoptera: Braconidae). *Eur. J. Entomol.* 95: 571–579.

Gerson, U., B. A. O'Connor and M. A. Houck. 1990. Acari. *In* D. Rosen (ed.). *Armored Scale Insects: Their Biology, Natural Enemies and Control.* Elsevier, Amsterdam. pp. 77–97.

Geus, A. 1969. Sporentierchen, Sporozoa: die Gregarinida der land- und susswasserbewohnenden Arthropoden Mitteleuropas. *In* F. Dahl (ed.). *Die Tierwelt Deutschlands und der angrenzenden Meeresteile nach ihren Merkmalen und nach ihrer Lebensweise.* Gustav Fisher Verlag, Jena. 608 pp.

Ghahari, H., P. A. Lehr, R. J. Lavigne, R. Hayat and H. Ostovan. 2007. New records of robber flies (Diptera, Asilidae) for the Iranian fauna with their prey records. *Far Eastern Entomologist.* 179: 1–9.

Ghorpade, K. D. 1977. On *Perilitus coccinellae* (Schrank) (Hymenoptera: Braconidae), an endoparasite of adult Coccinellidae (Coleoptera), in Karnataka. *Mysore J. Agric. Sci.* 11: 55–59.

Ghorpade, K. D. 1979. Further notes on *Perilitus coccinellae* (Hymenoptera: Braconidae) in India. *Curr. Res.* 8: 112–113.

Ginsberg, H. S., R. A. Lebrun, K. Heyer and E. Zhioua. 2002. Potential nontarget effects of Metarhizium anisopliae (Deuteromycetes) used for biological control of ticks (Acari: Ixodidae). *Environ. Entomol.* 31: 1191–1196.

Giroux, S., J-.C. Cote, C. Vincent, P. Martel and D. Coderre. 1994. Bacteriological insecticide M-One effects on predation efficiency and mortality of adult *Coleomegilla maculata lengi* (Coleoptera: Coccinellidae). *J. Econ. Entomol.* 87: 39–43.

Godeau, J-.F., J-.L. Hemptinne, A. F. G. Dixon and J-.C. Verhaeghe 2009. Reaction of ants to, and feeding biology of, a congeneric myrmecophilous and non-myrmecophilous ladybird. *J. Insect Behav.* 22: 173–185.

Gomiero, L. M., A. G. Manzatto and F. M. S. Braga. 2008. The role of riverine forest for food supply for the omnivorous fish *Brycon opalinus* Cuvier, 1819 (Characidae) in the Serra do Mar, Southeast Brazil. *Braz. J. Biol.* 68: 321–328.

Gonzales, G. 2006. Los Coccinellidae de Chile. http://www.coccinellidae.cl (accessed 27 July 2009).

Gourlay, E. S. 1930. Some parasitic Hymenoptera of economic importance in New Zealand. *N. Z. J. Sci. Tech.* 11: 339–343.

Graham, M. W. R. de V. 1969. Synonymic and descriptive notes on European Encyrtidae (Hym., Chalcidoidea). *Pol. Pismo Entomol.* 39: 211–319.

Haitlinger, R. 1998. New species *Chrysomelobia donati* and *Coccipolipus arturi* (Acari, Prostigmata, Podapolipidae) connected with insects from Cameroon and Sumatra. *Wiad. Parazytol.* 35: 161–164.

Hajek, A. 2004. *Natural Enemies: An Introduction to Biological Control.* Cambridge University Press, Cambridge, UK. pp. xv, 378.

Hajiqanbar, H., R. W. Husband, K. Kamali, A. Saboori and H. Kamali. 2007. *Ovacarus longisetosus* n. sp. (Acari: Podapolipidae) from *Amara (Paracelia) saxicoloa* Zimm. (Coleoptera: Carabidae) and new records of *Coccipolipus, Dorsipes, Eutarsopolipus* and *Tarsopolipus* from Iran. *Int. J. Acarol.* 33: 241–244.

Haldar, D. P., S. K. Ray and R. Bose. 1988. *Anisolobus indicus* sp.n. of septate gregarines (Apicomplexa: Sporozoea) from the beetle, *Coccinella septempunctata* (Coleoptera: Coccinellidae). *Acta Protozool.* 27: 165–170.

Hanks, M. L. and C. S. Sadof. 1990. The effect of ants on nymphal survivorship of *Coccus viridis* (Homoptera: Coccidae). *Biotropica.* 22: 210–213.

Hariri, G. 1965. Records of nematode parasites of *Adalia bipunctata* (L.) (Col., Coccinellidae). *Entomol. Mon. Mag.* 101: 132.

Harwood, J. D., C. Ricci, R. Romani et al. 2006a. Prevalence and association of the laboulbenialean fungus *Hesperomyces virescens* (Laboulbeniales: Laboulbeniaceae) on coccinellid hosts (Coleoptera: Coccinellidae) in Kentucky, USA. *Eur. J. Entomol.* 103: 799–804.

Harwood, J. D., C. Ricci, R Romani et al. 2006b. Historic prevalence of a laboulbenialean fungus infecting introduced coccinellids in the United States. *Antenna.* 30: 74–79.

Hemptinne, J. L. 1988. Ecological requirements for hibernating *Propylea quatuordecimpunctata* (L.) and *Coccinella septempunctata* (Col.: Coccinellidae). *Entomophaga.* 33: 505–515.

Hochmuth, R. C., J. L. Hellman, G. Dively and R. F. W. Schroder 1987. Effect of parasitic mite *Coccipolipus epilachnae* (Acari: Podapolipidae) on feeding, fecundity, and longevity of soybean-fed adult Mexican bean beetles (Coleoptera: Coccinellidae) at different temperatures. *J. Econ. Entomol.* 80: 612–616.

Hodek, I. 1973. *Biology of Coccinellidae.* Academia, Prague. 260 pp.

Hoffer, A. 1963. Descriptions of new species of the family Encyrtidae from Czechoslovakia (Hym., Chalcidoidea) I. *Acta Entomol. Mus. Nat. Pragae.* 35: 549–592.

Holloway, G. J., P. W. de Jong, P. M. Brakefield and H. de Vos. 1991. Chemical defence in ladybird beetles (Coccinellidae). I. Distribution of coccinelline and individual variation in defence in 7-spot ladybird (*Coccinella septempunctata*). *Chemoecology.* 2: 7–14.

Hoogendoorn, M. and G. E. Heimpel. 2002. Indirect interactions between an introduced and a native ladybird beetle species mediated by a shared parasitoid. *Biol. Contr.* 25: 224–230.

Hooker, M. E. and E. M. Barrows. 1992. Clutch size reduction and host discrimination in the superparasitizing gregarious endoparasitic wasp *Pediobius foveolatus* (Hymenoptera: Eulophidae). *Ann. Entomol. Soc. Am.* 85: 207–213.

Hoshide, K. 1980. Notes on the Gregarines in Japan 11. *Bull. Fac. Edu., Yamaguchi Univ., Nat. Sci.* 30: 49–60.

Houck, M. A. and B. M. O'Connor. 1991. Ecological and evolutionary significance of phoresy in the Astigmata (Acari). *Annu. Rev. Entomol.* 36: 611–636.

Howard, N. F. and B. J. Landis. 1936. Parasites and predators of the Mexican bean beetle in the United States. *U. S. Dep. Agric. Circular.* 418: 12.

Hubble, D. 2011. Kidney-spot ladybird *Chilocorus renipustulatus* (Scriba) (Coccinellidae), a new host for the parasitic fungus *Hesperomyces virescens* Thaxter (Ascomycetes: Laboulbeniales). *The Coleopterist.* 20: 135–136.

Hudon, M. 1959. First record of *Perilitus coccinellae* (Schrank) (Hymenoptera: Braconidae) as a parasite of *Coccinella novemnotata* Hbst. and *Coleomegilla maculata lengi* Timb. (Coleoptera: Coccinellidae) in Canada. *Can. Entomol.* 91: 63–64.

Hurst, L. D. 1991. The incidences and evolution of cytoplasmic male killers. *Proc. R. Soc. London (B).* 244: 91–99.

Hurst, G. D. D., M. E. N. Majerus and L. E. Walker. 1992. Cytoplasmic male killing elements in *Adalia bipunctata* (Linnaeus) (Coleoptera: Coccinellidae). *Heredity.* 69: 84–91.

Hurst, G. D. D., E. L. Purvis, J. J. Sloggett and M. E. N. Majerus. 1994. The effect of infection with male-killing *Rickettsia* on the demography of female *Adalia bipunctata* L. (two spot ladybird). *Heredity.* 73: 309–316.

Hurst, G. D. D., R. G. Sharpe, A. H. Broomfield et al. 1995. Sexually transmitted disease in a promiscuous insect, *Adalia bipunctata*. *Ecol. Entomol.* 20: 230–236.

Hurst, G. D. D., T. C. Hammarton, J. J. Obrycki et al. 1996. Male-killing bacterium in a fifth ladybird beetle, *Coleomegilla maculata* (Coleoptera: Coccinellidae). *Heredity.* 77: 177–185.

Hurst, G. D. D., M. E. N. Majerus and A. Fain. 1997a. Coccinellidae (Coleoptera) as vectors of mites. *Eur. J. Entomol.* 94: 317–319.

Hurst, G. D. D., T. M. Hammarton, C. Bandi et al. 1997b. The diversity of inherited parasites of insects: the male-killing agent of the ladybird beetle *Coleomegilla maculata* is a member of the Flavobacteria. *Genet. Res.* 70: 1–6.

Hurst, G. D. D., L. D. Hurst and M. E. N. Majerus 1997c. Cytoplasmic sex-ratio distorters. *In* S. L. O'Neill, A. A. Hoffmann and J. H. Werren (eds). *Influential Passengers.* Oxford University Press, Oxford. pp. 125–154.

Hurst, G. D. D., F. K. McMeechan and M. E. N. Majerus. 1998. Phoridae (Diptera) parasitizing *Coccinella septempunctata* (Coleoptera: Coccinellidae) select older prepupal hosts. *Eur. J. Entomol.* 95, 179–181.

Hurst, G. D. D., J. H. Graf von der Schulenburg, T. M. O. Majerus et al. 1999a. Invasion of one insect species, *Adalia bipunctata*, by two different male-killing bacteria. *Insect Mol. Biol.* 8: 133–139.

Hurst, G. D. D., C. Bandi, L. Sacchi, et al. 1999b. *Adonia variegata* (Coleoptera: Coccinellidae) bears maternally inherited Flavobacteria that kill males only. *Parasitology.* 118: 125–134.

Hurst, G. D. D., F. M. Jiggins, J. H. Graf von der Schulenburg et al. 1999c. Male-killing *Wolbachia* in two species of insect. *Proc. R. Soc. Lond. (B).* 266: 735–740.

Husain, T. and M. Y. Khan. 1986. Family Eulophidae. *In* B. R. Subba Rao and M. Hayat (eds). *The Chalcidoidea (Insecta: Hymenoptera) of India and the Adjacent Countries. Part II. A Catalogue of Chalcidoidea of India and the Adjacent Countries. Oriental Insects.* 20: 211–245.

Husband, R. W. 1972. A new genus and species of mite (Acarina: Podapolipidae) associated with the coccinellid *Cycloneda sanguinea*. *Ann. Entomol. Soc. Am.* 65: 1099–1014.

Husband, R. W. 1981. The African species of *Coccipolipus* with a description of all stages of *Coccipolipus solanophilae* (Acarina: Podapolipidae). *Rev. Zool. Afr.* 95: 283–299.

Husband, R. W. 1983. Distribution of African *Coccipolipus* with a description of two new species. *Rev. Zool. Afr.* 97: 158–171.

Husband, R. W. 1984a. New Central African *Coccipolipus* (Acarina: Podapolipidae). *Rev. Zool. Afr.* 98: 308–326.

Husband, R. W. 1984b. The taxonomic position of *Coccipolipus* (Acarina: Podapolipidae), a genus of mites which are parasites of ladybird beetles (Coccinellidae). *In* D. A. Griffith and C. E. Bowman (eds). *Acarology VI, vol 1.* E. Horward Publ., Chichester. pp. 328–336.

Husband, R. W. 1989. Two new species of *Coccipolipus* (Acari: Podapolipidae) parasites of *Chilocorus* spp. (Coccinellidae) from Vera Cruz and Morelos, Mexico and Florida and Wisconsin, USA. *Proc. Entomol. Soc. Wash.* 91: 429–435.

Ijichi, N., H. Shibao, T. Miura, T. Matsumoto and T. Fukatsu. 2005. Analysis of natural colonies of a social aphid *Colophina arma*: population dynamics, reproductive schedule, and survey for ecological correlates with soldier production. *Appl. Entomol. Zool.* 40: 239–245.

Inoue, T., K. Nakamura, S. Salmah and I. Abbas. 1993. Population dynamics of animals in unpredictably-changing tropical environments. *J. Biosci.* 18: 425–455.

Iperti, G. 1964. Les parasites des Coccinelles aphidiphages dans les Alpes-Maritimes et les Basses-Alpes. *Entomophaga.* 9: 153–180.

Iperti, G. 1966. Protection of coccinellids against mycosis. *In* I. Hodek (ed.). *Ecology of Aphidophagous Insects.* Academia, Prague and Dr W. Junk, The Hague. pp. 189–190.

Iperti, G. and D. van Waerebeke. 1968. Description, biologie et importance d'une novelle espece d'Allantonematidae (Nematode) parasite des coccinelles aphidiphages: *Parasitilenchus coccinellinae,* n. sp. *Entomophaga.* 13: 107–119.

Irwin, T. 2006. Female swarms of *Phalacrotophora* (Phoridae). *Diptera.info.* http://64.191.19.182/articles.php?article_id=9 (accessed 24 October 2008).

Ito, Y. 1989. The evolutionary biology of sterile soldiers in aphids. *Tr. Ecol. Evol.* 4: 69–73.

Izraylevich, S. and U. Gerson. 1993. Mite parasitisation on armoured scale insects: host suitability. *Exp. Appl. Acarol.* 17: 877–888.

Jackson, D. J. 1928. The biology of *Dinocampus (Perilitus) rutilus* Nees, a braconid parasite of *Sitona lineata* L. – Part I. *Proceedings of the General Meeting for Scientific Business of the Zoological Society of London,* 1928. pp. 597–630.

James, R. R., B. T. Shaffer, B. Croft and B. Lighthart. 1995. Field evaluation of *Beauveria bassiana*: its persistence and effects on the pea aphid and a non-target coccinellid in alfalfa. *Biocontrol Sci. Technol.* 5: 425–438.

Ji, L., U. Gerson and S. Izraylevich. 1994. The mite *Hemisarcoptes* sp. (Astigmata: Hemisarcoptidae) parasitizing willow oyster scale (Homoptera: Diaspididae) on poplars in northern China. *Exp. Appl. Acarol.* 18: 623–627.

Jiggins, F. M. and M. C. Tinsley. 2005. An ancient mitochondrial polymorphism in *Adalia bipunctata* linked to a sex-ratio-distorting bacterium. *Genetics.* 171: 1115–1124.

Jiggins, F. M., G. D. D. Hurst and M. E. N. Majerus. 1999. Sex ratio distorting *Wolbachia* causes sex role reversal in its butterfly host. *Proc. R. Soc. London (B).* 267: 69–73.

Johnson, N. F. 1992. Catalog of world Proctotrupoidea excluding Platygastridae. *Mem. Amer. Entomol. Inst.* 51: 1–825.

Johnstone, R. A. and G. D. D. Hurst. 1996. Maternally inherited male-killing microorganisms may confound interpretation of mtDNA variation in insects. *Biol. J. Linn. Soc.* 58: 453–470.

Joron, M. 2003. Aposematic colouration. *In* R. T. Carde and V. H. Resh (eds). *Encyclopedia of Insects.* Academic Press, New York. pp. 39–45.

Joshi, S. and C. A. Viraktamath. 2004. The sugarcane woolly aphid, *Ceratovacuna lanigera* Zehntner (Hemiptera: Aphididae): its biology, pest status and control. *Curr. Sci.* 87: 307–316.

Joudrey, P. and S. Bjornson. 2007. Effects of an unidentified microsporidium on the convergent lady beetle, *Hippodamia convergens* Guerin-Meneville (Coleoptera: Coccinellidae), used for biological control. *J. Invert. Pathol.* 94: 140–143.

Kadono-Okuda, K., H. Sakurai, S. Takeda and T. Okuda. 1995. Synchronous growth of a parasitoid, *Perilitus coccinellae,* and teratocytes with the development of the host, *Coccinella septempunctata. Entomol. Exp. Appl.* 75: 145–149.

Kadono-Okuda, K., F. Weyda and T. Okuda. 1998. *Dinocampus (= Perilitus) coccinellae* teratocyte-specific polypeptide: its accumulative property, localization and characterization. *J. Insect Physiol.* 44: 1073–1080.

Kamburov, S. S., D. J. Nadel and R. Kenneth. 1967. Observations on *Hesperomyces virescens* Thaxter (Laboulbeniales), a fungus associated with premature mortality of *Chilocorus bipustulatus* L. in Israel. *Israel J. Agric. Res.* 17: 131–134.

Kaneko, S. 2002. Aphid-attending ants increase the number of emerging adults of the aphid's primary parasitoid and hyperparasitoids by repelling intraguild predators. *Entomol. Sci.* 5: 131–146.

Kaneko, S. 2007. Larvae of two ladybirds, *Phymatosternus lewisii* and *Scymnus posticalis* (Coleoptera: Coccinellidae), exploiting colonies of the brown citrus aphid *Toxoptera citricidus* (Homoptera: Aphididae) attended by the ant *Pristomyrmex pungens* (Hymenoptera: Formicidae). *Appl. Entomol. Zool.* 42: 181–187.

Kato, T. 1968. Predatious behavior of coccidophagous coccinellid, *Chilocorus kuwanae* Silvestri in the hedge of *Euonymus japonicus* Thunberg. *Kontyu.* 36: 29–38. (In Japanese with English summary.)

Kerrich, G. J. 1960. A systematic study of the chalcidid genus *Uga* Girault (Hymenoptera: Chalcidoidea), parasites of epilachnine beetles. *Proc. R. Entomol. Soc. Lond. (B)* 29: 113–119.

Kerrich, G. J. 1963. A study of the Encyrtid genus *Aminellus* Masi with systematic notes on related genera. *Beitr. Entomol.* 13: 359–368.

Kerrich, G. J. 1973. A revision of the tropical and subtropical species of the eulophid genus *Pediobius* Walker (Hymenoptera: Chalcidoidea). *Bull. Br. Mus. nat. Hist. (Entomol.)* 29: 115–199.

Kimura, T. and H. Katakura. 1986. Life cycle characteristics of a population of the phytophagous ladybird *Henosepilachna pustulosa* depending on two host plants. *J. Fac. Sci. Hokkaido Univ. Ser. VI, Zool.* 24: 202–225.

Klausnitzer, B. 1967. Beobachtungen an Coccinellidenparasiten (Hymenoptera, Diptera). *Entomol. Abh. Staatl. Mus. Tierk. Dresden.* 32: 305–309.

Klausnitzer, B. 1969. Zur Kenntnis der Entomoparasiten mitteleuropaeischer Coccinellidae. *Abh. Ber. NaturkMus., Goerlitz.* 44(9): 1–15.

Knell, R. J. and K. M. Webberley. 2004. Sexually transmitted diseases of insects: distribution, evolution, ecology and host behaviour. *Biol. Rev.* 79: 557–581.

Kovalenko, T. K. 2002. On ecology of *Nothoserphus afissae* (Watanabe) (Hymenoptera, Proctotrupidae): a parasite of *Henosepilahna vigintioctomaculata* Motschulsky (Coleoptera,

Coccinellidae) in Primorskii krai. *A.I. Kurentsov's Annual Memorial Meeting.* 12: 38–42. (In Russian.)

Kovalenko, T. K. and V. N. Kuznetsov. 2005. Employment of the parasite *Nothoserphus afissae* (Watanabe) (Hymenoptera, Proctotrupidae) in biological control with *Henosepilachna vigintioctomaculata* Motschulsky (Coleoptera, Coccinellidae) in Primorskii krai. *A.I. Kurentsov's Annual Memorial Meeting.* 16: 73–80. (In Russian.)

Koyama, S. and M. E. N. Majerus. 2008. Interactions between the parasitoid wasp *Dinocampus coccinellae* and two species of coccinellid from Japan and Britain. *BioControl.* 53: 253–264.

Kristin, A. 1986. Heteroptera, Coccinea, Coccinellidae and Syrphidae in the food of *Passer montanus* L. and *Pica pica* L. *Biologia (Bratislava).* 41: 143–150. (In Slovak with English summary.)

Kristin, A., N. Lebedeva and J. Pinowski. 1995. The diet of nestling tree sparrows (*Passer montanus*). Preliminary report. *International Studies on Sparrows.* 20–21: 3–19.

Kulman, H. M. 1971. Parasitism of *Anatis quindecimpunctata* by *Homalotylus terminalis. Ann. Entomol. Soc. Am.* 64: 953–954.

Kundu, T. K. and D. P. Haldar. 1981. Observations on a new cephaline gregarine (Protozoa: Sporozoa), *Brustiospora indicola* gen. nov., sp. nov. from a coccinellid beetle, *Stethorus* sp. *Arch. Protistenk.* 124: 471–480.

Kurosu, U. and S. Aoki. 2003. *Tuberaphis owadai* (Homoptera), a new aphid species forming a large gall on *Styrax tonkinensis* in northern Vietnam. *Entomol. Sci.* 6: 89–96.

Kurosu, U., M. Kutsukake, S. Aoki et al. 2008. Galls of *Cerataphis bambusifoliae* (Hemiptera: Aphididae) found on *Styrax suberifolius* in Taiwan. *Zool. Stud.* 47: 191–199.

Kutsukake, M., H. Shibao, N. Nikoh et al. 2004. Venomous protease of aphid soldier for colony defense. *Proc. Natn. Acad. Sci.* 101: 11338–11343.

Kuznetsov, V. N. 1987. Parasites of coccinellids (Coleoptera Coccinellidae) in Far East. *In New Data on Systematics of Insects in Far East.* DVO AN SSSR, Vladivostok. pp. 17–22. (In Russian.)

Kuznetsov, V. N. 1997. *Lady Beetles of the Russian Far East.* Memoir no. 1, Center for Systematic Entomology, The Sandhill Crane Press, Inc., Gainesville, FL. 248 pp.

Laing, D. J. 1988. A comparison of the prey of three common web-building spiders of open country, bush fringe and urban areas. *Tuatara.* 30: 23–35.

Lal, B. 1946. Biological notes on *Pleurotropis foveolatus* Crawford: a larval parasite of *Epilachna vigintiocto-punctata* Fab. *Curr. Sci.* 5: 138–139.

Lall, B. S. 1961. On the biology of *Pediobius foveolatus* (Crawford) (Eulophidae: Hymenoptera). *Indian J. Entomol.* 23: 268–273.

LaMana, M. L. and J. C. Miller. 1996. Field observations on *Harmonia axyridis* Pallas (Coleoptera: Coccinellidae) in Oregon. *Biol. Contr.* 6: 232–237.

LaSalle, J. 1993. North American genera of Tetrastichinae (Hymenoptera: Eulophidae). *J. Nat. Hist.* 28: 109–236.

Laudeho, Y., R. Ormieres and G. Iperti. 1969. Les entomophages de Parlatoria blanchardi Targ. dans les palmeraies de l' Adar Mauritanien. II. Etude d'un parasite de Coccinellidae *Gregarina katherina* Watson. *Ann. Zool. Ecol. Anim.* 1: 395–406.

Laurent, P., J-.C. Braekman and D. Daloze. 2005. Insect chemical defense. *Topics in Curr. Chem.* 240: 167–229.

Leander, B. S., S. A. J. Lloyd, W. Marshall and S. C. Landers. 2006. Phylogeny of marine gregarines (Apicomplexa) – *Pterospora, Lithocystis* and *Lankesteria* – and the origin(s) of coelomic parasitism. *Protist.* 157: 45–60.

Lee, J. H., D. K. Reed, H. P. Lee and R. W. Carlson 1988. Parasitoids of *Henosepilachna vigintioctomaculata* Motschulsky (Coleoptera, Coccinellidae) in Kyonggido area, Korea. *Korean J. Appl. Entomol.* 27: 28–34.

Lee, S. C., N. Corradi, E. J. Byrnes III et al. 2008. Microsporidia evolved from ancestral sexual fungi. *Curr. Biol.* 18: 1675–1679.

Lengyel, G. D. 2009. *Phalacrotophora* species (Diptera: Phoridae) with four subequal scutellar setae and notes on the other European species. *Zootaxa.* 2172: 59–68.

Lengyel, G. D. 2011. A taxonomic discussion of the genus *Phalacrotophora* Enderlein, 1912 (Diptera: Phoridae), with the description of two new species from Southeast Asia. *Zootaxa.* 2913: 38–46.

Li, C-.D. and R-.S. Chai. 2008. A new species of *Cowperia* Girault (Hymenoptera: Encyrtidae) from China. *Entomotaxonomia.* 30: 196–198. (In Chinese.)

Liere, H. and I. Perfecto. 2008. Cheating on a mutualism: indirect benefits of ant attendance to a coccidophagous coccinellid. *Environ. Entomol.* 37: 143–149.

Lin, K. S. 1987. On the genus *Nothoserphus* Brues, 1940 (Hymenoptera: Serphidae) from Taiwan. *Taiwan Agric. Res. Inst., Spec. Publ.* 22: 51–66.

Lipa, J. J. 1967. Studies on gregarines (Gregarinomorpha) of arthropods in Poland. *Acta Protozool.* 5: 97–179.

Lipa, J. J. 1968a. *Nosema coccinellae* sp. n., a new microsporidian parasite of *Coccinella septempunctata, Hippodamia tredecimpunctata* and *Myrrha octodecimguttata. Acta Protozool.* 5: 369–374.

Lipa, J. J. 1968b. On gregarine parasites of Coccinellidae in California, USA. *Acta Protozool.* 6: 263–272.

Lipa, J. J. and V. P. Semyanov. 1967. The parasites of the ladybirds (Coleoptera, Coccinellidae) in the Leningrad region. *Entomol. Obozr.* 46: 75–80. (In Russian with English summary.)

Lipa, J. J. and E. A. Steinhaus. 1959. *Nosema hippodamiae* n. sp., a microsporidian parasite of *Hippodamia convergens* Guerin (Coleoptera, Coccinellidae). *J. Insect Pathol.* 1: 304–308.

Lipa, J. J., S. Pruszynski and J. Bartkowski. 1975. The parasites and survival of the lady bird beetles (Coccinellidae) during winter. *Acta Parasitol. Pol.* 23: 453–461.

Liu, G. 2001. *A Taxonomic Study of Chinese Phorid Flies (Part I) (Diptera: Phoridae)*. NEU Press, Shenyang. 292 pp. (In Chinese with English summary.)

Long, K. D. and S. A. Belokobylskij. 2003. A preliminary list of the Braconidae (Hymenoptera) of Vietnam. *Russian Entomol. J.* 12: 385–398.

Lotfalizadeh, H. 2010. Some new data and corrections on Iranian encyrtid wasps (Hymenoptera: Chalcidoidea, Encyrtidae) fauna. *Biharean Biol.* 4: 173–178.

Lotfalizadeh, H. and E. Ebrahimi. 2001. New report of *Homalotylus nigricornis* Mercet (Hym: Encyrtidae) in Iran. *J. Entomol. Soc. Iran.* 21: 115–116.

Lusis, J. J. 1947. Some aspects of the population increase in *Adalia bipunctata* L. 2. The strains without males. *Dokl. Akad. Nauk SSSR.* 57, 951–954. (In Russian.)

Mabbott, P. R. 2006. *Exochomus quadripustulatus* (L.) (Coleoptera: Coccinellidae) as a host of *Dinocampus coccinellae* (Schrank) (Hymenoptera: Braconidae). *Br. J. Entomol. Nat. Hist.* 19: 40.

Maeta, Y. 1969. Biological studies on the natural enemies of some Coccinellid beetles. I. On *Perilitus coccinellae* (Schrank). *Kontyu.* 37: 147–166. (In Japanese with English summary.)

Majerus, M. E. N. 1989. *Coccinella magnifica* (Redtenbacher): a myrmecophilous ladybird. *Br. J. Entomol. Nat. Hist.* 2: 97–106.

Majerus, M. E. N. 1994. *Ladybirds. No. 81, New Naturalist Series*. HarperCollins, London. 367 pp.

Majerus, M. E. N. 1997. Parasitization of British ladybirds by *Dinocampus coccinellae* (Schrank) (Hymenoptera: Braconidae). *Br. J. Entomol. Nat. Hist.* 10: 15–24.

Majerus, M. E. N. 2003. *Sex Wars: Genes, Bacteria, and Sex Ratios*. Princeton University Press, Princeton, NJ. 280 pp.

Majerus, M. E. N. 2006. The impact of male-killing bacteria on the evolution of aphidophagous coccinellids. *Eur. J. Entomol.* 103: 1–7.

Majerus, M. E. N. and G. D. D. Hurst. 1997. Ladybirds as a model system for the study of male-killing symbionts. *Entomophaga.* 42: 13–20.

Majerus, M. E. N. and T. M. O. Majerus. 1997a. Predation of ladybirds by birds in the wild. *Entomol. Mon. Mag.* 133: 55–61.

Majerus, M. E. N. and T. M. O. Majerus. 1997b. Cannibalism among ladybirds. *Bull. Amat. Entomol. Soc.* 56: 235–248.

Majerus, M. E. N. and T. M. O. Majerus. 2000. Female-biased sex ratio due to male-killing in the Japanese ladybird *Coccinula sinensis. Ecol. Entomol.* 25: 234–238.

Majerus, M. E. N., I. E. Geoghegan and T. M. O. Majerus. 2000a. Adaptive preferential selection of young coccinellid hosts by the parasitoid wasp *Dinocampus coccinellae* (Hymenoptera: Braconidae). *Eur. J. Entomol.* 97: 161–164.

Majerus, M. E. N., J. H. Graf von der Schulenburg and I. A. Zakharov. 2000b. Multiple causes of male-killing in a single sample of the 2 spot ladybird, *Adalia bipunctata* (Coleoptera: Coccinellidae) from Moscow. *Heredity.* 84, 605–609.

Majerus, M. E. N., J. J. Sloggett, J-.F Godeau and J-.L. Hemptinne. 2007. Interactions between ants and aphidophagous and coccidophagous ladybirds. *Popul. Ecol.* 49: 15–27.

Majerus, T. M. O. 2001. *The Evolutionary Genetics of Male-killing in the Coccinellidae*. Unpubl. PhD thesis, University of Cambridge, UK.

Majerus, T. M. O., M. E. N. Majerus, B. Knowles et al. 1998. Extreme variation in the prevalence of inherited male-killing microorganisms between three populations of the ladybird *Harmonia axyridis* (Coleoptera: Coccinellidae). *Heredity.* 81: 683–691.

Majerus, T. M. O., J. H. Graf von der Schulenburg, M. E. N. Majerus and G. D. D. Hurst. 1999. Molecular identification of a male-killing agent in the ladybird *Harmonia axyridis* (Pallas) (Coleoptera: Coccinellidae). *Insect Mol. Biol.* 8: 551–555.

Majerus, T. M. O. and M. E. N. Majerus 2010. Intergenomic arms races: detection of a nuclear rescue gene of male-killing in a ladybird. *PLOS Pathog.* 6(7): e1000987. doi:10.1371/journal.ppat.1000987.

Mandal, D., M. Rai, B. Pranhan, D. Gurung et al. 1986. On a new septate gregarine from *Coccinella septempunctata* L. of a tea garden, Darjeeling. *Arch. Protistenk.* 131: 299–302.

Marlatt, C. L. 1903. A chalcidid parasite of the Asiatic ladybird. *Proc. Entomol. Soc. Wash.* 5: 138–139.

Marples, N. M. 1993. Toxicity assays of ladybirds using natural predators. *Chemoecology* 4: 33–38.

Marples, N. M., P. M. Brakefield and R. J. Cowie 1989. Differences between the 7-spot and 2-spot ladybird beetles (Coccinellidae) in their toxic effects on a bird predator. *Ecol. Entomol.* 14: 79–84.

Matis, D. and A. Valigurova. 2000. Gregarines (Apicomplexa, Eugregarinida) of some invertebrates of Slovakia. *Folia Faun. Slovaka* 5: 11–22. (In Slovak with English summary.)

Matsuka, M., H. Hashi and I. Okada. 1975. Abnormal sex ratio found in the lady beetle *Harmonia axyridis* Pallas (Coleoptera: Coccinellidae). *Appl. Entomol. Zool.* 10: 84–89.

Matsura, T. 1976. Ecological studies of a coccinellid, *Aiolocaria hexaspilota* Hope. I. Interaction between field populations of *A. hexaspilota* and its prey, the walnut leaf beetle (*Gastrolina depressa* Baly). *Jpn. J. Ecol.* 26: 147–156. (In Japanese with English summary.)

McDaniel, B. and Morrill. W. 1969. A new species of *Tetrapolipus* from *Hippodamia convergens* from South Dakota (Acarina: Podapolipidae). *Ann. Entomol. Soc. Am.* 62: 1456–1458.

Miura, K. 2010. First records of two parasitoid species parasitizing the ladybird beetle, *Coccinella septempunctata* L. (Coleoptera: Coccinellidae) from Japan. *Entomol. News.* 121: 95–96.

Mizer, A. V. 1970. On eating of beetles from Coccinellidae family by birds. *Vest. Zool.* 6: 1970, 21–24. (In Russian with English summary.)

Moore, B. L., W. V. Brown and M. Rothschild. 1990. Methylalkylpyrazines in aposematic insects, their hostplants and mimics. *Chemoecology.* 1: 43–51.

Mracek, Z. and Z. Ruzicka. 1990. Infectivity and development of *Steinernema* sp. strain *Hylobius* (Nematoda, Steinernematidae) in aphids and aphidophagous coccinellids. *J. Appl. Entomol.* 110: 92–95.

Muesebeck, C. F. W. 1936. The genera of parasitic wasps of the braconid subfamily Euphorinae, with a review of the Nearctic species. *USDA Misc. Pub.* 241: 37 pp.

Muggleton, J. 1978. Selection against the melanic morphs of *Adalia bipunctata* (two-spot ladybird): a review and some new data. *Heredity.* 40: 269–280.

Murashevskaya, Z. S. 1969. Species of *Chilocorus. Zashch. Rast.* 14: 36–38. (In Russian.)

Myartseva, S. N. 1981. Species of *Homalotylus* Mayr (Hymenoptera, Encyrtidae) – coccinellides parazites (Coleoptera, Coccinellidae) in Turkmenistan. *Izv. Akad. Nauk Turkm. SSR, Ser. Biol. Nauk.* 6: 35–41. (In Russian.)

Nakamura, K. 1976. Studies on the population dynamics of the 28-spotted lady beetle, *Henosepilachna vigintioctopunctata* F. I. Analysis of life tables and mortality process in the field population. *Jpn. J. Ecol.* 26: 49–59. (In Japanese with English summary.)

Nakamura, K. 1983. Comparative studies on population dynamics of closely related phytophagous lady beetles in Japan. *Res. Popul. Ecol.* Suppl. 3: 46–60.

Nakamura, K. 1987. Population study of the large 28-spotted ladybird, *Epilachna vigintioctomaculata* (Coleoptera: Coccinellidae). *Res. Popul. Ecol.* 29: 215–228.

Nakamura, K. and T. Ohgushi. 1981. Studies on the population dynamics of a thistle feeding lady beetle, *Henosepilachna pustulosa* (Kôno) in a cool temperate climax forest. II. Life tables, key-factor analysis, and detection of regulatory mechanisms. *Res. Popul. Ecol.* 23: 210–231.

Nakamura, K., I. Abbas and A. Hasyim. 1988. Population dynamics of the phytophagous lady beetle, *Epilachna vigintioctopunctata*, in an eggplant field in Sumatra. *Res. Popul. Ecol.* 30: 25–41.

Nalepa, C. A. and K. A. Kidd. 2002. Parasitism of the multicolored Asian lady beetle (Coleoptera: Coccinellidae) by *Strongygaster triangulifer* (Diptera: Tachinidae) in North Carolina. *J. Entomol. Sci.* 37: 124–127.

Nalepa, C. A. and A. Weir. 2007. Infection of *Harmonia axyridis* (Coleoptera: Coccinellidae) by *Hesperomyces virescens* (Ascomycetes: Laboulbeniales): role of mating status and aggregation behavior. *J. Invert. Pathol.* 94: 196–203.

Narsi Reddy, Y. and P. Narayan Rao. 1984. First report of *Parasitylenchus coccinellae* Iperti and Waerebeke (1968) from India in two new hosts *Menochilus*

sexmaculatus (F.) and *Illeis indica* Timberlake. *Riv. Parassit.* 45: 133–137.

Nentwig, W. 1983. The prey of web-building spiders compared with feeding experiments (Araneae: Aranelidae, Linyphiidae, Pholcidae, Agelenidae). *Oecologia.* 56: 132–139.

Nentwig, W. 1986. Non-webbuilding spiders: prey specialists or generalists? *Oecologia.* 69: 571–576.

Neuenschwander, P., R. D. Hennessey and H. R. Herren. 1987. Food web of insects associated with the cassava mealybug, *Phenacoccus manihoti* Matile-Ferrero (Hemiptera: Pseudococcidae), and its introduced parasitoid, *Epidinocarsis lopezi* (De Santis) (Hymenoptera: Encyrtidae), in Africa. *Bull. Entomol. Res.* 77: 177–189.

New, T. R. 1975. An Australian host record for *Phalacrotophora* Enderlein (Diptera: Phoridae). *Austr. Entomol. Mag.* 2: 84.

Niijima, K. and K. Nakajima. 1981. Abnormal sex ratio in *Menochilus sexmaculatus* (Fabricus). *Bull. Fac. Agric. Tamagawa Univ.* 21: 59–67.

Nikol'skaya, M. 1934. List of chalcid flies (Hym.) reared in USSR. *Bull. Entomol. Res.* 25: 129–143.

Noyes, J. S. 2011. Universal Chalcidoidea Database. www.nhm.ac.uk/entomology/chalcidoids/index.html (last accessed 5 November 2011).

Noyes, J. S. and M. Hayat. 1984. A review of the genera of Indo-Pacific Encyrtidae (Hymenoptera: Chalcidoidea). *Bull. Br. Mus. nat. Hist. (Entomol.)* 48(3): 131–395.

Nyffeler, M. 1999. Prey selection of spiders in the field. *J. Arachnol.* 27: 317–324.

Obrycki, J. J. 1989. Parasitization of native and exotic coccinellids by *Dinocampus coccinellae* (Schrank) (Hymenoptera: Braconidae). *J. Kansas Entomol. Soc.* 62: 211–218.

Obrycki, J. J. and M. J. Tauber 1978. Thermal requirements for development of *Coleomegilla maculata* (Coleoptera: Coccinellidae) and its parasite *Perilitus coccinellae* (Hymenoptera: Braconidae). *Can. Entomol.* 110: 407–412.

Obrycki, J. J., M. J. Tauber and C. A. Tauber. 1985. *Perilitus coccinellae* (Hymenoptera: Braconidae): Parasitization and development in relation to host-stage attacked. *Ann. Entomol. Soc. Am.* 78: 852–854.

Ogloblin, A. 1913. Contribution a la biologie des Coccinelles. *Russ. Entomol. Obozr.* 13: 27–43. (In Russian.)

Ogloblin, A. 1924. Le role du blastoderme extraembryonnaire du *Dinocampus terminatus* Nees pendant l'etat larvaire. *Vest. Kral. Ces. Spolec. Nauk.* 2: 1–27.

O'Hara, J. E. 2009. Taxonomic and Host Catalogue of the Tachinidae of America North of Mexico. http://www.nadsdiptera.org/Tach/CatNAmer/Home/CatNAmerhom.htm (accessed 14 July 2009).

Ohgushi, T. 1986. Population dynamics of an herbivorous lady beetle, *Henosepilachna niponica*, in a seasonal environment. *J. Anim. Ecol.* 55: 861–879.

Okuda, T. and P. Ceryngier. 2000. Host discrimination in *Dinocampus coccinellae* (Hymenoptera: Braconidae), a solitary parasitoid of coccinellid beetles. *Appl. Entomol. Zool.* 35: 535–539.

Okuda, T. and K. Kadono-Okuda. 1995. *Perilitus coccinellae* teratocyte polypeptide: evidence for production of a teratocyte-specific 540kDa protein. *J. Insect Physiol.* 41: 819–825.

Olszak, R. W. 1986. The occurrence of *Propylea quatuordecimpunctata* in apple orchards in Poland and the mortality caused by mycosis and parasites. *In* I. Hodek (ed.). *Ecology of Aphidophaga.* Academia, Prague and Dr W. Junk, The Hague. pp. 531–535.

Ormond, E. L., J. K. Pell, A. P. M. Thomas and H. E. Roy 2006. Overwintering ecology of *Coccinella septempunctata, Beauveria bassiana* and *Dinocampus coccinellae. IOBC/wprs Bull.* 29: 85–88.

Ormond, E. L., A. P. M. Thomas, P. J. A. Pugh, J. K. Pell and H. E. Roy. 2010. A fungal pathogen in time and space: population dynamics of *Beauveria bassiana* in a conifer forest. *FEMS Microbiol. Ecol.* 74: 146–154.

Orr, C. J., J. J. Obrycki and R. V. Flanders. 1992. Host-acceptance behavior of *Dinocampus coccinellae* (Hymenoptera: Braconidae). *Ann. Entomol. Soc. Am.* 85: 722–730.

Osawa, N. 1992a. Effect of pupation site on pupal cannibalism and parasitism in the ladybird beetle *Harmonia axyridis* Pallas (Coleoptera, Coccinellidae). *Jpn. J. Entomol.* 60: 131–135.

Osawa, N. 1992b. A life table of the ladybird beetle *Harmonia axyridis* PALLAS (Coleoptera, Coccinellidae) in relation to the aphid abundance. *Jpn. J. Entomol.* 60: 575–579.

Osawa, N. 1993. Population field studies of the aphidophagous ladybird beetle *Harmonia axyridis* (Coleoptera: Coccinellidae): life tables and key factor analysis. *Res. Popul. Ecol.* 35: 335–348.

Osawa, N. and T. Nishida. 1992. Seasonal variation in elytral colour polymorphism in Harmonia axyridis (the ladybird beetle): the role of non-random mating. *Heredity.* 69: 297–307.

Otsu, Y., Y. Matsuda, H. Shimizu et al. 2003. Biological control of phytophagous ladybird beetles *Epilachna vigintioctopunctata* (Col., Coccinellidae) by chitinolytic phylloplane bacteria *Alcaligenes paradoxus* entrapped in alginate beads. *J. Appl. Entomol.* 127: 441–446.

Pal, A., M. M. Swain and S. Rath. 2007. Seasonal variation in the diet of the fan-throated lizard, *Sitana ponticeriana* (Sauria: Agamidae). *Herpetol. Conserv. Biol.* 2: 145–148.

Pankanin-Franczyk, M. and P. Ceryngier. 1999. On some factors affecting the population dynamics of cereal aphids. *Aphids and Other Homopterous Insects* 7: 289–295.

Park, H.-C., Y. C. Park, O. K. Hong and S. Y. Cho. 1996. Parasitoids of the aphidophagous ladybeetles, *Harmonia axyridis* (Pallas) (Coleoptera: Coccinellidae) in Chuncheon areas, Korea *Korean J. Entomol.* 26: 143–147. (In Korean with English summary.)

Parker, B. L., M. E. Whalon and M. Warshaw. 1977. Respiration and parasitism in *Coleomegilla maculata lengi* (Coleop-

tera: Coccinellidae). *Ann. Entomol. Soc. Am.* 70: 984–987.

Patnaik, H. P. and L. N. Mohapatra. 2004. Record on *Omphale* sp. (Hymenoptera: Eulophidae) as egg parasitoid of epilachna beetle, *Henosepilachna vigintioctopunctata* (Fabricius). *Insect Environ.* 10: 29–30.

Peck, O. 1963. A Catalogue of the Nearctic Chalcidoidea (Insecta: Hymenoptera). *Can. Entomol., Suppl.* 30: 1–1092.

Pena, G., J. Miranda-Rios, G. de la Riva et al. 2006. A *Bacillus thuringiensis* S-layer protein involved in toxicity against *Epilachna varivestis* (Coleoptera: Coccinellidae). *Appl. Environ. Microbiol.* 72: 353–360.

Ping, S., W. Qin-Ying, W. Hui-Xian, L. Xiu-Jun and W. Yong. 2008. Identification of the *cry* gene in *Bacillus thuringiensis* strain WZ-9 and its toxicity against *Henosepilachna vigintioctomaculata. Chin. J. Agric. Biotech.* 5: 245–250.

Poorani, J. 2008. Coccinellidae of the Indian Subcontinent. http://www.angelfire.com/bug2/j_poorani/ (accessed 13 July 2009).

Poorani, J., A. Slipinski and R. G. Booth 2008. A revision of the genus *Synona* Pope, 1989 (Coleoptera: Coccinellidae: Coccinellini). *Ann. Zool.* 58: 579–594.

Pope, R. D. 1979. Wax production by coccinellid larvae (Coleoptera). *Syst. Entomol.* 4: 171–196.

Poprawski, T. J., J. C. Legaspi and P. E. Parker. 1998. Influence of entomopathogenic fungi on *Serangium parcesetosum* (Coleoptera: Coccinellidae), an important predator of whiteflies (Homoptera: Aleyrodidae). *Environ. Entomol.* 27: 785–795.

Pschorn-Walcher, H. 1958. Zur Kenntnis der Proctotrupidae der *Thomsonina*-Gruppe (Hymenoptera). *Beitr. Entomol.* 8: 724–731.

Pschorn-Walcher, H. 1971. Heloridae et Proctotrupidae. *Insecta Helvetica* 4, *Hymenoptera.* Fotorotar, Zurich. 64 pp.

Quicke, D. L. J. 1997. *Parasitic Wasps.* Chapman and Hall, London. pp. xviii, 470.

Radwan, Z. and G. L. Lovei. 1982. Records of coccinellid parasites from apple orchard and corn fields. *Acta Phytopathol. Acad. Sci. Hung.* 17: 111–113.

Randerson, J. P., N. G. C. Smith and L. D. Hurst. 2000. The evolutionary dynamics of male-killers and their hosts. *Heredity.* 84: 152–160.

Rebolledo, R., J. Sheriff, L. Parra and A. Aguilera. 2009. Life, seasonal cycles, and population fluctuation of *Hippodamia variegata* (Goeze) (Coleoptera: Coccinellidae), in the central plain of La Araucania Region, Chile. *Chilean J. Agric. Res.* 6: 292–298.

Rehner, S. A. and E. P. Buckley. 2005. A *Beauveria* phylogeny inferred from nuclear ITS and EF1-alpha sequences: evidence for cryptic diversification and links to *Cordyceps* teleomorphs. *Mycologia.* 97: 84–98.

Rehner, S. A., F. Posada, E. P. Buckley, F. Infante, A. Castillo and F. E. Vega. 2006. Phylogenetic origins of African and Neotropical *Beauveria bassiana s.l.* pathogens of the coffee berry borer, *Hypothenemus hampei*. *J. Invert. Pathol.* 93: 11–21.

Reimer, R. J., M.-L. Cope and G. Yasuda. 1993. Interference of *Pheidole megacephala* (Hymenoptera: Formicidae) with biological control of *Coccus viridis* (Homoptera: Coccidae) in coffee. *Environ. Entomol.* 22: 483–488.

Rhamhalinghan, M. 1986a. Pathologies caused by *Coccinellimermis* Rubtzov (Nematoda: Mermithidae) in *Coccinella septempunctata* L. (Coleoptera: Coccinellidae). *Proc. Indian Natn. Sci. Acad.* B52: 228–321.

Rhamhalinghan, M. 1986b. Pathophysiological effects of *Coccinellimermis* Rubtzov (Nematoda: Mermithidae) on the respiration of *Coccinella septempunctata* L. (Coleoptera: Coccinellidae). *Proc. Indian Natn. Sci. Acad.* B52: 431–435.

Rhamhalinghan, M. 1986c. The behaviour of *Coccinella septempunctata* L. (Coleoptera: Coccinellidae) associated with the exit of *Coccinellimermis* Rubtzov (Nematoda: Mermithidae): 1. Prior to the emergence of the parasite. *Indian Zool.* 10: 95–97.

Rhamhalinghan, M. 1987a. Pathophysiological effects on the ovaries in *Coccinellimermis Rubtzov* (Nematoda: Mermithidae) infected *Coccinella septempunctata* L. (Coleoptera: Coccinellidae): A preliminary report. *Proc. Indian Natn. Sci. Acad.* B53: 31–34.

Rhamhalinghan, M. 1987b. Host parasite relationship in *Coccinella septempunctata* L. (Coleoptera: Coccinellidae) and *Coccinellimermis* Rubtzov (Nematoda: Mermithidae). *J. Adv. Zool.* 8: 94–99.

Rhamhalinghan M. 1987c. Seasonal variations in ovariolar output in *Coccinella septempunctata* L. (Coleoptera: Coccinellidae). *In* S. Palanichamy (ed.). *Proc. 5th Indian Symposium Invert. Reprod.* P.G. and Res. Dept. Zool., APA. College of Arts and Culture, Palani, India. pp. 149–157.

Rhamhalinghan. M. 1988. Behaviour of *Coccinella septempunctata* L. associated with the exit of *Coccinellimermis* Rubtzov. II. During and after the emergence of the parasite. *Meetchi.* 30: 1–5.

Rhule, E. L. and M. E. N. Majerus. 2008. The potential of the sexually-transmitted mite, *Coccipolipus hippodamiae*, to control the harlequin ladybird, *Harmonia axyridis*, in Britain. *Bull. Amat. Entomol. Soc.* 67: 153–160.

Rhule, E. L., M. E. N. Majerus, F. M. Jiggins and R. L. Ware. 2010. Potential role of the sexually transmitted mite *Coccipolipus hippodamiae* in controlling populations of the invasive ladybird *Harmonia axyridis*. *Biol. Contr.* 53: 243–247.

Richards, A. M. 1980. Defensive adaptations and behaviour in *Scymnodes lividigaster* (Coleoptera: Coccinellidae). *J. Zool. Lond.* 192: 157–168.

Richards, A. M. 1985. Biology and defensive adaptations in *Rodatus major* (Coleoptera: Coccinellidae) and its prey, *Monophlebus pilosior* (Hemiptera: Margarodidae). *J. Zool. Lond. (A)* 205: 287–295.

Richards, O. W. 1926. A note on a dipterous parasite of ladybirds. *Entomol. Mon. Mag.* 62: 99.

Richardson, M. L. and L. M. Hanks. 2009. Partitioning of niches among four species of orb-weaving spiders in a grassland habitat. *Environ. Entomol.* 38: 651–656.

Richerson, J. V. 1970. A world list of parasites of Coccinellidae. *J. Entomol. Soc. Brit. Columbia.* 67: 33–48.

Richerson, J. V. and C. J. DeLoach 1972. Some aspects of host selection by *Perilitus coccinellae*. *Ann. Entomol. Soc. Am.* 65: 834–839.

Richter, V. A. 1971. A list of Tachinidae (Diptera) from the Caucasus. I. Subfamily Exoristinae. *Entomol. Obozr.* 50: 587–597. (In Russian.)

Riddick, E. W. 2006. Influence of host gender on infection rate, density and distribution of the parasitic fungus, *Hesperomyces virescens*, on the multicolored Asian lady beetle, *Harmonia axyridis*. *J. Insect Sci.* 6: 1–15.

Riddick, E. W. 2010. Ectoparasitic mite and fungus on an invasive lady beetle: parasite coexistence and influence on host survival. *Bull. Insectol.* 63: 13–20.

Riddick, E. W. and P. W. Schaefer. 2005. Occurrence, density and distribution of parasitic fungus *Hesperomyces virescens* (Laboulbeniales: Laboulbeniaceae) on multicolored Asian lady beetle (Coleoptera: Coccinellidae). *Ann. Entomol. Soc. Am.* 98: 615–624.

Riddick, E. W., T. E. Cottrell and K. A. Kidd. 2009. Natural enemies of the Coccinellidae: parasites, pathogens, and parasitoids. *Biol. Control.* 51: 306–312.

Risbec, J. 1951. Les Chalcidoides d'A.O.F. *Mem. Inst. Fr. Afr. Noire.* 13: 1–409.

Romero-Nápoles, J., H. Bravo-Mojica and T. H. Atkinson. 1987. Biologia de los Epilachninae (Coleoptera: Coccinellidae) del estado de Morelos y su susceptibilidad al parasitismo por *Pediobius foveolatus* (Hymenoptera: Eulophidae). *Folia Entomol. Mex.* 71: 37–46.

Rossi, W. and E. Bergonzo. 2008. New and interesting Laboulbeniales from Brazil. *Aliso.* 26: 1–8.

Roy, H. E. and T. Cottrell. 2008. Forgotten natural enemies: interactions between coccinellids and insect-parasitic fungi. *Eur. J. Entomol.* 105: 391–398.

Roy, H. E. and J. K. Pell. 2000. Interactions between entomopathogenic fungi and other natural enemies: implications for biological control. *Biocontrol Sci. Techn.* 10: 737–752.

Roy, H. E., D. Steinkraus, E. Eilenberg, A. Hajek and J. K. Pell. 2006. Bizarre interactions and endgames: entomopathogenic fungi and their arthropod hosts. *Annu. Rev. Entomol.* 51: 331–357.

Roy, H. E., P. M. J. Brown, P. Rothery, R. L. Ware and M. E. N. Majerus. 2008. Interactions between the fungal pathogen *Beauveria bassiana* and three species of ladybird: *Harmonia axyridis*, *Coccinella septempunctata* and *Adalia bipunctata*. *BioControl.* 53: 265–276.

Roy, H. E., R. S. Hails, H. Hesketh, D. B. Roy and J. K. Pell. 2009. Beyond biological control: non-pest insects and their pathogens in a changing world. *Insect Conserv. Diver.* 2: 65–72.

Rubtsov, I. A. 1954. *Citrus Pests and Their Natural Enemies*. Izd. AN SSSR, Moscow-Leningrad. 260 pp. (In Russian.)

Rueckert, S. I. and B. S. Leander. 2008. Gregarina. Gregarines. Version 23 September 2008. *The Tree of Life Web Project*, http://tolweb.org/Gregarina/124806/2008.09.23.

Saito, T. and S. Bjornson. 2006. Horizontal transmission of a microsporidium from the convergent lady beetle, Hippodamia convergens Guerin-Meneville (Coleoptera: Coccinellidae), to three coccinellid species of Nova Scotia. *Biol. Contr.* 39: 427–433.

Saito, T. and S. Bjornson. 2008. Effects of a microsporidium from the convergent lady beetle, *Hippodamia convergens* Guerin-Meneville (Coleoptera: Coccinellidae), on three non-target coccinellids. *J. Invert. Pathol.* 99: 294–301.

Santamaría, S. 1995. New and interesting Laboulbeniales (Fungi, Ascomycotina) from Spain, III. *Nova Hedwigia.* 61: 65–83.

Santamaría. S. 2003. *Flora Mycologica Iberica. Vol. 5.Laboulbeniales, II. Acompsomyces–Ilyomyces*. Real Jardín Botanico Madrid and J. Cramer. 344 pp.

Santos, G. P. and A. C. Q. Pinto. 1981. Biologia de *Cycloneda sanguinea* e sua associacao com pulgao em mudas de mangueira. *Pesquisa Agropecuaria Brasileira.* 16: 473–476.

Schaefer, P. W., R. J. Dysart, R. V. Flanders, T. L. Burger and K. Ikebe. 1983. Mexican bean beetle (Coleoptera: Coccinellidae) larval parasite *Pediobius foveolatus* (Hymenoptera: Eulophidae) from Japan: field release in the United States. *Environ. Entomol.* 12: 852–854.

Schaefer, P. W. and V. P. Semyanov. 1992. Arthropod parasites of *Coccinella septempunctata* (Coleoptera: Coccinellidae); world parasite list and bibliography. *Entomol. News.* 103: 125–134.

Schroder, R. F. W. 1982. Effect of infestation with *Coccipolipus epilachnae* Smiley (Acarina: Podapolipidae) on the fecundity and longevity of the Mexican bean beetle. *Int. J. Acarol.* 8: 81–84.

Schroder, R. F. W. and P. K. Schroder 1989. Distribution and new host records of *Coccipolippus epilachnae* Smiley (Acari Podapolipidae) attacking the Mexican bean beetle in the United States, Central America and Mexico. *Int. J. Acarol.* 15: 55–56.

Schroeder, F. C., S. R. Smedley, L. K. Gibbons et al. 1998. Polyazamacrolides from ladybird beetles: Ring-size selective oligomerization. *Proc. Natl Acad. Sci. USA.* 95: 13387–13391.

Schulenburg, J. H. Graf von der. 2000. *The Evolution and Dynamics of Male-killing in the Two-spot Ladybird Adalia bipunctata L. (Coleoptera: Coccinellidae)*. Unpubl. PhD thesis, University of Cambridge, UK.

Schulenburg, J. H. Graf von der, M. Habig, J. J. Sloggett et al. 2001. The incidence of male-killing *Rickettsia* (alpha proteobacteria) in the 10-spot ladybird, *Adalia decempunctata* L. (Coleoptera: Coccinellidae). *Appl. Environ. Microbiol.* 67: 270–277.

Schwarz, E. A. 1890. Myrmecophilous Coleoptera found in temperate North America. *Proc. Entomol. Soc. Wash.* 1: 237–247.

Semyanov, V. P. 1986. Parasites and predators of *Coccinella septempunctata*. *In* I. Hodek (ed.). *Ecology of Aphidophaga*. Academia, Prague and Dr W. Junk, Dordrecht. pp. 525–530.

Semyanov, V. P. 1998. Parasites of ladybird *Menochilus sexmaculatus* (Fabr.) (Coleoptera, Coccinellidae) from southeastern China. *In The Problems of Entomology in Russia, vol. II.* Russian Academy of Sciences, Sankt-Petersburg. pp. 115–116. (In Russian.)

Sengupta, T. and D. P. Haldar. 1996. Three new species of septate gregarines (Apicomplexa: Sporozoea) of the genus *Gregarina* Dufour, 1828 from insects. *Acta Protozool.* 35: 77–86.

Shafee, S. A. and A. Fatma. 1984. Taxonomic notes on Indian species of *Echthroplexis* Forster (Hymenoptera: Encyrtidae), with descriptions of two species. *Mitt. Schweiz. Ent. Ges. / Bull. Soc. Ent. Suisse.* 57: 371–376.

Shapiro-Ilan, D. I. and T. E. Cottrell. 2005. Susceptibility of lady beetles (Coleoptera: Coccinellidae) to entomopathogenic nematodes. *J. Invert. Pathol.* 89: 150–156.

Shaw, M. R., I. E. Geoghegan and M. E. N. Majerus. 1999. Males of *Dinocampus coccinellae* (Schrank) (Hym.: Braconidae: Euphorinae). *Entomol. Rec. J. Var.* 111: 195–196.

Shepard, M. and G. T. Gale. 1977. Superparasitism of *Epilachna varivestis* (Col.: Coccinellidae) by *Pediobius foveolatus* (Hym.: Eulophidae): influence of temperature and parasitoid-host ratio. *Entomophaga.* 22: 315–321.

Shibao, H. 1998. Social structure and the defensive role of soldiers in a eusocial bamboo aphid, *Pseudoregma bambucicola* (Homoptera: Aphididae): a test of the defence-optimization hypothesis. *Res. Popul. Ecol.* 40: 325–333.

Shima, H., H.-Y. Han and T. Tachi. 2010. Description of a new genus and six new species of Tachinidae (Diptera) from Asia and New Guinea. *Zootaxa.* 2516: 49–67.

Shirai, Y. 1987. Ecological studies of phytophagous lady beetle, *Henosepilachna vigintioctomaculata* complex, (Coleoptera: Coccinellidae) in the Ina area, Nagano Prefecture. II. Population dynamics of the thistle-feeding *H. niponica. Jpn. J. Ecol.* 37: 209–218. (In Japanese with English summary.)

Shirai, Y. 1988. Ecological studies on phytophagous lady beetles, *Henosepilachna vigintioctomaculata* complex, (Coleoptera: Coccinellidae) in the Ina area, Nagano Prefecture. III. Population dynamics and life cycle characteristics of *H. yasutomii* feeding on deadly nightshade, *Scopolia japonica* (Solanaceae). *Jpn. J. Ecol.* 38: 111–119. (In Japanese with English summary.)

Shull, H. F. 1948. An all-female strain of lady beetles with reversions to normal sex ratio. *Am. Nat.* 82: 241–251.

Silva, R. B., I. Cruz, M. L. C. Figueiredo and A. M. Penteado-Dias. 2009. Ocorrencia e biologia de *Dinocampus coccinellae* (Schrank, 1802. (Hymenoptera; Braconidae: Euphorinae) em diferentes especies de Coccinellidae (Coleoptera). *Annais do IX Congresso de Ecologia do Brasil, 13-17 September 2009,* Sao Lourenco–MG. pp.1–3.

Silvestri, F. 1903. Contribuzioni alla conoscenza dei Mirmeco-fili. I. Osservazioni su alcuni Mirmecofili dei dintorni di Portici. *Ann. Mus. Zool. R. Univ. Napoli (Nuova Serie.)* 1(13): 5 pp.

Sloggett, J. J. 2010. Predation of ladybird beetles by the orb-web spider *Araneus diadematus*. *BioControl.* 55: 631–638.

Sloggett, J. J. and M. E. N. Majerus. 2000. *Myrrha octodecim-guttata* (L.) (Coleoptera: Coccinellidae), a newly recorded host of *Dinocampus coccinellae* (Schrank) (Hymenoptera: Braconidae). *Br. J. Entomol. Nat. Hist.* 13: 126.

Sloggett, J. J. and M. E. N. Majerus. 2003. Adaptations of *Coccinella magnifica*, a myrmecophilous coccinellid to aggression by wood ants (*Formica rufa* group). II. Larval behaviour, and ladybird oviposition location. *Eur. J. Entomol.* 100: 337–344.

Sloggett, J. J., A. Manica, M. J. Day and M. Majerus. 1999. Predation of ladybirds (Coleoptera: Coccinellidae) by wood ants, *Formica rufa* L. (Hymenoptera: Formicidae). *Entomol. Gaz.* 50: 217–221.

Sloggett, J. J., W. Völkl, W. Schulze, J. H. Graf von der Schulen-burg and M. E. N. Majerus. 2002. The ant-associations and diet of the ladybird *Coccinella magnifica* (Coleoptera: Coccinellidae). *Eur. J. Entomol.* 99: 565–569.

Sloggett, J. J., M. Webberley and M. E. N. Majerus. 2004. Low parasitoid success on a myrmecophilous host is maintained in the absence of ants. *Ecol. Entomol.* 29: 123–127.

Sloggett, J. J., R. A. Wood and M. E. N. Majerus. 1998. Adapta-tions of *Coccinella magnifica* Redtenbacher, a myrmecophil-ous coccinellid, to aggression by wood ants (*Formica rufa* group). I. Adult behavioral adaptation, its ecological context and evolution. *J. Insect Behav.* 11: 889–904.

Sluss, R. 1968. Behavioral and anatomical responses of the convergent lady beetle to parasitism by *Perilitus coccinellae* (Schrank) (Hymenoptera: Braconidae). *J. Invert. Pathol.* 10: 9–27.

Smiley, R. L. 1974. A new species of Coccipolipus parasitic on the Mexican bean beetle (Acarina: Podapolipidae). *J. Wash. Acad. Sci.* 64: 298–302.

Smirnoff, W. A. 1957. La cochenille du palmier dattier (*Parla-toria blanchardi* Targ.) en Afrique du Nord. Comportement, importance economique, predateurs et lutte biologique. *Entomophaga.* 11: 1–98.

Smith, B. C. 1960. Note on parasitism of two coccinellids, *Coccinella trifasciata perplexa* Muls. and *Coleomegilla macu-lata lengi* Timb. (Coleoptera: Coccinellidae) in Ontario. *Can. Entomol.* 92: 652.

Smith, O. J. 1953. Species distribution and host records of the braconid genera *Microctonus* and *Perilitus* (Hymenoptera: Braconidae). *Ohio J. Sci.* 53: 173–178.

Stathas, G. J. 2001. Ecological data on predators of *Parlatoria pergandii* on sour orange trees in southern Greece. *Phy-toparasitica.* 29: 207–214.

Stathas, G. J., P. A. Eliopoulos and G. Japoshvili. 2008. A study on the biology of the diaspidid scale *Parlatoria ziziphi* (Lucas) (Hemiptera: Coccoidea: Diaspididae) in Greece. *In* M. Branco, J. C. Franco and C. Hodgson (eds). *Proceedings of the XI International Symposium on Scale Insect Studies, Oeiras (Portugal), 24–27 September 2007.* ISA Press, Lisbon. pp. 95–101.

Steenberg, T. and S. Harding. 2009. Entomopathogenic fungi recorded from the harlequin ladybird, *Harmonia axyridis.* *J. Invert. Pathol.* 102: 88–89.

Steenberg, T. and S. Harding. 2010. Entomopathogenic fungi found in field populations of the harlequin ladybird, *Har-monia axyridis.* IOBC/wprs Bull. 58: 137–141.

Stern, D. L. 1998. Phylogeny of the tribe Cerataphidini (Homoptera) and the evolution of the horned soldier aphids. *Evolution.* 52: 155–165.

Stern, D. L. and W. A. Foster. 1996. The evolution of soldiers in aphids. *Biol. Rev.* 71: 27–79.

Stevens, L. M., A. L. Steinhauer and J. R. Coulson 1975a. Suppression of Mexican bean beetle on soybeans with annual inoculative releases of *Pediobius foveolatus. Environ. Entomol.* 4: 947–952.

Stevens, L. M., A. L. Steinhauer and T. C. Elden 1975b. Laboratory rearing of the Mexican bean beetle and the parasite *Pediobius foveolatus*, with emphasis on parasite lon-gevity and host–parasite ratios. *Environ. Entomol.* 4: 953–957.

Styrsky, J. D. and M. D. Eubanks. 2007. Ecological conse-quences of interactions between ants and honeydew-producing insects. *Proc. R. Soc. Lond. (B)* 274: 151–164.

Swaine, G. and D. A. Ironside. 1983. *Insect Pests of Crop Fields in Colour.* Queensland Department of Primary Industries, Information Series QI 8 3006, Brisbane. 138 pp.

Tachikawa, T. 1972. A pteromalid parasite, *Metastenus townsendi* (Ashmead), of the coccinellid in Japan (Hymenop-tera: Chalcidoidea). *Trans. Shikoku Entomol. Soc.* 11: 85–86.

Takizawa, T. and H. Yasuda. 2006. The effects of attacks by the mutualistic ant, *Lasius japonicus* Santschi (Hymenop-tera: Formicidae) on the foraging behavior of the two aphidophagous ladybirds, *Coccinella septempunctata brucki* Mulsant (Coleoptera: Coccinellidae) and *Propylea japonica* (Thunberg) (Coleoptera: Coccinellidae). *Appl. Entomol. Zool.* 41: 161–169.

Tavares, I. I. 1979. The Laboulbeniales and their arthropod hosts. *In* L. R. Batra (ed.). *Insect–Fungus Symbiosis, Nutri-tion, Mutualism, and Commensalism.* Wiley, New York. pp. 229–258.

Tavares, I. I. 1985. Laboulbeniales (Fungi, Ascomycetes). *Mycol. Mem.* 9: 1–627.

Telenga, N. A. 1948. *Biological Method of the Insect Pest Control (Predaceous Coccinellids and their Utilisation in USSR)* Izd. AN SSSR, Kiev. 120 pp. (In Russian.)

Thacker, J. R. M. 2002. *An Introduction to Arthropod Pest Control.* Cambridge University Press, Cambridge. 341 pp.

Thaxter, R. 1931 Contribution towards a monograph of the Laboulbeniaceae. Part V. *Mem. Am. Acad. Arts Sci.* 16: 1–435.

Timberlake, P. H. 1916. Note on an interesting case of two generations of a parasite reared from the same individual host. *Can. Entomol.* 48: 89–91.

Timberlake, P. H. 1918. Notes on some of the immigrant parasitic Hymenoptera of the Hawaiian Islands. *Proc. Haw. Entomol. Soc.* 3: 399–404.

Timberlake, P. H. 1919. Revision of the parasitic chalcidoid flies of the genera *Homalotylus* Mayr and *Isodromus* Howard, with description of two closely related genera. *Proc. U. S. Natn. Mus.* 56: 133–194.

Tinsley, M. C. 2003. *The Ecology and Evolution of Male-killing Bacteria in Ladybirds.* Unpubl. PhD thesis, University of Cambridge, UK.

Tinsley, M. C. and M. E. N. Majerus. 2006. A new male-killing parasitism: *Spiroplasma* bacteria infect the ladybird *Anisosticta novemdecimpunctata* (Coleoptera: Coccinellidae). *Parasitology.* 132: 757–765.

Toguebaye, B. S. and B. Marchand. 1984. Etude histopathologique et cytopathologique d'une microsporidiose naturelle chez la coccinelle des cucurbitacees d'Afrique, *Henosepilachna elaterii* (Col.: Coccinellidae). *Entomophaga.* 29: 421–429.

Townes, H. and M. Townes. 1981. A revision of the Serphidae (Hymenoptera). *Mem. Amer. Entomol. Inst.* 32: 1–541.

Triltsch, H. 1996. On the parasitization of the ladybird *Coccinella septempunctata* L. (Col., Coccinellidae). *J. Appl. Entomol.* 120: 375–378.

Triltsch, H. 1999. Another record of *Phalacrotophora delageae* Disney (Diptera: Phoridae) parasitizing the pupae of *Adalia bipunctata* L. (Coleoptera: Coccinellidae). *Studia Dipterologica.* 6: 233–234.

Trjapitzin, V. A. and S. V. Triapitsyn. 2003. A new species of *Homalotylus* (Hymenoptera: Encyrtidae) from Mexico, parasitoid of *Azya orbigera orbigera* (Coleoptera: Coccinellidae). *Entomol. News.* 114: 192–196.

Trjapitzin, V. A., E. Ya. Chouvakhina, E. Ruiz-Cancino and R. M. Thompson-Farfan. 1999. *Homalotylus mexicanus* (Hymenoptera: Encyrtidae), parasitoide de larvas de coccinelidos (Coleoptera: Coccinellidae) en San Luis Potosi, Mexico. *Acta Cient. Potos.* 14: 93–98.

Tsugawa, H. 1951. Studies on the gregarines from the Coleoptera in Japan. *Yamaguchi J. Sci.* 2: 93–106. (In Japanese with English summary.)

Usman, S. and T. S. Thontadarya. 1957. A preliminary note on the larval-parasites complex of *Epilachna sparsa* (Herbst) in Mysore. *Curr. Sci.* 26: 252–253.

Vakhidov, T. 1975. On species composition of the parasites of predatory coccinellids (Coleoptera, Coccinellidae) – destroyers of apple aphids. *In Ekologiya i Biologiya Zhivotnikh Uzbekistana.* Izd. 'Fan' Uzbek. SSR, Tashkent, pp. 93–96. (In Russian.)

Vega, F. E., M. S. Goettel, M. Blackwell et al. 2009. Fungal entomopathogens: new insights on their ecology. *Fungal Ecology.* 2: 149–159.

Venkatesha, M. G. 2006. Seasonal occurrence of *Henosepilachna vigintioctopunctata* (F.) (Coleoptera: Coccinellidae) and its parasitoid on ashwagandha in India. *J. Asia-Pacific Entomol.* 9: 265–268.

Volkl, W. 1995. Behavioral and morphological adaptations of the coccinellid, *Platynaspis luteorubra* for exploiting ant-attended resources (Coleoptera: Coccinellidae). *J. Insect Behav.* 8: 653–670.

Volkl, W. and K. Vohland. 1996. Wax covers in larvae of two *Scymnus* species: do they enhance coccinellid larval survival? *Oecologia.* 107: 498–503.

Walker, M. F. 1961. Some observations on the biology of the ladybird parasite *Perilitus coccinellae* (Schrank) (Hym., Braconidae), with special reference to host selection and recognition. *Entomol. Mon. Mag.* 97: 240–244.

Ware, R., L.-J. Michie, T. Otani, E. Rhule and R. Hall. 2010. Adaptation of native parasitoids to a novel host: the invasive coccinellid *Harmonia axyridis*. *IOBC/wprs Bull.* 58: 175–182.

Watson, M. E. 1915. Some new gregarine parasites from Arthropoda. *J. Parasitol.* 2: 27–36.

Watson, M. E. 1916. Studies on gregarines including descriptions of twenty-one new species and a synopsis of the eugregarine records from the Myriapoda, Coleoptera, and Orthoptera of the world. *Illinois Biol. Monogr.* 2 (3): 1–258.

Webberley, K. M. and G. D. D. Hurst. 2002. The effect of aggregative overwintering on an insect sexually transmitted parasite system. *J. Parasitol.* 88: 707–712.

Webberley, K. M., G. D. D. Hurst, R. W. Husband et al. 2004. Host reproduction and an STD: causes and consequences of *Coccipolipus hippodamiae* distribution on coccinellids. *J. Anim. Ecol.* 73: 1–10.

Webberley, K. M., J. Buszko, V. Isham and G. D. D. Hurst. 2006a. Sexually transmitted disease epidemics in a natural insect population. *J. Anim. Ecol.* 75: 33–43.

Webberley, K. M., M. Tinsley, J. J. Sloggett, M. E. N. Majerus and G. D. D. Hurst. 2006b. Spatial variation in the incidence of sexually transmitted parasites of the ladybird beetle Adalia *bipunctata*. *Eur. J. Entomol.* 103: 793–797.

Weinert, L. A., M. C. Tinsley, M. Temperley and F. M. Jiggins. 2007. Are we understanding the diversity and incidence of insect bacterial symbionts? A case study in ladybird beetles. *Biol. Lett.* 3: 678–681.

Weir A. 2002. The Laboulbeniales – an enigmatic group of arthropod-associated fungi. *In* J. Seckbach (ed.) *Symbiosis: Mechanisms and Model Systems.* Kluwer Acad. Publ., Dordrecht. pp. 613–620.

Weir, A. and G. W. Beakes. 1996. Correlative light- and scanning electron microscope studies on the developmental morphology of *Hersperomyces virescens*. *Mycologia* 88: 677–693.

Weir, A. and P. M. Hammond. 1997. Laboulbeniales on beetles: host utilization patterns and species richness of the parasites. *Biodivers. Conserv.* 6: 701–719.

Welch, V. L., J. J. Sloggett, K. M. Webberley and G. D. D. Hurst. 2001. Short-range clinal variation in the prevalence of a sexually transmitted fungus associated with urbanisation. *Ecol. Entomol.* 26: 547–550.

Werren, J. H., G. D. D. Hurst, W. Zhang et al. 1994. Rickettsial relative associated with male killing in the ladybird beetle (*Adalia bipunctata*). *J. Bacteriol.* 176: 388–394.

Wheeler, W. M. 1911. An ant-nest coccinellid (*Brachyacantha quadripunctata* Mels.). *J. N.Y. Entomol. Soc.* 19: 169–174.

Wieloch, M. 1975. Food of nestling house sparrows, *Passer domesticus* L. and tree sparrows, *Passer montanus* L. in agrocenoses. *Pol. Ecol. Stud.* 1: 227–242.

Wightman, J. A. 1986. *Sitona discoideus* (Coleoptera: Curculionidae) in New Zealand, 1975–1983: distribution, population studies, and bionomic strategy. *N. Z. J. Zool.* 13: 221–240.

Wright, J. E. 1978. Observations on the copulatory behaviour of *Perilitus coccinellae* (Hymenoptera: Braconidae). *Proc. Entomol. Soc. Ontario.* 109: 22.

Wright, E. J. and J. E. Laing. 1978. The effects of temperature on development, adult longevity and fecundity of *Coleomegilla maculata lengi* and its parasite, *Perilitus coccinellae. Proc. Entomol. Soc. Ontario.* 109: 33–47.

Wright, E. J. and J. E. Laing 1982. Stage-specific mortality of *Coleomegilla maculata lengi* Timberlake on corn in southern Ontario. *Environ. Entomol.* 11: 32–37.

Wylie, H. G. 1958. Observations on *Aphidecta obliterata* (L.) (Coleoptera: Coccinellidae), a predator of conifer-infesting Aphidoidea. *Can. Entomol.* 90: 518–522.

Yinon, U. 1969. The natural enemies of the armored scale lady-beetle *Chilocorus bipustulatus* (Col. Coccinellidae). *Entomophaga.* 14: 321–328.

Zakharov, I. A. and M. M. Eidelberg. 1997. Parasitic mite *Coccipolipus hyppodamia* McDaniel et Morrill (Tarsonemina, Podapolipidae) in populations of the two-spotted ladybird *Adalia bipunctata* L. (Coleoptera, Coccinellidae) *Entomol. Obozr.* 76: 680–683. (In Russian with English summary.)

Zakharov, I. A. and E. V. Shaikevich. 2001. The Stockholm populations of *Adalia bipunctata* (L.) (Coleoptera: Coccinellidae): a case of extreme female-biased population sex ratio. *Hereditas.* 134: 263–266.

Zeman, V. and J. Vanek. 1999. Hymenoptera (Braconidae, Ichneumonidae, Eurytomidae, Pteromalidae, Mymaridae, Proctotrupidae, Diapriidae, Scelionidae, Platygasteridae, Ceraphronidae, Megaspilidae, Crabronidae) in terrestrial traps in montane and sub-alpine zone in the Giant Mts. *Opera Corcontica.* 36: 171–179.

Chapter 9

COCCINELLIDS AND SEMIOCHEMICALS

Jan Pettersson

Department of Ecology, Swedish University of Agricultural Sciences, Box 7044, SE-750 07 Uppsala, Sweden

Ecology and Behaviour of the Ladybird Beetles (Coccinellidae), First Edition. Edited by I. Hodek, H.F. van Emden, A. Honěk.
© 2012 Blackwell Publishing Ltd. Published 2012 by Blackwell Publishing Ltd.

9.1 INTRODUCTION

Coccinellids are friendly, colourful insects, serving as valuable supporters of biological control of pests such as aphids and scales in important crops. However, they have developed a set of unpleasant and poisonous defence chemicals which together with striking colouration constitute an aposematic enemy barrier. The unique aposematic chemistry forms a basis for several prominent features of coccinellid population biology and behavioural ecology. Furthermore the ecological adaptability of coccinellids and their capacity to sense and respond to a broad set of information carried by stimuli from the environment is an important contribution to their ecological success (Hodek & Michaud 2008). Information mediated by semiochemicals contributes not only to different aspects of food and food search but also to social and ecological behavioural processes such as competition, reproduction and other important traits in behavioural ecology. Recently a valuable summary of coccinellid chemistry has been published by Durieux et al. (2010). This chapter summarizes semiochemical-supported mechanisms related to the main features of the population biology of coccinellids, their interaction with foes and friends, foraging and prey discrimination and social interaction.

9.2 APOSEMATISM AND REFLEX BLEEDING CHEMISTRY

9.2.1 Reflex bleeding

Coccinellids provide a typical case of aposematism i.e. striking colouration and, when provoked, promptly release of repellent and aggressive **defence substances** (Majerus 1994). Adults bleed from the tibio-femoral joints and larvae from dorsal glands. The **amount of reflex fluid** emitted by ladybirds in response to an attack can be very high (up to 20% of body fresh weight), and the alkaloid component can constitute several per cent of the weight of the fluid (Holloway et al. 1991, 1993). The exudates are yellowish/orange and are mostly toxic to other organisms and sometimes also distasteful with a strong flavour. Several studies have demonstrated the importance of these substances as a defence barrier against enemies (Marples 1993a, Majerus & Majerus 1997).

The aposematic chemicals play an important and complex role in the interaction between coccinellids and their enemies and other competitors. This relationship may involve finely tuned interactions as in the competition that occurs in the ant–coccinellid relationship.

9.2.2 Reflex bleeding substances

9.2.2.1 Experiments

The efficiency of the multiple-component defence secretion of *C. septempunctata* has been investigated in **feeding experiments** with Japanese quail, *Coturnix japonicus* (Marples et al. 1994). Even if the interaction between colour and taste was also important, colour pattern was the most significant deterrent for experienced birds. The birds detected the insect's smell but rarely used it as a cue to toxicity. No single element was sufficient to maintain avoidance comparable to that caused by the whole insect.

9.2.2.2 Identification

Aposematically active compounds have aroused a considerable amount of interest and a broad range from different coccinellid species have been **chemically identified**. So far 50 different **alkaloids** have been identified from 43 species (Daloze et al. 1994, Pasteels 2007). The **chemistry of** the **defensive substances** forms a specific framework for the semiochemistry of coccinellids (Schroeder et al. 1998, Laurent et al. 2005). Examples of aposematic substances are shown in Fig. 9.1. Most of them have negative effects on other organisms in that they are repellent and/or toxic. However, the insect pathogenic fungus, *Beauveria bassiana*, seems to overcome or tolerate them. *Beauveria bassiana* is a common mortality factor attacking hibernating coccinellids (Roy & Cottrell 2008; Chapter 8).

9.2.2.3 Sources

The **synthetic pathway** for most of the **autogenous** aposematic substances follows a common pattern, beginning with a hydrocarbon chain to which nitrogen from amino acids is added. The **site of synthesis** of coccinelline and adaline has been localized to cells in the **fat body** (Laurent et al. 2005)**.** The alkaloid content in the reflex bleeding is similar to that in the haemolymph (Holloway et al. 1991). Experimental results showing the occurrence of **haemocytes** in the

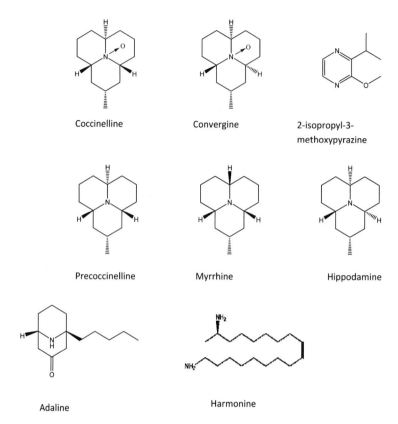

Coccinelline

Convergine

2-isopropyl-3-methoxypyrazine

Precoccinelline

Myrrhine

Hippodamine

Adaline

Harmonine

Figure 9.1 Chemical components in coccinellid aposematism and reflex bleeding systems.

reflex bleeding fluid may indicate shared origin of both the bleeding liquid and the haemolymph (Karystinou et al. 2004). Although most of the defensive alkaloids are synthesized by the coccinellids themselves, several cases have been reported where coccinellids **sequester toxins from** their aphid **prey**; the toxins originate from the host plant of the prey. Thus *C. undecimpunctata* sequester **cardenolides** from *Aphis nerii*, which in turn gets these by feeding on *Nerium oleander* (Rothschild et al. 1973). **Pyrrolizidines** are sequestered by *C. septempunctata* preying on *Aphis jacobaeae* feeding on *Senecio* (Witte et al. 1990).

In *C. septempunctata*, the fluid that is emitted at the tibio-femoral joints contains highly toxic alkaloids. One of them, **coccinelline** (Fig. 9.1), the N-oxide of the free base **precoccinelline**, has extremely high mammalian toxicity (Marples 1993b). Both compounds are released along with the volatile **pyrazine** (Fig. 9.1), which seems to serve a dual role in the

chemical ecology of *C. septempunctata* as an enemy repellent and a putative aggregation pheromone (Al Abassi et al. 1998).

9.2.2.4   Production cost

The production of aposematic compounds means **a cost for the coccinellid individual**. In a trade-off study with *Menochilus sexmaculatus*, the concentration of alkaloids present in reflex bleeding exudates in the larvae and adult females of different ages were determined and the cost of chemical defence to life-history parameters, viz. body weight, fecundity, egg weight, hatching success of eggs and longevity, was measured (Agarwala & Bhowmik 2007). The results showed that females which had been provoked to release chemicals had shorter longevity and produced smaller eggs by weight, and that the hatching success of their eggs was significantly reduced compared to eggs of unprovoked

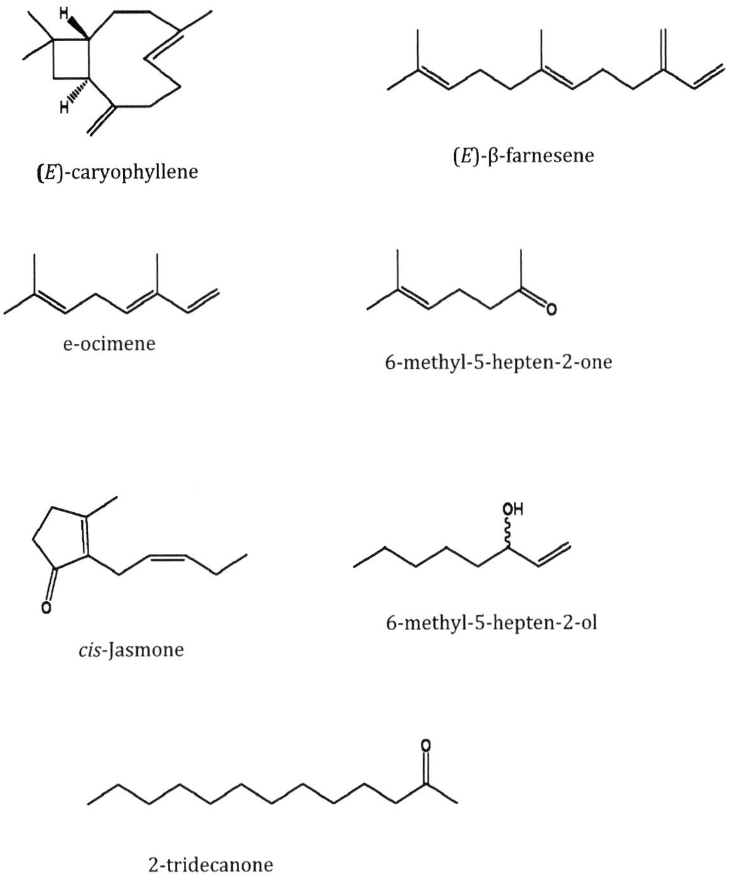

(E)-caryophyllene

(E)-β-farnesene

e-ocimene

6-methyl-5-hepten-2-one

cis-Jasmone

6-methyl-5-hepten-2-ol

2-tridecanone

Figure 9.2 Semiochemicals related to foraging and food searching behaviour in coccinellids.

control females. However, no effect was found on body weight, reproductive age and fecundity. Larvae that had been provoked developed into smaller females than unprovoked larvae. Results of exudate collection from individuals of different ages suggested that there was an **age-related variation** in the amount of defence fluid produced and the concentration of alkaloids, in line with what has been described for *Epilachna paenulata* mentioned above (Camarano et al. 2006).

9.2.2.5   Age and stage modifications

The composition of **toxins** (including autogenous aposematic compounds) may vary between developmental stages **in** some **phytophagous species**. The defensive chemistry of *Epilachna paenulata* was shown to

be a mixture of **piperidine**, **homotropane** and **pyrrolidine** alkaloids. Whole body extracts of adult beetles contain four major alkaloids, 1-(6-methyl-2,3,4,5-tetrahydro-pyridin-2-yl)-propan-2-one, 1-(6-methyl-2-piperidyl)-propan-2-one, 9-aza-1-methyl-bicyclo [3.3.1]nonan-3-one and 1- (2″- hydroxyethyl)-2-(12′-aminotridecyl)-pyrolidine. Comparisons of the composition of alkaloids in eggs, larvae, pupae and adults showed both qualitative and quantitative differences between the four life stages, and also varied during the adult stage, with optimum content during the oviposition period. Laboratory predation bioassays showed that **adults are better protected** than larvae and pupae against wolf spiders and, in field tests, the adult alkaloid extract also was deterrent to ants (Camarano et al. 2006).

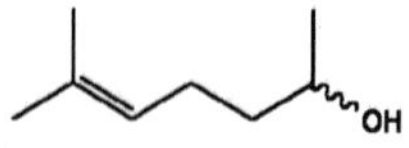

1- octen-3-ol

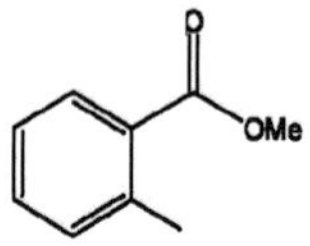

Methylsalicylate

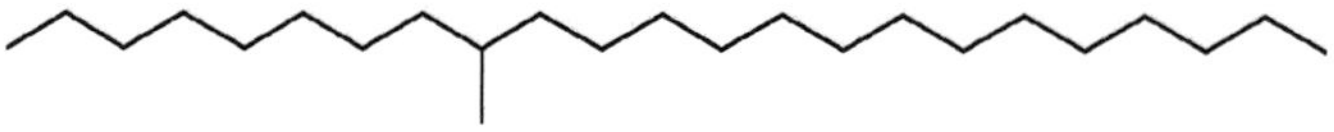

9-methyltricosane

N-pentacosane

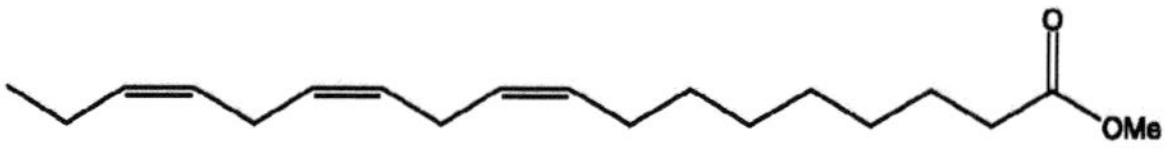

Methyl linolate

Figure 9.3 Examples of semiochemicals involved in kin recognition and mobility of coccinellids.

9.2.2.6   Species specific chemistry

Generally the composition of the reflex bleeding exudates for most coccinellid species seems to be species-specific and to contain only one or a few main components (review by Laurent et al. 2005). Thus it has been suggested that these alkaloids could be used as a monitoring **tool for estimation of intraguild predation** (IGP, Ch. 7; Hautier et al. 2008). Sloggett et al. (2009) used a gas chromatography-mass spectrometry (GC-MS) based approach for studies of predation of *Har. axyridis*. The alkaloid hippodamine from a single egg of *Hip. convergens* was detected in all 10 third instar larvae of another ladybird, *Har. axyridis*, for 12 hours after consumption. A comparison of the alkaloids of five ladybird species that co-occurred in the field study found that, in general, the alkaloids were sufficiently distinct to allow species identification of ladybirds that had been consumed by predators, although there was some overlap between species in alkaloid content. With the same analytical method, Kajita et al. (2010) found intraspecific variation of alkaloid content in egg clusters of *Har. axyridis* and *C. septempunctata*. The alkaloids affected egg consumption rates of adults and it was also found that *C. septempunctata* was more affected by alkaloids than

Har. axyridis. It was shown that this difference could be attributed to differences in alkaloid metabolism by the two species.

9.2.3   Relation to enemies and competitors

9.2.3.1   Parasitoids

The parasitoid, **Dinocampus coccinellae** (Ch. 8.3) causes mortality in adult *C. septempunctata* (9–72% in East Anglia) (Barron & Wilson 1998). The oviposition process is very rapid and seems to be initiated by a specific set of stimuli (Richerson & DeLoach 1972). Orr et al. (1992) compared the attack patterns and specificity of *D. coccinellae* with respect to four coccinellid species and concluded that orientation towards the host preceding final attack is an important step. It is suggested that volatile semiochemicals play a key role in this **location of adult** *C. septempunctata* by a *D. coccinellae* female. In laboratory experiments, it was demonstrated that the free base **precoccinelline** is attractive to the parasitoid and is highly active at the electrophysiological level (Al Abassi et al. 2001). Other naturally occurring ladybird alkaloids with similar structures e.g. **myrrhine** from the 18-spot ladybird, *Myrrha octodecimguttata*, and **hippodamine** from the convergent ladybird, *Hip. convergens*, showed significant biological activity on the parasitoid's behaviour. However, it remains to be seen to what extent other stimuli such as visual cues contribute to orientation of the wasp towards its host for the final attack and under natural conditions.

9.2.3.2   Ants

Ants compete with coccinellids for the **exploitation of aphids** and aphid products (honeydew) as food sources (5.4.1.6 and 8.2.4). Based on studies of aposematism and the interaction between *Lasius niger* and *C. septempunctata*, it has been suggested that alkaloids may also protect the ladybird from ant attack and thus increase its chances of consuming aphid prey (Marples 1993b). This form of protection seems to be better developed for *C. septempunctata* than for *A. bipunctata*, and is in agreement with the observation that *C. septempunctata* appears to use the trail pheromones of *Formica polyctena* in foraging (Bhatkar 1982). The specific relationship between ants and *C. septempunctata*, compared to that between ants and *Har. quadripunctata*, has also been observed for ant-attended aphid colonies on Scots pine, *Pinus sylvestris* (Sloggett & Majerus

2000). The results indicate that ant attendance prolongs the survival of aphid colonies, thus stabilizing the food supply for the ladybirds. Aqueous solutions of **convergine** (from *Hip. convergens*) and **coccinelline** (from *C. septempunctata*) have been tested for repellence to ants (*Formica rufa*) and quails (*Coturnix coturnix*) (Pasteels et al. 1973). Repellence to ants was consistent, while responses from individual quails showed variation. In a review of defensive chemistry in insects, Laurent et al. (2005) stress a wider ecological perspective on alkaloids, which are the most frequently encountered defensive compounds not only in coccinellids but also in other insects. This is also stressed by Majerus et al. (2007). Studies of the ovipositing behaviour of *A. bipunctata* showed that the presence of *L. niger* reduced (via semiochemicals) the number of eggs laid (Oliver et al. 2008). It was also found that the survival of eggs was reduced due to ants attacking but not consuming the coccinellid eggs.

Few cases of myrmecophily are known in coccinellids (8.2.4). However, Vantaux et al. (2010) showed that the larvae of the coccinellid *Diomus thoracicus* live safely inside the nests of the ant *Wasmannia auropunctata* and prey exclusively on the ant larvae. In contrast adults are always attacked. The tolerance of ants to the larvae in the ant nest is based on chemical mimicry of the ant cuticular chemistry.

9.2.3.3   Spiders

Coccinellids fall prey to spiders especially during the autumn migration to the hibernating sites. In feeding experiments, the European orb-web spider, *Araneus diadematus*, was offered *C. septempunctata* and *Har. axyridis* as prey (Sloggett 2010). Spiders were not deterred by the coccinellid alkaloids and no significant effect on spider development was observed. Several investigations on selectivity of spider foraging showed, with variable results (Nentwig 1983, 1986), that predator–prey combinations are important.

9.3   SEMIOCHEMICALS RELATED TO FOOD

9.3.1   Plant volatiles

9.3.1.1   Herbivore-induced plant volatiles

Informative volatile cues from herbivores and/or herbivore-infested plants shorten the searching time for the predator and increase foraging efficiency. There is experimental evidence for behavioural effects also of volatiles from plants on coccinellid food search having a short active range (Pasteels 2007) and they may be seen as part of the final step in foraging behaviour (9.3.4 and 5.4.1).

Semiochemicals indicating the presence of food can be produced directly by the prey/food and produced as plant responses to herbivore attack. Most coccinellids prey on a broad range of herbivores, and semiochemicals carrying information on **plant status**, i.e. **whether attacked or unattacked**, can be expected to be one of the leads in dispersal and foraging. The ecological importance of this kind of information has been well documented in mite predator–prey systems on lima bean where it was shown that predatory mites intensified their searching behaviour in contact with lima bean plants that had been attacked by mites (Dicke et al. 1998; Dicke 2000). Similar preferential responses have been shown in olfactometer experiments with *C. septempunctata* exposed to barley plants previously attacked by aphids, *Rhopalosiphum padi* (Ninkovic et al. 2001). Tea shoots attacked by aphids or mechanically damaged were shown to be more attractive to natural enemies and this was interpreted as a combined response to **volatiles produced by plants** (synomones) and by herbivores (kairomones) from the tea aphids (*Toxoptera aurantii*; Han & Chen 2002, 2005).

A **common volatile** released **from herbivore-stressed plants** is **methyl salicylate** (Walling 2000). However, Bi et al. (2007) showed that allelopathic responses in rice also included the production of methyl salicylate. Several investigations have shown that methyl salicylate is a messenger substance for coccinellids, releasing an **arresting behaviour**. Both in field and laboratory experiments, James and colleagues (James & Price 2004, James 2005) recorded such an arresting response in adult *C. septempunctata*. Similar results have been reported for other coccinellids (Zhu & Park 2005). Using sticky traps baited with herbivore-induced volatiles, James and colleagues (James & Price 2004, James 2005) showed that *Stethorus punctum picipes* was significantly attracted to sticky traps baited with a more complex mixture of volatiles, namely methyl salicylate, **cis-3-hexen-1-ol** and **benzaldehyde**.

So far only a few active substances have been identified that are exclusively linked to plant responses to herbivory. **cis-jasmone** ((Z)-jasmone) is a common plant volatile involved in a switch-on mechanism for **defence** metabolism **in plants**. In wind tunnel tests

C. septempunctata adults demonstrated attraction to (Z)-jasmone. When applied in the vapour phase to intact bean plants, (Z)-jasmone induced the production of volatile compounds, including the monoterpene (*E*)-ocimene, an attractant to the aphid parasitoid *Aphidius ervi* (Birkett et al. 2000, Bruce et al. 2008). Possibly, this is support for a general mechanism whereby predators and parasitoids of herbivores can trace plants under stress.

9.3.1.2   Adult receptors for food semiochemicals

Experiments with adult *Hip. convergens* indicate that receptor centres for food-related volatiles are located on the **antennal tips** (Hamilton et al. 1999). Responses of beetles whose antennae, maxillary palps or antennal tips had been amputated were compared in olfactometer experiments with non-injured beetles. Only beetles with their antennae or antennal tips intact responded positively both to the odour of radish leaves infested with peach–potato aphids, *Myzus persicae*, and to clean radish leaves. Scanning electron microscope studies of the distribution of different receptor types suggested that the primary function is executed by **trichoid sensilla** located **on the terminal antennal segment**. The synergistic effects of plant and prey volatiles have so far only been given moderate attention.

9.3.1.3   Responses of larvae

Most experiments on semiochemicals and food finding by coccinellids have been made with adults, but several reports also include the behaviour of larval stages. A common technique for laboratory studies of adult responses to semiochemicals is olfactometry, whilst experiments with **larvae** are usually based on recording **arresting responses** induced by volatiles in an arena. The searching behaviour of *Hip. convergens* larvae was studied as affected by volatile chemicals from tobacco aphids, *Myzus persicae nicotianae* (Jamal et al. 2001). Larvae (second, third and fourth instars) were exposed to three volatile sources: aphids alone, aphids on tobacco leaves and tobacco leaves previously exposed to aphids. The search path of larvae was traced in an arena and the different angles and velocities of movement were recorded. In general, older larvae were more efficient at searching than younger ones. In most cases significant effects attributable to olfactory cues were obtained. The results support the potential of olfactory cues to modify the behaviour of coccinellid larvae and call for more research on the semiochemicals involved.

9.3.1.4 Prey sex pheromones: predator kairomones

Rhyzobius spp. larvae responded to **sex pheromones** of pine bast scales *Matsucoccus*. The response of coccinellid larvae was tested in field trials using pine tree arenas baited with the sex pheromones of *M. josephi*, *M. feytaudi* and *M. matsumurae*. Both field and laboratory tests demonstrated a significant positive response to the sex pheromones of *M. feytaudi* and *M. matsumurae* (Branco et al. 2006).

It has been suggested that the foraging activity of coccinellids feeding/searching in aphid colonies also promotes formation of alate aphids (Dixon & Agarwala 1999). However, it is still open as to whether this is only due to disturbance of aphid feeding, or whether repellence of the volatile semiochemistry of the aggressive coccinellid also contributes to increasing the mobility of aphids that would promote alate offspring.

9.3.2 Prey alarm pheromones

9.3.2.1 Aphid alarm pheromone

When attacked by predators many aphids release a **secretion from the cornicles** that contains an alarm pheromone. Aphids that are nearby respond to the pheromone by pulling out their mouthparts and falling from the plant to escape (Nault et al. 1973). Similar observations on the behaviour of treehoppers (Membracidae) attacked by coccinellids have been reported (Nault et al. 1974). Since the main component of the aphid alarm pheromone, the sesquiterpene hydrocarbon **(E)-β-farnesene** (EBF), was identified (Nault et al. 1973) it has been given considerable attention both as an aphid repellent and as a **coccinellid attractant**. Experimental studies have shown that adult *C. septempunctata* and larvae of *A. bipunctata* are arrested/attracted by EBF and the odour of crushed aphids (Nakamuta 1991, Hemptinne et al. 2001b, Hatano et al. 2008).

9.3.2.2 Mechanism for modified responses

EBF is also a common substance in nature, occurring in several plants. This represents a problem, since a rigid positive response by ladybirds to EBF may cause them to stay longer in environments that are less favourable through being devoid of prey. Experiments have shown that the response of *C. septempunctata* to EBF is modified by another common plant substance, **(-)-β-caryophyllene** (Al Abassi et al. 2000). As long as the amount of EBF does not exceed a specific ratio that is maintained between it and the caryophyllene, no behavioural response of the coccinellids is elicited. Such is the case when a walking ladybird passes an undisturbed aphid colony on a plant that naturally manufactures EBF. However, if one of the aphid individuals in the colony responds to the threat of a predator and releases alarm pheromone, the proportion of EBF exceeds the critical ratio between the two active substances and the behaviour of the ladybird becomes aggressive. How far this mechanism also operates for other coccinellids is still unknown.

(-)-β-caryophyllene on its own elicited a positive **aggregation response** of adults of *Har. axyridis* in laboratory experiments by Verheggen et al. (2007), indicating that a certain semiochemical may have multiple roles depending on conditions and species involved.

9.3.2.3 EBF release: aphid individual risks

The release of alarm pheromone by an attacked aphid makes this individual more easily detectable. Thus the question has been raised as to whether it is a risk for aphids to reveal their presence by producing EBF. This topic has been studied for the *Har. axyridis–Acyrthosiphon pisum* relationship. However, the results did not show that this apparent **altruistic behaviour** would be costly for the individuals releasing the alarm pheromone. Its release did not have any negative effect for the producer (Mondor & Roitberg 2000).

Myzus persicae habituated to EBF had an increased reproduction rate compared to EBF sensitive aphids. However, predation by *Hip. convergens* significantly reduced the habituated aphid population and outweighed the initial increase in reproduction (de Vos et al. 2010).

9.3.3 Toxic substances in prey

The cabbage aphid, *Brevicoryne brassicae*, accumulates **glucosinolates** such as **sinigrin**, from its host plants (brassicas) and has an enzyme system, myrosinase (β-thioglucoside glucohydrolase), that breaks them down to behaviourally active compounds such as **isothiocyanates** (Kazana et al. 2007). When uptake of sinigrin from an artificial diet was studied, it was found that there was a significant **difference between winged and wingless** aphid morphs with regards to sequestration. Winged aphids excreted significantly higher amounts of glucosinolate in the honeydew than wingless aphids and the higher level of sinigrin in wingless aphids had a significantly stronger negative impact on survival of a ladybird predator. Larvae of *A. bipunctata* were unable to survive when fed on apterous aphids from a diet with 1% sinigrin. However, the larvae survived successfully when fed on winged aphids from the same diet.

Different coccinellid species may handle plant secondary metabolites in prey differently. In experiments with sinigrin, *B. brassicae* and larval stages of *A. bipunctata* and *C. septempunctata*, it was found that the first instar of *A. bipunctata* was unable to survive when fed with *B. brassicae* reared on *Brassica nigra* or artificial diets containing 0.2% **sinigrin**. By contrast, first instars of *C. septempunctata* were able to survive when fed with aphids reared on *B. nigra* or artificial diets containing up to 1% sinigrin, although the presence of sinigrin in the aphid diet reduced nymphal growth and increased the time the larvae took to reach second instar (Pratt et al. 2008).

Similar results were obtained in a series of experiments with *A. bipunctata* in which the effect of different aphid food plants was compared with special reference to **glucosinolates (GLS)** (Francis et al. 2001). The polyphagous aphid *M. persicae* was fed on broad bean (*Vicia faba*; GLS free), oilseed rape (*Brassica napus* subsp. *oleifera*; low GLS level) and white mustard (*Sinapis alba*; high GLS level). Both rape and mustard shortened the developmental time and increased the adult weight of *A. bipunctata*. No significant differences in mortality were observed, but rape-fed *M. persicae* caused higher egg production and larval emergence in the ladybird, while mustard-fed *M. persicae* caused lower fecundity and egg viability. A possible explanation for the differences in the effects on the coccinellids between the experiments with *B. brassicae* (Kazana et al. 2007) and *M. persicae* (Francis et al. 2001) may be that GLS metabolism differs between the polyphagous *M. persicae* and the cabbage feeder *B. brassicae*. (See also 5.2.6.)

9.3.4 Feeding stimulants for phytophagous coccinellids

Feeding experiments with *Epilachna admirabilis*, which mainly feeds on plants of the genus *Trichosanthes* (Cucurbitaceae), showed that the **cucurbitacins**, especially cucurbitacin E-glucoside, strongly stimulated both adults and larvae to feed (Abe & Matsuda 2000). In further studies of **phagostimulants** for *Henospilachna vigintioctomaculata* with potato *Solanum tuberosum*, an interesting synergism between different components from the host plant was shown. The feeding stimulants were isolated and identified as **methyl linolenate** from a lipid-soluble fraction, and glucose and fructose from an aqueous fraction. Although methyl linolenate alone was inactive, it acted positively synergistically with sugars. Methyl linolenate maximized the feeding activity on sugars at the concentration occurring in potato leaves. It is suggested that methyl linolenate plays an important role in host selection by *H. vigintioctomaculata* (Endo et al. 2004). The results indicate that host plant discrimination and feeding site acceptance have a semiochemically complex structure where a mix of specific and trivial plant compounds contribute to a complete message/cue.

9.3.5 Learning

Ladybirds use different sets of stimuli for food localization, and it has been suggested that **associative learning** may play an important role in increasing efficiency (Vet & Dicke 1992). However, conclusive experimental support for this hypothesis is as yet limited. Learned responses in terms of selectivity for prey were recorded in *A. decempunctata* in predation on *Hyalopterus pruni* (Dixon 1959). After previous feeding on *H. pruni*, *A. decempunctata* rejected this aphid when it was touched with the palps.

Adaptive changes in prey handling behaviour have been observed in the coccinellid *Anisolemnia tetrasticta* feeding on a plataspid heteropteran, *Caternaultiella rugosa*. It was found that the efficiency of attack behaviour improved with increasing experience of the prey (Dejean et al. 2003).

Ferrer et al. (2008) studied effects on fitness of adaptive food preferences (learning) in an experiment with *A. bipunctata* testing the effects on food preferences of feeding larvae on different food sources. One group of larvae were fed throughout their development on pea aphids, *A. pisum*, which is a high quality prey. The other group was fed on cowpea aphids (*Aphis craccivora*), considered a suboptimal prey on the basis that adults developing from larvae fed on this aphid were lighter and had fewer ovarioles and a lower overall fitness than individuals reared on pea aphid. When offered a choice, naïve first instar larvae of *A. bipunctata* from the cowpea aphid group more frequently attacked cowpea than pea aphids. However, older larvae did not show this preference and attacked the two species of prey irrespective of the aphids that were previously fed. This is different from what has been shown for parasitoids (van Emden et al. 2008) where the **selectivity** for prey **can develop late** during the larval–pupation phase and even not till emergence from the aphid mummy.

In laboratory experiments with fourth instar larvae of *Col. maculata* ssp. *lengi* Boivin et al. (2010) tested how previous experience affected prey rejection behaviour. It was found that the initial contact and prey discrimination was affected by previous experience but prey representing good food quality were always finally accepted. After 48 hours the learned behaviour appeared to be partially forgotten.

Finding food sources can be seen as a **two-step procedure** in which an initial patch or site preference phase is followed by a close range search process where several semiochemicals directly related to the food source change coccinellid behaviour from dispersal to feeding/attacking behaviour. This feeding response can be modified in an associative learning process by the presence of unique semiochemicals or specific ratios between trivial chemicals indicating presence of food sources. Olfactorily mediated associative learning has been demonstrated in experiments with adults of *C. septempunctata* and four cultivars of barley (Glinwood et al. 2011). *Coccinella septempunctata* did not prefer the odour of one aphid-infested barley cultivar over another. However, after feeding on aphids for 24 hours on a cultivar, adults preferred the odour of that particular cultivar in olfactometer experiments. After feeding experience on a different aphid infested cultivar this preference disappeared.

9.4   MATING AND SEX PHEROMONES

Knowledge about sex pheromones and other semiochemicals promoting coccinellid mating is still limited. However, experimental results indicate the involvement of a **sex pheromone** in the reproductive behaviour of *A. bipunctata* (Hemptinne et al. 1996). A putative **aggregation pheromone** identified in *C. septempunctata* (Al Abassi et al. 1998) also seems to be used by *A. bipunctata* (Hemptinne & Dixon 2000). Possibly the mating cue is separate from the set of volatiles involved in species discrimination where the pattern of aposematic substances may play a role (9.5.1).

9.4.1   Chemoreceptors on the antennae

Studies on *Cer. undecimnotata* showed a **difference** in terms of the presence of antennal receptors **between males and females** (Jourdan et al. 1995). Of 12 different types of antennal sensilla, two **cheatiform sensilla** on the male antenna are missing in the female antenna and one type is specific for the female antenna. Different responses to food and the opposite sex were shown by adult males and females of *A. bipunctata* (Hemptinne et al. 1996). While females mainly respond to prey density (aphids), the males primarily respond to females and mating opportunities. The suggested explanation for this difference in searching behaviour is that a **sexual difference in receptor presence** can be a complement to odour discrimination at the receptor level of the two sexes.

A significant **sexual difference in electroantennogram (EAG) responses** of male and female antennae was shown by Baker et al. (2003). The **male antennae** of *Col. maculata* responded to females but not to males, indicating that chemicals from females are involved in sexual communication. Hexane extracts from conspecific females but not those from males produced significant electrophysiological responses (EAG) from male antennae. The only identified chemical that corresponded to this pattern was **1-octen-3-ol**. A significant EAG response was also recorded to the extracts of fluids produced during 'reflex bleeding' and it may be hypothesized that these substances are part of a complex blend constituting the sex pheromone. Female *Col. maculata* antennae exhibited high thresholds in response to several compounds including α-terpineol, (Z)-3-**hexenol** and (4aS,7S,7aR)-**nepetalactone**. Field traps baited with **2-phenylethanol** and α-terpineol were highly attractive to adult *Col. maculata*.

9.4.2 Hydrocarbons on elytra

Bioassays indicated that the **female elytra** of *A. bipunctata* could be the source of the active semiochemicals (Hemptinne et al. 1996; Burns et al. 1998). The mating behaviour of *A. bipunctata* proceeds in steps: a male first palpates the female elytra with his maxillary palps, then mounts her and mates. **Elytra washed in** chloroform **failed to stimulate mating.** Analysis of the chloroform extracts of the elytra revealed that male and female ladybirds are coated with the same blend of hydrocarbons among which 9- and 7-**methyl tricosane** are dominant. It was suggested that visual stimuli, in particular movement, are necessary for a male to discriminate males from females while so far unidentified olfactory cues are important for species recognition. It may be hypothesized that again the set of species-specific aposematic compounds mentioned above play an active role.

9.4.3 Ultrastructure of the integumentary glands

Studies of adult *Cer. undecimnotata* showed that glands **with and without secretory ducts** are distributed over the head, thorax and abdomen. Glands without ducts are thought to release **volatile pheromones.** Such glands consist of a single cell and a secretory apparatus located within the thickness of the cuticle and equipped with a **cuticular cribriform plate.** This cribriform plate separates two superimposed cavities, and epicuticular filaments fill the lower cavity. The secretory products from glands with a duct are abundant and released on the surface of the cuticle in the shape of twisted cylinders, which are resistant to acetone treatment. However, their role in modifying behaviour is not yet known (Barbier et al. 1992).

To conclude, there is considerable evidence that sex attractants support mating of coccinellids. Although some active substances that may be components in a complex pheromone blend have been identified, conclusive information on the composition and role of the pheromones is still lacking. Behavioural experiments show that males and females can respond differently to mating partners. Sexual differences in the occurrence of olfactory receptors and behavioural responses to mating-related semiochemicals in behavioural responses to semiochemicals may explain the difference in behaviour between females and males.

9.5   OVIPOSITION

9.5.1 Oviposition deterrence: oviposition deterrence pheromones

Aphids constitute an unstable food supply and **cannibalism** on immobile/less mobile stages (eggs and young larvae) is a common trait in coccinellid biology. Unravelling the mechanistic regulation of this cannibalism via semiochemicals is an exciting ecological challenge as it deals with a complex contribution to the decisive importance of cannibalism for ecological fitness. (See also 5.4.1.3.)

An important requirement for the ovipositing female is to identify a site free from competitors and potential predators on her offspring. Hemptinne et al. (1993) suggested that contacts with coccinellid larvae by the female act as an oviposition inhibiting cue. In later work by Ruzicka (1994, 1997), the general mechanism of an oviposition deterrence pheromone (ODP) was described and defined in *C. septempunctata*. He demonstrated that gravid females utilized semiochemicals in **larval tracks to avoid oviposition** on the sites where conspecific larvae had already walked. ODPs have been shown to occur among coccinellids and are primarily a means for an ovipositing female to avoid sites where her offspring may later be exposed to predation by offspring from other females which had used the same site.

9.5.2 Species-specificity of oviposition deterrence pheromone substances

Species-specificity is illustrated by the fact that female *A. bipunctata* do not avoid tracks deposited by larvae of two other species, *A. decempunctata* and *C. septempunctata* (Hemptinne & Dixon 2000). The results of a study, in which the antennae and maxillary palps were amputated, indicate that females exclusively use contact chemoreceptors on their maxillary palps to detect oviposition-deterring tracks of conspecific larvae (Ruzicka 2003). Females of *Cer. undecimnotata* laid significantly smaller egg batches on paper strips with conspecific tracks than on clean paper strips. However, females of *Cycloneda limbifer* laid significantly larger batches of eggs on paper strips with **conspecific larval tracks** than on clean paper strips. This is the first evidence of an opposite effect in different species of conspecific oviposition-deterring larval tracks on egg clustering in aphidophagous coccinellids (Ruzicka 2003). The ecological background to these different behavioural responses is still an open question.

It has been suggested that **egg-surface** chemicals act as semiochemicals for the avoidance of intraguild (egg) predation between *C. septempunctata* and *A. bipunctata* (Hemptinne et al. 2001a). Behavioural experiments showing an interspecific relationship between *Har. axyridis* and *Calvia quatuordecimguttata* based on ODP are reported (Ware et al. 2008). The experiments are based on previous observations showing that eggs of *C. quatuordecimguttata* escape attack by *Har. axyridis*. The relationship between the deterring effects on *Har. axyridis* is discussed in relation to the palatability of the eggs as food and whether or not the semiochemicals located on the egg surface are really honest signals conveying information to the attacker.

The **persistence** of oviposition-deterring effects is variable and limited in time as shown in a series of experiments with different coccinellid species by Ruzicka (2001, 2002, 2006). Although the response to 10 day old conspecific larval tracks remained significant (see also Hemptinne et al. 2001a) it was considerably lower than the response to fresh tracks. In choice tests with fresh tracks of conspecifics, *Cycloneda limbifer, Cer. undecimnotata* and *Har. dimidiata* larvae, and 10 day old tracks of conspecific larvae, clutch sizes were smaller in the blank test without larval tracks than in tests with fresh tracks. Similarly, semiochemicals in the tracks of conspecific and heterospecific coccinellid larvae can contribute considerably to the spacing of *Menochilus sexmaculatus* offspring among prey of differing quality and that conspecific as well as heterospecific larval tracks can influence the **distribution of larvae. The persistence of intra- and interspecific** effects on the response of *M. sexmaculatus* to larval tracks has also been investigated. Fresh tracks of *M. sexmaculatus, Cycloneda limbifer* and *Cer. undecimnotata* larvae effectively deterred *M. sexmaculatus* females from ovipositing; larval tracks from two other ladybird species have also been shown to deter oviposition by females of *M. sexmaculatus*.

9.5.3 Active substances in oviposition deterrence pheromone tracks

The chemical nature of *A. bipunctata* semiochemicals has been investigated in relation to reproduction, oviposition deterrence and intraguild prey avoidance. **Alkanes** with chemical and structural similarity have been identified in each case, indicating that ladybirds exploit their natural product chemistry with parsimonious versatility (Hemptinne & Dixon 2000), similar to that described above for **pyrazine** (Al Abassi et al. 1998). Intraguild interactions between *Har. axyridis* and *P. japonica* are mediated by odour substances from the **faeces** (Agarwala et al. 2003). This mechanism is an interspecific complement to the intraspecific communication and reduces competition between these two coccinellid species.

Bioassays with larvae of *A. bipunctata* have shown that the anal disc of the 10th abdominal segment deposits ODP substances onto the substrate leaving tracks of these semiochemicals (Laubertie et al. 2006). The **chemical composition** of the larval tracks of *A. bipunctata* has been investigated (Hemptinne et al. 2001b) and found to be a complex mixture of around 40 chemically distinguishable alkanes. N-pentacosane is a major component (15.1%). The alkanes are likely to spread easily on the hydrophobic cuticle of plants and so leave a large signal. They are not quickly oxidized and can be long lasting. Observation showed that 10 day old larval tracks still significantly deterred oviposition. Klewer et al. (2007) identified **(Z)-pentacos-12-ene** as a key compound in the larval tracks of *Menochilus sexmaculatus* and **proved** its **oviposition deterring effects in bioassays**. Other compounds found in the tracks were tested in bioassays for oviposition deterrence but were not found to have a significant behavioural effect.

9.5.4 Relation to other aphid enemies

Elegant studies have shown that, to avoid **intraguild predation** (IGP; 7.8), females of the aphid parasitoid *Aphidius ervi* detect semiochemicals in fresh **adult and larval footprints** of *C. septempunctata*, thereby avoiding aphid colonies under attack from ladybirds (Nakashima et al. 2004, Nakashima & Akashi 2005). The response of three aphid parasitoid species, *Aphidius eadyi, A. ervi* and *Praon volucre* to footprint semiochemicals from adult *C. septempunctata* and *A. bipunctata* was investigated by Nakashima et al. (2006). Females of all three parasitoid species avoided leaves previously visited by *C. septempunctata* or *A. bipunctata* adults. *Praon volucre* avoided trails of both ladybird species to a similar degree but the avoidance of the *Aphidius* species was stronger to trails of *C. septempunctata* than to those of *A. bipunctata.* The footprint semiochemicals were identified and quantified. **n-pentacosane** and **n-heptacosane** occurred in significantly greater amounts in *C. septempunctata* trails than in those of *A. bipunctata* and the trails of the two species differed qualitatively in the other hydrocarbons present. The response of the three aphid parasitoids females was tested to three **footprint semiochemicals** occurring in both coccinellid species, n-tricosane ($C_{23}H_{48}$), n-pentacosane ($C_{25}H_{52}$), and n-heptacosane ($C_{27}H_{56}$). It was found that *A. eadyi* was more sensitive to **n-tricosane** than the other two species, only *Praon volucre* showed avoidance responses to n-heptacosane. All three species responded to n-pentacosane.

A similar mechanism has been described for the relationship between the aphid parasitoid, *Aphidius colemani* and *C. septempunctata*, *Har. axyridis* and *P. japonica*, where the number of eggs deposited in aphid colonies with coccinellid larvae was reduced in the presence of semiochemicals from the **larval tracks** of *C. septempunctata* (Takizawa et al. 2000). An important member of the aphidophagous guild *Chrysopa oculata* utilizes the footprint semiochemicals for intraspecific oviposition deterrence; these semiochemicals also have a similar role in *C. septempunctata* (Ruzicka 1997).

The examples given indicate that semiochemicals play a significant role in intraguild interactions between *C. septempunctata* and other aphid natural enemies (7.8).

9.5.5   Aphid abundance

So far most of the positive oviposition semiochemical mechanisms demonstrated for coccinellids in experiments are related to different aspects of food resources, and it has been shown experimentally that oviposition activity is positively affected by **aphid density itself** (Elliot 2000; Frechette et al. 2004). Investigations on the changes in volatiles that occur in aphid-attacked plants have shown that levels of **common plant volatiles** are elevated as discussed in the previous section. An interesting investigation by Oliver et al. (2006) shows that the trade-off between the repellence of larval tracks and increased quality of a certain oviposition site in favour of acceptance is positively related to aphid density, i.e. higher aphid density reduces the repellent effect of the larval tracks. Although three volatiles related to aphid density have been identified

(9.3.3) further messenger substances expressing aphid density, either from the aphids themselves or from the attacked plants, seem to be of importance in coccinellid feeding behaviour.

9.6   EGG AND PUPA PROTECTION

9.6.1   Protection of eggs

Cannibalism and interspecific predation are common coccinellid traits that are discussed in 5.2.7 and 5.2.8. The importance of eggs as a first food source for neonate larvae has been observed and discussed by several authors (Gagné et al. 2002; Santi et al. 2003; Michaud & Grant 2004; Michaud & Jyoti 2008). The level and ecological significance of cannibalism vary between species, but it seems common that **conspecific eggs are** identified and **preferred** to those of other coccinellid species and even to some preferred foods such as aphids.

The aposematic substances of coccinellids (9.2) show a species-specific pattern and so constitute a basis for discrimination between the eggs of different species. Thus Sloggett et al. (2009), in an elegant study, demonstrated the possibilities of gas chromatography–mass spectrometry (GC–MS) analysis of **prey alkaloids** to trace IGP in coccinellids. The amount of the alkaloid hippodamine in a single egg of *Hip. convergens* was enough for detection with this method in third instar larvae of *Har. axyridis* after 12 hours (and occasionally even after 36 hours) after consumption. Using an internal standard enabled this method to be used to make estimates of the number of eggs consumed.

9.6.2   Protection of pupae

Generally the pupal stage of holometabolous insects is a vulnerable phase in development, and **chemical defence** of coccinellid pupae has been reported. The pupal surface of *Subcoccinella vigintiquatuorpunctata* bears glandular hairs that produce a secretion consisting principally of three **polyazamacrolide alkaloids** that serve as a potent anti-predator defence; contact with it elicited pronounced cleaning activity by the predatory ant *Crematogaster lineolata.* Application of the secretion to palatable food items rendered them unacceptable to the ant (Smedley et al. 2002).

9.7 HIBERNATION AND AGGREGATION

Aggregation behaviour is discussed in detail in 6.3.1.4. The **intrinsic preference** to join conspecifics is an important mechanism in the formation of aggregations. The formation of denser aggregations seems to be a process affected by prevailing climatic conditions such as temperature and rainfall (Klausnitzer 1989). In the context of the specific conditions during the late autumn, any pheromone would need to have potent olfactory activity even if temperatures are low. To be active over a longer time it should also be chemically very stable. Indeed, a pheromone fulfilling these requirements was isolated and identified for *C. septempunctata* as **2-isopropyl-3-methoxypyrazine** (Al Abassi et al. 1997). In terms of properties, this compound is well suited for its expected functions i.e. high olfactory activity and low volatility and chemical stability under field conditions. Thus, it may function as a messenger substance even if aggregation itself is a slow process that is accentuated only on days with favourable temperatures.

9.8 HABITAT PREFERENCES: RESPONSES TO PLANTS AND PLANT VOLATILES

9.8.1 Habitat selection

Two different plant-derived semiochemical messages can be expected to contribute to **habitat/patch arrestment** of polyphagous coccinellids. One is information on **botanical composition** and possibly the presence of preferred plants (Schmid 1992). The second type of message would carry information on **plants under stress**, either from the presence of herbivores or caused by interaction on the plant–plant level. In summary there is experimental evidence that plant volatiles contribute to the patch preferences of several coccinellid species. Several of the recorded compounds are trivial plant volatiles, but some of them may also represent a stressed plant status, and there may be similarities between effects of herbivore (coccinellid prey) attacks and plant stress caused by plant–plant interactions. (See also Chapter 4 and 5.4.1)

Information related to both **prey density** and **plant status** contributes to the switch of an individual coccinellid from extensive foraging and dispersal to patch identification with intensified food-searching behaviour, and here a range of semiochemicals play an active role (Pettersson et al. 2008). The ecological importance of species-specific differences in the balance between the two types of behaviour is illustrated by a study comparing the searching behaviour efficiency of *Har. axyridis* and *Col. maculata* in **fragmented versus clumped landscapes** with clover patches infested with the pea aphid, *Acyrthosiphon pisum* (With et al. 2002). The two coccinellids did not differ in their search success within fragmented landscapes. It was only in clumped landscapes that *Har. axyridis* maximized search success and foraged within clover patches that had 2.5–3 times more aphids than those in which *Col. maculata* occurred. *Harmonia axyridis* was more efficient in finding aphid-infested clover cells, while *Col. maculata* made more efficient use of the aphids in such infested clover cells. A key factor for the difference between the two species seemed to be their difference in mobility and, therefore, different dependence on cues in the switch from dispersal to searching mode.

9.8.2 Plant stand traits

Coccinellid food can be classified as **essential food** that promotes development and propagation, and lower quality (**alternative**) **foods** that **only supports survival** (Hodek & Michaud 2008; 5.2.2. and 5.2.11). Thus volatile messenger substances carrying information on prey/food quality in terms of essential food and possibly longevity of the food source are important. Field investigations have shown a correlation between **increased botanical diversity** and the frequency of polyphagous predators (coccinellids), and different factors are suggested as contributing to this, including **microclimate** and **increased prey density** (Andow 1991; Vandermeer 1992). In a field site inventory study, *C. septempunctata* and *A. bipunctata* were dominating species and their distribution was significantly correlated with the percentage of ground cover of certain weed species (Leather et al. 1999). This is in line with studies by Schmid (1992) on plant discrimination by coccinellids. Of 73 plant species, most of which were common agricultural weeds, at least 20 were found to be highly attractive to coccinellids. Even moderate manipulations of the genetic homogeneity of a barley crop may be important in relation to habitat preferences. In a field experiment with pair-wise mixed and pure stands of different barley genotypes, it was found that, in the absence of prey, adult *C. septempunctata* preferred a specific genotype-pair mix (Ninkovic et al. 2011).

9.8.3 Avoided plants

Coccinellids have been shown to respond to species-specific **plant volatiles**. In laboratory experiments with **terpenoids** from catnip oil and from grapefruit seeds (Riddick et al. 2008), Z,E-dihydronepetalactone or E,Z-dihydronepetalactone caused a concentration-dependent avoidance behaviour in adult males and females of *Har. axyridis*. The avoidance behaviour was expressed by trying to fly, jumping back or turning away from the odour source. Other tested compounds, E,Z-iridomyrmecine and Z,E-**myrmecine**, were less effective. Finally, **nootkatone** and tetrahydronootka-tone were least effective. Based on the behavioural responses of the beetles, it is suggested that the substances should be tested as repellents to prevent aggregates of *Har. axyridis* at sites where they are unwanted. To what extent these preferences represent a permanent pattern of species-specific plant preference or an adaptive response remains to be tested. Investigations on effects of plant volatiles on coccinellid foraging responses have focussed on consumption patterns in *Aphidecta obliterata*, which as a conifer specialist was better adapted to spruce than the generalist *A. bipunctata* (Timms et al. 2008).

9.8.4 Plant–plant interactions/ attractive plants

A detectable plant stress indicator would be an important predator tool. The most common biotic challenge for a growing plant is sharing available resources with other plants and this has, in some cases, led to the development of advanced systems of plant–plant communication via semiochemicals. **Suppression of plant competitors** via semiochemicals defined as **allelopathy** by Molisch (1937) was later redefined by Rice (1984) as also involving effects related to microorganisms associated with plants. Advances in methods of chemical analysis and improved understanding of plant ecology have resulted in an increased scientific interest in allelopathy and its ecological importance.

However, as yet, the **intertrophic effects** of allelopathy (**allelobiosis** *sensu* Pettersson et al. 2003; Ninkovic et al. 2011), i.e. the effects upon herbivores and their natural enemies of plant responses in **plant–plant interactions**, have not been extensively studied. The efficiency of the searching behaviour of a polyphagous predator would benefit from a volatile cue allowing identification of plants stressed by herbivores.

Investigations of behavioural responses of adult *C. septempunctata* to barley and two common weeds have contributed to this topic. The distribution of *C. septempunctata* in a commercial barley crop with weeds indicates the importance of odour stimuli (Ninkovic & Pettersson 2003). Adults were significantly aggregated to patches with *Elytrigia repens* and *Cirsium arvense*, although no obvious food resource such as pollen, aphids or other small prey insects was abundant there. In olfactometer experiments, adult ladybirds showed no difference in orientation to either of the weeds. However, when barley and one of the two weeds were used together as the odour source, this was preferred compared to the odour of barley alone. Barley plants exposed to volatiles from *C. arvense* remained attractive even when the weed was taken away, whereas those exposed to *E. repens* lost their attractiveness in the absence of barley. This indicates that the positive effect of the barley–*E. repens* combination may merely be an effect of mixed volatiles, whereas the barley–*C. arvense* mixture is likely to represent a more complex mechanism involving **allelobiosis**.

A conclusion from these results is that **mixed plant stands** may have a stronger arresting effect on adult ladybirds than pure stands, if proximity of a different plant species makes a plant emit the volatile signature characteristic of general plant stress that would include herbivore-attacked plants. Hypothetically, an increased arresting effect could be a response merely to a complexity of plant volatiles. This would fit well with the *E. repens*–barley combination. However, the results from the tests of the *C. arvense*-barley combination suggest another possibility. Barley plants exposed to allelobiotic provocation from thistle volatiles become significantly less acceptable to aphids (Glinwood et al. 2004). Thus the positive ladybird response to *C. arvense*-exposed barley plants could be a response to a **plant-stress condition** upon herbivores and their natural enemies, which is similar whether induced by plant–plant interaction or aphid attack. This would be in line with the positive response of ladybirds to aphid-attacked barley plants (Ninkovic et al. 2001).

9.9 CONCLUSIONS AND FUTURE CHALLENGES

This chapter has summarized information on the chemical ecology of coccinellids. A considerable amount of experimentation has shown that coccinellids depend on semiochemical-mediated information to complete several of the important steps in their population ecology. However, so far there is only a limited understanding of the mechanisms involved and of how chemical ecology profoundly interacts with other stimuli regulating life processes.

Most of the behavioural responses of coccinellids to semiochemicals reported are related to major events in the life cycle. Thus the response to certain stimuli may vary depending on where in the **annual life cycle** the specific individual is. Several of the chemical cues related to mobility and foraging are commonly occurring substances, and one of the future challenges is to understand the subtle mechanisms that contribute to give this trivial set of information a precise meaning for the individual coccinellid. The integration of external information to be expressed in subsequent action is picked up by **chemoreceptors**, then processed at different sensory system levels and finally modified as a function of the status of the individual coccinellid. This status can be looked upon as an **endogenous response filter** that regulates the capability of the **coccinellid to respond to external stimuli**. Factors such as adaptive learning, mating and feeding status or different stages in the annual population cycle will modify the selectivity of this behaviour modifying filter (BMF). Basic elements controlling the dynamic variability of coccinellid behaviour such as **adaptive responses** and **learning** affected by a broad set of multiple blends of trivial chemicals such as green leaf volatiles and semiochemicals from other organisms are exciting fields where further research is needed.

The considerable knowledge on the chemical compounds that constitute the **aposematic chemistry** of coccinellids has created a challenging source for future studies on the behaviour of coccinellids in an ecological perspective. The metabolic costs of the powerful aposematic chemistry are high and further links to general ecological functions could be searched for. Increasing information of the aposematic chemistry in intra- and interspecific recognition stimulates further studies of the contributions of semiochemicals involved in aspects of the behavioural ecology of coccinellids such as territorial behaviour, cannibalism/foraging

and interaction with other behavioural stimuli such as colouration and tactile stimuli, where so far only very little is known for coccinellids. Some of the aposematic compounds are chemically complex and some of them (such as pyrazine) are behaviourally active in such small amounts that they challenge conventional methods for entrainment and chemical identification. Improved methods of chemical analysis will reduce these problems but also enable more sophisticated methods in studies of how coccinellids respond to semiochemicals. Further development of improved bioassay methods for studying coccinellid behaviour is needed in relation to processes such as kin recognition, aggressivity between individuals, courtship and mating and cues involved in migration.

Studies of the **olfactory receptor** systems on the antennae have shown sexual dimorphism. However, the behavioural consequences of this for different behavioural responses between males and females to different stimuli have so far not received much attention. Present knowledge on identified semiochemicals indicates that several biologically trivial compounds common both in several potential prey herbivores and their host plants convey important information to coccinellids. Responses to the same chemical can differ depending on time and place and the influence of the BMF mentioned above. Good examples of cues where these common **volatiles** may be behaviourally active are those **related to plant damage/stress**. Such volatiles form a vast set of active compounds that must be recognized by the olfactory receptors of coccinellids and, although the topic has often been discussed, there is only limited knowledge on how this challenge is met by different coccinellid species. Further studies of the principles for receptor function of coccinellids and its potential as a priming factor for specific chemical cues is therefore an interesting challenge.

With regards to foraging and feeding behaviour, the focus has been on **herbivore-induced volatiles** and other stress induced substances. Thus the borderline between stimuli expressed in food search and arresting responses in favourable patches is diffuse. Efficient use of a broad range of food sources means that identification of favourable habitats can be more profitable than searching for specific food sources. This creates a dynamic balance between investments in **localizing optimal habitats** where food is likely to be available, and in **finding specific** preferred food **resources**. This presupposes a pronounced adaptive capacity to meet a variable set of food sources.

Few insects can compete with coccinellids with regard to scientific challenges within the field of chemical ecology. Their own chemistry is expressed in complex aposematic compounds. They show an outstanding capacity to cope with different ecological requirements such as foraging, mating and responses to IGP, as well as to conspecific density. It is difficult to find another group of insects that is so close to human activities, and yet it must be admitted that conclusive data on the **identity and importance of semiochemicals** under natural conditions are still limited. It is to be expected that future studies will not only give results of general scientific value but also contribute to improved options for the different roles of coccinellids in human activities in biological production and nature conservation.

REFERENCES

Abe, M. and K. Matsuda. 2000. Feeding responses of four phytophagous lady beetle species (Coleoptera: Coccinellidae) to cucurbitacins and alkaloids. *Appl. Entomol. Zool.* 35: 257–264.

Agarwala, B. K. and P. J. Bhowmik. 2007. Effect of chemical defence on fitness of ladybird predator, Cheilomenes sexmaculata (Fabricius). *J. Biol. Control.* 21: 89–96.

Agarwala, B. K., H. Yasuda and Y. Kajita. 2003. Effect of conspecific and heterospecific feces on foraging and oviposition of two predatory ladybirds: Role of fecal cues in predator avoidance. *J. Chem. Ecol.* 29: 357–376.

Al Abassi, S., M. A. Birkett, J. Pettersson, J. A. Pickett, L. J. Wadhams and C. M. Woodcock. 1997. Ladybird beetle odour identified and found to be responsible for attraction between adults. *Cell. Mol. Life Sci.* 54: 876–879.

Al Abassi, S., M. A. Birkett, J. Pettersson et al. 1998. Ladybird beetle odour identified and found to be responsible for attraction between adults. *Cell. Mol. Life Sci.* 54: 876–879.

Al Abassi, S., M. A. Birkett, J. Pettersson et al. 2000. Response of the seven-spot ladybird to an aphid alarm pheromone and an alarm pheromone inhibitor is mediated by paired olfactory cells. *J. Chem. Ecol.* 26: 1765–1771.

Al Abassi, S., M. A. Birkett, J. Pettersson et al. 2001. Response of the ladybird parasitoid Dinocampus coccinellae to toxic alkaloids from the seven-spot ladybird, Coccinella septempunctata. *J. Chem. Ecol.* 27: 33–43.

Andow, D. A. 1991. Vegetational diversity and arthropod population response. *Annu. Rev. Entomol.* 36: 561–86.

Baker, T. C., J. J. Obrycki and J. W. Zhu. 2003. US Patent 6562332: *Attractants of Beneficial Insects,* issued 2003.

Barbier, R., A. Ferran, J. Le Lannic and M. R. Allo. 1992. Morphology and ultrastructure of integumentary glands of *Semiadalia undecimnotata* Schn. (Coleoptera: Coccinellidae). *Int. J. Insect Morphol. Embryol.* 21: 223–234.

Barron, A. and K. Wilson. 1998. Overwintering survival in the seven spot ladybird, *Coccinella septempunctata* (Coleoptera: Coccinellidae). *Eur. J. Entomol.* 95: 639–642.

Bhatkar A. P. (1982. Orientation and defence of ladybeetles (Coleoptera, Coccinellidae) following ant trail in search of aphids. *Folia Entomol. Mexicana* 53: 75–85.

Bi, H. H., R. S. Zeng, L. M. Su, M. An and S. M. Luo. 2007. Rice allelopathy induced by methyl jasmonate and methyl salicylate. *J. Chem. Ecol.* 33: 1089–103.

Birkett, M. A., C. A. M. Campbell, K. Chamberlain et al. 2000. New roles for cis-jasmone as an insect semiochemical and in plant defence. *Proc. Natl Acad. Sci. USA* 97: 9329–9334.

Boivin, G., C. Roger, D. Coderre and E. Wajnberg. 2010. Learning affects prey selection in larvae of a generalist coccinellid predator. *Entomol. Exp. Appl.* 135: 48–55.

Branco, M., J. C. Franco, E. Dunkelblum et al. 2006. A common mode of attraction of larvae and adults of insect predators to the sex pheromone of their prey (Hemiptera: Matsucoccidae). *Bull. Entomol. Res.* 96: 179–185.

Bruce, T. J. A., M. C. Matthes, K. Chamberlain et al. 2008. cis-Jasmone induces *Arabidopsis* genes that affect the chemical ecology of multitrophic interactions with aphids and their parasitoids. *Proc. Natl Acad. Sci. USA* 105: 4553–4558.

Burns, D. A., J. J. McDonnell, J. L. Hemptinne, G. Lognay and A. F. G. Dixon. 1998. Mate recognition in the two-spot ladybird beetle, *Adalia bipunctata*: role of chemical and behavioural cues. *J. Insect Physiol.* 44: 1163–1171.

Camarano, S., A. Gonzalez and C. Rossini. 2006. Chemical defence of the ladybird beetle *Epilachna paenulata*. *Chemoecology.* 16: 179–184.

Daloze, D., J. C. Braekman and J. M. Pasteels 1994. Ladybird defence alkaloids: structural, chemotaxonomic and biosynthetic aspects (Col.: Coccinellidae). *Chemoecology.* 5: 3–4.

Dejean, A., M. Gibernau, J. Lauga and J. Orivel. 2003. Coccinellid learning during capture of alternative prey. *J. Insect Behav.* 16: 859–864.

Dicke, M. 2000. Chemical ecology of host-plant selection by herbivorous arthropods: a multitrophic perspective. *Biochem. Syst. Ecol.* 28: 601–617.

Dicke, M., J. Takabayashi, M. A. Posthumus, C. Schütte and O. Krips. 1998. Plant–phytoseiid interactions mediated by herbivore induced plant volatiles: variation in production of cues and in responses of predatory mites. *Exp. Appl. Acarol.* 22: 311–333.

Dixon, A. F. G. 1959. An experimental study of the searching behaviour of the predatory Coccinellid beetle *Adalia decempunctata*. *J. Anim. Ecol.* 28: 259–281.

Dixon, A. F. G. and B. K. Agarwala 1999. Ladybird-induced life-history changes in aphids. *Proc. R. Soc. Lond.* (B) 266: 1549–1553.

Durieux, D., F. Verheggen, A. Vandereycken, E. Joie and E. Haubruge. 2010. Chemical ecology of ladybird beetles. *Biotechnologie, Agronomie, Société et Environnement* 14: 351–367.

Elliot, N. C. and R. W. Kieckhefer. 2000. Response by coccinellids to spatial variation in cereal aphid density. *Popul. Ecol.* 42: 81–90.

van Emden, H. F., A. P. Storeck, S. Douloumpaka et al. 2008. Plant chemistry and aphid parasitoids (Hymenoptera: Braconidae): Imprinting and memory. *Eur. J. Entomol.* 105: 477–483.

Endo, N., M. Abe, T. Sekine and K. Matsuda. 2004. *Appl. Entomol. Zool.* 39: 411–416.

Ferrer, A., A. F. G. Dixon and J. L. Hemptinne. 2008. Prey preference of ladybird larvae and its impact on larval mortality, some life-history traits of adults and female fitness. *Bull. Insect Ecol.* 61: 5–10.

Francis, F., E. Haubruge, P. Hastir and C. Gaspar. 2001. Effect of aphid host plant on development and reproduction of the third trophic level, the predator Adalia bipunctata (Coleoptera: Coccinellidae). *Environ. Entomol.* 30: 947–952.

Frechette, B., A. F. G. Dixon, H. Alauzet and J. L. Hemptinne. 2004. Age and experience influence patch assessment for oviposition by an insect predator. *Ecol. Entomol.* 29: 578– 583.

Gagné, I., D. Coderre and Y. Mauffette. 2002. Egg cannibalism by *Coleomegilla maculata lengi* neonates: preference even in the presence of essential prey. *Ecol. Entomol.* 27: 285–291.

Glinwood, R., V. Ninkovic, E. Ahmed and J. Pettersson. 2004. Barley exposed to aerial allelopathy from thistles (*Cirsium* spp.) becomes less acceptable to aphids. *Ecol. Entomol.* 29: 188–195.

Glinwood, R., E. Ahmed, E. Qvarfordt and V. Ninkovic. 2011. Olfactory learning of plant genotypes by a polyphagous insect predator. Oecologia DOI 10.1007/s00442-010-1892-x.

Hamilton, R. M., E. B. Dogane, G. B. Schaalje and G. M. Booth. 1999. Olfactory response of the lady beetle *Hippodamia convergens* (Coleoptera: Coccinellidae) to prey related odors, including a scanning electron microscopy study of the antennal sensilla. *Environ. Entomol.* 28: 812–822.

Han, B. Y. and Z. M. Chen. 2002. Behavioral and electrophysiological responses of natural enemies to synomones from tea shoots and kairomones from tea aphids, *Toxoptera aurantii*. *J. Chem. Ecol.* 28: 2203–2219.

Han, B. Y. and Z. M. Chen. 2005. Composition of the volatiles from intact and mechanically pierced tea aphid–tea shoot complexes and their attraction to natural enemies of the tea aphid. *J. Agric. Food Chem.* 50: 2571–2575.

Hatano, E., G. Kunert, J. P. Michaud and W. W. Weisser. 2008. Chemical cues mediating aphid location by natural enemies. *Eur. J. Entomol.* 105: 797–806.

Hautier, L., J. C. Gregoire, J. de Schauwers et al. 2008. Intraguild predation by *Harmonia axyridis* on coccinellids revealed by exogenous alkaloid sequestration. *Chemoecology.* 18: 191–196.

Hemptinne, J. L. and A. F. G. Dixon. 2000. Defence, oviposition and sex: Semiochemical parsimony in two species of ladybird beetles (Coleoptera, Coccinellidae)? A short review. *Eur. J. Entomol.* 97: 443–447.

Hemptinne J. L., A. F. G. Dixon, J. L. Doucet and J. E. Petersen 1993. Optimal foraging by hoverflies (Diptera: Syrphidae) and ladybirds (Coleoptera: Coccinellidae): Mechanisms. *Eur. J. Entomol.* 90: 451–455.

Hemptinne, J. L., A. F. G. Dixon and G. Lognay. 1996. Searching behaviour and mate recognition by males of the two-spot ladybird beetle, *Adalia bipunctata*. *Ecol. Entomol.* 21, 165–170.

Hemptinne, J. L., G. Lognay , C. Gauthier and A. F. G. Dixon. 2001a. Role of surface chemical signals in egg cannibalism and intraguild predation in ladybirds (Coleoptera: Coccinellidae). *Chemoecology.* 10: 123–128.

Hemptinne, J. L., G. Lognay, M. Doumbia and A. F. G. Dixon. 2001b. Chemical nature and persistence of the oviposition deterring pheromone in the tracks of the larvae of the two spot ladybird, *Adalia bipunctata* (Coleoptera, Coccinellidae). *Chemoecology.* 11: 43–47.

Hodek, I. and J. P. Michaud. 2008. Why is *Coccinella septempunctata* so successful? (A point-of-view). *Eur. J. Entomol.* 105: 1–12.

Holloway, G. J., P. W. De Jong, P. M. Brakefield, P. W. De Jong and H. De Vose. 1991. Chemical defense in ladybird beetles (Coccinellidae). I. Distribution of coccinelline and individual variation in defence in 7-spot ladybirds (*Coccinella septempunctata*). *Chemoecology.* 2: 7–14.

Holloway, G. J., P. W. De Jong and M. Ottenheim. 1993. The genetics and cost of chemical defence in the two-spot ladybird (*Adalia bipunctata* L.). *Evolution* 47: 1229–1239.

Jamal, E., C. Brown and C. Grayson. 2001. Orientation of *Hippodamia convergens* (Coleoptera: Coccinellidae) larvae to volatile chemicals associated with *Myzus nicotianae* (Homoptera: Aphidiidae). *Environ. Entomol.* 30: 1012–1016.

James, D. G. 2005. Further field evaluation of synthetic herbivore-induced plant volatiles as attractants for beneficial insects. *J. Chem. Ecol.* 31: 481–495.

James, D. G. and T. S. Price. 2004. Field testing of methyl salicylate for recruitment and retention of beneficial insects in grapes and hops. *J. Chem. Ecol.* 30: 1613–1628.

Jourdan, H., R. Barbier, J. Bernard and A. Ferran. 1995. Antennal sensilla and sexual dimorphism of the adult lady beetle Semiadalia undecimnotata Schn. (Coleoptera, Coccinellidae). *Int. J. Insect Morphol. Embryol.* 24: 307–322.

Kajita, Y., J. J. Obrycki, J. J. Sloggett and K.F. Haynes. 2010 Intraspecific alkaloid variation in ladybird eggs and its effects on con- and hetero-specific intraguild predators. *Oecologia* 163: 313–322.

Kazana, E., T. W. Pope, L. Tibbles et al. 2007. The cabbage aphid: a walking mustard oil bomb. *Proc. Biol. Sci.* 274: 2271–2277.

Karystinou, A., A. P. M. Thomas and H. E. Roy 2004. Presence of haemocyte-like cells in coccinellid reflex blood. *Physiol. Entomol.* 29: 94–96.

Klausnitzer, B. 1989. Aggregations of ladybirds on the Baltic coast (Col., Coccinellidae). *Entomol. Nachr. Ber.* 33: 189–195.

Klewer, N., Z. Ruzicka and S. Schulz. 2007. Z-pentacos-12-ene, an oviposition-deterring pheromone of *Cheilomenes sexmaculata*. *J. Chem. Ecol.* 33: 2167–2170.

Laurent, P., J. C. Braekman and S. Daloze. 2005. Insect chemical defense. *Topics Curr. Chem.* 240: 167–229.

Laubertie, E., X. Martini, C. Cadena et al. 2006. The immediate source of the oviposition-deterring pheromone produced by larvae of *Adalia bipunctata* (L.) (Coleoptera, Coccinellidae). *J. Insect Behav.* 19: 231–240.

Leather, S. R., R. C. A. Cooke, M. D. E. Fellowes and R. Rombe .1999. Distribution and abundance of ladybirds (Coleoptera: Coccinellidae) in non-crop habitats. *Eur. J. Entomol.* 96: 23–27.

Majerus, M. E. N. 1994. *Ladybirds.* Harper Collins, London. 367 pp.

Majerus, M. E. N. and T. M. O. Majerus, 1997. Predation of ladybirds by birds in the wild. *Entomol. Mon. Mag.* 133: 1592–1595.

Majerus, M. E. N., J. J. Sloggett, J. F. Godeau and J. L. Hemptinne. 2007. Interactions between ants and aphidophagous and coccidophagous ladybirds. *Popul. Ecol.* 49: 15–27.

Marples, N. M. 1993a. Toxicity assays of ladybirds using natural predators. *Chemoecology* 4: 33–28.

Marples, N. M. 1993b. Is the alkaloid in 2spot ladybirds (*Adalia bipunctata*) a defence against ant predation? *Chemoecology* 4: 29–32.

Marples, N. M., W. VanVeelen and P. M. Brakefield. 1994. The relative importance of colour, taste and smell in the protection of an aposematic insect *Coccinella septempunctata*. *Anim. Behav.* 48: 967–974.

Michaud, J. P. and A. K. Grant. 2004. Adaptive significance of sibling egg cannibalism in Coccinellidae: Comparative evidence from three species. *Ann. Entomol. Soc. Am.* 97: 710–719.

Michaud, J. P. and J. L. Jyoti. 2008. Dietary complementation across life stages in the polyphagous lady beetle *Coleomegilla maculata*. *Entomol. Exp. Appl.* 126: 40–45.

Mondor, E. B. and B. D. Roitberg. 2000. Has the attraction of predatory coccinellids to cornicle droplets constrained aphid alarm signalling behaviour? *J. Insect Behav.* 13: 321–329.

Molisch, H. 1937. *Der Einfluss einer Pflanze auf die andere-Allelopathie.* Gustaf Fischer Jena. 106 pp.

Nakashima, Y. and M. Akashi. 2005. Temporal and within-plant distribution of the parasitoid and predator complexes associated with *Acyrthosiphon pisum* and *A. kondoi* (Homoptera : Aphididae) on alfalfa in Japan. *Appl. Entomol. Zool.* 40: 137–144.

Nakashima, Y., M. A. Birkett, B. J. Pye, J. A. Pickett and W. Powell. 2004. The role of semiochemicals in the avoidance of the seven-spot ladybird, *Coccinella septempunctata* by the aphid parasitoid, *Aphidius ervi*. *J. Chem. Ecol.* 30: 1103–1116.

Nakashima, Y., M. A. Birkett, B. J. Pye, W. Powell. 2006. Chemically mediated intraguild predator avoidance by aphid parasitoids: Interspecific variability in sensitivity to semiochemical trails of ladybird predators. *J. Chem. Ecol.* 32(9):1989–1998.

Nakamuta, K. 1991. Aphid alarm pheromone component, (E)-beta-farnesene, and local search by a predatory lady beetle, *Coccinella septempunctata bruckii* Mulsant (Coleoptera: Coccinellidae). *Appl. Entomol. Zool.* 26: 1–7.

Nault, L. R., L. J. Edwards and W. E. Styer. 1973. Aphid alarm pheromones: secretion and reception. *Environ. Entomol.* 2: 101–105.

Nault ,L. R., T. K. Wood and A. M. Goff. 1974. Treehopper (Membracidae) alarm pheromones. *Nature* 249: 387–388.

Nentwig, W. 1983. The prey of web-building spiders compared with feeding experiments (Araneae: Araneidae, Linyphiidae, Pholcidae, Aegelenidae). *Oecologia* 56: 132–139.

Nentwig, W. 1986. Non web-building spiders: prey specialists or generalists? *Oecologia* 69: 571–576.

Ninkovic, V. and J. Pettersson. 2003. Plant/plant communication supports searching behaviour of the sevenspotted ladybird, *Coccinella septempunctata* (L.). *Oikos* 100: 65–70.

Ninkovic, V., S. Al Abassi and J. Pettersson. 2001. The influence of aphid-induced plant volatiles from barley on the searching behaviour of the seven spot ladybird, *Coccinella septempunctata*. *Biol. Control* 21: 191–195.

Ninkovic, V., S. Al Abassi, E. Ahmed, R. Glinwood and J. Pettersson. 2011. Effect of within-species plant genotype mixing on habitat preference of a polyphagous insect predator. *Oecologia* 166: 391–400.

Oliver, T. H., J. E. L. Timms, A. Taylor and S. Leather. 2006. Oviposition responses to patch quality in the larch ladybird *Aphidecta obliterata* (Coleoptera: Coccinellidae): effects of aphid density, and con- and heterospecific tracks. *Bull. Entomol. Res.* 96: 25–34.

Oliver, T. H., J. M. Cook and S. R. Leather. 2008. Avoidance responses of an aphidophagous ladybird, *Adalia bipunctata*, to aphid-tending ants. *Ecol. Entomol.* 33: 523–528.

Orr, C. J., J. J. Obrycki and R. V. Flanders 1992. Host acceptance behavior of *Dinocampus coccinellae* (Hymenoptera: Braconidae). *Ann. Entomol. Soc. Am.* 85: 722–730.

Pasteels, J. M. 2007. Chemical defence, offence and alliance in ants–aphids–ladybirds relationships. *Popul. Ecol.* 49: 5–14.

Pasteels, J. M., C. Deroe, B. Tursch et al. 1973. Distribution and activities of the defensive alkaloids of the Coccinellidae. *J. Insect Physiol.* 19: 1771–1784.

Pratt, C., T. W. Pope, G. Powell and J. T. Rossiter. 2008. Accumulation of glucosinolates by the cabbage aphid *Brevicoryne brassicae* as a defense against two coccinellid species. *J. Chem. Ecol.* 34: 323–329.

Pettersson, J., V. Ninkovic and R. Glinwood. 2003. Plant activation of barley by intercropped conspecifics and weeds: allelobiosis. *The BCPC International Congress: Crops Science and Technology.* 1135–1144.

Pettersson, J., V. Ninkovic, R. Glinwood et al. 2008. Chemical stimuli supporting foraging behaviour of *Coccinella septempunctata* L. (*Coleoptera: Coccinellidae*): volatiles and allelobiosis, *Appl. Entomol. Zool.* 43: 315–321.

Rice, E. L. 1984. *Allelopathy*, 2nd edn. Academic Press, New York.

Richerson, J. V. and C. J. DeLoach. 1972. Some aspects of host selection by *Perilitus coccinellae*. *Ann. Entomol. Soc. Am.* 65: 834–839.

Riddick, E. W., A. E. Brown and K. R. Chauhan. 2008. *Harmonia axyridis* adults avoid catnip and grapefruit-derived terpenoids in laboratory bioassays. *Bull. Insectol.* 61: 81–90.

Rothschild, M., J. V. Euw and T. Reichste. 1973. Cardiacglycosides in a scale insect (*Aspidiotus*), a ladybird (*Coccinella*) and a lacewing (*Chrysopa*). *J. Entomol. A* 48: 89–90.

Roy, H. and T. E. Cottrell. 2008. Forgotten natural enemies: interactions between coccinellids and insect-parasitic fungi. *Eur. J. Entomol.* 105: 391–398.

Ruzicka, Z. 1994. Oviposition-deterring pheromone in *Chrysopa oculata* (Neuroptera: Chrysopidae). *Eur. J. Entomol.* 91: 361–370.

Ruzicka, Z. 1997. Recognition of oviposition-deterring allomones by aphidophagous predators (Neuroptera: Chrysopidae, Coleoptera: Coccinellidae). *Eur. J. Entomol.* 94: 431–434.

Ruzicka, Z. 2001. Response of chrysopids (Neuroptera) to larval tracks of aphidophagous coccinellids (Coleoptera). *Eur. J. Entomol.* 98: 283–285.

Ruzicka, Z. 2002. Persistence of deterrent larval tracks in *Coccinella septempunctata*, *Cycloneda limbifer* and *Semiadalia undecimnotata* (Coleoptera: Coccinellidae). *Eur. J. Entomol.* 99: 471–475.

Ruzicka, Z. 2003. Perception of oviposition-deterring larval tracks in aphidophagous coccinellids *Cycloneda limbifer* and *Ceratomegilla undecimnotata* (Coleoptera: Coccinellidae). *Eur. J. Entomol.* 100: 345–350.

Ruzicka, Z. 2006. Oviposition-deterring effects of conspecific and heterospecific larval tracks on *Cheilomenes sexmaculata* (Coleoptera: Coccinellidae). *Eur. J. Entomol.* 103: 757–763.

Santi F, Burgio G and Maini S. 2003. Intra-guild predation and cannibalism of *Harmonia axyridis* and *Adalia bipunctata* in choice conditions. *Bull. Insectol.* 56, 207–210.

Schmid, A. 1992. Investigations on the attractiveness of agricultural weeds to aphidophagous ladybirds (Coleoptera, Coccinellidae). *Agrarökologie* 5: 122.

Schroeder, F. C., S. R. Smedley, L. K. Gibbons et al. 1998. Polyazamacrolides from ladybird beetles: Ring-size selective oligomerization. *Proc. Natl Acad. Sci. USA* 95: 13387–13391.

Sloggett, J. J. 2010. Predation of ladybirds by the orb-web spider *Araneus diadematus*. *BioControl* 55: 631–638.

Sloggett, J. J. and M. E. N. Majerus. 2000. Aphid-mediated coexistence of ladybirds (Coleoptera: Coccinellidae) and the wood ant *Formica rufa*: Seasonal effects, interspecific variability and the evolution of a coccinellid myrmecophile. *Oikos* 89: 345–359.

Sloggett, J. J., J. O. Obrycki and K. F. Haynes. 2009. Identification and quantification of predation: novel use of gas chromatography-mass spectrometric analysis of prey alkaloid markers. *Funct. Ecol.* 23: 416–426.

Smedley, S. R., K. A. Lafleur, L. K. et al. 2002. Glandular hairs: pupal chemical defense in a non-native ladybird beetle (Coleoptera: Coccinellidae). *Northeast. Naturalist* 9: 253–266.

Takizawa, T., H. Yasuda and B. K. Agarwala. 2000. Effect of three species of predatory ladybirds on oviposition of aphid parasitoids. *Entomol. Sci.* 3: 465–469.

Timms, J. E., T. H. Oliver, N. A. Straw and S. R. Leather. 2008. The effects of host plant on the coccinellid functional response: is the conifer specialist *Aphidecta obliterata* (L.) (Coleoptera: Coccinellidae) better adapted to spruce than the generalist *Adalia bipunctata* (L.) (Coleoptera: Coccinellidae)? *Biol. Control* 47: 273–281.

Vandermeer, J. 1992. *The Ecology of Intercropping*. Cambridge University Press, Cambridge, UK.

Vantaux, A., O. Roux, A. Magro et al. 2010. Host-specific myrmecophily and myrmecophagy in the tropical coccinellid *Diomus thoracicus* in French Guiana. *Biotropica* 42: 622–629.

Verheggen, F. J., Q. Fagel, S. Heuskin et al. 2007. Electrophysiological and behavioral responses of the multicolored Asian lady beetle, *Harmonia axyridis* Pallas, to sesquiterpene semiochemicals. *J. Chem. Ecol.* 33: 2148–2155.

Vet, L. E. M. and M. Dicke. 1992. Ecology of infochemical use by natural enemies in a tritrophic context. *Annu. Rev. Entomol.* 37: 141–172.

de Vos, M. and G. Jander. 2010. Volatile communication in plant–aphid interactions. *Curr. Opin. Plant Biol.* 13: 366–371.

Walling, L. L. 2000. The myriad of plant responses to herbivores. *J. Plant Growth Regulation* 19: 195–216.

Ware, R. L., F. Ramon-Portugal, A. Magro et al. 2008. Chemical protection of *Calvia quatuordecimguttata* eggs against intraguild predation by the invasive ladybird *Harmonia axyridis*. *BioControl* 53: 189–200.

With, K. A., D. M. Pavuk, J. L. Worchuck, R. K. Oates and J. L. Fisher. 2002. Threshold effects of landscape structure on biological control in agroecosystems. *Ecol. Appl.* 12: 52–65.

Witte, L., A. Ehmke and T. Hartmann. 1990. Interspecific flow of pyrrolizidine alkaloids from plants via aphids to ladybirds. *Naturwissenschaften* 77: 540–543.

Zhu, J. W. and K. C. Park. 2005. Methyl salicylate, a soybean aphid-induced plant volatile attractive to the predator *Coccinella septempunctata*. *J. Chem. Ecol.* 31: 1733–1746.

Chapter 10

QUANTIFYING THE IMPACT OF COCCINELLIDS ON THEIR PREY

J. P. Michaud[1] and James D. Harwood[2]

[1] Department of Entomology, Kansas State University, 1232 240th Ave., Hays, KS 67601, USA

[2] Department of Entomology, University of Kentucky, Lexington, KY 40546-0091, USA

10.1 INTRODUCTION

The importance of quantifying the impact of natural enemies in biological control programmes has often been emphasized, but the empirical measurement of impact in the field remains one of the most difficult challenges for ecological entomologists. This is especially true for predatory species such as coccinellids. The impact of coccinellids on their prey is a complex function of their local abundance, voracity and prey fidelity, combined with various indirect effects of their activities on prey survival, some mediated by other natural enemies and some by the host plant. Predation is difficult and time-consuming to observe directly in the field and often leaves no evidence of its occurrence. Since various coccinellid species frequently attack the same prey simultaneously, and may comprise only one fraction of a guild of predators, it can be challenging to partition impact among predator species even when rates of prey removal can be quantified. Despite such difficulties, efforts to quantify coccinellid predation directly are usually considered preferable to more indirect inferences derived from statistical correlations of predator and prey abundance. Too often, the collapse of large populations of **aphids** is attributed to the action of large numbers of coccinellids associated with them when, in actuality, **many biotic and abiotic factors interact to accelerate** such **population declines**. These include, but are not limited to, diminishing host plant quality, induced plant defence responses, development and emigration of winged forms of prey, fungal epizootics, high temperatures, wind and rain, etc. The respective contributions of these factors can be difficult to separate from the impact of predation. The study of *Diuraphis noxia* mortality conducted by Lee et al. (2005) illustrates some of the empirical challenges posed by such confounding factors and the fact that coccinellids, although abundant, may not necessarily suppress aphid populations below economic levels. Notwithstanding the challenges of attributing mortality to particular causes, the **important role of coccinellids** as agents of biological control is **undeniable in many cases**, even if it is difficult to quantify. For example, the mass exodus of spring generation *Hip. convergens* from mature wheat fields is observed with seasonal regularity on the High Plains of the USA and can only result from prodigious aphid consumption in the developing grain, without which more frequent and more extensive pesticide applications would almost certainly be necessary to preserve yields (JPM, pers. observ.). In contrast, Dixon (2005) concludes that, although natural enemies, including coccinellids, may occasionally inflict dramatic mortality on tree-dwelling aphids, it is the interactions between the aphids and their host plants that drive long-term cycles in the abundance of these species.

Coccidophagous coccinellids may match the reproductive rate of their prey, generation for generation, or in the case of *Rodolia cardinalis*, exceed it (Hodek 1973; Chapter 11), resulting in a highly effective numerical response. However, the numerical response of aphidophagous species cannot match the reproductive rate of most aphid species that can achieve multiple generations in the critical period of spring/early summer. Estimation of coccinellid impact on aphids is particularly challenging because of the ephemeral nature of aphid populations, their high reproductive rate (due to parthenogenesis and viviparity), and the sensitivity of biological control outcomes to initial conditions, especially the timing of predator arrival. Smith (1966) characterized five stages of the typical **aphid population cycle**: initiation, exponential growth, peak, collapse and scarcity. In the **initiation phase**, immigrant alatae form nuclear colonies on plants that have recently become suitable hosts; aphids remain at low density for some time because alate reproductive rate is low and many die prior to the maturation and reproduction of their apterous daughters. Although rapid recruitment by coccinellids and other aphid predators at this time can have great impact on the aphid population's growth potential, these predators may not respond in any numbers until aphid density reaches some threshold, and the timing of their arrival may hinge on the distance over which they must travel from other habitats or overwintering sites. It has been pointed out that any estimate of predator impact in established aphid colonies must be accompanied by some estimate of aphid reproduction during the same period in order to be meaningfully interpreted (Latham & Mills 2010). If natural enemy recruitment is either sparse or delayed, the aphid population enters a **phase of exponential growth** that coincides with the onset of reproduction by large numbers of maturing apterae. Soon thereafter, a threshold is surpassed beyond which the aphid replacement rate far exceeds the maximum rate of mortality that can be inflicted by natural enemies, regardless of their numbers, and biological control can be considered to have failed. Nevertheless, 'success' for coccinellids is not measured by decimation of the aphid population, but rather by achieving reproduction that is properly synchronized with the growth of the aphid colonies in order to ensure the survival of offspring (Kindlmann & Dixon 1993). Thus, females maximize their fitness by laying as many eggs as possible at the beginning of exponential growth phase, the so-called 'oviposition window' (Dixon 2000), so that their larvae can complete development during the peak phase when aphids are maximally abundant. This is followed by the decline or **collapse phase** when the last maturing aphids develop into alatae and disperse, and then finally the **period of aphid scarcity** that persists until a new population cycle is initiated, either on the same or alternative host plants.

From the **perspective of biological control**, three types of outcome are possible for each aphid population cycle. Coccinellids and other aphidophagous predators achieve maximum impact if they **arrive early and in sufficient numbers**, so that most aphid colonies are eliminated prior to the exponential growth phase. This is the anthropocentric objective, but has adverse consequences for the coccinellid population because many larvae starve, and those that survive must do so by resorting to cannibalism and intraguild predation. This scenario may well occur more often than is generally recognized, especially in non-agricultural systems, and thus contribute to a lack of recognition of coccinellid impact, simply because **aphid infestations eliminated** in such **early stages** often go **unnoticed**. At the other extreme, impact will be negligible if coccinellids arrive too late, regardless of their numbers, once a large proportion of aphid colonies are in the exponential growth phase. In many cases, the outcome lies somewhere between these extremes; the peak phase of the aphid population is suppressed to some degree, but perhaps not to the desired level. In these cases, whether impact is considered sufficient with respect to biological control hinges entirely on the economic threshold for aphids on the particular crop plant. Often, the threshold is too low to permit a realistic expectation of control through biological means alone, despite the fact that the potential for coccinellid impact is relatively high. Examples include cases of direct aphid contamination of a consumer product (e.g. the lettuce aphid, *Nasonovia ribisnigri*), and where aphids vector an important plant disease (e.g. *Myzus persicae* vectoring potato leaf roll

virus). Thus, from the perspective of pest management, coccinellid impact is relative; it may be substantial in terms of aphid population reduction and still fall short of economic requirements.

10.2 ASSAYS OF CONSUMPTION

There have been many efforts to **quantify the voracity** of coccinellid species as an indication of their potential impact. Bankowska et al. (1978) measured food consumption of coccinellids **in the laboratory** and then multiplied these values by the numbers of beetles observed in alfalfa fields to estimate their impact on aphid populations. Laboratory studies of coccinellid development and reproduction on particular prey species are useful for inferring their potential impact on prey populations (e.g. Michaud 2000, Uygun & Atlhan 2000, Michaud & Olsen 2004, Omkar & James 2004, Mignault et al. 2006) and for predicting functional and numerical responses (Mack & Smilowitz 1982, Atlhan & Ozgokce 2002, Butin et al. 2003). However, coccinellid **foraging efficiency** on a particular prey may vary as function of plant architecture (e.g. Clark & Messina 1998; 5.3.1.1) such that consumption rates measured off plants may yield unrealistic results (e.g. Corlay et al. 2007). Coccinellid **life tables** can be compiled from laboratory feeding data (e.g. Liu et al. 1997, Lopez et al. 2004) but their relevance to nature is low unless survival rates are known for different life stages in the field (e.g. Kirby & Ehler 1977). In general, **mortality** of coccinellid eggs and larval stages is likely much higher in the field than in the laboratory due to factors such as disease, intraguild predation, cannibalism, resource limitation, etc. Furthermore, their impact may be reduced by factors such as emigration, consumption of alternative prey, and climatic factors such as wind and rain that limit foraging activity. Since laboratory consumption assays are usually thought to generate upper limit estimates of impact potential, they complement direct field observations, but cannot substitute for them.

Latham and Mills (2009) **compared assays** of plum aphid, *Hyalopterus pruni*, consumption by various predators, both **in the laboratory and in the field**, and found significant discrepancies between the estimates generated. The authors reviewed possible sources of experimental error that could account for their results, in particular several that could have led to overestimates of consumption in the field. They also pointed out the importance of distinguishing between the **amount of biomass killed versus** the amount **consumed**. The latter value was significantly less than the former when the predator was *Chrysopa nigricornis*, but only slightly less in the case of *Har. axyridis*, leading the authors to infer that consumption assays in laboratory arenas were vulnerable to generating underestimates of predator impact, in contrast to the conventional assumption.

10.3 INDIRECT IMPACTS

Foraging coccinellids may disturb aphid colonies directly by bumping into aphids, and indirectly by stimulating release of the **aphid alarm pheromone** β-farnesene (5.3.1.2), causing adjacent aphids to abandon their feeding sites and incur costs in terms of growth and development (Nelson 2007). Whereas factors such as wind and rain can forcibly dislodge aphids from plants (Cannon 1986, Mann et al. 1995), an **aphid's propensity to drop** voluntarily in response to disturbance varies considerably depending on its species, growth stage, and feeding habits. Some species use quick dropping as an escape response (5.3.2), e.g. *Acyrthosiphum pisum* (Francke et al. 2008) and *Macrosiphum euphorbiae* (J.P.M., unpublished), whereas others may feed with their stylets so deeply imbedded in plant material that quick release of the plant is difficult, e.g. *Aphis fabae* and *Toxoptera citricidus*. Among species that drop readily in response to disturbance, it has been estimated that more aphids may be killed by disturbance than are actually consumed by the foraging beetle (Roitberg et al. 1979, Day et al. 2006). Aphids exposed to alarm pheromone may be reluctant to climb back onto plants (Klingauf 1976) and crop cultivars expressing antibiosis may increase the propensity of aphids to drop (e.g. Gowling & van Emden 1994), thus synergizing the impact of foraging coccinellids. Aphids dislodged from plants may succumb to the action of physical factors such as heat and desiccation on exposed soil (Roitberg & Myers 1979) or be consumed by other ground-dwelling predators before they can recover their host plant (Winder 1990, Losey & Denno 1998a). Thus, **laboratory measurements** of prey consumption rates **may** also **underestimate** various indirect impacts of coccinellids on prey populations in the field (McConnell & Kring 1990).

Historically, **classical biological control** programmes have often failed to adequately quantify the impact of released species following their establishment (Van Driesche & Hoddle 2000). Projects could be deemed successful on the basis of **indirect indicators** of economic benefit such as declining pest populations, reductions in pesticide usage and even diminishing numbers of public complaints about the pest. In actuality, such indirect measures can be driven by various factors other than the impact of specific natural enemies. Ideally, discrete populations of pest and predator should be monitored for several years at a variety of sites, including a series without coccinellid releases (e.g. Van Driesche et al. 1998). In **augmentation** programmes, some assumptions regarding the per capita impact of a coccinellid species are implicit since release rates and intervals must be selected. However, these are most often arrived at by trial and error rather than being interpolated from quantitative data (van Lenteren 2000).

10.4 TRADITIONAL APPROACHES TO THE STUDY OF PREDATION

Luck et al. (1988) outlined six general **techniques for evaluating the impact** of biological control agents in the field, pointing out the limitations of each and the merits of combining different tactics. (See also an earlier review, Hodek et al. 1972.) Of these traditional approaches, the most useful for assessing coccinellid impact have probably been **direct observation** and **exclusion**. Most coccinellid species are sufficiently conspicuous to make direct observation of their activities feasible, although this is not true in all cases. Whereas most aphidophagous species lay brightly coloured eggs in conspicuous clusters in the open, coccidophagous species typically lay eggs singly in concealed locations, making it more difficult to assess their reproductive activity in the field. van Emden (1963) developed a technique for re-visiting marked colonies of *Brevicoryne brassicae* to infer the intensity of mortality inflicted by different aphid natural enemies. A similar approach of re-iterative, non-destructive sampling was used by Michaud (1999) to infer the relative importance of various sources of mortality to *T. citricidus* colonies on citrus terminals.

Frazer and Gilbert (1976) maintained that all practical methods for **counting coccinellids in the field** substantially **underestimated** their **actual**

abundance. Actively foraging individuals are more likely to be observed and satiated individuals more likely to be resting in concealed locations and therefore overlooked (Frazer 1988). The efficacy of direct observation can also be greatly affected by the nature of the plant and the opportunities for concealment it offers, the time of day, prevailing weather conditions and any other factors affecting coccinellid activity. Different **sampling methods** may also give different results for different species and life stages within the same crop and the most efficient methods may vary between crops (Elliott & Michels 1997, Michels et al. 1997). Thus, efforts to sample coccinellids must be carefully tailored to suit a particular situation and take into account the structure of the prey-bearing plant, its stage of development, and differences in detectability among life stages. Since coccinellids are diurnal (5.4.1.1), nocturnal observations are not required to detect them, although they may be needed to determine the full range of predators contributing to prey mortality (Pfannenstiel & Yeargan 2002).

10.4.1   Selective exclusion

Selective exclusion is the method most often used for measuring coccinellid impact under field conditions, likely because many of their prey are comparatively sessile and thus well suited for field cage studies. It is often possible to select a mesh size to permit the selective entry or exclusion of specific natural enemies. Sticky barriers can also be judiciously applied to selectively exclude ants or other non-flying arthropods. Xiao and Fadamiro (2010) used a combination of barriers and exclusion cages to attribute mortality of citrus leafminer, *Phyllocnistis citrella*, in mandarin orange groves proportionally among various parasitoids and predators, including *Har. axyridis*. Exclusion studies can produce convincing data (e.g. Morris 1992), especially when one predatory species is clearly dominant, although data can be more ambiguous when multiple predator species are involved (Chambers et al. 1983, Hopper et al. 1995). Exclusion cages have often been successfully used on aphid colonies to generate estimates of coccinellid impact in situations where they are the dominant predators (Cherry & Dowell 1979, Kring et al. 1985, Liao et al. 1985, Rice & Wilde 1988, Nechols & Harvey 1998, Lee et al. 2005). For example, Wells et al. (1999) used a **combination of partial and total exclusion** treatments to

Figure 10.1 Exclusion cages mounted in an alfalfa field to determine impact of coccinellid species feeding on pea aphids, *Acyrthosiphon pisum* (photo: Edward Evans). (See colour plate.)

demonstrate the key role of *Scymnus* sp. in suppressing *Aphis gossypii* populations on cotton in the coastal plain of Georgia, USA. The study conducted by Winder (1990) used polyethylene barriers to exclude generalist epigeal predators, primarily carabids and linyphiid spiders, and estimate their contribution to mortality of *S. avenae* on wheat plants, separately from aphid-specific predators. Innovative traps were deployed to assess the rates at which dislodged aphids regained plants and demonstrate that polyphagous ground predators reduced their survival.

10.4.2   Field cages

Exclusion studies in field crops typically require the installation of large cages anchored to the ground and enclosing multiple plants (Fig. 10.1). In this manner, Costamagna and Landis (2006, 2007) used **direct observations in field cages** to quantify collective predation impact by coccinellid species on *Aphis glycines* relative to other predators in soybean. Selective exclusion of large coccinellid species (primarily *Har. axyridis* and *C. septempunctata* revealed that they exerted a disproportionate amount of pest suppression (Costamagna et al. 2008). On trees or shrubs, it may be sufficient to enclose infested branches or terminals. Michaud (2004) used exclusion cages to monitor the survival of immature Asian citrus psyllid, *Diaphorina citri*, on expanding citrus terminals (Fig. 10.2). This technique was effective in quantifying collective

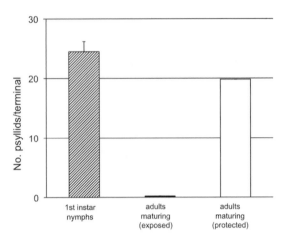

Figure 10.2 An exclusion cage designed for assessing impact of coccinellids and other natural enemies on aphids and psyllids infesting expanding grapefruit terminals. The cage is constructed from a clear plastic 2-litre soda bottle with mesh sleeves attached to either end with silicone and fastened to the branch with a length of wire secured to the inside of the bottle with packing tape. The sleeve mesh can be selected to permit the passage of particular insects (in this case, parasitoids), while excluding larger ones (coccinellids). The basal sleeve is attached tightly around the branch with a zip tie; the terminal sleeve is sufficiently long to permit unimpeded growth of the shoot for 2–3 weeks and can be untied to permit periodic access and counting of insects (photo: J.P. Michaud). (See colour plate.)

Figure 10.3 Mean numbers of first instar psyllid nymphs per grapefruit terminal and the numbers maturing to adulthood with and without exclusion of four coccinellid species (summarized from Michaud 2004).

predator impact caused primarily by four coccinellid species separately from that of the introduced parasitoid *Tamarixia radiata*. Although the various sources of juvenile psyllid mortality could not be fully partitioned, the comparative importance of ladybird species could be inferred from their relative abundance and the fact that their exclusion improved psyllid survival by a factor of 140 (Fig. 10.3). In addition, it was determined that parasitoid immatures suffered >95% mortality within their psyllid hosts as a consequence of intraguild predation by the coccinellids. Subsequently, Qureshi and Stansly (2009) applied a similar approach in south Florida and obtained similar results.

All forms of **enclosure modify** the **physical environment**, typically increasing temperature, reducing illumination and air flow, and providing some degree of unnatural protection from physical stresses such as wind and rain. This is sometimes addressed by including a **partial exclusion** treatment to reproduce the environmental effects of the cage while permitting the entry of natural enemies. However, physical enclosure, whether partial or complete, often results in significant changes in plant growth form, creating the potential for indirect effects on the prey population. The use of exclusion cages on aphids may also result in the **trapping of alatae** and thus lead to artificially elevated populations through forced re-infestation of the same plants (Lee et al. 2005). Studies that exclude coccinellids from plants naturally infested with their prey are always preferable to those where infestations are artificially established, although this is sometimes a necessity. **Artificial infestations** entail two significant risks; the plants infested manually may vary in quality in ways not evident to the researcher, and laboratory-raised insects may be unduly stressed when they are moved from protected rearing environments to the field. Exclusion of coccinellids can also be accomplished with insecticide treatments (e.g. Annecke et al. 1969) but these are often difficult to implement without affecting pest populations.

10.4.3   Cage inclusion

Cage inclusion, as opposed to exclusion, where predators are added to cages, has also been used to quantify coccinellid impact in the field. For example, Cudjoe et al. (1992) used selective cage inclusion and exclusion to quantify the relative impacts of the indigenous predator *Parexochomus troberti* and the exotic parasitoid *Anagyrus lopezi* on colonies of the cassava mealybug, *Phenacoccus manihoti*, in Ghana. Butin et al. (2003) demonstrated the proportionally greater impact on *Adelges tsugae* of *Scymnus ningshanensis* compared to *Sasajiscymnus (= Pseudoscymnus) tsugae* by caging colonies of the adelgid in the field with and without each of the two predators. Such studies generate inferences on potential coccinellid impact derived from relative comparisons of species-specific **functional and numerical responses** in a closed system, but they have few advantages over laboratory studies because they ignore searching efficiency and exclude important factors such as emigration and intraguild predation. However, they may be useful for assessing impact potential when the provision of prey is difficult to achieve in the laboratory, or when particular natural enemy interactions are of interest. For example, Straub and Snyder (2006) infested potato plants with *M. persicae* in field cages and then inoculated them with various predator combinations.

10.4.4 Manual removal

One alternative to exclusion cages is the manual removal of immature coccinellid life stages from a fraction of infested host plants to compare the subsequent survival and growth of prey colonies with and without predator removal. This technique was used effectively by Berthiaume et al. (2000) to demonstrate the impact of *Anatis mali* on *Mindarus abietinus* infesting balsam fir (*Abies balsamea*) in Christmas tree plantations. The feasibility of this approach may depend to some degree on host plant architecture and the ease of discovery of egg masses, but it has the advantage of high selectivity and not interfering with secondary sources of prey mortality. However, manual removal of immature stages will underestimate impact whenever predation by transient adults is significant. Rates of coccinellid recruitment to various host plant-prey combinations can also be assessed using **sentinel prey** on potted plants in a trap-line approach. A useful method for sampling potential parasites and predators is placing aphid colonies in the field and periodically collecting all natural enemies that arrive to attack them. Natural enemies of *Aphis fabae* were thus sampled in Germany (Müller 1966) and of *Schizaphis graminum* in the Czech

Republic (Starý & Gonzalez 1992). Although impact is not quantified with this technique, it is effective for assessing functional responses of coccinellid species and measuring the **speed of prey patch discovery**. Michaud and Belliure (2000) infested young citrus trees with various numbers of *T. citricidus* and revisited them daily for a period of weeks to tally the growth trajectories of aphid colonies and rates of natural enemy recruitment. Similarly, Noma et al. (2005) placed pots of wheat infested with *D. noxia* in a series of field locations for 2–7 day periods and then collected them to assess predation and parasitism. Although this approach was more effective for assessing parasitism, predatory larvae were also recovered. This approach could be used to estimate rates of prey discovery by coccinellids, but would more require frequent observations to detect transient visitation, or the use of trap plants with sufficiently large numbers of prey to stimulate oviposition by adult females.

10.5 STATISTICAL APPROACHES

Chambers and Aikman (1988) discussed the advantages and limitations of four different **analytical procedures** for estimating the extent to which predation was responsible for observed changes in aphid densities. Various regression techniques are often used to analyze patterns of coccinellid abundance in relation to that of their prey, both spatially and temporally (e.g. Rautapaa 1972, Elliott et al. 1999, Kriz et al. 2006). **Regression models** that incorporate plant growth stage and counts of predators and prey can be used to forecast whether the pest's economic injury threshold will be surpassed based on a current data sample from the field. This approach can be of practical value for guiding management decisions and providing indications of the collective impact of coccinellid assemblages on their prey at particular stages of crop growth or season. The limitation of such correlational analyses is that causal relationships are not necessarily implied and no information is generated with respect to underlying mechanisms.

It is possible to construct **life tables** for herbivores that partition mortality by life stage and then use key factor analysis to quantify the proportion of variance in abundance that is attributable to various submortalities, including coccinellid predation (Manly 1990, Dent 1997). This approach has not been used extensively to assess coccinellid impact, possibly because actual rates of prey removal can be difficult to assess and attribute to particular species, and without reliable estimates in this regard, the approach has little advantage over other regression techniques. Podoler et al. (1979) constructed **partial life tables** for black olive scale, *Saissetia oleae*, that revealed *Chil. bipustulatus* to be a key mortality factor for immature stages. Similarly, Hamilton et al. (1987) used a life table approach to infer significant impact of coccinellid predation on numbers of *S. graminum* on grain sorghum in central Missouri, USA. Life tables constructed for cereal leaf beetle, *Oulema melanopus*, (Shade et al. 1970) and citrus black fly, *Aleurocanthus woglumi*, (Cherry & Dowell 1979) also implicated coccinellid predation as key mortality factors for these pests.

10.6   RESOLVING COCCINELLID IMPACT WITHIN COMPLEX COMMUNITIES

Since coccinellids may serve as either **intraguild predators** or intraguild prey (Chapter 7), interactions between coccinellid species, and between coccinellids and other natural enemy taxa, may serve to either enhance or diminish herbivore suppression. For example, coccinellids often consume aphid parasitoids in their larval form (5.2.7) and this has been construed to impede, rather than assist, aphid control in systems where parasitism is inferred to be a key source of mortality (Colfer & Rosenheim 1995, Ferguson & Stiling 1996). In some cases, intraguild predation may have little or no effect on herbivore suppression (i.e. Snyder & Ives 2003) whereas in others, the role of coccinellids as key predators can more than counter their negative impact as intraguild predators of parasitoids (Michaud 2004, Gardiner & Landis 2007).

Tamaki and Weeks (1972) showed that the addition of *C. transversoguttata* improved control of *M. persicae* by predatory Hemiptera (*Geocoris bullatus* and *Nabis americoferus*) on caged sugar beet plants in a greenhouse, one of the first illustrations of how coccinellid predation can complement other biotic sources of mortality (Fig. 10.4). It has been pointed out that coccinellids are mobile predators that respond well to prey aggregations, whereas more **sedentary, resident predators** such as earwigs (*Forficula* spp.) are more likely to respond to aphid colonies early in the establishment phase when the proportional impact of

predation is much higher (Piñol et al. 2009). The latter authors used simulation modelling to explore the interactions of significant parameters affecting aphid population dynamics in a citrus grove that could give rise to the patterns of abundance observed.

10.6.1   Multi-species combinations

Many recent studies have sought to determine whether other multi-species combinations including one or more coccinellids either facilitate or impede the suppression of prey populations, and have yielded a wide range of results. Negative effects can arise via intraguild predation or foraging interference, whereas positive effects can be generated by spatial or temporal **niche partitioning** among predatory species. These include the differential use of prey life stages, preferences for foraging on different plant parts, or variation in diurnal or seasonal activity. Losey and Denno (1998b) used a combination of laboratory and field studies to show a synergistic impact of *C. septempunctata* and the carabid *Harpalus pennsylvanicus* on pea aphids that was double the sum of their individual predation impacts. Similarly, Cardinale et al. (2003) demonstrated a 'superadditive' effect of multiple natural enemies (*Har. axyridis*, *Nabis* sp. and *Aphidius ervi*) on populations of *A. pisum* caged on alfalfa. In contrast, Cardinale et al. (2006) inferred interference interactions among adult *Har. axyridis*, *C. septempunctata* and *Col. maculata* based on a lower level of aphid suppression when all three were present together than would be expected based on the sum of their effects when foraging alone. Straub and Snyder (2008) manipulated natural enemy diversity (*Aphidius matricariae*, *C. septempunctata*, *Hip. convergens* and *Nabis* sp.) in field cages of potatoes and collards infested with *M. persicae*. Their findings suggested that the identity of species within predator communities was more important for aphid control than species richness, per se, and that the coccinellid component (*C. septempunctata* and *C. transversoguttata*) had the greatest impact on aphid numbers. The authors concluded that increased natural enemy diversity **strengthened aphid suppression** because intraspecific competition was stronger than interspecific competition among predator species due to resource partitioning. Snyder (2009) reviewed predator diversity studies that included one or more coccinellid species and concluded that positive effects of increas-

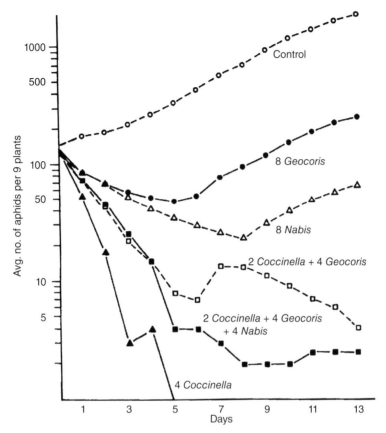

Figure 10.4 Effectiveness of three predators, used alone or in combination, against populations of the green peach aphid (after Tamaki and Weeks 1972).

ing predator diversity were generally more common than negative ones. Thus, coccinellids are emerging as potentially key predators that often act within a larger guild of natural enemies, the full complement of which may be required to provide acceptable levels of pest suppression.

10.6.2 Demographic-based estimates

In an effort to resolve the issue of aphid reproduction confounding the rate of removal by predators, Latham and Mills (2010) compared what they termed 'demographic-based' versus 'observation-based' estimates of **mealy plum aphid predation** in California. The latter represents a truly novel approach to

assessing predator impact, albeit one requiring a large investment in baseline observations. Periodic counts of aphids and predators throughout the season were used to estimate aphid abundance and predator densities, by species and life stage, that were then expressed in terms of leaf area on particular dates. Consumption rates were estimated for larval and adult predators of each species based on direct observation. Since the predators were diurnal (with the exception of adult *Chrysopa nigricornis*), their period of activity was assumed to correspond to actual day length. The proportion of time spent feeding was estimated from a series of point observations in the field, and the time required to consume an aphid estimated from laboratory observations. Collectively, this information was used to generate an estimate of aphid predation per unit of leaf area on all sampling dates that could be compared to the demographic estimate of aphid population dynamics. The latter was obtained from a simple model of aphid population growth using estimates of intrinsic rate of increase obtained from field cages from which predators were excluded. Despite the numerous sources of error inherent in this approach (thoroughly enumerated by the authors), the study confirmed an intuitive inference; the guild of predators (including *Har. axyridis*) were neither numerous enough nor voracious enough to effect directional changes in aphid population dynamics, nor suppress the aphids below economically damaging levels, although they were able to influence the rate of population growth or decline at some points in the cycle.

Our knowledge of coccinellid nutritional ecology has been greatly advanced by laboratory feeding studies that have tested the suitability of prey species for development and reproduction (5.2). However, our understanding of why coccinellids are and are not effective in suppressing prey populations would benefit from more complete knowledge of the full range of foods that species actually utilize under natural conditions (5.2.5, 5.2.9). Direct observation yields only fragmentary information in this regard, and one alternative is to dissect specimens and examine gut contents (Ricci et al. 1983, Mendel et al. 1985, Triltsch 1997; 5.2.1) or attempt to identify prey fragments in frass (Davidson & Evans 2010). Fortunately, newly developed molecular techniques have considerably enhanced our ability to identify prey types within gut contents beyond simple visual recognition of body parts. Since these techniques tend to yield only qualitative information, their value in aiding determinations of coccinellid impact on herbivores lies in complementing other techniques that yield more quantitative measures.

10.7 POST-MORTEM ANALYSIS OF PREDATION

Techniques for elucidating predator–prey interactions in the field have received considerable attention in recent years (reviewed by Symondson 2002, Sheppard & Harwood 2005, Harwood & Greenstone 2008, Weber & Lundgren 2009). These reviews have, to some extent, superseded earlier assessments of post-mortem techniques (e.g. Kiritani & Dempster 1973, Boreham & Ohiagu 1978, Washino & Tempelis 1983, Sunderland 1988, 1996, Greenstone 1996) due to the rapidly

advancing nature of molecular biology and biochemistry. However, the over-riding consensus has been that post-mortem analyses offer valuable insights, and sometimes advantages, for studying the ecology and impact of predator communities (including coccinellids) in complex environments. Despite a proliferation in the use of molecular approaches in ecology, the extent to which predation can be 'quantified' by such techniques remains controversial.

10.7.1 Antibody-based analysis of predation

Vertebrate antibodies have been used since the 1940s to qualify the presence (or absence) of target prey in predator guts (Brooke & Proske 1946). Brooke and Proske (1946) developed a **polyclonal antibody** in rabbits to identify trophic relationships between *Anopheles quadrimaculatus* and its natural enemies using a **precipitin test**. Although the specificity of the assay was too general to enable species-level elucidation of trophic relationships (the antiserum was reported to be 'order-specific'), this study provided the groundwork for the subsequent adoption of serology as the tool of choice for predator–prey research throughout the latter half of the twentieth century. Subsequent studies documented predation upon a diverse spectrum of prey with specificity reported at the level of order, family, genus and species (see Greenstone 1996 for a review). However, despite the plethora of studies diagnosing predation in the field using polyclonal antisera, it is surprising how few have focused on food relations of the Coccinellidae. In contrast, considerable attention has been directed towards examining ecological interactions between the Carabidae, Araneae and Hemiptera and their prey (see Boreham & Ohiagu 1978, Sunderland 1988, 1996, Greenstone 1996). **Enzyme-linked immunosorbent assay (ELISA)** technology (Fig. 10.5) relies of the binding of prey-specific antibodies to target antigen and subsequent visualization of the reaction using a spectrophotometer to signify the presence (or absence) of prey material within each ELISA plate well. The ELISA format offers a rapid, cost-effective and widely adopted approach for food web studies.

Early serological laboratory experiments illustrated their utility for evaluating coccinellid predation (Pettersson 1972), and Whalon and Parker (1978) demonstrated that a **polyclonal antiserum** developed in

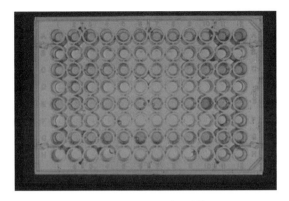

Figure 10.5 A 96-well microtitre plate following enzyme-linked immunosorbent assay. The amount of antigen–antibody binding is signified by the intensity of the reaction; absorbance is typically monitored by spectrophotometer to infer qualitative (and occasionally semi-quantitative) assessments of predation (photo: James Harwood). (See colour plate.)

rabbits could detect *Lygus lineolaris* proteins in *Col. maculata* for 42 hours after consumption. However, it was the study of Vickerman and Sunderland (1975) that provided the first insight into coccinellid feeding behaviour in the field. Traditionally, the adoption of gut dissection relied on the successful identification of indigestible remains of prey in coccinellid guts (Forbes 1883, Eastop & Pope 1969, Ricci 1986, Ricci & Ponti 2005, Ricci et al. 2005) or faecal samples (Honěk 1986), but Vickerman and Sunderland (1975) screened over 200 species of predators, including four coccinellids, collected by sweep netting and vacuum sampling in spring barley fields in the United Kingdom. High levels of predation upon *Metopolophium festucae* and/or *Sitobion avenae* were documented using a precipitin test (Fig. 10.6). The use of antisera to confirm consumption of specific prey types is thus a potentially useful tool for complementing direct observations of predation in studies that aim to assess coccinellid

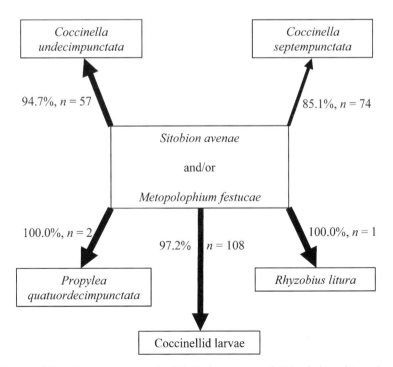

Figure 10.6 Adult coccinellid predation on the cereal aphids *Sitobion avenae* and *Metopolophium festucae* (summarized from Vickerman & Sunderland 1975). Size of arrow corresponds to strength of trophic pathway and the numeric value represents the percentage of predators screening positive for target prey.

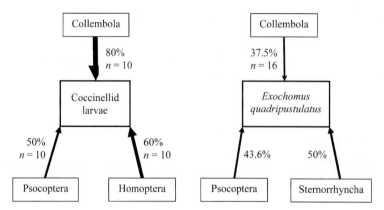

Figure 10.7 Adult and larval coccinellid predation on Homoptera, Collembola and Psocoptera (summarized from Turner 1984). Size of arrow corresponds to strength of trophic pathway and the numeric value represents the percentage of predators screening positive for target prey.

impact relative to other generalist predators, especially when these may be species that are nocturnal, cryptic or otherwise difficult to observe directly. For example, where estimates of predator abundance are used in models of herbivore suppression, these could be adjusted by an amount proportional to the percentage of each species testing positive for the focal prey.

Using a **modified Ouchterlony plate approach** (after Pickavance 1970), Turner (1984) was able to distinguish the consumption of insect prey from three orders (Psocoptera, Sternorrhyncha and Collembola) from the consumption of predatory arthropods from eight orders, including three coccinellid species (Fig. 10.7). In the same year, Leathwick and Winterbourn (1984) used the Ouchterlony approach to develop an antiserum targeting *Acyrthosiphon kondoi* and *A. pisum* from alfalfa in New Zealand, and reported that 70% of *C. undecimpunctata* tested positive for these aphids (*n* = 73). Although no cross reactivity was observed with *Sidnia kinbergi*, *Calocoris norvegicus* or *Philaenus spumarius*, there was a weak reaction to another hemipteran, *Nysius huttoni*. The characterization process of polyclonal and monoclonal antibodies typically requires extensive screening of the developed antibody against non-target organisms due to the possibility for cross-reactivity (whereby the 'pest-specific' antibody binds to the antigens of other invertebrates). Thus, the possibility of overestimating predation with this technique exists, especially when few non-target species are screened during characterization.

Hagley and Allen (1990) adopted an **immunoelectro-osmophoresis** approach on cellulose acetate membranes to examine the food relations of generalist predators of the green apple aphid, *Aphis pomi*, in apple orchards in Ontario. Although no cross-reactivity testing was reported and replication of field-collected specimens was low (*n* = 168), evidence was obtained of the importance of coccinellids as predators of the green apple aphid relative to other specialist and generalist aphid predators (Fig. 10.8). Such data can complement more conventional field-based studies designed to assess coccinellid effectiveness in pest suppression.

The principal limitation of polyclonal antisera has been an inherent lack of specificity that derives from their mode of production (simple immunization of a vertebrate animal, typically a rabbit, to produce polyclonal antisera targeting determinants common to many organisms). Therefore, potential 'false-positive' reactions generate interpretative errors that can lead to an overestimation of interaction pathway strength. Occasionally, cross-reacting material can be reduced through purification. For example, affinity chromatography can employ cross-reacting antibodies to bind the target antigen so that non-reacting material can be eluted from the column (Cuatrecasas 1970, Schoof et al. 1986). However, levels of sensitivity are sometimes compromised. Although species-specific (Dempster 1971, Pettersson 1972, Sunderland & Sutton 1980, Nemoto et al. 1985) and even stage-specific (Ragsdale

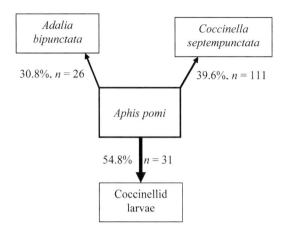

Figure 10.8 Adult (*Adalia bipunctata* and *Coccinella septempunctata*) and larval coccinellid predation on *Aphis pomi* in Ontario apple orchards (summarized from Hagley & Allen 1990). Size of arrow corresponds to strength of trophic pathway and the numeric value represents the percentage of predators screening positive for target prey.

et al. 1981) diagnostic systems have been developed using a polyclonal approach, the problems with specificity led to the proliferation of monoclonal antibody technology in food web biology during the 1990s. This approach relies on the immunization of a vertebrate animal, typically a mouse, with target material, and the subsequent extraction of lymphocytes that produce a hybridoma following fusion with myeloma cells. The hybridomas are cloned and single cells can be cultured *in vitro* to produce monoclonal antibodies of desired specificity (Köhler & Milstein 1975). Monoclonal antibodies have been used to target pests such as *Helicoverpa zea*, *Lygus hesperus*, *Bemisia tabaci*, *Pectinophora gossypiella*, *Heliothis virescens*, the slugs *Deroceras reticulatum*, *Arion hortensis* and *Tandonia budapestensis*, and aphids (reviewed by Greenstone 1996, Symondson 2002, Sheppard & Harwood 2005, Weber & Lundgren 2009). However, high initial development costs (Fournier et al. 2008, Harwood & Greenstone 2008) have driven monoclonal antibody-based research in the direction of pest management assessment and the development of pest-specific monoclonals. Many predators, including most coccinellids, have the ability to subsist on alternative prey types prior to colonizing pest-infested plants, with potentially important consequences for their physiological condition and the timing of their immigration into crop habitats. Thus,

the ability to recognize such consumption through the use of specific monoclonal antibodies could supplement our understanding of coccinellid effectiveness, or lack thereof, in particular contexts.

Large-scale field trials have been implemented to identify patterns of food web processes and have, in some cases, purported to quantify the role of natural enemies in biological control. In the largest study of its kind, Hagler and Naranjo (2005) utilized a whitefly-specific ELISA to assay 32,262 field-collected predators and assess the effect of insecticide regimes on the predator complex associated with *Bemisia tabaci*. It is perhaps surprising that such studies have yet to address role of coccinellids in agroecosystems, given the prior development of an aphid-specific monoclonal antibody.

Early-season predation has long been considered important in pest regulation by generalist predators (Settle et al. 1996, Landis & van der Werf 1997). Using an aphid-specific monoclonal antibody, Harwood et al. (2004) revealed that spider communities can impact aphids before their populations increase. By correlating the falling rates of aphids (and alternative prey) into webs with their consumption by the spiders, disproportionately high rates of feeding on *Sitobion avenae* were observed precisely when numbers were at their lowest. Despite the obvious power of monoclonal antibody technology, and its ability to yield somewhat quantitative inferences with respect to prey consumption, it has not yet been employed to any significant extent in studies addressing the impact of coccinellids on their prey. Hagler and Naranjo (1994) used monoclonal antibodies for *B. tabaci* and *P. gossypiella* to identify these species in the guts of 613 *Hip. convergens* adults, the only study to date that has used monoclonals to study coccinellid prey consumption. Whereas only 1.0% of *Hip. convergens* screened positive for eggs of pink bollworm, 38.0% contained whitefly, albeit with significant variation evident throughout the season, probably due to temporal variation in the abundance of whitefly and other (potentially preferred) prey.

10.7.2 Protein marking for predation analysis

Immunolabelling has been used for many years as a tool for marking insects and tracking dispersal (Hagler & Jackson 2001), but has recently been used to analyze predation (e.g. Hagler & Naranjo 2004, Hagler 2006).

This approach employs immunoglobulin labels that are applied either internally or externally to prey prior to their consumption, and then detected and visualized within the gut of a predator. A recent field assessment comparing the efficacy of protein labelling and monoclonal antibody-based ELISA (Mansfield et al. 2008) suggested superior levels of detection in the former. An extended period of detection was attributed to a longer persistence of rabbit protein markers in the predators (*Dicranolaius bellulus*, *Hip. variegata*, *C. leonina transversalis* and *Cheiracanthium* spp.) compared to antigen from *Helicoverpa armigera*. Post-consumption detection periods are critical in gut content analyses and can vary from a few hours to more than a week. Thus, immunolabelling incurs a compromise between the increased likelihood of detection over an extended period and the possibility of overestimating feeding events when prey remain detectable in guts for longer periods.

The frequency with which adult *Hip. convergens* fed on *L. hesperus*, *P. gossypiella* and *Trichoplusia ni* eggs was elucidated under varying light regimes in a field-cage system where the three pests were released after immunolabelling with rabbit immunoglobulin G (*T. ni*) or chicken immunoglobulin G (*L. hesperus*) (Hagler 2006). Consumption of *P. gossypiella* was detected with a *P. gossypiella* egg-specific monoclonal antibody. Predation on *T. ni* in the field occurred with the greatest frequency, with the peak proportion testing positive (~16%) between 0700–1300 hours, and lower frequencies of predation during the period before and after dusk (~11%) (3 hours light followed by 3 hours dark) and during the night (~11%), with little evidence of predation during the period before and immediately after dawn (3 hours dark followed by 3 hours light). Very low predation on *L. hesperus* and *P. gossypiella* was documented (<2% positive for prey proteins). Therefore, this technique can be used to discern not only **quantitative patterns** of coccinellid prey selection when multiple species are present, but also **diurnal cycles** of prey consumption. Although the low cost and ease of application of protein labels is appealing, the requirement for prey labelling and subsequent release into field plots for consumption by predators limits its utility and renders it impractical for assessing the consumption of prey that are naturally distributed in open-field conditions.

10.7.3   Detection of prey-specific DNA

Buoyed by increased access to molecular laboratories and equipment, ecologists have increasingly utilized **DNA-based gut content analysis** to examine the feeding relationships of terrestrial arthropods. This approach relies on the ability to visualize unique fragments of DNA in predator guts, typically the mitochondrial cytochrome oxidase I (COI) gene (Fig. 10.9),

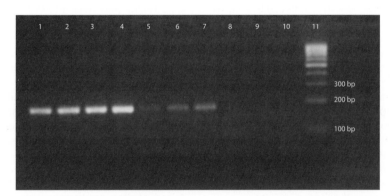

Figure 10.9 Agarose gel electrophoresis of PCR-amplified DNA. The presence of bands signifies the amplification of 'prey' DNA in predator samples, thus indicating the presence of prey in predator guts. This data can be used to identify the presence (or absence) of prey material and therefore infer the frequency of predation events (and thus impact of predators) in the field. Lanes 1–2, positive controls (target prey DNA) to ensure visualization of DNA on gel; lanes 3–7 field-collected predators screening positive for prey DNA; lanes 8–9, starved predators screening negative for prey DNA; lane 10, negative control; lane 11, 100 bp ladder to separate fragment sizes. Photograph courtesy of Eric G. Chapman (University of Kentucky).

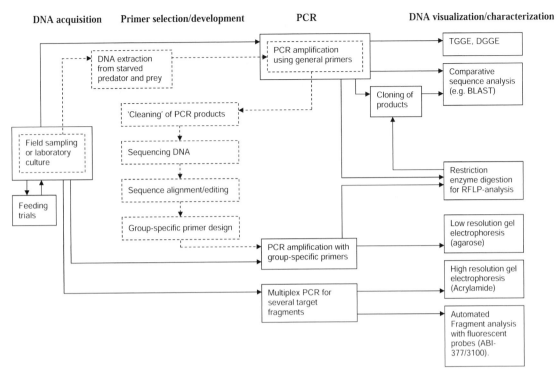

Figure 10.10 Decision flow diagram for the development of a gut-content analysis approach for detection of predation by PCR (from Sheppard & Harwood (2005), courtesy of Wiley-Blackwell).

although other targets have also been used. Traditionally, the approach was limited by an inability to detect DNA beyond a **short period after ingestion** (Asahida et al. 1997), thus reducing the likelihood of detecting foraging events under natural conditions. However, a reduction in target fragment size has enabled extended detection in both gut contents (Zaidi et al. 1999, Chen et al. 2000) and faecal material (Kohn & Wayne 1997, Farrell et al. 2000). For example, Zaidi and colleagues (1999) demonstrated that *Culex quinquefasciatus* DNA could be reliably detected in guts of *Pterostichus melanarius* up to 28 hours after digestion. Although the period of delectability is shorter than that of some monoclonal antibody-based assays, shorter detection periods may be advantageous when only the most recent foraging behaviour is of interest (Sheppard & Harwood 2005).

Although the details are beyond the scope of this chapter, DNA-based techniques utilize a series of protocols that enable the acquisition of DNA, the design

of molecular markers, the choice of a mechanism for amplifying target DNA, and the final visualization of results (Fig. 10.10). Although considerable care is required during the development and optimization of molecular detection systems, the relative ease of sequencing DNA and designing prey-specific primers has allowed cost-effective identification of a broad spectrum of food resources exploited by generalist predators. This advantage has been further enhanced by the **adoption of multiplex PCR** (Fig. 10.11) which amplifies prey DNA in a single reaction (e.g., Harper et al. 2005). The identification of diverse alternative prey is usually beyond the scope of monoclonal antibody-based analyses due to prohibitively high development costs, although these can be greatly offset in large-scale studies (e.g. Hagler & Naranjo 2005) due to the low per-sample screening cost of ELISA. Since PCR costs 15 to 20 times more than ELISA to screen each predator (Fournier et al. 2008, Harwood & Greenstone 2008), the trade-off between development cost

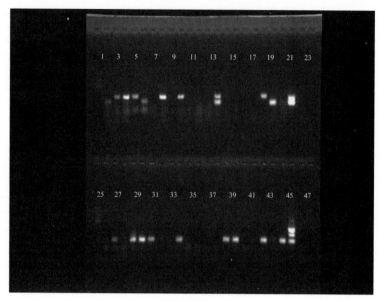

Figure 10.11 Agarose gel electrophoresis of multiplex-PCR-amplified DNA. Fluorescently labelled PCR primers allow the simultaneous amplification of prey DNA from within predator guts based upon variation in target fragment (e.g. lanes 5 and 6). Such technology even allows for the simultaneous amplification of multiple prey within a single sample (e.g. lane 14). Photograph courtesy of R. Andrew King (Cardiff University).

and screening cost influences the relative suitability of these two techniques for particular applications.

In the most comprehensive molecular gut-content study to date, Fournier et al. (2008) screened over 1200 predators (including *Hip. convergens* and *Har. axyridis*) by PCR against species-specific primers of the glassy-winged sharpshooter *Homalodisca vitripennis*. Although laboratory feeding trials revealed 100% successful detection of sharpshooter DNA for 8 hours after consumption, no field-collected coccinellids tested positive for the target prey, in contrast to spiders (18%), hemipterans (14%), mantids (13%), ants (12%) and neuropterans (10%). Similarly, very low levels of coccinellid predation on *Ostrinia nubilalis* were reported by Hoogendoorn and Heimpel (2002). Targeting the ITS1/18S region of nuclear DNA, the decline in detectability of the DNA of prey in predator guts followed a negative quadratic function over 12 hours and just 1% of field-collected *Col. maculata* and *Har. axyridis* were positive for these prey in a corn agroecosystem (Fig. 10.12).

Recently, Chacón et al. (2008) examined the food relations of four coccinellids along with *Chrysoperla carnea* and *Orius insidiosus* that were associated with the soybean aphid, *Aphis glycines*. Molecular gut-content analyses have providing valuable insights into the role of particular natural enemies in controlling this pest (Harwood et al. 2007, 2009, Chacón et al. 2008). For example, greater than 90% of *O. insidiosus*, a species previously reported to be an important predator of this aphid (Harwood et al. 2007, 2009), screened positive for soybean aphids, in contrast with previous reports that had used different molecular markers. Similarly, Zhang et al. (2007a) qualified *Bemisia tabaci* predation rates by a complex of predators, including Coccinellidae, in Chinese cotton agroecosystems (Fig. 10.12). We expect that molecular gut-content analyses will play an increasingly important role in evaluating food web dynamics and will serve to complement quantitative population monitoring and traditional field experiments in various ways to significantly advance our understanding of the relative

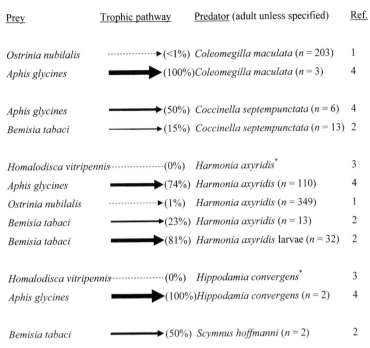

Prey	Trophic pathway	Predator (adult unless specified)	Ref.
Ostrinia nubilalis	(<1%)	*Coleomegilla maculata* (*n* = 203)	1
Aphis glycines	(100%)	*Coleomegilla maculata* (*n* = 3)	4
Aphis glycines	(50%)	*Coccinella septempunctata* (*n* = 6)	4
Bemisia tabaci	(15%)	*Coccinella septempunctata* (*n* = 13)	2
Homalodisca vitripennis	(0%)	*Harmonia axyridis*[*]	3
Aphis glycines	(74%)	*Harmonia axyridis* (*n* = 110)	4
Ostrinia nubilalis	(1%)	*Harmonia axyridis* (*n* = 349)	1
Bemisia tabaci	(23%)	*Harmonia axyridis* (*n* = 13)	2
Bemisia tabaci	(81%)	*Harmonia axyridis* larvae (*n* = 32)	2
Homalodisca vitripennis	(0%)	*Hippodamia convergens*[*]	3
Aphis glycines	(100%)	*Hippodamia convergens* (*n* = 2)	4
Bemisia tabaci	(50%)	*Scymnus hoffmanni* (*n* = 2)	2

Figure 10.12 Frequency of coccinellid predation in the field, revealed by molecular gut content analysis summarized from: 1, Hoogendoorn and Heimpel (2002); 2, Zhang et al. (2007a); 3, Fournier et al. (2008); 4, Chacón et al. (2008). Size of arrow corresponds to strength of trophic pathway and the numeric value represents the percentage of predators screening positive for target prey. *, exact sample size unspecified.

effectiveness of coccinellids as biological control agents in various ecological contexts.

10.8 CONCLUSIONS

Laboratory feeding studies remain a useful approach for determining prey acceptability and suitability for coccinellids and consumption assays produce estimates of potential impact, but the strength of laboratory studies usually lies in providing a base for interpreting field observations. Traditional methods of assessing the impact of coccinellids on focal prey species continue to be invaluable and the best insights are usually obtained through a combination of approaches selected for their suitability for particular prey–plant systems, as each has inherent limitations. Although molecular analyses have advanced our understanding of the trophic relations of predatory arthropods, including the Coccinellidae, they have yet to yield data that accurately **quantify predation** in the field. The better studies have linked population surveys with molecular gut-content analysis to strengthen inferences of trophic relationships and mechanisms of prey preference, as opposed to simply generating binary data on the presence/absence of target prey in predator guts. Semi-quantitative estimates of predation have been made using gut-content analysis by estimating the concentration of recognizable prey proteins in gut samples by ELISA (e.g. Symondson & Liddell 1993, Symondson et al. 2000, Harwood et al. 2004, Winder et al. 2005) or of prey DNA by quantitative PCR (Zhang et al. 2007b, Lundgren et al., 2009). Such studies provide valuable insights into the relative amount of recognizable material in the predator at a given time, but do not offer 'quantitative' estimates of predation because the number (and/or biomass) of prey consumed and the time elapsed since consumption cannot be inferred. Thus, although molecular gut-content analyses offer particular benefits for elucidating trophic interactions, they remain fraught with interpretive sources of error. Nevertheless, when used in combination with quantitative data from field experiments, they can generate unique insights that cannot be obtained with other approaches.

REFERENCES

Annecke, D. P., M. Karny and W. A. Burger. 1969. Improved biological control of the prickly pear, *Opuntia megacantha* Salm-Dyck, in South Africa through the use of an insecticide. *Phytophylactica*. 1: 9–13.

Asahida, T., Y. Yamashita and T. Kabayashi. 1997. Identification of consumed stone flounder *Kareius bicoloratus* (Basilewsky) from the stomach contents of sand shrimp, *Crangon affinis* (De Hahn) using mitochondrial DNA analysis. *J. Exp. Mar. Biol. Ecol.* 217: 153–163.

Atlhan, R. and M. S. Ozgokce. 2002. Development, fecundity and prey consumption of *Exochomus nigromaculatus* feeding on *Hyalopterus pruni*. *Phytoparasitica*. 30: 443–450.

Bankowska, R., W. Mikolajczyk, J. Palmowska and P. Trojan. 1978. Aphid–aphidophage community in alfalfa cultures (*Medicago sativa* L.) in Poland. Part 3. Abundance regulation of *Acyrthosiphum pisum* (Harr.) in a chain of ligophagous predators. *Ann. Zool., Warsaw.* 34: 39–77.

Berthiaume, R., C. Hebert and C. Cloutier. 2000. Predation on *Mindarus abietinus* infesting balsam fir grown as Christmas trees: the impact of coccinellid larval predation with emphasis on *Anatis mali*. *BioControl*. 45: 425–438.

Boreham, P. F. L. and C. E. Ohiagu. 1978. The use of serology in evaluating invertebrate predator–prey relationships: a review. *Bull. Entomol. Res.* 68: 171–194.

Brooke, M. M. and H. O. Proske. 1946. Precipitin test for determining natural insect predators of immature mosquitoes. *J. Natl Malaria Soc.* 5: 45–56.

Butin, E., J. Elkinton, N. Havil and M. Mongomery. 2003. Comparison of numerical response and predation effects of two coccinellid species on hemlock woolly adelgid (Homoptera: Adelgidae). *J. Econ. Entomol.* 96: 763–767.

Cannon, R. J. C. 1986. Summer populations of the cereal aphid *Metopolophium dirhodum* (Walker) on winter wheat – three contrasting years. *J. Appl. Ecol.* 23: 101–114.

Cardinale, B. J., C. T. Harvey, K. Gross and A. R. Ives. 2003. Biodiversity and biocontrol: emergent impacts of a multi-enemy assemblage on pest suppression and crop yield in an agroecosystem. *Eco. Lett.* 6: 857–865.

Cardinale, B. J., J. J. Wies, A. E. Forbes, K. J. Tilmon and A. R. Ives. 2006. Biodiversity as both a cause and consequence of resource availability: a study of reciprocal causality in a predator–prey system. *J. Anim. Ecol.* 75: 497–505.

Chacón, J. M., D. A. Landis and G. E. Heimpel. 2008. Potential for biotic interference of a classical biological control agent of the soybean aphid. *Biol. Control*. 46: 216–225.

Chambers, R. J. and D. P. Aikman. 1988. Quantifying the effects of predators on aphid populations. *Entomol. Exp. Appl.* 46: 257–265.

Chambers, R. J., K. D. Sunderland, I. J. Wyatt and G. P. Vickerman. 1983. The effects of predator exclusion and caging on cereal aphids in winter wheat. *J. Appl. Ecol.* 20: 209–224.

Chen, Y., K. L. Giles, M. E. Payton and M. H. Greenstone. 2000. Identifying key cereal aphid predators by molecular gut analysis. *Mol. Ecol.* 9: 1887–1898.

Cherry, R. and R. V. Dowell. 1979. Predators of citrus blackfly (Hom.: Aleyrodidae). *Entomophaga*. 24: 385–391.

Clark, T. L. and E. J. Messina. 1998. Plant architecture and the foraging success of ladybird beetles attacking the Russian wheat aphid. *Entomol. Exp. Appl.* 86: 153–161.

Colfer, R. F. and J. Rosenheim. 1995. Intraguild predation by coccinellid beetles on an aphid parasitoid *Lysiphlebus testaceipes*. *Proc. Beltwide Cotton Conf., San Antonio, January 1995* 2. National Cotton Council, Memphis. pp. 1033–1036.

Corlay, F., G. Boivin and G. Belair. 2007. Efficiency of natural enemies against the swede midge, *Contarinia nasturtii* (Diptera; Cecidomyiidae), a new invasive species in North America. *Biol. Control*. 43: 195–201.

Costamagna, A. C. and D. A. Landis. 2006. Predators exert top-down control of soybean aphid across a gradient of agricultural systems. *Ecol. Appl.* 16: 1619–1628.

Costamagna, A. C. and D. A. Landis. 2007. Quantifying predation on soybean aphid through direct field observations. *Biol. Control*. 42: 16–24.

Costamagna, A. C., D. A. Landis and M. J. Brewer. 2008. The role of natural enemy guilds in *Aphis glycines* suppression. *Biol. Control*. 45: 368–379.

Cuatrecasas, P. 1970. Protein purification by affinity chromatography. Derivatizations of agarose and polyacrylamide beads. *J. Biol. Chem.* 245: 3059–3065.

Cudjoe, A. R., P. Neuenschwander and M. J. W. Copland. 1992. Experimental determination of the efficiency of indigenous and exotic natural enemies of the cassava mealybug, *Phenacoccus manihoti* Mat.-Ferr. (Hom., Pseudococcidae), in Ghana. *J. Appl. Entomol.* 114: 77–82.

Davidson, L. N. and E. W. Evans. 2010. Frass analysis of diets of aphidophagous lady beetles (Coleoptera: Coccinellidae) in Utah alfalfa fields. *Environ. Entomol.* 39: 576–582.

Day, K. R., M. Docherty, S. R. Leather and N. A. C. Kidd. 2006. The role of generalist insect predators and pathogens in suppressing green spruce aphid populations through direct mortality and mediation of aphid dropping behavior. *Biol. Control*. 38: 233–246.

Dempster, J. P. 1971. The population ecology of the cinnabar moth, *Tyria jacobeae* L. (Lepidoptera, Arctiidae). *Oecologia.*. 7: 26–67.

Dent, D. R. 1997. Quantifying insect populations: estimates and parameters. *In* D. R. Dent and M. P. Walton (eds). *Methods in Ecological and Agricultural Entomology.* CAB International, Wallingford, UK. pp. 57–109.

Dixon, A. F. G. 2000. *Insect Predator–Prey Dynamics: Ladybird Beetles and Biological Control.* Cambridge University Press, Cambridge.

Dixon, A. F. G. 2005. *Insect Herbivore–Host Dynamics.* Cambridge University Press, Cambridge.

Eastop, V. F. and R. D. Pope. 1969. Notes on the biology of some British Coccinellidae. *Entomologist.* 102: 162–164.

Elliott, N. C. and G. J. Michels Jr. 1997. Estimating aphidophagous coccinellid populations in alfalfa. *Biol. Control.* 8: 43–51.

Elliott, N. C., R. W. Keickhefer, J. H. Lee and B. W. French. 1999. Influence of habitat and landscape structure related factors on aphid predator populations in wheat. *Landscape Ecol.* 14: 239–252.

van Emden, H. F. 1963. A field technique for comparing the intensity of mortality factors acting on the cabbage aphid, *Brevicoryne brassicae* (L.) (Hem. Aphididae) in different areas of a crop. *Entomol. Exp. Appl.* 6: 53–62.

Farrell, L. E., J. Roman and M. E. Sunquist 2000. Dietary separation of sympatric carnivores identified by molecular analysis of scats. *Mol. Ecol.* 9: 1583–1590.

Ferguson, K. I. and P. Stiling. 1996. Non-additive effects of multiple natural enemies on aphid populations. *Oecologia.* 108: 375–379.

Forbes, S. A. 1883. The food relations of the Carabidae and Coccinellidae. *Bull. Ill. St. Lab. Nat. Hist.* 1: 33–64.

Fournier, V., J. R. Hagler, K. Daane, J. de León and R. Groves. 2008. Identifying the predator complex of *Homalodisca vitripennis* (Hemiptera: Cicadellidae): a comparative study of the efficacy of an ELISA and PCR gut content assay. *Oecologia.* 157: 629–640.

Francke, D. L., J. P. Harmon, C. T. Harvey and A. R. Ives. 2008. Pea aphid dropping behavior diminishes foraging efficiency of a predatory ladybeetle. *Entomol. Exp. Appl.* 127: 118–124.

Frazer, B. D. 1988. Predators. *In* A. K. Minks and P. Harrewijn (eds). *Aphids, their Biology, Natural Enemies and Control,* vol. 2B. Elsevier, Amsterdam. pp. 217–230.

Frazer, B. D. and N. Gilbert. 1976. Coccinellids and aphids: a quantitative study of the impact of adult ladybirds (Coleoptera: Coccinellidae) preying on field populations of pea aphids (Homoptera: Aphididae). *J. Entomol. Soc. Brit.Columbia.* 73: 33–56.

Gardiner, M. M. and D. A. Landis. 2007. Impact of intraguild predation by adult *Harmonia axyridis* (Coleoptera: Coccinellidae) on *Aphis glycines* (Hemiptera: Aphididae) biological control in cage studies. *Biol. Control.* 40: 386–395.

Gowling, G. R. and H. F. van Emden. 1994. Falling aphids enhance impact of biological control by parasitoids on partially aphid resistant plant varieties. *Ann. Appl. Biol.* 125: 233–242.

Greenstone, M. H. 1996. Serological analysis of arthropod predation: past, present and future. *In* W. O. C. Symondson and J. E. Liddell (eds) *The Ecology of Agricultural Pests: Biochemical Approaches.* Chapman and Hall, London. pp. 265–300.

Hagler, J. R. 2006. Development of an immunological technique for identifying multiple predator–prey interactions in a complex arthropod assemblage. *Ann. Appl. Biol.* 149: 153–165.

Hagler, J. R. and C. G. Jackson. 2001. Methods for marking insects: current techniques and future prospects. *Annu. Rev. Entomol.* 46: 511–543.

Hagler, J. R. and S. E. Naranjo. 1994. Qualitative survey of two coleopteran predators of *Bemisia tabaci* (Homoptera: Aleyrodidae) and *Pectinophora gossypiella* (Lepidoptera: Gelechiidae) using multiple prey gut content ELISA. *Environ. Entomol.* 23: 193–197.

Hagler, J. R. and S. E. Naranjo. 2004. A multiple ELISA system for simultaneously monitoring intercrop movement and feeding activity of mass-released insect predators. *Int. J. Pest Manage.* 50: 199–207.

Hagler, J. R. and S. E. Naranjo. 2005. Use of a gut content ELISA to detect whitefly predator feeding activity after field exposure to different insecticide treatments. *Biocont. Sci. Technol.* 15: 321–339.

Hagley, E. A. C. and W. R. Allen 1990. The green apple aphid, *Aphis pomi* Degeer (Homoptera: Aphididae), as prey of polyphagous arthropod predators in Ontario. *Can. Entomol.* 122: 1221–1228.

Hamilton, G. C., R. L. Kirkland and I. D. R. Peries. 1987. Population ecology of *Schizaphis graminum* (Rondani) (Homoptera: Aphididae) on grain sorghum in Central Missouri. *Environ. Entomol.* 11: 618–628.

Harper, G. L., R. A. King, C. S. Dodd et al. 2005. Rapid screening of predators for multiple prey DNA targets. *Mol. Ecol.* 14: 819–828.

Harwood, J. D. and M. H. Greenstone. 2008. Molecular diagnosis of natural enemy–host interactions. *In* N. Liu (ed.). *Recent Advances in Insect Physiology, Toxicology and Molecular Biology.* Research Signpost, Trivandrum, India. pp. 41–57.

Harwood, J. D., K. D. Sunderland and W. O. C. Symondson. 2004. Prey selection by linyphiid spiders: molecular tracking of the effects of alternative prey on rates of aphid consumption in the field. *Mol. Ecol.* 13: 3549–3560.

Harwood, J. D., N. Desneux, H. Y. S. Yoo et al. 2007. Tracking the role of alternative prey in soybean aphid predation by *Orius insidiosus*: a molecular approach. *Mol. Ecol.* 16: 4390–4400.

Harwood, J. D., H. J. S. Yoo, M. H. Greenstone, D. L. Rowley and R. J. O'Neil. 2009. Differential impact of adults and nymphs of a generalist predator on an exotic invasive pest demonstrated by molecular gut-content analysis. *Biol. Invasions.* 11: 895–903.

Hodek, I. 1973. *Biology of Coccinellidae,* Academia, Prague and W. Junk, The Hague, 260 pp.

Hodek, I., K. S. Hagen and H. F. van Emden. 1972. Methods for studying effectiveness off natural enemies. *In Aphid Technology,* Acad. Press, London and New York. pp. 147–188.

Honěk, A. 1986. Production of faeces in natural populations of aphidophagous coccinellids (Col.) and estimation of predation rates. *J. Appl. Entomol.* 102: 467–476.

Hoogendoorn, M. and G. E. Heimpel. 2002. PCR-based gut content analysis of insect predators: using ITS-1 fragments from prey to estimate predation frequency. *In* R. G. Van Dreische (ed.). *Proc. 1 Int. Symp. Biol. Contr. Arthropods, Honolulu, January 2002.* US Department of Agriculture, Morgantown, WV. pp. 91–97.

Hopper, K. R., S. Aidara, S. Agret et al.1995. Natural enemy impact on the abundance of *Diuraphis noxia* (Homoptera: Aphididae) in wheat in southern France. *Environ. Entomol.* 24: 402–408.

Kindlmann, P. and A. F. G. Dixon 1993. Optimal foraging in ladybird beetles (Coleoptera: Coccinellidae) and its consequences for their use in biological control. *Eur. J. Entomol.* 90: 443–450.

Kirby, R. D. and L. E. Ehler. 1977. Survival of *Hippodamia convergens* in grain sorghum. *Environ. Entomol.* 6: 777–780.

Kiritani, K. and J. P. Dempster. 1973. Different approaches to the quantitative evaluation of natural enemies. *J. Appl. Ecol.* 10: 323–330.

Klingauf, F. 1976. Die Bedeutung der 'Stimmung' im Leben phytophager Insekten am Beispiel des Wirtswahl-Verhaltens von Blattläusen. *Z. Angew. Entomol.* 82: 200–209.

Köhler, G. and C. Milstein. 1975. Continuous culture of fused cells secreting antibody of predefined specificity. *Nature.* 256: 495–497.

Kohn, H. M. and R. K. Wayne 1997. Facts from faeces revisited. *Trends Ecol. Evol.* 12: 223–227.

Kring, T. J., F. E. Gilstrap and G. J. Michels. 1985. The role of indigenous coccinellids in regulating greenbug (Homoptera: Aphididae) on Texas grain sorghum. *J. Econ. Entomol.* 78: 269–273.

Kriz, J. C., S. D. Danielson, J. R. Brandle, E. E. Blankenship and G. M. Henebry. 2006. Effects of aphid (Homoptera) abundance and surrounding vegetation on the encounter rate of Coccinellidae (Coleoptera), Chrysopidae (Neuroptera), and Nabidae (Hemiptera) in alfalfa. *J. Entomol. Sci.* 41: 211–220.

Landis, D. A. and W. van der Werf. 1997. Early-season predation impacts the establishment of aphids and spread of beet yellows virus in sugar beet. *Entomophaga.* 42: 499–516.

Latham, R. D. and N. J. Mills. 2009. Quantifying insect predation: a comparison of three methods for estimating daily per capita consumption of two aphidophagous predators. *Environ. Entomol.* 38: 1117–1125.

Latham, R. D. and N. J. Mills. 2010. Quantifying aphid predation: the mealy plum aphid, *Hyalopterus pruni* in California as a case study. *J. Appl. Entomol.* 47: 200–208.

Leathwick, D. M. and M. J. Winterbourn. 1984. Arthropod predation on aphids in a lucerne crop. *N. Z. Entomol.* 8: 75–80.

Lee, J. H., N. C. Elliott, S. D. Kindler et al. 2005. Natural enemy impact on the Russian wheat aphid in southeastern Colorado. *Environ. Entomol.* 34: 115–123.

van Lenteren, J. C. 2000. Success in biological control of arthropods by augmentation of natural enemies. *In* G. Gurr and S. Wratten (eds.) *Biological Control: Measures of Success.* Kluwer, Dordrecht. pp. 77–104.

Liao, H. T., Harris, M. K., F. E. Gilstrap and F. Mansour. 1985. Impact of natural enemies on the blackmargined pecan aphid, *Monellia caryella* (Homoptera: Aphidae). *Environ. Entomol.* 14: 122–126.

Liu, T. X., P. A. Stansly, K. A. Hoelmer, L. S. Osborne. 1997. Life history of *Nephaspis oculatus* (Coleoptera: Coccinellidae), a predator of *Bemisia argentifolii* (Homoptera: Aleyrodidae). *Ann. Entomol. Soc. Am.* 90: 776–782.

Lopez, V. F., M. T. K. Kairo and J. A. Irish. 2004. Biology and prey range of *Cryptognatha nodiceps* (Coleoptera: Coccinellidae), a potential biological control agent for the coconut scale, *Aspidiotus destructor* (Hemiptera: Diaspididae). *Biocont. Sci. Technol.* 14: 475–485.

Losey, J. E. and R. F. Denno 1998a. Interspecific variation in the escape response of aphids: effect on risk of predation from foliar-foraging and ground-foraging predators. *Oecologia.* 115: 245–252.

Losey, J. E. and R. F. Denno 1998b. Positive predator–predator interactions: enhanced predation rates and synergistic suppression of aphid populations. *Ecology.* 79: 2143–2152.

Luck, R. F., B. M. Shepard and P. E. Kenmore. 1988. Experimental methods for evaluating arthropod natural enemies. *Annu. Rev. Entomol.* 33: 367–391.

Lundgren, J. G., M. E. Ellsbury and D. A. Prischmann. 2009. Analysis of the predator community of a subterranean herbivorous insect based on polymerase chain reaction. *Ecol. Appl.* 19: 2157–2166.

Mack, T. P. and Z. Smilowitz. 1982. Using temperature-mediated functional response models to predict the impact of *Coleomegilla maculata* (DeGeer) adults and 3rd-instar larvae on green peach aphids. *Environ. Entomol.* 11: 46–52.

Manly, B. F. J. 1990. *Stage-Structured Populations, Sampling, Analysis and Simulation.* Chapman and Hall, London, UK.

Mann, J. A., G. M. Tatchell, M. J. Dupuch et al. 1995. Movement of apterous *Sitobion avenae* (Homoptera: Aphididae) in response to leaf disturbances caused by wind and rain. *Ann. Appl. Biol.* 126: 417–427.

Mansfield, S., J. R. Hagler and M. E. A. Whitehouse. 2008. A comparative study on the efficacy of a pest-specific and prey-marking enzyme-linked immunosorbent assay for detection of predation. *Entomol. Exp. Appl.* 127: 199–206.

McConnell, J. A. and T. J. Kring. 1990. Predation and dislodgement of *Schizaphis graminum* (Homoptera: Aphididae), by adult *Coccinella septempunctata* (Coleoptera: Coccinellidae). *Environ. Entomol.* 19: 1798–1802.

Mendel, Z., H. Podoler and D. Rosen. 1985. A study of the diet of *Chilocorus bipustulatus* (Coleoptera: Coccinellidae) as evident from its midgut contents. *Israel J. Entomol.* 19: 141–146.

Michaud, J. P. 1999. Sources of mortality in colonies of the brown citrus aphid, *Toxoptera citricida. Biol. Control.* 44: 347–367.

Michaud, J. P. 2000. Development and reproduction of ladybeetles (Coleoptera: Coccinellidae) on the citrus aphids *Aphis spiraecola* Patch and *Toxoptera citricida* (Kirkaldy) (Homoptera: Aphididae). *Biol. Control.* 18: 287–297.

Michaud, J. P. 2004. Natural mortality of Asian citrus psyllid, *Diaphorina citri* (Homoptera: Psyllidae) in central Florida. *Biol. Control.* 29: 260–269.

Michaud, J. P. and B. Belliure. 2000. Consequences of foundress aggregation in the brown citrus aphid *Toxoptera citricida. Ecol. Entomol.* 25: 307–314.

Michaud, J. P. and L. Olsen. 2004. Suitability of Asian citrus psyllid, *Diaphorina citri* (Homoptera: Psyllidae) as prey for ladybeetles (Coleoptera: Coccinellidae). *Biol. Control.* 49: 417–431.

Michels, G. J. Jr, N. C. Elliott, R. L. Romero and W. B. French. 1997. Estimating populations of aphidophagous Coccinellidae (Coleoptera) in winter wheat. *Environ. Entomol.* 26: 4–11.

Mignault, M. P., M. Roy and J. Brodeur. 2006. Soybean aphid predators in Quebec and the suitability of *Aphis glycines* as prey for three Coccinellidae. *Biol. Control.* 51: 89–106.

Morris, W. F. 1992. The effects of natural enemies, competition, and host plant water availability on an aphid population. *Oecologia.* 90: 359–365.

Müller, H. J. 1966. Ueber mehrjaehrige Coccinelliden–Faenge auf Ackerbohnen mit hohem *Aphis fabae*-Besatz. *Z. Morph. Oekol. Tiere.* 58: 144–161.

Nechols, J. R. and T. L. Harvey. 1998. Evaluation of a mechanical exclusion method to assess the impact of Russian wheat aphid natural enemies. *In* S. S. Quisenberry and F. B. Peairs (eds). *Response Model for an Introduced Pest: The Russian Wheat Aphid.* Thomas Say Publications, ESA, Lanham, MD. pp. 270–279.

Nelson, E. H. 2007. Predator avoidance behavior in the pea aphid: costs, frequency, and population consequences. *Oecologia.* 151: 22–32.

Nemoto, H., Y. Sekijima, Y. Fujikura, K. Kiritani and S. Shibukawa. 1985. Application of an immunological method for the identification of predators of the diamond-back moth,

Plutella xylostella (L.) (Lepidoptera: Yponomeutidae). *Jpn J. Appl. Entomol. Zool.* 29: 61–66.

Noma, T., M. J. Brewer, K. S. Pike and S. D. Gaimari. 2005. Hymenopteran parasitoids and dipteran predators of *Diuraphis noxia* in the west-central Great Plains of North America: species records and geographic range. *BioControl.* 50: 97–111.

Omkar and B. E. James. 2004. Influence of prey species on immature survival, development, predation and reproduction of *Coccinella transversalis* Fabricius (Col., Coccinellidae). *J. Appl. Entomol.* 128: 150–157.

Pettersson, J. 1972. Technical description of a serological method for quantitative predator efficiency studies on *Rhopalosiphum padi* (L.). *Swedish. J. Agric. Res.* 2: 65–69.

Pfannenstiel, R. S. and K. V. Yeargan. 2002. Identification and diel activity patterns of predators attacking *Helicoverpa zea* (Lepidoptera: Noctuidae) eggs in soybean and sweet corn. *Environ. Entomol.* 31: 232–241.

Pickavance, J. R. 1970. A new approach to the immunological analysis of invertebrate diets. *J. Anim. Ecol.* 39: 715–724.

Piñol, J., X. Espadaler, N. Pérez and K. Beven. 2009. Testing a new model of aphid abundance with sedentary and non-sedentary predators. *Ecol. Model.* 229: 2469–2480.

Podoler, H., I. Bar-Zacay and D. Rosen. 1979. Population dynamics of the Mediterranean black scale, *Saissetia oleae* (Olivier), on citrus in Israel. I. A partial life-table. *J. Entomol. Soc. S. Afr.* 42: 257–266.

Qureshi, J. A. and P. A. Stansly. 2009. Exclusion techniques reveal significant biotic mortality suffered by Asian citrus psyllid *Diaphorina citri* (Hemiptera: Psyllidae) populations in Florida citrus. *Biol. Control.* 50: 126–136.

Ragsdale, D. W., A. D. Larson and L. D. Newsom. 1981. Quantitative assessment of the predators of *Nezara viridula* eggs and nymphs within a soybean agroecosystem using an ELISA. *Environ. Entomol.* 10: 402–405.

Rautapaa, J. 1972. The importance of *Coccinella septempunctata* L. (Col., Coccinellidae) in controlling cereal aphids, and the effect of aphids on the yield and quality of barley. *Ann. Agric. Fenn.* 11: 424–436.

Ricci, C. 1986. Seasonal food preferences and behaviour of *Rhizobius litura*. *In* I. Hodek (ed.). *Ecology of Aphidophaga.* Academia, Prague. pp. 119–123.

Ricci, C. and L. Ponti. 2005. Seasonal food of *Ceratomegilla notata* (Coleoptera: Coccinellidae) in mountain environments of northern Italian Alps. *Eur. J. Entomol.* 102: 527–530.

Ricci, C., G. Fiori and S. Colazza. 1983. Diet of the adult of *Tytthaspis sedecimpunctata* (L.) (Coleoptera Coccinellidae) in an environment with primary human influence: a meadow containing multiple plant species. *Proc. 13 Ital. Natl Congr. Entomol.* 691–698. (In Italian.)

Ricci, C., L. Ponti and A. Pires. 2005. Migratory flight and pre-diapause feeding of *Coccinella septempunctata*

(Coleoptera) adults in agricultural and mountain ecosystems of Central Italy. *Eur. J. Entomol.* 102: 531–538.

Rice, M. E. and G. E. Wilde. 1988. Experimental evaluation of predators and parasitoids in suppressing greenbugs (Homoptera: Aphidiidae) in sorghum and wheat. *Environ. Entomol.* 17: 836–841.

Roitberg, B. D. and J. H. Myers. 1979. Behavioural and physiological adaptations of pea aphids (Homoptera: Aphididae) to high ground temperatures and predator disturbance. *Can. Entomol.* 111: 515–519.

Roitberg, B. D., J. H. Myers and B. D. Frazer. 1979. The influence of predators on the movement of apterous pea aphids between plants. *J. Anim. Ecol.* 48: 111–112.

Schoof, D. D., S. Palchick and C. H. Tempelis. 1986. Evaluation of predator–prey relationships using an enzyme immunoassay. *Ann. Entomol. Soc. Am.* 79: 91–95.

Settle, W. H., H. Ariawan, E. T. Astuti et al. 1996. Managing tropical rice pests through conservation of generalist natural enemies and alternative prey. *Ecology.* 77: 1975–1988.

Shade, R. E., H. L. Hansen and M. C. Wilson. 1970. A partial life table of the cereal leaf beetle, *Oulema melanopus*, in northern Indiana. *Ann. Entomol. Soc. Am.* 63: 52–59.

Sheppard, S. K. and J. D. Harwood. 2005. Advances in molecular ecology: tracking trophic links through predator–prey food webs. *Funct. Ecol.* 19: 751–762.

Smith, R. F. 1966. Summing up section V. *In* I. Hodek (ed.). *Ecology of Aphidophagous Insects.* Academia, Prague and Dr. W Junk, The Hague. pp. 285–287.

Snyder, W. E. 2009. Coccinellids in diverse communities: Which niche fits? *Biol. Control.* 51: 323–335.

Snyder, W. E. and A. R. Ives. 2003. Interactions between specialist and generalist natural enemies: parasitoids, predators, and pea aphid biocontrol. *Ecology.* 84: 91–107.

Starý, P. and D. Gonzalez. 1992. Field acceptance of exposed exotic aphids by indigenous natural enemies (Homoptera: Aphidinea: Aphididae). *Entomol. Gener.* 17: 121–129.

Straub, C. S. and W. E. Snyder. 2006. Species identity dominates the relationship between predator diversity and herbivore suppression. *Ecology.* 87: 277–282.

Straub, C. S. and W. E. Snyder. 2008. Increasing enemy biodiversity strengthens herbivore suppression on two plant species. *Ecology.* 89: 1605–1615.

Sunderland, K. D. 1988. Quantitative methods for detecting invertebrate predation occurring in the field. *Ann. Appl. Biol.* 112: 201–224.

Sunderland, K. D. 1996. Progress in quantifying predation using antibody techniques. *In* W. O. C. Symondson and J. E. Liddell (eds). *The Ecology of Agricultural Pests: Biochemical Approaches.* Chapman and Hall, London. pp. 419–455.

Sunderland, K. D. and S. L. Sutton. 1980. A serological study of arthropod predation on woodlice in a dune grassland ecosystem. *J. Anim. Ecol.* 49: 987–1004.

Symondson, W. O. C. 2002. Molecular identification of prey in predator diets. *Mol. Ecol.* 11: 627–641.

Symondson, W. O. C. and J. E. Liddell. 1993. The detection of predation by *Abax parallelepipedus* and *Pterostichus madidus* (Coleoptera: Carabidae) on Mollusca using a quantitative ELISA. *Bull. Entomol. Res.* 83: 641–647.

Symondson, W. O. C., D. M. Glen, M. L. Erickson, J. E. Liddell and C. J. Langdon. 2000. Do earthworms help to sustain the slug predator *Pterostichus melanarius* (Coleoptera: Carabidae) within crops? Investigations using monoclonal antibodies. *Mol. Ecol.* 9: 1279–1292.

Tamaki, G. and R. E. Weeks. 1972. Efficiency of three predators, *Geocoris bullatus*, *Nabis americoferus*, and *Coccinella transversoguttata*, used alone or in combination against three insect prey species, *Myzus persicae*, *Ceramica pieta*, and *Mamestra configurata*, in a greenhouse study. *Environ. Entomol.* 1: 258–263.

Triltsch, H. 1997. Gut contents in field sampled adults of *Coccinella septempunctata* (Col.: Coccinellidae). *Entomophaga.* 42: 125–131.

Turner, B. D. 1984. Predation pressure on the arboreal epiphytic herbivores of larch trees in southern England. *Ecol. Entomol.* 9: 91–100.

Uygun, N. and R. Atlhan. 2000. The effect of temperature on development and fecundity of *Scymnus levaillanti*. *BioControl.* 45: 453–462.

Van Driesche, R. G. and M. S. Hoddle. 2000. Classical biological control: measuring success, step by step. *In* G. Gurr and S. Wratten (eds). *Biological Control: Measures of Success.* Kluwer, Dordrecht. pp. 39–76.

Van Driesche, R., K. Idoine, M. Rosec and M. Bryan. 1998. Evaluation of the effectiveness of *Chilocorus kuwanae* (Coleoptera: Coccinellidae) in suppressing euonymus scale (Homoptera: Diaspididae). *Biol. Control.* 12: 56–65.

Vickerman, G. P. and K. D. Sunderland. 1975. Arthropods on cereal crops: nocturnal activity, vertical distribution and aphid predation. *J. Appl. Ecol.* 12: 755–766.

Washino, R. K. and C. H. Tempelis. 1983. Mosquito host bloodmeal identification: methodology and data analysis. *Annu. Rev. Entomol.* 28: 179–201.

Weber, D. C. and J. G. Lundgren. 2009. Assessing the trophic ecology of the Coccinellidae: their roles as predators and as prey. *Biol. Control.* 51: 199–214.

Wells, L., J. R. Ruberson, R. M. McPherson and G. A. Herzog. 1999. Biotic suppression of the cotton aphid (Homoptera: Aphididae) in the Georgia coastal plain. *In* P. Duggar and D. Richter (eds). *Proc. Beltwide Cotton Conf., Orlando, January 1999* 2. National Cotton Council, Memphis, TN. pp. 1011–1014.

Whalon, M. E. and B. L. Parker. 1978. Immunological identification of tarnished plant bug predators. *Ann. Entomol. Soc. Am.* 71: 453–456.

Winder, L. 1990. Predation of the cereal aphid *Sitobion avenae* by polyphagous predators on the ground. *Ecol. Entomol.* 15: 105–110.

Winder, L., C. L. Alexander, J. M. Holland et al. 2005. Predatory activity and spatial pattern: the response of generalist carabids to their aphid prey. *J. Anim. Ecol.* 74: 443–454.

Xiao, Y. F. and H. Y. Fadamiro. 2010. Exclusion experiments reveal relative contributions of natural enemies to mortality of citrus leafminer, *Phyllocnistis citrella* (Lepidoptera: Gracillariidae) in Alabama satsuma orchards. *Biol. Control.* 54(3): 189–196.

Zaidi, R. H, Z. Jaal, N. J. Hawkes, J. Hemingway and W. O. C. Symondson 1999. Can the detection of prey DNA amongst the gut contents of invertebrate predators provide a new technique for quantifying predation in the field? *Mol. Ecol.* 8: 2081–2088.

Zhang, G. F., Z. C. Lü and F. H. Wan 2007a. Detection of *Bemisia tabaci* remains in predator guts using a sequence-characterized amplified region marker. *Entomol. Exp. Appl.* 123: 81–90.

Zhang, G. F., Z. C. Lü, F. H. Wan and G. L. Lövei 2007b. Real-time PCR quantification of *Bemisia tabaci* (Homoptera: Aleyrodidae) B-biotype remains in predator guts. *Mol. Ecol. Notes* 7: 947–954.

Chapter 11

COCCINELLIDS IN BIOLOGICAL CONTROL

J. P. Michaud

Department of Entomology, Kansas State University, 1232 240th Ave., Hays, Kansas, KS 67601, USA

Ecology and Behaviour of the Ladybird Beetles (Coccinellidae), First Edition. Edited by I. Hodek, H.F. van Emden, A. Honěk.

11.1 INTRODUCTION

Historically, the term biological control has had various meanings (DeBach 1964). **In the ecological sense**, it refers to the 'top-down' action of predators, parasites or pathogens on organisms occupying a lower trophic level, action that maintains their populations at lower levels than would occur in the absence of these natural enemies, i.e. a **natural process** that does not depend on any human intervention. **In the applied sense**, it refers to the **specific use or manipulation** of populations of organisms to achieve reduction or improved regulation of pest populations. In practice, this distinction is not always clear as modern integrated pest management programs usually rest on a foundation of 'background' natural control which occurs without intervention. However, cultural activities may be implemented or modified to facilitate natural biological control, or simply to spare beneficial species from adverse human impacts, including application of pesticides. The diverse approaches to achieving this end are now collectively termed **conservation biological control** (Barbosa 1998). In contrast, **classical biological control** involves importing and releasing exotic species with the goal of establishing a self-perpetuating population in a novel geographic region that exerts permanent suppression of a specific pest. Alternatively, **augmentation** has the goal of immediate, if temporary, reduction of a pest population through occasional or periodic releases of a natural enemy. Such releases may be **inundative** (large numbers released for immediate effect) or **inoculative** (fewer individuals released with some reproduction expected). Predatory coccinellids have been deployed and studied extensively in all these contexts. Furthermore, certain species, either through intentional or accidental introductions, have garnered attention as **invasive species** with significant impacts on biodiversity and human economies beyond their contributions to biological control.

11.2 THE ROLES OF COCCINELLIDAE IN BIOLOGICAL CONTROL

Dixon (2000) emphasized three attributes of successful biological control agents: specificity, voracity and a high rate of population increase. Whereas these attributes are certainly desirable for **rapid impact** when the target is increasing in abundance and approaching economic injury levels, many pest populations tend to be transient or seasonally cyclical. These same attributes may therefore become **handicaps for survival** of the agent when the pest is rare or absent. For example, the availability of alternative prey can be critical to coccinellid impact in biological control. This effect is illustrated by the case of *Cleobora mellyi*, which was introduced into New Zealand in 1977 for control of the eucalyptus tortoise beetle, *Paropsis charybdis*. Both larvae and adults of *C. mellyi* feed readily on tortoise beetle eggs and larvae, but rely heavily on psyllids to maintain their populations year-round in plantations of *Eucalyptus* and *Acacia* trees. Although establishment of the species on the North Island was initially successful, populations remained small and limited in distribution for about 25 years until the invasion of another tortoise beetle and several additional pysllid species led to a dramatic increase in the range and abundance of this 'forgotten ladybird' and its emergence as a significant biological control agent of this guild of arboreal defoliatiors (Withers & Berndt 2010).

11.2.1 Prey specificity

Prey specificity is desirable from the perspective of concentrating mortality on the pest and minimizing non-target impacts, but it can also be an impediment for predator survival when the preferred prey is at low density (Chang & Kareiva 1999). Likewise, **voracity** is not invariably an asset. Although high voracity is advantageous for maximizing the rate of conversion of prey biomass into predator biomass, it implies a greater food demand for successful development and reproduction of the predator. Thus coccinellids with lower voracity require a smaller critical intake of prey to complete development or remain actively foraging, making their populations more resilient to local extinctions when prey become scarce, and better able to track prey populations at low densities. Coccinellids exhibit a range of prey specificity (5.2), but species that dominate their communities tend to be **relative generalists**, though they often exhibit preferences for particular habitats (e.g. Colunga-Garcia et al. 1997). This is especially true of species that display good numerical responses to outbreaks of aphids, a prey that no natural enemies save fungal pathogens can match in reproductive rate. Adults of such species often survive for extended periods on **alternative prey** or **supplementary food** sources, even if these resources are inadequate to sustain their reproduction. Thus most ecologically dominant aphidophagous species tend to be opportunistic generalists with high voracity, traits adaptive for exploiting prey that are periodically abundant, but highly ephemeral and widely distributed. Although a few species do specialize on aphids associated with particular plants, most dietary specialists tend to exploit non-aphid prey that are more diffusely distributed, but more continuously available, in well-defined habitats, (e.g. coccids on trees).

Small ladybeetles of the genus *Stethorus* are specialized predators of phytophagous mites that can be important in biological control of tetranychid pests on a wide range of fruit crops (Biddinger et al. 2009). For example, Ullah (2000) found that *Stethorus vegans* was effective in locating and reproducing on *Tetranychus urticae* even at low densities, supplementing its diet with alternative prey as necessary. However, not all coccinellid species that specialize on one kind of prey are necessarily a cornerstone of its biological control. For example, ladybeetles of the genus *Nephaspis* are relatively specific to aleyrodid prey and *Clitostethus oculatus* is credited with reducing large infestations of spiralling whitefly, *Aleurodicus dispersus* in Hawaii (Waterhouse & Norris 1989) and India (Ramani et al. 2002). Nevertheless, this species does not provide adequate biological control of *A. dispersus* without the assistance of *Encarsia* spp. parasitoids that are better able to maintain the pest at low densities and are now considered to be solely responsible for the biological control of *A. dispersus* through much of Africa (Legg et al. 2003). Roy et al. (2005) concluded that predation by *Stethorus pusillus* (= *punctillum*) was complementary to that by *Amblyseius fallacis* in controlling *Tetranychus mcdanieli* in Quebec raspberry fields due to different patterns of seasonal activity. Another *Stethorus* sp. was observed to provide effective control of *Panonychus citri* in citrus when acting in concert with the predatory mite *Agistemus longisetus* (Jamieson et al. 2005). Similarly, Snyder et al. (2008) used field experiments to demonstrate that *C. septempunctata* and *Hip. convergens* provided more effective control of *Brevicoryne brassicae* and *Myzus persicae* on broccoli in combination with other natural enemies than when these were absent.

11.2.2   Generalist coccinellids

After multiplying their numbers on aphid outbreaks within particular habitats, many generalist coccinellids switch to feed on alternative prey species, a fact that warrants recognition in assessing their overall value as residents of agroecosystems. Many primarily aphidophagous species also contribute to biological control of pest species that serve only as supplementary food or non-essential prey. Through consumption of eggs and small larvae, *Col. maculata* may contribute to the control of *Ostrinia nubilalis* in corn (Musser & Shelton 2003), *Helicoverpa zea* in soybean and sweet corn (Pfannenstiel & Yeargan 2002), and *Leptinotarsa decemlineata* in potatoes (Groden et al. 1990). *Har. axyridis* may consume psyllids (Michaud 2004, Pluke et al. 2005), tetranychid mites (Lucas et al. 2002, Villanueva et al. 2004) and root weevil eggs (Stuart et al. 2002). Similarly, *C. septempunctata* prey on larvae of *Hypera postica* in alfalfa when aphids are scarce (Kalaskar & Evans 2001, Evans et al. 2004). By virtue of their ecology, such species tend to provide diffuse pest mortality in their favoured habitats, rather than being solely responsible for continuous suppression of a single pest. This is true in systems as diverse as the alfalfa fields of California (Neuenschwander et al. 1975) and the citrus groves of Puerto Rico (Michaud & Browning 1999).

11.2.3   Intraguild predation

Although large coccinellid species can supplement the effects of parasitoids in biological control (Bilu & Coll 2007) they may also be important agents of intraguild predation (Ferguson & Stiling 1996, Kaneko 2004, Michaud 2004; Chapter 7). Consequently, their role in suppressing a particular pest must be considered within the larger context of the entire guild of natural enemies that contribute to pest mortality. Because their interactions within and between trophic levels are complex, their net impacts in novel ecosystems are unpredictable. Thus, most conspicuous aphidophagous species are unsuited for foreign introduction even though they may be important agents of biological control in their native ecosystems. Once established, they may simply replace aphid mortality provided by local species, cause non-target impacts on other herbivores, or even displace autochthonous species from

Figure 11.2 The vedalia beetle *Rodolia cardinalis* with eggs and neonate larva on a mature cottony cushion scale (Jack Kelly Clark, courtesy UC Statewide IPM Program). (See colour plate.)

Figure 11.1 An assassin bug (Reduviidae) preying on an adult *Hippodamia convergens* (J.P. Michaud). (See colour plate.)

particular habitats. One the other hand, sometimes ladybirds, especially small species or immature stages, may have their effectiveness in biocontrol limited by intraguild predation from tertiary predators (e.g. Rosenheim et al. 2004; Fig. 11.1; Chapters 7 and 8).

11.3 SCALE INSECTS VERSUS APHIDS AS TARGETS OF EXOTIC INTRODUCTIONS

11.3.1 Coccidophagous coccinellids

In 1887 the vedalia beetle, **Rodolia cardinalis**, was introduced from Australia to combat the cottony cushion scale **Icerya purchasi** in the nascent California citrus industry (Fig. 11.2). The landmark success of this project became a textbook example of the great potential of classical biological control as a tactic for suppressing invasive pests (DeBach 1964). A total of 10,555 beetles were reared and released at a cost of $1500 and, within a year, the industry was saved from imminent destruction. The beetle was subsequently introduced to at least 29 other countries and control of the target was either complete or substantial in all locations. The success of the vedalia beetle, and the publicity surrounding its economic benefits, became a watershed event that catalyzed the introduction of many exotic coccinellids targeting other pests, although

rarely with the same degree of success (Caltagirone & Doutt 1989). Obrycki & Kring (1998) calculated that 18 ladybird species established in North America as a result of some 179 intentional introductions since 1900, an establishment rate of about 10%. In comparison, four exotic species became established in the UK over a similar period, although none were the result of intentional introductions (Majerus 1994). Neither does establishment equate to success in biological control. For example, the neotropical ladybirds *Diomus hennesseyi* and *Hyperaspis notata* were both established in a programme against the cassava mealybug, *Phenococcus manihoti*, in central and eastern Africa (Neuenschwander 2003), but neither became significant mortality factors. The high degree of specificity of *R. cardinalis* for *Icerya* species, in combination with a very short generation time, highly efficient detection of isolated host patches, and the ability of a larva to complete development on a single mature female scale are all thought to be key factors in the effectiveness of the vedalia beetle (Prasad 1990). Unfortunately, this ecological configuration is exceptional considering what is now known of the prey relationships of predatory Coccinellidae, and even atypical of other ladybird–coccid interactions. The importance of the rate of population increase of the predator relative to that of the prey has long been emphasized (Thorpe 1930, Hodek 1973, Hagen 1974), has recently been termed the 'generation time ratio' (Kindlmann & Dixon 1999, Dixon 2005) and is likely a key indicator of 'one on one' biocontrol potential.

Savoiskaya (1983) and Kuznetsov (1987) discussed examples of **coccinellid introductions across the Palearctic region** (Russia and Central Asia), many of which appeared to have failed because of climatic disparities between regions of collection and release. Similarly, introduction of a US strain of *Hip. convergens* to Kenya for control of *Schizaphis graminum* in 1911 failed despite a good history of the predator–prey association on wheat (Hunter 1909), probably because the source material was not adapted to tropical conditions (Greathead 1971). In contrast, many **introductions** of coccinellids **against scale** insects in tropical and subtropical habitats have been successful. DeBach (1964) listed introductions of nine coccidophagous coccinellids to which he attributed successful control of 11 coccid species other than *I. purchasi*. Of the 31 exotic species of Coccinellidae established in Hawaii, 17 are primarily scale-feeding species, five specialize on mealybugs and one on mites. The remainder feed on aphids or some combination of aphids, psyllids and other prey (Funasaki et al. 1988). Greathead (2003) reviewed classical programmes in Africa and listed nine coccinellid species successfully established, of which eight were scale-feeding species and the other the mealybug predator *Cryptolaemus montrouzieri*. In Fiji, successful control of the coconut scale, *Aspidiotus destructor*, was attributed to the introduction and establishment of *Cryptognatha nodiceps* Marshall from Trinidad (Singh 1976). Successful control of various scales in the subfamily Diaspidinae has been attributed to introductions of **Rhyzobius** (= *Lindorus*) **lophanthae** in Italy, the Black Sea coast of the Ukraine, and North Africa (Yakhontov 1960). This species is now considered an important introduced predator of citrus scale insects in both the United States (Flint & Dreisdadt 1998) and Australia (Smith et al. 1997), although it is less effective against heavily armoured species that are less vulnerable in later growth stages (Honda & Luck 1995).

Chilocorus bipustulatus is an effective predator of armoured scale insects and populations in the Middle East have been the source of successful introductions in North Africa (Iperti & Laudeho 1969), central Africa (Stansly 1984), Australia (Waterhouse & Sands 2001) and the USA (Huffaker & Doutt 1965). However, it failed to establish in New Zealand kiwi fruit orchards despite a vigorous release effort (Charles et al. 1995). Similarly, *Chil. nigritus* feeds on a variety of armoured scales and has been widely disseminated throughout tropical regions through both intentional and uninten-

tional introductions (Samways 1989). As early as 1940 it was introduced effectively against a complex of scale insects damaging coconuts in the Seychelles (Greathead 2003). This species is now mass-reared by various commercial suppliers of beneficial insects and widely used for augmentative biological control of armoured scales in glasshouses and interior plantscapes. *Chil. circumdatus*, a southeast Asian species, has been successfully introduced to Australia, South Africa, the USA and elsewhere (Samways et al. 1999) for control of citrus snow scale, *Unaspis citri*. It appears quite specific to its prey and is able to track it effectively at low densities, although in Hawaii it reportedly feeds on two additional species of Diaspididae (Funasaki et al. 1988). *Chil. circumdatus* is now considered to be the most important predator of *U. citri* in Australia (Smith et al. 1997) and appears to play a significant role in suppressing this pest in both Florida and California. Rosen (1986) reviewed natural enemies employed in biocontrol programmes against Diaspididae and concluded that, although useful, coccinellid species seldom provided adequate control without assistance from other natural enemies, especially parasitoids.

11.3.2   Aphidophagous coccinellids

Aphidophagous species have a long history of importation in classical biological control programmes, albeit with few recognized successes. As early as 1874, **C. undecimpunctata**, a relatively polyphagous species, was imported to New Zealand where it established to become an important predator of aphids and mealybugs in various fruit and forage crops (Dumbleton 1936). More recently, it established in various regions of North America (Wheeler & Hoebeke 1981) although its introduction there is thought to have been accidental. Dixon (2000) tallied 155 intentional introductions of coccinellid species worldwide that specifically targeted aphids and judged only one to be 'substantially successful'. The **high reproductive rate of aphids** is achieved through a combination of thelytoky and a telescoping of generations (live birth of pregnant daughters). Consequently, multiple aphid generations can be completed within the time required for a single generation of any coccinellid, forcing the numerical response of the predator to lag behind population growth of the prey. Borges et al. (2006) advanced the notion that the life history of ladybirds has evolved within constraints dictated by the ecology and distribution of their prey. They suggest that scale-feeding species have evolved a slow pace of life (slow development, low voracity and fecundity) in order to effectively exploit slowly developing prey that are continuously available, but widely distributed in small colonies. In contrast to scales, aphid colonies develop quickly and become much larger, but can be harder to find and more ephemeral in availability, attributes that favour their exploitation by voracious species that have faster development and higher fecundity. These disparities in life history, whether derived from predator–prey relations or not, suggest that aphidophagous coccinellids cannot be manipulated in biological control programmes by the same means, or for the same ends, as coccidophagous species, even though they may emerge as key sources of natural aphid mortality in field studies (Costamagna et al. 2008).

Various species of *Adelgidae* have been targeted with introduced coccinellids in **arboreal habitats**. The larch ladybird, *Aphidecta obliterata* is a species of European origin that was introduced to South Carolina in 1960 (Amman 1966) to control the balsam woolly adelgid, *Adelges piceae*, an invasive pest that arrived in North America around 1900. Later, *A. obliterata* was introduced to British Columbia, Canada (Harris & Dawson 1979) where it has emerged as an important biological control of both *A. piceae* and the hemlock woolly adelgid, *Adelges tsugae* (Humble 1994, Montgomery & Lyon 1996). The Asian species *Sasajiscymnus tsugae* was introduced from Japan to target *A. tsugae* and has been established in various regions of the eastern USA as a result of release programmes initiated in 1997 (Cheah et al. 2004). Although established, the impact of *S. tsugae* on the pest population was not as great as originally hoped, resulting in the importation and release of a Chinese species, *Scymnus ningshanensis*, that purportedly exhibits a stronger numerical response to increasing prey density (Butin et al. 2003).

Biological control programmes involving **introductions** have become increasingly **controversial** in recent decades as attention has been drawn to their potential **non-target impacts** (Howarth 1991) and other associated risks (van Lenteren et al. 2006). The wisdom of using natural enemy introductions as a first response to invasive arthropod pests has also been challenged (Wajnberg et al. 2001, Michaud 2002a). Regulatory restrictions on exotic species introductions have become increasingly stringent and most developed countries now require that entomophagous

species exhibit levels of prey specificity comparable to those previously required only for weed biocontrol agents (e.g. FAO 1996). Although few programmes currently seek to introduce exotic coccinellids into new regions, many displaced species (whether **intentionally or accidentally introduced**) now comprise a substantial proportion of the guild of aphidophagous insects in most Nearctic and Palearctic habitats. Many contribute significantly to control of pests in agricultural habitats, but others have been implicated in the **displacement of native species**. For example, the native species *Stethorus punctum* was replaced by the Palearctic *Stethorus pusillus* in Ontario around 1940 following its inadvertent establishment in Canada (Putman 1955).

11.3.3 Invasive coccinellids

Gordon (1985) listed over 150 species of Coccinellidae with records of intentional introduction to North America, mostly during the 20th century, with 19 successful establishments, and another 10 established apparently as a result of inadvertent introductions. However, species additional to these have been confirmed **established in the past 25 years**, including *P. quatuordecimpunctata*, *Har. axyridis*, *Har. quadripunctata* and *Hip. variegata* (Hoebeke & Wheeler 1996), *Curinus coeruleus* (Michaud 2002b), and probably others. Many of these species have gradually expanded their range through natural spread (e.g. *Hip. variegata*, Williams & Young, 2009) with mostly indeterminate ecological impacts. There is often a latent period before an introduced coccinellid assumes the status of an **invasive species**. Despite introductions to the USA as early as 1956, ***C. septempunctata*** was not confirmed established along the eastern seaboard until the late 1970s (Hoebeke & Wheeler 1980) and required another decade to expand its range westward to the Rocky Mountains. By the early 1990s, *C. septempunctata* had displaced the indigenous *Hip. convergens* as the dominant aphidophagous coccinellid in parts of the American Midwest and generated other non-target impacts (Horn 1991). Similar impacts were later attributed to *C. septempunctata* further west (Wheeler & Hoebeke 1995, Elliott et al. 1996, Hesler et al. 2005), although it has failed to displace *Hip. convergens* from dominance throughout much of the arid High Plains. Similarly, *A. bipunctata* remained confined to the Osaka region of Japan for about 10 years following its introduction to Japan, but is now expanding its range and utilizing a greater number of tree and aphid species (Toda & Sakuratani 2006). In contrast, the spread of *Har. axyridis* in North America was much more rapid, requiring only 10 years to colonize most of the Nearctic region from Florida to California in the south, and from Nova Scotia to British Columbia in the north. No case history is perhaps more remarkable than that of *Har. axyridis* and it has the ignominious distinction of being the first predatory coccinellid to be indexed in the Global Invasive Species Database (ISSG 2008) and have its worldwide distribution mapped as a plant pest (CABI 2007).

Harmonia axyridis, referred to as the harlequin ladybird in the UK and the Asian multi-colored lady-beetle in the USA, is a voracious, highly polyphagous species of Asian origin with a long history of introductions worldwide, partly because it is so easily reared on factitious diets, because it readily attacks a wide range of pests, and because its diapause is not obligatory (Chapter 6). It is has proved an excellent biological control agent of scales and aphids occurring on a multitude of plants including alfalfa, apple, citrus, maize, cotton, hops, pecan, pines and soybean. For example, With et al. (2002) showed that *Har. axyridis* was more effective than the indigenous *Col. maculata* in tracking pea aphids to low densities on patches of red clover in a structurally fragmented landscape. *Har. axyridis* has also has been commercially produced and released in augmentation programmes throughout western Europe since the 1980s with little apparent concern for the establishment of feral populations (Adriaens et al. 2003). In 1993 it was introduced into citrus orchards in Greece (Katsoyannos et al. 1997) and many hundreds of thousands of beetles have since been released in various orchard and urban settings there, although establishment remains uncertain as yet. Towards the end of the 1990s, *Har. axyridis* was released in Mendoza province, Argentina and by 2003 it had become the dominant aphid predator in walnut orchards in Buenos Aires province (Saini 2004). Range expansion across much of South America is now expected (Koch et al. 2006).

Between 1999 and 2002, a large scale programme to rear-and-release *Har. axyridis* in the Yucatan peninsula of Mexico was initiated by federal agricultural agencies in an effort to delay movement of the **brown citrus aphid, *Toxoptera citricidus***, into the primary citrus growing regions to the north (Lopez-Arroyo et al. 2008). Once established in the states of Yucatan and Quintana Roo, populations became increasingly abundant in other horticultural habitats. Additional releases were made for purposes of aphid control in pecan orchards in northern Mexico (Quinones & Tarango 2005). In 2002, *Har. axyridis* appeared unexpectedly in the mountains near Mexico City, far from any release site, and spontaneous colonization of various coastal regions has since been confirmed. In 2004, it appeared in southeast England and spread at a rate of 58–144 km per year over the next 2 years (Brown et al. 2007).

Intentional **North American introductions** of *Har. axyridis* began in the early 1900s in California (Gordon 1985) and the most recent US introduction was in 1985 in North Carolina (McClure 1987). Despite evidence of life cycle completion in the field, none of these original American introductions appear to have resulted in established populations. Similarly, *Har. axyridis* failed to establish in the Azores despite a series of releases in the 1990s and earlier (Borges et al. 2005). Soares et al. (2008) hypothesized extrinsic ecological or environmental factors as potentially responsible, but such failures may simply be due to founder effects in the source material – i.e. the introduction of small groups of individuals coincidentally lacking genetic composition immediately suited to local conditions. More recently, it appeared unexpectedly on the relatively barren Sable Island off the coast of Nova Scotia (Catling et al. 2009) and has been detected as a hitchhiker on ornamental plants imported into Norway from western Europe (Sathre et al. 2010). The first established North American populations of *Har. axyridis* were discovered in Louisiana and Georgia (Chapin & Brou 1991), hundreds of miles from any release site. It is now thought by some that *Har. axyridis* was accidentally introduced in cargo containers arriving at a Gulf of Mexico seaport (Day et al. 1994) and genetic studies seemingly point to a single founding population in North America (Krafsur et al. 1997). A detailed history of *Har. axyridis* introductions worldwide can be found in Soares et al. (2008).

11.3.4 Competitive displacement

Competitive displacement of native ladybirds has often been inferred from their declining abundance following establishment of *Har. axyridis* (Brown & Miller 1998, Colunga-Garcia & Gage 1998, Michaud 2002c, Mizell 2007) and *C. septempunctata* (Turnock et al.

2003). Numerous laboratory studies have revealed the advantages enjoyed by *Har. axyridis* in intraguild predation interactions with native species (Yasuda & Shinya 1997, Cottrell & Yeargan 1998, Michaud 2002c, Snyder et al. 2004, Burgio et al. 2005, Cottrell 2005). Nevertheless, *Har. axyridis* has coexisted for many years in Japan with its sibling species *Har. yedoensis*, a habitat specialist on pine trees (Osawa & Ohashi 2008). Although declines in the abundance of native species in regions invaded by these large exotic species have raised concern among conservationists, the complete elimination of autochthonous coccinellid species has not yet been documented and seems unlikely. Rather, native species may simply equilibrate to lower population densities in the cultivated or disturbed habitats dominated by invasive species, or retreat to more feral, ancestral habitats (Evans 2004, Acorn 2007, Harmon et al. 2007).

Aside from the apparent impact of **Har. axyridis** on biodiversity, it has gained the **status of a pest** in North America, something that might have been anticipated based on the behavior of this species in its native China (Wang et al. 2011). Its aggregative hibernation behaviour has raised the ire of home-owners when massive hordes of overwintering beetles invade houses (Fig. 11.3), foul living quarters, **trigger allergic reactions** in susceptible individuals (Albright et al. 2006), and even 'pinch' exposed skin with their mandibles (Yarbrough et al. 1999, Huelsman et al. 2002, Koch 2003). Large swarms of *Har. axyridis* were first observed in the American Midwest in 2001 (Huelsman et al. 2002) apparently as a direct result of a numerical

Figure 11.3 Aggregation of *Harmonia axyridis* attempting to enter a house under a door (courtesy of Marlin Rice). (See colour plate.)

response to the large populations of *Aphis glycines* that developed in soybean fields in the years following the aphid's invasion, and where the beetle continues to be important for biological control (Fox et al. 2004). Elsewhere, a propensity for **feeding on ripening fruit** has resulted in *Har. axyridis* damaging peaches and other soft-skinned fruits (Koch et al. 2004), and led to its emergence as a serious flavour contaminant in grapes (Pickering et al. 2007) to the point of requiring insecticide treatment (Galvan et al. 2006a). Consequently, the recent establishment of *Har. axyridis* in southeast England in 2004 (Roy et al. 2006) and in South Africa in 2006 (Stals & Prinsloo 2007) have been causes for alarm.

11.4 AUGMENTATION OF COCCINELLIDS

The term 'augmentation' refers to the **periodic release or inoculation** of a natural enemy in contexts where only the released insects or their immediate descendants are expected to exert biological control. Recent reviews of the augmentation approach continue to indicate that coccinellids are rarely utilized in such programmes (Collier & van Steenwyck 2004, Powell & Pell 2007). Two **limiting factors** are the cost of their production and the education required for their effective application by end users. Thus many demonstration projects have shown potential efficacy, but have failed to evolve into tactics that are attractive or economically viable pest control alternatives. For example, Baker et al. (2003) successfully released overwintered adults of *Cleobora mellyi* and *Har. conformis* against the leaf beetle *Chrysophtharta bimaculata* on eucalyptus trees in Australia, but concluded this approach was only economically feasible in small, environmentally sensitive areas where pesticide use was not acceptable. The pest population was reduced below economic threshold levels, but beetle numbers declined to pre-release levels within 7 days.

Coccinellid augmentation programmes may sometimes be **useful in developing countries** where labour costs associated with rearing and distribution are low, but are unlikely to prove a viable method of pest suppression on the large-scale commercial farms of the developed world, especially in low value row crops. For example, Heinz et al. (1999) tested augmentative releases of *Delphastus catalinae* against *Bemisia tabaci* (as *B. argentifolii*) on cotton in the

Imperial Valley of **California**. Although measurable reductions of whitefly populations were obtained in exclusion cages, open field evaluations revealed no significant effects of releasing adult beetles at rates of 3.5–5.5 beetles per plant. The authors implicated intraguild predation as one factor possibly limiting the impact of released *D. catalinae*. In contrast, smaller scale organic farms producing high value fruits and vegetables provide a more likely setting for coccinellid augmentation to succeed, especially in greenhouses that afford environmental control and some containment of released insects. Powell and Pell (2007) listed augmentation trials of ladybirds against aphids reporting the target species, crop, life stages released, and degree of success obtained. Notably, a number of positive results were obtained when parasitoids or other predators were released in conjunction with a coccinellid species.

11.4.1   The mealybug destroyer

Cryptolaemus montrouzieri is a relatively voracious predator with a long history of use in biological control making it probably more widely utilized in augmentation than any other coccinellid species. Introduced into California in 1892, it has been used in augmentation programmes against various mealybug pests (Pseudococcidae) for many years (Smith & Armitage 1920, 1931, Fisher 1963). Mealybug infestations are notoriously resilient to pesticide applications, a factor that has favoured the commercial production and sale of *C. montrouzieri* in many countries. The larvae of *C. montrouzieri* produce copious wax filaments that mimic those of mealybugs and enable them to forage effectively in the presence of tending ants that may attack other predators and parasitoids (Daane et al. 2007). Mani et al. (2004) reported that releases of *C. montrouzieri* provided complete control of *Rastrococcus invadens* on sapodilla, *Manikara zapota*, trees in a two month period with no other sources of mortality evident; similar results have been reported on mango (Mani & Krishnamoorthy 2001). Open field releases of *C. montrouzieri* have also been made in citrus (Smith & Armitage 1920, Moore & Hattingh 2004) and coffee (Hutton 2007), although not always with favourable results (Villalba et al. 2006).

   Cryptolaemus can also be useful for augmentation against soft scale species (Coccidae). Smith et al. (2004)

reported effective reduction of *Pulvinaria urbicola* on *Pisonia* trees following releases of *C. montrouzieri* in combination with three parasitoid species. Mani and Krishnamoorthy (2007) reported that *C. montrouzieri* provided effective control of the green shield scale, *Pulvinaria psidii*, on guava, *Psidium guajava*, when released at a rate of 10 adults per tree. This ladybird is often employed in greenhouse systems where mealybugs can be especially problematic, although it requires relatively warm temperatures to be effective (Shrewsbury et al. 2004). Typically, periodic augmentation is necessary for continued pest suppression because *C. montrouzieri* numbers crash abruptly as their prey comes under control. However, this is of little concern when **inundative releases** are used to assist with **rapid suppression** of large pest populations during periods of parasitoid introduction. For example, the pink hibiscus mealybug, *Maconellicoccus hirsutus*, has a very effective specialist parasitoid, *Anagyrus kamali*, but mass releases of *C. montrouzieri* can reduce large populations on the primary woody host plants much more rapidly, allowing many trees to recover (Villa Castillo 2005, Santiago-Islas et al. 2008). Chemical treatments are far less effective and would interfere with establishment of the parasitoid, which is subsequently able to maintain the pest at densities too low to sustain the predator in any numbers. However, Chong and Oetting (2007) recommended that parasitoid releases against citrus mealybug, *Planococcus citri*, should precede those of *C. montrouzieri* by a period sufficient to ensure substantial mummification of the next generation so that intraguild predation on parasitoid larvae would be minimized.

11.4.2   Redistribution of coccinellids

In the mountains of the Sierra Nevada, California, the aggregation of large numbers of **overwintering *Hip. convergens*** has, for almost a century, facilitated their collection for purposes of redistribution in vegetable and row crops (Fink 1915; Chapter 6). This practice has been criticized, largely because overwintered beetles tend to have a strong dispersal tendency that results in the **immediate emigration** of most individuals **from release sites** (Obrycki & Kring 1998), but also because of their low fecundity (Bjornson 2008) and the potential for distribution and transmission to other species of diseases such as microsporidia

(Saito & Bjornson 2006). Starks et al. (1975) explored night-time releases and the provision of shelter and water to retain imported *Hip. convergens* in sorghum fields to prey on *S. graminum* but these efforts failed to inhibit the dispersal of the coccinellids significantly. Furthermore, greenhouse trials indicated that a locally adapted strain was more effective in controlling the greenbug. Dreistadt and Flint (1996) provided over-wintered *H. convergens* with an opportunity to fly in screen cages as a pre-release conditioning treatment, but retention of the beetles on aphid-infested chrysanthemums was only marginally improved. Flint and Dreistadt (2005) calculated that effective control of rose aphids, *Macrosiphum rosae*, could be obtained with a release rate of about 2300 adult *Hip. convergens* per metre-squared of shrub-covered surface, a rate corresponding to double that normally recommended by commercial suppliers. The authors suggested that the approach was economically feasible given costs comparable to an insecticide treatment, but did not address the possible impact of such collections on local populations. The large scale collection and sale of *Hip. convergens* from the Sierra Nevada mountains continues until present times, but represents a commercial enterprise that exploits the public appeal of biological control, usually without delivering the results (e.g. Randolph et al. 2002).

11.4.3 Selection of source material for augmentation

Care should be taken in extrapolating control potential from laboratory feeding experiments, because the range of prey acceptable to coccinellids in confinement is often much broader than that in the field (Chapter 5). This can be true for various reasons, including prey specificity, plant and habitat preferences or simple refuge effects that are eliminated in simplified feeding situations. For example, Corlay et al. (2007) observed that larvae of the swede midge, *Contarinia nasturtii*, were acceptable prey to both *C. septempunctata* and *Har. axyridis* when presented in small containers, but completely escaped predation when presented on potted broccoli plants. The export of mass-collected beetles for augmentation in dissimilar habitats (where they are often released against unfamiliar prey) may be misguided, even when locally indigenous species are involved. Such export is based on the implicit assump-

tion that populations of a species are ecologically uniform across their range and ignores the likely importance of local adaptations in providing effective biological control.

11.4.4 Voltinism and diapause

Differences in voltinism and diapause behaviour have long been documented among European populations of **C. septempunctata** and other temperate species (Chapter 6) and such variation can pose a significant impediment to efficacy whenever augmentation programmes employ displaced material. In the New World, **Hip. convergens** is indigenous throughout tropical and sub-tropical America but has highly divergent ecology and prey relationships compared to populations inhabiting temperate regions.

11.4.5 Dietary requirements

The nutritional ecology of coccinellid species can also vary dramatically among geographically separated populations, probably reflecting underlying genetic divergence (Obrycki et al. 2001). *Curinus coeruleus* originates in Mexico and has been introduced to Florida, Hawaii and several Asian countries. Populations of *C. coeruleus* in Florida are unable to complete development on *T. citricidus* and rarely feed on this aphid, whereas in Hawaii they mature successfully on this prey and contribute substantially to its mortality on citrus trees (J.P. Michaud, unpublished). Similarly, Michaud (2000) was unable to obtain completed development of the introduced Eurasian species *Coelophora inaequalis* on *T. citricidus* using material collected in central Florida, despite observing successful maturation on this prey in Puerto Rico. Wang and Tsai (2001) obtained successful development of *Coelophora inaequalis* on *T. citricidus* using material from southern Florida and suggested this species had potential to control the pest. Intraspecific variation in prey utilization patterns probably arises from different founder effects in colonizing populations that are subsequently amplified by disruptive selection in local habitats. However widespread, such divergence of populations would seem to caution against augmentation efforts employing 'mail-order' beetles in favour of collecting and culturing

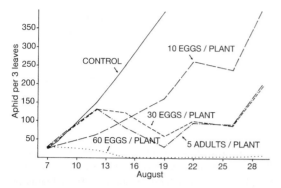

Figure 11.4 The effects of different rates of introduction of *Adalia bipunctata*, as either eggs or adults, on the numbers of *Myzus persicae* on sweet peppers in a greenhouse (from Hämäläinen 1977, with permission).

regionally adapted strains with proven local performance on the target.

11.4.6   Life stages for release

It is prudent to consider this in any particular coccinellid augmentation programme. Eggs are quite fragile and subject to a variety of mortality factors prior to hatching, especially predation and cannibalism, rendering them unsuitable for many applications. Nevertheless, Hämäläinen (1977) successfully used **egg masses** of *A. bipunctata* **for inundative control** of *M. persicae* on sweet peppers in a greenhouse (Fig. 11.4) and techniques have been developed for mass-producing eggs and applying them in the field (Shands et al. 1970). Kehrli and Wyss (2001) released *A. bipunctata* as eggs and **larvae** on apple trees over a range of dates in Switzerland and found that autumn releases of larvae could limit the deposition of overwintering eggs by aphids of the genus *Dysaphis*, leading to reductions in the numbers of fundatrices the following spring. By contrast, releasing egg masses was less effective due to the higher mortality. However, larvae are more costly to produce than eggs, especially if they require aphids for successful development, and they can suffer high levels of cannibalism in mass culture, depending on the species and the adequacy of the diet. Cannibalism is a particular problem in breeding strictly aphidophagous species and it is perhaps not surprising that most coccinellids mass-produced in commercial operations are species that can be reared on non-aphid prey or factitious diets. Pre-imaginal life stages are more practical for release in protected environments such as greenhouses and interior plantscapes where their survival can be enhanced by protection from natural enemies and control of the physical environment.

Adults are the most robust life stage for release. They are more **voracious** than larvae, but also the stage most prone to dispersal from release sites. Their effectiveness is enhanced if they can be induced to lay eggs prior to dispersal. Deng et al. (1987) claimed >90% reduction of sugar cane infestation by sugar cane woolly aphid, *Ceratovacuna lanigera*, by releasing field-collected adults of *Synonycha grandis* and *Coelophora biplagiata*. Yang (1985) reared and released between 600,000 and 800,000 *P. japonica* against *M. persicae* over a period of 4 years in cotton fields of Hubei, China. The study reported much lower aphid populations in release fields compared to control fields treated with insecticides, and at a lower cost. Ahmed et al. (2006) reported successful control of spider mites on cantaloupes in Egypt when releases of *Stethorus gilvifrons* were supplemented with applications of *Beauveria bassiana*, *Metarhizium anisopliae* (both entomopathogenic fungi) or insecticidal soap, none of which provided adequate control without the beetles. Presumably, the timing of field releases relative to pest population development is a critical factor influencing efficacy (10.1), as well as the elicitation of oviposition at release sites prior to emigration. Some progress in limiting emigration has been achieved by releasing flightless mutants, first discovered in *A. bipunctata* (Marples et al. 1993) and later mutagenically induced in *Har. axyridis* (Tourniaire et al. 2000). Weissenberger et al. (1999) showed that the flightless strain of *H. axyridis* has greater probability of laying eggs at a release site by virtue of its reduced dispersal capability. Unfortunately the flightless adults have significantly reduced fecundity (Ferran et al. 1998) and generally lower fitness. They appear more vulnerable to intraguild predation by tertiary predators such as Reduviidae (Heteroptera) and retain an urge to disperse, often wasting considerable time and energy on unsuccessful flight attempts (J.P. Michaud, unpublished).

11.5   CONSERVATION

In recent years, environmental concerns have given a renewed emphasis to the preservation and enhancement of indigenous coccinellid species in order to improve their biocontrol contributions in natural and managed ecosystems. There are a wide variety of tactics for conserving coccinellids in horticultural and agricultural settings. In general, these aim to encourage immigration, discourage emigration, mitigate mortality caused by cultural activities including the application of pesticides, or improve survival during critical periods by the provision of specific resources, usually in the form of particular plant species. (See also 11.6.4 below.)

11.5.1   Alternative or supplementary food

The rationale for providing **additional food** for coccinellids is that food can be limiting to coccinellid survival or retention in the crop during critical periods, or can be used to attract immigrant beetles into crop fields earlier in the season or in larger numbers. The assumption is that provision of food can **alter the spatial distribution** of coccinellids in the landscape so as to improve subsequent biological control of a pest that is initially present at very low levels, or one that arrives in a crop with some predictability with respect to season or plant growth stage. There exists some experimental evidence to support the effectiveness of this approach, but the potential has been largely ignored in commercial agriculture. (See also 5.2.10.)

Artificial honeydews have been employed **successfully** to **attract** a variety of aphid predators, including coccinellids, into cropping systems (Hagen et al. 1971, Nichols & Neel 1977). Evans and Swallow (1993) demonstrated that sucrose sprays were more effective than protein supplements for attracting coccinellids and chrysopids into alfalfa fields. Predator responses to sucrose persisted up to 7 days in the absence of rain and resulted in reduced densities of *Acyrthosiphum pisum* in sprayed plots. Mensah and Madden (1994) successfully employed sucrose sprays and sugar granules in feeding stations to retain adult *Cleobora mellyi*, in target areas. Alhmedi et al. (2010) showed that limonene was attractive to gravid *Har. axyridis* females in both laboratory and field trials and increased their oviposition on plants. More recently, **herbivore-induced plant volatiles** have been synthesized that demonstrated activity in attracting beneficial insects into hop orchards, including the mite predator *Stethorus punctum picipes* (James 2003).

The ecological significance of **herbivory by coccinellids** in agricultural landscapes has probably not received adequate attention (but 5.2.9). Both adult and larval coccinellids have been observed consuming tender plant foliage, although this has often been dismissed as simply a means to obtain water. Moser et al. (2008) have shown that larvae of *Har. axyridis* and *Col. maculata* readily feed on maize seedlings even in the presence of animal prey. Coccinellids regularly utilize the **extrafloral nectaries** on wild and cultivated sunflowers, *Helianthus annuus* as a source of hydration and supplementary nutrition on the arid High Plains of Kansas (Michaud & Qureshi 2005, 2006). In spring, large numbers of first generation coccinellids (mostly *Hip. convergens*, *Hip. sinuata*, *C. septempunctata* and *Col. maculata*) mature on the aphids infesting winter wheat and emigrate *en masse* from the crop just prior to harvest in mid June. These beetles face hot dry conditions, a scarcity of essential prey (Fig. 5.1), and often a complete lack of free water, save the occasional dew. By maintaining a reproductive diapause, the beetles are able to sustain themselves throughout summer months on the extrafloral nectar of sunflowers (Fig. 11.5) and the occasional supplementary prey item until aphids become available once again as cooler weather returns in the autumn. Thus the presence of abundant sunflowers in the region, including substantial cultivated acreage, represents a valuable resource

Figure 11.5 Close-up of an adult *Hippodamia convergens* drinking extra-floral nectar from the petiole of a sunflower plant *Helianthus annuus* (J.P. Michaud). (See colour plate.)

that enables the beetle population to survive adverse summer conditions and subsequently limit aphid establishment on emergent winter cereals in autumn. Coccinellid consumption of plant material other than pollen has received little attention and could provide some novel opportunities for area-wide conservation, just as the planting of suitable flowers can enhance parasitoid foraging.

11.5.2   Hibernation refuges

In temperate regions where overwintering mortality can be high, the planting of hibernation refuges such as '**beetle banks**' (Fig. 11.6) has successfully enhanced coccinellid survival in various agricultural contexts including sugar beet (Bombosch 1965), Brussels sprouts (van Emden 1965) and potatoes (Galecka 1966). Iperti (1966) developed traps that **simulated rocky crevices** to shelter overwintering *Ceratomegilla (= Semiadalia) undecimnotata*, reduce their infection by *Beauveria bassiana*, and facilitate their collection for redistribution. **Bandages on branches** (Nohara 1962) or metal bands on tree trunks (Tamaki & Weeks 1968) have proven useful as artifical hibernation refuges for coccinellids in arboral settings. Construction of artificial shelters may sometimes be feasible in high-value vegetable crops, but is probably cost-prohibitive in most large scale agricultural settings and current approaches to enhancing overwintering survival tend to rely on recognizing and preserving natural hibernation sites.

11.5.3   Habitat management

Conservation of coccinellid populations hinges on understanding the full range of resources utilized by all life stages of a coccinellid species throughout a complete annual cycle. Potentially important considerations include the preservation of natural shelter or overwintering sites, the diversification of plant communities to improve availability of food and shelter, and modifications to conventional agricultural practices that minimize impacts on coccinellid populations. These modifications include tactics such as strip-harvesting, intercropping, reduced tillage, restriction of pesticide applications to spot or strip treatments, and the use of pesticides with selective modes of action. Unfortunately, the scaling up of commercial

Figure 11.6 A 'beetle bank' comprising a strip of perennial grasses forming dense tussocks to serve as overwintering habitat for coccinellids and other beneficial insects (Otago Regional Council, New Zealand). (See colour plate.)

agriculture has generally resulted in larger fields of individual crops, reduced landscape complexity and increased habitat fragmentation for coccinellids, factors potentially reducing their ability to effectively track and control their prey (Elliott et al. 2002a, With et al. 2002). For example, Altieri and Todd (1981) showed that coccinellids remained **more concentrated in border rows** of soybean fields than in central parts. The floral composition of the surrounding landscape often explains a lot of observed variation in field-to-field coccinellid diversity and abundance (e.g. Elliott et al. 2002b). In addition, benefits observed in small-scale research plots may not be congruent with results obtained in larger scale commercial fields

(Corbett 1998). Thus, there is a need for continued study of coccinellid responses over various landscape scales if habitat management approaches to their conservation are to be successful.

11.5.3.1 Strip-harvesting

Strip-harvesting alfalfa (aka, lucerne) is one way to **prevent the post-harvest emigration** of local predator populations as a consequence of food deprivation, thus conserving natural enemies in fields and facilitating their rapid colonization of harvested areas when re-growth occurs. This approach has a long history of use in **alfalfa** production (Scholl & Medler 1947) where it functions to stabilize biological control of a range of pests and conserve coccinellids, among other beneficial species (Schlinger & Dietrick 1960, Hossain et al. 2001, Weiser et al. 2003). This approach has potential for conserving coccinellids in other perennial forage crops where aphids may cause losses. As with other forms of habitat management, even with proven results, the challenge in strip harvesting is one of implementation at farm level where the immediate convenience of conventional practices too often pre-empts consideration of future benefits that might be obtained with minor procedural modifications.

11.5.3.2 Floral diversity (non-crop plants)

Coccinellids demonstrate affinities for particular plants independent of prey availability but, although such preferences have been recognized, they have not been effectively exploited in biological control. For example, Schmid (1992) observed that coccinellids in Germany had consistent patterns of occurrence on particular **non-crop plant species**, mostly common weeds, and avoided others. These affinities were often independent of the presence of prey as fully 40% of the coccinellids were observed on plants without aphids. Lixa et al. (2010) demonstrated that six species of Coccinellidae were attracted to **aromatic species of Apiaceae** (dill, coriander and sweet fennel) particularly in their blooming seasons and Silva et al. (2010) found increased abundance of coccinellids and other beneficial insects in lemon orchards in response to ground cover vegetation.

Although the potential exists to enhance coccinellid biological control via management of vegetative diversity, it will always be difficult in practice to encourage tolerance by farmers of plants considered weeds in other contexts. Kranz and Sengonca (2001) tested a range of plants for their relative attractiveness to *C. septempunctata* and successfully used preferred plants (*Medicago sativa* and *Artemisia vulgaris*) to influence the distribution and abundance of beetles in the field. Harmon et al. (2000) found that a **pollen-bearing weed** (dandelion, *Taraxacum officinale*) interspersed in alfalfa attracted sufficient concentrations of *Col. maculata* to locally reduce *A. pisum* densities relative to alfalfa patches lacking the weed. The work of Grez et al. (2010) in Chile illustrates just how species-specific coccinellid responses can be to various types vegetation bordering agricultural fields. However, border vegetation planted to provide supplementary food resources or alternative prey for coccinellids may sometimes impede their timely dispersal into adjacent crops, a phenomenon often referred to as 'apparent competition' (Frere et al. 2007). For example, stinging nettle, *Urtica dioica*, has long been recognized as a reservoir plant for coccinellids (Perrin 1975), but such reservoirs cannot benefit biological control unless beetles leave them and enter crop fields. Thus, Alhmedi et al. (2007) suggested **cutting border strips of nettle** to encourage timely movement of coccinellids into neighbouring field crops in Belgium, since nettle aphids colonized earlier in the season than species infesting green pea and wheat.

11.5.3.3 Intercropping

Fye and Carranza (1972) showed that intercropping grain sorghum with cotton could increase populations of *Hip. convergens* in the cotton due to the abundance of greenbug in sorghum early in the growing season. Patt et al. (1997) showed that intercropping aubergine (egg plant) with dill or coriander improved coccinellid diversity and abundance and increased mortality of *L. decemlineata* eggs and larvae, largely due to the attractiveness and suitability of these flowers for both *Col. maculata* and *Chrysoperla carnea*. Unfortunately, as mentioned above, manipulation of vegetation can have both positive and negative effects on biological control in a particular crop, depending on the plant and insect species involved. Andow and Risch (1985) found that densities of *Col. maculata* remained higher in monocultures of maize than in polycultures where maize was intercropped with beans, squash or red clover. Predation on *O. nubilalis* eggs by *Col. maculata* was also higher in the monoculture, apparently because maize served as a better source of alternative foods (aphids and pollen) than did the other plants. Similarly, Seagraves & Yeargan 2006) used tomato as a companion plant to improve oviposition and egg survival of *Col. maculata* in corn plots, but this did not increase predation on *Helicoverpa zea* egg masses in the corn.

11.5.3.4   Reduced tillage

The widespread adoption of reduced tillage, or 'no-till' agriculture in the American High Plains over the past several decades was promoted to farmers for soil moisture conservation in this arid region. However, it also appears correlated with an area-wide reduction of cereal aphid problems in wheat and sorghum. Part of the effect may result from decreased rates of aphid colonization due to crop residues functioning as a mulch with reflective properties that reduce the 'apparency' of the crop plant (Burton & Krenzer 1985). There are also potential benefits for natural enemies such as coccinellids, the predators most often identified as responsible for biological control of cereal aphids in North America (Nechols & Harvey 1998, Michels et al. 2001). The structural complexity of habitats has been correlated with increased abundance and diversity of natural enemies, including coccinellids (Langellotto & Denno 2004). Crop residues in fields probably improve insect diversity by creating structural complexity on the soil surface, thus providing shelter for many species, as well as by providing food for detritivores and other non-pest insects, some of which may serve as alternative food for coccinellids. Although some studies have found no measurable effects of reduced tillage on coccinellid activity (Rice & Wilde 1991), others have noted favourable influences. Marti & Olson (2007) observed larger numbers of aphidophagous coccinellids in cotton fields under reduced tillage in Georgia, USA, although the treatment also increased populations of the red imported fire ant, *Solenopsis invicta*, a known antagonist of coccinellids and other aphid natural enemies (Eubanks et al. 2002). Although it is clear that coccinellids avoid tilled or bare soils, the extent to which they might utilize crop residues as overwintering sites has not been adequately explored.

Hesler and Berg (2003) observed that reduced tillage increased early season infestation by *Rhopalosiphum padi* in spring-sown cereals, an effect they inferred resulted from crop residues providing protection to the aphid which feeds preferentially on the lower stems of plants. However, this inference contrasts with the findings of Schmidt et al. (2004) who managed to improve biological control of *R. padi* in spring wheat using a **straw mulch treatment** that would have provided the aphids with physical protection similar to crop residues. These authors concluded that bare soil discouraged predators and rendered the crop more susceptible to pests. Given the similarity of the cropping systems, it seems unlikely that increased structural complexity alone can account for the contrasting results of these two studies, and the assemblages of natural enemies were also very different. Reduced tillage is also known to result in cooler soil temperatures and slower warming of the soil in spring, a factor that might impede the activity of coccinellids and other predators relative to aphids in specific situations. Thus it seems plausible that conditions may exist where the benefits of reduced tillage to coccinellids are offset by microclimatic effects.

Reports from **Mexico** suggest that adoption of no-till farming, known locally as 'direct seeding', has had a **long-term beneficial effect** on the abundance of aphidophagous coccinellids. In the Michoacán region of Mexico, summer crops of maize or sorghum are typically followed by winter crops of wheat or barley, a crop cycle similar to that seen across much of the Great Plains region to the north, albeit on an earlier seasonal schedule. Bahena and Fregoso (2007) sampled populations of beneficial insects annually in the winter crops of a number of fields that had been switched to continuous conservation tillage for periods as long as 10 years. Their data showed a trend toward increasing annual abundance of beneficial insects over time, particularly of *Hip. convergens* that comprised 80% of observed predators (Fig. 11.7). Reductions in the use of pesticides such as methyl parathion for aphids were also noted over this period although it is not clear to what extent they resulted from, or contributed to, the increase in *Hip. convergens* abundance.

11.6   ANCILLARY FACTORS INFLUENCING BIOLOGICAL CONTROL BY COCCINELLIDS

11.6.1   Ant-attendance of aphid colonies

Various ant species tend honeydew-producing insects and are notorious antagonists of coccinellids, repelling them from colonies of their prey (Cudjoe et al. 1993, Michaud & Browning 1999, Kaneko 2002) and even

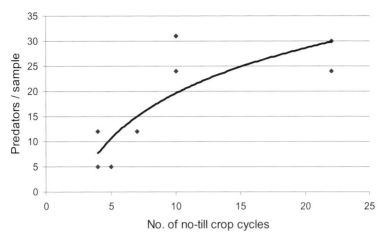

Figure 11.7 Mean numbers of predatory insects per sample (primarily *Hippodamia convergens*) in fields converted to reduced tillage cultivation in Michoacán, Mexico over a period of 10 years. The primary crop cycle was maize in summer followed by wheat in winter (courtesy of Bahena & Fregoso 2007). The logarithmic regression line obscures the fact that the maximum number of predators is actually reached in less than 10 cycles.

preying upon them (Sloggett et al. 1999). (See also 5.4.1.6 and 8.2.4) Where feasible, the **exclusion of ants** may improve biological control by coccinellids. For example, Itioka and Inoue (1996) obtained greatly improved control of citrophilus mealybug, *Pseudococcus cryptus*, by *Chil. kuwanae* when they excluded *Lasius niger* ants from citrus trees. In Hawaii, Reimer et al. (1993) obtained control of green scale, *Coccus viridis*, on coffee trees when they excluded tending *Pheidole megacephala* ants that preyed on larvae of four coccinellid species. Similarly, biological control of mealybugs on pineapples by parasitoids and predators, including the coccinellid *Nephus bilucernarius*, was substantially improved by the use of hydramethylnon **baits to reduce** populations of **tending ants**, primarily *P. megacephala* (Gonzalez-Hernandet et al. 1999). However, the larvae of some coccinellid species produce wax filaments that can provide protection against ant predation (e.g. *Scymnus lousianae*, Schwartzberg et al. 2010) and such species may actually benefit from ant attendance (e.g. *Azya orbigera*, Liere & Perfecto 2008).

11.6.2 Timing of arrival in annual crops

Timely immigration is often crucial to the impact of coccinellids on aphid populations and different species exhibit different seasonal cycles (Elliott & Kieckhefer 1990). These cycles are affected by many factors, especially weather. For example, a cold snap in spring can de-couple biological control of aphids in winter cereals by slowing the development and reproduction of coccinellids and other predators relative to the aphids, and by delaying crop development that in turn extends the period of plant vulnerability to aphid feeding. Early recognition of conditions potentially disruptive to prey tracking by coccinellids can permit a **timely intervention** and prevent economic losses.

The importance of seasonal cycles of crop colonization by coccinellids is well illustrated in **sorghum** production on the High Plains of the USA (Kring et al. 1985, Kring & Gilstrap 1986, Michels & Burd 2007). Early season infestations of *Rhopalosiphum maidis* attract large numbers of aphid predators, mostly coccinellids (*Hip.* spp., but also *C. septempunctata*, *Col. maculata* and other native species) and chrysopids into sorghum fields where they complete a generation on this prey. Although a number of *R. maidis* colonies escape control and become large, they cause no economic damage to the plants and largely disappear before the panicles emerge. The predators attracted to *R. maidis* also feed on *S. graminum*, a pest that is highly damaging to sorghum (Rice & Wilde 1988). Alates of *S. graminum* arrive later in mid-summer and confront a population of newly emerged, hungry coccinellid adults that exhaustively seek out and destroy greenbug colonies in their formative stages, usually preventing the development of economic populations. The generalist aphid parasitoid *Lysiphlebus testaceipes* is slower to respond, but eventually aids in maintaining the greenbug at low density. The risk of economic losses to greenbug is high only when *R. maidis* fails to colonize the crop in sufficient numbers to 'prime' the coccinellid population (Michels & Behle 1992). Thus, when relying on coccinellids to provide aphid control in annual field crops, agricultural producers should learn to **recognize and monitor the seasonal cycle of crop colonization** by these insects so that they can be prepared to intervene with supplementary controls in a timely manner in the event of any disruption.

11.6.3   Interaction of biological control by coccinellids with plant structure and chemistry

Since the plant is typically the theatre for biological control, plant architecture and chemistry may influence outcomes (5.4.1.1). Features such as **high structural complexity** (Grevstad & Klepetka 1992, Khan & Matin 2006) and **trichomes** (Belcher & Thurston 1982, Eisner et al. 1998, Heidari 1999) are known to impede coccinellid foraging ability, but may still be compatible with biocontrol. Shah (1982) demonstrated that varying the form and relative density of leaf trichomes had markedly divergent impacts on *A. bipunctata* larval foraging. Although larvae were unable to search leaves with dense upright or hooked hairs, searching efficiency was increased by widely scattered hairs that caused larvae to change direction frequently. In contrast, larval foraging on **waxy or highly glabrous leaves** was confined to edges and protruding veins. Obrycki et al. (1983) found that trichome density on potato cultivars was inversely correlated with aphid density when natural enemies (*Hip. convergens* and others) were excluded, but concluded that intermediate densities of trichomes would give the best combination of pest resistance and biocontrol in the field. Other plant features which discourage pests may also have unintended effects on coccinellid activity. Cotton cultivars lacking extrafloral nectaries reduce damage by *Lygus* spp. plant bugs, but such cultivars are also less attractive to natural enemies, including coccinellids (Scott et al. 1988). However, Katayama and Suzuki (2010) found that the presence of **extrafloral nectaries** on broad bean plants enhanced the survival of nuclear colonies of *Aphis craccivora* largely because of their greater attractiveness to two ant species that diminished the foraging activities of *C. septempunctata* larvae.

Certain plant characteristics may facilitate biological control by coccinellids. Kareiva and Sahakian (1990) found that a 'leafless' pea variety had higher than expected resistance to pea aphid in the field, simply because adult coccinellids were able to grasp the tendrils and forage more efficiently for the aphids than on **glossy leaves** from which they often slipped and fell. This tritrophic effect has been termed 'extrinsic resistance'. Eigenbrode et al. (1995) showed that a cabbage variety with glossy (as opposed to normal waxy) leaves reduced leaf mining by *Plutella xylostella* larvae, increasing their exposure to predation by *Hip. convergens* and other predators in addition to affording improved mobility for the predators. Similarly, Eigenbrode et al. (1998) showed that adult *Hip. convergens* foraged more effectively on pea cultivars with reduced epicuticular wax and Rutledge et al. (2003) found that such cultivars hosted lower pea aphid populations in field plots than isolines that lacked this trait. Comparing these isolines in cage and laboratory tests without predators revealed no differences between them in plant acceptance by aphids or their subsequent performance. The authors concluded that **higher foraging efficiency** was responsible for the greater coccinellid abundance and lower aphid populations on the reduced wax cultivar in the field.

The effectiveness of biocontrol is sometimes a function of plant susceptibility to the pest, creating the possibility for **synergism between plant resistance factors and natural enemies** such as coccinellids. In cereal breeding, much effort has been directed toward developing cultivars resistant to aphids, particularly *S. graminum* and the Russian wheat aphid *Diuraphis noxia*. Most sources of resistance available in commercial wheat, sorghum and barley cultivars express antibiosis that serves to impede aphid development and reproduction, coupled in some cases with a degree of antixenosis (van Emden 2007). Although such resistant cultivars ultimately succumb to uncontrolled aphid populations, they are able to survive longer than susceptible ones and effectively prevent yield losses when supported by the activities of natural enemies such as coccinellids. By slowing aphid growth and reproduction, they extend the period during which colonies remain small and vulnerable to elimination,

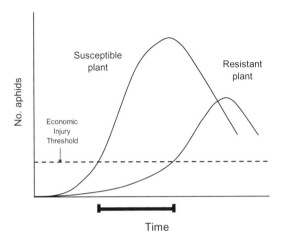

Figure 11.8 Hypothetical aphid colony growth trajectories on susceptible and resistant cereal plants in the absence of natural enemies. As the carrying capacity of the plant is reached, the plant begins to die and alate aphids develop and disperse. Since antibiosis delays aphid colony growth on the resistant plant, more time is required for the colony to reach the economic injury threshold (dashed line), and an incremental period (solid line) becomes available for the actions of coccinellids and other natural enemies to prevent economic damage.

and the time required for the economic injury level to be reached (Fig. 11.8). Wheat cultivars resistant to *D. noxia* are not selected simply for antibiosis, but also for resistance to leaf rolling, a plant response induced by aphid feeding that creates a physical refuge from foraging coccinellids and other natural enemies (Kauffman & LaRoche 1994). Farid et al. (1997) demonstrated that wheat **resistance to leaf rolling** by *D. noxia* was complementary to biological control by *Scymnus frontalis* because the suitability of *D. noxia* as prey was unaffected by resistance, while mortality of beetle larvae was lower on resistant plants due to the absence of rolled leaves that sometimes entrapped larvae on susceptible plants.

By definition, **plant antibiosis** has a negative impact on pest biology, leading to doubts about the suitability of prey fed on resistant plants. Most studies suggest only minor life history impacts of prey-mediated plant resistance factors on coccinellids that seem unlikely to significantly impact field populations or negate the benefits of such resistance in the field. Rice and Wilde (1989) fed *Hip. convergens* on *S.*

graminum reared on two resistant sorghum cultivars and found that larval survival was reduced by between 10 and 20% and development somewhat delayed. Martos et al. (1992) showed that survival and development of *Eriopis connexa* was reduced on a diet of *R. padi* raised on wheat seedlings containing the hydroxamic acid DIMBOA (also 5.2.6.1). Nevertheless, the deterrent flavour of the compound at higher concentrations causes beetles to avoid consuming aphids that contain a potentially lethal dose.

Aphids developing on resistant plants typically take longer to develop and have reduced reproductive rates (van Emden & Wearing 1965, Jyoti & Michaud 2005). They may move more and feed less, thus failing to achieve the size of those reared on susceptible plants. Smaller aphids are usually less profitable prey for coccinellids as they force a predator to spend more time searching and handling prey for a lower net energy return per unit foraging time. This could conceivably translate into delayed development for larvae or reduced clutch sizes for adult females. However, it is also possible for **reduced aphid size** to magnify the impact of coccinellid predation since a large number must be consumed before satiation occurs (Hassell et al. 1977). To date, there are no clear indications of negative effects of plant resistance on coccinellid abundance in the field. For example, Bosque-Perez et al. (2002) found no indications that *D. noxia*-resistant wheat had any adverse impact on the field abundance of coccinellids in Idaho.

With the widespread planting of **genetically modified** (GM) **insect-resistant crops** in many regions of the world, concern has been raised about possible impacts on coccinellids and other beneficial species. GM maize engineered to express *Bacillus thuringiensis* (Bt) toxins was considered a potential risk to coccinellids because the toxin was present in pollen (Harwood et al. 2007). Most studies have found no effect of Bt corn pollen consumption on fitness parameters of *Col. maculata* (Duan et al. 2002, Lundgren & Wiedenmann 2002) or other insect predators (Pilcher et al. 1997). Porcar et al. (2010) fed three types of solubilized Bt toxins directly to *A. bipunctata* and *Cryptolaemus montrouzieri* and obtained no increase in mortality compared to controls. However, (Moser et al. 2008) found that consumption of Bt maize leaf tissues by larvae resulted in a small delay in development relative to non-Bt maize. One study observed reductions in coccinellid abundance in Bt maize plots (Delrio et al. 2004), but McManus et al. (2005) found no reduction in abundance of *Col. maculata* in pre- or post-anthesis Cry3Bb1 maize fields expressing the Bt subsp. *kumamatoensis* toxin that is Coleoptera specific. Similarly, Riddick et al. (2000) looked for changes in abundance of generalist predators in fields of Cry3A-transgenic potatoes with resistance to Colorado potato beetle and found no impact on coccinellids or predatory Heteroptera. Others have suggested that Bt crops may benefit coccinellid populations via reductions in broadcast insecticide applications (Wadhwa & Gill 2007), and the results of a comprehensive multi-year study in Bt cotton (Head et al. 2005) appear to support this conclusion. Nevertheless, it should be noted that **aphids do not acquire the Bt toxin**, whereas other prey species such as spider mites may accumulate and concentrate it in their bodies. Thus Obrist et al. (2006) found high concentrations of Bt toxin in *Stethorus* sp. nymphs sampled in GM maize in Spain, although the impact on fitness of the beetle population was not determined.

11.6.4   Selective use of pesticides

This has long been recognized as an important tactic for conserving coccinellids, and efforts have been made in recent years to determine indices of relative toxicity that can be used to rank materials for **compatibility with ladybirds in IPM programmes**. In general, eggs and young larvae are the most sensitive life stages, and adults the least. As sessile stages, eggs and pupae may have a somewhat reduced risk of exposure in the field relative to actively foraging larvae and adults. One problem in assessing the impact of insecticides in field studies is that materials highly effective against the pest population will also reduce numbers of coccinellids as a consequence of eliminating their food supply, even if they lack direct toxicity to the beetles (Poehling & Dehne 1984, Mateeva et al. 2001). However, **careful timing of insecticide applications** relative to crop development can sometimes serve to synergize, rather than disrupt, pest control by a complex of natural enemies that includes coccinellids (Fagan et al. 2010).

A literature review by Croft and Brown (1975) reported 33 citations where coccinellids were rated **more tolerant of pesticides than their prey**. However, these cases did not include examples of resistant pests and tested mostly older generations of halogenated hydrocarbon and organophosphate insecticides. More recently, Gesraha (2007) found that

pirimicarb, imidacloprid and thiamethoxam were all less toxic to *C. undecimpunctata* than to its prey, *B. brassicae*, although only adult beetles were tested. Broad-spectrum pyrethroids tend to be more destructive to coccinellids than organophosphates (e.g. Kumar & Bhatt 2002), although there is evidence that populations of *Hip. convergens* can evolve substantial resistance to pyrethroids in agroecosystems with heavy usage patterns (Ruberson et al. 2007). Some materials that have shown moderate to low toxicity to coccinellids are listed in Table 11.1. Disparities in procedures, formulations and concentrations across studies make objective comparisons of materials difficult, but there is obvious variation among species in sensitivity to particular compounds. Some materials have a **selective mode of action** that can spare coccinellids from toxic exposure. For example, **spinosad** is generally benign to coccinellids, although Galvan et al. (2005) found sub-lethal impacts on development and reproduction. Because spinosad requires ingestion to induce toxicity, residue trials typically yield high safety ratings, whereas direct topical applications may result in some mortality, probably due to ingestion via grooming behaviour (e.g. Michaud 2002d). Although potentially hazardous to Hymenoptera, certain spinosad-based fly baits have tested relatively safe for coccinellids (Michaud 2003), possibly because the formulation did not stimulate consumption.

Pymetrozine selectively inhibits aphid feeding behaviour, has reputedly low toxicity to coccinellids, and has been used effectively in conjunction with *C. septempunctata* for control of *B. brassicae* on broccoli (Acheampong & Stark 2004). However, despite observing low acute toxicity of pymetrozine to *C. leonina transversalis*, Cole et al. (2010) found that less than 3% of exposed larvae survived to maturity. Insecticides toxic to coccinellids may also be **selective by virtue of formulation** or mode of application. Systemic materials applied as granules, soil drenches or planting time 'in-furrow' applications can give good control of aphids without causing direct mortality to coccinellids. Most recently, the modern **neonicotinoids** thiomethoxam and imidacloprid display good systemic activity in plants and have become increasingly available as **seed treatments** that are very effective against seedling pests, especially aphids and flea beetles, while sparing coccinellids and other predators from direct impacts, except when they engage in herbivory (Moser & Obrycki 2009). Another systemic neonicotinoid, dinotefuran, was shown to provide good control

Table 11.1 List of some insecticides potentially compatible with predatory coccinellids in integrated pest management. Risk levels were assigned for particular species based on the results of individual reports since different studies employed various exposure methods in field and laboratory, different material concentrations and tested different beetle life stages.

Material	Species tested	Type of test	Risk level	Reference
carbofuran	*Ceratomegilla undecimnotata*	laboratory: fed treated prey	sub-lethal effects	Papachristos & Milonas (2008)
diflubenzuron	*Scymnus spp.*	field	low	Matrangolo et al. (1987)
	Cycloneda sanguinea L. *Harmonia axyridis*	laboratory, topical/residue	moderate/low low/low	Michaud (2002d)
dimethoate	*Brumoides suturalis* *Menochilus sexmaculatus* *Coccinella septempunctata*	field	low-moderate low-moderate low-moderate	Sandhu (1986)
endosulfan	*Menochilus sexmaculatus*	semi-field, foliar	low	Jalali & Singh (2001)
	Menochilus sexmaculatus	laboratory, residue	low	Dhingra et al. (1995)
	Menochilus sexmaculatus	field	low	Sharma et al. (1991)
	Coccinella septempunctata		low	
	Micraspis discolor		low	
	Cycloneda sanguinea *Hippodamia convergens* *Olla v-nigrum*	laboratory: residue	moderate moderate moderate	Mizell & Schiffhauer (1990)
	Hippodamia convergens	laboratory, fed treated prey	high	Hurej & Dutcher (1994)
fenvalerate	*Menochilus sexmaculatus* *Coccinella septempunctata* *Micraspis discolor*	field	moderate moderate moderate	Sharma et al. (1991)
	Cycloneda sanguinea *Hippodamia convergens* *Olla v-nigrum*	laboratory: residue	moderate moderate moderate	Mizell & Schiffhauer (1990)
imidacloprid	*Coccinella undecimpunctata*	laboratory: topical	moderate	Gesraha (2007)
	Ceratomegilla undecimnotata	laboratory: fed treated prey	sub-lethal effects	Papachristos & Milonas (2008)
indoxacarb	*Curinus coeruleus* *Cycloneda sanguinea* *Harmonia axyridis* *Olla v-nigrum*	laboratory: topical	moderate low low moderate	Michaud & Grant (2003)
	Harmonia axyridis	laboratory: topical	moderate	Galvan et al. (2005)
	Harmonia axyridis	laboratory: topical/ residue/fed treated prey	moderate/low/ moderate	Galvan et al. (2006b)
methidathion	*Curinus coeruleus* *Cycloneda sanguinea* *Harmonia axyridis* *Olla v-nigrum*	laboratory: topical	moderate moderate moderate moderate	Michaud & Grant (2003)
neem oil	*Cycloneda sanguinea*	laboratory: topical	low	da Silva & Martinez (2004)
	Menochilus sexmaculatus	laboratory: topical	low	Krishnamoorthy et al. (2005)
	Coccinalla septempunctata	field	low	Dhingra et al. (2006)
	Harmonia axyridis	semi-field, foliar	low	Tenczar & Krischik (2006)
	Hippodamia convergens		low	
oxydemeton-methyl	*Brumoides suturalis* *Menochilus sexmaculatus* *Coccinella septempunctata*	field	low-moderate low-moderate low-moderate	Sandhu (1986)

(Continued)

Table 11.1 (*Continued*)

Material	Species tested	Type of test	Risk level	Reference
phosalone	*Coccinella septempunctata*	laboratory: residue	moderate	Hao et al. (1990)
	Cycloneda sanguinea	laboratory: residue	moderate	Mizell & Schiffhauer
	Hippodamia convergens		moderate	(1990)
	Olla v-nigrum		low	
phosmet	*Curinus coeruleus*	laboratory: topical	high	Michaud & Grant
	Cycloneda sanguinea		moderate	(2003)
	Harmonia axyridis		high	
	Olla v-nigrum		low	
	Hippodamia convergens	laboratory, fed treated prey	high	Hurej & Dutcher (1994)
phosphamidon	*Brumoides suturalis*	field	low-moderate	Sandhu (1986)
	Menochilus sexmaculatus		low-moderate	
	Coccinella septempunctata		low-moderate	
pirimicarb	*Adalia bipunctata*	laboratory: topical	low	Kalushkov (1982)
	Coccinella quinquipunctata		low	
	Cycloneda sanguinea		low	
	Coccinella septempunctata	laboratory: residue	low	Hao et al. (1990)
	Coccinella septempunctata	field	moderate	Mateeva et al. (2001)
	Coccinella undecimpunctata	laboratory: topical	moderate	Gesraha (2007)
pyriproxyfen	*Cycloneda sanguinea*	laboratory: topical/residue	moderate/low	Michaud (2002c)
	Harmonia axyridis		low/low	
	Rodolia cardinalis	field (selectively timed)	moderate	Grafton-Cardwell et al. (2006)
pymetrozine	*Coccinella septempunctata*	laboratory, field	low	Acheampong & Stark (2004)
spinosad	*Cycloneda sanguinea*	laboratory: topical/residue	moderate/low	Michaud (2002c)
	Harmonia axyridis		low/low	
	Coccinella septempunctata	laboratory, greenhouse	low	Miles & Dutton (2000)
	Hippodamia convergens		low	
	Menochilus sexmaculatus	laboratory, residue	low	Elzen & James (2002)
	Harmonia axyridis	laboratory, topical	moderate	Galvan et al. (2005)
	Harmonia axyridis	laboratory: topical/ residue/fed treated prey	low/low/low	Galvan et al. (2006b)
	Harmonia axyridis	semi-field, foliar	low	Tenczar & Krischik
	Hippodamia convergens		low	(2006)
Spinosad GF 120®	*Rodolia cardinalis*	laboratory: bait	low	Medina et al. (2004)
	Curinus coeruleus	laboratory: bait	low	Michaud (2003)
	Cycloneda sanguinea		low	
	Harmonia axyridis		low	
sucrose octanoate	*Curinus coeruleus*	laboratory: topical	low	Michaud & McKenzie
	Cycloneda sanguinea		low	(2004)
	Harmonia axyridis		low	
	Olla v-nigrum		low	
thiamethoxam	*Coccinella undecimpunctata*	laboratory: topical	moderate	Gesraha (2007)

of armored scales on Christmas trees when applied in a band around the base of the trunk, without impacting the foraging activities of the scale predators *Chil. stigma* and *Cybocephalus nipponicus* (Cowles 2010). However, **coccinellids** can be **impaired behaviourally** or reproductively by non-lethal concentrations of insecticides that they may acquire when feeding on contaminated pollen or nectar (Smith & Krischik 1999) or contaminated prey (Singh et al. 2004, Eisenback et al. 2010). **Insect growth regulators** (IGRs) such as buprofezen and pyriproxyfen generally lack acute toxicity to coccinellids, but may impair development (Hattingh and Tate 1995) and fecundity (Olszak et al. 1994). Nevertheless, good knowledge of insect ecology can sometimes enable the judicious use of these compounds without disrupting coccinellid populations. Despite the high sensitivity of *Rodolia cardinalis* to IGRs, and residual activity longer than 6 months under California conditions, IGRs can be used to control pyrethroid-resistant scale insects in citrus, provided that applications are delayed until after *R. cardinalis* has exerted control of *Icerya purchasi* in spring (Grafton-Cardwell et al. 2006).

11.7 CONCLUSIONS

The role of coccinellids in classical biological control programmes has diminished as various unanticipated ecological impacts of exotic species have come to light and certain large, dominant coccinellids have gained recognition as invasive species. Consequently, regulatory authorities are likely to permit only highly specialized, non-aphidophagous species for use in classical programmes in future. Certain species will remain important for augmentation in specialized contexts, but few novel augmentation applications for coccinellids been developed recently. The future appears brighter for improving **conservation** and **enhancing the efficacy** of naturally occurring species in open systems. Advances will hinge on improved holistic understanding of the ecological roles of coccinellids and their ability to complement other beneficial species. This information is critical for the development of novel approaches to habitat management that could improve the efficiency of established coccinellid guilds in particular agroecosystems and enhance their ability to track economically important prey species in time and space.

REFERENCES

Acheampong, S. and J. D. Stark. 2004. Can reduced rates of pymetrozine and natural enemies control the cabbage aphid, *Brevicoryne brassicae* (Homoptera, Aphididae), on broccoli? *Int. J. Pest Man.* 50: 275–279.

Acorn, J. 2007. Invasion of the seven-spotted certainty snatchers. *Am. Entomol.* 53: 192.

Adriaens, T., E. Branquart and D. Maes. 2003. The mulit-colored Asian Ladybird *Harmonia axyridis* Pallas (Coleoptera, Coccinellidae): a threat for native aphid predators in Belgium? *Belg. J. Zool.* 133: 201–202.

Ahmed, S. A., A. M. El-Adawy, Y. M. Ahmed, A. A. El-Sebae and M. M. Ibrahim. 2006. Release of the coccinellid predator *Stethorus gilvifrons* (Mulsant) and bio-rational pesticides to suppress the population of *Tetranychus urticae* Koch on cantaloupe plants in Egypt. *Egypt. J. Biol. Pest Control.* 16: 19–24.

Albright, D. D., D. Jordan-Wagner, D. C. Napoli et al. 2006. Multicolored Asian lady beetle hypersensitivity, a case series and allergist survey. *Ann. Allergy Asthma Immunol.* 97: 521–527.

Alhmedi, A., E. Haubruge, B. Bodson and F. Francis. 2007. Aphidophagous guilds on nettle (*Urtica dioica*) close to fields of green pea, rape and wheat. *Insect Sci.* 14: 419–424.

Alhmedi, A., E. Haubruge and F. Francis. 2010. Identification of limonene as a potential kairomone of the harlequin ladybird *Harmonia axyridis* (Coleoptera: Coccinellidae). *Eur. J. Entomol.* 107: 541–548.

Altieri, M. A. and J. W. Todd. 1981. Some influences of vegetational diversity on insect communities of Georgia soybean fields. *Prot. Ecol.* 3: 333–338.

Amman, G. D. 1966. *Aphidecta obliterata* (Coleoptera, Coccinellidae), an introduced predator of the balsam woolly aphid, *Chermes piceae* (Homoptera, Cherminae), established in North America. *J. Econ. Entomol.* 59: 506–508.

Andow, D. A. and S. J. Risch. 1985. Predation in diversified agroecosystems, relations between a coccinellid predator *Coleomegilla maculata* and its food. *J. Appl. Ecol.* 22: 357–372.

Bahena, J. F. and T. L. E. Fregoso. 2007. Entomophagous insects in fields under reduced tillage and conservation in the bay of México. *Proc. 15 Congr. Biol. Control: IOBC Symposium, Mérida, México, November 2007*, published on CD. (In Spanish.)

Baker, S. C., J. A. Elek, R. Bashford et al. 2003. Inundative release of coccinellid beetles into eucalypt plantations for biological control of chrysomelid leaf beetles. *Agric. Forest Entomol.* 5: 97–106.

Barbosa, P. (ed.). 1998. *Conservation Biological Control*. Academic Press, San Diego, CA. 396 pp.

Belcher, D. W. and R. Thurston. 1982. Inhibition of movement of larvae of the convergent lady beetle by leaf trichomes of tobacco. *Environ. Entomol.* 11: 91–96.

Biddinger, D. J., D. C. Weber and L. A. Hull. 2009. Coccinellidae as predators of mites: Stethorini in biological control. *Biol. Control.* 51: 268–283.

Bilu, E. and M. Coll. 2007. The importance of intraguild interactions to the combined effect of a parasitoid and a predator on aphid population suppression. *BioControl.* 52: 753–763.

Bjornson, S. 2008. Fecundity of commercially available convergent lady beetles, *Hippodamia convergens*, following shipment. *Biocont. Sci. Technol.* 18: 633–637.

Bombosch, S. 1965. Untersuchungen über die Disposition und Abundanz der Blattläusen und deren naturlichen Feinden. *Proc. 12 Int. Congr. Entomol., London, 1964.* pp. 578–580.

Borges, I., A. O. Soares and J. L. Hemptinne. 2006. Abundance and spatial distribution of aphids and scales select for different life histories in their ladybird beetle predators. *J. Appl. Entomol.* 130: 356–359.

Borges, P. A. V., R. Cunha, R. Gabriel et al. 2005. Description of the terrestrial Azorean biodiversity. *In* P. A. V. Borges, R. Cunha, R. Gabriel et al. (eds). *A List of the Terrestrial Fauna (Mollusca and Arthropoda) and Flora (Bryophyta, Pteridophyta and Spermatophyta) from the Azores.* Direcção Regional de Ambiente and Universidade dos Açores, Horta, Angra do Heroísmo and Ponta Delgada. pp. 21–68.

Bosque-Perez, N. A., J. B. Johnson, D. J. Schotzko and L. Unger. 2002. Species diversity, abundance, and phenology of aphid natural enemies on spring wheats resistant and susceptible to Russian wheat aphid. *BioControl.* 47: 667–684.

Brown, M. J., H. E. Roy, P. Rothery et al. 2007. *Harmonia axyridis* in Great Britain: analysis of the spread and distribution of a non-native coccinellid. *BioControl* 53: 55–67.

Brown, M. W. and S. S. Miller. 1998. Coccinellidae (Coleoptera) in apple orchards of eastern West Virginia and the impact of invasion by *Harmonia axyridis. Entomol. News.* 109: 136–142.

Burgio, G., F. Santi and S. Maini. 2005. Intra-guild predation and cannibalism between *Harmonia axyridis* and *Adalia bipunctata* adults and larvae, laboratory experiments. *Bull. Insectology.* 58: 135–140.

Burton, R. L. and E. G. Krenzer. 1985. Reduction of greenbug (*Schizaphis graminum*) (Homoptera, Aphididae) populations by surface residues in wheat tillage studies. *J. Econ. Entomol.* 78: 90–394.

Butin, E., J. Elkinton, N. Havil and M. Mongomery. 2003. Comparison of numerical response and predation effects of two coccinellid species on hemlock woolly adelgid (Homoptera, Adelgidae). *J. Econ. Entomol.* 96: 763–767.

CABI. 2007. *Distribution Maps of Plant Pests.* http://www.cababstractsplus.org/DMPP/ (accessed May 2008).

Caltagirone, L. E. and R. L. Doutt. 1989. The history of the vedalia beetle importation to California and its impact on the development of biological control. *Annu. Rev. Entomol.* 34: 1–16.

Catling, P., Z. O. E. Lucas and B. Freedman. 2009. Plants and insects new to Sable Island. *Can. Field Nat.* 123: 141–145.

Chang, G. C. and P. Kareiva. 1999. The case for indigenous generalists in biological control. *In* B. A. Hawkins and H. V. Cornell (eds). *Theoretical Approaches to Biological Control.* Cambridge University Press, Cambridge. pp. 103–115.

Chapin, J. B. and V. A. Brou. 1991. *Harmonia axyridis* (Pallas), the third species of the genus to be found in the U.S. (Coleoptera, Coccinellidae). *Proc. Entomol. Soc. Wash.* 93: 630–635.

Charles, J. G., M. G. Hill and D. J. Allan. 1995. Releases and recoveries of *Chilocorus* spp. (Coleoptera, Coccinellidae) and *Hemisarcoptes* spp. (Acari, Hemisarcoptidae) in kiwifruit orchards, 1987–93. *N. Z. J. Zool.* 22: 319–324.

Cheah, C. A. S. J., M. S. Montgomery, S. M. Salom et al. 2004. Biological control of the hemlock woolly adelgid. *USDA For. Ser. FHTET-2004-04,* FHTET, Fort Collins. 22 pp.

Chong, J. H. and R. D. Oetting. 2007. Intraguild predation and interference by the mealybug predator *Cryptolaemus montrouzieri* on the parasitoid *Leptomastix dactylopii. Biocontr. Sci. Technol.* 17: 933–944.

Cole, P. G., A. R. Cutler, A. J. Kobelt and P. A. Horne. 2010. Acute and long-term effects of selective insecticides on *Micromus tasmaniae* Walker (Neuroptera: Hemerobiidae), *Coccinella transversalis* F. (Coleoptera: Coccinellidae) and *Nabis kinbergii* Reuter (Hemiptera: Miridae). *Austral. J. Entomol.* 49: 160–165.

Collier, T. and R. van Steenwyck. 2004. A critical evaluation of augmentative biological control. *Biol. Control.* 31: 245–256.

Colunga-Garcia, M. and S. H. Gage. 1998. Arrival, establishment, and habitat use of the multicolored Asian lady beetle (Coleoptera, Coccinellidae) in a Michigan landscape. *Environ. Entomol.* 27: 1574–1580.

Colunga-Garcia, M., S. H. Gage and D. A. Landis. 1997. Response of an assemblage of Coccinellidae (Coleoptera) to a diverse agricultural landscape. *Environ. Entomol.* 26: 797–804.

Corbett, A. 1998. The importance of movement in the response of natural enemies to habitat manipulation. *In* C. H. Pickettt and R. L. Bugg (eds) *Enhancing Biological Control.* University of California Press, Berkeley, CA. pp. 25–444.

Corlay, F., G. Boivin and G. Belair. 2007. Efficiency of natural enemies against the swede midge, *Contarinia nasturtii* (Diptera; Cecidomyiidae), a new invasive species in North America. *Biol. Control.* 43: 195–201.

Costamagna, A. C., D. A. Landis and M. J. Brewer. 2008. The role of natural enemy guilds in *Aphis glycines* suppression. *Biol. Control.* 45: 368–379.

Cottrell, T. E. 2005. Predation and cannibalism of lady beetle eggs by adult lady beetles. *Biol. Control.* 34: 159–164.

Cottrell, T. E. and K. V. Yeargan. 1998. Intraguild predation between an introduced lady beetle, *Harmonia axyridis* (Coleoptera, Coccinellidae), and a native lady beetle, *Coleomegilla maculata* (Coleoptera, Coccinellidae). *J. Kans. Entomol. Soc.* 71: 159–163.

Cowles, R. S. 2010. Optimizing a basal bark spray of dinotefuran to manage armored scales (Hemiptera: Diaspididae) in Christmas tree plantations. *J. Econ. Entomol.* 103: 1735–1743.

Croft, B. A. and A. W. A. Brown. 1975. Responses of arthropod natural enemies to insecticides. *Annu. Rev. Entomol.* 20: 285–335.

Cudjoe, A. R., P. Neuenschwander and M. J. W. Copland. 1993. Interference by ants in biological control of the cassava mealybug *Phenacoccus manihoti* (Hemiptera, Pseudococcidae) in Ghana. *Bull. Entomol. Res.* 83: 15–22.

Daane, K. M., K. R. Sime, J. Fallon and M. L. Cooper. 2007. Impacts of Argentine ants on mealybugs and their natural enemies in California's coastal vineyards. *Ecol. Entomol.* 32: 583–596.

Day, W. H., D. R. Prokrym, D. R. Ellis and R. J. Chianese. 1994. The known distribution of the predator *Propylea quatuordecimpunctata* (Coleoptera, Coccinellidae) in the United States, and thoughts on the origin of this species and five other exotic lady beetles in eastern North America. *Entomol. News.* 105: 224–256.

DeBach, P. (ed.). 1964. *Biological Control of Insect Pest and Weeds.* Chapman and Hall, London. 844 pp.

Delrio, G., M. Verdinelli and G. Serra. 2004. Monitoring pest and beneficial insect populations in summer sown Bt maize. *OILB/SROP Bull.* 23: 43–48.

Deng, G. R., H. H. Yang and M. X. Jin. 1987. Augmentation of coccinellid beetles for controlling sugarcane wooly aphid. *Chin. J. Biol. Control.* 3: 166–168.

Dhingra, S., K. Murugesan and D. Sridevi. 1995. Insecticidal safety limits for the coccinellid, *Menochilus sexmaculatus* F. predating on different aphid species. *J. Entomol. Res.* 19: 43–47.

Dhingra, S., D. Sharma, S. Walia et al. 2006. Field appraisal of stable neem pesticide tetrahydroazadirachtin-A against mustard aphid (*Lipaphis erysimi*). *Indian J. Agric. Sci.* 76: 111–113.

Dixon, A. F. G. 2000. *Insect Predator–Prey Dynamics, Ladybird Beetles and Biological Control.* Cambridge University Press, Cambridge, UK. 257 pp.

Dixon, A. F. G. 2005. *Insect Herbivore–Host Dynamics, Tree-dwelling Aphids.* Cambridge University Press, Cambridge, UK. 199 pp.

Dreistadt, S. H. and M. L. Flint. 1996. Melon aphid (Homoptera, Aphididae) control by inundative convergent lady beetle (Coleoptera, Coccinellidae) release on chrysanthemum. *Environ. Entomol.* 25: 688–697.

Duan, J. J., G. Head, M. J. McKee et al. 2002. Evaluation of dietary effects of transgenic corn pollen expressing Cry3Bb1 protein on a non-target ladybird beetle, *Coleomegilla maculata*. *Entomol. Exp. Appl.* 104: 271–280.

Dumbleton, L. D. 1936. The biological control of fruit pests in New Zealand. *N.Z. J. Sci. Technol.* 18: 288–295.

Eigenbrode, S. D., S. Moodie and T. Castagnola. 1995. Predators mediate host plant resistance to a phytophagous pest in cabbage with glossy leaf wax. *Entomol. Exp. Appl.* 77: 335–342.

Eigenbrode, S. D., C. White, M. Rhode and C. J. Simon. 1998. Behavior and effectiveness of adult *Hippodamia convergens* (Coleoptera, Coccinellidae) as a predator of *Acyrthosiphon pisum* (Homoptera, Aphididae) on a wax mutant of *Pisum sativum*. *Environ. Entomol.* 27: 902–909.

Eisenback, B. M., S. M. Salom, L. T. Kok and A. F. Lagalante. 2010. Lethal and sublethal effects of imidacloprid on hemlock woolly adelgid (Hemiptera: Adelgidae) and two introduced predator species. *J. Econ. Entomol.* 103: 1222–1234.

Eisner, T., M. Eisner and E. R. Hoebeke. 1998. When defense backfires, detrimental effect of a plant's protective trichomes on an insect beneficial to the plant. *Proc. Natl Acad. Sci. USA* 95: 4410–4414.

Elliott, N. C. and R. Kieckhefer. 1990. Dynamics of aphidophagous coccinellid assemblages in small grain fields in Eastern South Dakota. *Environ. Entomol.* 19: 1320–1329.

Elliott, N. C., R. Kieckhefer and W. Kauffman. 1996. Effects of an invading coccinellid on native coccinellids in an agricultural landscape. *Oecologia* 105: 537–544.

Elliott, N. C., R. W. Kieckhefer and D. A. Beck 2002a. Effect of aphids and the surrounding landscape on the abundance of Coccinellidae in maize fields. *Biol. Control.* 3: 214–220.

Elliott, N. C., R. W. Kieckhefer, G. J. Michels Jr and K. L. Giles 2002b. Predator abundance in alfalfa fields in relation to aphids, within-field vegetation, and landscape matrix. *Environ. Entomol.* 31: 253–260.

Elzen, G. W. and R. R. James. 2002. Responses of *Plutella xylostella* and *Coleomegilla maculata* to selected insecticides in a residual insecticide bioassay. *Southwest. Entomol.* 27: 149–153.

van Emden, H. F. 1965. The role of uncultivated land in the biology of crop pests and beneficial insects. *Scient. Hort.* 17: 121–136.

van Emden, H. F. 2007. Host-plant resistance. *In* H. F. van Emden and R. Harrington (eds). *Aphids as Crop Pests*. CABI, Wallingford, UK. pp. 447–468.

van Emden, H. F. and C. H. Wearing. 1965. The role of the aphid host plant in delaying economic damage levels in crops. *Ann. Appl. Biol.* 56: 323–324.

Eubanks, M. D., S. A. Blackwell, C. J. Parrish, Z. D. Delamar and H. Hull-Sanders. 2002. Intraguild predation of beneficial arthropods by red imported fire ants in cotton. *Environ. Entomol.* 31: 1168–1174.

Evans, E. W. 2004. Habitat displacement of North American ladybirds by an introduced species. *Ecology* 85: 637–647.

Evans, E. W. and J. G. Swallow. 1993. Numerical responses of natural enemies to artificial honeydew in Utah alfalfa. *Environ. Entomol.* 22: 1392–401.

Evans, E. W., D. R. Richards and A. Kalaskar. 2004. Using food for different purposes, female responses to prey in the predator *Coccinella septempunctata* L. (Coleoptera, Coccinellidae). *Ecol. Entomol.* 29: 27–34.

Fagan, L. L., A. McLachlan, C. M. Till and M. K. Walker. 2010. Synergy between chemical and biological control in the IPM of currant-lettuce aphid (*Nasonovia ribisnigri*) in Canterbury, New Zealand. *Bull. Entomol. Res.* 100: 217–223.

FAO. 1996. *International Standards for Phytosanitary Meaasures. Part 1. Import Regulations. Code of Conduct for the Import and Release of Exotic Biological Control Agents. Publ. No. 3.* FAO, Rome. 21 pp.

Farid, A., J. B. Joyhnson and S. S. Quisenberry. 1997. Compatibility of a coccinellid predator with a Russian wheat aphid resistant wheat. *J. Kans. Entomol. Soc.* 70: 114–119.

Ferguson, K. I. and P. Stiling. 1996. Non-additive effects of multiple natural enemies on aphid populations. *Oecologia* 108: 375–379.

Ferran, A., L. Giuge, R. Tourniere, J. Gambier and D. Fournier. 1998. An artificial non-flying mutation to improve the efficiency of the ladybird *Harmonia axyridis* in biological control of aphids. *BioControl* 43: 53–64.

Fink, D. E. 1915. Control of injurious aphides by ladybirds in Tidewater Virginia. *Virginia Truck Exp. Stn Bull.* 16: 337–350.

Fisher, T. W. 1963. Mass culture of *Cryptolaemus* and *Leptomastix* natural enemies of citrus mealybug. *Bull. Calif. Agric. Exp. Stn* 797: 1–39.

Flint, M. L. and S. H. Dreisdadt (eds). 1998. *Natural Enemies Handbook. The Illustrated Guide to Biological Pest Control.* Publication no. 3386, University of California, Division of Agriculture and Natural Resources, CA. 154 pp.

Flint, M. L. and S. H. Dreistadt. 2005. Interactions among convergent lady beetle (*Hippodamia convergens*) releases, aphid populations, and rose cultivar. *Biol. Control.* 34: 38–46.

Fox, T. B., D. A. Landis, F. F. Cardoso and C. D. Difonzo. 2004. Predators suppress *Aphis glycines* Matsumura population growth in soybean. *Environ. Entomol.* 33: 608–618.

Frere, I., J. Fabry and T. Hance. 2007. Apparent competition or apparent mutualism? An analysis of the influence of rose bush strip management on aphid population in wheat field. *J. Appl. Entomol.* 131: 275–383.

Funasaki, G. Y., P. Y. Lai, L. M. Nakahara, J. W. Beardsley and A. K. Ota. 1988. A review of biological control introductions in Hawaii, 1890–1985. *Proc. Hawaiian Entomol. Soc.* 28: 105–160.

Fye, R. E. and R. L. Carranza. 1972. Movement of insect predators from grain sorghum to cotton. *Environ. Entomol.* 1: 790–791.

Galecka, B. 1966. The effectiveness of predators in control of *Aphis nasturtii* Kalt. and *Aphis frangulae* Kalt. on potatoes. *In* I. Hodek (ed.) *Ecology of Aphidophagous Insects.* Academia, Prague. pp. 255–258.

Galvan, T. L., R. Koch and W. D. Hutchison. 2005. Effects of spinosad and indoxacarb on survival, development, and reproduction of the multicolored Asian lady beetle (Coleoptera, Coccinellidae). *Biol. Control.* 34: 108–114.

Galvan, T. L., E. C. Burkness and W. D. Hutchison 2006a. Efficacy of selected insecticides for management of the multicolored Asian lady beetle on wine grapes near harvest. *Plant Health Prog.* October 2006: 1–5.

Galvan, T. L., R. Koch and W. D. Hutchison 2006b. Toxicity of indoxacarb and spinosad to the multicolored Asian lady beetle, *Harmonia axyridis* (Coleoptera, Coccinellidae), via three routes of exposure. *Pest Man. Sci.* 62: 797–804.

Gesraha, M. A. 2007. Impact of some insecticides on the Coccinellid predator, *Coccinella undecimpunctata* L. and its aphid prey, *Brevicoryne brassicae* L. *Egyp. J. Biol. Pest Control* 17: 65–69.

Gonzalez-Hernandez, H., M. W. Johnson and N. J. Reimer. 1999. Impact of *Pheidole megacephala* (F.) (Hymenoptera, Formicidae) on the biological control of *Dysmicoccus brevipes* (Cockerell) (Homoptera, Pseudococcidae). *Biol. Control.* 15: 145–152.

Gordon, R. D. 1985. The Coccinellidae (Coleoptera) of America north of Mexico. *J. N. Y. Entomol. Soc.* 93: 1–912.

Grafton-Cardwell, E. E., J. E. Lee, J. R. Stewart and K. D. Olsen. 2006. Role of two insect growth regulators in integrated pest management of citrus scales. *J. Econ. Entomol.* 99: 733–744.

Greathead, D. J. 1971. A review of biological control in the Ethiopian region. *Commonw. Inst. Biol. Control. Tech. Comm. No. 5.* CAB, Farnham-Royal. 162 pp.

Greathead, D. J. 2003. A historical overview of biological control in Africa. *In* P. Neuenschwander, C. Borgemeister and J. Langewald (eds). *Biological Control in IPM Systems in Africa.* CABI, Wallingford, UK. 414 pp.

Grevstad, F. S. and B. W. Klepetka. 1992. The influence of plant architecture on the foraging efficiencies of a suite of ladybird beetles feeding on aphids. *Oecologia* 92: 399–404.

Grez, A. A., C. Torres, T. Zaviezo, B. Lavandero and M. Ramirez. 2010. Migration of coccinellids to alfalfa fields with varying adjacent vegetation in Central Chile. *Cien. Inv. Agr.* 37: 111–121.

Groden, E., F. A. Drummond, R. A. Casagrande and D. L. Haynes. 1990. *Coleomegilla maculata* (Coleoptera, Coccinellidae), its predation upon the Colorado potato beetle (Coleoptera, Chrysomelidae) and its incidence in potatoes and surrounding crops. *J. Econ. Entomol.* 83: 1306–1315.

Hagen, K. S. 1974. The significance of predaceous Coccinellidae in biological and integrated control of insects. *Entomophaga* 7: 25–44.

Hagen, K. S., E. F. Sawall and R. L. Tassan. 1971. The use of food sprays to increase effectiveness of entomophagous insects. *Proc. Tall Timbers Conf. Ecol. Anim. Control Habit. Man.* 2: 59–80.

Hämäläinen, M. 1977. Control of aphids on sweet peppers, chrysanthemums and roses in small greenhouses using the ladybeetles *Coccinella septempunctata* and *Adalia bipunctata* (Col., Coccinellidae). *Ann. Agric. Fenn.* 16: 117–131.

Hao, X. C., F. Q. Hu and C. Y. Fang. 1990. Preliminary study on the reaction of *Coccinella septempunctata* to insecticides at different developmental stages. *Nat. Enem. Insects* 12: 62–65.

Harmon, J. P., A. R. Ives, J. E. Losey, A. C. Olson and K. S. Rauwald. 2000. *Coleomegilla maculata* (Coleoptera, Coccinellidae) predation on pea aphids promoted by proximity to dandelions. *Oecologia* 125: 543–548.

Harmon, J. P., E. Stephens and J. Losey. 2007. The decline of native coccinellids (Coleoptera, Coccinellidae) in the United States and Canada. *J. Insect Conserv.* 11: 85–94.

Harris, J. W. E. and A. F. Dawson. 1979. Predator release program for balsam wooly adelgid, *Adelges piceae*, (Homoptera, Adelgidae), in British Columbia, 1960–1969. *J. Entomol. Soc. B.C.* 16: 21–26.

Harwood, J. D., R. A. Samson and J. J. Obrycki 2007. Temporal detection of Cry1Ab-endotoxins in coccinellid predators from fields of *Bacillus thuringiensis* corn. *Bull. Entomol. Res.* 97: 643–648.

Hassell, M. P., J. H. Lawton and J. R. Beddington. 1977. Sigmoid functional responses by invertebrate predators and parasitoids. *J. Anim. Ecol.* 46: 249–262.

Hattingh, V. and B. Tate. 1995. Effects of field-weathered residues of insect growth regulators on some Coccinellidae (Coleoptera) of economic importance as biocontrol agents. *Bull. Entomol. Res.* 85: 489–493.

Head, G., W. Moar, M. Eubanks et al. 2005. A multiyear, large-scale comparison of arthropod populations on commercially managed Bt and non-Bt cotton fields. *Environ. Entomol.* 34: 1257–1266.

Heidari, M. 1999. Influence of host–plant physical defences on the searching behaviour and efficacy of two coccinellid predators of the obscure mealybug, *Pseudococcus viburni* (Signoret). *Entomologica* 33: 397–402.

Heinz, K. M., J. R. Brazzle, M. P. Parella and C. H. Pickett. 1999. Field evaluations of augmentative releases of *Delphastus catalinae* (Horn) (Coleoptera, Coccinellidae) for suppression of *Bemisia argentifolii* Bellows and Perring (Homoptera, Aleyrodidae) infesting cotton. *Biol. Control.* 16: 241–251.

Hesler, L. S. and R. K. Berg. 2003. Tillage impacts cereal-aphid (Homoptera, Aphididae) infestations in spring small grains. *J. Econ. Entomol.* 96: 1792–1797.

Hesler, L. S., R. W. Kieckhefer and M. M. Ellsbury. 2005. Abundance of coccinellids and their potential prey in field-crop and grass habitats in Eastern South Dakota. *Great Lakes Entomol.* 38: 83–96.

Hodek, I. 1973. *Biology of Coccinellidae.* Academia, Prague and Dr. W. Junk, The Hague. 260pp.

Hoebeke, E. R. and A. G. Wheeler. 1980. New distribution records of *Coccinella septempunctata* L. in the eastern United States (Coleoptera, Coccinellidae). *Coleop. Bull.* 34: 209–212.

Hoebeke, E. R. and A. G. Wheeler. 1996. Adventive lady beetles (Coleoptera, Coccinellidae) in the Canadian maritime provinces, with new eastern U.S. records of *Harmonia quadripunctata. Entomol. News* 107: 281–290.

Honda, J. Y. and R. F. Luck. 1995. Scale morphology effects on feeding behavior and biological control potential of *Rhyzobius lophanthae* (Coleoptera, Coccinellidae). *Ann. Entomol. Soc. Am.* 88: 441–450.

Horn, D. J. 1991. Potential impact of *Coccinella septempunctata* on endangered Lycaenidae (Lepidoptera) in NW Ohio, U.S.A. *In* L. Polgar, R. J. Chambers, A. F. G. Dixon and I. Hodek (eds). *Behavior and Impact of Aphidophaga.* SPB Academic Publishers, The Hague. pp. 159–162.

Hossain, Z., G. M. Gurr and S. D. Wratten. 2001. Habitat manipulation in lucerne (*Medicago sativa* L.): strip harvesting to enhance biological control of insect pests. *Int. J. Pest Man.* 47: 81–88.

Howarth, F. G. 1991. Environmental impacts of classical biological control. *Annu. Rev. Entomol.* 36: 485–509.

Huelsman, M. F., J. Kovach, J. Jasinski, C. Young and B. Eisley. 2002. Multicolored Asian lady beetle (*Harmonia axyridis*) as a nuisance pest in households in Ohio. *Proc. 4 Int. Conf. Urban Pests, Pocahontas Press, Blacksburg, July 2002*: 243–250.

Huffaker, C. B. and R. L. Doutt. 1965. Establishment of the coccinellid *Chilocorus bipustulatus* L. in California olive groves. *Pan-Pac. Entomol.* 41: 61–63.

Humble, L. M. 1994. Recovery of additional exotic predators of balsam wooly adelgid, *Adelges piceae* (Ratzeburg) (Homoptera, Adelgidae), in British Columbia, Canada. *Can. Entomol.* 126: 1101–1103.

Hunter, S. J. 1909. The green bug and its enemies. *Kans. Univ. Bull. No.* 137: 163 pp.

Hurej, M. and J. D. Dutcher. 1994. Indirect effect of insecticides on convergent lady beetle (Coleoptera, Coccinellidae) in pecan orchards. *J. Econ. Entomol.* 87: 1632–1635.

Hutton, A. F. 2007. The Australian mealy bug ladybird, *Crytolaemus montrouzieri. Antenna* 31, 40–43.

Iperti, G. 1966. Protection of coccinellids against mycosis. *In* I. Hodek (ed.). *Ecology of Aphidophagous Insects.* Academia. Prague and Dr. W. Junk, The Hague. pp. 189–190.

Iperti, G. and Y. Laudeho. 1969. Les entomophage de *Parlatoria blanchardi* Terg. dans les palmerales de l'adrar Mauritanien. *Ann. Zool. Ecol. Anim.* 1: 17–30.

ISSG. 2008. *Global Invasive Species Database.* http://www.issg.org/database/ (accessed May 2008).

Itioka, T. and T. Inoue. 1996. The role of predators and attendant ants in the regulation and persistence of a population of the citrus mealybug *Pseudococcus citriculus* in a satsuma orange orchard. *Appl. Entomol. Zool.* 31: 195–202.

Jalali, S. K. and S. P. Singh 2001. Residual toxicity of some insecticides to *Cheilomenes sexmaculata* (Fabricius). *J. Insect Sci., Ludhiana.* 14: 80–82.

James, D. G. 2003. Synthetic herbivore-induced plant volatiles as field attractants for beneficial insects. *Environ. Entomol.* 32: 977–982.

Jamieson, L. E., J. G. Charles, P. S. Stevens, C. E. McKenna and R. Bawden. 2005. Natural enemies of citrus red mite (*Panonychus citri*) in citrus orchards. *N. Z. Plant Prot.* 58: 299–305.

Jyoti, J. L. and J. P. Michaud. 2005. Comparative biology of a novel strain of Russian wheat aphid (Homoptera: Aphididae) on three wheat varieties. *J. Econ. Entomol.* 98: 1032–1039.

Kalaskar, A. and E. W. Evans. 2001. Larval responses of aphidophagous lady beetles (Coleoptera, Coccinellidae) to weevil larvae versus aphids as prey. *Ann. Entomol. Soc. Am.* 94: 76–81.

Kalushkov, P. 1982. The effect of five insecticides on coccinellid predators (Coleoptera) of aphids *Phorodon humuli* and *Aphis fabae* (Homoptera). *Acta Entomol. Bohemoslov.* 79: 167–180.

Kaneko, S. 2002. Aphid-attending ants increase the number of emerging adults of the aphid's primary parasitoid and hyperparasitoids by repelling intraguild predators. *Entomol. Sci.* 5: 131–146.

Kaneko, S. 2004. Positive impacts of aphid-attending ants on the number of emerging adults of aphid primary parasitoids and hyperparasitoids through exclusion of intraguild predators. *Jpn J. Entomol.* 7: 173–183.

Kareiva, P. and L. Sahakian. 1990. Tritrophic effects of a simple architectural mutation in pea plants. *Nature.* 345: 433–434.

Katayama, N. and N. Suzuki. 2010. Extrafloral nectaries indirectly protect small aphid colonies via ant-mediated interactions. *Appl. Entomol. Zool.* 45: 505–511.

Katsoyannos, P., D. C. Kontodimas, G. J. Stathas and C. T. Tsartsalis. 1997. Establishment of *Harmonia axyridis* on citrus and some data on its phenology in Greece. *Phytoparasitica.* 25: 182–191.

Kauffman, W. C. and S. L. LaRoche. 1994. Searching activities by coccinellids on rolled wheat leaves infested by the Russian wheat aphid. *Biol. Control.* 4: 290–297.

Kehrli, P. and E. Wyss. 2001. Effects of augmentative releases of the coccinellid, *Adalia bipunctata*, and of insecticide treatments in autumn on the spring population of aphids of the genus *Dysaphis* in apple orchards. *Entomol. Exp. Appl.* 99: 245–252.

Khan, M. R. and M. A. Matin. 2006. Effects of various host plant varieties on prey searching efficiency of *Coccinella septempunctata*. *J. Entomol. Res. Soc.* 8: 39–49.

Kindlmann, P. and A. F. G. Dixon. 1999. Strategies of aphidophagous predators, lessons for modelling insect predator–prey dynamics. *J. Appl. Entomol.* 123: 397–399.

Koch, R. L. 2003. The multicolored Asian lady beetle, *Harmonia axyridis*, A review of its biology, uses in biological control, and non-target impacts. *J. Ins. Sci., Madison* 3(32): http://www.insectscience.org/3.32 (accessed May 2008).

Koch, R. L., E. C. Burkness, S. J. W. Burkness and W. D. Hutchinson. 2004. phytophagous preferences of the multicolored Asian lady beetle (Coleoptera, Coccinellidae) for autumn-ripening fruit. *J. Econ. Entomol.* 97: 539–544.

Koch, R. L., R. C. Vennette and W. D. Hutchinson. 2006. Invasions by *Harmonia axyridis* (Pallas) (Coleoptera, Coccinellidae) in the western hemisphere: implications for South America. *Neotrop. Entomol.* 35: 421–434.

Krafsur, E. S., T. J. Kring, J. C. Miller et al. 1997. Gene flow in the exotic colonizing ladybeetle *Harmonia axyridis* in North America. *Biol. Control.* 8: 207–214.

Kranz, J. and C. Sengonca. 2001. Attractiveness of selected plant species for *Coccinella septempunctata* L. (Col., Coccinellidae) in the field. *Mitt. Deutsch. Gesell. Allge. Angew. Entomol.* 13: 189–192.

Kring, T. J. and F. E. Gilstrap. 1986. Beneficial role of corn leaf aphid in maintaining *Hippodamia* spp. (Coleoptera, Coccinellidae) in grain sorghum. *J. Crop. Prot.* 5: 125–128.

Kring, T. J., F. E. Gilstrap and G. J. Michels Jr. 1985. The role of indigenous coccinellids in regulating greenbug (Homoptera, Aphididae) on Texas grain sorghum. *J. Econ. Entomol.* 78: 269–273.

Krishnamoorthy, A., N. Rama and M. Mani. 2005. Toxicity of botanical insecticides to coccinellid predator *Cheilomenes sexmaculata* (Fabricius). *Insect Environ.* 11: 4–5.

Kumar, S. and R. I. Bhatt. 2002. Pyrethroid-induced resurgence of sucking pests in the mango ecosystem. *J. Appl. Zool. Res.* 13: 107–111.

Kuznetsov, V. N. 1987. The use of Far Eastern ladybeetles (Coleoptera, Coccinellidae) in biological control of plant pests. *Inf. Bull. EPS IOBC.* 21: 27–43. (In Russian.)

Langellotto, G. A. and R. F. Denno 2004. Responses of invertebrate natural enemies to complex-structured habitats, a meta-analytical synthesis. *Oecologia.* 139: 1–10.

Legg, J., D. Gerling and P. Neuenschwander. 2003. Biological control of whiteflies in Sub-Saharan Africa. *In* P. Neuenschwander, C. Borgemeister and J. Langewald (eds). *Biological Control in IPM Systems in Africa.* CABI, Wallingford, UK. pp. 87–100.

van Lenteren, J. C., J. Bale, F. Bigler, H. M. T. Hokkanen and A. J. M. Loomans. 2006. Assessing risks of releasing exotic biological control agents of arthropod pests. *Annu. Rev. Entomol.* 51: 609–634.

Liere, H. and I. Perfecto. 2008. Cheating on a mutualism: indirect benefits of ant attendance to a coccidophagous coccinellid. *Environ. Entomol.* 37: 143–149.

Lixa, A. T., J. M. Campos, A. L. S. Resende et al. 2010. Diversity of Coccinellidae (Coleoptera) using aromatic plants (Apiaceae) as survival and reproduction sites in agroecological system. *Neotrop. Entomol.* 39: 354–359.

Lopez-Arroyo, J. I., J. Loera-Gallardo, M. A. Rocha-Pena et al. 2008. Brown citrus aphid, *Toxoptera citricida* (Hemiptera: Aphididae). *In* H. C. Arredondo Bernal and L. A. Rodriguez del Bosque (eds). *Cases of Biological Control in Mexico.*

Mundi Presna México, SA de CV, Cuauhtemoc, México. pp. 279–292. (In Spanish.)

Lucas, E., I. Gagne and D. Coderre. 2002. Impact of the arrival of *Harmonia axyridis* on adults of *Coccinella septempunctata* and *Coleomegilla maculata* (Coleoptera, Coccinellidae). *Eur. J. Entomol.* 99: 457–463.

Lundgren, J. G. and R. N. Wiedenmann. 2002. Coleopteran-specific Cry3Bb toxin from transgenic corn pollen does not affect the fitness of a nontarget species, *Coleomegilla maculata* DeGeer (Coleoptera, Coccinellidae). *Environ. Entomol.* 31: 1213–1218.

Majerus, M. E. N. 1994. *Ladybirds*. Harper Collins, London. 367 pp.

Mani, M. and A. Krishnamoorthy. 2001. Evaluation of *Cryptolaemus montrouzieri* Muls. (Coleoptera, Coccinellidae) in the suppression of *Rastrococcus invadens* Williams on mango. *J. Insect Sci.. Ludhiana.* 14: 63–64.

Mani, M. and A. Krishnamoorthy. 2007. Recent trends in the biological suppression of guava pests in India. *Acta Hort.* 735: 469–482.

Mani, M., A. Krishnamoorthy and G. L. Pattar. 2004. Efficacy of *Cryptolaemus montrouzieri* Mulsant in the suppression of *Rastrococcus invadens* Williams on sapota. *J. Biol. Control.* 18: 203–204.

Marples, N. M., P. W. de Jong, M. M. Ottenheim, M. D. Verhoog and P. M. Brakefield. 1993. The inheritance of a wingless character in the 2-spot ladybird (*Adalia bipunctata*). *Entomol. Exp. Appl.* 69: 69–73.

Marti, O. G. and D. M. Olson. 2007. Effect of tillage on cotton aphids (Homoptera, Aphididae), pathogenic fungi, and predators in south central Georgia cotton fields. *J. Entomol. Sci.* 42: 354–367.

Martos, A., A. Givovich and H. M. Niemeyer. 1992. Effect of DIMBOA, an aphid resistance factor in wheat, on the aphid predator *Eriopis connexa* Germar (Coleoptera, Coccinellidae). *J. Chem. Ecol.* 18: 469–479.

Mateeva, A., M. Vassileva and T. Gueorguieva. 2001. Side effects of some pesticides on aphid specific predators in winter wheat. *IOBC/WPRS Bull.* 24(6): 139–142.

Matrangolo, E., L. A. Gavioli, S. Gravena, F. C. Moretti and N. K. Odake. 1987. Integration of diflubenzuron with naturally occurring arthropod predators of the cotton leafworm, *Alabama argillacea* (Huebner, 1818. (Lepidoptera, Noctuidae) *An. Soc. Entomol. Brasil.* 16: 5–18. (In Portuguese).

McClure, M. S. 1987. Potential of the Asian predator *Harmonia axyridis* Pallas (Coleoptera, Coccinellidae) to control *Matsucoccus resinosae* Bean and Godwin (Homoptera, Margarodidae) in the United States. *Environ. Entomol.* 16: 224–230.

McManus, B. L., B. W. Fuller, M. A. Boetel et al. 2005. Abundance of *Coleomegilla maculata* (Coleoptera, Coccinellidae) in corn rootworm-resistant Cry3Bb1 maize. *J. Econ. Entomol.* 98: 1992–1998.

Medina, P., I. Perez, F. Budia, A. Adan and E. Vinuela. 2004. Development of an extended-laboratory method to test bait insecticides. *OILB/SROP Bull.* 27: 59–66.

Mensah, R. K. and J. L. Madden. 1994. Conservation of two predator species for biological control of *Chrysophtharta bimaculata* (Col., Chrysomelidae) in Tasmanian forests. *Entomophaga.* 39: 71–83.

Michaud, J. P. 2000. Development and reproduction of lady-beetles (Coleoptera, Coccinellidae) on the citrus aphids *Aphis spiraecola* Patch and *Toxoptera citricida* (Kirkaldy) (Homoptera, Aphididae). *Biol. Control.* 18: 287–297.

Michaud, J. P. 2002a. Classical biological control: a critical review of recent programs against citrus pests in Florida. *Ann. Entomol. Soc. Am.* 95: 531–540.

Michaud, J. P. 2002b. Biological control of Asian citrus psyllid in Florida: a preliminary report. *Entomol. News.* 113: 216–222.

Michaud, J. P. 2002c. Invasion of the Florida citrus ecosystem by *Harmonia axyridis* (Coleoptera, Coccinellidae) and asymmetric competition with a native species, *Cycloneda sanguinea. Environ, Entomol.* 31: 827–835.

Michaud, J. P. 2002d. Relative toxicity of six insecticides to *Cycloneda sanguinea* and *Harmonia axyridis* (Coleoptera, Coccinellidae). *J. Entomol. Sci.* 37: 83–93.

Michaud, J. P. 2003. Toxicity of fruit fly baits to beneficial insects in citrus. *J. Insect Sci., Madison* 3:8. http://www.insectscience.org/3.8/ (accessed May 2008).

Michaud, J. P. 2004. Natural mortality of Asian citrus psyllid, *Diaphorina citri* (Homoptera, Psyllidae) in central Florida. *Biol. Control.* 29: 260–269.

Michaud, J. P. and H. W. Browning. 1999. Seasonal abundance of the brown citrus aphid, *Toxoptera citricida* (Kirkaldy), and its natural enemies in Puerto Rico. *Fla Entomol.* 82: 424–447.

Michaud, J. P. and A. K. Grant. 2003. IPM-compatibility of foliar insecticides for citrus: indices derived from toxicity to beneficial insects from four orders. *J. Insect Sci., Madison* 3:8. http://www.insectscience.org/3.18/ (accessed May, 2008).

Michaud, J. P. and C. L. McKenzie. 2004. Safety of a novel insecticide, sucrose octanoate, to beneficial insects in citrus. *Fla Entomol.* 87: 6–9.

Michaud, J. P. and J. A. Qureshi. 2005. Induction of reproductive diapause in *Hippodamia convergens* (Coleoptera, Coccinellidae) hinges on prey quality and availability. *Eur. J. Entomol.* 102: 483–487.

Michaud, J. P. and J. A. Qureshi. 2006. Reproductive diapause of *Hippodamia convergens* (Coleoptera, Coccinellidae) and its life history consequences. *Biol. Control.* 39: 193–200.

Michels, G. J., Jr and R. W. Behle. 1992. Evaluation of sampling methods for lady beetles (Coleoptera, Coccinellidae) in grain sorghum. *J. Econ. Entomol.* 85: 2251–2257.

Michels, G. J., Jr and J. D. Burd. 2007. IPM case studies: sorghum. *In* H. F. van Emden and R. Harrington (eds). *Aphids as Crop Pests.* CABI, Wallingford, UK. pp. 627–637.

Michels, G. J., Jr, N. C. Elliott, R. A. Romero, D. A. Owings and J. B. Bible. 2001. Impact of indigenous coccinellids on

Russian wheat aphids and greenbugs (Homoptera, Aphididae) infesting winter wheat in the Texas Panhandle. *Southwest. Entomol.* 26: 97–114.

Miles, M. and R. Dutton. 2000. Spinosad: a naturally derived insect control agent with potential for use in glasshouse integrated pest management systems. *British Crop Prot. Conf., Pests and Diseases, Brighton, November 2000.* 1: 339–344.

Mizell, R. F. 2007. Impact of *Harmonia axyridis* (Coleoptera: Coccinellidae) on native arthropod predators in pecan and crape myrtle. *Fla Entomol.* 90: 524–536.

Mizell, R. F. and D. E. Schiffhauer. 1990. Effects of pesticides on pecan aphid predators *Chrysoperla rufilabris* (Neuroptera, Chrysopidae, *Hippodamia convergens, Cycloneda sanguinea, Olla v-nigrum* (Coleoptera, Coccinellidae), and *Aphelinus perpallidus* (Hymenoptera, Encyrtidae). *J. Econ. Entomol.* 83: 1806–1812.

Montgomery, M. E. and S. M. Lyon. 1996. Natural enemies of adelgids in North America, their prospect for biological control of *Adelges tsugae* (Homoptera, Adelgidae). *Proc. 1 Hemlock Woolly Adelgid Review, Charlottesville, October 1995:* 89–102.

Moore, S. D. and V. Hattingh. 2004. Augmentation of natural enemies for control of citrus pests in South Africa: a guide for growers. *S. Afr. Fruit J.* 3(4): 45–47, 51, 53.

Moser, S. E., J. D. Harwood and J. J. Obrycki. 2008. Larval feeding on Bt-hybrid and non-Bt corn seedlings by *Harmonia axyridis* (Coleoptera, Coccinellidae) and *Coleomegilla maculata* (Coleoptera, Cocinellidae). *Environ. Entomol.* 37: 525–533.

Moser, S. E. and J. J. Obrycki. 2009. Non-target effects of neonicotinoid seed treatments; mortality of coccinellid larvae related to zoophytophagy. *Biol. Control.* 51: 487–492.

Musser, F. R. and A. M. Shelton. 2003. Predation of *Ostrinia nubilalis* (Lepidoptera, Crambidae) eggs in sweet corn by generalist predators and the impact of alternative foods. *Environ. Entomol.* 32: 1131–1138.

Nechols, J. R. and T. L. Harvey. 1998. Evaluation of a mechanical exclusion method to assess the impact of Russian wheat aphid natural enemies. *In* S. S. Quisenberry and F. B. Peairs (eds). *Response Model for an Introduced Pest: The Russian Wheat Aphid.* Thomas Say Publications, ESA, Lanham, MD. pp. 270–279.

Neuenschwander, P. 2003. Biological control of cassava and mango mealybugs in Africa. *In* P. Neuenschwander, C. Borgemeister and J. Langewald (eds). *Biological Control in IPM Systems in Africa.* CABI, Wallingford, UK. pp. 45–594.

Neuenschwander, P., K. S. Hagen and R. F. Smith. 1975. Predation on aphids in California's alfalfa fields. *Hilgardia* 43(2): 53–78.

Nichols, P. R. and W. W. Neel. 1977. The use of food wheast as a supplemental food for *Coleomegilla maculata* (DeGeer) (Coleoptera, Coccinellidae) in the field. *Southwest. Entomol.* 2: 102–105.

Nohara, K. 1962. On the overwintering of *Chilocorus kuwanae* Silvestri (Coleoptera: Coccinellidae). *Sci. Bull. Fac. Agric. Kyushu Univ.* 20: 33–39.

Obrist, L. B., A. Dutton, R. Albajes and F. Bigler. 2006. Exposure of arthropod predators to Cry1Ab toxin in Bt maize fields. *Ecol. Entomol.* 31: 143–154.

Obrycki, J. J. and T. J. Kring. 1998. Predaceous Coccinellidae in biological control. *Annu. Rev. Entomol.* 43: 295–321.

Obrycki, J. J., M. J. Tauber and W. M. Tingey. 1983. Predator and parasitoid interaction with aphid-resistant potatoes to reduce aphid densities: a two-year field study. *J. Econ. Entomol.* 76: 456–462.

Obrycki, J. J., E. S. Krafsur, C. E. Bogran, L. E. Gomez and R. E. Cave. 2001. Comparative studies of three populations of the lady beetle predator *Hippodamia convergens* (Coleoptera: Coccinellidae). *Fla Entomol.* 84: 55–62.

Olszak, R. W., B. Pawlik and R. Z. Zajac. 1994. The influence of some insect growth-regulators on mortality and fecundity of the aphidophagous coccinellids *Adalia bipunctata* L. and *Coccinella septempunctata* L. (Col-Coccinellidae). *J. Appl. Entomol.* 117: 58–63.

Osawa, N. and K. Ohashi. 2008. Oviposition strategies of sibling species *Harmonia yedoensis* and *H. axyridis* (Coleoptera, Coccinellidae), the roles of maternal investment through egg and sibling cannibalism on their sympatric coexistence. *Eur. J. Entomol.* 105: 445–454.

Papachristos, D. P. and P. G. Milonas. 2008. Adverse effects of soil applied insecticides on the predatory coccinellid *Hippodamia undecimnotata* (Coleoptera: Coccinellidae). *Biol. Control.* 47: 77–81.

Perrin, R. M. 1975. The role of the perennial stinging nettle, *Urtica dioica*, as a reservoir of beneficial natural enemies. *Ann. Appl. Biol.* 81: 289–297.

Patt, J. M., G. C. Hamilton and J. H. Lashomb. 1997. Impact of strip-insectary intercropping with flowers on conservation biological control of the Colorado potato beetle. *Adv. Hort. Sci.* 11: 175–181.

Pfannenstiel, R. S. and K. V. Yeargan. 2002. Identification and diel activity patterns of predators attacking *Helicoverpa zea* (Lepidoptera, Noctuidae) eggs in soybean and sweet corn. *Environ. Entomol.* 31: 232–241.

Pickering, G. J., K. Ker and G. J. Soleas. 2007. Determination of the critical stages of processing and tolerance limits for *Harmonia axyridis* for 'ladybug taint' in wine. *Vitus.* 46: 85–90.

Pilcher, C. D., J. J. Obrycki, M. E. Rice and L. C. Lewis. 1997. Preimaginal development, survival, and field abundance of insect predators on transgenic *Bacillus thuringiensis* corn. *Environ. Entomol.* 26: 446–454.

Pluke, R. W. H., A. Escribano, J. P. Michaud and P. A. Stansly. 2005. Potential impact of lady beetles on *Diaphorina citri* (Homoptera, Psyllidae) in Puerto Rico. *Fla Entomol.* 88: 123–128.

Poehling, H. M. and H. W. Dehne. 1984. Investigations on the infestation of winter wheat by cereal aphids under practical cultural conditions. II. Influence of insecticide treatments on the aphid populations and beneficial arthropods. *Meded. Fac. Landbouw. Rijksuniv. Gent* 49: 657–665. (In German.)

Porcar, M., I. Garcia-Robles, L. Dominguez-Escriba and A. Latorre. 2010. Effects of *Bacillus thuringiensis* Cry1Ab and Cry3Aa endotoxins on predatory Coleoptera tested through artificial diet-incorporation bioassays. *Bull. Entomol. Res.* 100: 297–302.

Powell, W. and J. K. Pell. 2007. Biological control. *In* H. F. van Emden and R. Harrington (eds). *Aphids as Crop Pests.* CABI, Wallingford. pp. 469–513.

Prasad, Y. K. 1990. Discovery of isolated patches of *Icerya purchasi* by *Rhodolia cardenalis*: a field study. *Entomophaga* 35: 421–429.

Putman, W. L. 1955. Bionomics of *Stethorus punctillum* Weise in Ontario. *Can. Entomol.* 87: 527–579.

Quinones, P. F. J. and R. S. H. Tarango. 2005. Development and survival of *Harmonia axyridis* Pallas (Coleoptera, Coccinellidae) in response to the prey species. *Agric. Tec. Mexico.* 31: 3–9. (In Spanish.)

Ramani, S. J. Poorani and B. S. Bhummanvar. 2002. Spiralling whitefly, *Aleurodicus dispersus*, in India. *Biocontr. News Inf.* 23: 55N–62N.

Randolph, T. L., M. K. Kroening, J. B. Rudolph, F. B. Peairs and R. F. Jepson. 2002. Augmentative releases of commercial biological control agents for Russian wheat aphid management in winter wheat. *Southwest. Entomol.* 27: 37–44.

Reimer, N. J., M. Cope and G. Yasuda. 1993. Interference of *Pheidole megacephala* (Hymenoptera, Formicidae) with biological control of *Coccus viridis* (Homoptera, Coccidae) in coffee. *Environ. Entomol.* 22: 483–488.

Rice, M. E. and G. E. Wilde. 1988. Experimental evaluation of predators and parasitoids in suppressing greenbugs (Homoptera, Aphididae) in sorghum and wheat. *Environ. Entomol.* 17: 836–841.

Rice, M. E. and G. E. Wilde. 1989. Antibiosis effect of sorghum on the convergent lady beetle (Coleoptera, Coccinellidae), a third-trophic level predator of the greenbug (Homoptera, Aphididae). *J. Econ. Entomol.* 82: 570–573.

Rice, M. E. and G. E. Wilde. 1991. Aphid predators associated with conventional- and conservation-tillage winter wheat. *J. Kans. Entomol. Soc.* 64: 245–250.

Riddick, E. W., G. Dively and P. Barbosa. 2000. Season-long abundance of generalist predators in transgenic versus nontransgenic potato fields. *J. Entomol. Sci.* 35: 349–359.

Rosen, D. 1986. Natural enemies of the Diaspididae and their utilization in biological control. *Boll. Lab. Entomol. Agrar. Fil. Silv.* 43: 189–194.

Rosenheim, J. A., T. E. Glik, R. E. Goeriz and B. Ramert. 2004. Linking a predator's foraging behavior with its effects on herbivore population suppression. *Ecology.* 85: 3362–3372.

Roy, M., J. Brodeur and C. Cloutier. 2005. Seasonal activity of the spider mite predators *Stethorus punctillum* (Coleoptera, Coccinellidae) and *Neoseiulus fallacis* (Acarina, Phyto-

seiidae) in raspberry, two predators of *Tetranychus mcdanieli* (Acarina, Tetranychidae). *Biol. Control.* 34: 47–57.

Roy, H. E., P. M. J. Brown, R. L. Ware and M. E. N. Majerus. 2006. Potential impacts of *Harmonia axyridis* on functional biodiversity. *OILB/SROP Bull.* 29: 113–116.

Ruberson, J. R., J. P. Michaud and P. M. Roberts. 2007. Pyrethroid resistance in Georgia populations of the predator *Hippodamia convergens* (Coleoptera, Coccinellidae). *Proc. Beltwide Cotton Conf., New Orleans, January 2007*: 361–365.

Rutledge, C. E., A. P. Robinson and S. D. Eigenbrode. 2003. Effects of a simple plant morphological mutation on the arthropod community and the impacts of predators on a principal insect herbivore. *Oecologia.* 135: 39–50.

Saini, E. D. 2004. Presence of *Harmonia axyridis* (Pallas) (Coleoptera, Coccinellidae) in the province of Buenos Aires. Aspects of biology and morphology. *Rev. Invest. Agropec.* 33: 151–160.

Saito, T. and S. Bjornson. 2006. Horizontal transmission of a microsporidium from the convergent lady beetle, *Hippodamia convergens* Guerin-Meneville (Coleoptera: Coccinellidae), to three coccinellid species of Nova Scotia. *Biol. Control.* 39: 427–433.

Samways, M. J. 1989. Climate diagrams and biological control, an example from the areography of the ladybird *Chilocorus nigritus* (F.) (Insecta, Coleoptera, Coccinellidae). *J. Biogeog.* 16: 345–351.

Samways, M. J., R. Osborn, H. Hastings and V. Hattingh. 1999. Global climate change and accuracy of prediction of species' geographical ranges, establishment success of introduced ladybirds (Coccinellidae, *Chilocorus* spp.) worldwide. *J. Biogeog.* 26: 795–812.

Sandhu, G. S. 1986. Chemical control of spotted alfalfa aphid *Therioaphis trifolii* (Monell) on lucerne with reference to conservation of coccinellid predators. *Indian J. Plant Prot.* 13: 125–127.

Santiago-Islas, T., A. Zamora-Cruz, E. A. Fuentes-Temblador, L. Valencia-Luna and H. C. Arredondo-Bernal. 2008. Pink hibiscus mealybug, *Maconellicoccus hirsutus* (Hemiptera: Pseudococcidae). *In* H. C. Arredondo-Bernal and L. A. Rodriguez del Bosque (eds). *Cases of Biological Control in Mexico.* Mundi Presna México, SA de CV. Cuauhtémoc, México. pp. 177–191. (In Spanish.)

Sathre, M. G., A. Staverl and E. B. Hagvar. 2010. Stowaways in horticultural plants imported from the Netherlands, Germany and Denmark. *Norweg. J. Entomol.* 57: 25–35.

Savoiskaya, G. I. 1983. *Coccinellids.* Izdatelstvo Nauka Kazachskoi SSR, Alma-Atin, 246 pp. (In Russian.)

Schlinger, E. I. and E. J. Dietrick. 1960. Biological control of insect pests aided by strip-farming alfalfa in experimental program. *Calif. Agric.* 14: 8–15.

Schmid, A. 1992. Untersuchungen zur Attraktivität von Ackerwildkräuter für aphidophage Marienkäfer (Coleoptera: Coccinellidae). *Agrarökologie.* 5: 1–122.

Schmidt, M. H, U. Thewes, C. Thies and T. Tscharntke. 2004. Aphid suppression by natural enemies in mulched cereals. *Entomol. Exp. Appl.* 113: 87–93.

Scholl, J. M. and J. T. Medler 1947. Trap strips control insects affecting alfalfa seed production. *J. Econ. Entomol.* 40: 448–450.

Schwartzberg, E. G., K. F. Haynes, D. W. Johnson and G. C. Brown. 2010. Wax structures of *Scymnus louisianae* attenuate aggression from aphid-tending ants. *Environ. Entomol.* 39: 1309–1314.

Scott, W. P., G. L. Snodgrass and J. W. Smith. 1988. Tarnished plant bug (Hemiptera, Miridae) and predaceous arthropod populations in commercially produced selected nectariless cultivars of cotton. *J. Entomol. Sci.* 23: 280–286.

Seagraves, M. P. and K. V. Yeargan. 2006. Selection and evaluation of a companion plant to indirectly augment densities of *Coleomegilla maculata* (Coleoptera: Coccinellidae) in sweetcorn. *Environ. Entomol.* 35: 1334–1341.

Shah, M. A. 1982. The influence of plant surfaces on the searching behaviour of coccinellid larvae. *Entomol. Exp. Appl.* 31: 377–380.

Shands, W. A., R. L. Holmes and G. W. Simpson. 1970. Improved laboratory production of eggs of *Coccinella septempunctata*. *J. Econ. Entomol.* 63: 315–317.

Sharma, R. P., R. P. Yadav and R. Singh. 1991. Relative efficacy of some insecticides against the field population of bean aphid (*Aphis craccivora* Koch.) and safety to the associated aphidophagous coccinellid complex occurring on *Lathyrus*, lentil and chickpea crops. *J. Entomol. Res.* 15: 251–259.

Shrewsbury, P. M., K. Bejleri and J. D. Lea-Cox. 2004. Integrating management practices and biological control to suppress citrus mealybug. *Acta Hort.* 633: 425–434.

Silva, E. B., J. C. Franco, T. Vasconcelos and M. Branco. 2010. Effect of ground cover vegetation on the abundance and diversity of beneficial arthropods in citrus orchards. *Bull Entomol. Res.* 100: 489–499.

da Silva, F. A. C. and S. S. Martinez. 2004. Effect of neem seed oil aqueous solutions on survival and development of the predator *Cycloneda sanguinea* (L.) (Coleoptera, Coccinellidae). *Neotrop. Entomol.* 33: 751–757.

Singh, S. 1976. Other insect pests of coconuts. *Cocomunity* 16: 18–24.

Singh, S. R., K. F. A. Walters, G. R. Port and P. Northing. 2004. Consumption rates and predatory activity of adult and fourth instar larvae of the seven spot ladybird, *Coccinella septempunctata* (L.), following contact with dimethoate residue and contaminated prey in laboratory arenas. *Biol. Control.* 30: 127–133.

Sloggett, J. J., A. Manica, M. J. Day and M. E. N. Majerus. 1999. Predation of ladybirds (Coleoptera, Coccinellidae) by wood ants, *Formica rufa* L. (Hymenoptera, Formicidae). *Entomol. Gaz.* 50: 217–221.

Smith, D., G. A. C. Beattie and R. Broadley (eds). 1997. *Citrus Pests and Their Natural Enemies: Integrated Pest Management in Australia*. Dept. of Primary Industries, Brisbane. 272 pp.

Smith, D., D. Papacek and M. Hallam. 2004. Biological control of *Pulvinaria urbicola* (Cockerell) (Homoptera, Coccidae) in a *Pisonia grandis* forest on North East Herald Cay in the coral sea. *Gen. Appl. Entomol.* 33: 61–68.

Smith, H. S. and H. M. Armitage. 1920. Biological control of mealybugs in California. *Calif. State Dept. Agric. Mon. Bull.* 9: 104–164.

Smith, H. S. and H. M. Armitage. 1931. The biological control of mealybugs attacking citrus. *Calif. Agric. Exp. Stn Bull. No. 509*, Univ. of California, Berkeley. 74 pp.

Smith, S. F. and V. A. Krischik. 1999. Effects of systemic imidacloprid on *Coleomegilla maculata* (Coleoptera, Coccinellidae). *Environ. Entomol.* 28: 1189–1195.

Snyder, G. B., D. L. Finke and W. E. Snyder. 2008. Predator biodiversity strengthens aphid suppression across single- and multiple-species prey communities. *Biol. Control.* 44: 52–60.

Snyder, W. E., G. M. Clevenger and S. D. Eigenbrode. 2004. Intraguild predation and successful invasion by introduced ladybird beetles. *Oecologia.* 140: 559–565.

Soares, A. O., I. Borges, P. A. V. Borges, G. Lanrie and E. Lucas. 2008. *Harmonia axyridis*: what will stop the invasion? *BioControl.* 53: 127–145.

Stals, R. and G. Prinsloo. 2007. Discovery of an alien invasive, predatory insect in South Africa, the multicoloured Asian ladybird beetle, *Harmonia axyridis* (Pallas) (Coleoptera, Coccinellidae). *S. Afr. J. Sci.* 103: 123–126.

Stansly, P. A. 1984. Introduction and evaluation of *Chilocorus bipustulatus* (Coleoptera, Coccinellidae) for control of *Parlatoria blanchardi* (Homoptera, Diaspididae) in date groves of Niger. *Entomophaga.* 29: 29–39.

Starks, K. J., E. A. Wood Jr, R. L. Burton and H. W. Somsen. 1975. *Behavior of Convergent Lady beetles in Relation to Greenbug Control in Sorghum. Observations and Preliminary Tests*. USDA, ARS, Stillwater, 53. pp. 1–10.

Stuart, R. J., J. P. Michaud, L. Olsen and C. W. McCoy. 2002. Lady beetles as potential predators of the citrus root weevil *Diaprepes abbreviatus* (Coleoptera, Curculionidae) in Florida citrus. *Fla Entomol.* 85: 409–416.

Tamaki, G. and Weeks R. E. 1968. Use of chemical defoliants on peach trees in integrated programs to suppress populations of green peach aphids. *J. Econ. Entomol.* 61: 431–435.

Tenczar, E. G. and V. A. Krischik. 2006. Management of cottonwood leaf beetle (Coleoptera, Chrysomelidae) with a novel transplant soak and biorational insecticides to conserve coccinellid beetles. *J. Econ. Entomol.* 99: 102–108.

Thorpe, M. A. 1930. The biology, post-embryonic development, and economic importance of *Cryptochaetum iceryae* (Diptera, Agromyzidae) parasitic on *Icerya purchasi* (Coccidae, Mono-phlebini). *Proc. Zool. Soc. London* 60: 929–971.

Toda, Y. and Y. Sakuratani. 2006. Expansion of the geographical distribution of an exotic ladybird beetle, *Adalia bipunctata* (Coleoptera, Coccinellidae), and its interspecific relationships with native ladybird beetles in Japan. *Ecol. Res.* 21: 292–300.

Tourniaire, R., A. Ferran, L. Giuge, C. Piotte and J. Gambier. 2000. A natural flightless mutation in the ladybird *Harmonia axyridis*. *Entomol. Exp. Appl.* 96: 33–38.

Turnock, W. J., I. L. Wise and F. O. Matheson. 2003. Abundance of some native coccinellines (Coleoptera, Coccinellidae) before and after the appearance of *Coccinella septempunctata*. *Can. Entomol.* 135: 391–404.

Ullah, I. 2000. *Aspects of the Biology of the Ladybird Beetle* Stethorus vagans *(Blackburn) (Coleoptera, Coccinellidae)*. PhD thesis, University of Western Sydney, Richmond, Australia. http://library.uws.edu.au/adt-NUWS/public/adt-NUWS20031103.132342 (accessed January 2008).

Villa Castillo, J. 2005. The use of biological control to control forest pests in Mexico. *Forestal XXI* 8, 11–12. (In Spanish.)

Villalba, M., N. Vila, C. Marzal and F. Garcia Mari. 2006. Influence of inoculative releases of natural enemies and exclusion of ants in the biological control of the citrus mealybug *Planococcus citri* (Hemiptera, Pseudococcidae), in citrus orchards. *Bol. San. Veg. Plagas.* 32: 203–213.

Villanueva, R., J. P. Michaud and C. C. Childers. 2004. Ladybeetles (Coleoptera, Coccinellidae) as predators of mites in citrus. *J. Entomol. Res.* 39: 23–29.

Wadhwa, S. and R. S. Gill. 2007. Effect of Bt-cotton on biodiversity of natural enemies. *J. Biol. Control.* 21: 9–16.

Wajnberg, E., J. K. Scott and P. C. Quimby (eds). 2001. *Evaluating Indirect Ecological Effects of Biological Control*. CABI, Wallingford. 261 pp.

Wang, J. J. and J. Tsai. 2001. Development and functional response of *Coelophora inaequalis* (Coleoptera, Coccinellidae) feeding on brown citrus aphid, *Toxoptera citricida* (Homoptera, Aphididae). *Agric. Forest Entomol.* 3: 65–69.

Wang, S., J. P. Michaud, X. L. Tan, F. Zhang and X. J. Guo. 2011. The aggregation behavior of *Harmonia axyridis* (Coleoptera: Coccinellidae) in Northeast China. *BioControl.* 56(2):193–206.

Waterhouse, D. F. and K. R. Norris. 1989. *Aleurodicus dispersus*. *Biological Control, Pacific Prospects – Supplement 1.* ACIAR monograph no.12. Australian Center for International Agricultural Research, Canberra. pp. 13–22.

Waterhouse, D. F. and D. P. A. Sands 2001. *Classical Biological Control of Arthropods in Australia*. ACIAR, Canberra. 559 pp.

Wheeler, A. G. and E. R. Hoebeke. 1981. A revised distribution of *Coccinella undecimpunctata* L. in eastern and western North America (Coleoptera, Coccinellidae). *Coleop. Bull.* 35: 213–216.

Wheeler, A. G. and E. R. Hoebeke. 1995. *Coccinella novemnotata* in northeastern North America, historical occurrence and current status (Coleoptera, Coccinellidae). *Proc. Entomol. Soc. Wash.* 97: 701–716.

Withers, T. and L. Berndt. 2010. Southern ladybeetle gets another chance. *New Zeal. Tree Grower.* 31: 37–38.

Weissenberger, A., J. Brun and C. Piotte. 1999. Comparison between the wild type and flightless type of the coccinellid *Harmonia axyridis* (Pallas) in the control of the damson hop aphid *Phorodon humuli* (Schrank). *Ann. 5 ANPP Conf, Int. Ravageurs Agric., Montpellier, December 1999*: 727–734. (In French.)

Weiser, L. A., J. J. Obrycki and K. L. Giles. 2003. Within-field manipulation of potato leafhopper (Homoptera, Cicadellidae) and insect predator populations using an uncut alfalfa strip. *J. Econ. Entomol.* 96: 1184–1192.

Williams, A. H. and D. K. Young. 2009. The alien *Hippodamia variegata* (Coleoptera: Coccinellidae) quickly establishes itself throughout Wisconsin. *Great Lakes Entomol.* 42: 100.

With, K. A., D. M. Pavuk, J. L. Worchuck, R. K. Oates and J. L. Fisher. 2002. Threshold effects of landscape structure on biological control in agroecosystems. *Ecol. Appl.* 12: 52–65.

Yakhontov, V. V. 1960. *Utilization of Coccinellids in the Control of Agricultural Pests* Izd. AN Uzbeksk, SSR, Tashkent. 85pp. (In Russian.)

Yang, J. H. 1985. Rearing and application of *Propylea japonica* (Coleoptera, Coccinellidae) for controlling cotton aphis. *Nat. Enem. Insects.* 7: 137–142.

Yarbrough, J. A., J. L. Armstrong, M. Z. Blumberg et al.1999. Allergic rhinoconjunctivitis caused by *Harmonia axyridis* (Asian lady beetle, Japanese lady beetle or lady bug). *J. Allergy Clin. Immunol.* 104: 704–705.

Yasuda, H. and K. Shinya. 1997. Cannibalism and interspecific predation in two predatory ladybirds in relation to prey abundance in the field. *Entomophaga.* 42: 155–165.

Chapter 12

RECENT PROGRESS AND POSSIBLE FUTURE TRENDS IN THE STUDY OF COCCINELLIDAE

Helmut F. van Emden[1] and Ivo Hodek[2]

[1] School of Biological Sciences, University of Reading, Whiteknights, Reading, RG6 6AS, UK
[2] Institute of Entomology, Academy of Sciences, CZ 370 05, České Budějovice, Czech Republic

A major objective of this book and this chapter is to stimulate new research; for this aim it is important not only to emphasize basic recent achievements, but also to question existing paradigms – and highlight the still unanswered problems.

Although in the Americas the coccinellids from the subfamily Epilachninae are important herbivorous pests and there represent the dominating economic focus, the worldwide emphasis is on Coccinellidae as biological control agents of sucking insects. The applied accent should not detract from the scientific interest and value of research on them that is not targeted directly to biological control. In any case, the economically-orientated material in this book needs setting in the wider context of coccinellids being only one component of a much larger complex of natural enemies discussed in Chapter 7. This makes their contribution to pest control very hard to isolate from that of the other taxa active in the relevant guilds (such as e.g. the aphidophagous guild).

In relation to the current increasing interest in conservation biological control, there are many attributes of coccinellids discussed in this book which should be assessed against the ecological and behavioural criteria needed for the different approaches which have been reviewed elsewhere (e.g. Barbosa 1998). With the relationships of ladybirds to habitats (Chapter 4) and food (Chapter 5), introducing single elements of diversity to improve biological control is probably more useful than increasing biodiversity *per se*.

In an even wider context, biological control is just one component of Integrated Pest Management (IPM). This began in the late 1950s as "integrated control" with increasing the ratio of natural enemies (including coccinellids) to aphids in alfalfa through reduced doses of insecticides. Biological control by coccinellids is likely to show positive synergism with both host plant resistance to the prey and insecticide, both of which are able to improve the numerical ratio of natural enemies to their prey (reviewed for IPM of aphids by van Emden 2007; see also Chapter 11.6.3 and Chapter 11.6.4).

The first two chapters, respectively on Phylogeny and Genetics, may appear out of place in relation to the title of this book. However, important recent progress has been made also in these fields, progress of which those interested in the biology and ecology of Coccinellidae should be aware as impacting on the interpretation of their results.

While the subfamilies of coccinellids are more or less worldwide in distribution, many tribes are restricted to particular biogeographical regions, and this has resulted in a number of alternative classifications (Vandenberg and Perez-Gelabert 2007). A comprehensive classification applicable to all geographical regions is needed. Some derived groups of species are often classified under a separate name, leaving the rest of the related species as a paraphyletic assemblage. Vandenberg and Perez-Gelabert (2007) propose that these sets of genera should be re-united or that paraphyletic genera be split to obtain a balanced classification. This can be only achieved by taxonomic revisions containing most described species. Phylogenetic relationships of genera, tribes and subfamilies and subsequent changes in higher classification must be obtained from analysis combining both morphological and molecular data. Mapping on cladograms of specific morphological and life history parameters, such as food specialization, possession of larval waxy exudations and defensive glands will allow us to understand which traits are homologies shared by descent and which are homoplasies achieved by parallel adaptations. Knowledge of the relationships of subfamilies and tribes within a family will allow better discrimination between general patterns and specific cases, as recommended by Sloggett (2005). This will enable us accurately to categorize groups between which comparisons are often made rather too facilely; for example, contrasts are often incorrectly proposed between aphidophagous (in fact almost exclusively Coccinellini) and coccidophagous coccinellids (in fact a polyphyletic assemblage from Chilocorinae, Scymninae and Coccidulinae). Contrasts between aphidophagous and coccidophagous ladybirds or other groups that differ in a specific trait should be analysed only between closely related species or genera (Chapter 1).

Methodological developments in the field of **molecular genetics and genomics**, including microarray technology and high-throughput sequencing, make molecular genetic studies of coccinellids much more achievable than they would have been even a few years ago. This applies equally to studies examining **gene expression** and to those using **molecular markers**. The wealth of data already available on coccinellid ecology and behaviour makes the group particularly well suited for research uniting both ecological and molecular approaches (Chapter 2). Aspects of coccinellid ecology that have already been the subject of intensive study from an evolutionary viewpoint, such as colour pattern polymorphism and male-killing, should particularly benefit from this approach. Our detailed knowledge of the **coccinellid genome** seems likely to increase by leaps and bounds in the next few years: it surely cannot be very long before the first whole coccinellid genome sequence becomes available.

What will be important is that the **research in molecular biology** does not end with the linking of genes with the insects' phenotypical characteristics, but **includes** the **biochemistry** by which the genes express their effect. Chapter 2 gives an account of some major areas where expansion of such research can make important new contributions. However, the presence of the DNA of **endosymbionts** (see below) can confuse the results and interpretation, and Chapter 2 explains the limits of using mitochondrial DNA from the body fluids of coccinellids. However, such genetic variation is still inherited and so subject to natural selection; this may well need to be included to explain the **genetic basis of behaviours** such as dietary specialization. It is when molecular techniques are targeted to identifying prey in gut contents or especially to determining phylogenetic relationships that techniques need to be sought that identify genetic markers on the nuclear chromatin of the particular species of prey or predator.

It seems very likely that **endosymbionts** are far more widely distributed in insects than we know at present, and it is time for a major assay of coccinellids to be carried out, both as regards their own symbionts and those of their prey. It could well be shown that some of the extraordinary **phenotypic plasticity** in the biology and behaviour (such as **food preferences** and **diapause behaviour**) may **not** have a **genetic basis** but reflect variation in the presence or titre of endosymbionts within genotypes. There are likely to be a great number of exciting new avenues of research in this emerging field of study; we should expect quite a lot of surprises.

One of the basic problems of the study of ladybird communities (Chapter 4) is that of **sampling**. Larvae are harder to spot than adults and are more nocturnal in activity, and so the many studies which count coccinellids visually at the same time as their prey collect a large number of zeros and singletons in distributions which are hard to define statistically and lead to dubious interpretation. Such dubious outcomes relate especially to the many studies seeking to compare populations in arable crops and more natural habitats. Two suggestions for future can be made. Firstly, **mark and recapture techniques** might be ideally suited

to coccinellids. They could give much more accurate population estimates and comparisons of population density in different habitats. Moreover, good estimates of vagility can be obtained by repeated marking and recapturing. The point is indeed made in Chapter 5.4.1.5 that the long-distance searching behaviour of ladybirds needs further study and it is clear that there is growing interest in studying coccinellid movements from "natural" to arable crop habitats. This trend in research has developed from the increasing priority being given to a consideration of more natural habitats as "ecosystem services". Good estimates of ladybird **immigration and emigration** therefore become essential and it is surprising that mark and recapture studies have still been rather neglected.

A second comment about sampling relates to an almost universal use of sampling which matches **catch per unit effort** (i.e. numbers in so many sweeps, on so many plants etc.). A far better approach for coccinellids, since a large proportion of samples often return a zero value, is probably to measure the **reciprocal**, the effort needed for a unit catch (i.e. number of plants examined to reach a target number of coccinellids). Density per unit of measurement can still be calculated, but with similar accuracy of the results across the whole range of densities occurring during a season. The ultimate improvement in this direction is to design a sequential sampling plan based on some preliminary sampling. This enables the decision to be made on each sampling occasion as to whether or not more samples need to be taken to determine population size to a given level of accuracy. Sequential sampling plans (Southwood 1966) do require quite a lot of preliminary work, but should be seriously considered for long-term studies in one agroecosystem.

There are two limitations of current work on coccinellid communities. Because of the difficulties in publishing **raw community data** in "high standard" journals, the majority end up in regional or local publications. It is a pity that this information is not collected and subjected to meta-analysis! The second drawback is that the quality of species determination is usually not very high in other groups than the tribe Coccinellini.

With food interrelations, probably most research progress has been made in the field of **behaviour during foraging** (Chapter 5.4). Fifteen years ago most authors considered the foraging movements as completely random, but there is now much laboratory evidence that ladybirds are guided both by **olfactory and visual cues** (Chapter 5.4.1.2). Also, the first studies on the relevant **receptors** have been published. Quite a breakthrough has been the discovery of **oviposition deterrence**, as this gives us insight into decision-making by female coccinellids, something that was previously quite a mystery. However, field assays on the action of volatiles, both from plants and prey, are still largely missing, together with the research on the mutual interference among volatiles (synergy v. inhibition).

It is stressed in chapters 5.2.2 and 5.2.11 that **food specificity** should be considered from the view of physiological/nutritional suitability: the food is either **essential**, enabling reproduction and development of larvae, or **alternative** that just prevents starvation. Although this principle difference has been accepted by leading coccinellid researchers, in some papers the authors limit themselves to the vague terms suitable v. non-suitable food and the distinction is wrongly based on acceptability. It has been shown here that ladybirds may feed also on toxic food.

It is now timely to revisit some of the results on coccinellids which were obtained many years ago. An unfortunate consequence of scientists relying almost exclusively on computer searches of the literature is that they are quite likely not to encounter highly relevant papers published in the 1950s and 1960s and even earlier. Thus studies demonstrating great differences between results at constant and **alternating conditions**, as for example found in the relationship between temperature and coccinellid **voracity** (Chapter 5.3.1), do not seem to be remembered today. The chapters in this book report many studies carried out in growth rooms controlled at constant temperature and humidity, but very few indeed where the experiments were conducted at fluctuating conditions, particularly reflecting night/day variation. As pointed out above, the effect of such fluctuating conditions on coccinellid voracity is quite large; what other attributes of coccinellids are similarly affected?

Although usually a very long phase of the life cycle, **diapause/dormancy**, has lately attracted much less research activity than other topics. This is unfortunate, because the timing of dormancy is essential for the modern ecological approaches to biological control (conservation, augmentation) and for IPM: we must know when the ladybirds need to be helped by providing attractants and alternative foods at their spring arrival on crops (Chapter 6.3.1.5), which includes the

period when the impact of coccinellids is often decided. Such information – also factors affecting survival during dormancy – are missing for most potentially effective coccinellid species. Apart from these applied aspects, diapause of coccinellids is a useful model for research in several fields (Chapter 6.5), e.g. the regulation of voltinism, i.e. annual number of generations (Chapter 6.2.16), or the ethological aspects of the flight to hibernation sites (Chapter 6.3.1.3).

Natural enemies of ladybirds (such as e.g. parasitoids) exert stronger and more persistent effects on the populations of phytophagous than on those of entomophagous Coccinellidae; also coccidophagous coccinellids may be more affected than aphidophagous ones. Phytophagous species, being typically **sedentary,** are more easily located by their enemies (especially by parasitoids) than are the **more mobile** predatory species of Coccinellidae. This explanation may also hold true for the greater susceptibility to natural enemies of coccidophages, since these are less mobile due to the relatively stable occurrence of their prey. Regardless of the coccinellid group considered, there is not much evidence for **top-down regulation** of their populations by natural enemies. Chapter 8 shows clearly that further studies are needed to clarify the role that natural enemies of ladybirds play in population dynamics of these predators. For ladybirds, as for their sternorrhynchan prey, parasitization has again proved easier to measure than predation.

Coccinellids depend on **semiochemical-mediated information** for important steps in their life history (Chapter 9). However, so far there is only a limited understanding of the mechanisms involved and of how chemical ecology interacts with other stimuli regulating life processes. Several of the chemical cues influencing mobility and foraging are commonly-occurring substances (such as green leaf volatiles and semiochemicals from other insects), and one of the future challenges is to understand the subtle mechanisms that contribute to the specificity of their activity for the individual coccinellid species.

This book presents the evidence, probably assembled for coccinellids for the first time, that the potential for **manipulating coccinellid behaviour with semiochemicals** has recently shown itself to have a practical future. However, as it seems unlikely that these chemicals can provide a grower-acceptable level of control on their own, it is important that development of techniques for using semiochemicals will be accompanied by the development of strategies for how they could be integrated with other approaches into an IPM package. Chapters 5.4.1 and 9 deal with active **behaviour-controlling molecules** as singletons, but the history of sex pheromone research has taught us that such compounds are often dramatically **synergized by triggering molecules,** which have little or no effect on their own. There is an example of this in alfalfa, with the attraction of chrysopid adults to a particular breakdown product of tryptophan in aphid honeydew; however, there is no response to the compound involved (indole acetaldehyde) unless the insects first perceive a volatile synomone from alfalfa (Hagen, 1986).

Although repeated evidence has accumulated on the effects of ladybirds in the natural control of several homopteran taxa, it may seem surprising that precise numerical data of their impact are hard to find, particularly on aphids in arable crops (Obrycki et al 2009). These problems are analysed in detail in Chapter 10 and 11, but three important ones are 1) that, as already mentioned at the start of this chapter, coccinellids are only one of over a dozen insect families which may be simultaneously predating on sucking insects in crops, 2) the technical difficulties of quantifying predation (as opposed to parasitism) and 3) the interaction in the field between predators and their prey is affected by several physical factors which vary with year.

Laboratory feeding studies (Chapter 10.2) remain a useful approach for determining prey acceptability and suitability for coccinellids, and consumption assays produce estimates of potential impact (Chapter 10.3). However, the strength of laboratory studies usually lies in providing a **base for interpreting field observations**. Traditional methods of assessing the impact of coccinellids on focal prey species (Chapter 10.4) continue to be invaluable and the best insights are usually obtained through a combination of approaches selected for their suitability for **particular prey-plant systems**, as each has inherent limitations. Although molecular analyses have advanced our understanding of the trophic relations of predatory arthropods, including the Coccinellidae, they have yet to yield data that accurately **quantify predation** in the field. The better studies have linked population surveys with molecular gut-content analysis to strengthen inferences about trophic relationships and mechanisms of prey specificity, as opposed to simply generating binary data on the presence/absence of target prey in predator guts. Semi-quantitative estimates of predation have been made using gut-content analysis by estimating

the concentration of recognizable prey proteins in gut samples by ELISA (Chapter 10.7) or of prey DNA by quantitative PCR. Such studies provide valuable insights into the relative amount of recognizable material in the predator at a given time, but **do not offer 'quantitative' estimates of predation** because the number (and/or biomass) of prey consumed and the time elapsed since consumption cannot be inferred. Thus, although **molecular gut-content analyses** offer particular benefits for elucidating trophic interactions, they remain fraught with interpretive sources of error. Nevertheless, when used in combination with quantitative data from field experiments, they can generate unique insights that cannot be obtained with other approaches.

In the end, however, the data from assays of predator voracity have to be included in a population dynamics exercise in the field. At present, even were we able to measure voracity independently for all the groups of aphid predators in the crop, we do not have a satisfactory numerical approach to analysing the population dynamics of the aphids themselves in the field. The emphasis for IPM needs to move from manipulating coccinellid density to manipulating coccinellid : aphid ratios, which are much more amenable to quantification than "predatory impact".

The role of introduced coccinellids in **classical** biological control programmes (Chapter 11.3.1 & 11.3.2) has **diminished** as unanticipated impacts of exotic species have come to light (Chapter 11.3.3 & 11.3.4). Consequently, regulatory authorities are likely to permit only highly specialized species for use in classical programmes. Certain species will remain important for **augmentation** in specialized contexts (Chapter 11.4), but few novel augmentation applications for coccinellids have as yet been developed. The future appears brighter for **improving conservation** and **enhancing the efficacy** of naturally-occurring species in the field (Chapter 11.5). Advances will hinge on **improved holistic understanding of the ecological roles** of coccinellids and their ability to complement other beneficial species. This information is critical for the development of novel approaches to **habitat management** (Chapter 11.5.3) that could improve the efficiency of established coccinellid guilds in particular agroecosystems and enhance their ability to track economically important prey species in time and space.

It is clear from the chapters in this book and Chapter 11.6.3 that **coccinellid/host plant resistance**

interactions are still a relatively blank canvas with great research opportunities, though as mentioned earlier there is already quite a lot of evidence that the **positive synergism** between ineffective biological control and ineffective host plant resistance can lead to dramatic reduction of aphid populations.

Finally, a concern raised in several of the chapters is that **differences in experimental protocols** of different researchers make it difficult to develop an overall picture when integrating results from the literature. We do not seek to select which approach should be uniformly adopted for each experimental situation. More realistically, a major contribution that this book can make is to identify where inconsistencies exist and provide the literature review material to allow workers in future to evaluate how best to proceed to make their work of maximum utility to their colleagues. This might even extend to taking data additional to those required for the aim of the experiment.

One thing we need to stress in conclusion is that, whatever merit there may or may not be in the suggestions we have made in this final chapter, the most exciting future research breakthroughs concerning coccinellids will, by definition, be matters we have not thought of!

ACKNOWLEDGEMENTS

We gratefully acknowledge that several authors of individual chapters have contributed ideas to this overview. The discussion in this chapter follows the order of the other chapters in the book.

REFERENCES

Barbosa, P. (ed.) 1998. *Conservation Biological Control.* Academic Press, San Diego, 396 pp.

van Emden, H. F. 2007. Integrated pest management of aphids and introduction to IPM case studies. *In* H. F. van Emden and R. Harrington (eds). *Aphids as Crop Pests.* CABI, Wallingford, pp. 537–548.

Hagen, K. S. 1986. Ecosystem analysis: plant cultivars (HPR), entomophagous species and food supplements. *In* D. J. Boethel and R. D. Eikenbary (eds). *Interactions of Plant Resistance and Parasitoids and Predators of Insects.* Ellis Horwood, Chichester, pp.138–197.

Obrycki, J. J., J.D. Harwood, T. J. Kring and R. J. O'Neil. 2009. Aphidophagy by Coccinellidae: Application of biological control in agroecosystems. *Biol. Contr.* 51: 244–254.

Sloggett, J. J. 2005. Are we studying too few taxa? Insights from aphidophagous ladybird beetles (Coleoptera: Coccinellidae). *Eur. J. Entomol.* 102: 391–398.

Southwood, T. R. E. 1966. *Ecological Methods.* Methuen, London 391 pp.

Vandenberg, N. J. and D. E. Perez-Gelabert. 2007. Redescription of the Hispaniolan ladybird genus *Bura* Mulsant (Coleoptera: Coccinellidae) and justification for its transfer from Coccidulinae to Sticholotidinae. *Zootaxa* 1586: 39–46.

APPENDIX: LIST OF GENERA IN TRIBES AND SUBFAMILIES

Oldrich Nedvěd[1] and Ivo Kovář[2]

[1] Faculty of Science, University of South Bohemia & Institute of Entomology, Academy of Sciences, CZ 37005 České Budějovice, Czech Republic
[2] Emer. Scientist of the National Museum, Prague; Current address: Zichovec, Czech Republic

COCCINELLIDAE LATREILLE, 1807

CHILOCORINAE MULSANT, 1846 [2:25]

Chilocorini Mulsant, 1846
Anisorcus Crotch, 1874
Arawana Leng, 1908
Axion Mulsant, 1850
Brumoides Chapin, 1965
Chilocorus Leach in Brewster, 1815
Chujochilus Sasaji, 2005
Cladia Mulsant, 1850
Curinus Mulsant, 1850
Egius Mulsant, 1850
Endochilus Weise, 1898
Exochomus Redtenbacher, 1843 (incl. *Brumus* Mulsant, 1850)
Halmus Mulsant, 1850
Harpasus Mulsant, 1850
Orcus Mulsant, 1850
Parapriasus Chapin, 1965
Parexochomus Barovsky, 1922
Phaenochilus Weise, 1895
Priasus Mulsant, 1850
Priscibrumus Kovář, 1997
Simmondsius Ahmad & Ghani, 1966
Trichorcus Blackburn, 1892
Xanthocorus Miyatake, 1970
Zagreus Mulsant, 1850
Telsimiini Casey, 1899
Hypocyrema Blackburn, 1892
Telsimia Casey, 1899

COCCIDULINAE MULSANT, 1846 [6:46]

Azyini Mulsant, 1850
Azya Mulsant, 1850
Bucolus Mulsant, 1850 (incl. *Bucolinus* Blackburn, 1892)
Cryptolaemus Mulsant, 1853
Pseudoazya Gordon, 1980
Coccidulini Mulsant, 1846
Adoxellus Weise, 1895
Apolinus Pope & Lawrence, 1990 (incl. *Platyomus* Mulsant, 1853 part)
Auladoria Brèthes, 1925
Botynella Weise, 1891
Coccidula Kugelann, 1798
Empia Weise, 1900
Epipleuria Fuersch, 2001
Erithionyx Blackburn, 1892
Eupalea Mulsant, 1850
Eupaleoïdes Gordon, 1994
Geodimmockius Chapin, 1930
Hazisia Weise, 1916
Hypoceras Chapuis, 1876
Iberorhyzobius Raimundo & Canepari, 2006
Lindorus Casey 1899 (incl. *Nothorhyzobius* Brèthes, 1925)
Microrhizobius Sicard, 1909
Mimoscymnus Gordon, 1994
Nat Ślipiński 2007
Nothocolus Gordon, 1994
Orbipressus Gordon, 1994
Planorbata Gordon 1994
Poorani Ślipiński 2007
Psorolyma Sicard, 1922
Rhyzobius Stephens, 1829
Robert Ślipiński 2007
Rodatus Mulsant, 1850
Scymnodes Blackburn, 1889 (incl. *Platyomus* Mulsant, 1853 part)
Sicara Strand, 1942
Stenadalia Weise, 1926
Stenococcus Weise, 1895
Syntona Weise, 1898
Wioletta Ślipiński 2007
Cranophorini Mulsant, 1850
Cleidostethus Weise 1885 (Arrow, 1929)
Cranophorus Mulsant, 1850
Cranoryssus Brèthes, 1921
Hoangus Ukrainsky, 2006 (syn. *Cassiculus* Weise, 1895)
Orynipus Brèthes, 1924
Paracranoryssus Hofmann, 1972
Monocorynini Sasaji 1989
Mimolithophilus Arrow, 1920
Monocoryna Gorham, 1885
Poriini Gordon, 1994
Poria Mulsant, 1850
Tetrabrachini Kapur, 1948 (syn. Lithophilini)
Tetrabrachys Kapur, 1948 (syn. *Lithophilus* Frölich 1799)

COCCINELLINAE LATREILLE, 1807 [5:94]

Coccinellini Latreille, 1807
Aaages Barovskij, 1926
Adalia Mulsant, 1846
Aiolocaria Crotch, 1871
Alloneda Iablokoff-Khnzorian, 1979
Anatis Mulsant, 1846 (incl. *Neopalla* Chapin, 1955)
Anegleis Iablokoff-Khnzorian, 1982
Anisolemnia Crotch, 1874
Antineda Iablokoff-Khnzorian, 1982
Aphidecta Weise, 1893
Archegleis Iablokoff-Khnzorian, 1984
Australoneda Iablokoff-Khnzorian, 1984
Bothrocalvia Crotch, 1874
Callicaria Crotch, 1871
Calvia Mulsant, 1846 (incl. *Eocaria* Timberlake, 1943)
Ceratomegilla Crotch, 1873
Cheilomenes Chevrolat, 1836
Chloroneda Timberlake, 1943
Cirocolla Vandenberg, 1992
Cleobora Mulsant, 1850
Clynis Mulsant, 1850
Coccinella Linnaeus, 1758 (incl. *Chelonitis* Weise, 1879)
Coelophora Mulsant, 1850 (incl. *Autotela* Weise, 1900, *Lemnia* Mulsant, 1850, *Microcaria* Crotch, 1871)
Cycloneda Crotch, 1871 (incl. *Coccinellina* Timberlake, 1943)
Cyrtocaria Crotch, 1874
Declivitata Fürsch, 1964
Docimocaria Crotch, 1874
Dysis Mulsant, 1850
Egleis Mulsant, 1850
Eoadalia Iablokoff-Khnzorian, 1977
Erythroneda Timberlake, 1943
Eumegilla Crotch, 1871
Harmonia Mulsant, 1850 (incl. *Leis* Mulsant, 1850)
Heterocaria Timberlake, 1943
Heteroneda Crotch, 1871
Hippodamia Chevrolat,1836 (syn. *Adonia* Mulsant, 1846)
Hysia Mulsant, 1850
Lioadalia Crotch, 1874
Megalocaria Crotch, 1871
Megillina Weise, 1909
Menochilus Timberlake, 1943
Micraspis Chevrolat, 1836
Microneda Crotch, 1871

Mononeda Crotch, 1874
Mulsantina Weise, 1906
Myrrha Mulsant, 1846
Myzia Mulsant, 1846
Neda Mulsant, 1850
Neocalvia Crotch, 1871
Neoharmonia Crotch, 1871
Nesis Mulsant, 1850
Oenopia Mulsant, 1850 (incl. *Pseudoenopia* Iablokoff-Khnzorian, 1986)
Oiocaria Iablokoff-Khnzorian, 1982
Olla Casey, 1899
Omalocaria Sicard, 1909
Palaeoneda Crotch, 1871
Paraneda Timberlake, 1943
Phrynocaria Timberlake, 1943 (incl. *Phrynolemnia*)
Procula Mulsant, 1850
Propylea Mulsant, 1846
Prototheа Weise, 1898 (incl. *Nedina* Hoang, 1983)
Pseudadonia Timberlake, 1943
Sospita Mulsant, 1846
Sphaeroneda Crotch, 1871
Spilindolla Vandenberg, 1996
Spiloneda Casey, 1908
Synona Pope, 1988
Synonycha Chevrolat, 1836
Xanthadalia Crotch, 1874

Discotomini Mulsant, 1850
Discotoma Mulsant, 1850
Euseladia Crotch, 1874
Pristonema Erichson, 1847
Seladia Mulsant, 1850
Vodella Mulsant, 1853

Halyziini Mulsant 1846 (syn. Psylloborini Casey, 1899)
Eothea Iablokoff-Khnzorian, 1986
Halyzia Mulsant, 1846
Illeis Mulsant, 1850 (incl. *Leptothea* Weise, 1898)
Macroilleis Miyatake, 1965
Metamyrrha Capra, 1945
Neohalyzia Crotch, 1871
Oxytella Weise, 1902
Psyllobora Chevrolat, 1836 (incl. *Thea* Mulsant, 1846)
Vibidia Mulsant, 1846

Singhikaliini Miyatake, 1972
Singhikalia Kapur, 1963

Tytthaspidini Crotch, 1874 (syn. Bulaeini)
Anisosticta Chevrolat, 1836
Bulaea Mulsant, 1850
Coccinula Dobzhanskiy, 1925

Coleomegilla Timberlake, 1920
Eonaemia Iablokoff-Khnzorian, 1982
Eriopis Mulsant, 1850
Isora Mulsant, 1850
Macronaemia Casey, 1899
Naemia Mulsant, 1850
Paranaemia Casey, 1899
Tytthaspis Crotch, 1874 (incl. *Barovskia* Iablokoff-Khnzorian, 1979)

EPILACHNINAE MULSANT, 1846 [4:24]

Cynegetini Thomson 1866 (syn. Madaini)
Cynegetis Chevrolat, 1836
Damatula Gordon, 1975
Figura Ukrainsky, 2006 (syn. *Bambusicola* Fürsch, 1986)
Lorma Gordon, 1975
Mada Mulsant, 1850
Malata Gordon, 1975
Megatela Weise, 1906
Merma Weise, 1898
Pseudodira Gordon, 1975
Tropha Weise, 1900

Epilachnini Mulsant, 1846
Adira (syn. *Dira* Mulsant, 1850)
Afidenta Dieke, 1947
Afidentula Kapur, 1958
Afilachna Bielawski, 1966
Afissula Kapur, 1958
Chnootriba Chevrolat, 1836
Epilachna Chevrolat, 1836
Henosepilachna Li, 1961
Macrolasia Weise, 1903
Subafissa Bielawski, 1963
Subcoccinella Huber, 1841 (Agassiz et Erichson, 1845)
Toxotoma Weise, 1899

Epivertini Pang & Mao, 1979
Epiverta Dieke, 1947

Eremochilini Gordon & Vandenberg 1987
Eremochilus Weise, 1912

EXOPLECTRINAE CROTCH, 1874 [2:24]

Exoplectrini Crotch, 1874
Ambrocharis Sicard, 1909

Anisorhizobius Hofmann 1972
Aulis Mulsant, 1850
Chapinula Ukrainsky, 2006 (syn. *Chapinella* Gordon, 1995)
Chnoodes Chevrolat, 1837
Coeliaria Mulsant, 1850
Cyrtaulis Crotch, 1874
Dapolia Mulsant, 1850
Dioria Mulsant, 1850
Exoplectra Chevrolat, 1837
Hovaulis Sicard, 1909
Iracilda Ślipiński 2007
Lucialla Ślipiński 2007
Neorhizobius Crotch, 1874
Oridia Gorham, 1895
Peralda Sicard, 1909
Sicardinus Ukrainsky, 2006 (syn. *Discoceras* Sicard, 1909; *Nurettinus* Özdikmen 2007)
Siola Mulsant, 1850
Sumnius Weise, 1892

Oryssomini
Neoryssomus Hofmann, 1972
Oryssomus Mulsant, 1850 (incl. *Gordonoryssomus* de Almeida & Lima 1995)
Pseudoryssomus Gordon, 1974
Rhizoryssomus Hofmann, 1972
Roger Ślipiński 2007

MICROWEISEINAE LENG, 1920 [4:23]

Carinodulini Gordon, Pakaluk a Ślipiński, 1989
Carinodula Gordon, Pakaluk a Ślipiński, 1989
Carinodulina Ślipiński & Jadwiszczak, 1995
Carinodulinka Ślipiński & Tomaszewska, 2002

Microweiseini Leng, 1920
Coccidophilus Brèthes, 1905 (incl. *Cryptoweisea* Gordon, 1970)
Gnathoweisea Gordon, 1970
Hong Ślipiński, 2007
Microcapillata Gordon, 1977
Microfreudea Fürsch, 1985
Microweisea Cockerell, 1903
Nipus Casey, 1899
Paracoelopterus Normand, 1936 (incl. *Diloponis* Pope, 1962)
Pseudosmilia Brèthes, 1924
Sarapidus Gordon, 1977
Stictospilus Brèthes, 1924

Serangiini
Delphastus Casey, 1899
Microscymnus Champion, 1913
Serangiella Chapin, 1940 (incl. *Microserangium* Miyatake, 1961)
Serangium Blackburn, 1889 (incl. *Catana* Chapin, 1940, *Catanella* Miyatake, 1961)
Sukunahikonini
Hikonasukuna Sasaji, 1967
Orculus Sicard, 1931
Paraphellus Chazeau, 1981
Pharellus Sicard, 1928
Scymnomorphus Weise 1897 (incl. *Scotoscymnus* Weise, 1901, syn. *Sukunahikona* Kamiya, 1960, syn. *Scymnomorpha* Blackburn, 1892)

ORTALIINAE MULSANT, 1850 [2:14]

Noviini Mulsant, 1850
Anovia Casey, 1908
Novius Mulsant, 1846
Rodolia Mulsant, 1850 (incl. *Eurodolia* Weise, 1895)
Vedalia Mulsant, 1850
Ortaliini Mulsant, 1850
Amida Lewis, 1896
Anortalia Weise, 1902
Azoria Mulsant, 1850
Ortalia Mulsant, 1850
Ortalistes Gorham, 1897
Paramida Sicard, 1909
Rhynchortalia Crotch, 1874
Scymnhova Sicard, 1909
Semra Özdikmen 2007 (syn. *Cinachyra* Gorham, 1899)
Zenoria Mulsant, 1850

SCYMNINAE MULSANT, 1846 [11:51]

Aspidimerini Mulsant, 1850
Acarinus Kapur, 1948
Aspidimerus Mulsant, 1850
Cryptogonus Mulsant, 1850
Pseudaspidimerus Kapur, 1948
Brachiacanthini Mulsant, 1850
Brachiacantha Dejean, 1837
Cyra Mulsant, 1850
Hinda Mulsant, 1850
Cryptognathini Mulsant, 1850
Cryptognatha Mulsant, 1850 (syn. Cryptognathus Mulsant, 1850)

Diomini Gordon, 1999
Andrzej Ślipiński 2007
Decadiomus Chapin, 1933
Dichaina Weise, 1926
Diomus Mulsant, 1850 (incl. *Amidellus* Weise, 1923)
Erratodiomus
Heterodiomus Brèthes, 1924
Magnodiomus
Hyperaspidini Mulsant 1846
Blaisdelliana Gordon, 1970
Helesius Casey, 1899
Hyperaspidius Crotch, 1873
Hyperaspis Chevrolat, 1836
Meltema Özdikmen 2007 (syn. *Corystes* Mulsant, 1850)
Thalassa Mulsant, 1850
Tiphysa Mulsant, 1850
Pentiliini Mulsant, 1850
Calloeneis Grote, 1873
Curticornis Gordon, 1971
Pentilia Mulsant, 1850
Platynaspidini Mulsant, 1846
Crypticolus Strohecker, 1953
Platynaspis Redtenbacher, 1843 (incl. *Paraplatynaspis* Hoang, 1983, *Phymatosternus* Miyatake, 1961, *Platynaspidius* Miyatake, 1961)
Scymnillini Casey, 1899
Viridigloba Gordon, 1978
Zagloba Casey, 1899
Zilus Mulsant, 1850 (incl. *Scymnillus* Horn 1895)
Scymnini Mulsant, 1846
Acoccidula Barowskij, 1931
Aponephus Booth 1991
Apseudoscymnus Hoang, 1984
Axinoscymnus Kamiya, 1963
Clitostethus Weise, 1885 (incl. *Nephaspis* Casey, 1899)
Cyrema Blackburn, 1889
Horniolus Weise, 1901
Keiscymnus Sasaji, 1971
Leptoscymnus Iablokoff-Khnzorian, 1978
Midus Mulsant, 1850
Nephus Mulsant, 1846 (incl. *Sidis* Mulsant, 1850, *Depressoscymnus* Gordon, 1976, *Geminosipho* Fürsch, 1987, *Parascymnus* Chapin, 1965, *Scymnobius* Casey, 1899)
Parasidis Brèthes, 1924
Propiptus Weise, 1901
Pullosidis Fürsch, 1987
Sasajiscymnus Vandenberg, 2004 (syn. *Pseudoscymnus* Chapin, 1962)

Scymniscus Dobzhanskiy, 1928
Scymnus Kugelann, 1794 (incl. *Didion* Casey, 1899, *Pullus* Mulsant, 1846)
Veronicobius Broun, 1893
Selvadiini Gordon 1985
Selvadius Casey, 1899
Stethorini Dobzhanskiy, 1924
Parastethorus Pang & Mao, 1975
Stethorus Weise, 1885

STICHOLOTIDINAE WEISE, 1901 [6:58]

Argentipilosini Gordon & de Almeida, 1991
Argentipilosa Gordon & de Almeida, 1991
Neojauravia Gordon & de Almeida, 1991
Cephaloscymnini
Aneaporia Casey, 1908
Cephaloscymnus Crotch, 1872
Neaporia Gorham, 1897
Prodilis Mulsant, 1850
Prodiloides Weise, 1922
Limnichopharini Miyatake, 1994
Limnichopharus Miyatake, 1994
Plotinini Miyatake, 1994
Ballida Mulsant, 1850 (incl. *Palaeoeneis* Crotch, 1874)
Buprestodera Sicard, 1910
Catanoplotina Kovář 1995
Haemoplotina Miyatake, 1969
Paraplotina Miyatake, 1969
Plotina Lewis, 1896
Protoplotina Miyatake, 1994
Sphaeroplotina Miyatake, 1969
Shirozuellini Sasaji, 1967
Ghanius Ahmad, 1973
Guillermo Ślipiński 2007
Medamatento Sasaji, 1989
Promecopharus Sicard, 1910
Sasajiella Miyatake, 1994
Shirozuella Sasaji, 1967
Sticholotidini Weise, 1901
Boschalis Weise, 1897
Bucolellus Blackburn, 1892 (incl. *Cycloscymnus* Blackburn, 1892)
Bura Mulsant, 1850
Chaetolotis Ślipiński 2004
Chilocorellus Miyatake, 1994
Coelolotis

Coelopterus Mulsant & Rey, 1852
Filipinolotis
Glomerella Gordon, 1977
Habrolotis Weise, 1895
Hemipharus Weise, 1897
Jauravia Motschulsky, 1858
Lenasa Gordon 1994
Lotis Mulsant, 1850
Mesopilo Duverger 2001
Mimoserangium
Neaptera Gordon 1991
Nelasa Gordon 1991
Neotina Gordon, 1977
Nesina Gordon, 1977
Nexophallus Gordon, 1969
Parajauravia Iablokoff-Khnzorian, 1972
Paranelasa Gordon 1991
Parinesa Gordon 1991
Pharopsis Casey, 1899
Pharoscymnus Bedel, 1906
Phlyctenolotis Sicard, 1912
Semiviride Gordon 1991
Sticholotis Crotch, 1874 (incl. *Gymnoscymnus* Blackburn, 1892, *Nesolotis* Miyatake, 1966, *Paranesolotis* Hoang, 1982)
Stictobura Crotch, 1874
Sulcolotis Miyatake
Synonychimorpha Miyatake, 1994
Trimallena Pope, 1962
Xamerpillus Sicard, 1912
Xanthorcus Weise, 1898
Xestolotis Casey, 1899

Figures in brackets following subfamily names are numbers of tribes and genera. After any taxon, in parentheses, are widely used or recently established synonyms.

Based on Iablokoff-Khnzorian (1982), Fuersch (1990), Gordon and de Almeida (1991), Vandenberg (1992), Duverger (2001), Ślipiński and Tomaszewska (2002), Vandenberg (2002), Poorani (2003), Ukrainsky (2006), Kovář (2007), Özdikmen (2007), Ślipiński (2007), and Synopsis of the described Coleoptera of the world (http://insects.tamu.edu/research/collection/hallan/test/Arthropoda/Insects/Coleoptera/Family/Coleoptera1.htm), http://www.scientific-web.com/en/Biology/Animalia/Arthropoda/Insects/Coccinellidae.html (accessed November 2010). Full citations are listed in Chapter 1.

SUBJECT INDEX

Note: Page numbers in **bold** face refer to tables, page numbers in *italic* face indicate figures and illustrations. Abbreviations used: LDT for lower development threshold; SET for sum of effective temperatures; SCP for super cooling point;

Aaron's rod see *Verbascum thapsus*
Abgrallaspis cyanophylli **60**, *87*, 91, **190**, **204**
Abies balsamea 471
abundance of coccinellids
    absolute and relative 111
    and character of landscape 127
    on crops **129**, **130**
    and habitat fragmentation 126–127
    link to aphid abundance 117, *118*
    methods of estimating 468–472
    reduced tillage 502, *503*
    sampling methods 112–114
    transgenic crops 122–123
    on trees **131**, **132**
Abutilon theophrasti 227, *227*
Acacia 154, 183, 489
Acalypha ostryaefolia 123, 126, 227, *227*
acarina
    Coccipolipus 411–414
    phoretic mites 345, 411
accepted food/prey 143, 144, 145, *146*
Aconitum 165
Acraea encedon 425
Acrocephalus schoenobaenus 379
Acutaspis umbonifera **190**
Acyrthosiphon caraganae **58**
Acyrthosiphon ignotum **58**
Acyrthosiphon kondoi 476
Acyrthosiphon nipponicum see *Neoaulacorthum nipponicum* 241

Acyrthosiphon pisum 57, **58–65**, 74–75, 78, 115, 118, 120, **145**, 147, 152–154, 157, **158**, *160*, 161, **161–163**, 164, 168–9, **168**, **170**, 171, *172*, 174, *182*, **188–195**, 196, 206, 208, 218, 220, 222–224, 234, 236, **237**, 238, *240*, 297, 345, 451, 457, 469, *534*
ant attendance 236, **237**
consumption rates *160*
essential prey **158**, **162**, *172*, **188–195**
larval period and pupal mass, effect on **161**
nutritive value 196–197
pea surface wax 217–218
ratio between disturbed and consumed *240*
visual ability 239
Adalia bipunctata 20, **21**, 23–32, *23*, *30*, 36, 37, *39*, 55–57, **58**, 68, 69, 72, 73, 77, 78, **79**, 82–85, *87*, 89, 90, 91, **93**, **96**, 113, 117, 119–123, *121*, 126, 128, 129, **129–132**, 144–146, 150, **151**, 151–154, **153**, 159, **162**, 163–169, *165*, **166**, 172–174, **173**, 179–180, **205**, 209, 211, 212, *215*, 216–217, 224, *230*, 231, 236, **291**, 305, **306**, 308, 309, 311, 318, 320–322, 324–326, 329,

332, 344, 356–358, 363, 376, 377, *378*, 391, **392**, **398**, **399**, 407, **407**, 408, **412**, 413, 414, 416, 417, **417**, **420**, **422**, 425, 449, 451, 454–458, *477*, 498, 504, 505, **508**
adaptation ability 157
aphid prey **153**
    capture of 238, **239**
    development on different **145**, **147**
    rejected/problematic 165–168
    toxic, effects of 161, 162, 452
augmentation 498
Coccipolypus hippodamiae, infection by *413*
colour pattern variation **21**, 22, *23*, 25
    assortative mating 27
    evolution of 26
    geographic variation 23–24
diapause regulation 290–291
dormancy behaviour 313–314
egg cannibalism 177–178, 233
essential foods **188**
expansion in Japan 493–494
inbreeding 40
intraguild predation *350*, 351–352, 361
larval attachment ability 218
larval food consumption 202, 203
    preference for 452–453

Ecology and Behaviour of the Ladybird Beetles (Coccinellidae), First Edition. Edited by I. Hodek, H.F. van Emden, A. Honěk.
© 2012 Blackwell Publishing Ltd. Published 2012 by Blackwell Publishing Ltd.